ADVANCED THERMODYNAMICS ENGINEERING

CRC Series in
COMPUTATIONAL MECHANICS and APPLIED ANALYSIS

Series Editor: J.N. Reddy
Texas A&M University

Published Titles

APPLIED FUNCTIONAL ANALYSIS
J. Tinsley Oden and Leszek F. Demkowicz

THE FINITE ELEMENT METHOD IN HEAT TRANSFER AND FLUID DYNAMICS, Second Edition
J.N. Reddy and D.K. Gartling

MECHANICS OF LAMINATED COMPOSITE PLATES: THEORY AND ANALYSIS
J.N. Reddy

PRACTICAL ANALYSIS OF COMPOSITE LAMINATES
J.N. Reddy and Antonio Miravete

SOLVING ORDINARY and PARTIAL BOUNDARY VALUE PROBLEMS in SCIENCE and ENGINEERING
Karel Rektorys

CRC Series in
COMPUTATIONAL MECHANICS and APPLIED ANALYSIS

ADVANCED THERMODYNAMICS ENGINEERING

Kalyan Annamalai
Department of Mechanical Engineering
Texas A & M University

Ishwar K. Puri
Department of Mechanical Engineering
University of Illinois at Chicago

CRC PRESS

Boca Raton London New York Washington, D.C.

Library of Congress Cataloging-in-Publication Data

Annamalai, Kalyan.
Advanced thermodynamics engineering / Kalyan Annamalai & Ishwar K. Puri.
p. cm. — (CRC series in computational mechanics and applied analysis)
Includes bibliographical references and index.
ISBN 0-8493-2553-6 (alk. paper)
1. Thermodynamics. I. Puri, Ishwar Kanwar, 1959- II. Title. III. Series.

TJ265 .A55 2001
621.402′1—dc21 2001035624

Direct all inquiries to CRC Press LLC, 2000 N.W. Corporate Blvd., Boca Raton, Florida 33431.

International Standard Book Number 0-8493-2553-6
Library of Congress Card Number 2001035624
Printed in the United States of America 1 2 3 4 5 6 7 8 9 0
Printed on acid-free paper

KA dedicates this text to his mother Kancheepuram Pattammal Sundaram, who could not read or write, and his father, Thakkolam K. Sundaram, who was schooled through only a few grades, for educating him in all aspects of his life. He thanks his wife Vasanthal for companionship throughout the cliff–hanging journey to this land of opportunity and his children, Shankar, Sundhar and Jothi for providing a vibrant source of "energy" in his career.

IKP thanks his wife Beth for her friendship and support and acknowledges his debt to his sons Shivesh, Sunil, and Krishan, for allowing him to take time off from other pressing responsibilities, such as playing catch. His career has been a fortunate journey during which his entire family, including his parents Krishan and Sushila Puri, has played a vital role.

PREFACE

We have written this text for engineers who wish to grasp the *engineering physics* of thermodynamic concepts and *apply* the knowledge in their field of interest rather than merely digest the abstract generalized concepts and mathematical relations governing thermodynamics. While the fundamental concepts in any discipline are relatively invariant, the problems it faces keep changing. In many instances we have included physical explanations along with the mathematical relations and equations so that the principles can be relatively applied to real world problems.

The instructors have been teaching advanced thermodynamics for more than twelve years using various thermodynamic texts written by others. In writing this text, we acknowledge that debt and that to our students who asked questions that clarified each chapter that we wrote. This text uses a "down–to–earth" and, perhaps, unconventional approach in teaching advanced concepts in thermodynamics. It first presents the phenomenological approach to a problem and then delves into the details. Thereby, we have written the text in the form of a self–teaching tool for students and engineers, and with ample example problems. Readers will find the esoteric material to be condensed and, as engineers, we have stressed applications throughout the text. There are more than 110 figures and 150 engineering examples covering thirteen chapters.

Chapter 1 contains an elementary overview of undergraduate thermodynamics, mathematics and a brief look at the corpuscular aspects of thermodynamics. The overview of microscopic thermodynamics illustrates the physical principles governing the macroscopic behavior of substances that are the subject of classical thermodynamics. Fundamental concepts related to matter, phase (solid, liquid, and gas), pressure, saturation pressure, temperature, energy, entropy, component property in a mixture and stability are discussed.

Chapter 2 discusses the first law for closed and open systems and includes problems involving irreversible processes. The second law is *illustrated* in Chapter 3 rather than presenting an axiomatic approach. Entropy is introduced through a Carnot cycle using ideal gas as the medium, and the illustration that follows considers any reversible cycle operating with any medium. Entropy maximization and energy minimization principles are illustrated. Chapter 4 introduces the concept of availability with a simple engineering scheme that is followed by the most general treatment. Availability concepts are illustrated by scaling the performance of various components in a thermodynamic system (such as a power plant or air conditioner) and determining which component degrades faster or outperforms others. Differential forms of energy and mass conservation, and entropy and availability balance equations are presented in Chapters 2 to 4 using the Gauss divergence theorem. The differential formulations allow the reader to determine where the maximum entropy generation or irreversibility occurs within a unit so as to pinpoint the major source of the irreversibility for an entire unit. Entropy generation and availability concepts are becoming more important to energy systems and conservation groups. This is a rapidly expanding field in our energy–conscious society. Therefore, a number of examples are included to illustrate applications to engineering systems. Chapter 5 contains a postulatory approach to thermodynamics. In case the reader is pressed for time, this chapter may be entirely skipped without loss of continuity of the subject.

Chapter 6 presents the state equation for real gases including two and three parameter, and generalized equations of state. The Kessler equation is then introduced and the methodology for determining $Z^{(0)}$ and $Z^{(1)}$ is discussed. Chapter 7 starts with Maxwell's relations followed by the development of generalized thermodynamic relations. Illustrative examples are presented for developing tables of thermodynamic properties using the Real Gas equations. Chapter 8 contains the theory of mixtures followed by a discussion of fugacity and activity. Following the methodology for estimating the properties of steam from state equations, a methodology is presented for estimating partial molal properties from mixture state equations. Chapter 9 deals with phase equilibrium of multicomponent mixtures and vaporization and boiling. Applications to engineering problems are included. Chapter 10 discusses the regimes of stable and metastable states of fluids and where the criteria for stability are violated. Real gas state equations are used to identify the stable and unstable regimes and illustrative examples with physical explanation are given.

Chapter 11 deals with reactive mixtures dealing with complete combustion, flame temperatures and entropy generation in reactive systems. In Chapter 12 criteria for the direction of chemical reactions are developed, followed by a discussion of equilibrium calculations using the equilibrium constant for single and multi-phase systems, as well as the Gibbs minimization method. Chapter 13 presents an availability analysis of chemically reacting systems. Physical explanations for achieving the work equivalent to chemical availability in thermodynamic systems are included. The summary at the end of each chapter provides a brief review of the chapter for engineers in industry.

Exercise problems are placed at the end. This is followed by several tables containing thermodynamic properties and other useful information.

The field of thermodynamics is vast and all subject areas cannot be covered in a single text. Readers who discover errors, conceptual conflicts, or have any comments, are encouraged to E–mail these to the authors (respectively, kannamalai@tamu.edu and ikpuri@uic.edu). The assistance of Ms. Charlotte Sims and Mr. Chun Choi in preparing portions of the manuscript is gratefully acknowledged. We wish to acknowledge helpful suggestions and critical comments from several students and faculty. We specially thank the following reviewers: Prof. Blasiak (Royal Inst. of Tech., Sweden), Prof. N. Chandra (Florida State), Prof. S. Gollahalli (Oklahoma), Prof. Hernandez (Guanajuato, Mexico), Prof. X. Li. (Waterloo), Prof. McQuay (BYU), Dr. Muyshondt. (Sandia National Laboratories), Prof. Ochterbech (Clemson), Dr. Peterson, (RPI), and Prof. Ramaprabhu (Anna University, Chennai, India).

KA gratefully acknowledges many interesting and stimulating discussions with Prof. Colaluca and the financial support extended by the Mechanical Engineering Department at Texas A&M University. IKP thanks several batches of students in his Advanced Thermodynamics class for proofreading the text and for their feedback and acknowledges the University of Illinois at Chicago as an excellent crucible for scientific inquiry and education.

Kalyan Annamalai, College Station, Texas
Ishwar K. Puri, Chicago, Illinois

ABOUT THE AUTHORS

Kalyan Annamalai is Professor of Mechanical Engineering at Texas A&M. He received his B.S. from Anna University, Chennai, and Ph.D. from the Georgia Institute of Technology, Atlanta. After his doctoral degree, he worked as a Research Associate in the Division of Engineering Brown University, RI, and at AVCO-Everett Research Laboratory, MA. He has taught several courses at Texas A&M including Advanced Thermodynamics, Combustion Science and Engineering, Conduction at the graduate level and Thermodynamics, Heat Transfer, Combustion and Fluid mechanics at the undergraduate level. He is the recipient of the Senior TEES Fellow Award from the College of Engineering for excellence in research, a teaching award from the Mechanical Engineering Department, and a service award from ASME. He is a Fellow of the American Society of Mechanical Engineers, and a member of the Combustion Institute and Texas Renewable Industry Association. He has served on several federal panels. His funded research ranges from basic research on coal combustion, group combustion of oil drops and coal, etc., to applied research on the cofiring of coal, waste materials in a boiler burner and gas fired heat pumps. He has published more that 145 journal and conference articles on the results of this research. He is also active in the Student Transatlantic Student Exchange Program (STEP).

Ishwar K. Puri is Professor of Mechanical Engineering and Chemical Engineering, and serves as Executive Associate Dean of Engineering at the University of Illinois at Chicago. He received his Ph.D. from the University of California, San Diego, in 1987. He is a Fellow of the American Society of Mechanical Engineers. He has lectured nationwide at various universities and national laboratories. Professor Puri has served as an AAAS-EPA Environmental Fellow and as a Fellow of the NASA/Stanford University Center for Turbulence Research. He has been funded to pursue both basic and applied research by a variety of federal agencies and by industry. His research has focused on the characterization of steady and unsteady laminar flames and an understanding of flame and fire inhibition. He has advised more than 20 graduate student theses, and published and presented more than 120 research publications. He has served as an advisor and consultant to several federal agencies and industry. Professor Puri is active in international student educational exchange programs. He has initiated the Student Transatlantic Engineering Program (STEP) that enables engineering students to enhance their employability through innovative international exchanges that involve internship and research experiences. He has been honored for both his research and teaching activities and is the recipient of the UIC COE's Faculty Research Award and the UIC Teaching Recognition Program Award.

NOMENCLATURE*

Symbol	*Description*	*SI*	*English*	*Conversion SI to English*
A	Helmholtz free energy	kJ	BTU	0.9478
A	area	m^2	ft^2	10.764
a	acceleration	$m\ s^{-2}$	$ft\ s^{-2}$	3.281
a	specific Helmholtz free energy	$kJ\ kg^{-1}$	$BTU\ lb_m^{-1}$	0.4299
a	attractive force constant			
$\bar{a}$	specific Helmholtz free energy	$kJ\ kmole^{-1}$	$BTU\ lbmole^{-1}$,	0.4299
$\bar{b}$	body volume constant	$m^3\ kmole^{-1}$	$ft^3\ lbmole^{-1}$	16.018
c	specific heat	$kJ\ kg^{-1}\ K^{-1}$	BTU/lb R	0.2388
COP	Coefficient of performance			
E	energy, (U+KE+PE)	kJ	BTU	0.9478
E_T	Total energy (H+KE+PE)	kJ	BTU	0.9478
e	specific energy	$kJ\ kg^{-1}$	$BTU\ lb_m^{-1}$	0.4299
e_T	methalpy = h + ke + pe	$kJ\ kg^{-1}$	$BTU\ lb_m^{-1}$	0.4299
F	force	kN	lb_f	224.81
f	fugacity	kPa(or bar)	$lb_f\ in^{-2}$	0.1450
G	Gibbs free energy	kJ	BTU	0.9478
g	specific Gibbs free energy (mass basis)	$kJ\ kg^{-1}$	$BTU\ lb_m^{-1}$	0.4299
g	gravitational acceleration	$m\ s^{-2}$	$ft\ s^{-2}$	3.281
g_c	gravitational constant			
$\bar{g}$	Gibbs free energy (mole basis)	$kJ\ kmole^{-1}$	$BTU\ lbmole^{-1}$	0.4299
$\hat{g}$	partial molal Gibb's function,	$kJ\ kmole^{-1}$	$BTU\ lbmole^{-1}$	0.4299
H	enthalpy	kJ	BTU	0.9478
h_{fg}	enthalpy of vaporization	$kJ\ kg^{-1}$	$BTU\ lb_m^{-1}$	0.4299
h	specific enthalpy (mass basis)	$kJ\ kg^{-1}$	$BTU\ lb_m^{-1}$	0.4299
$h_{o,h}$*	ideal gas enthalpy	$kJ\ kg^{-1}$	$BTU\ lb_m^{-1}$	0.4299
I	irreversibility	kJ	BTU	0.9478
I	irreversibility per unit mass	$kJ\ kg^{-1}$	$BTU\ lb_m^{-1}$	0.4299
I	electrical current	amp		
J	Joules' work equivalent of heat	(1 BTU = 778.14 ft lb_f)		
J_k	fluxes for species, heat etc	$kg\ s^{-1}$, kW	$BTU\ s^{-1}$	0.9478
J_k	fluxes for species, heat etc	$kg\ s^{-1}$, kW	$lb\ s^{-1}$	0.4536
K	equilibrium constant			
KE	kinetic energy	kJ	BTU	0.9478
ke	specific kinetic energy	$kJ\ kg^{-1}$	$BTU\ lb_m^{-1}$	0.4299
k	ratio of specific heats			
L	length, height	m	ft	3.281
l	intermolecular spacing	m	ft	3.281
l_m	mean free path	m	ft	3.281
LW	lost work	kJ	BTU	0.9478
LW	lost work	kJ	ft lb_f	737.52
M	molecular weight, molal mass	$kg\ kmole^{-1}$	$lb_m\ lbmole^{-1}$	
m	mass	kg	lb_m	2.2046

* Lower case (lc) symbols denote values per unit mass, lc symbols with a bar (e.g., $\bar{h}$) denote values on mole basis, lc symbols with a caret and tilde (respectively, $\hat{h}$ and $\tilde{h}$) denote values on partial molal basis based on moles and mass, and symbols with a dot (e.g. $\dot{Q}$) denote rates.

Y	mass fraction			
N	number of moles	kmole	lbmole	2.2046
N_{Avag}	Avogadro number	molecules $kmole^{-1}$	molecules $lbmole^{-1}$	0.4536
n	polytropic exponent in PV^n = constant			
P	pressure	$kN\ m^{-2}$	kPa $lb_f\ in^{-2}$	0.1450
PE	potential energy	kJ	BTU	0.9478
pe	specific potential energy			
Q	heat transfer	kJ	BTU	0.9478
q	heat transfer per unit mass	$kJ\ kg^{-1}$	$BTU\ lb^{-1}$	0.4299
q_c	charge			
R	gas constant	$kJ\ kg^{-1}\ K^{-1}$	$BTU\ lb^{-1}\ R^{-1}$	0.2388
$\overline{R}$	universal gas constant	$kJ\ kmole^{-1}\ K^{-1}$	$BTU\ lbmole^{-1}\ R^{-1}$	0.2388
S	entropy	$kJ\ K^{-1}$	$BTU\ R^{-1}$	0.5266
s	specific entropy (mass basis)	$kJ\ kg^{-1}\ K^{-1}$	$BTU\ lb^{-1}\ R^{-1}$	0.2388
$\overline{s}$	specific entropy (mole basis)	$kJ\ kmole^{-1}\ K^{-1}$	$BTU\ lbmole^{-1}\ R^{-1}$	0.2388
T	temperature	°C, K	°F, °R	(9/5)T+32
T	temperature	°C, K	°R	1.8
t	time	s	s	
U	internal energy	kJ	BTU	0.9478
u	specific internal energy	$kJ\ kg^{-1}$	$BTU\ lb^{-1}$	0.4299
$\overline{u}$	internal energy (mole basis)	$kJ\ kmole^{-1}$	$BTU\ lbmole^{-1}$	0.4299
V	volume	m^3	ft^3	35.315
V	volume	m^3	gallon	264.2
V	velocity	$m\ s^{-1}$	$ft\ s^{-1}$	3.281
v	specific volume (mass basis)	$m^3\ kg^{-1}$	$ft^3\ lb_m^{-1}$	16.018
$\overline{v}$	specific volume (mole basis)	$m^3\ kmole^{-1}$	$ft^3\ lbmole^{-1}$	16.018
W	work	kJ	BTU	0.9478
W	work	kJ	$ft\ lb_f$	737.5
w	work per unit mass	$kJ\ kg^{-1}$	$BTU\ lb^{-1}$	0.4299
w	Pitzer factor			
ω	specific humidity	$kg\ kg^{-1}$	$lb_m\ lb_m^{-1}$	
x	quality			
x_k	mole fraction of species k			
Y_k	mass fraction ofspecies k			
z	elevation	m	ft	3.281
Z	compressibility factor			

Greek symbols

$\hat{\alpha}_k$	activity of component k, /f_k			
β_P, β_T,	compressibility	K^{-1}, atm^{-1}	R^{-1}, bar^{-1}	0.555, 1.013
β_s		atm^{-1}	bar^{-1}	1.013
γ_k	activity coefficient, $\hat{\alpha}_k/\hat{\alpha}_k^{id}$			
$\hat{\phi}_k/\phi_k$	Gruneisen constant			
λ	thermal conductivity	$kW\ m^{-1}\ K^{-1}$	$BTU\ ft^{-1}\ R^{-1}$	0.1605
η	First Law efficiency			
η_r	relative efficiency			

ω	specific humidity			
ρ	density	kg m^{-3}	$1b_m$ ft^{-3}	0.06243
ϕ	equivalence ratio, fugacity coefficient			
ϕ	relative humidity,			
Φ	absolute availability(closed system)	kJ	BTU	0.9478
Φ'	relative availability or exergy	kJ kg^{-1}	BTU lb^{-1}	0.4299
ϕ	fugacity coefficient			
J_T	Joule Thomson Coefficient	K bar^{-1}	°R atm^{-1}	1.824
μ	chemical potential	kJ $kmole^{-1}$	BTU $lbmole^{-1}$	0.4299
ν	stoichiometric coefficient			
σ	entropy generation	kJ K^{-1}	BTU R^{-1}	0.2388
Ψ	absolute stream availability	kJ kg^{-1}	BTU lb^{-1}	0.2388
Ψ'	relative stream availability or exergy			

Subscripts

a	air
b	boundary
c	critical
chem	chemical
c.m.	control mass
c.v.	control volume
e	exit
f	flow
f	saturated liquid (or fluid)
f	formation
fg	saturated liquid (fluid) to vapor
g	saturated vapor (or gas)
H	high temperature
I	inlet
inv	inversion
id	ideal gas
iso	isolated (system and surroundings)
L	low temperature
max	maximum possible work output between two given states (for an expansion process)
m	mixture
min	minimum possible work input between two given states
net	net in a cyclic process
p	at constant pressure
p,o	at constant pressure for ideal gas
R	reduced, reservoir
rev	reversible
r	relative pressure, relative volume
s	isentropic work, solid
sf	solid to fluid (liquid)
sh	shaft work
Th	Thermal
TM	Thermo-mechanical
TMC	Thermo-mechanical-chemical
w	wet mixture

v	at constant volume
v,o	at constant volume for ideal gas
v	vapor (Chap. 5)
0 or o	ambient, ideal gas state

Superscripts

(0)	based on two parameters
(1)	Pitzer factor correction
α	alpha phase
β	beta phase
id	ideal mixture
ig	ideal gas
P	liquid
g	gas
l	liquid
res	residual
sat	saturated
o	pressure of 1 bar or 1 atm
-	molal property of k, pure component
^	molal property when k is in a mixure

Mathematical Symbols

$\delta(\ )$	differential of a non-property, e.g., δQ, δW, etc.
d ()	differential of property, e.g., du, dh, dU, etc.
Δ	change in value

Acronyms

CE	Carnot Engine
c.m.	control mass
c.s	control surface
c.v	control volume
ES	Equilibrium state
HE	Heat engine
IPE,ipe	Intermolecular potential energy
IRHE	Irreversible HE
KE	Kinetic energy
ke	kinetic energy per unit mass
LHS	Left hand side
KES	Kessler equation of state
MER	Mechanical energy reservoir
mph	miles per hour
NQS/NQE	non-equilibrium
PC	piston cylinder assembly
PCW	piston cylinder weight assembly
PE	Potential energy
pe	potential energy per unit mass
PR	Peng Robinson
RE, re	Rotational energy
RHE	Reversible HE
RHS	Right hand side
RK	Redlich Kwong

RKS	Redlich Kwong Soave
QS/QE	Quasi-equilibrium
ss	steady state
sf	steady flow
TE, te	translational
TER	Thermal energy reservoir
TM	thermo-mechanical equilibrium
TMC	Thermo-mechanical-chemical equilibrium
uf	uniform flow
us	uniform state
VE,ve	Vibrational energy
VW	Van der Waals

Laws of Thermodynamics in Lay Terminology

First Law: It is impossible to obtain something from nothing, but one may break even

Second Law: One may break even but only at the lowest possible temperature

Third Law: One cannot reach the lowest possible temperature

Implication: It is impossible to obtain something from nothing, so one must optimize resources

The following equations, sometimes called the accounting equations, are useful in the engineering analysis of thermal systems.

Accumulation rate of an extensive property B: dB/dt = rate of B entering a volume ($\dot{B}_i$) – rate of B leaving a volume ($\dot{B}_e$) + rate of B generated in a volume ($\dot{B}_{gen}$) – rate of B destroyed or consumed in a volume ($\dot{B}_{des/cons}$).

Mass conservation: $dm_{cv}/dt = \dot{m}_i - \dot{m}_e$.

First law or energy conservation: $dE_{cv}/dt = \dot{Q} - \dot{W} + \dot{m}_i e_{T,i} - \dot{m}_e e_{T,e}$,

where $e_T = h + ke + pe$, $E = U + KE + PE$, $\delta w_{rev,\ open} = -v\ dP$, $\delta w_{rev,\ closed} = P\ dv$.

Second law or entropy balance equation: $dS_{cv}/dt = \dot{Q}/T_b + \dot{m}_i s_i - \dot{m}_e s_e + \dot{\sigma}_{cv}$,

where $\dot{\sigma}_{cv} > 0$ for an irreversible process and is equal to zero for a reversible process.

Availability balance: $d(E_{cv} - T_o S_{cv})/dt = Q(1 - T_0/T_R) + \dot{m}_i \psi_i - \dot{m}_e \psi_e - \dot{W} - T_o \dot{\sigma}_{cv}$,

where $\psi = (e_T - T_0 s) = h + ke + pe - T_0 s$, and $E = U + KE + PE$.

Third law: $S \to 0$ as $T \to 0$.

CONTENTS

Chapter 1

1. INTRODUCTION

A. IMPORTANCE, SIGNIFICANCE AND LIMITATIONS

Thermodynamics is an engineering science topic,which deals with the science of "motion" (*dynamics*) and/or the transformation of "heat" (*thermo*) and energy into various other energy–containing forms. The flow of energy is of great importance to engineers involved in the design of the power generation and process industries. Examples of analyses based on thermodynamics include:

The transfer or motion of energy from hot gases emerging from a burner to cooler water in a hot–water heater.

The transformation of the thermal energy, i.e., heat, contained in the hot gases in an automobile engine into mechanical energy, namely, work, at the wheels of the vehicle.

The conversion of the chemical energy contained in fuel into thermal energy in a combustor.

Thermodynamics provides an understanding of the nature and degree of energy transformations, so that these can be understood and suitably utilized. For instance, thermodynamics can provide an understanding for the following situations:

In the presence of imposed restrictions it is possible to determine how the properties of a system vary, e.g.,

The variation of the temperature T and pressure P inside a closed cooking pot upon heat addition can be determined. The imposed restriction for this process is the fixed volume V of the cooker, and the pertinent system properties are T and P.

It is desirable to characterize the variation of P and T with volume V in an automobile engine. During compression of air, if there is no heat loss, it can be shown that $PV^{1.4} \approx$ constant (cf. Figure 1).

Inversely, for a specified variation of the system properties, design considerations may require that restrictions be imposed upon a system, e.g.,

A gas turbine requires compressed air in the combustion chamber in order to ignite and burn the fuel. Based on a thermodynamic analysis, an optimal scenario requires a compressor with negligible heat loss (Figure 2a).

During the compression of natural gas, a constant temperature must be maintained. Therefore, it is necessary to transfer heat, e.g., by using cooling water (cf. Figure 2b).

It is also possible to determine the types of processes that must be chosen to make the best use of resources, e.g.,

To heat an industrial building during winter, one option might be to burn natural gas while another might involve the use of waste heat from a power plant. In this case a thermodynamic analysis will assist in making the appropriate decision based on rational scientific bases.

For minimum work input during a compression process, should a process with no heat loss be utilized or should one be used that maintains a constant temperature by cooling the compressor? In a later chapter we will see that the latter process requires the minimum work input.

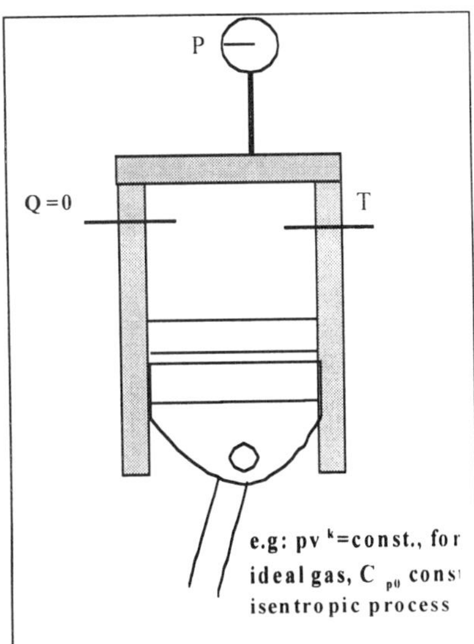

Figure 1: Relation between pressure and volume

The properties of a substance can be determined using the relevant state equations. Thermodynamic analysis also provides relations among nonmeasurable properties such as energy, in terms of measurable properties like P and T (Chapter 7). Likewise, the stability of a substance (i.e., the formation of solid, liquid, and vapor phases) can be determined under given conditions (Chapter 10).

Information on the direction of a process can also be obtained. For instance, analysis shows that heat can only flow from higher temperatures to lower temperatures, and chemical reactions under certain conditions can proceed only in a particular direction (e.g., under certain conditions charcoal can burn in air to form CO and CO_2, but the reverse process of forming charcoal from CO and CO_2 is not possible at those conditions).

B. LIMITATIONS OF THERMODYNAMICS

It is not possible to determine the rates of transport processes using thermodynamic analyses alone. For example, thermodynamics demonstrates that heat flows from higher to lower temperatures, but does not provide a relation for the heat transfer rate. The heat conduction rate per unit area can be deduced from a relation familiarly known as Fourier's law, i.e.,

$$\dot{q}'' = \text{Driving potential} \div \text{Resistance} = \Delta T/R_H, \quad (1)$$

where ΔT is the driving potential or temperature difference across a slab of finite thickness, and R_H denotes the thermal resistance. The Fourier law cannot be deduced simply with knowledge of thermodynamics. Rate processes are discussed in texts pertaining to heat, mass and momentum transport.

1. Review

a. *System and Boundary*

A system is a region containing energy and/or matter that is separated from its surroundings by arbitrarily imposed walls or boundaries.

A *boundary* is a closed surface surrounding a system through which energy and mass may enter or leave the system. *Permeable* and process boundaries allow mass transfer to occur. Mass transfer cannot occur across *impermeable* boundaries. A *diathermal* boundary allows heat transfer to occur across it as in the case of thin metal walls. Heat transfer cannot occur across the *adiabatic* boundary. In this case the boundary is impermeable to heat flux, e.g., as in the case of a Dewar flask.

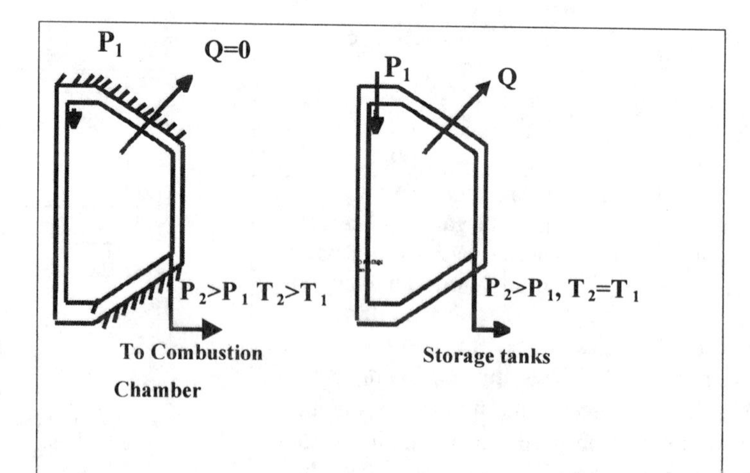

Figure 2: (a) Compression of natural gas for gas turbine applications; (b) Compression of natural gas for residential applications.

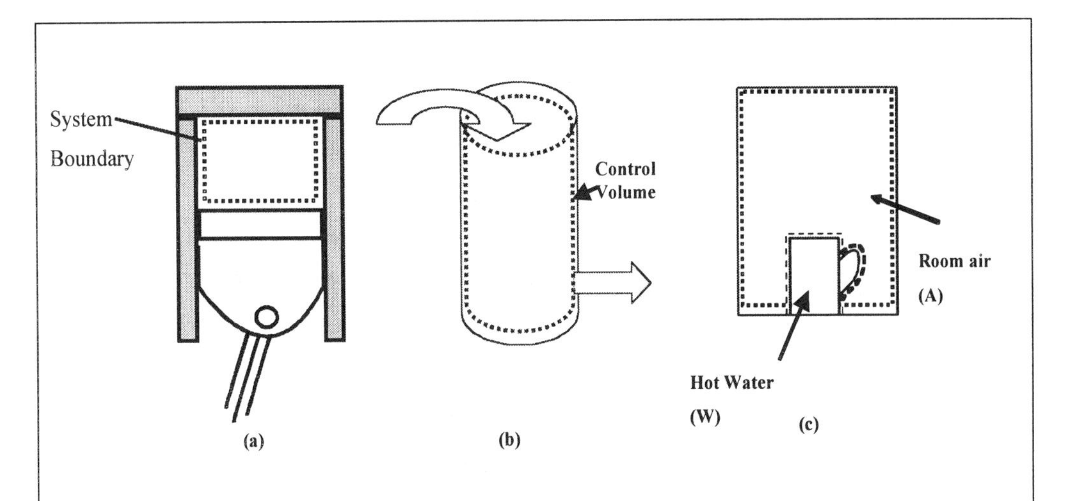

Figure 3. Examples of: (a) Closed system. (b) Open system (filling of a water tank with drainage at the bottom). (c) Composite system.

A *moveable/deforming* boundary is capable of performing "boundary work".
No boundary work transfer can occur across a *rigid* boundary. However energy transfer can still occur via shaft work, e.g., through the stirring of fluid in a blender.
A *simple* system is a homogeneous, isotropic, and chemically inert system with no external effects, such as electromagnetic forces, gravitational fields, etc.
Surroundings include everything outside the system (e.g. dryer may be a system; but the surroundings are air in the house + lawn + the universe)
An *isolated* system is one with rigid walls that has no communication (i.e., no heat, mass, or work transfer) with its surroundings.
A *closed* system is one in which the system mass cannot cross the boundary, but energy can, e.g., in the form of heat transfer. Figure 3a contains a schematic diagram of a closed system consisting of a closed–off water tank. Water may not enter or exit the system, but heat can . A philosophical look into closed system is given in Figure 4a.
An *open* system is one in which mass can cross the system boundary in addition to energy (e.g., as in Figure 3b where upon opening the valves that previously closed off the water tank, a pump now introduces additional water into the tank, and some water may also flow out of it through the outlet).
A *composite* system consists of several subsystems that have one or more internal constraints or restraints. The schematic diagram contained in Figure 3c illustrates such a system based on a coffee (or hot water) cup placed in a room. The subsystems include water (W) and cold air (A)

b. *Simple System*

A *simple* system is one which is macroscopically homogeneous and isotropic and involves a single work mode. The term macroscopically homogeneous implies that properties such as the density ρ are uniform over a large dimensional region several times larger than the mean free path (ℓ_m) during a relatively large time period, e.g., 10^{-6} s (which is large compared to the intermolecular collision time that, under standard conditions, is approximately 10^{-15} s, as we will discuss later in this chapter). Since,

$$\rho = \text{mass} \div \text{volume}, \tag{2}$$

where the volume $V \gg \ell_m^3$, the density is a macroscopic characteristic of any system.

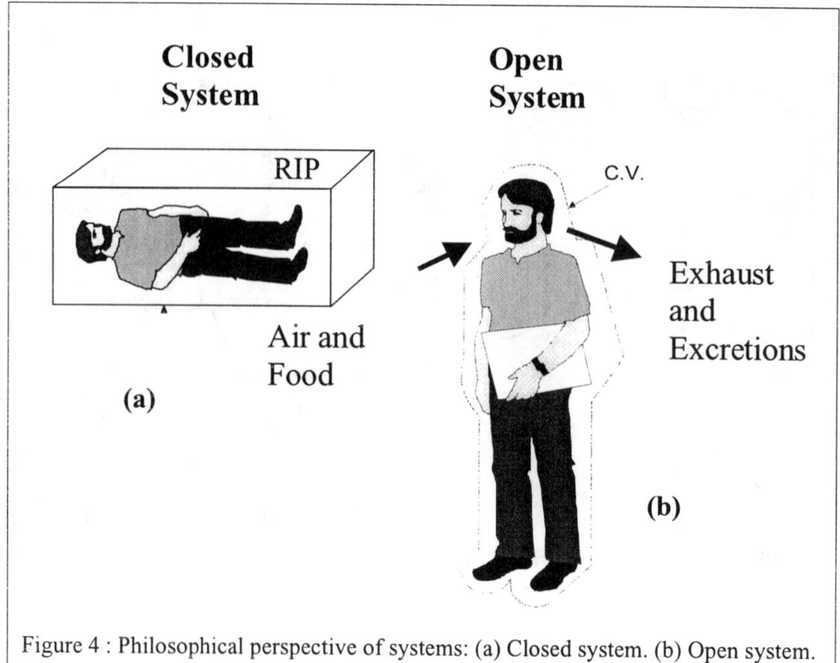

Figure 4 : Philosophical perspective of systems: (a) Closed system. (b) Open system.

An *isotropic* system is one in which the properties do not vary with direction, e.g., a cylindrical metal block is homogeneous in terms of density and isotropic, since its thermal conductivity is identical in the radial and axial directions.

A simple *compressible* system utilizes the work modes of compression and/or expansion, and is devoid of body forces due to gravity, electrical and magnetic fields, inertia, and capillary effects. Therefore, it involves only volumetric changes in the work term.

c. *Constraints and Restraints*

Constraints and restraints are the barriers within a system that prevent some changes from occurring during a specified time period.

A *thermal* constraint can be illustrated through a closed and insulated coffee mug. The insulation serves as a thermal constraint, since it prevents heat transfer.

An example of a *mechanical* constraint is a piston–cylinder assembly containing compressed gases that is prevented from moving by a fixed pin. Here, the pin serves as a mechanical constraint, since it prevents work transfer. Another example is water storage behind a dam which acts as a mechanical constraint. A *composite* system can be formulated by considering the water stores behind a dam and the low–lying plain ground adjacent to the dam.

A *permeability* or *mass* constraint can be exemplified by volatile naphthalene balls kept in a plastic bag. The bag serves as a non–porous impermeable barrier that restrains the mass transfer of naphthalene vapors from the bag. Similarly, if a hot steaming coffee mug is capped with a rigid non–porous metal lid, heat transfer is possible whereas mass transfer of steaming vapor into the ambient is prevented.

A *chemical* constraint can be envisioned by considering the reaction of the molecular nitrogen and oxygen contained in air to form NO. At room temperature N_2 and O_2 do not react at a significant rate and are virtually inert with respect to each other, since a chemical constraint is present which prevents the chemical reaction of the two species from occurring. (Non–reacting mixtures are also referred to as inert mixtures.) The chemical constraint is an activation energy, which is the energy required by a set of reactant species to chemically react and form products. A substance which prevents the chemical reaction from occurring is a chemical restraint, and is referred to as an anti–catalyst, while catalysts (such as platinum in a catalytic converter which converts carbon monoxide to carbon dioxide at a rapid rate) promote chemical reactions (or overcome the chemical restraint).

d. *Composite System*

A *composite* system consists of a combination of two or more subsystems that exist in a state of constrained equilibrium. Using a cup of coffee in a room as an analogy for a composite system, the coffee cup is one subsystem and room air another, both of which might exist at different temperatures. The composite system illustrated in Figure 3c consists of two subsystems hot water (W) and air (A) under constraints, corresponding to different temperatures.

e. *Phase*

A region within which all properties are uniform consists of a distinct *phase*. For instance, solid ice, liquid water, and gaseous water vapor are separate phases of the same chemical species. A portion of the Arctic Ocean in the vicinity of the North Pole is frozen and consists of ice in a top layer and liquid water beneath it. The atmosphere above the ice contains some water vapor. The density of water in each of these three layers is different, since water exists in these layers separately in some combination of three (solid, liquid, and gaseous) phases. Although a vessel containing immiscible oil and water contains only liquid, there are two phases present, since $\rho_{oil} \neq \rho_{water}$. Similarly, in metallurgical applications, various phases may exist within the solid state, since the density may differ over a solid region that is at a uniform temperature and pressure.

In liquid mixtures that are miscible at a molecular level (such as those of alcohol and water for which molecules of one species are uniformly intermixed with those of the other), even though the mixture might contain several chemical components, a single phase exists,

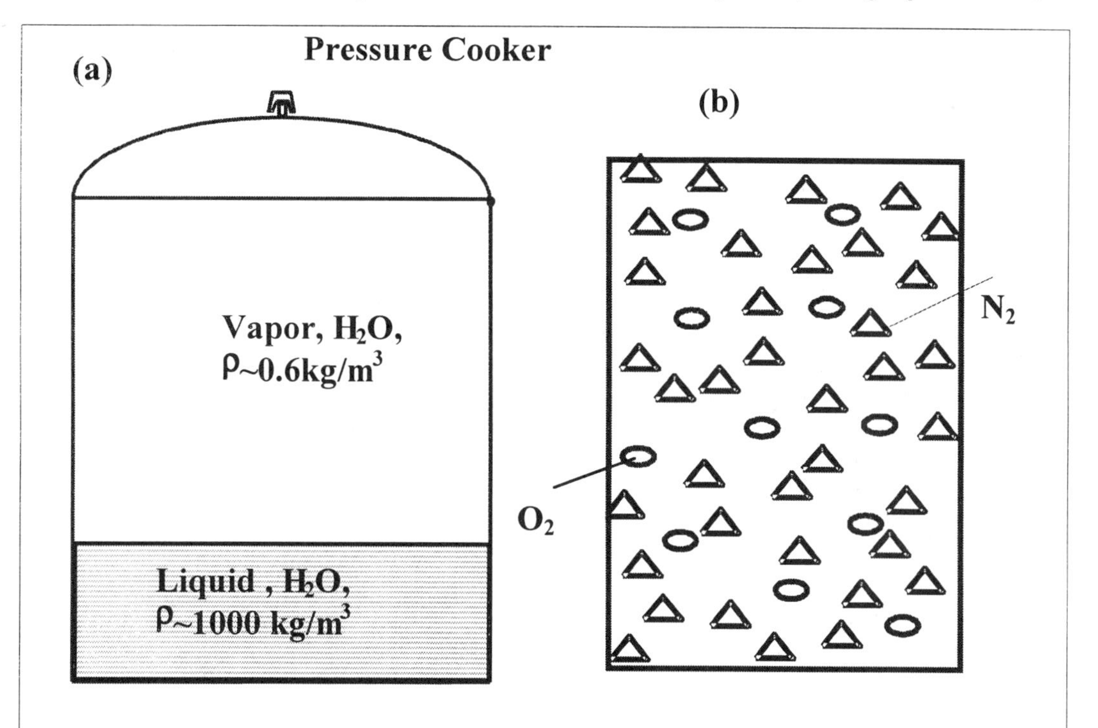

Figure 5 : (a) Pure substance illustrated by the presence of water and its vapor in a pot; (b) A homogeneous system in which each O_2 molecule is surrounded by about four N_2 molecules.

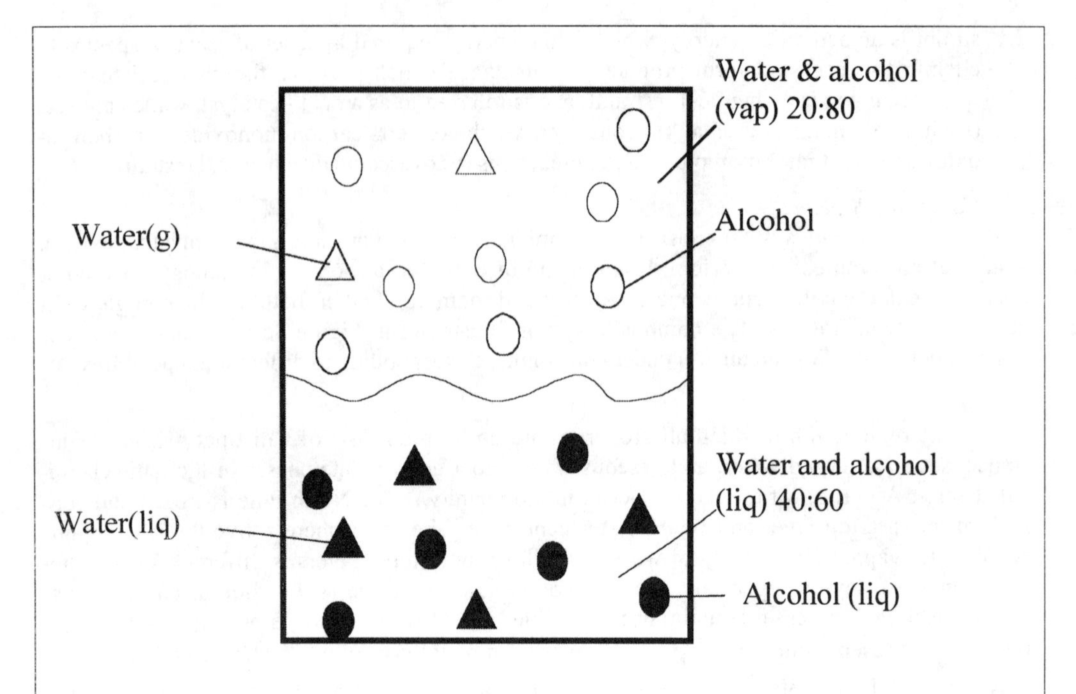

Figure 6: A heterogeneous system consisting of binary fluid mixtures. The liquid phase contains a water–alcohol mixture in the ratio 40:60, and the vapor phase water and alcohol are in the ratio 20:80.

since the system properties are macroscopically uniform throughout a given volume. Air, for example, consists of two major components (molecular oxygen and nitrogen) that are chemically distinct, but constitute a single phase, since they are well–mixed.

f. Homogeneous

A system is *homogeneous* if its chemical composition and properties are macroscopically uniform. All single–phase substances, such as those existing in the solid, liquid, or vapor phases, qualify as homogeneous substances. Liquid water contained in a cooking pot is a homogeneous system (as shown in Figure 5a), since its composition is the same everywhere, and, consequently, the density within the liquid water is uniform. However, volume contained in the entire pot does not qualify as a homogeneous system even though the chemical composition is uniform, since the density of the water in the vapor and liquid phases differs.

The water contained in the cooker constitutes two phases, liquid and vapor. The molecules are closely packed in the liquid phase resulting in a higher density relative to vapor, and possess lower energy per unit mass compared to that in the vapor phase.

Single–phase systems containing one or more chemical components also qualify as homogeneous systems. For instance, as shown in Figure 5b, air consists of multiple components but has spatially macroscopic uniform chemical composition and density.

g. Pure Substance

A *pure* substance is one whose chemical composition is spatially uniform. At any temperature the chemical composition of liquid water uniformly consists of H_2O molecules. On the other hand, the ocean with its salt–water mixture does not qualify as a pure substance, since it contains spatially varying chemical composition. Ocean water contains a nonuniform fraction of salt depending on the depth. Multiphase systems containing single chemical components consist of pure substances, e.g., a mixture of ice, liquid water, and its vapor, or the liquid water and vapor mixture in the cooking pot example (cf. Figure 5a). Multicomponent single–phase systems also consist of pure substances, e.g., air (cf. Figure 5b).

Heterogeneous systems may hold multiple phases (e.g., as in Figure 5a with one component) and multicomponents in equilibrium (e.g., Figure 6 with two components). Well–mixed single–phase systems are simple systems although they may be multicomponent, since they are macroscopically homogeneous and isotropic, e.g., air. The vapor–liquid system illustrated in Figure 6 does not qualify as a pure substance, since the chemical composition of the vapor differs from that of the liquid phase.

h. Amount of Matter and Avogadro Number

Having defined systems and the types of matter contained within them (such as a pure, single phase or multiphase, homogeneous or heterogeneous substance), we will now define the units employed to measure the amount of matter contained within systems.

The *amount of matter* contained within a system is specified either by a molecular number count or by the total mass. An alternative to using the number count is a mole unit. Matter consisting of 6.023×10^{26} molecules (or Avogadro number of molecules) of a species is called one kmole of that substance. The total mass of those molecules (i.e., the mass of 1 kmole of the matter) equals the molecular mass of the species in kg. Likewise, 1 lb mole of a species contains its molecular mass in lb. For instance, 18.02 kg of water corresponds to 1 kmole, 18.02 g of water contains 1 gmole, while 18.02 lb mass of water has 1 lb mole of the substance. Unless otherwise stated, throughout the text the term mole refers to the unit kmole.

i. Mixture

A system that consists of more than a single component (or species) is called a mixture. Air is an example of a mixture containing molecular nitrogen and oxygen, and argon. If N_k denotes the number of moles of the k–th species in a mixture, the mole fraction of that species X_k is given by the relation

$$X_k = N_k/N, \qquad (3)$$

where $N = \Sigma N_k$ is the total number of moles contained in the mixture. A mixture can also be described in terms of the species mass fractions mf_k as

$$Y_k = m_k/m, \qquad (4)$$

where m_k denotes the mass of species k and m the total mass. Note that $m_k = N_k M_k$, with the symbol M_k representing the molecular weight of any species k. Therefore, the mass of a mixture

$$m = \Sigma N_k M_k.$$

The molecular weight of a mixture M is defined as the average mass contained in a kmole of the mixture, i.e.,

$$M = m/N = \Sigma N_k M_k/N = \Sigma X_k M_k \qquad (5)$$

---

a. Example 1

Assume that a vessel contains 3.12 kmoles of N_2, 0.84 kmoles of O_2, and 0.04 kmoles of Ar. Determine the constituent mole fractions, the mixture molecular weight, and the species mass fractions.

Solution

Total number of moles N = 3.12 + 0.84 + 0.04 = 4.0 kmoles

$x_{N_2} = N_{N_2}/N = 3.12/4 = 0.78$. Similarly, $x_{N_2} = 0.21$, and $x_{Ar} = 0.01$.

The mixture molecular weight can be calculated using Eq. 5, i.e.,

$M = 0.78\times28 + 0.21\times32 + 0.01\times39.95 = 28.975$ kg per kmole of mixture.

The total mass m = 3.12×28.02 + 0.84×32 + 0.04×39.95 = 115.9 kg, and mass fractions are:

$Y_{N2} = m_{N_2}/m = 3.12\times28.02/115.9 = 0.754$. Similarly $Y_{O2} = 0.232$, and $Y_{Ar} = 0.0138$.

Remark

The mixture of N_2, O_2, and Ar in the molal proportion of 78:1:21 is representative of the composition of air (see the Appendix to this chapter).

When dealing specifically with the two phases of a multicomponent mixture, e.g., the alcohol–water mixture illustrated in Figure 6, we will denote the mole fraction in the gaseous phase by $X_{k,g}$ (often simply as X_k) and use $X_{k,\ell}$ $X_{k,s}$ to represent the liquid and solid phase mole fraction, respectively.

At room temperature (of 20°C) it is possible to dissolve only up to 36 g of salt in 100 g of water, beyond which the excess salt settles. Therefore, the mass fraction of salt in water at its solubility limit is 27%. At this limit a one–phase saline solution exists with a uniform density of 1172 kg m^{-3}. As excess salt is added, it settles, and there are now two phases, one containing solid salt (ρ = 2163 kg m^{-3}) and the other a liquid saline solution (ρ = 1172 kg m^{-3}). (Recall that a phase is a region within which the properties are uniform.)

Two liquids can be likewise mixed at a molecular level only within a certain range of concentrations. If two miscible liquids, 1 and 2, are mixed, up to three phases may be formed in the liquid state: (1) a miscible phase containing liquids 1 and 2 with $\rho = \rho_{mixture}$, (2) that containing pure liquid 1 ($\rho = \rho_1$), and (3) pure liquid 2 ($\rho = \rho_2$). A more detailed discussion is presented in Chapter 8.

j. Property

Thus far we have defined systems, and the type and amount of matter contained within them. We will now define the properties and state of matter contained within these systems.

A *property* is a characteristic of a system, which resides in or belongs to it, and it can be assigned only to systems in equilibrium. Consider an illustration of a property the temperature of water in a container. It is immaterial how this temperature is reached, e.g., either through solar radiation, or electrical or gas heating. If the temperature of the water varies from, say, 40°C near the boundary to 37°C in the center, it is not single–valued since the system is not in equilibrium, it is, therefore, not a system property. Properties can be classified as follows:

Primitive properties are those which appeal to human senses, e.g., T, P, V, and m.

Derived properties are obtained from primitive properties. For instance, the units for force (a derived property) can be obtained using Newton's second law of motion in terms of the fundamental units of mass, length and time. Similarly, properties such as enthalpy H, entropy S, and internal energy U, which do not directly appeal to human senses, can be derived in terms of primitive properties like T, P and V using thermodynamic relations (Chapter VII). (Even primitive properties, such as volume V, can be derived using state relations such as the ideal gas law V = mRT/P.)

Intensive properties are independent of the extent or size of a system, e.g., P (kN m^{-2}), v (m^3 kg^{-1}), specific enthalpy h (kJ kg^{-1}), and T (K).

Extensive properties depend upon system extent or size, e.g., m (kg), V (m^3), total enthalpy H (kJ), and total internal energy U (kJ).

An *extrinsic* quantity is independent of the nature of a substance contained in a system (such as kinetic energy, potential energy, and the strength of magnetic and electrical fields).

An *intrinsic* quantity depends upon the nature of the substance (examples include the internal energy and density).

Intensive and extensive properties require further discussion. For example, consider a vessel of volume 10 m^3 consisting of a mixture of 0.32 kmoles of N_2, and 0.08 kmoles of O_2 at 25°C (system A), and another 15 m^3 vessel consisting of 0.48 kmoles of N_2 and 0.12 kmoles of O_2 at the same temperature (system B). If the boundary separating the two systems is removed, the total volume becomes 25 m^3 containing 0.8 total moles of N_2, and 0.2 of O_2. Properties which are additive upon combining the two systems are extensive, e.g., V, N, but intensive properties such as T and P do not change. Likewise the mass per unit volume (density) does not change upon combining the two systems, even though m and V increase. The kinetic energy of two moving cars is additive $m_1 V_1^2/2 + m_2 V_2^2/2$ as is the potential energy of two masses at different heights (such as two ceiling fans of mass m_1 and m_2 at respective heights Z_1 and Z_2 with a combined potential energy $m_1 g Z_1 + m_2 g Z_2$). Similarly, other forms of energy are additive.

An extensive property can be converted into an intensive property provided it is distributed uniformly throughout the system by determining its value per unit mass, unit mole, or unit volume. For example, the specific volume v = V/m (in units of m^3 kg^{-1}) or V/N (in terms of m^3 $kmole^{-1}$). The density ρ = m/V is the inverse of the mass–based specific volume. We will use lower case symbols to denote specific properties (e.g.: v, $\bar{v}$, u, and $\bar{u}$, etc.). The overbars denote mole–based specific properties. The exceptions to the lower case rule are temperature T and pressure P. Furthermore we will represent the differential of a property as d(*property*), e.g., dT, dP, dV, dv, dH, dh, dU, and du. (A mathematical analogy to an exact differential will be discussed later.)

k. State

The condition of a system is its *state*, which is normally identified and described by the observable primitive properties of the system. The system state is specified in terms of its properties so that it is possible to determine changes in that state during a process by monitoring these properties and, if desired, to reproduce the system. For example, the normal state of an average person is usually described by a body temperature of 37°C. If that temperature rises to 40°C, medication might become necessary in order to return the "system" to its normal state. Similarly, during a hot summer day a room might require air conditioning. If the room temperature does not subsequently change, then it is possible to say that the desired process, i.e., air conditioning, did not occur. In both of the these examples, temperature was used to described an aspect of the system state, and temperature change employed to observe a process. Generally, a set of properties, such as T, V, P, N_1, N_2, etc., representing system characteristics define the state of a given system.

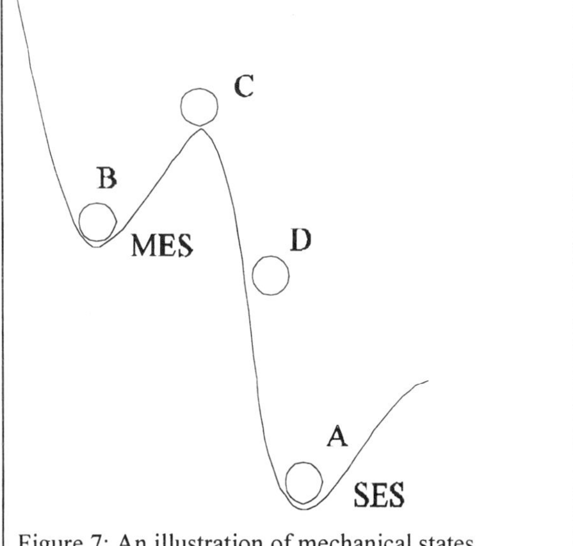

Figure 7: An illustration of mechanical states.

Figure 7 illustrates the mechanical analogy to various thermodynamic states in a gravitational field. *Equilibrium states* can be characterized as being stable, metastable, and unstable, depending on their response to a perturbation. Positions A, B and C are at an equilibrium state, while D represents a non-equilibrium position. Equilibrium states can be classified as follows:

A *stable equilibrium state* (SES), is associated with the lowest energy, and which, following perturbation, returns to its original state (denoted by A in Figure 7). A closed system is said to achieve a state of stable equilibrium when changes occur in its properties regardless of time, and which returns to its original state after being subjected to a small perturbation. The partition of a system into smaller sub–systems has a negligible effect on the SES.

If the system at state B in Figure 7 is perturbed either to the left or right, it reverts back to its original position. However, it appears that a large perturbation to the right is capable of lowering the system to state A. This is an example of a *metaequilibrium state* (MES). It is known that water can be superheated to 105°C at 100 KPa without producing vapor bubbles which is an example of a metastable state, since any impurities or disturbances introduced into the water can cause its sudden vaporization (as discussed in Chapter 10).

A slight disturbance to either side of an *unstable* equilibrium *state* (UES) (e.g., state C of Figure 7) will cause a system to move to a new equilibrium state. (Chapter 10 discusses the thermodynamic analog of stability behavior.)

The system state cannot be described for a *nonequilibrium* (NE) position, since it is transient. If a large weight is suddenly placed upon an insulated piston–cylinder system that contains an ideal compressible fluid, the piston will move down rapidly and the system temperature and pressure will continually change during the motion of the piston. Under these transient circumstances, the state of the fluid cannot be described.

Furthermore, various equilibrium conditions can occur in various forms:

Mechanical equilibrium prevails if there are no changes in pressure. For example, helium constrained by a balloon is in mechanical equilibrium. If the balloon leaks or bursts open, the helium pressure will change.

Thermal equilibrium exists if the system temperature is unchanged.

Phase equilibrium occurs if, at a given temperature and pressure, there is no change in the mass distribution of the phases of a substance, i.e., if the physical composition of the system is unaltered. For instance, if a mug containing liquid water is placed in a room with both the liquid water and room air being at the same temperature and the liquid water level in the mug is unchanged, then the water vapor in the room and liquid water in the mug are in phase equilibrium. A more rigorous definition will be presented later in Chapters 3, 7, and 9.

Chemical equilibrium exists if the chemical composition of a system does not change. For example, if a mixture of H_2, O_2, and H_2O of arbitrary composition is enclosed in a vessel at a prescribed temperature and pressure, and there is no subsequent change in chemical composition, the system is in chemical equilibrium. Note that the three species are allowed to react chemically, the restriction being that the number of moles of a species that are consumed must equal that which are produced, i.e., there is no net change in the concentration of any species (this is discussed in Chapter 12).

The term *thermodynamic state* refers only to equilibrium states. Consider a given room as a system in which the region near the ceiling consists of hot air at a temperature T_B due to relatively hot electrical lights placed there, and otherwise cooler air at a temperature of T_A elsewhere. Therefore, a single temperature value cannot be assigned for the entire system, since it is not in a state of thermal equilibrium. However, a temperature value can be specified separately for the two subsystems, since each is in a state of internal equilibrium.

l. Equation of State

Having described systems, and type and state of matter contained within them in terms of properties, we now explore whether all of the properties describing a state are independent or if they are related.

A thermodynamic state is characterized by macroscopic properties called *state variables* denoted by $x_1, x_2, \ldots, x_n$ and F. Examples of state variables include T, P, V, U, H, etc. It has been experimentally determined that, in general, at least one state variable, say F, is not independent of $x_1, x_2, \ldots, x_n$, so that

$$F = F(x_1, x_2, \ldots, x_n). \quad (6)$$

Equation (6) is referred to as a *state postulate* or *state equation.* The number of independent variables x_1, x_2, ... ,x_n (in this case there are n variables) is governed by the laws of thermodynamics. Later, in Chapter 3, we will prove this generalized state equation. For example, if x_1 = T, x_2 = V, x_3 = N, and F = P, then

$$P = P\,(T, V, N).$$

For an ideal gas, the functional form of this relationship is given by the ideal gas law, i.e.,

$$P = N\bar{R}T/V, \qquad (7)$$

where $\bar{R}$ is known as the universal gas constant, the value of which is 8.314 kJ kmole^{-1} K^{-1}. The universal gas constant can also be deduced from the Boltzmann's constant, which is the universal constant for one molecule of matter (defined as $k_B = \bar{R}/N_{Avog}$ = 1.38×10^{-28} kJ molecule^{-1} K^{-1}). Defining the molar specific volume = V/N, we can rewrite Eq. (7) as

$$P = \bar{R}T/\bar{v}. \qquad (8)$$

Equation (8) (stated by J. Charles and J. Gay Lussac in 1802) is also called *an intensive equation of state*, since the variables contained in it are intensive. The ideal gas equation of state may be also expressed in terms of mass units after rewriting Eq. (7) in the form

$$P = (m/M)\bar{R}T/V = mRT/V \qquad (9)$$

where $R = \bar{R}/M$. Similarly,

$$P = RT/v \qquad (10)$$

Equation (10) demonstrates that P = P (T,v) for an ideal gas and is known once T and v are prescribed. We will show later that this is true for all single–component single–phase fluids.

Consider the composite system containing separate volumes of hot and cold air (assumed as ideal gas) at temperatures T_A and T_B, respectively. We cannot calculate the specific volume for the entire system using Eq. (10), since the temperature is not single valued over the entire system. For a nonequilibrium system, state equations for the entire system are meaningless. However, the system can be divided into smaller subsystems A and B, with each assumed to be in a state of internal equilibrium. State equations are applicable to subsystems that are in local equilibrium.

m. Standard Temperature and Pressure

Using Eq. (8), it can be shown that the volume of 1 kmole of an ideal gas at *standard temperature and pressure* (*STP*), given by the conditions T = 25°C (77°F) and P = 1 bar (≈ 1 atm) is 24.78 m^3 kmole^{-1} (392 ft^3 lb mole^{-1},) This volume is known as a standard cubic meter (SCM) or a standard cubic foot (SCF). See the attached tables for the values of volume at various STP conditions.

n. Partial Pressure

The equation of state for a mixture of ideal gases can be generalized if the number of moles in Eq. (7) is replaced by

$$N = N_1 + N_2 + N_3 + ... = \Sigma N_k, \qquad (11)$$

so that Eq. (7) transforms into

$$P = N_1\bar{R}T/V + N_2\bar{R}T/V + ... \qquad (12)$$

The first term on the right hand side of Eq. (12) is to be interpreted as the component pressure (also called the partial pressure for an ideal gas mixture, this is the pressure that would have been exerted by component 1 if it alone had occupied the entire volume). Therefore,

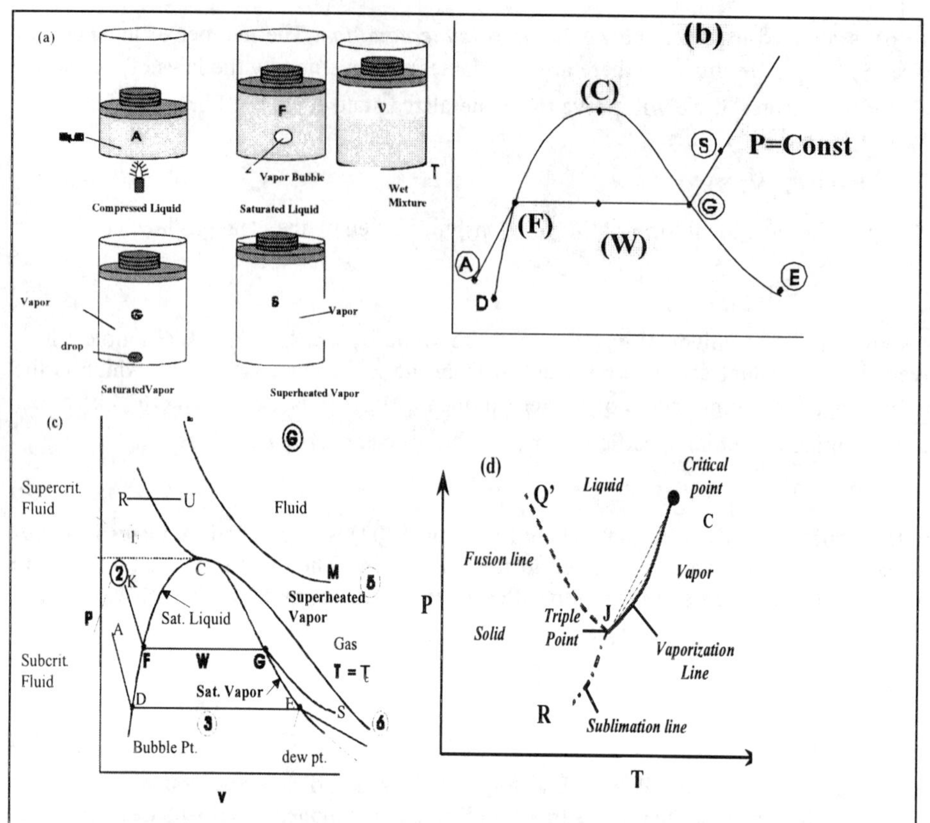

Figure 8: a. Isobaric heating of a fluid; b. Pressure–volume diagram; c. Some terminology used to describe liquid–vapor regimes; d. A schematic illustration of a generalized phase diagram.

$$p_l = N_l \bar{R}T/P = X_l N \bar{R}T/V = X_l P. \quad (13)$$

Assuming air at a standard pressure of 101 kPa to consist of 21 mole percent of molecular oxygen, the pressure exerted by O_2 molecules alone p_{O_2} is 0.21×101 = 21.21 kPa. Further details of mixtures and their properties will be discussed in Chapters 8 and 9.

o. Process

A process occurs when a system undergoes a change of state (i.e., its properties change) with or without interaction with its surroundings. A *spontaneous* process changes the state of a system without interacting with its environment. For instance, if a coffee cup is placed in an insulated rigid room, the properties of the composite system (e.g., T_{air}, T_{coffee}) change with time even though there is no interaction of the room with the outside environment through work or non–work (e.g., heat.) energy transfer.

During an *isothermal* process there are no temperature changes, i.e., dT = 0. Likewise, for an *isobaric* process the pressure is constant (dP = 0), and volume remains unchanged during an isometric process (dV = 0). Note that if the temperature difference during a process $\Delta T = T_f - T_{in} = 0$, this does not necessarily describe an isothermal process, since it is possible that the system was heated from an initial temperature T_{in} to an intermediate temperature T_{int} (> T_{in}) and cooled so that the final temperature $T_f = T_{in}$. An *adiabatic* process is one during which there is no heat transfer, i.e., when the system is perfectly insulated.

If the final state is identical to the initial system state, then the process is *cyclical*. Otherwise, it is *noncyclical*.

p. *Vapor–Liquid Phase Equilibrium*

Having defined systems, matter, and some relations among system properties (including those for ideal gases), we now discuss various other aspects of pure substances. Consider a small quantity of liquid water contained in a piston–cylinder assembly as illustrated in Figure 8a.

Assume the system temperature and pressure (T,P) to be initially at standard conditions. State A shown in Figure 8a is the compressed liquid state (illustrated on the pressure–volume diagram of Figure 8b) corresponding to sub–cooled liquid. If the water is heated, a bubble begins to form once the temperature reaches 100°C (at the bubble point or the saturated liquid state, illustrated by point F in Figure 8b). This temperature is called the saturation temperature or boiling point temperature at the prescribed pressure. The specific volume of the liquid at this saturated state is denoted by v_f. As more heat is added, the two liquid and vapor phases coexist at state W (in the two–phase or wet region, $v > v_f$). The ratio of vapor (subscript g) to total mass m is termed quality ($x = m_g/m$). As more heat is added, the liquid completely converts to vapor at state G which is called the saturated vapor state or the dew point. Upon further heat addition at the specified pressure, the system temperature becomes larger than the saturation temperature, and enters state S which is known as the superheated vapor state.

In the context of Figure 8a and b, the symbol A denotes the subcooled liquid or compressed liquid state; F the saturated liquid state (it is usual to use the subscript f to represent the system properties of the fluid at that state) for which the quality $x = 0$. W is the wet state consisting of a mixture of liquid and vapor, G a saturated vapor state (denoted with the subscript g, $x = 1$), and S represents the superheated vapor

The curve AFWGS in Figure 8b describes an isobaric process. If the system pressure for the water system discussed in the context of Figure 8a is changed to, say, 10, 100, 200, and 30,000 kPa, these pressures correspond to different saturation temperatures, liquid and vapor volumes, and isobaric process curves. Joining all possible curves for saturated liquid and vapor states it is possible to obtain the saturated liquid and vapor curves shown in Figure 8b which intersect at the critical point C that corresponds to a distinct critical temperature and pressure T_c and P_c.

The Table A-1 contains critical data for many substances while Tables A-4 contain information regarding the properties of water along the saturated vapor and liquid curves, and in the superheated vapor regions and Table A-5 contains same information for R-134a. A representative P–v diagram and various liquid–vapor regime terminology are illustrated in Figure 8c as follows:

If the vapor temperature $T > T_c$, and its pressure $P < P_c$ the vapor is called a *gas*. A gas is a fluid that, upon isothermal compression, does not change phase (i.e., from gas to liquid as in Curve LM, $T > T_c$). Otherwise, the fluid is called a vapor (a fluid in a vapor state may be compressed to liquid through a process such as along the curve SGWFK).

Substances at $P>P_c$ and $T>T_c$ are generally referred to as fluid (e.g., point U of Figure 8c).

If a supercritical fluid is isobarically heated there is no change of phase (e.g., line RU of Figure 8c).

Above the critical point, i.e., when $P>P_c$, and $T>T_c$ the vapor is called a fluid which exists in a *supercritical state*. If $P > P_c$, and $T < T_c$ it is referred to as a *supercritical fluid* (region (1), Figure 8c).

A *subcritical fluid* is one for which $P < P_c$, $T < T_c$, and $v < v_f$.

Both liquid and vapor are contained in the *two–phase dome* where $P < P_c$, $T < T_c$, and $v_f < v < v_g$.

If $P < P_c$, $T < T_c$, and $v > v_g$, the vapor is in *superheated state*.

The saturated liquid line of Figure 8c joins those states that have been denoted by the subscript F in the context of Figure 8a, e.g., points D, F, etc. Likewise, the saturated vapor line joins the states represented by the subscript g, i.e., the points G, E, etc. of Figure 8c. At the critical point C the saturated liquid and vapor states are identical. Upon plotting the pressure with respect to the saturation temperatures T_{sat} along the saturated curve, the phase diagram of Figure 8d is obtained. In that figure the vaporization curve is represented by JC, JQ is the melting curve for most solids (JQ′ is the analogous melting curve for ice), and JR represents the sublimation curve. The intersection J of the curves JC and JQ′ is called the triple point at which all three phases co–exist. For water, this point is characterized by P = 0.0061 bar (0.006 atm) and T = 273.16 K (491.7 R), whereas for carbon the analogous conditions are T ≈ 3800 K, P ≈ 1 bar. For triple points of other substances see Table A-2.

C. MATHEMATICAL BACKGROUND

Thus far, we have discussed the basic terminology employed in thermodynamics. We now briefly review the mathematical background required for expressing the conservation equations in differential form (that will be discussed in Chapters. 2, 3 and 4), equilibrium criteria (Chapter 3), conversion of the state equations from one form to another (Chapter 5), Maxwell's relations (Chapter 7), the Euler equation (Chapters 3 and 8) stability behavior of fluids (Chapter 10 and entropy maximization and Gibb's function minimization (Chapters 3 and 12).

1. Explicit and Implicit Functions and Total Differentiation

If P is a known function of T and v, the explicit function for P is

$$P = P(v,T), \tag{14}$$

and its total differential may be written in the form

$$dP = \left(\frac{\partial P}{\partial v}\right)_T dv + \left(\frac{\partial P}{\partial T}\right)_v dT. \tag{15}$$

Consider the P, T, v relation

$$P = RT/(v-b) - a/v^2, \tag{16}$$

where a and b are constants. Equation (16) is explicit with respect to P, since it is an explicit function of T and v. On the other hand, v cannot be explicitly solved in terms of P and T, and, hence, it is an implicit function of those variables. The total differential is useful in situations that require the differential of an implicit function, as illustrated below.

---

b. Example 2

If state equation is expressed in the form

$$P = RT/(v-b) - a/v^2, \tag{A}$$

find an expression for $(\partial v/\partial T)_P$, and for the isobaric thermal expansion coefficient $\beta_P = (1/v)(\partial v/\partial T)_P$.

Solution

For given values of T and v, and the known parameters a and b, values of P are unique (P is also referred to as a point function of T and v). Using total differentiation

$$dP = (\partial P/\partial v)_T\, dv + (\partial P/\partial T)_v\, dT. \tag{B}$$

From Eq. (A)

$$(\partial P/\partial v)_T = -RT/(v-b)^2 + 2a/v^3, \text{ and} \tag{C}$$

$$(\partial P/\partial T)_v = R/(v - b) \quad \text{(D)}$$

Substituting Eqs.(C) and (D) in Eq. (B) we obtain

$$dP = (-RT/(v - b)^2 + 2a/v^3)_T\, dv + (R/(v - b))_v\, dT. \quad \text{(E)}$$

We may use Eq. (E) to determine $(\partial v/\partial T)_P$ or $(\partial v/\partial P)_T$. At constant pressure, Eq. (E) yields

$$0 = (-RT/(v - b)^2 + 2a/v^3)_T\, dv + (R/(v - b))_v\, dT, \quad \text{(F)}$$

so that

$$(\partial v_P / \partial T_P) = (\partial v/\partial T)_P = -(R/(v - b))/(-RT/(v - b)^2 + 2a/v^3), \quad \text{(G)}$$

and the isobaric compressibility

$$\beta_P = 1/v\, (\partial v/\partial T)_P = -R/(v(-RT/(v-b) + 2a(v-b)/v^3)). \quad \text{(H)}$$

Remarks

It is simple to obtain $(\partial P/\partial T)_v$ or $(\partial P/\partial v)_T$ from Eq. (A). It is difficult, however, to obtain values of $(\partial v/\partial T)_P$ or $(\partial v/\partial P)_T$ from that relation. Therefore, the total differentiation is employed.

Note that Eqs. (C) and (D) imply that for a given state equation:

$$(\partial P/\partial T)_v = M(T,v), \text{ and} \quad \text{(I)}$$

$$(\partial P/\partial v)_T = N(T,v), \text{ and} \quad \text{(J)}$$

Since,

$$dP = M(T,v)\, dv + N(T,v)\, dT, \quad \text{(K)}$$

Differentiating Eq. (C) with respect to T,

$$\partial/\partial T\, (\partial P/\partial v) = (\partial M/\partial T)_v = -R/(v - b)^2. \quad \text{(L)}$$

Likewise, differentiating Eq. (D) with respect to v,

$$\partial/\partial v\, (\partial P/\partial T) = (\partial N/\partial v)_T = -R/(v - b)^2. \quad \text{(M)}$$

From Eqs. (L) and (M) we observe that

$$\partial M/\partial T = \partial N/\partial v \text{ or } \partial^2 P/\partial T\partial v = \partial^2 P/\partial v\partial T. \quad \text{(N)}$$

Eq. (N) illustrates that the order of differentiation does not alter the result. The equation applies to all state equations or, more generally, to all point functions (see next section for more details).

From Eq. (B), at a specified pressure

$$(\partial P/\partial v)_T dv + (\partial P/\partial T)_v dT = M(T,v)\, dv + N(T,v)\, dT = 0.$$

Therefore,

$$(\partial v/\partial T)_P = -M(T,v)/N(T,v) = -(\partial P/\partial T)_v/(\partial P/\partial v)_T. \quad \text{(O)}$$

Eq. (O) can be rewritten in the form

$$(\partial v/\partial T)_P\, (\partial T/\partial P)_v\, (\partial P/\partial v)_T = -1, \quad \text{(P)}$$

which is known as the cyclic relation for a point function.

2. Exact (Perfect) and Inexact (Imperfect) Differentials

If the sum Mdx+Ndy (where M = M (x,y) and N = N (x,y)) can be written as d(sum), it is an exact differential which can be expressed in the form

$$dZ = M(x,y)\,dx + N(x,y)\,dy. \quad (17)$$

If the sum cannot be written in the form d(sum), it is more properly expressed as

$$\delta Z = M(x, y)\,dx + N(x, y)\,dy, \quad (18)$$

For instance, the expression xdy + ydx is an exact differential, since it can be written as d(xy), i.e., dZ = xdy + ydx = d (xy), where Z = xy + C. A plot of the function Z versus x and y is a surface. The difference Z_2-Z_1 = x_2 y_2-x_1 y_1 depends only on the points (x_1, y_1), (x_2, y_2) and not on the path connecting them. But the expression x^2dy + ydx cannot written as d(xy) and hence the sum x^2dy + ydx denoted as $\delta Z = x^2dy + ydx$.

Exact differentials may also be defined through simple integration. Consider a differential expression that is equal to $9x^2y^2dx + 6x^3ydy$, the integration of which can be problematic. It is possible to specify a particular path (Say path AC in Figure 9), e.g., by first keeping x constant while integrating the expression with respect to y, we obtain $Z = 3\,x^3\,y^2$ (for a moment let us ignore the integration constant). On the other hand we can keep y constant and integrate with respect to x and obtain $Z = 3\,x^3\,y^2$, which is same as before. Only exact differentials yield such identical integrals. Hence, the the sum $9x^2y^2dx + 6x^3ydy$ is an exact differential. Consider $6x^2ydx + 6x^3ydy$. If we adopt a similar procedure we get different results (along constant x, $Z = 3x^3y^{\,2}$ and along constant y, $Z = 2\,x^3y$). Instead of integrating the algebraic expressions, we can also integrate the differentials between two given points. For an exact differential of the form dZ = Mdx + Ndy the integrated value between any two finite points (x_1,y_1) and (x_2,y_2) is path independent. If the integration is path dependent, an inexact differential (of the form δZ = Mdx + Ndy) is involved.

Consider the exact differential dZ = xdy + ydx. The term xdy represents the elemental area bounded by the y–axis (GKJH) in Figure 10 and ydx is the corresponding area bounded by the x–axis (KLMJ). The total area xdy + ydx is to be evaluated while moving from point F(x_1,y_1) to point C(x_2,y_2). An arbitrary path "a" can be described joining the points F and C. Integrating along "a" from F to C, the integrated area $\int(xdy) + \int(ydx)$ is a sum of the areas EFaCD + AFaCB. Using another path "b" results in exactly the same area, since regardless of the path that traversed to connect F and C, the integrated value is the same, i.e., Z_2–Z_1 = (x_2y_2–x_1y_1). Therefore, $\int(xdy + ydx)$ is path independent.

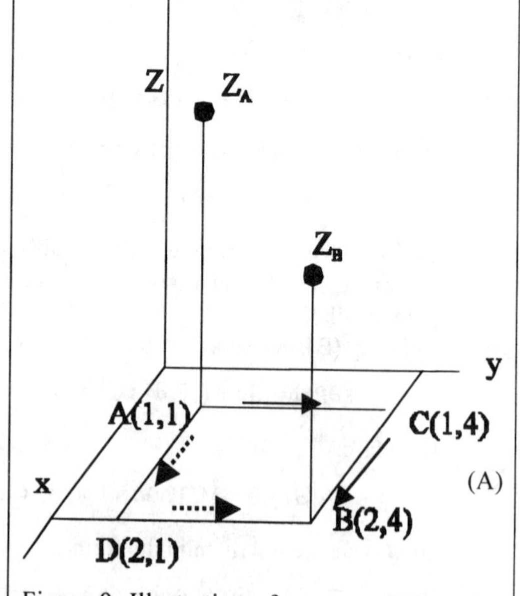

Figure 9: Illustration of an exact differential using path integration.

c. *Example 3*

Is the function

$$9x^2y^2dx + 6x^3ydy$$

an exact or inexact differential? Prove or disprove by adopting path integration (Figure 9).

Solution

The difference Z_A–Z_B can be determined

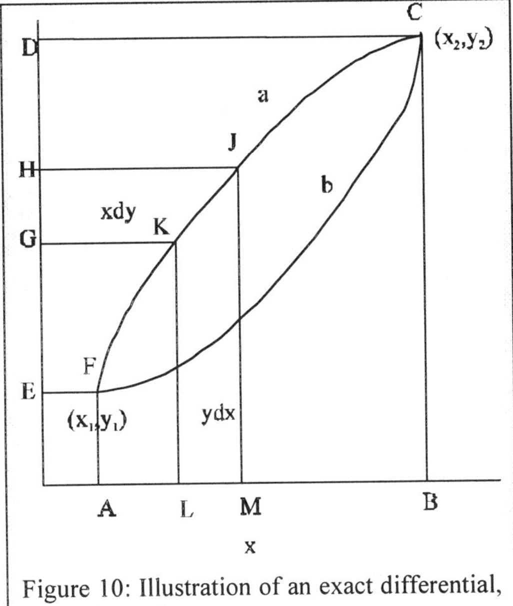

Figure 10: Illustration of an exact differential, dz = xdy + ydx.

by moving along paths ACB or ADB, as illustrated in Figure 9. Consider the path ACB along AC for which x = 1. Integrating the relation while keeping x constant,

$$Z_C - Z_A = (6x^3y^2/2)_{1,1}^{1,4} = 48 - 3 = 45. \quad \text{(A)}$$

Along the path CB, y is held constant (y = 4). Integrating the relation at constant y,

$$Z_B - Z_C = (9x^3y^2/3)_{1,4}^{2,4} = 336, \text{ i.e.,} \quad \text{(B)}$$

$$Z_B - Z_A = \text{Eq. (A)+Eq. (B)} = 381. \quad \text{(C)}$$

The integration can also be performed along path ADB, i.e., along AD, keeping y at a constant value of 1. Using the relation

$$Z_D - Z_A = \int 9x^2y^2dx = (9x^3y^2/3)_{1,1}^{2,1} = 21. \quad \text{(D)}$$

Similarly, x is constant (x = 2) along DB, so that

$$Z_B - Z_D = (3x^2y^2)_{2,1}^{2,4} = 360. \quad \text{(E)}$$

From Eqs. (D) and (E), $Z_B - Z_A = 381$.

The integral is the same for paths ACB and ADB. Thus the differential is an exact differential.

Remarks

If the integration is performed along the path ACB and continued from B to A along BDA, the cyclic integral $\oint dz = \int_{ACB} dz + \int_{BDA} dz = 381 - 381 = 0$.

The difference ($Z_B - Z_A$) is independent of the path selected to reach point B from point A, since Eq. (A) is an exact differential. The function Z is a point function, since it only depends upon the selected coordinates.

In the context of Figure 9, the value of Z_B-Z_A via path C will be the same as via path D. Thus if we take a cyclic process from A-B via path C and then from B-A via path D, there will be no net change in Z, i.e $\oint dZ = 0$.

d. Example 4

Determine if $6x^2y^2dx + 6x^3ydy$ is an exact or inexact differential.

Solution

Consider the path ACB along which $Z_C - Z_A = (6x^3y^2/2)_{1,1}^{1,4} = 45$ and $Z_B - Z_C = (6x^3y^2/3)_{1,4}^{2,4} = 224$ so that $(Z_B - Z_A) = 269$.

Likewise, following the path ADB $Z_D - Z_A = (2x^3y^2)_{1,1}^{2,1} = 14$ and $Z_B - Z_D = (3x^3y^2)_{2,1}^{2,4} = 360$ so that $(Z_B - Z_A) = 374$.

The value of $(Z_B - Z_A)$ along the path ADB does not equal that along path ACB. Consequently, the expression for Z is not a property, since it is path dependent, and is, therefore, an inexact differential.

Remark

If the integration is first performed along the path ACB and continued from B back to A along BDA, the integrated value is 269 (ACB)–374 (BDA) = –105. If the integration is first performed along the path ACB and continued from B back to A along BDA, the integrated value is $\oint \delta Z \neq 0$ since 269 (ACB)-374 (BDA) = -105. In general, the cyclic integral of an inexact differential is nonzero.

a. Mathematical Criteria for an Exact Differential

i. Two Variables (x and y)

The path integration procedure helps determine whether a differential is exact or inexact. However, the mathematical criterion that will now be discussed avoids lengthy path integration and saves time. Example 2 shows that a point function of the form $P = P(T,v)$ may be written as

$$dP = M(T,v)\,dv + N(T,v)\,dT, \tag{19}$$

and it possesses the property

$$\partial M/\partial T = \partial N/\partial v, \tag{20}$$

i.e., $\partial^2P/\partial v\partial T = \partial^2P/\partial T\partial v$. Substituting for $x = v$, $y = T$, and $Z = P$,

$$dZ = (\partial Z/\partial x)_y\,dx + (\partial Z/\partial y)_x\,dy = M(x,y)\,dx + N(x,y)\,dy. \tag{21}$$

The function M is called the conjugate of x, and N is the conjugate of y. The necessary and sufficient condition for Z to be a point function is given by Eq. 20, namely,

$$\partial^2Z/\partial y\,\partial x = \partial^2Z/\partial x\partial y \text{ or } (\partial M/\partial y)_x = (\partial N/\partial x)_y. \tag{22}$$

This is another criterion describing an exact differential, and it is also referred to as the condition of integrability. A differential expression of the form $M(x,y)\,dx + N(x,y)\,dy$ is said to be in the linear differential (or Pfaffian) form. A differential expression derived from a point function or a scalar function, such as $P = P(T,v)$, in the Pfaffian form satisfies the criterion for being an exact differential.

e. Example 5

Consider the expression $-(Ry/x^2)dx + (R/x)\,dy$, where R is a constant. Is this an exact differential? If so, integrate and determine "Z".

Solution

$M = -R\,y/x^2$, and $N = R/x$. Therefore,

$(\partial M/\partial y)_x = -R/x^2$, and $(\partial N/\partial x)_y = -R/x^2$, i.e., the criterion for being an exact differential is satisfied by the expression. Therefore,

$$dZ = -Ry/x^2\,dx + R/x\,dy, \text{ i.e.,} \tag{A}$$

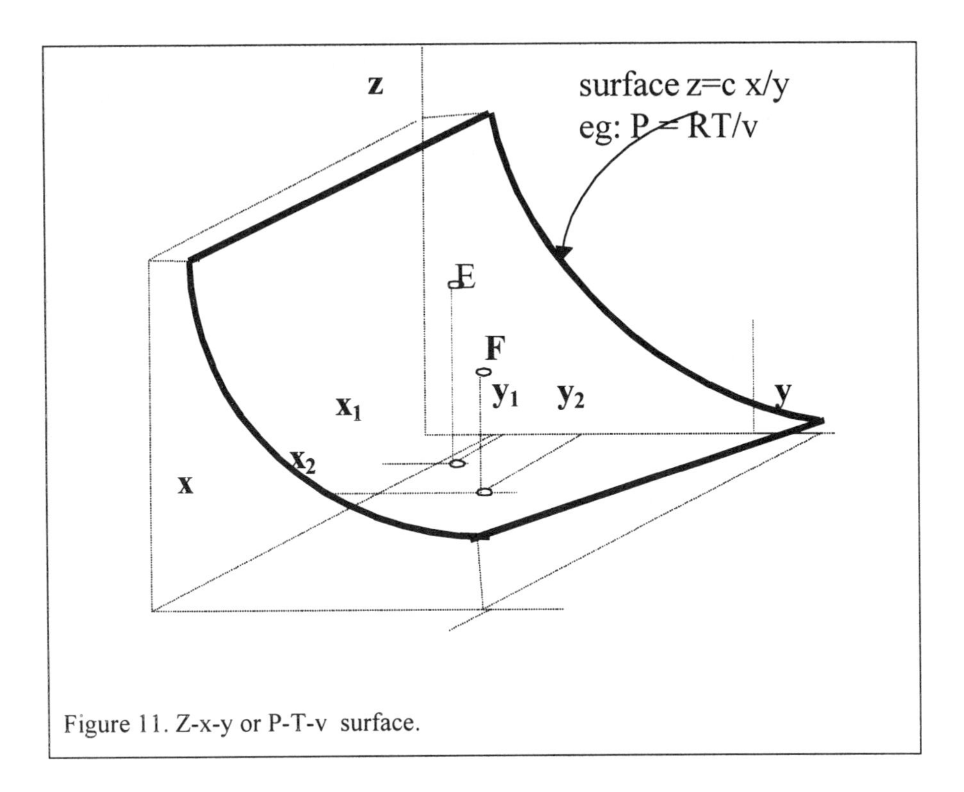

Figure 11. Z-x-y or P-T-v surface.

$\partial Z/\partial y = R/x = N$, and $\partial Z/\partial x = -Ry/x^2 = M$. (B)

To determine Z, we can integrate along constant x to obtain

$Z = Ry/x + f(x)$. (C)

Upon differentiating Eq. (C) with respect to x

$\partial Z/\partial x = -Ry/x^2 + f'(x)$. (D)

However, from Eqs. (A) and B, $(-Ry/x^2) = M$ so that $f'(x) = 0$, i.e., f(x) is a constant, and

$Z = Ry/x + C$. (E)

Remarks

Assume that C = 0 and R = 8. Once x and y are specified (say, respectively with values of 2 and 4) in Eq. (E), the value of Z is fixed (= 16) irrespective of the path along which the point (x = 2, y = 4) is reached. In this case Z is called a point function.

If C = 0, y = T, x = v, and Z = P, then using Eq. (E) the point function that is obtained is of the form

$P = RT/v$,

which is the familiar ideal gas equation of state. A plot of P with respect to T is presented in Figure 12 while a plot of P versus both T and v describes a surface (Figure 11) Starting a process at point (T_1, v_1) (i.e., point A, of Figure 12), the pressure P_2 can be determined at a point (T_2,v_2) (i.e., Point B of the figure) using either of the paths ACB or ADB.

From the preceding discussion we note that $\oint dz = 0$ if Z is a point function. Therefore, if Z denotes the temperature T, then $\oint dT = 0$. However, if Z denotes the heat transfer Q which is not a point function, $\oint \delta Q = 0$. Point functions such as $T = T(P,v) = Pv/R$ can be specified for only those systems that are in equilibrium (or which have a uniform property distribution within them). If a system exists in a nonequilibrium state, it can be shrunk and continually made smaller until a uniform property domain is reached. At that stage point function relations can be applied to determine the properties of the smaller system.

ii. Three or More Variables

The exactness criteria can be generalized to systems involving more than two variables. Consider that Z is described by three variables x_1, x_2, and x_3, i.e.,

$$Z = Z(x_1,x_2,x_3). \quad (23)$$

The total differential of Z is

$$dZ = \partial Z/\partial x_1\, dx_1 + \partial Z/\partial x_2\, dx_2 + (\partial Z/\partial x_3)\, dx_3. \quad (24)$$

Since dZ is exact

$$\partial Z/\partial x_1 = \partial Z/\partial x_2,\ \partial Z/\partial x_2 = \partial Z/\partial x_3, \text{ and } \partial Z/\partial x_3 = \partial Z/\partial x_1. \quad (25)$$

We now have three conditions in terms of all three variables. Generalizing these expressions for k variables when

$$Z = Z(x_1,x_2,x_3,\ldots,x_k). \quad (26)$$

The total differential may be written in the form

$$dZ = \sum_{i=1}^{k} (\partial Z/\partial x_i)dx_i, \quad (27)$$

and by analogy the criteria describing an exact differential for this case are

$$\frac{\partial}{\partial x_j}\left(\frac{\partial Z}{\partial x_i}\right) = \frac{\partial}{\partial x_i}\left(\frac{\partial Z}{\partial x_j}\right), j \neq 1. \quad (28)$$

When Z is a function of two variables alone, i.e., $Z(x_1,x_2)$, one criterion describes an exact differential. If more than two variables are involved, i.e., $Z(x_1,x_2,x_3,\ldots,x_k)$ it is possible to write the following equations in terms of x_1, namely, $\partial^2/\partial x_1\partial x_2 = \partial^2/\partial x_2\partial x_1$, $\partial^2/\partial x_1\partial x_3 = \partial^2/\partial x_3\partial x_1$,..., etc., and generate (k–1) equations for the k variables. Likewise, in terms of $\partial^2/\partial x_2\partial x_1 = \partial^2/\partial x_1\partial x_2$, $\partial^2/\partial x_2\partial x_3 = \partial^2/\partial x_3\partial x_2$, ..., etc. However, $\partial^2/\partial x_1\partial x_2 = \partial^2/\partial x_2\partial x_1$, which appears in both equations so that only ((k–1)–1) equations can be generated in terms of x_2. Similarly, ((k–1)–2) criteria can be generated for x_3, and so on resulting in ((k–1) + ((k–1)–1) + ((k – 1) – 2) + ((k – 1) – 3)+....) criteria. Simplifying, k(k–1) – (1 + 2 + 3 + ... + k–1) = k(k–1) – (1/2)(k–1)k = k(k–1)/2. Therefore, the number of criteria describing an exact differential of $Z(x_1,x_2,x_3,\ldots,x_k)$ are k(k–1)/2. When a point function involves more than two variables i.e. $Z = Z(x_1,x_2,x_3,\ldots,x_k)$ a hypersurface is produced, e.g., the plot of $P = N_1\bar{R}T/V + N_2\bar{R}T/V + .. = P(T,V,N_1,N_2,N_3,\ldots)$.

3. Conversion from Inexact to Exact Form

It is possible to convert an inexact differential into an exact differential by using an integrating factor.

---

f. *Example 6*

Consider the following inexact differential

$$\delta q_{rev} = c_{v,o}(T)dT + ((RT/(v - b))dv, \quad (A)$$

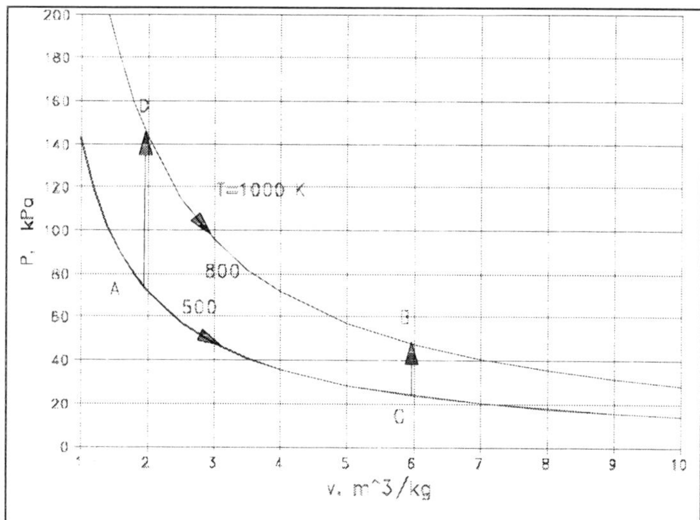

Figure 12: Plot of pressure vs. volume with temperature as a parameter.

where $c_{v,o}$ denotes the specific heat at constant volume, R the gas constant, and b a virial constant. Show that the differential is inexact. If Eq. (A) is throughout divided by T, is q_{rev}/T an exact differential?

Solution

Considering $M = c_{v,o}(T)$ and $N = RT/(v-b)$, then

$$\partial M/\partial v = 0, \text{ and } \partial N/\partial T = R/(v-b).$$

Since, $\partial M/\partial v \neq \partial N/\partial T$, q_{rev} is inexact. Dividing Eq. (A) by T

$\delta q_{rev}/T = c_{v,o}(T)dT/T + ((R/(v-b))dv$,

Now, consider $M = c_{v,o}(T)/T$, and $N = R/(v-b)$ so that

$$\partial M/\partial v = \partial N/\partial T = 0.$$

In this case, $\delta q_{rev}/T$ is an exact differential.

Remarks

In Chapter 3 we will discuss that q_{rev} is termed the reversible heat transfer, which is not a system property, but $q_{rev}/T = ds$, where s is a system property called the entropy.

We can similarly show that q_{rev}/v is inexact.

4. Relevance to Thermodynamics

We will now discuss the relevance of exact differentials to thermodynamic analyses.

a. *Work and Heat*

Both work transfer W and heat transfer Q are path dependent, while properties such as P and T are path independent. If a gas that is initially at state 1 (cf. Figure 13) corresponding to the conditions $T_1 = 500$ K, $v_1 = 2$ m^3 kg^{-1} (i.e., $P_1 = 71.8$ kPa from Figure 13) is isothermally expanded (process AC) to $v_2 = 6$ m^3 kg^{-1}, following which heat addition occurs at fixed volume so that the gas temperature rises to 1000 K, the pressure at state 2, i.e., P_2, is found to be 47.8 kPa. The same end state may be achieved by first adding heat at constant volume (process AD) to raise the temperature to 1000 K, and then expanding isothermally to $v_2 = 6$ m^3 kg^{-1}. The final pressure following the latter process will still be 47.8 kPa.

In a closed system containing an ideal gas, the incremental work $\delta W = P\,dV$ (this will be discussed more thoroughly in Chapter 2, with the total work W for a process being given by the area under the corresponding P–v curve (e.g., Figure 12). For the path ACB the work W_{ACB} is the area under the curve ACB, while for a process along ADB (Figure 12), W_{ADB} is the area under that curve. It is apparent that $W_{ACB} \neq W_{ADB}$ even though the initial and final pressures, temperatures and volumes (all of which are properties) are the same. Therefore, we can determine v for given values of T and P, but not the work done, since it is path dependent. So "v" could be tabulated at given T and P as in Steam and R 134a Tables (Tables A-4 and A-5) but work cannot be tabulated. The differentials of path dependent quantities are inexact differentials (e.g., δW, δQ etc.), and their cyclic integrals $\oint \delta Q \neq 0$ and $\oint \delta W \neq 0$. In general, the heat transfer between any two states 1 and 2,

$$\int \delta Q \neq Q_2 - Q_1. \tag{29}$$

b. Integral Over a Closed Path (Thermodynamic Cycle)

Over a cycle for which the final and initial states are identical

$$\oint dT = \oint dP = \oint du = \ldots = 0. \tag{30}$$

In general, for a process occurring between two distinct states 1 and 2, the property change

$$\int_{T_1}^{T_2} dT = T_2 - T_1,\ \int_{P_1}^{P_2} dP = P_2 - P_1,\ \text{etc.} \tag{31}$$

The internal energy can be expressed as an exact differential by the relation $du = T\,ds - P dv$, i.e., $u = u(s,v)$, $M(s,v) = T$, and $N(s,v) = -P$. The exact differential criterion $(\partial T/\partial v)_s = -(\partial P/\partial s)_v$ for this case is also referred to as a Maxwell relation, details of which are given in Chapter 7. In general, all system properties, e.g., T, P, V, v, u, U, etc., are path independent and point functions, and, therefore, form exact differentials. Consider the exact differential form $du = c_{v,0}dT + (a/T)v^2dv$ where "a" and c_{v0} are constants. If the internal energy difference is to be determined between states A and B (cf. Figure 12) either of paths ACB (isothermal conditions along AC, and constant volume along CB), or ADB (v constant along AD and T

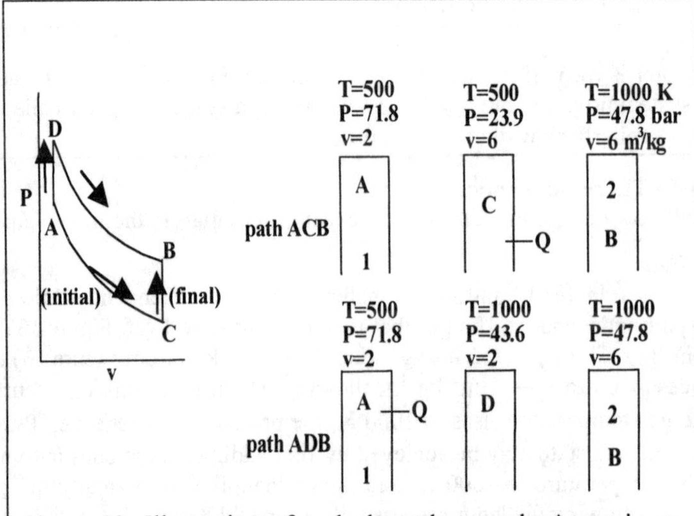

Figure 13: Illustration of path dependent work. A gas is expanded to state B using paths ACB or ADB.

unchanged along DB) can be used to integrate the expression, with the difference (u_B–u_A) being the same regardless of path.

The Appendix contains several relations between irrotational scalar fields that are useful in fluid mechanics, criteria for exact differentials, and thermodynamic properties.

5. Homogeneous Functions

Homogeneous functions possess certain mathematical characteristics and the term homogeneous must not be confused with the thermodynamic definition of homogeneity. The total energy U in the air contained in a vessel is readily determined if the state (say, number of moles, temperature, and pressure) is known. If three identical vessels containing air at the same conditions are combined, these will contain three times as many moles, and, therefore, three times as much energy (since U is an extensive property). The combined internal energy U_c = $U_c(T,P, 3\times0.3 \text{ moles } O_2, 3\times1.2 \text{ moles } N_2) = 3\, U(T, P, 0.3 \text{ moles } O_2, 1.2 \text{ moles } N_2)$. Mathematically,

$$U(a,b, \lambda N_{O_2}, \lambda N_{N_2}) = \lambda^1 U(a,b, N_{O_2}, N_{N_2}).$$

The function U is called a homogeneous function of degree 1, λ is a multiplier, and a and b are constants (which in this case are fixed values of T and P). If the three vessels are combined in an equilibrium state, the density, which is an intensive property, does not change. Therefore,

$$\rho(T,P, \lambda N_{O_2}, \lambda N_{N_2}) = \lambda^0 \rho\, (a,b, N_{O_2}, N_{N_2}),$$

where ρ is a homogeneous function of degree zero. The definition of homogeneous function can be generalized as follows:

In general, a function $F(a,b,x_1,x_2,x_3,\ldots,x_k)$ is a homogeneous function of degree m if

$$F(a,b, \lambda x_1, \lambda x_2, \lambda x_3,\ldots, \lambda x_k) = \lambda^m F(a,b,x_1,x_2,x_3,\ldots,x_k), \qquad (32)$$

where a and b are constants. Homogeneous functions for which m = 1 describe extensive properties, and those with m = 0 describe intensive properties.

For instance, consider the function

$$F(a,b,x_1,x_2) = a x_1^2 x_2^3 / (b^2 x_3). \qquad (33)$$

Assuming $x_{1,new} = \lambda x_1$, $x_{2,new} = \lambda x_2$,...

$$F(a,b,x_{1,new},x_{2,new},,\ldots) = F(a,b, \lambda x_1, \lambda x_2,\ldots) = a x_{1,newl}^2 x_{2,new}^3 / (b^2 x_{3,new})$$

$$= a\lambda^2 x_1^2 \lambda^3 x_2^3 / (b^2 \lambda x_3) = a\lambda^4 x_1^2 x_2^3 / (b^2 x_3) = \lambda^4\, F(a,b,x_{1,new},x_{2,new},,\ldots).$$

Therefore, F is a homogeneous function of degree 4. If a=b=1, x_1=1, x_2=2, and x_3=1, F(1,2,1) = 8. Furthermore, if λ = 2, then F($2x_1$, $2x_2$,...) = F(2,4,2), and using Eq. (32)

$$F\,(2,4,2) = 24\, F(1,2,1) = 16\times8 = 128.$$

This result may be checked using the above values for the variables in Eq. (33) so that $x_{1,new} = \lambda x_1 = 2x_1 = 2$, $x_{2,new} = \lambda x_2 = 2x_2 = 4$, and $x_{3,new} = \lambda x_3 = 2x_3 = 2$. In that case as well, the function F = 128.

Consider the following homogeneous functions: $F_1(x,y) = \sin^2(x/y)$ is a function of degree 0, since its phase is unchanged by λ; $F_2(x,y) = x^{-\pi}\sin(x/y) + xy^{-\pi-1}\ln(y/x)$ is one of degree m = $-\pi$, and $F_3(x,y) = 3x^3/y^2$ of degree m = 1.

A necessary and sufficient condition for F to be homogeneous and of degree m, is that the Euler equation

$$\Sigma_{k=0}^{K} x_k (\partial F / \partial x_k) = mF. \quad (34)$$

holds, the proof for which is contained in the Appendix to this chapter.

g. *Example 7*

Prove Euler's equation with the function

$$Z(x,y) = ax^2y + 2bxy^2. \quad (A)$$

Solution

$$F = Z(x,y) \quad (B)$$

is a homogeneous function of degree m = 3.
We must prove that

$$x(\partial Z/\partial x) + y(\partial Z/\partial y) = 3Z.$$

Differentiating Eq. (A) with respect to x and then y, the resultant expression is

$$= x(2axy + 2by^2) + y(ax^2 + 4bxy) = 2ax^2y + 2bxy^2 + ax^2y + 4bxy^2$$

$$= 3ax^2y + 6bxy^2 = 3(ax^2y + 2bxy^2) = 3Z.$$

A function F is oftentimes not homogeneous with respect to all of the variables. If F is partly homogeneous in terms of j among k variables so that

$$F(a,b,x_1,x_2,x_3,\ldots,x_k) = F(a,b,\lambda x_1,\lambda x_2,\lambda x_3,\ldots,\lambda x_j,x_{j+1},\ldots,x_k), \quad (35)$$

the Euler equation Eq. (34) assumes the form

$$\Sigma_{i=0}^{j} x_i (\partial F / \partial x_i) = mF. \quad (36)$$

h. *Example 8*

Is the function

$$F(a,b,x,y,t) = ax^3y/t + x^2y^2/t^3 + bxy^3/t^7$$

a homogeneous function, a and b being constants. What is the Euler equation?

Solution

The function F is not fully homogeneous, since $F(a,b,x,y,t) \neq \lambda^m F(a,b,x,y,t)$. If the powers of x, y, and t are added, the first term on the RHS of the expression yields 3, the second term 1, and the third term –3. However, if t is excluded, the sum of the powers of x and y for each term is 4. Therefore, the function is partly homogeneous (with respect to x, and y) so that $F(a,b,\lambda x,\lambda y,t) = \lambda^4 F(a,b,x,y,t)$.

The Euler equation assumes the form

$$x\partial F/\partial x + y\partial F/\partial y = 4F.$$

If a function F = F (a,b,x,y) is homogeneous and of degree m, λ can be specified equal to 1/x so that

$$F(a,b,\lambda x,\lambda y) = F(a,b,1,y/x) = (1/x)^m F(a,b,x,y). \quad (37)$$

Therefore,

$$F(a,b,x,y) = x^m F(a,b,y/x) \quad (38)$$

i. *Example 9*

Consider the function

$$Z(a,b,x,y) = ax^3 + bxy^2, \quad (A)$$

and show that $Z(a,b,x,y) = x^3 Z(a,b,y/x)$.

Solution

Equation (A) may be written in the form

$$Z(a,b,x,y) = x^3(a + b(y/x)^2), \quad (B)$$

where the terms in the brackets correspond to $Z(a,b,y/x)$. Therefore,

$$Z(a,b,x,y) = x^3 Z(a,b,y/x).$$

We now summarize the properties of homogeneous functions:

$$F(a,b, \lambda x_1, \lambda x_2, \lambda x_3,\ldots, \lambda x_k) = \lambda^m F(a,b,x_1,x_2,x_3,\ldots,x_k),$$

$$\Sigma_{k=0}^{K} x_k (\partial F/\partial x_k) = mF.$$

$$F(a,b,x,y) = x^m F(a,b,y/x)$$

a. *Relevance of Homogeneous Functions to Thermodynamics*

i. Extensive Property

A thermodynamic variable or property F is *extensive* if it is a homogeneous function of the first degree with respect to all of its extensive parameters in a functional relation. Mathematically, F is an extensive property if $m = 1$ in Eq. (32), namely

$$F(\lambda x_1, \lambda x_2,\ldots) = \lambda F(x_1, x_2,\ldots),$$

where x_1, x_2,... are all extensive properties.

If F is taken to represent the internal energy $U = U(S,V,\lambda N_{O_2},\lambda N_{N_2},\lambda N_{Ar})$ of air contained in a vessel of volume V and of entropy S, where N_{O_2}, N_{N_2}, and N_{Ar} denote moles of oxygen, nitrogen and argon. The entropy is an extensive property (that is discussed in greater detail in Chapter 3 which has the units of kJ K^{-1}. If λ identical vessels are combined into a system, the internal energy of the composite system is λU, and the volume is λV contains λN_i moles of each of the species i. Therefore,

$$U(\lambda S, \lambda V, \lambda N_1, \lambda N_2, \lambda N_3) = \lambda U(S,V,N_1,N_2,N_3). \quad (39)$$

For sake of illustration assume each vessel to be at S = 2 kJ K^{-1}, with volume of 5 m^3 containing 1 kmole of argon (N_1), 78 kmole of nitrogen (N_2) and 21 kmole of oxygen (N_3). Assume that the internal energy in each vessel is 500 kJ. If $\lambda = 3$, three vessels have been combined and the volume and number of moles increases threefold. Using the notation of Eq. (39)

$$U(3\times2,3\times5,3\times1,3\times78,3\times21) = 3\times U(2,5,1,78,21).$$

Therefore, $m = 1$, and we confirm once again that U is an extensive variable.

ii. Intensive Property

A thermodynamic variable or property F is said to be intensive if it is a homogeneous function of zero degree with respect to all of its extensive parameters. In mathematical terms F is intensive when $m = 0$ in Equation (32) or if

$$F(\lambda x_1, \lambda x_2,\ldots) = F(x_1, x_2,\ldots),$$

We can define $T = \partial U/\partial S$. Since U is a function of S, V, and of the number of moles of various species, as discussed above, $\partial U/\partial S$ is also a function of those variables. Therefore,

$$T = \partial U/\partial S = T(S,U, N_1,N_2,...). \quad (40)$$

If the energy of each vessel in the above discussion is increased by, say, dU = 3 kJ and, for sake of illustration, the corresponding change in dS = 0.01 kJ K^{-1}, the temperature inside the vessels must be

$$T = \partial U/\partial S = 3 \text{ kJ}/0.01 \text{ kJ K}^{-1} = 300 \text{ K}.$$

If three vessels are combined, the volume, number of moles, energy, and entropy all triple, i.e., dU = 3×3, dS = 3×0.01. Therefore, the temperature of the combined system is still

$$T = 9/0.03 = 300 \text{ K},$$

as expected. More rigorously,

$$T = (\partial U/\partial S)_{V,N_1,N_2,...} = T(S,V,N_1,N_2,...).$$

Since for the combined system $U_c = \lambda U$, and $S_c = \lambda S$,

$$T_c = (\partial U_c/\partial S_c)_{V,N_1,N_2,...} = \partial(\lambda U)/\partial(\lambda S) = \partial U/\partial S = T.$$

We, therefore, conclude that intensive properties are unchanged upon addition of identical systems, i.e.,

$$T(\lambda S, \lambda U, \lambda N_1, \lambda N_2,...). = \lambda^0 T(\lambda S, \lambda U, \lambda N_1, \lambda N_2,...). \quad (41)$$

Additional applications will be discussed in Example 10 and Chapters 3, 5 and 8.

iii. Partly Homogeneous Function

The volume given by the ideal gas law V = NRT/P where V = V(T,P,N) is a partly homogeneous function of the number of moles N. Consider a vessel containing air at a temperature of 298 K and pressure of 1 bar. If three identical vessels are combined into another system, the values of V and N triple, although T and P are unaffected. Therefore,

$$V(T,P,\lambda N_{O_2},\lambda N_{N_2},\lambda N_{Ar}) = \lambda V(T,P, N_{O_2}, N_{N_2},N_{Ar}), \quad (42)$$

which shows that V is a partly homogeneous function of degree 1 with respect to N_{O_2}, N_{N_2}, and N_{Ar}.

iv. Conversion of Extensive Into Intensive Properties

We have shown that U = U(S,V,N) is a homogeneous function of degree 1, namely,

$$U = U(\lambda S, \lambda V, \lambda N) = \lambda U(S,V,N)$$

Using a value of $\lambda = 1/N$, U (S/N,V/N,1) = (1/N) U(S,V,N), or

$$U(\bar{s},\bar{v},1) = (1/N)U(S,V,N) \text{ or } N\bar{u}(\bar{s},\bar{v}) = U(S,V,N).$$

---

j. Example 10

Consider the following state equation for the entropy of an electron gas

$$S(N,U,V) = C\, N^{1/6}U^{1/2}V^{1/3}, \quad (A)$$

Show that S is a homogeneous function of degree 1 (i.e., it is extensive).

Assuming $T = (\partial U/\partial S)_{V,N}$, show that T is a homogeneous function of degree 0 (i.e., it is intensive).

Solution

$$S(\lambda N,\lambda V,\lambda U) = C(\lambda N)^{1/6}(\lambda V)^{1/3}(\lambda U)^{1/2} = \lambda CN^{1/6}V^{1/3}U^{1/2} = \lambda S(N,U,V). \quad \text{(B)}$$

Therefore, S is homogeneous function of degree m = 1, S being an extensive property. From Eq. (A),

$$dS_{V,N} = CN^{1/6}V^{1/3}((1/2)U^{-1/2}dU_{V,N}) \text{ and}$$

$$T(N,U, V) = \partial U_{V,N}/\partial S_{V,N} = 2U^{1/2}/(CN^{1/6}V^{1/3}). \quad \text{(C)}$$

The temperature

$$T(\lambda N,\lambda V,\lambda U) = 2(\lambda U)^{1/2}/(C(\lambda N)^{1/6}(\lambda V)^{1/3}) = \lambda^0 2U^{1/2}/(CN^{1/6}V^{1/3}),$$

that proves that T is a homogeneous function of degree 0 which cannot be altered by increasing or decreasing the system size (or λ).

Remarks

The entropy S is an extensive property (m = 1), whereas the temperature T is an intensive property (m = 0).

Since m = 1, Euler's equation for S(U, V, N) assumes the form

$$U(\partial S/\partial U) + V(\partial S/\partial V) + N(\partial S/\partial N) = S. \quad \text{(D)}$$

We will show in Chapter 3 that

$$\partial S/\partial U = 1/T,\ \partial S/\partial V = P/T, \text{ and } \partial S/\partial N = -\mu/T. \quad \text{(E)}$$

where μ is called the chemical potential. If S is expressed in units of J K^{-1} and U in J, $\partial S/\partial U$ is in units K^{-1}. (Similarly, you may verify that $\partial S/\partial V$ can be expressed in units of N m^{-2} K^{-1}.) Using Eqs. (D) and (E),

$$U/T + V(P/T) - \mu/T = S, \text{ i.e., } U + P V - T S = \mu N.$$

k. *Example 11*

The internal energy U is an extensive property, since it is a homogeneous function of degree m = 1. In general, $U = U(S,V,N_1,N_2,...,N_k)$ so that k+2 extensive properties are required to determine U for a k–component simple compressible system. Show that $\bar{u}$, which is an intensive property, is a function only of k+1 intensive variables.

Solution

Select $\lambda = 1/N$, where N denotes the total number of moles in the system so that

$$U(S/N,V/N,N_1/N,N_2/N,...,N_k/N) = (1/N)U(S,V,N_1,N_2,...,N_k), \text{ or}$$

$$U(S,V,N_1,N_2,...,N_k) = N\,\bar{u}\,(\bar{s}, \bar{v}, x_1,x_2,...,x_k),$$

where x_i represents the mole fraction of the i–component in a gaseous system (we can replace x_i with $x_{l,i}$ for a system containing a liquid mixture). Therefore,

$$\bar{u} = U/N = \bar{u}\,(\bar{s}, \bar{v}, x_1,x_2,...,x_k).$$

Since $N_1+N_2+ \cdots +N_k = N$, then

$$N_1/N+N_2/N+ \cdots +N_k/N = 1, \text{ or } x_1+x_2+ \cdots +x_k = 1.$$

Therefore, $x_k = 1-x_1-x_2-...-x_{k-1}$, and $\bar{u}(\bar{s}, \bar{v}, x_1,x_2,...,x_{k-1})$ is an intensive property which is a function of only k–1+2 = k+1 intensive variables.

6. Taylor Series

The value of a function w(x) at neighboring point x+δx, namely, w(x+δx), can be determined in terms of its value at x by using a Taylor series as follows:

$$w(x+\delta x) = w(x)+|dw/dx|_x\delta x+(1/2!)|d_2w/dx^2|_x(\delta x)^2 +(1/3!)|d^3w/dx^3|_x(\delta x)^3+...+(1/n!)|d^nw/dx^n|_x(\delta x)^n+R', \quad (43)$$

where 2!=2×1, 3!=3×2×1,..., n!=n×(n–1)×(n–2)×...×4×3×2×1, and R′ denotes the remainder.

If w = w(x,y,z), then

$$w(x+\delta x,y+\delta y,z+\delta z) = w(x,y,z) + (\partial/\partial x\ \delta x + \partial/\partial y\ \delta y + \partial/\partial z\ \delta z)w + (1/2!)(\partial/\partial x\ \delta x + \partial/\partial y\ \delta y + \partial/\partial z\ \delta z)^2\ w+ (1/3!)\ (...) + ..., \text{ or}$$

$$w(x+\delta x,y+\delta y,z+\delta z) = w(x,y,z) + \delta w + \delta^2 w + ... + \delta^n w + R', \text{ where} \quad (44)$$

$$\delta^2 w = (\partial/\partial x\ \delta x + \partial/\partial y\ \delta y + \partial/\partial z\ \delta z)^2\ w$$
$$= \partial w/\partial x^2\ \delta x^2 + \partial^2 w/\partial y^2\ \delta y^2 + \partial^2 w/\partial z^2\ \delta z^2 + 2\ \partial^2 w/\partial x\partial y\ \delta x\delta y + 2\ \partial^2 w/\partial y\partial z\ \delta y\delta z + 2\ \partial^2 w/\partial z\partial x\ \delta z\delta x, \text{ and}$$
$$\delta^n w = (\partial/\partial x\ \delta x + \partial/\partial y\ \delta y + \partial/\partial z\ \delta z)^n\ w.$$

The Taylor series expansion will be used to derive conservation equations in Chapter 2, the entropy balance equation in Chapter 3, the availability balance equation in Chapter 4 and stability criteria in Chapter 9. In place of a Taylor series, Callen uses the expression

$$w(x+dx, y+dy,z+dz) = \exp((\partial/\partial x\ \delta x + \partial/\partial y\ \delta y + \partial/\partial z\ \delta z)\ w(x,y,z)),$$

where the term related to the exponential is treated as a small quantity.

7. LaGrange Multipliers

The LaGrange multiplier method allows us to optimize (i.e., either maximize or minimize) a function u = u(x,y,z), say, subject to the conditions g(x,y,z) = 0, and h(x,y,z) = 0. The method involves the following steps:

1. A function F is formed such that

$$F(x,y,z,\lambda^1,\lambda^2) = u(x,y,z) + \lambda^1 g(x,y,z) + \lambda^2 h(x,y,z). \quad (45)$$

2. Since F is to be optimized, Eq. (45) is differentiated specifying

$$\partial F/\partial x = 0,\ \partial F/\partial y = 0, \text{ and } \partial F/\partial z = 0.$$

3. x, y, z, λ^1, and λ^2 are solved using the constraints and Eq. (45) at the optimum condition.

We will use the LaGrange multiplier method later to determine the equilibrium conditions for multicomponent and multiphase systems.

---

l. Example 12

Use the LaGrange Multiplier method and optimize the function

$$G(A,B,x,y) = x(A+ \ln(x/(x+y))) + y\ (B+\ln(y/(x+y))) \quad (A)$$

subject to the condition that

$$2x + y = N, \quad (B)$$

where A, B and N are constants. Obtain a numerical solution for x, y, and G when A = –30.27, B = –12.95, and N = 4.

Solution

Using Eq. (45) we form the function

$$F = G(A,B,x,y) + \lambda(2x + y - N) = 0, \text{ where} \quad (C)$$

$$\partial F/\partial x = 0, \partial F/\partial y = 0$$

Using Eqs. (A) and (C), and differentiating the latter with respect to x and y,

$$\partial F/\partial x = x(1/x - 1/(x+y)) + (A + \ln (x/(x+y)) + y(-1/(x+y)) + 2\lambda = 0.$$

Upon simplification,

$$A + \ln (x/(x+y)) + 2\lambda = 0, \text{ and} \quad (D)$$

$$\partial F/\partial y = y (1/y - 1/(x+y)) + (B + \ln(y/(x+y))) + x(-1/(x+y)) + \lambda = 0 \text{ so that}$$

$$B + \ln(y/(x+y)) + \lambda = 0. \quad (E)$$

Multiplying Eq. (E) by 2 and subtracting it from Eq. (D)

$$A + \ln(x/(x+y)) - 2B - 2\ln(y/(x+y)) = 0, \text{ or}$$

$$\ln(y/(x+y))^2 - \ln (x/(x+y)) = \exp(A-2B), \quad (F)$$

where x and y are obtained from the condition $2x + y = N$. Using $y = N - 2x$ in Eq. (F),

$$(N-2x)^2/(x(N - x)) = K, \quad (G)$$

or $x^2-xN + N^2/(4 + K) = 0$, where

$$K = \exp(A-2B). \quad (H)$$

Equation (G) is a quadratic in terms of x and can therefore be solved if it is optimized for a particular value of x. Using A = –30.27, B = –12.95, N = 4 in Eq.(H), K = 0.0127. Substituting these data in Eq. (G) we obtain $(4-2x)^2 = 0.0127 \text{ x } (4-x)$. Solving for x and selecting the root for which x>0, and y>0,

$$x = 1.8875, \text{ and } y = N - 2x = 4 - 2\times1.8875 = 0.225, \text{ and } G(A,B,x,y) = -60.78$$

Remark

In Chapter 12, an example illustrates that A, B and K are functions of T and P, and Eq. (F) corresponds to a chemical equilibrium condition when G is minimized for a problem in which 4 moles of oxygen are admitted to a reactor while x moles of O_2 and y moles of O–atoms leave the reactor. The solution for x corresponds to the chemical equilibrium composition.

8. Composite Function

Consider the ideal gas law v = RT/P to apply for a process during which a gas expands in a two–dimensional nozzle. If we travel downstream with the gas, as a consequence of the expansion, the pressure and temperature typically decrease and the specific volume v occupied by 1 kg mass of the gas typically increases according to the ideal gas law. An alternative way of looking at the problem is to consider the entire nozzle domain in an x–y dimensional plane (called an Eulerian frame of reference) in which P and T (and, therefore, v) are functions of x and y. The specific volume v(x,y) can be determined using the state equation. Functions

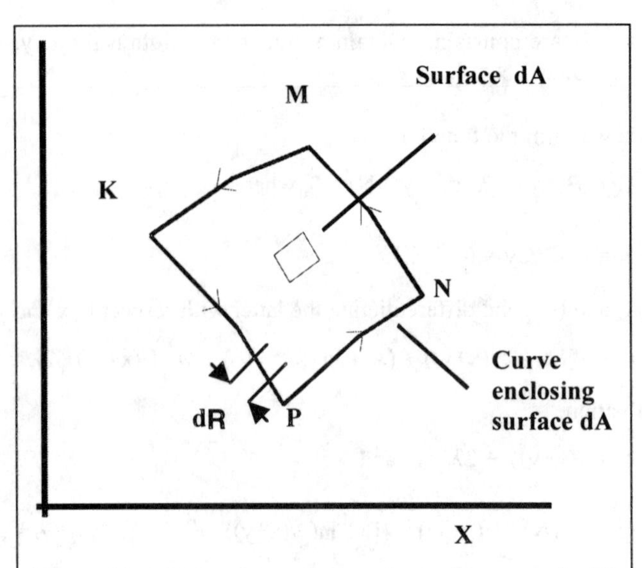

Figure 14: (a) Line integral around an arbitrary path. (b) Line integral around a circular path.

such as v = v(T,P), with T and P being themselves functions of x and y, are called composite functions. In order to determine $|\partial v/\partial x|_y$, i.e., how the specific volume changes with respect to displacements y from the nozzle centerline, v = v(T,P) and, therefore, T(x,y) and P(x,y) must be known. From the state equation,

$$dv = (\partial v/\partial T)dT + (\partial v/\partial P)dP \quad (46)$$

However, since T = T(x,y) and P = P(x,y),

$$dT = (\partial T/\partial x)\,dx + (\partial T/\partial y)\,dy, \text{ and } dP = (\partial P/\partial x)\,dx + (\partial P/\partial y)\,dy. \quad (47)$$

Substituting from Eqs. (47) in Eq. (46)

$$dv = (\partial v/\partial T)((\partial T/\partial x)dx + (\partial T/\partial y)dy)) + (\partial v/\partial P)((\partial P/\partial x)dx + (\partial P/\partial y)dy),$$

and the variation of v along x for fixed y

$$|dv/dx|_y = (\partial v/\partial T)(\partial T/\partial x) + (\partial v/\partial P)(\partial P/\partial x). \quad (48)$$

Using the relation v = RT/P

$$|dv/dx|_y = (R/P)(\partial T/\partial x) + (-RT/P^2)(\partial P/\partial x). \quad (49)$$

Similar arguments apply if T = T(x,y,z), and P = P(x,y,z).

9. Stokes and Gauss Theorems

These theorems are required in order to convert equations from integral to differential forms as will be shown in Chapters 2 to 4 for energy conservation, and the entropy and availability balance equations. A brief overview of vector calculus which covers dot and cross products, the gradient of scalar, curl of a vector, and the relationship between thermodynamic properties and scalar fields is presented in the Appendix.

a. *Stokes Theorem*

The relation between the line integral (KMNP) and the surface integral (illustrated in Figure 14) is given by the relation

$$\oint \vec{F} \cdot d\vec{s} = \oint_{cs} (\vec{\nabla} \times \vec{F}) \cdot d\vec{A}, \tag{50}$$

where $\vec{\nabla} \times \vec{F}$ is the curl of $\vec{F}$. The area vector $d\vec{A}$ is the outer normal perpendicular to the surface (e.g., for a closed curve lying in the x–y plane, the area lies in the x–y plane and, if the integration for the line integral is performed in a counter–clockwise direction, the area vector faces outward similar to a screw moving out of a surface). The integral over the area simplifies to $(F_x y - F_y x)$.

b. Gauss–Ostrogradskii Divergence Theorem

As illustrated in Figure 15, the relation between surface and a volume integral is given by

$$\oint_{cs} \vec{F} \cdot d\vec{A} = \oint_{cv} (\vec{\nabla} \cdot \vec{F}) \cdot dV, \tag{51}$$

If $\vec{F} = \rho\vec{v}$, the Gauss divergence theorem yields $\oint_{cs} (\rho\vec{v}) \cdot d\vec{A} = \oint_{cv} (\vec{\nabla} \cdot (\rho\vec{v})) \cdot dV$, where ρ denotes density, $\vec{v}$ velocity, and $\rho\vec{v}$ the mass flux per unit area. The control surfaces cs (comprising surfaces ABCD, EFGH, BFGC, CDHG, HDAE, and ABFE) enclose the control volume cv. If the total mass flux leaving the volume from all of the surfaces is known, that flux must equal the flux $\vec{\nabla} \cdot (\rho\vec{v})$ leaving a small elemental volume integrated over the entire volume.

c. The Leibnitz Formula

If the gas contained within a balloon is discharged, the balloon volume shrinks, and the mass contained in it decreases. The rate of change of the mass can be determined by applying the Leibnitz Formula, i.e.,

$$\frac{\partial}{\partial t} \iiint_{V(t)} \rho dV = \iiint_{V(t)} \frac{\partial \rho}{\partial t} dV + \iint_{A(t)} \rho \vec{v}_d \cdot d\vec{A}, \tag{52}$$

where $\vec{v}_d(t)$ denotes the instantaneous deformation velocity of the balloon. This formula is useful for solving problems related to the material covered in Chapter 2 that involve deformable control volumes. In the case of a balloon releasing gas, the balloon shrinks and the $\vec{v}_d(t)$ vector is inward, while the area vector is outward, and $\vec{v}_d \cdot d\vec{A} < 0$. On the other hand, if gases are pumped into the balloon, it expands, so that $\vec{v}_d \cdot d\vec{A} > 0$.

D. OVERVIEW OF MICROSCOPIC THERMODYNAMICS

In order to understand the physical processes governing behavior in thermodynamic systems, such as the variations in energy and temperature with work and heat input; the relations between pressure and temperature in gases, liquids, and solids; the directions of heat and mass transfer and chemical reactions; the relation between the saturation pressure and temperature, etc., we must understand the microscopic behavior of molecules constituting the matter of those systems. This understanding is also useful in interpreting many classical thermodynamic relations. A detailed treatment of microscopic thermodynamics is beyond the scope of this text, and, therefore, only a brief overview of the subject is presented herein.

1. Matter

Feynman describes matter as follows: "...all things are made of atoms – little particles that move around in perpetual motion attracting each other when they are a little distance apart, but repelling upon being squeezed into one another." Atoms are of the order of 1–2 Angstroms (i.e., $1–2\times10^{-10}$ m) in radius. The water molecule, H_2O, is a heteronuclear molecule consisting of two atoms of H (located apart by 105°) separated from one atom of O by a distance of about 1 Å (Figure 16). Adjacent water molecules are separated by an intermolecular distance ℓ. The variation of the intermolecular force F between molecules as a function of ℓ is illustrated in Figure 17. In a piston–cylinder–weight assembly, this distance can be varied by varying the

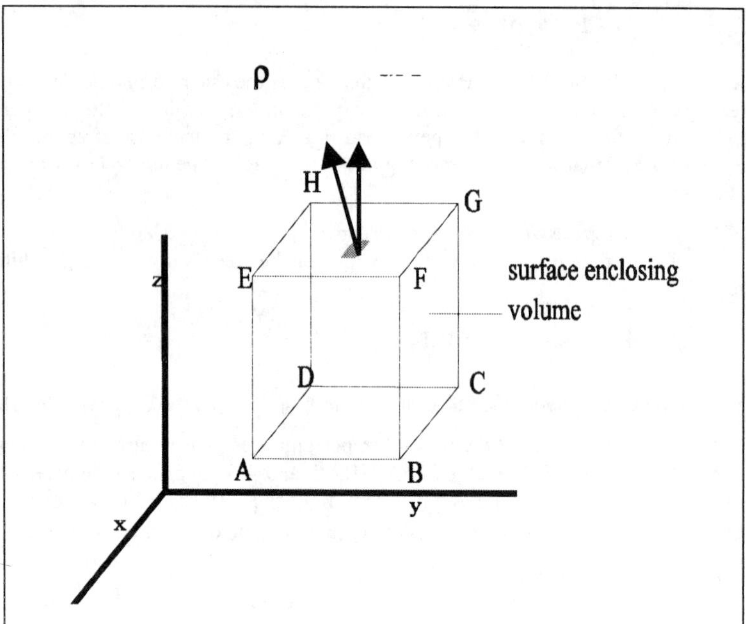

Figure 15: Surface and volume integrals used in the Gauss divergence theorem.

volume through the addition or removal of weights. The intermolecular force is negative when attractive, i.e., it attempts to draw molecules closer together, while positive forces correspond to closer intermolecular spacing and are repulsive, i.e., they attempt to move the molecules away from each other. The distances ℓ_i, ℓ_m, ℓ_0 and σ that are illustrated in Figure 17 will be described later.

2. Intermolecular Forces and Potential Energy

Consider the earth's mass m_E (whose origin is at its center). Newton's law of gravitation states that the force F exerted by the earth towards its origin on another mass m located at a distance r is given by the relation $F = C\ mm_E/r^2$, where C is the gravitational constant. In vector form

$$\vec{F}(r) = C\ mm_E\ \vec{r}/|r^3|, \tag{53}$$

where $C = 6.67\times10^{-8}$ N m^2 kg^{-2} , and $\vec{F}(r)$ in units of N. The force exerted on a unit mass by the earth, i.e., its gravitational acceleration, $\vec{g}(r) = C\ m_E\ \vec{r}/|r^3|$. (If $\vec{F}(r)$ is an attractive force, it carries a negative sign, since it acts towards the origin. Typically, $\vec{g}<0$, since it is attractive towards the earth, and, in order to move a mass away from the earth through a distance $d\vec{r}$, work must be done to overcome the earth's attractive force.) Therefore, the work done upon a mass m, i.e., the work input to raise that mass, is given by

$$\delta\phi = \delta W = -\vec{F}(r)\cdot d\vec{r}. \tag{54}$$

We see from Eq. (54) that $\delta W/d\vec{r} = -\vec{F}(r)$. Using the relation for the gravitational acceleration, the work performed to raise a unit mass is

$$w = W/m = \phi_g = -Cm_E/r + C_1,$$

where ϕ_g is known as the gravitational potential. As $r\to\infty$, . $\phi_g \to C_1$ so that $C_1=0$. Therefore,

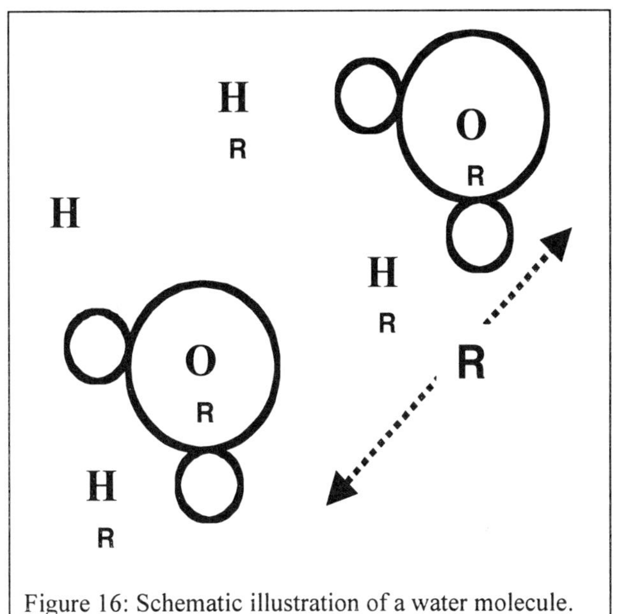

Figure 16: Schematic illustration of a water molecule.

$$\phi_g = -Cm_E/r, \qquad (55)$$

and $d\phi_g/d\vec{r}$ represents the gravitational force exerted on a unit mass. The energy stored in a mass under the influence of the earth's gravitational field grows with an increase in the distance r. This gravitational potential energy is similar to the energy contained within a raised weight that induces it to fall unless it is constrained. Similarly, work must be performed to move in a charge of Q_c coulombs through an electrical potential.

Likewise, if a molecule A is located at an origin and molecule B is situated at a distance ℓ removed from it, the potential energy stored within the molecule can be determined if the characteristics of the force field are known. Alternatively, if the potential is known, the force exerted by a molecule on another can be determined (as illustrated above by the derivative $-d\phi_g/d\vec{r}$). The Lennard–Jones' (LJ) (6-12 law) empirical approach for like molecular pairs, such as the homonuclear molecular pair O_2–O_2, furnishes the intermolecular potential energy in the form

$$\Phi(\ell) = 4\varepsilon\,((\ell_0/\ell)^{12} - (\ell_0/\ell)^6), \qquad (56)$$

where ε represents the characteristic interaction energy between molecules, i.e., the maximum attraction energy or minimum potential energy Φ_{min} ($\varepsilon = \Phi_{min} \approx 0.77\ k_B\ T_c$, with k_B denoting the Boltzmann constant and T_c the critical temperature), ℓ_0 represents the distance at which the potential is zero (cf. Figure 17) and is approximately equal to the characteristic or collision diameter σ of a molecule at which the potential curve shown in Figure 17 is almost vertical. Tables A-3 tabulate σ and ε/k_B (in K) for many substances. The term k_B is called Boltzmann constant ($= \bar{R}/N_{Avog} = 1.33\times10^{-26}$ (kJ /molecule K)). In order to calculate the minimum potential energy ℓ_{min}, Eq. (56) can be differentiated with respect to ℓ and set equal to zero. From this exercise $\ell_{min}/\ell_0 = 2^{1/6} = 1.1225$, and the corresponding value of $\Phi_{min} = \varepsilon$. Hence,

$$\Phi(\ell)/\ \Phi_{min} = 4((\ell_0/\ell)^{12} - (\ell_0/\ell)^6), \qquad (57)$$

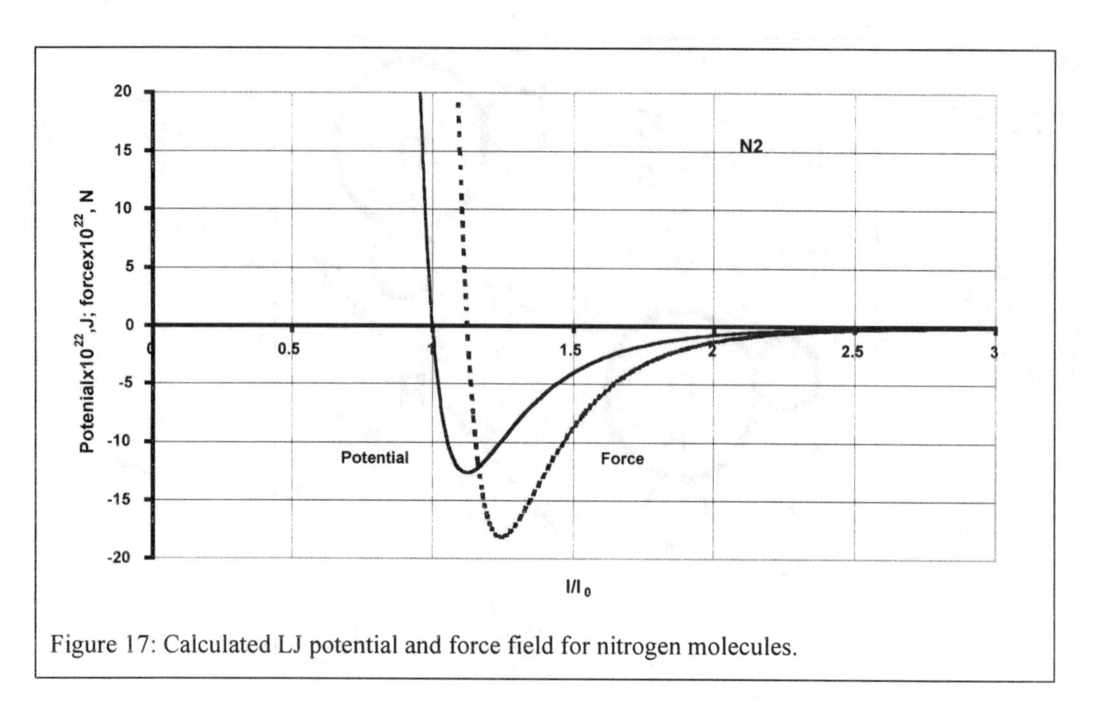

Figure 17: Calculated LJ potential and force field for nitrogen molecules.

Figure 18 presents a plot of the nondimensional intermolecular potential with respect to ℓ/ℓ_0. If we approximate an ideal gas as that gas where $|\phi\ (\ell_{limit})\ | \approx 0.01|\ \phi_{min}|$, then we obtain ℓ_{limit}/ℓ_0 = 3.075 (from Eq. (57)) . Note that we are comparing attractive potential of gases with those of maximum potential (i.e of liquids/solids). A better definition will be given in Chapter 6. The interaction force between the molecules is given by the relation $F(\ell) = -d\Phi/d\ell$, so that $F(\ell)/\ \Phi_{min} = (4/\ell_0)(12(\ell_0/\ell)^{13} - 6(\ell_0/\ell)^7)$. The maximum attractive force occurs at $\ell_{max}/\ell_0 = 1.2445$, and the corresponding force $|F_{max}| = 2.3964\ |\Phi_{min}|\ell_0$. Therefore,

$$F(\ell)/\ |F_{max}| = -0.599(-12(\ell_0/\ell)^{13} + 6\ (\ell_0/\ell)^7). \tag{58}$$

It is seen from Eqs. (58) and (53) that gravitational forces are proportional to masses (independent of the chemical composition) and inverse of distance square while the LJ forces are inversely proportional $1/l^7$, and depends upon the chemical composition of the masses. Assuming ℓ_0 to equal σ, for molecular nitrogen $\sigma = \ell_0 = 3.681$ Å and $\varepsilon/k_B = 91.5$ K. Using the value of $k_B = 1.38\times10^{-23}$ J molecule^{-1} K^{-1}, Φ and F can be determined for given values of ℓ_0/ℓ. Results are presented for molecular nitrogen in Figure 17.

If the molecules are spaced relatively far apart, the attractive force is negligible. Ideal gases fall into this regime. As the molecules are brought closer together, although the attractive forces increase, the momentum of the moving molecules is high enough to keep them apart. As the intermolecular distance is further decreased, the attractive forces become so strong that the matter changes phase from gas to liquid. Upon decreasing this distance further, the forces experienced by the molecules become negligible (i.e., $d\Phi/d\ell = 0$ or Φ is maximized), and the matter is now a solid in which the molecules are well–positioned.

From Eq. (58) we see that the attractive force $F(l) \propto (\ell^3)^{-7/3}$ has units of approximately (volume)$^{-2}$. This concept can be used in developing van der Waals' equation of state (see Chapter 6). The LJ relation assumes the force field to be spatially symmetric around the molecule, an assumption which is valid over a wide range of conditions for gases such as O_2, N_2, and He and the other noble gases. However, this is not necessarily true for polar molecules such as H_2O (cf. Figure 16 in which H–atoms are positively charged and O–atom is negatively charged, since the O–atom pulls electrons away from H–atoms due to its heavier mass) and NH_3. For the sake of illustration we will assume the LJ relation to also hold for polar gases.

3. Internal Energy, Temperature, Collision Number and Mean Free Path

a. *Internal Energy and Temperature*

At low pressures and high temperatures the intermolecular spacing in gases is usually large and the molecules move incessantly over a wide range of velocities. The molecules also vibrate and rotate. The total energy possessed by them is due to these translational, rotational, and vibrational modes (Figure 19).

For the sake of illustration, consider H_2O vapor–phase molecules at a pressure of 1 bar and a temperature of 200°C. Typically, these molecules move with an average velocity of 350 m s^{-1} at temperatures around 300 K. Since $\ell \gg 3\ell_{imit}$, attractive forces can be ignored. As the water vapor is compressed, the intermolecular distance decreases and attractive forces become significant as the gas reaches a certain volume (or pressure). Upon further compression, the attractive forces become so strong that the vapor changes phase to become liquid. According to liquid cell theory, each molecule is confined to a small cell of volume v′ (which is the total volume divided by the number of molecules contained in it). If the molecular diameter is small compared to the cell volume, a molecule is free to move within its cell without interacting with its nearest neighbors. Therefore, the translational energy of that molecule decreases, although it possesses the same rotational and vibrational energies. As the liquid is further compressed it becomes a solid. The interactions of a molecule with its neighbors are strongest when motion is restricted to conditions corresponding to the minimum potential energy, i.e., when $\ell = \ell_{min}$. At this state the molecules possess most of their energy in the vibrational mode. The relative position of molecules (or their configuration) is fixed in solids. Gases correspond to the other extreme and contain a chaotic molecular distribution and motion. Liquids fall in a regime intermediate between gases and solids, since their molecular kinetic energies are comparable to the maximum potential energies. Therefore, the molecular energy changes significantly with compression and phase change.

The position of an atom within a molecule can be fixed by three spatial coordinates (say, x,y and z). A polyatomic molecule containing δ atoms requires 3δ coordinate values in order to fix the atomic positions, and, consequently, has 3δ degrees of freedom. Molecules can have three translational energy modes. A monatomic gas (δ =1) has three translational energy modes, and a linear molecule such as CO_2, which has all of its atoms arranged in a straight line, possesses two rotational degrees of freedom (since rotation about its own axis is negligible) while H_2O, which is a nonlinear molecule, possesses three rotational degrees of freedom. Therefore, the number of vibrational energy modes for a nonlinear molecule must be equal to the difference between the total degrees of freedom and the sum of the translational and rotational energy modes, i.e., (3δ–6). Since a linear molecule possesses three translational and two rotational modes, its vibrational energy modes must number (3δ–5). The total energy associated with a molecule $u' = e'_T + e'_R + e'_V$ is known as the molecular internal energy, where e'_T, e'_R, and e'_V, respectively, represent the total translational, rotational, and vibrational energies of that molecule.

b. *Collision Number and Mean Free Path*

Molecules contained in matter travel a distance ℓ_{mean} before colliding with another molecule. Consider a molecule A that first collides with another molecule after traveling a distance ℓ_{mean}, then undergoes another collision after moving a distance of $2\ell_{mean}$, and so on, until colliding for the Nth time with another molecule after having moved along a distance $N\times\ell_{mean}$. If these N collisions occur in one second, the molecule A is said to undergo N collisions per unit time (also known as the collision number).

If the molecular diameter of a molecule is σ (also called the collisional diameter), this is the closest distance at which another molecule can approach it. At this distance the repulsive force between the two molecules is infinitely large as shown in Figure 17. Assume that the average molecular velocity V_{avg} is the distance through which the molecule travels in one second. Now consider a geometrical space shaped in the form of a cylinder of radius σ and length V_{avg}. There are $n'\pi\sigma^2 V_{avg}$ molecules within this cylinder where n′ denotes the number of molecules per unit volume. A molecule traveling through the cylinder will collide with all of the molecules contained within it, since the cylinder radius equals σ. Therefore, the number of collisions occurring per unit time Z_{coll} is $n'\pi\sigma^2 V_{avg}$, and the time taken for a single collision is the inverse of this quantity. The average distance traveled by the molecule during this time is called its mean free path ℓ_{mean}, where

$$\ell_{mean} = V_{avg}/(n'\pi\sigma^2 V_{avg}) = 1/(n'\pi\sigma^2).$$

Another relation for the mean free path is

$$\ell_{mean} = 1/(2^{1/2}\pi n'\sigma^2).$$

Typically, the number of collisions is of the order of $10^{39} m^{-3} s^{-1}$. The mean free path is also the average distance between adjacent molecules. For instance, consider a room consisting of N_2 molecules at 298K, 1 bar. Then $n' = 2.43 \times 10^{25}$ molecules/m^3, σ= 3.74 Å, and ℓ_{mean} = 0.0662 μm or 662 Å or 66.2 nm.

All of the molecules do not travel at the average velocity. The typical velocity distributions (also called the Maxwellian distributions) of helium molecules at different temperatures are illustrated in Figure 20. The typical velocity distributions can be determined from the expression

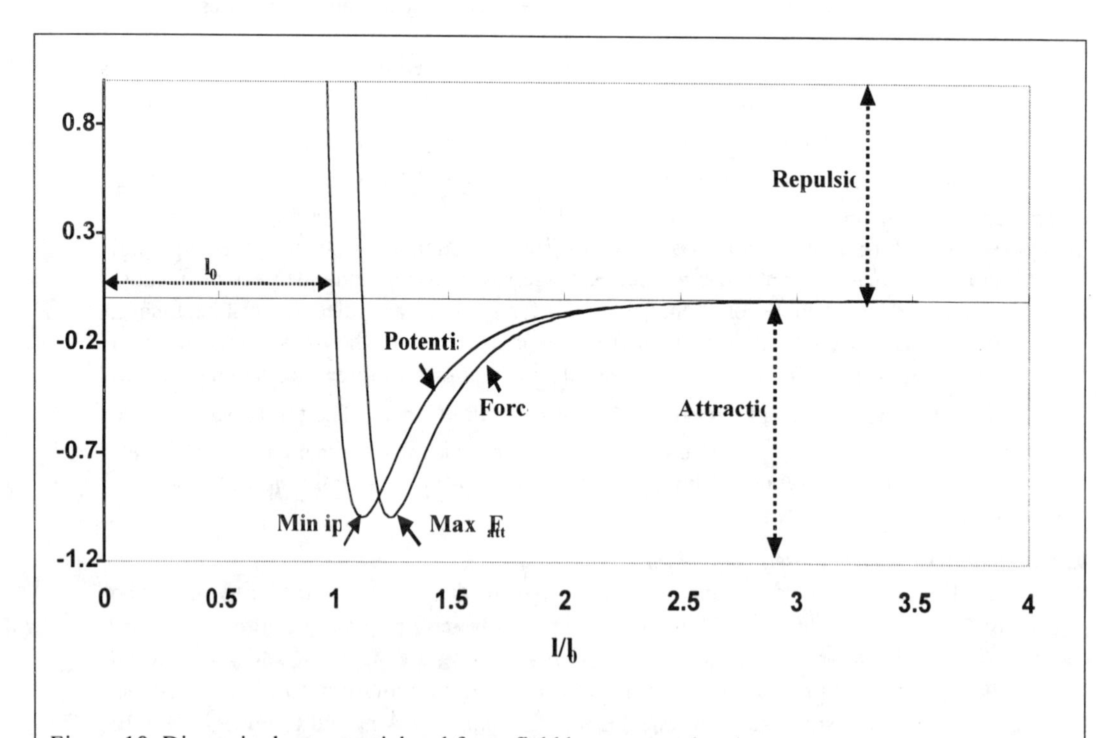

Figure 18. Dimensionless potential and force field between molecules.

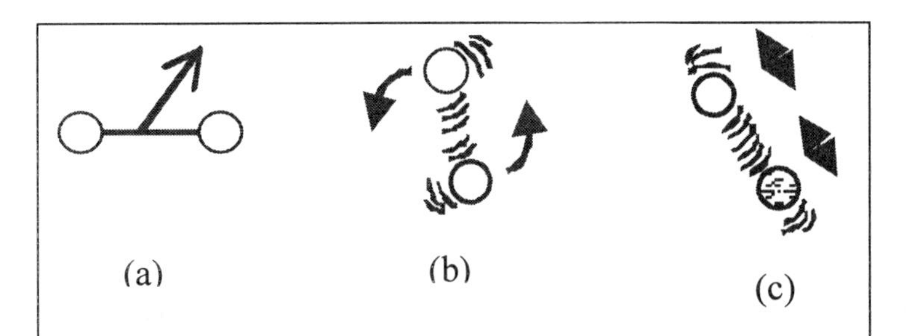

Figure 19. Illustration of the energy modes associated with a diatomic molecule. (a) Translational energy (TE). (b) Rotational energy (RE). (c) Vibrational energy (VE).

$$(1/N)(dN'_V/dV) = 4\pi^{-1/2}\,(m/(2k_BT))^{3/2} V^2 \exp(-(1/2)mV^2/(k_BT))) \quad (59)$$

where N'_V represents the number of molecules moving with a velocity in the range V and V+dV, N the total number of molecules, m the molecular mass (= M/N_{Avog}), with M denoting the molecular weight). Therefore, the translational energy varies among the molecules, and integration of Eq. (59) between the limits V = 0 and ∞ results in a number fraction of unity. Microscopically, the molecules are in state from which the average energy is subject to perturbations of varying strengths. In Chapter 10 we will learn that these perturbations cause certain states to become stable, metastable, or unstable.

Equation (59) can be rewritten in terms of the energy $e = mV^2/2$ and integrated to obtain the fraction of molecules possessing energy in the range from E to ∞, i.e.,

$$N'_E/N = 2\pi^{-1/2}((E/(\bar{R}T))^{1/2}\exp(-(E/(\bar{R}T)) + (1 - \mathrm{erf}((E/\bar{R}T)^{0.5})), \quad (60)$$

where $E = e'N_{Avog} = M\,V^2/2$, M denotes the molecular weight (or the mass of 1 kmole), and $\bar{R} = k_B\,N_{Avog}$ is the universal gas constant. As E→0, so does the error function and the first term in Eq. (60), and, therefore, as is logical, the term $(N_{0\le E\ge\infty}/N)\to 1$. This term becomes negligibly small as E→∞, since the volume fraction of molecules associated with extremely large energies normally approaches zero. Since $E/\bar{R}T$ is typically large, the value of the last term on the RHS of Eq. (60) is negligibly small. Hence, the fraction of molecules with a velocity in the range V to ∞ (or E≤E≤∞) may be expressed as

$$N'_V/N = 2\pi^{-1/2}(E/(\bar{R}T)^{1/2}\exp(-(E/(\bar{R}T)). \quad (61)$$

Equation (61) indicates that the fraction of molecules associated with an energy of value E and greater is proportional to $\exp(-(E/(\bar{R}T))$. Chemical reactions between reactant molecules occur when the energy E exceeds the minimum activation value, which is required to overcome the molecular bond energies, thereby allowing the atoms to be rearranged in the form of products.

The average molecular speed V_{avg} is

$$V_{avg} = (8/(3\pi)]^{1/2}\, V_{rms} = (8\,k_BT/(\pi m)\}^{1/2} = (8\,\bar{R}T/(M\,\pi)\}^{1/2} \quad (62)$$

Where m is the mass of molecule and the expression for the most probable speed is

$$V_{mps} = (2/3)^{1/2}\, V_{rms} = (2k_BT/m)^{1/2} = (2\,\bar{R}T/M)^{1/2}. \quad (63)$$

The root mean square speed V_{rms} can be expressed as

$$V_{rms} = (3k_BT/m)^{1/2} = (3\,\overline{R}\,T/M)^{1/2}. \quad (64)$$

where $V_{rms}^2 = V_x^2 + V_y^2 + V_z^2$ is based on the three velocity components, and

$$mV_{rms}^2 = m(V_x^2 + V_y^2 + V_z^2)/2 = (3/2)k_BT.$$

From Eq. (64) note that average te per molecule $3k_B T/2$ where $k_B = \overline{R}/N_{Avog}$. It is customary to assume three velocity components to equal each other in magnitude, i.e., each translational degree of freedom contributes energy equivalent to $(1/2)k_BT$ to the molecule. At standard conditions $V_{rms} \approx 1770$, 470, and 440 m s^{-1}, respectively, for H_2, N_2 and O_2, and is typically of the same magnitude as the sound speed in those gases. Recall that for an ideal gas the sound speed $c = (k\overline{R}T/M)^{1/2}$, where $1 \le k \le 5/3$. For gaseous N_2 and H_2, respectively, at standard conditions $V_{avg} \approx 475$ and 1770 m s^{-1}; $m = 4.7\times10^{-26}$ kg and 0.34×10^{-26} kg; $\sigma = 3.74$ Å and 2.73 Å; $\ell = 650$ Å and 1230 Å; and $Z_{coll} = 7\times10^9$ and 14.4×10^9 collisions s^{-1}. Recall that for an ideal gas the sound speed $c = \sqrt{k\overline{R}T/M}$, where $1 < k < 5/3$. The sound speed is comparable to average molecular velocity.

i. Monatomic Gas

The only molecular energy mode in monatomic gases is translational. Helium, argon, and other noble gases are examples of monatomic gases. The energy per molecule u′ in a monatomic gas is

$$u' = e'_T = (3/2)k_BT. \quad (65)$$

where energy per degree of freedom is given by $(1/2)(k_B T)$ and at 298 K energy per degree of freedom is given as $0.5*1.38\text{x}10^{-26}$ kJ/molec. K * 298 = $2.05\text{x}10^{-24}$ kJ/molec. Monatomic gas has 3 degrees of freedom. For a mass containing Avogadro's number of molecules N_{Avog},

$$\bar{u} = (3/2)N_{Avog}\,k_BT = (3/2)\,\overline{R}\,T\text{, i.e.,} \quad (66)$$

$T = 2/3(\bar{u}/\overline{R})$. If an ideal monatomic ideal gas is placed in a rigid container and heated, the intermolecular spacing remains unchanged and, as shown in Figure 18, the potential energy is still negligible. However, due to a rise in the translational energy, the internal energy increases.

ii. Diatomic Gas

There are three translational and two rotational modes for a diatomic gas. At low temperatures the vibrational modes can be neglected so that

$$u' = e'_T + e'_R = (5/2)k_BT. \quad (67)$$

At higher temperatures there are $(3n-5) = 1$ vibrational modes. If a diatomic molecule is visualized as two atoms attached by a spring, each vibrational mode for this combination has two degrees of freedom, i.e., due to the potential energy (that is similar to the energy stored in a spring), and to the kinetic energy of the atoms with respect to the center of mass. Each degree of freedom contributes an energy equivalent to $(1/2)k_BT$, and

$$e'_{V,diatomic} = 2(1/2)k_BT = k_BT. \quad (68)$$

At higher temperatures, since $u' = (e'_T + e'_R) + e'_V$, its value equals $(7/2)k_BT$. Therefore, for diatomic gases

$$\bar{u} = (7/2)N_{Avog}\,k_BT = (7/2)\,\overline{R}\,T\text{, i.e.,} \quad (69)$$

$$T = 2/7(\bar{u}/\overline{R}).$$

Comparing Eqs. (66) and (69) it is seen that for similar increase in $\bar{u}$, the temperature change for the diatomic molecule gas is smaller compared to a monatomic gas due to the higher energy storage capacity of the diatomic molecule.

iii. Triatomic Gas

We have seen that each vibrational mode has two degrees of freedom for linear molecules containing δ number of atoms. Therefore, linear triatomic molecules each have (3+2+(3δ −5)×2), i.e., (6δ–5) degrees of freedom, while nonlinear molecules have (3+3+(3δ–6)×2) or (6δ –6) degrees of freedom. Each mode contributes $(1/2)k_BT$ of energy. The molecular energy in a linear polyatomic molecule is

$$u' = (6\delta-5)\,(1/2)k_BT, \text{ i.e., } \bar{u} = (6\delta-5)\,(1/2)\bar{R}\,T. \quad (70)$$

Likewise, for a nonlinear molecule

$$u' = (6\delta-6)\,(1/2)k_BT, \text{ i.e., } \bar{u} = (6\delta-6)\,(1/2)\bar{R}\,T. \quad (71)$$

This simplified theory suggests that the internal energy per mole is proportional to the temperature. The translational energy $e'_T \approx 0$ for liquids, while for solids both e'_T and e'_R are negligible.

4. Pressure

When a racquetball is thrown against a wall it bounces back after impact. If several balls are thrown against the wall periodically, the impact due to the balls becomes regular and, at a high enough frequency, can be considered as a force. Similarly, the pressure that we experience is due to a continuum of matter that strikes us incessantly (as shown in Figure 21). In the case of gases, large numbers of molecules travel at high speeds at standard conditions and impinge on surfaces, thereby creating pressure. Under atmospheric conditions, the force exerted by impinging air molecules (due to their change in momentum as they strike a surface) is equivalent to placing a weight of 10^5 N on each m^2 of the surface (i.e., 100 KPa). A relatively

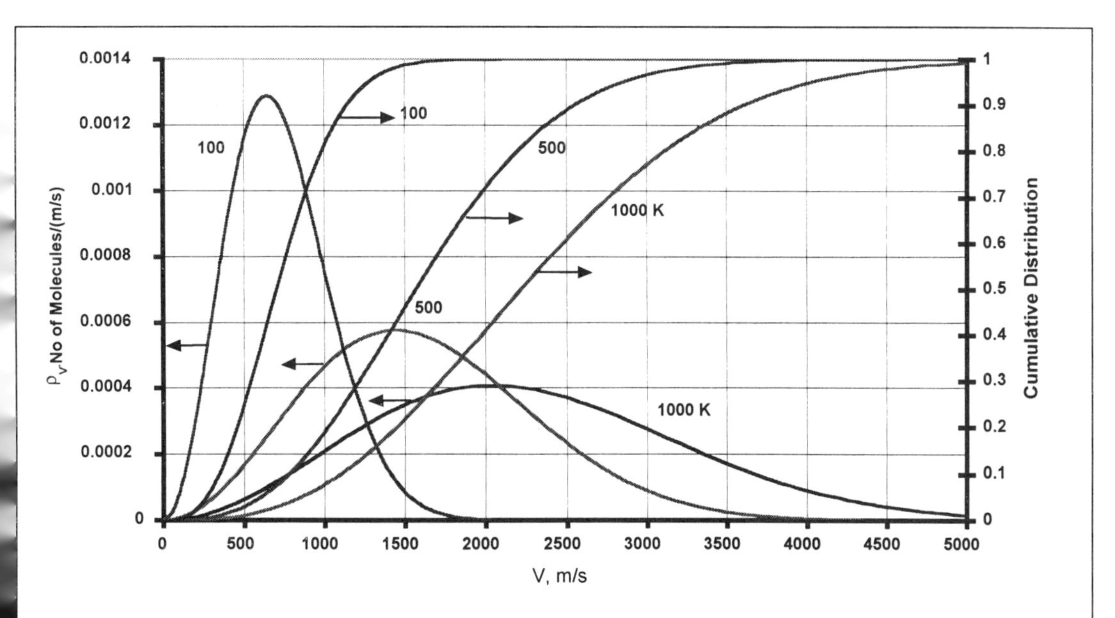

Figure 20. Maxwellian distribution of the absolute velocity in helium, which is a perfect gas. (Helium with $m = 6.65\times 10^{-23}$ g).

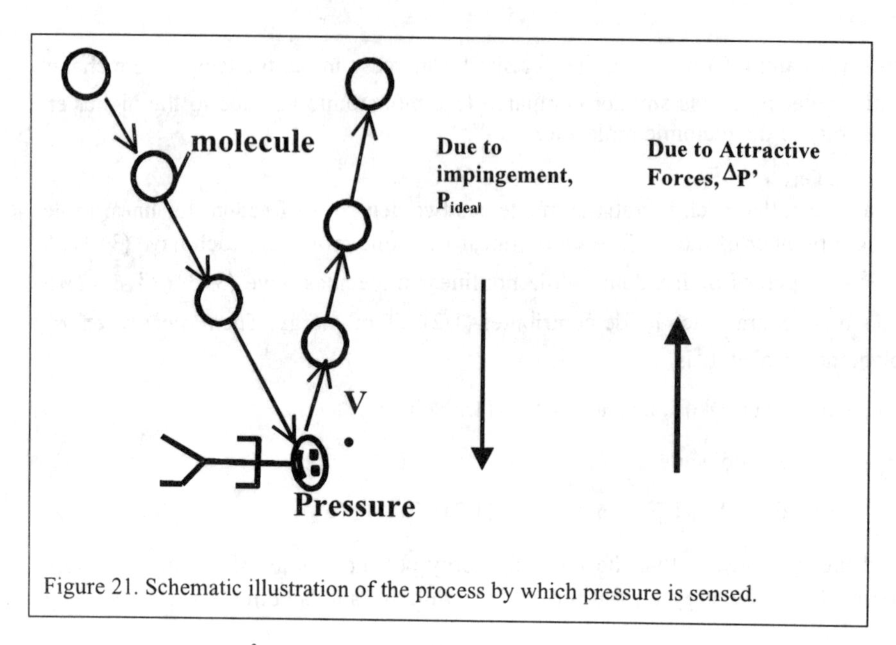

Figure 21. Schematic illustration of the process by which pressure is sensed.

small volume of 1 mm^3 can contain 4×10^{15} molecules of air at 298 K that travel with an average molecular velocity of 350 m s^{-1}, and which create a pressure of 100 KPa. If the number of molecules is doubled, but the molecular velocity and volume are held constant (i.e., the translational energy and temperature are unchanged), the pressure will double to 200 KPa. If the original number of molecules is retained within the original volume but the molecular velocity is raised, i.e., the translational energy and, thereby, temperature are increased, the pressure will also rise. Therefore, the pressure in a container can be altered by changing the molecular velocity (hence, temperature), the number of molecules per unit volume (n′), or the number of moles per unit volume (n).

a. Relation between Pressure and Temperature

Ideal Gas: We have seen that for gases the exchange of momentum is related to pressure. Consider a surface of area ℓ^2 upon which molecules impinge and apply pressure. A molecule of mass m traveling with a velocity V_y imparts a momentum mV_y to the surface followed by other similar molecules. Since ℓ also denotes the intermolecular distance, the time interval between successive collisions t_{coll} on the surface is ℓ/V_y. Therefore, the momentum impingement rate is $m\ V_y/(\ell/V_y)$ or mV_y^2/ℓ, where $V_y^2 = (1/3)V^2$. The momentum rate per unit area or pressure on the $\ell\times\ell$ area is $mV^2/3\ell^3$, provided there are no attractive forces between the gas molecules and the surface. If the molecules are assumed placed at the corners of a cube of dimension $\ell\times\ell\times\ell$, the number of molecules per unit volume n′ is approximately $1/\ell^3$. Recalling that $n' = N'/V$, where V is the volume,

$$P = n'mV^2/3 = 2N'(TE)/3V = (2/3)(3/2)N'k_BT/V = N\bar{R}T/V, \quad (72)$$

where $N = n'/N_{Avog}$ is the mole number, TE denotes the translational energy per molecule, and $\bar{R} = N_{Avog}k_B$. Equation (74) is a statement of the ideal gas law.

Air, which is a mixture, can be assumed to contain 79% nitrogen and 21% oxygen by volume. Therefore, for approximately four collisions due to N_2 molecules on a surface that is adjacent to air, one collision is due to an O_2 molecule with the consequence that 79% of the pressure felt by the surface is due to nitrogen and 21% due to oxygen. The contribution of each species to the total pressure is called its partial pressure. The partial pressure exerted by the

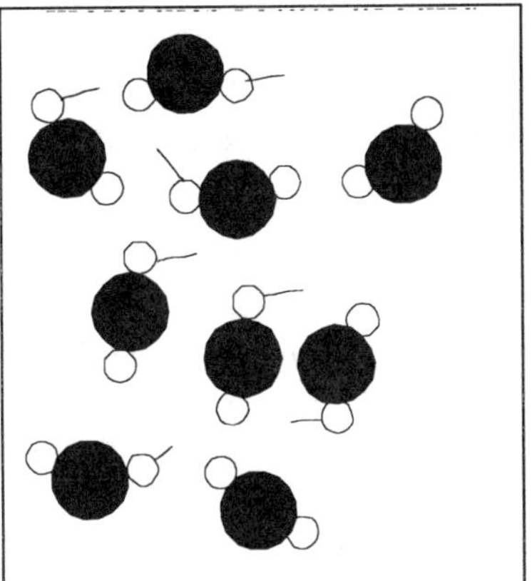

Figure 22. Illustration of the molecules contained in liquid water that are in motion relative to each another.

nitrogen in air p_{N_2} on a surface at standard conditions is 0.79 bar, whereas the partial pressure due to oxygen p_{O_2} is 0.21 bar. In general, for any species k, $p_k = N_k \bar{R} T/V$, where p_k is the component pressure due to the k–th species. Hence, the general state equation for an ideal gas mixture is given by

$$P = \Sigma p_k = \Sigma N_k \bar{R} T/V = N \bar{R} T/V, \text{ or}$$

$$p_k/P = N_k/N = X_k. \tag{73}$$

where X_k is the mole fraction of the k–th component in the mixture.

Real Gas: The relation given by Eq. (74) for the ideal gas pressure ignores the attractive forces between molecules. Consider the interior of any system (e.g., the interior of room air). There is no net attractive force between the interior molecules since the intermolecular forces cancel out. However the attractive forces between a molecule at the boundary (Fig. 19a) and the interior molecules causes a net attractive force or pressure ΔP_{attr} toward the interior, thereby reducing the ideal gas pressure (P_{ig}) caused by the exchange of momentum. The pressure of a real gas $P = P_{ig} - \Delta P_{attr}$. Recall that the ideal gas pressure is proportional to the number of molecules per unit volume (n′) and the momentum exchange by each molecule. Similarly ΔP_{attr} is also proportional to n′ and the attractive force between experienced by each molecule within n′ and all the interior molecules per unit volume (n′). Therefore, $\Delta P_{attr} \propto n'^2 \propto 1/v^2 = \mathbf{a}/v^2$ where **a** is a constant. The real gas pressure $P \approx RT/v - \mathbf{a}/v^2$. The ideal gas and real gas regimes can be delineated by comparing the pressure reduction due to attractive forces with the ideal gas pressure (details of this will be found in Chapter 6). The ideal gas assumption is valid if

$$(\bar{v}/\bar{b}) \gg 0.9/T_R^{\ 0.43}$$

where T_R denotes the reduced temperature T/T_c and $\bar{b}$ the body volume of all molecules per kmole of the substance ($\approx N_{Avog}\, \pi\, \sigma^3/6$). Further discussion is provided in Chapter 6.

Knudsen Number: The pressure relations are valid only when the surface on which a molecule impinges has a dimension much larger than the mean free path l. Consider a small particle of the order of say 0.01 μm surrounded by N_2 gas. We wish to determine the pressure exerted on this small particle. Say that the mean free path of N_2 molecule is 0.1 μm . Hence, it is possible that molecules located 0.1 μm apart may not collide at all on the surface of the particle. For such cases the pressure cannot be calculated through continuum equations. The Knudsen number is defined as

$$Kn = \ell_{mean}/d,$$

where d denotes the particle diameter. This number is useful in defining continuum properties such as the pressure, thermal conductivity coefficient, etc. If Kn «l, the continuum approximation is valid.

5. Gas, Liquid, and Solid

When matter is compressed, its molecules exist closer to each other. As the intermolecular distance is reduced, the attractive force between adjacent molecules becomes large enough to reduce the molecular velocity. Through this process gas molecules slow down to a state at which the matter changes phase and becomes liquid. The atoms (that are part of molecules) in liquids can vibrate, and molecules can rotate around each other to assume any configuration as shown in Figure 22. This rotational capability of the molecules disallows their placement at particular positions, and is a characteristic of a fluid. Liquid molecules contain negligible translational energy. The sum of their rotational and vibrational energies defines their warmth or "heat". In general, the sum of the translational, rotational, and vibrational, energies for fluids are comparable to the minimum potential energy with the consequence that fluids are mobile.

As liquids are compressed, the intermolecular distance ℓ decreases further, and the net force on the molecules (i.e., the maximum attractive potential) declines to eventually become negligibly small. Therefore, molecules cease to move around each other with the consequence that the rotational energy tends to zero, although the vibrational energy is still finite with the vibrations occurring about a fixed position ℓ_{min} (cf. Figure 23). The cessation of rotation "glues" the molecules to definite positions as shown in Figure 23, and at this fixed configuration matter becomes solid. As solids are compressed $\ell < \ell_{min}$, although individual atoms contained in the various molecules vibrate, the intermolecular forces are repulsive. Upon stretching solids, $\ell > \ell_{min}$, and the intermolecular force becomes attractive, thereby bringing the molecular configuration to its original state. If the solid temperature is raised, the molecular vibrational energy increases. The consequent rise in the vibration amplitude tends to stretch the molecules over greater distances although $\ell<\ell_{min}$. Since intermolecular attractive forces increase weakly as compared to repulsive forces, molecules can be spaced farther apart at higher temperatures, leading to their thermal expansion.

---

m. Example 13

Water is contained inside a piston-cylinder assembly.

Assuming the water to be gaseous, determine both the rms and average velocities and internal energies of the molecules at 293 K and 3000 K.

If 1 kmole of water is contained in a piston–cylinder–weight assembly, the volume of which is either 1041.5 m^3 or 0.0805 m^3, determine the average volume around each molecule.

Assuming these volumes to be spheres of radius r′, determine the sphere radii and the intermolecular spacing for the two cases.

If the collision diameter ($\approx\ell_0$) of water molecules is ≈2.56 Å (1 Å = 10^{-10} m), determine ℓ/ℓ_{max} for each of the two cases. Express the answers in terms of ℓ/ℓ_{max}. If it is assumed that 1 kmole of H_2O behaves as an ideal gas at 1041.5 m^3, determine the mean free path at 293 K. Comment on the results.

Solution

The Boltzmann constant,

$$k_B = \bar{R}/N_{Avog} = 8314 \text{ J K}^{-1} \text{ kmole}^{-1}/(6.023\times10^{26} \text{ molecule kmole}^{-1})$$

$$= 1.38\times10^{-23} \text{ J molecule}^{-1} \text{ K}^{-1}.$$

The molecular mass

$$m = M/N_{Avog} = 18.02 \text{ kg kmole}^{-1}/(6.023\times10^{26} \text{ molecule kmole}^{-1})$$

$$= 2.99\times10^{-26} \text{ kg molecule}^{-1}.$$

$$V_{rms} = (3k_BT/m)^{1/2} = (3\times1.38\times10^{-23}\times293/2.99\times10^{-26})^{1/2} = 637 \text{ m s}^{-1}.$$

$$V_{avg} = (2\pi^{-1/2})\ V_{rms} = 718.5 \text{ m s}^{-1}.$$

The energy per molecule,

$$u' = (1/2)m(V_{rms})^2 = (1/2)\ 2.99\times10^{-26}\times637^2 = 6.066\times10^{-21} \text{ J molecule}^{-1}, \text{ and}$$

$$\bar{u} = 6.066\times10^{-21} \text{ J molecule}^{-1}\times6.023\times10^{26} \text{ molecule kmole}^{-1}=3654 \text{ kJ kmole}^{-1}.$$

At 3000 K,

$V_{rms} = (3\times1.38\times10^{-23}\times 3000/2.99 \times10^{-26})^{1/2} = 2037.4 \text{ m s}^{-1}$, and $V_{avg} = 2299 \text{ m s}^{-1}$.

$u'=7.901\text{x}10^{-21}$ J/molecule , u= 4760 kJ/kmole

For a volume of 1041.5 m^3,

$$v' = \bar{v}/N_{Avog} = 1041.5 \text{ m}^3 \text{ kmole}^{-1}/(6.023\times10^{26} \text{ molecule kmole}^{-1})$$

$$= 1.73\times10^{-24} \text{ m}^3 \text{ molecule}^{-1}.$$

$$r' = (3v'/4\pi)^{1/3} = 74.46\times10^{-10} \text{ m or } 74.46 \text{ Å}.$$

$$\ell = 2r' = 2\times74.46 \text{ Å} = 148.9 \text{ Å}.$$

For a volume of 0.0805 m^3,

$v' = 13.4\times10^{-29}$ m^3 molecule^{-1}.

$r' = 3.172\times10^{-10}$ m or 3.172 Å.

$\ell = 6.34$ Å.

Furthermore, since $\ell_{min}/\ell_0=1.1225$, $\ell_{max}/\ell_0=1.2445$, and $\ell_0\approx\sigma = 2.56$ Å, $\ell_{max} = 3.19$ Å, and

$\ell/\ell_{max} = 148.91/3.19 = 46.68$ at $\bar{v} = 1041.5$ m^3 kmole^{-1}, and

$\ell/\ell_{max} = 3.85/3.19 = 1.21$ at $\bar{v} = 0.018304$ m^3 kmole^{-1}.

Finally $l_{mean}=1/(n'\pi\sigma^2$ }, n' $=6.023\text{x}10^{26}/1041.5$ $= 5.783 \text{ x}10^{23}$, $l_{mean} =1/(5.783\text{x}10^{23}$ $*\pi*(2.53\text{x}10^{-10})^2$ }

$=8.6\text{x}10^{-06}$ m or 86000 Å!

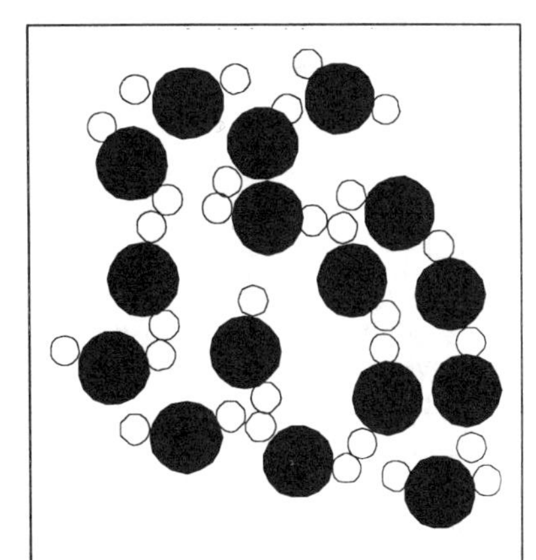

Figure 23. Illustration of ice molecules that exist in a fixed configuration with respect to each other.

Remarks

Attractive forces are negligible for the larger volume (ℓ/ℓ_{max}= 46.68) and, hence, the water molecules behave as in an ideal gas. However, upon compression to the smaller volume ℓ/ℓ_0= 1.34, and attractive forces become strong enough for the water to exist as either liquid or a solid (ice).

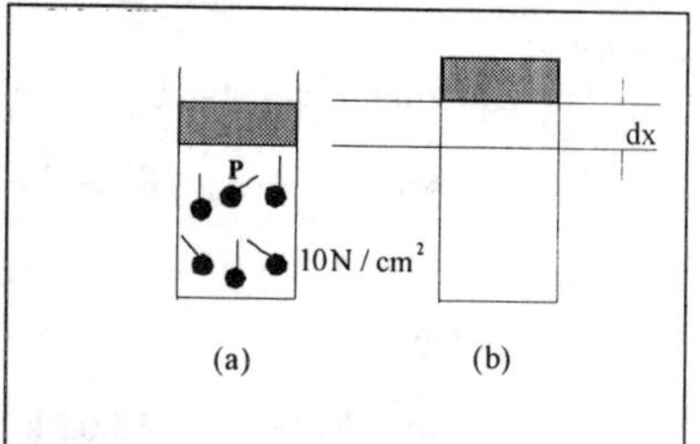

Figure 24. Schematic illustration of work being done.

If velocity of a hypothetical ideal gas tends to zero, so does u′, and, consequently, u=0. In reality, as the molecular momentum becomes negligibly small, matter is drawn together due to the intermolecular attractive forces, thereby condensing it into a liquid or a solid.
The molecules are located farther apart in the larger volume at a specified temperature or average molecular velocity. Consequently, the number of molecules per unit volume is lower than in the smaller volume, resulting in a lower pressure. Using the ideal gas law, the pressure exerted by the larger volume at T = 293 K is 0.023 bar. However, this law cannot be applied once the molecules are relatively closely spaced as in a liquid or solid, and cannot be used to predict the pressure under these conditions, since attractive forces are not considered in its development. In Chapter 6 we will discuss real gas equations of state which consider the effect of attractive forces on pressure.

6. Work

Gas molecules contained in a piston–cylinder assembly at a specified temperature move with a certain average velocity. The impact of these molecules (or the gas pressure) on the piston induces a net force on it as shown in Figure 24. This force will cause the piston to move upwards unless it is constrained by an equal force. If the constraining force is smaller, the piston will move some distance, say dx so that dV = Adx, with V and A, respectively denoting volume and area until the force exerted by the gas on the piston reduces sufficiently to equal the constraining force. Therefore, a force or pressure difference causes a volumetric change dV. The work done to accomplish this change is

$$W = F\,dx = PA\,dx = P\,dV, \tag{74}$$

where P is the pressure exerted by the gases within the cylinder at the end of expansion, and is equal to the external pressure P_{ext} exerted by the imposed constraining force. The work done is assumed reversible (or performed in quasi equilibrium such that at each step during the volume change $P_{ext} \approx P$), and the process is itself mechanically reversible (i.e., a positive or negative fluctuation in pressure within the cylinder can cause the piston to move in either direction).

Consider a piston–cylinder arrangement containing an isothermal gas that is internally divided by a partition into two unequal sections A and B, as shown in Figure 25. Assume that the smaller section A away from the piston contains a larger number of molecules per unit volume as compared to section B. Since the temperatures in both sections are identical, the molecules everywhere travel with the same average velocity. There is a pressure differential across the partition, since section A is denser than B (i.e., $P_A>P_B$). If the internal partition is removed, molecules in section A will migrate and collide with those in B. However, during a short initial period, molecules in B that are adjacent to the piston will be unaware of that migration, and the pressure exerted by them will remain unchanged at P_B. After this initial period the migrating molecules will reach all portions of B and the pressure will become uniform everywhere within the cylinder. As a result the piston will move unless constrained. In this example work is performed as a result of a nonuniform pressure difference within the system during the short ini-

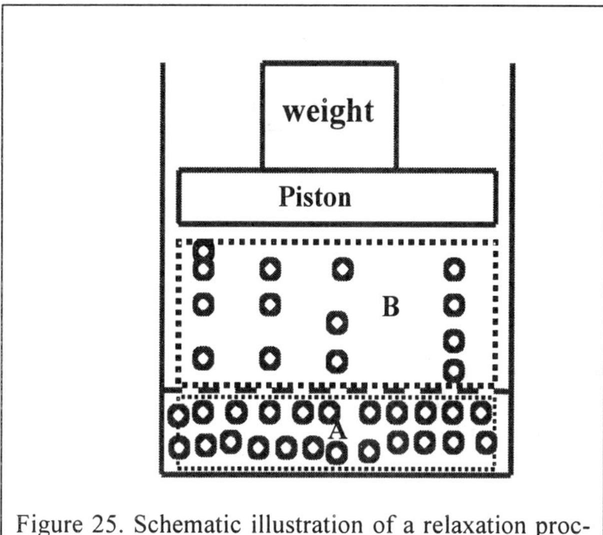

Figure 25. Schematic illustration of a relaxation process where $P_A>P_B$, and $T_A>T_B$.

tial mixing period. Hence a single value of P cannot be assigned for the whole system during the relaxation process. After this relaxation time the pressure is uniform and work ceases.

Work is a result of organized motion. Therefore, when a piston moves, gas molecules contained in a cylinder also move in the same direction. A pressure difference may be employed to accelerate these molecules in a particular direction, consistent with Newton's laws of motion. As a result of the acceleration, gas molecules can acquire a higher kinetic energy, resulting in higher temperatures.

7. Heat

Next, we discuss the concepts of thermal equilibrium and reversible heat transfer through the following example. Assume a vessel to be partitioned into two sections A and B by a hypothetical permeable surface (Figure 26). Gas is heated in section A and cooled in B. Since they are at a higher temperature (being heated), the molecules contained in section A possess greater energy than those in section B. Further, assume that at any given time N molecules from A cross the partition and randomly move into section B, while a similar number migrate from B to A so that there is no net mass transfer.

However, energy transfer occurs from A to B, since molecules migrating from section A have greater energy compared to those in B. This energy transfer occurs as heat transfer that is due to a temperature gradient serving as the driving potential. Heat transfer causes the random motion of molecules to increase in all directions, regardless of phase. If the heating of matter in section A and its cooling in section B are ceased, eventually the molecules contained in both sections will move at the same average velocity, i.e., they will be at the same temperature and, thus, have the same kinetic energy. At this state thermal equilibrium has been reached. Molecules in a system at a uniform temperature have, on average, the same translational energy. Therefore, although there is a microscopic molecular velocity distribution in the matter; there is no net exchange of energy. Under these conditions any heat transfer is reversible since at any time an equal number of molecules with the same energy crosses in either direction.

Consider a vessel containing hot liquid water placed in a room under atmospheric conditions. The water molecules possess energy in their vibrational and rotational modes, with each mode contributing energy equivalent to $(1/2)k_BT$. The water molecules transfer energy from these modes to the gas molecules in air that impinge on the liquid. In turn, these gas molecules transfer energy to other gas molecules that are farther removed from the liquid surface, and so on. Eventually, the water cools and air heats until the liquid and gas are in thermal equilibrium, i.e., they exist at the same temperature. At this state since not all molecules within the liquid or gas have the same energy, energy (heat) transfer still occurs between the low and high energy molecules, although there is no net energy exchange. Heat transfer under these conditions is reversible.

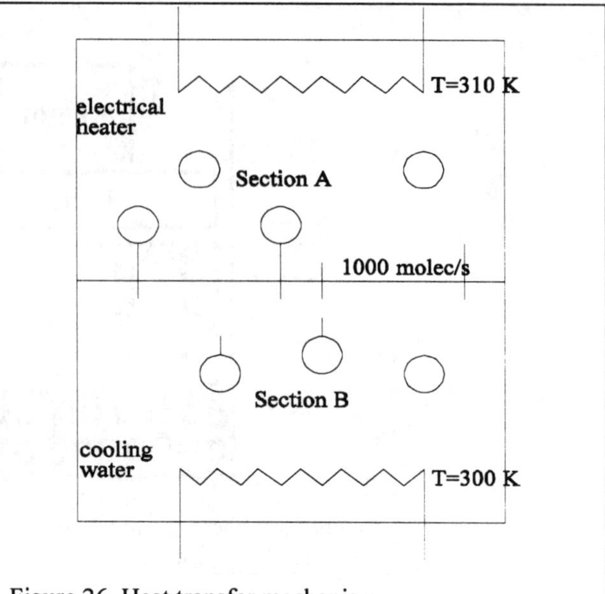

Figure 26. Heat transfer mechanism.

8. Chemical Potential

The chemical potential drives mass (or species) transfer in a manner similar to the thermal potential that drives heat transfer from higher to lower temperatures.

a. Multicomponent into Multicomponent

Consider a vessel divided into two sections C and D (as shown in Figure 27) that initially contains oxygen throughout, and in which charcoal is spread over the floor of section D. Assume that as the charcoal is burned, sections C and D consist of two components: oxygen and CO_2. Further, consider a specific time at which the mole fraction of O_2 in section C (say, x_{O_2}= 80%) is larger compared to that in section D (say, X_{O2} = 30%).

Since molecules move randomly, for every 1000 molecules that migrate from C into D through the section Y–Y, 1000 molecules will move from D into C. Consequently, 800 molecules of O_2 will move into D while only 300 molecules of this species will migrate to C from D, so that there is net transfer of 500 molecules of O_2 from section C into D. Simultaneously, there is a net transfer of 500 molecules of CO_2 across the Y–Y plane from section D into C. The oxygen transfer enables continued combustion of the charcoal. This mass transfer (or species transfer) due to random molecular motion is called diffusion.

The chemical potential μ for ideal gases is related to the species concentrations (hence, their mole fractions). A higher species mole fraction implies a higher chemical potential for that species. For instance, the chemical potential of O_2, μ_{O_2} is higher in section C compared to D, thereby inducing oxygen transfer from C to D. If the charcoal is extinguished, CO_2 production (therefore, O_2 consumption) ceases, and eventually a state of species equilibrium is reached. At this state the chemical potential of each species or its concentration is uniform in the system.

b. Single Component into Multicomponent

Consider the following scenario. A vessel is divided into two sections E and F by a porous membrane, as shown in Figure 28a. Section E initially contains a single component (denoted by o) at a lower pressure, and Section F contains a multicomponent gas mixture at the same temperature, but at double the pressure. Assume that the mole fraction of o molecules in section F is initially $x_{o,F}$ = 0.2, and that there are 50 molecules per unit volume contained in section E and 100 molecules per unit volume in section F. Further, assume the porosity of the membrane to be selective such that it allows only o molecules to be transferred through its pores (i.e., it is a semipermeable membrane). Assuming 200 molecules s^{-1} of o to migrate from E into F, 400 molecules of all species will attempt to transfer into E from F due to the higher pressure in that section. However, the semipermeable membrane allows only o molecules to

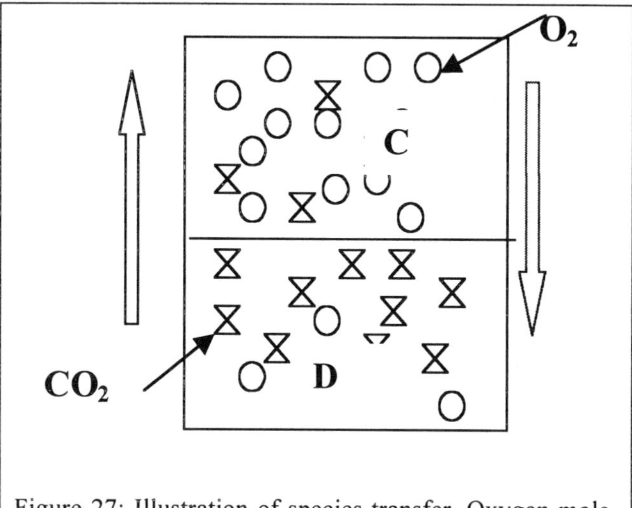

Figure 27: Illustration of species transfer. Oxygen molecules are denoted by o and CO_2 molecules by x.

transfer from F, so that of these 400 only the 80 molecules of o move from F into E. Therefore, there is net flow equal to (200–80)=120 molecules s^{-1} from E into F. If the pressure in section F is increased eightfold, molecules of species o can no longer be transferred into it, since of the 1600 molecules that now attempt to migrate every second, the membrane allows only the 320 which are of o to move into section E (cf. Figure 28b). The net motion is 320 – 200 = 120 molecules s^{-1} into section E from F.

Therefore, by adjusting the pressure in section F, we can control the direction of species transfer, or prevent it altogether by maintaining chemical equilibrium. For example, if under these conditions, the pressure in section F is five times that in E, 1000 molecules s^{-1} attempt to migrate from F to E, but only 200 molecules s^{-1} of o actually do, balancing the transfer of the same amount from E to F. The chemical potential of species o becomes uniform across the membrane at this state. Altering the pressure from this condition will change the chemical potential. In general, the larger the pressure, the higher the chemical potential.

9. Boiling/Phase Equilibrium

a. *Single Component Fluid*

Consider an open vessel that is partly filled with liquid water and placed in the atmospheric, as shown in Figure 29a. If the water is heated, its molecular energy and intermolecular distances increase. While the molecules in the interior of the liquid are surrounded in all directions by molecules exerting very strong attractive forces, those near the surface are partially unbalanced being somewhat weakly attached. Upon further heating, the intermolecular spacing keeps increasing, and the rotating molecules near the water surface attain sufficient rotational energy to overcome the attractive forces. At this point these molecules move (or escape) into the space occupied by air and/or water vapor. This process, whereby molecules are removed from the liquid mass into the vapor space is called evaporation. Likewise, water vapor molecules can approach the liquid surface and be pulled (or captured) into the liquid phase by the strong attractive forces exerted by the liquid molecules. This process is called condensation.

Consider a closed evacuated vessel into which a small quantity of water is injected and then heated to a temperature T, as shown in Figure 29b. Upon heating, the pressure in the vapor/gas phase increases as the initially liquid molecules transform into it. Initially, due to the sparse population of gas-phase molecules the return rate to the liquid will be lower compared to the escape rate into gas phase. As the pressure of the vapor/gas phase rises, the return rate to

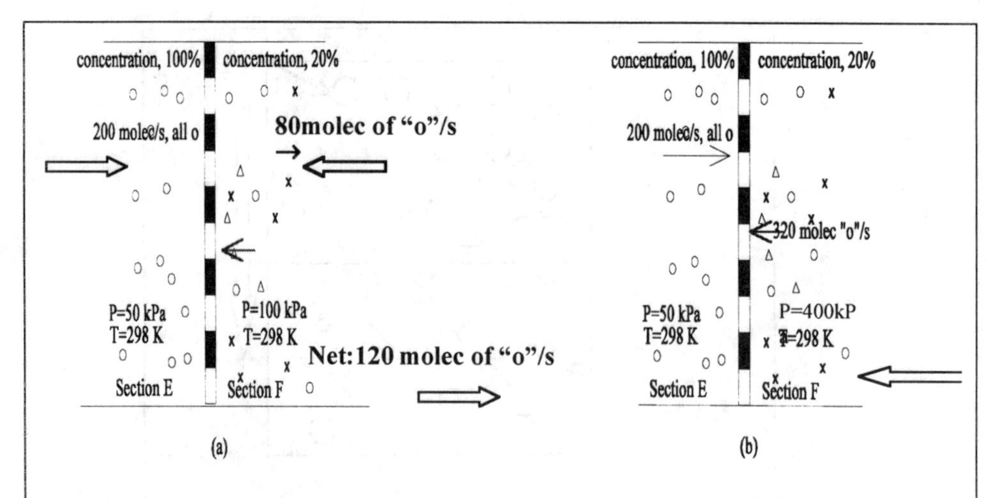

Figure 28: Illustration of a semipermeable membrane that allows species transfer from (a) Section E to F; (b) Section F to E.

the liquid phase also increases. Eventually a condition will be reached at which the return and escape rates equal each other. At this state of phase equilibrium the water level will remain unchanged over time. For any species k at a specific temperature, this pressure is called the saturation pressure $P^{sat}{}_{k}$. In a microscopic sense, the net evaporation rate = escape rate to the vapor phase return rate to the liquid and hence if there is evaporation, there is no absolute phase equilibrium.

In a microscopic sense, (the net evaporation rate) = (escape rate to the vapor phase) – (return rate to the liquid) and, hence, in the case of evaporation there is no absolute phase equilibrium. However, since there may be trillions of molecules crossing the interface at any time, a few million molecules evaporating and condensing per unit time will not significantly affect the phase equilibrium properties. As the liquid temperature is raised, the molecular energy increases and, therefore, more molecules escape the liquid into the gas phase. If phase equilibrium is to be maintained at this stage, the vapor pressure should be increased such that the capture rate of vapor molecules into liquid equals the escape rate of liquid molecules into vapor. The saturation pressure increases with a rise in temperature (as seen in the Steam Tables A-4).

Sublimation of solids into vapor occurs when the vibrational energy is high enough to overcome the intermolecular forces within the solid. As with the liquid–vapor interface discussed above, at the same time vapor molecules strike the solid surface and are captured into the solid phase. Phase equilibrium is achieved when the escape rate from the solid equals the capture rate of impinging gaseous molecules.

b. Multiple Components

At a temperature of 50°C, water molecules have stronger attractive forces as compared to ethanol molecules, since ethanol is highly volatile. Therefore, water attains phase equilibrium at a relatively lower vapor pressure (the saturation pressure of water at 50°C is 10 kPa while that of ethanol is 40 kPa). Consider the system illustrated in Figure 30 that contains a 90% water and 10% ethanol mixture at 50°C. Assume that the escape and capture rates both equal 1000 molecules s^{-1} cm^{-2} for water, and 4000 molecules s^{-1} cm^{-2} for ethanol. Typically, the capture rate is proportional to the saturation pressure. Due to mixing at the molecular level, of each cm^2 of surface area 0.1 cm^2 corresponds to ethanol molecules and the rest, i.e., 0.9 cm^2, to those of water. Therefore over each square cm of mixture surface, the number of ethanol molecules that escape equals 4000×0.1 = 400. In order for phase equilibrium to prevail the ethanol condensation rate should also equal 400 molecules cm^{-2}. Equilibrium with respect to water requires a condensation rate of 900 molecules cm^{-2}. The gas phase consists of both species in some proportion (say, x_w and x_e that, respectively, denote the vapor mole fractions of water and ethanol). Therefore, phase equilibrium at 50°C requires that a total of 1300 molecules condense per cm^2, whereas each of these species alone would have condensed at the rate of 4000 molecules/cm^2 (for ethanol) and 1000 molecules/cm^2 (for water). The capture rate is proportional to the vapor pressure or mole fraction. The capture rate is proportional to the vapor pressure through effects due to the molecular density and energy. Through this example we see that molecularly mixed multicomponent substances have two effects on phase equilibrium:

The ethanol mole fraction in the gas phase x_e is different compared to the liquid mole fraction $X_{e,l}$. Whereas $X_{e,l} = 0.1$, $X_e = 400/1300 = 0.3$ with the consequence that the gas–phase mole fraction of the volatile component is higher compared to its liquid phase mole fraction. The gas-phase mixture pressure P at equilibrium is greater than that for water p_w^{sat}, but lower than that for ethanol p_e^{sat}. The partial vapor pressures exerted by the two species are $X_{e,l}\, p_e^{sat}$ and $x_{w,l}\, p_w^{sat}$. This relation for partial pressure is known as Raoult's law (see Chapter 9). The total gas-phase pressure P is a sum of these partial pressures given by $P = X_{e,l}\, p_e^{sat} + X_{w,l}\, p_w^{sat} = 0.1 \times 40 + 0.9 \times 10 = 13$ kPa, where $p_w^{sat} < P < p_e^{sat}$. Note that $X_e = (.1 * 40)/13 = 0.31$ and $X_w = 0.69$.

Therefore, for this example, phase equilibrium exists at a pressure of 13 kPa at 50°C at which the vapor phase ethanol mole fraction is 30%. If the vapor pressure is suddenly lowered to 12 kPa, keeping the temperature and liquid mixture composition the same, the capture rates for both species will be lower, implying that the escape rate from the liquid must be reduced in order to restore phase equilibrium. This may be accomplished by reducing the liquid temperature (without altering the composition) so that fewer molecules escape into the vapor phase, or by changing the composition, but maintaining the same temperature. That composition can be determined by applying Raoult's Law, i.e., $(1-x_e)\times 10 + x_e \times 40 = 12$ so that $x_e = 6.7\%$. Likewise, $x_{e,l} = x_e \times 40/P = 0.067 \times 40/12 = 023$, and its equilibrium value will reduce as the vapor pressure is lowered. A more detailed discussion is presented in Chapter 9.

10. Entropy

Molecules undergo random motion as shown in Figure 31a. The energy of random motion is indicated by the temperature (e.g., $T \propto m V^2$ where V denotes the molecular velocity; a random velocity distribution is provided by a Maxwellian law) while "ipe" depends upon the volume V of the system. In addition to "te", molecules contain energy in the form of "ve" and "re" at various rotational speeds. As seen in Figure 31a, the random motion which occurs in all

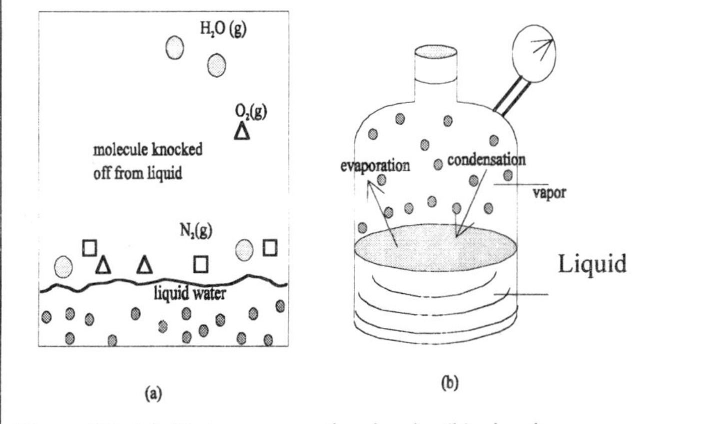

Figure 29: (a) Water evaporation in air; (b) simultaneous evaporation and condensation.

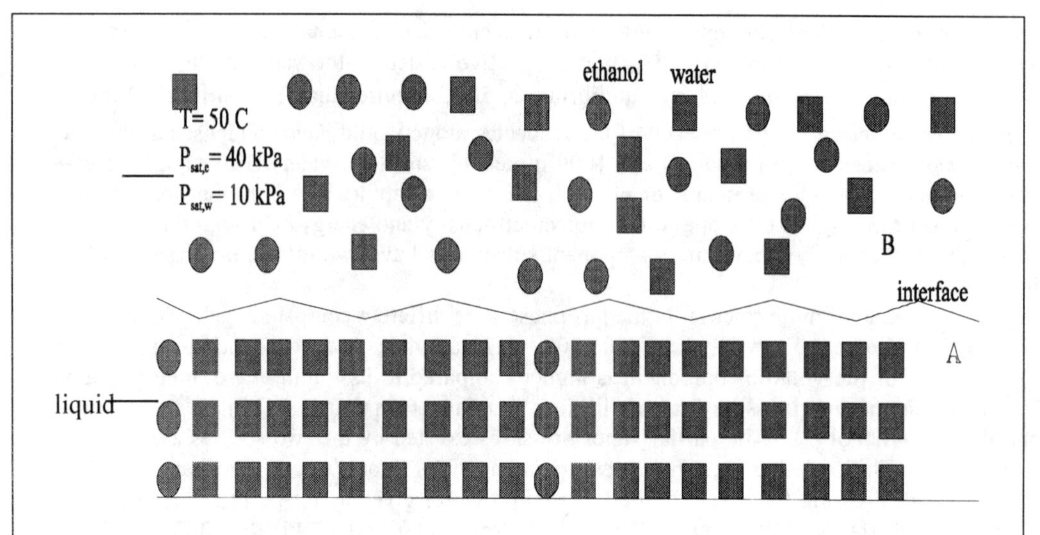

Figure 30: Boiling and condensation of a multicomponent solution of water (w) and ethanol (e).

directions and in all velocities cannot be converted directly into work. If the motion is organized (cf. Figure 31b - water flow from a dam) with macroscopic flow and kinetic energy, the work capability is improved. Now consider blending a fluid in a blender and its temperature increases; here, organized shaft work is converted into thermal (random) energy of the matter. Entropy is a measure of the number of "random" states in which molecules store energy, just as there are several ways to store physical items in a cabinet (depending upon the number of items and the shelves in the cabinet). Assuming all of these states to be equally probable, the entropy S is defined as a quantity proportional to the logarithm of the number of macro states Ω, in which energy is stored i.e., $S \propto \ln\Omega$.

The probability of predicting a particular macro state out of all possible macro states is low, and the entropy is also defined as a quantity proportional to ln (1÷probability). Just as it may be possible to rearrange the items in the cabinet among the various shelves over a very short duration, it may be possible to reorganize the energy among the macro states of a system.

We now illustrate how the energy is distributed. Consider a monatomic gas. The total translational energy of an individual particle is given as

$$\varepsilon_{ijk} = (1/2)\ m\ V_{ijk}2 = (\ h_P^2/(8mV^{2/3}))(i^2+j^2+k^2), \tag{75}$$

where ε_{ijk} denotes the energy at a quantum number, h_P the Planck's constant (=6.623x10^{-37} kJ-s/molecule), V volume, and where i, j, and k are quantum numbers in the x, y, and z directions, and $V_{ijk}^2 = V_i^2 + V_j^2 + V_k^2$. Note that as volume is decreased, the energy per quantum state is increased. A crude explanation is that the molecules have frequent collisions within a smaller volume thereby maintaining narrower velocity distribution or more molecules/atoms having higher quanta of energy. For monatomic gas the energy is mostly translational and hence has three degrees of freedom (i, j, and k). For a diatomic gas, the additional terms to be included are

$$\varepsilon_\ell = (h_P\nu)(\ell+(1/2)),\ \text{and} \tag{76}$$

$$\varepsilon_r = (\ h_P^2 r/(8\pi^2 I))(r+1), \tag{77}$$

where ε_ℓ denotes the vibrational quantum number, ν the frequency, r the rotational quantum number, and I the moment of inertia. As we have previously discussed, diatomic molecules have 3 translational quantum numbers, and one vibrational and one rotational quantum number with five consequent degrees of freedom (i.e., we can assign 5 numbers, e.g., i, j, k, ℓ, r for a diatomic gas). As an illustration of energy quanta, consider the emission of light occurs due to the excitation of electrons from a lower to a higher energy level (ε_1) followed by decay to a ground state (ε_0); the frequency of light emitted (ν) by a single photon is given by the expression

$$h_P \nu = (\varepsilon_1 - \varepsilon_0).$$

The number of photons depends upon the number of electrons undergoing similar processes.

Consider the example of water molecules contained within a rigid vessel of fixed volume. The energy stored in the molecules exists in various forms. Each particle has energy at various quantum levels. Therefore, the macroscopic energy of a group of particles consists of, say, for the sake of illustration, particle A at a hypothetical quantum state i=2, j=3, k=5 (cf. Eq. (77)), particle B at i= 5, j=7, k=14, etc. Figure 32a and b illustrates two possible quantum or macro states (particle A at 0, B at 1, C at 2, D at 3 and A at 0, B at 2,C at 2, D at 2 or the 0,1,2,3 and 0,2,2,2 states) for a group having 4 particles with 6 units of total energy. The other 3 possible quantum arrangements are (3,3,0,0; 2,2,1,1; 3,1,1,1). Considering millions of particles with a total energy of U, there exists an immense number of arrangements in quantum states for the same group containing fixed energy. The entropy S is defined as $S \approx k \ln\Omega$, (also known as the Boltzmann Law) where Ω denotes the total number of quantum states and all macro states are considered to be equally probable. Hence, $S \approx -k$ ln (probability of a particu-

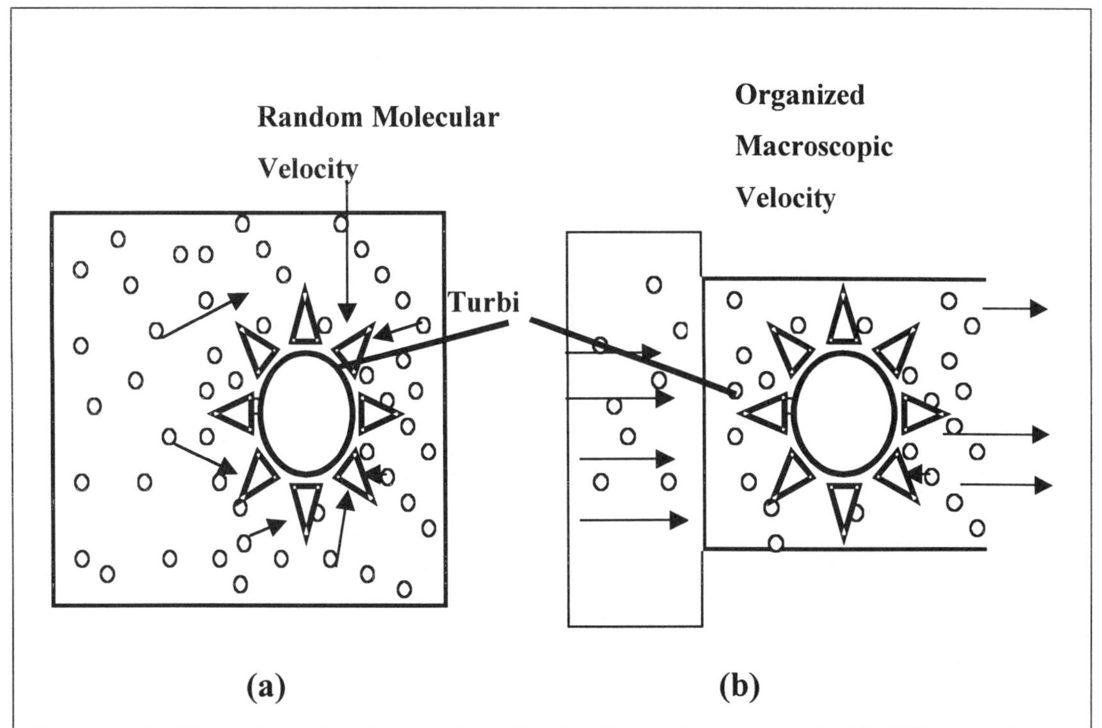

Figure 31 (a) Illustration of random motion of molecules; molecules travel with different velocities (in steps of quanta); work capability is less. (b) Illustration of macroscopic motion of molecules (e.g.: water from a dam), work capability is high.

lar macro state).

One can question what happens to the number of states as the volume is changed. The number of quantum states depends upon the free (or available) space between molecules and the larger the volume for a set number of molecules, the greater the number of quantum states within which energy storage is possible. Therefore, the energy per quantum state in the smaller volume is higher (cf. Eq. (77)). Upon compression, the molecules of a gas come closer together in a smaller volume and while the number of particles at each state is fixed, each particle contains energy at a higher energy state. Therefore, the entropy decreases as the volume is decreased (or the pressure increases) at a fixed energy level. The energy value for quantum levels ε_i is proportional to the inverse of the intermolecular spacing or $(\text{volume})^{2/3}$. Therefore, quantum levels ε_0 (ground energy), ε_1, ε_2, etc., can contain a higher energy value in a smaller volume. For example, if in a larger volume ε_0 corresponds to 2 units of energy, upon compression to 1/8th of that volume ε_0 contains 8 energy units. Therefore, upon compression, the total number of macro states reduces (or the energy per macro state increases) if the total energy and total number of molecules are fixed.

It is seen from Eq. (77) and Figure 32a and b that when the energy contained in matter U is increased at a specified value of V, there are more quantum numbers or more states in which molecules store energy and, hence, entropy. On the other hand, given the same energy U but at a reduced volume V, the energy per quantum state increases (cf. Eq. (77)) and, hence, the entropy declines. Therefore, the entropy is a function of the energy and volume or S= S(U, V). It is a property and a measure of the number of ways (macrostates) molecules store energy. This relation will be rigorously discussed in Chapter 3 using classical thermodynamics.

Consider the entropy of a crystalline solid at 0 K that contains negligible energy. The probability of molecules in the zero–energy macro state is unity, and S = 0. As the substance temperature is increased, its molecules move apart (which increases the specific volume) and it expands, and the entropy increases due to the increase in energy and change in volume. Upon further heating, the solid changes phase to become liquid, and molecules are allowed to move around other molecules. At this stage $S_{liquid}>S_{olid}$. Upon further heating, the liquid can vaporize so that the molecules translate at a relatively high speed (which, for water, is approximately 400 m s^{-1} at its boiling point). As a consequence, there is an increasing number of quantum states within the translational energy mode, and the entropy becomes larger. This entropy increase continues as the vapor is further heated.

We now discuss the relation between entropy and energy increase at fixed volume. We will show later that dS/dU (=change in entropy or change in number of quantum states/ change in energy) = 1/T (cf. Chapter 3). The greater the number of molecules with high "te" values (i.e. with larger molecular velocities) in a monatomic gas, the smaller is the increase in entropy with an increase in the value of U at a specified volume. This implies a smaller increase in the number of energy states, since most molecules possess energy at higher quantum number. An analogy is having only five bills of 100 dollars each with a total worth of 500 dollars. Thus, if you are given another 100 dollars in a single bill, the total number of bills becomes six (much like a small increase in entropy). On the other hand if one has 500 one dollar bills and is given another 100 dollars in like bills, the total number of bills is 600 (much like a large increase in entropy). A large amount of energy transfer may be required at higher temperatures to create a similar increase in entropy as compared to a system at a lower temperature.

Consider the energy $U = \Sigma N_i'\varepsilon_i$ so that

$$dU = \Sigma dN_i'\varepsilon_i + \Sigma d\varepsilon_i\, N_i'. \tag{78}$$

The term "dU" represents the change in energy brought out by a process. For example, the compression process (e.g. the work input) and heating process (heat flow into the matter) increase the energy by " dU" (Figure 32 c and e). The right hand side represents the mode in which the molecules store the energy. The first summation in the above equation can be interpreted as the increase in the number of molecules dN_i' at a given quantum energy state. This is caused by the heat addition to the mass at fixed volume (i.e., in the absence of a compression process, or at a fixed number of quantum states) that increases the internal energy which moves a few molecules from lower energy levels to higher energy levels (Figure 32c). Now, with more energy and a fixed number of particles, a larger number of arrangements is possible with a consequent increase in entropy. The second summation term represents the storage for the same number of molecules due to the increased energy level at the same quantum state. The latter occurs during compression in the absence of heat i.e. the molecules remain at the same quantum level after the magnitude of that energy level has increased. This is denoted as a bigger step size at the same quantum level (Figure 32e) and is known as the volume effect or the PdV work effect. The second term will not change with the number of macro states, and hence does not alter entropy.

We will show in Chapter 2 that (Net energy gain, dU = Energy gain due to heat transfer, δQ – (Energy decrease due to reversible expansion work transfer, δW) where $\delta W_{rev} = PdV$. The term "rev" will be explained in Chapters 2 and 3. *The energy gain due to heat transfer results in an entropy increase*. In Chapter 3 we will see that $dS = \delta Q_{rev}/T$. It can be shown that $\delta Q = \Sigma\, dN_i'\, \varepsilon_I$, $\delta W = -\Sigma d\varepsilon_i\, N_i'$.

Entropy increases with heat transfer only but not due to PdV work. When PdV work is performed, a group of molecules are exerted on with a force. This accelerates the x–wise

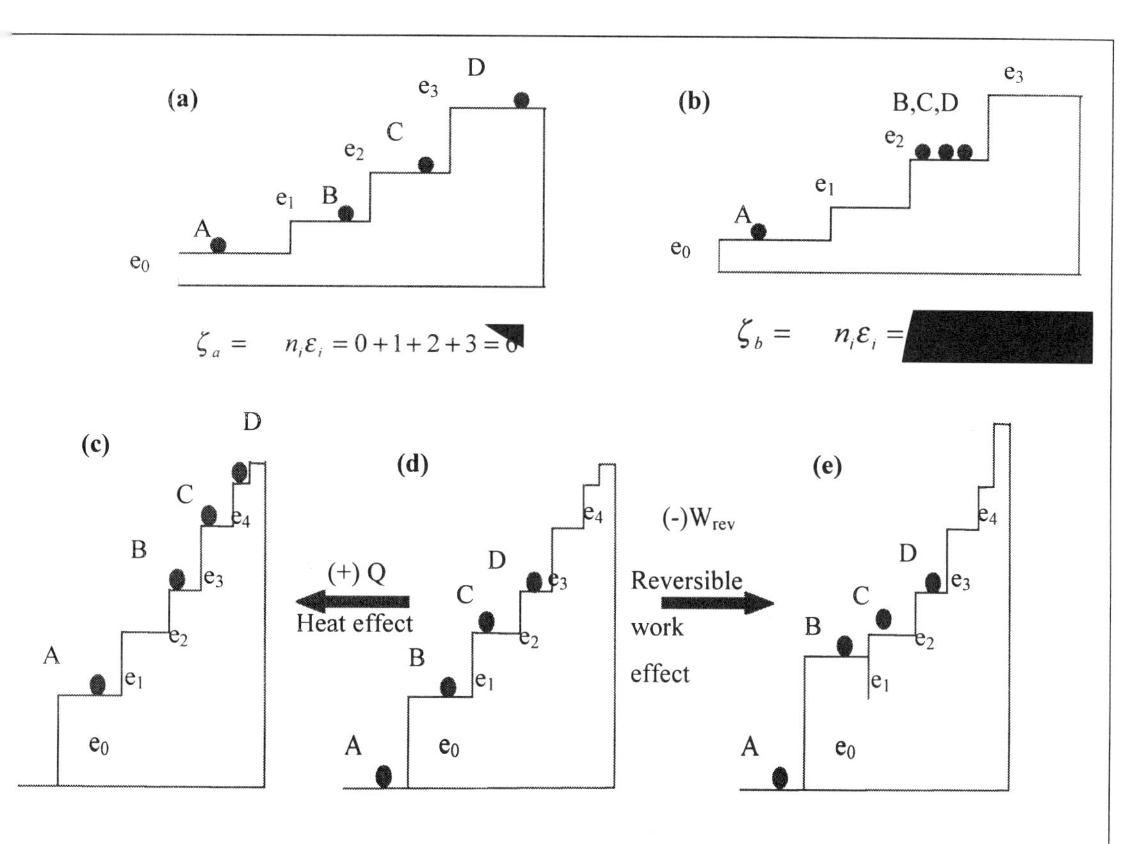

Figure 32a and b. Two possible quantum states (a) and (b). The system in (c) after it has received heat (re from system (d) and system in (e) after it has received reversible work (from Sussman and redrawn; Sussman, M.V., *Elementary General Thermodynamics*, Addison-Wesley Publishing Co., Reading, Mass., 1972. p 191 and 209. With permission).

component of the molecular velocity *V* that increases the "te". The energy level of this group of molecules is raised as shown in Figure 32e. Thus, the total number of states do not change even though the energy level for each group has increased due to work input. Now consider the energy transfer due to heat (i.e. due to temperature difference) through solid walls into a gas with the solid being at a higher temperature. The molecules within a group of gas molecules impinging on the wall pick up the energy randomly and these can be placed at different energy levels as shown in Figure 32c. The energy transfer through heat results in an entropy increase while energy transfer through work does not. In Chapter 3 we will see that $dS = \delta Q_{rev}/T$ (but not $\delta W_{rev}/T$ or PdV/T).

The entropy increases as two different species are mixed. This can be illustrated through the example of two adjacent adiabatic containers of volumes V_1 and V_2 at the same temperature that, respectively, contain nitrogen and oxygen. If the partition between them is removed, then N_2 and O_2 gases have a new set of quantum states due to extension of volume from V_1 and V_2 to V_1+V_2. This increases the entropy of each species. Hence mixing causes an increase in entropy, and, consequently the system entropy. In this instance, mixing causes the entropy to increase even though total energy of nitrogen and oxygen is unchanged due to mixing.

11. Properties in Mixtures – Partial Molal Property

A kmole of any substance at standard conditions contains 6.023×10^{26} molecules known as Avogadro number. The molecular energy is in the form of vibrational, rotational, and translational energy, and the molecules are influenced by the intermolecular potential energy (ipe). At the standard state, the energy of pure water $\bar{u}$ is 1892 kJ/kmole (the bar at the top indicates pure property on a kmole basis). If a kmole of water is mixed at the molecular level at standard conditions with 2 kmoles of ethanol, each H_2O molecule is now surrounded by 2 molecules of ethanol. Since the temperature is unchanged, the intermolecular distance is virtually unaltered before and after mixing. The attractive forces due to the water-ethanol molecules are different from those between water-water molecules (this is true of non-ideal solutions and will be discussed in Chapter 8) and, consequently, the potential energy is different for the two cases. Therefore, the combined energy contribution to the mixture by a kmole (or 6×10^{26} molecules) of water in the mixture $\hat{u}_{H2O,}$ is different from that of a kmole of pure water $\bar{u}_{H2O}$. The heat at the top of $\bar{u}_{H2O}$ indicates property when the component is inside the mixture. Here, $\hat{u}_{H2O}$ denotes the partial molar internal energy. Similarly, the enthalpy and entropy of the water are different in the mixture from its unmixed condition. This is further discussed in Chapter 8.

If the solution were ideal, i.e., if the ethanol-ethanol intermolecular attractive forces were the same as those for water-water molecules, the water-ethanol attractive forces would equal those in the pure states. In that case $\hat{u}_{H2O} = \bar{u}_{H2O}$, for an ideal gas mixture and $\bar{\mu}_k = \bar{\mu}_k$ since attractive forces do not influence the property. However, even then, $\hat{s}_{H2O}$ would not equal $\bar{s}_{H2O}$, since the water molecules would be spread over greater distances in the mixture with the result that the number of quantum states for water molecules would increase.

E. SUMMARY

We have briefly reviewed various systems (such as open, closed, and composite), mixtures of substances, exact and inexact differentials and their relation to thermodynamic variables, homogeneous functions and their relation to extensive and intensive variables, Taylor series, the LaGrange multiplier method for optimization, and the Gauss and Stokes theorems. The background material and mathematical concepts will be used through a quantitative language useful to engineers involved with the design and optimization of thermodynamic systems. We have also briefly covered the nature of intermolecular forces and potential, the physical meanings of energy, pressure; of temperature, and of thermal, mechanical, and species equilibrium; boiling and saturation relations; and, finally, entropy. These concepts are useful in the physical interpretation of various thermodynamic relations that are presented in later chapters.

F. APPENDIX

1. Air Composition

Species	Mole %	Mass %
Ar	0.934	1.288
CO_2	0.033	0.050
N_2	78.084	75.521
O_2	20.946	23.139
Rare gases	0.003	0.002

Molecular Weight: 28.96 kg kmole^{-1}.

2. Proof of the Euler Equation

Assume that our objective is to determine a system property F, where

$$F(\lambda x_1, \lambda x_2,...) = \lambda^m F(x_1, x_2,...), \text{ and} \quad (79a)$$

$x_{1,new} = \lambda x_1$, $x_{2,new} = \lambda x_2$,.... Differentiating Eq. (80) with respect to λ (and treating it as a variable),

$$(\partial F/\partial(\lambda x_{1,new}))(\partial x_{1,new}/\partial\lambda)+(\partial F/\partial(\lambda x_{2,new}))(\partial x_{2,new}/\partial\lambda)+\ldots = m\lambda^{m-1}F(x_1, x_2,...).(81b)$$

Since $\partial x_{1,new}/\partial\lambda = x_1$, $\partial x_{2,new}/\partial\lambda = x_2$, ..., Eq. (81b) assumes the form

$$(\partial F/\partial(\lambda x_{1,new}))x_1 + (\partial F/\partial(\lambda x_{2,new}))x_2 + \ldots = m\lambda^{m-1}F(x_1, x_2,...).$$

Multiplying both sides of the above equation by λ, and noting that

$$\lambda^m F(x_1, x_2,...) = F(x_{1,new}, x_{2,new},...),$$

we have the relation

$$(\partial F/\partial(\lambda x_{1,new}))x_{1,new} + (\partial F/\partial(\lambda x_{2,new}))x_{2,new} + \ldots = mF(x_{1,new}, x_{2,new},...). \quad (80)$$

If m = 1,

$$\Sigma_{k=0}^{K} x_k(\partial F / \partial x_k) = mF. \quad (81)$$

3. Brief Overview of Vector Calculus

a. Scalar or Dot Product

i. Work Done to Move an Object

Consider a surfboard being dragged over water along an elemental path $d\vec{s}$ by a power boat that applies a force of $\vec{F}$ on the board. The work done is given as

$$\delta W = \vec{F}\cdot d\vec{s} = F \cos\theta \; ds,$$

where θ denotes the angle between the force and the elemental path.

ii. Work Done to Move an Electrical Charge

Similarly if an electrical charge of strength Q is located at an origin, the force $\vec{F}$ exerted by it on another charge of strength q situated at distance $\vec{r}$ removed from the origin is

$$\vec{F} = (\varepsilon q Q \, \vec{r})/|r^3|,$$

where ε denotes the Coulomb constant. If the product (qQ) > 0 (i.e., the two are like charges), the force is repulsive. In case (qQ) < 0 (i.e., the charges are unlike) the force is one of attraction. The work done to move charge q away from Q

$$\delta W = \vec{F} \cdot d\vec{r}.$$

b. *Vector or Cross Product*

The area $\vec{A}$ due to a vector product

$$\vec{A} = \vec{x} \times \vec{y}, \tag{82}$$

can be written in the form

$$\vec{A} = \vec{k} \,|\, x \,||\, y \,|\sin\theta, \tag{83}$$

where $\vec{k}$ denotes the unit vector in a plane normal to that containing the vectors $\vec{x}$ and $\vec{y}$, and θ the angle between these two vectors. The vector product yields an area vector in a direction normal to the plane containing the two vectors.

Consider the circular motion of an object around an origin in a plane. The force due to that object in the plane

$$\vec{F} = \vec{i}\,F_x + \vec{j}\,F_y = \vec{i}\,F\cos\theta + \vec{j}\,F\sin\theta, \tag{84}$$

where θ denotes the angle between the force and an arbitrary x–wise coordinate at any instant, and $\vec{i}$ and $\vec{j}$ denote unit vectors in the x– and y– directions, respectively. The torque exerted about the center

$$\vec{B} = \vec{F} \times \vec{r} = (\vec{i}\,F\cos\theta + \vec{j}\,F\sin\theta) \times (\vec{i}\,x + \vec{j}\,y) = \vec{k}\,(y\,F\cos\theta + x\,F\sin\theta), \tag{85}$$

where $\vec{i} \times \vec{i} = 0$, $\vec{i} \times \vec{j} = \vec{k}$, and $\vec{j} \times \vec{i} = -\vec{k}$.

When a screw is loosened from a flat surface by rotating it in the counter clockwise direction, it emerges outward normal to the surface, say, in the z–direction. To place the screw back into the surface, it must be rotated in the clockwise direction, i.e., it may be visualized as moving towards the origin of the z–direction. The rotation is caused by an applied torque that is a vector. If the term (F cos θ y – Fsinθ x) = 0 in Eq.(87), then there is no rotation around the z–axis. In general, a force has three spatial components, i.e.,

$$\vec{F} = \vec{i}F_x + \vec{j}F_y + \vec{k}F_z, \tag{86}$$

and the torque is described by the relation

$$\vec{B} = \vec{F} \times \vec{r} = \vec{i}(F_y z + F_z y) + \vec{j}(F_z x + F_x z) + \vec{k}(F_x y + F_y x), \text{ i.e.,} \tag{87}$$

there are rotational components in the x– and y– directions also. If $\vec{F}$ and $\vec{r}$ are parallel to each other, e.g., $\vec{F} = \vec{i}F_x$ and $\vec{r} = \vec{i}x$, then

$$\vec{B} = \vec{F} \times \vec{r} = 0.$$

c. *Gradient of a Scalar*

Consider a one–dimensional heat transfer problem in which the temperature T is only a function of one spatial coordinate, say, y, i.e., T = T(y). In this case T(y) is a point or scalar function of y, since its value is fixed once y is specified. In general, the gradient of T is defined as

$$\vec{\nabla}T = (\vec{i}\partial/\partial x + \vec{j}\partial/\partial y + \vec{k}\partial/\partial z)T, \quad (88)$$

which for the one–dimensional problem assumes the form

$$\vec{\nabla}T = \vec{j}\partial T/\partial y, \quad (89)$$

The x–z plane contains isotherms, since $T \neq T(x,z)$, and $\vec{\nabla}T$ is a vector along normal to the isotherms in the y–direction.

Consider, now, the temperature profile in an infinite cylindrical rod. Assume that the temperature is constant along the axial direction z, once a cross–sectional location (x,y) is specified, i.e., $T=T(x,y)$, and $T \neq T(z)$. Assume an axisymmetric problem for which the isotherms are circular in the x–y plane and form cylindrical surfaces. In this case,
$\vec{\nabla}$

$$dT = (\partial T/\partial x)dx + (\partial T/\partial y)dy = \vec{\nabla}T \cdot d\vec{s}, \quad (90)$$

where , $\vec{\nabla}T = \vec{i}\partial T/\partial x + \vec{j}\partial T/\partial y$. Therefore,

$$dT/ds = \vec{\nabla}T \cdot d\vec{s}/ds, \text{ i.e.,} \quad (91)$$

the gradient dT/ds varies, depending upon the direction of the gradient between any two isotherms. Along any circular isotherm $\vec{\nabla}T \cdot d\vec{s} = 0$ according to Eq. (93), since $\vec{\nabla}T$ and $d\vec{s}$ are normal to each other.

In general, if $T=T(x,y,z)$ then isotherms form surfaces that lie in all three (x,y,z) coordinates, and, at any location, $\vec{\nabla}T$ represents a vector that lies normal to a scalar surface on which T is constant.

d. Curl of a Vector

Consider a vector

$$\vec{\nabla} \times \vec{F} = (\vec{i}\partial/\partial x + \vec{j}\partial/\partial y + \vec{k}\partial/\partial z) \times (\vec{i}F_x + \vec{j}F_y + \vec{k}F_z)$$

$$= \vec{i}(\partial F_y/\partial z - \partial F_z/\partial y) + \vec{j}(\partial F_z/\partial x - \partial F_x/\partial z) + \vec{k}(\partial F_x/\partial y - \partial F_y/\partial x) \quad (92)$$

The LHS of Eq. (94) is a vector called curl $\vec{F}$. If $\vec{\nabla} \times \vec{F} = 0$, then the two are parallel to each other, i.e., the vector field is irrotational. Assume that

$$\vec{F} = \vec{\nabla}T. \quad (93)$$

Now assume that instead of a spatial coordinate system, x denotes pressure P, y denotes the specific volume v, and z represents x_1 (i.e., the mole fraction of component 1 in a binary mixture), i.e.,

$$\begin{aligned}\vec{\nabla} \times \vec{\nabla}T = &\ \vec{i}(\partial/\partial x_1(\partial T/\partial v) - \partial/\partial v(\partial T/\partial x_1)) \\ &+ \vec{j}(\partial/\partial P(\partial T/\partial x_1) - \partial/\partial x_1(\partial T/\partial P)) \\ &+ \vec{k}(\partial/\partial v(\partial T/\partial P) - \partial/\partial P(\partial T/\partial v)).\end{aligned} \quad (94)$$

The vector $\vec{\nabla}T$ lies in a direction normal to the isothermal surface T, and $\vec{\nabla} \times \vec{\nabla}T$ lies normal to the plane containing $\vec{\nabla}$ and $\vec{\nabla}T$. This implies that $\vec{\nabla} \times \vec{\nabla}T$ is a vector that lies back on the isothermal scalar surface T, and, therefore, $\vec{\nabla} \times \vec{\nabla}T = 0$. Note that the terms in the brackets satisfy the criteria for exact differentials and the RHS of Eq. (96) equals zero. All thermodynamic properties satisfy the irrotationality condition. Functions such as $T=T(P,v,x_1)$ are known as properties, point functions, scalar functions, or scalar potentials. Terms in exact differential form, such as $dT = \partial T/\partial P\, dP + \partial T/\partial v\, dv + \partial T/\partial x_1\, dx_1$, are called Pfaffians.

Chapter 2

2. FIRST LAW OF THERMODYNAMICS

A. INTRODUCTION

Chapter 1 contains an introduction to thermodynamics, provides some basic definitions, a microscopic overview of thermodynamic properties and processes, and briefly reviews the necessary mathematics. We will use that material to formulate thermodynamic laws based either on a generalization of experimental observations, or in terms of four mathematical postulates that are not necessarily based on these experimental results. The laws of thermodynamics are presented in Chapters 2 and 3, and the postulate concepts are addressed in Chapter 5.

The *thermodynamic laws* are simply restrictions on the transformation of energy from one form into another. For example,

If the thermal energy content of a given mass of steam is 100,000 kJ, it is impossible to obtain a work output of 150,000 kJ from it in the absence of another energy input. Here, the First Law of Thermodynamics provides a restriction.

If that same mass of steam containing the same energy content exists at a temperature, it is impossible to obtain a work output of 90,000 kJ from steam at 1000 K. In this case, the restriction is due to the second law of thermodynamics that constrains the degree of conversion of heat energy.

In this chapter we will briefly discuss the zeroth and first laws that deal with energy conservation, examine problems involving reversible and irreversible, and transient and steady processes; and, finally, present the formulation of the conservation equations in differential form. The second law and its consequences will be considered in Chapter 3.

1. Zeroth Law

The Zeroth law forms the basis for the concept of thermal state (or temperature). Consider the body temperature of two persons (systems P_1 and P_2) read using an oral thermometer (system T). If the systems P_2 and T are in thermal equilibrium, and so are systems T and P_1, then systems P_2 and P_1 must exist at the same thermal state. Therefore, both persons will manifest the same body temperature. Similarly, if the hot gas inside an electric bulb is in thermal equilibrium both with the electrical filament and the glass wall of the bulb, the glass wall is necessarily in thermal equilibrium with the filament.

2. First Law for a Closed System

We will present the First law of thermodynamics for a closed system, and illustrate applications pertaining to both reversible and irreversible processes.

a. Mass Conservation

For closed systems the mass conservation equation is simply that the mass

$$m = \text{Constant}, \tag{1}$$

In the field of atomic physics, mass and energy E are considered convertible into each another and, taken together, are conserved through the well–known Einstein relation $E = mc^2$, where c denotes the light speed. However, in the field of thermodynamics it is customary to assume that the conversion of mass and energy into each other is inconsequential and, therefore, either is separately conserved.

b. Energy Conservation

An informal statement regarding energy conservation is as follows: "Although energy assumes various forms, the total quantity of energy is constant, with the consequence that when energy disappears in one form, it appears simultaneously in others".

i. Elemental Process

For a closed system undergoing an infinitesimally slow process, (Figure 1a) during which the only allowed interactions with its environment are those involving heat and work, the first law can be expressed quantitatively as follows

$$\delta Q - \delta W = dE, \tag{2}$$

where δQ denotes the elemental (heat) energy transfer across the system boundaries due to temperature differences (Figure 1a), δW the elemental (work) energy in transit across the boundaries (e.g., the piston weight lifted due to the expansion of the system), and dE the energy change in the system. The "E" includes internal energy U (=TE+VE+RE etc.) which resides in the matter, kinetic energy KE and potential energy PE. Note that Q and W are transitory forms of energy and their differentials are written in the inexact forms δQ and δW (see Chapter 1) while differential of resident energy E is written as an exact differential. Dividing Eq. (2) by m,

$$\delta q - \delta w = de, \tag{3a}$$

where q denotes the heat transfer per unit mass Q/m, w the analogous work transfer W/m, and, likewise, e = E/m.

It is customary to choose a sign convention for the work and heat transfer that follows common sense. In the absence of work transfer, i.e., $\delta W = 0$, addition of heat causes an increase in energy. Therefore, it is usual to accord a positive sign for heat transfer into a system. For an adiabatic system ($\delta Q = 0$), if the work done by the system is finite and conferred with a positive sign ($W > 0$), then, from Eq. (2), $dE < 0$. This is intuitively appropriate, since in order to perform work, the system must expend energy. On the other hand, if the system of Fig. 2 is adiabatically compressed, work is done on the system (so that $W < 0$), and the stored energy in the system increases ($dE > 0$).

The system energy consists of the internal, potential, and kinetic energies. Equation (2) may be rewritten for a static system in the form

$$\delta Q - \delta W = dU. \tag{3b}$$

ii. Internal Energy

At a microscopic level the internal energy is due to the molecular energy which is the sum of the (1) molecular translational, vibrational and rotational energies (also called the thermal portion of the energy), (2) the molecular bond energy (also called the chemical energy), and the (3) intermolecular potential energy, ipe (cf. Chapter 1). At a given temperature the energy depends upon the nature of a substance and, hence, is known as an *intrinsic* form of energy.

iii. Potential Energy

The potential energy of a system is due to the work done on a system to adiabatically move its center of gravity through a force field. The potential energy of a system whose center of gravity is slowly raised vertically (so as not to impart a velocity to it) in the earth's gravity field through a distance of dz increases by a value equal to mgdz. The first law

$$\delta Q - \delta W = dU + d(PE) + d(KE),$$

where PE and KE denote the potential and kinetic energies, can be applied after noting that for this case $\delta Q = dU = d(KE) = 0$, so that

$$0 - \delta W = 0 + d(PE) + 0. \tag{4}$$

Now, $\delta W = -F\, dz$. The negative sign arises since work is done on the system by a force F that lifts it through a distance dz. In raising the mass, the direction of the force is vertically upward. In the absence of any acceleration of the mass, this force is also called a body force. Using the relation $F = mg$ for the force with g denoting the local gravitational acceleration, the work done $W = -mg\, dz$, and using Eq. (4) $d(PE) = -\delta W = mg\, dz$. Integrating this expression across a vertical displacement that extends from z_1 to z_2, the potential energy change is given as

$$\Delta PE = mg(z_2 - z_1).$$

The potential energy per unit mass due to the gravitational acceleration at a location z above a stipulated datum is also called the gravitational potential pe, i.e.,

$$pe = gz. \quad (5)$$

In SI units, pe can be expressed in J kg^{-1} or in units of $m^2\ s^{-2}$, where

$$pe \text{ (in units of kJ kg}^{-1}) = g(\text{in units of m s}^{-2})\ z(\text{in units of m})/1000.$$

In English units, pe can be expressed as BTU $lb^{-1} = g(ft\ s^{-2})\ z(ft) \div (g_c\mathbf{J})$, where $g_c = 32.174$ (lb ft $s^{-2}lbf^{-1}$) is the gravitational constant, and **J** denotes the work equivalent of heat of value 778.1(ft lbf BTU^{-1}).

iv. Kinetic Energy

In order to move a mass along a level frictionless surface, a boundary or surface force must be exerted on it. Applying the first law, namely, $\delta Q - \delta W = dU + d(PE) + d(KE)$, the adiabatic work due to these forces can be expressed as

$$0 - \delta W = 0 + 0 + d(KE). \quad (6)$$

The work performed in moving the center of gravity of a system through a distance dx is ($-$ F dx), where the force $F = m\, d\mathcal{V}/dt$, the velocity $\mathcal{V} = dx/dt$, and t denotes time. In order to be consistent with the standard sign convention, the work done on the system is considered negative. Therefore, $d(KE) = m\ (d\mathcal{V}/dt)\times(\mathcal{V}\ dt) = m\mathcal{V}d\mathcal{V}$. Upon integration, the kinetic energy change of the system as it changes its velocity from a value $\mathcal{V}_1$ to $\mathcal{V}_2$ is

$$\Delta KE = (1/2)m(\mathcal{V}_2^2 - \mathcal{V}_1^2).$$

The kinetic energy per unit mass ke is

$$ke = 1/2\ \mathcal{V}^2. \quad (7)$$

In SI units, ke is expressed in J kg^{-1}, namely, $ke(J\ kg^{-1}) = (1/2)\mathcal{V}^2(m^2\ s^{-2})$. Often, it is preferable to express ke as

$$ke\ (kJ\ kg^{-1}) = (1/2000)V^2.$$

In English units, $ke(BTU\ lb^{-1}) = \mathcal{V}^2(ft^2\ s^{-2}) \div (g_c\mathbf{J})$. The kinetic and potential energies are independent of the nature of the matter within a system, and are known as *extrinsic* forms of energy.

v. Integrated Form

Integrating Eq. (2) between any two thermodynamic states (1) and (2) we have

$$Q_{12} - W_{12} = E_2 - E_1 = \Delta E. \quad (8)$$

The heat and work transfers are energy forms in transit and, hence, do not belong to the matter within the system with the implication that neither Q nor W is a property of matter. Therefore, while it is customary to write the energy change for a process $E_{12} = E_2 - E_1$, we cannot write Q_{12}

$= (Q_2–Q_1)$ or $W_{12} = (W_2–W_1)$. Since for a cycle the initial and final states are identical, $\oint \delta Q = \oint W=0$.

Writing Eq. (8) on a unit mass basis

$$q_{12} - w_{12} = e_2 - e_1 = \Delta e. \tag{9}$$

The application of the first law to systems require these to be classified as either coupled systems in which the transit energy modes, namely, Q and/or W, affect particular storage forms of energy, or as uncoupled systems if the heat and/or work transfer affect more than one mode of energy as illustrated below.

vi. Uncoupled Systems

Consider an automobile that is being towed uphill on a frictionless road during a sunny summer afternoon from initial conditions $Z_1 = V_1 = U_1 = 0$ to an elevation Z_2, velocity V_2 and energy U_2. Taking the automobile as a system, the heat transfer Q_{12} from the ambient to the car is determined by applying Eq. (8), i.e.,

$$Q_{12} - W_{12} = E_2 - E_1 = \Delta U + \Delta PE + \Delta KE,$$

so that $Q_{12} = \Delta U$. Therefore, the heat transfer across the boundary increases the system internal energy by ΔU which changes the static state of the system. The work performed to tow the automobile is

$$- W_{12} = \Delta\ PE + \Delta\ KE,$$

which influences the dynamic state of the system.

a. Example 1

A car of mass 2000 kg is simultaneously accelerated from a velocity $V = 0$ to 55 mph (24.6 m s^{-1}) and elevated to a height of 100 m. Determine the work required. Treat the problem as being uncoupled.

Solution

$$Q_{12} - W_{12} = = U_2 - U_1 + KE_2 - KE_1 + PE_2 - PE_1.$$

$$0 - W_{12} = (0+(2000\div2)(24.6\text{ m s}^{-1})^2–0+2000\times(9.81\times100)–0)\div1000 = 2568\text{ kJ}.$$

Remark

All of the work can be recovered if the car is made to slide down on a frictionless road to ground level (i.e., to zero potential energy) so that the potential energy is completely converted into kinetic energy. Upon impact against a spring the vehicle kinetic energy is further transformed into the spring potential energy, thereby recovering the work. Hence, the process is uncoupled.

vii. Coupled Systems

In coupled systems two or more interactions across the system boundary (e.g., heat and work) influence the same energy mode. For example, if a tow truck pulls a car on a rough high-friction road, the work performed is higher than that for an uncoupled system, since additional work is required in order to overcome the external friction. Frictional heating can cause the internal energy of the car tires to increase, and if the tires do not serve as good insulators, heat transfer to the road can occur. Therefore, the work is coupled with both internal energy and heat transfer. This is illustrated in the following example.

b. Example 2

A car of mass 2000 kg that is simultaneously accelerated from a velocity $V = 0$ to 55 mph (24.6 m s^{-1}) and elevated to a height of 100 m requires a work input of 3000 kJ. If the car is well insulated, what is the change in the internal energy of the car?

Solution

$$Q_{12} - W_{12} = = U_2 - U_1 + KE_2 - KE_1 + PE_2 - PE_1.$$

$$0+3000=U_2-U_1+((2000\div2)(24.6ms^{-1})^2-0+2000\times(9.81\times100)-0)\div1000=2568 \text{ kJ},$$

i.e.,

$$U_2 - U_1 = 3000 - 2568 = 432 \text{ kJ}.$$

Remarks

The work input is more than ΔKE and ΔPE. Thus additional work is used to overcome friction. Frictional work results in heating. If the tires (which are part of the car) are well insulated, their internal energy increases by 432 kJ. In this case the work is coupled to changes in the internal, kinetic, and potential energies of the system.
Dividing the work into intrinsic and extrinsic contributions

$$W_{12} = W_{12,int} + W_{12,ext},$$

we find that $W_{12,int}$ = 432 kJ, which results in the change in U, and that $W_{12,ext}$ = 3000 – 432 = 2568 kJ, which results in a change in the kinetic and potential energies, (diathermic) and the tire remains at fixed temperature. Then there is no change in the internal energy. Hence heat must be lost from the tires, i.e.,

$$Q_{12} - W_{12} = \Delta U + \Delta PE + \Delta KE = 0 + \Delta PE + \Delta KE,$$

where Q_{12} = –432 kJ, and W_{12} = –3000 kJ. In this case the work is coupled with the heat transfer. The heat transfer affects the intrinsic energy by changing U.
When the car moves at a high velocity, frictional drag due to the atmosphere can cause its body to heat, thereby increasing the internal energy. The work done on the car also increases its potential and kinetic energies, and the process becomes coupled. The work cannot be recovered, since the car will contain a higher internal energy even after impacting it against the spring, as illustrated in the previous example.

The heating of matter offers another example involving a coupled system. Consider constant pressure heating that causes a system of gases to expand and lift a weight of 100 kg through a distance of 2 m, if Q_{12} = 10 kJ, W_{12} = 1.96 kJ. (Here we neglect any change in the center of gravity of the matter contained in the system.) If there is no change in the system kinetic energy, from Eq. (8)

$$Q_{12} - W_{12} = U_2 - U_1 = \Delta U. \quad (10)$$

In this case $Q_{12} \neq \Delta U$. As a result of the work and heat interaction, ΔU = 10 – 1.96 = 8.04 kJ. If the system is confined to include only the moving boundary and the lifted weight, and these are considered adiabatic, then $(-W_{12}) = \Delta(PE)$, so that the work performed alters the system potential energy.

The illustrations of coupled and uncoupled systems demonstrate that it is necessary to understand the nature of a problem prior to applying the mathematical equations.

c. *Systems with Internal Motion*

Consider a mass of warm water contained in a vessel. If it is stirred, the entire effort imparts kinetic energy to that mass in the absence of frictional forces, and the center of gravity of each elemental mass of water moves with a specific kinetic energy ke. If the kinetic energy distribution is uniform throughout the system, its total value equals m x ke. Such a situation exists in an automobile engine when fresh mixture is admitted or when the exhaust valve opens. Oftentimes, the kinetic energy is destroyed due to internal frictional forces between the system walls and moving matter, which converts the kinetic energy into internal energy, as in coupled systems.

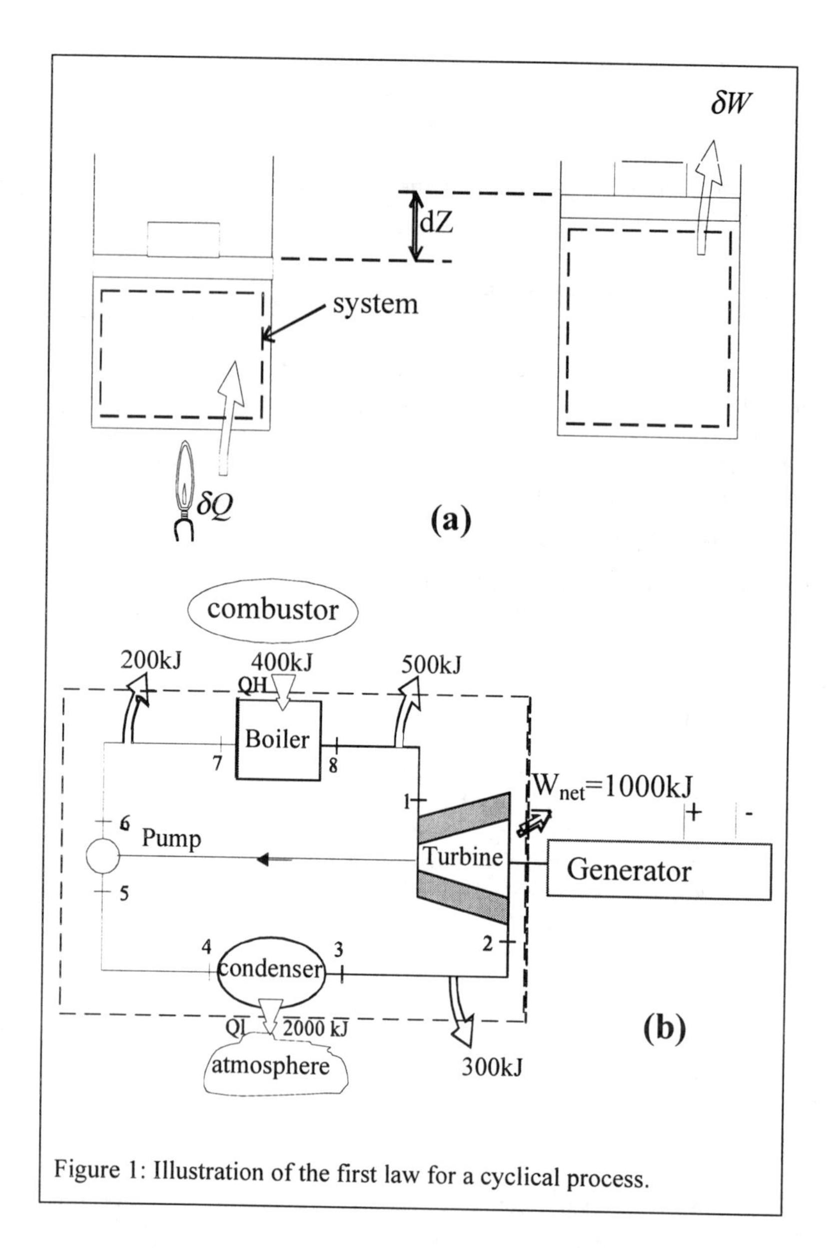

Figure 1: Illustration of the first law for a cyclical process.

viii. Adiabatic Work and Caratheodary Axiom I

The work performed during all adiabatic processes ($Q_{12} = 0$) between two given states is the same. Applying Eq. (8)

$$W_{12} = E_2 - E_1 = \Delta E.$$

This statement is called the *Caratheodary Axiom I* (see Postulate III of Chapter 5). For example, the electrical work (= voltage × charge current) used to heat a fluid adiabatically between two temperatures is identical to the mechanical work (e.g., performed using a pulley–paddle assembly to stir the water) required for a similar adiabatic heating process.

d. Cyclical Work and Poincare Theorem

ix. Cyclical Work

For a closed system undergoing a thermodynamic cyclical process,

$$\oint dE = 0. \tag{11}$$

Hence, the first law (Eq. (2)) yields,

$$\oint \delta Q = \oint \delta W \ , \tag{12}$$

and Eq. (12) implies that if $\oint \delta Q \neq 0$, then $\oint \delta W \neq 0$. On a unit mass basis

$$\oint \delta q = \oint \delta w \ .$$

Figure 1b illustrates the cyclical process in a steam power plant for which the heat transfer during the various processes is indicated. Applying Eq. (12) for all processes, i.e., 1–2, 2–3, 3–4,..., and 8–1,

$$\oint \delta Q = Q_{12} + Q_{23} + \cdots + Q_{81} = \oint \delta W = W_{12} + W_{23} + \cdots + W_{81},$$

so that

$$\oint \delta W = 0 - 300 - 2000 + 0 + 0 - 200 + 0 + 4000 - 500 = 1000 \text{ kJ}.$$

Therefore, by considering the net heat transfer for this cyclical process, the net work output of the plant can be determined.

x. Poincare Theorem

Consider an adiabatic system containing water and a mechanical stirrer. Work transfer through the stirrer is used to raise the water temperature from a quiescent state 1 to another motionless state 2. Since $Q_{12} = 0$, the change in internal energy can be obtained by applying the first law (Eq. (10)), and $U_2 - U_1 = \Delta U = |W_{12}|$. Next, if the insulation is removed and the water allowed to cool to its initial state, Eq.(10) can again be used to determine the heat flow Q_{21}. In this case $Q_{21} = |U_2 - U_1| = |W_{21}|$ as a consequence of the *Poincare theorem* of thermodynamics, which states that during a cyclical process the net heat interactions equal the net work interactions. While the Caratheodary axiom states the First Law in context of a single adiabatic process, the Poincare theorem expresses it for a cyclical process.

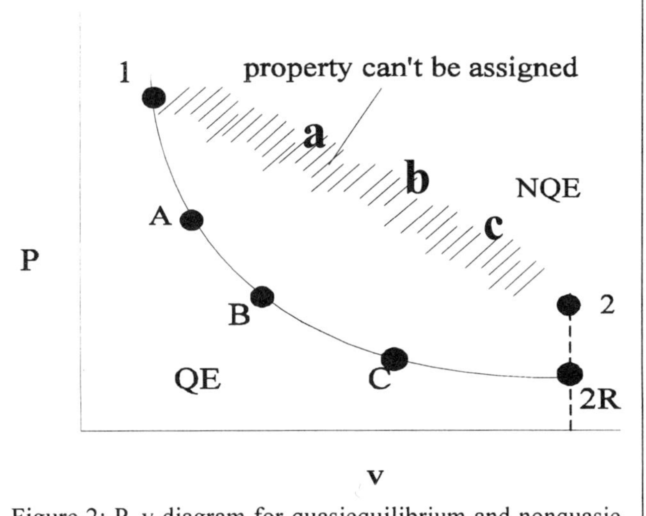

Figure 2: P–v diagram for quasiequilibrium and nonquasiequilibrium processes.

xi. Rate Form

Equation (2) can be used to express the change in state over a short time period δt (i.e., $\delta Q = \dot{Q}\delta t$, and $\delta W = \dot{W}\delta t$ to obtain the First Law in rate form, namely,

$$\dot{Q} - \dot{W} = dE/dt. \tag{13}$$

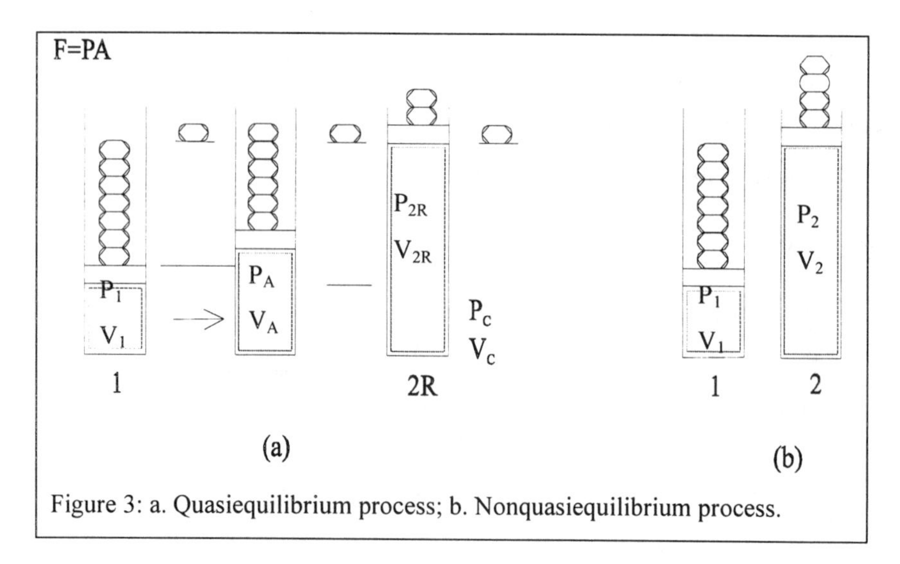

Figure 3: a. Quasiequilibrium process; b. Nonquasiequilibrium process.

The rate of work $\dot{W}$ is the energy flux crossing the boundary in the form of macroscopic work (e.g., due to the system boundary motion through a distance dz as illustrated in Figure 2). The heat flux $\dot{Q}$ is a consequence of a temperature differential, and does not itself move the boundary, but alters the amplitude of molecular motion that manifests itself in the form of temperature.

We see that energy conservation can be expressed in various forms (e.g., Eqs. (2), (3), (8), (9), (12) and (13)). The laws of thermodynamics are constitutive equation independent. It is possible to determine dE/dt accurately if $\dot{W}$ and $\dot{Q}$ are measured. Calculations of $\dot{Q}$ and/or $\dot{W}$ may require constitutive relations. In the context of the relation $\dot{Q} = -\lambda\nabla T$, a constitutive equation for heat transfer is employed with $\dot{W} = 0$. Therefore, the value of dE/dt depends upon the accuracy of the Fourier law and can differ from actual experimental data.

e. *Quasiequilibrium Work*

Consider an adiabatic frictionless piston–cylinder assembly on which infinitesimal weights are placed as illustrated in Figure 3a and Figure 3b. If the small weights are slowly removed, the system properties remain almost uniform throughout the removal process. Therefore, at any instant following the removal of an infinitesimal weight, if the system is isolated, it is in an equilibrium state (i.e., its properties are invariant with respect to time). Since the intensive state can be determined during any part of the process involving the successive removal of weights, the path along which the process proceeds can be described (e.g., as illustrated in Figure 2 for a quasiequilibrium process that moves the system from state 1 to 2R along the path ABC). Due to their nature, quasiequilibrium processes are also termed quasistatic.

However, not all quasistatic processes are at quasiequilibrium. Consider the example of a gasoline–air mixture (system) contained in a piston–cylinder assembly. At the end of a compression process, spark is initiated, hot region develops around the spark plug, while the remainder of the mixture is much colder. Even though the piston moves slowly (i.e., it is quasistatic) during this process, the spark initiation results in a non–equilibrium state, since the temperature distribution is nonuniform, and it is not possible to assign a single system temperature.

The consequences of the quasiequilibrium processes illustrated in Figure 3 are as follows:

If the infinitesimal weights are slowly removed, then at any time the force of the weights F $\approx$ PA, where A denotes the piston surface area. A force of P×A is exerted by the system. Therefore, the infinitesimal work W performed by the system as the individual weights are removed, and the piston moves infinitesimally through a displacement dx, is

$$\delta W = F\,dx = PAdx = PdV. \quad (14)$$

Consequently, the work done during the process 1–2 is

$$W_{12} = \int_1^2 PdV. \quad (15)$$

This is an illustration of reversible work.
The work performed by the system results in a p\otential energy gain for the remaining weights that are placed in the environment outside the system.
Since energy is transferred to the environment, according to the first law the system loses internal energy. The process can be reversed by slowly placing the weights back on the piston. This action will push the piston inward into the system, reduce the potential energy of the weights placed in the system environment, and restore the system to its initial state. A quasiequilibrium process is entirely reversible, since the initial states of both the system and environment can be completely restored without any additional work input or heat interaction.
We will see later that a totally reversible process is always a quasiequilibrium process.
The work done on or by the system PdV is due to the matter contained within it. The sign convention follows, since it is positive for expansion (when work is done by the system), and negative during compression (when work is performed on the system).
It can be mathematically shown that W is an inexact differential. Equation (14) may be written in the form $\delta W = PdV + 0 \times dP$. Using the criteria for exact differentials (discussed in Chapter 1) with M = P, and N = 0, it is readily seen that $\partial M/\partial P = 1$, and $\partial N/\partial V = 0$. Therefore,

$$\partial M/\partial P \neq \partial N/\partial V.$$

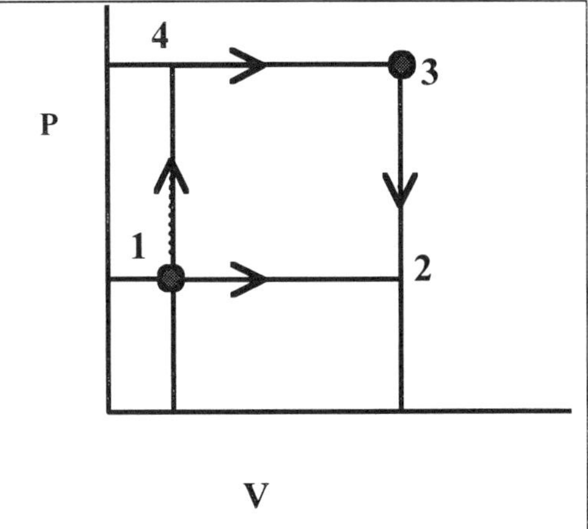

Figure 4: P–v diagram with P expressed in units of bar and v in $m^3\ kg^{-1}$.

c. *Example 3*

Air is isobarically expanded from state 1 (P_1 = 1 bar, v_1 = 1 $m^3\ kg^{-1}$), to state 2 (P_2 = 1 bar, v_2 = 3 $m^3\ kg^{-1}$), and then compressed isometrically to state 3 (P_3 = 3 bar, v_3 = 3 $m^3\ kg^{-1}$). Determine the final temperature and the net work.
Air is isometrically compressed from state 1 (P_1 = 1 bar, v_1 = 1 $m^3\ kg^{-1}$), to state 4 (P_4 = 3 bar, v_4 = 1 $m^3\ kg^{-1}$), and then expanded isobarically to state 3 (P_3 = 3 bar, v_3 = 3 $m^3\ kg^{-1}$). Determine the final temperature and the net work.

Solution

The P–v diagram for this example is illustrated in Figure 4. The final temperature T_3 is independent of the work path, and

$$T_3 = P_3 v_3/R = 300 \times 3 \div 0.287 = 3136 \text{ K}.$$

The work along the two paths

$$w_{123} = P_1 (v_2 - v_1) = 1 \times 100 \times (3 - 1) = 200 \text{ kJ kg}^{-1}, \text{ and} \quad (A)$$

$$w_{143} = P_4 (v_3 - v_2) = 3 \times 100 \times (3 - 1) = 600 \text{ kJ kg}^{-1}. \quad (B)$$

Remarks

The net work in the second case, i.e., w_{143}, is larger compared to W_{123}. The temperature represents the state of the system, and its functional form, e.g., $T_3 = P_3v_3/R$, is *independent of the path* selected to reach that state. However, the work expressions w_{123} and w_{143} (Eqs. A and B) depend upon the path selected to reach the same final state, even though the expressions for work (contain variables that only represent properties. Therefore, the final temperature is path independent, but the net work is not.

The inexact differential W integrated between two identical states along dissimilar paths 1–2–3 and 1–4–3 yields different results. An inexact differential can only be integrated if its path is known.

f. Nonquasiequilibrium Work

In the context of Figure 3, the initial pressure in the system is such that $P_1A = F_1$, where F_1 denotes the combined weight of the piston and the aggregate weights placed upon it. If all of the weights are abruptly removed, rather than slowly as discussed previously, the force exerted on the system near the piston will be much smaller than F_1. The difference between these two forces results in an acceleration of the piston due to Newton's law, and the piston mass acquires kinetic energy. Thereupon, the system pressure in the vicinity of the piston rapidly decreases. The translational energy of these molecules decreases with the pressure reduction.However, molecules further removed from the piston still possess their initial velocities (i.e. higher T, higher P), and the system is in an internally nonequilibrium state. The matter adjacent to the piston also acquires kinetic energy (e.g., Section A in Figure 5) while that removed from it does not (e.g., Section B). Hence at any instant the system properties are nonuniform and, consequently, the process is not at quasiequilibrium.

The time taken for the system to equilibrate, also called its relaxation time t_{relax}, is of the order of the distance divided by average molecular velocity (that approximately equals the sound speed V_s). Typically, the sound speed in air at room temperature is 350 m s^{-1}. It follows that if L = 10 cm, t_{relax} = 3H10^{-3} s. Consequently, a disturbance near the piston, such as a decreased pressure or decreased molecular velocity is communicated through random molecular motion to molecules located 10 cm away after roughly 0.3 ms. If the piston is displaced by 10

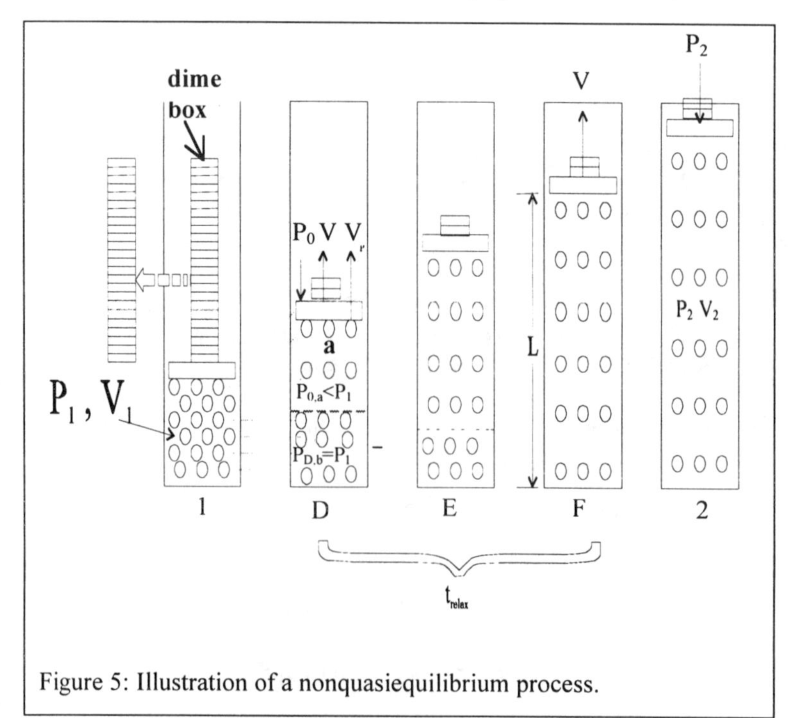

Figure 5: Illustration of a nonquasiequilibrium process.

cm every 0.3 ms, the disturbance perpetuates, and a non-equilibrium condition continually prevails.

This behavior is similar to that of a disturbance due to stones dropped into a placid pond. The disturbance is always present unless the time interval between two sequentially dropped stones is long. If the rate of stones being dropped is fast enough, the disturbance strengthens. In the piston–cylinder example, if the piston moves with a velocity of 1 m s^{-1} (which is much lower than the sound speed), the typical time scale involving motion through a 10 cm displacement ($=L/V_P$) is 100 ms, which is much larger than the relaxation time. In this case, the pressure rapidly conforms to a uniform value within the whole system. Therefore, a process may be assumed to be in quasiequilibrium as long as its relaxation timescale ($=L/V_S$) is considerably smaller than the process timescale ($=L/V_P$) responsible for the property gradients that are the source of nonequilibrium system conditions. If $V_P = 350$ m s^{-1}, quasiequilibrium cannot be assumed, and the system properties (i.e., its state) along the process path cannot be described. Therefore, an uncertain path is used to illustrate such a process in Figure 2.

During the quasiequilibrium process 1–2R described in Figure 2, the system performs more work than any corresponding non-equilibrium process 1–2 (path D-E-F-2), since part of the non-equilibrium work imparted to the piston in the form of kinetic energy is converted into thermal energy. As a consequence, even if expanded to the same final volume, the temperature at the end of a non-equilibrium process is higher, and applying the ideal gas law $P_2 > P_{2R}$ (state 2). This may also be understood by envisioning the frictional effects that dissipate and raise the system internal energy (therefore, temperature) during non-equilibrium processes. These processes are irreversible, since the original system state cannot be reverted to its original state by simply reversing the work transfer. An additional amount of work is required to overcome the effects due to friction.

Placing the system boundary immediately around the piston and the external weights (that are respectively, of mass m_p and m_w), the force experienced by the system is

$$m\, dV_P/dt = P_bA - P_RA, \tag{16a}$$

where dV_P/dt denotes the piston acceleration, P_b the pressure at the system boundary, $P_R = P_o + mg/A$ is the sum of the ambient pressure and the pressure due to the piston weight, and $m = m_p + m_w$. The work performed to move the accelerating mass m through a displacement dz is the difference between the work performed by the system and that performed to overcome the resistance to its motion. Multiplying Eq. (16a) by dZ

$$m\,(dV_P/dt)\, dz = \delta W - (P_odV + mg\, dz),$$

where the boundary work $\delta W = P_bdV$. Therefore,

$$\delta W = m\,(dV_P/dt)\, dz + P_odV + mg\, dz. \tag{16b}$$

Using the relation $dz = V_Pdt$ in Eq. (16a), and integrating appropriately,

$$W = m\, \Delta ke + P_o\Delta V + m\Delta pe, \tag{17}$$

where $P_oke = V_P^2/2$ and $\Delta pe = gz$.

If the system pressure is uniform (e.g., $t_{relax} \ll L/V_p$), then $P_b = P$ and the work performed by system

$$\delta W \approx PdV,$$

which requires a functional relation between P and V for the matter contained in the system. The following example illustrates a nonquasiequilibrium process.

---

d. Example 4:

A mass of air is contained in a cylinder at P = 10 bar, and T = 600 K. A mass of 81.5 kg is placed on the piston of area 10 cm^2 and the piston is constrained with a pin. If

the pin is removed, assuming the piston mass and atmospheric pressure to be negligible:
Determine the piston acceleration just after the pin is released.
Write an expression for the work performed on the surroundings.
Write an expression for the work done by the system matter if it exists at a uniform state.
Why is there a difference between the answers to the questions above?
What are the effects of a frictional force of 0.199 kN? (See also Figure 6.)

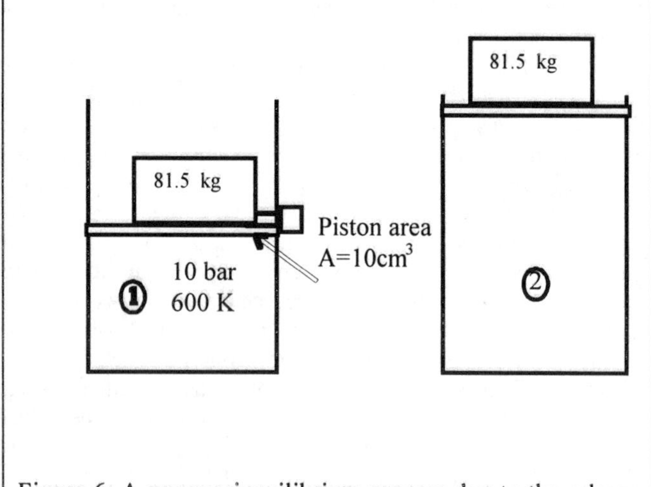

Figure 6: A nonquasiequilibrium process due to the release of a mass accelerated by a pressurized system.

Solution

The force due to mass of 81.5 kg placed on the piston equals 81.5×9.81÷1000 = 0.8 kN. The pressure due to a weight of 0.8 kN equals 800 kPa (or 8 bar). Since the system pressure is 10 bar, there is a force imbalance equivalent to 2 bars, and the mass is accelerated. The force F = m×a = m$d\mathcal{V}/dt$, i.e.,

$$F = (10-8)\text{bar}\times 100\ \text{kN m}^{-2}\ \text{bar}^{-1}\times 10\ \text{cm}^2\times 10^{-4}\ \text{m}^2\ \text{cm}^{-2}\times 1000\ \text{N kN}^{-1} = 200\ \text{N}.$$

Hence, the initial acceleration $d\mathcal{V}/dt = F/m = 2.45\ \text{m s}^{-2}$.

The work $\delta W = 800\times dV$, which changes the potential energy of the mass.

If the process is internally reversible, the matter is internally in a quasiequilibrium state, and $\delta W = PdV$.

The difference between the work performed by the system and that transmitted to the weight in the form of potential energy increases the kinetic energy of the weight. If the imparted kinetic energy is zero (or $d\mathcal{V}/dt = 0$), the work done by the system equals that done on its surroundings, i.e., there are no losses.

In the case of a frictional force of 0.199 kN, the resistance force F = 0.8 + 0.199 = 0.999 kN so that the resistance pressure $P = 0.999\ \text{kN}/10^{-3}\ \text{m}^2 = 999$ kPa, which is virtually identical to the system pressure. Therefore, the force imbalance is negligible, and $m\ d\mathcal{V}/dt \approx 0$. If the process is *internally* reversible, the work done by the system $\delta W_{system} = P\ dV$, and that done on the weight $\delta W_W \approx 800\times dV$. Hence, the frictional work

$$W_F = P\ dV - 800\times dV = (P - 800)\times dV.$$

e. *Example 5*

A mass of 50 kg is placed on a 10 cm^2 area weightless piston (cf. Figure 7). The ambient is a vacuum, i.e., the pressure is zero in it. The initial gas pressure is 100 bar, and the initial volume is 10 cm^3. The cylinder height is 10 cm. A pin, constraining the piston in place is suddenly released.

Consider the gases in the piston–cylinder assembly to constitute a system A. If the process in system A is internally reversible and isothermal, determine the work output of the gas.
Let system B be such that it includes the piston, weight, and ambient, but excludes the gases. What is the velocity of the piston when its position is at the cylinder rim? Assume system B to be adiabatic.

Solution

System A delivers work to system B during the process 1–2.

$V_1 = 10\ cm^3$, $V_2 = 10\ cm \times 10\ cm^2 = 100\ cm^3$.

The work done by system A is:

$$W_A = \int PdV = \int(mRT/V)dV = mRT \ln(V_2/V_1) = P_1V_1 \ln(V_2/V_1) \quad (A)$$

$$\therefore W_A = 100\ bar \times 100\ kN\ m^{-2}\ bar^{-1} \times 10\ cm^3 \times 10^{-6}\ m^3\ cm^{-3} \ln(100/10)$$

$$= 0.230\ kJ.$$

The work input from system A into system B results in an increase of the kinetic and potential energies of the weight. The initial and final heights of the piston in the cylinder are:

$$Z_1 = V_1/A = 10\ cm^3 \div 10\ cm^2 = 1\ cm,\ Z_2 = 10\ cm. \quad (B)$$

Applying Eq. (8) to system B, i.e.,

$$Q_{12} - W_{12} = E_2 - E_1 = \Delta U + \Delta PE + \Delta KE, \quad (C)$$

where $\Delta PE = 50\ kg \times 9.81\ m\ s^{-2} \times (10-1) cm \times 0.01\ m\ cm^{-1} \div (1000\ J\ kJ^{-1}) = 0.044\ kJ$, and

$\Delta U = 0.$

Using this result and Eq. (A) in Eq. (C),

$0 - (-0.230) = 0 + \Delta KE + 0.044$, i.e.,

$\Delta KE = (1/2)m(V_2^2 - V_1^2) = 0.230 - 0.044 = 0.186\ kJ.$

Since the initial velocity V_1 is zero, $(1/2)mV_2^2/1000 = 0.186\ kJ$, and substituting m=50 kg,

$V_2 = 2.73\ m\ s^{-1}$.

Remarks

Instead of the 50 kg weight, a projectile of very small mass can be similarly used. If the projectile were fired from the chamber using, say, gunpowder, the gases would expand, although the

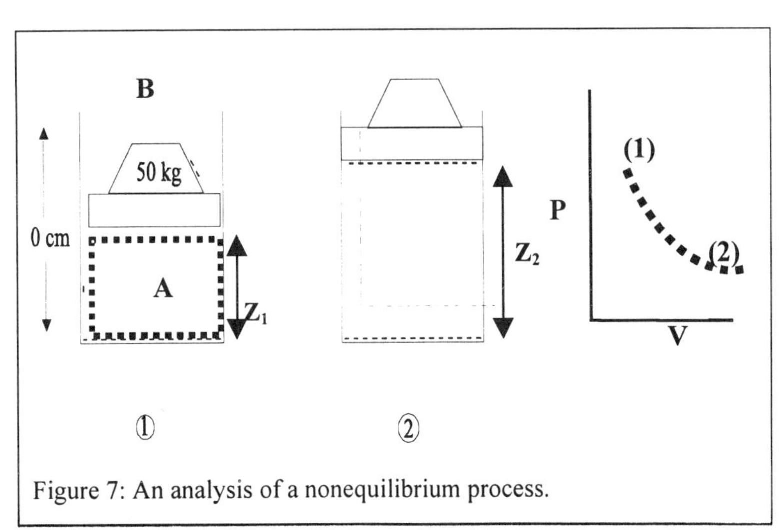

Figure 7: An analysis of a nonequilibrium process.

high temperature would remain unchanged over the period of interest due to the combusting powder. In that case, the projectile velocity can be determined using the above example.
Since the velocity in the example is of the order of 2.73 m s^{-1}, which is much slower than the room temperature molecular velocity of 350 m s^{-1}, one can assume rapid equilibration within the system. However, at lower temperatures, the quasiequilibrium assumption is invalid, since the molecular velocity can approach the process velocity.
If the ambient pressure P_o is 1 bar, the work transmitted to the matter, which is also called useful work, is given by the relation

$$W_u = \int PdV - \int P_o dV = \int (P - P_o)dV = \int PdV - P_o(V_2 - V_1),\text{ i.e.,}$$

$$W_u = 0.230 - 1 \times 100 \times 90 \times 10^{-6} = 0.221 \text{ kJ}.$$

Therefore, the kinetic energy change is

$$\Delta KE = (0.221 - 0.044) = 0.177 \text{ kJ, and}$$

$$\mathcal{V}_2 = (2 \times 1000 \times 0.177 \div 50)^{1/2} = 2.66 \text{ m s}^{-1}.$$

---

g. *First Law in Enthalpy Form*

If the kinetic and potential energies are neglected, Eq. (2) transforms into

$$\delta Q - \delta W = dU.$$

The enthalpy can replace the internal energy in this equation. The enthalpy of any substance is defined as

$$H = U + PV, \text{ or} \quad (18)$$

$$h = u + Pv.$$

For ideal gases PV = mRT and, hence, H = U + mRT. Substituting Eq. (18) in Eq. (3')

$$\delta Q - \delta W = d(H - PV).$$

For a quasiequilibrium process $\delta W = PdV + \delta W_{other}$. Therefore,

$$\delta Q - PdV - \delta W_{other} = d(H - PV).$$

Simplifying, this expression

$$\delta Q + VdP - \delta W_{other} = dH. \quad (19)$$

If $\delta W_{other} = 0$

$$\delta Q + VdP = dH.$$

The First Law can be written in the form

$$\delta Q - \delta W' = dH,$$

where for a reversible process

$$\delta W' = -VdP.$$

For a quasiequilibrium process at constant pressure

$$\delta Q_P = dH. \quad (20)$$

If an electric resistor is used to heat a gas contained in an adiabatic piston–cylinder–weight assembly, as shown in Figure 8b, the constant pressure electrical work

$$-\delta W_{elec,P} = dH. \quad (21)$$

The constant volume work (cf. Figure 8a) is

$$-\delta W_{elec} = dU. \quad (22)$$

Note that the First Law is valid whether a process is reversible or not. However, once the equality $\delta W = P\ dV$ is accepted, a quasiequilibrium process is also assumed.

xii. Internal Energy and Enthalpy

Experiments can be performed to measure the internal energy and enthalpy using Eqs. (21) and (22). For instance, electrical work can be supplied to a fixed volume adiabatic piston cylinder assembly (cf. Figure 8a), and Eq. (22) used to determine the internal energy change dU or du. Alternately, using a constant pressure adiabatic assembly (cf. Figure 8b), the electric work input equals the enthalpy change, and Eq. (21) can be utilized to calculate dH or dh.

The internal energy is the aggregate energy contained in the various molecular energy modes (translational, rotational, vibrational) which depend upon both the temperature and the intermolecular potential energy which is a function of intermolecular spacing or volume (see Chapter 1). Therefore, $u = u(T,v)$ or $u = u(T,P)$, since the specific volume is a function of pressure. While differences in internal energy can be determined, its absolute values cannot be obtained employing classical thermodynamics. However, we are generally interested in differences. For tabulation purposes a reference condition is desired. If the initial condition $u_1 = u_{ref}$ is the reference condition and $u_2 = u$ during a process 1–2, the difference

$$\Delta u = u(T,P) - u_{ref}(T_{ref},P_{ref}).$$

We normally set $u_{ref} = 0$ at the reference temperature and pressure T_{ref}, and P_{ref}, which characterize the reference state. For example, for tabulation of steam properties, the triple point ($T_{tp} = 0.01°C$, $P_{tp} = 0.006$ bar) is used as the reference state. Once u is calculated with respect to the reference condition $u_{ref} = 0$, Eq. (18) can be used to determine h. From the relation

$$h_{ref} = u_{ref} + P_{ref}\, v_{ref} = 0 + P_{ref}\, v_{ref}.$$

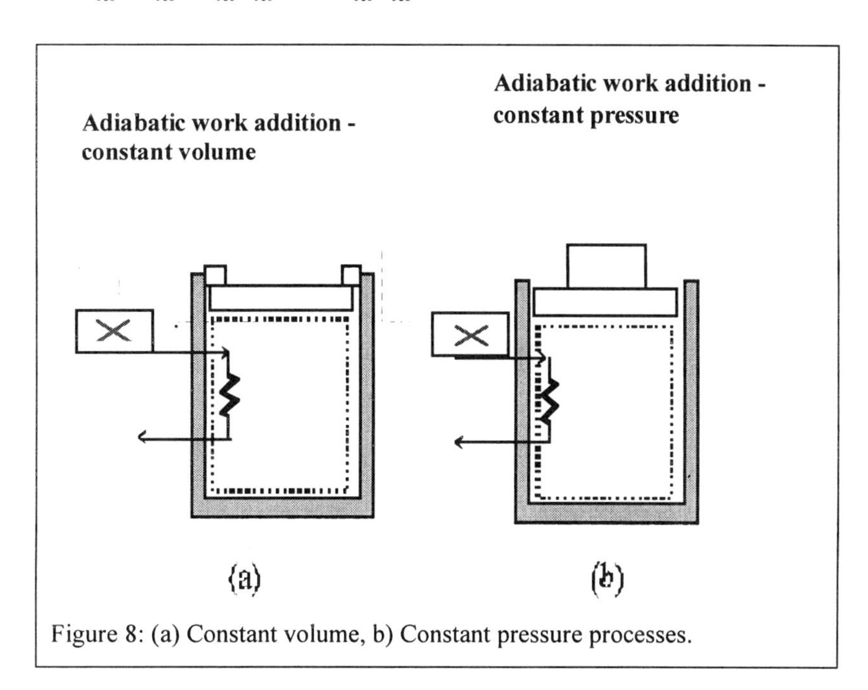

Figure 8: (a) Constant volume, b) Constant pressure processes.

we note that $h_{ref} \neq 0$ even though $u_{ref} = 0$. However, a separate reference condition can be used for the enthalpy so that

$$\Delta h = h(T,P) - h_{ref}(T_{ref},P_{ref}).$$

The internal energy can be separately calculated at this reference state. Property tables for many substances set $h_{ref} = 0$ at (T_{ref},P_{ref}) (Steam tables usually use $T_{ref} = 0.01$ C and $P_{ref} = 0.0061$ bar for liquid water).

---

f. Example 6

One kilogram of water at a temperature $T = T_{ref} = T_{tp} = 0.01°C$ is contained in an adiabatic piston cylinder assembly. The assembly resides in an evacuated chamber and a weight is placed on top of the piston such that $P = P_{ref} = 0.61$ kPa. At these reference conditions, the specific volume $v(T_{ref},P_{ref}) = 0.001\ m^3\ kg^{-1}$ is assumed to be independent of temperature. During an isobaric process, a current of 0.26 A provided at a potential of 110 V over a duration of 60.96 min raises the water temperature to 25°C. Determine the enthalpy of water at that state if $h_{ref} = 0$.

Solution

We will use the energy conservation equation

$\delta Q - \delta W = dU$

and select the water mass as the system. In general, the work term will include a volumetric change component in addition to the electrical work so that

$\delta Q - PdV - \delta W_{elec} = dU.$

At constant pressure, $\delta Q_P - \delta W_{elec,P} = dU + PdV = dH$, and on a unit mass basis

$\delta q_P - \delta w_{elec,P} = du + Pdv = dh.$

Recalling that the system is adiabatic ($q_P = 0$), and integrating the latter expression

$$-w_{elec,P} = h - h_{ref}.$$

Now, $W_{elec} = 0.26 \times 110 \times 60.96 \times 60 = 104.6$ kJ. Therefore, $-(-104.6) = h - 0$, and

$$h(25°C, 0.61\ kPa) = 104.6\ kJ\ kg^{-1}.$$

Furthermore,

$$u = h - Pv = 104.6 - 0.61 \times 0.001 \approx 104.6\ kJ\ kg^{-1}.$$

Remarks

The experiments may be repeated at different pressures for the same temperature range, and the enthalpy tabulated as a function of pressure. If the specific volume is known, applying the relation $u = h - Pv$, the internal energy can also be tabulated, as is done in the Steam tables.

Through experiments performed on ideal gases, it is found that $h = h(T)$ which is independent of the pressure, e.g., the enthalpy of air at 25°C and 1 bar is identical to that at 25°C and 10 bar ($\approx 300\ kJ\ kg^{-1}$).

Denoting the enthalpy of an ideal gase by h(T),

$$u = h(T) - Pv = h(T) - RT = u(T). \quad (23)$$

(Later in this text, ideal gas properties will be denoted as u_0, h_0, etc.). In general, for any substance $u = u(T,v)$. However, when an ideal gas is isothermally heated in a piston–cylinder assembly, the molecular translational, rotational, and vibrational energies remain constant, while the gas expands, thereby increasing the intermolecular spacing. Under these conditions, the intermolecular potential energy for ideal gases is also unchanged, since intermolecular attractive forces are absent. Therefore, the internal energy of an ideal gas is a function of temperature alone. A more detailed discussion of this is contained in Chapters 6 and 7.

xiii. Specific Heats at Constant Pressure and Volume

As the matter contained within a system is heated, the temperature and internal energy change. Applying the First Law to a constant volume closed system $\delta q_v = du_v$. The specific heat at constant volume c_v is defined as

$$c_v = (\partial u/\partial T)_v = \delta q_v/dT_v. \quad (24)$$

If instead of heating, electrical work is supplied to an adiabatic system (as in Figure 8)

$$c_v = (\partial u/\partial T)_v = (|\delta w_{elec\ v}|/dT)_v.$$

If the matter contained in a piston–cylinder–weight assembly that ensures isobaric processes is likewise heated (as illustrated in Figure 8b), the constant pressure specific heat c_p is defined as

$$c_p = (\partial h/\partial T)_P = (|\delta w_{elec\ v}|/dT)_p. \quad (25)$$

For any substance, the values of the properties c_p and c_v can be experimentally measured. In general, incompressible liquids and solids are characterized by a single specific heat c which is a function of the temperature alone, i.e., $c_p \approx c_v = c(T)$. A more detailed discussion is contained in Chapters 3 and 7. The enthalpy at a given pressure can be determined as a function of temperature by integrating Eq. (25), namely,

$$dh_p = c_p\, dT. \quad (26)$$

The ratio of the two specific heats $k = c_p/c_v$ is an important thermodynamic parameter. Typically the value of k is 1.6 for monatomic gases (such as Ar, He, and Ne), 1.4 for diatomic gases (such as CO, H_2, N_2, O_2) and 1.3 for triatomic gases (CO_2, SO_2, H_2O).

---

g. *Example 7*

Consider an electron gas, the enthalpy of which is $h = 3CT^6/P^2$. Obtain an expression for c_p.

Solution

$$c_p = (\partial h/\partial T)_P = 18CT^5/P^2 = f(T,P).$$

Remarks

Although the differentiation is carried out at constant pressure, c_p is a function of both pressure and temperature.

If water is isobarically heated at 100 kPa from 25 to 60°C its specific heat at constant pressure averaged over that temperature range is measured to be 4.184 kJ kg^{-1} K^{-1}. If the water is isobarically heated at 2 bars (e.g., in a pressure cooker) over the same temperature range, the average value of c_p is 4.17 kJ kg^{-1} K^{-1}, illustrating that the specific heat varies with pressure within the same temperature range.

For ideal gases, since u and h are functions of temperature alone, so are the two specific heats, rendering the subscripts somewhat meaningless, i.e.,

$$c_{vo} = du/dT = c_{vo}(T), \text{ and } c_{po} = (dh/dT) = c_{po}(T). \quad (27)$$

For ideal gases the subscript v is to be interpreted as differentiation of u with respect to T, while the subscript P may be interpreted as differentiation of h with respect to T. Substituting Eq. (23) in Eq. (27)

$$c_{po} = c_{vo} + R. \quad (28)$$

Table A-6F presents relations for $c_{po}(T)$ for many ideal gases while Table A-6C provides c_{po} values at specific temperatures. The internal energy and enthalpy of an ideal gas can be calculated using Eq. (27), i.e.,

$$h = \int_{T_{ref}}^{T} c_{p,o}(T)\,dT, \text{ and } u = \int_{T_{ref}}^{T} c_{v,o}(T)\,dT,$$

where $h_{ref} = u_{ref} = 0$. Once either the enthalpy or internal energy is known, the other property can be calculated from the ideal gas relation $u = h - RT$. For instance, if $c_{po}(T)$ is specified (Tables A-6F), one can generate h and u tables for ideal gases (Tables A-7 for air and A-8 to A-19 for many other ideal gases).

---

h. Example 8

In order to determine c_p for an unknown ideal gas, 0.1 kg of its mass is deposited into an adiabatic piston–cylinder–weight assembly and electrically heated (cf. Figure 8b) by a current of 0.26 A at 110V for a duration of 30 seconds. The resultant temperature rise is measured to be 10°C. Calculate c_p, assuming it to be constant.

The experiment is repeated by removing the weight, but constraining the assembly with a pin so that the volume is kept constant (cf. Figure 8a). For the same temperature rise of 10°C to occur, the current must now be applied for 23 seconds. Determine c_v.

Determine the molecular weight of the unknown gas from the measured specific heats assuming the gas to be ideal.

Solution

$$\delta W_{elec} - P\,dV = dU, \text{ or } -\delta W_{elec} = d(H - PV) + P\,dV$$

Since the pressure is held constant,

$$-\delta W_{elec} = dH = m\,c_p\,dT.$$

Assuming $c_p \approx$ constant in the narrow temperature range, and integrating

$$-W_{elec} = m\,c_p\,(T_2 - T_1).$$

Substituting for $(T_2 - T_1) = 10°C$, and using a negative sign for the electrical work transfer to the system

$$c_p = (30\times0.26\times110\div1000)\div(0.1\times10) = 0.85 \text{ kJ kg}^{-1}\text{ K}^{-1}.$$

Since V = constant, $-\delta W_{elec} = dU = m\,c_v\,dT$, and

$$c_v = (23\times0.26\times110\div1000)\div(0.1\times10) = 0.65 \text{ kJ kg}^{-1}\text{ K}^{-1}.$$

With the ideal gas assumption $c_p = c_{po}$, and $c_v = c_{vo}$, using the relation,

$$c_{po} - c_{vo} = R = \bar{R}/M,$$
$$R = 0.85 - 0.65 = 0.2 \text{ kJ kg}^{-1}\text{ K}^{-1}, \text{ and}$$
$$M = 8.314\div0.2 = 42 \text{ kg kmole}^{-1}.$$

Remarks

An alternative method to determine the molecular weight of an unknown gas is by charging a known mass of that gas into a bulb of known volume, measuring the temperature and pressure, and employing the relation $M = m\bar{R}T\div(PV)$.

If the gas molecular weight is known, c_{vo} can be determined if c_{po} is known, and vice versa, since $c_{vo} = c_{po} - R$.

At higher pressures, close to critical pressure, the intermolecular spacing becomes small, and the effects of intermolecular potential energy on u and h, and, therefore, c_p and c_v, become significant for gases. This is discussed in Chapter 7.

The temperature remains constant for a liquid being vaporized at a fixed pressure. Since, according to the First Law, the heat transfer per unit mass of liquid equals its latent heat of vaporization, namely, $q_p = h_{fg}$, the enthalpy change is finite while $dT = 0$. Therefore, $c_p = (\partial h/\partial T)_p \rightarrow \infty$ during vaporization. (Although c_p for both the liquid and vapor phases has a finite value, that value is infinite during phase change. Therefore, c_p is discontinuous during phase change.)

xiv. Adiabatic Reversible Process for Ideal Gas with Constant Specific Heats

For any reversible process, $\delta w_{rev} = P\ dv$. For an ideal gas $du = c_{v0}\ dT$. Hence, for an adiabatic reversible process involving ideal gases

$$0 - P\ dv = c_{v0}\ dT$$

Using ideal gas law $P = RT/v$ and simplifying with the relations $R = c_{p0} - c_{v0}$ and $k = c_{p0}/c_{v0}$

$$-(k-1)dv/v = dT/T$$

Assuming constant specific heats and integrating,

$$-(k-1)\ln v = \ln T + B',\ \text{i.e., } \ln T + (k-1)\ln v = C',\ \text{or } \ln T\ v^{k-1} = C'.$$

Therefore,

$$T\ v^{k-1} = C'', \quad (29a)$$

where $C'' = \exp(C')$. Using the relation $T = Pv/R$, we find that $(Pv/R)\ v^{k-1} = C$, or

$$Pv^k = C. \quad (29b)$$

For air $c_{p0} = 1$, $c_{v0} = 0.714$, i.e., $k = 1 \div 0.714 = 1.4$.

Note that if a gas is compressed adiabatically and reversibly from state 1 to 2 and then expanded back adiabatically and reversibly from state 2 to 1, the net cyclic work is zero. For the cyclic work to be finite, one must add heat at the end of the adiabatic compression process; since the expansion line is parallel. In this case, the cycle cannot be closed unless heat is rejected after the reversible expansion, which is manifest through the Second Law (cf. Chapter 3).

We now discuss why the temperature increases during adiabatic compression. Consider a 1 kg mass that is compressed for which $\delta q - \delta w = du$, where $\delta w = Pdv$. If the system is adiabatic, $\delta q = 0$. The deformation or boundary work (which is an organized form of energy with motion in a specified direction) is used to raise the internal energy of the 1 kg mass, thereby raising the internal energy (manifest through the random energy of molecules that equals te+ve+re) and, hence, temperature. For an adiabatic process, if $\delta w = 0$ then the internal energy is unchanged, i.e., $u = u(T)$ (as for an incompressible substance). The temperature does not change during the adiabatic compression of an incompressible substance.

xv. Polytropic Process

In practical situations, processes may not be adiabatic. It is possible to determine the relation between P and v and find for most substances that

$$Pv^n = C$$

where n may not necessarily equal k. Note that $n = 1$ for an isothermal process involving an ideal gas and $n = 0$ for an isobaric process.

---

i. Example 9

Air is contained in an adiabatic piston cylinder assembly at $P_1 = 100$ kPa, $V_1 = 0.1$ m^3, and $T_1 = 300$ K. The piston is constrained with a pin, and its area A is 0.01 m^2. Vacuum surrounds the assembly. A weight Wt of 2 kN is rolled on to the piston, and the pin is released. Assuming that k_o ($=c_p/c_v$) = 1.4, and $c_{v0} = 0.7$ kJ kg^{-1} K^{-1},
Is the process 1–2 reversible or irreversible?
What are the final pressure, volume, and temperature?

Solution

We will select our system to include both the air and the weight rather than the air alone because the sudden process by which it changes state cannot be completely characterized. The process is clearly irreversible, since the system cannot be restored to its initial state unless the weight is lifted back to its original position, which requires extra work.

$$P_2\ A = Wt,\ \text{or } P_2 = Wt/A. \quad (A)$$

With $Wt = 2$ kN, $P_2 = 2 \div 0.01 = 200$ kPa.

Applying the First Law to the system,

$$Q_{12} - W_{12} = 0 = E_2 - E_1, \text{ or } E_2 = E_1.$$

Neglecting the kinetic energy,

$$E_2 = U_2 + Wt\, Z_2, \text{ and } E_1 = U_1 + Wt\, Z_1. \quad \text{(B)}$$

Substituting Eq. (A) in (B), since $E_2 = E_1$,

$$U_2 - U_1 = Wt\,(Z_2 - Z_1) = Wt\,(V_2 - V_1) \div A = P_2\,(V_1 - V_2), \text{ or}$$

$$m\,(u_2 - u_1) = Wt\,(V_1 - V_2) \div A. \quad \text{(C)}$$

Treating the air as an ideal gas, Eq. (C) may be written in the form

$$m\, c_{vo}\,(T_2 - T_1) = Wt\,(V_1 - V_2) \div A. \quad \text{(D)}$$

The two unknowns in Eq. (D) are T_2, V_2, so that an additional equation is required to solve the problem. Invoking the ideal gas law for the fixed mass

$$P_1V_1/RT_1 = P_2V_2/RT_2, \quad \text{(E)}$$

Equations (D) and (E) provide the solution for V_2 and T_2. Substituting for V_2 from Eq. (E) in (D), we obtain a solution for T_2/T_1, namely,

$$T_2/T_1 = (P_2/P_1 + c_{vo}/R)/(1 + c_{vo}/R). \quad \text{(F)}$$

Using $R = \overline{R}/M = 8.314 \div 28.97 = 0.287$ kJ kg^{-1} K^{-1},

$$T_2/T_1 = (200 \div 100 + 0.7 \div 0.287) \div (1 + 0.7 \div 0.287) = 1.29, \text{ or}$$

$$T_2 = 387 \text{ K}.$$

Substituting this result in Eq. (E),

$$V_2/V_1 = (1 + (c_{vo}/R)(P_1/P_2))/(1 + c_{vo}/R). \quad \text{(G)}$$

$$\therefore\ V_2/V_1 = (1 + 0.7 \times 100 \div (0.287 \times 200)) \div (1 + 0.7 \div 0.287) = 0.65, \text{ and}$$

$$V_2 = 0.65 \times 0.1 = 0.065 \text{ m}^3.$$

Remarks

The potential energy of the weight is converted into thermal energy in air.

Once P_2 and T_2 are known, it is possible to determine k_o (= c_{po}/c_{vo}) for an ideal gas using Eq. (F). Furthermore, employing the identity $\overline{R} = \overline{c}_{p,o} - \overline{c}_{v,o}$ it is possible to calculate the molar specific heats. The gas molecular weight is required in order to ascertain the mass–based specific heats.

If the ambient pressure is finite, then Eq. (F) and (G) remain unaffected, but $P_2 = W/A + P_o$.

A machine that violates the first law of thermodynamics is termed a perpetual motion machine of the first kind (PMM1) (e.g., the "magician" David Copperfield lifting a man and, thus, changing potential energy without performing any work). Such a machine cannot exist.

We have thus far presented the First Law in the context of closed systems containing fixed masses. This analysis is applicable, for example, to expansion and compression processes within automobile engines, and the heating of matter in enclosed cooking pots. Most of the practical systems involve open systems such as compressors, turbines,

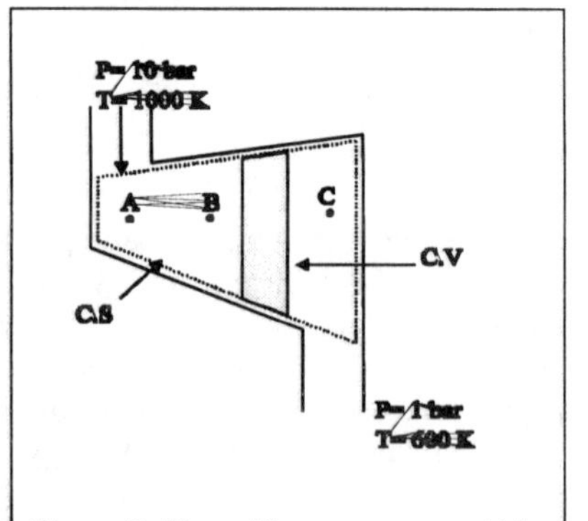

Figure 9: Nonuniform property within a control volume.

heat exchangers, biological species, etc. In the next section we will examine the derivation of the first law for an open system.

3. First Law For an Open System

In open systems, mass crosses the system boundary (also known as the control surface cs which encloses a control volume cv). In addition to heat and work interactions with the environment, interactions also occur through an exchange of constituent species between the system enclosure and its environment. Consequently the mass contained within the system may change. Examples of open systems include turbines which have a rigid boundary, thereby implying a fixed control volume (as in Figure 9) or automobile engine cylinders in which the cs deforms during the various strokes (as illustrated in Figure 10) We will initially restrict our analysis to situations for which boundary deformation occurs only in that part of the c.v. in which mass does not enter or exit the system (e.g., the portion H in Figure 10).

In general, the system properties are spatially nonuniform within the control volume, e.g., in the turbine illustrated in Figure 9, $T_A \neq T_B \neq T_C$ so that internal equilibrium for the entire system mass cannot be assumed and hence a single property cannot be assigned for the whole control volume. However, the c.v. can be treated as though each elemental volume dV within it is internally in a state of quasiequilibrium, and constitutes a subsystem of the open composite system. The mass contained in any elemental volume (cf. Figure 9) is dV/(v(T,P)).

An open system energy conservation equation is equivalent to that for a closed system if the energy content of an appropriate fixed mass in the open system is temporally characterized using the Lagrangian method of analysis. However, the problem becomes complicated if the matter contains multiple components. It is customary to employ an Eulerian approach that fixes the control volume, and analyzes the mass entering and leaving it. We now formulate the Eulerian mass and energy conservation equations, and illustrate their use by analyzing various flow problems. At the end of the chapter, we will also develop differential forms of these equations that are useful in problems involving fluid mechanics, heat transfer, and chemically reacting flows.

a. Conservation of Mass

An elemental mass δm_i awaits entry through the inlet port of an open system (such as the automobile cylinder illustrated in Figure 10) at time t. The cs enclosing the open system is marked by the boundary FBHCDE. Another boundary AGFEDCHBA (called the control mass surface c.m.s) includes both the mass within the c.v. and the elemental mass δm_i. During an infinitesimal time period δt (does not necessarily denote an inexact differential), while the mass δm_i enters the c.v, another elemental mass δm_e exits it. Thus if mass in-flow rate is $\dot{m}_i$ (say 0.2 kg/s) and if time period is δt (say 2 ms) then mass waiting outside the c.v. is $\delta m_i = \dot{m}_i \delta t$ (i.e 100 g). Thus every 2 ms, a slug of 100 g will enter our c.v. We will be concerned with mass and energy conservation equations within δt first. The piston moves simultaneously performing deformation work δW_d. As the mass δm_i moves into the c.v., the boundary of the c.m.s moves from AG to BF and extends to LK, i.e., the c.m.s moves from AGFEDCHBA to BFEDLKCHB in such a manner that it contains the same mass at both times t and t $+\delta t$. In summary, m

Quantity	At time t	At time t+δt
mass in c.v	mc.v,t	$m_{c.v,t+\delta t}$
mass outside c.v	δm_i	δm_e
mass within c.m.s	$m_{c.v,t} + \delta m_i$	$m_{c.v,t+\delta t} + \delta m_e$

Since the mass enclosed within the c.m.s does not change during the time t,

$$m_{c.v,t} + \delta m_i = m_{c.v,t+\delta t} + \delta m_e. \quad (30a)$$

Applying a Taylor series expansion (see Chapter 1) at time t+δt to the RHS of Eq. (29),

$$m_{c.v,t+\delta t} = m_{c.v,t} + (dm_{c.v}/dt)\ \delta t + (1/2!)(d^2m_{c.v}/\delta\tau^2)_t(\delta t)^2 + \ldots, \quad (30b)$$

where $(dm_{c.v.}/dt)_t$ denotes the time rate of change of mass in the c.v. at time t. Substituting Eq. (30b) in Eq. (30a)

$$m_{c.v,t} + \delta m_i = m_{c.v,t} + (dm_{c.v}/dt)_t\delta t + (1/2!)(d^2m_{c.v}/dt^2)_t(\delta t)^2 + \ldots + \delta m_e.$$

Simplifying, and dividing throughout by δt,

$$\delta m_i/\delta t = (dm_{c.v}/dt)_t + (1/2!)(d^2m_{c.v}/dt^2)_t(\delta t) + \ldots + \delta m_e/\delta t. \quad (31)$$

xvi. Nonsteady State

In the limit $\delta t \rightarrow 0$, the higher order terms in Eq. (31), namely $d^2m_{c.v}/dt^2$ and so on vanish. Therefore,

$$dm_{c.v}/dt = \dot{m}_i - \dot{m}_e, \quad (32)$$

i.e., the rate of mass accumulation within the c.v. equals the difference between the mass flow into it and that out of it. In Eq. (32) $\dot{m}_i$ and $\dot{m}_e$, respectively, denote the mass flow rate crossing the system boundary at its inlet and its exit, and its LHS the rate of change of mass within the c.v. The relation is derived for a c.v. containing a single inlet and exit.

The mass in the control volume can be evaluated in terms of density and the volume. The density may be spatially nonuniform (Figure 9). Considering an elemental volume dV, and evaluating the elemental mass as ρdV, an expression for $m_{c.v.}$ can be obtained in the form

$$m_{c.v.} = \int_{cv}\rho dV, \quad (33)$$

Substituting in Eq. (32),

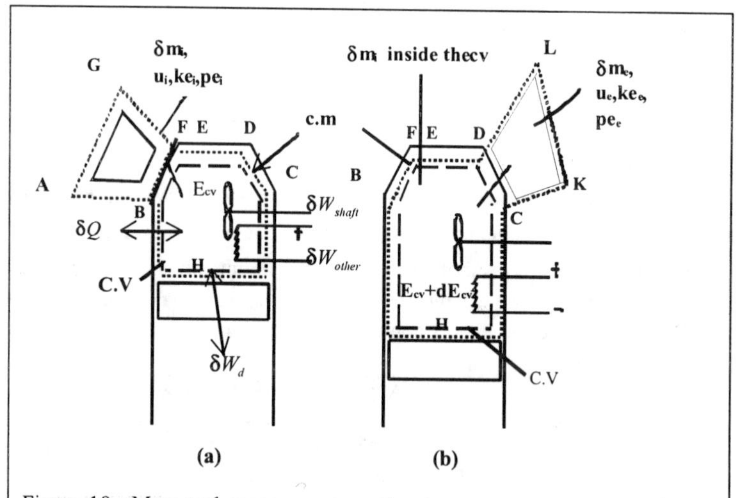

Figure 10 : Mass and energy conservation in an open system at: a: time t; and b: time t+δt.

$$d(\int_{cv}\rho dV)/dt = \dot{m}_i - \dot{m}_e. \quad (34)$$

xvii. Elemental Form

For a small infinitesimal time period t, Eq. (32) may be written in the form

$$dm_{c.v.} = dm_i - dm_e, \quad (35)$$

where $dm_i = \dot{m}_i d\,t$ denotes the elemental mass entering the c.v. during the time period dt, $dm_e = \dot{m}_e dt$ is the mass that exits during that time, and $dm_{c.v.}$ is the elemental mass that accumulates within the c.v. over the same period.

xviii. Steady State

Steady state prevails when the system properties and characteristics are temporally invariant. (Property gradients within the system may exist at steady state, e.g., the spatial non–uniformities in a turbine even though the local property values within the turbine are invariant over time.) Therefore,

$$dm_{c.v.}/dt = 0, \quad (36)$$

and Eq. (32) implies that $\dot{m}_i = \dot{m}_e$, since, at steady state, $m_{c.v.}$ = constant. The mass within the c.v. is time independent in a steady flow open system. Although the steady flow open system exchanges mass with its environment, while the closed system does not, both systems contain constant mass.

xix. Closed System

Since mass cannot cross the system boundary $\dot{m}_i = \dot{m}_e = 0$, and, hence at steady state, once again Eq. (34) applies so that $m_{c.v.}$ = constant. Equation (36) implies that the mass within c.v. is time independent even in a steady flow open system. Note that the steady flow open system exchanges mass with its environment even though it has constant mass within c.v., while the closed system does not allow mass to cross the boundaries.

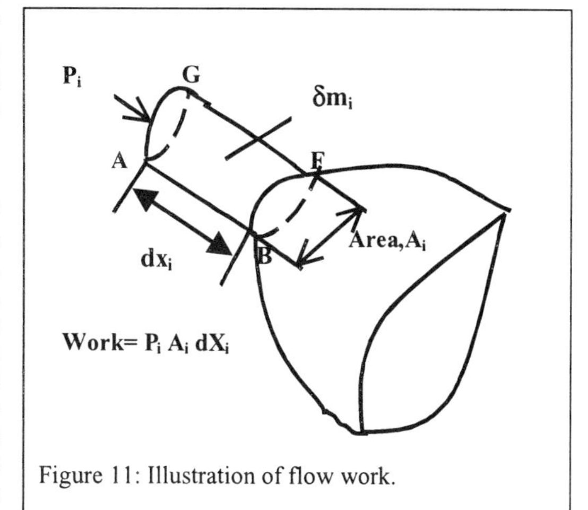

Figure 11: Illustration of flow work.

b. *Conservation of Energy*

The specific energy "e" of a mass of matter δm_i entering the inlet port of an open system during an arbitrary time interval δt is due to its kinetic, potential, and internal energies, ke_i, pe_i, and u_i (e.g., as in the automobile engine illustrated in Figure 10). Applying the First Law to the c.m.s,

$$\delta Q_{c.m.s} - \delta W_{c.m.s} = dE_{c.m.s}, \quad (2)$$

where $\delta Q_{c.m.s}$ and $\delta W_{c.m.s}$ refer to the heat and work transfer across the c.m.s boundary. The $\delta W_{c.m.}$ includes electrical work δW_{elec}, shaft work δW_{shaft}, deformation work δW_d (=PdV) etc..The energy accumulation within the c.m.s during δt is

$$dE_{c.m.s} = E_{c.m.s,t+\delta t} - E_{c.m.s,t}. \quad (37)$$

The energy in the system at time t

$$E_t = E_{c.v,t} + \delta m_i\, e_i,$$

where $E_{c.v,t}$ denotes the energy within the c.v, $\delta m_i e_i$ the energy due to the mass m_i, and e_i is the specific energy of the inlet mass. The specific energy

$$e = u + ke + pe. \tag{38}$$

At time t = t + δt, the energy content of the c.m.s is

$$E_{c.m.s,t+\delta t} = E_{c.v,t+\delta t} + \delta m_e e_e, \tag{39}$$

and the subscript e refers to the exit conditions. Using Eqs. (2), (37), and (39),

$$\delta Q_{c.m.s} - \delta W_{c.m.s} = E_{c.v,t+\delta t} + \delta m_e e_e - E_{c.v,t} - \delta m_i e_i. \tag{40}$$

Expanding $E_{c.v,t+\delta t}$ using a Taylor's series,

$$E_{c.v,t+\delta t} = E_{c.v,t} + (dE_{c.v}/dt)_t\delta t + (1/2)(d^2E_{c.v}/dt^2)_t(\delta t)^2 + \ldots,$$

and using this result in Eq. (40), and simplifying

$$\delta Q_{c.m.s} - \delta W_{c.m.s} = (dE_{c.v}/dt)_t\delta t + (1/2)(d^2E_{c.v}/dt^2)_t(\delta t)^2 + \ldots + \delta m_e e_e - \delta m_i e_i. \tag{41}$$

In general, the work interaction through the c.m. boundary (analogous to closed system work) can involve the shaft work δW_{shaft}, the c.v. boundary deformation work δW_d (e.g., at boundary H in Figure 10), flow work δW_f (a kind of boundary work involving the deformation of c.m. boundary; for example, the boundary AG in Figure 11 is pushed in by applying inlet pressure P_i so that the mass δm_i can enter the c.v. within dt, and due to exit pressure P_e that pushes out the mass δm_e within dt),and other work forms δW_{other} (e.g., electrical work).

$$\delta W_{c.m.s} = \delta W_{shaft} + \delta W_f + \delta W_d + \delta W_{other}. \tag{42}$$

xx. Flow Work

The control mass boundary deforms at the inlet and exit due to the mass entering and leaving the c.v. Therefore,

$$\delta W_f = \delta W_{f,i} + \delta W_{f,e}. \tag{43}$$

At the inlet, the boundary AG is pushed towards BF as illustrated in Figure 11. Work is performed on the control mass by pushing it through a distance dx_i during time δt, i.e.,

$$\delta W_{f,i} = -P_iA_idx_i.$$

The distance $dx_i = V_i\delta t$, where V_i denotes the inlet velocity. Since P_iA_i have positive values, the negative sign is added in order to satisfy the sign convention for work input into the system. The previous expression may be written as

$$\delta W_{f,i} = -P_iA_iV_i\delta t = -P_i\,\dot{m}_i v_i\delta t. \tag{44}$$

Similarly, at the exit the work done by the system in pushing the mass dm_e out is

$$\delta W_{f,e} = -P_eA_eV_e\delta t = -P_e\,\dot{m}_e v_e\delta t.$$

Therefore,

$$\delta W_f = P_e \dot{m}_e v_e \delta t - P_i \dot{m}_i v_i \delta t. \tag{45}$$

Further, substituting Eqs. (42), and (45) in Eq. (41) and dividing throughout by δt

$$\delta Q/\delta t - \delta W_{shaft}/\delta t + (P_i \dot{m}_i v_i - P_e \dot{m}_e v_e) - \delta W_d/\delta t - \delta W_{other}/\delta t$$

$$= (dE_{c.v}/dt)_t \delta t + (1/2)(d^2E_{c.v}/dt^2)_t(\delta t)^2 + \ldots + (\delta m_e/\delta t)e_e - (\delta m_i/\delta t)e_i. \tag{46}$$

xxi. Nonsteady State

In the limit $\delta t \to 0$ in the context of Eq. (46), δm_i shrinks to an infinitesimally small volume (see the remarks below), but the ratio $\delta m_e/\delta t$ is still finite, the boundary AG approaches BF (cf. Figure 11 and Figure 10), and the boundary LK approaches DC. Since the c.m.s is virtually identical with the cs, $\delta Q_{c.m.s}/\delta t = \dot{Q}_{cv}$, $\delta W_{shaft}/\delta t = \dot{W}_{shaft}$, $\delta W_d/\delta t = \dot{W}_d$, $\delta W_{other}/\delta t = \dot{W}_{other}$. Using the expression $\delta W_{c.v.} = \delta W_{shaft} + \delta W_d + \delta W_{other}$, and the definition $\dot{w}_{c.v.} = \dot{w}_{shaft} + \dot{w}_d + \dot{w}_{other}$, Eq. (46) assumes the form Eq. (46) assumes the form

$$\dot{Q}_{cv} - \dot{W}_{cvt} = (dE_{c.v}/dt) + \dot{m}_e e_e - \dot{m}_i e_i - P_i \dot{m}_i v_i + P_e \dot{m}_e v_e. \tag{47}$$

Simplifying this expression

$$\dot{Q}_{cv} - \dot{W}_{cv} = (dE_{c.v}/dt) + \dot{m}_e e_{T,e} - \dot{m}_i e_{T,i}, \tag{48}$$

where $e_T = h + ke + pe$ is called the methalpy or total enthalpy. The Pv term in the enthalpy is due to the work flow of matter into and out of the c.v. Equation (48) may be rewritten in the form

$$(dE_{c.v.}/dt) = \dot{Q}_{cv} - \dot{W}_{cv} + \dot{m}_i e_{T,i} - \dot{m}_e e_{T,e}, \tag{49}$$

the physical meaning of which is as follows: The energy accumulation rate = Energy added through the c.s. by heat transfer – Energy transfer through work interactions + Methalpy addition by advection – Methalpy expulsion through advection. The term $E_{c.v.}$ must be evaluated for the entire open system in which property gradients may exist. For instance, in a steam turbine (Figure 10) as steam is admitted in order to start it, the turbine c.v. is warmed up and $E_{c.v.}$ increases over time. However, the temperature and pressure near its inlet are higher than at the exit so that the specific energy varies within the c.v. We can evaluate $E_{c.v.}$ by using the relation

$$(dE_{c.v.}/dt) = d/dt(\int_{cv} \rho e dV) = \dot{Q}_{cv} - \dot{W}_{cv} + \dot{m}_i e_{T,i} - \dot{m}_e e_{T,e}. \tag{50}$$

For a nondeformable c.v., such as a turbine, $\dot{w}_d = 0$ so that $\dot{w}_{c.v.} = \dot{w}_{shaft} + \dot{w}_{other}$.
The term $\dot{w}_{c.v.}$ does not include the flow work which is already accounted for in the enthalpy term (i.e., h = u + Pv =internal energy + flow work).
The elemental mass δm_i in Figure 10 is the mass waiting outside control volume that will subsequently enter the c.v. within the duration δt. As $\delta t \to 0$, δm_i and its volume $\to 0$, and δm_e and its volume $\to 0$. In this case the c.m.s. $\to$ c.v., and the two boundaries merge.
The heat transfer across the c.v. boundary $\dot{Q}_{c.v.} = \delta Q/\delta t \neq 0$ as $\delta t \to 0$, although $\delta Q_{c.m.} \to 0$.
The term $E_{c.v.} = (U + KE + PE)_{c.v.}$
The terms $dE_{c.v}/dt$, $\dot{Q}$, and $\dot{W}$ are expressed in similar units and the dot over symbols $\dot{Q}$, $\dot{W}$, and $\dot{m}$ is used to indicate heat, work and mass transfer across the c.s., and time differentials, e.g., $dE_{c.v}/dt$, indicate accumulation of properties within the c.v.

Equations (49) and (50) can be applied to various cases such as steady ($\partial/\partial t = 0$), adiabatic ($\dot{Q}_{c.v}= 0$), closed systems ($\dot{m}_i= 0$, $\dot{m}_e= 0$), and heat exchange devices like boilers ($\dot{W}c.v. = 0$).

xxii. Elemental Form

Upon multiplying Eq. (49) by δt we obtain

$$dE_{c.v.} = \dot{Q}_{cv}dt - \dot{W}_{cv}dt + \dot{m}_i e_{T,i}dt - \dot{m}_e e_{T,e}dt, = \delta Q_{c.v}dt - \delta W_{c.v}dt + \dot{m}_i e_{T,i}dt - \dot{m}_e e_{T,e}dt. \quad (51)$$

In Eq. (51) $dE_{c.v.}$ denotes the energy accumulation, and $\delta Q_{c.v.}$ and $\delta W_{c.v.}$ the heat and work transfer over a small time period dt.

xxiii. Steady State

Open systems, e.g., turbines, compressors, and pumps, often operate at steady state, i.e., when $dE_{c.v.}/dt = 0$, $dm_{c.v.}/dt = 0$, and $\dot{m}_i = \dot{m}_e = 0$. Hence,

$$\dot{Q}_{cv} - \dot{W}_{cv} + \dot{m}_i e_{T,i} - \dot{m}_e e_{T,e} = 0. \quad (52)$$

xxiv. Rate Form

Consider the special case of a single inlet and exit with no boundary work. At steady state, properties within the c.v. do not vary over time, although spatial variations may exist. Therefore, Eq. (49) simplifies to the form

$$\dot{Q}_{cv} - \dot{W}_{cv} = \dot{m}\Delta e_T, \quad (53)$$

where $\Delta e_T = e_{T,i} - e_{T,e}$.

xxv. Unit Mass Basis

The unit mass–based equation may be obtained by dividing Eq. (53) by mass, i.e.,

$$\dot{q}_{cv} - \dot{w}_{cv} = \Delta e_T. \quad (54a)$$

where $q_{c.v.} = \dot{Q}_{cv}/\dot{m}$, and $w_{c.v} = \dot{W}_{c.v}/\dot{m}$. For an elemental section of a turbine Eq. (54a) can be written as,

$$\delta q_{c.v.} - \delta w_{c.v.} = de_T.$$

If the KE and PE are neglected,

$$\delta q_{c.v} - \delta w_{c.v} = dh. \quad (54b)$$

In a Lagrangian reference frame, a unit mass enters a turbine (as illustrated in Figure 12) with an inlet energy $e_{T,i}$ which decreases due to heat loss to the ambient (i.e., $\delta q_{c.v.} < 0$) and work output ($\delta w_{c.v.} > 0$). At the same time the unit mass undergoes deformation due to changes in volume. If one travels with the mass, then

$$\delta q_{c.m.} - \delta w_{c.m} = du_{c.m.} \quad (54c)$$

where $\delta w_{c.m}$ is the work involved within c.m. But $\delta w_{c.m} = \delta w_{Pdv}$ for a reversible process. Since the internal energy change $du = dh - P\,dv - v\,dP$, then the equation (54c) becomes

$$\delta q + v\,dP = dh,$$

where the subscript c.m. has been omitted. Upon comparison with Eq. (54c) with Eq. (54b), the reversible shaft work

$\delta w_{c.v,rev} = - vdP.$

xxvi. Elemental Form

Over a small time period δt during which a mass dm both enters and leaves a steady state open system, Eq. (52) yields

$$\delta Q_{c.v.} - \delta W_{c.v.} = dm(\Delta e_T). \quad (55)$$

This expression may also be obtained by multiplying Eq. (53) by dm.

xxvii. Closed System

Equation (36) is also an expression of closed system mass conservation. The energy conservation expression of Eq. (50) can be applied to closed systems. Since $\dot{m}_i = \dot{m}_e = 0$,

$$(dE_{c.v.}/dt) = \dot{Q} - \dot{W}. \quad (56)$$

The subscript c.v. has been omitted for the closed system. The elemental form of Eq. (56), namely, $\delta Q - \delta W = dE$, is identical to Eq. (2).

xxviii. Remarks

For a nondeformable c.v, such as a turbine, $W_d = 0$ so that $W_{c.v.} = W_{shaft} + W_{other}$. The term $W_{c.v.}$ does not include the flow work which is accounted for through the enthalpy term (h = u + Pv = internal energy + flow work). The elemental mass δm_i in Figure 10 is the mass waiting outside control volume that will subsequently enter the c.v. within the duration δt. As $\delta t \rightarrow 0$, δm_i and its volume $\rightarrow 0$, and δm_e and its volume $\rightarrow 0$. However, $\delta m_i/\delta t \neq 0$, and $\delta m_e/\delta t \neq 0$. In this case the c.m.s. $\rightarrow$ c.s., and the two surfaces merge.

The heat transfer across the c.v. boundary $\dot{Q}_{cv} = \delta Q/\delta t \neq 0$ as $t \rightarrow 0$, although $\delta Q_{c.m.s} \rightarrow 0$. The control volume energy $E_{c.v.} = (U + KE + PE)_{c.v.}$.

The dot over symbols Q, W, and m is used to indicate heat, work and mass transfer across the cs and time differentials, e.g., $dE_{c.v}/dt$, indicate accumulation.

Equations (49) and (50) can be applied to various cases such as steady ($\partial/\partial t = 0$), adiabatic ($\dot{Q}_{cv} = 0$), closed systems ($\dot{m}_i = 0$, $\dot{m}_e = 0$), and heat exchange devices such as boilers ($\dot{W}_{cv} = 0$).

xxix. Steady State Steady Flow (SSSF)

Steady flow need not necessarily result in steady state, e.g., during the mixing of a hot and cold fluid. Likewise, during intensive steady state, i.e., when the properties are temporally uniform, a system may not experience steady flow, e.g., as a fluid is drained from a vessel.

j. Example 10

As liquid water flows steadily through an adiabatic valve the pressure decreases from $P_1 = 51$ bar to $P_2 = 1$ bar. If the inlet water temperature is 25°C, what is the exit temperature? Assume that the specific volume of water is temperature independent and equal to 0.001 m^3/kg, and that u = cT, where c = 4.184 kJ kg^{-1} K^{-1}. Neglect effects due to the kinetic and potential energies.

Solution

Mass conservation implies that $dm_{c.m.}/dt = 0$. Therefore, $\dot{m}_i = \dot{m}_e = \dot{m}$. Furthermore, $dE_{c.v}/dt = 0$, and $\dot{Q}_{cv} = \dot{W}_{cv} = 0$. Applying Eq. (50), $\dot{m}\Delta e_T = 0$.

Since $e_T = h + ke + pe$, this implies that $h_2 = h_1$. A process during which the enthalpy is unchanged (i.e., $h_2 = h_1$) is called a throttling process. Furthermore, since $v_2 = v_1 = v$, and $u_2 + P_2v_2 = u_1 + P_1v_1$ (as a consequence of $h_2 = h_1$),

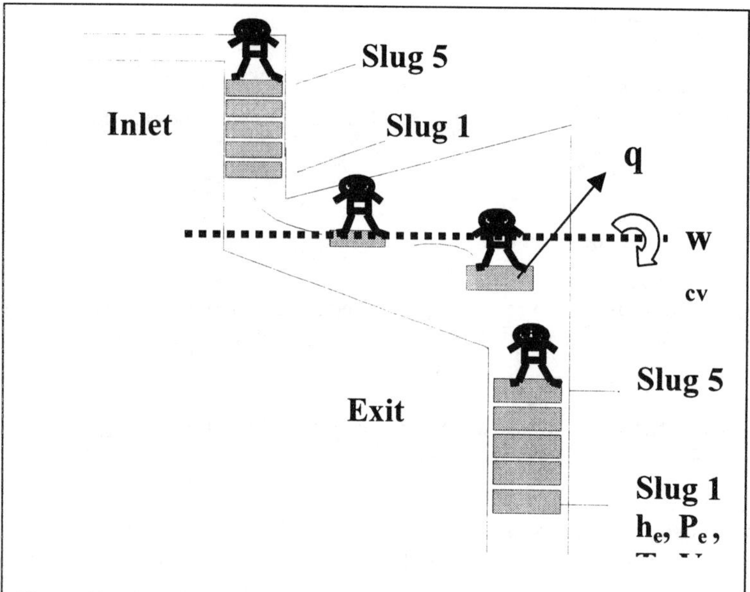

Figure 12: First Law applied to a steady state open system in a Lagrangian reference frame.

$c\,(T_2 - T_1) = -v(P_2 - P_1)$, i.e.,

$T_2 - T_1 = (0.001 \times (51-1)$ bar $\times$ 100 kPa bar$^{-1}) \times (4.184$ kJ kg^{-1}K$^{-1})$ = 1.2 K, and $T_2 = (298 + 1.2) = 299.2$ K.

Remarks

The temperature increases during the adiabatic throttling of liquids. We saw that when a fluid of fixed mass is compressed adiabatically in a closed system, the boundary or deformation work raises the internal energy and, hence, the temperature. If the fluid is incompressible, liquid deformation work is absent and u and T cannot change. On the other hand, if there is some flow *inside* the closed system, e.g., incompressible water around inside a piston–cylinder assembly, applying the relation $\delta q - \delta w = du + d(ke) = 0 - 0 = du + d(ke)$. If the water slows down due to friction effects at the walls, du increases while d (ke) decreases, i.e., the temperature changes for an incompressible substance occur only due to friction. The internal energy and temperature of a compressible fluid changes during an adiabatic process due to Pdv work during compression due to frictional heating.

We will now consider a throttling process. Water at the inlet must overcome an inlet pressure over a cross-sectional area A. Consider an elemental mass having dimensions of area A and distance L. This mass is pushed into the control volume with a pressure of P. Therefore, the work done by applying a force P A to move along a distance L is equal to P A L. Mathematically, P A L = P V = P v m (which is the inlet flow work). Using the subscript 1 to denote conditions at the inlet, the total energy of the mass after it enters the inlet (that includes the work required to push it) is $u_1 + P_1 v_1$ which will increase the c.v. internal energy $u_{c.v.}$ (see also Example 14). In order to maintain steady state conditions, a similar mass must be pushed out of the exit by flow work equal to $P_2 v_2$ m. However, $P_2 < P_1$, and the outlet flow work is lower than the inlet flow work. Consequently, energy starts accumulating within the control volume heating the water. Therefore, the mass leaving the c.v. must be at a higher tem-

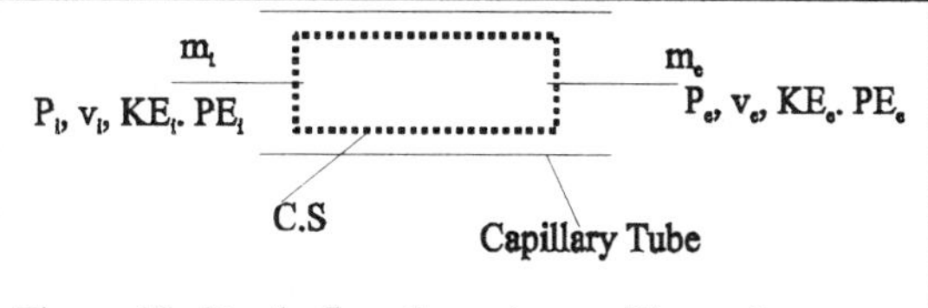

Figure 13: Steady flow through a capillary tube.

perature compared to the mass at the inlet in order to maintain the steady state condition. Chapter 7 contains further discussion regarding the temperature change in a gas that accompanies a pressure drop during a sudden expansion process that is also called a throttling process. In that case $v_2 > v_1$.

k. Example 11

A fluid flows through a capillary tube with an inlet velocity V_i and exit velocity V_e. Apply the mass and energy conservation equations for steady flow and simplify the expression.(Figure 13)

Solution

Mass conservation implies that $\dot{m}_i = \dot{m}_e = \dot{m}$. From energy conservation $dE_{c.v}/dt = 0$, and $\dot{W}_{cv} = 0$ so that Eq. (50) implies that

$$\dot{Q}_{cv} - \dot{m}\Delta e_T = 0, \text{ i.e., } q_{c.v} - \Delta e_T = 0. \quad (A)$$

For an adiabatic system, Eq. (A) assumes the form

$$\Delta(h + ke) + \Delta pe = 0, \text{ or } (h + ke) + pe = \text{Constant}. \quad (B)$$

The sum (h + ke) is called the stagnation enthalpy and is commonly used in fluid dynamics analyses. If u is a function of temperature alone, e.g., as for an ideal gas or incompressible liquid, $\Delta u = 0$ and $\Delta h = \Delta (Pv)$. In this case,

$$(Pv + ke) + pe = \text{Constant, or } P/\rho + V^2/2 + gZ = \text{Constant}. \quad (C)$$

Remarks

Suppose eq. (C) is applied to an incompressible fluid between inlet and exit. Then

$$u_i + P_i v + ke_i + pe_i = u_e + P_e v + ke_e + pe_e. \quad (D)$$

Rewriting this relation, we have

$$P_i/\rho + V_i^2/2 + gz_i - P_e/\rho + V_e^2/2 + gz_e = u_i - u_e. \quad (E)$$

The term $e_m = P/\rho + V^2/2 + gz$ is the mechanical portion of the energy. Rewriting Eq. (E),

$$e_{m,i} - e_{m,e} = u_i - u_e. \quad (F)$$

From Eq. (E),

$$P_{m,i} - P_{m,e} = (u_i - u_e)\rho \quad (G)$$

where $P_m = P + \rho V^2/2 + \rho\, gz$ denotes the mechanical pressure, or

$$H_{m,i} - H_{m,e} = (u_i - u_e)/g \quad (H)$$

where, $H_m = P/\rho g + V^2/2g + Z$. The difference $H_{m,i}$ - $H_{m,e}$ is the mechanical head loss $H_{m,L}$, i.e.,

$$H_{m,L} = H_{m,i} - H_{m,e} = (u_i - u_e)/g.$$

If there is no frictional loss for an incompressible fluid, $u_i = u_e$, and Eq. (E) yields

$$Pv + ke + pe \text{ or } P/\rho + V^2/2 + gZ = e_m = \text{Constant}. \quad \text{(I)}$$

Eq. (I) is the Bernoulli energy equation which is well known in the field of fluid mechanics. Therefore,

$$e_{m,i} = e_{m,e};\ P_{m,i} = P_{m,e};\ H_{m,i} = H_{m,e}.$$

Instead of a capillary tube with constant mass flow, consider natural gas flow through a pipeline of variable cross section. The velocity distribution across the pipe may be spatially nonuniform, and the density also may vary axially as the flow proceeds. If an imaginary capillary tube of very small cross section is inserted into the pipe, such that velocity across the capillary tube is spatially invariant at both inlet and outlet, the energy change for a non-adiabatic elemental mass flow through it is given by Eq. (B), and Eq. (D) if adiabatic. Such a tube is called a stream tube, and if we imagine ourselves to be situated on top of a unit mass travelling through the stream tube, the energy of the mass is governed by Eqs. (B), (D) or (I). An infinite number of stream tubes can theoretically be inserted across a cross section, and the total energy change can be calculated.

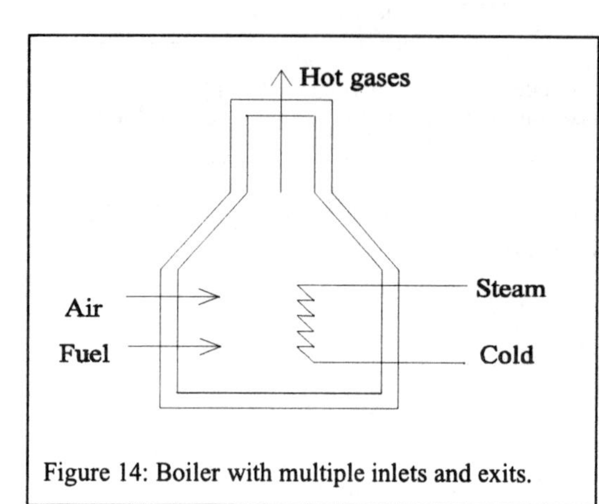

Figure 14: Boiler with multiple inlets and exits.

c. *Multiple Inlets and Exits*

For the mass boiler illustrated in Figure 14,

$$dm_{c.v}/dt = \sum \dot{m}_i - \sum \dot{m}_e, \text{ and} \quad \text{(57a)}$$

$$(dE_{c.v.}/dt) = \dot{Q}_{cv} - \dot{W}_{cv} + \sum \dot{m}_i e_{T,i} - \sum \dot{m}_e e_{T,e}, \quad \text{(57b)}$$

d. *Nonreacting Multicomponent System*

xxx. Mass Conservation

For multicomponent nonreacting systems having a single inlet and exit, Eq. (32) may be written in terms of each component, namely,

$$dm_{k,c.v}/dt = \dot{m}_{k,i} - \dot{m}_{k,e}. \quad \text{(58)}$$

The mass flow rate of a component is the product of the component molar flow rate multiplied by its molecular weight, i.e., $\dot{m}_k = \dot{N}_k M_K$. The mole balance equation for each species is written in the form

$$dN_k/dt = \dot{N}_{k,i} - \dot{N}_{k,e}. \quad \text{(59)}$$

and the overall mass balance assumes the form

$$dm_{c.v}/dt = d\sum m_{k,cv}/dt = -\sum \dot{m}_e. \quad \text{(60)}$$

The mass conservation equation will be extended to reacting systems in Chapter 11.

xxxi. Energy Conservation

The energy conservation may be written in the molal form as

$$(dE_{c.v}/dt) = \dot{Q}_{c.v.} - \dot{W}_{c.v.} + \Sigma \dot{N}_{k,i}\, \hat{e}_{kTi} - \Sigma \dot{N}_{k.e}\, \hat{e}_{k,T,e}.$$

The advection energy for mixtures is defined as

$$\Sigma \dot{m}_i e_{T,i} = \Sigma \dot{N}_{k,i}\, \hat{e}_{kTi} ,$$

and the advection enthalpy $\dot{m}h$ for a mixture

$$\dot{m}h = \Sigma_k \dot{N}_k \hat{h}_k ,$$

where $\hat{e}_{kTi}$ and $\hat{h}_k$ denote the energy and enthalpy of the k–the component in the mixture (cf. Chapter 1). If we assume that $\hat{e}_{kTi} = \bar{e}_{kTi}$ and $\hat{h}_k = \bar{h}_k$ (which is the enthalpy of component k in its pure form at the same temperature and pressure – discussed in greater detail in Chapter 8),

$$(dE_{c.v}/dt) = \dot{Q}_{c.v.} - \dot{W}_{c.v.} + \Sigma \dot{N}_{k,i}\, \bar{e}_{kTi} - \Sigma \dot{N}_{k.e}\, \bar{e}_{kTe}. \quad (61)$$

4. Illustrations

The simplest class of problems pertains to those concerned with spatially uniform properties within a c.v., also known as uniform state problems. However, the system characteristics can change over time.

a. *Heating of a Residence in Winter*

A transient analysis must be employed to predict the temperature as a residential space is heated from a colder to a warmer temperature.

l. Example 12

A rigid residential space is at a temperature of 0°C (which equals the ambient temperature). A gas heater is used to warm it up to 20°C. Heat is lost from the walls, floor and ceiling of the space to the ambient at a rate of 3 kW. The net air equivalent mass inside the house is 400 kg. What is the required blower capacity so that the space can be warmed to the desired temperature in 15 minutes. Assume warm air to be available at 40°C, the air mass in the space to be constant, and the air exhausts at the space temperature. Assume also that $c_{vo} = 0.71$ kJ kg^{-1} K, and R =0.287 kJ kg^{-1} K^{-1}. As we heat the space and if P = constant, then mass (=PV/RT) will decrease inside the control volume and mass conservation requires that mass leaving must be more compared to mass entering. Then mass must be decreasing. However, in order to simplify the problem, assume that mass in the residential space is constant.

Solution

We start by writing the generalized mass and energy conservation equations for open systems. Using the mass conservation relation, namely, Eq. (32), at steady state

$$\dot{m}_i = \dot{m}_e = \dot{m}. \quad (A)$$

Neglecting the potential and kinetic energies, $E_{c.v.} = U$, and Eq. (50) assumes the form

$$d/dt(\textstyle\int_{cv} \rho e dV) = \dot{Q}_{cv} - \dot{W}_{cv} + \dot{m}_i h_i - \dot{m}_e h_e. \quad (B)$$

Substituting Eq. (A) in Eq. (B), and assuming constant specific heat and uniform temperature throughout the space

$$m_{c.v} c_{vo} dT/dt = \dot{Q}_{cv} - \dot{W}_{cv} + c_{po}(T_i - T_e). \quad (C)$$

Since $\dot{W}_{cv} = 0$ and the exit temperature T_e equals the house temperature T, Eq. (C) may be written in the form

$$\frac{m_{cv} c_{vo} dT}{\dot{Q}_{cv} - \dot{m} c_{po} (T - T_i)} = dt. \quad \text{(D)}$$

After integration and simplifying,

$$\ln(A - B\,T) = -Bt + C. \quad \text{(E)}$$

The constants

$$A = BT_i + \dot{Q}_{cv}/(m_{c.v}c_{vo}), \text{ and} \quad \text{(F)}$$

$$B = \dot{m}c_{po}/(m_{c.v}c_{vo}). \quad \text{(G)}$$

Applying the initial condition $T = T_o$ at time $t = 0$, using Eq. (E)

$$\ln(A - BT_o) = C. \quad \text{(H)}$$

Employing Eq. (E)–(H),

$$((\dot{Q}_{cv}/\dot{m}c_{po}) + (T_i - T)) / ((\dot{Q}_{cv}/\dot{m}c_{po}) + (T_i - T_e)) = \exp(-(t/t_c)), \quad \text{(I)}$$

where the characteristic time associated with heating the house

$$t_c = (m_{c.v}c_{vo})/(\dot{m}c_{po}) = m_{c.v}/(\dot{m}k) = (284\ \dot{m}^{-1}) \quad \text{(J)}$$

where $k = c_{po}/c_{vo}$. The mass heat capacity of the house C_{vo} equals $m_{c.v}c_{vo}$. Using the desired temperature $T = 293$ K in the allotted time $t = 900$ s, with the initial and inlet temperatures $T_e = 273$ K, $T = 293$ K and $T_i = 313$ K, the heat loss $\dot{Q}_{cv} = -3$ kJ s^{-1}, $m_{c.v.} = 400$ kg, and the properties $c_{vo} = 0.71$ kJ kg^{-1} K^{-1} and $c_{po} = 1$ kJ kg^{-1} K^{-1}, Eq. (I) yields

$$(-3\ \dot{m}^{-1} + 313 - 293) \div (-3\ \dot{m}^{-1} + 313 - 273) = \exp(-3.17\ \dot{m}).$$

Solving iteratively $\dot{m} = 0.327$ kg s^{-1}. Then from Eq.(J), $t_c = 284\ \dot{m}^{-1} = 869$ s.

Remarks

The characteristic time t_c is a useful time scale. The higher the house heat capacity, the larger is the value of t_c, implying that it takes longer to heat a space with the same flow rate. Typically, the mass to be heated depends upon the building area. For most residential buildings, the air equivalent mass is 50 kg per m^2 of heated area (10 lb ft^{-2}), and for larger houses and commercial buildings this mass is 150-350 50 kg m^{-2}. The example uses a low air mass equivalent.

b. Thermodynamics of the Human Body

There is perpetual heat loss from the human body and, yet, normally the body temperature remains virtually unchanged.

m. Example 13

Consider a spherical tank of radius 0.38 m (radius R). We wish to pack electric bulbs each of radius 0.01 m (radius a). The power to each bulb is adjusted such that the surface temperature of the tank is maintained at 37°C. The heat transfer coefficient is about 4.63 W/m^2 K. Assume steady state. Determine a) heat loss from the tank (= $h_H A\ (T - T_0)$) for $T_0 = 25$°C, b) number of bulbs you can pack in the tank, c) amount of electrical power required for each bulb so that the tank surface is always at 37 C ; d) what are the answers for (a) to (c) if the tank radius is reduced to 0.19 m but the surface temperatureis still maintained at 37°C and the bulb size is fixed?

Solution

$\dot{Q} = h_H\, A\, (T - T_0) = 4.63 \times 4\,\pi \times 0.38^2 \times (37\text{-}25) = 101$ W.

Since the volume of each bulb $V = 4/3 \times\pi \times R^3$, the number of bulbs $R^3/a^3 = 54872$.

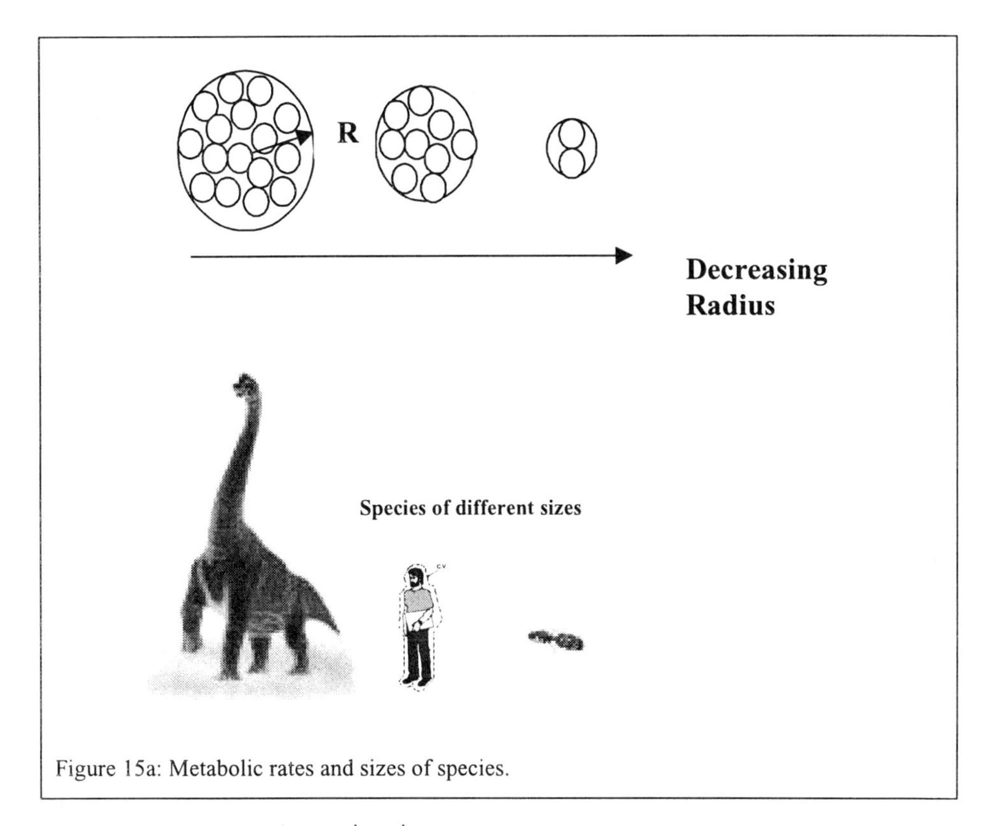

Figure 15a: Metabolic rates and sizes of species.

At steady state $dE_{c.v}/dt = \dot{Q} - \dot{W} + \dot{m}_i e_{T,i} - \dot{m}_e e_{T,e} = 0$, $\dot{m}_i = \dot{m}_e = 0$, i.e.,
$\dot{Q} - \dot{W} = 0$.

Therefore, the electrical work $\dot{W} = \dot{Q}$, and

$\dot{W}/N = \dot{Q}/N = 101/54872 = 0.00184$ W/bulb.

The heat transfer

$\dot{Q}/N = hA(T-T_0)/(R^3/a^3) = h_H\, 4\, \pi(T- T_0)\, a^3/R = 0.0007/R$.

Therefore, by reducing the radius R, the electrical work per bulb increases, and the power per bulb doubles to 0.00368 W.

Remarks

As an analogy, the cells in a human or animal body can be thought of as replacing the bulbs in the above example. The electrical power can then be replaced by the slow metabolism of fuel (glucose and fat). As the size of a species decreases (Figure 15a), there is a smaller number of cells (which decrease in proportion to the length scale R^3), while the surface area decreases more slowly (according to R^2). Thus, a larger amount of fuel metabolism is required in each cell.

We can now obtain scaling groups. The heat loss from an organism $\dot{Q}_L = \dot{Q}_L'' A = hA(T_b - T)$, where A denotes the organism body area. The heat ransfer rate h is constant for most mammals. We note that $\dot{Q}_L \propto m_b^{2/3}$. Experiments yield that $\dot{Q}_L = 3.552\, m_b^{0.74}$.

The metabolic rate during the human lifetime keeps varying with the highest metabolic rate being for a baby and the lowest being for a relatively senior citizen.

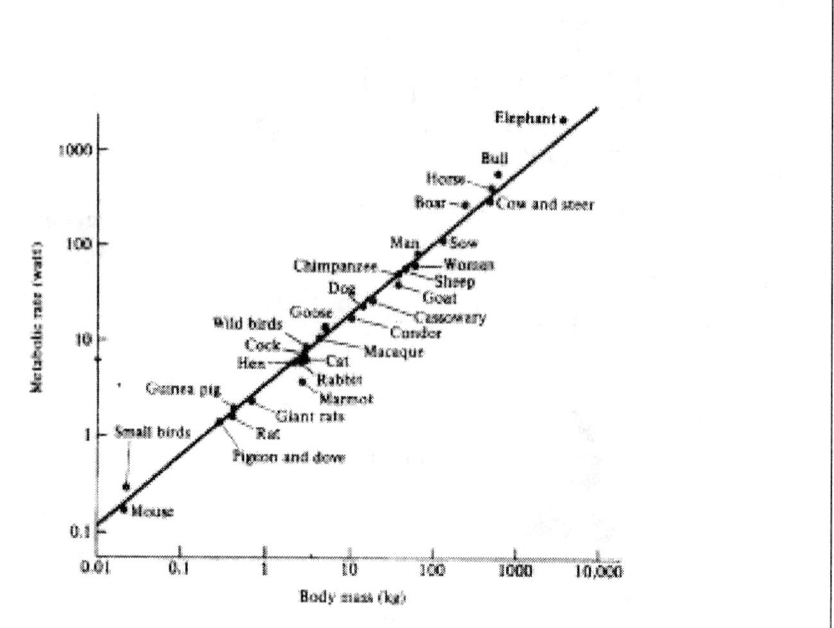

Figure 15b: Metabolic rates of different species. (Adapted from Scaling: *Why is animal size so important.* K.S. Nielsen, Cambridge University Press, p 57, 1984. With permission).

The minimum metabolic rate required for maintaining bodily functions is of the order of 1 W. The open system energy balance under steady state provides the relation $\dot{Q}_{c.v.} = \dot{m}_e h_e - \dot{m}_i h_i = \dot{m}(h_e - h_i)$.

If the body temperature rises, (e.g., during fever), then $dE_{c.v}/dt \neq 0$.

c. *Charging of Gas into a Cylinder*

Compressed gas cylinders, containing high–pressure gases, are commonly used to supply gas for welding torches and fire extinguishers. The following problem considers the time required to charge gases up to a specified pressure into a cylinder of known volume.

n. *Example 14*

A rigid cylinder is charged with an ideal gas through a pressurized line, and the flow is choked. Determine:

The enthalpy in the tank at a time $t \gg 0$, assuming $m_{t=0} = 0$.

A relation between the cylinder and line temperatures.

The cylinder temperature, pressure, and mass as a function of time.

Solution

For this problem, $\dot{m}_e = 0$ so that

$$dm_{c.v}/dt = \dot{m}_i. \quad \text{(A)}$$

Assuming that the gas charging occurs over a short duration, heat losses can be ignored, i.e., $\dot{Q}_{cv} = 0$. Furthermore, the kinetic and potential energies, and boundary and shaft work can also be assumed negligible, and, using Eq. (50),

$$(dU_{c.v}/dt) = \dot{m}_i h_i. \quad \text{(B)}$$

Assuming a uniform state in the c.v, $U_{c.v.} = \int u\rho dV = u\rho\int dV = um$, and Eq. (B) may be written in the form

$$(d(mu)/dt) = \dot{m}_i h_i, \text{ or} \tag{C}$$

$$m\, du/dt + u\, dm/dt = \dot{m}_i h_i. \tag{D}$$

From Eqs. (A) and (D)

$$dm/dt\,(h_i - u) = m\, du/dt. \tag{E}$$

Assuming the inlet state to be at steady state, i.e., $h_i \neq h(t)$, Eq. (E) may be written as

$$dm/m = du/(h_i - u), \tag{F}$$

which upon integration, using the initial conditions $u = u_o$, $m = m_o$, at $t = 0$, assumes the form

$$m/m_o = (h_i - u_o)/(h_i - u). \tag{G}$$

Using Eq. (G),

$$m(h_i - u) = m_o(h_i - u_o). \tag{H}$$

Simplifying with $m = m_0 + m_i$,

$$u = h_i - (m_o\ /(m_0 + m_i))\,(h_i - u_o). \tag{I}$$

Equation.(I) is valid whether the matter is an ideal gas or not. Since $dh = c_{po}dT$ and $du = c_{vo}dT$, and c_{po} and c_{vo} remain constant, $h_i = c_{po}T_i$ and $u = c_{vo}T$. Using these relations in Eq. (I) we obtain the relation

$$T= (m_o\, T_0 + m_i\, k_0\, T_i)/(m_0 + m_i), \tag{J}$$

where $k_o = c_{po}/c_{vo}$. The pressure in the tank $P(t)= m(t)\, RT/V$, and

$$P(t) = (m_oT_0 + m_ik_0\, T_i)\, R\,/V \tag{K}$$

If $m_o = 0$, from Eq. (H) $m(h_i - u) = 0$. Since $m \neq 0$, for this initial condition

$$h_i = u. \tag{L}$$

Further with constant c_{po} and c_{vo}, $h_i = c_{po}T_i$ and $u = c_{vo}T$. Using these relations in Eq. (L) we obtain,

$$T = k_oT_i, \tag{M}$$

where $k_o = c_{po}/c_{vo}$. When the line pressure is large, gas dynamic considerations indicate that the flow is choked. Therefore, the mass flow rate depends only upon the line pressure and temperature. For fixed line conditions, $\dot{m}_i$ is a constant irrespective of the downstream cylinder pressure. Integrating Eq. (A) with the initial condition $m(t = 0) = m_o = 0$,

$$m = \dot{m}_i t + m_o = \dot{m}_i t. \tag{N}$$

From the ideal gas law $m(t) = P(t)V/RT$. Using Eqs. (M) and (N) we can solve for pressure

$$P(t) = \dot{m}_i t\, R\, k_oT_i/V = \alpha t. \tag{L}$$

Remarks

In the initial period m_i is comparable to m_0, and T will increase as more mass is added (cf. Eq. (J). Since $m_i \gg m_o$, the temperature reaches a constant value as indicated by Eq.(M), which is a form of Eq. (J) obtained when $m_0 = 0$.

The temperature in the c.v. is higher due to the conversion of flow work (i.e., the pumping work performed in pushing the mass into the cylinder) into thermal energy in the form of u. The mass entering the cylinder contains an enthalpy h which is converted into u, i.e., the Pv (or flow work) at the inlet is converted into internal energy.

Eq. (L) states that $P(t)/t = \alpha$. If $\dot{m}_i$, T_i, and R are known, α may be determined by examining the time gradient of P(t). Once α is known, the ratio k_o can be calculated, Since $R = c_{po} - c_{vo}$, than c_{vo} and c_{po} can be determined.

The kinetic energy of the fluid entering the tank has been neglected in this analysis. If the inlet line has a large diameter, the flow velocity is relatively low, and the enthalpy of the fluid in the line is at its stagnation enthalpy (since, $h_{stg} = h + V^2/2$ where h is called static enthalpy). At the throat of the tank, the enthalpy $h_{throat} < h$,. but $V_{throat} \gg V$ and h_{stg} is the same as in the line. Once the fluid enters the cylinder (of far larger diameter than the line), assuming the c.v. to be situated several throat diameters downstream, $V^2/2 = 0$. Therefore, in this case the enthalpy of fluid entering the c.v. is the same as the stagnation enthalpy that in a large diameter line.

If the cylinder is charged with a reciprocating compressor, the mass flow will depend upon the cylinder pressure, and the mass flow may vary with time.

o. *Example 15*

The tank volume in an automobile is 757 liters. This volume is to be charged with methane at a station where the line pressure is considerably higher than the tank pressure, and the line temperature is 300 K. Assume that methane is an ideal gas (with $\bar{c}_{po}$ = 36 kJ kmole^{-1} K^{-1}, $\bar{R}$= 8.314 kJ kmole^{-1}). What should the pressure in the tank be so that it can contain a heating value equivalent to 20 gallons of gasoline? The heat release due to 1 gallon of gasoline is the same as that released from 2.4 kg of CH_4.

Solution

Mass of methane = 20 × 2.4 = 48 kg of CH_4.

Number of methane moles N_{CH_4} = mass ÷ molecular weight, i.e.,

$$N_{CH_4} = 48 \text{ kg} \div 16.05 \text{ kg kmole}^{-1} = 2.99 \text{ kmole}.$$

The temperature of gas inside the tank is higher than the line temperature due to the flow work performed on the mass. The constant volume specific heat

$$\bar{c}_{vo} = \bar{c}_{po} - \bar{R} = 36 - 8.314 = 27.69 \text{ kJ kmole}^{-1} \text{ K}^{-1}, \text{ i.e.,}$$

$$k_o = \bar{c}_{po}/\bar{c}_{vo} = 36 \div 27.69 = 1.3.$$

Using Eq. (J) of Example 14, $T = k_o T_i$, i.e.,

$$T = 1.3 \times 300 = 390 \text{ K}.$$

Therefore, the tank pressure

$$P = N_{CH_4} \bar{R}T/V = 2.99 \times 8.314 \times 390 \div (757 \text{ l} \times 10^{-3} \text{ m}^3 \text{ l}^{-1}), \text{ or}$$

$$P = 12{,}807 \text{ kPa or } 128 \text{ bars}.$$

Remarks

If the tank is cooled during charging to 300 K, the tank pressure required to charge the same mass of methane can be reduced to 128 × 300 ÷ 390 = 98.46 bars. If the tank is pressurized to 128 bars at 300 K, a larger number of moles of CH_4, i.e., N_{CH_4} = 12,807 × 757 × 10^{-3} ÷ (8.314 × 300) = 3.89 kmole (or 62.4 kg) can be charged.

d. Discharging Gas from Cylinders

Gases are discharged from cylinders for use in welding torches as well as other applications. Here, the time variation in cylinder pressure as the gases are discharged becomes consequential.

---

p. Example 16

Gas is discharged from a pressurized rigid cylinder. Determine the change in pressure in a rigid cylinder as the specific volume of the gas ($v = V/m$) contained in it changes.

Solution

For this problem, $\dot{m}_i = 0$, and $m_{c.v.} = m$, and Eq. (32) simplifies to the form

$$(dm/dt) = -\dot{m}_e. \qquad (A)$$

For a relatively short time period δt

$$dm = -dm_e. \qquad (B)$$

For a rigid cylinder $\delta W_d = 0$. Since there is no shaft work, and the potential energy is negligible, Eq. (50) assumes the form

$$(dE_{c.v}/dt) = \dot{Q}_{cv} - \dot{m}_e (h_e + ke_e).$$

For the duration δt

$$dE_{c.v.} = \delta Q_{c.v.} - dm_e (h_e + ke_e). \qquad (C)$$

Note that as matter is discharged, the cylinder temperature and pressure may vary so that h_e can change over time.

When gas leaves the tank its enthalpy h_e differs from that of the stagnant gas in the tank. For the mass near the exit, the specific energy e includes the energies of the stagnant and moving gas (i.e., $u + ke$). The energy balance between a unit mass of exiting gas and stagnant gas yields $h_e + ke_e = h$ (see Example 11). Omitting the subscript c.v, Eq. (C) may be written as

$$d(em) = \delta Q_{c.v.} - dm_e\, h. \qquad (D)$$

The specific energy is not uniform in the c.v. However, the mass containing kinetic energy adjacent to the exit is small as compared to the rest of the mass of stagnant gas. Therefore, the assumption $e \approx u$ or $E_{c.v.} = U_{c.v.} = m \times u$ is a good approximation. Expanding the LHS of Eq. (D) and using Eq. (B) to eliminate dm_e,

$$m\, du + u\, dm = \delta Q + dm\, h.$$

If the system is adiabatic, $\delta Q = 0$, and

$$dm\, (h - u) = m\, du. \qquad (E)$$

Since $m = V/v$, $\ln(m) = \ln(V) - \ln(v)$ so that

$$dm/m = -\, dv/v. \qquad (F)$$

Using Eqs. (E) and (F)

$$-(dv/v)(h - u) = du, \text{ or } -dv\,((Pv)/v) = du, \text{ i.e., } du + P\, dv = 0. \qquad (G)$$

The relation in Eq. (G) is independent of the nature of the system. Assuming the gas to be ideal, $du = c_{vo}\, dT$ and $P = RT/v$, i.e., $R = c_{po} - c_{vo}$. Using these relations in Eq. (G),

$$(c_{vo}/T)\, dT = -((c_{po} - c_{vo})/v)\, dv. \qquad (H)$$

Therefore,

$dT/T = -(k_o - 1)(dv/v)$, that, upon integration, results in (I)

$Tv^{k-1} = \text{Constant}.$ (J)

Using the ideal gas law to replace T (with Pv/R) in Eq. (J), and simplifying

$Pv^k = \text{Constant}.$ (K)

As the specific volume increases due to a decrease in mass, the pressure also decreases. If the value of k is known, both temperature and pressure can be predicted.

Remarks

The above step-by-step procedure illustrates the process of simplification while using valid assumptions.

The Washburn experiment (discussed later in Chapter 7) involves gas discharge from a tank into the atmosphere with the pressurized tank and its valves immersed in an isothermal bath. The gas leaving the tank is always at the bath temperature. In that case, Pv = constant if gas is ideal, i.e., k = 1.

e. Systems Involving Boundary Work

As air is blown into a balloon the c.v. deforms due to deformation work. In the following example we discuss the consequent change in balloon energy. The resulting expression will be used later in Chapter 3 to illustrate the dependence of internal energy U on properties, such as S, V and N in an open system.

q. Example 17

Obtain an expression for dU when dN moles of single component (e.g., N_2) fluid is pumped into a balloon. Neglect the kinetic and potential energies. Assume the ambient pressure to be slightly below the balloon pressure.

Solution

The balloon expands during the filling process resulting in deformation work. Employing the mole balance equation

$$dN_{c.v}/dt = \dot{N}_i - \dot{N}_e. \quad (59)$$

where the subscripts i and e, refer to the inlet and exit respectively. Neglecting the kinetic and potential energy in the energy conservation equation,

$$dU_{c.v}/dt = \dot{N}_i h_i - \dot{N}_e h_e + \dot{Q}_{cv} - \dot{W}_{cv}. \quad (A)$$

For a small time period δt, since $N_e = 0$, Eqs. (59) and (A) yield

$$dN_{c.v.} = dN_i, \text{ and} \quad (B)$$

$$dU = dN_i h_i - \delta W_{c.v.} + \delta Q. \quad (C)$$

Since there is no shaft work, $\delta W_{c.v.} = \delta W_d = PdV$ (as the process is in quasiequilibrium). Furthermore, since $dN_i = dN$, and $h_i = h$, Eq. (C) may be written in the form

$$dU = \bar{h}dN - PdV + \delta Q, \quad (D)$$

where dN denotes the number of moles accumulated within the balloon. Internal energy in the balloon increases due to energy input into the balloon along with gas inflow, work performed and heat added.

Remarks

In the context of Eq. (D) it is incorrect to write U = U(N,V,Q), since the differential of Q is inexact, i.e., Q is not a property. For a closed system, since dN = 0, Eq. (D) yields,

$$\delta Q = dU + PdV \text{ or } \delta Q = dU - (- PdV) \quad \text{(E)}$$

Equation (D) will be used later in Chapter 3 for an open system to obtain a relation for energy in terms of entropy, moles accumulated and the volume.

Consider a wet carpet drying over a period of time. If a c.v. is selected around the carpet, the system is open, since the liquid vaporizes (here dN denotes the change in the number of moles of fluid contained in the carpet) due to heat transfer (δQ supplied by the ambient air) that supplies latent heat to the fluid. Equation (D) is still applicable to this case.

r. *Example 18*

Suppose pressurized air is admitted into a pneumatic piston–cylinder assembly that jacks up an automobile. As the cylinder of cross-sectional area A is pressurized by the air, the force cA on the piston exceeds the weight of the car, thereby lifting it against gravity. Determine the volume and temperature, V(t) and T(t), for any given mass m(t).

Solution

For this problem, $\dot m_e = 0$, and $m_{c.v.} = m$, and Eq. (32) simplifies to the form

$$(dm/dt) = -\dot m_i . \quad \text{(A)}$$

Assume the system to be adiabatic, and the kinetic and potential energies to be negligible. Therefore, h_i is time independent. Since the c.v. is deformed, deformation work $\dot W_d$ is performed, and from Eq. (50)

$$(dUc.v/dt) = -\dot W_d + \dot m_i hi, \text{ where} \quad \text{(B)}$$

$$\dot W_d = P\, dV/dt. \quad \text{(C)}$$

If the weight of the automobile is W, then P= W/A. Initially the pressure is less than P and as such volume will not change until $P \ge P_0$. If the gas is admitted over a small duration δt, multiplying Eq. (B) by that period we have

$$dU = - P\, dV + d\, m_i h_i. \quad \text{(D)}$$

If h_i is invariant, then integrating Eq.(D),

$$U - U_0 = \int_{V_0}^{V(t)} PdV + m_i\, h_i\, .$$

In the initial periods when the value of PA is lower than the weight, dV= 0. Therefore,

$$U - U_0 = m_i\, h_i, \text{ and} \quad \text{(E)}$$

$$(U - U_o) = - P_W (V-V_0) + m_i h_i,\ P \ge P_0, \quad \text{(F)}$$

where the subscript 0 represents the initial conditions. Eq. (E) presents results for the charging problem (Example 14). Assuming uniform properties within the c.v. U = mu,, Eq.(F) implies that

$$(m\, u - m_0 u_o) = - P(V-V_0) + m_i h_i. \quad \text{(G)}$$

Here V_o, m_o, and u_o denote the initial volume, mass and specific internal energy. Note that when $V= V_0$, (i.e., piston does not move), Eq. (G) converts to the charging problem solved in Example 14. Employing the ideal gas law

$$m = P(t)V(t) /RT, \text{ and } m_o = P_0V_o/RT_o. \tag{H}$$

Using Eqs. (G) and (H),

$$(PV/RT)\, c_{vo}\, R\, T\, (t) - (P_oV_o/RT_0)\, c_{vo}\, T_0 + P(V-V_o) = m_ih_i. \text{ i.e.,} \tag{I}$$

$$(PV - P_oV_o)(c_{vo}/R) + P_W(V-V_o)\,\Omega = m_ih_i. \tag{J}$$

Using the relation $R = c_{po}- c_{vo}$ and solving for mass m(t) at any time, based on the inlet mass flow rate

$$m(t) = m_o + \dot{m}_i\, t = m_o + m_i\,(t) \tag{K}$$

Once $\dot{m}_i$ is known, m (t) can be determined using Eq. (K) (and for known P, V(t) can be calculated using Eq. (J)). Using the ideal gas law,

$$T(t)/T_0 = (PV/(P_oV_o))/(1 + (PV\, k/(P_oV_o) - (1+ (k-1)\, P/ P_o))\, (T_o/ kT_i)) \tag{L}$$

Remarks

Charging Period:

During the initial period before the piston starts to move, $V = V_0$. Here, the air is charging the cylinder as in Example 14, increasing the temperature. For the condition $V\rightarrow V_o$ in Eq. (L), we recover the charging solution

$$T/T_o = (P/P_o)/(1 + (P\, k/P_o - (1+ (k-1)\, P/ P_o))\, (T_o/ kT_i)), \tag{M}$$

which yields the expression for the gas temperature as a function of the gas pressure in the cylinder. For an initial pressure $P_o\approx 0$, Eq. (M) reduces to the form $T = k\, Ti$ (as shown in Example 14).

Lifting Period:

If the initial pressure $P_0 = P$, then Eq.(L) implies that

$$T/T_o = (V(t)/V_o)/(1 + (V(t)/V_o -1)(T_o/ T_i)),, \; V(t) \geq V_0, \tag{N}$$

which is independent of gas medium pumped in. It then follows that

$$m_i = P\, V(t)/(R\, T(t)) = P\, k/(k-1)(V-V_o)/h_i. \tag{O}$$

For the case $V \gg V_0$, Eq. (N) simplifies to the form $T = T_i$ and Eq. (O) yields $m_i = m = PV/(c_{p0}\, T_i)$. In this example flow work is employed to lift a weight so that $T = T_i$.

f. Charging of a Composite System

The following example illustrates the case of a composite system involving two phases.

s. Example 19

An insulated gasoline tank must be replaced in an automobile with the system shown in Figure 16. The tank is filled to 90% of capacity with the initial pressure above the liquid pressure $P_{ambient}$. Valve A is then opened, the gas is throttled to pressure P_2, and then admitted through valve B so that the gas fills the space above the liquid fuel in the rigid tank, and increases the pressure to P_2. When the pressure reaches a value P_2, valve C is opened, the fuel is fed into the engine for combustion. As the liquid level drops, the pressure above the liquid space is maintained at P_2 by adjusting the valve A. As an engineer, you are asked to analyze the process of charging the tank and draining the liquid gasoline. The tank volume is V and initial fuel temperature is $T_{F,0}$.

Assume that the gas is ideal and the line temperature is T_1. However, draining may involve a longer time.

What will be the temperature T_2 after throttling by the valve A. Assume that the charging of gas occurs so rapidly that the process in the tank can be assumed as adiabatic. Also assume that the space above the liquid initially contains a negligible amount of matter.

Determine the maximum temperature within the c.v.

Analyze the problem of draining by employing the mass and energy conservation equations. Assume that mass flow through valve B $\dot{m}_{e,F} = A\,(2\rho(P-P_{ambient}))^{1/2}$. How will you obtain the liquid volume in the tank over time in terms of P? An explicit answer is not required.

Solution

Throttling in valve A is adiabatic, i.e., for an ideal gas $T_2 = T_1$.

Since the charging occurs rapidly, the tank can be assumed to be adiabatic. We know from Example 14 that the gas temperature above the liquid surface (cf. c.v. G in Figure 16)

$$T_g = k\,T_i = 1.4\,T_2.$$

Therefore, $T = k\,T_1$. This result is reasonable as long as there is insufficient time for heat transfer to occur to the liquid.

One can determine amount of charged gas in the tank $m_{g,charge}$ using the results of Example 14.

Consider the gas space. The mass and energy conservation equations are

$$(dm_{c.v}/dt) = \dot{m}_i - \dot{m}_e, \text{ and}$$

$$(dE_{c.v}/dt) = \dot{Q}_{c.v.} - \dot{W}_{c.v.} + \dot{m}_i e_{T,i} - \dot{m}_e e_{T,e}.$$

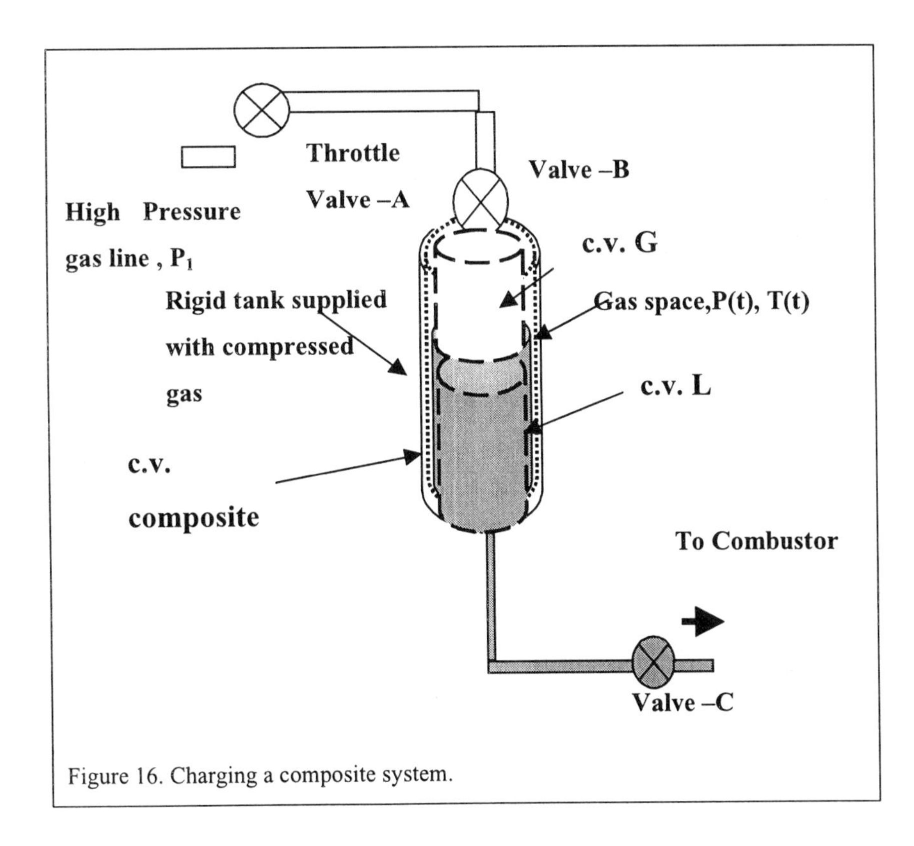

Figure 16. Charging a composite system.

For this problem, $\dot{m}_e = 0$ and $m_{c.v.} = m$ so that $(dm/dt) = \dot{m}_i$
Taking the control volume around the liquid only (cf. c.v. L in Figure 16) and assuming that there is no heat transfer from the gas to liquid, and recalling that work is done on the liquid,

$$(u_F\ dm_F/dt) + (m_F\ du_F/dt) = 0 - P\ dV/dt - \dot{m}_{e,F}\ h_{e,F}. \tag{A}$$

Applying mass conservation,

$$(1/v_F)\ dV/dt = -\ \dot{m}_{e,F} \tag{B}$$

where $\dot{m}_{e,F} = A(2\rho(P-P_{ambient}))^{1/2.}$ For the gas space,

$$(1/\ v_g)\ dV/dt = \dot{m}_{i,g}, \text{ and} \tag{C}$$

$$\dot{m}_{i,g}/\ \dot{m}_{e,F} = v_F/v_g. \tag{D}$$

The temperature of the gas space remains constant. The volume of gas space increases for a constant pressure. Since P and T are constant, the specific volume of gas must remain constant during the draining process. With the flow rate maintained constant, the gas flow rate must also be constant. Therefore, the mass of gas admitted during the draining process $m_{g,drain}$ can be determined.
Eq. (A) then implies that

$$\dot{m}_{e,F}\ P\ v_F = -\ \dot{m}_{e,F}\ (u_F - h_F) + ((m_F c_F)\ dT/dt). \tag{E}$$

Simplifying, dT/dt =0 or T is constant.

Remarks

The gas temperature during charging increases. First, the gas transfers heat to the liquid. Any subsequent mass admitted during draining will have the temperature T_2. It is more likely that final temperature of liquid and gas T_f will be such that $m_F\ c_F\ T_f + m_g\ c_{v,g,}\ T_f = m_F\ c_F\ T_{F,0} + m_{g,charge}\ c_g\ \ k\ T_2 + m_{g,drain}\ c_g\ T_2$ where $m_g = m_{\ g,charge} + m_{g,drain.}$

B. INTEGRAL AND DIFFERENTIAL FORMS OF CONSERVATION EQUATIONS

1. Mass Conservation

a. *Integral Form*

If the rigid boundary around a turbine is demarcated, the properties within the c.v. vary spatially from inlet to exit. Therefore, the mass within the c.v. can only be determined by considering a small elemental volume dV, and integrating therefrom over the entire turbine volume. Since the velocity distribution also varies spatially (Figure 17), the inlet and exit mass flow rates can be represented as

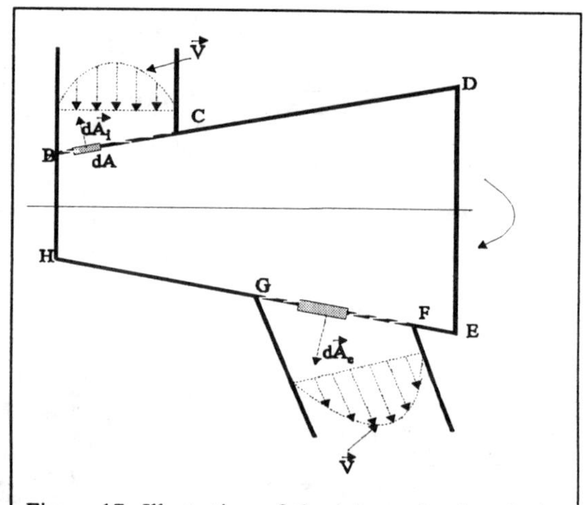

Figure 17: Illustration of the inlet and exit velocity vectors for a turbine.

$$\dot{m}_i = -\int_{A_i} \rho\vec{V}\cdot d\vec{A},\ \dot{m}_e = \int_{A_e} \rho\vec{V}\ d\vec{A}. \tag{62}$$

The negative sign in Eq. (62) is due to the velocity vector of the entering mass that points towards the elemental area dA, while the area vector always points outward normal to the c.s. (cf. Figure 17). Hence, the dot product of the integral in Eq. (62) evaluated at the inlet is negative. According to our previously stated convention, the mass entering the turbine must carry a positive sign. This is satisfied by providing the negative sign to the equation. Using Eqs. (62) and (34)

$$d/d\int_{cv}(\rho dV) = -\int_{A_i}\rho\vec{V}\cdot d\vec{A} + \int_{A_e}\rho\vec{V}\cdot d\vec{A}\text{, i.e.,} \quad (63)$$

$$d/d\int_{cv}(\rho dV) = -\oint\rho\vec{V}\cdot d\vec{A}. \quad (64)$$

The cyclical integration implies that the mass is tracked both in and out throughout the c.s. of the system (e.g., the surface BCDEFGH in Figure 17).

b. Differential Form

Applying the Gauss divergence theorem to the RHS of Eq. (64) (cf. Chapter 1),

$$\oint\rho\vec{V}\cdot d\vec{A} = -\oint\vec{\nabla}\cdot\rho\vec{V}\,dV. \quad (65)$$

(As mentioned in Chapter 1, the x–wise component of the RHS is $(\partial(\rho v_x)/\partial y)dy(dxdz)$.) If the control volume is time independent, i.e., it has rigid boundaries, the LHS of Eq. (64) may be written in the form

$$d/dt(\int\rho dV) = \int(\partial\rho/\partial t)dV. \quad (66)$$

Using Eqs. (65) and (66), we can rewrite Eq. (64) as

$$\int_{cv}(\partial\rho/\partial t + \vec{\nabla}\cdot\rho\vec{V})dV = 0. \quad (67)$$

Since the c.v. is arbitrarily defined, if it is shrunk to a small volume, Eq. (67) still holds, and

$$\partial\rho/\partial t + \vec{\nabla}\cdot\rho\vec{V} = 0. \quad (68)$$

2. Energy Conservation

a. Integral Form

In a steam turbine the term $E_{c.v.}$ must be evaluated for the entire turbine in which the properties are spatially nonuniform. Therefore, the energy varies within the c.v. The methalpy crossing the cs is

$$-\int_{in}\rho e_T\vec{V}\cdot d\vec{A}, \quad (69)$$

where $\vec{V}\cdot d\vec{A} < 0$ for the incoming flow and $\vec{V}\cdot d\vec{A} > 0$ for the exiting flow. A negative sign is added to the value in order to be consistent with our sign convention for the mass inflow and outflow. Assuming the c.v. boundary to be rigid, $\dot{W}_d = 0$. Therefore,

$$\dot{W}_{cv} = \int_{cv} w'_{cv}dV, \quad (70)$$

where w'_{cv} denotes the work done per unit volume. The heat crossing the system boundary $\dot{Q}_{cv}$ is given by

$$\dot{Q}_{cv} = -\int_{cv}q''\cdot d\vec{A}, \quad (71)$$

The negative sign associated with Eq. (71) is explained as follows. The vector $d\vec{A}$ is outward normal to the surface, i.e., $d\vec{A} = \vec{n}dA$. The incoming heat flux vector q″ enters the surface in a direction that is opposite to the outward normal vector $\vec{n}$. Therefore, the dot product of the heat flux vector and the area vector is negative, implying that the sign for heat input is also negative. Since our stated sign convention in First Law states that heat input in context of the First Law is positive, a negative sign is placed in Eq. (71). Using the relationships of Eqs. (69)–(71) in Eq. (50), the integral form of the energy conservation equation for a rigid c.v. assumes the form

$$d/d\int_{cv}(\rho e dV) = \oint q'' \cdot d\vec{A} - \int w'_{cv} dV - \oint \rho e_T \vec{V} \cdot d\vec{A}. \qquad (72)$$

The term d/dt($\int\rho e dV$) in Eq. (72) denotes the rate of change of energy in the entire control volume.

b. Differential Form

Applying the Gauss divergence theorem to Eq. (72), and converting the surface integral into a volume integral, we obtain

$$d/d\int_{cv}(\rho e dV) + \int \vec{\nabla} \cdot (\rho e_T \vec{V}) dV = -\int \vec{\nabla} \cdot \vec{q}'' dV - \int w'_{cv} dV. \qquad (73)$$

Simplifying the result

$$\partial(\rho e)/\partial t + \vec{\nabla} \cdot (\rho e_T \vec{V}) = -\vec{\nabla} \cdot \dot{\vec{Q}}'' - w'_{cv}. \qquad (74)$$

If the heat transfer in the c.v. occurs purely through conduction and the Fourier law applies,

$$\dot{\vec{Q}}'' = -\lambda \vec{\nabla} T. \qquad (75)$$

Furthermore, if no work is delivered, and the kinetic and potential energies are negligible (namely, $e_T = h$, and $e = u$), Eq. (74) may be expressed in the form

$$\partial(\rho u)/\partial t + \vec{\nabla} \cdot (\rho \vec{V} h) = \vec{\nabla} \cdot (\vec{\nabla} T). \qquad (76)$$

Using the relation $u = h - P/\rho$, Eq. (76) may be written as

$$\partial(\rho h)/\partial t + \vec{\nabla} \cdot (\rho \vec{V} h) = \partial P/\partial t + \vec{\nabla} \cdot (\vec{\nabla} T). \qquad (77)$$

These differential forms of the energy conservation equation are commonly employed in analyses involving heat transfer, combustion, and fluid mechanics.

c. Deformable Boundary

Examples of a deforming boundary include the surface of a balloon while it is being filled, and the leakage of gases past a piston. The above formulations have accounted for deformation work, but assumed that there is no flow at the deforming boundary. If mass flow occurs at the deforming boundary, the input and exit mass and energy flows will be influenced. Consider the leakage of air past a piston in addition to the mass otherwise entering and leaving a cylinder. If the absolute velocity of the leaking fluid adjacent to the piston is $\vec{V}$ and deformation velocity is $\vec{V}_d$, then leakage flow rate will be zero if $\vec{V} = \vec{V}_d$. If $\vec{V} > \vec{V}_d$, then the leakage flow rate of fluid past the deforming boundary is

$$\dot{m} = \int \rho \vec{V}_r \cdot d\vec{A}, \qquad (78)$$

where the relative velocity $\vec{V}_r = \vec{V} - \vec{V}_d$. The RHS of Eq. (34), written in terms of the mass flow rates, can now be expressed in terms of the relative velocity. Furthermore, the integrals

$$d/dt(\int \rho dV) \rightarrow \oint (\partial\rho/\partial t)\, dV, \text{ and } d/dt(\int \rho e dV) \rightarrow \oint (\partial(\rho e)/\partial t)\, dV. \qquad (79)$$

Therefore, the mass and energy conservation equations can be written in the forms

$$\oint (\partial\rho/\partial t)\, dV = \oint \rho\vec{V}_r \cdot d\vec{A}, \text{ and } \oint (\partial(\rho e)/\partial t)\, dV = \dot{Q}_{cv} - \dot{W}_{cv} + \oint \rho e_T \vec{V}_r \cdot d\vec{A}. \quad (80)$$

Rigorous proof of the formulation of Eqs. (80) is contained in the Appendix to this chapter.

If a balloon releases gas (as shown in Figure 18) at an absolute velocity of 8 m s^{-1} and it shrinks at the rate of –2 m s^{-1}, the magnitude of the relative velocity V_r = 8–(–2) = 10 m s^{-1}. If the cross-sectional area through which leakage occurs is 1 mm^2, assuming the density of air to be 1.1 kg m^{-3}, the mass flow exiting the deforming balloon equals $10\times1\times10^{-9}\times1.1$ = 1.1×10^{-8} kg s^{-1}.

C. SUMMARY

In this chapter we have discussed the conservation equations in the context of closed and open systems. Problems pertaining to quasiequilibrium and nonquasiequilibrium problems in closed systems have been discussed, and several applications of the conservation equations expressed in transient form were illustrated. The conservation equations have been expressed both in differential and integral forms.

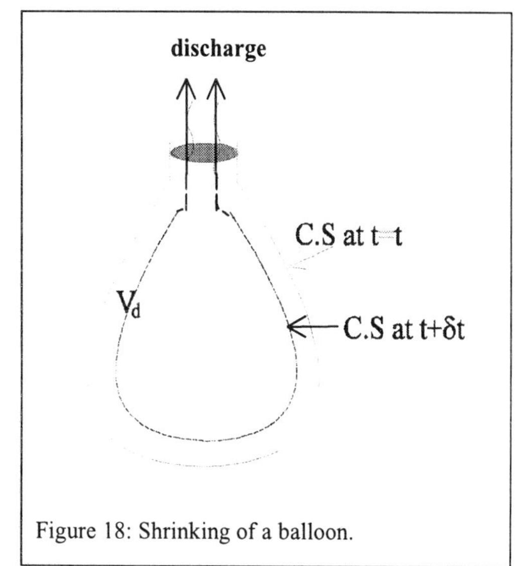

Figure 18: Shrinking of a balloon.

We have learned that the general method to solve problems pertaining to thermodynamic systems is as follows:

Select the system and determine whether it is closed or open

Determine the transactions across the boundary (i.e., the heat, work, and mass transfer across the boundary).

Determine the nature of matter contained within the system (e.g., ideal gas, incompressible liquid, etc.)

Determine the known properties (e.g., using the ideal gas assumption, or the relevant property tables).

Write the mass and energy conservation equations in a dimensionally conforming manner and following a consistent sign convention.

Characterize the processes that occur within the system (e.g., isothermal, adiabatic, isobaric, etc.).

Make reasonable assumptions in order to simplify the problem, and solve the problem.

D. APPENDIX

1. Conservation Relations for a Deformable Control Volume

In the context of Eq. (50) we focus attention on the term $d/dt(\int_{cv}\rho e dV)$. For a small time period δt, the accumulation of energy within a deforming c.v. is $d(\int_{cv}\rho e dV)$, i.e., the change in energy within the original volume plus the change in energy within an incremental deformed volume. Therefore,

$$d(\int_{cv}\rho e dV) = -d\oint \rho e\, dV + \oint \rho e\, d\vec{x}_d \cdot d\vec{A},$$

where $d\vec{x}_d$ is the incremental deformation length. Dividing the entire equation by δt, in the limit δt→0,

$$d/dt \int_{cv}(\rho e dV) = \int \partial(\rho e)/\partial t\, dV + \oint \rho e \vec{V}_d \cdot d\vec{A}, \tag{A}$$

where $\vec{V}_d$ denotes the deformation velocity of the volume. Likewise, the last two terms in Eq. (50)

$$\dot{m}_i e_{T,i} - \dot{m}_e e_{T,e} = \oint \rho e_T \vec{V} \cdot d\vec{A}. \tag{B}$$

We have assumed that the c.v. performs boundary work W_b in addition to other forms of work transfer, i.e., $W_{c.v.} = W_b + W_{shaft} + W_{other}$. For a small time period, the boundary work

$\delta W_b = PdV + P_{surr}dV$, i.e.,

it equals the work done by the flow as it enters and leaves the c.v. (represented by the first term on the RHS) in addition to the work done by the boundary as it is subjected to the surrounding pressure P_{surr} (i.e., the second term). Therefore,

$$\delta W_b = \int_{A_f} P\, d\vec{x}_d \cdot d\vec{A} + \int_A P_{surr}\, d\vec{x}_d \cdot d\vec{A}.$$

Again, dividing the entire equation by δt, in the limit $\delta t \to 0$,

$$\dot{W}_b = \int_{A_f} P\, \vec{V}_d \cdot d\vec{A} + \oint P_{surr}\, \vec{V}_d \cdot d\vec{A}. \tag{C}$$

Manipulating Eq. (50), and Eqs. (A), (B), and (C), we obtain

$$\int \partial(\rho e)/\partial t\, dV = \dot{Q}_{cv} - (\dot{W}_{shaft} + \dot{W}_{other} + \int_{A_f} P\vec{V}_d \cdot d\vec{A} + \oint P_{surr}\, \vec{V}_d \cdot d\vec{A})$$

$$+ \oint \rho e_T \vec{V} \cdot d\vec{A} - \oint \rho e \vec{V}_d \cdot d\vec{A}. \tag{D}$$

Using the relation

$$\oint \rho e_T \vec{V}_d \cdot d\vec{A} = \oint \rho e \vec{V}_d \cdot d\vec{A} + \int_{A_f} P\vec{V}_d \cdot d\vec{A}, \tag{E}$$

Eq. (D) assumes the form

$$\int \partial(\rho e)/\partial t\, dV = \dot{Q}_{cv} - (\dot{W}_{shaft} + \dot{W}_{other} + \oint P_{surr}\, \vec{V}_d \cdot d\vec{A}) + \oint \rho e_T \vec{V}_r \cdot d\vec{A}. \tag{F}$$

where the relative velocity $\vec{V}_r = \vec{V} - \vec{V}_d$.

Chapter 3

3. SECOND LAW AND ENTROPY

A. INTRODUCTION

The First law of thermodynamics does not limit the degree of conversion of cyclical heat input into cyclical work output, as discussed in Chapter 2. The Second law establishes this limit and, for instance, it prevents heat engines from converting their entire heat input into work. Consider the example of a car. A gallon of gasoline releases 120,000 kJ of thermal energy. If the work transfer to the wheels of the car is only 40,000 kJ then the remaining 80,000 kJ must be accounted for. Assume that the heat loss from the automobile radiator accounts for 40,000 kJ while the exhaust accounts for 40,000 KJ. The ratio between work and heat (40,000/120,000) is the efficiency η. For the above example η = Energy Sought/Energy Bought = 1/3. In no engines is all of the heat absorbed converted into work, i.e., $\eta \neq 1$. An upper limit on the efficiency can, however, be obtained by applying the Second law.

This chapter presents the statements of the Second law and expressions for maximum possible efficiency; defines entropy; introduces the concept of entropy generation and its relation to work loss; and summarizes the relations with which the entropy of substances can be evaluated. Entropy balance equations are also presented in integral and differential forms. Finally entropy maximum and energy minimum principles are illustrated with examples. However, prior to discussing the Second law any further, some pertinent concepts are first presented.

1. Thermal and Mechanical Energy Reservoirs

A *thermal energy reservoir* is a large repository of heat that acts as a source or sink. Heat exchange can occur with the reservoir (or repository system) without changing its temperature. It also acts as a reversible heat source, since it contains no temperature gradients within itself. Upon heat addition, only the thermal energy content of the reservoir changes, indicating that it is implicitly rigid. Therefore, by the First law, $dU = \delta Q$. The reservoir temperature is virtually unaffected during heat transfer, since it has a large mass.

Examples of thermal energy reservoirs include the atmosphere and oceans. If the mass of an ocean is 10^{10} kg and its specific heat c = 4.184 kJ/kg K, then 10^6 kJ of heat addition to it will produce a temperature rise of only 2.4×10^{-5} K (= $10^6 \div 4.184\times10^{10}$), i.e., the temperature is virtually unchanged. The internal energy change dU is nonzero, since, even though the temperature increment is negligibly small, the system is massive. Therefore, $dU = mcdT = 4.184\times10^{10} \times 2.4\times10^{-5} = 10^6$ kJ, which equals the transferred heat energy.

A *mechanical energy reservoir* is a large body acting as a source or sink with which work can be exchanged without affecting its characteristics. Examples include a large flywheel containing a large amount of kinetic energy, or a large pressure reservoir consisting of a gas contained in a piston–cylinder assembly that has an infinitely large weight placed on the piston. If work is exchanged with the pressure reservoir, its pressure is unaffected. Likewise, work transfer to the flywheel does not alter its kinetic energy.

Thermodynamic Cycle: Final thermodynamic state after a (cyclic) process is the same as initial state, e.g., in a steam power plant.

Mechanical Cycle: Final mechanical position is the same as its initial position (e.g., the position of piston at top of a cylinder returns after one revolution of a crank).

Closed Cycle: The working fluid undergoes a series of processes and returnes to its original state, i.e., fluid is retained in the system.

Open Cycle: The working fluid is different from that at its initial state and is discarded at the conclusion of the cycle.

Unless otherwise specified, a cycle refers to thermodynamic cycle

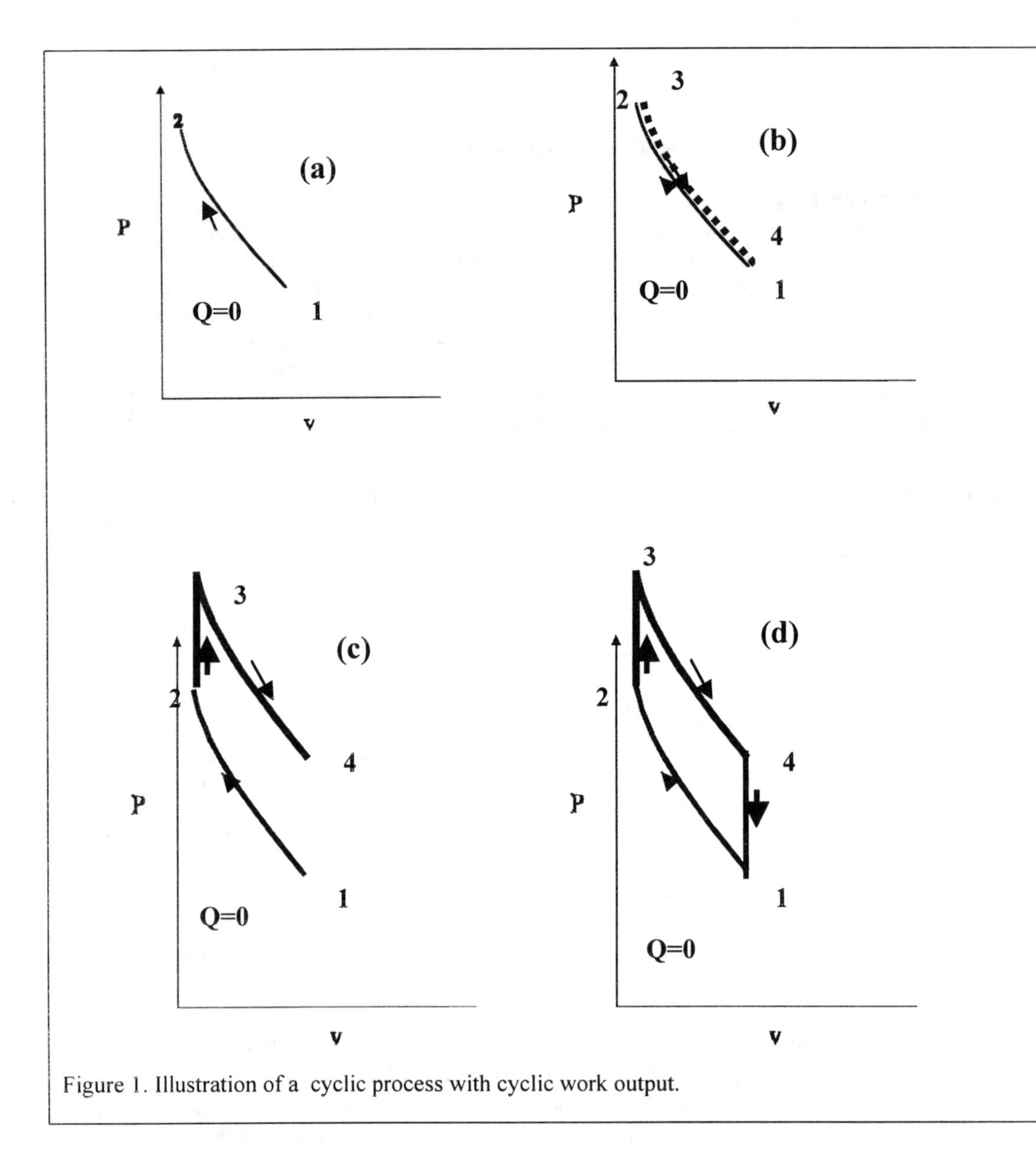

Figure 1. Illustration of a cyclic process with cyclic work output.

a. Heat Engine

A heat engine is a cyclic device in which the heat interactions with higher and lower temperature thermal energy reservoirs are converted into work interactions with mechanical energy reservoirs. Heat engines can operate either with vapor (e.g., steam) or gas (e.g., air) as the medium of fluid. Consider Figure 1 of Chapter 2, which is known as the Rankine cycle. In this cycle steam is produced in a boiler through heat input to water. The steam is subsequently expanded in a turbine, thereby producing work, and later condensed back into the liquid phase through heat rejection from the spent steam in a condenser. The water is then pumped back to the boiler.

b. Heat Pump and Refrigeration Cycle

Systems in which work interactions with mechanical energy reservoirs result in heat transfer from lower– to higher–temperature thermal energy reservoirs are either heat pumps or refrigeration devices.

B. STATEMENTS OF THE SECOND LAW

If a gas is adiabatically compressed from a state characterized by P_1 and v_1 to a state (P_2,v_2) through work transfer w_{12} (Figure 1), adiabatic expansion from state 3 (which is the same as state 2) to (P_4,v_4) (where state 4 is the same as state 1) provides work output $|w_{34}| = |w_{12}|$, i.e., $w_{cyc}= 0$. If a higher pressure is required at the end of the expansion process then heat must be supplied at state 2 (e.g., at constant volume so that $v_3= v_2$ and $P_3 > P_2$). The expansion to state 4 (for which $v_4= v_1$) produces $P_4 > P_1$. In this case the adiabatic curves 3-4 and 1-2 are almost parallel and, consequently, $T_4 > T_1$. We make the point that heat rejection must close the cyclic process. The Otto cycle used in automobiles is schematically illustrated through part (d) of Figure 1 in the form of a P–v diagram. It consists of four processes: adiabatic quasiequilibrium compression (1–2), constant volume heat addition (2–3), adiabatic quasiequilibrium expansion (3–4), and constant volume heat rejection (4–1).

The simplest statement of Second law is that *for heat input a cyclic process requires heat rejection.*

1. Informal Statements

We present two informal statements of the Second law.

Statement 1: "The efficiency of a heat engine is less than unity."

The efficiency η of a thermodynamic cycle is defined as the ratio of the work output to the thermal input, i.e.,

$$\eta = = \text{Sought/Bought} = W_{cycle}/Q_{in}. \quad (1)$$

For a cyclical process, the First law states that

$$\oint \delta Q = Q_{in} - Q_{out} = W_{cycle} = \oint \delta W. \quad (2)$$

Using Eqs. (1), (2) and the informal statement,

$$\eta = (1 - Q_{out}/Q_{in}) < 1. \quad (3)$$

Note that by using the subscripts "in" and "out", the sign convention has already been accounted for. For a finite work output, Eq. (3) implies that Q_{out} is always nonzero. The ratio of the actual mechanical work to cycle work is called the system mechanical efficiency, and the product of the cycle and mechanical efficiencies is sometimes referred to as the thermal efficiency.

Statement 2: "An isolated system initially in a state of nonequilibrium will spontaneously achieve an equilibrium state."

A spontaneous process is one that occurs without outside intervention (i.e., without work or heat transfer). Consider a cup of warm water placed in a room made of rigid, insulated, and nonpermeable walls. Heat transfer occurs from the water and cup to room air until all three are at the same temperature. At this state equilibrium is reached. It is impossible to reheat the water and cup back to their initially higher temperature without external intervention, i.e., reheating is impossible by using the room air alone, since the final state consisting of the higher–temperature water (and cup) placed in lower–temperature air would be in a state of nonequilibrium, and in violation of the Second law. We will illustrate this concept using Examples 1 and 2.

---

a. Example 1

A stirrer is used to warm 0.5 kg of water that is contained in an insulated vessel. The water is initially at a temperature of 40°C. A pulley–weight assembly rotates the stirrer through a gear mechanism such that a 100 kg mass falls through a height of 4220 cm. Is this process really possible? What is the final temperature? Assume that u = cT (with T expressed in °C), and c = 4.184 kJ kg^{-1} K^{-1}.

Solution

The potential energy of the falling weight changes, and the consequent work input into the system heats the water. The potential energy change is

$$\Delta PE = 100\times9.81\times42.2\div1000 = 42 \text{ kJ}.$$

The First law states that $Q_{12} - W_{12} = U_2 - U_1$. At 40°C,
$U_1 = 0.5\times40\times4.184 = 84$ kJ.
Since $Q_{12} = 0$, $\Delta PE = -(-42) = U_2 - U_1$, or
$U_2 = 42+84 = 126$ kJ.
Therefore, $u_2 = 126\div0.5 = 252$ kJ kg^{-1}, and $T_2 = 252\div4.184 = 60$°C.
The process is possible and all the work is converted into "heat" (thermal energy)

Remarks

Even though the internal energy cannot be measured directly, the pulley system enables us to implicitly calculate its values. Note that in this case all of the work can be converted into the thermal (heat) energy of water.
Another example involves running a blender filled with ice cream over a long time period. Again, in this case, work input is converted into "heat" and the ice cream will melt.
One can complete a cyclic process by rejecting heat to the ambient reservoir and cooling the water from 60 °C to 40 °C. In this cyclic process, the First law states that $\int\delta Q - \int\delta W = 0$ or $\int\delta W = W_{cyc} = \int\delta Q < 0$, i.e., a cyclic process can be completed with heat rejection to a single reservoir but with work input.

b. *Example 2*

42 kJ of heat energy are transferred from a thermal energy reservoir that exists at a temperature of 27°C and are converted into work using a hypothetical heat engine undergoing a cyclical process. The process raises a 100 kg weight through a 4220 cm height. The weight is then allowed to fall (as in Example 1) and the work is used to heat 0.5 kg of water that is initially at a temperature of 40°C. Is such a scenario possible? What is the final water temperature? Assume $u = cT$, and $c = 4.184$ kJ kg^{-1} K^{-1}.

Solution

In this scenario, the combined system that includes the heat engine and weight performs no net work, but can continuously transfer heat from the lower temperature thermal energy reservoir at 27°C to warmer water that exists at 40°C. This cannot be done through direct contact alone. Therefore, the combined system within the boundary E can be made to proceed further towards nonequilibrium, which is counter intuitive and the process is impossible.
However, this process can still be analyzed on the basis of the First law alone, i.e.,

$$-(-42) = U_{2W} - U_{1W}, \tag{A}$$

where, $U_{1W} = 0.5\times4.184\times40 = 84$ kJ. From Eq. (A),

$$U_{2W} = 42 + 84 = 126 \text{ kJ},\ u_{2W} = 252 \text{ kJ};\ T_{2W} = 60°\text{C}.$$

Remark

Recall that work can be converted into "heat" and hence the problem lies in the conversion of all of the "heat" into work.

a. *Kelvin (1824-1907) – Planck (1858-1947) Statement*

"It is impossible to devise a machine (i.e., a heat engine) which, operating in a cycle, produces no effect other than the extraction of heat from a thermal energy reservoir and the performance of an equal amount of work."

The First law conserves energy, and the Second law prohibits the complete conversion of thermal energy into work during a cyclical process. Using the First law for a closed system undergoing a cyclical process in a heat engine, $(\oint \delta Q = \oint \delta W) > 0$. Note that the Kelvin-Planck statement does not preclude the condition $\int\delta W \approx 0$ (cf. Example 1). Consider an adiabatic mass of water being heated by the action of a frictionless stirrer that raises the water temperature. If the insulation and stirrer are removed, it is possible to cool the water to its initial temperature by losing an amount of thermal energy that is equal to the stirrer work input. In this case, the water undergoes a cyclical process, but still rejects heat to a single thermal energy reservoir. Now, consider the following possibility: Instead of rejecting heat to the reservoir after the insulation is removed, the stirrer is retained, and the water cooled to its initial temperature by converting all the thermal energy (i.e. heat) into work. Such a process is impossible, since it violates the Kelvin–Planck statement, and no such cyclical device can be designed. Once work is converted into heat, all of the heat cannot be converted back into work. Therefore, heat energy (i.e., $Q = U_2 - U_1$) possesses a lower quality than an equal amount of work energy, since it is capable of a smaller amount of useful work.

b. *Clausius (1822-1888) Statement*

Heat cannot flow from a lower to higher temperature. However, heat can be transferred from a lower-temperature thermal energy reservoir to a higher-temperature reservoir in the presence of work input. The Clausius statement (due to Rudolf Clausius, 1822–1888) regarding this is as follows:

"It is impossible to construct a device that operates in a thermodynamic cycle and produces no effect other than the transfer of heat from a cooler to a hotter body."

An air conditioner transfers heat from the lower temperature indoor space of a house to a higher–temperature ambient during the summer. A heat pump delivers heat from a lower–temperature ambient to a higher–temperature system (such as a house). For example, using a heat pump, if 150 kJ of heat is transferred from cold air at, say, 0°C in conjunction with a work input of 100 kJ, the pump is capable of delivering 250 kJ of heat to a space. Rather than the efficiency, a *refrigeration cycle* is characterized by a Coefficient of Performance, namely

$$COP_{refrigeration} = \text{Energy Sought/Energy Bought}$$

$$= \text{Heat transferred from the lower temperature system/Work input.} \quad (4)$$

In the case of a *heat pump*,

$$COP_{heat\ pump} = \text{Energy Sought/Energy Bought}$$

$$= \text{Heat transferred to the higher temperature system/Work input.} \quad (5)$$

Heat pumped at the rate of 3.516 kW (200 BTU/min or 211 kJ/mim) from a system constitutes a unit that is referred to as one ton of capacity. The physical implication is derived from the cooling of water, i.e., if 3.516 kW of thermal energy is removed from 1 ton of liquid water at 0°C, transforming it completely into ice at 0°C over a duration of 24 hours. Instead of $(COP)_{cooling}$ which is dimensionless, industries use a unit called HP/ton of refrigeration (1 HP = 550 ft lb_f/s = 0.7457 kW= 42.42 BTU/min, HP/Ton = 4.715/COP). Employing Eqs. (2), (4), and (5)

$$COP_{refrigeration} = Q_{in\ from\ lower\ T\ system} \div |Q_{in\ from\ lower\ T\ system} - Q_{out\ to\ higher\ T\ system}| \quad (6)$$

$$COP_{heat\ pump} = Q_{out\ to\ higher\ T\ system} \div |(Q_{in\ from\ lower\ T\ system} - Q_{out\ to\ higher\ T\ system}| \quad (7)$$

Consider a system undergoing a cyclical process and pumping heat from a lower-temperature thermal energy reservoir to a higher-temperature thermal energy reservoir. From the First law ($\oint \delta Q = \oint \delta W$) < 0, since there is work input into such a process. Therefore, for a refrigeration cycle, such as one for an air conditioner $(- Q_H + Q_L) < 0$, where Q_H denotes the heat leaving a system and entering a warmer thermal energy reservoir (e.g., the ambient), and Q_L is that entering the system from a cooler thermal energy reservoir (e.g., a cooled residential space).

i. Perpetual Motion Machines

A machine that obeys the First law but violates the Second law of thermodynamics is known as a perpetual motion machine of the second kind (PMM2). Other terms for it include anti–Clausius machine and anti–Kelvin machine.

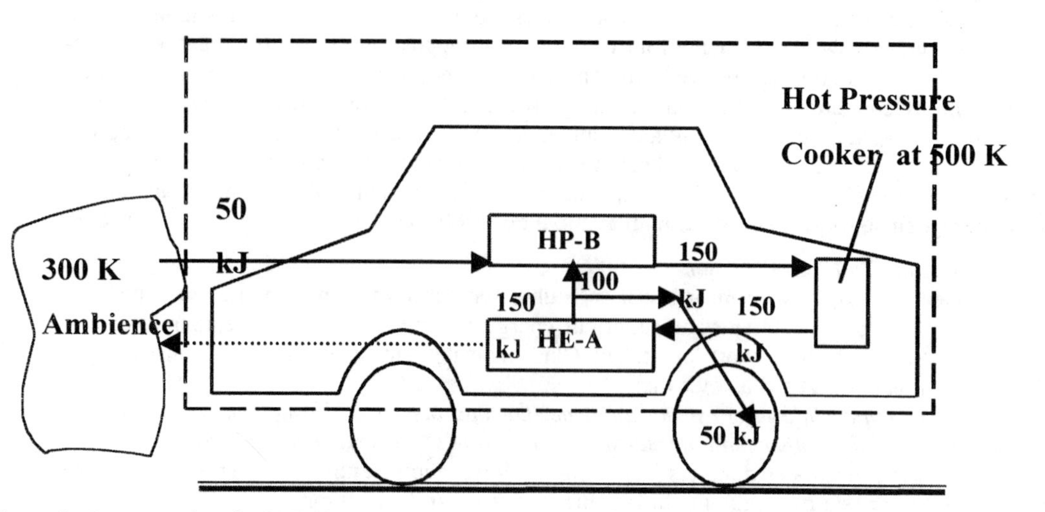

Figure 2: An example of a PMM2.

c. Example 3

Consider the following hypothetical scenario based solely on the First law. An insulated pressure vessel contains superheated steam. The thermal energy contained in the steam is converted into work in order to run a heat engine A ($Q_{in,A}$, $Q_{out,A}$, $W_{cycle,A}$). However, this decreases the steam temperature, and a heat pump B ($Q_{in,B}$, $Q_{out,B}$, $W_{cycle,B}$) is employed to pump thermal energy into the vessel to raise the temperature to its initial value, such that $Q_{out,B} = Q_{in,A}$. Note that the numbers carry a positive sign for all the symbols. Now, $W_{cycle,A} = Q_{in,A} - Q_{out,A}$, and $W_{cycle,B} = Q_{out,B} - Q_{in,B} = Q_{in,A} - Q_{in,B}$. Therefore, $W_{wheels} = W_{cycle,A} - W_{cycle,B} = Q_{in,B} - Q_{out,A}$. If $Q_{in,B} > Q_{out,A}$, we can harness the difference $W_{cycle,A} - W_{cycle,B}$ in order to run an automobile. Is it possible to operate an automobile perpetually in this manner without consuming fuel and, thereby, emitting no pollutants or greenhouse gases? (See Figure 2.)

Solution

Consider the case $Q_{out,A} = 0$. This is a violation of the Second law. A consequence is that with $Q_{in,A} = 150$ kJ, $W_{cycle,A} = 150$ kJ, $Q_{in,B} = 50$ kJ, $Q_{out,B} = Q_{in,A} = 150$ kJ, $W_{cycle,B} = 100$ kJ, $W_{cycle,A} - W_{cycle,B} = Q_{in,B} - Q_{out,A} = 50$ kJ. The question is whether one can use the energy from the ambient to run an automobile. (Note that frictional work transferred at the tires can be used to supply energy back into the ambient.) This is an example of a perpetual motion machine of the second kind. The heat engine A violates the Kelvin–Planck statement of the Second law, and its existence is not possible.

If $Q_{out,A} = 45$ kJ, the machine A is not in violation of the Second law. In this case $W_{wheels} = Q_{in,B} - Q_{out,A} = 5$ kJ. However, for the dashed boundary, 5 kJ of energy enters from the ambient and it is all converted into work. Again, this violates the Kelvin–Planck statement of the Second law.

Thus, the system within the dashed boundary in Figure 2 remains unchanged during both of these processes, and the entire heat crossing the boundary is converted into work. This is impossible.

C. CONSEQUENCES OF THE SECOND LAW

1. Reversible and Irreversible Processes

Water may be heated using direct heat (e.g., using an electrical heater) or work (e.g., using a stirrer producing mechanical work that is converted into internal energy by mechanical frictional processes). The stirring process is irreversible, since all the increased energy contained in the water cannot be converted back to obtain a cyclic work output that is equivalent to the work input. Hence conversion of all mechanical work into thermal energy is an irreversible process as a consequence of the Second law. In general, irrversibilities are caused by frictional processes and property gradients within systems. All processes are not irreversible. For example, if a gas is adiabatically compressed under quasiequilibrium process, its internal energy and pressure increase. However, upon expansion, the initial state can be retrieved without any work transfer across the system boundary.

Is irreversibility undesirable? Irreversibility is required in order to force a process, since energy cannot be extracted from a substance in equilibrium with its surroundings. Water stored behind a dam is in thermal and chemical equilibrium and, if constrained by the dam, is also in mechanical equilibrium. The mechanical constraint must be removed to create a nonequilibrium state in order to extract the energy. Similarly, high temperature steam contained in an insulated boiler drum is not in thermal equilibrium with its environment. The resulting temperature difference is used to transfer heat and extract work using a heat engine, such as a steam turbine. A thermal process cannot be executed without creating a thermal potential difference. However, Second law restrictions result in irreversibilities during such a process.

2. Cyclical Integral for a Reversible Heat Engine

We will show that for any reversible heat engine involving an ideal gas as the medium the cyclical integral

$$\oint \delta Q/T = 0.$$

For such a heat engine using the First law

$$\delta q - \delta w = du.$$

If the processes within it are in quasiequilibrium, then $\delta q_{rev} - \delta w_{rev} = du$. Since $\delta w_{rev} = Pdv$, and for an ideal gas $du = c_{v0}\, dT$,

$$\delta q_{rev} - P\, dv = c_{v0}\, dT, \text{ i.e., } \delta q_{rev}/T - (R/v)dv = c_{v0}\, dT/T.$$

Integrating over a cycle, we find that the RHS of these relations is zero, i.e.,

$$\oint \delta q_{rev}/T = \int R\, dv/v + \int c_{v0}\, dT/T = 0. \quad (8)$$

The Carnot cycle (due to Sadi Carnot, 1796–1832) uses ideal gas as its working fluid and consists of four quasiequilibrium processes, as illustrated in Figure 3: adiabatic compression, isothermal heat addition Q_H from a higher–temperature reservoir at a temperature T_H, adiabatic expansion, and heat rejection Q_L to a lower–temperature reservoir at T_L. From Eq. (8) it is evident that

$$Q_L/Q_H = T_L/T_H \quad (9)$$

for a Carnot cycle. Hence the efficiency of a Carnot cycle is given by the expression

$$\eta = W_{cycle}/Q_H = 1 - Q_L/Q_H = 1 - T_L/T_H. \quad (10)$$

If the P–v–T relationship for ideal gases were of the form $P\, v = R\, f(\Theta)$ where $T = f(\Theta)$, in that case $du = c_{v0}\, f'(\Theta)\, d\,\Theta$, and $\oint \delta Q/T = \oint \delta Q/f(\Theta) = 0$. Therefore,

$$Q_L/Q_H = f(\Theta_H)/f(\Theta_L), \text{ i.e.,} \quad (11)$$

$$Q_L/Q_H = f(\Theta_L, \Theta_H). \quad (12)$$

From Eqs. (10) or (12) it is obvious that all Carnot heat engines running between the reservoirs at the same high and low temperatures, and using an ideal gas as a medium have the same efficiency. This is known as Carnot's Second Corollary. Let us now state Carnot's corollaries:

First: The thermal efficiency of an irreversible power cycle is always less than the thermal efficiency of a reversible power cycle when each operates between the same two reservoirs.

Second: All reversible power cycles with any medium of fluid operating between the same two thermal reservoirs must have the same thermal efficiencies.

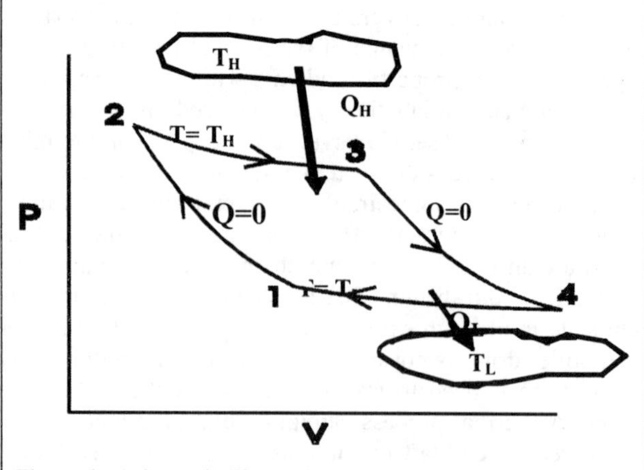

Figure 3. Schematic illustration of a Carnot cycle.

For instance, a Carnot cycle using steam as a medium must have the same efficiency as one using ideal gas as the working fluid if they operate between the same thermal reservoirs. The Kelvin–Planck statement of the Second law is violated if Carnot's Second Corollary is violated as illustrated in the next example. We will prove this I) for a cycle with ideal gas as medium and II) then with steam as medium. Also in general, for any Carnot cycle operating with any medium $\oint \delta Q/T = 0$.

d. Example 4

An automobile engine consists of an Carnot cycle–based heat engine using steam as its medium, and operating between a thermal reservoir at 500 K and ambient air at 300 K. Assume its efficiency to be 50%. A similar heat engine which uses ideal gas as its medium is operated in reverse in order to replenish the heat lost by the higher–temperature reservoir. This tandem operation of steam heat engine and ideal gas heat pump occurs perpetually in the absence of any fuel input to power the automobile. Is this possible? (See Figure 4.)

Solution

The efficiency of a Carnot cycle–based engine using ideal gas as its working fluid is $\eta_G = 1 - 300 \div 500 = 0.4$, i.e., $Q_L = 0.6\ Q_H$. We are asked to assume that for the steam engine $\eta_s = 0.5$, $Q_L = 0.5\ Q_H$. Since the engines are reversible, the ideal gas–based machine can be operated as a heat pump. Using both the steam heat engine and ideal gas heat pump in tandem, it is possible to obtain a net energy production (as work) equal to $0.1Q_H$ or 10 kJ if $Q_H = 100$ kJ while the entire heat lost by the higher–temperature reservoir is replenished. The combined system within the dashed boundary removes 10 kJ of thermal energy from the lower–temperature reservoir at 300 K and converts it completely into work with 100% efficiency in violation of the Kelvin–Planck statement of the Second law. Clearly, this is impossible.

Remarks

An assumption was made that $\eta_s > \eta_G$. Instead, if we assume $\eta_s < \eta_G$, the tandem operation of an ideal gas engine and a similar steam heat pump can be hypothesized to prove that this is not possible. Thus, the only realistic scenario occurs when $\eta_s = \eta_G$.

The conclusion from this example is that the efficiency of a Carnot cycle using steam as its medium is the same as for a cycle employing air as the medium as long as they operate between the same TERs. In other words the Carnot efficiency is independent of the medium used in the cycle or constitutive relation (e.g. ideal gas law) of the medium. Therefore, Eqs. (8)–(10) are applicable to any Carnot cycle utilizing any medium, and $Q_L/Q_H = T_L/T_H$ for all Carnot cycles. To assume otherwise would violate the Second law. If two Carnot cycles are operated between the same higher–temperature reservoirs at T_H, but different lower–temperature reservoirs, the higher efficiency will belong to one operating at the lowest temperature T_L.

---

e. Example 5

Steam is generated at a temperature of 1000 K. It is possible to transfer 2000 kW from it to a Carnot heat engine. Calculate the work done if the engine is used in:
A desert where the ambient temperature is 47°C?
A polar region where the ambient temperature is –13°C?

Solution

Q_H = 2000 kW. Therefore, heat rejection from the engine at 47°C (320 K) is

$$Q_L = -Q_H\, T_L/T_H = -2000 \times (320 \div 1000) = -640 \text{ kW}.$$

$$W = \oint \delta Q = Q_H - Q_L = 2000 - 640 = 1360 \text{ kW, and}$$

$$\eta = 1 - 320 \div 1000 = 0.68.$$

At the lower temperature of –13°C (260 K), the heat rejection

$$Q_L = -Q_H\, T_L/T_H = -2000 \times (260 \div 1000) = -520 \text{ kW},\ W = 2000 - 520 = 1480 \text{ kW, and}$$

$$\eta = 1 - 260 \div 1000 = 0.74.$$

A larger amount of work is possible with the same thermal input if the temperature of the lower–temperature reservoir is reduced.

---

f. Example 6

What is the work required to run a Carnot heat pump that provides 2000 kW of thermal energy to a 1000 K high-temperature reservoir in a desert that has an ambient temperature of 47°C.

Solution

The COP = $Q_H/W = T_H/(T_H - T_L)$ = 1000 ÷ (1000–320) = 1.47. Therefore,

$$W = 2000 \div 1.47 = 1360 \text{ kW}.$$

This work input equals the output of the heat engine discussed in Example 5 above.

---

3. Clausius Theorem

The Clausius theorem proves that for any reversible cycle (using any medium)

$$\oint \delta Q/T = 0. \qquad (13)$$

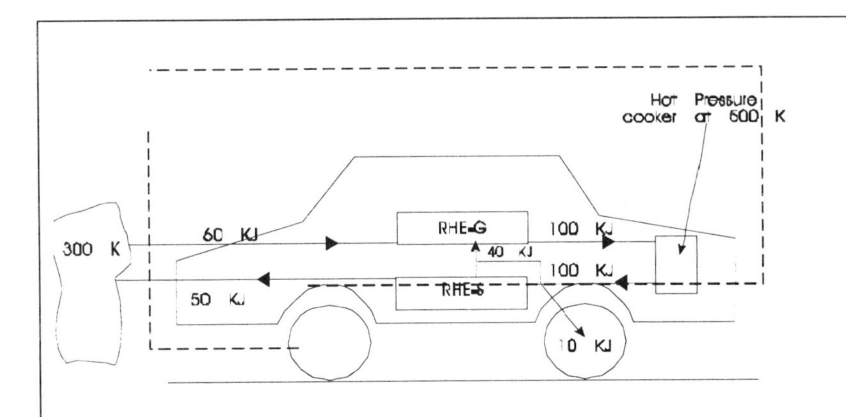

Figure 4: An automobile using ambient air as its energy source.

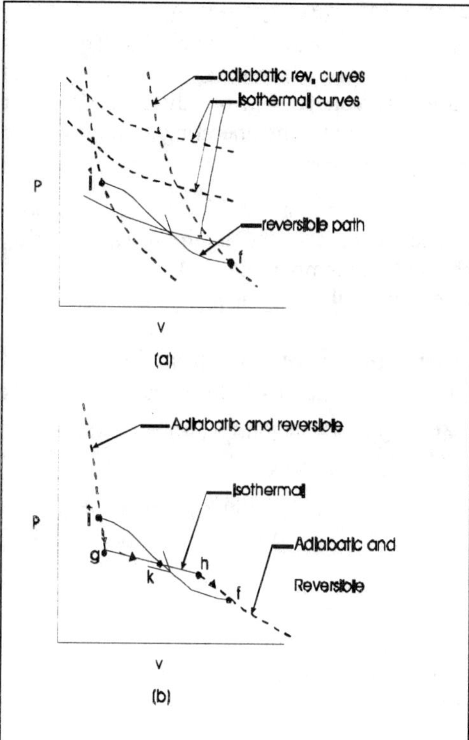

Figure 5: a. A reversible process "if"; b. Replacement of the reversible process with adiabatic and isothermal reversible paths.

The theorem converts a reversible cycle into the equivalent of an aggregate of a series of Carnot cycles. Consider a single reversible process i-f depicted in Figure 5a. This process can be replaced by a sum of adiabatic reversible (i-g), isothermal (g-h), and adiabatic reversible (h-f) processes (as shown in Figure 5b) such that the area under the P-v curve for process Ai-f@ equals that under the path i-g-h-f.

Applying the First law to the process i–f

$$q_{if} - w_{if} = u_{if} = u_f - u_i. \tag{14}$$

We wish to prove that the path "i–f" can be replaced by i–g, g–h, and h–f as long as

$$w_{ighf} = w_{if}. \tag{15}$$

To do so, select the state g such that Eq. (15) is satisfied (or the area under reversible path is the same as that under those due to the isothermal and adiabatic reversible processes). Applying the First law to the process i–g–h–f

$$q_{ighf} - w_{ighf} = u_{ighf} = (u_f - u_i), \text{ i.e.,} \tag{16}$$

$$q_{ighf} - w_{if} = u_{if}. \tag{17}$$

From Eqs. (14) and (17)

$$q_{if} = q_{ighf} = q_{ig} + q_{gh} + q_{hf}. \tag{18}$$

However, $q_{ig} = q_{hf} = 0$. Therefore,

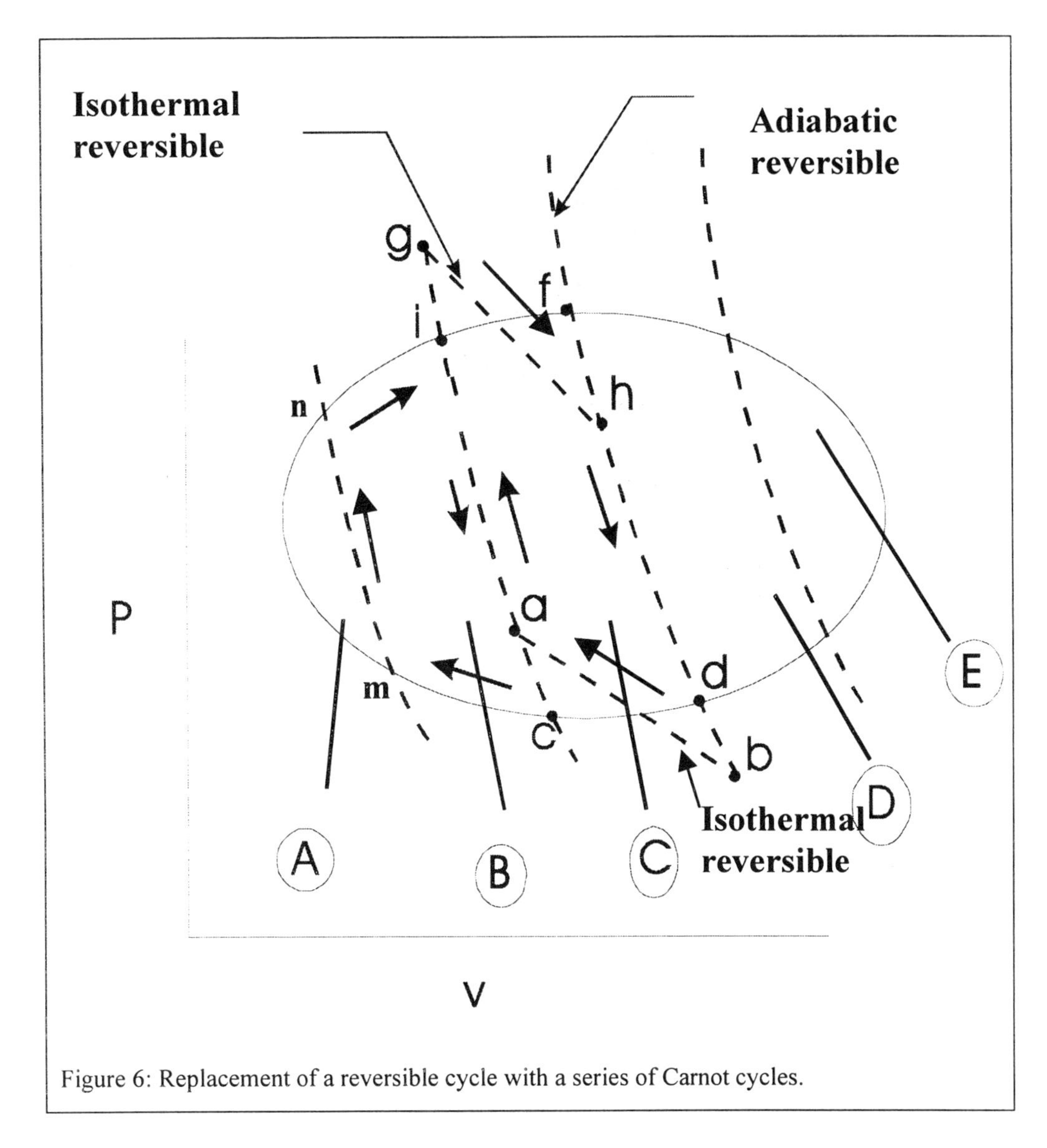

Figure 6: Replacement of a reversible cycle with a series of Carnot cycles.

$$q_{if} = q_{ighf} = q_{gh} \tag{19}$$

This discussion illustrates that the heat interaction during a reversible process which is a part of an arbitrary reversible cycle, e.g., along the path i-f, can be replaced by isothermal processes (such as gh) and adiabatic processes (e.g., i-g and h-f) both of which are part of Carnot cycles. For instance, consider cycle j-i-k-d-c-j as illustrated in Figure 6,. We can draw adiabatic reversible lines as shown in the figure, and the integral $\oint \delta Q/T$ can be evaluated by dividing the entire cycle j-i-f-k-d-c-j into a series of cycles A, B, C, D, and E. For instance, cycle C is along paths m-n, n-i, i-c and c-m.

Consider the cycle i-f-d-c-i in which the processes c-i and f-d are adiabatic and reversible. Using the Clausius theorem, we can replace the path i-f (which is a part of the reversible cycle C) by processes i-g, g-h, h-f which are part of a Carnot cycle. The work transfer $w_{if} = w_{ighf}$, and heat transfer $q_{if} = q_{ighf}$. Similarly, the process d-c can be replaced by the path d-b, b-a, and a-c. Therefore, the cycle i-f-d-c-i is equivalent to the sum of the processes i-f ($\int$(i-g+(g-h)+(h-f)), f-d, d-c ($\int$(d-b)+(b-a)+(a-c)), and c-i so that it can be replaced by the equivalent Carnot cycle a-g-h-b-a. Consequently,

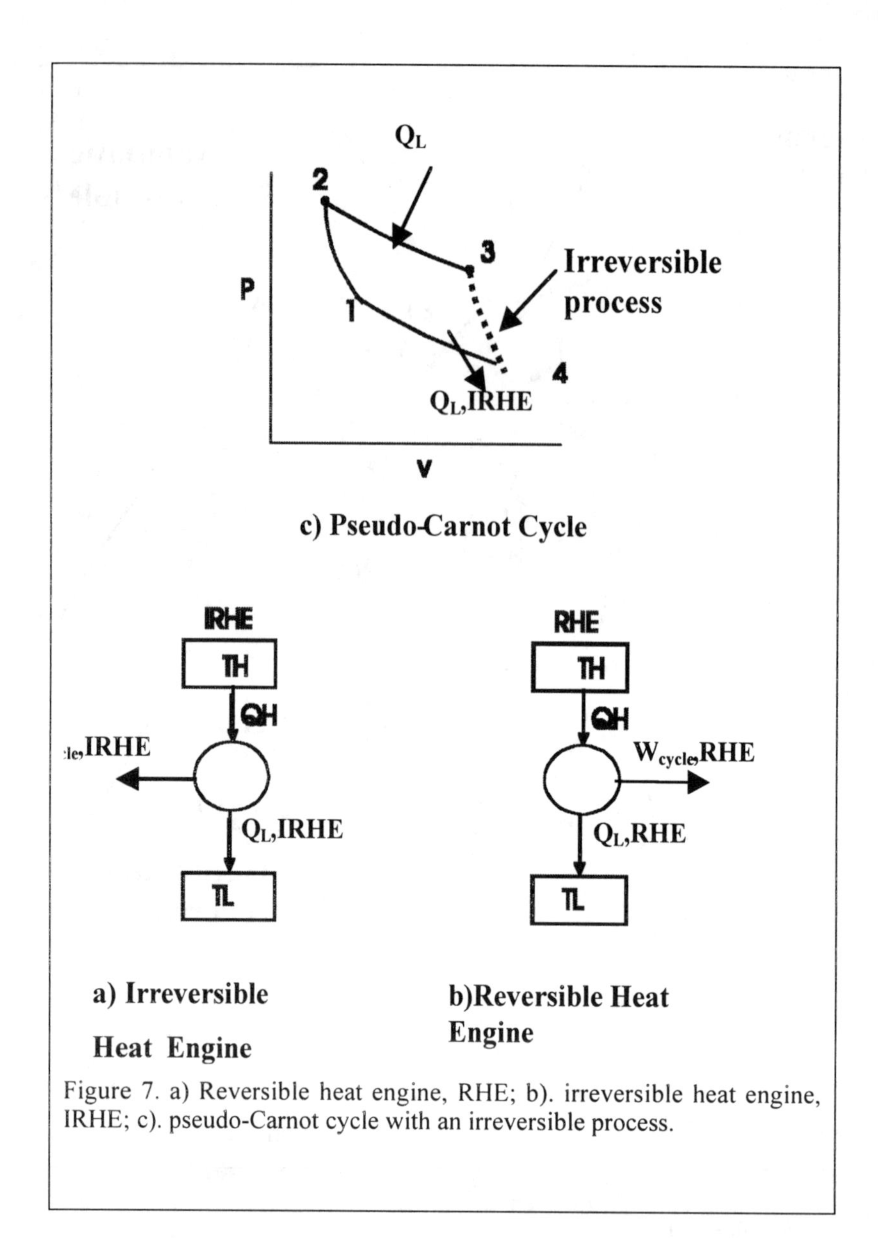

Figure 7. a) Reversible heat engine, RHE; b). irreversible heat engine, IRHE; c). pseudo-Carnot cycle with an irreversible process.

$$\oint_{if+fd+dc+ci} \delta Q / T = \delta Q_{gh}/T_{gh} + 0 + \delta Q_{ba}/T_{ba} + 0. \qquad (20)$$

Once the reversible cycle is split into an infinite number of Carnot cycles $T_{gh} \rightarrow T_{if}$, and the local Carnot efficiency for cycle C may be expressed as $\eta = 1 - (T_{dc}/T_{if})$. Equation (20) can be rewritten in the form

$$\Sigma \frac{\dot{Q}}{T} = \frac{\dot{Q}_{if}}{T_{if}} + \frac{\dot{Q}_{dc}}{T_{dc}} = 0, \text{ or } \oint \delta Q / T = \delta Q_{if}/T_{if} + \delta Q_{dc}/T_{dc} = 0, \text{ i.e.,}$$

(the integral of the local heat transfer for a reversible process) ÷ (local temperature of the system) = 0. This relation is valid for any reversible cycle, i.e.,

$$\oint \delta Q / T = 0. \qquad (21a)$$

4. Clausius Inequality

Consider two heat engines, one irreversible (IRHE) and the other reversible (RHE). The efficiency of the irreversible heat engine is lower than that of a reversible heat engine operating between identical higher– and lower–temperature thermal reservoirs, (Carnot's First Corollary) i.e.,

$$\eta_{IRHE} < \eta_{RHE}.$$

Consider a Carnot cycle that involves reversible processes (e.g., a reversible heat engine) and a pseudo-Carnot cycle (the irreversible heat engine shown in Figure 7b). The pseudo-Carnot cycle involves a single irreversible process depicted by the irreversible expansion path 3-4 in Figure 7c. The irreversible path 3-4 creates frictional heating, which requires more heat rejection to complete the cycle. Recall that for any heat engine $\eta = 1 - Q_{out}/Q_{in}$, and that for a Carnot engine $Q_{out}/Q_{in} = Q_L/Q_H = T_L/T_H$, which implies that for the same value of Q_H

$$Q_{L,IRHE} > Q_{L,RHE}.$$

For an irreversible cycle

$$\oint (\delta Q/T)_{IRHE} = Q_H/T_H - Q_{L,IRHE}/T_L = (Q_H/T_H - Q_{L,RHE}/T_L) + Q_{L,RHE}/T_L - Q_{L,IRHE}/T_L.$$

The expression contained in the parenthesis equals zero, since $Q_{L,RHE}/Q_H = T_L/T_H$. Furthermore, $Q_{L,IRHE} > Q_{L,RHE}$, and

$$\oint \delta Q/T = 0 + (Q_{L,RHE} - Q_{L,IRHE})/T_L, \text{ i.e.,}$$

$$\oint \delta Q/T < 0. \qquad (21b)$$

This mathematical statement is known as the Clausius inequality.

In a manner similar to that used for a closed reversible cycle, an irreversible cycle may be represented by an infinite number of pseudo–Carnot cycles (involving one irreversible process). By doing so, it becomes possible to prove that for any cycle involving irreversible processes, $\oint \delta Q/T < 0$.

For the same of illustration, consider a realistic automobile engine running on an Otto cycle. Due to irreversible frictional processes, the engine must reject more heat to the cooling water than an analogous reversible engine so that the cyclic process is achieved. In this case, $\delta Q_{out}/T_{out} > \delta Q_{in}/T_{in}$ leading to the inequality of Eq. (21b). The medium (gaseous combustion products) in the engine can exist at a temperature different from that of the reservoirs. Therefore, the relevant temperatures in the integral of Eq. (21b) may differ from the reservoir temperatures. In subsequent sections we will generalize the Clausius inequality, and refer to medium temperatures during the cyclical process rather than the reservoir temperatures.

5. External and Internal Reversibility

A Carnot cycle is illustrated in Figure 8a. Although during the process 2–3, the cycle medium temperature is 1000 K, the temperature of the corresponding thermal energy reservoir $T_H' = 1200$ K. Likewise, the process 4–1 occurs at a medium temperature of 400 K, while the thermal energy reservoir is colder with $T_L' = 300$ K. In this case, irreversibilities occur between the cylinder wall and the hot (at T_H') and cold (at T_L') reservoirs. Assuming uniform gas temperatures within the system during these processes (e.g., $T_A = T_B = 1000$ K as the process 2–3 proceeds, as shown in Figure 8b), it is clear that while the closed system is internally reversible; it is externally irreversible. This spatial property uniformity causes the process to be internally reversible.

The efficiency of the Carnot cycle 1–2–3–4 equals $1 - T_L/T_H$, where $T_L = 400$ K; and

T_H = 1000 K. This efficiency is based on the internal temperatures of the system. Therefore, Eq. (21) can be written as

$$\oint(\delta Q/T)_{int\,rev} = 0, \quad (22)$$

where the T denotes the uniform internal system temperature.

6. Entropy

In Chapter 1, entropy is defined as a measure of the number of states in which energy is stored. The calculation of entropy requires knowledge of energy states of molecules. Now using classical thermodynamics, a mathematical definition will be given for estimating the entropy in terms of macroscopic properties.

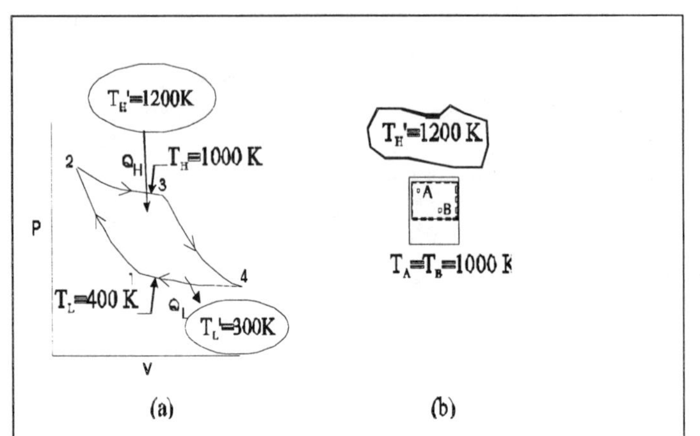

Figure 8 a. Carnot cycle with a thermal energy reservoir at T_H' and T_L; b. piston-cylinder assembly with a temperature different from the thermal energy reservoir at T_H'.

a. Mathematical Definition

For any cycle involving internally reversible processes, $\oint(\delta Q/T)_{int\,rev} = 0$. Since the cyclical integral for any property is also zero, e.g., $\oint du = \oint dh = \oint du = \oint du = 0$, we can define $(\delta Q/T)_{rev}$ in terms of the entropy which is a property. Therefore,

$$(\delta Q/T)_{rev} = dS. \quad (23)$$

The subscript "int" is omitted hereon for the sake of convenience. The absolute entropy can be expressed in units of kJ K^{-1} or BTU R^{-1}. On a unit mass basis $(\delta q/T)_{rev} = ds$ (in units of kJ Kg^{-1} K^{-1} or BTU lb^{-1} R^{-1}). Similarly, on a mole basis, $(\delta \bar{q}/T)_{rev} = d\bar{s}$ (expressed in units of kJ $kmole^{-1}$ K^{-1} or BTU lb $mole^{-1}$ R^{-1}). The absolute entropy S is an extensive property as are the absolute internal energy and enthalpy, and volume, and can be converted into its intensive form s or $\bar{s}$.

b. Characteristics of Entropy

The entropy is a measure of the energy distribution within the constituent molecules of the matter contained in a system. The larger the number of ways that energy can be distributed in a system, the greater the entropy. The classical theory suggests that the entropy change can be evaluated by the relation $dS = \delta Q_{rev}/T$ rather than using the energy distribution approach. For a reversible process, it is seen from Eq. (23) that

$$TdS = \delta Q_{rev} \text{ or } T\,ds = \delta q_{rev}. \quad (24)$$

Processes can now be depicted on a T–S diagram (as shown in Figure 9). The area under a process path 1-2 represents the reversible heat transfer. If a process is reversible and adiabatic, $\delta Q_{rev} = 0$, implying that the entropy remains unchanged during it (in this case the process is termed as being isentropic). The T–S diagram for a Carnot cycle operating between fixed temperature reservoirs forms a rectangle as illustrated in Figure 10. For this cycle the entropy change during the heat absorption process ΔS_H equals that during the heat rejection process ΔS_L. The entropy change of a composite system (e.g., containing two subsystems) ΔS is simply the sum of the entropy changes in both systems. The proof of this statement is contained in the Appendix. Therefore, $\Delta S_{1+2} = \Delta S_1 + \Delta S_2$, or $S_{1+2} = S_1 + S_2$.

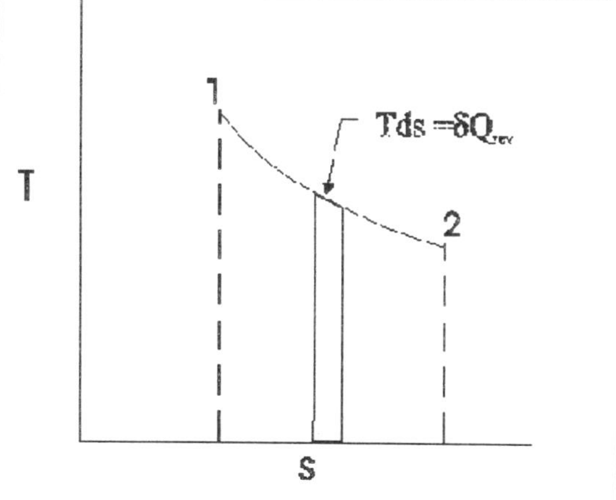

Figure 9: A process represented on the temperature-entropy (T-S) diagram.

g. *Example 7*

A house, initially at a temperature T_1 during a hot summer day, must be cooled to a temperature T_2, while the ambient temperature is T_0.
Obtain an expression for the minimum work required.
If T_0 = 310 K, T_1 = 310 K, T_2 = 294 K, c_{v0} = 0.718 kJ kg^{-1} K^{-1}, determine the minimum work required to cool a house containing a living area of 200 m^2 with equivalent mass of 50 kg m^{-2} of living area.

Solution

An air–conditioning cycle which absorbs heat at a temperature T, and rejects heat to ambient at T_0 is used (see Figure 11). The temperature of the house decreases as progressively more heat is absorbed from the house (1-2), and discarded to the ambient. The heat transfer decreases the entropy of the house, and the ambient gains entropy (line K–L).

We assume the air–conditioning to occur through a Carnot cycle GHCFG that consists of a series of elemental reverse Carnot cycles that operate at the same high temperature T_0, but their lower–temperature reservoirs have different temperatures ranging from T_1 to T_2 (or T_G to T_H). Consider one such elemental cycle A–B–C–D which absorbs heat δQ_{in} during the process A–B from the house which is at temperature T. Applying the First law to the reversed Carnot engine,

$$\delta W = \delta Q_{in} - \delta Q_{out}, \text{ and} \quad \text{(A)}$$

$\delta Q_{out}/\delta Q_{in} = T_0/T$. Therefore,

$$\delta W = \delta Q_{in}(1 - T_0/T). \quad \text{(B)}$$

Note that the heat transfer to the reversed Carnot cycle

$$\delta Q_{in} = -\delta Q_H, \quad \text{(C)}$$

where δQ_H is the heat transfer from the house. From Eqs. (C) and (B)

$$\delta W = -\delta Q_H(1 - T_0/T) = -\delta Q_H + T_0\, dS.$$

Applying the First law to the house

$$\delta Q_H - \delta W_H = dU_H.$$

Since the work transfer to the rigid house $\delta W_H = 0$,

$$\delta Q_H = dU_H. \quad \text{(D)}$$

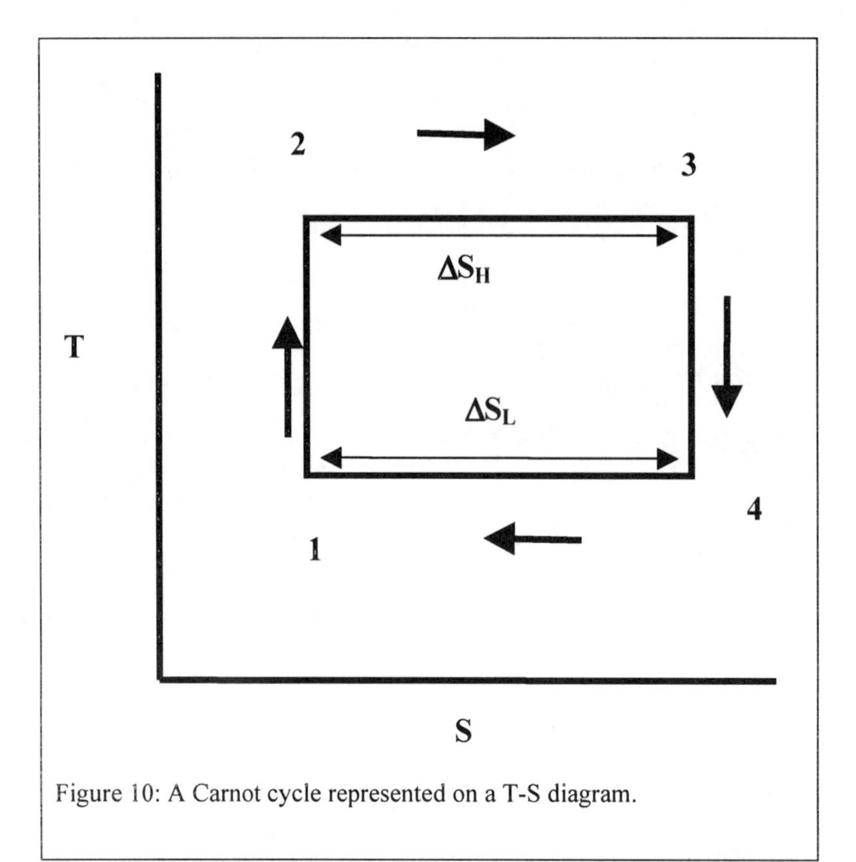

Figure 10: A Carnot cycle represented on a T-S diagram.

Furthermore, from Eqs. (D) and (C)

$$\delta W = -\,dU_H + T_0\,dS_H. \tag{E}$$

Equation (E) can be integrated to obtain

$$W_{min} = W = -(U_2 - U_1) + T_0(S_2 - S_1), \tag{F}$$

where U and S refer to properties of the house. (The availability concepts introduced in Chapter 4 will yield similar results.) Per unit mass of the house,

$$w_{min} = W/m = w = -(u_2 - u_1) + T_0\,(s_2 - s_1). \tag{G}$$

$T_1 = 310$ K, $T_2 = 294$ K, $c_{v0} = 0.718$ kJ kg^{-1} K^{-1}.

$w = -0.718 \times (294 - 310) + 310 \times (0.718 \ln (294/310) - R \ln (v_2/v_1))$.

However, $v_2 = v_1$, since the house is rigid, and

$w = -\,0.718 \times (294 - 310) + 310 \times 0.718 \ln (294/310) = -0.3071$ kJ kg^{-1}. m = 200 × 50 = 10,000 kg and hence,

$W = -0.30701 \times 10{,}000 = -3070.1$ kJ

Remarks

The overall cycle diagram for the combined Carnot cycle involving several elemental cycles is depicted in Figure 11b as the dashed line E–F–G–H.

Figure 11b illustrates the change in entropy of the house and ambient air. For the Carnot cycle operating between the variable temperature reservoir and the ambient, the T–S diagram is no longer a rectangle (area E–F–G–H in Figure 11b). The area under the lower–temperature path 1–2 represents the heat absorbed from the house by the medium in the Carnot cycle.

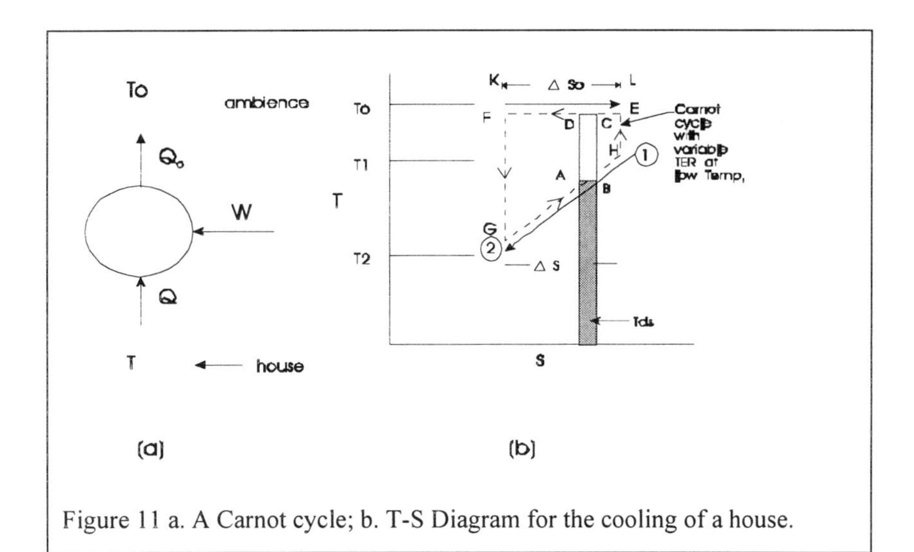

Figure 11 a. A Carnot cycle; b. T-S Diagram for the cooling of a house.

In Chapter 4 we will discuss the concept of availability. There, Eq. (F) will reappear in the relation for optimum work.
If the expression in Eq. (F) is used in the context of a thermal energy reservoir at a temperature T_H, $W = (U_1 - U_2)(1 - T_0/T_H)$, where $U_1 - U_2$ of the reservoir equals Q_{in}.
If the house is initially at a temperature of $T_{H,1}$, and a sudden cold front at T_0 moves in, how much work must be supplied to cool the house to $T_{H,2}$?
One can show that Eq.(G) reduces to

$$w\times = (w_{min}/(c_{v0}\, T_{H,1})) = -(\theta_2 - 1) + \theta_0 \ln(\theta_2),\ \theta = T/T_{H,1} \tag{H}$$

where $w\times$ is a maximum when $\theta 2 = \theta_0$.

Figure 11c illustrates the variation of $w\times$ as a function of θ_2. It is possible to produce work when cooling the house (e.g., the house supplies heat to Carnot heat engine which rejects heat to the ambient). The sensible energy of a warm house can be used to produce work during a cooling process. However, the work produced decreases once the house temperature falls below the ambient temperature since a part of the produced work is used as a work input in a Carnot heat pump. When the house temperature reaches a certain value, then

$$-(\theta_2 - 1) + \theta_0 \ln(\theta_2) = 0 \text{ or } \theta_2 = 1 - \theta_0 \ln(1/\theta_2), \text{ i.e., } w = 0,$$

(see the straight line on the plot). If the house temperature decreases further, then external work input is necessary. When $T_0 = 298$ K, $T_{house\text{-}2} = 298$ K, $w = -0.718 \times (298\text{–}310) + 298 \times 0.718 \ln(298/310) = 0.169$ kJ kg^{-1} while at $T_{house\text{-}2} = 285$ K, $w\times = -0.718 \times (285\text{–}310) + 298 \times 0.718 \ln(285/310) = -0.041$ kJ kg^{-1}.

h. Example 8

Discuss an optimal path to heat 1 kg of water from 300 K and 1 bar (point K of Figure 12) to 780 K, with the final pressure being arbitrary. A pump that is used to compress the fluid consumes virtually negligible work.

Solution

Two different paths may be selected: (1) the path K–F–G–H at a pressure of 1 bar, or (2) the path K–D–E along which the final pressure is the same as critical pressure. The area under the path S on the T–S diagram represents the amount of heat required

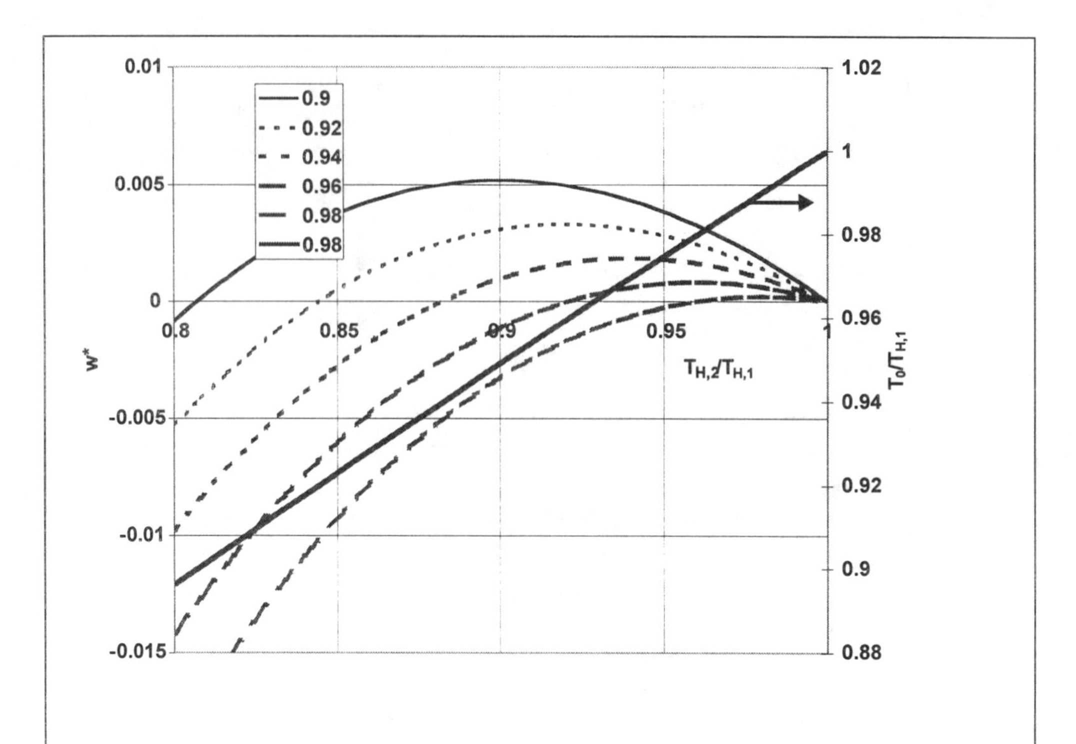

Figure 11c: Variation of work ($w\times = w_{min} /(c_{v0}\ T_{H,1})$)) as a function of the nondimensional house temperatures ($T_{H,2}/T_{H,1}$) with nondimensional ambient temperatures ($T_0/T_{H,1}$) as parameters.

under reversible conditions. For the second path the area A–K–D–E–B is lower compared to that for the first path, i.e., area A–K–F–G–H–C. Regardless of the choice of final pressure (within the constraint that the fluid is not supercritical), the path along the critical pressure represents the least heat input.

Remark

For path (2), the water pressure must be raised from 1 bar to its critical pressure P_c = 220.9 bar. Since water is a liquid (hence, incompressible), the pump work required is negligible.

7. Relation between ds, δq and T During an Irreversible Process

Assume that a large aggregate of infinitesimal weights is suddenly placed on top of a piston of a piston–cylinder assembly, thereby compressing the gas contained adiabatically, but irreversibly (path 1-2 in Figure 13 and Figure 14a). The abrupt action of placing the weight on the piston causes a higher–velocity macroscopic movement of molecules and closer spacing of molecules near the piston surface, while molecules farther away from this surface, still spread far apart, are macroscopically virtually motionless. This results in the formation of property gradients and frictional heating of the gases through the destruction of macroscopic kinetic energy, and in a temperature rise at the conclusion of the compression process. After reaching the final volume V_2, if the infinitesimal weights are slowly removed, not all of the weights must be taken away in order to regain the initial volume V_1 (path 2-3 in Figure 13 and Figure 14b), since the initial compression created too high a temperature due to rapid motion of the piston where kinetic energy of molecules is eventually converted into heat. Therefore, the expansion which now occurs through a quasiequilibrium process back to the initial volume cannot have the same initial temperature.

The two processes are illustrated through the P–v diagram in Figure 13. The irreversible compression process is represented by the dashed line 1–2, while the quasiequilibrium expansion is depicted by the solid line 2–3. Since $T_3 > T_1$, the pressure P_3 following expansion is greater than the initial pressure P_1, although the corresponding volumes are identical. The processes can be organized into a cycle by adding a quasistatic equilibrium heat rejection process at constant volume. Thus, the cycle 1-2-3-1 involves three processes: (1) irreversible adiabatic compression 1-2 , path A ($Q_{12} = 0$), (2) reversible adiabatic expansion 2-3 ($Q_{23} = 0$), and (3) constant volume heat rejection 3-1 ($Q_{31} < 0$). Let us lump the processes 2-3 and 3-1 as path B. Applying the First law to the processes, since $\oint \delta Q = \oint \delta W$ and $\oint \delta Q < 0$, then $\oint \delta W < 0$ implying work input into the cycle.

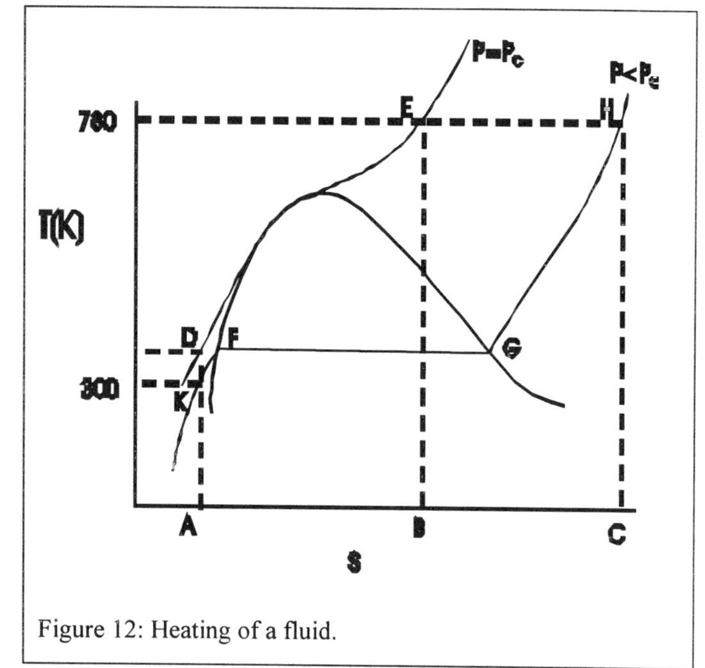

Figure 12: Heating of a fluid.

The Clausius inequality must be satisfied for any cycle involving irreversible processes so that

$$\int_1^2 \left(\frac{\delta Q}{T}\right) + \int_2^3 \left(\frac{\delta Q}{T}\right)_{rev} + \int_3^1 \left(\frac{\delta Q}{T}\right) < 0.$$

Lumping the reversible processes 2-3 and 3-1 together as a single reversible process 2-1 (path B),

$$\int_1^2 (\delta Q / T)_A + \int_2^1 (\delta Q / T)_{rev,B} < 0.$$

The process 2-3-1 is reversible, and the term $\delta Q/T$ can be replaced by dS. Therefore, $(S_2\text{-}S_1) < 0$. Since the value of $(S_2\text{-}S_1)$ along path B is the same as that along the irreversible path A, then for any irreversible process

$$S_2 - S_1 > \int_1^2 (\delta Q/T). \quad (25)$$

Equation 25 implies that the entropy change between two equilibrium states 1 and 2 exceeds the entropy change induced by the heat transfer process or the transit entropy alone due to the irreversibility.

The entropy transfer due to heat transfer across a boundary $\delta Q/T$ will be termed as the transit entropy (abbreviated as tentropy). It is a not a property. The transit entropy $\oint \delta Q/T = 0$ only if $Q = 0$. (e.g., the irreversible process 1–2 illustrated in Figure 14a). It is also possible to reach state 2 through a combination of reversible processes. For instance, in a first quasiequilibrium process, the infinitesimal weights can be slowly placed on the piston (Figure 14b) to reach state 2R at which $V_{2R} = V_2$, but $T_{2R} < T_2$. Following this, state 2 can be reached from 2R through constant volume heating. The entropy change along path 1–2R–2 can be determined by employing the expression $dS = \delta Q_{rev}/T$ to obtain $S_2 - S_1$ (recall that S is a property, and independent of the process path). This is the procedure that we will adopt in the next section for evaluation of entropy.

As an example, assume that 2000 kJ of heat crosses a system boundary at 999 K, and that this temperature subsequently increases to 1001 K. Since the average temperature is 1000 K, the tentropy equals 2 kJ K^{-1}. If the entropy change in the system dS is measured to be 2.5 kJ K^{-1}, then the above inequality given by Eq. (25) is satisfied. Thus, the process is possible.

As another example, consider a 2 kg mass of air contained in an insulated piston–cylinder assembly at 25°C and 100 bar. Stirring work (= 14 kJ) is performed to raise the air temperature to 35°C (Figure 15a) at constant volume. It is impossible to convert all of the 14 kJ of thermal energy back into work since, for the 14 kJ of heat extraction, the Carnot work is 0.46 kJ (cf. Eq. F in Example 8). Therefore, a work capability of 13.54 kJ is lost. However, if in the first instance, the air was adiabatically and quasistatically compressed to 35°C utilizing 14 kJ of work (as illustrated in Figure 15b), the gas could have been expanded to its original state to recover the entire amount of work. The latter process is reversible while the former is irreversible. In the former process, moving the stirrer causes viscous dissipation (which is a frictional process converting work into "heat") to occur, and the Second law prevents the conversion of the entire amount of heat into work. For that process, since $\delta Q = 0$, using the relation $dS > \delta Q/T$, $dS > 0$. Since the latter process is reversible, $dS = \delta Q/T$. Furthermore, since $\delta Q = 0$, by implication $dS = 0$ implying that S = constant.

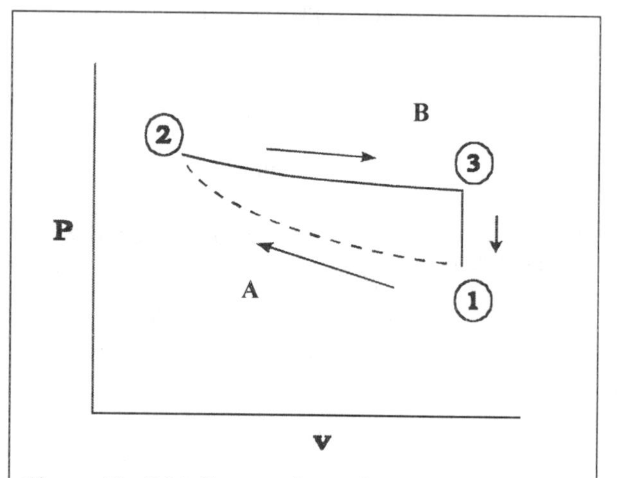

Figure 13: P-V diagram for a thermodynamic cycle consisting of an irreversible process.

a. Caratheodary Axiom II

The first Caratheodary axiom has been previously discussed in Chapter 2. We will illustrate the second axiom through the following example: Consider the adiabatic, but irreversible, compression process (from V_1 to V_2) depicted in Figure 14(c). The same state (2) can be reached by adiabatic reversible compression 1-2R and then via heat transfer at constant volume $V_{2R} = V_2$. Is $T_{2R} > T_2$ or $T_{2R} < T_2$? Although $V_2 = V_{2R}$, the Caratheodary axiom II postulates that T_{2R} must always be lower than T_2.

ii. Proof

Assume the axiom to be true. Since the process 1–2R–2 is reversible, the cycle 1–2–2R–1 (which contains an irreversible process 1–2) is possible. For the cycle,

$$\oint \delta Q/T = (\int \delta Q/T)_{1-2} + (\int \delta Q/T)_{2-2R} + (\int \delta Q/T)_{2R-1} = 0 + (\text{negative number}) + 0,$$

and the Clausius inequality is satisfied.

Now, assume that the axiom is incorrect, and that state 2R lies above 2 on the P–V diagram (cf. Figure 14d). For the cycle 2–2R–1–2,

$$\oint \delta QT = (\int \delta Q/T)_{1-2} + (\int \delta Q/T)_{2R} + (\int \delta Q/T)_{2R} = 0 + (\text{positive number}) + 0.$$

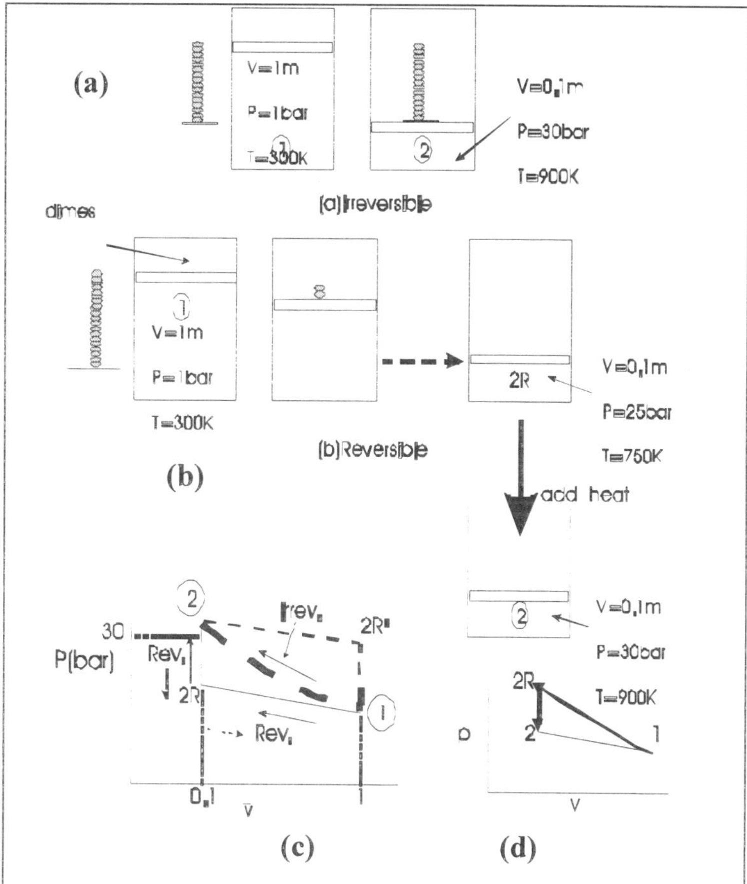

Figure 14: Illustration of reversible and irreversible processes that reach the same final state – a. Irreversible compression by suddenly placing a large system of infinitesimal weights; b. Reversible compression by placing one infinitesimal weight at a time; c. P-V diagram for the processes illustrated in Figures a and b; d. Proof of the Caratheodary axiom for the example.

The RHS of this expression is positive, which violates the Clausius inequality. Therefore, some states cannot be reached through an adiabatic process once the final volume is fixed. This is the essence of the Caratheodary axiom II.

D. ENTROPY BALANCE EQUATION FOR A CLOSED SYSTEM

The Clausius inequality states that the entropy change is always larger than the transit entropy for irreversible processes. This statement can be expressed in a balanced form that is similar to the energy and mass conservation equations, except for the fact that entropy is not a conserved quantity. Entropy balance is an important tool in designing and optimizing heat exchangers, heat engines and pumps, and various other thermodynamic systems. It provides quantitative information regarding the operating conditions and the extent of inefficiency of a device or system.

1. Infinitesimal Form

a. Uniform Temperature within a System

The entropy generated is the difference between the entropy change and the transit entropy during a process. For a closed system, the differential relation of Eq. (25) may be rewritten to explicitly include the entropy generation s, i.e.,

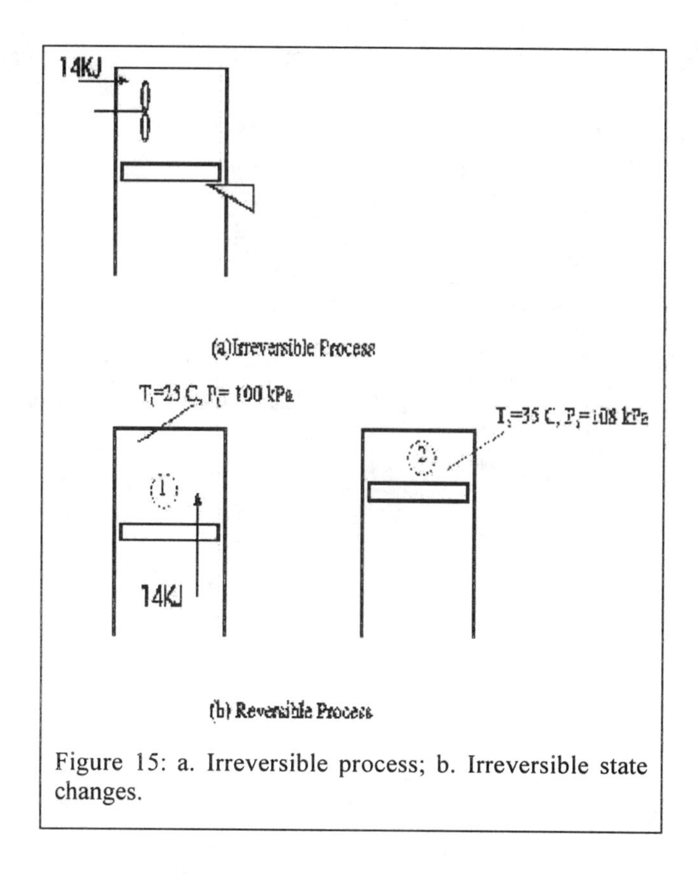

Figure 15: a. Irreversible process; b. Irreversible state changes.

$$dS = \delta Q/T + \delta\sigma. \tag{26}$$

This relation is also known as Gibbs' equation. The entropy generation $\delta\sigma > 0$ for internally irreversible processes and is zero for internally reversible processes. Although the boundary temperature may be uniform, thereby indicating thermal reversibility, other irreversibilities, such as those due to chemical reactions, can contribute to σ, as will be discussed in Chapter 11.

---

i. *Example 9*

Assume a large primary system to consist of a vessel containing warm water at a system temperature of 350.001 K (T_1). An infinitesimal amount of its heat (1050 J) is transferred to a secondary system consisting of room air at a temperature of 300 K (T_2). Consequently, the water temperature drops to 349.999 K. What is the entropy generation:

If the system is cooled in air (as illustrated through process (a) (as shown in Figure 16a)?

If the heat removed from the primary system is used to run a Carnot engine (process (b)), and that rejected by the engine is transferred to the secondary system (cf. Figure 16b)?

Solution

Assuming an internally reversible cooling and heating process for the water and air, using Eq. (26), $\delta\sigma = 0$ for both systems. Therefore,

$$dS = \delta Q/T \tag{A}$$

Since the temperature is approximately constant ($\approx$350 K), upon integrating Eq. (A)

$$\Delta S_1 = -1050 \div 350 = -3 \text{ J K}^{-1}. \quad \text{(B)}$$

which is represented by path AB in Figure 16c. Similarly,

$$\Delta S_2 = 1050 \div 300 = 3.5 \text{ J K}^{-1}. \quad \text{(C)}$$

The entropy gain for the air is represented by the path C–D–E in Figure 16c. If a boundary is placed around systems 1 and 2, as illustrated by the dashed line in Figure 16a, there is no heat or work transfer across the composite isolated system. However, an irreversible process occurs within the composite system, and

$$dS - (0/T) = \delta\sigma, \text{ i.e., } dS_1 + dS_2 = \delta\sigma, \text{ or} \quad \text{(D)}$$

$$\Delta S_1 + \Delta S_2 - 0 = \sigma. \quad \text{(E)}$$

Employing Eqs. (B) and (C),

$$\sigma = -3.0 + 3.5 - 0 = 0.5 \text{ J K}^{-1}.$$

The path D–E, shown in Figure 16c illustrates the net entropy gain for the isolated system that occurs since entropy is generated due to irreversible heat transfer between the two subsystems 1 and 2.

The second scenario is illustrated in Figure 16d. In this case subsystem 2 undergoes the same entropy change as does subsystem 1 so that

$$\Delta S_1 = -3.0 \text{ J K}^{-1}, \text{ and } \Delta S_2 = Q_2/T_2. \quad \text{(F)}$$

Therefore, $Q_L/Q_H = Q_2/Q_1 = T_2/T_1 = 300 \div 350 = 0.857$.
Since $Q_2 = 1050 \times 0.857 = 900$ J, and

$$\Delta S_2 = 900 \div 300 = +3.0 \text{ J K}^{-1}. \quad \text{(G)}$$

For the composite system $\Delta S_1 + \Delta S_2 - 0 = \sigma$. Employing Eqs. (F) and (G),

$$-3.0 + 3.0 - 0 = \sigma = 0.$$

In this particular case $|\Delta S_1| = |\Delta S_2|$.

Remarks

For the first case 1050 J of thermal energy is transferred from the warm water to the ambient and $\Delta U_1 = -\Delta U_2 = -1050$ J. It is impossible to transfer the 1050 J back from the ambient to restore the water to its original state. Further, $\Delta U = \Delta U_1 + \Delta U_2 = 0$

For the second case work is produced and used to lift a weight (e.g., lift an elevator) with the consequence that a smaller amount of heat is rejected to the ambient while accomplishing the same change in state of the water. In this case, $\Delta U_1 = -1050$ J, and $\Delta U_2 = +900$ J so that ΔU ($=\Delta U_1 +$

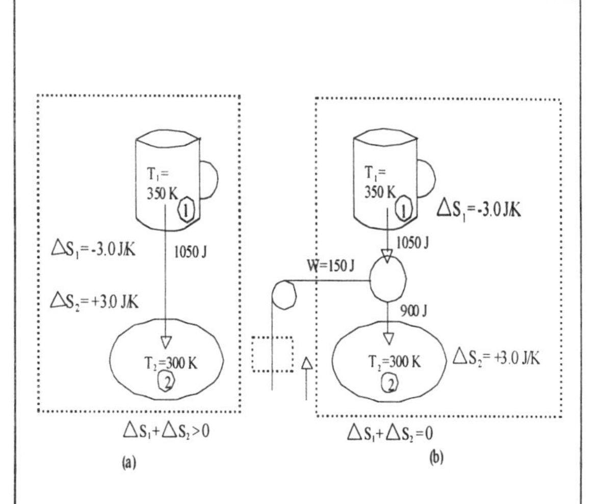

Figure 16: a). Illustration of a) direct cooling; b). cooling using a Carnot engine.

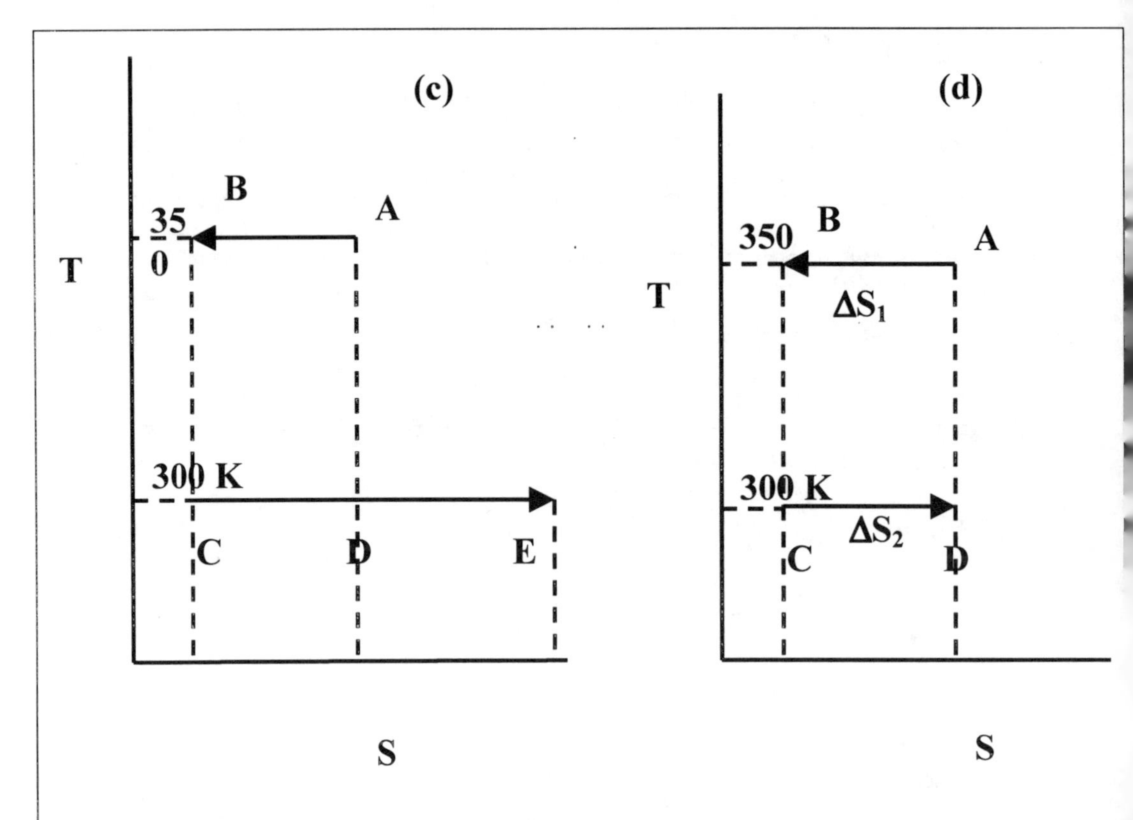

Figure 16 c) T-S diagram for direct cooling; d) T-S diagram for cooling via Carnot engine.

ΔU_2) = –150 J which equals the potential energy change in the weight.

For the second case the ambient gains 150 J in potential energy due to the work done in raising the weight. In order to reverse the process, i.e., to heat the water by heat transfer from the ambient, work input is required. If the weight is lowered to provide work for a heat pump cycle Figure 17, $Q_H/Q_L = T_H/T_L = 350 \div 300 = 1.17$. Therefore, $Q_L = 1050 \div 1.17 = 900$ J, and the absolute value of the work |W| = |Heat absorbed – Heat rejected| = $|Q_L - Q_H| = 150$ J. By lowering the weight to its original position, it is possible to supply 150 J of work to the heat pump to heat the water. In this manner both the water and ambient are restored to their initial states. Processes during which $\sigma = 0$ are entirely reversible.

In order to restore the water back to its original state for the first case one can extract 900 J of heat from the ambient with an external work input of 150 J so that it is possible to pump 1050 J of thermal energy. However, we now require an additional 150 J of work. For the first system $\Delta U_1 = 0$. For the second system $\Delta U_2 = +150$ J and the ambient loses a work equivalent of 150 J (which equals $T_0\sigma = 300 \times 0.5 = 150$ J). Ideally, heat engines should operate in a manner similar to the second case.

b. Nonuniform Properties within a System

Thus far, we have considered uniform-temperature systems such that $dS - \delta Q/T = \delta\sigma$. This relation must be modified to account for nonuniform system temperatures.

Consider a system that changes irreversibly from state 1 to 2 due to the rapid compression induced by the sudden inward movement of a piston in a cylinder (cf. Figure 18a and b). At the final state, the system is a composite of two subsystems A and B. Subsystem A contains molecules adjacent to the piston that are more closely packed than those in B (cf. Figure

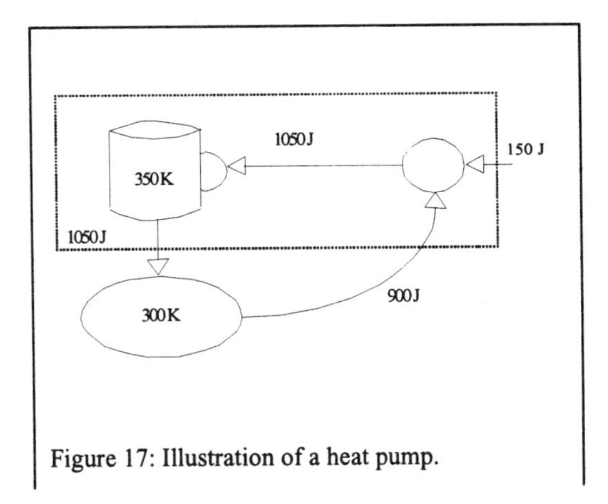

Figure 17: Illustration of a heat pump.

18b). Consequently, the two subsystems exchange different amounts of heat (respectively, δQ_A and δQ_B) through the system boundary. Heat transfer occurs internally between the two subsystems in the amount $\delta Q'$, where $\delta Q' < 0$ for A, and has a positive value for B. Assuming these subsystems to be internally reversible, applying Eq. (26),

$$dS_A - \delta Q_A/T_A - \delta Q'/T_A = 0, \text{ and } dS_B - \delta Q_B/T_B + \delta Q'/T_B = 0.$$

Adding the two relations,

$$(dS_A + dS_B) - (\delta Q_A/T_A + \delta Q_B/T_B) = \delta Q' \,(1/T_A - 1/T_B), \text{ or} \quad (27a)$$

$$(dS_A + dS_B) - (\delta Q_A/T_A + \delta Q_B/T_B) = \delta\sigma. \quad (27b)$$

Note that $\delta Q' < 0$ if $T_A > T_B$, and is positive if $T_A < T_B$. The RHS of Eq. (27a) is always a positive number that represents the entropy generated due to irreversible heat transfer within the composite system. Equation (27b) may be interpreted as follows: (The combined entropy change of the two subsystems within the composite system or a system where gradients exist) – (The transit entropy across the boundary calculated using the composite system boundary temperatures and heat fluxes) = (The entropy generated due to internal gradients).

Figure 18c illustrates the processes occurring in a composite system that consists of three subsystems. Generalizing Eq. (27b) to a system containing several subsystems (that is uniquely defined by its properties),

$$\Sigma dS_j = \Sigma \delta Q_j/T_{b,j} + \delta\sigma \quad (27c)$$

The term dS_j denotes the entropy change in the j–th subsystem, δQ_j is the heat flux across that subsystem at the subsystem boundary temperature $T_{b,j}$, and $\delta\sigma$ the entropy generated for the entire system as a result of irreversible interactions within the various subsystems. If the boundary temperature T_b is uniform across the system boundary, but differs from the system temperature, Eq. (27) can be simplified into the form

$$dS = \delta Q/T_b + \delta\sigma. \quad (28)$$

Equations (27c) and (28) are called entropy balance equations for specified mass or closed systems. Equation (28) expresses the entropy change dS within a system at a boundary temperature T_b. This change is caused by the transit entropy $\delta Q/T_b$ and the entropy generated due to irreversibilities. All processes (including chemical changes) must satisfy Eq. (28).

If the control surface is slightly extended to lie outside a system so that $T_b = T_0$ (cf. Figure 18d), all irreversibilities lie within the system, and Eq. (28) assumes the form

$$dS = \delta Q/T_o + \delta\sigma. \quad (29)$$

As an example, approximate the engine walls of an automobile engine to be adiabatic. The compressed "cold" gasoline–air mixture in the engine exists at a temperature ≈600 K, which, after burning, is converted into hot gases ≈2000 K (cf. Figure 19). Upon ignition, the vicinity of the spark plug is "warmer" than other locations such that the system is a composite of (A) hot spots and (B) cold spots. For the two subsystems, $\Delta S_A + \Delta S_B > 0$, since the entropy increases due to "internal equilibration". For the composite system illustrated in Figure 19,

$$dS_A + dS_B = \delta Q_A/T_A + \delta Q_B/T_B + \delta\sigma.$$

Frictional heating between moving gases and the fixed walls offers another example of an irreversible process. The friction causes a temperature differential near the wall that subsequently transfers heat towards the system interior, thereby generating entropy.

iii. Simple rule

If the properties of a system are uniform throughout (i.e., the system contains no property gradients), in that case processes are "internally" reversible. When temperature, pressure, or kinetic energy gradients are created within a system, processes involving it become internally irreversible, and, consequently, entropy is generated.

2. Integrated Form

The integrated form of Eq. (28) is

$$S_2 - S_1 = \int\delta Q/T_b + \sigma. \quad (30)$$

If a process satisfies Eq. (30) (e.g., with $\sigma \geq 0$), there is no assurance that the end state (2) is

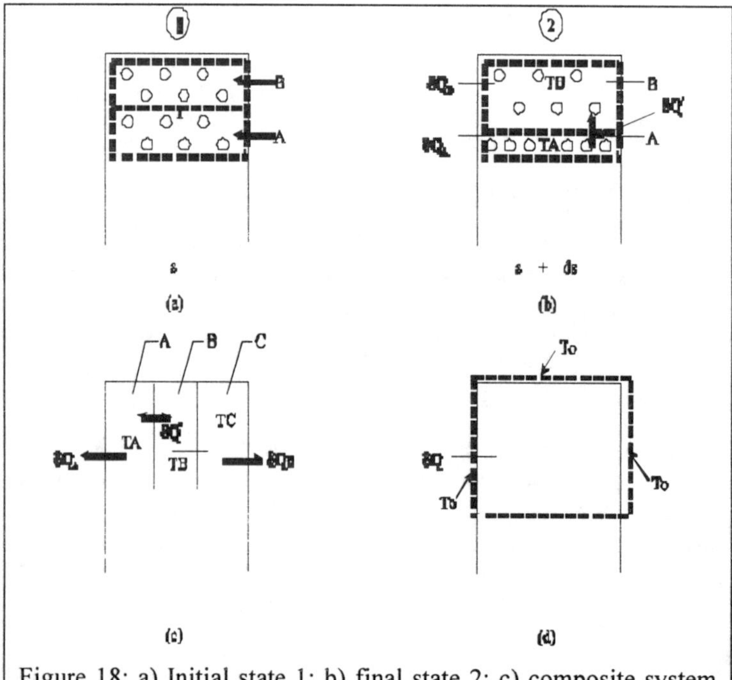

Figure 18: a) Initial state 1; b) final state 2; c) composite system with three sub-systems; d) system with $T_b = T_0$.

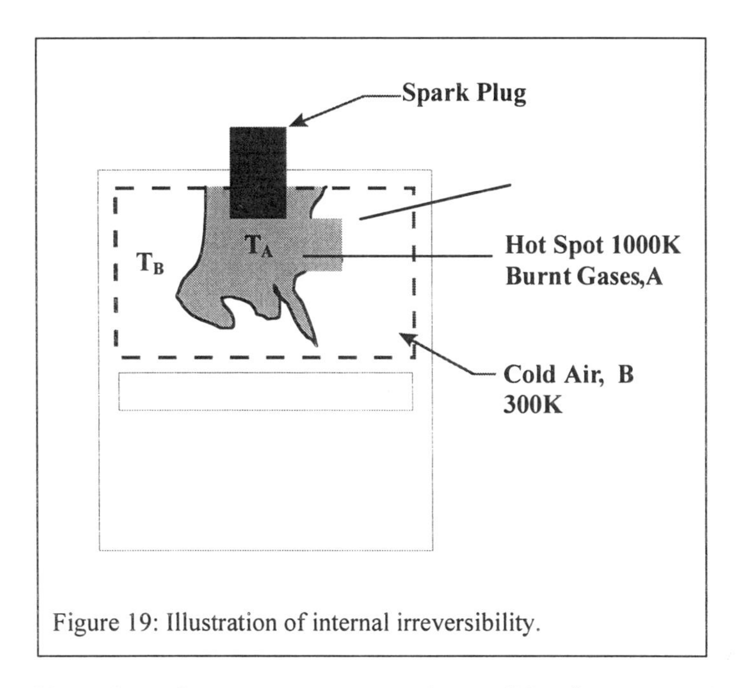

Figure 19: Illustration of internal irreversibility.

realized. As we will see later, for a process to occur, the condition $\delta\sigma \geq 0$ must be satisfied during each elemental part of the process. Therefore, Eq. (28) is more meaningful than Eq. (30).

3. Rate Form

The time derivative of Eq. (28) returns the rate form of that relation, i.e.,

$$dS/dt - \dot{Q}/T_b = \dot{\sigma}. \quad (31)$$

For instance, if a blender containing water is turned on or when a coffee pot with an immersed electrical heating coil is switched on, work is destroyed. Equation (31) is convenient to use to determine the entropy generation rate. The thermodynamic laws are constitutive equation independent, but can be used to validate (or invalidate) any constitutive equation. For instance, it is possible to determine $\dot{\sigma}$ accurately if $\dot{Q}$ is known or can be accurately measured (cf. Examples 1 and 2) and dS/dt is also known (e.g., from property tables such as Tables A-4, A-5, etc. or from basic measurements using pulley-weight assembly systems). On the other hand calculations of $\dot{Q}$ and/or dS/dt may require application of a constitutive relation. For instance, by applying $\dot{Q}/T = -(\lambda\nabla T)/T$, we have used a constitutive equation for heat transfer. If it is possible to show that $\dot{\sigma}<0$ with a constitutive relation, and since the entropy balance equation follows from the Clauisus inequality (which is a mathematical form of the Second law), then the constitutive relation is inaccurate.

4. Cyclical Form

Integrating Eq. (28) over a cyclical process, $0 = \oint \delta Q/T_b + \sigma_{cycle}$. Since $\sigma_{cycle} > 0$,

$$\oint \delta Q/T_b < 0. \quad (32)$$

This relation is a restatement of the Clausius inequality with the temperature replaced by T_b.

The entropy generation concept is a powerful tool to determine the extent of irreversibilities occurring during cyclical processes, which result in increased heat rejection and lead to lower efficiencies. Evaluation of the entropy generation during the individual processes constituting a cycle allows the determination of their relative irreversibilities, and quantifies those due to heat transfer, the destruction of mechanical work, etc., as illustrated in Example

12. Most idealized cyclical processes (e.g., the Rankine, Otto, and Brayton cycles) assume $\sigma = 0$. Based upon the values of σ_{cycle} or $\oint \delta Q/T_b$, practical cyclical heat engines can be assigned a rating of 0 (σ_{cycle} highest with work production of zero) to 1 ($\sigma_{cycle} = 0$, i.e., idealized cycles with maximum work production). Cycles may deteriorate over time due to hardware problems, with the consequence that σ_{cycle} increases.

5. Irreversibility and Entropy of an Isolated System

Since $\delta Q = 0$ for isolated systems, and $\delta\sigma > 0$ for irreversible processes, Eq. (28) yields that $dS > 0$. When warm water is exposed to ambient air, as illustrated in Example 9, the system "drifts" in the absence of internal constraints towards an equilibrium state. We saw from Example 9 that for a composite system consisting of (1) warm water directly losing heat to air (cf. Figure 16a) and (2) that water being supplied with heat equal to the lost value through a Carnot heat pump, a net work loss occurred. This loss is called the irreversibility I of the composite system. In the case discussed in Example 9,

$$I = T_0\,\sigma.$$

A rigorous proof of this equality is contained in Chapter 4.

j. *Example 10*

An uninsulated coffee pot is maintained at a temperature of 350 K in a 300 K ambient by supplying 1050 W of electrical work. The heat transfer coefficient is 0.2 kW m^{-2} K^{-1}, and heat transfer occurs over a pot surface area of 0.5 m^2. Determine the entropy generated:

In the system contained within the boundary cs_1, as illustrated in Figure 20a (i.e., for only the coffee within the pot), assuming the pot boundary temperature to be 350 K.

The matter contained within cs_2, as illustrated in Figure 20b (i.e., for the system including both the coffee and pot)

For the system containing the coffee, pot, and the ambient (i.e., bounded by the surface cs_3 illustrated in Figure 20c) for which $T_b = T_0$ which is the ambient temperature.

Solution

Selecting the control surface internally, and applying the First law, $\dot{Q} - \dot{W}_{elec} = dE/dt$. At steady state, $dE/dt = 0$ so that $\dot{Q} = \dot{W}_{elec} = -1050$ W. Applying the entropy balance equation in rate form $dS/dt - \dot{Q}/T_b = \dot{\sigma}$, we obtain

$$0 - (-1050/T_b) = \dot{\sigma}.$$

Since $T_b = 350$ K,

$$\dot{\sigma} = 1050 \div 350 = 3 \text{ W K}^{-1}.$$

Selecting the control surface cs_2 to be flush with the pot walls, the boundary temperature T_b must be determined. Applying the convection heat transfer relation

$$h\,A\,(T_b - T_0) = \dot{Q} = 1050 \text{ W},$$

the boundary temperature is determined as,

$T_b = 1.05 \div (0.2 \times 0.5) + 300 = 310.5$ K, and $\dot{\sigma} =$

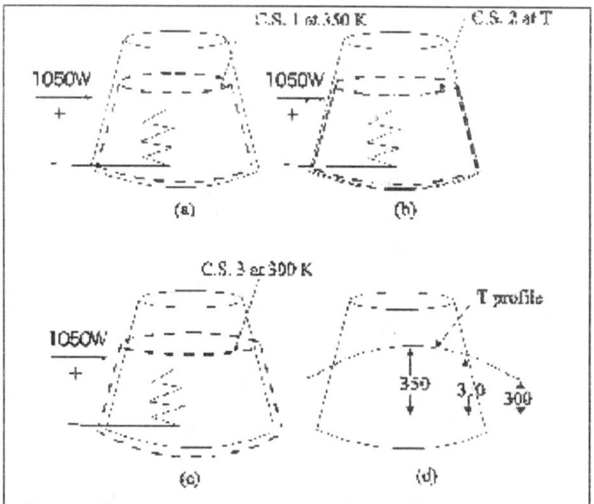

Figure 20: Entropy generation within a coffee pot.

$-(-1050 \div 310.5) = 3.382$ W K^{-1}.

Upon comparison with the previous solution, we find that irreversible heat transfer between the coffee and pot walls causes an entropy generation of $3.382 - 3 = 0.382$ W K^{-1}.

Selecting the control surface cs_3 such that the boundary exists outside the pot, $T_b = T_0$,

$$0 - (-1050 \div 300) = \sigma, \text{ i.e., } \dot{\sigma} = +3.5 \text{ W K}^{-1}.$$

No irreversibilities exist outside the boundary of the control surface cs_3. The entropy change in this composite system (using $T_b = T_0$) equals the entropy change in an isolated system, since there is no entropy production within the ambient.

Remarks

For the matter contained within the surfaces cs_2 and cs_3 which include the pot wall,

$$\dot{\sigma} = 0 - (1050 \div 310.5) - (-1050 \div 300) = 0.118 \text{ W K}^{-1}$$

due to the heat transfer between the ambient and pot walls.

By a suitable choice of the boundary, we are able to determine contributions to overall σ. The major contribution is due to destruction of electrical work into heat called electrical frictional work.

The change in entropy due to:

destruction of electrical work within the coffee pot = 3 W K^{-1}.

irreversible heat transfer between coffee and pot walls = 0.38 W K^{-1}.

irreversible heat transfer between pot walls and ambient = 0.12 W K^{-1}.

The change in entropy of the isolated system = 3.5 W K^{-1}.

k. *Example 11*

An uninsulated coffee pot is maintained at a temperature of 350 K in a 300 K ambient. Instead of supplying electrical work, we can compensate for the heat loss by placing a heat pump between the coffee pot and ambient, as shown in Figure 21. What is the electrical work required to operate the heat pump?

Solution:

COP = $350 \div (350-300) = 7$, i.e., $\dot{W}_{elec} = 1050 \div 7 = 150$ W.

The pot can be maintained at 350 K by providing 150 W of electrical power to a heat engine, rather than directly supplying 1050 W as in the previous example.

6. Degradation and Quality of Energy

Consider a Carnot cycle operating between thermal energy reservoirs at the high and low temperatures T_H and T_L, respectively. The term $(1- T_0/T_H)$ represents the quality of the energy or the work potential per unit energy that can be extracted in the form of heat from a thermal energy reservoir at a temperature T_H. Therefore, the heat Q_H extracted from the thermal energy reservoir at T_H has the potential to perform work equal to $Q_H \times$ quality = Q_H $(1- T_0/T_H)$ where quality of energy at T_H is given by $(1- T_0/T_H)$. It is seen that at a specified temperature, the quality equals the efficiency of a Carnot engine that is operated between TERs at temperatures T and T_0.

Assume that it is possible to remove 100 kJ of heat from hot gases that are at 1000 K (and which constitute a thermal energy reservoir) using a Carnot engine operating between 1000 K and the ambient temperature of 300 K (cf. Figure 22a). Using the engine under these conditions, it is possible to produce a work output of

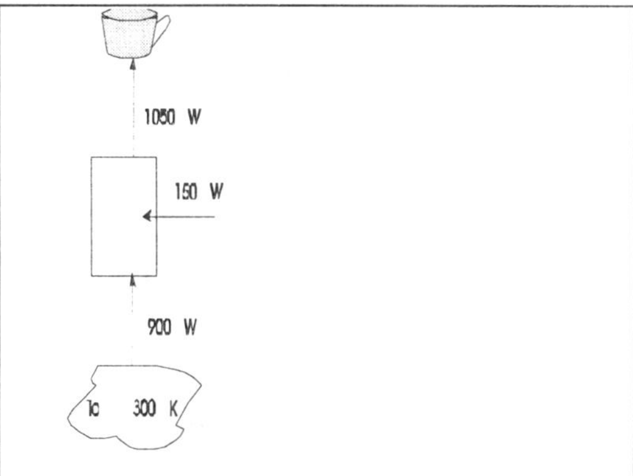

Figure 21: The heating of a coffee pot using a Carnot heat pump.

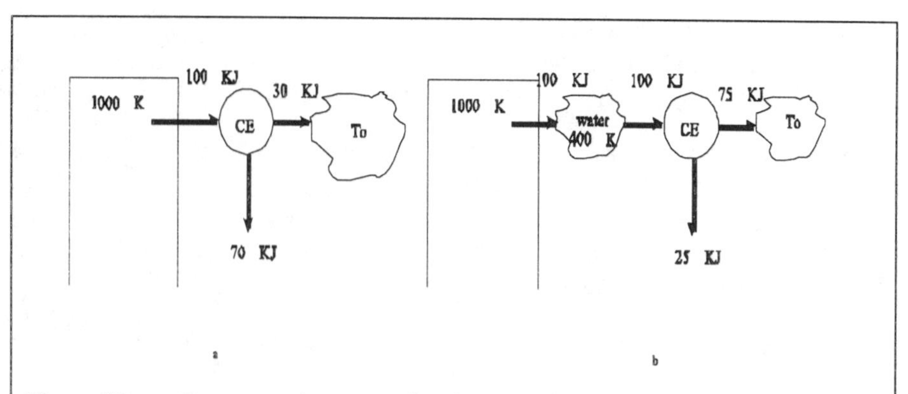

Figure 22: a. Carnot engine operating between hot gases and the ambient; b. Carnot engine operating between water and the ambient.

100×(1 – 298÷1000) ≈ 70 kJ, as illustrated in Figure 22a. Therefore, the quality of energy at 1000 K is 70%. The entropy change of the hot gases is –0.1 kJ K^{-1} (= –100 ÷ 1000), while the entropy gain for the ambient is 0.1 kJ K^{-1} (i.e., 30 kJ÷300 K).

Alternatively, we can cool the hot gases using water to transfer the 100 kJ of energy. Assume that during this process the water temperature rises by 2 K from 399 K to 401 K (with the average water temperature being 400 K, as shown in Figure 22b). If a Carnot engine is placed between the water at 400 K and the ambient at 300 K, then for the same 100 kJ of heat removed from radiator water, we can extract only 100×(1 – 298÷400) ≈ 25 kJ, and 75 kJ is rejected to the ambient. In this case, the energy quality is only 25% of the extracted heat. Figure 23 illustrates the processes depicted in Figure 22a and b using a T–S diagram. The cycle A–B–C–D–A in Figure 23 represents the Carnot engine (CE) of Figure 22a, while area ABJIA and EGKIE represent heat transfer from engine and to hot water, respectively, for Figure 22b, while the area C–D–I–H represents the rejected heat of CE for the first case, and the area D-C-H–K–J–I–D that for the latter case. Since more heat is rejected for the second case, the work potential or the quality of the thermal energy is degraded to a smaller value at the lower temperature. This is due to the irreversible heat transfer or the temperature gradients between hot gases and radiator water (as shown in Figure 22b). In general property gradients cause entropy generation.

Now, one might ask about the Maxwell-Boltzmann distribution of molecular velocities. Consider a monatomic gas within a container with rigid adiabatic walls. A "pseudo" temperature distribution exists for the monatomic gas. The question is whether with collision and transfer of energy, there can be degradation of energy or generation of entropy. First, temperature is a continuum property and the temperature cannot be associated with a group of molecules. Secondly, after frequent collisions, at that location where frequent transfers occur, the intensive state is not altered over a time period much larger than collision time. Thus, no gradient exists and there is no entropy generation.

a. Adiabatic Reversible Processes

Recall that for any process within a closed or fixed mass system, $dS = \delta Q/T_b + \delta\sigma$. For any reversible process $\delta\sigma = 0$ so that $dS = \delta Q/T$. For an adiabatic reversible process, $\delta Q = \delta\sigma = 0$, so that

$$dS = 0.$$

Consequently, the entropy remains unchanged for an adiabatic reversible process. These processes are also known as *isentropic* processes.

E. ENTROPY EVALUATION

The magnitude of heat transfer can be determined through measurements or by applying the First law. Thereupon, in the context of Eq. (28), if the entropy change is known, $\delta\sigma$ may be determined for a process. The entropy is a property that depends upon the system state and is evaluated at equilibrium.

Consider the irreversible process illustrated in Figure 24 involving the sudden compression of a gas contained in a piston–cylinder assembly with a large weight. The dashed curve in Figure 24 depicts the accompanying irreversible process. Applying the First law to the process we obtain

$$Q_{12} - W_{12} = U_2 - U_1.$$

The change in state due to an irreversible process can also be achieved through a sequence of quasiequilibrium processes as described by the path A in Figure 24. Applying the First law to this path, we obtain the relation

$$Q_{1-2R-2} - W_{1-2R-2} = U_2 - U_1.$$

Integrating Eq. (28) (with $\delta\sigma = 0$) along this path A ,

$$S_2 - S_1 = \int_1^2 \delta Q_R/T. \qquad (33)$$

since $\delta\sigma = 0$. The infinitesimal heat transfer δQ_R along the path A is obtained from the First law for a sequence of infinitesimal processes occurring along the reversible path 1–2R–2, i.e.,

$$\delta Q_R = dU + \delta W_R = dU + P\,dV.$$

Therefore, Eq. (33) may be written in the form

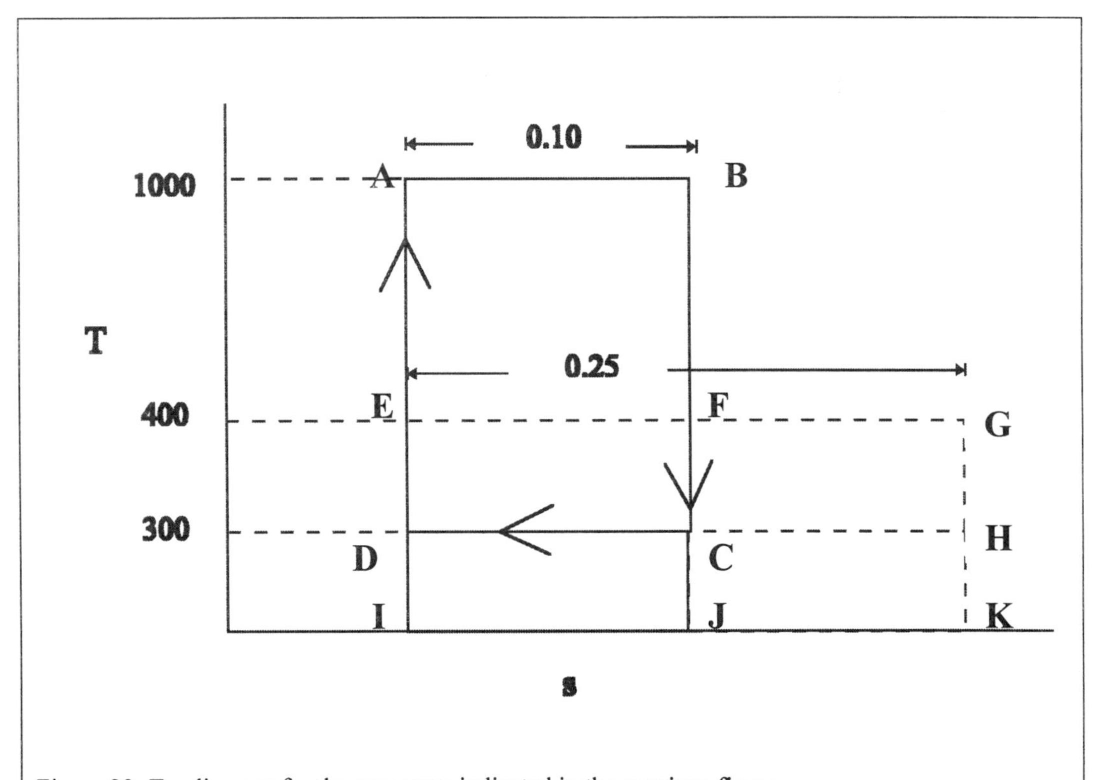

Figure 23: T-s diagram for the processes indicated in the previous figure.

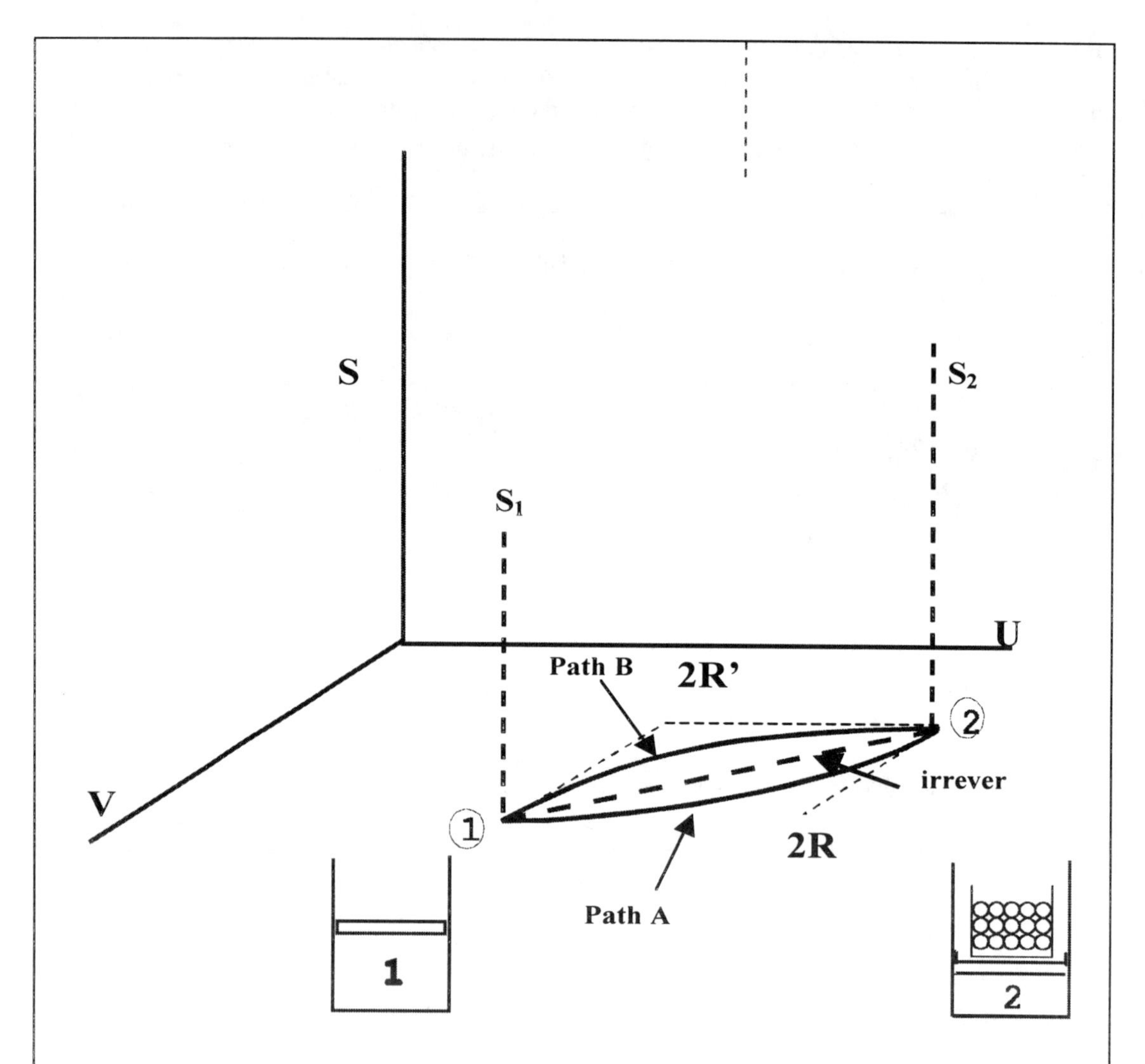

Figure 24: An irreversible process depicted on a U-V-S diagram. An illustration of estimating s by reversible path.

$$S_2 - S_1 = \int(dU + PdV)/T. \tag{34}$$

Integrating this relation between the initial and final equilibrium states

$$S_2 - S_1 = \int_{U_1}^{U_2} T^{-1}dU + \int_{V_1}^{V_2} PT^{-1}dU. \tag{35}$$

The values of pressure and temperature along the path 1–2R–2 in Eq.(35) are different from those along the dashed line 1–2, except at the initial (T_1, P_1) and final (T_2, P_2) states. If the state change is infinitesimal

$$dS = (dU/T) + (P/T)\, dV, \text{ or} \tag{36}$$

$$TdS = dU + PdV, \tag{37}$$

which is also known as the TdS relation. Equation (37) results from a combination of the First and Second laws applied to closed systems.

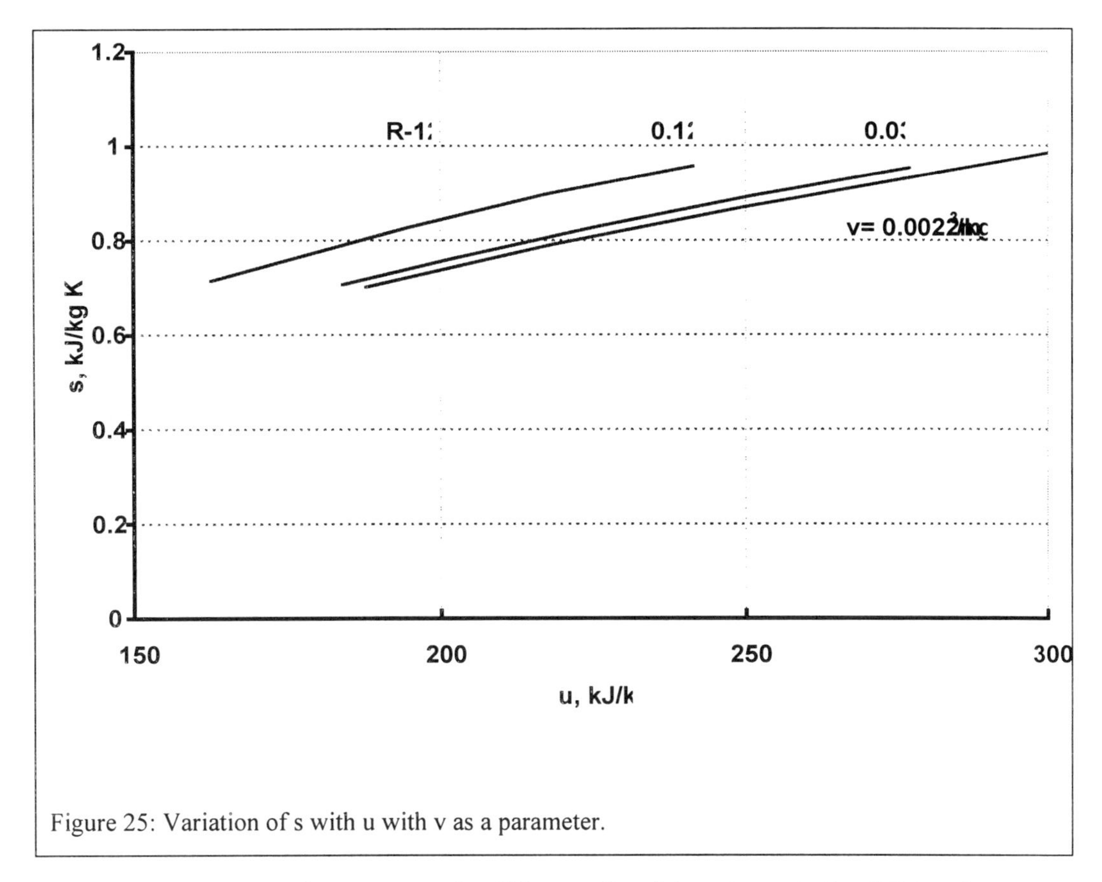

Figure 25: Variation of s with u with v as a parameter.

Since the entropy is a property, the difference ($S_2- S_1$) as shown in Eq. (35) is a function of only the initial (U_1,V_1) and final (U_2,V_2) states, i.e., for a closed system S = S(U,V). For example, if the initial and the final pressures and volumes are known, the temperature difference $T_2 - T_1$ can be determined using the ideal gas relation $T_2 = P_2 V_2/(mR)$ and $T_1 = P_1 V_1/(mR)$, even though the final state is reached irreversibly, i.e., the functional relation for $T_2 - T_1$ is unaffected. Likewise, to determine the final functional form for the difference (S_2–S_1), any reversible path A or B may be selected, since its value being path–independent depends only upon the initial and final states. (This is also apparent from Eq. (36) from which it follows that dS = 0 if dU = dV = 0.) For the processes being discussed, the internal energy change assumes the form

$$dU = T\,dS - P\,dV. \quad (38)$$

For an infinitesimal process, Eq. (38) represents the change of internal energy between two equilibrium states with the properties U and U+ dU, S and S +dS, and V and V+dV.

Recall from Chapter 1 that the higher the energy, the greater the number of ways by which molecules distribute energy. In confirmation, according to Eq.(36), as the internal energy increases in a fixed mass and volume system, the entropy too must increase. Therefore, the entropy is a monatomic function of the internal energy for a given volume and mass. The gradient of the entropy with respect to the internal energy is the inverse of the temperature T^{-1}. If the internal energy is fixed, Eq.(36) implies that as the volume increases, so does the entropy (which confirms the microscopic overview outlined in Chapter 1). This is to be expected, since more quantum states are available due to the increased intermolecular spacing.

Upon integrating Eq. (36), functional relation for the entropy is

$$S = S(U,V) + C, \text{ or } U = U(S,V) + C. \quad (39a \text{ and } b)$$

The latter is also known as the Gibbs fundamental relation for systems of fixed matter. If the composition of a system is known, it is possible to evaluate the constant C which is a function of the number of moles of the various species (N_1, N_2,..., etc.) or their masses (m_1, m_2,..., etc.) that are contained in the closed system of fixed total mass m. If the composition of the system is fixed, i.e., if the number of species moles N_1, N_2,..., etc. are fixed, then S = S(U,V) which is also known as the fundamental equation in entropy form.

On a unit mass basis Eq. (36) may be written in the form

$$ds = du/T + Pdv/T, \tag{40}$$

so that for a closed system of fixed mass

$$s = s\,(u,v). \tag{41}$$

Figure 25 contains an experimentally–determined relationship between s and u with v as a parameter for the refrigerant R–12.

Since dU = dH – d (PV), Eq. (36) assumes the form

$$dS = dH/T - VdP/T. \tag{42}$$

It is apparent from Eq. (42) that S = S(H,P). Writing Eq.(42) on unit mass basis

$$ds = dh/T - v\,dP/T, \text{ i.e.,} \tag{43}$$

$$s = s(h,P). \tag{44}$$

Note that only for exact differentials or differentials of properties can one give the functional relation like Eq. (44). On the other hand consider the example of electrical work supplied to a piston–cylinder–weight assembly resulting in gas expansion. In that case. the work

$$\delta W = P\,dV - E_{elec}\,\delta q_c, \tag{45}$$

where δq_c denotes the electrical charge and E_{elec} the voltage. It is not possible to express W = W(V,q_c), since δW is an inexact differential (so that W is not a point function).

1. Ideal Gases

Substituting for the enthalpy dh = c_{p0} (T) dT, Eq. (43) may be written in the form

$$ds = c_{po}\,dT/T - R\,dP/P. \tag{46}$$

a. *Constant Specific Heats*

Integrating Eq. (46) from (T_{ref},P_{ref}) to (T,P)

$$s(T,P) - s(T_{ref},P_{ref}) = c_{po} \ln(T/T_{ref}) - R \ln(P/P_{ref}). \tag{47a}$$

Selecting P_{ref} = 1 atmosphere and letting s(T_{ref},1) = 0, we have

$$s(T,P) = c_{p0} \ln(T/T_{ref}) - R \ln(P(atm)/1(atm)). \tag{47b}$$

Selecting an arbitrary value for T_{ref}, and applying Eq. (47b) at states 1 and 2,

$$s(T_2,P_2) - s(T_1,P_1) = c_{po} \ln(T_2/T_1) - R \ln(P_2/P_1). \tag{47c}$$

For an isentropic process $s_2 = s_1$. Consequently,

$$c_{po} \ln(T_2/T_1) = R \ln(P_2/P_1). \tag{47d}$$

Since $R = c_{po} - c_{vo}$, applying Eq. (47d)

$$T_2/T_1 = (P_2/P_1)^{k/(k-1)}, \text{ or } P_2/P_1 = (T_2/T_1)^{(k-1)/k}, \tag{47e and f}$$

where $k = c_{p0}/c_{v0}$. Finally, upon substituting for T = Pv/R in Eq. (47e), we obtain the relation

$$Pv^k = \text{Constant}. \quad (47g)$$

b. Variable Specific Heats

Consider an ideal gas that changes state from (T_{ref},P_{ref}) to (T,P). Integrating Eq. (46) and setting $s(T_{ref},P_{ref}) = 0$ we have

$$s(T,P) = \int_{T_{ref}}^{T} (c_{p0}(T)/T)dT - R \ln (P/P_{ref}).$$

For ideal gases, the first term on the right is a function of temperature alone. Setting $P_{ref} = 1$ atm, the entropy

$$s(T,P) = s^0(T) - R \ln (P(atm)/1(atm)), \text{ where} \quad (48a)$$

$$s^0(T) = \int_{T_{ref}}^{T} (c_{p0}(T)/T)dT. \quad (48b)$$

If data for the specific heat $c_{p0}(T)$ are available (Tables A-6F), Eq. (48b) can be readily integrated. In general, tables listing $s^0(T)$ assume that $T_{ref} = 0$ K(Tables A-7 toA-19). Therefore,

$$s(T,P) = s^0(T) - R \ln (P/1), \quad (49)$$

where the pressure is expressed in units of atm or bars. If P = 1 atm or approximately bar, $s(T, 1) = s^o(T)$ which is the entropy of an ideal gas at a pressure of 1 bar and a temperature T. The second term on the RHS of Eq. (49) is a pressure correction. Applying Eq. (49) to states 1 and 2,

$$s(T_2,P_2) - s(T_1, P_1) = s^0(T_2) - s^0(T_1) - R \ln (P_2/P_1). \quad (50a)$$

For an isentropic process

$$s^0(T_2) - s^0(T_1) - R \ln (P_2/P_1) = 0. \quad (50b)$$

Therefore, for an isentropic process, if the initial and final pressures, and T_1 are specified, $s^0(T_2)$ can be evaluated. Using the appropriate tables for $s^0(T)$ (e.g.: Tables A-7 to A-19), T_2 may be determined.

For processes for which the volume change ratios are known, it is useful to replace the pressure term in Eq. (50b) using ideal gas law:

$$s^0(T_2) - s^0(T_1) - R \ln ((RT_2/v_2)/(RT_1/v_1)) = 0.$$

Simplifying this relation,

$$s^0(T_2) - s^0(T_1) - R \ln (T_2/T_1) + R \ln (v_2/v_1) = 0. \quad (50c)$$

For a known volume ratio and temperature T_1, Eq. (50c) may be used to solve for T_2 iteratively. In order to avoid the iterative procedure, relative pressures and volumes, P_r and v_r, may be defined using Eq. (50b) as follows (further details are contained in the Appendix to this chapter)

$$P_r(T) = \exp(s^0(T)/R)/\exp(s^0(T_{ref}')/R), \text{ and} \quad (50d)$$

$$v_r = (T/\exp(s^0(T)/R))/(T_{ref}'/\exp(s^0(T_{ref}')/R)), \quad (50e)$$

where T_{ref}' is an arbitrarily defined reference temperature. For air, T_{ref}' is taken to be 273 K, and

$$P_r = 0.00368 \exp(s^0(T)/R)$$

Equations (50b) and (50c) can also be written in the form

$$P_2/P_1 = P_{r2}/P_{r1}, \text{ and} \quad (50f)$$

$$v_2/v_1 = v_{r2}/v_{r1}. \quad (50g)$$

The value of v_r in SI units is based on the relation

$$v_r = 2.87\ T/P_r$$

Tabulations for P_r and v_r particularly for solution of isentropic problems were necessary in the past due to the nonavailability of computers. Since their advent, the system properties at the end of isentropic compression or expansion are readily calculated.

The isentropic and nonisentropic processes can now be explained as follows. Consider a monatomic gas. When an adiabatic reversible compression process occurs in a closed system the work input is converted into a translational energy increase (e.g., due to increased molecular velocity ($V_x^2 + V_y^{\ 2} + V_z^2$) because of a force being applied in a specific direction, say "x" which increases V_x). Thus, the total number of macro-states cannot change. A crude way to interpret is that dS = dU/T + P dV/T so that S generally increases with increased energy U but decreases due to a decrease in volume V. The entropy first increases due to increased U because of work input (the first term on the RHS) but decreases due to the reduced volume (as the second term, due to the intermolecular spacing, is reduced and, consequently, the number of states in which energy can be stored also decreases). The second term counteracts the entropy rise due to the increased internal energy, and the entropy is unchanged.

---

l. Example 12

Air is adiabatically and reversibly compressed from P_1 = 1 bar, and T_1 = 300 K to P_2 = 10 bar. Heat is then added at constant volume from a reservoir at 1000 K (T_R) until the air temperature reaches 900 K (T_3). During heat addition, about 10% of the added heat is lost to the ambient at 298 K. Determine:

The entropy generated σ_{12} in kJ kg^{-1} K^{-1} for the first process 1–2;

The net heat added to the matter;

The heat supplied by the reservoir;

The entropy generated in an isolated system during the process from (2) to (3).

Solution

$$S_2 - S_1 - \int \delta Q/T_b = \sigma_{12}. \quad \text{(A)}$$

Since the process is reversible,

$$\sigma_{12} = 0, \quad \text{(B)}$$

which implies that no gradients exist within the system. Therefore,

$$T_b = T. \quad \text{(C)}$$

Using Eqs. (A), (B), and (C)

$$S_2 - S_1 = \int \delta Q/T. \quad \text{(D)}$$

Since the process is adiabatic $\delta Q = 0$, and $S_2 = S_1$ or $s_2 = s_1$. At state 1, from the air tables (Tables A-7), p_{r1} = 1.386, u_1 = 214.07, and h_1 = 300.19. Therefore,

$$s_1 = s^0(T_1) - R \ln P/1 = 1.702 - 0 = 1.702 \text{ kJ kg}^{-1} \text{ K}^{-1}.$$

For the isentropic process

$$p_{r2}(T_2)/p_{r1}(T_1) = p_2/p_1 = 10. \text{ Hence, } p_{r2} = p_{r1}\ 10 = 1.386\times10 = 13.86 \text{ so that}$$

$$T_2 = 574 \text{ K}, u_2 = 415 \text{ kJ kg}^{-1}, h_2 = 580 \text{ kJ kg}^{-1}, \text{ and } s_2 = s_1 = 1.702 \text{ kJ kg}^{-1} \text{ K}^{-1}.$$

Temperature gradients can develop inside a system during heat addition from a thermal reservoir or heat loss to the ambient, thereby making a process internally irreversible. In this example, the final states are assumed to be at equilibrium. Applying the First law to the constant volume process, the heat added to the system can be evaluated as follows

$$q_{23} = u_3 - u_2 = 674.58 - 415 = 260 \text{ kJ kg}^{-1}.$$

If q_R denotes the heat supplied by reservoir, the heat added $q_{23} = 0.9\ q_R$, i.e.

$q_R = 288.88$ kJ kg^{-1}.

The heat loss to the ambient is $q_0 = 288.88 - 260 = 28.88$ kJ kg^{-1}.

Since we must determine the entropy of an isolated system, assuming that there are no gradients outside that system, and selecting the system boundaries to include the reservoir at T_R and the ambient at T_0, it follows that

$$s_3 - s_2 - q_R/T_R - q_0/T_0 = \sigma.$$

Now, $P_3/P_2 = T_3/T_2 = 900 \div 574$, i.e., $P_3 = 15.68$ atm, and $s_3 = 2.849 - 0.287 \ln (15.68 \div 1) = 2.059$ kJ kg^{-1} K^{-1}. Therefore,

$$2.059 - 1.702 - (289/1000) - (-29/298) = \sigma \text{ so that}$$

$$\sigma = 0.165 \text{ kJ kg}^{-1} \text{ K}^{-1}.$$

Remarks

It is possible to tabulate p_r values for a particular gas using Eq. (50c).

2. Incompressible Liquids

For incompressible liquids and solids, the specific volume v is constant. Since u = u(T,v), for incompressible substances it follows that u = u(T). The intermolecular spacing in incompressible liquids is constant and, consequently, the intermolecular potential energy is fixed so that the internal energy varies only as a function of temperature. Since, h = u + Pv, for incompressible liquids

$$h(T,P) = u(T) + P\,v.$$

Differentiating with respect to the temperature at fixed pressure,

$$c_P = (\partial h/\partial T)_P = (\partial u/\partial T)_P.$$

Since u = u(T), it follows that

$$c_P = (\partial u/\partial T)_P = du/dT = c_v = c,$$

and for incompressible substances

$$du = c dT. \tag{51}$$

The values of c for liquids and solids are tabulated in Tables A-6A and A-6B. Using Eq. (40), ds = du/T + 0, so that

$$ds = c dT/T. \tag{52}$$

Therefore, the entropy is a function of temperature alone. For any substance s = s(T,v) so that if v = constant, s = s(T). Note that Eq. (43) cannot be used since h = h(T,P). Equations (51) and (52) are applied to evaluate the internal energy and entropy of compressed liquids.

For example, water at 25°C and 1 bar exists as compressed liquid, since $P > P^{sat}(25°C)$. The Steam tables (Tables A-4A) tabulate values of u(T) and s(T) as a function of temperature for saturated water. If the entropy of liquid water is desired at 25°C and 1 bar, since $u(T,P) \approx u(T,P^{sat}) = u_f(T)$, and $s(T,P) \approx s(T,P^{sat}) = s_f(T)$, the respective tabulated values are 104.9 kJ kg^{-1} K^{-1} and 0.367 kJ kg^{-1} K^{-1}. Likewise, the enthalpy at that state is

$$h = u + P\,v = 104.9 + 1{\times}100{\times}0.001 = 105 \text{ kJ kg}^{-1}.$$

An incompressible substance with constant specific heat is also called a perfect incompressible substance. For these substances, integration of Eq. (52) between T and reference temperature T_{ref} yields

$$s - s_{ref} = c \ln(T/T_{ref}), \tag{53a}$$

or between two given states

$$s_2 - s_1 = c \ln(T_2/T_1). \tag{53b}$$

When an incompressible liquid undergoes an isentropic process, it follows from Eq. (53b) that the process is isothermal.

3. Solids

Equation (53), which presumes constant specific heat, is also the relevant entropy equation for incompressible solids. However as $T \to 0$, Eq. (53) becomes implausible, forcing us to account for the variation of the specific heat of solids at very low temperatures. At these temperatures

$$c_v(T) = 3 R(1 - (1/20)(\theta_D/T)^2), \text{ where } T \gg (\theta_D = 3 R (4 \pi^4/5) (T/\theta_D)^3), \tag{54}$$

where θ_D is known as the Debye temperature. A solid that behaves according to Eq. (54) is called a Debye solid.

Another pertinent relation is the Dulong–Petit law that states that

$$c_v \approx 3 R.$$

This is based on the presumption that a mole of a substance contains N_{avag} independent oscillators vibrating in three directions, with each molecule contributing an amount $(3/2)k_BT$ to the energy. Molecules contribute an equal amount of potential energy, i.e., $(3/2)k_BT$. At low temperatures, the Dulong–Petit constant specific heat expression leads to erroneous results, and a correction is made using the Einstein function $E(T_{Ein}/T)$, i.e.,

$$c_v \approx 3 R\, E(T_{Ein}/T), \text{ where}$$

$$E(T_{Ein}/T) = (T_{Ein}/T)^2 \exp(T_{Ein}/T)/(\exp(T_{Ein}/T)-1)^2.$$

Here, T_{Ein} denotes the Einstein temperature. (For many solids, $T_{Ein} \approx 200$ K.) As $T \to 0$, $E(T_{Ein}/T) \to T^2$. For coals,

$$c_v \approx 3 R((1/3)E(T_{Ein,1}/T) + (2/3)\, E\, (T_{Ein,2}/T)),$$

where $T_{Ein,2}$ denotes the second Einstein temperature.

4. Entropy During Phase Change

Consider the case of a boiling liquid. Since the pressure and temperature are generally unchanged during a phase transformation, applying Eq. (43),

$$ds = dh/T - vdP/T = dh/T. \tag{55}$$

Integrating the expression between the saturated liquid and vapor states

$$s_g - s_f = (h_g - h_f)/T = h_{fg}/T. \tag{56}$$

Generalizing for any change from phase α to β,

$$s_\alpha - s_\beta = h_{\alpha\beta}/T \tag{57}$$

---

m. Example 13

The entropy of water at T_{tp} = 0°C , P_{TP} = 0.611 kPa, is arbitrarily set to equal zero, where the subscript tp refers to the triple point. Using this information, determine:
s(liquid, 100°C) assuming c = 4.184 kJ kg^{-1} K^{-1}. Compare your results with values tabulated in the Steam tables (Tables A-4).
s(sat vapor, 0°C, 0.611 kPa) assuming h_{fg} = 2501.3 kJ^{-1} kg^{-1} K^{-1}.
The entropy generated if the water at 0°C and 0.611 kPa is mechanically stirred to form vapor at 0°C in an adiabatic blender.

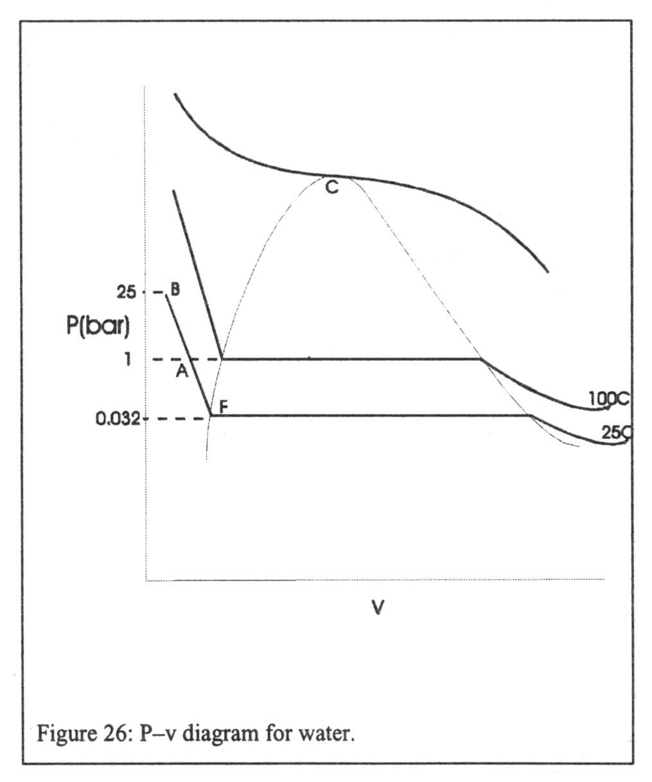

Figure 26: P–v diagram for water.

s(393 K, 100 kPa) assuming $c_{p,0}$ = 2.02 kJ kg^{-1} K^{-1} and that steam behaves as an ideal gas.

Solution

Applying Eq. (53),

$$s(373) - s(273) = 4.184 \ln (373/273) = 1.306 \text{ kJ kg}^{-1} \text{ K}^{-1}.$$

Since s(0°C) = 0, s(100°C) =1.306 kJ kg^{-1} K^{-1}.

From the Table A-4A, s(100°C) = 1.3069 kJ kg^{-1} K^{-1}, which is very close.

Applying Eq. (56) to the vaporization process at the triple point,

$s_g - s_f = 2501.3 \div 273 = 9.16$ kJ kg^{-1} K^{-1}. Since,

$$s_f (273 \text{ K}, 0.611 \text{ kPa}) = 0,\ s_g (273 \text{ K}, 0.611 \text{ kPa}) = 9.16 \text{ kJ kg}^{-1} \text{ K}^{-1}. \quad \text{(A)}$$

$ds - \delta q/T_b = \delta\sigma$. Since $\delta q = 0$, $ds = \delta\sigma$. Integrating this expression,

$$s_g - s_f = \sigma. \quad \text{(B)}$$

Using Eq. (A) and (49b), with s_f (0°C, 0.611 kPa) = 0

$$s_g - s_f = 9.16 \text{ kJ kg}^{-1} \text{ K}^{-1} = \sigma. \quad \text{(C)}$$

s(393 K, 100 kPa) – s(273 K, 0.611) =

$$2.02 \ln (393 \div 273) - (8.314 \div 18.02) \ln (100 \div 0.611) = -1.616 \text{ kJ kg}^{-1} \text{ K}^{-1}, \text{ or}$$

$$s(393, 100 \text{ kPa}) = 9.16 - 1.616 = 7.54 \text{ kJ kg}^{-1} \text{ K}^{-1}.$$

Conventional Steam tables (e.g., Table A-4A) yield a value of 7.467 kJ kg^{-1} K^{-1}.

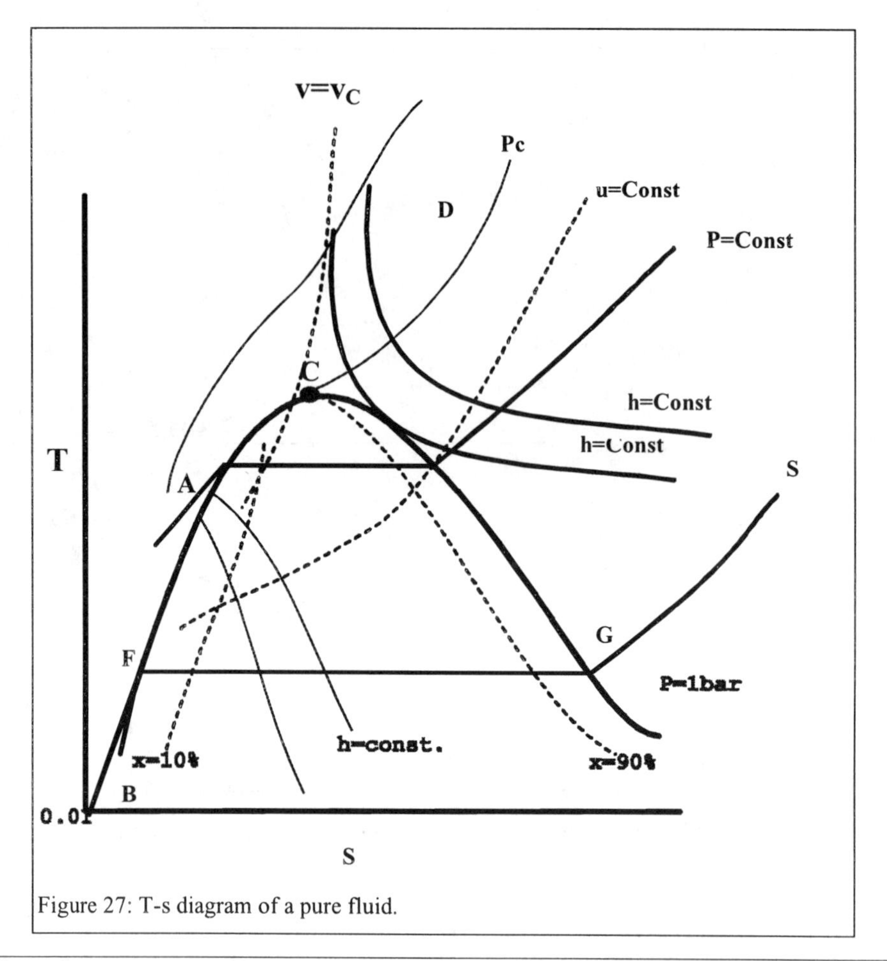

Figure 27: T-s diagram of a pure fluid.

Remarks

For estimating entropy in vapor phase at low pressures, one can use ideal gas tables (Tables A-12) also, e.g., s(393,100) – s (273,0.611)= (s^0 (393) – R ln (100/100)) – (s^0 (273) – R ln (0.611/100)) where P_{ref} = 100 kPa

The stirring process is irreversible. Therefore, viscous dissipation converts mechanical energy into thermal energy. The heat vaporizes the liquid, and increases the entropy.

---

n. *Example 14*

Determine the enthalpy of water at 25°C and 1 bar (i.e., at point A of Figure 26) if the enthalpy of saturated liquid (at point F) at that temperature is known.

Solution

From the Steam tables (A-4A) P^{sat} = 0.03169 bar and h_f = 104.89 kJ kg^{-1} for saturated liquid water at 25°C (at point F). Since c = constant for incompressible liquids,

$$ds = c\ dT/T \qquad (A)$$

Along the 25°C isotherm (curve FA), Eq. (A) illustrates that ds = 0, and the process is isentropic. Since, dh = Tds + vdP, for this case

$$dh = v\ dP.$$

Upon integrating between points F and A,

h_A (25 C, 1 bar) – h_F (25°C, 0.03169 bar) = $\int v\, dP \approx v_f$ (25°C, 0.03169) (P_A – P_F) = $1.0029\times10^{-3} \times (100 - 3.169) = 0.09711$ kJ kg^{-1}, and

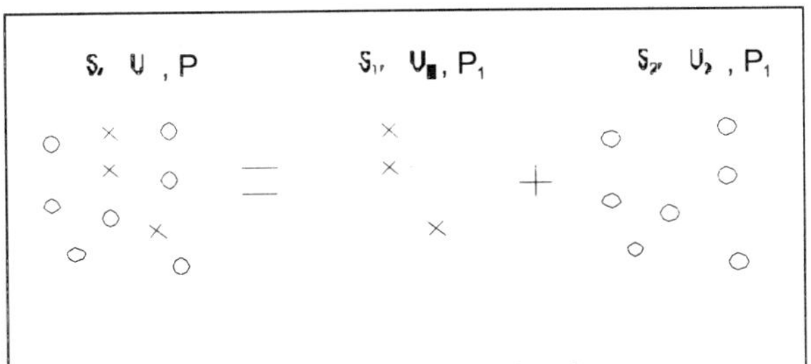

h_A (25 C, 1 bar) = 104.89 + 00.09711= 104.987 kJ kg^{-1}.

Figure 28: Illustration of the Gibbs-Dalton law.

Remarks

Since the enthalpy values are virtually insensitive to pressure, one can assume that $h_A \approx h_F$, i.e., the enthalpy of a compressed liquid at given temperature and pressure is approximately that of the saturated liquid at that temperature.

If pressure at point A is 25 bar, h_A = 104.89 + 2.504 = 107.394 kJ kg^{-1}. Use of Table A-4 yields a value of 107.2 kJ kg^{-1}, which is very close, with the difference being due to the assumption of constant specific volume.

a. T–s Diagram

We are now in a position to discuss the representation of the states of a pure fluid on a T–s diagram. For instance, we may arbitrarily assign a zero entropy to liquid water at its triple point (i.e., point B of Figure 27). For incompressible liquids, we can assume that s(0.01°C, 1 bar) ≈ s(0.01°C,0.006 bar) = 0. If the water is again heated from 0.01°C to 100 C at 1 bar, it is possible to evaluate the values of s, and those of s_f(100°C, 1 bar), (point F) using Eq. (53), and s_g (100°C, 1 bar) (point G) using Eq. (56). If the vapor behaves as an ideal gas (which is generally true at lower pressures) the entropy may be evaluated using either of Eqs. (47c) or (50a) (Point S) . In this manner, the behavior of a substance can be characterized at lower pressures on the T–s diagram, as illustrated by the curve BFGS in Figure 27 at 1 bar. Thereafter, by changing the pressure, entropy values can be obtained at higher pressures. Since the ideal gas assumption is flawed at elevated pressures, Eqs. (47c) or (49) must be modified. This will be discussed further in Chapters 6 and 7. As is apparent from the path A–C–D, an inflection occurs in the slope of the isobar (at the critical pressure P_c) at the critical point C, i.e., $(\partial T/\partial s)_{Pc}$ = 0 at this point. Also illustrated on the diagram are isometric, isenthalpic and isoquality lines.

5. Entropy of a Mixture of Ideal Gases

a. Gibbs–Dalton's law

The application of the Gibbs–Dalton law to characterize a multicomponent gaseous mixture is illustrated in Figure 28. Two components species are hypothetically separated, and the component pressures P_1 and P_2 are obtained. Thereby, the component pressure P_k is determined as though component k alone occupies the entire volume (i.e., no other components are present) at the mixture temperature. Thereafter, using the component pressures, the entropy is evaluated, i.e.,

$$S\,(T,P,N) = \Sigma S_k\,(T,p_k,N_k) = N_k\,\bar{s}_k(T,\,p_k). \qquad (58)$$

For ideal gases, the component pressure for a species is identical to its partial pressure, namely,

$$p_k' = X_k P = p_k, \qquad (59)$$

where p_k' denotes the partial pressure of species k in the mixture. For ideal gases $p_k' = p_k$. This subject is discussed in greater detail in Chapter 8 on mixtures.

b. Reversible Path Method

A general method to determine the mixture entropy using the relation $dS = \delta Q_{rev}/T$ is derived in the Appendix.

---

o. Example 15

A piston–cylinder assembly contains a 0.1 kmole mixture consisting of 40% CO_2 and 60% N_2 at 10 bars and 1000 K (state 1). The mixture is heated to 11 bars and 1200 K (state 2). The work output from the assembly is 65.3 kJ. Evaluate the entropy change $S_2–S_1$ and σ_{12} for the following cases:

The boundary temperature T_b equals that of the gas mixture.

T_b is fixed and equals 1300 K during heat up.

Solution

$$S = \sum N_k \bar{s}_{k(T,p)} \quad \text{(A)}$$

For the mixture

$$S = N_{CO_2}\bar{s}_{CO_2}(T, P_{CO_2}) + N_{N_2}\bar{s}_{N_2}(T, P_{N_2}), \quad \text{(B)}$$

$$S_1 = (N_{CO_2}\bar{s}_{CO_2}(T, P_{CO_2}) + N_{N_2}\bar{s}_{N_2}(T, P_{N_2}))_1, \quad \text{(C)}$$

$$S_2 = (N_{CO_2}\bar{s}_{CO_2}(T, P_{CO_2}) + N_{N_2}\bar{s}_{N_2}(T, P_{N_2}))_2, \text{where} \quad \text{(D)}$$

$N_{CO_2} = 0.4 \times 0.1 = 0.04$ kmole, and $N_{N_2} = 0.6 \times 0.1 = 0.06$ kmole.

Now, $\bar{s}_{CO_2}(T,P_k) = \bar{s}^0_{CO_2}(T) - \bar{R}\ln(P_{CO_2}/1)$, where

$(P_{CO_2})_1 = 0.4 \times 10 = 4$ bar, $(P_{CO_2})_2 = 0.4 \times 11 = 4.4$ bar, and

$(P_{N_2})_1 = 0.6 \times 10 = 6$ bar, $(P_{N_2})_2 = 0.6 \times 11 = 6.6$ bar.

Therefore, at conditions 1 and 2, respectively,

$\bar{s}_{CO_2}(1200\text{K}, 4.4 \text{ bar}) = \bar{s}^0_{CO_2}(1200 \text{ K}) - \bar{R}\ln(4.4 \div 1)$

$= 234.1 - 8.314 \times \ln(4.4 \div 1) = 221.8$ kJ kmole^{-1} K^{-1}, and,

$\bar{s}_{CO_2}(1000\text{K}, 4 \text{ bar}) = \bar{s}^0_{CO_2}(1000 \text{ K}) - \bar{R}\ln(4 \div 1) = 216.6$ kJ kmole^{-1} K^{-1}.

Likewise,

$\bar{s}_{N_2}(1200\text{K}, 6.6 \text{ bar}) = \bar{s}^0_{N_2}(1200 \text{ K}) - \bar{R}\ln(6.6 \div 1)$

$= 279.3 - 8.314 \times \ln(6.6 \div 1) = 263.6$ kJ kmole^{-1} K^{-1}, and

$\bar{s}_{N_2}(1000\text{K}, 6 \text{ bar}) = 269.2 - 8.314 \times \ln 6 = 254.3$ kJ kmole^{-1} K^{-1}.

Using Eqs. (C) and (D)

$S_1 = 0.04 \times 216.6 + 0.06 \times 254.3 = 23.92$ kJ K^{-1},

$S_2 = 0.04 \times 221.8 + 0.06 \times 263.6 = 24.69$ kJ K^{-1}, and

$S_2 - S_1 = 24.69 - 23.92 = 0.77$ kJ K^{-1}.

$$S_2 - S_1 - Q_{12}/T_b = \sigma_{12}. \quad \text{(E)}$$

Applying the First law, $Q_{12} = U_2 - U_1 + W_{12}$. Therefore,

$U_2 = 0.04 \times 43871 + 0.06 \times 26799 = 3362.8$ kJ,

$U_1 = 0.04 \times 34455 + 0.06 \times 21815 = 2687.1$ kJ, and

$Q_{12} = 3362.8 - 2687.1 + 65.3 = 741$ kJ.

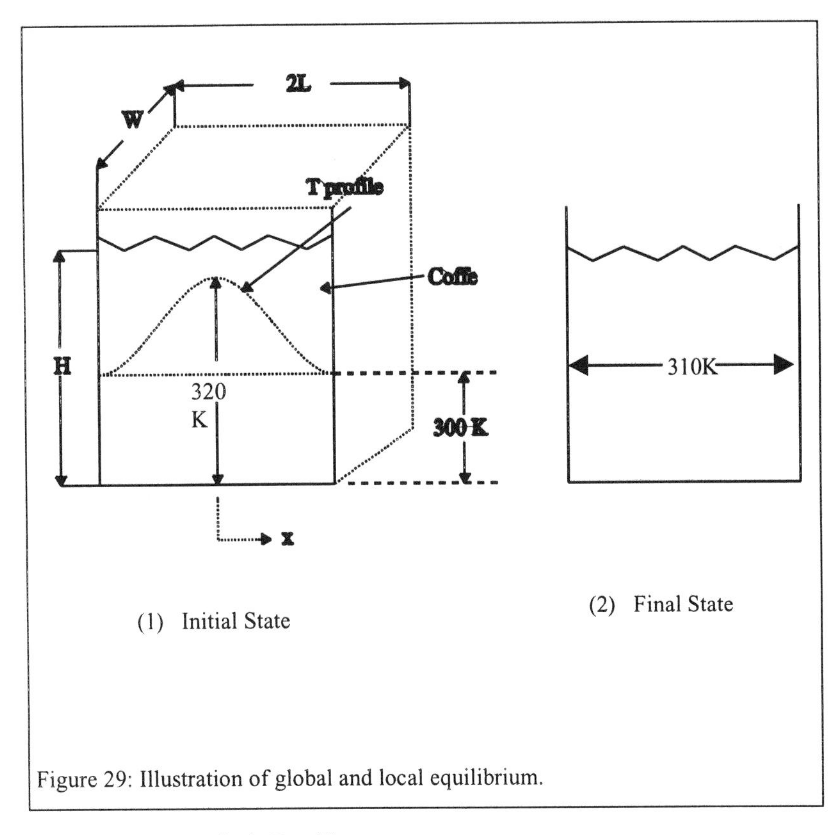

Figure 29: Illustration of global and local equilibrium.

Using these results in Eq. (E)

$$\sigma_{12} = 0.77 - 741/1300 = 0.2 \text{ kJ K}^{-1}.$$

F. LOCAL AND GLOBAL EQUILIBRIUM

A system exists in a state of thermodynamic equilibrium if no changes occur within the system in the absence of any interactions (of mass or energy). The entropy cannot be evaluated for a system that contains internal temperature gradients. However, it is possible to determine the entropy for a small elemental mass with the assumption of local equilibrium. Summing the local entropy over all the elemental masses contained in a system, the system entropy can be determined. However, the concept of intensive system entropy is meaningless in this case.

---

p. Example 16

Consider a container of length 2L, width W, and height H that is filled with water. Due to cooling, at a specified time t, the water temperature at the center of the container is 320 K, while that adjacent to the walls is 300 K (cf. Figure 29). The initial temperature profile follows the relation

$$T = T_{max} - (T_{max} - T_0)\, x/L \qquad (A)$$

Assuming local equilibrium initially, obtain an expression for S_1 at time t.

The pool is now insulated and the entire pool is allowed to reach equilibrium. What are the final system and specific entropies? Assume the water mass to equal 1000 kg. What is the entropy generated for the above process?

Solution

Assuming local equilibrium for an elemental mass $dm = W\ H\ dx\ \rho$,

$$S_1 = 2\int_{x=0}^{x=L} (c \ln (T/T_{ref}))\ (W\ H\ dx\ \rho). \quad \text{(B)}$$

Employing Eqs. (A) and (B), with $T_{ref} = T_0$ we have

$$S_1/(2L\ W\ H\ \rho) = c(T_0/(T_{max}-T_0))((T_{max}/T_0) \ln(T_{max}/T_0)-T_{max}/T_0+1). \quad \text{(C)}$$

$S_1/(2LWHP)$=0.1365 kJ/kgk

Noting that 2LWHP is the pool mass of 1000 kg.

Therefore, S_1 = 136.5 kJ K^{-1}.

Therefore, S_1 = 1159 kJ K^{-1}.
At the final state $S_2/(2\ L\ W\ H\ \rho) = c \ln (T_2/T_0)$. Using values for c = 4.184 kJ kg^{-1} K^{-1}, T_{max} = 320 K, T_0 = 300 K, mass m = 2LWHP= 1000 kg, and using a linear temperature profile that yields T_2 = 400 K we have,

$$S_2 = 137.2 \text{ kJ K}^{-1}, \text{ and } s_2 = 0.1372 \text{ kJ kg}^{-1} \text{ K}^{-1}.$$

Therefore, $(S_2 - S_1)$ =0.7 kJ K^{-1}.

Although the internal energy, volume, and mass remain unchanged, the entropy changes during the irreversible equilibration process.

Since the process is adiabatic, $s = S_2 - S_1$ = 0.7 kJ K^{-1}.

Remarks

Assume that the universe was formed from a highly condensed energy state during a big bang that resulted in temperature gradients (e.g., formed by the temperatures at the surface of the sun and the earth). With this description the universe is currently in the process of approaching an equilibrium state, and, consequently, its entropy is continually increasing. Once the equilibrium state is reached no gradients will exist within the universe, and the entropy will reach a maximum value.

G. SINGLE–COMPONENT INCOMPRESSIBLE FLUIDS

For incompressible fluids the internal energy and entropy may be written in the forms $u = c\ (T-T_{ref})$, and $s = c \ln(T/T_{ref})$. Manipulating the two relations

$$s = c \ln ((u/c + T_{ref})/T_{ref}), \text{ i.e.,} \quad \text{(60a)}$$

$s = s(u)$. This relation is called the entropy fundamental equation for an incompressible single–component fluid. Likewise, expressing the internal energy as a function of the entropy,

$$u = c\ T_{ref} (\exp (s/c) - 1), \text{ i.e.,} \quad \text{(60b)}$$

it follows that $u = u(s)$. This relation is called the energy fundamental equation for an incompressible single–component fluid. Similarly, the enthalpy fundamental equation may be expressed in the form

$$h = u + Pv_{ref} = c\ T_{ref}\ (\exp(s/c) -1) + Pv_{ref} = h\ (s,P).$$

This is discussed further in Chapter 7.

q. *Example 17*

Consider a vapor–liquid mixture in a closed system that is adiabatically and quasistatically compressed. Is the process isentropic at low pressures? Assume that the mixture quality does not change significantly (cf. Figure 30).

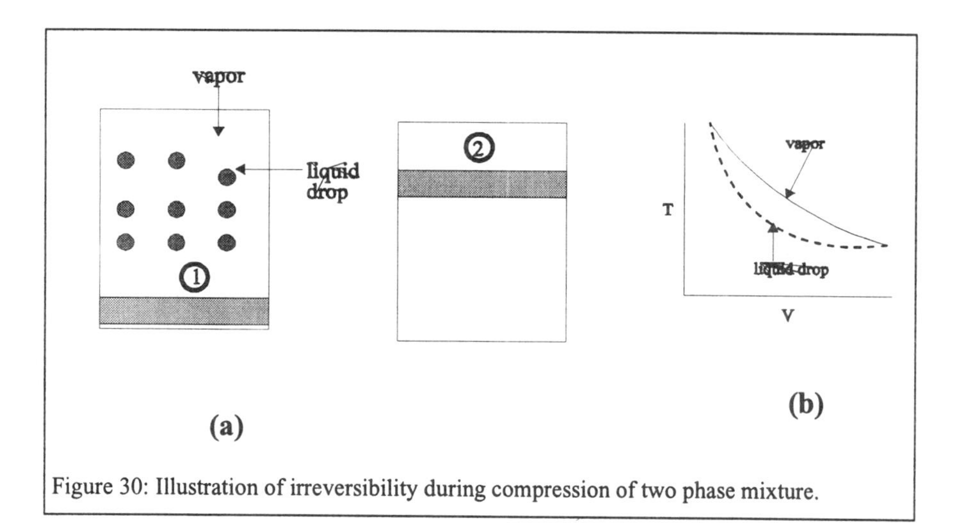

Figure 30: Illustration of irreversibility during compression of two phase mixture.

Solution

During the process the vapor is more readily compressed, which, in turn, compresses the liquid droplets. If the process is to be isentropic, there should be no temperature difference between the vapor and liquid drops. However isentropic compression of incompressible drops cannot create a temperature rise, while it can do so for vapor. Thus the vapor must heat the drops. Therefore, even though the process is quasistatic, it is not a quasiequilibrium process, since internal temperature gradients exist during compression, which cause irreversible heat transfer between the vapor and liquid drops. Applying the First law,

$$-P\,(dV_v + dV_l) = dU_v + dU_l, \tag{A}$$

where the subscripts v and l, respectively, denote vapor and liquid. Upon compression, the increased vapor temperature causes the liquid drops to heat up, and it is usual that the liquid temperature lags behind the vapor temperature. Finally, the system equilibrates so that $T_{v,2} = T_2 = T_{l,2}$. In order to simplify the problem, we assume that there is no vaporization during compression, i.e., first the vapor is compressed as though the drops are insulated from it. Since the liquid has a specific volume of 0.001 $m^3\ kg^{-1}$ while the vapor specific volume is of the order of 1 $m^3\ kg^{-1}$ we can neglect the small change in drop volume. Assuming ideal gas behavior for the vapor, we can show using Eq. (A) (or assuming an isentropic process for the vapor) that

$$T_{v,2}/T_{v,1} = (V_1/V_2)^{(k-1)}. \tag{B}$$

Following compression the liquid and vapor reach the equilibrium temperature T_2 without any change in their respective volumes. Applying the First law,

$$m_v\, c_{v0}\, (T_{v2} - T_2) = m_l\, c_l\, (T_2 - T_1). \tag{C}$$

Solving for T_2/T_1

$$T_2/T_1 = (x\, c_{v0}\, (T_{v,2}/T_{v,1}) + (1-x)\, c_l)/(x\, c_{v0} + (1-x)\, c_l) \tag{D}$$

where $x = m_v/(m_l + m_v)$ denotes the mixture quality. Applying the entropy balance equation $S_2 - S_1 - \int\delta Q/T_b = \sigma$,

$$s_2 - s_1 = \sigma/m = x\,(c_{v0} \ln (T_2/T_1 + R \ln (V_2/V_1)) + (1-x)\, c_l \ln (T_2/T_1) \tag{E}$$

Using values for the compression ratio $V_1/V_2 = 2$, $c_{v0} = 1.5$, $c_l = 4.184$, and $R = 0.46$, a plot of σ with respect to x with r_v as a parameter can be generated (Figure 31). When x = 1, the mixture is entirely vapor, and the process is reversible. When x = 0, the mixture only contains liquid, and the process is again reversible. The entropy generation term σ reaches a maxima (which can be found by differentiating Eq. (E) with respect to x, and, subsequently, setting $d\sigma/dx = 0$) in the vicinity of x = 0.6.

Remarks

We have ignored the influence of phase equilibrium and vaporization in the above analysis. As the vapor is compressed, its temperature and pressure increase according to the relation $T_{v,2}/T_1 = (P_2/P_1)^{(k-1)/k}$. Phase equilibrium effects induce the liquid temperature $T^{sat}(P)$ to increase with the pressure according to a different relationship compared to variation of vapor temperature during compression. If local phase equilibrium near the drop surface is assumed during the compression process, invariably a temperature difference exists causing irreversibility.

r. *Example 18*

Air (for which k = 1.4) is contained in an insulated piston–cylinder assembly under the conditions P = 100 kPa, V = 0.1 m^3 and T = 300 K. The piston is locked with a pin and its area is 0.010 m^2. A weight of 2 KN is rolled onto the piston top and the pin released.
Is the process reversible or irreversible?
Does the relation Pv^k = constant, which is valid for an isentropic process, describe the process?
Determine the final state.
Evaluate the entropy change $s_2 - s_1$.

Solution

When the pin is released, the molecules adjacent to the piston are immediately compressed making the local gas hotter while those farther away are not. Therefore, the pressure near the piston top is higher than the cylinder bottom. This effect continues as the piston moves inward, and the system is not at a uniform state. Thus a pressure and temperature gradients are established. The process is irreversible.

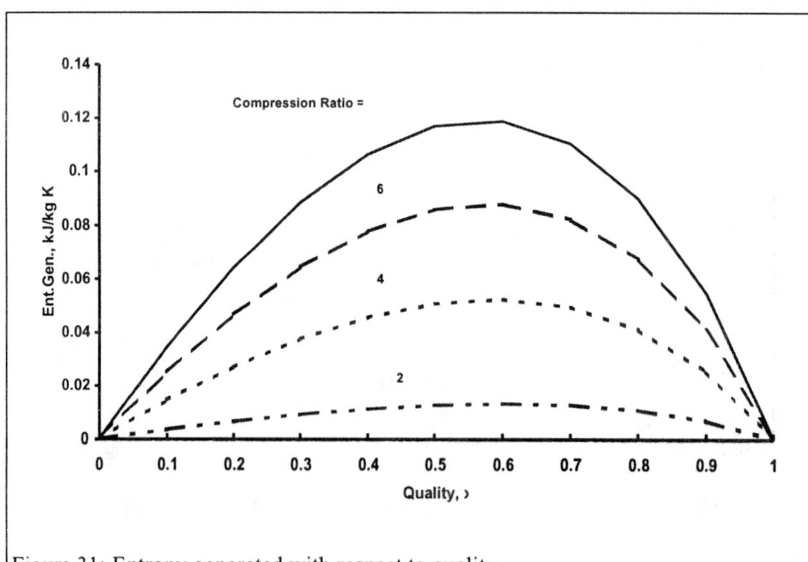

Figure 31: Entropy generated with respect to quality.

The relation Pv^k = constant does not apply, since the process is not isentropic.

Using the method of Example 9 of Chapter 2, $P_2 = 2/0.010 = 200$ kPa, $T_2 = 386$ K, and $V_2 = 0.065$ m^3. Therefore,

$s_2 - s_1 = c_{p0} \ln (T_2/T_1) - R \ln(P_2/P_1) = \ln (386 \div 300) - 0.286 \ln(200 \div 100)$

$= 0.0538$ kJ kg^{-1} K^{-1}.

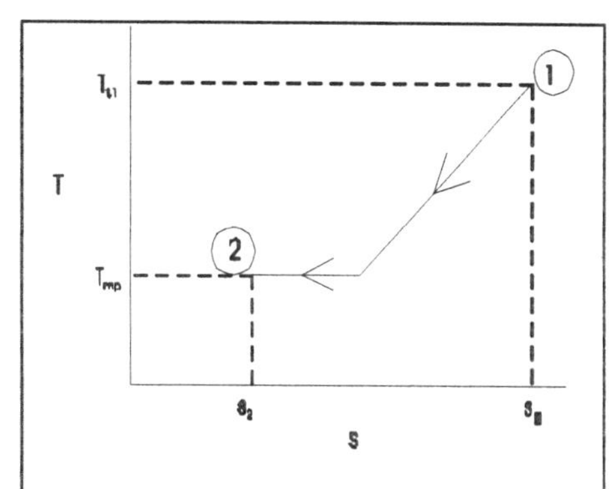

Figure 32: T-s diagram illustrating water in a storage tank.

s. *Example 19*

Consider an idealized air conditioning cycle used for storage tank applications. The objective is to cool the water stored in a tank from 25°C ($T_{t,1}$) to make ice at 0°C ($T_{t,2}$) by circulating cold Freon inside the tank. The ambient temperature is 25°C (T_0). Determine the minimum work required for every kg of water contained in the storage tank. The heat of fusion for water ($h_{sf} \approx u_{sf}$) is 335 kJ kg^{-1} (cf. Figure 32).

Solution

This example is similar to that contained in Example 7. We assume a Carnot refrigeration cycle discarding heat to a variable low temperature reservoir. The cycle operates at a fixed higher temperature T_0. First, the water is cooled from the initial state (state 1) to the melting point (MP) of ice, and then frozen at that temperature (state 2). As shown in Example 7,

$W_{min} = -(U_{t,2} - U_{t,1}) + T_0 (S_{t,2} - S_{t,1})$,

$W_{min} = -(mc(T_2 - T_1) - m\, u_{sf}) + T_0 (mc \ln (T_2/T_1) - m\, u_{sf}/T_{freeze})$, or

$w_{min} = W_{min}/m = -(4.184\times(0-25)-335)+298\times(4.184\times\ln(273\div298)-335\div273)$

$= 439.6 - 298 \times 1.594 = -35.41$ kJ kg^{-1}.

Remarks

Figure 32 contains a representation of the process on a T–s diagram. The area under the path in the figure represents the reversible heat absorbed from the tank. We have assumed that the low temperature (of Freon) during the refrigeration cycle is exactly equal to the storage tank temperature. However, in practice it is not possible to transfer heat in the absence of a temperature gradient without inducing some irreversibility between the Freon and water contained in the tank.

t. *Example 20*

0.5 kg of coffee is contained in a cup at a temperature of 370 K. The cup is kept in an insulated room containing air at a temperature of 300 K so that, after some time, it cools to 360 K. The air mass is 100 kg. Assume the properties of coffee to be the same as those of water, and determine the following:

The change in internal energy dU ($= dU_{coffee} + dU_{air}$).

The initial entropy of the coffee and air if it is assumed that both subsystems exist at an equilibrium state.

The heat transfer across the cup boundary δQ.

The temperature change of the air.

The entropy change of coffee dS_{coffee}.

The entropy change of air dS_{air}.

The entropy generated.

Solution

Consider an isolated composite system consisting of two subsystems, i.e., coffee and air. In the absence of external interactions, isolated systems attain a stable equilibrium state. The internal energy change dU = 0 by applying the First law for a combined system, i.e.,

$$dU_{coffee} = -dU_{air}. \quad \text{(A)}$$

$$S_{coffee} = m\,s = 0.5 \times s_{f,370\,K} = 0.5 \times 1.25 = 0.625 \text{ kJ K}^{-1}. \quad \text{(B)}$$

$$S_{air} = 100 \times s_{300\,K} = 100 \times 1.7 = 170 \text{ kJ K}^{-1}. \quad \text{(C)}$$

Therefore,

$$S = 170.625 \text{ kJ K}^{-1}. \quad \text{(D)}$$

For the coffee $\delta Q - \delta W = dU_{coffee} = m\,c\,dT_{coffee} = 0.5 \times 4.184 \times (-10) = -20.92$ kJ.

Since there is no volume change, $\delta W = 0$, and

$$\delta Q = -20.92 \text{ kJ}. \quad \text{(E)}$$

Applying the First law $dU_{coffee} = -dU_{air}$ (see Eq. (A)), since $dU_{coffee} = -20.92 = -dU_{air} = -100 \times 1.0 \times dT_{air}$,

$$dT_{air} = 0.21 \text{ K}. \quad \text{(F)}$$

The air temperature does not rise significantly.

Assuming the coffee temperature to be uniform within the cup, we will select the system so as to exclude boundaries where temperature gradients exist. Using the entropy balance equation for closed systems, and for internally reversible processes,

$$dS_{coffee} = \delta Q/T = -20.92 \div ((360 + 370) \times 0.5) = -0.0573 \text{ kJ K}^{-1}. \quad \text{(G)}$$

Employing Eq. (B),

$S_{coffee} = 0.625 - 0.0573 = 0.5677$ kJ K^{-1}.

$dS_{air} = 20.92 \div ((300 + 300.21) \times 0.5) = +0.0697$ kJ K^{-1}, and

$S_{air} = 170.0697$ kJ K^{-1}.

Forming a combined system that includes both coffee and air, $\delta Q = 0$. Applying the entropy balance equation $dS - 0 = \delta\sigma$,

$dS = dS_{coffee} + dS_{air} = 0.0573 + 0.0697 = 0.0124$ kJ K^{-1} so that

$\delta\sigma = 0.0124$ kJ K^{-1}.

You will also find that $\delta\sigma \to 0$ when the coffee temperature almost equals that of the air.

Remarks

In this case, $\delta\sigma > 0$ during the irreversible process in the isolated system.

$\delta\sigma \to 0$ when the coffee temperature almost equals the air temperature (or as reversibility is approached).

Vaporization of water into the air has been neglected.

u. *Example 21*

A gas undergoes an expansion in a constant diameter horizontal adiabatic duct. As the pressure decreases, the temperature can change and the velocity increases, since the gas density decreases. What is the maximum possible velocity?

Solution

From mass conservation

$$d(V/v) = 0, \tag{A}$$

where V denotes velocity. Therefore, $dV/v + V\, d(1/v) = 0$. Applying energy conservation

$$d(h + V^2/2) = 0, \text{ and} \tag{B}$$

utilizing the entropy equation

$$dh = T\, ds + v\, dP, \text{ or } ds = dh/T - v\, dP/T. \tag{C}$$

Using Eqs. (A) and (B),

$$dh = -V\, dV = -V^2\, dv/v. \tag{D}$$

From Eqs. (D) and (C),

$$ds = -V^2\, dv/(T\, v) - v\, dP/T. \tag{E}$$

For an adiabatic duct, $ds = \delta\sigma$. Since $\delta\sigma \geq 0$ for an irreversible process,

$$ds \geq 0. \tag{F}$$

Using Eqs. (F) and (E), $V^2 \leq -v^2 (\partial P/\partial v)_s$. Typically $(\partial P/\partial v)_s < 0$. Hence $V^2 > 0$. For a reversible (i.e., isentropic) process,

$$V^2 = -v^2 (\partial P/\partial v)_s \tag{G}$$

which is the velocity of sound in the gas.

Remarks

In Chapter 7, we will discuss use of the enclosed software to determine the sound speed in pure fluids.

H. THIRD LAW

The Third law states that a crystalline solid substance at an absolute temperature of zero (i.e., 0 K) possesses zero entropy. This implies that the substance exists in a state of perfect order at that temperature in the absence of energy, a condition that is not particularly useful, since it is no longer possible to extract work from it. In other words, the entropy of any crystalline matter tends to zero as $\partial U/\partial S \to 0$. We will see in Chapter 7 that s(0 K) is independent of pressure, i.e., s(0 K, P =1 bar) = s(0 K, P). Entropy values are tabulated for most substances using the datum s = 0 at 0 K.

In general, substances at low temperatures exist in the condensed state so that for an incompressible substance

$$ds = c_s\, dT/T.$$

At very low temperatures the specific heat–temperature relation for a solid, $c_s = \alpha T^m$ can be applied, so that

$$(s - s_{ref}(0)) = \alpha T^m/m,\ m \neq 0 \tag{61a}$$

For Debye solids m = 3, and

$$\alpha = (1944/\theta_D^3)\text{ kJ kmole}^{-1}\text{ K}^{-4}, \text{ at } T < 15\text{ K}, \tag{61b}$$

where θ_D is a constant dependent upon the solid.

In summary, according to the Third law $s = s_{ref}(0) = 0$ at an absolute temperature of zero.

---

v. Example 22

The specific heat of a Debye solid (for temperatures less than 15 K) is represented by the relation $\bar{c}_s = (1944\ T^3/\theta^3)$ kJ kmole^{-1} K^{-1}. Obtain a relation for the entropy with respect to temperature for cyclopropane C_3H_6 for which $\theta = 130$. What are the values of the entropy and internal energy at 15 K. (cf. also Figure 33).

Solution

The molar specific entropy (0 K, 1 bar) = 0 kJ kmole^{-1} K^{-1} (i.e., point A in Figure 33a). Since $\bar{s} = \int \bar{c}_s d\,T/T$,

$\bar{s} = (1944\ T^3/\theta^3)/3$ kJ kmole^{-1} K^{-4}.

At 15 K, $\bar{s}$(15 K, 1 bar) = $(1944 \times 15^3/130^3)/3 = 0.995$ kJ kmole^{-1} K^{-1}.

The internal energy $\bar{u} = \int \bar{c}_s dT$ so that at 15 K,

$\bar{u} = (1944 \times 15^4 \div 130^3) \div 4 = 11.2$ kJ kmole^{-1}.

---

w. Example 23

At a pressure of 1 bar, evaluate the entropy and internal energy of cyclopropane C_3H_6 when it exists as (a) saturated solid; (b) saturated liquid; and (c) saturated vapor given that the specific heat c_s follows the Debye equation with $\theta_D = 130$ K and m = 3 when T < 15 K, $\bar{c}_s$ = 28.97 kJ kmole^{-1} K^{-1} for 15 K < T < T_{MP}, where the melting point T_{MP} = 145.5 K at P = 1 bar, h_{sf} = 5442 kJ k mole^{-1}, and the normal boiling point T_{BP} = 240.3 K; and for the liquid $\bar{c}_l$ = 76.5 kJ kmole^{-1} K^{-1}, and $\bar{h}_{fg}$ = 20,058 kJ kmole^{-1}.

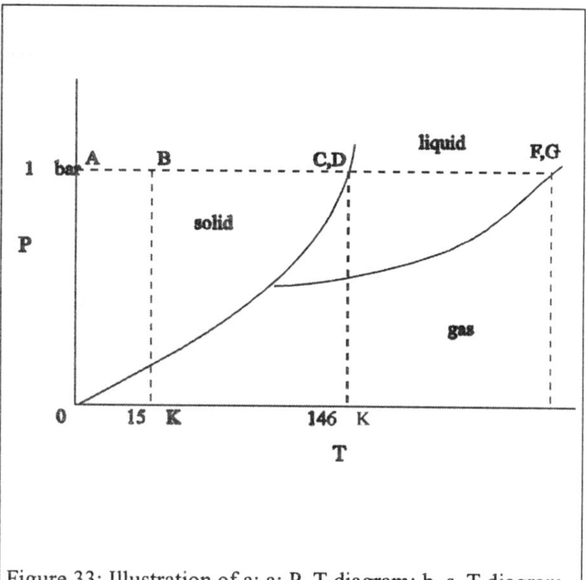

Figure 33: Illustration of a: a: P–T diagram; b. s–T diagram.

Solution

From Example 22, $\bar{s}$(s, 15 K) = 0.99 kJ kmole^{-1} K^{-1} (at point B in Figure 33). Hence, $\bar{s}$(s, 145.5) − $\bar{s}$(s,15) = 28.97 ln (145÷15) (i.e., point C, saturated solid in Figure 33a and b which represents a saturated solid that is ready to melt). Therefore,

$\bar{s}$(s, 145.5) = 65.72 + 0.99 = 66.71 kJ kmole^{-1} K^{-1}, and

$\bar{u}$ (s, 145.5) = 28.97 × (145 − 15) + 11.2 = 377.7 kJ kmole^{-1}.

The liquid entropy may be evaluated as follows:

$\bar{s}(\ell$, 145.5) − s(s,145.5) = 5442÷145.5 = 37.40 kJ kmole^{-1} K^{-1}.

Therefore,

$\bar{s}$ (ℓ, 145.5) = 37.4 + 66.71 = 104.11 kJ kmole^{-1} K^{-1} (point D in Figure 33) so that

$\bar{s}(\ell,240.3) - s(\ell,145.5) = 76.5 \ln(240.3 \div 145.5) = 38.4$ kJ kmole^{-1} K^{-1}, and

$\bar{s}(\ell,240.3) = 38.4 + 104.11 = 142.51$ kJ kmole^{-1} K^{-1} (point F in Figure 33).

Since h = u + Pv, and for solids and liquids Pv « u, it follows that for these substances $h \approx u$ or $h_{sf} \approx u_{sf}$ and $h_{fg} \approx u_{fg}$. Hence,

$u(\ell,145.5) = 377.7 + 5442 = 5819.9$ kJ kmole^{-1}, and

$u(\ell,240.3) = 5819.9 + 76.5 \times (240.3 - 145.5) = 13132$ kJ kmole^{-1}.

In the gaseous state

$\bar{s}(g, 240.3) - \bar{s}(1,240.3) = 20{,}058 \div 240.3 = 83.5$ kJ kmole^{-1} K^{-1}, i.e.,

$\bar{s}(g,240.3) = 83.5 + 142.51 = 226.01$ kJ kmole^{-1} K^{-1} (point G in Figure 33).

Therefore,

$\bar{s}(g,240.3) = 226$ kJ kmole^{-1} K^{-1}, and

$h(g, 240.3) = 13132 + 20.058 = 33{,}190$ kJ kmole^{-1}.

$u(g, 240.3) = 33{,}190 - 8.314 \times 240.3 = 31{,}192$ kJ kmole^{-1}.

Remarks

If the reference condition for the entropy is selected at the saturated liquid state (i.e., at point D), we can arbitrarily set s = 0 there. Therefore, at point C, saturated solid, $\bar{s}_C = -37.40$ kJ kmole^{-1} K^{-1}, and at point B, $\bar{s}_B = -37.40 - 65.72 = -103.12$ kJ kmole^{-1} K^{-1}. Such a procedure is generally used for water, since the reference condition with respect to its entropy is based on the saturated liquid state at its triple point temperature of 0.01°C. At this state it is usual to set $s = h \approx u = 0$. Methods for evaluating at any pressure and temperature will be discussed in Chapter 7.

Recall from the First law that $\delta Q - \delta W = dU$. Therefore, if a process involves irreversibility, then $dS = \delta Q/T_b + \delta\sigma$ so that

$$dS = dU/T_b + \delta W/T_b + \delta\sigma.$$

In case the process is mechanically reversible, then the entropy balance equation for a closed system can also be written as

$$dS = dU/T_b + P\,dV/T_b + \delta\sigma.$$

where $\delta\sigma > 0$ for irreversible processes and equals zero for reversible processes.

I. ENTROPY BALANCE EQUATION FOR AN OPEN SYSTEM

We have presented the entropy balance equation Eq. (28) for a closed system and obtained relations describing the entropy of a fixed mass. We will now formulate the entropy balance for an open system in an Eulerian reference frame.

1. General Expression

The entropy balance equation for a closed system containing a fixed control mass assumes the form

$$dS_{c.m.} - \delta Q/T_b = \sigma_{c.m.}, \tag{28}$$

where the subscript c.m. denotes the control mass. Work does not explicitly enter into this expression. For an open system which exchanges mass, heat and work with its ambient (cf. Figure 34), the derivation of the corresponding balance equation is similar to that of the energy conservation equation. The entropy change within a fixed mass over an infinitesimal time δt can be written in the form

$$dS_{c.m.} = S_{c.m.,t+dt} - S_{c.m.,t}. \tag{62a}$$

The control mass at time t (illustrated within the dashed boundary in Fig. 25(a)) includes both the control volume mass and a small mass dm_i waiting to enter the control volume. After an

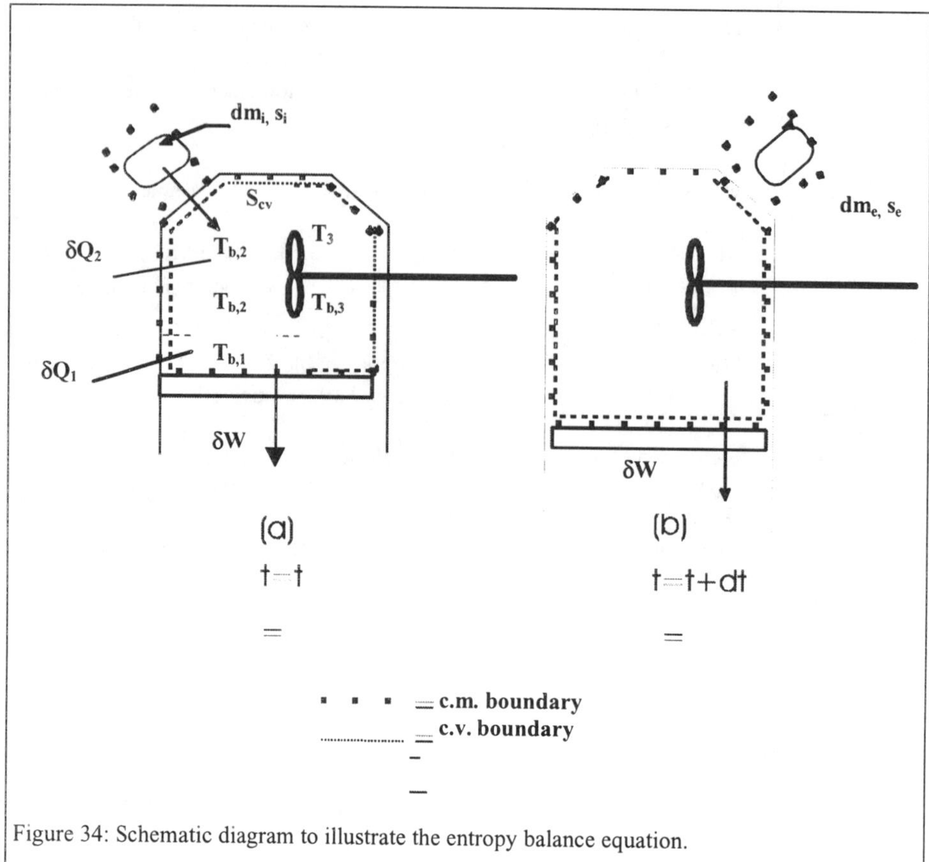

Figure 34: Schematic diagram to illustrate the entropy balance equation.

infinitesimal time δt as the mass dm_i enters at the inlet, a small mass dm_e leaves through the control volume exit, and for the control volume Eq. (62a) may be expressed in the form

$$dS_{c.m.} = (S_{c.v.,t+dt} + dm_e\, s_e) - (S_{c.v.,t} + dm_i s_i). \tag{62b}$$

Heat transfer can occur across the control volume boundary. In general, the boundary temperature at its inlet (but within the dashed boundary) is different from the corresponding temperature at the exit, and the heat transfer rate δQ may vary from the inlet to exit. For the same of analysis we divide the boundary into sections such that at any section j the boundary temperature is $T_{b,j}$ and the heat transfer rate across the boundary is $\dot{Q}_j$. For an infinitesimally small time period δt, the term δQ/T is given as

$$(\delta Q/T_b) = \Sigma_j \dot{Q}_j \delta t/T_{b,j}. \tag{63}$$

Using Eqs. (28), (62), and (63), expanding $S_{c.v.,t+\delta t}$ in a Taylor series around time t, and dividing the resultant expression by δt, and letting when $\delta t \to 0$ so that the control mass and control volume boundaries merge (i.e., c.m. $\to$ c.v.), we obtain

$$dS_{c.v.}/dt = \dot{m}_i s_i - \dot{m}_e s_e + \Sigma_j \dot{Q}_j/T_{b,j} + \dot{\sigma}_{cv}. \tag{64}$$

On a mole basis Eq. (64) may be written in the form

$$dS_{c.v.}/dt = \dot{N}_i\bar{s}_i - \dot{N}_e\bar{s}_e + \Sigma_j \dot{Q}_j/T_{b,j} + \dot{\sigma}_{cv}, \quad (65)$$

and generalizing for multiple inlets and exits

$$dS_{c.v.}/dt = \Sigma\, \dot{m}_i s_i - \Sigma\, \dot{m}_e s_e + \Sigma_j \dot{Q}_j/T_{b,j} + \dot{\sigma}_{cv}. \quad (66)$$

where $\dot{Q}_j$ denotes the heat interaction of the system with its surroundings at section j with a boundary temperature $T_{b,j}$. The first term on the RHS of Eq. (66) represents the input entropy through the various inlets, the second term the outlet entropy, and the third term the transit entropy due to heat transfer. Equation (66) may be interpreted as follows: The rate of entropy accumulation in the control volume = entropy inflow through advection – entropy outflow through advection + change in the transit entropy through heat transfer + entropy generated due to irreversible processes. The various terms are also illustrated in Figure 35.

Equations (64) and (65) may be, respectively, rewritten in the form

$$dS_{c.v.} = dm_i\, s_i - dm_e\, s_e + \Sigma_j \delta Q_j/T_{b,j} + d\sigma, \text{ and} \quad (67a)$$

$$dS_{c.v.} = dN_i\, \bar{s}_i - dN_e\, \bar{s}_e + \Sigma_j \delta Q_j/T_{b,j} + d\sigma. \quad (67b)$$

Recall that for closed system $TdS = \delta Q + d\sigma$. Equation (67) is the corresponding equation for the open system. In case of elemental reversible processes in the presence of uniform control volume properties (e.g., in an isothermal swimming pool)

$d\sigma = 0$, $T_{b,j} = T$ so that

$TdS = \delta Q + dm_i\, Ts_i - dm_e\, Ts_e$.

For an open system with property gradients within the control volume, as in a turbine, the system may be divided into small sections to apply the relation

$TdS = \delta Q + dm_i\, Ts_i - dm_e\, Ts_e$

to each subsystem within which the temperature is virtually uniform.

For a closed system, Eq. (64) reduces to Eq. (31). If the process is reversible, Eq.(66) becomes

$$dS_{c.v.}/dt = \Sigma\, \dot{m}_i s_i - \Sigma\, \dot{m}_e s_e + \Sigma_j \dot{Q}_j/T_{b,j}. \quad (68)$$

Equation (66) provides information on the rate of change of entropy in a control volume. If it becomes difficult to evaluate $\Sigma_j \dot{Q}_j/T_{b,j}$, the system boundary may be drawn so that $T_{b,j} = T_0$, i.e., all irreversibilities are contained inside the selected control volume. For instance, if in Figure 34 the boundary is selected just outside the control volume, then $T_{b,j} = T_0$.
At steady state Eq. (66) becomes

$$\Sigma\, \dot{m}_i s_i - \Sigma\, \dot{m}_e s_e + \Sigma_j \dot{Q}_j/T_{b,j} + \dot{\sigma}_{cv} = 0. \quad (69)$$

For single inlet and exit at steady state

$$s_i - s_e + \Sigma q_j / T_{bj} + \sigma_{c.v} = 0$$

Where σ_m, entropy generates per unit mass. If process is reversible $\sigma_{c.v} = 0$, $T_b = T$ and hence

$$s_i - s_e + \Sigma q_j / T = 0$$

or

$$ds = \Sigma_I\, \{\delta q_j)_{rev}/T$$

which is similar to relation for an open system

In case of a single inlet and exit, but for a substance containing multiple components, the relevant form of Eq. (66) is

$$dS_{c.v.}/dt = \Sigma_k \dot{m}_{k,i}\bar{s}_{k,i} - \Sigma_k \dot{m}_{k,e}\bar{s}_{k,e} + \Sigma_j \dot{Q}_j/T_{b,j} + \dot{\sigma}_{cv}. \quad (70)$$

where $\bar{s}_k$ denotes the entropy of the k–th component in the mixture. For ideal gas mixtures $\bar{s}_k(T,P) = \bar{s}_k^0(T) - \ln(PX_k)$.

---

x. *Example 24*

Water enters a boiler at 60 bar as a saturated liquid (state 1). The boiler is supplied with heat from a nuclear reactor maintained at 2000 K while the boiler interior walls are at 1200 K. The reactor transfers about 4526 kJ of heat to each kg of water. Determine the entropy generated per kg of water for the following cases assuming the system to be in steady state:

For control surface 1 shown in Figure 36 with P_2 = 60 bars and T_2 = 500°C.

For control surface 2 which includes the nuclear reactor walls with P_2 = 60 bars and T_2 = 500°C.

For control surface 1 with P_2 = 40 bars and T2 = 500°C.

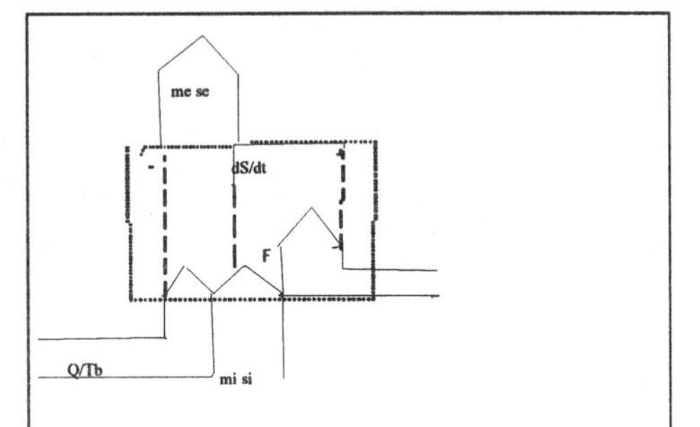

Figure 35: Illustration of the entropy band diagram.

Solution

At steady state

$$\dot{m}(s_i - s_e) + \dot{Q}_j/T_{b,boiler} + \dot{\sigma} = 0, \text{ i.e.,} \quad (A)$$

$$\sigma = (s_2 - s_1) - q_{12}/T_{b,boiler}. \quad (B)$$

Using the standard Steam tables (A-4A) s_1 = 3.03 kJ kg^{-1} at P_1 = 60 bar for the saturated liquid and s_2 = 6.88 kJ kg^{-1} at P_2 = 60 bars for steam at 500°C. The heat transfer q_{12} = 4526 kJ kg^{-1} of water and $T_{b,boiler}$ = 1200 K. Substituting the data in Eq.(B), σ = (6.88 – 3.03) – 4526÷1200 = 0.07833 kJ kg^{-1} of water.

For control surface 2, Eq.(B) may be written in the form

$$\sigma = s_2 - s_1 - q_{12}/T_{b,reactor}. \quad (C)$$

Using the Steam tables (A-4) data

σ = (6.88 – 3.03) – 4526 ÷ 2000 = 1.587 kJ kg^{-1} of water.

For this case s_2 = 7.0901 at P_2 = 40 bars and T_2 = 500 C, and

σ = 7.09 – 3.03 – 4526 ÷ 1200 = 0.288 kJ kg^{-1} of water.

Remarks

The difference in the value of s between cases (a) and (b), (i.e., 1.587 – 0.0783 = 1.509 kJ kg^{-1} of water) is due to the irreversible heat transfer between the two control surfaces 1 and 2. Therefore,

$$\sigma_q = q_{12}(1/T_{b,boiler} - 1/T_{b,reactor}),$$

where σ_q denotes the entropy generated in the thin volume enclosed within control surfaces 1 and 2 due to the irreversible heat transfer.

Power plant systems are designed to minimize the generated entropy.

If the heat transfer q_{12} for the first case is given as 5000 kJ, $\sigma = -0.32$ kJ kg^{-1}. Is this possible or is the heat transfer value incorrect?

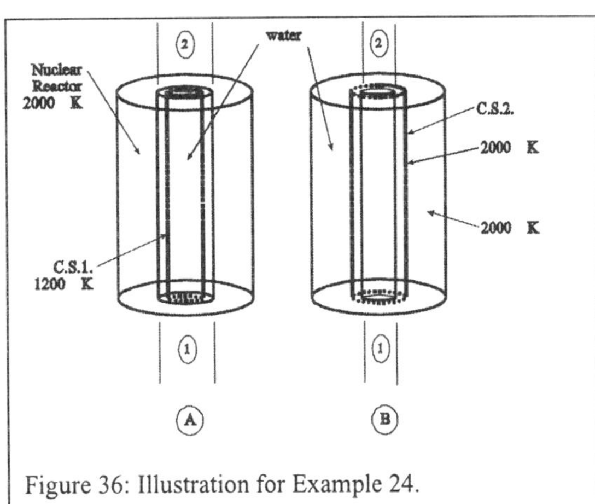

Figure 36: Illustration for Example 24.

y. *Example 25*

Consider a human being who weighs 70 kg. At 37°C the typical heat loss is 100 W. The person is injected with glucose (s = 1.610 kJ kg^{-1} K^{-1}) at the rate of 0.54 kg day^{-1}. Air at 27°C is inhaled at the rate of 0.519 kg hr^{-1}. Assume steady state, no excretion, and for the products to possess the same properties as air. The products are exhausted through the nose at 37°C. Select the c.v. boundary so that it lies just below human skin.

a. Write the mass conservation and entropy balance equations for the system and simplify the equations.

b. Determine the exhaust mass flow rate (e.g., through the human nose).

c. What is the entropy generation rate per day?

d. What is the entropy generation rate per day per unit mass?

e If the entropy generation during a species' life cannot exceed 10,000 kJ kg^{-1} K^{-1}, what is this human being's life span?

Solution

$dm_{cv}/dt = \dot{m}_i - \dot{m}_e$, and

$dS_{cv}/dt = \Sigma \dot{Q}_{cv,j}/T_{b,j} + \Sigma \dot{m}_i s_i - \Sigma \dot{m}_e s_e + \dot{\sigma}_{cv}$.

At steady state

$\dot{m}_i - \dot{m}_e = 0$, and

$\Sigma \dot{Q}_{cv,j}/T_{b,j} + \Sigma \dot{m}_i s_i - \Sigma \dot{m}_e s_e + \dot{\sigma}_{cv} = 0$.

From mass conservation

$\dot{m}_i = \dot{m}_e = \dot{m} = 0.519$ kg hr^{-1}.

Therefore,

$$\dot{\sigma}_{cv} = -100\times3600\times24/(310\times1000)+0.54\times1.610+12.456\times1.702-(0.54+0.519\times24)\times1.735 = 28.35 \text{ kJ K}^{-1} \text{ day}^{-1}.$$

Per unit mass

$\dot{\sigma}_{cv} = 28.35/70 = 0.405$ kJ kg^{-1} day^{-1}.

The life span is 10000/(365×0.405) = 68 years.

Remarks

In Chapter 11, the irreversibility due to metabolism will be considered.

From Example 12 in Chapter 2 we see that $\dot{q}_G \propto m_b^{-0.33}$ while empirical results suggest that $\dot{q}_G$ (kW/kg) = $0.003552 m_b^{-0.26}$. Part (e) of the problem can be mathematically expressed as $\approx \dot{\sigma}/m = \dot{\sigma}_m \approx (\dot{q}_G/T_b)$ equals the specific meatbolic rate $\div T_b$.

Likewise, from part (c), the life span of a species $\approx CT_b/\dot{q}_G$ where $C \approx$ 10,000 kJ kg^{-1} K^{-1}.

The metabolic rate during the lifetime of an organism varies, with the highest metabolic rate being for a baby and the lowest for an older person. The minimum metabolic rate for maintaining bodily functions is of the order of 1 W. The expression in part d is based on an average metabolic rate.

The entropy change in the environment can be obtained by considering the atmosphere as the system.

There are no gradients in the environment outside of the skin. Therefore, the entropy generation is zero. The entropy growth rate in the environment is

$$dS_{env}/dt = 0 + \dot{Q}/T_\infty + 0 = 100/300 = 0.33 \text{ W K}^{-1} \text{ per human being.}$$

---

2. Evaluation of Entropy for a Control Volume

Recall that for closed systems we evaluate entropy by connecting a reversible path between two given states and then use $dS = dU/T + PdV/T$ along the reversible path in order to determine the entropy change. We need to obtain a corresponding relation for a control volume. For instance, let us say that we wish to find the entropy change of the air in a tire when air is pumped into it. Assume that the initial (T_1,P_1,V_1,N_1) and final states (T_2, P_2, V_2, N_2) are known. The difference $N_2–N_1$ represents the number of moles that are pumped into the tire. The process may or may not be reversible. We will now show that for a single component undergoing a reversible process in an open system

$$dS = dU/T - PdV/T + (\mu/T)dN. \tag{71}$$

where the chemical potential $\mu = g = h - Ts$.

---

z. *Example 26*

Consider a balloon that is charged with gaseous nitrogen. The balloon is kept in a room whose pressure can be closely matched to that of the balloon. The moles within the balloon and the volume increase. Assume the process to be reversible. Show that the entropy change when dN moles of a pure component are pumped into it is represented by Eq.(71).

Solution

Boundary work is performed as the balloon expands. Heat transfer may also occur across the boundary. Since matter enters the balloon but does not leave it during the charging process, over a small time period

$$dE = dU = dN_i\,\bar{h}_i + \delta Q - \delta W. \tag{A}$$

The form of energy conservation has been previously discussed in Chapter 2. For entropy balance, we can apply Eq. (65) with a uniform boundary temperature, i.e.,

$$dS = dN_i\,\bar{s}_i + \delta Q/T_b + \delta\sigma_{c.v.}. \tag{B}$$

The mass or mole conservation relation is

$$dN = dN_i. \tag{C}$$

Since the processes in the balloon are internally reversible, the pressure of the gas in the balloon is almost equal to the outside pressure so that the deformation work W_d = PdV is reversible and $\delta\sigma_{c.v.} = 0$, implying that $T_b = T$. Dropping the subscript i in Eqs. (A) and (B)

$$dU = dN\,\bar{h} - PdV + \delta Q, \text{ and} \tag{D}$$

$$dS = dN\,\bar{s} + \delta Q/T. \tag{E}$$

Equation (E) reveals that even if an open system is adiabatic and reversible dS can be nonzero since moles enter the system. Eliminating δQ from Eqs. (D) and (E), the resulting expression is (in mass form) dU = dNh − PdV + TdS − TsdN, where $\mu = g = (h - Ts)$ is the chemical potential or Gibbs function of a species entering the system. Therefore, U = U(S,V,N). Since δW = PdV, the third term on the right hand side of the resultant equation, i.e., μdN, can be viewed as the reversible chemical work, $\delta W_{chem.rev} = -\mu dN$. The negative sign occurs, since the chemical work input when moles are added to a system is negative according to sign convention for work.

Therefore, the change of internal energy of the matter in the balloon equals the energy transfer into the control volume due to reversible heat addition less the energy outflow through the deformation work, but in addition to the energy transfer due to the chemical work.

If a gas is pumped into a rigid volume (e.g., rigid tank) then P dV = 0, and if the entropy is kept fixed (e.g., if the volume is cooled while matter is introduced into it so that $\bar{s}_{final}N_{final} = \bar{s}_{initial}N_{initial}$), the energy change $dU = -\delta W_{chem,rev} = \mu dN$ is the chemical work performed to adiabatically pump the incremental number of moles dN.

Pressure potential causes the volume to change and perform deformation (boundary) work, temperature potential causes the entropy change through the heat transfer process, and the chemical potential causes the matter to move into or out of the control volume. Rewriting Eq. (F) in the form

$$dS = dU/T + (P/T)dV + (\mu/T)dN. \quad \text{(H)}$$

which represents the change of entropy along a reversible path. As before, S = S(U,V,N). Note that for a closed system containing inert matter, N is fixed and hence S = S(U,V).

Remarks

In Example (26), we determine dS for a control volume. With this expression we can integrate Eq. (G) between two states to obtain the resultant entropy change.

If a multi–component gas mixture (e.g., air) is reversibly pumped into a uniform temperature volume, Eqs. (A), (B), and (D) are modified as (with subscripts k denoting species)

$$dU = \Sigma dN_k \bar{h}_k + \delta Q - P\,dV, \quad \text{(A')}$$

$$dS = \Sigma dN_k \bar{s}_k + \delta Q/T, \text{ and} \quad \text{(B')}$$

$$dU = \Sigma dN_k(\bar{h}_k - T\bar{s}_k) - PdV + TdS, \text{ or} \quad \text{(D')}$$

$$dU = TdS - PdV + \Sigma \mu_k\, dN_k. \quad \text{(72a)}$$

Here, $\mu_k = \bar{g}_k = (\bar{h}_k - T\bar{s}_k)$ (a more detailed discussion is contained in Chapter 8). The first term on the RHS of Eq. (72a) represents the reversible heat transfer and the last term is the accumulation due to the exchange of matter. In a closed system containing inert components there is no change in the number of moles, and Eq. (72a) can be written in the form dU = TdS − PdV.

Adding d(PV) to both sides of Eq. (72a) we have

$$dH = dU + d(PV) = TdS + VdP + \Sigma \mu_k dN_k. \quad \text{(72b)}$$

Subtracting d(TS) from both sides of Eq. (72a)

$$dA = dU - d(TS) = -SdT - PdV + \Sigma \mu_k dN_k \quad \text{(72c)}$$

where $A = U - TS$ is the Helmholtz function. Subtracting d(TS) from both sides of Eq. (72c)

$$dG = dH - d(TS) = - SdT + VdP + \Sigma\mu_k dN_k, \quad (72d)$$

where $G = H-TS$ is the Gibbs function.
It is clear from the energy fundamental equation

$$U = U(S,V,N_1,N_2, \ldots ,N_k), \quad (73)$$

that k + 2 properties are required to determine the *extensive* state of an open system. For the same of illustration, assume a 9 m^3 room containing 0.09 kmole of O_2 (component 1) and 0.36 kmole of N_2 (component 2). Fresh warm air is now pumped into the room ($N_{1,i}$ = 0.1 kmole and $N_{2,i}$ = 0.376 kmole) and some air leaves the room ($N_{1,e}$ = 0.05 and $N_{2,e}$ = 0.188). Hence, N_1 = 0.09 + 0.1 – 0.05 = 0.14 kmole, N_2 = 0.36 + 0.376 – 0.188 = 0.548 kmole. The entropy increases due to the temperature rise as well as the larger number of moles of gas in the room, while the gas volume remains the same.
If the room is divided into three equal parts A, B, and C, then VA = VB = VC = V/3 = 9/3 m^3, and likewise $S_A = S_B = S_C = S/3$, $N_{O_2,A} = \ldots = N_{O_2}/3$, and $U_A = U_B = U_C = U/3$. Therefore, $U(S_A,V_A, \ldots) = U(S/3,V/3, N_{O_2}/3, \ldots) = 1/3\ U(S,V, N_{O_2}, \ldots)$. In general, if the room is divided into λ′ parts,

$$U(S/\lambda', V/\lambda', \ldots = 1//\lambda'\ U(S,V, \ldots),$$

or if $1/\lambda' = \lambda$ then $U(\lambda S, \lambda V, ..) = \lambda U(S,V, \ldots)$ which is a homogeneous function of degree 1. Note that intensive properties are independent of the extent of the system and are homogeneous functions of degree 0. Since $T = \partial U/\partial S$, and the property in any section is $1/\lambda'$ that of the original extensive property, i.e.,

$$T = \partial U_A/\partial S_A = \partial(U/\lambda')/\partial(S/\lambda') = \partial U/\partial S.$$

Since the internal energy is an extensive property (or a homogeneous function of degree 1), it must satisfy the Euler equation (cf. Chapter 1), namely,

$$\partial U/\partial S + V\ \partial U/\partial V + N_1\ \partial U/\partial N_1 + \ldots = 1 \quad (74a)$$

However, $\partial U/\partial S = T$, $\partial U/\partial V = - P$, and $\partial U/\partial N_1 = \mu_1$ so that Eq. (74a) assumes the form

$$U = ST - PV + \mu_1 N_1 + \mu_2 N_2 + \ldots, \quad (74b)$$

which, upon simplification, may be written as

$$U + PV - TS = H - TS = G = \Sigma\mu_k N_o. \quad (74c)$$

Likewise, applying the Euler equation to evaluate the property $H(S,P,N_1,..)$ (cf. Eq. (72b)),

$$H\ (S,P,N_1\ ..) = TS + \Sigma\mu_k N_o, \quad (74d)$$

Similarly, the Helmholtz and Gibbs functions can be written in the forms

$$A\ (T,V,N_1, \ldots) = -PV + \Sigma\mu_k N_o, \text{ and} \quad (74e)$$

$$G\ (T,P,N_1, \ldots) = \Sigma\mu_k N_o. \quad (74f)$$

Equation (73) implies that

$$dU = TdS + SdT - PdV - VdP + \mu_1 dN_1 + N_1 d\mu_1 + \ldots . \quad (75a)$$

Subtracting Eq. (72a) from Eq. (75a) we obtain

$$SdT - VdP + N_1 d\mu_1 + N_2 d\mu_2 + \ldots = 0. \quad (75b)$$

Equation (75b), which is also known as the Gibbs–Duhem equation, implies that

$$T = T(P,,\mu_1,\mu_2, \ldots, \mu_k). \quad (76)$$

The temperature, which is an *intensive* property, is a function k+1 intensive variables. Equation (76) is also known as the intensive equation of state.
Applying Eq. (72a) to examine the fluid discharge from a rigid control volume we obtain the relation

$$dU = T\, dS - \Sigma\mu_k dN_k, \quad (77a)$$

which describes the internal energy change due to the change in the number of moles (i.e., dN_k). Even if $\bar{s}_k$ is constant, the change in entropy can be nonzero, since in this case $dS = d(\Sigma N_k \bar{s}_k) = \Sigma \bar{s}_k dN_k$ so that

$$dU = T\Sigma \bar{s}_k dN_k - \Sigma\mu_k dN_k = T\Sigma \bar{s}_k dN_k - \Sigma(\bar{h}_k - T\bar{s}_k)\, dN_k = -\Sigma\, \bar{h}_k\, dN_k. \quad (77b)$$

Therefore, the internal energy change when matter is isentropically discharged equals the enthalpy of the matter leaving the system (Example 16, Chapter 2, gas discharge from tank).

We may rewrite Eq. (72a) in the form of the entropy fundamental equation,

$$dS = dU/T + (P/T)dV - \Sigma\, \{\mu_k/T)\, dN_k.$$

i.e., $S = S(U,V,N_1,\ldots)$.

aa. Example 27

Nitrogen is pumped into a 0.1 m^3 rigid tank. The initial state of the gas is at 300 K and 1 atm. Determine the chemical work done to isentropically pump 0.016 kmole of the gas into the tank to a 10 bar pressure.

Solution

Applying the ideal gas law,

$N_1 = 1 \times 0.1 \div (0.08314 \times 300) = 0.004$ kmole.

Therefore, the entropy change $dS = d(\bar{s}N) = 0$, i.e., $d\bar{s}/\bar{s} = dN/N$, or

$\bar{s}_2/\bar{s}_1 = N_1/N_2 = 0.004 \div 0.020 = 0.2$.

Now, $\bar{v}_2 = 0.1 \div 0.02 = 5\ m^3\ kmole^{-1}$, and $\bar{v}_1 = 0.1 \div 0.004 = 25\ m^3\ kmole^{-1}$ so that

$\bar{s}_2/\bar{s}_1 = (\bar{c}_{v0}\ln(T_2/T_{ref}) + \bar{R}\ln(\bar{v}_2/\bar{v}_{ref}))/(\bar{c}_{v0}\ln(T_1/T_{ref}) + \bar{R}\ln(\bar{v}_1/\bar{v}_{ref}))$

Using the values $T_{ref} = 273$ K, $\bar{v}_{ref} = 1\ m^3\ kmole^{-1}$, $\bar{c}_{v0} = 20$ kJ $kmole^{-1}\ K^{-1}$, and $\bar{R} = 8.314$ kJ $kmole^{-1}\ K^{-1}$,

$T_2 = 186$ K.

The chemical work, $W_{chem,rev} = -\int\mu dN = -(U_2 - U_1)$. Now,

$U_2 - U_1 = 0.02 \times 20 \times (186 - 273) - 0.004 \times 20 \times (300 - 273) = -36.96$ kJ, i.e.,

$W_{chem,rev} = +36.96$ kJ.

a. Example 28

Determine the chemical potential of pure O_2 at T = 2000 K and P = 6 bar, and O_2 present in a gaseous mixture at T = 2000 K and P = 6 bar, and $X_{O2} = 0.3$, assuming the mixture to behave as an ideal gas.

Solution

$\mu = \bar{g} = (\bar{h} - T\bar{s})$, where for ideal gases $\bar{s} = \bar{s}^0 - R\ \ln(P/P_{ref})$. Using values from tables (Table A-19),

$\mu_{O_2} = 67881$ kJ kmole^{-1} – 2000 K × (268.655 kJ kmole^{-1} K^{-1} – 8.314 kJ kmole^{-1} K^{-1} × ln(6 bar ÷1 bar)) = – 439,636 kJ kmole^{-1}.

For an ideal gas mixture according to the Gibbs–Dalton law,

$\bar{s}_k(T,p_k) = \bar{s}_k^0(T) - \ln(p_k/P_{ref})$ = (268.655 kJ kmole^{-1} K^{-1} – 8.314 kJ kmole^{-1} K^{-1} × ×ln ((6 bar × 0.3) ÷1 bar) = 272.747 kJ kmole^{-1} K^{-1}.

Since $\mu_k = (\bar{h}_k - T\bar{s}_k)$, μ_{O_2} = – 477,613 kJ (kmole O_2 in the mixture)$^{-1}$.

Remarks

The chemical potential of O_2 decreases as its concentration is reduced from 100% (pure gas) to 30%.

You will later see that the chemical potential plays a major role in determining the direction of chemical reactions (in Chapter 10) and of mass transfer (in this chapter), just as temperature determines the direction of heat transfer.

3. Internally Reversible Work for an Open System

It has been shown in Chapter 2 that for a steady flow open system $\delta q - \delta w = de_T$, where $e_T = h + ke + pe$. For an internally reversible process that occurs between two static states

$$Tds - \delta w_{c.v.,rev} = de_T. \qquad (78)$$

Since $dh = Tds + vdP$,

$$Tds - \delta w_{c.v.,rev} = Tds + vdP + d(ke + pe), \text{ i.e.,}$$

$$\delta w_{c.v.,rev} = -vdP - d(ke + pe).$$

Neglecting the kinetic and potential energies, the work delivered by the system is represented through the relation

$$w_{c.v.,rev} = -\int_{P_i}^{P_e} vdP. \qquad (79)$$

bb. Example 29

The systolic (higher) and diastolic (lower) blood pressures are measured to be, respectively, 120 mm and 70 mm of mercury for a healthy person. Determine the work performed by the heart per kilogram of blood that is pumped. Assume that blood has the same properties as water.

Solution

$$w_{c.v.,rev} = -\int_{P_i}^{P_e} vdP.$$

Therefore,

$w_{c.v.,rev} = -\ v_{blood}(P_e - P_i)$ = 0.001 m^3 kg^{-1} (120–70) (mm Hg)×(100 ÷760) kPa (mm Hg)$^{-1}$ = 0.0063 kJ kg^{-1}.

Remarks

Blood is contained within a finite volume of blood vessels. The body maintains the amount of water to a fixed concentration. If the two pressures simultaneously increase

(true for non–exercising persons), but the term $(P_e - P_i)$ remains unchanged (e.g., P_e = 190 mm and P_i = 140 mm), the work performed by the heart may not change. However, the blood vessels may now become stressed and fail at the higher pressures.
In Chapter 9 we will discuss that the amounts of dissolved CO_2, N_2, and O_2 in the blood rise as the pressure increases, but do so disproportionately, depending upon their boiling points.

---

cc. *Example 30*

Pressurized gas tanks are employed in space power applications. As the gas contained in the tanks is used, the tank pressure falls (say, from $P_{t,1}$ to $P_{t,2}$) so that the work done per unit mole can vary. Determine the work that can be done if a 2 m^3 turbine and the tank are kept in an isothermal bath, P_e = 1 bar, $P_{t,1}$ = 50 bars, and $P_{t,2}$ = 1 bar.

Solution

For an open system,

$$\delta \bar{w}_{shaft,rev} = -\bar{v}dP = -(\bar{R}T/P)dP, \text{ i.e.,} \quad (A)$$

$$\bar{w}_{shaft,rev} = -\bar{R}T\ln(P_e/P_i). \quad (B)$$

The pressure P_i varies as the gas is progressively withdrawn from the tank. From Eq. (B)

$$\delta \bar{W}_{shaft,rev} = \bar{w}_{shaft,rev}dN_i = -\bar{R}T\ln(P_e/P_i)dN_{i,turbine}, \text{ where} \quad (C)$$

$$dN_{tank} = -dN_{i,turbine}. \quad (D)$$

Since $P_{tank}V = N_{tank}\bar{R}T$, $dP_{tank} = dN_{tank}\bar{R}T/V$, and

$$dN_{tank} = dP_{tank}V/\bar{R}T. \quad (E)$$

Therefore, using Eqs. (C) and (E)

$$\delta \bar{W}_{shaft,rev} = --\bar{R}T\ln(P_e/P_i)\,dP_{tank}V/\bar{R}T = -V\ln(P_e/P_i)\,dP, \quad (F)$$

where $P_i = P_{tank}$. Hence,

$$\delta \bar{W}_{shaft,rev} = -VP_e\ln(P_e/P_{tank})\,P_{tank}/P_e, \text{ and}$$

$$\bar{W}_{shaft,rev} = VP_e((P_{tank,2}/P_e)\ln(P_{tank,2}/P_e) - (P_{tank,1}/P_e)\ln(P_{tank,1}/P_e)). \quad (G)$$

Using the values V = 2 m^3, P_e = 1 bar, $P_{tank,1}$ = 50 bars, $P_{tank,2}$ = 1 bar,

$$\bar{W}_{shaft,rev} = 2\times 1\times 100\,(1 \div 1\ln(1\div 1) - (50\div 1)\ln(50\div 1)) = 39120 \text{ kJ}$$

Remarks

If we select the control volume to include both the turbine and the tank, assuming that there is no accumulation of energy or entropy in the turbine,

$$dU_{tank} = \delta Q - PdV - \delta\bar{W}_{shaft,rev} - \bar{h}_e dN_e. \quad (H)$$

Applying the entropy balance equation to the tank,

$$dS_{tank} = \delta Q/T - dN_e\bar{s}_e. \quad (I)$$

From Eqs. (H) and (I),

$$\delta \overline{W}_{shaft,rev} = TdS_{tank} - dU_{tank} - PdV + T\, dN_e \bar{s}_e - \bar{h}_e dN_e. \qquad (J)$$

Since $dN_e = -dN_{tank}$, $dV = 0$, $dS_{tank} = \bar{s}_{tank}\, dN_{tank} + N_{tank} d\bar{s}_{tank}$, $\bar{h}_e(T) = \bar{h}_{tank}(T) = \bar{u}_{tank} + \bar{R}T$, $dU_{tank} = \bar{u}_{tank}\, dN_{tank} + N_{tank}\, \bar{u}_{tank}$, and $d\,\bar{u}_{tank} = 0$,

$$\delta W_{shaft,rev} = T\,(\bar{s}_{tank}\, dN_{tank} + N_{tank}\, d\,\bar{s}_{tank}) - d(dN_{tank}\,\bar{u}_{tank} + N_{tank}\, d\,\bar{u}_{tank}) - 0 - T\, dN_{tank}\,\bar{s}_e + (\bar{u}_{tank} + RT)\, dN_{tank}.$$

Simplifying this relation

$$dW_{shaft,rev} = dN_{tank}T(s_{tank} - s_e) = -\, dN_{tank}TR\, \ln(P_{tank}/P_e) = -V\, dP_{tank}\ln(P_{tank}/P_e). \qquad (K)$$

If the process within the control volume is adiabatic and reversible,

$$dU_{tank} = 0 - 0 - \delta\overline{W}_{shaft,rev} - \bar{h}_e(T)\, dN_e, \text{ or} \qquad (H)$$

$$\delta\overline{W}_{shaft,rev} = -\,\bar{u}_{tank}\, dN_{tank} - N_{tank}\, d\,\bar{u}_{tank} + (\bar{u}_{tank} + \bar{R}T(t))dN_{tank}$$

$$= -\,N_{tank}\,\bar{c}_{v0}dT_{tank} + \bar{R}T\,(t)dN_{tank}$$

$$= -\,N_{tank}\,\bar{c}_{v0}dT_{tank} + VdP - \bar{R}N_{tank}dT_{tank}$$

$$-\,dT_{tank}N_{tank}(\bar{c}_{v0} + \bar{R}) + VdP = -\,dT_{tank}\,N_{tank}\,\bar{c}_{p0} + VdP.$$

Therefore the relationship between the temperature and N_{tank} is of the form $T_{tank}/T_{tank,1} = (P_{tank}/P_{tank,1})^{(k-1)/k} = (N_{tank}T_{tank}/N_{tank,1}T_{tank,1})^{(k-1)/k}$, i.e.,

$$(T_{tank}/T_{tank,1})^{(1/k)} = (N_{tank}/N_{tank,1})^{(k-1)/k}, \text{ and}$$

$$\delta\overline{W}_{shaft,rev} = -\,dT_{tank}\,N_{tank,1}(T_{tank}/T_{tank,1})^{(1/(k-1))}\,\bar{c}_{p0} + VdP.$$

The efficiency of heat engines can oftentimes be improved by increasing the peak temperature in the relevant thermodynamic cycle. However, materials considerations impose a restriction on the peak temperature. Materials may be kept at a desired safe temperature by providing sufficient cooling. In that case entropy is generated through cooling which must be compared with the work loss by reducing the peak temperature. When heat exchangers and cooling systems are designed for work devices, information on possible work loss should be provided.

---

dd. Example 31

Consider a 1 meter long turbine that operates steadily and produces a net power output of 1000 kW. Gases enter the turbine at 1300 K (T_i) and 10 bar (P_i), and leave at 900 K (T_e) and 1 bar (P_e). The turbine walls are insulated, but its blades are cooled. The cooling rate per unit area of the blade is given by the relation $h(T_{avg} - T_{blade})$ where $T_{avg} = (T_i + T_e)/2$. The Nusselt number ($Nu = hC/\lambda$) on the gas side of the blade is 1000, where h denotes the convective heat transfer coefficient (kW m^{-2} K^{-1}), C the chord length (which is 15 c.m. along axial direction), and λ the thermal conductivity of the hot gases (= 70_10^{-6} kW m^{-2} K^{-1}). The blade A= C _ blade height which is assumed to be the same as the chord length. There are approximately 40 blades for each rotor and 3 rotors for every meter of length. Assume that $c_p = 1.2$ kJ kg^{-1} K^{-1} and $R = 0.287$ kJ kg^{-1} K^{-1}.

Write the generalized overall energy conservation equation.
What is the heat loss rate if the blade temperature $T_{blade} = 900$ K.
Determine the gas mass flow rate.
Write the entropy balance equation and simplify it for this problem.
Determine the entropy generation rate.

Solution

The energy equation can be written in the form

$dE/dt = \dot{Q} - \dot{W} + \dot{m}_i(h + ke + pe)_i - \dot{m}_e(h + ke + pe)_e$.

At steady state, $dE_{c.v.}/dt = 0$, $ke = 0$, $pe = 0$, $dm_{c.v.}/dt = 0$, hence $\dot{m}_i = \dot{m}_e = \dot{m}$, i.e.,

$$0 = \dot{Q} - \dot{W} + \dot{m}(h_i - h_e). \quad (A)$$

$$\dot{Q} = h\,A\,(T_{avg} - T_{blade}), \text{ and} \quad (B)$$

$$A = 0.15 \times 01.5 \times 40 \times 3 = 2.7 \text{ m}^2. \quad (C)$$

Since, $h\,C/\lambda = 1000$,

$$h = 1000 \times 70 \times 10^{-6} \div 0.15 = 0.467 \text{ kW m}^{-2} \text{ K}^{-1}. \quad (D)$$

Using Eqs. (B), (C), and (D),

$$\dot{Q} = 0.462 \times 2.3 \times (1100 - 900) = 252 \text{ kW}. \quad (E)$$

$$h_e - h_i = c_{p0}(T_i - T_e) = 1.2 \times (1300 - 900), \text{ and} \quad (F)$$

$$\dot{W} = 1000 \text{ kW}. \quad (G)$$

Using Eqs. (A), (E), (F), and (G),

$$0 = 252 - 1000 - \dot{m} \times 1.2 \times (1300 - 900), \text{ i.e.,}$$

$$\dot{m} = 2.6 \text{ kg s}^{-1}. \quad (H)$$

The entropy balance equation at steady state is

$$dS_{c.v.}/dt = 0 = \dot{Q}/T_{b,j} + \dot{m}(s_i - s_e) + \dot{\sigma}_{cv}. \quad (I)$$

$$(s_e - s_i) = c_{p0} \ln(T_e/T_i) - R \ln(P_e/P_i). = 1.2 \ln(900 \div 1300) - 0.287 \ln(1/10) \quad (J)$$

$$= -0.441 + 0.66 = 0.220 \text{ kJ kg}^{-1} \text{ K}^{-1}. \quad (K)$$

Using Eqs. (H)–(K),

$$\dot{\sigma}_{c.v.} = -(-252/900) - 2.6 \times (-0.22) = 0.28 + 0.57 = 0.85 \text{ kW K}^{-1}.$$

Remark

The entropy generation due to blade cooling results in a work loss of $T_o\,\dot{\sigma}_{c.v.} = 298 \times 0.85 = 250$ kW (this is discussed further in Chapter 4). However, the increased gas temperature at the turbine inlet results in a larger work output.

4. Irreversible Processes and Efficiencies

Adiabatic expansion and compression processes prevent energy loss related to heat transfer. Idealized adiabatic processes are also isentropic. However, actual processes may not occur under quasi–equilibrium conditions and may, therefore, be adiabatic, but irreversible Figure 37. In this case it is useful to compare various adiabatic devices operating at identical pressure ratios for either expansion or compression (e.g., turbines and compressors) or over the same expansion or compression ratios (e.g., the compression and expansion strokes in automobile engines). The resulting term represents the adiabatic (or isentropic) efficiency η_{ad} that is

η_{ad} = actual work output ÷ isentropic work output = w/w_s (expansion processes) or

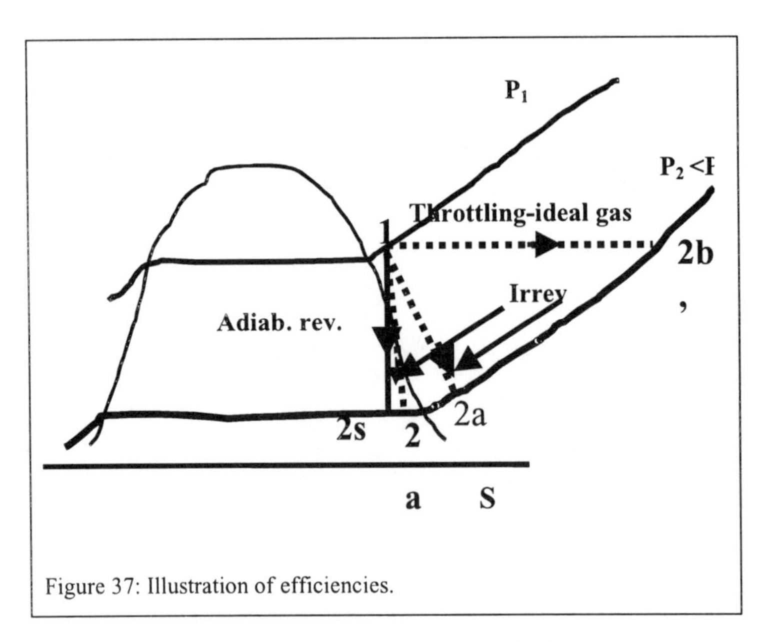

Figure 37: Illustration of efficiencies.

η_{ad} = isentropic work input ÷ actual work input = w_s/w (compression processes). For these processes, $w = |h_1 - h_2|$, and $w_s = |h_1 - h_{2s}|$.

Compressors and turbines are also designed for isothermal processes. The work input can be reduced even for isothermal processes. The isothermal efficiency η_{iso} is defined as

$\eta_{iso} = w/w_{iso}$ (expansion processes) or w_{iso}/w (compression processes).

These efficiencies are also sometimes referred to as First law efficiencies. The work $w_{rev} = \int Pdv$ or $-\int vdP$, respectively, for closed and open systems.

5. Entropy Balance in Integral and Differential Form

We have previously determined the entropy generation by assuming the system properties to be spatially uniform, except at the boundary. We now present the appropriate balance equations for irreversible processes that occur in continuous systems containing spatial non–uniformities. The methodology is similar to that used to present the differential forms of the mass and energy conservation equations in Chapter 2.

a. *Integral Form*

Equation (64) may be expressed in integral form as

$$d/dt(\int_{c.v.} \rho s dV) + \oint \rho \vec{V} s \cdot d\vec{A} = -\oint (\vec{Q}''/T)\, d\vec{A} + \int_{c.v.} \dot{\sigma}'\, dV. \quad (80)$$

The relevant density and temperature are those at the boundary of the control surface and $\vec{Q}''$, $\vec{V}$, and $\vec{A}$, respectively, denote the heat flux, velocity, and area vectors.

b. *Differential Form*

The entropy balance equation can be written in a differential form to evaluate the entropy generation in a control volume. Applying the Gauss Divergence theorem to Eq. (80) we obtain

$$d/dt(\rho s) + \vec{\nabla}\cdot\rho\vec{V}s = -\vec{\nabla}\cdot(\vec{Q}''/T) + \dot{\sigma}'''_{cv}. \tag{81}$$

The control surface is shrunk to a surface around an infinitesimally small volume dV and the boundary temperature becomes the elemental volume temperature. Employing the mass conservation equation

$$d\rho/dt + \nabla.\rho\vec{V} = 0,$$

Eq. (81) can be simplified into the form

$$\rho ds/dt + \rho\vec{V}\cdot\vec{\nabla}s = -\vec{\nabla}\cdot(\vec{Q}''/T) + \dot{\sigma}'''_{cv}. \tag{82}$$

Recall that the Fourier law is represented by the relation $\vec{Q}'' = -\lambda\vec{\nabla}T$. For isotropic materials (that have uniform properties in all spatial directions) the law involves a linear relation between the two vectors $\vec{Q}''$ and $\vec{\nabla}T$, where the direction $\vec{Q}''$ depends upon that of the temperature gradients that are normal to the isotherms. That the two vectors must be parallel to each other is known as Curie's principle. For a one–dimensional problem

$$\vec{Q}'' = -\lambda\,\vec{i}(dT/dx).$$

The heat flux vector has a positive direction with respect to x when heat flows from a higher temperature to a lower temperature, and the introduction of the negative sign satisfies the Second law. We will later prove that this sign in the Fourier law can be determined using the criterion that $\dot{\sigma}'''_{cv} > 0$ for irreversible heat transfer for all materials having a positive thermal conductivity coefficient. Applying the Fourier law to Eq. (82),

$$\rho\partial s/\partial t + \rho\vec{V}.\nabla s = \vec{\nabla}\cdot(\lambda\vec{\nabla}T/T) + \dot{\sigma}'''_{cv}. \tag{83}$$

We have assumed that the local volume is in thermodynamic equilibrium and that entropy generation occurs due to irreversibilities between the various local volumes.

6. Application to Open Systems

The entropy generation $\dot{\sigma}'''_{cv}$ can be determined as a function of spatial location within a volume by solving Eq. (83) provided that the temperature and pressure are known. In Chapter 4 we will discuss that the work lost due to irreversibilities is given by the product $T_0\dot{\sigma}'''_{cv}$.

Fins are used in heat exchangers to increase the heat transfer rate. We have discussed that a hot body can be cooled either by directly transferring heat to the ambient (by generating entropy) or by using a heat engine to produce reversible work (without producing entropy). The fins are entropy generators at steady state for which Eq. (83) yields

$$\rho\vec{V}\cdot\vec{\nabla}s = -\vec{\nabla}\cdot(\lambda\vec{\nabla}T/T) + \dot{\sigma}'''_{cv}. \tag{84}$$

Since there is no convection heat transfer within the solid fins,

$$-\vec{\nabla}\cdot(\lambda\vec{\nabla}T/T) + \dot{\sigma}'''_{cv} = 0.$$

a. Steady Flow

Consider the one–dimensional steady flow of a fluid with negligible temperature and velocity gradients. For this case Eq. (83) simplifies to the form

$$\rho v\, ds/dx = \dot{\sigma}'''_{cv}.$$

For an ideal gas

$$ds = c_p\, dT/T - R\, dP/P$$

Since the temperature is virtually constant. Therefore,

$$\rho v\,(-R/P)\, dP/dx = \dot{\sigma}'_{cv}.$$

In this case $dP/dx \le 0$. For a fixed mass flow rate, integrating between the inlet ($P = P_i$) and any other section where $P < P_i$,

$$\dot{\sigma}'_{cv} = (R\rho v/x)\ln(P_i/P).$$

Friction causes pressure losses so that. $\dot{\sigma}'_{cv} > 0$.

b. Solids

The energy conservation for solids can be written in the form

$$\rho c\,\partial T/\partial t = -\vec{\nabla}\cdot\dot{\vec{Q}}'' \tag{85a}$$

Since $s = c\, dT/T$, manipulating Eq. (83)

$$(\rho c/T)\partial T/\partial t = -\vec{\nabla}\cdot(\dot{\vec{Q}}''/T) + \dot{\sigma}'_{cv}. \tag{85b}$$

Dividing Eq. (85a) by the temperature and subtracting the result from Eq. (85b) we obtain

$$(\vec{\nabla}\cdot\dot{\vec{Q}}'')/T + \vec{\nabla}\cdot(\dot{\vec{Q}}''/T) + \dot{\sigma}'_{cv} = 0\text{, i.e.,}$$

$$\dot{\sigma}'_{cv} = (1/T^2)\dot{\vec{Q}}''\vec{\nabla}\cdot T\text{, i.e.,}$$

Since $\dot{\sigma}'_{cv} > 0$,

$$\dot{\vec{Q}}''\cdot\vec{\nabla}T < 0.$$

The important implication is that $\dot{\vec{Q}}'' > 0$ if $\vec{\nabla}T < 0$, i.e., heat can only flow in a direction of decreasing temperature.

---

ee. Example 32

Consider a 5 mm thick infinitely large and wide copper plate, one surface of which is maintained at 100°C and the other at 30°C. The specific heat and thermal conductivity of copper are known to be, respectively, 0.385 kJ kg^{-1} K^{-1} and 0.401 kW m^{-1} K^{-1}. Determine the entropy at 100°C and 30°C, and the entropy production rate.

Solution

$$s = c\ln(T/T_{ref}). \tag{A}$$

At 100°C, $s = 0.385\ln(373 \div 273) = 0.1202$ kJ kg^{-1} K^{-1}.
Likewise, at 30°C $s = 0.0401$ kJ kg^{-1} K^{-1}.
Using Eq. (85b), for the one-dimensional conduction problem,

$$\dot{\sigma}'_{cv} = -d/dx(\lambda/T\, dT/dx). \tag{B}$$

From energy conservation,

$$-\lambda\, dT/dx = \dot{q}'' = \text{Constant}. \tag{C}$$

Therefore, integrating from the boundary condition, namely, $T = T_0$ at $x = 0$,

$$T_0 - T = \dot{q}''x/\lambda \tag{D}$$

Dividng Eq. (C) by T,replacing T with Eq.(D), and then using the result in Eq. (B)

$\dot\sigma'_{cv} = d/dx(\dot q''/(T_0 - \dot q'' x/\lambda)$

Multiplying by dx and integrating within the limits x=0 and x= L,

$$\dot\sigma''_{c.v.}(x) = \frac{\dot q''}{T_0 - \dot q'' x/\lambda} - \frac{\dot q''}{T_0} \quad \text{(E)}$$

For the given problem $\dot q''$ = 0.401 × (373 – 303) ÷ 0.005 = 5614 kW m^{-2}, and

$\dot\sigma_{c.v.}''(L)$ =(614×(1÷(373–5614×0.005÷0.401)–(÷(373))== 3.48 kW m^{-2} K^{-1}.

Even though entropy is produced at a rate of 3.48 kW m^{-2} K^{-1} within the plate, the entropy at the edges (across the plate thickness) will not increase, since the entropy is a property that depends only on the local temperatures. In addition, the entropy that is produced is flushed out in the form of transit entropy at x = 0, i.e., through thermal conduction. This statement can be verified by employing the entropy balance equation for a control volume around the entire plate,

$dS/dt - \int \delta \dot Q''/T_b = \dot\sigma'_{cv}$.

Since the temperature at any location is constant, the entropy cannot accumulate at steady state and dS/dt = 0. Therefore, $\int d\vec{\dot Q}''/T_b = -\dot q''$ (1÷373 – 1÷303) = $\dot\sigma'_{cv}$ = 3.48 kW m^{-2} K^{-1}.

A similar example involves electric resistance heating for which the coil temperatures can be maintained at steady state by transferring heat out of the coils as fast as it is produced. The entropy flux $d\vec{\dot Q}''/T_b$ acts in the same manner in order to maintain constant entropy.

Summarizing this section on the entropy balance,

$\Delta S = \sigma$ for an isolated system.

$dS_{c.v.}/dt = \dot\sigma_{c.v.}$, in rate form for an isolated system.

$dS_{c.v.}/dt = \dot\sigma_{c.v.}$, in rate form for an adiabtic closed system.

$dS_{c.v.}/dt = \dot Q_{c.v.}/T_b + \dot\sigma_{c.v.}$, in rate form for any system.

$dS_{c.v.}/dt = \dot Q_{c.v.}/T_b + \dot m_i s_i - \dot m_e s_e + \dot\sigma_{c.v.}$, in rate form for an open system.

J. MAXIMUM ENTROPY AND MINIMUM ENERGY

The concept of mechanical states is illustrated in Figure 38 using the example of balls placed on a surface of arbitrary topography. Position A represents a nonequilibrium condition, whereas B, C, D, and F are different equilibrium states. A small disturbance conveyed to the ball placed at position C will cause it to move and come to rest at either position B or D. Therefore, position C is an unstable equilibrium state for the ball. Small disturbances that move the ball to positions B and D will dissipate, and the ball will return to rest. If the ball is placed at B, a large disturbance may cause it to move to position D (which is associated with the lowest energy of all the states marked in the figure). If the ball is placed at position D and then disturbed, unless the disturbance is inordinately large, it will return to its original (stable) position. Position D is an example of a stable equilibrium state, whereas position B represents a metastable equilibrium state. Also it is noted that the change

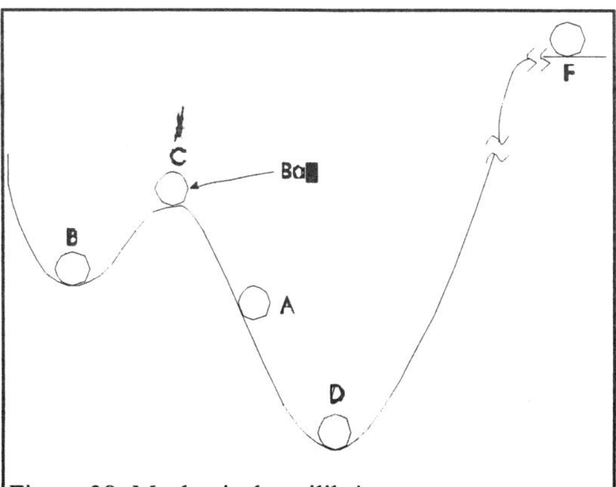

Figure 38: Mechanical equilibrium states.

of states say from (C) to (B) or (D) are irreversible. A criteria for "stability"can also be described based on the potential energy associated with the various states depicted in Figure 38. If the potential energy decreases (i.e., $\delta(PE) < 0$) as a system is disturbed from its initial state, that state is unstable (state C). On the other hand, if the potential energy increases once the system is disturbed ($\delta(PE) > 0$), that initial state is stable (states B and D).

Similarly, the stability of matter in a system can be described in terms of its thermodynamic properties. A composite system containing two subsystems is illustrated in Figure 39. The first subsystem consists of an isolated cup of warm coffee or hot water (W), while the room air surrounding it is the other subsystem (A). The internal constraints within the composite system are the insulation (which is an adiabatic constraint) around the coffee mug and the lid (which serves as a mechanical constraint) on the cup. The two subsystems will eventually reach thermal equilibrium state, once the constraints are removed. The magnitude of the equilibrium temperature will depend upon the problem constraints. For example, if the walls of the room are rigid and insulated, the temperature of the room air will increase as the coffee cools. Consequently, the air pressure will increase, but the internal energy of the combined system will not change.

If the mechanical constraint is still in place, it is only possible to reach thermal equilibrium. If the mechanical constraint is removed, say, by using an impermeable but movable piston placed on top of the water, then thermo-mechanical (TM) equilibrium is achieved. If the piston is permeable, in that case the water may evaporate and also reach phase (or chemical) equilibrium. Therefore, the conditions of a system depend upon the constraints that are imposed. If the walls of the composite system are uninsulated, heat losses may occur from the system, thereby reducing the internal energy. The equilibrium temperature will be lower for this case. Therefore, equilibrium may be reached in a variety of ways so that various scenarios may be constructed, depending on the constraints, as follows.

The room may be insulated, impermeable, and rigid so that the composite system is isolated. In this case entropy generation will occur due to irreversible processes taking place inside the system.

The room may be diathermal, rigid and impermeable. Interactions with the environment (which serves as a thermal energy reservoir at a temperature T_0) are possible. In this case the combined entropy of the coffee and room air may not change as the two subsystems undergo the equilibration process. The coffee will transfer heat to the room air, which in turn will transfer it to the environment. Consequently, the internal energy of the composite system will decrease.

The room might be diathermal with a flexible ceiling allowing the pressure to be constant during the process. In this case, we can show that the enthalpy decreases while the entropy, pressure, and mass are fixed.

1. Maxima and Minima Principles

a. Entropy Maximum (For Specified U, V, m)

The isolated system shown within the dotted boundary in Figure 39 contains two subsystems say hot water (W) and air (A). That isolated system contains two subsystems (W) and (A). The total internal energy

$$U = U_W + U_A. \tag{86}$$

Similarly, under a constrained equilibrium state the entropy is additive, i.e.,

$$S = S_W + S_A. \tag{87}$$

Once the constraints are removed, the composite system enters a nonequilibrium state (analogous to the ball at position A in Figure 38). Thus an irreversible occurs. We can apply Eq. (28) to the system within the dotted boundary,

$$dS = \delta Q/T_b + \delta\sigma,$$

For an irreversible process, $\delta\sigma > 0$. Since the composite system is adiabatic $\delta Q = 0$. Hence $dS = dS_w + dS_A > 0$ or S keeps increasing as long the irreversibility exists. The internal energy ($U = U_W + U_A$) is conserved, but the entropy increases until an equilibrium state is reached at which dS = 0 and the entropy is at a maxima as shown in Figure 40 (analogous to the ball at position D in Figure 38). The final equilibrium state is achieved when the entropy of the isolated system reaches a maximum. Note that constitutive rate equations (e.g., the Fourier law, Newton's laws etc.) are not required in order to determine the entropy generation as long as S is known as a function of U and V from the basic pulley-weight type of experiments. Even though the composite system is isolated, the local temperature within it is time varying, as are other local system properties. Although, dU = dV = 0, since an irreversible process occurs, entropy is generated, since, $\delta\sigma > 0$.

Recall from earlier derivation that at equilibrium.

$$S = S(U, V, N_1, \ldots N_n) \tag{88}$$

while during an irreversible process occurring in the isolated system

$$dS \neq 0, \; U,V, m \text{ fixed (implying dU=0, dV=0 during the process).} \tag{89}$$

Then, at the entropy maxima, (cf. Figure 40)

$$d^2S < 0, , U,V, m \text{ fixed} \tag{90}$$

Intuitively, equilibrium is reached as the temperatures of the two subsystems W and A approach each other. However, in a chemically reacting system, the temperature alone cannot describe equilibrium, since the composition may change and S-max principle can be invoked to describe the equilibrium.

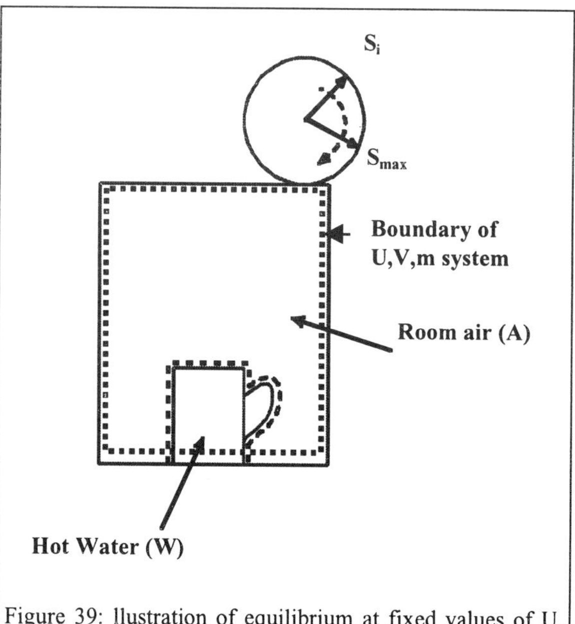

Figure 39: llustration of equilibrium at fixed values of U, V, m.

Recall the entropy balance equation for a closed system $dS = dU/T + P\,dV/T + \delta\sigma$. If this expression were to be used in the context of the composite system of Figure 39 (for which $dU = dV = 0$), then $dS = 0$. (By implication, an adiabatic fixed-volume system is incapable of an entropy change.) This relation cannot be applied to a composite system that is in a non-equilibrium state. For example, one cannot assign a single temperature to the composite system illustrated in Figure 39. However, it may be applied separately for each subsystem to properly analyze the problem if each sub-system in internal equilibrium with irreversibility confined to a thin boundary between the two subsystems W and A . i.e., $dS_W = dU_W/T_W + P_W\,dV_W/T_W$ with $\delta\sigma_w = 0$ and $dS_A = dU_A/T_A + P_A\,dV_A/T_A$ with $\delta\sigma_A = 0$. We will use such a procedure to illustrate the S_{max} principle with a simple example of cooling of coffee in room air.

---

ff. Example 33

The initial temperature of a completely covered and insulated coffee cup containing 1 kg of coffee is 350 K. Assume the properties of coffee to be the same as those of water ($c = 4.184$ kJ kg^{-1} K^{-1}). The ambient air ($c_{v0,A} = 0.713$ kJ kg^{-1} K^{-1}) is contained in a rigid and insulated room at a temperature of 290 K. The air mass is 0.4 kg. If the insulation is removed, but the covering retained (to allow heat transfer but no mass transfer) from the coffee, and the cup is cooled gradually, determine the following (ignoring evaporation and assuming $u_W = c_W(T_W - 273)$, $u_A = c_{v0,A}(T_a - 273)$, $s_W = c\ln(T_W/273)$, and $s_A = c_{v0,A}\ln(T_A/273) + R\ln(v/v_{ref})$):

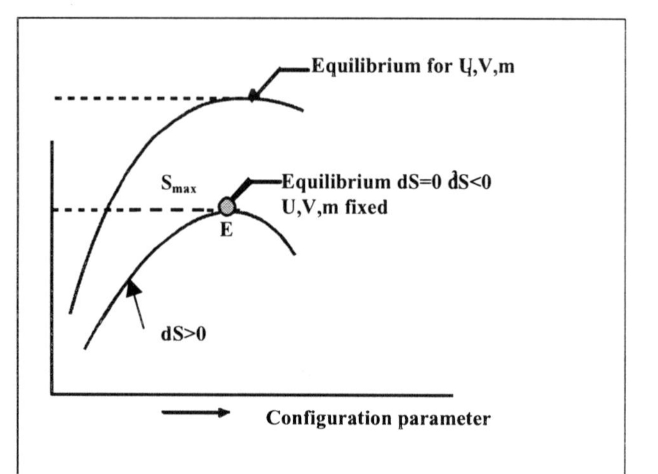

Figure 40 : Entropy maximum S_{max} principle for specified values of U, V, m.

The heat removed from the water for each 0.5 K drop in the coffee temperature.
The temperature of air T_A when the coffee cools to 349.5 K.
The entropy of the water (S_W) at $T_W = 350$, and 349.5 K.
The entropy of the air (S_A) when $T_W = 350$, and 349.5 K.
The total entropy $S(U, V, m) = S_A + S_W$ when $T_W = 350$, and 349.5 K.
Repeat these steps for $T_W = 349, 348.5, \ldots$ and so on, and plot the entropy with respect to T_A and U_A.

Solution

$$U_W = m_W c_W (T_W - 273), \qquad \text{(A)}$$

$$U_A = m_A c_{v0A} (T_A - 273), \qquad \text{(B)}$$

$$V_W = m_W/\rho_W = 1 \div 1000 = 0.001\ \text{m}^3,$$

$$\rho_A = P_A/RT_A = 100 \div (0.287 \times 290) = 1.2\ \text{kg m}^{-3},$$

$V_A = m_A / \rho_A = 0.4 \div 1.2 = 0.333 \text{ m}^3$, and the total volume

$V = V_W + V_A = 0.001 + 0.333 = 0.334 \text{ m}^3$.

The volume of the coffee cup V_W =0.001 m^3, and it contains internal energy U_W = 322.2 kJ. The room air contains internal energy U_A = 13.3 kJ, and volume V_A = 0.333 m^3. The total internal energy U = 335.5 kJ, total volume V = 0.334 m^3, and total mass m = 1.4 kg. Given this information, we must determine the possible states that the system can attain if the internal constraints are removed. One such state results when the two subsystem temperatures equal each other. (This technique will not always work as will be seen later in the case of chemical reactions in Chapter 10.) Another possibility is to examine those states as the system entropy increases to a maximum from its initial value.

If the coffee cools through a temperature decrement $dT_W = -0.5$ K, using the First law,

$$\delta Q_W - \delta W_W = dU_W, \text{ so that} \tag{C}$$

$$\delta Q_W - 0 = dU_W = m_W\, c\, dT = 1 \times 4.184 \times (-0.5) = -2.092 \text{ kJ}.$$

Heat is transferred resulting in a temperature rise in the air.

Applying the First law to the room air,

$$\delta Q_A - \delta W_A = dU_A, \text{ i.e., } 2.092 - 0 = m_A\, c_A\, dT_A \text{ so that} \tag{D}$$

$$dT_A = 2.092 \div (0.4 \times 0.713) = 7.335 \text{ K}.$$

Therefore, $T_A = 290 + 7.335 = 297.335$ K.

Since the composite system is in a constrained equilibrium state, the total entropy can be obtained by summing up the subsystem entropies. Using Eq. (53) with $T_2 = T_W$ and

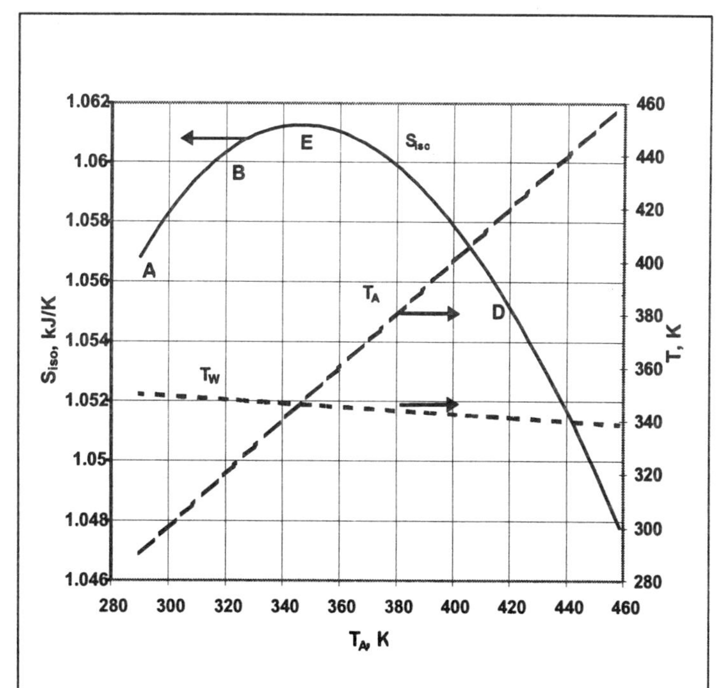

Figure 41 : The entropy with respect to T_A For Specified U, V, and m values.

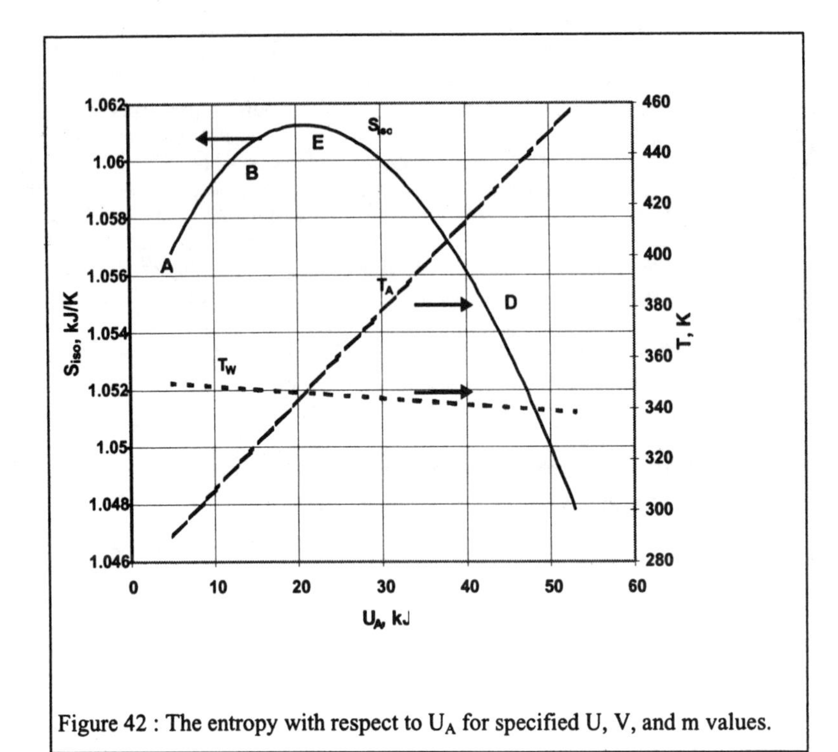

Figure 42 : The entropy with respect to U_A for specified U, V, and m values.

$T_1 = 273$ K,

$s_W = c_W \ln (T_W/273)$, so that

$$S_W = m_W s_W = m_W c_W \ln (T_W/273) \quad \text{(E)}$$

$= 1$ (kg) $\times$ 4.184 (kJ kg^{-1} K^{-1}) $\times$ ln(350 $\div$273) = 1.0395 kJ K^{-1}.

Similarly, at temperatures of 349.5, 349, 348.5, ..., K, S_W = 1.0336, 1.0276, ..., kJ K^{-1}.
For a 0.5 K temperature drop dS_W = 1.0336 – 1.0395 = –0.0059 kJ K^{-1}.
Since the volume of air does not change,

$S_A = m_A c_{vA} \ln (T_A/273) + R \ln (v/v_{ref})$. Let $v_{ref} = v = (V/m_A)$, so that

$$S_A = m_A c_{vA} \ln (T_A/273). \quad \text{(F)}$$

Therefore, at 290 K, S_A = 0.4 (kg) $\times$ 0.713 (kJ kg^{-1} K^{-1}) $\times$ ln(290/273) = 0.01722 kJ K^{-1}, and at 297.335 K, S_A = 0.02435 kJ K^{-1}.
For a 7.335 K temperature rise, dS_A = 0.02435 – 0.01722 = 0.00713 kJ K^{-1}.
At 350 K, S = $S_W + S_A$ = 1.0395 + 0.01722 = 1.0568 kJ K^{-1}. At 349.5 K, S = 1.0579 kJ K^{-1}. Therefore,

$dS = dS_W + dS_A$ = – 0.0059 + 0.00713 = 0.00173 kJ K^{-1}.

The combined system is a closed and adiabatic system so that $dS - \delta Q/T = \delta\sigma$.

Since $\delta Q = 0$, $\delta\sigma = dS$ = 0.0014 kJ K^{-1} for the infinitesimal process during which dT_W = – 0.5 K. As entropy is generated due to system irreversibilities the entropy of the composite system changes.

Figure 41 illustrates the change in the composite system entropy with respect to the temperature T_A of subsystem A. Likewise Figure 42 presents the variation of the composite system entropy with respect to the internal energy U_A contained in air at the

constrained equilibrium state (i.e., by allowing the coffee to lose heat, then placing the insulation back around the coffee mug and preventing any further heat transfer). The entropy increases at the fixed U, V, and m values so that the entropy generation $\delta\sigma > 0$ along the branch ABE. At point E there are no constraints within the system, the temperatures T_A and T_W are equal, and the entropy reaches a maxima.
Therefore, equilibrium is that state at which the entropy is the highest of all possible values after considering all of the constrained equilibrium states (for specified values of U, V and m). This is called the highest entropy principle.

Remarks

Thermal equilibrium: The case of a homogeneous-single component system.
Now, consider the case when the total internal energy, volume, and mass are held constant, but the initial mass of cold system can be changed. Figure 43 contains two curves for molecular nitrogen corresponding to the same values of U,V, and m. The curve AEC has been generated for a cold system mass equal to 0.2 kg and a hot system mass of 0.4 kg. The hot and cold conditions are T_{cold} = 290 K, V_{cold} = 0.13 m^3, U_{cold} = 43 kJ, m_{hot} = 0.4 kg, T_{hot} = 350 K, V_{hot} = 0.26 m^3, U_{hot} = 104 kJ. The total values (for the system) are U = 147 kJ, V = 0.36 m3, and m = 0.6 kg. The curve BED corresponds to a cold mass of 0.25 kg with T_{cold} = 290 K, U_{cold} = 54 kJ; and m_{hot} = 0.35 kg, T_{hot} = 359 K, U_{hot} = 93 kJ (for the same total values for U, V and m as for curve AEC). Both sets of initial conditions reach the same value for S_{max} at equilibrium, i.e., the maximum entropy is a function of only U, V, and m.

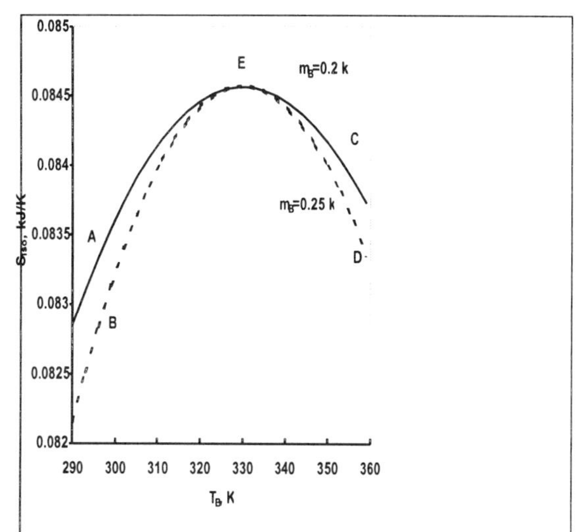

Figure 43 : The entropy with respect to the temperature T_B during the interaction of hot and cold nitrogen.

The case of irreversibility during spontaneous processes.
In the presence of constraints each subsystem (e.g. A and W) is in equilibrium. Once the constraints are removed, the entropy reaches a maximum following a spontaneous process. In order to reverse that process, the entropy of the composite system should decrease which is impossible according to the Second law. The use of a Carnot heat pump in this case would require external work input. Secondly, the entropy of the coffee increases ($\Delta S_{coffee} > 0$) with heat addition. As the air cools back to its initial temperature, its entropy S_A decreases ($\Delta S_A < 0$). However, $|\Delta S_A| = |\Delta S_{coffee}|$, and the overall entropy remains unchanged. Therefore, it is not possible to restore the entropy of an isolated system back to its initial state.

The S-U-V surface.
In our analyses, we have assumed the various subsystems to undergo quasiequilibrium processes. Each of these processes within each subsystem can be mapped in S–U–V space. Since the composite system is in a nonequilibrium state, the process that it undergoes cannot be mapped in this S–U–V space (Figure 44). However, if the third coordinate is elected as the deformation coordinate (e.g., internal energy of wa-

ter in the current example increasing away from the origin "O"), the process ABE in Figure 41 is mapped as A–K–2U in Figure 44.. During the process A-K-2U, one could have stopped the process at state K by covering the water with an insulated lid. However dS at this point is not yet zero, i.e., the entropy has not yet reached a maximum value. This state K could be construed as constrained equilibrium states of each subsystem constituting the composite system The final equilibrium state can be located in that space (points 2S,2U, Figure 44) where dS=0. Note that the process D-L-2U is possible only if the air is warmer than the water, i.e., equilibrium is approached towards increasing entropy.

Equilibrium.

At the equilibrium point E (Figure 41), at a microscopic level at $S = S_{max}$ a few energetic molecules of water (at a slightly higher temperature than of the air taken as a whole) transfer heat to air. An equal number of energetic air molecules transfer heat back to the water resulting in reversible heat transfer. Although the system is constantly disturbed at the equilibrium state, the state is overall stable. At the stable equilibrium state dS = 0, so that $\delta\sigma = 0$ i.e $T_W \approx T_A = 346.2$ K, with the implication that the entropy is constant adjacent to its peak value. The processes are reversible, i.e., the air can supply a differential amount of heat to the coffee and *vice versa*. The process A–B–E illustrated in Figure 41 is possible but process E–D is not. However, if the initial state of the system lies at point D (i.e., with hot air and cold coffee), the process D–E is possible, and $\delta\sigma > 0$.

Using Eqs. (A), (B), (E), and (F) to eliminate T_W and T_A in terms of U_W and U_A we obtain

$$S = S_W + S_A$$

$$= m_W c_W \ln(U_W/(m_W c_W \times 273) + 1) + m_A c_{v0,A} \ln(U_A/(m_A c_{v0,A} \times 273) + 1), \quad (G)$$

where $U = U_A + U_W$ is fixed. Therefore, the entropy is a function of U_A alone if U, m_W, and m_A are fixed. In order for the entropy to reache a maxima, the necessary conditions are

$$\partial S/\partial U_a = 0 \text{ , and } \partial^2 S/\partial U_a^2 < 0 \text{ at equilibrium, } U, V.m \text{ fixed}$$

By differentiating Eq. (G) it is possible to determine U_A or T_A when $S = S_{max}$. Inversely, if a system is in equilibrium, dS = 0 and $d^2S < 0$ for any small disturbance. (This is called the stability criteria, and is discussed later in Chapter 10.)

If the subsystem masses and total volume are kept constant, but the initial temperatures and internal energies are changeable, the equilibrium value S_{max} changes, since S = S(U,V,m). Similarly, if only the volume is changed, the pressure also changes, and the value of S_{max} is different. Changes in the total mass or number of moles have a like effect, since S = S(U, V, N). Differentiating the entropy near equilibrium,

$$dS = \partial S/\partial U\, dU + \partial S/\partial V\, dV + \partial S/\partial N\, dN, \quad (H)$$

so that

$$dS = dU/T - P\, dV/T + \mu\, dN. \quad (I)$$

This expression is only valid between two equilibrium states, namely, S(U,V,N) and S(U+dU,V+dV,N+dN).

Heterogeneity and equilibrium

That equilibrium exists between the coffee and air does not imply that the pressure and internal energy are uniform. If the internal energy is everywhere the same, a system exists in a homogeneous state (or phase). Generally, when two subsystems that are initially in a nonequilibrium state reach equilibrium with each other, heterogeneous states (or two phases) may exist. Differences in the system density and internal energy describe these phases.

A stability test for equilbrium at specified values of U, V and m.

Recall that the system at microscopic level is incessantly dynamic. Hence disturbance occurs continuously. Let us consider the impact of small disturbance on the system at fixed values of the internal energy, volume, and mass. It is possible for a group of air molecules to exist at a temperature $T' > T_{equil}$. This implies that a group of water molecules must correspond to a temperature $T'' < T_{equil}$. Such a situation can arise if a few air molecules gain some energy dU_A. At that state the entropy of air $S_A(U_A + dU_A) = S(U_A) + dS/dU_A\, dU_A + d^2S/dU_A{}^2\, dU_A{}^2 + \ldots$. The water can lose an equivalent amount of energy dU_W (since the total energy is unchanged), and its entropy $S_W(U_W + dU_W) = S + dS/dU(dU_W) + d^2S/dU_W{}^2(dU_W)^2$. Therefore, $S_A(U_A + dU_A) + S_W(U_W + dU_W) = S_A + S_W + 1/T(dU_A + dU_W) + (d^2S/dU^2)\,(dU_A{}^2 + dU_W{}^2) + \ldots$. Since the energy $U = U_A + U_W$ is fixed, $dU_A = -\, dU_W$ and $dU_A{}^2 = dU_W{}^2$. However, in this case the disturbed state entropy is lower than the maximum value, and $S_A(U + dU) + S_W(U_W + dU_W) - (S_A(U_A, V_A) + S_W(U_W, V_W)) < 0$. This implies that $(d^2S/dU^2)\,(dU_A)^2 < 0$ if the initial state is at the "maximum entropy" or stable equilibrium state (which is also known as the stability condition, Chapter 10). In this example, the small group of hotter air molecules will attempt to equilibrate after contact with the water molecules. Such a process, where the system self–adjusts to a disturbance, is said to follow Le Chatelier's principle.

An application.

Assume that at the end of the compression stroke of an Otto cycle a gasoline–air mixture reaches a gas–phase temperature of 600 K. tIn the presence of constraints each subsystem is in equilibrium. Once the constraints are removed, the entropy reaches a maximum following a spontaneous process. In order to reverse that process, the entropy of the composite system should decrease which is impossible according to the Second law. Can we use Carnot heat pump. First, it requires external work input. Secondly, the entropy of the coffee increases ($\Delta S_{coffee} > 0$) with heat addition. As the air cools back to its initial temperature, its entropy S_A decreases ($\Delta S_A < 0$). However, $|\Delta S_A| = |\Delta S_{coffee}|$, and the overall entropy remains unchanged. Therefore, it is not possible to restore the entropy of isolated system back to its initial state that time, a spark initiates combustion. After about say 2 ms, half of the chamber is filled with hot gases at 2000 K, the unburned side is still at 600 K, and the reaction is frozen. At this time the system is insulated and the piston locked (i.e., U, V, and m are fixed). Equilibrium is achieved for this system when the entropy reaches a maxima (or when $T_A = T_B$).

We have so far dealt with the equilibrium conditions for isolated systems that (1) have no interactions with the environment, or (2) undergo spontaneous processes. In the atmosphere the temperature and pressure are approximately constant. However, irreversible processes continue to occur within the atmosphere. Therefore, the question arises as to the criteria for equilibrium? Does the entropy continue to increase to a maximum value at fixed T and P or does it decrease to a minimum in this case? Are there other extensive properties which reach a maxima or minima if we change the constraints from constant internal energy and volume to, for instance, specified values of T and P, or S and V? In the following sections we will discuss the various equilibrium conditions when different parameters are held constant.

b. Internal Energy Minimum (for specified S, V, m)

Recall the previous example. Near the maximum entropy state, at fixed internal energy $(\partial S/\partial UA)_U = 0$ and $(\partial S/\partial U_W)_U = 0$. Therefore, $(\partial S/\partial U_A)\,(\partial U_A/\partial U)\,(\partial U/\partial S) = -1$ near equilibrium, i.e., $(\partial U/\partial U_A)_S = -(\partial S/\partial U_A)T = 0$. Since $(\partial S/\partial U_A)_U = 0$, $(\partial U/\partial U_A)_S = 0$ implying that the internal energy must be extremized Figure 45 with respect to U_A at a given value of entropy as discussed below.

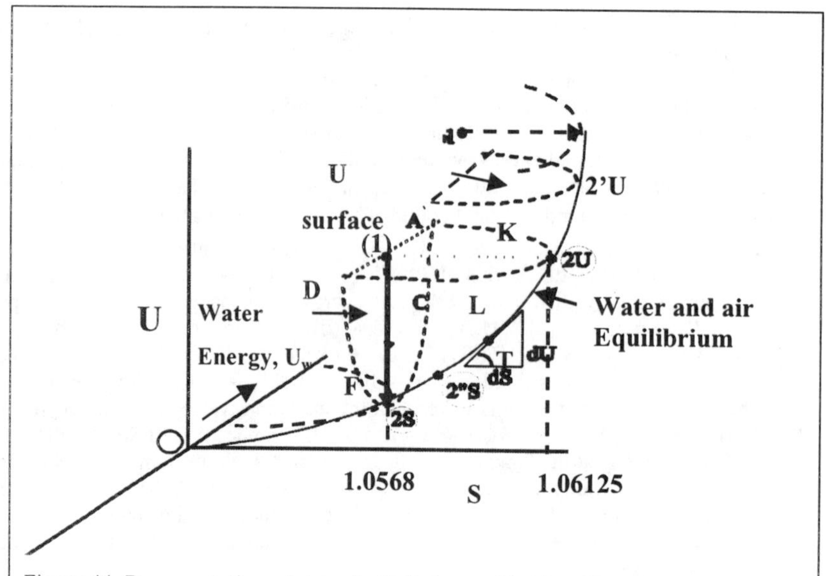

Figure 44: Representation of states in U-S plane. $U= U_W+ U_A$, $S = S_W + S_A$,

State (1): Composite system: W +A, water at higher T; Path A-K-2U: direct cooling; Path A-C-2S: cooling via Carnot engine; Path A-K-2U-2S: direct cooling followed by heat loss to ambient.

Systems may, in general, interact with heat, work, and mass reservoirs. One method to cool coffee isentropically is to connect a Carnot heat engine between the coffee and the (ambient) room air. By this method we can remove, say, δQ_{coffee} amount of heat from the coffee to produce work δW, and reject heat ($\delta Q_{coffee} - \delta W$) to the ambient air. Using the First law, we know that the internal energy of the system must decrease by an amount $dU = \delta Q_{coffee} - \delta W$. Therefore, the internal energy U decreases if the values of S, V, and m are unchanged. When the coffee temperature equals that of the room air, it is no longer possible to extract work from it. In this case, the internal energy of the composite system has reached a minimum value. The mechanical analogy is a coin lying with one face down. It has the lowest possible energy in this position and is more stable compared to a coin standing on its edge.

Consider the expression of the First law

$$\delta Q - \delta W = dU, \tag{91a}$$

where the work

$$\delta W = \delta W_b + \delta W_{other} = P\, dV + \delta W_{other}. \tag{91b}$$

The subscript b refers to the system boundary work. For a process to occur within a fixed mass system,

$$dS = \delta Q / T_b + \delta\sigma, \text{ or } \delta\sigma \geq 0. \tag{89}$$

Using Eqs. (89), (91a), and (91b) to eliminate δQ

$$T_b\, dS = dU + P\, dV + \delta W_{other} + T_b\, \delta\sigma. \quad (92)$$

Since $\delta\sigma \geq 0$, the following restatements of Eq. (92) apply, i.e.,

At constant U, V, m, and $\delta W_{other} = 0$ (i.e., for an isolated system), or

$$dS = \delta\sigma \geq 0. \quad (93a)$$

At constant U, V, m, and $\delta W_{other} \neq 0$, $dS = \delta W_{other}/T_b + \delta\sigma$, or

$$dS \geq \delta W_{other}/T_b. \quad (93b)$$

For a composite system if S, V, m are fixed, and $\delta W_{other} = 0$, $dU = -T_b\, \delta\sigma$.

For a composite system at fixed S, V, m and $\delta W_{other} \neq 0$, $dU \leq -\delta W_{other} - T_b\, \delta\sigma$, or

$$dU \leq 0, \text{ S,V, m fixed}. \quad (94a)$$

For a composite system at fixed S, V, m and $\delta W_{other} \neq 0$, $dU \leq -\delta W_{other} - T_b\, \delta\sigma$, or

$$dU \leq -\delta W_{other}. \quad (94b)$$

For a Carnot–engine operating at constant S, V, and m, Eq. (94b) is applicable as an equality, since no irreversible processes occur, and

$$dU = -\delta W_{other}, \quad \text{S,V, m fixed implying dS=0, dV=0 during the process} \quad (95)$$

where δW_{other} is the work leaving the composite system. Therefore, the internal energy continually decreases as work is extracted from the engine and reaches a minimum at fixed S,V and m (Figure 45).

In place of a Carnot engine, we can use the room air to transfer heat to the ambient so that the combined entropy of the coffee and air remain constant. For this irreversible process, $\delta W_{other} = 0$ so that

$$dU \leq 0, \text{ S,V, m specified}. \quad (96)$$

gg. Example 34

One kg of hot coffee at a temperature of 350 K ($T_{W,0}$) is kept in an adiabatic room that contains 0.4 kg of air at a temperature of 290 K ($T_{A,0}$). The cup is initially insulated. A Carnot engine is used to cool the coffee, lift a weight, and reject heat to the room air until the coffee and air temperatures equilibrate (cf. Figure 46). During the equilibration process:

How does the internal energy of the composite system change?

How does the entropy of the coffee change?

How does the entropy of the room air change?

How does the entropy of the composite system change?

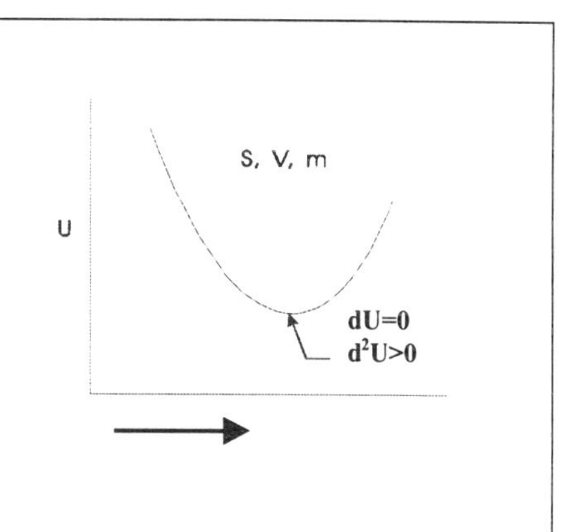

Figure 45: Energy minimum U_{min} principle for specified S, V, m.

Assuming coffee temperatures to be 340 K, 339.5 K, ... K, calculate T_A such that $dS_W + dS_A = 0$.
Determine the initial internal energy of the composite system and plot the internal energy as a function of air temperature as the coffee cools.

Solution

Using the First law $\delta Q - \delta W = dU$. Since, $\delta Q = 0$ and $\delta W > 0$, $dU < 0$. Therefore, the internal energy will decrease, since energy is converted into work.
The entropy of the coffee decreases, since heat is transferred from it.
The entropy of the air increases, since heat is transferred to it.
There is no entropy change in the composite system, since a Carnot engine is used for which the entropy changes in its source (coffee) and sink (air) are equal.
The total entropy change $dS_A + dS_W = dS = 0$. Therefore, $dS_A = -dS_W$ so that $dS_W = m_W c_W\, dT_W/T_W$. If the coffee cools by 0.5 K, $dT_w = -0.5$ K, and

$$dS_W = 1 \times 4.184 \times (-0.5 \div 350) = -0.00598 \text{ kJ K}^{-1}.$$

$$dS_A = m_A\, c_{vA}\, dT_A/T_A = -dS_W = 0.00598 \text{ kJ K}^{-1}.$$

$$dT_A = 0.00598 \times 290 \div (0.4 \times 0.713) = 6.078 \text{ K}.$$

$U = U_W + U_A = m_W\, c_W\, (T_W - 273) + m_A\, c_{vA}\, (T_A - 273) = 327.02$ kJ.
After cooling by 0.5 K $U = U_W + U_A = m_W\, c_W\, (T_W - 273) + m_A\, c_{vA}\, (T_A - 273) = 326.66$ kJ.
The internal energy decreases as work is delivered. These calculations may be repeated for the other temperatures, i.e., $T_w = 389$, 388.5 K, ...

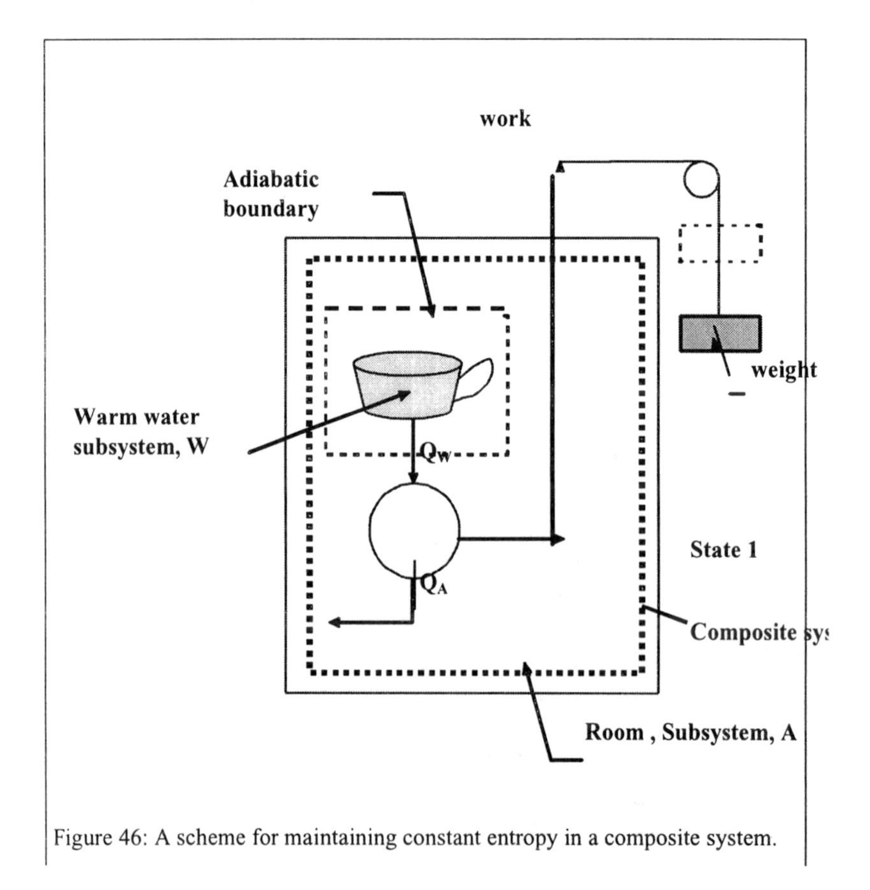

Figure 46: A scheme for maintaining constant entropy in a composite system.

Remarks

Minimum U for a reversible path at specified values of S, V and m.

The internal energy reaches a minimum value of 325.47 kJ at $T_A = T_W = 345.8$ K. Figure 47 presents the variation of the composite system internal energy U as a function of U_A. The entropy, volume, and mass of the composite system are fixed, but its internal energy decreases as work is delivered. At equilibrium dU = 0, and the work delivered is the maximum possible. The entropy of the water decreases, but that for room air increases, whereas in Example 33, $\Delta S_A > \Delta S_W$. In this example, the Carnot engine is connected between subsystems W and A with the condition that $|\Delta S_w| = |\Delta S_A|$. We can reverse the process by operating the Carnot engine as a heat pump that lowers a weight. The process at constant S,V and m is represented by A-C-2S in Figure 44. If we define a larger system that includes the environment and the two subsystems, the total energy E (=U + PE) remains constant, although the internal energy of subsystem W decreases. The energy that leaves the coffee partly heats the air, and partly raises the weight.

Consider the line where the entropy surface S_1 intersects the surface U_1. State 1 is in constrained equilibrium, but does not lie on the curve (2S)-(2'S)-(2U)-(2'U), since the initial state of the composite system is not an equilibrium state. The dashed curve A-C- (2S)-F or D-F-2S (in case air is hotter) is the path along which the composite system internal energy is minimized at constants. Equilibrium can be defined as that state at which the internal energy (=U_A+U_W) is the lowest among all possible values for specified values of S, V and m. This is called the lowest energy principle. The final temperature $T_W = T_A = 345.8$ K is obtained from the slope of the internal energy with respect to the entropy at the point (2S). (Analogously, a mechanical equilibrium state in Figure 37 is defined as that with the lowest energy among all possible states, state D in Figure 37).

Minimum U for an irrversible path but not at fixed S, V and m.

Consider warm water and cool air that mix spontaneously. As in example 33 consider the equilibrium state 2U that is reached ireversibly via the path A-K-2U as shown in Figure 44. Now consider the case when the subsystem W is always in thermal equilib-

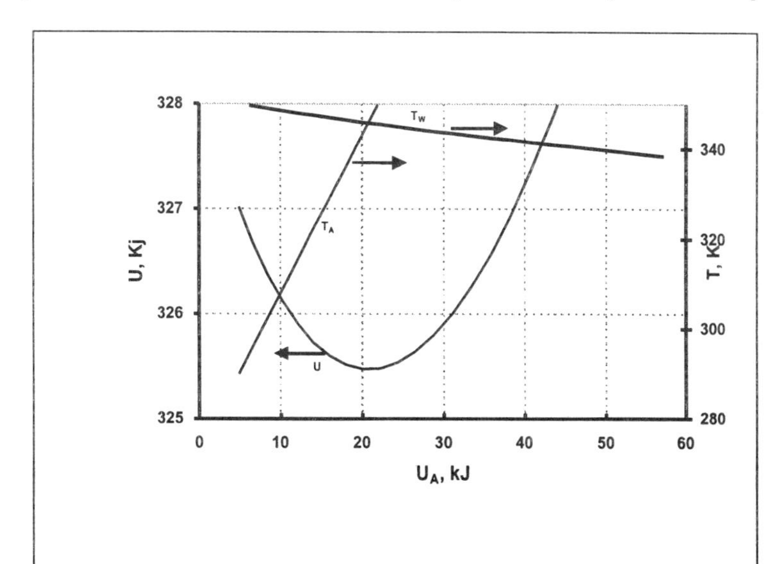

Figure 47: The total internal energy U as a function of the internal energy of the air U_A for specified values of S, V and m.

rium with subsystem A (e.g., state 2U in Figure 44). If we transfer heat from both W and A to the ambient in such a manner that $T_W = T_A$, then $S = (m_W c_W + m_A c_{vA}) \ln (T/273)$ and $U = (m_W c_W + m_A c_{vA}) (T-273)$. Eliminating the temperature from these two equations, $U = 273 (m_W c_W + m_A c_{vA})(\exp (S/(m_W c_W + m_A c_{vA})) - 1)$. This relation for U(S) is plotted in Figure 44 as a curve (2S)-(2'S)-(2U)-(2'U). The entropy along this curve is a single valued function once U, V, and m are prescribed, since there is a single stable equilibrium state. The slope of the curve $(\partial U/\partial S)_V$ is the temperature of the $T_W = T_A = T$. surface. The slope at 2S (fixed values of S, V, and m) is lower than that at 2U (fixed values of U, V, and m), since, in the current example, energy is transferred in order to maintain constant entropy.
From Figure 44 we also note that the slope $dU/dS = T$ is positive. As the temperature increases, the internal energy increases and, consequently, the entropy increases. Since $T = dU/dS$, then the convex nature of the curve requires that the temperature (or slope) must increase with increasing entropy. (At $T = 0$ K, $dU/dS = 0$.) The nature of this curve will be discussed further in Chapter 10.

---

hh. Example 35

Two kg of hot water (subsystem A), initially at a temperature of 600 K, is mixed with 1 kg of cold water (subsystem B) that is initially at 300 K.
What are the equilibrium temperature and entropy if both A and B are isolated subsystems.
Now assume that the two subsystems are not isolated. Once the composite system reaches equilibrium, heat is removed so that the final entropy value of A+B equals the initial entropy of A+B. What is the final temperature?

Solution

Using the First law $U_{1A} + U_{1B} = U_{2A} + U_{2B}$, or $2 \times (T_2 - 600) + 1\ (T_2 - 300) = 0$, i.e.,

$T_2 = 500$ K. Therefore,

$S_2 - (S_{1A} + S_{1B}) = 2 \times 4.184 \times \ln (T_2/600) + 1 \times 4.184 \times \ln (T_2/300) = 0.612$, i.e.,

$S_2 = 0.612$ kJ K^{-1}.

In order that $S_3 = S_1$, the entropy increase must be countered. Therefore,

$3 \times 4.184 \times \ln (T_3/500) = -0.612$, i.e.,

$T_3 = 476.2$ K

When the entropy is maintained constant for specified values of volume and mass, the initial internal energy

$U_1 = U_{1A} + U_{1B} = 2 \times 4.184 \times (600 - 273) + 1 \times 4.184 \times (300 - 273) =$ 358.65 kJ,

and at the final state

$U_2 = U_{2A} + U_{2B} = 2 \times 4.184 \times (476.2 - 273) + 1 \times 4.184 \times (476.2 - 273) =$ 255.1 kJ.

The internal energy reaches a minimum value (Note that intitial S and final S are fixed but during the process $dS \neq 0$) even though the isolated system entropy increases.

---

c. Enthalpy Minimum (For Specified S, P, m)

Rewriting Eq. (92)

$$dU = T_b\, dS - P\, dV - (\delta W_{other}) - T_b\, \delta\sigma \quad (97)$$

Using the relation

$$dU = dH - d(PV),$$

$$dH = T_b dS + VdP - \delta W_{other} - T_b\, \delta\sigma, \quad (98)$$

$$dH \le T_b\, dS + V\, dP - \delta W_{other}.,\qquad (99)$$

$$dS = dH/T_b - (V/T_b)\, dP + (\delta W_{other})/T_b + \delta\sigma.\qquad (100)$$

$$\text{At constant H, P, and m, } \delta W_{other} = 0, \text{ and } dS \ge 0\qquad (101)$$

(Equation (100) is particularly useful for adiabatic reacting flows occuring in open systems, cf. Chapters 11 and 12.)

At constant H, P, and m, when $\delta W_{other} \neq 0$, and $dS \ge \delta W_{other}$.

At constant (Figure 48a) S, P, and m, when $\delta W_{other} = 0$, and $dH \le 0$. (102)

At constant S, P, and m, when $\delta W_{other} \neq 0$, and $dH \le -\delta W_{other}$.. (103)

---

ii. Example 36

The pressure (P = 100 kPa), entropy ($S = S_o$), and mass in the problem of Example 33 are maintained constant. Obtain an expression for the variation of the enthalpy with respect to the air temperature T_A).

Solution

The entropy change $dS = dS_A + dS_W = 0$. At constant pressure, $dS_A = m_A\, c_{pA}\, dT_A/T_A$ so that

$$m_A\, c_{pA}\, dT_A/T_A = -\, m_W\, c_W\, dT_W/T_W.\qquad (A)$$

Integrating Eq. (A) we obtain

$$\ln T_A = -\, m_W\, c_W/m_A\, c_{pA} \ln T_W + C.$$

Eliminating the integration constant by using the initial condition

$$T_A/T_{A0} = (T_W/T_{W0})^{m_W c_W/(m_A c_{pA})},\qquad (B)$$

the temperatures T_W, T_A can be evaluated. When T_W = 349 K,

$$T_A = 290 \times (349/350)^{(-1\times 4.184/(0.4\times 1.0))} = 298.8 \text{ K}.$$

Therefore, it becomes possible to evaluate the enthalpy change

$$dH = dH_A + dH_W = m_a\, c_{pA}\, dT_A + m_W\, c_W\, dT_W.\qquad (C)$$

Fixing the reference state at 273 K, and integrating

$$H = m_A\, c_{pA}\, (T_A - 273) + m_W\, c_W\, (T_W - 273).\qquad (D)$$

If water temperature of 349 K and an air temperature of 298.8 K are assumed, H = 328 kJ. The net heat transfer across the boundary of the room can be calculated from the relation

$$\delta Q_b = dH = dH_A + dH_W = m_A\, c_{pA}\, dT_A + m_W\, c_W\, dT_W, \text{ i.e.,}$$

$$Q_b = H - H_0 = m_A\, c_{pA}\, (T_A - T_{A0}) + m_W\, c_W\, (T_W - T_{W0}).$$

Remarks

Explicit solution for T at H_{min}.

Using Eqs. (B) and (D), and differentiating H with respect to T_W, the water temperature at the minimum enthalpy H_{min} can be obtained, i.e., through the relation

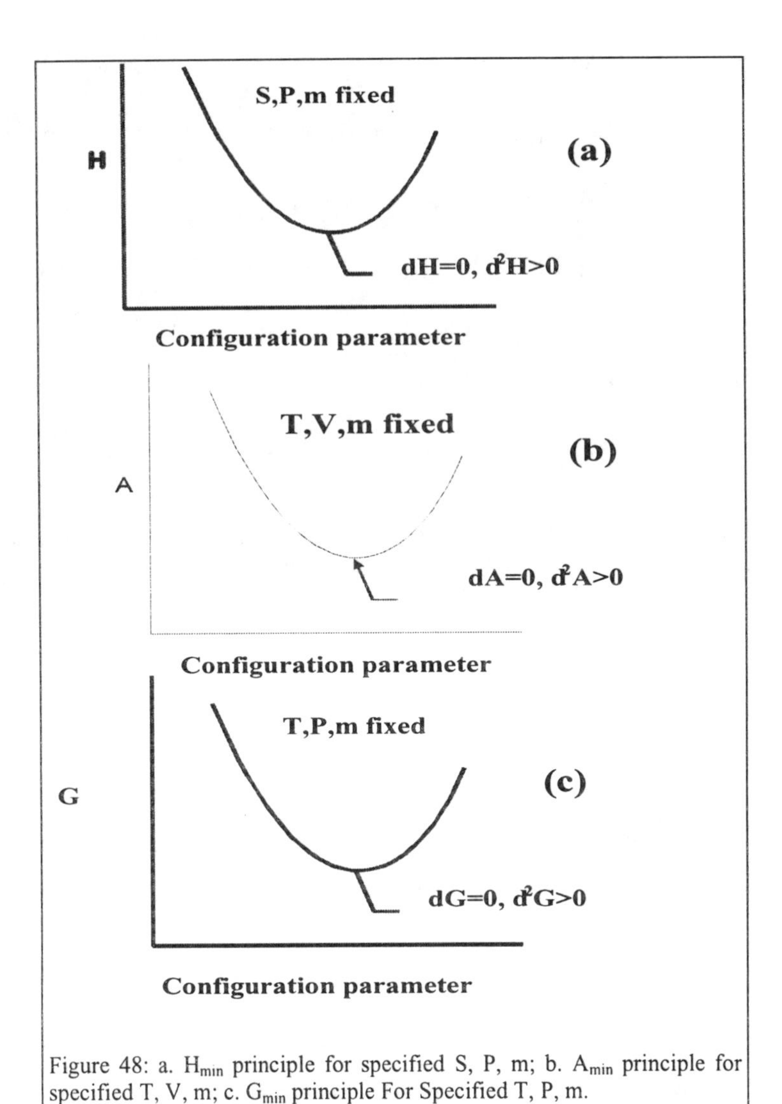

Figure 48: a. H_{min} principle for specified S, P, m; b. A_{min} principle for specified T, V, m; c. G_{min} principle For Specified T, P, m.

$$m_A c_{pA} T_{A0} (-m_W c_W/(m_A c_{pA}))(T_W/T_{W0})^{(m_W c_W/(m_A c_{pA})-1)} (1/T_{W0}) + m_W c_W = 0, \text{ or}$$

$$(T_{A0}/T_{W0})(T_W/T_{W0})^{(m_W c_W/(m_A c_{pA})-1)} - 1 = 0.$$

Therefore, $T_W/T_{W0} = (T_{W0}/T_{A0})^{(m_W c_W/(m_A c_{pA})+1)^{-1}} = 0$, so that

$$T_w = 0.9837 \times 350 = 344.3 \text{ K, and } T_A = 344.3 \text{ K.}$$

The two temperatures still equal one another, but the equilibrium temperature at H_{min} is lower than that at S_{max} (which was calculated in the previous example). This is to be expected, since work is delivered in order to maintain the pressure constant as the air is heated during the process.

An application.

Recall from Eq. (100) that at constant H,P,m, $dS \geq 0$. Combustion occurs at almost constant pressure during a diesel cycle. During such a hypothetical cycle, a mass of air is compressed until it reaches a temperature of 600 K, and diesel is injected into its center. Assume that combustion is initiated at 2 ms later when half of the chamber is filled with hot gases at 2000 K while the unburned mass is at the 600 K temperature (cf. Figure 49). Further, assume that the reaction is frozen at this instant. As heat is transferred from the burnt to unburned gases we allow the piston to move in order to maintain constant pressure. The chamber walls are insulated. Irreversible heat transfer between the hot and cold gases causes the entropy to increase. Equilibrium for the multicomponent system is achieved when the entropy reaches a maximum at fixed H,P, m while the Otto cycle is an example involving the maximization of entropy at fixed U, V,and m.

---

d. *Helmholtz Free Energy Minimum (For Specified T, V, m)*

Oftentimes, the internal energy and entropy cannot be directly measured. It becomes useful to fix the temperature, volume, and mass in order to examine the change of state from a nonequilibrium to equilibrium state. Using the relation Eq.(97)

$$dU = T_b\, dS - P\, dV - (\delta W_{other}) - T_b\, \delta\sigma \tag{92}$$

Since

$$dU = dA + d(TS).$$

then at constant temperature, volume, and mass (with $T_b = T$),

$$dA = -SdT - P\, dV - \delta W_{other} - T_b\, \delta\sigma \tag{104}$$

If $\delta W_{other} = 0$,

$$dA \leq 0,\ T,\ V \text{ and m fixed. Implying } dT=0, dV=0 \tag{105}$$

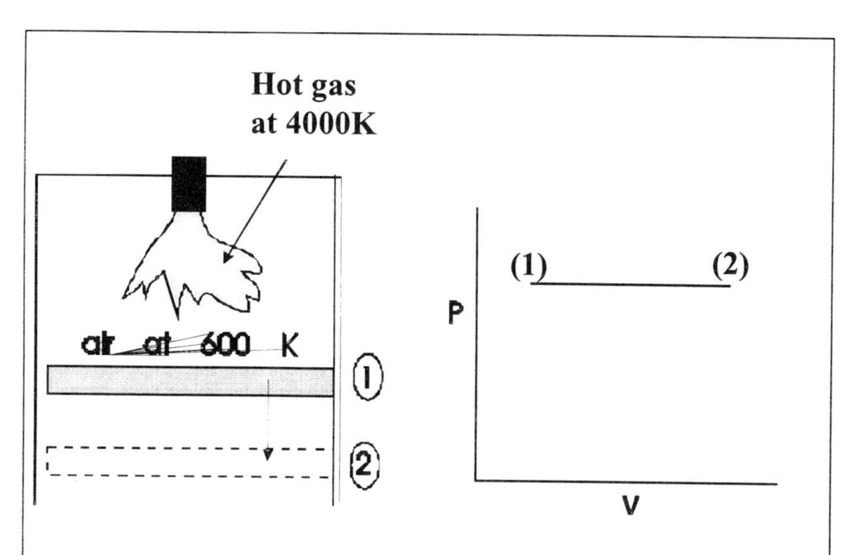

Figure 49 : An application of the entropy maximization principle at fixed H,P and m.

Thus A is minimized at fixed T, V and m (Figure 48b). In Chapter 7 we will show that we can determine the saturation pressure at any given temperature using this principle. For example, we can pour liquid water at 50° C into a rigid evacuated vessel of volumeV immersed in an isothermal bath at T. As vaporization proceeds at constant T, V, m, the sum of A of H_2O (g) and H_2O (ℓ) decreases and vaporization stops once A is minimized or phase equilibrium is reached. If

$$\delta W_{other} \neq 0,\ dA \leq -\delta W_{other}. \tag{103b}$$

Consider molecular nitrogen and molecular oxygen at the same temperature and pressure in two adjacent containers separated by a partition. Even though at thermal and mechanical equilibrium, once the partition is removed, the composition of the composite system changes until the Helmholtz Free Energy reaches a minimum value.

e. Gibbs Free Energy Minimum (For Specified T, P, m)

Again, using the relation Eq.(99)

$$dH = T_b dS + V\,dP - \delta W_{other} - T_b\,\delta\sigma \tag{98}$$

Simplifying Eq.(98)

$$dG = dH - d(TS). \tag{106}$$

Assume that there are no thermal and mechanical irreversibilities in the system (i.e., $T_b = T$, and P = uniform) using Eqs.(98) and (104)

$$dG = -S\,dT + V\,dP - \delta W_{other} - T_b\,\delta\sigma \tag{107}$$

Simplifying this relation,

$$dG \leq -S\,dT + V\,dP - \delta W_{other}. \tag{108}$$

At constant P, T, and m, if $\delta W_{other} = 0$, $dG \leq 0$. (109)

See Figure 48c. Equation (109) has applications to phase change problems (Chapters 7), mixing problems (Chapters 8) and chemically reacting systems (Chapters 11 and 12). We will show in Chapter 7 that it is possible to determine the saturation pressure of a fluid at any temperature using the G_{min} principle.

At constant P, T, and m, if $\delta W_{other} \neq 0$, $dG \leq -\delta W_{other}$. (110)

---

jj. Example 37

Consider section (A) in a constant–pressure device to consist of a 10 m^3 volume that contains molecular oxygen at 25°C and 100 kPa. Section (B) in the same device consists of the remaining volume of 15 m^3 which contains molecular nitrogen at the same temperature and pressure. When the partition is removed, molecules of both species diffuse into one another. The molecules are instantaneously distributed throughout the section they diffuse into. Plot the relationship of G_{A+B} with respect to $Y_{N2,A}$. What is the value of G for the combined system at equilibrium? Assume the two specific heats $c_{pN2} = 1.04$ kJ kg^{-1} K^{-1}, and $c_{P,O2} = 0.92$ kJ kg^{-1} K^{-1} (cf. Figure 50).

Solution

$N_A = N_{O2} PV/\bar{R}T = 100 \times 10/(298 \times 8.314) = 0.404$ kmole.

Similarly, $N_B = N_{N2} = 0.605$ kmole.

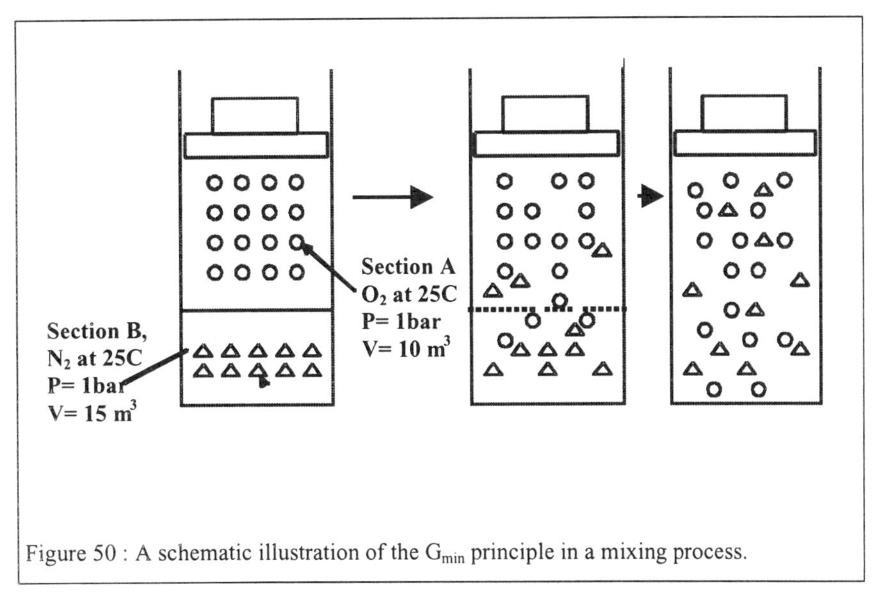

Figure 50 : A schematic illustration of the G_{min} principle in a mixing process.

$\bar{h}_{N_2}$ = 1.04×28×298 = 8678 kJ kmole^{-1}, and, $\bar{h}_{O_2}$ = 0.92×32×298 = 8773 kJ kmole^{-1}. Now,

$$G = N_{O2,A}\, \bar{g}_{O_2} + N_{N2,B}\, \bar{g}_{N_2}, \text{ and} \tag{A}$$

$\bar{g}_{O_2,A} = \bar{g}_{O_2} - T\ \bar{s}_{O_2,A}(T,p_{O2,A})$, where

$\bar{s}_{O_2,A}(T,p_{O2,A}) = \bar{c}_{p,O_2} \ln(T/T_{ref}) - \bar{R} \ln(p_{O2,A}/p_{ref})$.

The reference state is selected to be 298 K and 1 bar. Since $p_{O2,A} = Y_{O2,A}\, P$,

$$\bar{g}_{O_2,A} = \bar{g}_{O_2}(T,P) + \bar{R}T \ln Y_{O2,A}. \tag{B}$$

Similarly,

$$\bar{g}_{N_2,B} = \bar{g}_{N_2}(T,P) + \bar{R}T \ln Y_{N2,B}, \text{ i.e.,} \tag{C}$$

$\bar{g}_{O_2,A}$ = 8773 + 8.314 × 298 ln 1 = 8773 kJ kmole^{-1}, and $\bar{g}_{N_2,B}$= 8678 kJ kmole^{-1}.

Initially,

G_1 = 0.404 × 8773 + 0.605 × 8678 = 8794 kJ.

Assume that 10% of the oxygen molecules (i.e., 0.0605 kmole) cross into section B after the partition is removed and 10% of the N_2 molecules (i.e., 0.0404 kmole) likewise cross into section A (at the same temperature and pressure, i.e., 298 K and 1 bar). Following the molecular crossover, the number of moles contained in Section A are

$N_{N2,A}$ = 0.10 × 0.605 = 0.0605 kmole, and

$N_{O2,A}$ = 0.404 – 0.1 × 0.404 = 0.364 kmole.

Similarly the number of moles contained in section B are

$N_{N2,B}$ = 0.605 – 0.1 × 0.605 = 0.545 kmole, and

$N_{O2,B}$ = 0.1 × 0.4 = 0.0404 kmole.

Therefore,

$Y_{N2,A}$ = 0.0605 ÷ (0.0605 + 0.364) = 0.143, and

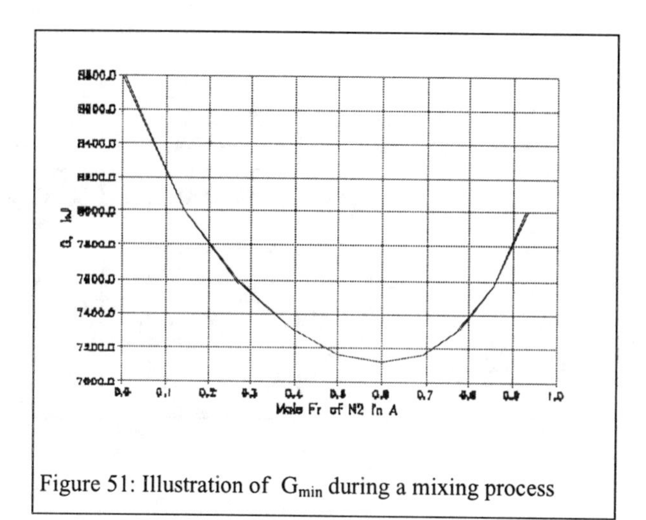

Figure 51: Illustration of G_{min} during a mixing process

$Y_{O2,A} = 0.364 \div (0.0605 + 0.364) = 0.857.$

Similarly in Section B,

$Y_{N2,B} = (0.545) \div (0.545 + 0.0404) = 0.931$, and

$Y_{O2,B} = 0.0404 \div (0.0404 + 0.545) = 0.069.$

Now,

$G = N_{O2,A}\ \bar{g}_{O_2,A} + N_{N2,A}\ \bar{g}_{N_2,A} + N_{O2,B}\ \bar{g}_{O_2,B} + N_{N2,B}\ \bar{g}_{N_2,B}$, where

$\bar{g}_{N_2,A} = 8678 + 8.314 \times 298 \times \ln 0.143 = 3859$ kJ kmole^{-1}.

Similarly,

$\bar{g}_{N_2,B} = 8678 + 8.314 \times 298 \times \ln 0.931 = 8501$ kJ kmole^{-1},

$\bar{g}_{O_2,A} = 8773 + 8.314 \times 298 \times \ln 0.857 = 8391$ kJ kmole^{-1}, and

$\bar{g}_{O_2,B} = 8773 + 8.314 \times 298 \times \ln 0.069 = 2149$ kJ kmole^{-1}, so that

$G = 0.364 \times 8391 + 0.0605 \times 3859 + 0.0404 \times 2149 + 0.545 \times 8501 = 8007$ kJ.

These calculations can be repeated for 20%, 30%, etc., of each species diffusing into each other. Figure 51 contains a plot of G with respect to $Y_{N2,A}$. Note that if all of the molecular nitrogen diffuses into Section A and all of the O_2 diffuses into Section B the value of G reverts to G_1. The minimum value is reached when $Y_{N2,A} = 0.6$.

At equilibrium

$Y_{O2} = 0.404 \div (0.404 + 0.605) = 0.4$, $Y_{N2} = 0.6$,

$\bar{g}_{O_2} = 8773 + 8.314 \times 298 \times \ln 0.4 = 6503$ kJ, and

$\bar{g}_{N_2} = 8678 + 8.314 \times 298 \times \ln 0.6 = 7412$ kJ.

Hence,

$G_2 = 0.404 \times 6503 + 0.605 \times 7412 = 7111$ kJ.

$dG_{T,P} < 0$ when mixing (which is irreversible) occurs. We will discuss the G_{min} principle further in Chapters 7 and 8.

Remarks

Explicit solution for equilibrium concentration with LaGrange method

Applying Eq.(107) at constant, T,P,m and zero other work

$$dG_{T,P,m} = -\ T_b\ \delta\sigma \qquad (111)$$

we need to evaluate either σ or G. The Gibbs energy cannot be evaluated for Sections A and B together since an irreversible process occurs between them. However G can be evaluated if we assume the process within each section to occur reversibly. In that case, $dG_{T,P,m}$ = (dG_A +dG_B) <0; each section acts as an open system. In Chapters 4, 11, and 12, we will show that for the open system, $dG_A = dN_{O2,A}\ \bar{g}_{O_2,A} + dN_{N2,A}\ \bar{g}_{N_2,A}$, where $dN_{O2,A}$ is the change in Section A for O_2 moles due to transfer from Section B ; $dG_B = dN_{O2,B}\ \bar{g}B + dN_{N2,B}\ \bar{g}_{N_2,B}$. The concentration at G = G_{min} can be obtained from Eqs. (A), (B), and (C) by minimizing G(T, P, N_{N2}, N_{O2}) subject to $N_{O2} = N_{O2,0}$ (the initial value of O_2) and $N_{N2} = N_{N2,0}$ (the initial value of N_2). Using the LaGrange multiplier method,

$$F = G_A(T,P,N_{N2,A},N_{O2,A}) + G_B(T,P,N_{N2,B},N_{O2,B}) +$$

$$\lambda_1(N_{O2,A} + N_{O2,B} - N_{O2,0}) + \lambda_2(N_{N2,A} + N_{N2,B} - N_{N2,0}). \quad \text{(D)}$$

where λ is the LaGrange multiplier. Therefore,

$$G(T,P,N_{N2},N_{O2}) = N_{O2}(\bar{g}_{O_2}(T,P) + \bar{R}T \ln (N_{O2}/(N_{N2} + N_{O2}))) +$$

$$N_{N2}(\bar{g}_{N_2}(T,P) + \bar{R}T \ln(N_{O2}/(N_{N2} + N_{O2}))). \quad \text{(E)}$$

Using the relations $\partial F/\partial N_{O2,A} = 0$, $\partial F/\partial N_{O2,B} = 0$, $\partial F/\partial N_{N2,A} = 0$, $\partial F/\partial N_{N2,B} = 0$, and differentiating Eq. (E),

$$N_{O2,A}(\bar{R}T/(N_{O2,A} - (N_{N2,A}+N_{O2,A}))^{-1}) +$$

$$\bar{g}_{O_2}(T, P) + \bar{R}\ \ln(N_{O2,A}/(N_{N2,A} + N_{O2,A})) + \lambda_1 = 0.$$

Simplifying the equation

$$\bar{R}T\ N_{N2,A}/(N_{N2,A} + N_{O2,A}) + \bar{g}_{O_2}(T, p_{O2,A}) + \lambda_1 = 0, \text{ or}$$

$$Y_{N2,A} + \bar{g}_{O_2}(T, p_{O2,A}) = -\lambda_1' \quad \text{(G)}$$

where $\lambda_1' = \lambda/RT$. Similarly,

$$Y_{N2,B} + \bar{g}_{O_2}(T, p_{O2,A}) = -\lambda_1'. \quad \text{(H)}$$

It is seen from Eqs. (G) and (H) that mole fractions in both Sections A and B must be the same at equilibrium (i.e., at G_{min}) ,i.e.,

$Y_{N2,A} = Y_{N2,B}$ or $p_{O2,A} = p_{O2,B}$.

---

2. Generalized Derivation for a Single Phase

We have thus far obtained the conditions for thermal (Example 33) and chemical (Example 37) equilibrium. Consider, once again, hot coffee contained in a rigid cup with a firm lid as in Example 33. If the coffee is replaced with warm nitrogen at 350 K, heat transfer will occur from the cup. As the room air warms, the room pressure increases, while that of the gas in the cup decreases. Thermal equilibrium will be reached at the maximum entropy state $(S_{max})_{TE}$, but with a mechanical constraint in place. If the mechanical constraint is removed, i.e., the lid is replaced with a nonpermeable and moveable piston, then equilibrium will be achieved at another maximum entropy state $(S_{max})_{TM}$. Finally, if the impermeable piston is replaced with a permeable piston, the system pressure and temperature may not change, but the nitrogen mole fraction in the room will change until chemical or species equilibrium is reached at yet another maximum entropy state $(S_{max})_{TMC}$. In this case $(S_{max})_{TMC} > (S_{max})_{TM} > (S_{max})_{Thermal\ Equil}$, i.e., "equilibrium" is reached when the entropy attains the highest possible value when all the constraints (thermal, mechanical and chemical) are removed. In this section we will discuss a generalized analysis for such an equilibration process.

Assume that two subsystems A and B contain two species, namely, species 1 and 2, as illustrated in Figure 52(a). Assume that subsystem A has a slightly higher pressure, a slightly higher temperature, and contains a slightly larger number of moles of species 1 as compared to subsystem B so that the two subsystems are infinitesimally apart from equilibrium with one another. There are three initial constraints: a rigid plate that is nonporous, is a good thermal energy conductor, and serves as a chemical constraint; a porous rigid insulation which serves as a thermal constraint allowing only mass transfer when subsystems A and B are at same temperature; and a pin holding the rigid plate firmly in place (serving as a mechanical constraint) which, when removed, allows work transfer. When all of the constraints are removed, assuming the combined system to be insulated, rigid, and impermeable, changes in U, V, N_1 and N_2 occur only in each of the subsystems. Therefore, the entropy of each subsystem changes subject to the condition $U = U_A + U_B$, or

$$dU = 0 = dU_A + dU_B \tag{112}$$

Similarly,

$$dV = dV_A + dV_B = 0, \tag{113}$$

$$dN_{A1} + dN_{B1} = 0, \text{ and} \tag{114}$$

$$dN_{A2} + dN_{B2} = 0. \tag{115}$$

Since each subsystem is initially in a state of equilibrium, then

$$S_A = S_A(U_A, V_A, N_{A1}, N_{A2}), \text{ and} \tag{116}$$

$$S_B = S_B(U_B, V_B, N_{B1}, N_{B2}), \tag{117}$$

the entropy of subsystems A and B changes as soon as the constraints are removed. Employing a Taylor series expansion for the relevant expressions around each initial subsystem state,

$$S_A + dS_A = S_A(U_A, V_A, N_{A1}, N_{A2}) + ((\partial S/\partial U)_A\, dU_A + (\partial S/\partial V)_A\, dV_A +$$

$$\partial S_A/\partial N_{A1}\, dN_{A1} + \partial S_A/\partial N_{A2}\, dN_{A2}) + \ldots. \tag{118}$$

Similarly,

$$S_B + dS_B = S_B(U_B, V_B, N_{B1}, N_{B2}) + ((\partial S/\partial U)_B\, dU_B + (\partial S/\partial V)_B\, dV_B +$$

$$(\partial S/\partial N_{B1})\, dN_{B1} + \partial S_B/\partial N_{B2}\, dN_{B2}) + \ldots. \tag{119}$$

Since

$$dS = dS_A + dS_B = \delta\sigma \tag{120}$$

considering only the first-order derivatives,

$$dS_A = (\partial S/\partial U)_A\, dU_A + (\partial S/\partial V)_A\, dV_A + (\partial S_A/\partial N_{A1})\, dN_{A1} + (\partial S_A/\partial N_{A2})\, dN_{A2}, \tag{121}$$

$$dS_B = (\partial S/\partial U)_B\, dU_B + (\partial S/\partial V)_A\, dV_B + (\partial S_B/\partial N_{B1})\, dN_{B1} + (\partial S_B/\partial N_{B2})\, dN_{B2}. \tag{122}$$

Recall that for a closed system $dS = dU/T + P/T\, dV - \Sigma(\mu_k/T)\, dN_k$. Therefore,

$$dS_A = dU_A/T_A + P_A/T_A\, dV_A - \Sigma(\mu_k/T)_A\, dN_{k,A},$$

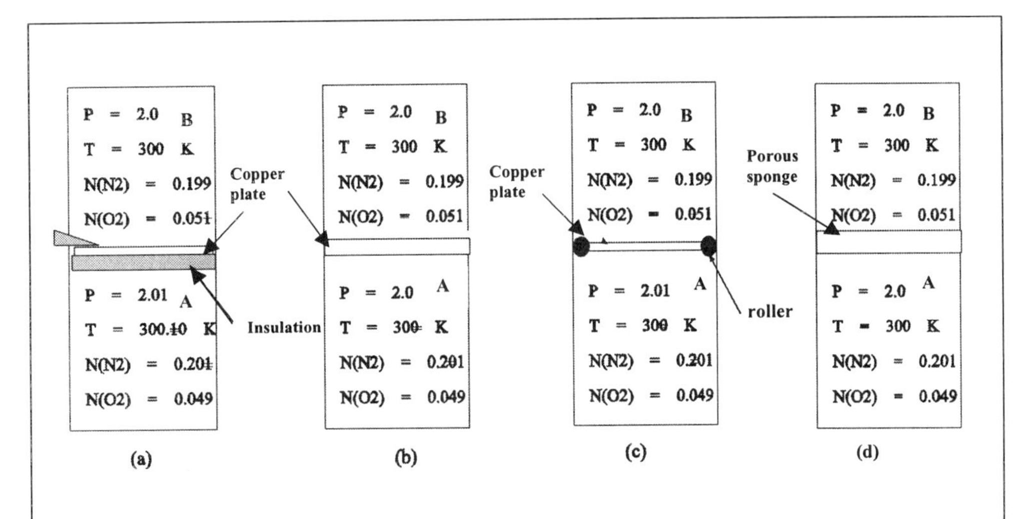

Figure 52 a. Initial state of a composite system; b. thermal equilibration with a copper plate in contact; c. mechanical equilibration with a movable partition; d. chemical equilibration with a porous partition.

$$(\partial S_A/\partial U_A)_{VA,NA1,NA2} = 1/T_A, \text{ and} \quad (123)$$

$$(\partial S/\partial V) = P_A/T_A. \quad (124)$$

We can also define

$$\partial S_A/\partial N_{A1} = -\mu_{A1}/T, \text{ and } \partial S/\partial N_{A2} = -\mu_{A2}/T. \quad (125)$$

The process within the adiabatic and isolated composite system is irreversible so that $dS = \delta\sigma$. Therefore, the entropy increases due to heat, work, and mass transfer, and $dS > 0$.

Using Eqs. (121) to (125) we can express Eq. (120) in the form

$$dS = \delta\sigma = (1/T_A - 1/T_B)\, dU_A + (P_A/T_A - P_B/T_B)\, dV_A +$$

$$(\mu_{B1}/T_B - \mu_{A1}/T_A)\, dN_{A1} + (\mu_{B2}/T_B - \mu_{A2}/T_A)\, dN_{A2} \geq 0. \quad (126)$$

This relation can be expressed in rate form that is valid during the entropy transfer, i.e.,

$$\delta\sigma/dt = (1/T_A - 1/T_B)\, dU_A/dt + (P_A/T_A - P_B/T_B)\, dV_A/dt +$$

$$(\mu_{B1}/T_B - \mu_{A1}/T_A)\, dN_{A1}/dt + (\mu_{B2}/T_B - \mu_{A2}/T_A)\, dN_{A2}/dt \geq 0. \quad (127)$$

a. Special Cases

iv. No Thermal Constraint

In this case, consider the rigid impermeable plate and pin in place, but with the porous insulation removed (cf. Figure 52(b)) so that the two subsystems are each rigid and impermeable, i.e., $dV_A = 0$, $dV_B = 0$, $dN_1 = 0$. $dN_2 = 0$. Using Eq. (126),

$$\delta\sigma = (1/T_A - 1/T_B)\, dU_A, \text{ or}$$

$$\delta\sigma/dt = (1/T_A - 1/T_B)\, dU_A/dt. \quad (128)$$

We can also express (128) in terms of heat flux. From the First law $\delta W_A/dt = 0$, since the pin is in place. Therefore, the rate of change in internal energy dU_A/dt equals the conduction heat flux Q_A (= J_Q) from subsystem A, and the relation

$$\delta\sigma/dt = J_Q (1/T_A - 1/T_B).$$

If $dU_A/dt < 0$ (e.g., heat loss from A and $J_Q < 0$) and $\delta\sigma/\delta t > 0$, $T_A > T_B$. At thermal equilibrium $\delta\sigma = 0$. Then, using Eq. (128)

$$T_A = T_B.$$

v. No Mechanical Constraint

Now assume that, initially, both temperatures are equal. Once the pin and insulation are removed, heat and work transfer is possible, but mass transfer is not (cf. Figure 52(c)). Thereupon, as subsystem A expands, subsystem B will be compressed. The temperature in both sections will equilibrate with the consequence that the first term on the RHS of Eq. (124) will equal zero, since $T_A = T_B$. In this case

$$dS/dt = \delta\sigma/dt = (P_A/T_A - P_B/T_B)\, dV_A/dt = ((P_A - P_B)/T)\, J_V, \quad (129)$$

where J_V is the deformation rate flux as a result of the generalized deformation force $(P_A - P_B)/T$. If $dV_A > 0$ and $\delta\sigma > 0$, $P_A > P_B$. At the mechanical equilibrium condition, $\delta\sigma = 0$ so that $P_A = P_B$.

vi. No Chemical Constraint

If the initial temperatures and pressures are equal in the subsystems (as illustrated in Figure 52(d)), but the boundary is permeable Eq. (127) takes the form

$$(\mu_{B1}/T_B - \mu_{A1}/T_A) dN_{A1}/dt + (\mu_{B2}/T_B - \mu_{A2}/T_A)\, dN_{A2}/dt \geq 0. \quad (130)$$

If $dN_{A1} < 0$ (i.e., there is net transfer of species 1 from subsystem A), since $\delta\sigma > 0$, $\mu_{A1}/T_A > \mu_{B1}/T_B$. Furthermore, since $T_A = T_B$,

$$\mu_{A1} > \mu_{B1},$$

and Eq. (130) may be written in the form

$$\delta\sigma = ((\mu_{B1} - \mu_{A1})/T)\, J_{NA1} + ((\mu_{B2} - \mu_{A2})/T)\, J_{NA2} \geq 0.$$

The term dN_{A1}/dt is the species flux J_{N1} crossing the boundary as a result of the generalized (species–1) flux force $(\mu_{B1} - \mu_{A1})/T$ that is conjugate to J_{N1}. At chemical or species equilibrium

$$\mu_{A1} = \mu_{B1}.$$

Similarly, $\mu_{A2} = \mu_{B2}$.

Consider the example of pure liquid water at a temperature of 100°C and pressure of 1 bar (subsystem A) and water vapor at the same temperature and pressure contained in subsystem B. When both subsystems are brought into equilibrium $T_A = T_B$, $P_A = P_B$, and $\mu_{1A} = \mu_{1B}$. This is also known as the phase equilibrium condition.

vii. Other Cases

Oftentimes in a mixture, the higher the concentration of a species, the higher is its chemical potential (cf. Chapter 8) and hence the species transfers from regions of higher concentration to those at lower concentration through molecular diffusion.

Since there are four independent variables dU_A, dV_A, dN_{A1} and dN_{A2} contained in Eq. (126), when $\delta\sigma > 0$ each term in the relation must be positive. Consequently $T_A > T_B$ if $dU_A < 0$, $P_A > P_B$ if $dV_A > 0$, and $\mu_{B1} > \mu_{A1}$ if $dN_{A1} > 0$, i.e., heat flows from higher to lower temperatures, work flows from higher to lower pressures, and species likewise flow from higher to lower chemical potentials.

In case of multiphase and multicomponent mixtures, the derivation of the equilibrium condition is not simple. In this case the LaGrange multiplier scheme can be used to maximize the entropy subject to constraints, such as fixed internal energy, volume, and mass. This method is illustrated in the Appendix.

If a system initially in equilibrium (i.e., $T_A = T_B$, $P_A = P_B$, $\mu_{A1} = \mu_{B1}$, $\mu_{A2} = \mu_{B2}$) is disturbed, its entropy decreases, i.e., $dS < 0$. A Taylor series expansion around the equilibrium state that includes second order derivatives reveals that

$$dS = 0, \text{ and } d^2S < 0,$$

which are conditions for entropy maximization indicating the initial state to be stable. This will be further discussed in Chapter 10.

A process (or effect) occurs only if a nonequilibrium state (or cause) exists. For instance, fluid flows in a pipe due to a pressure differential, and heat transfer only occurs if a temperature differential exists. A stable equilibrium state will not support the occurrence of any process.

---

kk. Example 38

Consider a mixture of O_2 and N_2 (in a volumetric ratio of 40:60) contained in chamber A at a 10 bar pressure as shown in Figure 53. Chamber B contains only O_2 at the same temperature, but at a pressure P_B such that chemical equilibrium is maintained for O_2 between chambers A and B that are separated by a semipermeable membrane that is permeable to only O_2. Determine P_B. (A semipermeable membrane is a device that allows the transfer of specific chemical specie.)

Solution

At chemical equilibrium, there is no net flow of molecular oxygen, i.e.,

$$\bar{\mu}_{O2,A} = \mu_{O2,B}, \text{ and}$$

$$h_{O2}(T) - T\,(s_{O2,A}^0(T) - R \ln p_{O2,A}/1) = h_{O2}(T) - T\,(s_{O2,B}^0(T) - R \ln p_B/1).$$

Upon simplification

$$p_B = 0.4 \times 10 = 4 \text{ bars}.$$

Remark

Note that there is a mechanical imbalance of forces across the semipermeable membrane so that the membrane must be able to withstand a pressure difference of 10 – 4 = 6 bars.

K. SUMMARY

Chapter 3 introduces concepts on Second law of thermdodynamics, entropy and entropy generation. While Chapter 1 defined entropy in terms of the number of states in which energy is stored, Chapter 3 yields an expression for entropy in terms of properties of the substance of the system. Design of thermal systems should minimize entropy generation so that useful work output is maximized. The entropy maximum, and energy minimum principles are illustrated with simple examples. Finally the driving potentials for heat, work and species transfer are defined using entropy generation concepts.

L. APPENDIX

1. Proof for Additive Nature of Entropy

Consider a well–insulated coffee cup (1) at 320 K that is closed with an insulated rigid lid and placed in a room that is at a temperature of 300 K. We are asked to determine the entropy of composite system (1+2). Assume that there are two Carnot engines, one connected

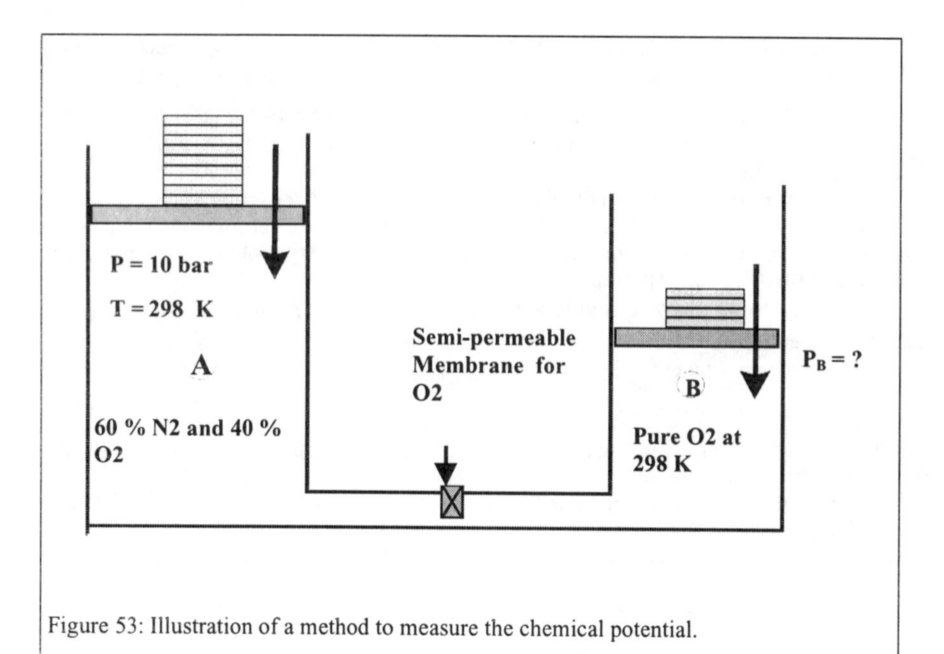

Figure 53: Illustration of a method to measure the chemical potential.

between a large sauna that exists at 350 K and the coffee cup (that is at 320 K), and another engine that is placed between the sauna and the room air (that is at 300 K). If $Q_{1,H}$ denotes the heat absorbed from the sauna by the engine between the sauna and the coffee, and $Q_{2,H}$ represents the heat absorbed by the engine placed between the sauna and the room air, then the change of entropy in the sauna is

$$\Delta S_H = (Q_{1,H} + Q_{2,H})/T_H = Q_{1,H}/T_H + (Q_{2,H})/T_H \tag{131}$$

However, this change in the sauna entropy must equal the sum of the changes in the entropy of the coffee and the room air. Therefore,

$$\Delta S_L = \Delta S_{1+2} = Q_{1,L}/T_{coffee} + Q_{2,L}/T_{room} = \Delta S_1 + \Delta S_2. \tag{132}$$

Therefore, the entropy change in the composite system is the sum of the entropy change in each of the subsystems constituting the composite system.

2. Relative Pressures and Volumes

Recall that

$$s(T_2,P_2) - s(T_1, P_1) = s^0(T_2) - s^0(T_1) - R \ln (P_2/P_1). \tag{133}$$

For *isentropic processes*, $s(T_2,P_2) = s(T_1,P_1)$, and Eq. (133) can be used to determine the ratio

$$P_2/P_1 = \exp(s^0(T_2) - s^0(T_1))/R = \exp(s^0(T_2)/R)/\exp(s^0(T_1)/R). \tag{134}$$

Changing the notation so that $P_2 = P$, $T_2 = T$, $P_1 = P_{ref}'$ and $T_1 = T_{ref}'$,

$$P_r(T)= P/P_{ref}' = \exp (s^0(T)/R)/\exp(s^0(T_{ref}')/R), \tag{135}$$

where $P_r(T)$ is referred to as the relative pressure. It represents a fictitious pressure ratio for an isentropic process during which the temperature changes from T_{ref} to T. In the air tables (Tables A-7) T_{ref}' is generally set equal to 273 K. We can determine the values of P_r as a function of temperature for an ideal gas by applying Eq. (50d). With values for $s^0(273) = 1.6073$ kJ kg^{-1}

K^{-1}, $R = 0.287$ kJ kg^{-1} K^{-1}, $s^0(1000) = 2.9677$ kJ kg^{-1} K^{-1}, $P_r(1000K) = 114$. Thus Eq.(50d) may be written as (cf. Tables A-7)

$$P_r(T) \approx 0.00368 \exp(s^0(T)/R)$$

Irrespective of the reference condition, for an isentropic process Eq. (134) may be expressed in the form

$$P_2/P_1 = P_{r2}(T_2)/P_{r1}(T_1). \quad (136)$$

Using the ideal gas law, Eq. (134) can be expressed in terms of volume, i.e.,

$$(RT_2/v_2)/(RT_1/v_1) = \exp(s^0(T_2)/R)/\exp(s^0(T_1)/R). \quad (137)$$

The ratio

$$v_2/v_1 = (T_2/\exp(s^0(T_2)/R)) / (T_1/\exp(s^0(T_1)/R)) \text{ or } v_2\, P_{r2}/T_2 = v_1\, P_{r1}/T_1 \quad (138)$$

The relative volume $v_r'' = TR/P_r$ which is a dimensional quantity. The dimensionless analog

$$v_r = v/v_{ref} = (T/\exp(s^0(T)/R)) / (T_{ref}/\exp(s^0(T_{ref})/R)) = (T/T_{ref})/P_r. \quad (139)$$

For an isentropic process, Eq. (50g) assumes the form

$$v_2/v_1 = v_{r2}/v_{r1} = (T_2/P_{r2})/(T_1/P_{r1}). \quad (140)$$

Expressing, $v_r'' = T\,R/P_r$, and using the value $R = (53.3 \text{ ft lb}_f/\text{ lb}_m R)/(144 \text{ in}^2/\text{ft}^2)$, then $v_r''(1800 \text{ R}) = 1800\times 53.3/(144\times 114) = 5.844 \text{ ft}^3 \text{ lb}_f/(\text{lb}_m \text{ in}^2)$ which is same as the tabulated value. In general,

$$v_r'' = 2.87\, T/P_r \text{ in SI units, and } v_r'' = 0.37\, T/P_r \text{ in English units.}$$

We can eliminate the units once we define

$$v_r = v/v_{ref} = (T/\exp(s^0(T)/R)) / (T_{ref}/\exp(s^0(T_{ref})/R)), \text{ i.e.,} \quad (141)$$

$$v_r = (T/T_{ref})/P_r. \quad (142)$$

If we select $T_{ref} = 273$ K for air, then at $T = 1000$ K, $P_r = 114$, $v_r = 0.0321$ (however, in this case, the tables provide a value of 25.17).

3. LaGrange Multiplier Method for Equilibrium

a. U, V, m System

One can use the LaGrange multiplier method to maximize the entropy. In case an analysis involves several nonreacting subsystems containing several species so that

$$S = S^{(1)}(U^{(1)}, V^{(1)}, N_1^{(1)}, N_2^{(1)}, \ldots) + S^{(2)}(U^{(2)}, V^{(2)}, N_1^{(2)}, N_2^{(2)}) + \ldots, \quad (143)$$

the entropy may be maximized subject to the constraint that

$$U = U^{(1)} + U^{(2)} + \ldots = \text{Constant}, \quad (144)$$

$$V = V^{(1)} + V^{(2)} + \ldots = \text{Constant, and} \quad (145)$$

$$N^1 = N_1^{(1)} + N_1^{(2)} + \ldots = \text{Constant}. \quad (146)$$

Using the LaGrange multiplier method and Eqs. (143) to (146),

$$S = S^{(1)}(U^{(1)}, V^{(1)}, N_1^{(1)}, N_2^{(1)}, \ldots) + S^{(2)}(U^{(2)}, V^{(2)}, N_1^{(2)}, N_2^{(2)}) + \ldots +$$

$$\lambda_U(U - (U^{(1)} + U^{(2)} + \ldots)) + \lambda_V(V - (V^{(1)} + V^{(2)} + \ldots)) + \lambda^{N1}(N^1 - (N_1^{(1)} + N_1^{(2)} + \ldots)). \quad (147)$$

The maximization process requires that $\partial S/\partial U^{(1)}= 0$, $\partial S/\partial U^{(2)} =0$, ...Differentiating Eq. (147),

$$\partial S/\partial U^{(1)} = \partial S^{(1)}/\partial U^{(1)} - \lambda_U = 0 \text{ or } \partial S^{(1)}/\partial U^{(1)} = 1/T^{(1)} = \lambda_U. \quad (148)$$

$$\partial S/\partial U^{(2)} = \partial S^{(2)}/\partial U^{(2)} + \lambda_U = 0, \text{ or } \partial S^{(2)}/\partial U^{(2)} = 1/T^{(2)} = \lambda_U. \quad (149)$$

...

Therefore, $T^{(1)} = T^{(2)} =$....., that represents the thermal equilibrium condition. Since

$$\partial S/\partial V^{(1)} = \partial S^{(1)} /\partial V^{(1)} - \lambda_V = 0 \text{ or } \partial S^{(1)} /\partial V^{(1)} = P^{(1)}/T^{(1)} = \lambda_V. \quad (150)$$

$$\partial S/\partial V^{(2)} = \partial S^{(2)} /\partial V^{(2)} - \lambda_V = 0, \text{ or } \partial S^{(2)} /\partial V^{(2)} = P^{(2)}/T^{(2)} = \lambda_V. \quad (151)$$

Since $T^{(1)} = T^{(2)} = \ldots$, $P^{(1)} = P^{(2)} = \ldots$, mechanical equilibrium condition between different phases. Furthermore,

$$\partial S/\partial N_1^{(1)} = \partial S^{(1)}/\partial N_1^{(1)} - \lambda_{N1} = 0 \text{ or } \partial S^{(1)}/\partial N_1^{(1)} = \mu_1^{(1)}/T^{(1)} = -\lambda(N_1). \quad (152)$$

$$\partial S/\partial N_2^{(1)} = \partial S^{(1)}/ \partial N_2^{(1)} + \lambda_{N2} = 0 \text{ or } \partial S^{(1)}/\partial N_2^{(1)} = \mu_2^{(1)}/T^{(1)} = -\lambda(N_2). \quad (153)$$

Since the temperatures within the subsystems are identical, $\mu_1^{(1)} = \mu_2^{(1)} = \ldots$, i.e., phase 1 is in equilibrium and no chemical reaction occurs. Repeating the process for the other subsystems,

$$\partial S/\partial N_1^{(2)} = \partial S^{(2)}/\partial N_1^{(2)} - \lambda_{N1} = 0 \text{ or } \partial S^{(2)}/\partial N_1^{(2)} = \mu_1^{(2)}/T^{(2)}= -\lambda(N_1). \quad (154)$$

$$\partial S/\partial N_2^{(2)} = \partial S^{(2)}/\partial N_2^{(2)} + \lambda_{N2} = 0 \text{ or } \partial S^{(2)}/\partial N_2^{(2)} = \mu_2^{(2)}/ T^{(2)}= -\lambda(N_2). \quad (155)$$

Furthermore, we assume identical values of λ so that , $T^{(1)} = T^{(2)} = \ldots$, and $\mu_1^{(1)} = \mu_1^{(2)}, \ldots$, $\mu_2^{(1)} = \mu_2^{(2)}$, etc. Therefore, in the nonreacting subsystems, the equilibrium condition requires the temperatures, pressures, and chemical potentials in all of the subsystems to, respectively, equal one another.

A similar procedure can be adopted to determine the equilibrium condition at given T, P and N.

b. T, P, m System

viii. One Component

Consider N moles of a pure substance (say, H_2O) kept at 0°C and constant pressure P (say, 0.6 kPa, the triple point pressure). The substance attains equilibrium in multiple phases (e.g., the water forms three – $\pi = 3$ – phases: solid ,liquid and gas). In general, the number of moles in each phase is different (say, $N^{(\ell)}$, $N^{(g)}$, $N^{(s)}$,) may change, and the Gibbs energy is minimized at equilibrium. For the composite system that includes all the phases

$$G = G^{(1)} (T,P,N^{(1)}) + G^{(2)} (T,P,N^{(2)}) + \ldots = G(T,P, N^{(1)}, N^{(2)} \ldots N^{(\pi)}), \quad (156)$$

which is to be minimized subject to the constraint $N = \Sigma N^{(j)} =$ constant. We again use the La-Grange multiplier method and form the function

$$F = G + \lambda(\Sigma N^{(j)} - N), \text{ so that} \quad (157)$$

$$\partial F/\partial N^{(1)} = \partial G/\partial N^{(1)} + \lambda = \bar{g}^{(1)} + \lambda = 0. \quad (158)$$

Similarly, for the other phases

$$\partial F/\partial N^{(2)} = \partial G/\partial N^{(2)} + \lambda = \bar{g}^{(2)} + \lambda = 0. \quad (159)$$

Therefore,

$$\bar{g}^{(1)} = \bar{g}^{(1)} = \ldots = -\lambda, \quad (160)$$

implying that at equilibrium the molal Gibbs function is identical for all species.

ix. Multiple Components

The Gibbs energy

$$G = G^{(1)}(T,P,N_1^{(1)},N_2^{(1)},\ldots,N_K^{(1)}) + G^{(2)}(T,P,N_1^{(2)},N_2^{(2)},\ldots,N_K^{(2)}) + \ldots +$$

$$G^{(\pi)}(T,P,N_1^{(\pi)},N_2^{(\pi)},\ldots,N_K^{(\pi)}) \quad (161)$$

$$= G(T,P,N_1^{(1)},N_2^{(1)},\ldots,N_K^{(1)}, N_1^{(2)},N_2^{(2)},\ldots,N_K^{(2)}, N_1^{(\pi)}, N_2^{(\pi)},\ldots,N^{(\pi)}). \quad (162)$$

We must minimize G subject to the constraints

$$N_1 = N_1^{(1)} + N^{(2)} + \ldots + N_1^{(\pi)}. \quad (163)$$

$$N_2 = N_2^{(1)} + N_2^{(2)} + \ldots + N_1^{(\pi)}.$$

$$\ldots$$

$$N_K = N_K^{(1)} + N_K^{(2)} + \ldots N_K^{(\pi)}.$$

Therefore,

$$F = G + \lambda_1(N_1^{(1)} + N_1^{(2)} + \ldots + N_1^{(\pi)} - N_1) +$$

$$\lambda_2(N_1^{(1)} + N_1^{(2)} + \ldots + N_1^{(\pi)} - N_2) + \ldots +$$

$$\lambda_K(N_1^{(1)} + N_1^{(2)} + \ldots + N_1^{(\pi)} - N_K);\text{ and} \quad (164)$$

$$\partial F/\partial N_1^{(1)} = 0 = \hat{g}_1^{(1)} + \lambda_1, \quad (165)$$

$$\partial F/\partial N_1^{(2)} = 0 = \hat{g}_1^{(2)} + \lambda_1,$$

so that

$$\hat{g}_1^{(1)} = \hat{g}_1^{(2)} = \ldots .$$

Likewise,

$$\hat{g}_2^{(1)} = \hat{g}_2^{(2)} = \ldots .$$

The partial molal Gibbs function for each component must be identical in all of the phases at equilibrium.

Chapter 4

4. AVAILABILITY

A. INTRODUCTION

The Second law illustrates that the energy contained in a system in the form of thermal or internal energy cannot be entirely converted into work in a cyclic process even though the system may exist at a higher temperature than its ambient. On the other hand, if an equivalent amount of energy is contained in the same system in the form of potential energy, that energy can be entirely converted into work. Therefore, 1000 kJ of thermal energy contained in a system at a temperature of 1000 K that interacts with an ambient at 300 K can at most potentially provide only 700 kJ of electrical work (through a Carnot heat engine) while 1000 kJ of potential energy in the same system can produce possibly 1000 kJ of electrical work.

For obvious reasons, it is desirable to convert the entire amount of energy in applications (e.g., the chemical energy of gasoline) into work. This is potentially possible in fuel cells (cf. Chapter 13) in which the chemical energy of a fuel can be almost fully converted into electrical energy. However, if the same amount of fuel is burned and the chemical energy contained in it is converted into thermal energy by the production of hot combustion gases in an adiabatic reactor, the conversion of heat into work is limited by the system and ambient temperatures. Therefore, it is useful to develop a method to determine the availability (or work potential) of energy in its various forms (such as heat, chemical, work, advection or flow energy).

Availability is a measure of the work potential of energy. Availability concepts enable the continuous monitoring of the work potential of thermodynamic systems and the associated work losses as they undergo changes in their respective states.

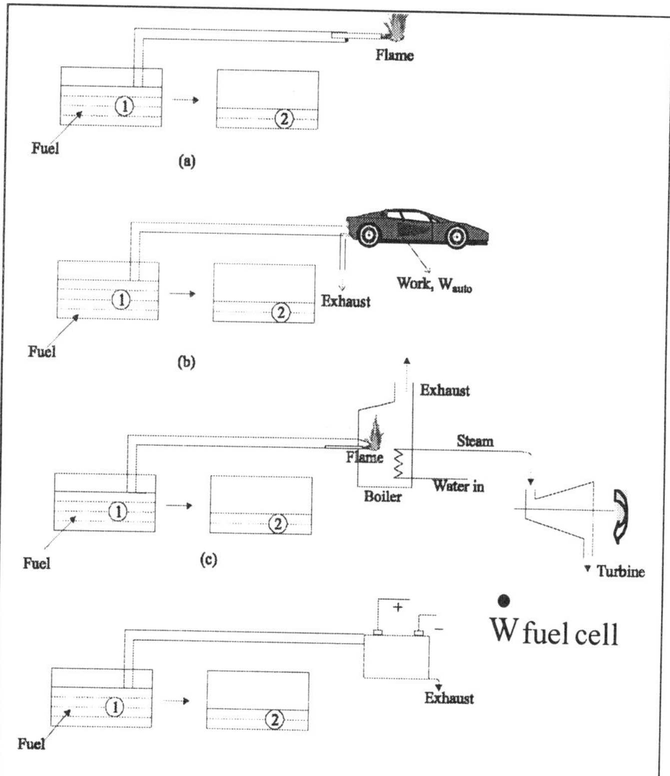

Figure 1: a. simple combustion; b. automobile engine; c. steam power plant; and d. fuel cell.

The work output from a process must satisfy both the First and Second laws of thermodynamics. For instance, if δW and dU are known, δQ can be determined using the First law ($\delta Q - \delta W = dU$), fixed mass and then used in the context of the Second law ($dS - \delta Q/T_b = \delta\sigma$), fixed mass to examine if the Second law is satisfied by the relation $\delta\sigma \geq 0$. When the availability concept is used, the compliance of both laws is ensured. Application of the concept leads to the best use of resources for a prescribed state change.

Suppose, for the sake of illustration, that a cup of hot coffee at 100°C is to be cooled to 30°C. One option is to transfer heat to the ambient atmosphere. Then, if the coffee is to be reheated to its original temperature, a heat pump may be used, however, it involves external work input. Alternatively, the coffee can be cooled by transferring heat to a Carnot heat engine that produces work, and converts it into potential energy by, say, lifting a small elevator. For this process the coffee may be reheated by lowering the elevator and supplying work to a reversed Carnot cycle (or Carnot heat pump) to heat the coffee. The latter process makes the best use of the energy resources and the combined cooling via Carnot engine and reheating process is potentially reversible while the direct cooling to atmosphere is irreversible. Other examples are as follows.

The fuel contained in the gas tank of an automobile can be supplied to a burner and simply burned (cf. Figure 1a) or used to idle an automobile (cf. Figure 1b). For both processes the chemical energy of the fuel is simply wasted. Useful work can be obtained by utilizing the automobile for transportation, but part of the energy will still remain unused and will escape with the exhaust and radiator water. The fuel may also be used to fire a boiler, produce steam, or run a turbine that delivers work (cf. Figure 1c). Finally, the same amount of fuel can be supplied to a fuel cell that converts the chemical energy of the fuel into electrical energy and, thereby, electrical work (cf. Figure 1d). Even though the fuel consumption is the same for all of these cases, the largest amount of work is typically delivered by fuel cells. This leads to the question: What is the maximum work possible from a particular device, given a specific amount of fuel? The answer to this question is provided by availability analyses.

B. OPTIMUM WORK AND IRREVERSIBILITY IN A CLOSED SYSTEM

Consider an initial equilibrium state for a uniform temperature and pressure (say, 2000 K and 20 atm) air mass contained in an insulated and locked piston–cylinder assembly that is placed in a cooler, lower–pressure ambient (at, say, 298 K and 1 atm). Upon removal of the insulation and the locking pin, the hot air will cool and the piston will move and produce work. During the cooling and work delivery phase, realistically, the system temperature will most likely be nonuniform during an initial phase (for the sake of illustration, say, 2000 K at the center and 400 K at boundaries) so that the process is irreversible. At this time the system will contain a certain internal energy $U_{1'}$ and entropy $S_{1'}$. If the nonuniform temperature piston–cylinder system is once again restrained and insulated, another uniform temperature and pressure equilibrium state will eventually be reached (again, for the sake of illustration at, say, 1900 K and 16 atm for a fixed U, V, and m process). At this final equilibrium state, although the internal energy $U_2 = U_{1'}$, the entropy $S_2 > S_{1'}$ due to heat transfer among the various nonuniform temperature masses prior to equilibration (Chapter 3). The First Law applied to the system states that

$$Q - W = U_2 - U_1. \tag{1}$$

For the irreversible process that occurs between the two equilibrium states,

$$S_2 - S_1 = \int_1^2(\delta Q/T_b) + \sigma \tag{2}$$

where S_1 is the initial entropy (at 2000 K and 20 atm) and S_2 the entropy at the final state (at 1900 K and 16 atm).

Realistically, the boundary temperature T_b will also most likely vary along the system boundary during the process and assume different values as the state changes. If T_b is maintained constant (e.g., by using a water jacket around the cylinder) during the irreversible process, the Second law may be simplified into the following form

$$S_2 - S_1 = Q/T_b + \sigma. \tag{3}$$

Eliminating Q in Eqs. (1) and (4) we have

$$S_2 - S_1 = (U_2 - U_1)/T_b + W/T_b + \sigma. \tag{4}$$

Therefore, the work

$$W = -(U_2 - U_1) + T_b(S_2 - S_1) - T_b\,\sigma. \quad (5)$$

Now consider an elemental process between the equilibrium states (U,S) and (U + dU, S + dS) so that $S_2 - S_1 \rightarrow dS$, and $U_2 - U_1 = dU$. (Note that the condition of constant T_b is more appropriate for the elemental process.) For the elemental process Eq. (5) may be written as

$$\delta W = -dU + T_b\,dS - T_b\,\delta\sigma. \quad (6)$$

1. Internally Reversible Process

An internally reversible system contains no temperature gradients during any part of the process so that T_b equals the system temperature T and, within the system, $\sigma = 0$. For an internally reversible process, Eq. (6) yields

$$TdS - PdV = dU,$$

i.e., we recover a combination of the First and Second Law expressions.

2. Useful or External Work

The work expressed in Eq. (5) is delivered by the *matter* of the system and it crosses the system boundary. However, the piston rod does not receive the entire amount of work, since a part of that work (say, W_0) is used to push against the ambient gases that exist at a pressure P_0 adjacent to the piston and which resist the piston motion. Therefore, the net available external or useful work

$$W_u = W - W_0 \quad (7)$$

is delivered through the piston rod, while the total work

$$W_0 = P_0(V_2 - V_1). \quad (8)$$

Therefore, with eqs.(5) and (8) , Eq. (7) may be expressed in the form

$$W_u = -(U_2 - U_1) + T_b(S_2 - S_1) - P_0(V_2 - V_1) - T_b\,\sigma. \quad (9)$$

3. Internally Irreversible Process with no External Irreversibility

Recall that the temperature T_b in Eqs. (5) and (6) may be unknown during a process. However, if the system boundary is slightly extended outside of the system into the uniform ambient, the boundary temperature $T_b \rightarrow T_0$, and its boundary pressure $P \rightarrow P_0$. In this case all of the irreversibilities occur inside the system since there are no temperature and pressure gradients outside it. Hence, the generated entropy is the same as that generated in an isolated system. Rewriting Eq. (9) with $T_b = T_0$,

$$W_u = -(U_2 - U_1) + T_0(S_2 - S_1) - P_0(V_2 - V_1) - T_0\,\sigma. \quad (10)$$

If, the state of a system changes reversibly during a process, the work produced is optimized. Since no internal gradients are created during an idealized reversible process so that no entropy is generated, losses are eliminated while delivering the same amount of work as for an irreversible process. With $\sigma = 0$ in Eq. (10) the optimum work is expressed as

$$W_{u,opt} = -(U_2 - U_1) + T_0(S_2 - S_1) - P_0(V_2 - V_1). \quad (11)$$

The availability concept is based on Eq. (11) and we will later discuss a method to achieve processes for which $\sigma = 0$.

a. Irreversibility or Gouy–Stodola Theorem

The difference between $W_{u,opt}$ and W_u (from Eqs. (10) and (11)) is called the irreversibility or lost work, i.e.,

$$I = W_{u,opt} - W_u = T_0\sigma. \quad (12)$$

Equation (12) is also known as Gouy–Stodola theorem.

4. Nonuniform Boundary Temperature in a System

If, during a process, the boundary temperatures are nonuniform in a system* then $Q = \Sigma\delta Q_j$ and the term $\delta Q/T_b$ in Eq. (2) must be replaced by $\Sigma\delta Q_j/T_{b,j}$ where $T_{b,j}$ denotes the boundary temperature of the j–th infinitesimal element of the control surface surrounding the system. In that case

$$dS = \Sigma\delta Q_j/T_{b,j} + \delta\sigma \quad (13)$$

Expanding,

$$dS = \delta Q_0/T_0 + \delta Q_1/T_{b,1} + \delta Q_2/T_{b,2} + \ldots + \delta\sigma, \quad (14)$$

where δQ_0 denotes the heat exchange between the system and its environment (that exists at a temperature T_0). The First law may be expressed in the differential form

$$\delta Q_0 + \delta Q_1 + \delta\delta Q_2 + \ldots - \delta W = dU. \quad (15)$$

Eliminating δQ_0 between (14) and (15)

$$\delta W = -dU + T_0\, dS + \delta Q_1(1 - T_0/T_{b,1}) + \delta Q_2(1 - T_0/T_{b,2}) + \ldots - T_0\,\delta\sigma \quad (16)$$

If the different boundary temperatures remain unaltered during the process, Eq. (16) can be integrated

$$W_u = -(U_2 - U_1) - P_0(V_2 - V_1) + T_0(S_2 - S_1) + Q_1(1 - T_0/T_{b,1})$$

$$+ Q_2(1 - T_0/T_{b,2}) + \ldots - T_0\sigma \quad (17)$$

We will see later that oftentimes $T_{b,1}$, $T_{b,2}$, ... denote the boundary temperatures of thermal energy reservoirs. By setting $\Phi = 0$ in Eq. (17), the optimum work is obtained as

$$W_{u,opt} = -(U_2 - U_1) - P_0(V_2 - V_1) + T_0(S_2 - S_1) + Q_1(1 - T_0/T_{b,1})$$

$$+ Q_2(1 - T_0/T_{b,2}) + \ldots \quad (18)$$

C. AVAILABILITY ANALYSES FOR A CLOSED SYSTEM

We will now discuss the maximum possible work or optimum work given the initial and final states of a system. Equation (11) provides an expression that combines the First and Second laws to determine the optimum work. In this section we present an idealized scheme to achieve that optimum work. In addition we will also define useful work (W_u), and present the availability functions for closed systems.

1. Absolute and Relative Availability Under Interactions with Ambient

Consider a piston–cylinder assembly in which steam (at an initial equilibrium state (P_1,T_1)) expands to a final equilibrium state (P_2,T_2) during an irreversible process, i.e., the system properties may be non-uniform and the temperatures at its boundary and in the ambient could be different during the process (cf. Figure 2a). It thereby loses heat Q_0 to its ambient and produces work.

---

* e.g., in an automobile engine the temperature at the cylinder walls and heads is different from that at the piston surface.

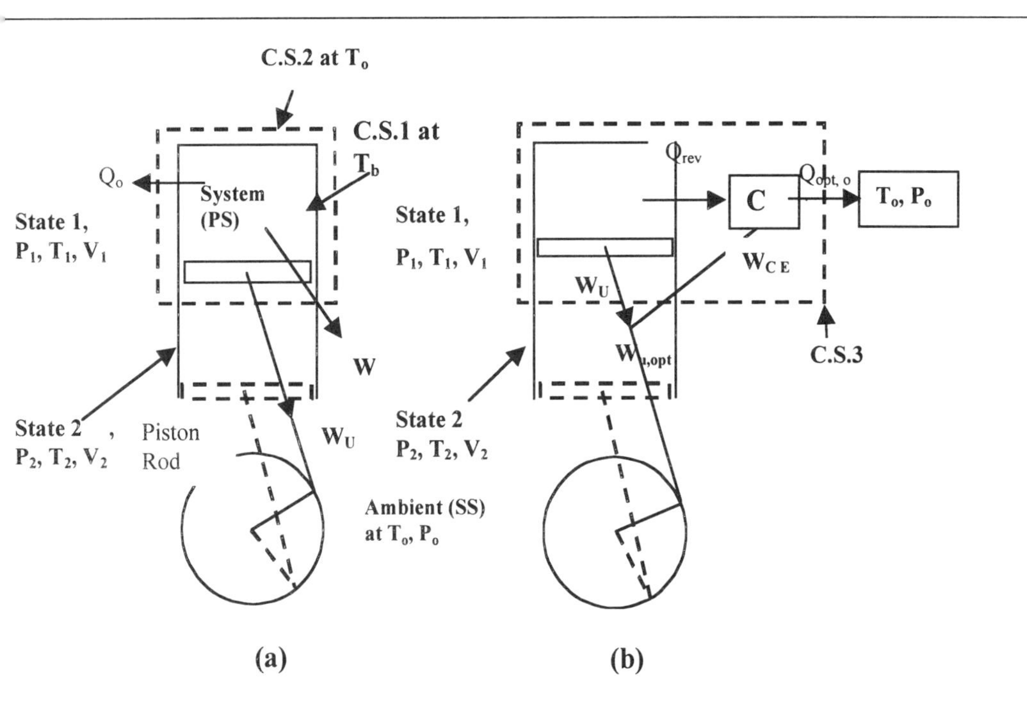

Figure 2: (a) Irreversible process (for control surface 1 $T_b \neq T_0$, whereas for control surface 2 $T_b = T_0$). (b) Reversible process with uniform properties within the system

We denote the matter within the c.s.1 of the piston–cylinder assembly as M and that in the ambient as A, and allow M to undergo an arbitrary change in state from (U_1,V_1) to (U_2,V_2) so that the energy in A changes from $U_{1,0}$ to $U_{2,0}$, and the corresponding volumetric change (in A) is $V_0 = -(V_2 - V_1)$. The total energy E = U (of M) + U_0 (of A) + PE_0 (of A), total volume V = V (of M) + V_0 (of A), and the total mass of the isolated system consisting of both M and A are unchanged, but irreversible processes within the isolated system result in entropy generation. A reversible process (cf. Figure 2b) that involves work transfer $W_{re,}$ heat transfer Q_{rev} across the boundaries of M which is then used to run a Carnot engine that, in turn, rejects Q_0 (amount of heat to the ambient) and produces a work of W_{CE}, can also change the initial state of M to the same final state as shown in Figure 2a, but in this case without altering the entropy of the isolated system. For the latter case, the reversible work done by M, i.e., W_{rev}, and the work delivered by the Carnot engine W_{CE} can be combined so that $W_{opt} = W_{rev} + W_{CE}$.

As the Carnot engine absorbs heat from M (Figure 2b), the temperature of M changes and, consequently, the Carnot efficiency continually changes. Hence consider an infinitesimal reversible process:

$$\delta W_{rev} = \delta Q_{rev} - dU. \tag{19a}$$

If δW_{rev} = 50 kJ, dU = -100 kJ, δQ_{rev} will be -50 kJ; if dU is fixed at –100 kJ, δW_{rev} = 0 kJ, then δQ_{rev} = -100 kJ. The higher the work delivered by the mass, the lower the amount of heat transfer for the same value of dU. The heat δQ_{rev} is supplied from M to the Carnot engine. Since the heat gained by the Carnot engine is $(-\delta Q_{rev})$, the work done by the engine is

$$\delta W_{CE} = -\delta Q_{rev}(1 - T_0/T). \tag{19b}$$

Furthermore, since entropy change of matter M, $dS = \delta Q_{rev}/T$, the above equation assumes the form

$$\delta W_{CE} = -\delta Q_{rev} + T_0\, dS. \quad (20)$$

Adding Eqs. (19a) and (19b) and considering an infinitesimal state change,

$$\delta W_{opt} = -\, dU + T_0\, dS. \quad (21)$$

The higher the work δW_{rev} delivered by the matter, the lower the amount of heat transfer for the same value of dU and the lower the value of δW_{CE}. However, $\delta W_{opt} = \delta W_{CE} + \delta W_{rev}$ remains independent of how much work δW_{rev} is delivered by the matter M within the system. Integrating Eq. (21) between initial and final states, respectively, denoted as 1 and 2,

$$W_{opt} = U_1 - U_2 - T_0\,(S_1 - S_2). \quad (22)$$

This is the net work delivered by the matter through the heat transfer Q_{rev} (i.e., W_{CE}) and Wrev during change of state from state 1 to 2. However, the work through the piston rod is less since a part of W_{opt} is used to overcome atmospheric resistance ($P_0\,(V_2 - V_1)$). The useful or external optimum work during the process is represented by the relation

$$W_{u,opt} = W_{opt} - W_0 = W_{opt} - P_0\,(V_2 - V_1), \text{ and} \quad (23)$$

$$W_{u,opt} = (U_1 - U_2) - T_0\,(S_1 - S_2) + P_0\,(V_2 - V_1). \quad (24)$$

This is the same expression as Eq. (11) and represents the optimum useful work delivered by the matter M. It is more appropriate to refer to $W_{u,opt}$ as the external work delivered rather than as the useful work, since for compression processes the term "useful" can be confusing to readers. For a compression process $W_{u,opt}$ is the external work required to compress the fluid. Based on a unit mass basis

$$w_{u,opt} = (u_1 - u_2) - T_0\,(s_1 - s_2) + P_0\,(v_1 - v_2). \quad (25)$$

For an expansion process, the term $w_{u,opt}$ represents the maximum useful work for the same initial and final states of M. Both processes are represented on the T–s diagram contained in Figure 3. The term $T_0(s_1 - s_2)$ in Eq. (25) is the unavailable portion of the energy (represented by the hatched area DEGF in Figure 3).

We denote ϕ as the absolute closed system availability, i.e.,

$$\phi = u - T_0\, s + P_0\, v, \text{ so that} \quad (26)$$

$$w_{u,opt} = \phi_1 - \phi_2. \quad (27)$$

(The term ϕ is known as "availability" in the European literature.) The availability is expressed in units of kJ kg^{-1} in the SI system and in BTU lb^{-1} in the English system. The term ϕ_1 represents the potential to perform work in a closed system. It is not a property, since it also depends upon the environmental conditions surrounding a system. If, during a process, the state of a system is known, the availability can be determined and the optimum work can be compared with the actual work being produced.

If state 2 at which the temperature and pressure of M approach T_0 and P_0 (state 0 in Figure 3) thereby achieving thermo–mechanical equilibrium (called "restricted dead state") with the environment , Eq. (25) assumes the form

$$\phi_1' = w_{u,opt,0} = (u_1 - u_0) - T_0\,(s_1 - s_0) + P_0\,(v_1 - v_0) = \phi_1 - \phi_0. \quad (28)$$

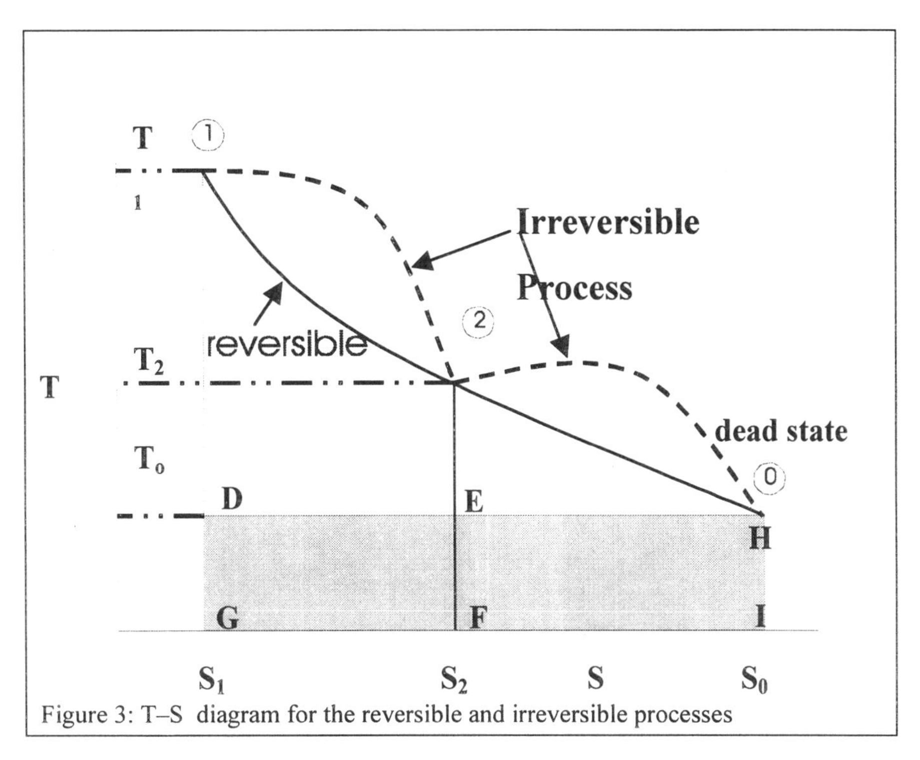

Figure 3: T–S diagram for the reversible and irreversible processes

We denote ϕ_1' as closed system relative availability at state 1, (also known as closed system exergy in Europe or availability in the US). The relative stream availability or exergy has a clearer physical meaning in comparison to the absolute specific flow availability. Recall that the properties for u, s, etc., are tabulated assuming that u=0, s=0 at a prescribed reference state (cf. Chapters 2 and 3). Thus, the values for u and s differ depending upon the reference state and hence absolute availability will change depending upon this choice. However, the exergy, ϕ' is unaffected by these changes so that it is a truer representation of the potential to perform work.

In other words, the closed system exergy ϕ_1' = the energy of matter relative to the dead state $(u_1 - u_0)$ – the unavailable portion of energy $T_0\,(s_1 - s_0)$ (area DHIG shown in Figure 3) – the work to be performed to overcome the ambient (or atmospheric) resistance $P_0\,(v_1 - v_0)$. We now illustrate the physical implication of the relative availability or exergy by considering a thermodynamic system that contains steam at high pressure and temperature (state 1). If the steam is reversibly expanded until it reaches a "dead state", e.g., when it is fully converted into liquid water, the energy that is extracted in the form of useful work is also the relative availability at the initial state. Eq. (27) can be expressed as

$$w_{u,opt} = \phi_1' - \phi_2'. \tag{29}$$

So far we have only considered expansion processes that involve interactions with the environment. During totally reversible compression $w_{u,opt}$ denotes the minimum external work input ($w_{u,min}$) that must be exerted in order to achieve the state change. In this case, the ambient may actually aid in the compression process. Consider a gas compression process at sea level and the same process conducted at the top of a high mountain. The higher ambient pressure at sea level will lead to a lower work input in comparison with that at higher altitudes.

2. Irreversibility or Lost Work

During a reversible process (cf. Figure 2b) for the same change in state as for an irreversible process (cf. Figure 2) the work $W = W_u = W_{u,opt}$. The irreversibility I or lost work LW are both defined as the energy that is unavailable for conversion to work as a result of irreversibilities, i.e.,

$$LW = I = (W_{opt} - W), \text{ or} \quad (30)$$

$$LW = I = (W_{u,opt} - W). \quad (31)$$

Applying the First and Second Laws to the system shown within the control surface cs2 in Figure 2a, $W = Q_0 - (U_2 - U_1)$ and the entropy generation $\sigma = S_2 - S_1 - Q_0/T_0$. Eliminating Q_0 from these two relations ,

$$W = (U_2 - U_1) - T_0 (S_2 - S_1) - T_0 \sigma = W_{opt} - T_0 \sigma, \quad (32)$$

which has a form similar to Eq. (5) with $T_b = T_0$. In the context of Eqs. (30) and (31)

$$LW = I = W_{opt} - W = W_{u,opt} - W = T_0 \sigma. \quad (33)$$

Since $\sigma > 0$, $LW > 0$ and $W_{opt} \geq W$.

a. Comments

For the reversible process $Q_{opt,,0} - W_{opt} = \Delta U$ while for irreversible process $\Delta U = Q_0 - W$. Therefore, $W_{opt} - W = LW = Q_{opt,,0} - Q_0$. Since $LW > 0$, $Q_{opt,,0} > Q_0$. Due to our convention, the heat rejection carries a negative sign. For example, if $Q_{,0}$ = -100 kJ, $Q_{opt,,0}$ = -50 kJ then $Q_{opt,,0} > Q_0$ criterion is satisfied. The implication is that an irreversible process rejects a larger amount of heat and further raises the internal energy of the ambient A due to the irreversibility it must overcome.

Recall that in Chapter 3 we have discussed the entropy maximum principle at conditions corresponding to fixed U_{iso}, V_{iso} and m_{iso}, and the energy minimum principle at fixed S_{iso}, V_{iso}, and m_{iso}. Now, consider the hot matter M as it interacts with the ambient A. If M is cooled by A, the combined entropy of M and A,$S_{iso} = S_{M+A} = S + S_0$, increases at fixed U_{iso} $(= U + U_0)$, V_{iso} $(= V + V_0)$, and $m_{iso} = m + m_0$, and eventually reaches equilibrium at state 2U by travelling along the path 1–A–B–2U that is shown in Figure 4. The point 1 represents the initial state of the composite system in a U_{M+A}–S_{M+A}–U coordinate system. If matter M delivers work and transfer heat to its ambient then the irreversible process is represented by 1–2. If, at state 2, the system is insulated and its piston is restrained from moving (i.e., a constrained equilibrium state is established), S_{M+A} will increase, U_{M+A} will decrease (since work is delivered), while V_{M+A} and m_{M+A} will remain unchanged. If the process is continued until the "dead state" is reached (0,I) , U will have decreased at fixed V and m along the path 1–2–O_I.

Now let us consider the optimum process 1-2opt for which no entropy is generated during a totally reversible process of the composite system. Consequently, S_{M+A} is unchanged while U_{M+A} decreases with V_{M+A} and m_{M+A} fixed as depicted by the path 1–2_{opt} in Figure 4. If a constraint is placed between M and A once the composite system reaches the state 2_{opt}, this, once again, results in a constrained equilibrium condition. The entropy change for the process 1–2_{opt} $dS_{M+A} = dS_M + dS_0 = 0$. Even if the state of M is identical at points 2 and 2 $_{opt}$ so that $U_{M,2} > U_{M2_{opt}}$,$U_{M+A,2} > U_{M+A2_{opt}}$, since $U_{A,2} > U_{A2_{opt}}$ due to more heat rejection in the irreversible process. With S, V and m for the composite system being fixed, the maximum optimum work is obtained when the system M reaches a dead state along the path 1–2_{opt}–0 at which thermo-mechanical equilibrium is achieved within the ambient. At this point the energy U_{M+A} the minimum and no more work can be delivered by M.

Consider a constant volume system, e.g., a car battery if $T_0 = T_1 = T_2 = T$ then

$$W_{u,opt} = (U_1 - U_2) - T (S_1 - S_2), \text{ i.e.,} \quad (34)$$

$$W_{u,opt} = (U_1 - T S_1) - (U_2 - T S_2), \text{ or}$$

$$W_{u,opt} = (A_1 - A_2), \quad (35)$$

where $A = U - T S$ denotes the Helmholtz function or free energy. The magnitude of A represents the capability of a closed system to deliver work.

---

a. *Example 1*

Air is expanded to perform work in a piston–cylinder assembly. The air is initially at $P_1 = 35$ bar and $T_1 = 2000$ K, and the expansion ratio r_v ($= v_2/v_1$) is 7. The ambient temperature $T_0 = 298$ K and pressure $P_0 = 1$ bar.
Determine the useful work that is delivered for an isentropic process.
If process is nonadiabatic and $P_2 = 2.5$ bar, what are the absolute closed system availabilities at the initial and final states? Determine the optimum and useful optimum work.
If a dynamometer measures the useful work to be 0.8 kJ for non-adiabatic process and the initial volume V_1= 0.000205 m^3, determine the heat loss, irreversibility (or lost work), and the entropy generated during the process.

Solution

From the tables for air (Table A-7) at 2000 K, u_1 =1679 kJ kg^{-1}, s_1^0 = 3.799 kJ kg^{-1} K^{-1}, v_{r1} = 2.776, and P_{r1} = 2068.
For the isentropic processes,
$v_{2s}/v_1 = v_{r2s}/v_{r1} = 7$. $\therefore$ v_{r2s} = 19.43, and from the tables P_{r2} = 161, T_2 = 1090 K, and u_{2s} = 835 kJ kg^{-1}. Hence,
$P_2/P_1 = P_{r2}/P_{r1}$ = 161/2068 = 0.0779, and P_2 = 2.725 bar,
and the isentropic work
$w_s = u_1 - u_{2s}$ = (1679-835) = 844 kJ kg^{-1}.
The specific volumes
$v_1 = RT_1/P_1$ = 0.164 m^3 kg^{-1}, and v_2 = 1.148 m^3 kg^{-1} (using the expansion ratio).
Hence, the useful work
$w_{u,s} = 844 - 100 \times (1.148 - 0.164) = 745.6$ kJ kg^{-1}.
Since the cylinder mass is constant, applying the ideal gas law, $T_2 = T_1 P_2 v_2/(P_1 v_1)$, i.e.,
$T_2 = 2.5 \times 7 \times 2000 \div 35 = 1000$ K. The initial entropy
$s_1 = s_1^0 - R \ln (P_1/P_0) = 3.799 - 0.287 \ln (35/1) = (3.799 - 1.020) = 2.779$ kJ kg^{-1} K^{-1},
and the final entropy
$s_2 = s_2^0 - R \ln (P_2/P_0) = 2.97 - 0.287 \ln (2.5/1) = 2.71$ kJ kg^{-1} K^{-1}.
The absolute availabilities
$\phi_1 = u_1 - T_0 s_1 + P_0 v_1 = 1679 - 298 \times 2.779 + 100 \times 0.164 = 867.23$ kJ kg^{-1}, and
$\phi_2 = u_2 - T_0 s_2 + P_0 v_2 = 758.9 - 298 \times 2.71 + 100 \times 1.148 = 66.1$ kJ kg^{-1}. Therefore,
$w_{u,opt} = \phi_1 - \phi_2 = 867.2 - 66.1 = 801.1$ kJ kg^{-1}. Now,
$w_{opt} = w_{u,opt} + P_0 (v_2 - v_1) = 801.1 + 100 \times (1.148 - 0.164) = 899.5$ kJ kg^{-1}.
Hence, the optimal heat transfer
$q_{opt} = (u_2 - u_1) + w_{opt} = 758.9 - 1679 + 899.5 = -20.6$ kJ kg^{-1}.
Applying the relation $V_2/V_1 = v_2/v_1 = 7$, since V_1 = 0.000205 m^3,
$V_2 = 7 \times 0.000205 = 0.00144$ m^3
The mass $m = P_1V_1/RT_1 = 35 \times 10^2 \times 0.000205/(0.287\times2000) = 0.00125$ kg.
Since the total useful work W_u = 0.8 kJ, on a mass basis w_u = 0.8/0.00125 = 640 kJ kg^{-1}.
Therefore, the work $w = 640 + 100 \times (1.148 - 0.164) = 738.4$ kJ kg^{-1}.

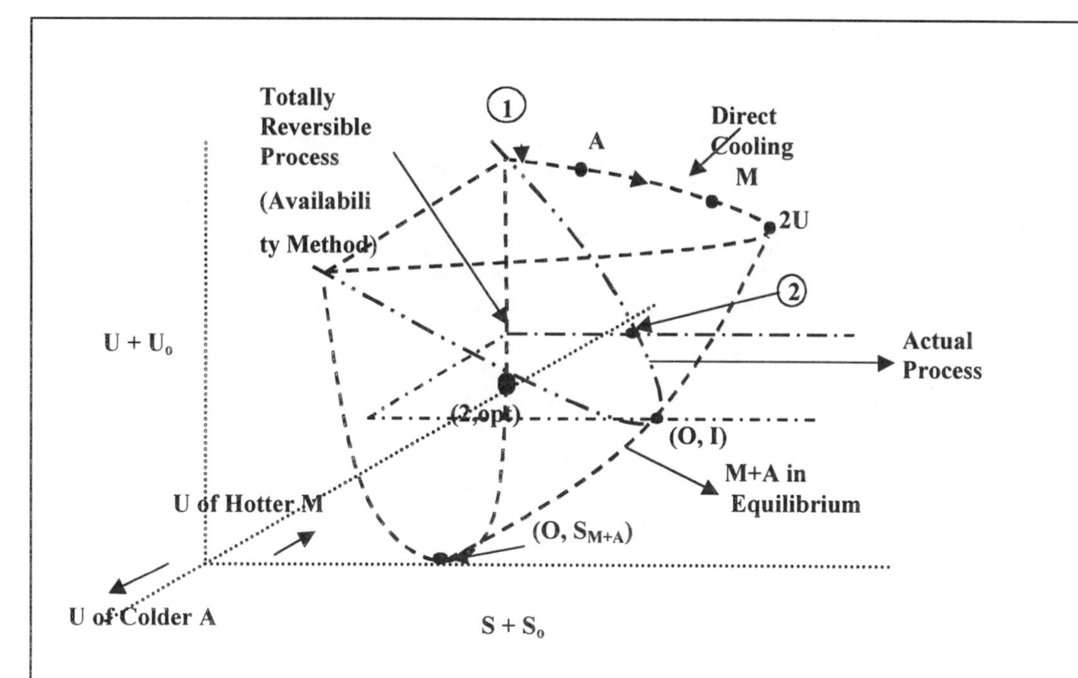

Figure 4: Illustration of reversible and irreversible processes for M and A. The optimum path is along A–D–E whereas a realistic path might be along A–B–C.

The heat transfer $q = (u_2 - u_1) + w = (758.9 - 1679) + 738.4 = -181.7$ kJ kg^{-1}.
The irreversibility $i = w_{u,opt} - w_u = 801.2 - 640 = 161$ kJ kg^{-1} (or 0.201 kJ).
The entropy generation $\sigma = I/T_0 = 161/298 = 0.0007$ kJ K^{-1} (or 0.54 kJ kg^{-1} K^{-1}).

Remarks

For the last part of the problem $\sigma = 161.2/298 = 0.54$ kJ kg^{-1} K^{-1}. The entropy change in the system $s_2 - s_1 = -0.07$ kJ kg^{-1} K^{-1}, the atmospheric entropy change $s_0 = -q/T_0 = 182/298 = 0.61$ kJ kg^{-1} K^{-1} so that $\sigma = -0.07 + 0.61 = 0.54$ kJ kg^{-1} K^{-1} (as before).

The entropy change can be negative for nonadiabatic processes, since the entropy is lowered when a system is cooled.

We note that $|q_{opt}| < |q|$.

We will now illustrate the optimum process. The work $w_{u,opt}$ could have been obtained for same initial and final system states through a totally reversible process which generated no entropy in the isolated system (for which S, V, and m are fixed).

At the initial states for the system M and the ambient A, $U_{1,M} = 1679 \times 0.00125 = 2.1$ kJ, (cf. Point 1 on Figure 4) and we assume that $U_{1,A} = 0$. Therefore, the combined isolated system energy is $U_{1,M+A} = 2.1$ kJ.

At their final states

$U_{2,M} = 758.9 \times 0.00125 = 0.95$ kJ, i.e., the energy change $\Delta U_M = -1.15$ kJ.

Applying the First law

$Q_A - W_A = U_{2,A} - U_{1,A}$, where

$W_A = P_0 \Delta V_A = -0.062$ kJ. Now,

$Q_A = 0.00125 \times 20.6 = 0.026$ kJ. Therefore,

$U_{2,A} - U_{1,A} = 0.026 - (-0.062) = 0.088$ kJ, i.e., $U_{2,A} = 0.088$ kJ, and

$U_{2,M+A} = 0.95 + 0.088 = 1.038$ kJ.

Assume that the optimum work is used to raise a weight so that the ambient potential energy increases by an amount

$\Delta PE = 801.2 \times 0.00125 = 1$ kJ (based on the useful work).

The internal energy change in the isolated system

$\Delta U_{M+A} = U_{2,M+A} - U_{1,M+A} = 1.038 - 2.1 = -1.06$ kJ.

The internal energy of the combined isolated system is lowered (cf. Path $1\text{–}2_{opt}$ on Figure 4), although S, V, and m are held constant.

For the actual process considered in part five of the problem $U_{2,A} - U_{1,A} = Q_A$. With $Q_A = 0.227$ kJ,

$U_{2,A} = 0.227 + 0.062 = 0.289$ kJ, and

$U_{2,M+A} = 0.95 + 0.289 = 1.239$ kJ (cf. Point 2 on Figure 4).

The actual process increases the isolated system entropy (cf. Path 1–2 on Figure 4). The combined system energy $U_{2,M+A}$ is higher in this case than for the corresponding reversible process, since a larger amount of heat is rejected to environment. While the increased heat transfer raises the ambient internal energy, in case a weight is lifted, the ambient gains a lower amount of potential energy ΔPE due to the smaller amount of work that is done on it.

In case the expansion is continued to the dead state, T_0= 25°C, P_0= 1 bar, using the properties of air at that state, it is possible to determine that

$\phi_0 = u_0 - T_0 s_0 + P_0 v_0 = 213 - 298 \times 1.677 + 100 \times 0.855 = -201.2$ kJ kg^{-1},

$w_{u,opt,0} = \phi_1 - \phi_0 = 867.3 - (-201.2) = 1068.5$ kJ kg^{-1},

$w_{opt,0} = w_{u,opt,0} + P_0 (v_0 - v_1) = 1068.5 + 100 \times (0.855 - 0.164) = 1137.6$ kJ kg^{-1}, and

$q_{opt} = (u_0 - u_1) + w_{opt,0} = 213 - 1679 + 1137.6 = -328.5$ kJ kg^{-1}, i.e., the heat entering the ambient under this condition $Q_{opt,0} = 0.374$ kJ. Furthermore,

$U_{0,A} = 0.374 + 0.062 = 0.436$ kJ and $U_{0,M} = 213 \times 0.00125 = 0.266$ kJ so that

$U_{0,M+A} = 0.436 + 0.266 = 0.702$ kJ (cf. Point 0 on Figure 4). Therefore,

$U_{0,M+A} - U_{1,M+A} = U_{0,M} - U_{1,M} + U_{0,A} - U_{1,A} = 0.266 - 2.1 + 0.436 - 0 = -1.398$ kJ.

This is the lowest possible energy value that can be contained in the combined isolated system keeping the combined S, V and m constant. In terms of the potential energy change, at this point a lifted weight will reach its highest elevation.

For the actual process (cf. to Point 2 on Figure 4) a lifted weight will rise to a lower elevation and the internal energy and entropy of the combined isolated system will have higher values.

Removing all restraints between M and at the dead state A will result in no change in state of the isolated system (if both M and A consist of air). However, if the gas composition is different in M and A, removing the restraints will result in mixing. This will be discussed later in the context of chemical availability.

---

b. *Example 2*

Two kg of water are to be heated from a temperature $T_1 = 25$°C to $T_2 = 100$°C.

Natural gas heaters with an 85% efficiency are used for the purpose. How much natural gas is required, assuming that it can release 18,400 kJ m^3 of heat.

How much electrical work is required by an electrical range to do so?(Figure 5a)

Determine the minimum work required using an availability analysis (Figure 5b)

If the cost of natural gas is \$4 per MJ and the electricity cost is 4¢ per kW–hr, determine the costs associated with the problem.

Solution

Assume that the ambient temperature $T_0 = 298$ K and that the water specific heat c = 4.184 kJ kg^{-1} K^{-1}.

Applying the First law

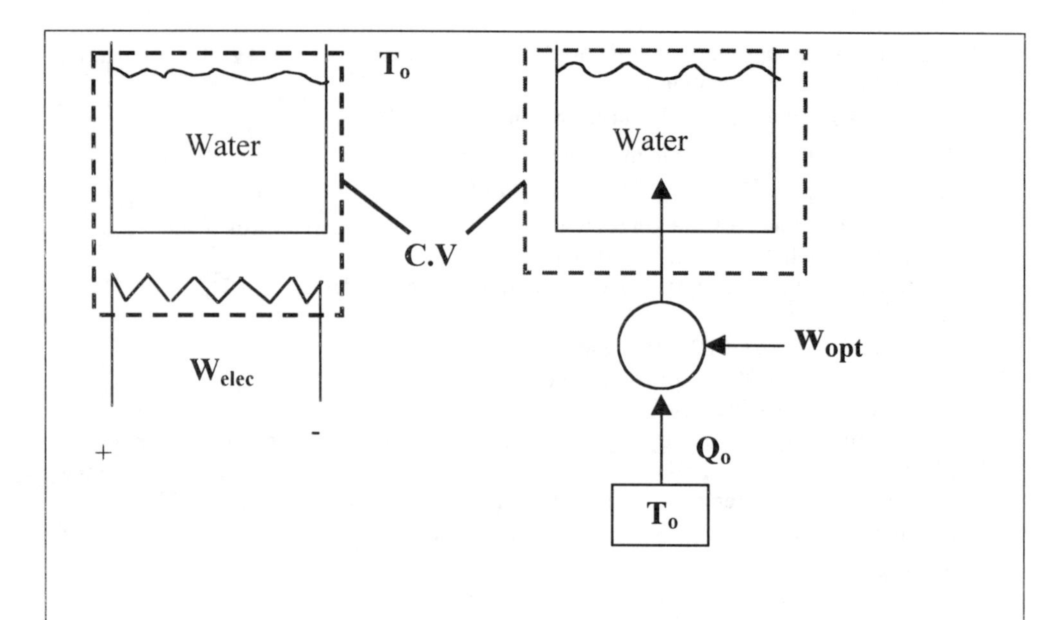

Figure 5: a. Direct heating of water with an electrical range; b. heating of water with a heat pump where we use ambient Q_0 to supply a part of the heat.

$q - 0 = u_2 - u_1 = c\,(T_2 - T_1) = 4.184 \times (100 - 25) = 313.8$ kJ kg^{-1}.

The actual heat input that is required

$q_{in} = 313.8/0.85 = 369.2$ kJ kg^{-1} or $Q_{in} = 2 \times 369.2 = 738.4$ kJ.

The volumetric amount of natural gas required = $738.4 \div 18400 = 0.04$ m^3.

Assuming that the electrical work is 100% efficient

$W_{elec} = -313.8 \times 2 = -627.6$ kJ.

The optimum work

$w_{opt} = u_1 - u_2 - T_0\,(s_1 - s_2) + P_0\,(v_1 - v_2)$.

Since $v_1 \approx v_2$ for liquids due to negligible expansion,

$w_{opt} = c\,(T_2 - T_1) - T_0\,(c \ln (T_2/T_1)) = -4.184 \times 75 + 298 \times 4.184 \times \ln (373/298) =$

$-313.8 + 279.9 = -33.9$ kJ kg^{-1}. Therefore,

$W_{opt} = -33.9 \times 2 = -67.8$ kJ, and

$W_{opt} = W_{min}$, the minimum work required in order to heat the water.

The cost comparisons to heat the 2 kg of water are as follows:

Natural gas heating:	$\$4 \times 10^{-6} \times 738.4 = 0.296$¢.
Electrical heating:	4¢ kWh^{-1} $\times$ 313.8 = 0.7¢.
Heat pump:	4¢ kWh^{-1} $\times$ 33.9 = 0.076¢.

Remarks

Using a heat pump, the electrical bill for heating the water can be reduced by 89%. Availability analyses provide the information on how best to achieve desired end states with minimum work input or by obtaining maximum work output. The only allowed interactions are the ones with the environment.

The heat pump can be run by using 33.9 kJ kg^{-1} of electrical work to run it. In turn the heat pump will accept 279 kJ kg^{-1} from the ambient air at 25°C and deliver 313.8 (= 279.9 + 33.9) kJ kg^{-1} of heat to the water. The heat pump must be operated between a

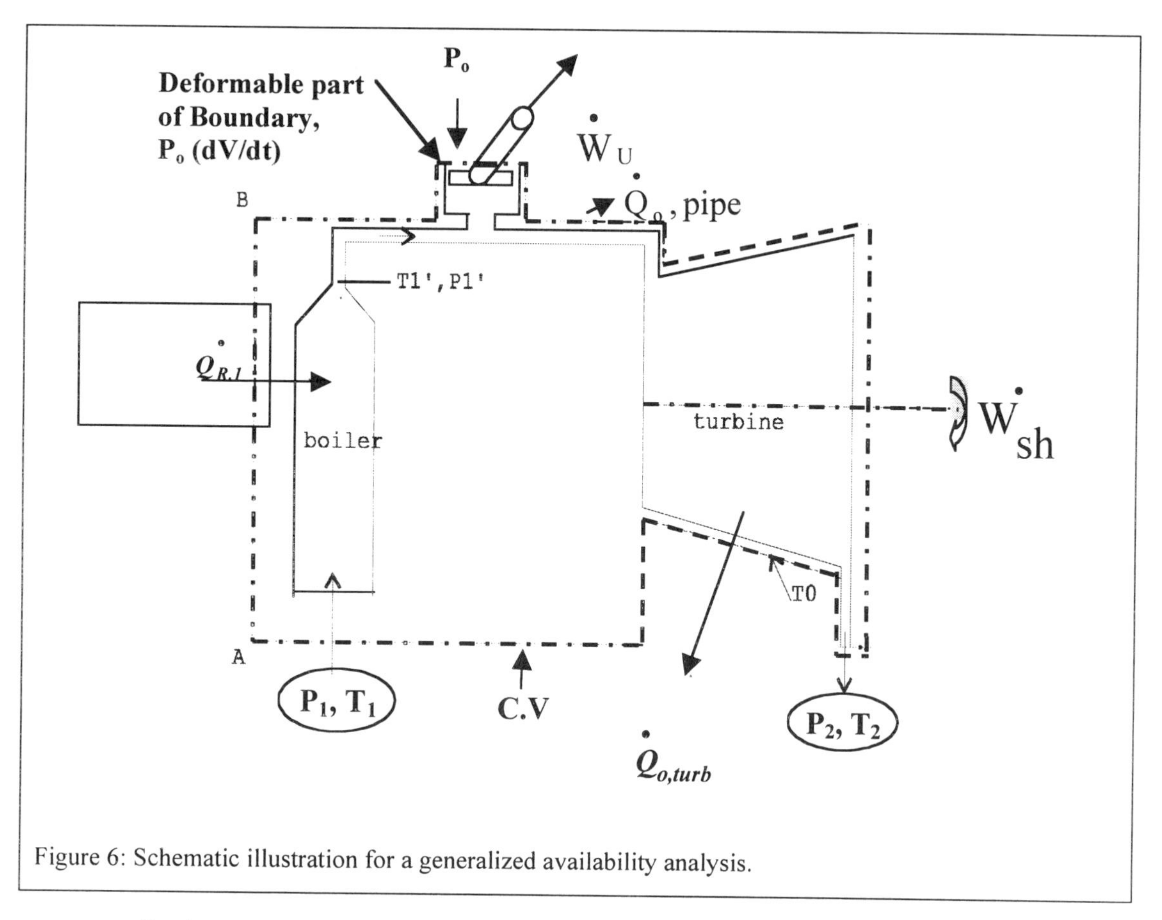

Figure 6: Schematic illustration for a generalized availability analysis.

fixed temperature thermal reservoir at a temperature T_0 =25°C and the variable–temperature hot water reservoir.
This example is pertinent also to domestic heating applications.

D. GENERALIZED AVAILABILITY ANALYSIS

The previous section considered a closed system undergoing expansion or compression processes with simultaneous heat exchange with a constant–temperature environment. Many practical applications (e.g., automobiles, steam power plants, and gas turbines) involve open systems that interact with heat sources that are thermal energy reservoirs (such as boilers, reheaters, and combustors), and reject heat to the ambient (e.g., condensers in steam power plants, through heat losses from steam pipes, and automobile and turbine exhaust). These systems may also operate in an unsteady mode. Therefore, a more generalized availability analysis is required. We will start by discussing a generalized analysis for an open system and then simplifying it to specific systems.

1. Optimum Work

Consider water at high pressure P_1 and low temperature T_1 (cf. Figure 6) that is heated to produce steam at (P_1', T_1') using a large thermal reservoir (such as a boiler) that exists at a constant temperature $T_{R,1}$. The reservoir transfers heat at a rate $\dot{Q}_{R,1}$ to the water. The steam first delivers useful work $\dot{W}_u$ through a deformable piston–cylinder assembly followed by shaft work $\dot{W}_{Shaft}$ through a steam turbine. The cylinder boundary deforms, producing deformation or boundary work $P_0\ dV_{cyl}/dt$ against the atmosphere. The boundary is selected in such a manner that no property gradients exist outside it, and the boundary temperature is T_0 (as illustrated in Figure 6). Therefore, there are no irreversibilities outside of the system. We assume that there is no entropy generation in either the reservoir inside the control volume or the environment outside it. Since heat is transferred to the water and steam from the hot gases in the boiler, a temperature gradient exists inside the boiler. Other (temperature and pressure) gradients exist inside the pipes, the cylinder and turbine, and other components. For instance, the turbine blades may have rusted, resulting in frictional heat generation. Boundary layer effects on pipe and boiler walls and turbulent viscous dissipation over the turbine blades can add heat to the flow. Realistically, the process from state 1 to state 2 will be irreversible so that entropy must have been generated within the control volume and the net actual work output rate for such a process is

$$\dot{W}_{cv} = \dot{W}_u + \dot{W}_{sh} + \dot{W}_0.$$

where $\dot{W}_0$ denotes the atmospheric work $P_o dV_{cyl}/dt$.

In this context we will now determine the optimum work for a process that occurs between the same inlet (T_1,P_1) and outlet (T_2,P_2) states and which withdraws an identical amount of heat from the reservoir, $\dot{Q}_{R,J}$. In our analysis we will consider the presence of several thermal energy reservoirs at various temperatures, i.e., $T_{R,1}$, $T_{R,2}$, $T_{R,3}$,...etc. with all irreversibilities being maintained within the system control volume. Applying the First law for an open system,

$$\frac{dE_{cv}}{dt} = \dot{Q}_{cv} - \dot{W}_{cv} + \dot{m}_i e_{T,i} + - \dot{m}_e e_{T,e}, \tag{36}$$

where the control volume work expression $\dot{W}_{cv}$ includes $\dot{W}_u$, $\dot{W}_{shaft}$, $P_0 dV_{cyl}/dt$, and any other work forms. The total heat $\dot{Q}_{cv}$ transferred from the control volume includes the various heat interactions $\dot{Q}_{R,1}$, $\dot{Q}_{R,2}$, $\dot{Q}_{R,3}$,... with the thermal energy reservoirs, and that with the environment $\dot{Q}_0$. Therefore,

$$\frac{dE_{cv}}{dt} = \dot{Q}_0 + \dot{Q}_{R,1} + \dot{Q}_{R,2} + \dot{Q}_{R,3} + \ldots\ldots - \dot{W}_{cv} + \dot{m}_i e_{T,i} - \dot{m}_e e_{T,e} \tag{37}$$

Typically $\dot{Q}_0 < 0$. If turbine blades get rusted over a period of time, more energy is used to overcome friction than to produce work; thus, frictional heating will occur which will cause turbine exit temperature T_e to increase. In order to maintain the same T_e, the heat loss to the environment $\dot{Q}_{0,turb}$ must have to be increased so that under steady state operation a smaller amount of work is delivered.

We have assumed that temperature gradients only exist within the boundaries of the control volume and, consequently, entropy is generated only within it. Applying the entropy balance equation (Chapter 3),

$$\frac{dS_{cv}}{dt} = \frac{\dot{Q}_0}{T_0} + \frac{\dot{Q}_{R,1}}{T_{R,1}} + \frac{\dot{Q}_{R,2}}{T_{R,2}} + \frac{\dot{Q}_{R,3}}{T_{R,3}} + \cdots - \dot{m}_e s_e + \dot{m}_i s_i + T_0 \dot{\sigma}_{cv}, \tag{38}$$

We observe that the higher the loss $|\dot{Q}_0|$, the higher $\dot{\sigma}_{cv}$ for a steady state operation for fixed s_i and s_e. From energy balance, the higher the heat loss, the lower is the work output. Expressing $\dot{Q}_0$ in terms of $\dot{\sigma}_{cv}$ Eq. (37) can be expressed in the form

$$\begin{aligned} \dot{W}_{cv} &= T_0\left(\frac{dS_{cv}}{dt} - \Sigma_{j=1}^{N} \frac{\dot{Q}_{R,j}}{T_{R,j}} + \dot{m}_e s_e - \dot{m}_i s_i\right) + \\ &\Sigma_{j=1}^{N} \dot{Q}_{R,j} - \frac{dE_{cv}}{dt} - \dot{m}_e e_{T,e} + \dot{m}_i e_{T,i} - T_0 \dot{\sigma}_{cv} \end{aligned} \tag{39}$$

Rewriting Eq. (39),

$$\dot{W}_{cv} = -d(E_{cv}-T_0S_{cv})/dt + \Sigma_{j=1}^{N}\dot{Q}_{R,j}\left(1-T_0/T_{R,j}\right) - \dot{m}_e\psi_e + \dot{m}_i\psi_i - T_0\dot{\sigma}_{cv}, \quad (40)$$

The absolute specific stream availability or the absolute specific flow or stream availability ψ is defined as

$$\psi(T,P,T_0) = e_T(T,P) - T_0\, s(T,P) = (h(T,P) + ke + pe) - T_0\, s(T,P). \quad (41)$$

where the terms ψ_i and ψ_e denote the absolute stream availabilities, respectively, at the inlet and exit of the control volume. They are not properties of the fluid alone and depend upon the temperature of the environment. The optimum work is obtained for the same inlet and exit states when $\dot{\sigma}_{cv} = 0$. In this case, Eq. (40) assumes the form

$$\dot{W}_{cv,opt} = -d(E_{cv}-T_0S_{cv})/dt + \Sigma_{j=1}^{N}\dot{Q}_{R,j}\left(1-T_0/T_{R,j}\right) - \dot{m}_e\psi_e + \dot{m}_i\psi_i. \quad (42)$$

where the term $\dot{Q}_{R,j}(1-T_0/T_{R,j})$ represents the availability in terms of the quality of heat energy or the work potential associated with the heat transferred from the thermal energy reservoir at the temperature $T_{R,j}$. When the kinetic and potential energies are negligible,

$$\psi = h - T_0\, s. \quad (43)$$

For ideal gases, $s = s^0 - R \ln(P/P_{ref})$, where the reference state is generally assumed to be at P_{ref} = 1 bar. Therefore,

$$\psi = \psi^0 + R\, T_0 \ln(P/P_{ref}), \quad (44)$$

and $\psi^0 = h^0 - T_0 s^0$. The enthalpy $h(T) = h^0(T)$ for ideal gases, since it is independent of pressure.

If the exit temperature and pressure from the control volume is identical to the environmental conditions T_0 and P_0, i.e., the exit is said to be at a restricted dead state, in that case $\dot{W}_{cv,opt} = \left.\dot{W}_{cv,opt}\right|_0$ and the exit absolute stream availability at dead state may be expressed as

$$\psi_{e,0} = e_{T,e,0}(T_0,P_0) - T_0\, s_{e,0}(T_0,P_0). \quad (45)$$

Note that $e_{T,e,0} = h_0$ since ke and pe are equal to zero at dead state.

2. Lost Work Rate, Irreversibility Rate, Availability Loss

The lost work is expressed through the lost work theorem, i.e.,

$$LW = \dot{I} = \dot{W}_{cv,opt} - \dot{W}_{cv} = T_0\dot{\sigma}_{cv}. \quad (46)$$

The terms $\dot{W}_{cv}, \dot{W}_{cv,opt} > 0$ for expansion processes and $\dot{W}_{cv}, \dot{W}_{cv,opt} < 0$ for compression and electrical work input processes. The lost work is always positive for realistic processes.

The availability is completely destroyed during all spontaneous processes (i.e., those that occur without outside intervention) that bring the system and its ambient to a dead state. An example is the cooling of coffee in a room.

3. Availability Balance Equation in Terms of Actual Work

We will rewrite Eq. (40) as

$$d(E_{cv}-T_0S_{cv})/dt = \dot{m}_i\psi_i + \Sigma_{j=1}^{N}\dot{Q}_{R,j}\left(1-T_0/T_{R,j}\right) - \dot{m}_e\psi_e - \dot{W}_{cv} - \dot{I}. \quad (47)$$

The term on the LHS represents the availability accumulation rate within the control volume as a result of the terms on the RHS which represent, respectively: (1) the availability flow rate into the c.v.; (2) the availability input due to heat transfer from thermal energy reservoirs; (3) the availability flow rate that exits the control volume; (4) the availability transfer through ac-

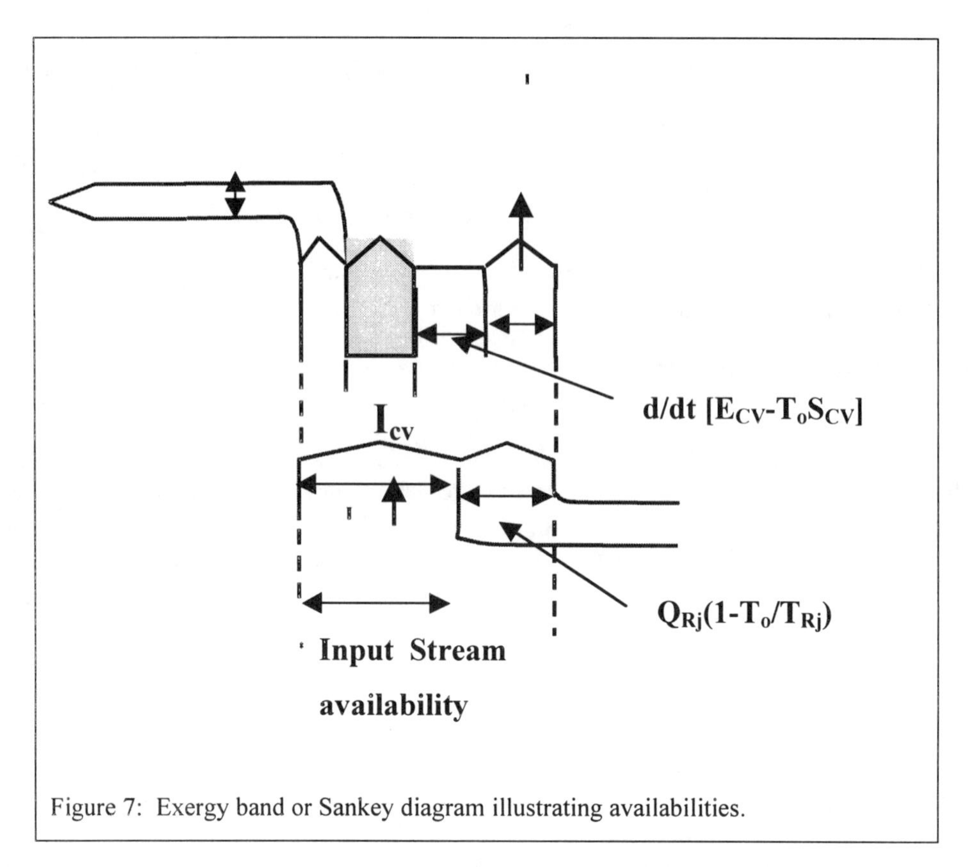

Figure 7: Exergy band or Sankey diagram illustrating availabilities.

tual work input/output; and (5) the availability loss through irreversibilities. The Band or Sankey diagram illustrated in Figure 7 employs an accounting procedure to describe the availability balance. This includes irreversibility due to temperature gradients between reservoirs and working fluids, such as water in a boiler with external gradients, the irreversibilities in pipes, turbines, etc.

a. Irreversibility due to Heat Transfer

We can separate these irreversibilities into various components. For instance consider the boiler component. Suppose the boiler tube is enclosed by a large Tubular TER. For a single TER the availability balance equation is given as

$$d(E_{cv}-T_0S_{cv})/dt = \dot{Q}_b(1-T_0/T_{RJ})+\dot{m}_i\psi_i-\dot{m}_e\psi_e+\dot{I}_{b+R}, \tag{48}$$

We have seen that an irreversibility can arise due to both internal and external processes. For instance, if the boundary AB in Figure 6 is selected so as to lie just within the boiler, and the control volume encloses the gases within the turbine, then Eq. (47) becomes

$$\frac{d(E_{cv}-T_0S_{cv})}{dt} = \dot{Q}_b(1-T_0/T_{w,b})+\dot{m}_i\psi_i-\dot{m}_e\psi_e+\dot{I}_b, \tag{49}$$

where $T_{w,b}$ denotes the water temperature just on the inside surface of the boiler (assumed uniform) and $\dot{Q}_{R,1} = \dot{Q}_b$, the heat transfer from the reservoir to the water. The irreversibility $\dot{I}_b$ arises due to temperature gradients within the water. Subtracting (48) from (49), the irreversibility that exists to external temperature gradient between reservoir and wall temperature alone can be expressed as

$$\dot{I}_{b+R}-\dot{I}_b = \dot{Q}_bT_0(1/T_{w,b}-1/T_{RJ}) \tag{50}$$

Recall the entropy generation $\dot{\sigma}_{cv} = \dot{I}_{cv}/T_0$. Thus, the entropy generated due to gradients existing between a TER and a boiler tube wall

$$\dot{\sigma} = \frac{\dot{I}_{b+R} - \dot{I}_b}{T_o} = \dot{Q}_b\left(\frac{1}{T_{w,b}} - \frac{1}{T_{R,l}}\right) \quad (51)$$

4. Applications of the Availability Balance Equation

We now discuss various applications of the availability balance equation.

An unsteady situation exists at startup when a turbine or a boiler is being warmed, and the availability starts to accumulate. Here,

$$d(E_{cv} - T_0 S_{cv})/dt \neq 0.$$

If a system has a nondeformable boundary, then

$$W_{cv} = W_{shaft},\ P_0 dV_{cyl}/dt = 0,\ \dot{W}_u = 0$$

When a system interacts only with its ambient (that exists at a uniform temperature T_0), and there are no other thermal energy reservoirs within the system, the optimum work is provided by the relation

$$\dot{W}_{cv,opt} = \dot{W}_{cv} + \dot{I} = \dot{m}_i \psi_i - \dot{m}_e \psi_e - d(E_{cv} - T_0 S_{cv})/dt. \quad (52)$$

For a system containing a single thermal energy reservoir (as in the case of a power plant containing a boiler, turbine, condenser and pump, (Figure 8) or the evaporation of water from the oceans as a result of heat from the sun acting as TER), omitting the subscript 1 for the reservoir,

$$d(E_{cv} - T_0 S_{cv})/dt = \dot{m}_i \psi_i + \dot{Q}_R(1 - T_0/T_R) - \dot{m}_e \psi_e - \dot{W}_{cv} - \dot{I}. \quad (53)$$

For a steady state steady flow process (e.g., such as in power plants generating power under steady state conditions), mass conservation implies that $\dot{m}_i = \dot{m}_e = \dot{m}$. Furthermore, if the system contains a single inlet and exit, the availability balance assumes the form

$$\dot{m}(\psi_i - \psi_e) + \Sigma_{j=1}^{N} \dot{Q}_{R,j}(1 - T_0/T_{R,j}) - \dot{W}_{cv} - \dot{I} = 0. \quad (54)$$

On unit mass basis

$$\psi_i - \psi_e + \Sigma\, q_{R,j}(1 - T_0/T_{R,j}) - w_{cv} - i = 0, \quad (55)$$

where $q_{R,j} = \dot{Q}_{R,j}/\dot{m}, w_{cv} = \dot{W}_{cv}/\dot{m}, i = \dot{I}/\dot{m}$. When a system interacts only with its ambient at T_0 and there are no other thermal energy reservoirs within the system, the optimum work is given by the relation

$$\dot{W}_{cv,opt} = \dot{m}_i \psi_i - \dot{m}_e \psi_e - d(E_{cv} - T_0 S_{cv})/dt. \quad (56)$$

In case the exit state is a restricted dead state, (e.g., for H_2O, dead state is liquid water at 25°C 1 bar)

$$\dot{W}_{cv,opt} = \dot{m}\psi' + \Sigma_{j=1}^{N} \dot{Q}_{R,j}(1 - T_0/T_{R,j}) \quad (57)$$

where $\psi' = \psi - \psi_0$ is the specific stream exergy or specific-relative stream availability (i.e., relative to the dead state). Since $\psi_0 = h_o - T_o s_o$ in the absence of kinetic and potential energy at the dead state, as $T_0 \to 0$, $\psi_0 \to 0$, and the relative and absolute stream availabilities become equal to each other.

For a system containing multiple inlets and exits the availability equation is

$$d(E_{cv} - T_0 S_{cv})/dt = \sum_{inlets} \dot{m}_i \psi_i + \sum_{j=1}^{N} \dot{Q}_{R,j}(1 - T_0/T_{R,j}) - \sum_{exits} \dot{m}_e \psi_e - \dot{W}_{cv} - \dot{I}. \quad (58)$$

For a single inlet and exit system containing multiple components the expression can be generalized as

$$d(E_{cv} - T_0 S_{cv})/dt = \sum_{species} \dot{m}_{k,i} \psi_{k,i} + \sum_{j=1}^{N} \dot{Q}_{R,j}(1 - T_0/T_{R,j}) - \sum_{species} \dot{m}_{k,e} \psi_{k,e} - \dot{W}_{cv} - \dot{I} \quad (59)$$

where $\psi_k = h_k(T,P,X_k) - T_0\, s_k(T,P,X_k)$ denotes the absolute availability of each component, and X_k the mole fraction of species k. For ideal gas mixtures,

$$\psi_k = h_k - T_0\,(s_k^0 - R \ln (p_k/P_{ref})),$$

since the partial pressure of the k–th species in the ideal gas mixture $P_k = X_k\, P$.

Consider an automobile engine in which piston is moving and at the same time mass is entering or leaving the system (e.g., during the intake and exhaust strokes). In addition to the delivery of work through the piston rod $\dot{W}_u$, atmospheric work is performed during deformation, i.e., $\dot{W}_0 = P_0\, dV/dt$. Therefore,, $\dot{W}_{cv} = \dot{W}_u + P_0\, dV_{cyl}/dt$ and the governing availability balance equation is

$$d(E_{cv} - T_0 S_{cv})/dt = \sum_{species} \dot{m}_{k,i} \psi_{k,i} + \sum_{j=1}^{N} \dot{Q}_{R,j}(1 - T_0/T_{R,j}) - \sum_{species} \dot{m}_{k,e} \psi_{k,e} - \dot{W}_u - P_0 dV/dt - \dot{I}$$

Simplifying.

$$d(E_{cv} - T_0 S_{cv} + P_0 dV/dt)/dt = \sum_{species} \dot{m}_{k,i} \psi_{k,i} + \sum_{j=1}^{N} \dot{Q}_{R,j}(1 - T_0/T_{R,j}) - \sum_{species} \dot{m}_{k,e} \psi_{k,e} - \dot{W}_u - \dot{I}$$

For steady cyclical processes the accumulation term is zero within the control volume, and $\psi_i = \psi_e$. Therefore,

$$\dot{W}_{cv,cycle} + \dot{I} = \sum_{j=1}^{N} \dot{Q}_{R,j}(1 - T_0/T_{R,j}).$$

---

c. *Example 3*

Steam enters a turbine with a velocity of 200 m s^{-1} at 60 bar and 740°C and leaves as saturated vapor at 0.2 bar and 80 m s^{-1}.The actual work delivered during the process is 1300 kJ kg^{-1}. Determine inlet stream availability, the exit stream availability, and the irreversibility.

Solution

$\psi_i = (h_1 + v^2/2g) - T_0\, s_1 = 3989.2 + 20 - 298 \times 7.519 = 1769$ kJ kg^{-1}. Likewise,

$\psi_e = 2609.7 + (80^2 \div 2000) - 298 \times 7.9085 = 256$ kJ kg^{-1}. Therefore,

$w_{opt} = 1769 - 256 = 1513$ kJ kg^{-1}, and

$I = 1513 - 1300 = 213$ kJ kg^{-1}. The entropy generation

$\sigma = 213 \div 298 = 0.715$ kJ kg^{-1} K^{-1}.

Remarks

The input absolute availability is 1769 kJ kg^{-1}.
The absolute availability outflow is 256 kJ kg^{-1}.
The absolute availability transfer through work is 1300 kJ kg^{-1}.
The availability loss is 213 kJ kg^{-1}.
The *net* outflow is 1769 kJ kg^{-1}.

d. *Example 4*

This example illustrates the interaction between a thermal energy reservoir, its ambient, a steady state steady flow process, and a cyclical process. Consider the inflow of water in the form of a saturated liquid at 60 bar into a nuclear reactor (state 1). The reactor temperature is 2000 K and it produces steam which subsequently expands in a turbine to saturated vapor at a 0.1 bar pressure (state 2). The ambient temperature is 25°C. The reactor heat transfer is 4526 kJ per kg of water. Assume that the pipes and turbines are rigid.

What is the maximum possible work between the two states 1 and 2?

If the steam that is discharged from turbine is passed through a condenser (cf. Figure 8) and then pumped back to the nuclear reactor at 60 bar, what is the maximum possible work under steady state cyclical conditions? Assume that the inlet condition of the water into the pump is saturated liquid.

Solution

If the boundary is selected through the reactor, for optimum work I = σ = 0. Under steady state conditions time derivatives are zero, and, since the body does not deform $W_u = 0$, so that $W_{cv} = W_{shaft}$ and $\dot{m}_i = \dot{m}_e = \dot{m}$. Therefore,

$$\dot{m}(\psi_i - \psi_e) + \dot{Q}_{RJ}(1 - T_0/T_{RJ}) - \dot{W}_{cv,opt} = 0. \qquad (A)$$

Dividing Eq. (A) throughout by the mass flow rate,

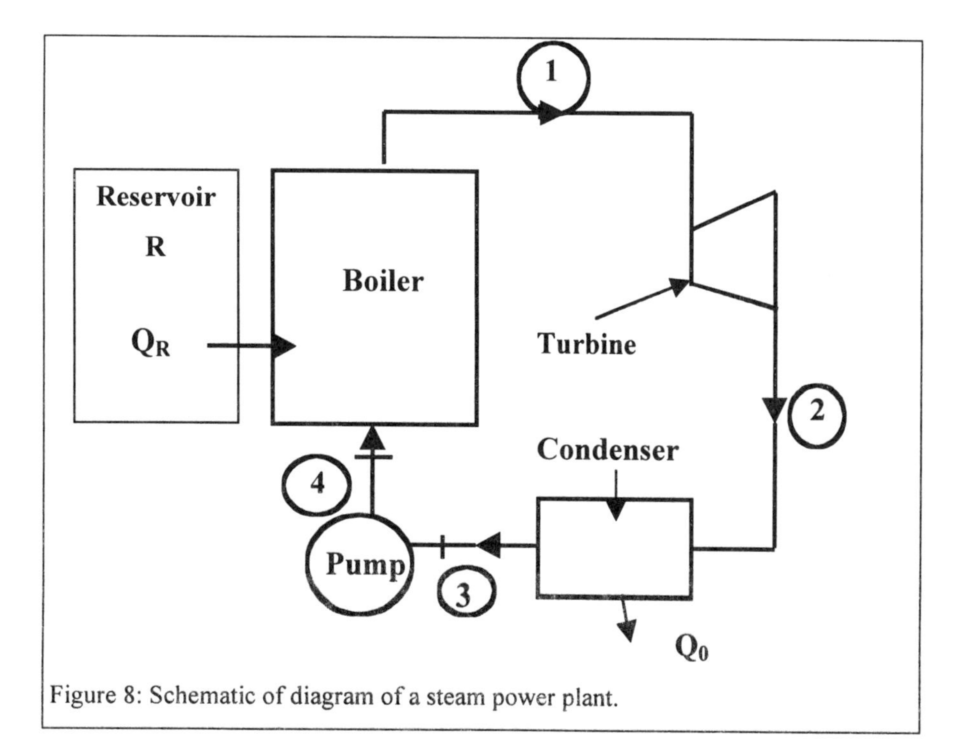

Figure 8: Schematic of diagram of a steam power plant.

$$w_{cv,opt} = (\psi_i - \psi_e) + q_{R,1}(1 - T_0 / T_{R,1}). \quad (B)$$

Using the steam tables (A-4A)

$$\psi_i = 1213.4 - 298 \times 3.03 = 310.5 \text{ kJ kg}^{-1}, \text{ and} \quad (C)$$

$$\psi_e = 2584.7 - 298 \times 8.15 = 156.0 \text{ kJ kg}^{-1}. \quad (D)$$

Therefore,

$$q_{R,1} = 4526 \text{ kJ kg}^{-1}, \text{ and} \quad (E)$$

$$w_{cv,opt} = 4526(1 - 298 \div 2000) + (310.5 - 156) = 4006 \text{ kJ kg}^{-1}. \quad (F)$$

For the cycle, $\psi_i = \psi_e$. Therefore,

$$w_{cv,opt,cycle} = q_{R,1}\,(1 - T_0/T_{R,1}) = 3852 \text{ kJ kg}^{-1}.$$

Note that this work is identical to that of a Carnot cycle with an efficiency of $(1-T_0/T_{R,1})$.

Remarks

A realistic cyclical process contains inherent irreversibilities due to irreversible heat transfer and internal irreversibilities so that

$$w_{cv,cycle} < w_{cv,opt,cycle}.$$

The work $w_{cv,cycle}$ usually deteriorates over time, since internal irreversibilities in the cycle increase. Once the state is known, it is possible to ascertain ψ at various points during a process to determine $w_{cv,opt}$ and w_{cv}, and to calculate $\sigma_{cv} = (w_{cv,opt} - w_{cv})/T_0$.

It is seen from Eq. (40) that the higher the entropy generation, the larger the mount of lost work and lesser the work output. Entropy generation occurs basically due to internal gradients and frictional processes within a device; it can also originate due to poor design, such as through irreversible heat transfer between two systems at unequal temperatures as in heat exchangers. For instance, consider a parallel flow heat exchanger in which hot gases enter at 1000 K and are cooled to 500 K by cold water that enters at 300 K (cf. Figure 9a). The water can, at most, be heated to 500 K. A large temperature difference of 700 K exists at the inlet resulting in large entropy generation during the process. In a counter flow heat exchanger the temperature difference can be minimized to reduce σ (cf. Figure 9b). Therefore, it is important to consider entropy generation/availability concepts during the design of thermal systems.

e. Example 5

Hot air at a temperature of 400°C flows into an insulated heat exchanger (the fire tube boiler shown in Figure 10) at a rate of 10 kg s^{-1}. It is used to heat water from a saturated liquid state to a saturated vapor condition at 100°C. If the air exits the heat exchanger at 200°C, determine the water flow in kg s^{-1} and the irreversibility. Assume that $c_p = 1$ kJ kg^{-1} K^{-1}, $h_{fg} = 2257$ kJ kg^{-1}, and $T_0 = 298$ K.

Solution

The energy required to heat the water is obtained by applying the First Law, namely,

$$dE_{c.v.}/dt = \dot{Q}_0 + \dot{Q}_{R,1} + \dot{Q}_{R,2} + \dot{Q}_{R,3}.. - \dot{W}_{c.v.} + \Sigma \dot{m}_i e_{T,i} - \Sigma \dot{m}_e e_{T,e}.$$

Since the fire tube boiler is assumed to be an adiabatic, steady and non-work producing device, this relation assumes the form

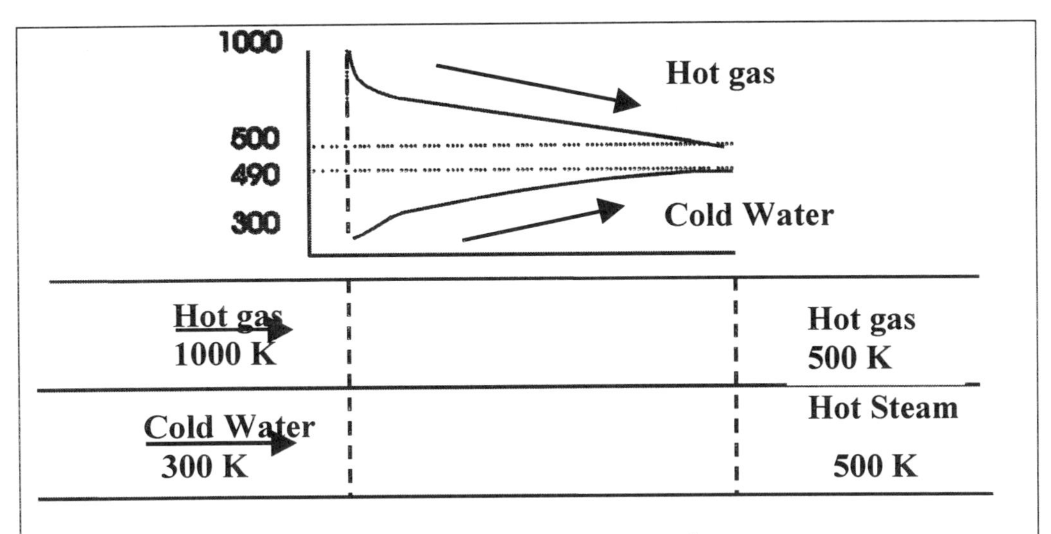

(a) Parallel flow heat exchanger

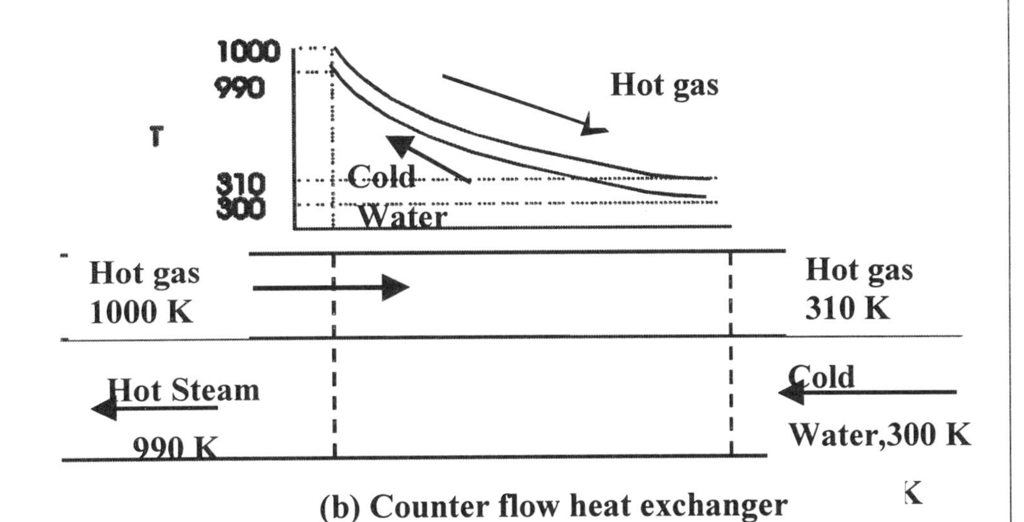

(b) Counter flow heat exchanger

Figure 9. Schematic illustration of: (a) parallel flow heat exchanger; and (b) counterflow heat exchanger.

$$0 = + \dot{m}_a (h_{a,i} - h_{a,e}) + \dot{m}_w (h_f - h_g), \text{ or}$$

$$\dot{m}_a c_p (T_2 - T_1) = \dot{m}_w h_{fg},$$

where the subscripts a and w, respectively, refer to the air and water. Therefore,

$$\dot{m}_w = 10 \times 1 \times 200 \div 2257 = 0.886 \text{ kg s}^{-1}.$$

The optimum work

$$\dot{W}_{cv,opt} = (\dot{m}_a \psi_{a,i} + \dot{m}_w \psi_{w,i}) - (\dot{m}_a \psi_{a,e} + \dot{m}_w \psi_{w,e}), \text{ i.e.,} \quad (A)$$

$$\psi_{a,i} = h_{a,i} - T_0 s_i = 1 \times 673 - 298 \times (1 \times \ln(673/298)) = 430.2 \text{ kJ kg}^{-1},$$

$$\psi_{a,e} = 473 - 298 \times 1 \times \ln(473/298) = 335.3 \text{ kJ kg}^{-1},$$

$$\psi_{w,i} = 419 - 298 \times 1.31 = 28.6 \text{ kJ kg}^{-1}, \text{ and}$$

$$\psi_{w,e} = 2676.1 - 298 \times 7.35 = 485.8 \text{ kJ kg}^{-1}.$$

Therefore,

$$\dot{W}_{cv,opt} = 10\times430.2 + 0.89\times28.6 - (10\times335.3 + 0.89\times485.8) = 544 \text{ kW, and}$$

$$I = \dot{W}_{cv,opt} - \dot{W} = 544 - 0 = 544 \text{ kW}.$$

Remarks

Hot combustion products enter the fire tubes of fire tube boilers at high temperatures and transfer heat to the water contained in the boiler drum. The water thereby evaporates, producing steam. This example reveals the degree of irreversibility in such a system.

The irreversibility exists due to the temperature difference between the hot gases and the water. An alternative method to heat the water would be to extract work by running a Carnot engine that would operate between the variable–temperature hot gases and the uniform–temperature ambient. A portion of the Carnot work can be used to run a heat pump in order to transfer heat from the ambient to the water and generate steam. The remainder of the work would be the maximum possible work output from the system. However, such a work output is unavailable from conventional heat exchangers in which the entire work capability is essentially lost.

We now discuss this scenario quantitatively. Assume that the air temperature changes from $T_{a,i}$ to $T_{a,e}$ as it transfers heat to the Carnot engine. For an elemental amount of heat $\delta\dot{Q}$ extracted from the air, the Carnot work.

$$\delta\dot{W}_{CE} = \delta\dot{Q}\,(1 - T_0/T). \quad \text{(B)}$$

Since

$$\delta\dot{Q} = \dot{m}_a\, c_p\, dT, \quad \text{(C)}$$

$$\delta\dot{W}_{CE} = -\dot{m}_a\, c_p\, dT\,(1 - T_0/T).$$

Upon integration,

$$\dot{W}_{CE} = \dot{m}_a c_p\,((T_e - T_i) - T_0 \ln(T_e/T_i)). \quad \text{(D)}$$

$$= 10 \times 1 \times (200 - 298 \times \ln(473/673)) = 949 \text{ kW}.$$

This is the Carnot work obtained from the transfer of heat from the air. Now, a portion of this work will be used to run a heat pump operating between constant temperatures T_0 and T_w (100°C) in order to supply heat to the water. The heat pump COP is given by the expression

$$COP = \dot{Q}_H/\dot{W}_{in,heat\ pump} = T_w/(T_w - T_0) = 4.97.$$

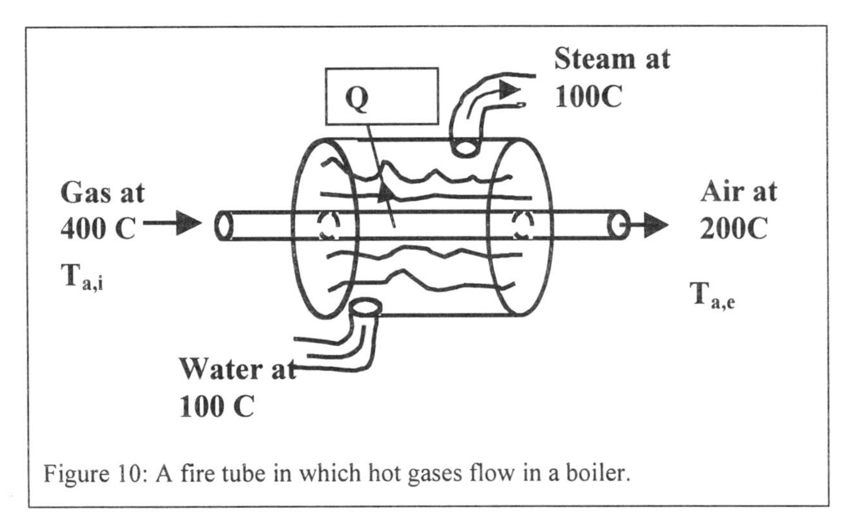

Figure 10: A fire tube in which hot gases flow in a boiler.

Since the heat transfer $\dot{Q}_H = 2257 \times 0.89 = 2009$ kW, The work input $\dot{W}_{in,heat\ pump} = 2009 \div 4.97 = 404$ kW. Therefore, the net work that is obtained

$$\dot{W}_{cv,opt} = 949 - 404 = 545 \text{ kW}.$$

This is identical to the answer obtained for the irreversibility flux using the availability analysis. Due to the high cost of fabricating such a system, conventional heat exchangers are instead routinely used.

We now examine the feasibility of installing a Carnot engine between the hot gases and the water that exists at 100°C so that heat could be directly pumped into water. You will find that it is impossible to achieve the same end states as in the heat exchanger while keeping $\sigma_{cv} = 0$ without any interaction with the environment.

5. Gibbs Function

Assume that a system is maintained at the ambient temperature T_0 (a suitable example is a plant leaf that is an open system in which water enters through the leaf stem and evaporates through the leaf surface). In this case, the absolute stream availability can be expressed as

$$\psi = h - T s = h - T_0 s = g, \tag{60}$$

where g denotes the Gibbs function or Gibbs free energy. It is also referred to as the chemical potential of a single component and is commonly used during discussions of chemical reactions (e.g., as in Chapter 11). The product (Ts) in Eq. (60) is the unavailable portion of the energy. Therefore, the Gibbs function of a fixed mass is a measure of its potential to perform optimum work in a steady flow reactor. We recall from Chapter 3 that a system attains a stable state when its Gibbs function reaches a minimum value at given T and P. This tendency to reach a stable state is responsible for the occurrence of chemical reactions during non-equilibrium processes (Chapter 12).

6. Closed System (Non–Flow Systems)

In this section we will further illustrate the use of the availability balance equation Eq. (47), particularly the boundary volume changes resulting in deformation work (Figure 11).

a. *Multiple Reservoirs*

For closed systems, $\dot{m}_i = \dot{m}_e = 0$, and the work $\dot{W}_{cv} = \dot{W}_{shaft} + \dot{W}_u + P_0 dV_{cyl}/dt$. For a closed system containing multiple thermal energy reservoirs, the balance equation assumes the form

$$d(E_{cv}-T_0S_{cv})/dt = \Sigma_{j=1}^{N}\dot{Q}_{R,j}\left(1-T_0/T_{R,j}\right)-\dot{W}_{cv}-I. \quad (61)$$

If this relation is applied to an automobile piston–cylinder assembly with negligible shaft work (δW_{shaft}=0) and with the inlet and exhaust valves closed, the useful optimum work delivered to the wheels over a period of time dt is $\delta W_{u.}$ The work

$$\delta W_u = \left(-dE + T_0 dS\right) - P_0 dV + \Sigma\,\delta Q_R\left(1-T_0/T_{R,j}\right)w - \delta I\text{, i.e.,} \quad (62)$$

$$W_u = \left(-\Delta E + T_0\Delta S\right) - P_0\Delta V + \Sigma\,Q_R\left(1-T_0/T_{R,j}\right) - I \quad (63)$$

where $\Delta E=E_2-E_1$, $\Delta S=S_2-S_1$, $\Delta V= V_2-V_1$. Dividing the above relation by the mass m,

$$w_u = \left(-\Delta e + T_0\Delta s\right) - P_0\Delta v + \Sigma\,q_R\left(1-T_0/T_{R,j}\right) - i. \quad (64)$$

b. *Interaction with the Ambient Only*

With values for q_R = 0, i = 0, and e = u, Eq. (64) simplifies as

$$w_{u,opt} = \phi_1 - \phi. \quad (27)$$

When $\phi_2 = \phi_0$,

$$w_{u,opt,0} = \phi_1' = \phi_1 - \phi_0.$$

The term ϕ' is called closed system exergy or closed system relative availability. Consider the cooling of coffee in a room, which is a spontaneous process (i.e., those that occur without outside intervention). The availability is completely destroyed during such a process that brings the system and its ambient to a dead state. Thus, w_u = 0 and $i = w_{u,opt,0} = \phi_1 - \phi_0$.

c. *Mixtures*

If a mixture is involved, Eq. (63) is generalized as,

$$\begin{aligned} W_u &= \left(-\Sigma(N_{k,2}\hat{e}_{k,2} - N_{k,1}\hat{e}_{k,1}) + T_0(\Sigma N_{k,2}\hat{s}_{k,2} - N_{k,1}\hat{s}_{k,1})\right) \\ &\quad - P_0\Sigma(N_{k,2}\hat{v}_{k,2} - N_{k,1}\hat{v}_{k,1}) + \Sigma\,Q_R\left(1-T_0/T_{R,j}\right) - I \end{aligned}, \quad (65)$$

where, typically, $\hat{e} \approx \bar{e} \approx \bar{u}$, and, $\hat{s}_k = \bar{s}_k^0 - \bar{R}\ln p_k/P_{ref}$ for a mixture of ideal gases and P_{ref} = 1 bar

---

f. *Example 6*

This example illustrates the interaction of a closed system with its ambient. A closed tank contains 100 kg of hot liquid water at a temperature T_1 = 600 K. A heat engine transfers heat from the water to its environment that exists at a uniform temperature T_0 = 300 K. Consequently, the water temperature changes from T_1 to T_0 over a finite time period. What is the maximum possible (optimum) work output from the engine? The specific heat of the water c = 4.184 kJ kg^{-1} K^{-1}.

Solution

Consider the combined closed system to consist of both the hot water and the heat engine. Since there are no thermal energy reservoirs within the system and, for optimum work, I = 0,

$$d(E_{cv}-T_0S_{cv})/dt = \dot{W}_{cv,opt}\text{, or} \quad (A)$$

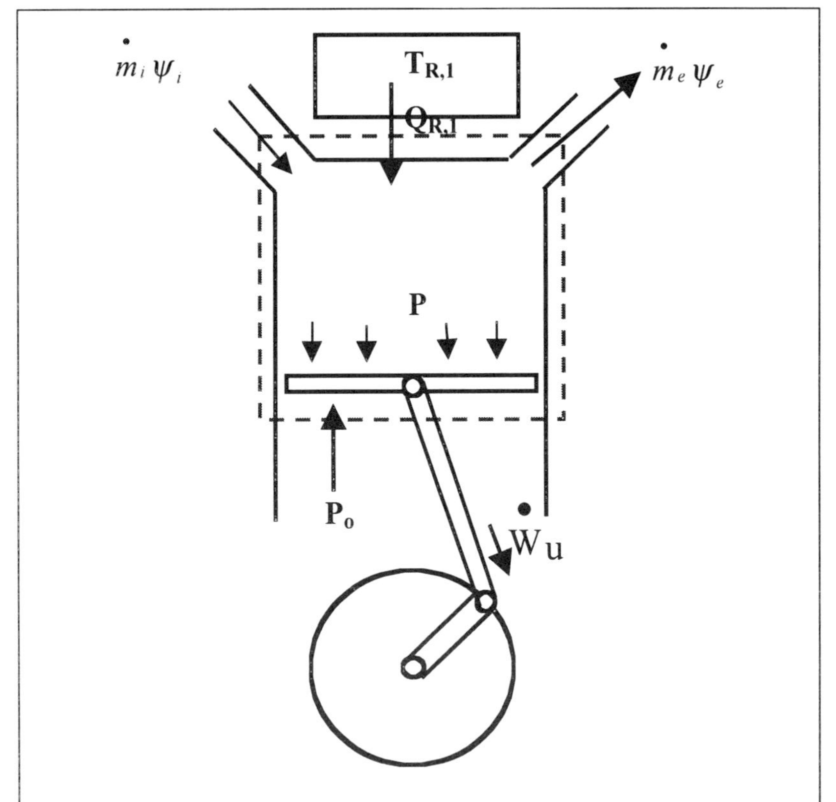

Figure 11. Application of the availability balance for a piston-cylinder assembly.

$W_{cv,opt} = (E_{cv} - T_0\, S_{cv})_1 - (E_{cv} - T_0\, S_{cv})_2$, where (B)

$(E_{cv})_1 = U_1 = m\, c\, T_1$, $(E_{cv})_2 = U_2 = m\, c\, T_2$, and $(S_{cv})_1 - (S_{cv})_2 = m\, c\, \ln(T_1/T_0)$. (C)

Substituting Eq. (C) into Eq. (B), we obtain

$W_{cv,opt} = m\, c\, (T_1 - T_0) - T_0\, m\, c\, \ln(T_1/T_0) =$

$100 \times 4.184 \times (600 - 300 - 300 \times \ln(600/300)) = 38520$ kJ.

Remarks

If only the heat engine is considered to be part of the system, it interacts with both the hot water and the ambient. In this case the hot water is a variable–temperature thermal energy reservoir. Since the heat engine and, therefore, the system, is a cyclical device, there is no energy accumulation within it. Therefore, for an infinitesimal time period

$$\delta Q_{R,w}\, (1 - T_0/T_{R,w}) = \delta W_{cv,opt}, \quad \text{(D)}$$

where the hot water temperature $T_{R,w}$ decreases as it loses heat. Applying the First and Second laws to the variable–temperature thermal energy reservoir, $\delta Q_{R,w} = -dU_{R,w} = -m_w\, c_w\, dT_{R,w}$ and $\delta Q_{R,w}/T_{R,w} = dS_{R,w}$. Using these relations in the context of Eq. (D) we obtain the same answer as before.

7. Helmholtz Function

In case a closed rigid system exists at its ambient temperature T_0, its absolute availability can be expressed as

$$\phi = u - T\, s = u - T_0 = a, \quad (66)$$

where a denotes the Helmholtz function or the Helmholtz free energy. The Helmholtz function is another measure of the potential to perform work using a closed system. Consider, for instance, an automobile battery in which chemical reactions occur at room temperature and produce electrical work, and the chemical composition of the battery changes with time. The optimum work for such a situation is given by the expression

$$W_{cv,opt} = (\Sigma N_k\ \hat{\phi}_k)_{initial} - (\Sigma N_k\ \hat{\phi}_k)_{final}. \quad (67)$$

Since we have assumed that $T = T_0$,

$$W_{cv,opt} = (\Sigma N_k\ \hat{a}_k)_{initial} - (\Sigma N_k\ \hat{a}_k)_{final}.$$

The Helmholtz function of a closed system represents its potential to perform work.

---

g. Example 7

This example considers an air–conditioning cycle. In some areas the cost of electricity is higher during the day than at night, making it expensive to use air conditioning. The following scheme is proposed to alleviate the cost. The air conditioner is to be operated during the night in order to cool water in a storage tank to its freezing temperature. During the day a fan is to be used to blow ambient air over the water tank, thereby cooling the air and circulating it appropriately. Use a generalized availability analysis and derive an expression for the minimum work that is required in order to produce ice from the water if its initial temperature is 300 K.

Solution

Consider a closed control volume that encloses the tank and the air conditioner, but excludes the ambient. The minimum work

$$\dot{W}_{cv,opt} = d(E_{cv} - T_0 S_{cv})/dt.$$

Integrating over the time period required to convert the water into ice,

$$W_{cv,min} = (E_{tank,1} - E_{tank,2}) - T_0\ (S_{tank,1} - S_{tank,2}).$$

Assuming the energy for the process E = U, since the mass contained in the tank is unchanged,

$$w_{cv,min} = (u_{tank,1} - u_{tank,2}) - T_0\ (s_{tank,1} - s_{tank,2}).$$

Note that state 1 is liquid while state 2 is ice; thus, sensible energy must be removed to reduce T_1 to T_{freeze} and then latent energy to form ice. Assuming constant properties for water,

$$u_{tank,2} = u_{tank,1} - c_w\ (T_1 - T_{freeze}) - u_{fs}, \text{ and } s_{tank,2} = s_{tank,1} - c_w \ln (T_{freeze}/T_1) - s_{fs}.$$

Therefore,

$$w_{cv,min} = c_w\ (T_1 - T_{freeze}) + u_{fs} - T_0\ (c_w \ln (T_{freeze}/T_1) + s_{fs}) \quad (A)$$

Since $s_{fs} = h_{fs}/T_{freeze} = (u_{fs} + P\ v_{fs})/T_{freeze} \approx u_{fs}/T_{freeze}$, we have

$$w_{cv,min} = c_w\ (T_1 - T_{freeze}) + u_{fs}\ (1 - T_0/T_{freeze}) - T_0\ c_w \ln (T_1/T_{freeze}). \quad (B)$$

Using the values $c_w = 4.184$ kJ kg^{-1} K^{-1}, $u_{fs} = 334.7$ kJ kg^{-1}, and $T_0 = 300$ K,

$w_{cv,min} = 4.184\ (300 - 273) + 335\ (1- 300/273) - 300 \times 4.184 \times \ln (300/273)$

$= -38.54$ kJ kg^{-1} of ice made.

In case $T_1 = T_2 = T_{freeze} = 273$K, using Eq. (B)

$w_{cv,min} = u_{fs}\ (1 - T_0/T_{freeze}) = |\text{Heat removed} \div COP_{Carnot}|$, where

$COP_{Carnot} = T_{freeze}/(1 - T_0/T_{freeze}) = 10.11$, so that

$|\ w_{cv,min}| = 335/10.11 = 33.1$ kJ kg^{-1} of ice.

Remarks

Practical air conditioning systems involve a throttling process which is irreversible and, therefore, $\sigma > 0$ during air–conditioning cycles. Although the actual work will be greater than 38.54 kJ/kg of ice that is made, the design goal should be to approach this value. In an ideal air–conditioning cycle, isentropic expansion in a turbine may be used rather than using a throttling device in order to eliminate entropy generation.

E. AVAILABILITY EFFICIENCY

Availability analyses help to determine the work potential of energy. As the energy of systems is altered due to heat and work interactions, their work potential or availability changes. The analyses lead to the maximization of work output for work–producing systems (heat engines, turbines, etc.) and to the minimization of work input for work–absorbing systems (heat pumps, compressors, etc.) so as to achieve the same initial and end states. Under realistic conditions systems may produce a lower work output or require more work input as compared to the results of availability analyses. In that case it is pertinent to evaluate how close the actual results are compared to their optimum values. The analyses also allow us to evaluate irreversibilities of heat exchangers that are neither work–producing nor work–absorbing devices. This section presents a method of evaluating the performance of heat engines, heat pumps, turbines, compressors, and heat exchangers using availability concepts.

1. Heat Engines

a. *Efficiency*

Different heat engines employ various cyclical processes (e.g., the Rankine, Brayton, and Otto cycles) that first absorb heat and then reject it to the environment in order to produce work. The efficiency η = Sought/Bought = work output ÷ heat input = $W/Q_{in} = (Q_{in}-Q_{out})/Q_{in}$ (Figure 3a) presents an energy band diagram for a heat engine operating between two fixed–condition thermal energy reservoirs. The Carnot efficiency of an ideal heat engine $\eta_{CE} = 1 - T_L/T_H$, where T_L and T_H, respectively, are the low and high temperatures associated with the two reservoirs. We note that even for idealized cycles involving isothermal energy reservoirs and internally reversible processes, $\eta_{CE} < 1$ due to Second law implications, and availability analysis tells us that work potential of heat is equal to $Q_{in}\ (1- T_0/T_H)$. A part of Q_{in} is converted to $W_{opt,cyc}$ and $Q_{0,opt,cyc}$ is rejected to the ambient under ideal conditions. y.

b. *Availability or Exergetic (Work Potential) Efficiency*

In power plants based on the Rankine cycle, heat is transferred from hot boiler gases to cooler water in order to form steam. A temperature difference exists between the gases and the water, thereby creating an external irreversibility even though the plant may be internally reversible. Therefore its work output W_{cyc} is lower than the maximum possible work output $W_{opt,cyc}$ for the same heat input, and hot gas, ambient, and cold water temperatures. The *Availability Efficiency* is defined as

$$\eta_{Avail} = W/W_{opt,cyc} = W/W_{max,cyc}, \text{ where} \quad (68)$$

$$W_{opt,cyc} = W_{max,cyc} = (W_{cyc} + I_{cyc}), \text{ and} \quad (69)$$

$I_{cyc} = T_0\,\sigma_{cyc}$. Note that σ_{cyc} refers to entropy generation in isolated system during a cyclic process. Exergetic efficiency for a cycle is a measure of deviation of an actual cycle from an ideal reversible cycle. Equation (68) can be used to compare different cycles that operate between similar thermal energy reservoirs. For a cycle operating between fixed–temperature thermal energy reservoirs

$$W_{opt,cyc} = W_{max,cyc} = Q_{R,1}\,(1 - T_0/T_{R,1}), \text{ and} \quad (70)$$

the optimum cyclic process rejects a smaller amount of heat $Q_{0,opt,cy}$ as compared to a realistic process. The difference $Q_{0,opt,cyc} - Q_{0,cyc} = I = T_0\,\sigma_{cyc}$ is the irreversibility. Figure 12a and b illustrate the energy and availability band diagrams for a heat engine. The term $Q_R\,(1\text{-}T_0/T_R)$ is the availability associated with heat Q_R, W_{cyc} is the availability transfer through work, and I_{cyc} is the availability loss in the cycle.

---

h. Example 8

A nuclear reactor transfers heat to water in a boiler that is at a 900 K temperature, thereby producing steam at 60 bar, and 500ºC. The steam exits an adiabatic turbine in the form of saturated vapor at 0.1 bar. The vapor enters a condenser where it is condensed into saturated liquid at 0.1 bar and then pumped to the boiler using an isentropic pump.

Determine:

The optimum work.

The availability efficiency.

The overall irreversibility of the cycle.

The irreversibility in the boiler, turbine, and condenser.

Solution

Analyzing the Rankine cycle:

The ***turbine*** work is

$$q_{12} - w_{12} = h_2 - h_1, \text{ i.e.,} \quad (A)$$

$$w_{12} = 2585 - 3422 = 837 \text{ kJ kg}^{-1}.$$

The heat rejected in the ***condenser***

$$q_{23} - w_{23} = h_3 - h_2, \text{ i.e.,}$$

$$q_{23} = q_{out} = 192 - 2585 = -2393 \text{ kJ kg}^{-1}.$$

Likewise, in the ***pump***

$$q_{34} - w_{34} = h_4 - h_3 \approx v_3\,(P_4 - P_3), \text{ or} \quad (B)$$

Since properties for the liquid state at 4 may be unavailable, they can be otherwise determined. The work

$$w_{34} = -0.001 \times (60 - 0.1) \times 100 = -6 \text{ kJ kg}^{-1}.$$

From Eq. (B) $h_3 = 192$ kJ kg^{-1} (sat liquid at 0.1), and

$$h_4 = 192 + 6 = 198 \text{ kJ kg}^{-1}.$$

In the ***boiler***

$$q_{in} = q_{41} - w_{41} = h_1 - h_4 = 3422 - 198 = 3224 \text{ kJ kg}^{-1}.$$

Therefore, the cyclical work

$$w_{cyc} = w_t - w_p = 837 - 6 = q_{in} - q_{out} = 3224 - 2393 = 831 \text{ kJ kg}^{-1}.$$

The efficiency

$$\eta = w_{cyc}/q_{in} = 831/3224 = 0.26.$$

Integrating the general availability balance equation over the cycle

$$w_{cyc,opt} = q_{in}\,(1 - T_0/T_b) = 3224\,(1 - 298/900) = 2156 \text{ kJ kg}^{-1}.$$

The actual Rankine cycle work = 831 kJ kg^{-1} and the actual cycle efficiency η = 0.26. The Carnot work is 2156 kJ kg^{-1} and the Carnot efficiency η_{Carnot} = 0.67. The relative efficiency is

$$\eta_{Avail} = w_{cyc}/w_{cyc,opt} = \eta/\eta_{Carnot} = 831/2156 = 0.39.$$

The overall irreversibility of the cyclical process is

$$I = w_{cyc,opt} - w = 2156 - 831 = 1325 \text{ kJ kg}^{-1}.$$

An availability analysis can be performed on the various system components as follows:
For the turbine,
$s_1 = 6.88$ kJ kg^{-1} K^{-1}, and $s_2 = 8.15$ kJ kg^{-1} K^{-1} so that $s_2 > s_1$. Furthermore,
$\psi_1 = h_1 - T_0\, s_1 = 3422 - 298 \times 6.88 = 1372$ kJ kg^{-1}, and, likewise,
$\psi_2 = 2585 - 298 \times 8.15 = 156.3$ kJ kg^{-1}. Therefore,
$w_{t,opt} = \psi_1 - \psi_2 = 1372 - 156.3 = 1216$ kJ kg^{-1}, and
$I_t = 1216 - 837 = 379$ kJ kg^{-1}.
For the condenser,
$\psi_3 = 191.8 - 298 \times 0.649 = -1.6$ kJ kg^{-1}, and
$w_{cond,opt} = \psi_2 - \psi_3 = 156.3 - (-1.6) = 157.9$ kJ kg^{-1}. Therefore,
$I_{cond} = 157.9$ kJ kg^{-1}.
For the pump,
$s_4 = s_3 = 0.649$ kJ kg^{-1} K^{-1}, and
$\psi_4 = h_4 - T_0\, s_4 = 198 - 298 \times 0.649 = 4.6$ kJ kg^{-1}. Consequently,
$w_{p,opt} = \psi_3 - \psi_4 = -1.6 - 4.6 = -6.2$ kJ kg^{-1}.
Since $w_p = -6.2$ kJ kg^{-1}, $I_p = 0$ kJ kg^{-1}.
For the boiler,
$W_{b,opt} = = q_b\,(1 - T_0/T_b) + \psi_4 - \psi_1 = 2156 + 4.6 - 1372 = 789$ kJ kg^{-1}, and
$I_b = = W_{b,opt} - W_b = 789 - 0 = 789$ kJ kg^{-1}.
The total irreversibility = 379+158+0+789 = 1326 kJ kg^{-1} is the same as that calculated above.
The availability input at the boiler inlet $\psi_4 = 4.6$ kJ kg^{-1}, and

$$\psi_0 = h_0 - T_0\, s_0 \approx h_f^{sat}\,(25\text{ C}) - 298\times s_f^{sat}\,(25\text{ C}) = 104.89 - 298\times 0.3674 = -4.6 \text{ kJ kg}^{-1}.$$

The availability input through heat transfer in the boiler $\psi_{41} = q_b\,(1 - T_0/T_b) = 3224 \times (1 - 298/2000) = 2156$ kJ kg^{-1}. Figure 13 illustrates the exergy band diagram for the cyclic process.

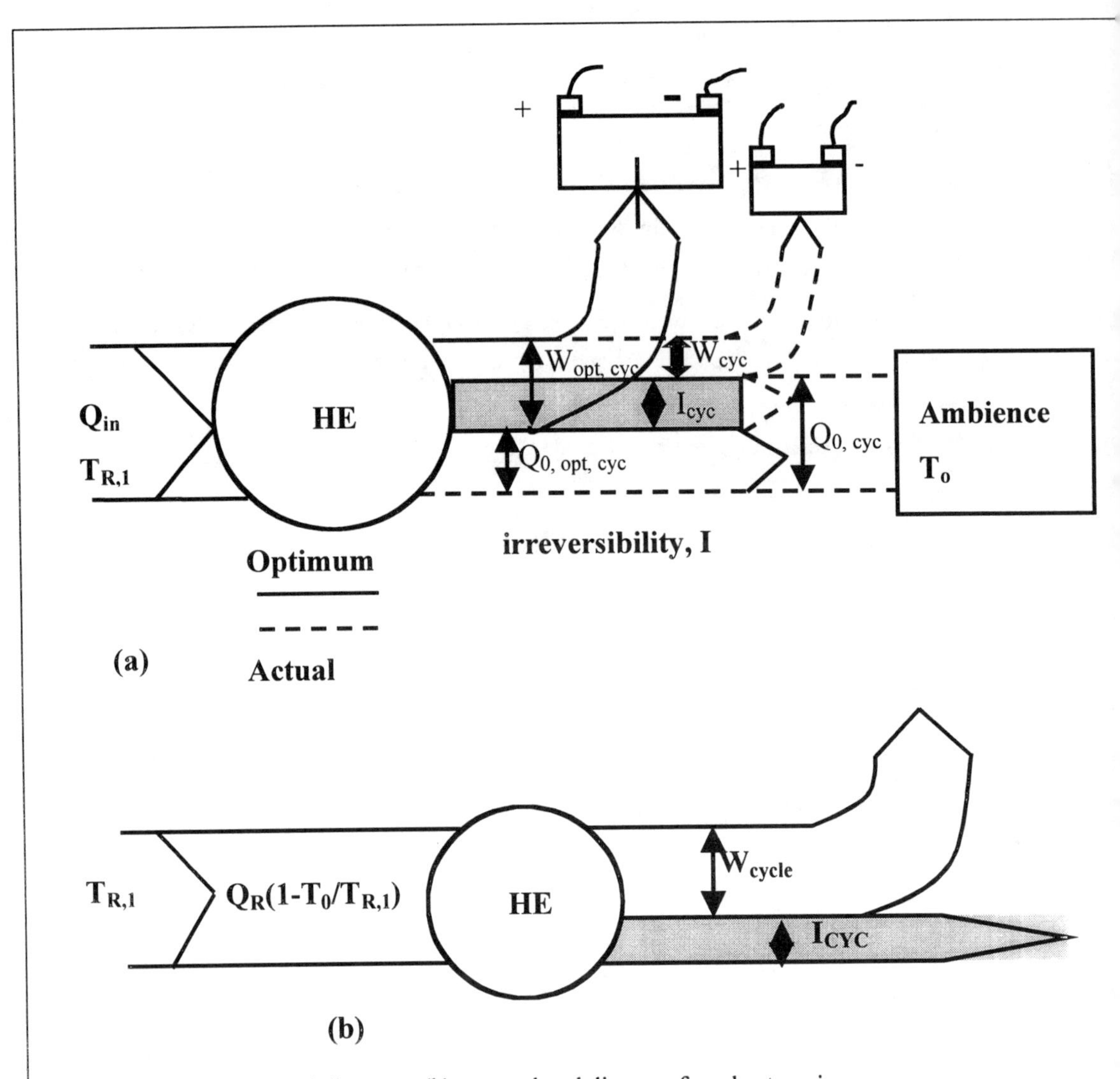

Figure 12(a): Energy band diagram , (b) exergy band diagram for a heat engine.

These results are summarized in tabular form below.

State	T,C	P, bar	x	H, kJ kg^{-1}	s, kJ kg^{-1} K^{-1}	q, kJ kg^{-1}	w, kJ kg^{-1}	ψ, kJ kg^{-1}	i, kJ kg^{-1}	$\psi' = \psi - \psi_0$
1	500	60	-	3422	6.8	0		1372		1376.6
2		0.1	1.0	2585	8.15	0	837	156.3	379	160.9
3		0.1	0.0	191.8	0.65	-2393	0	-1.6	158	3.0
4		60	-	198	0.65	-	-6	4.6	0	9.2
1		60	-	3422	6.8	3224	-	1372	789	1376.6

Remarks

In this example, the processes comprising the Rankine cycle are all reversible. The irreversibility arises due to the temperature difference between the thermal energy reservoir and the boiler. In this case the maximum work output $w_{cyc,opt}$ can be obtained by placing two Carnot heat engines, one between the reactor and the boiler (to supply heat to the boiler), and the second between the condenser and its ambient (to reject heat to the ambient).

The boiler accounts for 24.5% of the total irreversibility.

Many practical systems do not interact with a fixed–temperature reservoir, e.g., in a coal– or oil–fired power plant. In that case

$$\eta_{Avail} = w_{cyc}/(\text{inlet stream exergy into a system } \psi'),$$

where $\psi' = (h - T_0\, s) - (h^0 - T_0\, s_0)$. The definition of h^0 (involving the chemical energy of a species) will be discussed in Chapter 11.

2. Heat Pumps and Refrigerators

a. Coefficient of Performance

Heat pumps and refrigerators are used to transfer heat through work input and are characterized by a coefficient of performance COP (= heat transfer ÷ work input) instead of an efficiency. For a heat pump

$$COP_H = Q_H/\text{Work input} = Q_{out}/(Q_{in} - Q_{out}), \qquad (71)$$

and for a refrigerator

$$COP_R = Q_L/\text{Work input} = Q_{in}/(Q_{in} - Q_{out}). \qquad (72)$$

For a Carnot heat pump $Q_{out}/Q_{in} = Q_H/Q_L = T_H/T_L$. Therefore,

$$COP_H = T_H/(T_H - T_L). \qquad (73)$$

Likewise, for Carnot refrigerators

$$COP_R = T_L/(T_H - T_L). \qquad (74)$$

Figure 14a contains an energy band diagram for a heat pump and refrigerator that interacts with fixed–temperature thermal energy reservoirs. Figure 14b illustrates the corresponding availability. Both the Carnot COPs $\to \infty$ as $T_H \to T_L$, and approach either zero (in case of COP_R) or unity (in case of COP_H) as the difference $(T_H - T_L)$ becomes very large. The availability COP

$$COP_{avail} = |W_{cyc,opt}|/|W_{cyc}|, \text{ where} \qquad (75)$$

$W_{cyc,opt} = W_{cyc,min} = W_{cyc\_} - T_0\, \sigma_{cyc}$.

---

i. Example 9

A Carnot heat pump delivers heat to a house maintained at a 25°C temperature in a 0°C ambient. The temperature of the evaporator in the heat pump is –10°C, while the condenser temperature is 35°C so that external irreversibilities exist. Determine:

The COP based on the evaporator and the condenser temperatures.

The COP based on the house and the ambient temperatures.

The minimum work input that is required.

The availability efficiency.

The irreversibility.

Solution

Using the relation

$$COP_{Carnot} = |Q_{condenser}|/|\text{Work Input}| = |Q_{condenser}|/(|Q_{condenser}| - |Q_{evaporator}|), \qquad (A)$$

$$Q_{condenser}/Q_{evaporator} = T_{condenser}/T_{evaporator}. \qquad (B)$$

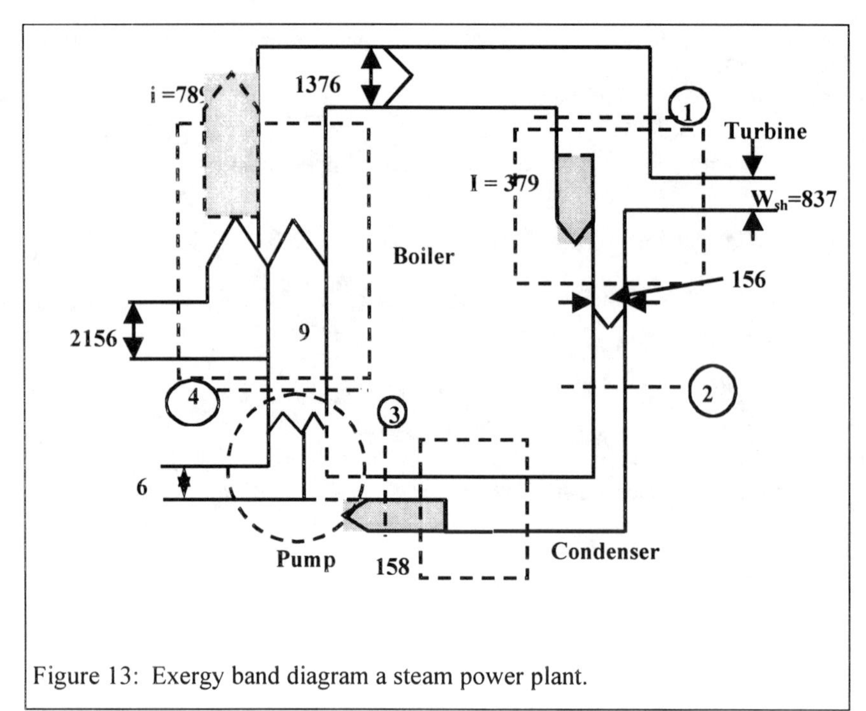

Figure 13: Exergy band diagram a steam power plant.

Therefore,

$$COP_{Carnot} = T_{condenser}/(T_{condenser} - T_{evaporator}) = (308)/(308 - 263) = 6.844.$$

In the absence of external irreversibilities, the evaporator and ambient temperatures should be identical, as should the condenser and the house temperatures. Therefore $COP_{Carnot,ideal} = |Q_{house}|/|\text{Work Input}| = T_{house}/(T_{house} - T_{ambient}) = 11.92$, and $W_{Carnot} = |Q_{house}|/COP = 1/11.92 = 0.0839$ kJ per kJ of heat pumped into the house. Consider the generalized availability equation

$$\dot{W}_{cv} = \sum_{inlets} \dot{m}_i \psi_i + \Sigma_{j=1}^{N} \dot{Q}_{R,j}(1 - T_0/T_{R,j}) - \sum_{exits} \dot{m}_e \psi_e - d(E_{cv} - T_0 S_{cv})/dt.$$

For a steady state cyclical process with one inlet and exit, $d/dt = 0$, $\dot{m}_i = \dot{m}_e$(steady), and $\psi_i = \psi_e$ (cyclical). Therefore,

$$W_{cyc,min} = |Q_{house}|(1 - T_0/T_{house})/T_{house} = |Q_{house}|/COP_{Carnot}, \text{ and} \qquad (C)$$

$W_{cyc,min} = 1/11.92 = 0.0839$ kJ per kJ of heat pumped in.
The availability COP
$COP_{avail} = 0.08389/0.146 = 0.57$.
We considered an internally reversible process with external irreversibilities existing at the evaporator and condenser due to the temperature differences between these reservoirs and the ambient and the house, respectively. In that case
$W_{cyc} = |Q_{house}|/COP = 1/6.844 = 0.146$ kJ per kJ of heat pumped into the house.
The overall irreversibility associated with every kJ of heat that is pumped into the environment is
$I = W_{cyc,min} - W_{cyc} = T_0\, \sigma_{cyc} = -0.08389 - (-0.146) =$
 0.06211 kJ for every kJ of heat pumped into the house.
Since the ambient temperature is 0°C,

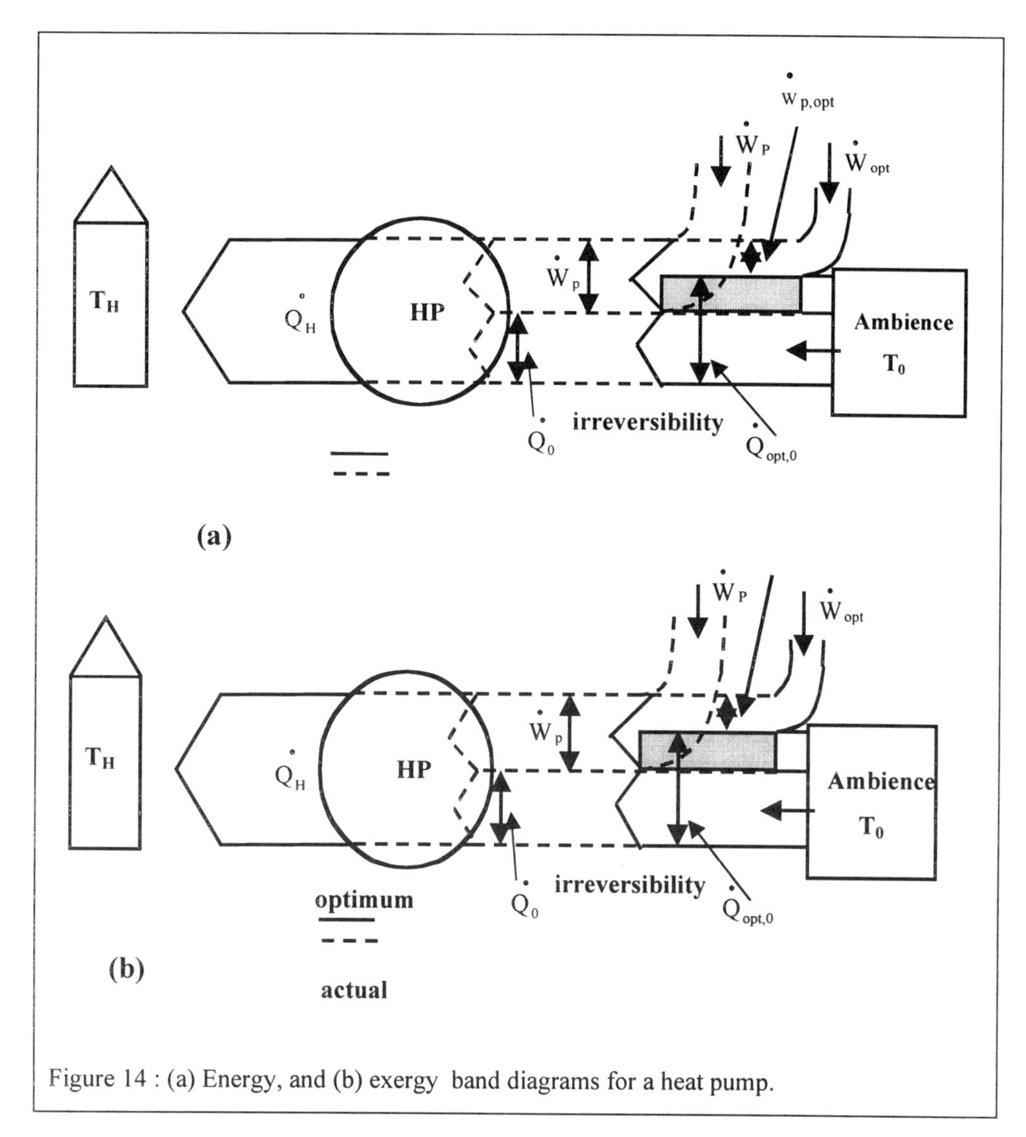

Figure 14 : (a) Energy, and (b) exergy band diagrams for a heat pump.

σ_{cyc} = 0.0621/273 = 0.00023 kJ K^{-1} for every kJ of heat pumped into the house.

Remarks

The overall irreversibility can also be obtained by considering entropy balance equation for the system by including the thermal reservoirs at 25°C and 0°C, i.e.,
Entropy change in the isolated system = Entropy change in the house (at temperature T_H) + Entropy change in the ambient (at temperature T_L) + Entropy change in the control volume of interest during the cyclical process due to internal irreversibilities. Therefore,

$\sigma = \Delta S_H + \Delta S_L + 0 = \Delta S_H + \Delta S_L$.

Based on each unit heat transferred to the house, the two entropy changes are

$\Delta S_H = Q_H/T_H = 1/298 = 0.00336$ kJ K^{-1}, and

$\Delta S_L = - Q_L/T_L$.

Since $W_{cyc} = 0.146$ kJ per kJ of heat pumped into the house,
$Q_L = 1 - 0.146 = 0.854$ kJ per kJ of heat pumped into the house,
$\Delta S_L = - 0.854/273 = -0.00313$ kJ K^{-1}, and

$\sigma = 0.00336 - 0.00313 = 0.00023$ kJ per kJ of heat pumped into the house.

This is the same answer as that obtained in the solution.

3. Work Producing and Consumption Devices

In order to change a system to a desired end state from a specified initial state, energy must be transferred across its boundaries. In work producing or absorbing devices, this energy transfer is in the form of work. Since W (for zero irreversibility) differs from W (for realistic processes), the value of the availability efficiency η_{avail} is instructive in assessing the overall system design.

a. Open Systems

For a work producing device $W_{opt} = W_{max}$ and

$$\eta_{avail} = W/W_{max}. \tag{76}$$

Availability or exegetic efficiency is a measure of deviation of an actual process from an ideal reversible process for the prescribed initial and final states.The maximum value of the availability efficiency is unity and the presence of irreversibilities reduces that value. The overall irreversibility

$$I = W_{max} - W = W_{max}(1 - W/W_{max}) = W_{max}(1 - \eta_{avail}). \tag{77}$$

If the end state of a working fluid emanating from a work–producing device, e.g., a gas turbine, is at a higher temperature or pressure than that of its ambient, the fluid still contains the potential to perform work. Therefore, it is useful to define the availability efficiency considering the optimum work (which is based on the assumption that the optimal end state is a dead state). In that case $W_{max,0} = W_{opt,0}$, and

$$\eta_{avail,0} = = W/W_{max,0} = (W/W_{max})(W_{max}/W_{max,0}) = \eta_{avail}\,(W_{max}/W_{max,0}). \tag{78}$$

Note that $W_{max} \le W_{max,0}$ and hence $\eta_{avail1,0} \le \eta_{avail.}$ For a work–consuming device such as a compressor,

$$\eta_{avail} = W_{min}/W. \tag{79}$$

If the exit state from a work-producing device is the dead state, then the availability efficiency is. This ratio informs us of the extent of

$$\eta_{avail,0} = (\text{work output}) \div (\text{input exergy}), \tag{80}$$

and the conversion of the input exergy into work, but gives no indication as to whether the exergy is lost as a result of irreversibility or with the availability leaving along with the exit flow.

For a work–consuming device such as a compressor

$$\eta_{avail} = |W_{min}|/|W|,\ \eta_{re\ 1,0} = |W_{min,0}|/|W|$$

b. Closed Systems

For processes involving work output from a closed system (which is usually expansion work such as that obtained during the gas expansion in an automobile engine) $W_{u,opt} = W_{u,max}$, and

$$\eta_{avail} = |W_u|/|W_{u,max}|. \tag{81}$$

Likewise, for processes during which work is done on a closed system (which is usually compression work, e.g., air compression in a reciprocating pump) $W_{u,opt} = W_{u,min}$ so that

$$\eta_{avail} = |W_u|/|W_{u,min}|, \text{ and} \quad (82)$$

$$\eta_{avail,0} = \eta_{avail}|W_u|/|W_{u,min,0}|. \quad (83)$$

The isentropic efficiency is not same as the availability efficiency, since isentropic work can involve an end state that is different from a specified end state, while the determination of optimum work is based on the specified end state. These differences are illustrated in the example below.

---

j. *Example 10*

An adiabatic steady–state, steady–flow turbine expands steam from an initial state characterized by 60 bar and 500°C (State 1) to a final state at 10 kPa at which the quality x= 0.9 (state 2).
Is the process possible?
Determine the turbine work output.
What *would have been* the quality x_{2s} at the exit and isentropic work output for the same initial conditions for the same P_2 = 10 kPa?
Determine the work output if the final state is to be reached through a combination of a reversible adiabatic expansion process that starts at the initial state followed by reversible heat addition until the final state is reached.
Determine the maximum possible (optimum) work.
Calculate the availability efficiency based on the actual inlet and exit states and availability efficiency based on the optimum work.

Solution

Applying the generalized entropy balance equation

$$dS_{cv}/dt = \dot{m}_i s_i - \dot{m}_e s_e + \dot{Q}/T_b + \dot{\sigma}. \quad (A)$$

Under adiabatic steady state steady flow conditions, d/dt = 0, $\dot{m}_i = \dot{m}_e$(steady), and $\dot{Q}$=0. Therefore, Eq. (A) assumes the form

$$s_2 - s_1 = \sigma.$$

The specific entropies s_1(60 bar, 773 K) = 6.88 kJ kg^{-1} K^{-1}, and s_2(0.1 bar, x = 0.9) = 0.1 × 0.65 + 0.9 × 8.15 = 7.4 kJ kg^{-1} K^{-1} so that σ = 0.52 kJ kg^{-1} K^{-1}. The process is irreversible, since $\sigma > 0$.

Applying the energy conservation equation for an adiabatic (q = 0) steady–state, steady–flow process

$$-w = h_2 - h_1. \quad (C)$$

The specific enthalpies h_1(60 bar, 773 K) = 3422.2 kJ kg^{-1}, h_2(0.1 bar, x = 0.9) = 0.1 × 191.83 + 0.9 × 2584.7 = 2345.4 kJ kg^{-1} so that w = –(2345.4 – 3422.2) = 1076.8 kJ kg^{-1}.

For an isentropic process the end state $s_{2s} = s_1$ (= 6.88 kJ kg^{-1} K^{-1}) with the final pressure $P_{2s} = P_2$ (although the quality of the steam differs at these two states). Therefore,

$$s_{2s} = 6.88 = (1-x_{2s}) \times 0.65 + x_{2s} \times 8.15, \text{ i.e., } x_{2s} = 0.83.$$

Applying the First law to the process 1–2s,

$$-w_{12s} = h_{2s} - h_1, \text{ i.e.,} \quad (D)$$

$$h_{2s} = 0.17 \times 191.83 + 0.83 \times 2584.7 = 2177.9 \text{ kJ kg}^{-1}, \text{ and}$$

$w_{12s} = 3422.2 - 2177.9 = 1244.3$ kJ kg^{-1}.

We see that $x_2 > x_{2s}$, since the irreversible (frictional) process generates heat and, consequently, the steam leaves the turbine with a relatively higher enthalpy at the conclusion of process 1–2. Therefore, $w_{12} < w_{12s}$.
The adiabatic or isentropic efficiency is

$$\eta = w_{12}/w_{12s} = 0.865.$$

The infinitesimal enthalpy change dh = $\delta q - \delta w$. One could react state 2 by using an isentropic process first to P2= 10 kpa and x2s =0.83 and then adding heat at constant T_2 to state 2 to obtain the quality x_2= 0.9. Since the paths 1–2s and 2s–2 are reversible, δq = T ds. Hence,

$$T\,ds - w = dh. \tag{E}$$

Integrating the equation appropriately, we have

$$\int_{s_1}^{s_{2s}} T ds + \int_{s_{2_s}}^{s_2} T ds - w_{1-2_s-2} = h_2 - h_1.$$

The path 1–2_s involves no entropy change so that

$$T\,(s_2 - s_{2s}) - w_{12} = h_2 - h_1.$$

Hence, $- w_{1-2s-2} = 2345.4 - 3422.2 - 318.8 \times (7.4 - 6.88) - 1242.6$ kJ kg^{-1}. Since work is path–dependent and the paths 1–2 and 1–2s–2 are different, it is incorrect to write w_{1-2s-2} as w_{12}. The work w_{1-2s-2} is larger than the answer obtained in part 0 of the solution, since the process 1–2_s–2 is reversible. During the process 2_s–2 the reversible heat added $q_{2s-2,rev} = T\,(s_2 - s_{2s}) = 165.8$ kJ kg^{-1}. A portion of this heat is converted into additional work. We have not, however, given any information on what source is used to add the heat. The heat addition process involves an interaction with a source other than the ambient.
We will now use an availability analysis to determine the maximum work output that is possible in the absence of entropy generation while maintaining the same initial and final states. Simplifying the availability balance equation for this situation, the optimum work

$$w_{opt} = \psi_1 - \psi_2, \tag{F}$$

where $\psi_1 = h_1 - T_0\,s_1 = 3422.2 - 298 \times 6.88 = 1371.9$ kJ kg^{-1}. Likewise, $\psi_2 = 2345.4 - 298 \times 7.4 = 140.2$ kJ kg^{-1}, and $w_{opt} = 1231.7$ kJ kg^{-1}.
The availability efficiencies

$$\eta_{avail} = w_{12}/w_{opt} = 0.874, \text{ and}$$

$$\eta_{avail,0} = w_{12}/w_{opt,0}, \text{ where}$$

$w_{opt,0}= \psi_1 - \psi_0$ and $\psi_0 = h_0 - T_0\,s_0$. The dead state for the working fluid is that of liquid water at 298 K, 1 bar, so that h_0 = 104.89 kJ kg^{-1} and s_0 = 0.3674 kJ kg^{-1} K^{-1}, and $\psi_0 = 104.89 - 298 \times 0.3674 = -4.6$ kJ kg^{-1}.
Consequently, $w_{opt,0}$= 1376.5 kJ kg^{-1} and $\eta_{avail,0} = 0.78$.

Remarks

We see that $w_{12} < w_{opt} < w_{1-2s-2} < w_{opt,0}$. For the optimum work the only outside interaction that occurs is with the ambient that exists at the temperature T_0 while $W_{1\text{-}2s\text{-}2}$ is

achieved with an external heat input from an unknown source. However, heat for path 2s-2 can be pumped without using an external heat source; instead we use a heat pump. We can first employ the isentropic expansion process 1–2s to produce a work output of w_{1-2s} = 1244.3 kJ kg^{-1}. Then we can use a portion of this work to operate a Carnot heat pump that absorbs heat from the ambient (at T_0) and adds 165.8 kJ kg^{-1} of heat to the process 2s–2. The Carnot heat pump must operate between 319 K and 298 K. Therefore, the Carnot COP = 318.8 ÷ (318.8 – 298) = 15.3. Since 165.8 kJ kg^{-1} of heat is required, a work input of 165.8 ÷ 15.3 = 10.8 kJ kg^{-1} is necessary. This 10.8 kJ kg^{-1} of work is subtracted from w_{1-2s} and, consequently, w_{opt} = 1244.3 – 10.8 = 1233.5 kJ kg^{-1}, which is essentially the same answer as that based on the above availability analysis. For the optimum process 1-2s-2, entropy generation is zero.
The availability calculation does not explain why a final state x_2 = 0.9 at P_2 = 0.1 bar is reached instead of the state x_{2s} = 0.83 at P_2 = 1 bar. It only provides information as to what the optimum work could have been had the inlet and exit states been fixed. In a cyclic process, all the states are normally fixed. In a power plant all the states are normally fixed to maintain a steady state. A power plant operator must monitor the exit conditions, optimum work, and the entropy generation as the plant equipment degrades over time.

4. Graphical Illustration of Lost, Isentropic, and Optimum Work

The previous example illustrates the differences between isentropic, actual, and optimum work for a steady state steady flow process. These differences and those between the adiabatic and availability or exergetic efficiencies can now be graphically illustrated. This is done in Figure 15 that contains a representative T–s diagram for gas expansion in a turbine from a pressure P_1 to P_2. The solid line 1–2s represents the isentropic process, while the dashed curve 1–2 (for which the actual path is unknown) represents the actual process. The isenthalphic curve h_1 intersects the isobaric curves P_1 at (1) and P_2 at the point K. The isentropic and actual work represent the areas that lie, respectively, under the lines 2s–K (i.e., B–2s–K–D) and 2–k (i.e., C–2–K–D). The proof follows.

For any adiabatic process, w = dh. For the isentropic process 1–2s

$$-w_{1-2s} = h_{2s} - h_1,$$

while for the actual process 1–2

$$-w_{1-2} = h_2 - h_1.$$

The work loss during the irreversible adiabatic process is

$$w_{1-2s} - w_{1-2} = h_2 - h_{2s}.$$

Consider the relation

$$T\,ds + v\,dP = dh$$

which is valid for a fixed mass of simple compressible substances. At constant pressure

$$T\,ds = dh,$$

at a constant pressure P_2

$$\int_{2_s}^{1} T ds = \int_{h_{2_s}}^{h_1} dh.$$

Consider an ideal gas as an example (Figure 15). The constant temperature lines are same as constant enthalpy lines. For illustration consider the expansion process 1-2 with $P_2<P_1$; the area under 2s-K along constant P_2 line represents the work output for isentropic process.

Similarly, the area under 2–K (C–2–K–D) represents the actual work. The area under 2s–2 (i.e., B–2s–2–C), therefore, represents the difference between isentropic and non-

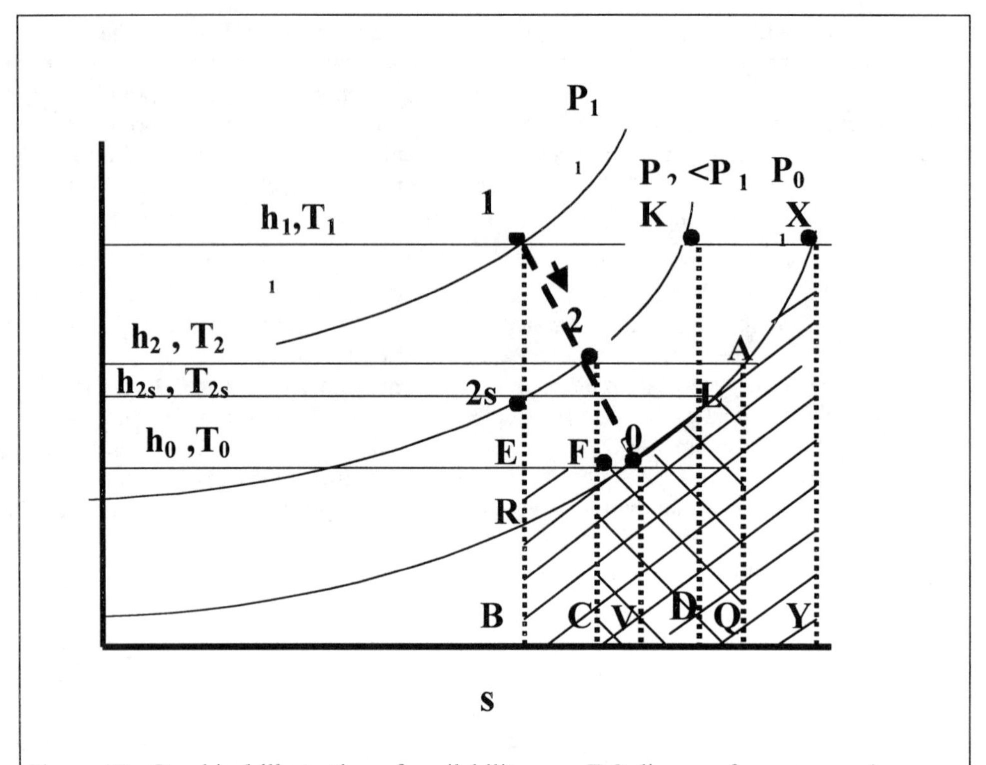

Figure 15 : Graphical illustration of availability on a T-S diagram for an expansion process.

isentropic processes. Similarly, if fluid is expanded from state (1) to dead state, (say $P_2 = P_0$, $T_2 = T_0$) then work is given by area V-0-X-Y and availability stream availability at state 1 ($h_1 - h_0 - T_0 (s_1 - s_0)$) is given by area B-E-0-V+V-0-X-Y and at state 2 by area (E-F-0-V+V-0-A-Q) thus, the area E-F-C-B-+Q-A-X-Y represents w_{opt}. The irreversibility is given by area BEFC + QAXY.

Since the actual work is given by area 2KDC+QAXY, the consequent availability loss $T_0 (s_2 - s_1)$ that is represented by the area BEFC which is smaller than the work loss area B–2s–2–C. The reason for this is that the end–state conditions are maintained identical for the availability calculations, i.e., part of the isentropic work is used to pump heat so that state 2 is reached from the state 2s. Hence, the optimum work $w_{opt} < w_{1-2s-2}$ so that $(w_{opt} - w_{12}) < (w_{12s} - w_{12})$. This difference is represented by the area E–2s–2–F. The adiabatic or isentropic efficiency is provided by the relation

$$\eta = \text{actual work} \div \text{isentropic work} = (\text{Area C2KD}) \div (\text{Area B2S}_s\text{KD}).$$

The availability efficiency

$$\eta_{Avail} = \text{actual work} \div \text{maximum work} = (\text{Area C2KD} \div (\text{Area C2KD+EFCB}).$$

Similar diagrams can be created for compression processes for which

$$\eta = \text{isentropic work} \div \text{actual work} = (\text{Area B2sKD}) \div (\text{Area C2KD}), \text{ and}$$

$$\eta_{Avail} = \text{actual work} \div \text{maximum work} =$$

$$(\text{Area C2ID}) \div (\text{Area C2KD}) \div (\text{Area C2KD + Area EFCB}).$$

5. Flow Processes or Heat Exchangers

Heat exchangers are used to transfer heat rather than to *directly* produce work. Therefore, the definition for availability efficiency that is just based on work is unsuitable for heat exchangers. Hence the availability efficiency for a heat exchanger must be defined in terms of its capability to maintain the work potential after heat exchange. Hence $\eta_{Avail,f}$ = (Exergy leaving the system) ÷ (Exergy entering the system). A perfect heat exchange will have $\eta_{avail,f}=1$

Since the stream exergy leaving a system equals that entering it minus the exergy loss in the system,

$$\eta_{Avail,f} = 1 - ((\text{Exergy loss in the system}) \div (\text{Exergy entering the system})). \quad (84)$$

a. Significance of the Availability or Exergetic Efficiency

For instance, heat is transferred in a boiler from hot gases to water in order to produce steam. However, the steam may be used for space heating and/or to produce work, and the higher the η_{Avail} value in the boiler, the higher will be the potential of the steam to perform work in a subsequent work–producing device. The availability efficiency represents the ratio of the exiting exergy to the entering exergy.

Assume that a home is to be warmed by a gas heater to a 25°C temperature during the winter when the ambient temperature is 0°C. Assume, also, that the heater burns natural gas as fuel and produces hot combustion gases at a temperature of say 1800 K. These hot gases are used to heat cooler air in a heat exchanger. The flow through the house is recirculated through the heater in which the cold air enters at a temperature of say 10°C and leaves at 25°C. Consequently, the hot gases transfer heat to the colder air and leave the heat exchanger at a 500 K temperature. Extreme irreversibilities are involved. Typically η_{Avail} is very low indicating a large loss in work potential.

"Smart" engineering systems can be designed to heat the home and at the same time provide electrical power to it for the same conditions as in the previous gas heater arrangement. Assume that the hot product gases at 1800°K are first cooled to the dead state (at 273 K) using a Carnot engine to produce work equal to $\Psi_{g,1800}$. The cold air at 283 K can also be cooled to the dead state to run another Carnot engine that produces work, $\Psi_{a,283}$. The work produced from both engines $\Psi_{g,1800} + \Psi_{a,283}$ can then be used to run a heat pump that raises the temperature of the air from the dead state to the desired temperature (298 K) and, consequently, increases the exergy contained in the air and raises the temperature of the product gases from the dead state to the exiting gas temperature (500 K). We will still be left with a potential to do work (= exergy of hot gases and cooler air entering the heat exchanger – exergy due to the cooled gases and heated air leaving the heat exchanger) which can be used to provide electricity to the home.

b. Relation Between $\eta_{Avail,f}$ and $\eta_{Avail,0}$ for Work Producing Devices

If the exit state from a work producing device is the dead state, then the availability efficiency is $\eta_{Avail,0}$ = (work output) ÷ (input exergy). This ratio informs us of the extent of the conversion of the input exergy into work, but gives no indication as to whether the exergy is lost as a result of irreversibility or with the exit flow. The flow availability efficiency $\eta_{Avail,f}$, which compares the exergy ratio leaving a system to that entering it, is able to convey that information.

F. CHEMICAL AVAILABILITY

Our discussion thus far has considered systems for which the dead state is in thermo–mechanical (TM) equilibrium. For instance, consider compressed dry air that is contained in a piston–cylinder assembly that is placed in an ambient under standard conditions. The air may be expanded to its dead state and, in the process, produce work. At the dead state the air exists in thermo–mechanical equilibrium with its environment. If the constraint, i.e., the piston, between the air and the ambient is removed, no change of state occurs within the cylinder or the ambient. Now, assume that the cylinder initially contains a mixture consisting of 40% N_2 and 60% O_2, while the ambient still contains dry air (consisting of 79% N_2 and 21% O_2). Although thermo–mechanical equilibrium is achieved when the gas is fully expanded to restricted dead state conditions (thermo-mechanical equilibrium), mass transfer occurs when the constraint is removed, i.e., the composition within the cylinder changes irreversibly. Recall that chemical potential of species k is the same as Gibb's function which depends upon species concentration. Thus, the difference in concentration between the gas in the system and air in the ambient leads to difference in Gibb's function and hence irreversible mass transfer of species k (Chapter 3). The ambient gains O_2 molecules, trying to alter its partial pressures in the environment. The overall composition of the combined isolated system (piston–cylinder and ambient) is not the same as it was before implying that the entropy of the isolated system must have increased. Therefore, even if a system exists in thermo–mechanical equilibrium, this does not assure a zero entropy increase when the constraints upon it are removed.

Similarly, consider a turbine in which compressed air is expanded to the dead state which exists at standard conditions. Upon discharge to the dead state, it exits the turbine with negligible kinetic energy and its state does not change, since the air exists in thermo–mechanical equilibrium with the ambient. On the other hand, if compressed nitrogen is expanded through the turbine, its state will change from pure nitrogen to an air–nitrogen mixture as it mixes with the ambient air. In this section we will discuss a methodology to determine the optimum work in cases where thermo–mechanical equilibrium exists, but irreversible mixing occurs. If somehow, the N_2 is released at pressure equal to ambient partial pressure of N_2, then there is no irreversible mixing and chemical equilibrium now exists in addition to TM equilibrium. We wish to derive relations for the optimum work when matter reaches thermo-mechanical-chemical (TMC) equilibrium. Before doing so, we will briefly describe semi-permeable membranes. These membranes are permeable to specific species only, e.g., if dirty water is filtered through a charcoal bed, the bed can be designed to be permeable mostly to water, but impermeable to any particulate matter that it carries. Similarly, semipermeable membranes can be designed to separate water (solvent) and salt (solute), and gas mixtures.

1. Closed System

We now discuss thermo–mechanical–chemical equilibrium. In the following we present a methodology of achieving TMC equilibrium followed by brief derivation. Consider a cylinder containing an ideal gas mixture consisting of 40% N_2 and 60% O_2 by volume which has expanded from initial state to the final restricted dead state (Figure 16), i.e., in thermo-mechanical equilibrium with the ambient. Once in the TM state the cylinder can be divided into two chambers A and B by a partition and constrained by two pistons placed on either side of the partition that can move independently, as shown in Figure 16. The partial pressures of oxygen and nitrogen in the cylinder are $p_{O2,0}$= 0.6 and $p_{N2,0}$ = 0.4 bars, respectively, while the corresponding ambient pressures are $p_{N2,\infty}$ = 0.79 and $p_{O2,\infty}$ = 0.21 bars. Next, a semi-permeable membrane that is only permeable to O_2 replaces the partition. One piston, say A, is then moved so as to decrease the pressure in chamber A without altering its temperature. Therefore, the partial pressure of O_2 in chamber A decreases below its corresponding value in chamber B and, consequently, oxygen molecules migrate across the membrane from B to A (Chapter 1 and Chapter 3). By maneuvering the piston to achieve very low pressure, virtually all of the oxygen can be transferred from chamber B into A. Next, the O_2–permeable membrane can be replaced with one permeable to nitrogen and all N_2 molecules can be transferred from A into chamber B. By so manipulating the two pistons and the semipermeable membranes, the two components can be separated so that chamber A consists of only O_2 and chamber B consists of only N_2

Once the separation process is completed semipermeable membrane is replaced with a partition impermeable to either of the species, the pistons A and B can be moved so that $p_{N2,B}$

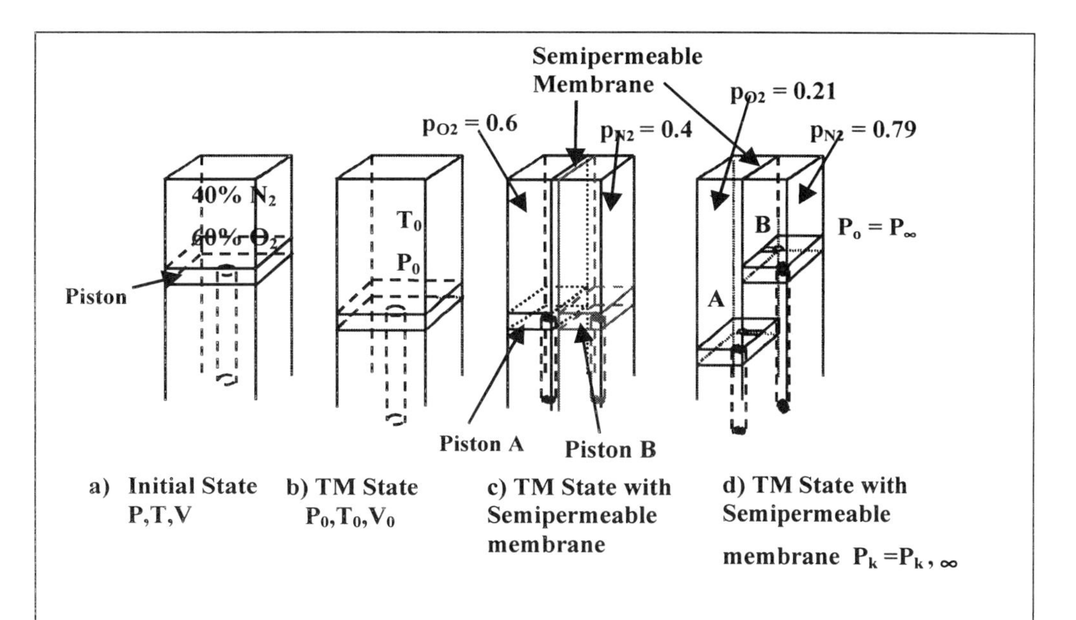

Figure 16: Illustration of a device that may be used to achieve thermo–mechanical–chemical equilibrium of a system with its environment. a) Gas mixture at initial state; b): gas mixture at dead state; c) partition with semi-permeable membrane with two separate pistons; d) separation of components and adjustments to partial pressures of ambient.

$= p_{N2,\infty}$ and $p_{O2,A} = p_{O2,\infty}$ (Figure 16d), i.e., same as the partial pressures of the two species in the environment. Note that $T_0 = T_\infty$ in this section of the chapter. Isothermal work called chemical work $W_{ch,02}$ is obtained from chamber A, since $p_{O2,\infty} < p_{O2,0}$ (the initial partial pressure of oxygen), but compression work must be performed on chamber B for which $p_{N2,\infty} > p_{N2,0}$. The pistons A and B can now be removed and replaced with rigid semipermeable membrane pistons that are, respectively, permeable to O_2 and N_2. There can not be transfer of species across this membrane since partial pressures are the same. In this manner chemical equilibrium will be achieved in the combined isolated system consisting of the chambers A and B and the ambient.

Thus, the work can be obtained through a two–step process consisting of (1) the work obtained during expansion from the initial state to that in the thermo–mechanical equilibrium $W_{u,0}$, and (2) the chemical work obtained in proceeding from the thermo–mechanical state to the thermo–mechanical–chemical equilibrium state W_{ch}.

Therefore, for a closed system in TMC equilibrium,

$$W_{u,opt,\infty} = W_{u,opt,0} + W_{ch}, \text{ where} \tag{85}$$

$$W_{ch} = (U_0-U_\infty) - T_0(S_0-S_\infty) + P_0(V_0-V_\infty) = (H_0-H_\infty) - T_0(S_0-S_\infty) == G_0-G_\infty. \tag{86}$$

Where properties U_∞, S_∞, V_∞ are determined for O_2 and N_2 contained within the sections A and B at $T_0 = T_\infty$, $p_{N2,\infty}$ and $p_{O2,\infty}$. The work $w_{u,opt,0}$ can be obtained using Eq. (28). On a unit mass basis,

$$w_{ch} = (h_0 - h_\infty) - T_0\,(s_0 - s_\infty) = g_0 - g_\infty.$$

It is seen that the chemical work per unit mass is also the same as the difference in stream availability between the restricted dead state, i.e., TM to TMC equilibrium state (see next section) and the chemical work can be represented as the change in the Gibbs free energy due to the change in state from TM to TMC states.

For ideal gas mixtures $\bar{g}_0 = \Sigma X_k \bar{g}_k (T_0, p_{k,0})$, where $p_k = X_k P$, and

$$\bar{g}_k(T_0, p_k) = \bar{h}_k(T_0) - T_0(\bar{s}^0(T) - \bar{R}\ln(p_k/1) = \bar{g}_k(T_0, P_0) + \bar{R}T_0 \ln(p_k/1).$$

Therefore,

$$\bar{w}_{ch} = \Sigma_k X_k \ \phi_{k,0}(T_0, p_{k,0}) - \Sigma X_k \phi_{k,\infty}(T_0, p_{k,\infty}). \quad (87)$$

and writing in terms of g

$$\bar{w}_{ch} = \Sigma X_k \bar{g}_{k,0}(T_0, p_{k,0}) - \Sigma X_{k,} \bar{g}_{k,\infty}(T_0, p_{k,\infty}). \quad (88)$$

$$\bar{w}_{ch} = \bar{\phi}_0 - \bar{\phi}_\infty, \text{ where} \quad (89)$$

$$\bar{\phi}_0 = \Sigma \ X_{k,0} \bar{g}_{k,0}(T_0, P_0),\ X_k = X_{k,0} = X_{k,e}, \text{ and} \quad (90)$$

$$\bar{\phi}_\infty = \Sigma \ X_{k,0} \bar{g}_{k,\infty}(T_\infty, P_\infty). \quad (91)$$

---

k. *Example 11*

Determine the maximum work that can be performed if a gas mixture consisting of 40% O_2 and 60% N_2 is expanded from 2000 K and 60 bars to the dead state at which it is at thermo–mechanical–chemical equilibrium.

Solution

The maximum work

$$\bar{w}_{max,\infty} = \bar{w}_{max,0} + \bar{w}_{ch}, \text{ where} \quad (A)$$

$$\bar{w}_{max,0} = \bar{\phi} - \bar{\phi}_0.$$

Now, $\bar{\phi} = \bar{u} - T_0 \bar{s} + P_0 \bar{v}$, where $\bar{v} = 0.08314 \times 2000 \div 60 = 2.77$ m^3 kmole^{-1}. The specific entropy, $\bar{s}_{N_2} = \bar{s}^0_{N_2} - \bar{R}$ ln (p_{N2}/1) = 251.6 – 8.314 (ln(0.4×60)/1) = 225.2 kJ kmole^{-1} K^{-1}, and $\bar{s}_{O_2} = \bar{s}^0_{O_2} - \bar{R}$ ln (P_{O2}/1) = 268.7 – 8.314 (ln(0.6×60)/1) = 238.9 kJ kmole^{-1} K^{-1}. The mixture initial specific entropy and internal energy are

$\bar{s} = (0.4\ \bar{s}_{N_2} + 0.6\ \bar{s}_{O_2}) = 233.4\ 9$ kJ kmole^{-1} K^{-1}, and

$\bar{u} = (0.4 \times 48{,}181 + 0.6 \times 51{,}253) = 50{,}024$ kJ kmole^{-1}.

Therefore,

$\bar{\phi} = 50{,}024 - 298 \times 233.4 + 100 \times 2.77 = -19252.2$ kJ kmole^{-1},

$\bar{u}_0 = 0.4 \times 6190 + 0.6 \times 6203 = 6197.8$ kJ kmole^{-1},

$\bar{s}_0 = 4(191.52 - 8.314 \times \ln(0.4/1)) + 0.6(205.0 - 8.314 \times \ln(0.6/1)) = 205.2$ kJ kmole^{-1} K^{-1},

$\bar{v}_0 = 0.08314 \times 298/1 = 24.78$ m^3 kmole^{-1},

$\bar{\phi}_0 = 6197.8 - 298 \times 205.2 + 100 \times 24.78 = -52473.8$ kJ kmole^{-1}. Therefore,

$w_{max,0} = -19252.2 - (-52473.8) = 33{,}221.6$ kJ kmole^{-1}. Since

$w_{ch} = \bar{\phi}_0 - \bar{\phi}_\infty$ and

$\bar{\phi}_\infty = \Sigma X_k\ \bar{\phi}_{k,\infty}$

$\bar{\phi}_{O_2,\infty}(T_0, P_{O2,\infty}) = \bar{u}_{O_2}(T_0, p_{O2,\infty}) - T_0\ \bar{s}_{O_2}(T_0, p_{O2,\infty}) + P_0\ \bar{v}_{O_2}(T_0, p_{O2,\infty})$

$= 6203 - 298 \times (205.0 - 8.314 \times \ln (0.21/1)) + 100 \times (0.08314 \times 298/0.21)$

$= -46981$ kJ kmole^{-1}.

Likewise,

$\bar{\phi}_{N_2,\infty}(T_0, p_{N2,\infty})$

$= 6190 - 298 \times (191.5 - 8.314 \times \ln (0.79/1)) + 100 \times (0.8314 \times 298/0.79)$

$= 6190 - 298 \times 193.5 + 100 \times 31.4 = -48{,}333$ kJ kmole^{-1}.

Therefore,

$\bar{\phi}_\infty = 0.6 \times (-46981) + 0.4 \times (-48333) = -47521.8$ kJ kmole^{-1}, and

$w_{ch} = -52473.8 - (-47521.8) = -4952$ kJ kmole^{-1}.

$w_{u,\,max,\,\infty} = 33{,}221.6 + (-4952) = 28{,}269.6$ kJ/mole

---

Remarks

The chemical work per kmole of mixture is negative, since a larger work input is necessary to compress 0.4 kmoles of N_2 from 0.4 bars to 0.79 bars as compared to the work output obtained from the expansion of 0.6 kmoles of O_2 from 0.6 bar to 0.21 bar.

It will be shown later that the expression for w_{ch} for an open system is similar to that for a closed system. Several cases involving water, air–vapor mixtures, and product gas mixtures will be dealt with later in the context of open systems.

2. Open System

As before, we assume that the availability can be characterized by a two–step process. First, the mixture is brought from a specified state to a dead state considering only thermomechanical equilibrium (i.e., there is no change in the system composition), and then allowing the system components to arrive at phase and/or chemical equilibrium with the environment (which is a mixture of specified composition, e.g., containing O_2, N_2, H_2O, CO_2, Ar, etc.).

a. Ideal Gas Mixtures

Assume that you travel with an unit mass (or a kmole) of an ideal gas mixture as it enters a turbine and leaves it for the environment in a state as shown in Figure 17. Consequently, the composition within the unit mass at X_k changes irreversibly when it enters the environment at $X_{k,\infty}$. In order to achieve TMC equilibrium we adopt the scheme illustrated in Figure 18. The exiting gases (at the state (T_0, P_0) are passed through semipermeable membranes in order to separate the mixture into its pure components. These components subsequently enter a chemical turbine so that the exhaust pressure of component k matches its partial in the ambient. Using the generalized availability relation

$$d(E_{cv} - T_0 S_{cv})/dt = \sum_{inlet} \dot{N}_k \hat{\psi}_k (T,P,x_k) - \sum_{exit} \dot{N}_k \hat{\psi}_k (T,P,x_k) + \Sigma_{j=1}^{N} \dot{Q}_{R,j}\left(1 - T_0/T_{R,j}\right) - \dot{W}_{cv} - \dot{I},$$

where $\hat{\psi}_k$ denotes the absolute stream availability of the k–th species in the mixture. For a steady state of the system without any reservoir and any irreversibility,

$$\dot{W}_{opt,0} = \sum_{inlet} \dot{N}_k \psi_k (T,P,X_k) - \sum_{turb.exit} \dot{N}_k \psi_{k,0}(T_0,P_0,X_{k,0}),\ \text{where} \quad (92a)$$

$$\hat{\psi}_k (T_0,P_0,x_k) = \hat{h}_k (T_0,P_0,x_k) - T_0 \bar{\bar{s}}(T_0,P_0,X_k) = \hat{g}_k (T_0,P_0,X_{k,0}). \quad (92b)$$

Recall that Eq. (92a) is still valid when chemical reactions occur where moles change (cf. Chapter 13). If a system is nonreacting, then $\dot{N}_{k,in} = \dot{N}_{k,exit}$, $\dot{N}_{in} = \dot{N}_{exit}$. Recognizing that

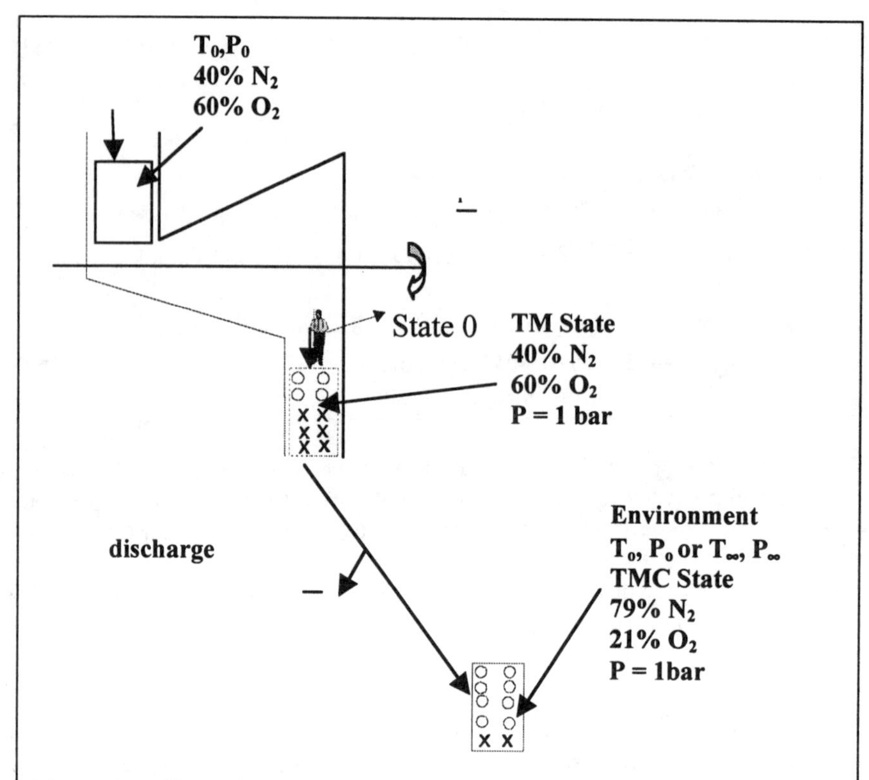

Figure 17: Illustration of equilibrium with the environment at a restricted (thermo–mechanical) dead state.

$$\dot{W}_{opt,\infty} = \dot{W}_{opt,0} + \dot{W}_{ch}\text{, where} \tag{93a}$$

$$\dot{W}_{ch} = \sum_{turb.exit} \dot{N}_k \psi_{k,0}(T_0,P_0,X_{k,0}) - \sum_{exits} \dot{N}_k \psi_{k,\infty}(T_\infty,P_\infty,X_{k,\infty}) = \sum \dot{N}_k \overline{w}_{ch}\text{, and} \tag{93b}$$

$$\hat{\psi}_k(T_\infty,P_\infty,X_{k,\infty}) = \hat{h}_k(T_\infty,P_\infty,x_{k,\infty}) - T_0\bar{\bar{s}}(T_\infty,P_\infty,X_{k,\infty}) = \hat{g}_{k,\infty}(T_\infty,P_\infty,X_{k,\infty}). \tag{93c}$$

Note that $T_0 = T\infty$, $P_0 = P_\infty$, but $X_{k,0} \neq X_{k,\infty}$.

Using Eqs. (92a) and (93b)

$$\dot{W}_{opt,\infty} = \sum_{inlet} \dot{N}_k \hat{\psi}_k(T,P,X_k) - \sum \dot{N}_k \hat{\psi}_{k,\infty}(T_\infty,P_\infty,X_{k,\infty})\text{, and} \tag{93d}$$

Eq. (93b) can be rewritten in the form

$$\dot{W}_{ch} = \dot{N}\overline{w}_{ch} = \sum \dot{N}_k \hat{w}_{ch,k} = \sum \dot{N}_k [\hat{g}_{k,0}(T_0,P_0,X_{k,0}) - \hat{g}_{k,\infty}(T_0,P_0,X_{k,\infty}), \tag{94a}$$

$$\hat{w}_{ch,k} = \hat{g}_{k,0}(T_0,P_0,X_{k,0}) - \hat{g}_{k,\infty}(T_0,P_0,X_{k,\infty}). \tag{94b}$$

Dividing Eq. (94a) throughout by $\dot{N}$

$$\overline{w}_{ch} = \frac{\dot{W}_{ch}}{\dot{N}} = \sum X_{k,0}\hat{w}_{ch,k} = \sum X_k[\hat{g}_{k,0}(T_0,P_0,X_{k,0}) - \hat{g}_{k,\infty}(T_0,P_0,X_{k,\infty}). \tag{94c}$$

For an ideal gas mixture, Eq. (94b) has the form

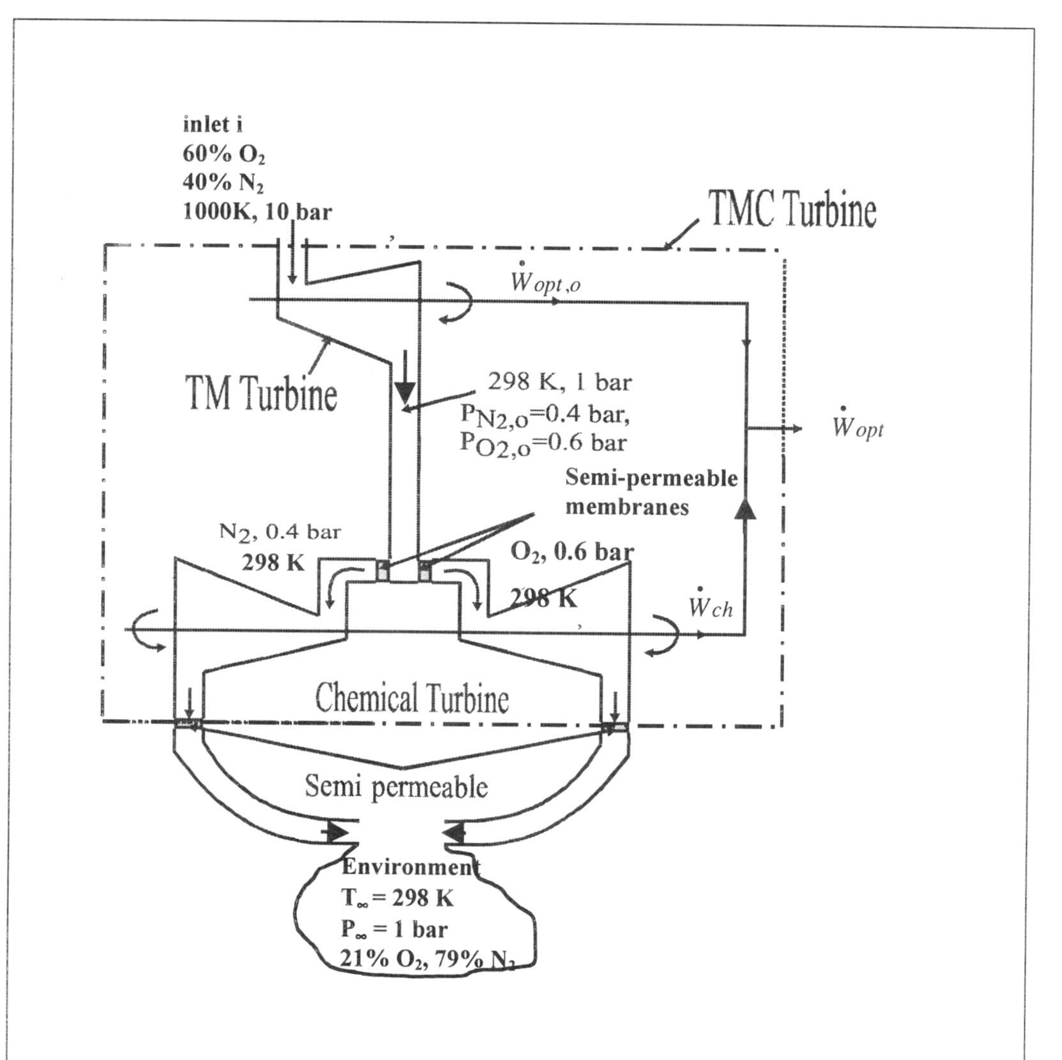

Figure 18: Illustration of a method to achieve thermo–mechanical–chemical equilibrium for a mixture of ideal gases.

$$\hat{w}_{ch,k} = \overline{R}T_0 X_{k,0} \ln(X_{k,0} / X_{k,\infty}). \tag{95}$$

For example if a turbine operates with a single component, say pure nitrogen so that $X_{k,0} = 1$ and $X_{k,\infty} = 0.79$ in the ambient (air), in this case

$$\hat{w}_{ch} = -RT_0 \ln X_{N_2,\infty}.$$

---

l. Example 12

What is the minimum work required to separate oxygen from air if the oxygen exits the separation system at (1) $p_{O2} = 0.21$ bar and 298 K; and (2) $p_{O2} = P = 1$ bar, 298 K. N_2 leaves the system separately at 0.79 bar in both the cases.

Solution

Air is supplied to a device at 298 K and 1 bar to separate O_2 at 0.21 bar and 298 K, while N_2 will be discharged by the device to the atmosphere in chemical equilibrium at a 0.79 bar pressure. Employing the availability relation,

$$w_{opt} = \psi_{air}(T_0,P_0) - (\psi_{O_2,0}(T_0,P_0,X_{O_2,0} = 0.21) + \psi_{N_2,0}(T_0,P_0,X_{O_2,0} = 0.79)). \quad (A)$$

Since $\psi = h - T_0\, s$, by simplifying Eq. (A) it may be shown that $w_{opt} = 0$. No work is required, since the sum

$\psi_{O_2,0}(T_0,P_0,X_{O_2,0} = 0.21) + \psi_{N_2,0}(T_0,P_0,X_{O_2,0} = 0.79) = \psi_{air,0}(T_0,P_0)$.

If the separated oxygen exits the system at a pressure of 1 bar, then work must be done in order to compress it from 0.21 bar to the higher pressure, i.e., $X_{O2,\infty} = 1$, $X_{O2,0} = 0.21$. Using Eq. (95) or using isothermal compression at 298 K to raise the pressure from 0.21 bar to 1 bar,

$$\overline{w}_{ch} = RT_0 \ln\frac{0.21}{1.0} = -3867 \text{ kJ kmole}^{-1} \text{ (of } O_2).$$

Remarks

Since the nitrogen must be discharged in a state of chemical equilibrium, one way to achieve this is to separate the oxygen in a relatively small quantity δN_{O2} and to discharge the remaining mixture (consisting mostly of air) directly to the atmosphere. Applying the availability relation,

$$\delta W_{opt} = (N_{O2,0}\, \hat{\psi}_{O2,0}(T_0, P_0, X_{O2,0}) + N_{N2,0}\, \hat{\psi}_{N2,0}(T_0, P_0, X_{N2,0}))$$

$$- \delta N_{O2}\, \hat{\psi}_{O2}(T_0, P_0) - (N_{O2,\infty}\, \hat{\psi}_{O2,\infty}(T_0, P_0, X_{O2,\infty}) + N_{N2,\infty}\, \hat{\psi}_{N2,\infty}(T_0, P_0, X_{N2,\infty})).$$

Since, $N_{O2,\infty} = N_{O2,0} - \delta N_{O2}$, for small values of δN_{O2}, $\hat{\psi}_{O2,0} \approx \hat{\psi}_{O2,\infty}$ and $\hat{\psi}_{N2,0} \approx \hat{\psi}_{N2,\infty}$, so that

$$\delta W_{opt}/\delta N_{O2} = \hat{\psi}_{O2,\infty}(T_0,P_0,X_{O2,\infty}) - \hat{\psi}_{O2}(T_0,P_0) = \hat{\psi}_{O2,0}(T_0,P_0,X_{O2,\infty}) - \hat{\psi}_{O2}(T_0,P_0).$$

Therefore,

$$\delta W_{opt}/\delta N_{O2} = RT_0 \ln X_{O2,\infty} = -3867 \text{ kJ kmole}^{-1} \text{ (of } O_2).$$

b. *Vapor or Wet Mixture as the Medium in a Turbine*

Oftentimes, turbines run on a single component that is condensable when cooled. For instance, when steam is expanded to a dead state at 25°C and 1 bar (at which it exists in thermo–mechanical equilibrium with the environment) it liquefies, and the availability at the turbine exit is

$$\psi(T_0,P_0) = \psi_0 = \psi_{H_2O(\ell)} = g_0 = g_{H_2O(\ell)} = (h_{H_2O(\ell)} - T_0 s_{H_2O(\ell)}). \quad (96)$$

The properties for the liquid state can be obtained using compressed liquid tables (Tables A-4A for H_2O and A-5A for R 134A) or we can select the properties at the saturated liquid states at specified temperatures as an approximation.

Moreover, the liquid from a turbine cannot be discharged into the ambient wet air where the water vapor partial pressure is typically $p_{H2O} = 0.02$ bars, since the liquid water so discharged into wet air will partially vaporize and mix with the atmosphere irreversibly. In order to avoid the irreversibility we can first expand the steam from its initial state T, P all the way to $P_{H2O}=0.02$ and $T_o=298$K and subsequently release the steam after passing it through a semipermeable membrane so that it is in equilibrium with the environment (cf. Figure 20). As mentioned above, it is useful to consider expansion as the work performed during this two–step process: $W_{opt,\infty} = W_{opt,0} + W_{ch}$, (cf. Eq. (93a)).

However, there are some difficulties in using the tables (A-4C) for the enthalpy and entropy of superheated vapors at pressures as low as 2 kPa and temperatures of 25°C. At low pressures ideal gas behavior for the vapor can be assumed. Since the enthalpy of ideal gases does not depend upon the pressure,

$$h(T_0,P_{H2O(g),\infty}) = h(T_0,P^{sat}). \quad (97)$$

Therefore, the enthalpy of superheated vapors at low pressures can be assumed to be the same as that of saturated vapor at the same temperature. However, the entropy of the superheated vapor does depend on pressure

$$s(T_\infty,p_{H2O,\infty}) = s(T_\infty,P^{sat}) - R \ln (p_{H2O,\infty}/P^{sat}(T_\infty)), \text{ i.e.,}$$

$$\psi_\infty = h^{sat}_{H_2O}(T_\infty) - T_\infty\, s(T_\infty,p_{H2O,\infty}) = h^{sat}_{H_2O}(T_\infty) - T_\infty\, s(T_\infty,P^{sat}) + RT_\infty \ln(p_{H2O,\infty}/P^{sat}(T_\infty)), \text{ or}$$

$$\psi_\infty = g^{sat}_{H_2O}(T_\infty) + RT_0 \ln (P_{H2O,\infty}/P^{sat}(T_\infty)) = g^{sat}_{H_2O}(T_\infty) + RT_0 \ln(RH) \quad (98)$$

where the ratio $(P_{H2O,\infty}/P^{sat}(T_\infty))$ is the *relative humidity* of water in air.

c. *Vapor–Gas Mixtures*

Consider a mixture containing water vapor ($X_{H2O,0} = 0.2$) and nitrogen ($X_{N2,0} = 0.8$) that expands to 1 bar and 298 K, a state that is in thermo–mechanical equilibrium (Figure 19). The expected partial pressure of water vapor at this dead state is 0.2 bar if it remains entirely in the form of vapor. However, at $p_{H2O} = 0.2$ bar the required temperature for water to remain as vapor is 333 K. Therefore, the vapor will partially condense. For the vapor–N_2 mixture to be in restricted dead state requires that $p_{H2O(g)} = p_{H2O(g)}{}^{sat}$ (298 K) = 0.032 bar. The partly liquid water and remaining gaseous N_2 may form a wet mixture. On the other hand typical wet air in ambient may consist of vapor at $p_{H2O(g)} = 0.02$ bars. For a TMC, system, therefore, the vapor must be further expanded from 0.032 to 0.02 bar and all the liquid must be vaporized and N_2 must be discharged at partial pressures that are the same as the partial pressure in the ambient.

Figure 19 illustrates a device in which the vapor–gas mixture can be first expanded to a state of thermo–mechanical equilibrium (liquid water, vapor and gaseous N_2) following which the two components are separated using semipermeable membranes. Each component is expanded in a chemical turbine and then released into the ambient in thermo–mechanical–chemical equilibrium. The total chemical work

$$\overline{w}_{ch} = X_{N2,0}\ \hat{w}_{ch,N2} + X_{H2O,0}\ \hat{w}_{ch,H2O}, \quad (99a)$$

and from Eq. (95)

$$\hat{w}_{ch,N2} = \overline{R}T_0 \ln (p_{N_2,0}/p_{N_2,\infty}). \quad (99b)$$

Since a wet mixture may exist at T_0, we must determine the molal Gibbs function of the mixture. The molal Gibbs function of any species k at the saturated liquid state, its saturated vapor state and for a wet mixture are identical. (Recall from Chapter 3 that $dG = -S\,dT + VdP$ for a closed system. When boiling occurs at fixed pressure, the temperature is also fixed, and $dg = 0$, i.e., g is constant during phase change. Further details follow in Chapters 7 and 9.) Thus,

$$\hat{w}_{Ch,\,H2O} = \hat{g}_{H2O,0}{}^{sat}(T_0) - \hat{g}_{H2O,\infty}(T_0, P_0, X_{H2O,\infty}), \quad (99c)$$

where $\hat{g}_{H2O,\infty}(T_0, P_0, X_{H2O,\infty}) = \overline{g}^{sat}{}_{H2O,0}(T_\infty) + \overline{R}T_0 \ln(p_{H2O,\infty}/P^{sat}(T_0))$ if the vapor is an ideal gas. Simplifying

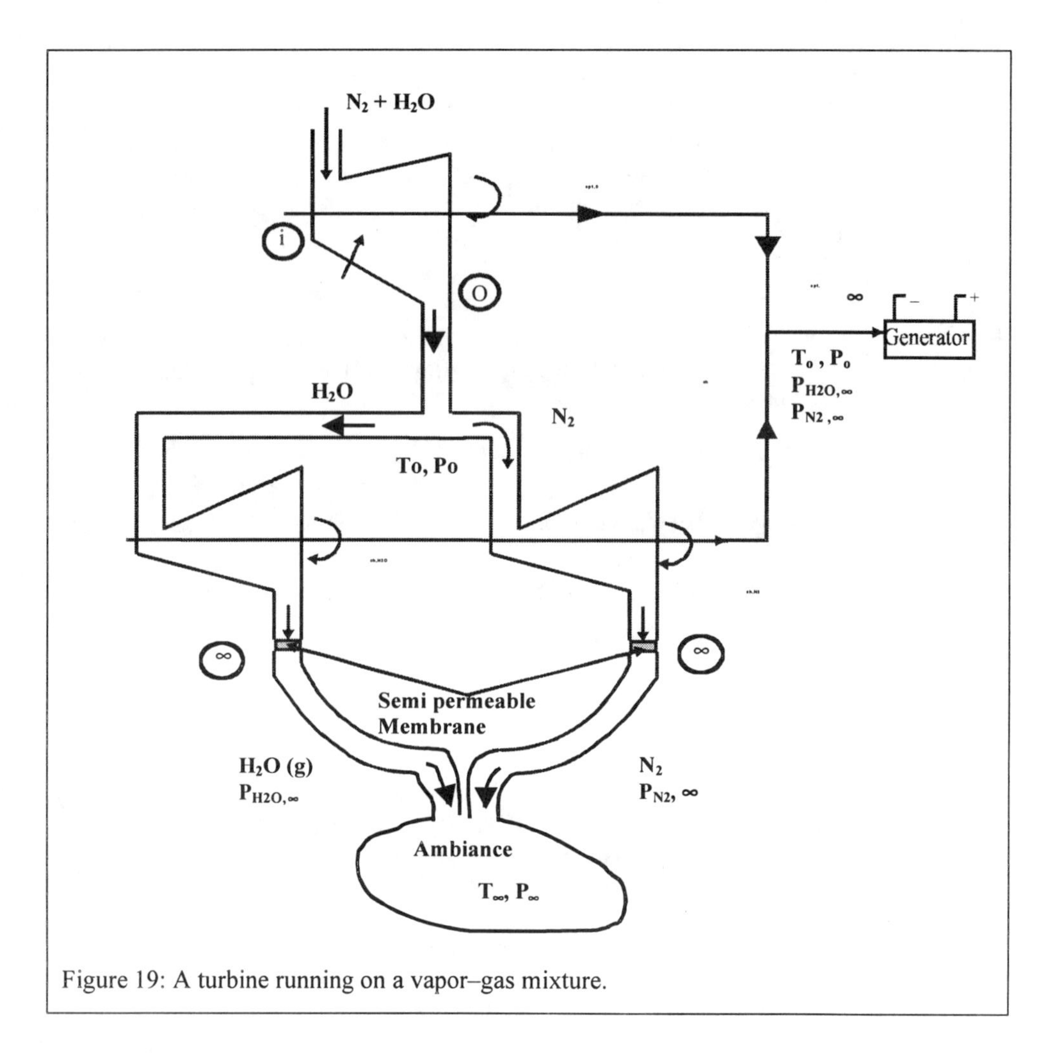

Figure 19: A turbine running on a vapor–gas mixture.

$$\hat{w}_{Ch,H2O,} = \hat{g}^{sat}{}_{H2O,0}(T_0) - \hat{g}_{H2O,\infty}(T_0, P_0, X_{H2O,\infty}), \text{ i.e.,}$$

$$\hat{w}_{ch,H2O} = -\bar{R}T_0 \ln (p_{H2O,\infty}/P^{sat}(T_0)) = -\bar{R}T_0 \ln(RH). \quad (99d)$$

d. Psychometry and Cooling Towers

Moist air containing both dry air and water vapor is characterized by the following parameters:

The humidity ratio or specific humidity or mixing ratio $w = m_v/m_a$, where m_v denotes the mass of vapor and m_a the dry air mass. In a specified volume V, $m_v = (p_vV)/(R_vT)$ and $m_a = (p_aV)/(R_aT)$ where P_v and P_a are the partial pressures of vapor and air, respectively. Further $R_k = \bar{R}/M_k$ where the symbol M_k denotes the molecular weight of species k (k= v and a). Therefore,

$$w = (M_v/M_a)(P_v/P_a) = 0.622\,(p_v/p_a). \quad (100)$$

The degree of saturation is the ratio of the vapor mass actually present at a specified temperature and pressure to the maximum possible mass that could have been present without condensation m_v^{sat} at the same temperature and pressure. Hence,

$$\mu = m_v(T,P)/m_v^{sat}(T,P) = N_v(T,P)/N_v^{sat}(T,P). \quad (101)$$

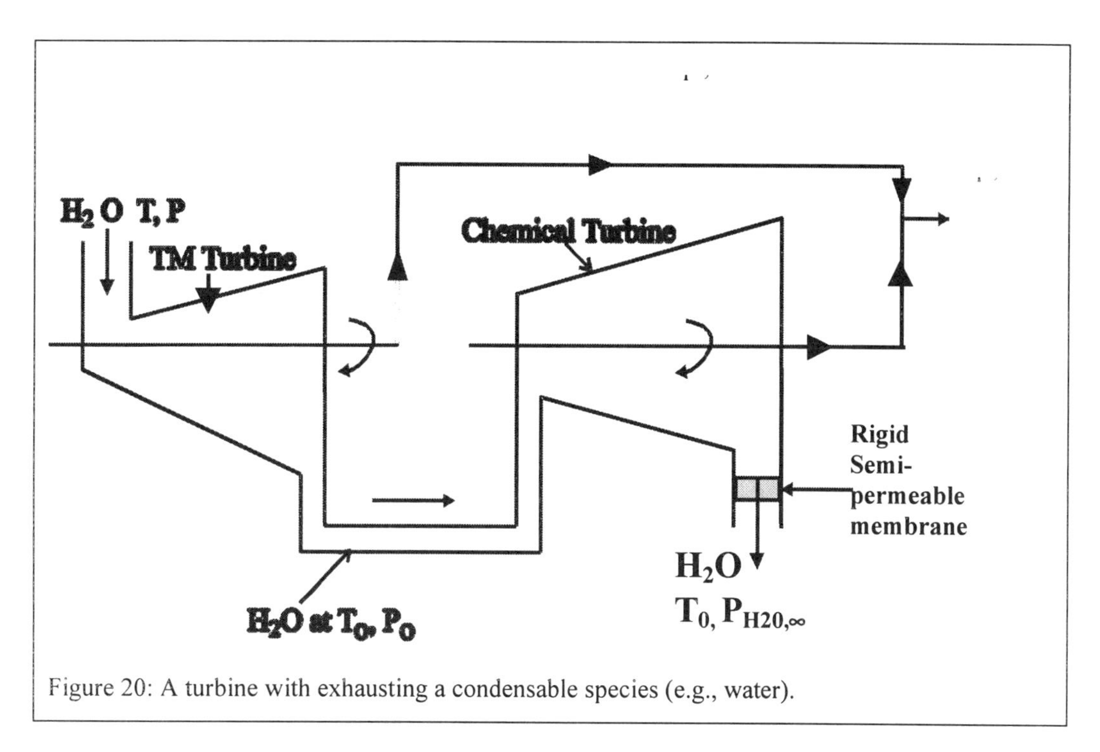

Figure 20: A turbine with exhausting a condensable species (e.g., water).

The relative humidity RH is defined as the ratio of the water vapor mole fraction at a specified temperature and pressure to the mole fraction that would exist under saturated conditions at the same temperature and pressure, i.e.,

$$RH = X_v(T,P)/ X_v^{sat}(T,P)= (N_v(T,P)/N(T,P))/(N_v^{sat}(T,P)/N^{sat}(T,P)). \quad (102)$$

In terms of partial pressures

$$RH = P_v(T)/ P_v^{sat}(T). \quad (86')$$

Since $N^{sat}(T,P) = N(T,P) + N_v^{sat}(T,P) - N_v(T,P)$ where N denotes the total number of molecules, the relative humidity can also be expressed as

$$RH = \mu\ (1 - X_v) + X_v = \mu/(1 - X_v^{sat}(1 - \mu)).$$

Using Psychometric charts (Appendix Figure B.1) for given T_{DB} (dry bulb temperature) an T_{WB} (wet bulb) the *relative humidity* can be determined. Replacing N_2 with air in Figure 19 and from Eq. (95)

$$\hat{w}_{ch,air} = \overline{R}T_O \ln(\frac{P_{air,o}}{P_O}), \hat{w}_{ch,H2O} = -\overline{R}T_O \ln\frac{p_{H_2O,\infty}}{p_{H_2O}^{sat}(T_0)}$$

or

$$\hat{\psi}_{ch,air} = X_{air}\ \overline{R}T_O \ln(p_{air,\,0}/P_o) - X_{H2O}\ \overline{R}\,T_0 \ln(RH) \quad (103)$$

where $RH = p_{H_2O,\infty} / p_{H_2O}^{sat}(T_O)$

m. Example 13

A wet cooling tower is used to cool the water discharged from the condenser of a power plant. (Figure 21). Water enters the tower at 45°C (state 1) and leaves at 25°C (state 2). Additional makeup water enters the tower at 25°C. Air enters the tower at

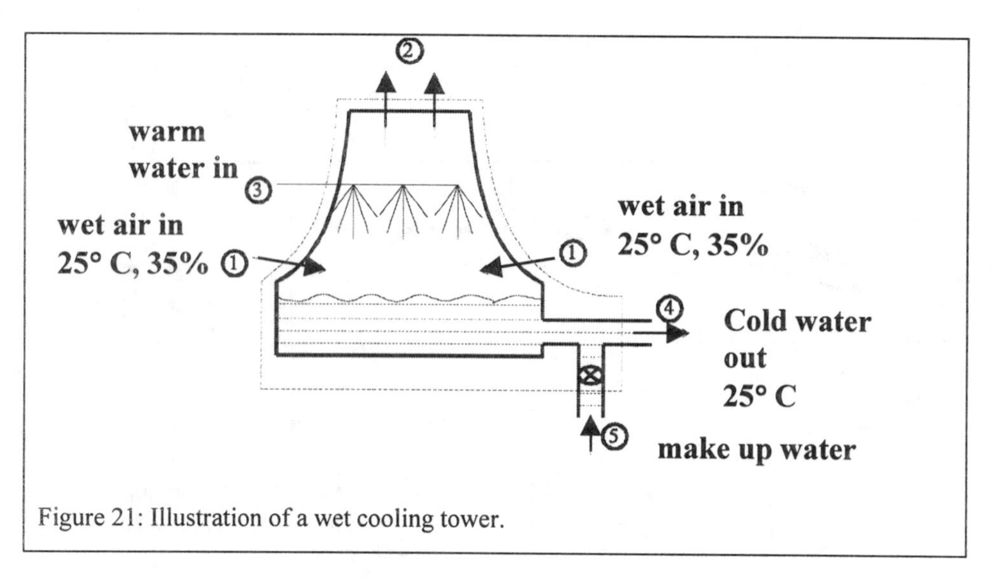

Figure 21: Illustration of a wet cooling tower.

20°C and 35% relative humidity, and leaves it at 35°C and 70% relative humidity. Assume $c_{p,v}$ = 0.603 kJ kg^{-1} K^{-1}. Determine:

The mass flow rate of dry air; and

The optimum work.

Solution

With p_1^{sat} (20C) = 0.02339 bars and p_1^{sat} (35C) = 0.05628 bars. Using Eqs. (86′) and known values of RH, p_{v1} =0.0082 bars, p_{v2} = 0.0394 bars. Hence $p_{a,1}$ = 1-0.008 = 0.992 bar, and $p_{a,2}$ = 0.961 bar. Then from Eq. (100), w_1 = 0.0051, w_2 = 0.025 (Alternately use psychrometric charts in Appendix Fig B-1). Using constant specific heats $h_{a,1} = c_{pa}$ T (in C) = 20 kJ kg^{-1}, $h_{a,2}$ = 35 kJ kg^{-1}, and from the Table A-4, h_{v1} = 2547.2, h_{v2} = 2565.3, s_g (25C) = 8.558, with R= 8.314/18.02=0.461 kJ/kgk, $s_{g1}=s_g^{sat}$ (25C) - Rln(p_{v1}/P_v^{sat} (T_1))= 8.6672 kJ/kg K, s_{g2} = 8.3531 kJ/kg K, $h_{f,3}$ = 188.45 kJ kg^{-1}, $s_{f,3}$ = 0.6387 kJ kg^{-1} K^{-1}, $h_{f,4}$= $h_{f,5}$= 104.89 kJ kg^{-1} K^{-1}, $s_{f,4}$ = 0.3674 kJ kg^{-1} K^{-1}.

From the mass balance for dry air

$$\dot{m}_{a1} = \dot{m}_{a1} = \dot{m}_a.$$

For water

$$dm_{water}/dt = \dot{m}_{v1} + \dot{m}_{f3} + \dot{m}_{f5} - \dot{m}_{v2} - \dot{m}_{f4}, \text{ where}$$

$$\dot{m}_{f5} = \dot{m}_{v1} - \dot{m}_{v2}$$

Dividing throughout by m_a,

$$\dot{m}_{f5}/\dot{m}_a = \dot{m}_{v2}/\dot{m}_a - \dot{m}_{v1}/\dot{m}_a = w_2 - w_1, \text{ and}$$

$$\dot{m}_{f5}/\dot{m}_a = 0.025 - 0.005 = 0.020 \text{ kg of water per kg dry air.}$$

Through an energy balance

$$dE/dt = \dot{Q} - \dot{W} + \dot{m}_{v1} h_{v1} + \dot{m}_{a1} h_{a1} + \dot{m}_{f3} h_{f3} + \dot{m}_{f5} h_{f5}$$
$$- \dot{m}_{v2} h_{v2} - \dot{m}_{a2} h_{a2} - \dot{m}_{f4} h_{f4}.$$

Dividing by m_a and assuming steady state,

$$0 = \dot{Q}/m_a - \dot{W}/\dot{m}_a + \dot{m}_{v1} h_{v1}/\dot{m}_a + \dot{m}_{f3} h_{f3}/\dot{m}_a + \dot{m}_{f5} h_{f5}/\dot{m}_a$$
$$- \dot{m}_{v2} h_{v2}/\dot{m}_a - \dot{m}_{f4} h_{f4}/\dot{m}_a.$$

Since $\dot{Q} = 0$, $\dot{W} = 0$, $\dot{m}_{f4} = \dot{m}_{f3}$, using the mass balance equation,

$$\dot{m}_{f3}/\dot{m}_a = (w_2 h_{v2} + h_{a2}) - (w_1 h_{v1} + h_{a1}) - (w_2 - w_1) h_{f5})/(h_{f3} - h_{f4})$$
$$= ((0.025\times2565.3+35)-(0.005\times 2538.1+20)$$
$$-(0.02)\times104.9)/(188.45 - 104.9)$$
$$= 0.794 \text{ kg of water per kg dry air.}$$

The optimum work

$$\dot{W}_{opt} = \dot{m}_a\,\psi_{a,1} + \dot{m}_{v1}\,\psi_{v,1} + \dot{m}_{f,3}\,\psi_{f,3} + \dot{m}_{f,5}\,\psi_{f,5} - \dot{m}_a\,\psi_{a,2} - \dot{m}_{v,2}\,\psi_{v,2} - \dot{m}_{f,4}\,\psi_{f,4},\text{ i.e.,}$$

$$w_{opt} = ((h_{a,1} - h_{a,2} - T_0(s_{a,1} - s_{a2})) + w_1\,(h_{v,1} - T_0\,s_{v,1}) - w_2\,(h_{v,2} - T_0\,s_{v,2}) + \dot{m}_{f3}/\dot{m}_a\,(h_{f,3} - h_{f,4} - T_0(s_{f,3} - s_{f,4})) + (w_2 - w_1\,\}\psi_{f,5}.$$

$\psi_{f,5} = h_{f,5} - T_0 s_{f,5} = 104.89 - 298\times\ 0.3674 = -4.595$ kJ/kg

$$w_{opt} = (-15 - 298 \times (1\ \ln(293\div308) - 0.287\ \ln(0.992\div0.961)))$$
$$+ 0.005\ \ (2538.1 - 298\times(8.6672\ -0.461\ \ln\ (0.0082/0.02339)))$$
$$- 0.025 \times (2565.3 - 298\times(8.3531\ -0.461\ \ln\ (0.0394/0.05628)))$$
$$+ 0.794\ (188.45\text{-}104.89 - 298 \times (0.6387\text{-}0.3674)) + 0.02\times(-5.595)$$
$$= (2.594 + 0.005\times(-188.722) - 0.025\times 27.0909 + 0.794\times 2.713 - 0.112$$
$$= 3.015\text{ kJ kg}^{-1}\text{ (of dry air).}$$

Expressing in kJ per kg of water pumped,

$w_{opt} = 3.015/0.791 = 3.811$ kJ kg^{-1} of water pumped to cooling tower.

For air we use reference temperature as 273.15 K, and reference pressure as 1 bar.

$$\psi_2 = 35 - 298\times(1\ \ln(308/273.15) - 0.287\ \ln(0.961/1)) + 0.025(2565.3 - 298\times(8.3531\ -0.461\ \ln\ (0.0394/0.05628))$$
$$= -3.79 + 0.025 \times (27.09) = -3.11\text{ kg of mix at 2 per kg of dry air.}$$

Now,

$\psi_2' = \psi_2 - \psi_0$, where

$$\psi_0 = 25 - 298\times(1\ \ln(298/273.15) - 0.287\ \ln(0.961/1)) + 0.025(2547.2 - 298 \times (8.558\ -0.461\ \ln\ (0.05628/0.03169))$$
$$= -4.60\ + 0.025 \times 75.82 = -2.71\text{ kJ of mixture per kg of dry air.}$$

$$\psi_2' = \psi_2 - \psi_0 = -3.11 + 2.71 = -0.04\text{ kJ kg}^{-1}\text{ dry air.}$$

$$\psi_\infty = 25 - 298\times(1\ \ln(298/273.15) - 0.287\ \ln(0.992/1)) + 0.025(2547.2 - 298\times(8.558\ -0.461\ \ln\ (0.0082/0.03169))$$
$$= -1.8\ + 0.025 \times (-188.9) = -6.523\text{ kJ kg}^{-1}\text{ dry air.}$$

$$w_{ch} = \psi_0 - \psi_\infty = -2.71 + 6.523 = 3.81\text{ kJ kg}^{-1}\text{ dry air.}$$

Remarks

The results are provided per kg of dry air since dry air mass flow remains constant during cooling or heating of wet mixtures.

The optimum work is lost in the power plant. The design could have used (1) a Carnot engine to obtain work from the warm water, (2) run a heat pump to heat the air, (3) vaporize some water, and (4) add vapor to the exiting air stream.

The same method can be used to determine the work required to separate water vapor from air so that the vapor does not condense on the evaporator coils in air conditioning devices (and, consequently, does not reduce the heat transfer rate).

G. INTEGRAL AND DIFFERENTIAL FORMS

1. Integral Form

Consider a control volume for which mass with an availability Θ leaves the control surface of an elemental area dA with velocity a $\vec{v}$. The surface is at a temperature T and the availability $\dot{Q}$ (1- T_0/T) associated with the heat ($\vec{\dot{Q}}'' \bullet d\vec{A}$) leaves the elemental control surface $d\vec{A}$. The appropriate formulation in the integral form is

$$\int_{cv}\rho(e - T_0 s)dV = -\int_{cs}\rho\psi\vec{v}\cdot d\vec{A} - \int_{cs}\dot{q}''(1 - T_0/T)\cdot d\vec{A} - \int_{cv}\dot{w}'''dV - \int_{cv}\dot{i}'''dV. \quad (104)$$

This relation implies that the accumulation rate = -availability leaving the c.v - availability exiting with heat - availability exiting with work – availability loss rate. Note that when the mass, heat and work enter the c.v., the first three terms on the right are positive due to the vectorial notation.

2. Differential Form

The differential form of this relation can be obtained using the Gauss divergence theorem, i.e.,

$$\rho\partial(e - T_0 s)/\partial t = -\vec{\nabla}\cdot\rho\psi\vec{v} - \vec{\nabla}\cdot\dot{q}''(1 - T_0/T) - \dot{w}''' - \dot{i}''', \tag{105}$$

where e denotes the energy and the availability loss per unit volume $\dot{i}''' = T_0\dot{\sigma}'''$. Applying the mass conservation equation and the Fourier heat conduction law $\dot{q}'' = -\lambda\vec{\nabla}T$ to Eq. (105),

$$\rho\partial(e - T_0 s)/\partial t + \rho\vec{v}\cdot\vec{\nabla}\psi = \vec{\nabla}\cdot((\lambda\vec{\nabla}T)(1 - T_0/T)) - \dot{w}''' - \dot{i}''' \tag{106}$$

the availability equations can also be obtained by coupling the entropy balance and energy conservation equations in the manner ((availability balance) = (energy conservation) – T_0(entropy balance)). Mixing and velocity, temperature, and species concentration gradients cause a loss in availability.

3. Some Applications

Using Eq. (90), the availability of the hot gases within a boiler can be mapped if the local state data is available. Locations where the availability loss rate is large can be identified.

At steady state, neglecting both the kinetic and potential energies, e = h and Eq. (106) simplifies to the form

$$\rho\vec{v}\cdot(\vec{\nabla}h - T_0\vec{\nabla}s) = \vec{\nabla}\cdot((\lambda\vec{\nabla}T)(1 - T_0/T)) - \dot{w}''' - \dot{i}''' \tag{107}$$

In the absence of internal temperature gradients, for a reversible process

$$\rho\vec{v}\cdot\vec{\nabla}\psi = w'''_{opt}, \text{ i.e.,} \tag{108}$$

the local value of ψ is a measure of reversible work for a steady state adiabatic system containing negligible kinetic and potential energy.

For an isothermal turbine performing reversible work, Eq. (90) may be expressed as

$$\rho\partial a/\partial t + \rho\vec{v}\cdot\vec{\nabla}g = -w''', \tag{109}$$

where a = u – Ts, and g = h – Ts. Therefore, changes in the Helmholtz and Gibbs functions are measures of work in reversible systems. In a closed system $\rho\vec{v}\cdot\vec{\nabla}g = 0$ so that $\partial a/\partial t = \dot{w}'''/\rho = \dot{w}$ which is the work rate per unit mass. Since, the system is internally reversible, i.e., there are no internal gradients, i.e., $A_2 - A_1 = -W$.

On the other hand in a steady open reversible system

$$\rho\vec{v}\cdot(\nabla g) = -\dot{w}'''. \tag{110}$$

The advective term is absent in nonflow systems and, consequently, Eq. (106) simplifies to the form

$$\rho\partial(e - T_0 s)/\partial t = \nabla\cdot(\lambda\nabla T(1 - T_0/T)) - \dot{w}''' - \dot{i}'''. \tag{111}$$

For instance, if a large pot of coffee is cooled such that it is internally isothermal at all instants, this relation can be expressed as

$$\rho\partial u/\partial t - T_0\partial s/\partial t = \rho c(1 - T_0/T)\partial T/\partial t = -\dot{i}''', \tag{112}$$

since for this case de = du = c dT, and ds = c dT/T. The variation of the coffee temperature with time is related to the availability loss rate per unit volume. Similar analyses are valid for the cooling of solids, such as hot steel billets and associated loss in availabilities.

---

n. Example 14

A turbine blade of length L is subjected to hot gases at a temperature T_∞. The blade surface temperature T_w is maintained by wall cooling. The blade can be simulated as a flat plate subjected to laminar flow of hot gases at T_∞ so that a boundary layer $\delta(x)$ grows over the plate as shown in Figure 22. At steady state, the temperature profile within the boundary layer is given by the expression

$$\Theta = 1 - 2\xi - + 2\xi^3 - \xi^4, \text{ where} \quad (A)$$

$$\xi = y/\delta(x), \quad (B)$$

where $\delta(x) = C\,x^{1/2}$, C is a constant and

$$\Theta = (T - T_\infty)/(T_w - T_\infty). \quad (C)$$

Using the differential form of the generalized availability balance, obtain an expression for the availability loss rate per unit volume at the wall.

Obtain an expression for the availability loss rate per unit volume in the free stream.

If T_∞ = 1300 K, T_w = 900 K, the thermal conductivity $\lambda = 70\times10^{-6}$ kW m^{-2} K^{-1}, and $\delta(x)$ = 0.005 m, determine the availability loss rate per unit volume at the wall.

What is the availability transfer rate associated with the heat flow at the wall?

Solution

For a steady non–work–producing two–dimensional process, the simplified differential form of availability balance equation is

$$\rho v_x \frac{\partial\psi}{\partial x} + \rho v_y \frac{\partial\psi}{\partial y} = \frac{\partial}{\partial y}\left[\left\{\lambda\frac{\partial T}{\partial y}\right\}\left\{1 - \frac{T_0}{V}\right\}\right] - i'''. \quad (D)$$

At the wall $v_x = v_y = 0$ so that

$$i''' = \frac{\partial(\lambda(\partial T/\partial y)(1 - T_0/T))}{\partial y} = \lambda\frac{\partial^2 T}{\partial y^2}(1 - \frac{T_0}{T})) + \lambda\frac{T_0}{T^2}\left(\frac{\partial T}{\partial y}\right)^2. \quad (E)$$

At the blade wall $T = T_w$ so that

$$i''' = \lambda\left.\frac{\partial^2 T}{\partial y^2}\right|_{y=0}(1 - \frac{T_0}{T_w}) + \lambda\frac{T_0}{T_w^2}\left.\left(\frac{\partial T}{\partial y}\right)^2\right|_{y=0}. \quad (F)$$

Using Eqs. (B) and (C)

$$(\partial T/\partial y)_x = (dT/d\xi)(\partial\xi/\partial y)_x = d\Theta/d\xi\,(T_w - T_\infty)/\delta(x). \quad (G)$$

At the wall

$$(\partial T/\partial y)_x|_{y=0} = d\Theta/d\xi\,|_{\xi=0}\,(T_w - T_\infty)/\delta(x). \quad (H)$$

From Eq. (A),

$$d\Theta/d\xi = -2 + 6\xi^2 - 4\xi^3, \text{ and} \quad (I)$$

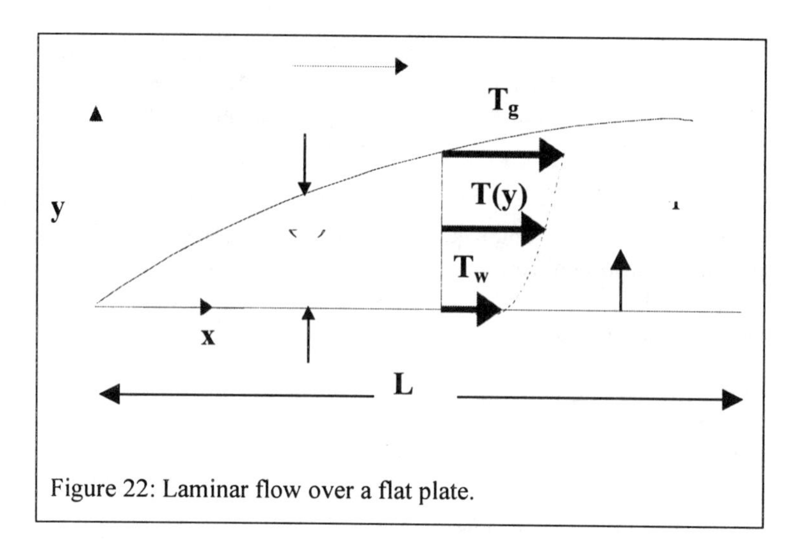

Figure 22: Laminar flow over a flat plate.

$d\Theta/d\xi(\xi=0) = -2.$

Using this result in Eq. (H),

$$(\partial T/\partial y)_x|_{y=0} = -2\,(T_w - T_\infty)/\delta(x). \quad \text{(J)}$$

Similarly,

$$(\partial^2 T/\partial y^2)_x = (d/d\xi)((d\Theta/d\xi)(T_w - T_\infty)/\delta(x)) = (d^2\Theta/d\xi^2)(T_w - T_\infty)/(\delta\,(x))^2. \quad \text{(K)}$$

Since $d^2\Theta/d\xi^2(\xi=0) = 0$,

$$(\partial^2 T/\partial y^2)_x|_{y=0} = 0. \quad \text{(L)}$$

Using Eqs. (F), (J), and (L),

$$i'''_{y=0} = \lambda(T_0(T_w - T_\infty)^2 / T_w^2)(4/(\delta(x))^2). \quad \text{(M)}$$

The availability loss rate is always a positive quantity. In the free stream, $\partial T/\partial y = 0$, and $\partial^2 T/\partial y^2 = 0$. Therefore for the free stream

$$i''' = 0 \quad \text{(N)}$$

Using Eq. (M),

$$i''' = 70\times10^{-6} \times 298 \times ((900 - 1300)^2 \div 900^2) \times (4 \div 0.005)^2 = 2637\ \text{kW m}^{-3}.$$

For forced convection $\delta(x) = C\,x^{1/2}$ where C denotes a constant so that $\delta = C^2x$. Therefore, using Eq. (M), the *dimensionless* availability loss rate is

$$i'''(0)\,C^2x/(\lambda\,T_0) = 4(1- T_\infty/T_w)^2 = 4(1- 1300 \div 900)^2 = 0.79.$$

The local availability flow rate into the plate due to the heat transfer is

$$\dot q''_{y=0}\,(1-T_0/T_w) = -\lambda(\partial T/\partial y)_{y=0}(1-T_0/T_w)$$
$$= -\lambda\, d\Theta/d\xi(\xi=0)\,(T_w - T_\infty)\,(1-T_0/T_w)/\delta(x)$$
$$= 70\times10^{-6} \times 2 \times (-400) \times (1- 298 \div 900) \div 0.005 = -7.492\ \text{kW m}^{-2}.$$

Remarks:

Assuming a 2.7m^2 of total blade area, there is a net 20 kW loss in availability. This transfer should be compared with the work gain due to the higher operational temperature. For instance, $(1-T_0/T_g)$ is the availability transferred from the gas at T_g in the absence of

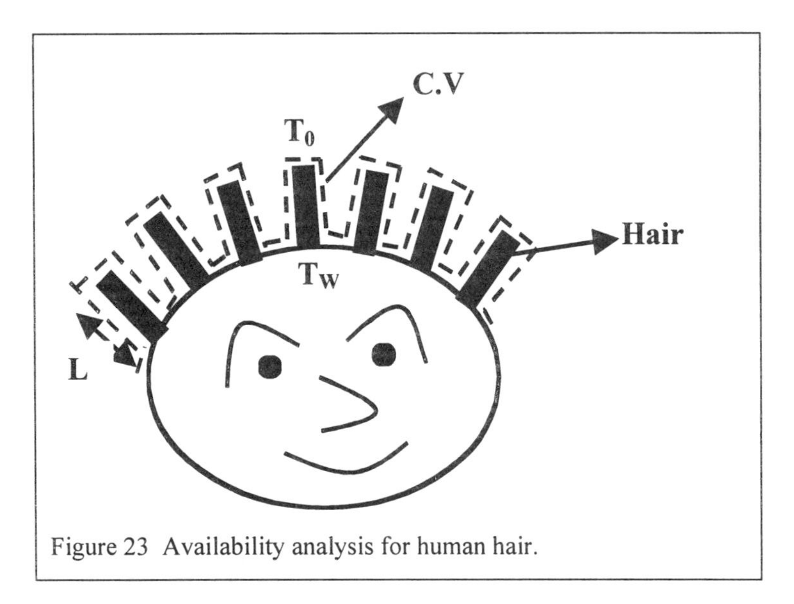

Figure 23 Availability analysis for human hair.

cooling. This availability is increased if cooling is used since operating gas temperature is increased to $T_g' = T_g + \Delta T_g$ and, consequently, the availability increase is $((1-T_0/T_g')-(1-T_0/T_g)) \approx (\Delta T_g T_0/T^2_g)$ per unit amount of heat transferred from the hot gases. The availability loss due to the blade cooling is $2A_b\lambda(T_w-T_g)(1-T_0/T_w)/\delta(x)$, where A_b denotes the blade area. If the availability gain is to be larger than the loss, then $|\dot{Q}''|\Delta T_g T_0/T^2_g > (2\lambda(T_w-T_g)(1-T_0/T_w)/\delta(x))$, where $|\dot{Q}''|$ is the heat or energy loss from the hot gases per unit blade area.

---

o. *Example 15*

Determine the irreversibility loss for a human hair of length L. The hair has a surface temperature of T_w at one end and an adiabatic tip at the other end, and is exposed to an ambient temperature T_0. The literature informs us that the heat transfer rate from the base of a fin of length L,

$$\dot{Q}_w = \lambda A_f(T_w - T_0) \tanh \alpha L,$$

where $\alpha = (h_H C/\lambda A_f)^{1/2}$, C denotes the hair circumference, A_f is the cross-sectional area of the fin (hair), and h_H is the convective heat transfer coefficient. Use the values $\rho = 165$ kg m^{-3}, $\lambda = 0.036$ W m^{-1} K^{-1}, the hair diameter d = 0.1 mm, and $h_H = 0.1$ W m^{-2} K^{-1}. Note that $C/A_f = 4/d$ for a circular cross section.

What is the maximum possible power that can be developed using a "hot head"?

Solution

Selecting the control volume around a single hair,

$$I = \dot{Q}_w(1 - T_0/T_w) - Q_0(1 - T_0/T_0) = \dot{Q}_w(1 - T_0/T_w),$$

where $T_w = 310.2$ K (average body temperature). Now,

$$\dot{Q}_w = \lambda A_f(T_w - T_0) \tanh \alpha L, \text{ where } L = 10^{-2} \text{ m},$$

$\alpha = (h_H C/\lambda A_f)^{1/2} = (0.1 \times 4 \div (0.0001 \times 0.036))^{1/2} = 333.3$ m^{-1}. Therefore,

$$\dot{Q}_w = 0.036\times(\pi(1\times10^{-4})^2\div4)\ (310.2 - 298) \times \tanh(333.3\times0.1\times10^{-4})$$

$$= 0.036\times3.1427\times0.1^2\div4 \times 10^{-4}\ (310.2-298) \times 1 = 3.7\times10^{-7} \text{ W}$$

$$I = 3.45\times10^{-7} \times (1-298\div310.2) = 1.36\times10^{-8} \text{ W}$$

The maximum possible work rate equals the actual work rate plus the irreversibility rate. Since in this case no work is done, the irreversibility is the maximum possible work.

Remarks

Shorter hair will have a lower heat loss.

In order to reduce this irreversibility, you can couple a Carnot engine to each strand of hair and use the work so obtained to propel yourself!

Since the human body is warm, body heat loss through our skin occurs at the rate of about 1 kW. Again, a Carnot engine may be used to extract work. Since the Carnot efficiency will be 1 – 298/310 = 0.039, the power developed would be roughly 3.9 W which is of the order of power of a night lamp.

H. SUMMARY

We have discussed *mass* and *energy* conservation in Chapter 2, the *entropy balance* equation in Chapter 3, and the *availability balance* equation in this chapter. Most thermodynamic systems can be efficiently designed with two conservations and two balance concepts. One can perform a coupled thermodynamic and economics analysis which provides a cost and efficiency basis to account for the lost energy in each component of a larger system. However, in order to use these relations and perform analyses, thermodynamic properties (e.g., h, s, u, v, etc.) must be known. These can either be directly obtained from experiments or by using real gas state equations or ideal gas properties which are considered next in Chapters 6 and 7. Properties of matter that consist of a mixture of gases are discussed in Chapter 8. Chapter 9 will present a stability analysis which will explain the formation of multiples phases of a single component, while Chapter 10 will present the enthalpy and entropy relations for reacting species. Chapter 5 considers the subject from the perspective of thermodynamic postulates without using the more conventional laws that we have discussed thus far in Chapters 2 and 3. However, Chapter 5 is not essential for following the material presented in subsequent chapters.

Chapter 5

5. POSTULATORY (GIBBSIAN) THERMODYNAMICS

A. INTRODUCTION

In the previous chapters we discussed the thermodynamics laws by employing a classical approach. In this chapter we will discuss the subject using a set of postulates or rules, fundamental state equations, and other mathematical tools such as Legendre transforms. We will first establish the classical rationale behind such an approach and relate it to some postulates (without invoking any laws). Thereafter, we will introduce the Legendre transform using which it is possible to transfer an equation from one coordinate system to others (e.g., from an entropy–based coordinate to a temperature–based coordinate). Next, we will relate the energy to work. We will discuss the postulates in a mathematical context, present the entropy and energy fundamental equations, and describe intensive and extensive properties by using the properties of homogeneous functions. Finally, we will derive the Gibbs–Duhem relation using fundamental and Euler equations.

B. CLASSICAL RATIONALE FOR POSTULATORY APPROACH

We have seen that a Stable Equilibrium State (SES) is achieved when the entropy reaches a maximum value for fixed values of U, V and m (or for fixed number of moles N_1, N_2, ...). The internal energy of an open system that exchanges mass with its surroundings is represented by the relation

$$U = U(S, V, N_1, N_2, N_3, ...), \tag{1}$$

which is also known as the energy fundamental equation. The internal energy is a single valued function of S,V, N_1, N_2, N_3, ..., since there is a single stable equilibrium state for a specified set of conditions. Upon differentiating Eq. (1) we obtain the relation

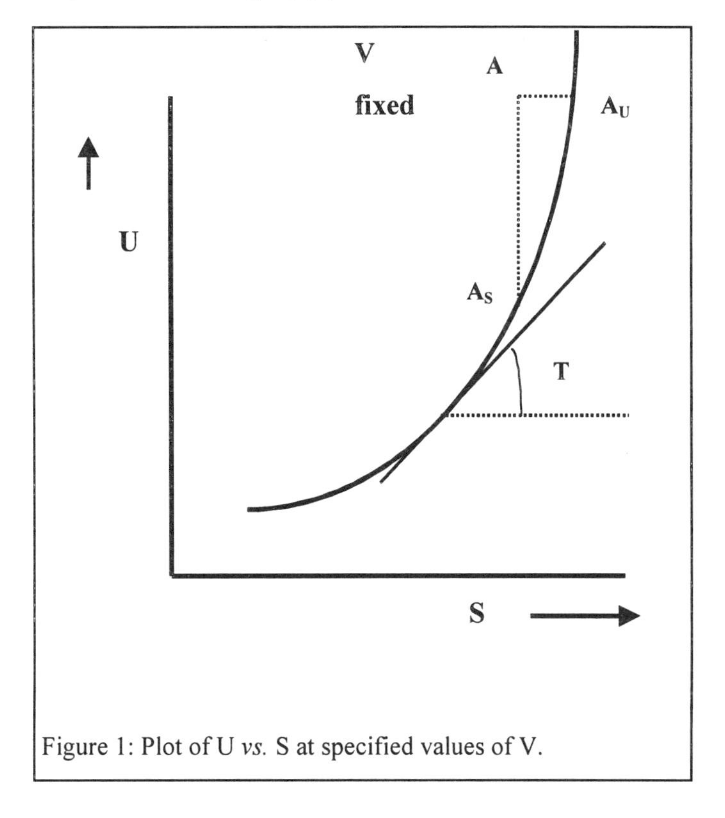

Figure 1: Plot of U *vs.* S at specified values of V.

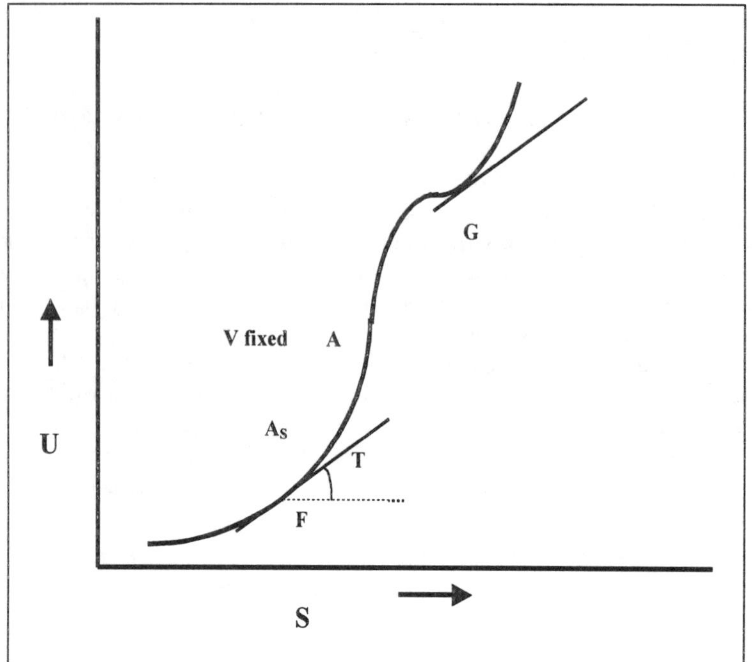

Figure 2: A plot of U *vs*. S plot at specified values of V showing similar temperatures (slopes) at two different values of S.

$$dU = TdS - PdV + \Sigma\mu_k dN_k, \text{ where} \tag{2}$$

$\partial U/\partial S = F_T = T$ is the affinity or force driving heat transfer (cf. Figure 1), $\partial U/\partial V = F_P = -P$ is the affinity driving mechanical work, and $\partial U/\partial N_k = F_{m,k} = \mu_k$ is the affinity that drives mass transfer (say, during a chemical reaction or a phase transition). In general, the partial derivative $\partial U/\partial \xi_j$ represents the force driving a parameter ξ_j. For instance, if $\xi_i = N_i$, i.e., the number of moles N_i of the i–th species in the system, then $\partial U/\partial N_i = \mu_i$ represents the chemical potential of that species.

Rewriting Eq.(2),

$$dS = (1/T)dU + (P/T)dV - \Sigma(\mu_i/T)dN_i, \text{ or } S = S(U, V, N_1, N_2, ...). \tag{3}$$

Equation (3) is called the entropy fundamental equation. It implies that the equilibrium states are described by the extensive set of properties $(U, V, N_1, ..., N_n)$, which is also known as Postulate I.

We have seen that the entropy is a single valued function (for prescribed values of U, V, and m), since there is a unique stable equilibrium state. The fundamental equation (cf. Eq. (3)) is written in terms of extensive parameters. Each sub–system in a composite system can be described by the fundamental equation. However, the equation cannot be applied to the composite system itself. (This is also called Postulate II.) Equations (1) and (3) are, respectively, the energy and entropy representations of the fundamental equation, and these are valid only in a positive coordinate system in which the values of the variables $U, V, N_1, ..., S > 0$. We can define $(\partial S/\partial U)_{V, N_i} = F_T = 1/T$, $T(\partial S/\partial V)_{U, N_i} = F_P = P$, etc. for any system that changes from one equilibrium state to another. This definition fails for systems that do not exist in equilibrium states.

It is possible to generalize Eq. (3) in the rate form as

$$dS/dt = \Sigma(dS/dx_k)(dx_k/dt) = \Sigma F_k\, J_k. \tag{4}$$

where x_k 's are U, V, N_1, N_2 etc.. For a single component system, Eq. (4) yields S = S(U,V,N) or S = S(U,V,M) so that s = s(u,v), i.e.,

$$1/T = (\partial s/\partial u)_v, \text{ and } P/T = (\partial s/\partial v)_u. \tag{5}$$

These relations are valid only in the octant where (u,v > 0), and 1/T and P/T represent tangents to the S–surface in the (u,v) plane. It is possible to have identical values of P/T and 1/T for various combinations of values of s, u and v. Examples of this include the saturated liquid and saturated vapor states for a specified pressure. Even though these states correspond to identical P and T, the values of s, u and v differ for the two states (cf. Figure 2).

1. Simple Compressible Substance

Since Eq.(1) is a first order homogeneous equation, the application of Euler equation (cf. Chapter 1) yields

$$S\frac{\partial U}{\partial S} + V\frac{\partial U}{\partial V} + \sum N_i \frac{\partial U}{\partial N_i} = U. \tag{6}$$

Using the partial derivatives described by Eqs.(5),

$$TS - PV + \Sigma\mu_i N_i = U. \tag{7}$$

The total differentiation of Eq. (7) and use of Eq.(2) yields the Gibbs–Duhem (G–D) equation, namely,

$$SdT - VdP + \Sigma N_i\, d\mu_i = 0 \tag{8}$$

Equation (8) gives the intensive equation of state and it is apparent that

$$T = T(P, \mu_1, \mu_2, \ldots). \tag{9}$$

Dividing Eq.(8) by N and solving for dμ (for, say, species 1),

$$x_1\, d\mu_1 = -(S/N)\, dT + (V/N)dP - x_2\, d\mu_2 - x_3\, d\mu_3 - \ldots - x_n\, d\mu_n. \tag{10}$$

Since $x_1 = 1 - x_2 - x_3 - \ldots - x_n$, this results in an intensive equation of state that is the zeroth order homogeneous function

$$\mu_1 = f(T, P, x_2, x_3, x_4, \ldots, x_n). \tag{11}$$

As before, we see that (n+1) intensive properties describe the intensive state of an n–component simple compressible substance.

C. LEGENDRE TRANSFORM

1. Simple Legendre Transform

Consider the relation

$$y = 2x^2 + 5 \tag{12}$$

We call y as the basis function, $y = y^{(0)}$, i.e.,

$$y^{(0)} = 2x^2 + 5. \tag{13}$$

We can use a series of points to draw the curve ABCD (through point geometry) as shown in Figure 3. The slope

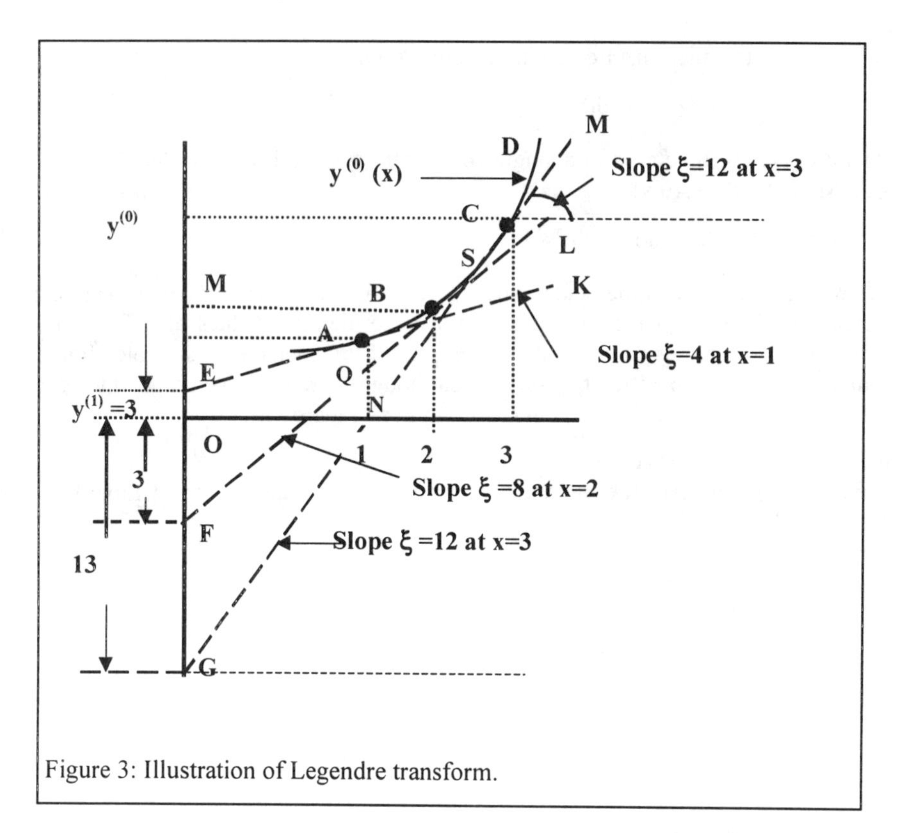

Figure 3: Illustration of Legendre transform.

$$y'(x) = \partial y^{(0)}/\partial x = \xi = 4x, \text{ i.e.,} \tag{14}$$

$$y^{(0)} = (1/8)\xi^2 + 5. \tag{15}$$

Can we draw the same curve $y^{(0)}$ vs x in the $y^{(0)}$ - x plane which we drew with Eq. (13), but just by using Eq. (15)? Given slope, we cannot place $y^{(0)}$ in $y^{(0)}$-x plane. Equation (15) is a differential equation, and by replacing x with the slope information regarding the value of x is lost. From Eq. (15),

$$y^{(0)} = f(\xi), \tag{16}$$

which is a relation that does not provide the same information as the expression

$$y^{(0)} = f(x). \tag{17}$$

Therefore, we require a different function $y^{(1)}(\xi)$ that describes the curve $y^{(0)}(x)$. In the context of Eq. (14) $\xi = 4$at $x = 1$, $\xi = 8$ at $x = 2$, and so on. A tangent drawn at point A ($x = 1$) intersects the $y^{(0)}$ axis at point E ($y^{(0)} = 3$) (Figure 3) and one at point B ($x=2$) intersects the $y^{(0)}$ axis at point F ($y^{(0)} = -3$). The intercepts will be denoted as $y^{(1)}$. The locus tangent to all the lines is the curve $y^{(0)}$. Thus, we can construct the curve $y^{(0)}(x)$. Hence, a curve may be drawn either along points (point geometry, Eq. (12)) or with a series of slopes or lines (line geometry). Line geometry requires a functional relation between the intercept $y^{(1)}(\xi)$ and ξ to draw the curve as $y^{(0)}(x)$ while point geometry requires a func-

x	0	1	2	3
$y^{(0)}$	5	7	13	23
ξ	0	4	8	12
$y^{(1)}$	5	3	–3	–13

Table 1: Values of x, $y^{(0)}$, ξ and $y^{(1)}$ for the basis function $y^{(0)} = 2x + 5$.

tional relation between the ordinate $y^{(0)}$ and x. The line geometry relation is

$$y^{(1)} = y^{(0)} - \xi\, x. \quad (18)$$

Replacing $y^{(0)}$ in Eq. (18) using Eqs. (15) and (14), we obtain the relation

$$y^{(1)}(\xi) = 5 - \xi^{2}/4. \quad (19)$$

Equations (18) and (19) provide the same information as does Eq. (12). The functional relation $y^{(1)}(\xi)$ is known as the first Legendre transform of $y^{(0)}$ with respect to x. For the above example (cf. Eqs. (12)–(19)), Table 1 presents values of x, $y^{(0)}$, ξ and $y^{(1)}$ for the basis function ($y^{(0)} = 2x^2 + 5$).

It is apparent from Eq. (18) or (19) that the first Legendre transform slope of ξ is an independent variable and $y^{(1)}$ is a dependent variable. In the context of the basis function x is the independent variable and $y^{(0)}$ is the dependent variable. Differentiating Eq.(18)

$$dy^{(1)} = dy^{(0)} - dx\,\xi - x\, d\xi\, y^{(1)}. \quad (20)$$

From Eq. (17),

$$dy^{(0)} = \xi\, dx, \text{ i.e.,} \quad (21)$$

Using the result in Eq.(20)

$$dy^{(1)} = \xi\, dx - dx\,\xi - x\, d\xi = -x\, d\xi, \text{ or } x = -\partial y^{(1)}/\partial\xi. \quad (22)$$

a. Relevance to Thermodynamics

We cannot measure the entropy directly, so that it must be derived from other measurements. We can use the relation U = U(S) and measure isometric temperatures, which represent the slope $\xi = (\partial U/\partial S)_V$. Thereafter, the function $A = y^{(1)}(\xi)$ can be generated to construct the U(S) curve.

---

a. Example 1

The following measurements are made of $u^{(1)}$ and $T = (\partial u/\partial s)_v$.

T, K	275	280	290	300	320	340
$u^{(1)}$, kJ kg^{-1}	–0.030	–0.4	–2.3	–5.5	–16.2	–32.0

Can you plot $u^{(0)}$ vs s ; It is known for an incompressible liquid that

$$y^{(0)} = u^{(0)} = 273\, c\, (\exp(s/c) - 1), \quad (A)$$

where the specific heat c = 4.184 kJ kg^{-1} K^{-1}. Also apply the first Legendre transform for Eq. (A) to obtain an expression for $u^{(1)}(T)$.

Solution

At specified values of ξ the intercepts $u^{(1)}$ are known. Hence, the $u^{(0)}(s)$ curve can be constructed from the above data.

b) Using the first Legendre transform,

$$u^{(0)} = u^{(1)} - (\partial u/\partial s)_{v,s} = u^{(1)} - Ts. \quad (B)$$

Differentiating Eq. (A)

$$(\partial u^{0}/\partial s)_v = 273 \exp(s/c) = T, \text{ or } s/c = \ln (T/273). \quad (C)$$

Using Eq. (C) in (A)

$$u^{(0)} = cT - 273\ c, \text{ and } u^{(1)} = c(\xi - 273) - \xi c \ln(\xi/273), \text{ where } \xi = T. \quad (D)$$

2. Generalized Legendre Transform

Consider the generalized expression for a basis function involving more than one variable, i.e.,

$$y^{(0)} = y^{(0)}(x_1, x_2, x_3, ..., x_n) \text{ so that } y^{(1)} = y^{(0)} - \xi_1 x_1. \quad (23)$$

The function $y^{(1)} = y^{(1)}(\xi_1)$ for prescribed $x_2, ..., x_n$, can also be used to describe $Y^{(0)}$. The first Legendre transform with respect to x_1 is expressed in the form

$$y_{x_1}{}^{(1)} = y^{(1)}(\xi_1, x_2, ..., x_n) \quad (24)$$

If $x_1, x_3, ..., x_n$ are held constant, the first Legendre transform with respect to x_2

$$Y_{x_2}{}^{(1)} = Y^{(1)}(x_1, \xi_2, x_3, ..., x_n). \quad (25)$$

Since

$$\partial y^{(1)}/\partial x_2 = \partial y^{(0)}/\partial x_2 - 0 = \xi_2, \quad (26)$$

the second Legendre transform is

$$y^{(2)} = y^{(1)} - (\partial y^{(1)}/\partial x_2)x_2 = y^{(1)} - \xi_2 x_2, \text{ i.e.,} \quad (27)$$

$$y^{(2)} = y^{(1)} - \xi_2 x_2, \text{ or} \quad (28)$$

$$y^{(2)}(\xi_1, \xi_2, x_3, x_4...) = y^{(0)}(x_1, x_2, x_3..) - \xi_1 x_1 - \xi_2 x_2. \quad (29)$$

Generalizing Eq. (29) to the m–th Legendre transform (where $m<n$),

$$y^{(m)} = y^{(0)} - \Sigma_{i=1,m}\, \xi_i x_i, \text{ and} \quad (30)$$

the n–th Legendre transform

$$y^{(n)} = y^{(0)} - \Sigma_{i=1,n}\, \xi_i x_i. \quad (31)$$

In the context of the second Legendre transform,

$$dy^{(2)} = dy^{(0)} - d\xi_1 dx_1 - \xi_1 dx_1 - d\xi_2 x_2 - \xi_2 dx_2 =$$

$$\xi_1 dx_1 + \xi_2 dx_2 + \xi_3 dx_3 + ... - d\xi_1 x_1 - \xi_1 dx_1 - d\xi_2 x_2 - \xi_2 dx_2, \text{ i.e.,}$$

$$dy^{(2)} = \xi_3 dx_3 + \xi_4 dx_4 + \ldots - d\xi_1 x_1 - d\xi_2 x_2, \quad (32)$$

for m–th Legendre transform

$$dy^{(m)} = \Sigma_{j=m+j,n}\, \xi_j dx_j - \Sigma_{j=1,m}\, d\xi_j x_j,\ m < n, \quad (33)$$

and the n–th Legendre transform

$$dy^{(n)} = -\Sigma_{j=1,n}\, d\xi_j x_j. \quad (34)$$

If, in the context of Eq. (32) $\xi_2, x_3, x_4, \ldots, x_n$ are held constant, then

$$(dy^{(2)})_{\xi 2, x3, x4...} = -d\xi_1\, x_1, \text{ i.e.,} \quad (35)$$

$$(\partial y^{(2)}/\partial \xi_1)_{\xi_2, x_3, x_4, \ldots} = -x_1. \tag{36}$$

Similarly,

$$(\partial y^{(1)}/\partial \xi_1)_{x_2, x_3, x_4, \ldots} = -x_1.$$

Likewise,

$$(dy^{(2)})_{\xi_1, \xi_2, x_4, x_5, \ldots} = \xi_3 dx_3, \text{ i.e.,} \tag{37}$$

$$(\partial y^{(2)}/\partial x_3)_{\xi_1, \xi_2, x_4, x_5, \ldots} = \xi_3 = (\partial y^{(1)}/\partial x_3)_{\xi_1, x_2, x_4, \ldots} = (\partial y^{(0)}/\partial x_3)_{x_1, x_2, x_4, \ldots}, \text{ and} \tag{38}$$

$$(\partial y^{(2)}/\partial x_4)_{\xi_1, \xi_2, x_3, x_5, \ldots} = \xi_4 = (\partial y^{(1)}/\partial x_4)_{\xi_1, x_2, x_3, x_5, \ldots} = (\partial y^{(0)}/\partial x_4)_{x_1, x_2, x_3, x_5, \ldots}. \tag{39}$$

If the variables in the basis function are extensive, then the Euler equation must be satisfied so that

$$\Sigma_{i=1,n}\, x_i\, \partial y^{(0)}/\partial x_i = \Sigma_{i=1,n}\, \xi_i x_i = y^{(0)}. \tag{40}$$

Since, $y^{(n)} = y^{(0)} - \Sigma_{i=1,n}\, \xi_i x_i$, $y^{(n)} = 0$, and

$$\Sigma_{i=1,n}\, x_i\, d\xi_i = 0, \tag{41}$$

which is known as the Gibbs–Duhem relation

---

b. *Example 2*

Consider a basis function involving five variables. Obtain the first and third Legendre transforms.

Solution

The basis function is

$$y^{(0)}(x_1, x_2, \ldots. x_5), \text{ i.e.,} \tag{A}$$

$$y^{(3)} = y^{(0)} - \xi_1 x_1 - \xi_2 x_2 - \xi_3 x_3 \text{ (cf. Eq. 29),} \tag{B}$$

$$dy^{(3)} = -(\xi_4 dx_4 + \xi_5 dx_5) - (x_1 d\xi_1 + x_2 d\xi_2 + x_3 d\xi_3) \text{ (cf. Eq. 32),} \tag{C}$$

$$dy^{(2)} = -(\xi_3 dx_3 + \xi_4 dx_4 + \xi_5 dx_5) - (x_1 d\xi_1 + x_2 d\xi_2), \tag{D}$$

$$dy^{(1)} = -(\xi_2 dx_2 + \xi_3 dx_3 + \xi_4 dx_4 + \xi_5 dx_5) - (x_1 d\xi_1), \text{ and} \tag{E}$$

$$dy^{(0)} = -(\xi^1 dx^1 + \xi_2 dx_2 + \xi_3 dx_3 + \xi_4 dx_4 + \xi_5 dx_5). \tag{F}$$

Employing Eqs. (C)–(E),

$$(\partial y^{(3)}/\partial \xi_1)_{\xi_2, \xi_3, x_4, x_5} = (\partial y^{(2)}/\partial \xi_1)_{\xi_2, x_3, x_4, x_5} = (\partial y^{(1)}/\partial \xi_1)_{x_2, x_3, x_4, x_5} = -x_1, \text{ where}$$

$$y^{(1)} = f(\xi_1, x_2, x_3, \ldots), \text{ i.e., } \partial y^{(1)}/\partial \xi_1 = -x_1.$$

Differentiating with respect to x_2,

$$\partial y^{(1)}/\partial x_2 = y_2^{(1)} = \xi_2, \text{ and . } (\partial/\partial \xi_2)(\partial y^{(1)}) = 1.$$

More generally,

$$\partial y^{(m)}/\partial x_k = y_k{}^{(m)} = \xi_k,\ k > m, \text{ and } \partial/\partial\xi_l(y_k{}^{(m)}) = \partial\xi_k/\partial\xi_l.$$

For instance,

$$(\partial y^{(3)}/\partial\xi_2)_{\xi_1,\xi_3,x_4,x_5} = (\partial y^{(2)}/\partial\xi_2)_{\xi_1,\xi_3,x_4,x_5} = -x_2, \text{ and}$$

$$(\partial y^{(3)}/\partial\xi_3)_{\xi_1,\xi_3,x_4,x_5} = -x_3.$$

More generally,

$$\partial y^{(m)}/\partial\xi_k = \partial y^{(m-1)}/\partial\xi_k = \ldots = \partial y^{(k)}/d\xi_k = -x_k,\ k < m, \text{ and } m \neq n.$$

Employing Eqs. (C)–(F),

$$(\partial y^{(3)}/\partial x_5)_{\xi_1,\xi_2,\xi_3,x_4} = (\partial y^{(2)}/\partial x_5)_{\xi_1,\xi_2,x_3,x_4} = (\partial y^{(1)}/\partial x_5)_{\xi_1,x_2,x_3,x_4} = (\partial y^{(0)}/\partial x_5)_{x_1,x_2,x_3,x_4} = -\xi_5,$$

$$(\partial y^{(3)}/\partial x_4)_{\xi_1,\xi_2,\xi_3,x_5} = (\partial y^{(2)}/\partial x_4)_{\xi_1,\xi_2,x_3,x_5} = (\partial y^{(1)}/\partial x_4)_{\xi_1,x_2,x_3,x_5} = (\partial y^{(0)}/\partial x_4)_{x_1,x_2,x_3,x_5} = -\xi_4,$$

$$(\partial y^{(2)}/\partial x_5)_{\xi_1,\xi_2,x_3,x_4} = (\partial y^{(1)}/\partial x_5)_{\xi_1,x_2,x_3,x_4} = (\partial y^{(0)}/\partial x_5)_{x_1,x_2,x_3,x_4} = -\xi_5,$$

$$(\partial y^{(2)}/\partial x_4)_{\xi_1,\xi_2,x_3,x_5} = (\partial y^{(1)}/\partial x_4)_{\xi_1,x_2,x_3,x_5} = (\partial y^{(0)}/\partial x_4)_{x_1,x_2,x_3,x_5} = -\xi_4,$$

$$(\partial y^{(2)}/\partial x_3)_{\xi_1,\xi_2,x_4,x_5} = (\partial y^{(1)}/\partial x_3)_{\xi_1,x_2,x_4,x_5} = (\partial y^{(0)}/\partial x_3)_{x_1,x_2,x_4,x_5} = -\xi_3,$$

$$(\partial y^{(1)}/\partial x_5)_{\xi_1,x_2,x_3,x_4} = (\partial y^{(0)}/\partial x_5)_{x_1,x_2,x_3,x_4} = -\xi_5,$$

$$(\partial y^{(1)}/\partial x_4)_{\xi_1,x_2,x_3,x_5} = (\partial y^{(0)}/\partial x_4)_{x_1,x_2,x_3,x_5} = -\xi_4,$$

$$(\partial y^{(1)}/\partial x_3)_{\xi_1,x_2,x_4,x_5} = (\partial y^{(0)}/\partial x_3)_{x_1,x_2,x_4,x_5} = -\xi_3,$$

$$(\partial y^{(1)}/\partial x_2)_{\xi_1,x_3,x_4,x_5} = (\partial y^{(0)}/\partial x_2)_{x_1,x_3,x_4,x_5} = -\xi_2,$$

$$(\partial y^{(0)}/\partial x_5)_{x_1,x_2,x_3,x_4} = -\xi_5,$$

$$(\partial y^{(0)}/\partial x_4)_{x_1,x_2,x_3,x_5} = -\xi_4,$$

$$(\partial y^{(0)}/\partial x_3)_{x_1,x_2,x_4,x_5} = -\xi_3,$$

$$(\partial y^{(0)}/\partial x_2)_{x_1,x_3,x_4,x_5} = -\xi_2, \text{ and}$$

$$(\partial y^{(0)}/\partial x_1)_{x_2,x_3,x_4,x_5} = -\xi_1.$$

In general,

$$\partial y^{(m)}/\partial x_k = \partial y^{(m-1)}/\partial x_k \ldots \partial y^{(0)}/\partial x_k = -\xi_k,\ m < k, \text{ and } m \neq n, \text{ and}$$

$$y_{11}{}^{(0)} = \partial^2 y^{(0)}/\partial x_1{}^2,\ y_{1k}{}^{(0)} = (\partial/\partial x_1)(\partial y^{(0)}/\partial x_k).$$

---

3. Application of Legendre Transform

We will now relate the Legendre transform methodology to various thermodynamic relations.

---

c. *Example 3*

Obtain the first Legendre transform with respect to S, the second Legendre transform with respect to S and V, and the (n+2) Legendre transform for the basis function

$$U = U(S, V, N_1, N_2, ..N_n). \quad \text{(A)}$$

Show that $G = \Sigma\mu_k N_k$.

Solution

The first Legendre transform

$$U^{(1)} = U^{(0)} - S\,(\partial U^{(0)}/\partial S)_{V,N_k}, \text{ i.e.,} \quad \text{(C)}$$

$$U^{(1)} = U^{(0)} - TS. \quad \text{(D)}$$

where $T = \partial U^{(0)}/\partial S$ is the thermodynamic potential. Since $(U - TS) = A$,

$$A = U^{(1)} = U^{(0)} - TS = U^{(1)}(T, V, N_1, N_2,..N_n). \quad (E)$$

The function $U^{(1)} = A(T,V,N_1,N_2...)$ is as fundamental equation, which is the intercept of the curve $U^{(0)}(S)$. We can draw a series of lines with this intercept relation, the loci of which yield $U^{(0)}(S)$. Likewise,

$$U^{(2)} = U^{(0)} - S(\partial U^{(0)}/\partial S)_{V, N_k} - V(\partial U^{(0)}/\partial V)_{S, N_k} = U^{(0)} - TS - V(-P) = H - TS = G, \text{ i.e.,}$$

$$U^{(2)} = G(T, P, N_1, N_2,..N_n). \quad (F)$$

The term $U^{(2)}$ is the intercept of the $U^{(1)}(V)$ curve, and $-P$ is the slope of the $U^{(1)}(V)$ at specified values of T or of $U^{(0)}(V)$ at specified values of S, i.e.,

$$\partial G/\partial(\xi_1) = \partial G/\partial T = S, \text{ and } \partial G/\partial(-P) = V. \quad (G)$$

where $\xi_1 = T = (\partial U^{(0)}/\partial S)_V$

Further, the (n+2) Legendre transform

$$U^{(n+2)} = 0 = U^{(0)} - TS + PV - N_1\, \partial U^{(0)}/\partial N_1 - N_2\, \partial U^{(0)}/\partial N_2 - ...$$

$$U^{(n+2)} = 0 = G - \mu_1 N_1 - \mu_2 N_2 - ... , \text{ i.e.,}$$

$$G = \Sigma \mu_k N_k \quad (B)$$

D. GENERALIZED RELATION FOR ALL WORK MODES

Recall that for a multicomponent system,

$$dU = \delta Q - \delta W + \Sigma \mu_k dN_k. \quad (42)$$

where $\delta W = P\, dV$ for simple compressible system involving one reversible work mode. In the following more work modes will be discussed.

1. Electrical Work

In case of electrical work a current I flows as a result of a potential E over a period dt and the electrical work performed

$$\delta W = -E\, I\, dt = -E\, dq_c, \quad (43)$$

where E is the electromotive potential. The charge provided over time δq_c, = I dt. For instance, consider an electrical coil (system) which heats up a bowl of water. The volume of the coils slightly increases by dV due to heating. Then the work expression is given as

$\delta W = P\, dV - E\, dq_c$ and $dU = \delta Q - \delta W = T\, dS - P\, dV + E\, dq_c$ and

hence $U = U(S, V, q_c)$ for this example.

2. Elastic Work

When an anchored rectangular rod of length L on a plate is stretched in the x–wise direction due to a force F_x, the work done by surroundings on the rod

$$\delta W = -F_x\, dx = -P_s\, A\, dx = -P_s\, AL\, dx/L = -P_s\, V\, d\varepsilon, \quad (44)$$

where the stress $P_s = F_x/A$, A denotes the cross-sectional area, V the volume, and the linear strain $\varepsilon = dx/L$. Ideally, once the force is removed, the rod will retain its original length and the

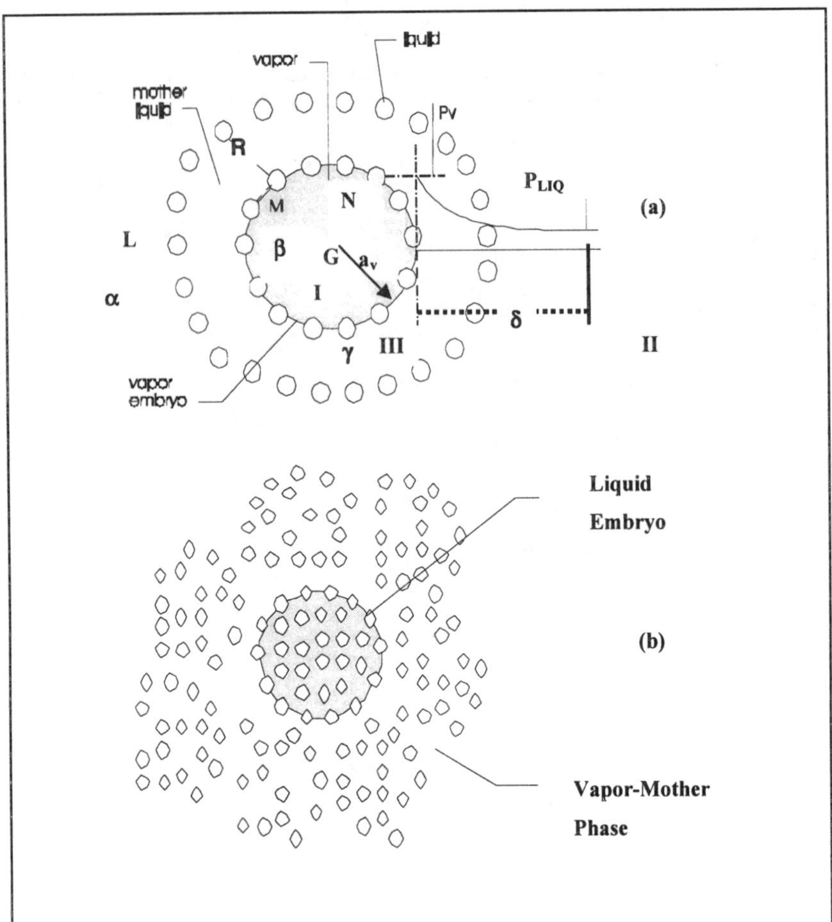

Figure 4: Vapor embryo surrounded by mother liquid. (b) Liquid drop embryo surrounded by mother vapor phase.

work will be recovered. (The Young's modulus $E = (\partial P_s/\partial \varepsilon)_T$ and the isentropic Young's modulus $E_S = (\partial P_s/\partial \varepsilon)_S$.)

3. Surface Tension Effects

Surface tension can be illustrated through Figure 4. Figure 4a shows a vapor bubble surrounded by boiling liquid (during evaporation), while Figure 4b illustrates a drop surrounded by vapor (during condensation). The fluid molecules at the surface of the bubble in Figure 4a are under tension. We will call the bubble the embryo phase and the boiling liquid the bulk mother phase. Beyond the interface MN, the molecules are faced by liquid molecules (L) on one side (i.e., molecules at closer intermolecular spacing with stronger intermolecular forces), while on the other side they are faced by vapor molecules (G) which are at a larger intermolecular spacing with weaker intermolecular forces. Thus, the molecules at the surface of vapor embryo are pulled towards L due to the very strong intermolecular forces. However, such a pull results in an increase in the intermolecular spacing along MN that decreases the liquid density at surface along MN, resulting in stretching forces. These effects (i.e., the decreased liquid density) persist over a small distance δ. Three regions are formed, namely, (1) vapor of uniform density separated from (2) liquid of uniform density by (3) a layer of nonuniform density of thickness δ. Tensile forces exist on the surface that lies normal to this thickness. At larger distances from the interface, liquid molecules are surrounded by other like molecules and surface tension effects vanish. Here the and pressure equals the liquid pressure P_{liq}. In the small thickness δ, the tensile force varies with distance r and hence P = P(r).

Consider a vapor bubble of radius a_v in a liquid. The net tensile force exerted by the vapor at the mid plane of the bubble normal to the area $A = \pi a_v^2$ equals $(P_v - P_{liq})A$. This force pulls the molecules against the attractive forces within the layer δ′. At equilibrium,

$$(P_v - P_{liq})A = \sigma' 2\pi a_v \delta, \text{ i.e., } \sigma' \delta' = (P_v - P_{liq})\, A/C, \quad (45)$$

where σ′ denotes the attractive force per unit area that counterbalances the pressure forces and C the circumference. As $\delta' \to 0$, $\sigma' \to \infty$. The thickness $\delta' \to 0$ is a surface discontinuity and $\sigma'\delta' = \sigma$, which denotes the surface tension. Therefore,

$$(P_v - P_{liq}) = A/(C\sigma), \text{ i.e.,} \quad (46)$$

$$(P_v - P_{liq}) = 2\,\sigma/a_v. \quad (47)$$

which is known as the Laplace equation.

This discussion has considered the mother phase to be liquid enclosing a vapor embryo phase. We can also develop the relations for the other scenario, e.g., for a condensing water droplet, in which case the mother phase is vapor having its molecules farther apart with weaker intermolecular forces, while the liquid embryo molecules exist at closer intermolecular spacing. Mechanical equilibrium exists if the embryo (liquid) phase pressure is higher than that of the mother phase (converse to the above example, cf. Eqs. (46) and (47)). For a condensing drop of radius a_{liq} that is surrounded by vapor at P_v, at equilibrium,

$$(P_{liq} - P_v) = 2\,\sigma/a_{liq}. \quad (48)$$

More generally $(P_{embroyo} - P_{mother}) = 2\sigma/a$ where a denotes the embryo radius.

In expanding a bubble or increasing drop radius, surface tension work must be performed. For e.g., consider a film of liquid contained within a rectangular wire whose sides are L and W. If one of the the sides of width is pulled by dX , the film the film area A increases by L dx. The work done on the film

$$\delta W = -F\, dx = -\sigma\, L\, dx.$$

Likewise, as a bubble of radius "a" expands, work is performed to stretch its film surface from $A = 4\pi a^2$ to A+ dA.

$$\delta W = -\sigma\, dA. \quad (49)$$

where $dA = 8\pi a\, da$ or $dA = 2\, dV/a$ where $V = (4/3)\pi a^3$

4. Torsional Work

In case of torsional work,

$$\delta W = \tau\, d\theta \quad (50)$$

where τ denotes the torque that causes an angular deformation dθ.

5. Work Involving Gravitational Field

Consider the energy change in the earth as it revolves around the sun

$$dE = dU + d(PE) + d(KE). \quad (51)$$

For a closed system $du = TdS - PdV + \Sigma\mu_k N_k$, i.e.,

$$dE = T\,dS - P\,dV + \Sigma\mu_k N_k + d(PE) + d(KE)$$

$$= T\,dS - P\,dV + \Sigma(\mu_k + M_k\,\phi_k + M_k\,v_k^2/2)\,dN_k, \quad (52)$$

where M_k denotes the molecular weight of the k–th species available on the earth, and ϕ_k is the potential energy of that species.

6. General Considerations

In general,

$$U = U(S, V, q_c, E, A, \ldots, N_1, N_2, \ldots, N_m) \quad (53)$$

The number of independent variables equals the number of all work modes plus number of species. In the context of Eq. (52), excluding any one thermodynamic property, the configuration parameters include S, V, q_c, E, A, … , and m constituents. In this case, if there are n configurational parameters and m constituents, there are (m+n+1) independent variables that describe the system. Generalizing Eq (52),

$$U = U(S, x_1, x_2 \ldots x_n, N_1 \ldots N_p), \quad (54)$$

where $x_1 = V$, $x_2 = q_c$, … . In addition, with all other properties and components held unchanged,

$$\partial U/\partial S = T,\ \partial U/\partial x_1 = -P,\ \partial U/\partial x_2 = E \ldots,\ \partial U/\partial N_1 = \mu_1,\ \partial U/\partial N_2 = \mu_2, \ldots \quad (55)$$

The m th Legendre transform of Eq. (54) such that $1 < m < n$,

$$U^{(m)} = U - TS - \Sigma_{i=1,m} f_i x_i, \quad (56)$$

where $f_k = \partial U/\partial x_k$ denotes the driving force. For instance, x_j could denote deformation due to the application of a force f_j. such as the deformation in V due to a pressure P. Taking (K+J+1) th transform

$$U^{(J+K+1)} = U + PV - TS - \Sigma_{i=1,J-1+K} f_i Z_i \quad (57)$$

E. THERMODYNAMIC POSTULATES FOR SIMPLE SYSTEMS

We will use four thermodynamic postulates (or rules) instead of stating thermodynamics laws.

1. Postulate I

Equilibrium states of simple systems exist that are characterized completely by the internal energy U, the volume V, and the mole numbers $N_1, N_2, \ldots N_n$ of the chemical constituents. Thus, (n+2) variables are required to fix the state of a simple system. The internal energy U for a closed system is defined by the relation

$$dU = \delta Q - \delta W. \quad (58)$$

Postulate I is a state equation describing a system at equilibrium and can be mathematically expressed in the form

$$U = U(S, V, N_1, N_2, \ldots, N_n),$$

which is the energy fundamental equation for an n–component mixture. There are no unique fundamental relations for all substances.

2. Postulate II

Processes that do not influence the environment change all systems with specified internal restraints in such a manner that they approach a stable equilibrium state for each simple subsystem contained within them. In a limiting condition, the entire system is in equilibrium. Postulate II resembles the Second Law, since it states that all systems in a nonequilibrium state eventually reach equilibrium.

There exists a function S (called the entropy of the extensive parameters U, V, N_1,N_2, ... , N_n) of any isolated composite system that is defined for all equilibrium states having those properties. The values assumed by the extensive parameters in the absence of an internal constraint are those that maximize S over the allowed constrained equilibrium states.

The equilibrium state is described by the relation

$$S = S(U,V, N_1,N_2, \ldots , N_n), \tag{59}$$

Equation (59) is the entropy fundamental equation and is similar to the combined First and Second Law in the engineering approach.

3. Postulate III

The entropy of a composite system is additive over its subsystems. The entropy is continuous and differentiable and is a monotonically increasing function of the internal energy. This postulate defines the entropy as an extensive property.

4. Postulate IV

Postulate IV is also known as the Nernst Postulate. For any system state

$$(\partial U/\partial S)_{V,N_j} \to 0 \text{ as } S \to 0. \tag{60}$$

This postulate implies that the value of U increases with an increase in the entropy. The postulate is similar to the Third law of thermodynamics. Since $(\partial U/\partial S)_{V,N_j} = T$, $T \to 0$ as $S \to 0$.

The entropy is an extensive property and is a homogeneous function of degree 1. If the values of U, and V are doubled, that of S is also doubled. However, the differentiation in Eq. (60) accounts for the size increase in both U and S so that the temperature remains unchanged. Therefore, T is a homogeneous function of U and V of degree 0.

F. ENTROPY FUNDAMENTAL EQUATION

The fundamental entropy equation described by Postulate II involves certain restrictions, namely,

The equation cannot involve derivatives, since integration will lead to the presence of unknown constants.

The entropy is continuous and differentiable with respect to all of its arguments, since the derivatives will yield intensive properties of matter (e.g., for a simple compressible substance, $dS/dU = 1/T$ and $dS/dV = -P/T$).

S is a homogeneous function of the first degree (i.e., it is an extensive property).

For all appropriate values of U and S,

$$1/T = (\partial S/\partial U)_{V,N_j} \geq 0, \text{ or} \tag{61}$$

$$T = (\partial S/\partial U)_{V,N_j} \geq 0. \tag{62}$$

---

d. Example 4

The state equation for the entropy of an electron gas in a metal is

$$S = C_1 N^{1/6} V^{1/3} (U-U_o)^{1/2}, \tag{A}$$

where

$$C_1 = 2^{3/2}\,\pi^{4/3}\,k_B\,m^{1/2}/(3^{1/3}\,h), \tag{B}$$

$k_B = 1.3804\times10^{-26}$ kJ K^{-1} denotes the Boltzmann constant, $h = 6.62517\times10^{-37}$ kJ s the Planck constant, $m = 9.1086\times10^{-31}$ kg the electron mass, and N the number of electrons (say, mole number × Avogadro number). The energy at 0 K $U_o = 3/5\ N\ \mu_o$, where the Fermi energy or chemical potential at 0 K

$$\mu_o = C_2\,(N/V)^{2/3}, \text{ and} \tag{C}$$

$$C_2 = 3^{2/3}\,h^2/(8\,\pi^{2/3}\,m). \tag{D}$$

Show that equation (A) is an entropy fundamental equation.

Solution

$$S = C_1\,N^{1/6}\,(V^{2/3}\,U - (3/5)\,C_2\,N^{5/3})^{1/2} = S(U,V,N). \tag{E}$$

Equation (E) and, therefore, Eq. (A) are forms of the entropy fundamental equation. Both equations imply that U = U(S,V,N), which is the energy fundamental equation.

G. ENERGY FUNDAMENTAL EQUATION

The functions S and U are monotonically related. For a single phase

$$S = S(U, V, N_1, N_2,..., N_n) \tag{63}$$

Since S is a monotonic function of U, a single valued solution for U can be obtained, i.e.,

$$U = U(S, V, N_1, N_2,..., N_n), \text{ i.e.,} \tag{64}$$

$$dU = (\partial U/\partial S)_{V,N_j}\,dS + (\partial U/\partial V)_{S,N_j}\,dV + \Sigma_{k=1,n}(\partial U/\partial N_k)_{S,V,N_{j\neq k}}\,dN_k, \text{ where} \tag{65}$$

$$(\partial U/\partial S)_{V,N_j} = T,\ (\partial U/\partial V)_{S,N_j} = -P, \text{ and } (\partial U/\partial N_k)_{S,V,N_{j\neq k}} = \mu_k. \tag{66}$$

Therefore,

$$dU = TdS - PdV + \Sigma_{k=1,n}\mu_k dN_k. \tag{67}$$

Equation (67) describes the differential change between adjacent stable equilibrium states U and (U + dU). Rearranging Eq. (67),

$$dS = dU/T + PdV/T \text{ z- } \Sigma_{k=1,n}(\mu_k/T)dN_k. \tag{68}$$

In the context of these relations, there are (2+K) independent functions that specify either U or S. State variables in the fundamental equations are extensive, and the properties T, P, and μ_k are functions of S, V, and N_k. For a single component system

$$U=U(S, V, N). \tag{69}$$

three extensive properties fix the extensive state. In its differential form Eq.(69) has the form

$$dU = TdS - PdV + \mu dN, \text{ or for a closed system } dU = TdS - PdV. \tag{70}$$

This is a representation of the combined First and Second Laws of Thermodynamics.

H. INTENSIVE AND EXTENSIVE PROPERTIES

Consider a thermodynamic property

$$\phi = \phi(x_1,x_2...x_n). \tag{71}$$

The function ϕ depends upon the extensive variables x_i. If it is an intensive property, it is homogeneous function of degree zero. On the other hand, ϕ is extensive if it is homogeneous function of degree one. If extensive, the derivatives of ϕ with respect to extensive variables are intensive, since the degree of the derivatives is zero.

---

e. Example 5

The simplified form of a fundamental equation for an electron gas is

$$S = C_1 N^{1/6} V^{1/3} U^{1/2}. \quad (A)$$

Show that S is an extensive property and that the temperature is an intensive property.

Solution

In case the system is doubled, then $N_{new} = 2N$, $V_{new} = 2V$, and $U_{new} = 2U$. In that case,

$$S(U_{new}, V_{new}, N_{new}) = C_1 (2N)^{1/6} (2V)^{1/3} (2U)^{1/2} = 2C_1 N^{1/6} V^{1/3} U^{1/2} = 2S.$$

Therefore, S is doubled, which implies that it is an intensive property.
Since, $T = (dU/dS)_{V,N}$, from the relation

$$U = S^2/(C_1 N^{1/3} V^{2/3}), \quad (B)$$

$$T = (dU/dS)_{V,N}, = 2\, S/(C_1 N^{1/3} V^{2/3}). \quad (C)$$

Applying Eqs. (B) and (C),

$$T = 2\, U^{1/2} /(C_1 N^{1/6} V^{1/3}), \text{ i.e.,} \quad (D)$$

$$T = T\,(U, V, N) \quad (E)$$

This shows that T is a homogeneous function of U, V and N, but does not indicate the degree. However, you will see that by doubling U, N and V, the value of T is unchanged. Therefore, T is a homogeneous function of degree 0, which implies that it is an intensive property. Table 2 presents a summary of thermodynamic relations and potentials.

Table 2: A summary of thermodynamic relations and potentials.

Potential and independent variables	Definition	State equations	Integrated form	Gibbs–Duhem equation
Entropy S(U,V,N)		$1/T = (\partial S/\partial U)_{V,N}$, $P/T = (\partial S/\partial V)_{U,N}$, $\mu/T = (\partial S/\partial N)_{U,V}$	$S = U/T + PV/T - \mu N/T$	$Ud(1/T) + Vd(P/T) - Nd(\mu/T) = 0$
Internal energy U(S,V,N)		$T = (\partial U/\partial S)_{V,N}$, $P = -(\partial U/\partial V)_{S,N}$, $\mu = (\partial U/\partial N)_{S,V}$	$U = TS - PV + \mu N$	$SdT - VdP + Nd\mu$
Enthalpy H(S,P,N)	$H = U + PV$	$T = (\partial H/\partial S)_{P,N}$, $V = (\partial H/\partial P)_{S,N}$, $\mu = (\partial H/\partial N)_{S,P}$	$H = TS + \mu N$	$SdT - VdP + Nd\mu$
Helmholtz F(T,V,N)	$A = U - TS$	$T = (\partial A/\partial T)_{V,N}$, $P = -(\partial A/\partial V)_{T,N}$, $\mu = (\partial A/\partial N)_{T,V}$	$A = -PV + \mu N$	$SdT - VdP + Nd\mu$
Gibbs G(T,P,N)	$G = U + PV - TS$	$T = -(\partial G/\partial T)_{P,N}$, $V = (\partial G/\partial P)_{T,N}$, $\mu = (\partial G/\partial N)_{T,P}$	$G = \mu N$	$SdT - VdP + Nd\mu$
Massiew J(1/T,V,N)	$J = S - U/T$	$P/T = -(\partial J/\partial(1/T))_{V,N}$, $P/T = -(\partial J/\partial V)_{T,N}$, $\mu/T = -(\partial J/\partial N)_{T,V}$	$J = PV/T - \mu N/T$	$Ud(1/T) + Vd(P/T) - Nd(\mu/T) = 0$
Planck Y(1/T,P,N)	$Y = S - U/T - P/T$	$H = -(\partial Y/\partial(1/T))_{P,N}$, $V/T = -(\partial Y/\partial P)_{T,N}$, $\mu/T = -(\partial Y/\partial N)_{T,P}$	$Y = -\mu N/T$	$Hd(1/T) + (V/T)dP - Nd(\mu/T) = 0$

In the Table, the first row contains the entropy representation and the second the energy representation of the appropriate fundamental equation. The table contains only simple forms of the relations, i.e., $N_j = N$. For multicomponent systems the term μN should be replaced by $\Sigma\mu_j N_j$, and $Nd(\mu/T)$ by $\Sigma N_j\, d(\mu_j/T)$. The enthalpy function is partly homogeneous.

I. SUMMARY

A postulatory approach is given to describe thermodynamics without invoking thermodynamic laws. Then using Legendre transforms, various thermodynamic functions, such as U, H, A, G, etc., are defined.

Chapter 6

6. STATE RELATIONSHIPS FOR REAL GASES AND LIQUIDS

A. INTRODUCTION

In the previous chapters we have discussed the First and Second laws of thermodynamics. In order to analyze the performance and conduct availability analyses of thermodynamic systems (such as power plants and heat pumps) and devices (e.g., turbines and compressors), and to design high pressure vessels we require the thermodynamic properties and the constitutive state equations for the matter in the various systems. Chapters 6 and 7 discuss the methods that can be used to determine the system properties by using state equations. In this chapter we will discuss the state equations for real gases (i.e., gases at relatively high pressures) and fluids in terms of measurable properties, such as pressure, volume, and temperature.

In case experimental data is available for P, v, and T (e.g., in the steam or refrigerant tables), it is possible to develop empirical relations that relate these properties. Otherwise, the relations between these properties must be derived using physical principles. First, we will discuss the simplest equation of state, i.e., the ideal gas relation. This relation will be extended to consider real gases and other fluids in terms of two– and three–parameter equations of state (for which spreadsheet software is also available). Finally, approximate equations of state for liquids and solids are discussed.

B. EQUATIONS OF STATE

The ideal gas equation of state is also considered to be a thermally or mechanically perfect state equation. In Chapter 1 we presented a simple derivation of this equation by using microscopic thermodynamic considerations and neglecting intermolecular forces and the molecular body volume. The resulting relation was

$$P\bar{v}_0 = \bar{R}T. \quad (1)$$

The subscript 0 implies that the gas is ideal at the given conditions. If the measured gas volume at given P and T values is identical to that calculated by using Eq. (1), the gas is said to be an ideal gas. However, if the measured specific volume at the same pressure and temperature differs from that determined using Eq. (1), the gas is considered to be a real gas. The simplest way to present the real gas equation of state is by introducing a correction to the specific volume by defining the compressibility factor Z, i.e.,

$$Z(T,P) = \bar{v}(T,P)/\bar{v}_0(T,P). \quad (2)$$

The actual specific volume $\bar{v}(T,P)$ can be determined from experiments while theoretical volume can be determined using ideal gas law. From Eqs. (1) and (2) we obtain the relation

$$P\bar{v}(T,P) = Z(T,P)\bar{R}T, \quad (3)$$

which is called the real gas or imperfect equation of state. If Z = 1, Eq. (3) reduces to the ideal gas equation of state. Equation (3) can be represented using reduced properties. Applying Eq. (2) at the critical point

$$Z(T_c,P_c) = \bar{v}(T_c,P_c)/\bar{v}_0(T_c,P_c), \quad (4)$$

where T_c and P_c, respectively, denote the critical temperature and pressure. Table A-1 tabulates these values for many substances. Applying Eq. (3) at the critical point, we obtain the following relation:

$$P_c\bar{v}_c = Z(T_c,P_c)\bar{R}T_c. \quad (5)$$

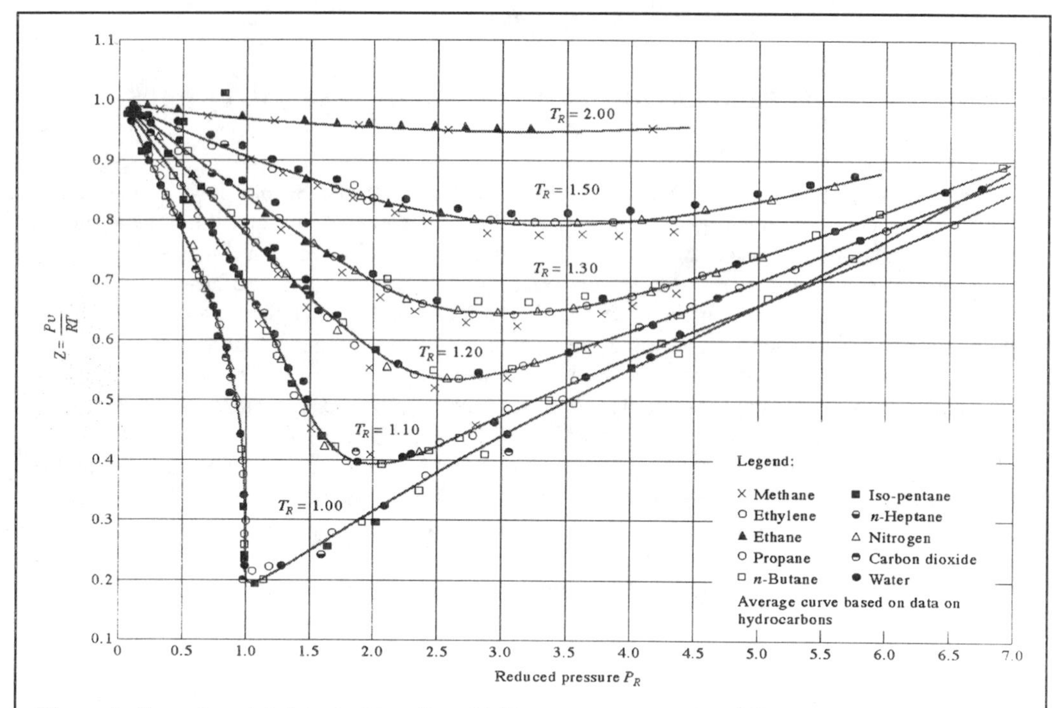

Figure 1: Experimental data for Z *vs.* P_R with T_R as a parameter for different gases (from G.J. Su, "Modified Law of Corresponding States," *Ind. Eng. Chem.*, 38, 803, 1946. With permission.).

From Eqs. (3) and (5), we can express the compressibility factor

$$Z(T_R,P_R) = P_R\,\bar{v}(T_R,P_R)\,Z(T_c,P_c)/T_R = f(P_R,T_R)\,Z_c, \tag{6}$$

where P_R denotes the reduced pressure P/P_c, and T_R the reduced temperature T/T_c. According to Van Der Waals equation of state (later sections), Z_c = 3/8 and is same for all substances. Then it is apparent from Eq. (6) that Z is only a function of T_R, v_R. Fig. B.2a shows the compressibility chart. In general, values of Z_c lie in the range from 0.2–0.3.

Figure 1 contains experimental data for Z *vs.* P_R with T_R as a parameter for different gases. Compressibility charts (Chart B.2a) to determine the value of Z can be used at the appropriate reduced pressures and temperatures in order to ascertain whether a gas is real or ideal under specified conditions. Experiments can also be conducted to determine which equation of state the gas observes and to measure the compressibility factor. It is also possible to obtain an approximate criterion for real or ideal gas behavior using the intermolecular force potential diagram presented in Chapter 1. When $\ell \gg 3\ell_0$, the gas molecules move randomly in the absence of intermolecular attractive forces. If the specific volume of a solid $\bar{v}_s$ or (liquid $\bar{v}_f$) are known, the molecular number density is $n' = N_{Avog}/\bar{v}_s$ (or $= N_{Avog}/\bar{v}_f$), and $\ell \approx n'^{-1/3}$.

C. REAL GASES

According to ideal gas law, the product $P\bar{v}$ must be constant at specified temperature. Figure 2 presents the variation of $P\bar{v}$ vs. P for the temperature range 150 K<T<300 K for nitrogen. It is apparent that experimental data does not support ideal gas law.

1. Virial Equation of State

When experimental data is represented by a polynomial expansion of $P\bar{v}$ in terms of P, the resulting equation of state is called the virial (i.e., force) equation of state.

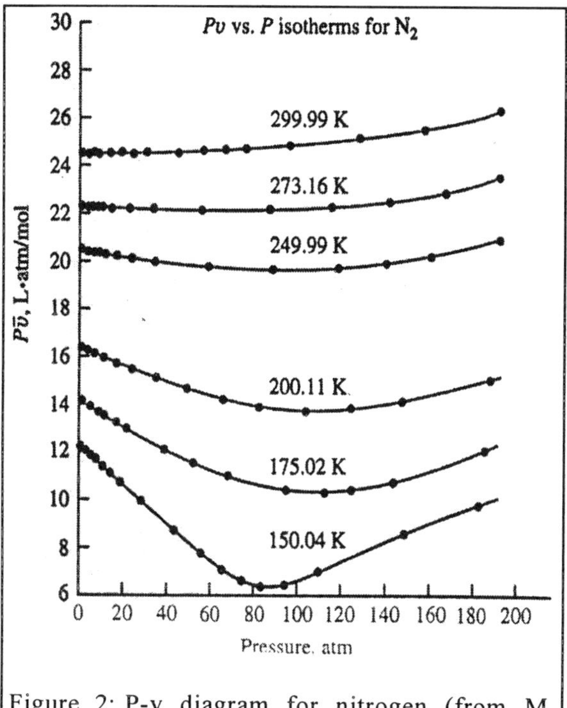

Figure 2: P-v diagram for nitrogen (from M. Zemansky, 4th Ed., McGraw Hill, 1957).

a. Exact Virial Equation

The exact virial equation of state has the form

$$P\bar{v} = A_1'(T) + B_1'(T)P + C_1'(T)P^2 + D_1'(T)P^3 + \cdots, \tag{7}$$

where A_1', B_1' (and so on) are, respectively, called the first, second (and so on) virial coefficients. As $P\rightarrow 0$, $P\bar{v}\rightarrow A_1'(T)$. Since the ideal gas state is approached as $P\rightarrow 0$,

$$A_1'(T) = \bar{R}T. \tag{8}$$

Dividing Eq.(7) by $\bar{R}T$,

$$Z = 1 + B_1(T)\,P + C_1(T)\,P^2 + \ldots, \text{ where} \tag{9}$$

$$B_1(T) = B_1'(T)/\bar{R}T,\ C_1(T) = C_1'(T)/\bar{R}T, \ldots \tag{10}$$

For nitrogen in the pressure range $0 < P < 200$ atm, at $T = 298$ K the second through ninth virial coefficients in Eq. (9) are, respectively, -10.281 atm^{-1}, 0.065189 atm^{-2}, 0, 5.1955×10^{-7} atm^{-4}, 0, -1.3156×10^{-11} atm^{-6}, 0, and 1.009×10^{-16} atm^{-8}.

Alternatively the polynomial expansion can be performed in terms of $\bar{v}^{-1}$ (as $P\rightarrow 0$, $\bar{v}^{-1}\rightarrow 0$), i.e.,

$$P\bar{v} = A'(T) + \frac{B'(T)}{\bar{v}} + \frac{C'(T)}{\bar{v}^2} + \frac{D'(T)}{\bar{v}^3} + \cdots. \tag{11}$$

As $\bar{v}\rightarrow\infty$, $A'(T) = \bar{R}T$ and

$$Z = 1 + B(T)\,\overline{v}^{-1} + C(T)\,\overline{v}^{-2} + \ldots, \tag{12}$$

where

$$B(T) = B'(T)/\overline{R}T,\ C(T) = C'(T)/\overline{R}, \ldots \tag{13}$$

b. Approximate Virial Equation

As P→0, terms of the order of P^2 and higher can be neglected, and Eq. (9) can be reduced to a first order approximation of the form

$$P\overline{v}/\overline{R}T = Z = 1 + B_1(T)P. \tag{14}$$

For nitrogen, the value of B_1 is given by the relation

$$B_1(T)\ \overline{R}T = \overline{b}(T) = (0.0395 - 10\,T^{-1} - 1084\,T^{-2})\ m^3\ kmole^{-}. \tag{15}$$

Equation (14) suggests that Z follows a linear law with P at low pressures and at specified temperatures. The low-pressure experimental data contained in Figure 1 confirms such a relation at specified T_R. Abbott has suggested the following empirical relation

$$Z = 1 + B_1(T_R)\,P_R,\ \text{for } v_R > 2, \text{ where} \tag{16}$$

$$B_1(T_R) = T_R(0.083 - 0.422\,T_R^{-1.6})\,T_R^{-1}. \tag{17}$$

The value of Z is greater than unity at larger T_R and lower than unity for smaller values of T_R. According to the first order approximation of the vrial equation (i.e., Eq. (14), the value of Z can be larger or smaller than unity, depending upon the sign of $B_1(T_R)$. Equation (17) implies that the slope $\partial Z/\partial P_R = B_1(T_R)$ is finite as $P_R \to 0$. At low pressures and $T_R \approx 2.76$,

$$B_1(T_R) = 0,$$

so that $\partial Z/\partial P_R = 0$ and $Z = 1$, i.e., the real gas behaves like an ideal gas. This reduced temperature is also known as the Boyle temperature. For CO_2, $T_c = 304$ K so that its Boyle temperature is 839 K.

2. Van der Waals (VW) Equation of State

We now develop a rational approach to develop a real gas equation of state. Later we will use various equations of state in determining the stability characteristics (Chapter 10). Prior to the presentation of VW equation, Clausius I equation of state will be described because of its relevance to VW equation.

a. Clausius–I Equation of State

Consider N moles of a gas that occupy a volume V in a container at some pressure and temperature. The gas molecules undergo random motion and pressure forces exist due to their impact on the walls of the container. For instance, at room temperature the force due to impact of air molecules on any surface is 10 N cm^{-2}. This is the average pressure experienced by the surface due to the impact of molecules that travel at velocities of approximately 350 m s^{-1}. If b′ denotes the volume of each molecule and N′ the number of molecules in the volume V, then the total body volume of the molecules is given by the product N′b′. If the volume N′b′ is insignificant compared to geometrical volume V, the molecules can be assumed to be point masses, an approximation that is valid at low pressures. The volume V then denotes the free volume in which these point mass molecules can move. In case intermolecular forces are negligible (except upon impact) one can derive the ideal gas or perfect gas equation of state (cf. Chapter 1) in the form

$$PV = N\overline{R}T,$$

where $N = N'/N_{Avog}$.

In case the geometrical volume is reduced while keeping the number of molecules unchanged so that the volume N′b′ is comparable to V, the ideal gas equation of state must be modified. In this case V–N′b′ denotes the free space available for equivalent point mass molecules to move randomly.

Assume, for sake of illustration, that N′ = 8 and consider a cube of side 2σ (where σ denotes the molecular diameter) and volume V. The volume of each molecule (assumed to be spherical) $b' = \pi\sigma^3/6$ and, if the eight molecules are tightly packed inside the cube, the free volume available to them to move around is $V - 8(\pi\sigma^3/6)$. However, since the intermolecular distance in the cube is 2σ and adjacent molecules touch each other, the empty space between any two molecules is unavailable for movement. Therefore, knowing the molecular diameter alone is insufficient information regarding the free volume. The shortest possible distance between any two molecules at which there is contact is σ (= $\sigma/2 + \sigma/2$). No other molecule can be included within the volume defined by the radius σ (or diameter 2σ), and the "forbidden volume" per pair of molecules is $\pi(2\sigma)^3/6$. For a single molecule the forbidden volume is $\pi(2\sigma)^3/12$, and for N′ molecules it is

$$b' = N'\pi(2\sigma)^3/12 = 4\,N'\pi\sigma^3/6 = N'\ (\text{collisional volume} \div 2), \tag{18}$$

where the collisional volume is defined as $4\pi\sigma^3/3$.

The ideal gas equation can be corrected by subtracting the product N′b′ from the geometrical volume V. The free volume V–N′b′ is the volume occupied by equivalent point masses in an ideal gas and N′b′ is considered to be the *apparent* body volume of the molecules contained in the volume V. The reduced free volume increases the number of molecules per unit free volume, which, in turn, increases the number of collisions per unit time. This results in a higher pressure. Based on the reduced free volume, the ideal gas equation may be modified into the form

$$P = N\bar{R}T/(V - N'b') = \bar{R}T/(\bar{v} - \bar{b}), \tag{19a}$$

where

$$\bar{b} = N'b'/N = (2/3)N_{Avog}\pi\sigma^3. \tag{19b}$$

The product $\pi\sigma^3$ is the forbidden volume per kmole of the gas, and is four times the body volume. The number of molecules and moles are related by the expression

$$N' = N_{Avog}N.$$

Rewriting Eq. (19a) (which is known as the Clausius–I equation of state),

$$\bar{v} = \bar{R}T/P + \bar{b} = \bar{v}_0 + \bar{b}. \tag{20}$$

Note that $\bar{v}_0$ would have been the volume had the gas been ideal.

The compressibility factor

$$Z = \bar{v}(T,P)/\bar{v}_0\,(T,P) = P\bar{v}/\bar{R}T = 1 + P\bar{b}/\bar{R}T, \text{ or} \tag{21}$$

$$Z - 1 = \bar{b}(P/\bar{R}T). \tag{22}$$

Equation (22) is also called the deviation function for the compressibility factor. The deviation function tends to zero as $P \to 0$ at a specified temperature. According to the Clausius–I equation of state, the geometrical volume at a specified state (i.e., fixed T and P) is equal to ideal gas volume (predicted by ideal gas equation) plus a correction for the molecular body volume. Upon comparing Eq. (21) with Eq. (14), it is clear that

$$\bar{b}(T) = B_1(T)(\bar{R}T). \tag{23}$$

Although Eq. (21) implies that Z>1 at all conditions, experimental data indicates that Z<1 at intermediate pressures.

b. VW Equation

The Clausius–I equation of state (Eq. (19)) does not account for the intermolecular attraction forces that are significant when the molecular spacing is relatively close (e.g., at high pressures). Therefore, Eq. (19) must be appropriately modified.

Let us denote the pressure that would exist in absence of attractive forces as P′. In order to understand the effects of attractive forces on the pressure P′ we use the analogy of the gas molecules represented by groups of five particles. Each analogous group consists of a single particle labeled as "M" surrounded by four other particles that are geometrically located 90° apart. Assume that a group travels at a velocity of 350 m s^{-1}. There is no net attractive force on the particle contained in the interior of the group, since the attractive forces due to the four exterior particles cancel each other out. On the other hand when the particle "M' (and similar particles) impinges on a surface so as to create a pressure, the particle that was originally in the interior is now surrounded by only three particles (since one particle has already struck the surface). Therefore, at this time, the particle group will exert a lower pressure than 10^5 N m^{-2}, since there is now a net attraction force exerted on the interior particle "M".

In order to determine the reduction in pressure, we need a functional form for the attractive force exerted between a pair of molecules separated by an intermolecular distance ℓ. Such a relation can either be represented by empirical relations (e.g., by using the Lennard–Jones (LJ) empirical intermolecular potential energy $\Phi(\ell)$ between a pair of molecules) that was discussed in Chapter 1, or it can be deduced through a phenomenological approach. Applying the LJ approach for a like molecular pair, the intermolecular force function (cf. Chapter 1)

$$F(\ell) = -(4\varepsilon/\sigma)(12(\sigma/\ell)^{13} - 6(\sigma/\ell)^7, \tag{24}$$

where ε denotes the characteristic energy of interaction between the molecules (which corresponds to the maximum attraction energy $\approx$ 0.77 k_B T_c, k_B is the Boltzmann constant), σ the characteristic or collision diameter of the molecule (= 2σ′), and ℓ the intermolecular distance. The first term on the RHS of Eq. (24) represents the repulsion force between a molecular pair, and the second term arises due to the intermolecular attraction between the two molecules. The repulsive force is only of interest if the substance is a solid or a liquid. The reduction in pressure ΔP_{attr} due to attractive forces is derived in the Appendix, and is

$$\Delta P = 2.667\pi\,\varepsilon\sigma^3 n'^2 \approx 3\pi\,\varepsilon\sigma^3 n'^2 \tag{25}$$

where n′ denotes the number of molecules per unit volume= N_{Avog} n, and n the number of moles per unit volume. By using Eqs. (19a) and (25) we obtain the Van der Waals (VW) equation of state (named after Johannes Diderik Van der Waals, 1837–1923) in the form

$$P = \bar{R}T/(\bar{v} - \bar{b}) - (\bar{a}/\bar{v}^2), \tag{26}$$

where $\bar{a}$ = 2.667 $\pi\varepsilon\sigma^3 N_{Avog}^2$. The units for $\bar{b}$ and $\bar{v}$ are identical (in m^3 $kmole^{-1}$), while those for $\bar{a}$ are atm m^6 $kmole^{-2}$. According to a more rigorous derivation based on the potential

$$\bar{a} = 2.667\pi\varepsilon\sigma^3\ N_{Avog} \tag{27}$$

The first term in Eq.(26) is the pressure exerted due to collision and bouncing off an imaginary plane and is proportional to thermal part of energy (i.e te+ve+re etc) of all the molecules within unit volume of free space. The second term is the reduction in force due to attractive force exerted on those molecules by neighboring molecules. A typical experimentally determined P–v diagram is illustrated in Figure 3. The saturated liquid state occurs along the line FAC, saturated vapor along GBC. A few isotherms are also included in Figure 3. If the pressure is fixed, the temperature is constant during vaporization. As the critical point is approached, vaporization occurs at single point with the result that the variation of pressure with volume must show an inflection for the critical isotherm (at Point C in Figure 3 at which $T = T_c$, $P = P_c$, and $\bar{v} = \bar{v}_c$). Thus, one must select values for "a" and "b" such that the following inflexion condition is satisfied, i.e.,

$$(\partial P/\partial \bar{v})_{T=T_c} = (\partial^2 P/\partial \bar{v}^2)_{T=T_c} = 0. \quad (28)$$

Therefore, in context of Eq. (26)

$$(\partial P/\partial \bar{v})_{T=T_c} = -\bar{R}T_c/(\bar{v}_c - \bar{b})^2 + 2\bar{a}/\bar{v}_c^3 = 0, \text{ and} \quad (29)$$

$$(\partial^2 P/\partial \bar{v}^2)_{T=T_c} = 2\bar{R}T_c/(\bar{v}_c - \bar{b})^3 + 6\bar{a}/\bar{v}_c^4 = 0. \quad (30)$$

Hence,

$$\bar{a} = (\bar{R}T_c/(\bar{v}_c - \bar{b})^2)\bar{v}_c^3/2, \text{ and} \quad (31)$$

$$\bar{b} = \bar{v}_c/3. \quad (32)$$

Finally, combining Eqs. (31) and (32) we obtain

$$\bar{a} = (9/8)\bar{R}T_c\bar{v}_c. \quad (33)$$

If critical data on T_c and $\bar{v}_c$ are available then the constants $\bar{a}$ and $\bar{b}$ can be determined from Eqs. (32) and (33) and the critical pressure can then be determined from the state equation (27) at the critical point, i.e., by using the result for $\bar{a}$ and $\bar{b}$ from Eqs. (32) and (33)

$$P_c = (\bar{R}T_c/\bar{v}_c)(3/8). \quad (34)$$

Therefore,

$$Z_c = P_c\bar{v}_c/\bar{R}T_c = 3/8. \quad (35)$$

The critical temperature and volume for water are, respectively, $T_c = 647.3$K and $\bar{v}_c = 0.0558$ m^3 kmole^{-1} (Table A-1) Thereafter, using Eq. (32), $\bar{b} = 0.0186$ m^3 kmole^{-1}, and from Eq. (33) $\bar{a} = 60.54$ bar m^6 kmole^{-2}. (The calculated value of P_c from Eq. (34) is 362 bar, but the measured value is 220.9 bars, i.e., the experimental data for the critical pressure of water at specified T_c and $\bar{v}_c$ deviates far from values obtained from the VW equation of state when $Z_c \neq 3/8$. For the moment we will presume that this equation of state is accurate at the critical point and proceed to express $\bar{a}$ and $\bar{b}$ in terms of P_c and T_c using Eqs. (32)–(34). Therefore,

$$\bar{a} = (27/64)\ \bar{R}^2T_c^2/P_c = c_1\bar{R}^2T_c^2/P_c, \text{ and } \bar{b} = \bar{R}T_c/(8P_c) = c_2\bar{R}T_c/P_c, \quad (36)$$

where the constants $c_1 = 27/64 = 0.4219$ and $c_2 = 1/8 = 0.125$. If measured data for all three critical properties P_C, T_C, $\bar{v}_C$ are available, one has a choice of either Eqs. (32) and (33) or Eqs. (36). Use of eqs. (36) is recommended for better comparison with experimental data. One may tabulate "a" and "b" values for a few substances using Eqs. (36) as is done in many texts. Table A-1 lists the VW values of $\bar{a}$ and $\bar{b}$ for many substances.

A possible experiment to measure the critical properties of a substance (say, water) can be constructed in the following manner. First, pour the water into a quartz made piston–cylinder assembly that is maintained at a constant pressure. Then, slowly heat the water until a small bubble appears. Determine the specific volume of the water just as bubble begins

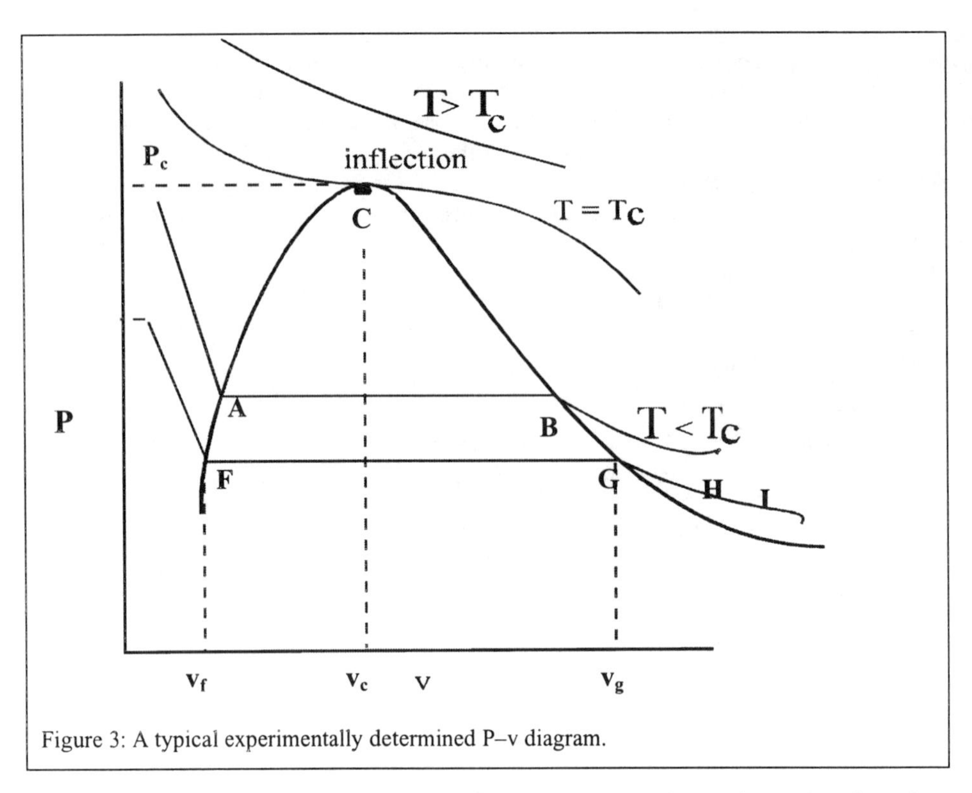

Figure 3: A typical experimentally determined P–v diagram.

to appear (i.e, v_f, saturated liquid). Thereafter, determine the gas–phase volume when the entire mass of the water has been evaporated (i.e., v_g for saturated vapor). Also measure the boiling temperature of water under the specified conditions. Repeat the experiment several times after incrementally increasing the system pressure and plot the response of v_f to the pressure (i.e., curve FAC in Figure 3), and of v_g to P (curve GBC in the Figure 3). Two distinct phases will be observed until the critical point is reached (where the curves for both v_f and v_g vs. P intersect). The pressure at which we cannot observe a clear demarcation between the liquid and vapor (i.e., v_f being equal to v_g) is the critical pressure.

i. Comments

Equation (26) can be rewritten in the form

$$\bar{v}^3 + \bar{v}^2 (-\bar{b}P - \bar{R}T)/P + \bar{v}\,(\bar{a}/P)\,(-\bar{a}\,\bar{b}/P) = 0.$$

which is a cubic equation in terms of $\bar{v}$. Therefore, for a specified pressure and temperature, there are one real and two imaginary solutions, and/or three real positive solutions for $\bar{v}$.

The constants a and b are related to the critical properties. The appendix presents explicit solutions for cubic equations.

Recall that $\bar{b} = (2/3)N_{Avog}\pi\sigma^3$, so that the collision diameter and, hence, the molecular size can be determined once we know $\bar{b}$ from critical properties.

Instead of using the inflection approach for evaluation of $\bar{a}$ and $\bar{b}$, the constants c_1 and c_2 (in Eq. (36)) can be selected to minimize the difference between the experimental data and theoretical results obtained from Eq. (26).

If we assume that v » b (i.e., a point mass approximation is used), $P \approx \bar{R}T/\bar{v} - \bar{a}/\bar{v}^2$, which is a quadratic equation. In this case, the compressibility factor $Z = P\bar{v}/(\bar{R}T) = 1 -$

$\bar{a}/(\bar{v}\bar{R}T)$. Therefore, $Z < 1$ if one considers only attractive forces. When $v \approx \bar{b}$, the first term in Eq. (26) dominates and hence $P \approx \bar{R}T/(\bar{v}-\bar{b})$ which is the Clausius I equation of state. The compressibility factor $Z > 1$ when body volume effects are more significant. Thus when both effects are included $Z>1$ or $Z<1$

For a closed system the reversible work $\bar{w} = \text{I}P(T, \bar{v})\, dv$. If a real gas is involved, one must use the real gas equation of state Eq. (26) for $P(T, \bar{v})$. Similarly, for an open system the reversible specific work $\bar{w} = -\text{I}\ \bar{v}(T,P)\, dP$.

Once $\bar{v}$ is specified, $\partial P/\partial T = R/(v-b)$, i.e., isochoric curves are linear in P-T plane for a fluid following VW equation of state.

Since $\bar{a} = (27/64)\ \bar{R}^2T_c^2/P_c$ and $\bar{b} = \bar{R}T_c/(8P_c)$, then using Eqs. (19b) and (27), we can obtain the following relations:

$$\sigma = 0.3908 k_B^{1/3}(T_c/P_c)^{1/3}$$

and $\varepsilon = (27/64\ R^2T_c^2/P_c)/(3\ N_{Avag}^2 \pi\sigma^3)$. Simplifying

$$\varepsilon = 0.75\ k_BT_c,$$

which is close the value of $0.77\ k_BT_c$ cited in the literature.

---

a. *Example 1*

Determine $\bar{v}$ for $H_2O(g)$ at $P = 140$ bars and $T = 673$ K using the ideal gas equation, the compressibility chart, the steam tables and the VW equation of state. What is the size of a single molecule?

Solution

Ideal gas.

$\bar{v} = \bar{R}T/P = 0.08314$ bar m^3 kmole^{-1} K^{-1} $\times$ 673 K $\div$ 140 bar $= 0.4$ m^3 kmole^{-1}.

Compressibility chart.

$P_R = P/P_c = 140$ bar $\div$ 220.9 bar $= 0.634$,

$T_R = T/T_c = 673$ K $\div$ 647.3 K $= 1.04$. From the compressibility chart, $Z = 0.78$.

Since $P\bar{v} = Z\bar{R}T$,

$\bar{v} = 0.78 \times 0.08314$ bar m^3 kmole^{-1} K^{-1} $\times$ 673 K$\div$140 bar $= 0.312$ m^3 kmole^{-1}.

Tables.

$\bar{v} = 17.22$ cm^3 g^{-1} $\times$ 10^{-6} m^3 cm^{-3} $\times$ 10^3 g kg^{-1} 18.02 kg kmole^{-1} $= 0.31$ m^3 kmole^{-1}.

Van der Waals equation.

$\bar{a} = 27\bar{R}^2\ T_c^2/64\ P_c = 27\times0.08314^2$ bar^2 m^6 kmole^{-2} K^{-2}$\times$647.3^2 K^2$\div$(64 $\times$ 220.9 bar), i.e.,

$\bar{a} = 5.531$ bar m^6 kmole^{-2}.

Likewise, $\bar{b} = \bar{R}T_c/(8P_c) = (0.08314$ bar m^3 kmole^{-1} K^{-1} $\times$ 647.3 K$\div$(8 $\times$ 220.9 bar), i.e.,

$\bar{b} = 0.0305$ m^3 kmole^{-1}.

However, using Eq. (27), at 140 bar, T= 673 K, $\bar{v} = 0.31$ m^3 kmole^{-1}, according to VW equation. (Note that use of this equation produces a closer prediction to the tabulated value than was obtained using the ideal gas law.)

Molecular diameter

Recall that $\bar{b}$ represents the finite volume or 4 $\times$ (body volume) of the molecules. A kmole contains 6.023 $\times$ 10^{26} molecules. Therefore, the body volume of a single molecule

≈ 0.0305 m^3 kmole^{-1} $\div$ (4 $\times$ 6.023 $\times$ 10^{26} molecules) $= 1.264 \times 10^{-28}$ m^3.

The molecular volume

$\pi d^3/6 = 1.264\times10^{-29}$ m^3, i.e., $d = 2.89\times10^{-10}$ m $= 2.89$ Å.

Remarks

The term $-\bar{a}/\bar{v}^2$ in the VW equation corresponds to the reduction in pressure due to intermolecular forces. When $\bar{v}$ = 0.31 m^3 kmole^{-1}, $-\bar{a}/\bar{v}^2$ = –57.6 bar, while the first term $\bar{R}T/(\bar{v}-\bar{b})$ = 199.5 bar. This implies that the frequent molecular collisions create a pressure of 199.5 bars, but due to the strong attraction forces in the dense gas phase the molecules are pulled back together with a pressure equivalent of 57.6 bar. Therefore, the net pressure is 141.9 bar.

Since $\sigma = ((6/\pi)(\bar{b}/4)/N_{Avog})^{1/3}$, and $\bar{b} = (1/8)\bar{R}T_c/P_c$ using the VW equation of state, $\sigma = ((6/32\pi)\bar{R}T_c/(P_cN_{Avog}))^{1/3} = 0.391(k_BT_c/P_c)^{1/3}$. Using the value $k_B = 1.3804 \times 10^{-26}$ kJ K^{-1} molecule^{-1}, $\sigma = = 2.02(T_c(K)/P_c)^{1/3}$ Å with P_c expressed in bar. For instance, if T_c = 647 K and P_c = 221 bar, σ = 2.89 Å.

---

3. Redlich–Kwong Equation of State

We will now discuss the Redlich–Kwong (RK) equation of state that is a more accurate representation of a gaseous state, particularly for $T>T_c$, and is also easier to use. This equation will be repeatedly used in later chapters in context of the thermodynamic properties u, h, and s. The RK equation is of the form

$$P = \bar{R}T/(\bar{v}-\bar{b}) - \bar{a}/(T^{1/2}\,\bar{v}(\bar{v}+\bar{b})). \quad (37)$$

Attractive forces are modeled differently by the RK equation of state. The two parameters $\bar{a}$ and $\bar{b}$ can be evaluated using the inflection conditions as outlined while developing the VW equation. Since $(\partial P/\partial v) = 0$ and $\partial^2P/\partial v^2 = 0$ at the critical point, it can be shown that

$$(\bar{b}/\bar{v}_c)^3 - 3(\bar{b}/\bar{v}_c)^2 - 3(\bar{b}/\bar{v}_c) + 2 = 0, \text{ and} \quad (38)$$

$$\bar{a}/(T_c^{1/2}\bar{v}_c) = (1 + (\bar{b}/\bar{v}_c)^2)/((1-(\bar{b}/\bar{v}_c)^2)\,(2 + (\bar{b}/\bar{v}_c))). \quad (39)$$

From. Eqs. (38) and (39),

$$\bar{b}/\bar{v}_c = 0.26, \text{ and } \bar{a}/(RT_c^{3/2}\,\bar{v}_c) = 1.2844.$$

Using these values in Eq. (37) at the critical point

$$Z_c = 1/3. \quad (40)$$

Therefore,

$$\bar{a} = c_3\,\bar{R}^2T_c^{2.5}/P_c, \text{ and } \bar{b} = c_4\,\bar{R}T_c/P_c. \quad (41)$$

where c_3 = 0.4275, c_4 = 0.08664. Table A-1 lists the RK values of $\bar{a}$ and $\bar{b}$ for many substances using their critical properties.

Instead of using the inflection approach for determining the values of $\bar{a}$ and $\bar{b}$, the constants c_3 and c_4 (in Eq. (41)) can be selected so as to minimize the differences between experimental data for P, v, T and theoretical results obtained by applying Eq. (37). Then, it is possible to solve for T_c and P_c by employing Eqs. (41), e.g.,

$$T_{c,mod} = ((a\,c_4/(c_3b))^{2/3},\ P_{c,mod} = c_4\,RT_c/b = ((Ra\,c_4^{5/3}/(c_3^{2/3}b^{5/3}).$$

Dimensionless charts based on the RK equation can be developed (cf. Chapters 6, 7, 8, and 10), e.g.,

$$(T_{c,mod}/T_c) = ((a\,c_4/(c_3b))^{2/3})/((a\times0.08664)/(0.4275\,b))^{2/3} = 2.8983\ (c_4/c_3)^{2/3}.$$

Similarly,

$$P_{c,mod}/P_c = 0.3345(c_4^{5/3}/c_3^{2/3}),$$

and using the modified critical properties in the charts, the ratios $(T_{c,mod}/T_c)$ and $P_{c,mod}/P_c = 1$ when $c_3 = 0.4275$ and $c_4 = 0.08664$.

One can prove that for gases whose behavior is modeled by the RK equation of state, isochoric curves (i.e., of P with respect to T at a specified value of $\bar{v}$) are no longer linear. Equation (37) can also be expressed in the form

$$\bar{v}^3 + \bar{v}^2(-\bar{R}T/P) + \bar{v}\,(\bar{a}/(PT^{1/2}) - \bar{b}\,\bar{R}T/P - \bar{b}^2) + (-\bar{a}\,\bar{b}/(PT^{1/2})) = 0.$$

This cubic equation can be explicitly solved for $\bar{v}$ in terms of T and P.

4. Other Two–Parameter Equations of State

Table in the Appendix section (end of this chapter) contains descriptions of other two–parameter equations of state. We comment on one of these, namely, the Dietrici equation of state that is represented by the expression

$$P = (\bar{R}T/(\bar{v} - \bar{b}))\exp(-\bar{a}/(\bar{R}T\bar{v})).$$

This equation predicts a reasonable value of Z_c (= 0.271) and does not yield negative values of pressures. Its performance is superior in the neighborhood of the critical point.

b. Example 2

Plot the variation of pressure with respect to $\bar{v}$ at a constant temperature of 593 K for H_2O using the ideal gas equation of state and the RK equation of state.

Solution

Ideal gas equation.

$$P = (\bar{R}T)/\bar{v} = 0.08314\times593/\bar{v} = 49.3/\bar{v}. \quad (A)$$

This hyperbolic behavior is illustrated by the curve QRS in Figure 4. As $\bar{v} \to 0$, P $\to$

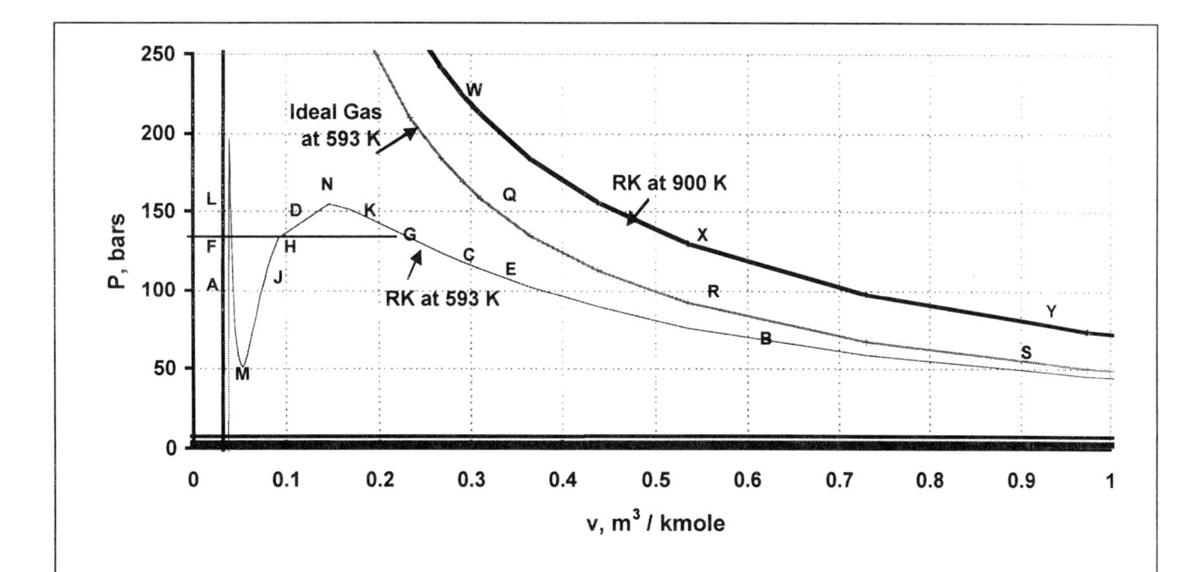

Figure 4: The response of pressure to $\bar{v}$ at two temperatures (593 K and 900 K) using the RK equation, and of P vs. $\bar{v}$ employing the ideal gas equation at 593 K.

∞.

RK equation.

$\bar{a} = 0.4275\,\bar{R}^2 T_c^{2.5}/P_c$, and $\bar{b} = 0.08664\,\bar{R}\,T_c/P_c$. $T_c = 647.3$ K and $P_c = 220.9$ bar.

$\bar{a} = 142.64$ bar m^6 K$^{0.5}$ kmole^2; $\bar{b} = 0.0211$ m^3 kmole^{-1}, and

$P = (49.30/(\bar{v} - 0.0211))\ (5.86/(\bar{v} + 0.0211))$.

The curve BECGKNDHJMAFL in Figure 4 describes this behavior at T = 593 K.

Remarks

For ideal gases, at a specified temperature and pressure $P \propto 1/\bar{v}$. In general, $\bar{v} = \bar{v}(P,T)$ is a single–valued function of P and T.

The real gas equation is a cubic equation in $\bar{v}$ at specified P and T. When $T > T_c$, at given T and P the relation yields two complex roots and one real root for the volume. The pressure is a monotonic function of $\bar{v}$ at $T>T_c$ (e.g., illustrated by curve WXY in Figure 4) and hence v= v(T,P), a single valued function. A plot of the pressure with respect to volume is a hyperbola at $T \gg T_c$. As $T \rightarrow T_c$ deviations from hyperbolic behavior occurs around $T=T_c$. Non–monotonic behavior occurs at $T<T_c$ inside the vapor dome (e.g., curve BECGKNDHJMAFL). This is a pressure–explicit type of equation of state, i.e., P = P(T, $\bar{v}$) is a single–valued function of $\bar{v}$ and T, but it is not a monotonic function of T and $\bar{v}$, at a specified temperature $T < T_c$, P does not increase monotonically as $\bar{v}$ decreases. As the volume decreases at constant temperature, first the pressure increases along BEC (according to the relation $P \propto \bar{R}T/\bar{v}$), and as $\bar{v}$ is further reduced (along CGKN), attractive forces become stronger so that the rate of pressure increase is reduced. As the volume is reduced further, the attractive forces become so strong that the pressure starts to decrease (along curve NDHJM). Near the point M, the volume is so small that $\bar{v} \approx \bar{b}$ and, hence, $P \approx \bar{R}T/(\bar{v} - \bar{b})$. The free volume available for molecular movement becomes very small, with the result that there are frequent collisions that result in an increase in the pressure (along the curve MAFL with the attractive forces holding the matter in a liquid state).

At specified pressures and temperature, multiple solutions for $\bar{v}$ can occur. Therefore, $\bar{v}$ is not a single–valued function of P and T at T < Tc for certain ranges of pressures. For instance, at $50<P<155$ bars, three real equilibrium solutions are possible for $\bar{v}$ while at 50 and 155 bars, only two solutions exist. For P >155, and P < 50, only one solution (or v is single valued function of T and P) is possible. However, all of the solutions are not at stable equilibrium states. This can be illustrated in the following manner. Consider water contained at state E in Figure 4 (i.e., 100 bar and 593 K) in a piston–cylinder–weight (PCW) assembly. Upon pushing the piston down slightly and then releasing it, the decreased volume results in the increase of the fluid pressure and the fluid comes back to original state. On the other hand if the fluid were initially at state H (i.e., 133 bar and 593 K), the same disturbance test causes a slight decrease in volume first; but the fluid pressure is decreased to a value less than the external pressure. Therefore, the fluid would be compressed past state M, finally reaching an equilibrium state in the vicinity of F. State E is stable while state H is unstable. Further discussion regarding the stability of the three possible states is contained in Chapter 10. The single–valued and monotonic characteristics are very important while discussing the fundamental relations and thermodynamic postulates (e.g., in Chapter 5).

At large specific volumes, the vapor behaves like an ideal gas ($\bar{v} \gg \bar{b}$, $\bar{a}/\bar{v}^2 \approx 0$) and $P \propto 1/\bar{v}$. Alternately, if we let $\bar{a} = \bar{b} = 0$ the ideal gas equation of state can be obtained from the real gas equation.

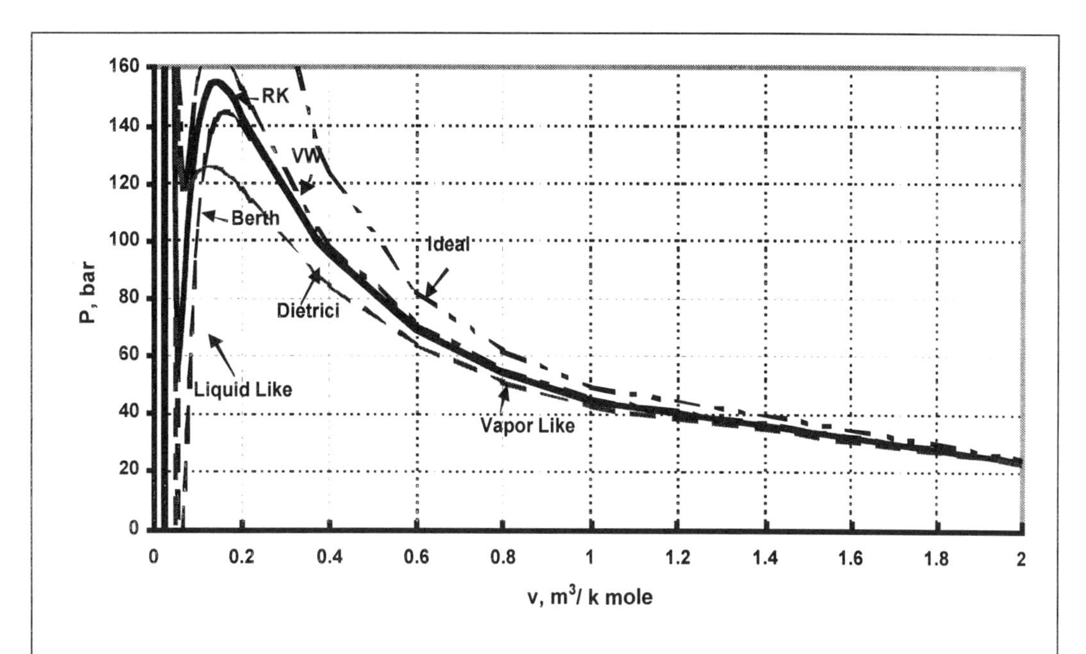

Figure 5: The variation of pressure with specific volume for water using several state equations at 593 K.

Since the real gas state equation is cubic in terms of specific volume, explicit algebraic expressions are available for solving $\bar{v}(P,T)$. Normally there are three real solutions for $\bar{v}$ at specified P and T for $T < T_c$ (e.g., the points F, H, and G in Figure 4 at T = 320°C, P = 113 bar). The smallest value (point F) corresponds to a solution in the liquid phase while the largest value (point G) corresponds to a solution in the vapor phase. The middle value (point H) has no physical meaning. (These liquid–like and vapor–like solutions will be used later for determining P^{sat} *versus* T or for drawing phase diagram.) Points N and M are called spinodal vapor and liquid points, and will be discussed in Chapter 10. Note that sometimes as $\bar{v}$ becomes very small, it approaches liquid–like behavior due to strong attractive forces ($= \bar{a}/\bar{v}^2$), and the pressure becomes negative (a liquid under tension).

Eqs. (26) and (37) are also known as two parameter equations of state since they involve two parameters, $\bar{a}$ and $\bar{b}$. The variation of pressure with specific volume using several state equations is illustrated in Figure 5.

The ratio $P_c \bar{v}_c/T_c = 3/8$ for VW gas while $Z_c = 0.333$ for RK gas. Some representative experimentally obtained values of Z_c are as follows:

Non–polar gases	(e.g., Ar, He, Ne, N_2, etc.)	$0.29 \le Z_c \le 0.3$.
Hydrocarbons	(e.g., C_nH_m)	$0.26 \le Z_c \le 0.29$.
Polar Gases	(e.g., H_2O)	$0.22 \le Z_c \le 0.26$.

Isochoric curves (i.e., $\bar{v}$ = constant or P vs. T curves) obtained from the VW equations are linear, while those obtained from the RK equation are not necessarily so.

The ideal gas equation is applicable when $\bar{v} \gg \bar{b}$ and $\bar{a}/(T^{1/2}\bar{v}^2) \ll \bar{R}T/\bar{v}$. In terms of reduced variables, $v_R' \gg 0.4275/T_R^{3/2}$ and, using the state equation, $P_R \ll T_R^{5/2}/0.4275$.

At 100°C and1 bar, the value $\bar{v}$ can be obtained by applying the RK equation (0.0264 m^3 $kmole^{-1}$), as can the collision pressure ($= \bar{R}T/(\bar{v}-\bar{b}) \approx 5894$ bar) and the pressure reduction due to attractive force ≈ 5893 bar, which indicates that attractive forces cannot be ignored in comparison with collision forces and, hence, the ideal gas equation is not applicable. On the other hand, the ideal gas equation can be used when $\bar{v} \gg \bar{b}$ and if the pressure reduction due to attractive force is far less than the ideal gas pressure. Therefore, for ideal gas behavior to apply $\bar{a}/(T^{1/2}\,\bar{v}^2) \ll \bar{R}T/\bar{v}$. Since $\bar{v} = N_{Avog}\,\ell^3$, $\bar{v} \gg \bar{a}/(\bar{R}T^{3/2})$, i.e.,

$$\ell \gg (C_3 k_B T_C / T_R^{1.5} P_C)^{1/3}, \text{ or } \ell \gg (C_3 \bar{v}_C' / N_{Avag} T_R^{1.5})^{1/3}, \text{ or } \ell / \ell_c' \gg 3.75 / T_R^{1/2}.$$

which implies that

$$\frac{\ell}{\ell_c} = \frac{\text{molecular spacing at any T and P}}{\text{molecular spacing at the critical point}} \gg \frac{5.41}{T_R^{1/2}}.$$

Here, $\bar{v}_c' = N_{Avog}(\ell_c')^3$, $\bar{v}_c = N_{Avog}\,\ell_c^3$, and $\ell_c = 0.693\,\ell_c'$, since $\bar{v}_c/\bar{v}_c' = 1/3$. In terms of reduced variables, $v_R' \gg 0.4275/T_R^{3/2}$ and, using the state equation, $P_R \ll T_R^{5/2}/0.4275$. The volume at minimum potential is related to critical volume. We will see later that $v/v_c \gg 0.7/T_R^{0.5}$ for ideal gas behavior to occur.

The term $\bar{a}$ represents a measure of attractive forces between the molecules. The higher the $\bar{a}$, higher the energy required to detach the molecules from liquid phase, the higher is the saturation temperature. Table A-1 lists boiling and melting points of several substances. Figure 6 plots the boiling and melting points of several substances with respect to $\bar{a}$. Can you tell which substances may be more volatile: methane or water at specified T? The values of "a" are respectively 32.22 and 142.6 (bar m^6 $K^{0.5}$)/kmole^2 for the RK equation.

---

c. *Example 3*

If the number of molecules per unit volume $n' \approx 1/\ell^3$, where ℓ denotes the average distance (or mean free path between molecules), determine the value of ℓ for water vapor at 320°C and 100 bar for the following cases: (a) when the ideal gas law is applicable, and (b) when the RK equation is applicable, and

Compare the answer from part (b) with the molecular diameter obtained from the $\bar{b}$ value.

Using the LJ potential function, determine the ratio of attractive potential to minimum attractive potential.

Using Tables A-4 for the saturated properties of water, determine the intermolecular spacing for the saturated liquid at 320°C

Determine the intermolecular spacing for saturated liquid and vapor at T_{TP}, and at the critical point

---

Solution

The temperature T = 593 K.

$n = 100 \div (0.08314 \times 593) = 2.028$ kmole m^{-3}, and

$n' = 2.028 \times 6.023 \times 10^{26} = 1.22 \times 10^{27}$ molecules m^{-3},

$\ell = 935.44 \times 10^{-12}$ m = 9.35 Å.

$P = \bar{R}T/(\bar{v} - \bar{b}) - \bar{a}/(T^{1/2}\,\bar{v}\,(\bar{v} + \bar{b}))$

$100 = 0.08314 \times 593/(\bar{v} - 0.0210) - 142.64/(593^{1/2}\,\bar{v}(\bar{v} + 0.0210)$

Solving iteratively, $\bar{v} = 0.375$ m^3 $kmole^{-1}$, $n = 2.667$ kmole m^{-3}.

$n' = 1.606 \times 10^{27}$ molecules m^{-3}, $\ell = 853.94 \times 10^{-12}$ m or 8.54 Å.

Note that the first term in P expression is 139.3 bars due to collision of molecules at 593 K with the wall while the pressure of attraction/intermolecular forces is determined as 39.4 bars

$d \approx ((6/4\pi)\,0.0210)^{1/3} = 0.000255$ m or 2.55 Å, $\ell/d = 3.35$, $d/\ell = 0.299$.

Using the LJ function we find that (intermolecular force/maximum attractive force) $\approx (0.299^{13} - 2 \times 0.299^7) = 0.0014$.

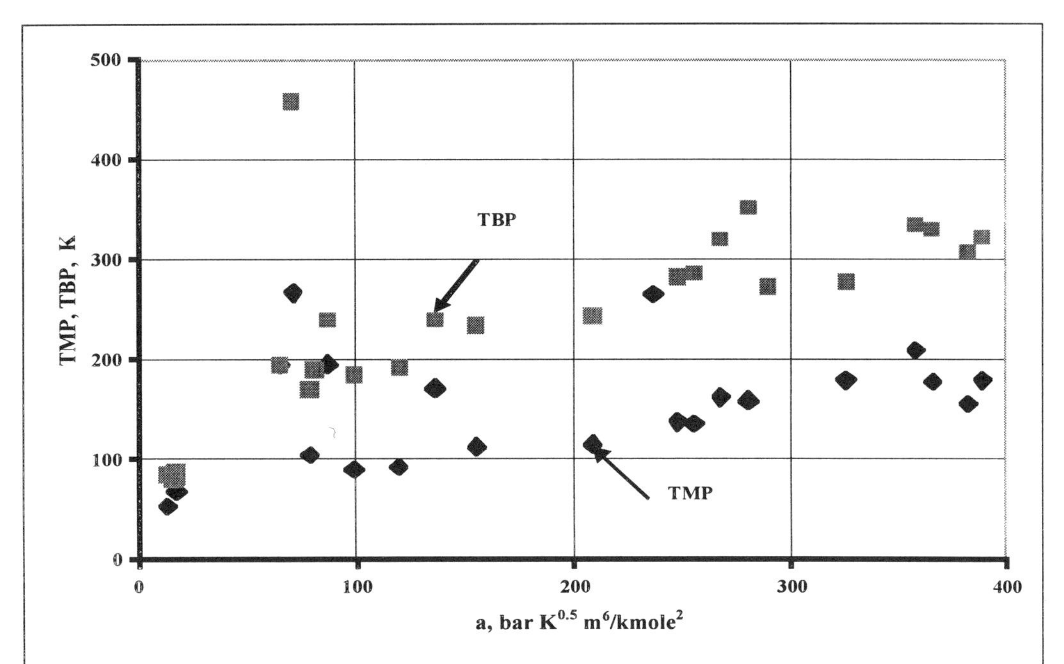

Figure 6: The boiling and melting points of several fluids with respect to $\bar{a}$: solid diamond: MP, solid square: BP.

$v_f = 1.4988\times10^{-3}$ m^3 kg^{-1} or 0.0270 m^3 kmole^{-1}, n = 37.03 kmole m^{-3} or 22.30x10^{27} molecules m^{-3}, $\ell = 355\times10^{-12}$ m, i.e., 0.000355 μm or 3.55 Å which is about half the spacing for vapor at 320°C, and 100 bar. Therefore, ℓ/d ≈ 1.39 and some space remains between the water molecules. The intermolecular potential is 25.8% of the minimum potential, indicating strong attractive forces.

At T_{TP}, v_f = 0.001 m^3 kg^{-1} or $\bar{v}_f$ = 0.0180 m^3 kmole^{-1}, n = 55.55 kmole m^{-3} or 33.46×10^{27} molecules m^{-3}, and $\ell_{f,TP}$ = 3.1 Å, Similarly, v_g = 206.136 m^3 kg^{-1}, n = 0.000269 kmole m^{-3}, or 1.62×10^{21} molecules, and $\ell_{g,TP}$ = 183 Å. At the critical point, v_c = 0.003155 m^3 kg^{-1}, or 0.568 m^3 kmole^{-1}, n = 1.762 kmole m^{-3} or 1.06×10^{27} molecules m^{-3}, and $\ell_{f,C} = \ell_{g,C}$ = 9.8 Å. At lower pressures, for vapor states the molecules are separated at farther distances, while they are more closely packed in the liquid state.

d. Example 4

Determine an expression for the reversible work done by a Van der Waals gas for:
A closed system.
An open system.
Assume the processes to be isothermal.

Solution

The work done during the isothermal compression of a closed system can be written in the form

$$w = \int_{v_1}^{v_2} P dv = \int_{v_1}^{v_2} (RT/(v-b) - a/v^2) dv$$

$$= RT \ln((v_2-b)/(v_1-b)) + a(1/v_2 - 1/v_1)$$

The work done during isothermal compression in a steady state steady flow device is:

$$w = -\int_{v_1}^{v_2} v dP = -(Pv)_{v_1}^{v_2} + \int_{v_1}^{v_2} P dv$$

$$= P_1v_1 - P_2v_2 + RT \ln((v_2-b)/(v_1-b)) + a(1/v_2 - 1/v_1)$$

Remarks

Even though the work expression has properties at states 1 and 2, the functional form changes if we choose a different path for the same final state. Thus, w is a path dependent quantity.

5. Compressibility Charts (Principle of Corresponding States)

We now illustrate how compressibility charts can be constructed by employing Eq. (3) and the RK equation of state, Eq. (37). We define

$$v_R' = v/v_c', \text{ where } v_c' = RT_c / P_c. \tag{42}$$

Using Eq. (42) in Eq. (37) and writing in terms of reduced variables we obtain the expression:

$$P_R = \frac{T_R}{(v_R' - 0.08664)} - \frac{0.42748}{T_R^{0.5} v_R' (v_R' + 0.08664)}. \tag{43}$$

Therefore, for given values of P_R and T_R, v_R' can be obtained. Thereafter, the compressibility factor is determined from the relation $Z = P\bar{v}/\bar{R}T$ by inserting the reduced variables, since

$$Z = P_R P_c v_R' v_c'/(\bar{R} T_R T_c) = P_R v_R'/T_R. \tag{44}$$

It is useful to eliminate v_R' in Eqs. (43) and (44) and obtain a relation for Z_{RK} (based on the RK equation of state) in terms of P_R and T_R, i.e.,

$$P_R = \frac{T_R}{Z_{RK} T_R / P_R - 0.08664} - \frac{0.42748}{Z_{RK} T_R^{1.5} / P_R (Z_{RK} T_R / P_R + 0.08664)} \tag{45}$$

This relation can be simplified in order to obtain Z (= Z_{RK}) in terms of P_R and T_R, namely, $Z^3 - Z^2 + (a' - b'^2 - b') Z - a' b' = 0$, where

$$a' = 0.4275 P_R/T_R^{2.5}, \tag{46}$$

$$\text{and } b' = 0.08664 P_R/T_R. \tag{47}$$

The appendix presents explicit solutions for the three roots of Z. As an illustration, Figure 6a contains a compressibility chart for Z vs P_R with T_R as a parameter for an equation of state for which Z_C=0.2801. Figure 7 presents a compressibility chart for Z vs P_R with T_R as a parameter for an equation of state for which Z_C=0.2801.

For sake of illustration, we consider water at a pressure of 250 bar, a temperature of 873 K. Therefore, P_R = 250÷220.9 = 1.132, T_R = 873÷647.3 = 1.349, so that a′ = 0.329 and b′ = 0.0727. Consequently, the value of $(a' - b'^2 - b')$ is 0.151, and a′ b′ = 0.01665, and a single real root exists for Eq. (46), i.e., Z = 0.845 (point A, Fig. 6a). Likewise, if P = 133 bars and T = 593 K, P_R = 0.601, T_R = 0.916, a′ = 0.320, b′ = 0.0569, $(a' - b'^2 - b')$ = 0.260, and a′ b′ = 0.01824. There are now three roots for the equation, i.e.,, Z_1 = 0.115 (for a liquid–like solution, point L), Z_2 = 0.249 (unstable solution, point M; see Chapter 10), and Z_3 = 0.632 (vapor–like solution, point V).

e. *Example 5*

What is the value of Z_{RK} at P_R = 1.5 and v_R' = 0.45?

What is the value of Z_{RK} at P_R = 1.5 and T_R = 1.15?

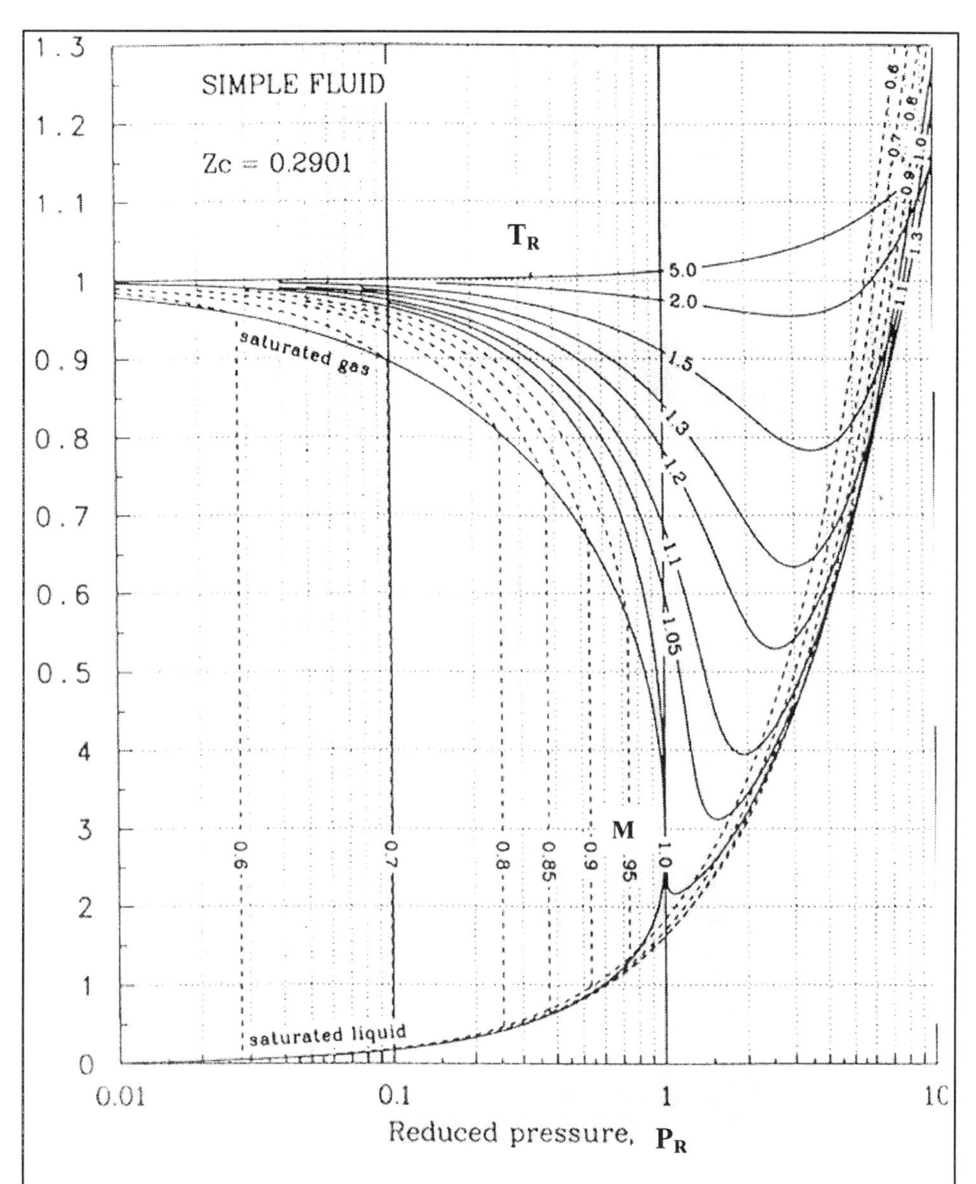

Figure 7: A compressibility chart for Z *vs.* P_R with T_R as a parameter for an equation of state for which Z_C=0.2801 (from R. Sonntag, C. Borgnakke, and G. J. Wiley, *Fundamentals of Classical Thermodynamics*, 5th Ed. John Wiley &Sons, 1998, p 763. With permission.).

What is the value of Z for CH_4 at T = 219.3 K and P = 69.6 bars according to the RK equation?

What is the value of Z for N_2 at T = 145.1 K and P = 50.85 bars according to the RK equation?

Solution

Consider Eqs. (37) and (41), and try Z_{RK} = 0.6. You will find that

$1.5 \neq (0.45\times1.5 \div (0.6 \times (0.45-0.08664)) - 0.42748 \times 0.6^{0.5}/((1.5^{0.5} \times 0.45 \div .5)(0.45 + 0.08664) = 3.0961 - 1.6689 = 1.42)$.

Next, try Z_{RK} = 0.58, and in this case the RHS

= 3.2029 – 1.6409 = 1.56.

By interpolating we find that Z_{RK} = 0.59. The compressibility charts give the same result.

For the second problem assume Z_{RK} = 0.6.

The LHS = 1.5.

The RHS = 1.15 ÷ (0.6 × 1.15 ÷ 1.5 – 0.08664) – 0.42748 × 1.5 ÷ ($1.15^{1.5}$ × 0.6 × (0.6 × 1.15 ÷ 1.5 + 0.08664)) = 3.0801 – 1.5853 = 1.494.

Hence, the LHS ≈ RHS.

P_R = 69.6 ÷ 46.4 = 1.5, T_R = 219.3 ÷ 190.7 = 1.15, and, therefore, Z_{RK} = 0.6.

P_R = 50.85 ÷ 33.9 = 1.5, T_R = 145.1 ÷ 126.2 = 1.15, and, therefore, Z_{RK} = 0.6.

Remarks

It is seen that at specified P_R and T_R Z_{RK} is equal for all real gases.

f. Example 6

Determine the value of v and compare it with that obtained from the steam tables, the value of v'_R, and that of Z_{RK} for water at a pressure of 250 bars and a temperature of 600°C.

Solution

For water P_c = 220.9, and T_c = 647.3 K. Using Eq. (41)

$\bar{a} = 0.42748\ R^2T_c^{2.5}/P_c = 0.42748\times(0.0814\text{ bar m}^3\text{ kmole}^{-1}\text{ K}^{-1})^2\times(647.3\text{ K})^{2.5}\div 220.9$ bar

= 142.59 bar m^6 $K^{1/2}$ $kmole^{-2}$

$\bar{b} = 0.08664\ RT_c/P_c = 0.08664 \times 0.08314$ bar m^3 $kmole^{-1}$ K^{-1} × 647.3 K ÷ 220.9 bar

= 0.0211 m^3 $kmole^{-1}$.

Therefore, for RK equation (37)

250 bar = (0.08314 bar m^3 $kmole^{-1}$ K^{-1} × 873 K) ÷ ($\bar{v}$ – 0.0211)

– 142.59 bar m^6÷($873^{1/2}$ $\bar{v}$ ($\bar{v}$ + 0.0211)).

Solving for $\bar{v}$, we obtain three real solutions. Selecting the largest of the three values, which corresponds to a vapor–like solution, $\bar{v}$ = 0.246 m^3 $kmole^{-1}$.

∴ v = 0.246 m^3 $kmole^{-1}$ ÷ 18.02 kg $kmole^{-1}$ = 0.01361 m^3 kg^{-1}.

The steam tables give a value of 0.014137, a difference of –3.7%.

$v'_c = T_c/P_c$ = 0.08314 bar m^3 $kmole^{-1}$ K^{-1} × 647.3 K ÷ 220.9 bars = 0.244 m^3 $kmole^{-1}$.

Since, $v'_R = \bar{v}/v'_c$ = 0.246 ÷ 0.244 = 1.008, and P_R = 1.132, T_R = 1.349, using Eq. (43), v'_R = 1.008.

Now, $(P\bar{v})_{RK} = Z_{RK}\bar{R}T$. Since, $\bar{v}_{RK}$ = 0.246 m^3 $kmole^{-1}$,

Z_{RK} = 250 bar × 0.246 m^3 $kmole^{-1}$ ÷ (0.08314 bar m^3 $kmole^{-1}$ K^{-1} × 873 K) = 0.845.

Remarks

The reduced parameters $P_R = P/P_c$ = 250 ÷ 220.9 = 1.132, and $T_R = T/T_c$ = 873 ÷ 647.3 = 1.349. A value of v'_R = 1.007 can be obtained using Eq. (43). Thereafter, since v'_c = 0.244, $\bar{v}$ = 0.246 m^3 $kmole^{-1}$.

Equation (46) is a representation of the principle of corresponding states, which states that the compressibility factor for all gases is the same at specified values of P_R and T_R. The factor evaluated with two parameter equation of state is normally denoted as $Z^{(0)}$.

Using Eq. (48) we can now plot Z_{RK} *vs.* P_R using v'_R as a parameter. At fixed temperatures, as the pressure of a gas increases from very low values, its volume decreases, the product Pv ≈ constant, and, hence, initially Z ≈ 1. As the volume de-

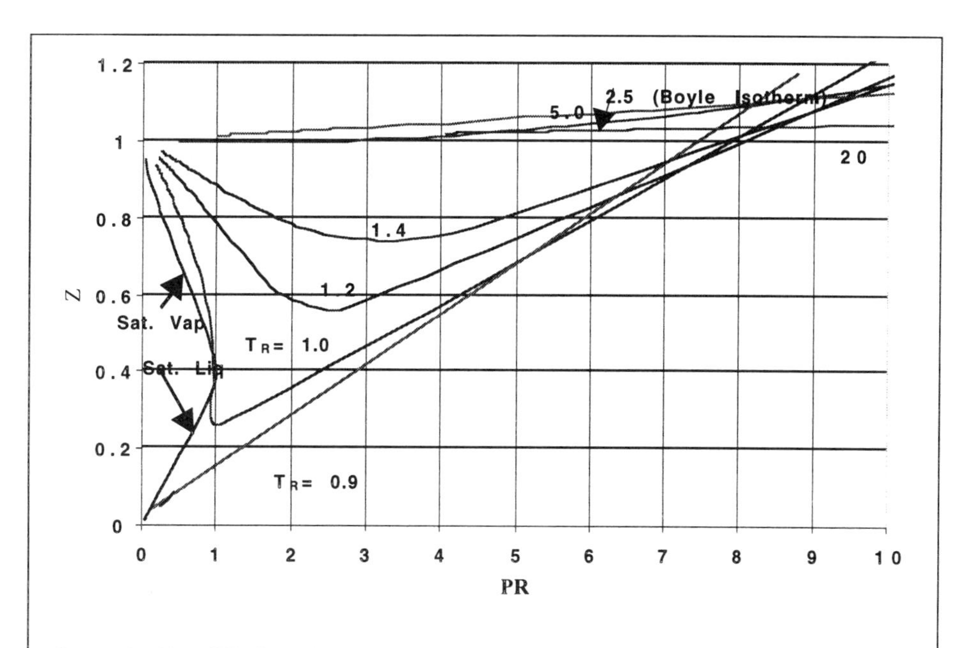

Figure 8: Simplified Z chart illustrating various regimes. Most of the plots were generated using RK equation.

creases, the pressure increases due to frequent molecular collisions of molecules. However, the intermolecular attractive forces reduce the pressure so that (Pv) < $(Pv)_{ideal\ gas}$ and, consequently, the value of Z decreases. Beyond a certain pressure, further pressure increments produce smaller and smaller reductions in the volume (i.e, liquid volumes) so that the value of Pv (or Z) again increases. Therefore, the compressibility factor passes through a minimum value with respect to P_R (e.g., point B in Figure 8 at T_R=1.2), since the finite body volume dominates pressure effects at high pressures and overwhelms the intermolecular attractive forces. At low pressures, the variation in the values of Z with pressure has both negative and positive slopes. The temperature T_R at which $(\partial Z/\partial P_R) = 0$ is known as the Boyle temperature. At higher pressures Z = 1 at a particular value of P_R, once T_R is fixed, i.e., the decrease in pressure due to attractive forces at this condition equals the increase in pressure due to more frequent collisions within a smaller free volume available for molecular motion. The Z–minimum condition, the Boyle temperature, and the Z = 1 condition are discussed below.

Table 1 contains a comparison of values of Z from charts and other state equations.

Table 1: Comparison of values of Z from charts and other state equations.

		Chart	***VW***		***RK***		***BWR***	
Pr	***Tr***	***Z***	***Z***	***% error***	***Z***	***% error***	***Z***	***% error***
2	1	0.31	0.43	38.7	0.25	12.9	0.32	3.2
2	1.3	0.7	0.65	-7.1	0.7	0	0.69	-1.4
2	3	0.96	0.92	-4.2	0.95	-1	0.95	-1
4	1	0.56	0.74	32.1	0.59	5.4	0.58	3.6
4	1.3	0.69	0.75	8.7	0.7	1.4	0.66	-4.3
4	2	0.96	0.91	-5.2	0.95	-1	0.96	0
6	1	0.78	0.96	23.1	0.82	5.1	0.81	3.8
6	1.3	0.82	0.97	18.3	0.84	2.4	0.82	0

		Chart	*VW*		*RK*		*BWR*	
Pr	*Tr*	*Z*	*Z*	*% error*	*Z*	*% error*	*Z*	*% error*
6	2	1.01	1.05	4	0.99	-2	1	-1
Av. % Diff.				15.7		3.5		2

6. Boyle Temperature and Boyle Curves

a. Boyle Temperature

Using the RK equation, the Boyle temperature can be determined as follows. First, multiply Eq. (37) by v to obtain

$$Pv = RTv/(v-b) - a/(T^{1/2}(v+b)). \quad (48a)$$

Dividing by RT,

$$Z = Pv/RT = v/(v-b) - a/(RT^{3/2}(v+b)). \quad (48b)$$

Since $Z = Pv/RT$, $\partial Z/\partial P = (1/RT)\ \partial(Pv)/\partial P$. Therefore, if $\partial Z/\partial P = 0$, it implies that $\partial(Pv)/\partial P = 0$, and

$$\partial(Pv)/\partial P = (RT/(v-b) - RTv/(v-b)^2 + a/(T^{1/2}(v+b)^2))\ (\partial v/\partial P)$$

$$= (a/(T^{1/2}(v+b)^2 - RTb/(v-b)^2)\ (\partial v/\partial P). \quad (49)$$

Since

$$\partial v/\partial P = 1/((a(2v+b)/T^{1/2}\ v^2\ (v+b)^2) - RT/(v-b)^2) \quad (50)$$

As $P \to 0$, the volume becomes large, and Eq (50) assumes the form

$$\partial(Pv)/\partial P \to (b - a/RT^{3/2}),\ \text{i.e.,}\ \partial(Z)/\partial P \to (b/RT - a/R^2T^{5/2}). \quad (51)$$

Expressed in terms of reduced variables.

$$\partial Z/\partial P_R = 0.08664/T_R - 0.4275/T_R^{5/2}\ \text{as}\ P_R \to 0. \quad (52)$$

Recall from Abbott's correlation, (Eq. 16) that $\partial Z/\partial P_R = B_1(T_R) = 0.083/T_R - 0.4275/T_R$. Hence, using the RK equation, when $T_R = 1$, the slope $\partial Z/\partial P_R = -0.3409$. This slope tends to zero when $T_R^{3/2} \to 2.8983$. The corresponding temperature is called the Boyle temperature. (This result compares well with the value of 2.76 that is obtained using the empirical virial expressions.) The slope is negative when $T_R < 2.8983$ and is positive when $T_R > 2.8983$. It can be shown that the slope has a maximum value of 0.009737 at $T_R = 5.33873$ and decreases slowly to 0.004093 as $T_R \to 20$, indicating that the value of $Z \approx 1$ even at high pressures and temperatures. For instance, in diesel engines pressures as high as 80 bars are attained but the gaseous mixtures behave like an ideal gas, since for many combustion–related species $T_R > 20$.

b. Boyle Curve

If $T_R < 2.8983$, Z approaches a minimum value at a particular reduced pressure. The loci of all the minima of $Z = Z_{minimum}$ constitute the Boyle curve along which the gas behavior is similar over a wide pressure range. The Boyle curve may be characterized as follows. The $Z_{minimum}$ condition occurs where the product (Pv) has a minimum value at a specified temperature. Applying Eq. (49a),

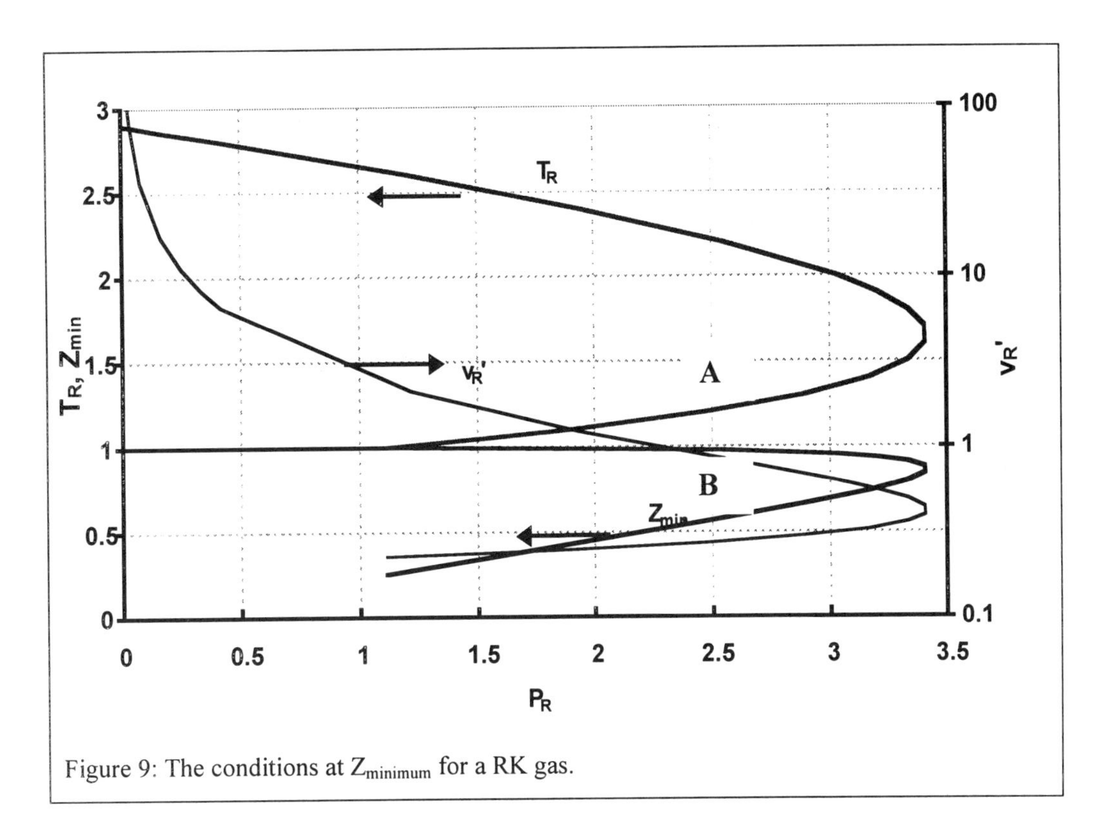

Figure 9: The conditions at $Z_{minimum}$ for a RK gas.

$$(\partial(Pv)/\partial P) = (a/(T^{1/2}(v+b)^2 - R\,T\,b/(v-b)^2)\,(\partial v/\partial P), \tag{53}$$

at the minima $\partial(Pv)/\partial P = 0$. Therefore, at this condition, either $\partial v/\partial P = 0$, which implies that there is no volumetric change during compression (an unrealistic supposition), or

$$(R\,T/(v-b) - R\,T\,v/(v-b)^2 + a/(T^{1/2}(v+b)) = 0. \tag{54}$$

Using Eq. (55) to solve for T,

$$T^{3/2} = a\,(v-b)^2/(R\,b(v+b)^2).$$

In terms of reduced variables

$$T_R^{3/2} = 4.9342\,(v_R' - 0.08664)^2/(v_R' + 0.08664)^2. \tag{55}$$

Further, solving for v_R',

$$v_R' = 0.08664\,(1 + 0.4502\,T_R^{3/4})/(1 - 0.4502\,T_R^{3/4}) \tag{56}$$

Therefore, at specified values of T_R, v_R' can be determined using Eq. (57) and the result substituted into Eq. (45) to determine P_R at $Z_{minimum}$. There is no solution for v_R' and $Z_{minimum}$ for values of $T_R > 2.898$ (i.e., above the Boyle temperature), since Z always increases. These results are illustrated in Figure 9.

c. *The Z = 1 Island*

For specified values of T_R, the corresponding values of Z first decrease below unity as the pressure is increased, pass through a minima, and then increase to cross over Z = 1. On a pressure–temperature graph there is restricted regime or an island on which Z = 1. Here the real gas behaves as an ideal gas. Using Eq. (49b) at this condition, we obtain the relation

$$Pv/RT = Z = 1/(1 - b/v) - a/(R\,T^{3/2}\,v\,(1 + b/v)) = 1, \text{ i.e.,} \tag{57}$$

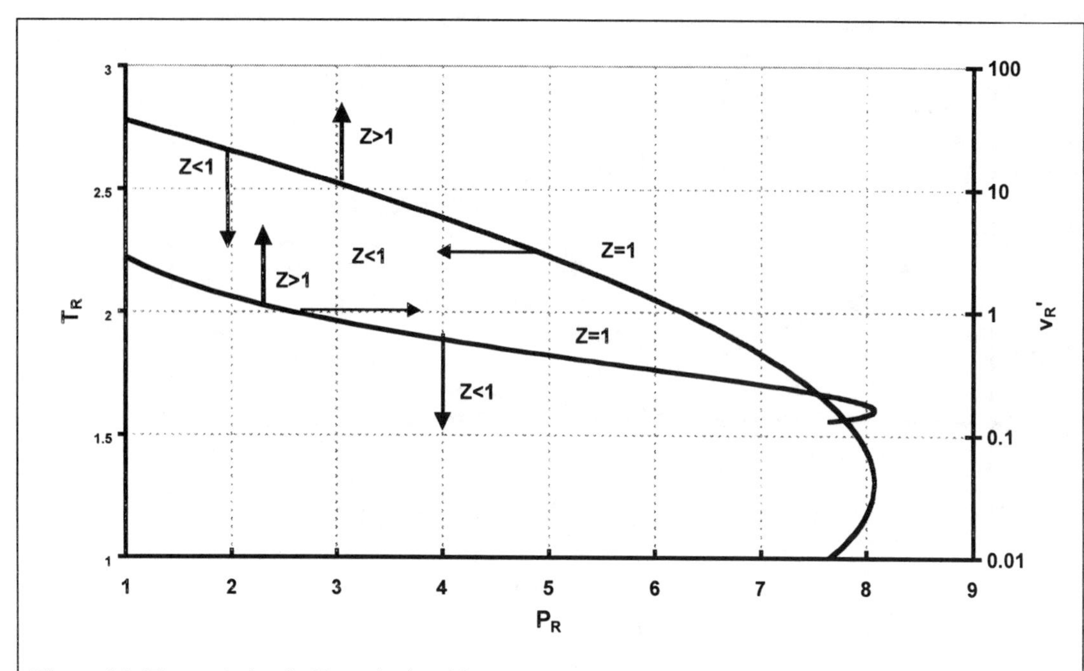

Figure 10: The variation in T_R and v'_R with respect to P_R at the condition Z = 1.

$$(v\, R\, T^{3/2}/a) = (1 - b/v)/(1 + b/v), \tag{58}$$

which yields a relation between v and T at Z = 1. Solving for the product (bv^{-1}), we obtain

$$b/v = (A - 1)/(A + 1), \tag{59}$$

where $A = (a/vRT^{3/2})$. Substituting Eq. (60) into the RK equation of state,

$$P = (RT/(2\,A\,b))\,(A - 1) - (a/(2\,b^2\,T^{1/2}\,A(1 + A)))\,(1 - A)^2. \tag{60}$$

In the terms of reduced parameters, $A = 4.9342/T_R^{3/2}$, and $v'_R = 0.08664\,(A + 1)/(A - 1)$, so that

$$P_R = 5.77101\,T_R\,(A - 1) - 28.47536(A - 1)^2/(T_R^{1/2}\,A\,(1 + A))). \tag{61}$$

These results are illustrated in Figure 10.

7. Deviation Function

The function $(v - v_0) = (v - RT/P)$ is called the deviation function for volume. It provides a measure of the deviation of the volume of a real gas from that of an ideal gas under the same T and P. The generalized RK equation of state can be expressed as

$$P = RT/(v - b) - a/(T^n\,v\,(v + c)) \tag{62}$$

where n=1/2, c= b for a RK fluid, n=0, c=0 for VW fluids, and n=1, c=0 for Berthelot fluids.

As $P \to 0$, v tends to large values, and b/v becomes smaller. Under these conditions Eq. (63) assumes the form

$$P = (RT/v)\,(1 - b/v)^{-1} - (a/T^{1/2})\,(1/v^2)\,(1 + c/v)^{-1}.$$

Upon expansion, $(1 + b/v)^{-1} = 1 - b/v + (b/v)^2 + \ldots$, and neglecting terms of the order v^3 and higher,

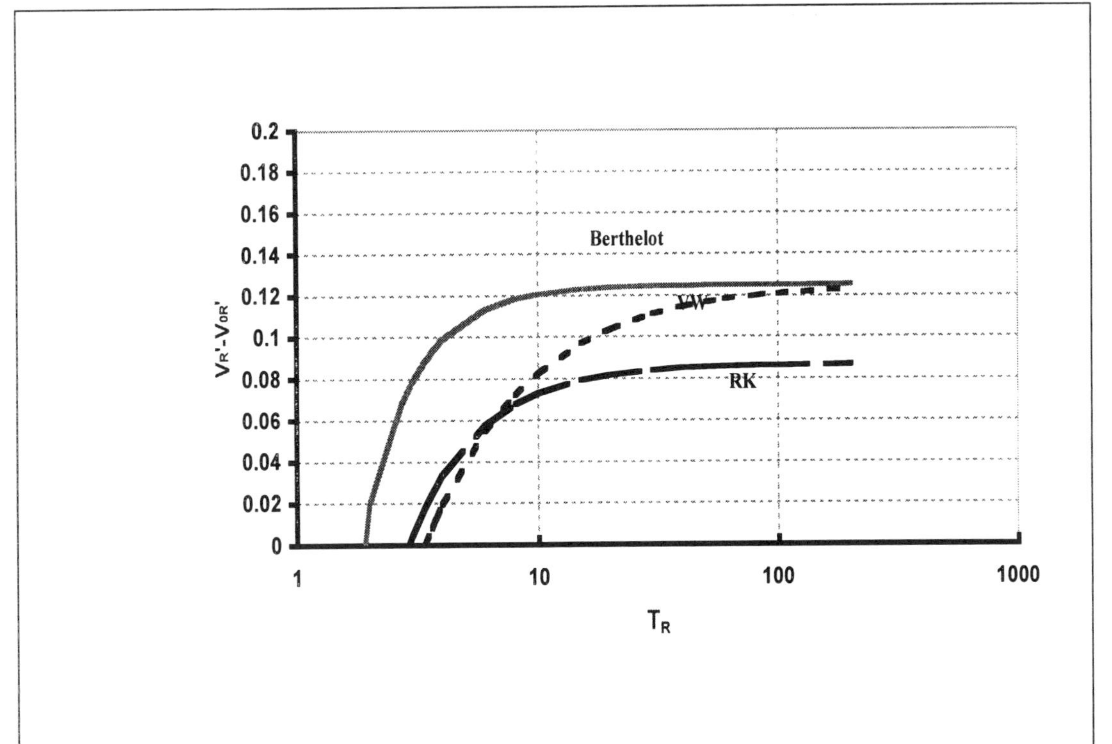

Figure 11: The deviation function for Berthelot, RK,and VW gases, $P_R \to 0$

$$P = (RT/v) + (1/v^2)\,(bRT - a/T^n). \tag{63}$$

Therefore,

$$Pv/RT = Z\,(v,T) = (1 + b/v) - a/(vRT^{(n+1)}) = 1 + B(T)/v, \tag{64}$$

Where $B(T) = b - a/RT^{(n+1)}$. As $P \to 0$, $v \to \infty$, and $Z \to 1$.

Solving for v from Eq. (64),

$$v = RT/2P\,(1 \pm (1 + (4\,P/RT)(b - a/RT^{\,n+1}))^{1/2}).$$

As $P \to 0$,

$$v \approx RT/2P\,(1 \pm (1 + 2\,P/RT\,(b - a/RT^{n+1}))),$$

only positive values of which are acceptable. Therefore,

$$v = RT/P + (b - a/RT^{n+1}),$$

Since $RT/P = v_0$,

$$v = v_0 + b - a/(RT^{n+1}),\ \text{for RK, VW and Berthelot} \tag{65}$$

Therefore, at lower pressures, $(v - v_0)_{P\to 0} = b - a/RT^{n+1}$, where $n \geq 0$. For example, if $n = 0$, the volume deviation function has a value equal to $(b - a/RT)$, and is a function of temperature. In this case, the real gas volume never approaches the ideal gas volume even when $T \to \infty$. In dimensionless form

$$v'_R - v'_{o,R} = a\times - b\times/T_R^{(n+1)},\ \text{for RK, VW and Berthelot fluids.}$$

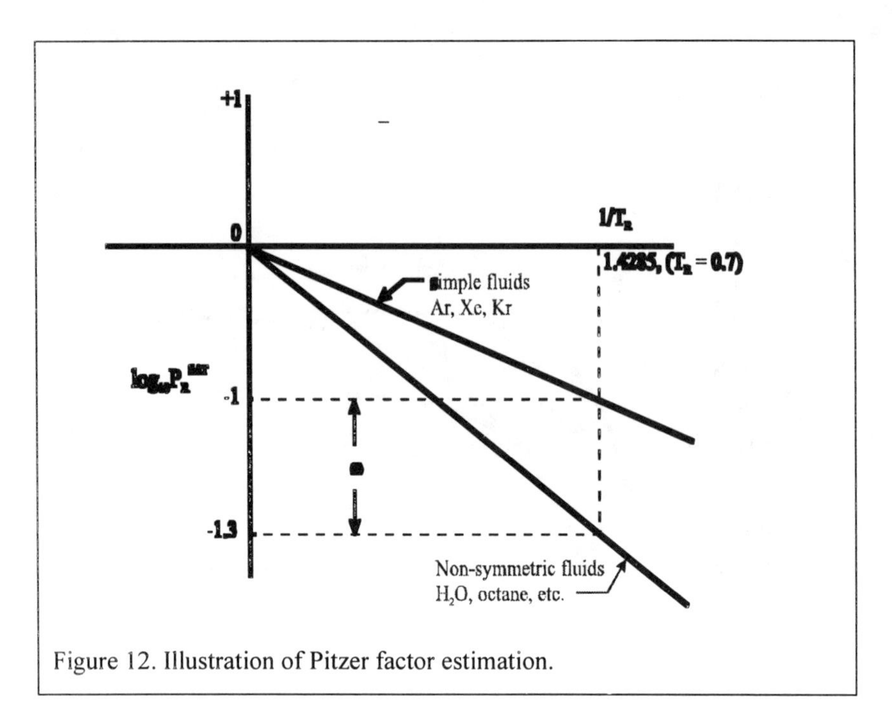

Figure 12. Illustration of Pitzer factor estimation.

where n = 0, 1, 1/2, ax = (27/64), (27/64), 0.4275, and bx = 0.125, 0.125, 0.08664, respectively, for the Van der Waals, Berthelot and RK equations. The difference between v'_R and v'_{oR} are illustrated with respect to T_R for these equations as $P_R \to 0$ in Figure 11.

If $\bar{a} = \bar{b} = 0$ in any of the real gas equations of state, these equations are identical to the ideal gas state equation.

8. Three Parameter Equations of State

If $v = v_c$ (i.e., along the critical isochore), employing the Van der Waals equation,

$$P = RT/(v_c - b) - a/v_c^2,$$

which indicates that the pressure is linearly dependent on the temperature along that isochore. Likewise, the RK equation also indicates a linear expression of the form

$$P = RT/(v_c - b) - a/(T^{1/2} v_c (v_c + b)).$$

However, experiments yield a different relation for most gases. Simple fluids, such as argon, krypton and xenon, are exceptions. The compressibility factors calculated from either the VW or RK equations (that are two parameter equations) are also not in favorable agreement with experiments. One solution is to increase the number of parameters.

a. Critical Compressibility Factor (Z_c) Based Equations

Clausius developed a three parameter equation of state which makes use of experimentally measured values of Z_c to determine the three parameters, namely

$$P = \bar{R}T/(\bar{v} - \bar{b}) - \bar{a}/(T(\bar{v} + \bar{c})^2). \quad (66)$$

where the constants can be obtained from two inflection conditions and experimentally known value of Z_C, critical compressibility factor.

b. Pitzer Factor

The polarity of a molecule is a measure of the distribution of its charge. If the charge it carries is evenly or symmetrically distributed, the molecule is non–polar. However, for some chemical species, such as water, octane, toluene, and freon, the charge is separated across the molecule, making it uneven or polar. The compressibility factors for nonsymmetric or polar fluids are found to be different from those determined using two parameter equations of state. Therefore, a third factor, called the Pitzer or acentric factor ω has been added so that the empirical values correspond with those obtained from experiments. This factor was developed as a measure of the structural difference between the molecule and a spherically symmetric gas (e.g., a simple fluid, such as argon) for which the force–distance relation is uniform around the molecule. In case of the saturation pressure, all simple fluids exhibit universal relations for P_R^{sat} with respect to T_R (as illustrated in Figure 14). In Chapter 7 we can derive such a relation using a two parameter equation of state. For instance, when $T_R = 0.7$, all simple fluids yield $P_R^{sat} \approx 0.1$, but polar fluids do not. The greater the polarity of a molecule, the larger will be its deviation from the behavior of simple fluids. Figure 14 could also be drawn for $\log_{10} P_r^{sat}$ vs. $1/T_R$ as illustrated in Figure 12. The acentric factor ω is defined as

$$\omega = -1.0 - \log_{10} (P_R^{sat})_{TR=0.7} = -1 - 0.4343 \ln (P_R^{sat})_{TR=0.7}. \quad (67)$$

Table A-1 lists experimental values of ω" for various substances. In case they are not listed, it is possible to use Eq. (68).

i. Comments

The vapor pressure of a fluid at $T_R = 0.7$, and its critical properties are required in order to calculate ω. For simple fluids ω = 0.

For non-spherical or polar fluids, a correction method can be developed. If the compressibility factor for a simple fluid is $Z^{(0)}$, for polar fluids $Z \neq Z^{(0)}$ at the same values of T_R and P_R.

We assume that the degree of polarity is proportional to ω. In general, the difference $(Z - Z^{(0)})$ at any specified T_R and P_R increases as ω becomes larger (as illustrated by the line SAB in Figure 13).

With these observations, we are able to establish the following relation, namely.

$$(Z\ (\omega,T_R,P_R) - Z^{(0)}\ (P_R, T_R)) = \omega Z^{(1)}\ (T_R, P_R). \quad (68)$$

Evaluation of $Z(\omega,T_R,P_R)$ requires a knowledge of $Z^{(1)}$, w and $Z^{(0)}\ (P_R, T_R)$.

c. *Evaluation of Pitzer factor,* ω

i. Saturation Pressure Correlations

The function $\ln(P^{sat})$ varies linearly with T^{-1}, i.e.,

$$\ln P^{sat} = A - B\ T^{-1}. \quad (69)$$

Using the condition $T = T_c$, $P = P_c$, if another boiling point T_{ref} is known at a pressure P_{ref}, then the two unknown parameters in Eq. (70) can be determined. Therefore, the saturation pressure at $T = 0.7T_c$ can be ascertained and used in Eq. (69) to determine ω.

ii. Empirical Relations

Empirical relations are also available, e.g.,

$$\omega = (\ln P_R^{sat} - 5.92714 + 6.0964/T_{R,BP} + 1.28862 \ln T_{R,BP} - 0.16935\ T_{R,BP})/$$

$$(15.2578 - 15.6875/T_{R,NBP} + 0.43577\ T_{R,NBP}), \quad (70)$$

where P_R denotes the reduced vapor pressure at normal boiling point (at P = 1 bar), and $T_{R,NBP}$ the reduced normal boiling point.

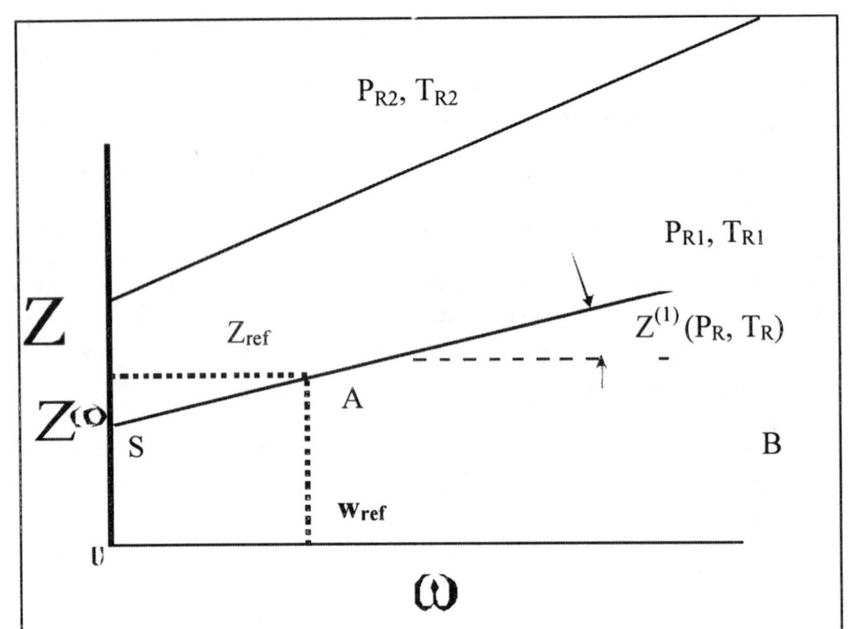

Figure 13: An illustration of the variation in the compressibility factor with respect to the acentric factor.

An alternative expression involves the critical compressibility factor, i.e.,

$$\omega = 3.6375 - 12.5\, Z_c. \tag{71}$$

Another such relation has the form

$$\omega = 0.78125/Z_c - 2.6646. \tag{72}$$

9. Other Three Parameter Equations of State

Other forms of the equation of state are also available.

a. One Parameter Approximate Virial Equation

For values of $v_R > 2$ (i.e., at low to moderate pressures),

$$Z = 1 + B_1(T_R)\, P_R, \tag{73}$$

where $B_1(T_R) = B^{(0)}(T_R) + \omega B^{(1)}(T_R)$, $B^{(0)}(T_R) = (0.083\ T_R^{-1}) - 0.422\ T_R^{-2.6}$, and $B^{(1)}(T_R) = 0.139 - 0.172\ T_R^{-5.2}$.

b. Redlich–Kwong–Soave (RKS) Equation

Soave modified the RK equation into the form

$$P = RT/(v-b) - a\,\alpha\,(\omega,T_R)/(v(v+b)), \tag{74}$$

where $a = 0.42748\ R^2T_c^2P_c^{-1}$, $b = 0.08664\ RT_cP_c^{-1}$, and $\alpha\,(\omega,T_R) = (1 + f(\omega)(1 - T_R^{0.5}))^2$, which is determined from vapor pressure correlations for pure hydrocarbons. Thus, $f(\omega) = (0.480 + 1.574\ \omega - 0.176\ \omega^2)$.

c. Peng–Robinson (PR) Equation

The Peng–Robinson equation of state has the form

$$P = (RT/(v - b)) - (a\,\alpha(\omega,T_R)/((v + b(1 + 2^{0.5}))(v + b(1 - 2^{0.5}))), \tag{75}$$

where $a = 0.45724\ R^2T_c^2P_c^{-1}$, $b = 0.07780\ RT_cP_c^{-1}$ and $\alpha(\omega,T_R) = (1 + f(\omega)(1 - T_R^{0.5}))^2$, $f(\omega) = 0.37464 + 1.54226\ \omega - 0.26992\ \omega^2$. Equation (75) can be employed to predict the variation of P^{sat} with respect to T, and can be used to explicitly solve for T(P,v).

10. Generalized Equation of State

Various equations of state (e.g., VW, RK, Berthelot, SRK, PR, and Clausius II) can be expressed in a general cubic form, namely,

$$P = RT/(v-b) - a\alpha(\omega,T_R)/(T^n(v+c)(v+d)). \tag{76}$$

In terms of reduced variables this expression assumes the form

$$P_R = T_R/(v'_R - b') - a'\alpha(\omega,T_R)/(T_R^n(v'_R + c')(v'_R + d')), \tag{77}$$

where $a' = a/(P_c v_c'^2 T_c^n)$, $b' = b/v'_c$, $c' = c/v'_c$, and $d' = d/v'_c$. Tables are available for the parameters a′ to d′. Using the relation $Z = P_R v'_R/T_R$, we can obtain a generalized expression for Z as a function of T_R and P_R, i.e.,

$$Z^3 + Z^2((c' + d' - b')P_R/T_R - 1) + Z(a'\alpha(\omega,T_R)P_R/T_R^{2+n} -$$

$$(1 + b'P_R/T_R)(c' + d')P_R/T_R + c'd'P_R^2/T_R^2) -$$

$$(a'\alpha(\omega,T_R)b'P_R^2/T_R^{(3+n)} + (1 + P_R b'/T_R)(c'd'P_R^2/T_R^2)) = 0. \tag{78}$$

Writing this relation in terms of v'_R,

$$v_R'^3 P_R + v_R'^2((c' + d' - b')(P_R/T_R) - 1) + v'_R((c'd' - b'c' - b'd')P_R - (c' + d')T_R$$

$$+ a\times/T_R^n) - P_R b'c'd' - a'\alpha(\omega,T_R)b'/T_R - T_R c'd' = 0. \tag{79}$$

Using this equation along with the relation $T_R = P_R v'_R/Z$, the compressibility factor can be obtained as a function of P_R and v'_R, i.e.,

$$Z^{(3+n)}(a'\alpha(w, T_R)/(v_R'^{(2+n)}P_R^{(1+n)}))(1 - b'/v'_R) +$$

$$Z^3(1 + (c' + d' - b')/v'_R - (b'/v_R'^2)(c' + d' - d'/v_R)) - Z^2(1 + (c' + d')/v'_R - c'd'P_R/v'_R +$$

$$d'/v_R'^2) - Z(b'c'/v'_R) - c' = 0.. \tag{80}$$

where T_R in $\alpha(w, T_R)$ expression must be replaced by $P_R v_R'/Z$. Table 2 tabulates values of α, n, a′, b′, c′, and d′ for various equations of state.

Table 2: Constants for the generalized real gas equation of state.

	Berthelot	*Clausius II*	*PR*	*PR with w*	*RK*	*SRK*	*VW*
a′	0.421875	0.421875	0.45724	0.4572	0.42748	0.42748	0.421875
b′	0.125	-0.02	0.0778	0.0778	0.08664	0.08664	0.125
c′	0	0.145	0.187826	0.187826	0.08664	0.08664	0
d′	0	0.375	-0.03223	-0.03223	0	0	0
n	1	1	0	0	0.5	0	0
f(ω), H_2O				0.873236		1.000629	

Note that Z_c is required for Clausius II while ω is required for RKS, PR, f(ω) for H_2O with ω = 0.344

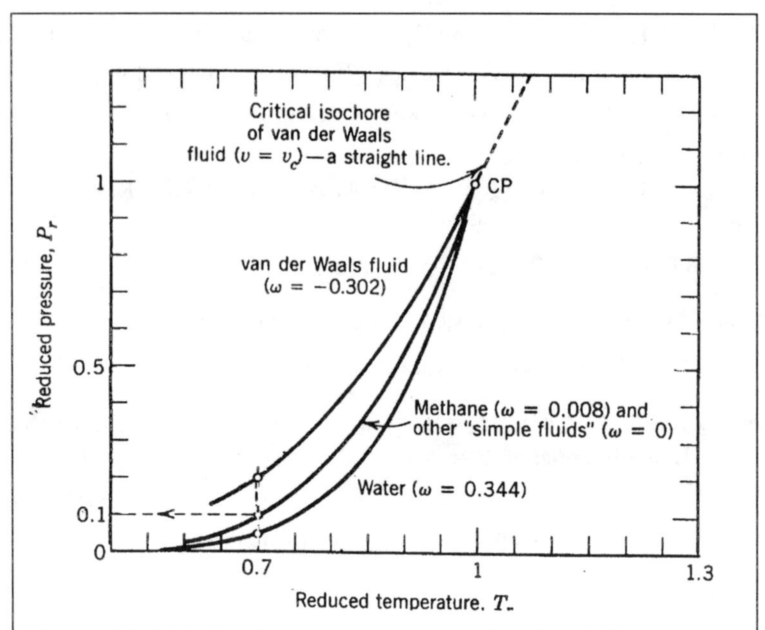

Figure 14: Relation between pressure and volume for compression/expansion of air (from A. Bejan, *Advanced Engineering Thermodynamics*, John Wiley and Sons., 1988, p 281).

11. Empirical Equations Of State

These equations accurately predict the properties of specified fluid; however, they are not suitable for predicting the stability characteristics of a fluid (Chapter 10).

a. *Benedict–Webb–Rubin Equation*

The Benedict Webb Rubin (BWR) equation of state which was specifically developed for gaseous hydrocarbons, has the form

$$P = RT/v + (B_2RT - A_2 - C_2/T^2)/v^2 + (B_3RT - A_3)/v^3 + A_3C_6/v^6$$

$$+ (D_3/(v^3T^2))(1+E_2/v^2)\exp(-E_2/v^2) \qquad (81)$$

The eight constants in this relation are tabulated in the literature. This equation is not recommended for polar fluids. Table A-20A lists the constants.

b. *Beatie – Bridgemann (BB) Equation of State*

This equation is capable representing P-v-T data in the regions where VW and RK equations of state fail particularly when $\rho < 0.8\,\rho_c$. It has the form

$$P\,\bar{v}^2 = \bar{R}T\,(\bar{v} + B_0\,(1-(\bar{b}/\bar{v}))\,(1-c/(\bar{v}T^3)) - (A_0/\bar{v}^2)(1-(a/\bar{v})).$$

Table A-20B contains several equations and constants.

c. *Modified BWR Equation*

The modified BWR equation is useful for halocarbon refrigerants and has the form

$$P = \sum_{n=1}^{9} A_n(T)/v^n + \exp(-v_c^{\,2}/v^2)\sum_{n=10}^{15} A_n(T)/v^{(2n-17)}. \qquad (82)$$

d. Lee–Kesler Equation of State

This is another modified form of the BWR equation which has 12 constants and is applicable for any substance. This relation is of the form

$$P_R = (T_R/v'_R)(1+A/v'_R+B/v'^2_R+C/v'^5_R+(D/v'_R)(\beta+\gamma/v'^2_R)\exp(-\gamma/v'^2_R)), \quad (83a)$$

$$Z = P_R v'_R/T_R = 1+A/v'_R+B/v'^2_R+C/v'^5_R+(D/v'_R)(\beta+\gamma/v'^2_R)\exp(-\gamma/v'^2_R), \quad (83b)$$

where $A = a_1 - a_2/T_R - a_3/T_R^2 - a_4/T_R^3$, $B = b_1 - b_2/T_R + b_3/T_R^3$, $C = c_1 + c_2/T_R$, and $D = d_1/T_R^3$. The constants are usually tabulated to determine $Z^{(0)}$ for all simple fluids and $Z^{(ref)}$ for a reference fluid, that is usually octane (cf. Table A-21). Assuming that

$$Z^{(ref)} - Z^{(0)} = \omega Z^{(1)}, \quad (83c)$$

$Z^{(1)}$ can be determined.

A general procedure for specified values of P_R and T_R is as follows: solve for v_R' from Eq. (83a) with constants for simple fluids and use in Eq. (83b) to obtain $Z^{(0)}$. Then repeat the procedure for the same P_R and T_R with different constants for the reference fluid, obtain $Z^{(ref)}$, and determine $Z^{(1)}$ from Eq.(83c). The procedure is then repeated for different sets of P_R and T_R. A plot of $Z^{(0)}$ is contained in the Appendix and tabulated in Table A–23A. The value of $Z^{(1)}$ so determined is assumed to be the same as for any other fluid. Tables A-23A and A-23B tabulate $Z^{(0)}$ and $Z^{(1)}$ as function of P_R and T_R.

e. Martin–Hou

The Martin–Hou equation is expressed as

$$P = RT/(v-b) + \sum_{j=2}^{5} F_j(T)/(v-b)^j + F_6(T)/e^{Bv}, \quad (84)$$

where $F_i(T) = A_i + B_i T + C_i \exp(-KT_R)$, b, B and F_j are constants (typically $B_4 = 0$, $C_4 = 0$ and $F_6(T) = 0$). This relation is accurate within 1 % for densities up to 1.5 ρ_c and temperatures up to $1.5T_c$.

12. State Equations for Liquids/Solids

a. Generalized State Equation

The volume v = v (P,T), and $dv = (\partial v/\partial P)_T dP + (\partial v/\partial T)_P dT$, i.e.,

$$dv = (\partial v/\partial P)_T dP + (\partial v/\partial T)_P dT. \quad (85a)$$

We define

$$\beta_P = (1/v)(\partial v/\partial T)_P, \quad (85b)$$

$$\beta_T = -(1/v)(\partial v/\partial P)_T, \quad (85c)$$

$$\kappa_T = 1/(\beta_T P) = (-v/P)(\partial P/\partial T)_T \quad (85d)$$

where β_P, β_T and κ_T are, respectively, the isobaric expansivity, isothermal compressibility, and isothermal exponent. The isobaric expansivity is a measure of the volumetric change with respect to temperature at a specified pressure. We will show in Chapter 10 that $\beta_T>0$ for stable fluids. Upon substituting these parameters in Eq. (86a),

$$dv = v\beta_P\, dT - v\beta_T\, dP, \text{ or } d(\ln v) = \beta_P dT - \beta_T dP.$$

If β_P and β_T are constant, the general state equation for liquids and solids can be written as

$$\ln(v/v_{ref}) = \beta_P (T - T_{ref}) - \beta_T (P - P_{ref}). \quad (86)$$

This relation is also referred to as the explicit form of the thermal equation of state. In terms of pressure, the relation

$$P = P_{ref} + (\beta_P/\beta_T)(T - T_{ref}) - \ln (v/v_{ref})/(\beta_T v_{ref}), \quad (87)$$

is an explicit, although approximate, state equation for liquids and solids. Both Eqs. (87) or (88) can be approximated as

$$(v - v_{ref})/v_{ref} = \beta_P (T - T_{ref}) - \beta_T (P - P_{ref}). \quad (88)$$

Solving the relation in terms of pressure

$$P = P_{ref} + (\beta_P/\beta_T) (T - T_{ref}) - (v - v_{ref})/(\beta_T v_{ref}), \quad (89)$$

which is an explicit, although approximate, state equation for liquids and solids.

The pressure effect is often small compared to the temperature effect. Therefore, Eq. (89) can be approximated in the form

$$\ln(v/v_{ref}) \approx \beta_P (T - T_{ref}). \quad (90)$$

In case $\beta_P (T - T_{ref}) \ll 1$, then

$$v/v_{ref} = (1 + \beta_P (T - T_{ref})). \quad (91)$$

which is another explicit, although approximate, state equation for liquids and solids

Copper has the following properties at 50°C: v, β_P, and β_T are, respectively, 7.002×10^{-3} m^3 kmole^{-1}, 11.5×10^{-6} K^{-1}, and 10^{-9} bar^{-1}. Therefore, heating 10 kmole of the substance from 50 to 51°C produces a volumetric change that can be determined from Eq.(87) as $7.002\times10^{-3} \times 10 \times 11.5\times10^{-6} = 805$ cm^3. If a copper bar containing 10 kmole of the substance is vertically oriented and a weight is placed on it such that the total pressure on the mass equals 2 bar, the volume of the copper will reduce by a value equal to $-7.002\times10^{-3} \times 10 \times 0.712\times10^{-9} = -0.05$ mm^3. Therefore, changing the state of the 10 kmole copper mass from 50°C and 1 bar to 51°C and 2 bars, will result in a volumetric change that equals 805 – 0.05 = 804.95 mm^3.

For solids β_P is related to the linear expansion coefficient α. The total volume $V \propto L^3$, and

$$\beta_P = 1/V(\partial V/\partial T)_P = 1/L^3\, \partial(L^3)/\partial T = (3/L)\, \partial L/\partial T = 3\alpha, \quad (92)$$

where $\alpha = (1/L)\,(\partial L/\partial T)_P$.

---

g. *Example 7*

Water is compressed isentropically from 0.1 bar and 30°C to 60 bar. Determine the change in volume, and work required to compress the fluid. Treat water as a compressible substance, and assume that at 30°C, $\beta_P = 2.7\times10^{-4}$ K^{-1}, $\beta_T = 44.8\times10^{-6}$ bar^{-1}, $v = 0.00101$ m^3 kg^{-1}, and $c_p = 4.178$ kJ kg^{-1} K^{-1}.

Solution

Since $\beta_P = 44.8\times10^{-6}$ bar^{-1} and $dv = -\beta_P\, dP\, v$, $\ln v_2/v = -\beta_T (P_2 - P_1) = -0.00268$, i.e., $v_2/v = 0.997$.

Now, $v_2 = 0.997\times0.00101 = 0.001007$ m^3 kg^{-1}, so that

$v_2 - v = 0.001007 - 0.001010 = 0.000997$ m^3 kg^{-1}.

$\delta w = -v dP$ (for a reversible process in an open system).

$\therefore \delta w = -v (dP/dv) dv = (1/\beta_T) dv$.

Integrating this expression,

$w = (1/\beta_T)(v_2 - v_.) = 100$ kPa bar^{-1}×(0.001007–0.00101)÷44.8×10^{-6} = –6.76 kJ kg^{-1}.

---

h. Example 8

A defective radiator does not have a pressure relief valve and there is no drainage provision for the reservoir. The radiator water temperature increases from 25°C to 90°C. Assuming that the radiator is rigid, what is the final water pressure in the radiator?

Solution

Since $d \ln v = \beta_P dT - \beta_T dP$ and the volume is constant,

$dP/dT = \beta_P/\beta_T = 2.7\times10^{-4}$ K^{-1}/44.8×10^{-6} bar^{-1} = 6.03 bar K^{-1}.

Assuming that β_T and β_P are constants,

$\Delta P = 6.03\times65 = 391$ bar.

---

b. Murnaghan Equation of State

If we assume that the isothermal bulk modulus B_T (= $1/\beta_T$) is a linear function of the pressure, then

$$B_T(T,P) = (1/\beta_T) = -v(\partial P/\partial v)_T = B_T(T,0) + \alpha P \quad (93)$$

where $\alpha = (\partial B_T/\partial P)_T$. Therefore,

$$\partial P\ \beta_T(1,0)/(1 + \alpha P\ \beta_T(1,0)) = -dv/v. \quad (94)$$

Integrating, and using the boundary condition that as $P \to 0$, $v \to v_0$, we obtain the following relation

$$v/v_0 = 1/(1 + (\alpha P \beta_T(1,0))^{(1/\alpha)}, \text{ i.e.,} \quad (95)$$

$$P(T,v) = ((v_0/v)^\alpha - 1)(1/\alpha\beta_T(1,0)). \quad (96)$$

c. Racket Equation for Saturated Liquids

The specific volume of saturated liquid follows the relation given by the Racket equation, namely,

$$v_f = v_c Z_c^{(1-T_R^{0.2857})}. \quad (97)$$

d. Relation for Densities of Saturated Liquids and Vapors.

If ρ_f denotes the saturated liquid density, and ρ_g the saturated vapor density, then

$$\rho_{Rf} = \rho_f/\rho_c = 1 + (3/4)(1 - T_R) + (7/4)(1 - T_R)^{1/3}, \text{ and} \quad (98)$$

$$\rho_{Rg} = \rho_g/\rho_c = 1 + (3/4)(1 - T_R) - (7/4)(1 - T_R)^{1/3}. \quad (99)$$

These relations are based on curve fits to experimental data for Ne, Ar, Xe, O_2, CO, and CH_4. It is also seen that

$$\rho_{Rf} - \rho_{Rg} = (7/2)(1 - T_R)^{1/3}. \quad (100)$$

At low pressures,

$$\rho_{Rf} \approx (7/2)(1 - T_R)^{1/3} \text{ since } \rho_{Rf} >> \rho_{Rg}$$

In thermodynamics, $\rho_{Rf} - \rho_{Rg}$ is called order of parameter. If ρ_{Rg} is known at low pressures (e.g., ideal gas law), then ρ_{Rf} can be readily determined. Another empirical equation follows the relation

$$\rho_{R,f} = 1 + 0.85(1-T_R) + (1.6916 + 0.9846\psi)(1-T_R)^{1/3} \quad (101)$$

where $\psi \approx \omega$.

e. Lyderson Charts (For Liquids)

Lyderson charts can be developed based on the following relation, i.e.,

$$\rho_R = \rho/\rho_c = v_c/v. \quad (102)$$

The appendix contains charts for ρ_R *vs.* P_R with T_R as a parameter. In case the density is known at specified conditions, the relation can be used to determine P_c, T_c and ρ_c. Alternatively, if the density is not known at reference conditions, the following relation, namely,

$$\rho/\rho_{ref} = v_{ref}/v = \rho_R/\rho_{R,ref} \quad (103)$$

can be used.

f. Incompressible Approximation

Recall that liquid molecules experience stronger attractive forces compared to gases due to the smaller intermolecular spacing. The molecules are at conditions close to the lowest potential energy where the maximum attractive forces occur. Therefore, any compression of liquids results in strong repulsive forces that produce an almost constant intermolecular distance. This allows us to use the incompressible approximation, i.e., v = constant.

D. SUMMARY

This chapter describes how some properties can be determined for liquids, vapors, and gases at specified conditions, e.g., the volume at a given pressure and temperature. Compressibility charts can be constructed using the provided information and fluid characteristics, such as the Boyle temperature, can be determined. The relations can be used to determine the work done as the state of a gas is changed. Various methods to improve the predictive accuracy are discussed, e.g., by introducing the Pitzer factor. State equations for liquids and solids are also discussed.

E. APPENDIX

1. Cubic Equation

One real and three imaginary solutions are obtained for Z when $T_R>1$. However, when $T_R<1$, we may obtain one to three real solutions.

The following method is used in spreadsheet software to determine the compressibility factor. Consider the relation

$$Z^3 + a_2 Z^2 + a_1 Z + a_0 = 0.$$

Furthermore, let

$$\alpha = a_2^2/9 - a_1/3,\ \beta = -a_2^3/27 + a_1 a_2/6 - a_0/2, \text{ and } \gamma = \alpha^2 - \beta^3.$$

a. *Case I: $\gamma > 0$*

i. Case Ia: $\alpha > 0$

There is one real root for this case, i.e.

$$Z = Z = (\alpha + \gamma^{0.5})^{1/3} + (\alpha - \gamma^{0.5})^{1/3} + 1/3.$$

ii. Case Ib: $\alpha < 0$

Again, only one real root exists. If $\tan \varphi = (-p)^{1.5}/q$, $\tan \theta = (\tan(\varphi/2))^{1/3}$ if $\varphi > 0$, and $-(\tan(-\varphi/2))^{1/3}$ if $\varphi<0$, then

$$Z = (-2)\,(-\alpha)^{0.5}/\tan(|2\theta|) + 1/3.$$

b. *Case II: $\gamma < 0$*

Three real roots exist for this case. If $\cos \phi = \beta/\alpha^{1.5}$, then

$$Z_1 = 2\alpha^{1/2} \cos(\phi/3) + 1/3,$$

$$Z_2 = 2\alpha^{1/2} \cos(\phi/3 + 4\pi/3) + 1/3, \text{ and}$$

$$Z_3 = 2\alpha^{1/2} \cos(\phi/3 + 8\pi/3) + 1/3.$$

---

i. Example 9

Consider the RK equation of state. Determine Z at $T_R = 1.5$ and $P_R = 1.2$, and at $T_R = 1.2$ and $P_R = 10$, and Z_1, Z_2, Z_3 at $T_R = 0.9161$ and $P_R = 0.602$.

Solution

$a^\times = 0.4275 P_R/T_R^{2.5} = 0.1862$, and $b^\times = 0.08664 P_R/T_R = 0.06931$

Using Eqs. (46) and (106), $a_2 = -1$, $a_1 = a^\times - b^{\times 2} - b^\times = 0.1121$, $a_0 = -a^\times b^\times = -0.01291$.

Therefore,

$\alpha = a_2^2/9 - a_1/3 = 1/9 - 0.1121/3 = 0.07374$,

$\beta = -a_2^{\,3}/27 + a_1 a_2/6 - a_0/2 = 1/27 - 0.1121/6 + 0.01291/2 = 0.02481$, and

$\gamma = \beta^2 - \alpha^3 = 0.0002145$.

Since, $\gamma > 0$ and $\alpha > 0$, Case Ia is applicable, and

$Z = (\alpha + \gamma^{0.5})^{1/3} + (\alpha - \gamma^{0.5})^{1/3} + 1/3$

$= (0.02481 + 0.0002145^{0.5})^{1/3} + (0.02481 - 0.0002145^{0.5})^{1/3} + 1/3$

$= 0.340 + 0.2166 + 0.333 = 0.8899$.

In the second case, $a^\times = 2.7109$, $b^\times = 0.722$, $a_1 = 1.4668$, $a_0 = -1.9569$ so that

$\alpha = -0.3778$, $\beta = 0.7709$, and $\gamma = 0.6482$.

Hence, Case Ib is applicable.

$\tan \phi = (-\alpha)^{1.5}/\beta = 0.3012$, i.e., $\phi = 16.76$.

$\tan \theta = (\tan(\phi/2))^{1/3} = 0.4366$, i.e., $\theta = 27.84$. Therefore,

$Z = 2\,(-\alpha)^{0.5}/\tan(2\theta) + 1/3 = 2 \times 0.3778^{0.5}/\tan(2\times27.84) + 1/3$

$= 0.8392 + 0.333 = 1.1725$.

For the third case, $a^\times = 0.3204$, $b^\times = 0.05693$, $a_1 = .2602$, $a_0 = -0.01824$, and

$\alpha = 0.02437$, $\beta = 0.002789$, and $\gamma = -6.78\times10^{-6}$.

Case II applies, and there are three roots to the equation.

$\cos\phi = \beta/\alpha^{1.5} = 0.7329$, i.e., $\phi = 42.87$.

$Z_1 = 2\alpha^{0.5}\cos(\phi/3) + 1/3 = 2 \times 0.02437^{0.5}\cos(42.87/3) + 1/3$

$= 0.3122 \times 0.9691 + 1/3 = 0.3025 + 0.333 = 0.6359$.

$Z_2 = 2\alpha^{0.5}\cos(\phi/3 + 120) + 1/3 = 0.3122 \times \cos(134.29) + 0.333$

$= -0.2180 + 0.333 = 0.1153$.

$Z_3 = 2\alpha^{0.5}\cos(\phi/3 + 240) + 1/3 = 0.3122 \times \cos(254.29) + 0.33333$

$= -0.08453 + 0.3333 = 0.2488$

The spreadsheet software uses this methodology for solving the cubic equation.

---

2. Another Explanation for the Attractive Force

The net force acting on the molecules on a wall is proportional to the number of surrounding molecules that exert an attraction force. The net force on each molecule near the wall equals the force exerted on the wall by collision minus the attraction force. Therefore, for n molecules on the wall, (n × the force exerted on the wall by collisions per molecule) – (n × attractive force per molecule) = (n × net force on the wall per molecule). Since the attraction force per molecule ∝ n of surrounding the system, then (n × force exerted on the wall by collision per molecule) – (n × n × constant) = (n × net force). Therefore, the net pressure equals the pressure that would have been exerted in the absence of attraction forces minus the term (n^2 × constant), i.e.,

$$P = (RT/(V - b') - \text{attraction force (which is} \propto n^2)$$

$$= RT/(V - b') - \text{attraction force} \propto N^2/V^2 = RT/(V - b') - a'/V^2.$$

If we compare the attractive force component to the LJ force function (cf. Chapter 1), the attractive force $\propto 1/\ell^6$, i.e., the attractive force being proportional to the n^2 seems to be the reason that the exponent is 6 in the attractive force relation. (In the context of the gravitational law $F = G\, m_E m'/r^2$, where $G = 6.67\times10^{-14}$ kN m^2 kg^{-2}, since $g = 9.81$ m s^{-2} at $r = r_E$, $F = Gm_E/r_E^2$. If the radius of the earth is known, then its mass can be determined. This derivation also enables a simplistic relation for the pressure due to inter-planetary forces between planets in the universe.)

3. Critical Temperature and Attraction Force Constant

Consider an $\ell\times\ell$ cross section of a wall containing a single molecule M. Other molecules that collide with M impart a momentum to it due to their velocity V. The momentum transfer rate to M is $mV^2/3\ell$. The molecule M also experiences attraction forces. The attraction force between a molecular pair is $4(\varepsilon/\sigma)\sigma^7/r^7$ according to the LJ model.

Now consider a semicircular segment characterized by the dimensions dr and dθ located at a radial distance r from M. There are $\pi r n' r d\theta dr$ molecules within that shell pulling M away from the wall in the radial direction. The net force on the molecules in that direction is $(r^2\, dr\, d\theta \cos\theta\, \pi n'\, 24(\varepsilon/\sigma)\, \sigma^7)/r^7)$. Assuming the force field to be continuous and integrating this expression over $r = \sigma$ to ∞ and $\theta = 0$ to $\pi/2$, the net force on M equals $3\pi n'(\varepsilon\sigma^2)$. We must subtract this force from the momentum transfer rate. Dividing by the area ℓ^2, the pressure equals $mV^2/3\ell^3 - \pi n'(3\varepsilon\sigma^2)/\ell^2$. Since $N' = \ell^{-3}$, the pressure

$$n'mV^2/3 - \pi n'^2(3\varepsilon\sigma^2)\,\ell \propto \bar{R}T/(\bar{v} - \bar{b}) - a/\bar{v}^2.$$

Therefore, $a = N_{Avag}^{2} 3\pi\varepsilon\sigma^{2}\ell$ is a weak function of the intermolecular spacing. In case $\ell \approx \sigma$, $a = N_{Avag}^{2} 3\pi\varepsilon\sigma^{3}$. A more rigorous derivation based on the potential gives the relation $a = 2.667 N_{Avag}^{2}$.

Chapter 7

7. THERMODYNAMIC PROPERTIES OF PURE FLUIDS

A. INTRODUCTION

In this chapter, we will make use of the properties of ideal gases, the critical properties of substances, and the state equations that can be applied to describe their behavior in order to determine the thermodynamic properties of pure fluids.

B. IDEAL GAS PROPERTIES

The molecules of ideal gases can be considered to be point masses that are uninfluenced by intermolecular attractive forces, and follow the state relationship

$$P\,v = R\,T. \tag{1}$$

The molecular energy of an ideal gas u_o can be determined if the molecular structure and velocity are known. (The subscript o is taken to denote ideal gas properties, which can be interpreted as the condition $P \to 0$). The value of u_o depends only upon the temperature. Using the relation $h_o(T) = u_o(T) + (Pv)_o = u_o(T) + RT$, the internal energy may be expressed as:

$$u_o(T) = h_o(T) - RT. \tag{2}$$

For ideal gases, $c_{v,o} = c_{p,o} - R$, where $c_{p,o} = dh_o/dT$ and $c_{v,o} = du_o/dT$. Therefore,

$$h_o(T) - h_{o,ref}(T_{ref}) = \int_{T_{ref}}^{T} c_{p,o}(T)dT, \tag{3}$$

where the difference $h_0(T)$-$h_0(T_{ref})$ is called thermal enthalpy. If $h_{o,ref} = 0$ at $T_{ref} = 0$ K,

$$h_o(T) = \int_0^T c_{p,o}(T)\,dT. \tag{4}$$

If $c_{p0}(T)$ = constant, then Eq. (4) states that $h_0(T) = c_{p0}\,T$. Thereafter, $u_o(T)$ can be determined. Similarly, using the relation developed in Chapter 3,

$$s_o(T,P) = s^o(T) - R\ln(P/P_{ref}). \tag{5}$$

Usually, P_{ref} is taken as 1 atm, and

$$s^o(T) = \int_{T_{ref}}^{T} c_{p,o}(T)\,dT/T. \tag{6}$$

Any substance, whether solid, liquid or real gas can be "converted" into a hypothetical ideal gas by removing the attractive forces and reducing the body volume of molecules to a "point volume"(Chapter 6).

C. JAMES CLARK MAXWELL (1831–1879) RELATIONS

Maxwell provided relations for several nonmeasurable properties in terms of measurable properties (e.g., T, v and P). The basis for the derivation of relations is as follows. If

$$dz = (M(x,y)\,dx + N(x,y)\,dy)$$

is an exact differential, it must then satisfy the exactness criterion, i.e.,

$$(\partial M/\partial y)_x = (\partial N/\partial x)_y.$$

The variable M (x,y) is called the conjugate of x and N(x,y) is the corresponding conjugate of y. If the exactness criterion is satisfied, the sum (M(x,y) dx + N(x,y) dy) = dZ, the integration of which yields a point (or state) function Z(x,y) that is a property (Chapter 1). Inversely if Z is a property, then dZ is exact and, since dZ equals the aforementioned sum, the criterion for an exact differential is satisfied.

1. First Maxwell Relation

The First law for a process occurring in a closed system can be expressed in the form

$$\delta Q - \delta W = dU. \tag{7}$$

For a process occurring along an internally reversible path

$$\delta Q_{rev} = TdS \text{ and } \delta W_{rev} = PdV,$$

so that Eq. (7) can be written as

$$dU = TdS - PdV. \tag{8}$$

For a unit mass, the corresponding relation is

$$du = Tds - Pdv. \tag{9}$$

a. Remarks

Equation (9) is an expression of the combined First and Second laws for a closed system. We observe that once s and v are fixed, du = 0. The relation u = u(s,v) is an intensive state equation that is expressed in terms of intensive variables.

Assume that you are to visit a planet on which only s and v can be measured, but for some reason not T and P. Equation (9) can be written in the form

$$du = T(s,v)\, ds - P(s,v)\, dv \tag{10}$$

The slopes of u at specified s and v are

$$(\partial u/\partial s)_v = u_s = T(s,v), \text{ and } (\partial u/\partial v)_s = u_v = -P(s,v). \tag{11}$$

The temperature T is the conjugate of s and (–P) the conjugate of v. It is noted that $(\partial u/\partial s)_v \rightarrow 0$ as $T \rightarrow 0$ and hence $u = u(v)$ as $T \rightarrow 0$. Obtaining total differential of T(s,v) and using Eq. (10),

$$dT = (\partial T/\partial s)_v\, ds + (\partial T/\partial v)_s\, dv, = u_{ss}\, ds + u_{sv}\, dv, \tag{12}$$

where $u_{ss} = \partial^2 u/\partial s^2$, $u_{sv} = \partial^2 u/\partial s \partial v$. Similarly,

$$-dP = u_{sv}\, ds + u_{vv}\, dv. \tag{13}$$

These relations are useful in stability analyses (cf. Chapter 10).

At constant volume, i.e., along an isometric curve Eq. (9) yields the expression

$$Tds_v = du_v, \text{ or } T(\partial s/\partial T)_v = (\partial u/\partial T)_v = c_v. \tag{14}$$

Using the first of these two relations, the area under the resulting curve on a T–s diagram represents the internal energy change for the isometric process. Rewriting Eq. (10) in the form, we obtain the fundamental relation in entropy form s = s(u,v), i.e.,

$$ds = (1/T(s,u))\, du + (P(s,u)/T(s,u))\, dv.$$

Since du is an exact differential, Eq. (10) must satisfy the corresponding criterion, namely,

$$(\partial T/\partial v)_s = -(\partial P/\partial s)_v, \tag{15}$$

which is known as the First relation. Table 1 summarizes the relations.

2. Second Maxwell Relation

Adding the term d(Pv) to Eq. (9) and simplifying, we obtain the expression

	Differential	Conjugate	Maxwell Relation	Remarks
u	$du = Tds - Pdv$	T, –P	$\partial T/\partial v = -\partial P/\partial s$	$u = u(s,v)$
h	$dh = Tds + vdP$	T, v	$\partial T/\partial P = \partial v/\partial s$	$h = h(s,P)$
a	$da = -sdT - Pdv$	–s, –P	$\partial s/\partial v = \partial P/\partial T$	$a = a(T,v)$
g	$dg = -sdT + vdP$	–s, v	$-\partial s/\partial P = \partial v/\partial T$	$g = g(T,P)$
j	$dj = -P/T\, dv + u/T^2\, dT$	$-P/T$, u/T^2	$\partial j/\partial v = -P/T$, $\partial j/\partial T = u/T^2$	$j = -a/T = s - u/T$; Massieu function, $j = j(T,v)$
r	$dr = (v/T)\, dP + (h/T^2)\, dT$	v/T, h/T^2	$\partial r/\partial P = v/T$, $\partial r/\partial T = h/T^2$	$r = -g/T = s - h/T$; Planck function $r = r(T,P)$

Table 1: Summary of relations

$$dh = Tds + v\, dP. \quad (16)$$

a. Remarks

Equation (16) is a form of the state equation $h = h(s,P)$, and

$$(\partial h/\partial s)_P = T(s,P), \text{ and } (\partial h/\partial P)_s = v(s,P). \quad (17a)$$

We see that $(\partial h/\partial s)_P \to 0$ as $T \to 0$, i.e., $h = h(P)$ as $T \to 0$. Furthermore,

$$dT = h_{ss}\, ds + h_{sp}\, dP, \text{ and } dv = h_{sp}\, ds + h_{pp}\, dP.$$

Using Eq. (16), $ds = dh/T(s,P) - v(s,P)\, dP/T(s,P)$, i.e., $s = s(h,P)$. At constant pressure,

$$T\, ds_P = dh_P.$$

Therefore, for an isobaric process, the area under the corresponding curve on a T–s diagram represents the enthalpy. In addition,

$$T(\partial s/\partial T)_P = (\partial h/\partial T)_P = c_P. \quad (17b)$$

At the critical point, $\partial T/\partial s = 0$ (cf. Chapter 3), and $c_p \to \infty$.
The second relation has the form

$$(\partial T/\partial P)_s = (\partial v/\partial s)_P \quad (18)$$

---

a. Example 1

Verify the Nernst Postulate, namely, $c_v \to 0$ and $c_p \to 0$ as $T \to 0$.

Solution

Consider the relation

$$(\partial s/\partial T)_v = c_v/T.$$

Since the Third law states that $s \to 0$ as $T \to 0$, three possibilities exist for the slope $(\partial s/\partial T)_v$, namely,

$(\partial s/\partial T)_v \to 0$.

$(\partial s/\partial T)_v$ is finite.

$(\partial s/\partial T)_v \to \infty$ as either $T \to 0$ or $s \to 0$.

For the first two of these three cases as $T \to 0$, $c_v/T \to 0$ or has finite values. Therefore, in either case $c_v \to 0$ as $T \to 0$. For the third case, since $(\partial s/\partial T)_v \to \infty$, we will

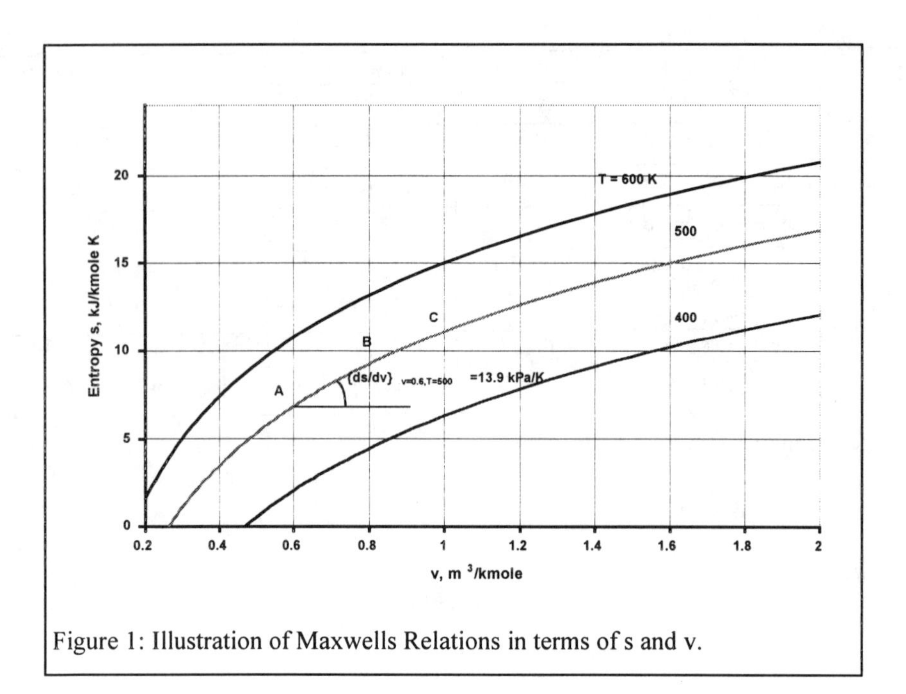

Figure 1: Illustration of Maxwells Relations in terms of s and v.

use the result from Chapter 3 that $c \propto T^3$ at low temperatures. Thereafter, assuming $c = c_v$, $c_v/T \approx T^2$. Therefore, for all three cases, $c_v \rightarrow 0$ as $T \rightarrow 0$.

Similarly, using the relation $(\partial s/\partial T)_P = c_P/T$, we can show that $c_P \rightarrow 0$ as $T \rightarrow 0$.

3. Third Maxwell Relation

The Helmholtz function is defined as

$$a = u - Ts, \text{ and } da = du - d\,(Ts). \tag{19}$$

The entropy is a measure of how energy is distributed. The larger the number of quantum states at a specified value of the internal energy, the larger the value of the entropy. Therefore, if two systems that exist at the same temperature and internal energy, the Helmholtz function is lower for the system that has a larger specific volume. Substituting from Eq. (9) for du in Eq. (19),

$$da = -P\,dv - s\,dT, \text{ and } a = a(v,T). \tag{20}$$

a. Remarks

Equation (20) implies that

$$da = -P(v,T)\,dv - s(v,T)\,dT, \text{ where}$$

$$(\partial a/\partial T)_v = -\,s(T,v) \text{ and } (\partial a/\partial v)_T = -\,P(T,v). \tag{21}$$

Using the differentials of Eq.(21)

$$-ds = a_{TT}\,dT + a_{Tv}\,dv, \text{ and } -dP = a_{vT}\,dT + a_{vv}\,dv.$$

Using Eq. (20), the third relation is derived as

$$(\partial P/\partial T)_v = (\partial s/\partial v)_T. \tag{22}$$

Equation (22) provides a relation for s in terms of the measurable properties P, v, and T. (The value of the LHS of the equation is measurable while the RHS value is nonmeas-

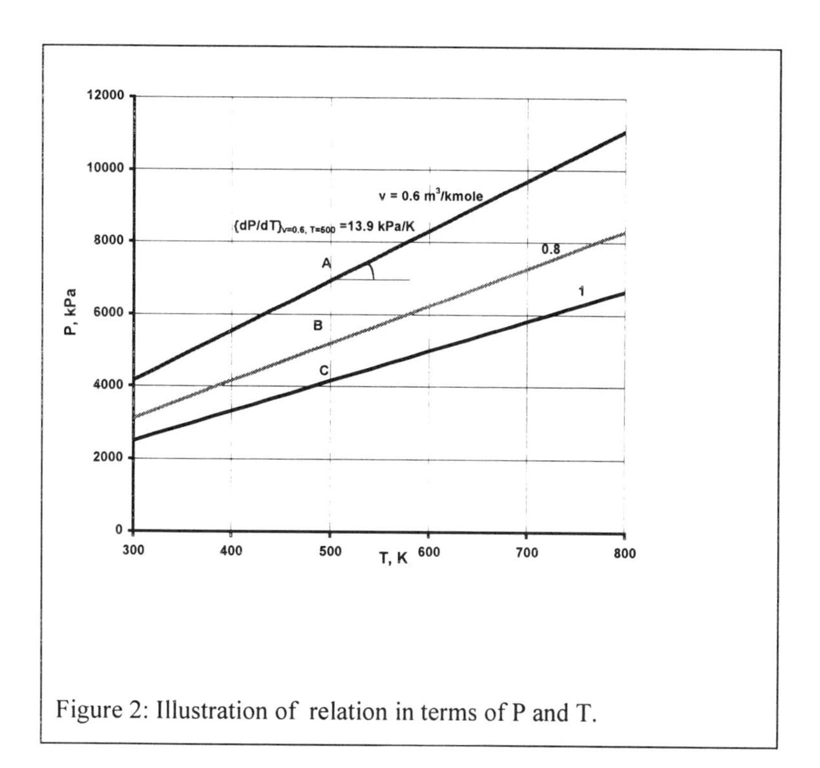

Figure 2: Illustration of relation in terms of P and T.

urable.) relations are illustrated in Figure 1 and Figure 2. Since $a = u - Ts$ and $s = -(\partial a/\partial T)_v$, $a/T = u/T + (\partial a/\partial T)_v$. Therefore,

$$\partial((a/T)/\partial T) = (1/T)\ \partial u/\partial T - u/T^2 - (\partial s/\partial T)_v = c_v/T - u/T^2 - (\partial s/\partial T)_v.$$

From the fundamental relation in entropy form, $T(\partial s/\partial T)_v = (\partial u/\partial T)_v = c_v$, so that

$$\partial((a/T)/\partial T) = - u/T^2 \text{ or } \partial((a/T)/\partial(1/T)) = u. \quad (23)$$

Furthermore, since

$$da_T = P\ dv_T,$$

the area under an isotherm on a P–v diagram represents the Helmholtz function. The work transfer during an isothermal process results in a change in the Helmholtz function. Recall that "a" is a measure of the availability in a closed system. Knowing P=P (v,T), one can obtain Helmholtz function "a".

The Massieu function j is defined as

$$j = -a/T = s - u/T, \text{ i.e.,}$$

$$dj = -da/T + a/T^2dT = (1/T^2)(PT\ dv - u\ dT) = j(T,v).$$

---

b. *Example 2*

The fundamental relation for the entropy of an electron gas can be approximated as

$$S(U,V,N) = B\ N^{1/6}\ V^{1/3}\ U^{1/2}, \text{ where} \quad (A)$$

$$B = 2^{3/2}\pi^{4/3}k_B m^{1/2} N_{avag}{}^{1/6}/(3^{1/3}h_P). \quad (B)$$

Here, k_B denotes the Boltzmann constant that has a value of $\overline{R}/N_{Avag} = 1.3804\times10^{-26}$ kJ K^{-1}, h_P is the Planck constant that has a value of 6.62517×10^{-37} kJ s, m denotes the electron mass of 9.1086×10^{-31} kg, N the number of kmoles of the gas, V its volume in m^3, and U its energy in kJ. Determine $\bar{s}$, T, and P when $\bar{u}$ = 4000 kJ k $mole^{-1}$, and $\bar{v}$ = 1.2 m^3 $kmole^{-1}$.

Solution

The value of B = 5.21442 $kg^{1/2}$ k $mole^{1/6}$ s K^{-1}. From Eq. (A),

$$\bar{s} = S/N = (B/N)\, N^{1/6}(\bar{v}N)^{1/3}(\bar{u}N)^{1/2} = B\,\bar{v}^{1/3}\,\bar{u}^{1/2}, \text{ i.e.,} \quad (C)$$

$$\bar{s} = 5.21442\ (kg^{1/2}\ K^{-1}\ Kmole^{1/6}\ s)(1.2\ m^3\ k\ mole^{-1})^{1/3}\ (4000\ kJ\ kmole^{-1})^{2}.$$

Recalling that the units kg (m/s^2) m ≡ J.

$$\bar{s} = 350\ kg^{1/2}\ m\ kJ^{1/2}\ kmole^{-1}\ K^{-1}. = 350.45\ kJ\ kmole^{-1}\ K^{-1}.$$

From the entropy fundamental equation

$$1/T = (\partial\bar{s}/\partial\bar{u})_{\bar{v}}.$$

Differentiating Eq. (C) with respect to $\bar{u}$ and using this relation,

$$1/T = (1/2)\, B\, \bar{v}^{1/3}/\bar{u}^{1/2} = 0.04381 \text{ or } T = 22.8\ K. \quad (D)$$

Similarly, since

$$P/T = (\partial\bar{s}/\partial\bar{v})_u,$$

Upon differentiating Eq. (C) and using the above relation,

$$P/T = (1/3)\, B\, \bar{u}^{1/2}/\bar{v}^{2/3} = 94.35\ kPa\ K^{-1}. \quad (E)$$

Using the value for T = 22.83 K, the pressure P = 2222.4 kPa.. The enthalpy

$$\bar{h} = \bar{u} + P\bar{v} = 4000 + 2222.4 \times 1.2 = 6666.9\ kJ\ kmole^{-1}.$$

Remarks

Eq. (C) can be expressed in the form

$$\bar{u}(\bar{s},\bar{v}) = \bar{s}^2/(B^2\ \bar{v}^{2/3}). \quad (F)$$

Equation (F) is referred to as the energy representation of the fundamental equation (cf. Chapter 5).
Rewriting Eq. (D)

$$\bar{u}\ (T,\bar{v}) = (1/4)\, B^2\ \bar{v}^{2/3}T^2. \quad (G)$$

Differentiating this relation with respect to T we obtain the result

$$c_v = (\partial u/\partial T)_v = (1/2)\, B^{2/3}\ \bar{v}^{2/3}\, T. \quad (H)$$

Dividing Eq. (E) by Eq. (D) we obtain the expression

$$\bar{u}(P,\bar{v}) = (3/2)\, P\,\bar{v}. \quad (I)$$

Likewise, using the entropy fundamental state equation (Eq. (A)), we can also tabulate other nonmeasurable thermodynamic properties such as $\bar{a}$ (= $\bar{u} - T\bar{s}$) and $\bar{g}$ (= $\bar{h} - T\bar{s}$).
Eliminating $\bar{u}$ in Eqs. (D) and (E) we obtain the state equation P = P(T, $\bar{v}$) for an electron gas in terms of measurable properties, i.e.,

$$P(T, \bar{v}) = (B^2/6)\, T^2/\bar{v}^{1/3}. \quad (J)$$

If this state equation (in terms of P, T and $\bar{v}$) is known, it does not imply that $\bar{s}$, $\bar{u}$, $\bar{h}$, $\bar{a}$, and $\bar{g}$ can be subsequently determined. This is illustrated by considering the temperature and pressure relations

$$T = \partial \bar{s}/\partial \bar{u}, \text{ and } P/T = \partial \bar{s}/\partial \bar{v}. \quad (K)$$

One can use Eq. (J) in (K). These expressions indicate that Eqs. (K) are differential equations in terms of $\bar{s}$ and, in order to integrate and obtain $\bar{s} = \bar{s}(T, \bar{v})$, an integration constant is required which is unknown. Therefore, a fundamental relation is that relation from which all other properties at equilibrium (e.g., T, P, $\bar{v}$, $\bar{s}$, $\bar{u}$, $\bar{h}$, $\bar{a}$, $\bar{g}$, c_p, and c_v) can be directly obtained by differentiation alone. While the Eq. (A) represents a fundamental relation, we can see that the relation Eq. (J) does not.

---

c. *Example 3*

An electron gas follows the state equation

$$\bar{a}(T, \bar{v}) = -(1/4)\, B^2\ \bar{v}^{2/3} T^2, \quad (A)$$

where B = 5.21442 $kg^{1/2}\ K^{-1}\ kmole^{1/6}\ s$. Determine the functional relations for properties such as $\bar{s}$, P, $\bar{u}$, and $\bar{h}$.

Solution

Using Eq. (21), we obtain the relation

$$(d\bar{a}/dT)_v = -\bar{s} = -1/2\ B^2 T\, \bar{v}^{2/3}. \quad (B)$$

The pressure is obtained from the expression

$$(d\bar{a}/d\bar{v})_T = -P = -(1/6)\, B^2 T^2/\bar{v}^{1/3}. \quad (C)$$

Since $\bar{u} = a + T\bar{s}$, using Eqs. (A) and (B), we obtain

$$\bar{u} = -(1/4)\, B^2\, \bar{v}^{2/3}\ T^2 + (1/2)\, B^2 T^2\, \bar{v}^{2/3} = 1/4\ B^2 T^2\, \bar{v}^{2/3}. \quad (D)$$

Differentiating Eq. (D), we obtain an expression for the constant volume specific heat, i.e.,

$$c_v = (\partial u/\partial T)_v = (1/2)\, B^2 T\, \bar{v}^{2/3}.$$

Furthermore, $\bar{h} = u + P\bar{v}$ so that

$$\bar{h} = (1/4)\, B^2 T^2\ \bar{v}^{2/3} + (1/6)\, B^2 T^2\, \bar{v}^{2/3} = (5/12)\, B^2 T^2\, \bar{v}^{2/3}, \text{ and} \quad (E)$$

$$c_P = (\partial h/\partial T)_P = (5/12)\, B^2 (2T\, \bar{v}^{2/3} + (2/3)\, T^2\, \bar{v}^{-1/3} (\partial \bar{v}/\partial T)_P). \quad (F)$$

The value of $(\partial \bar{v}/\partial T)_P$ can be obtained from Eq. (C).

Remarks

Alternatively, one can use Eq. (23) and get $\bar{u}$ shown in Eq. (D) directly.

The Gibbs energy is a measure of the chemical potential, and

$$\bar{g} = \bar{h} - T\bar{s} = (5/12)\, B^2 T^2\ \bar{v}^{2/3} + (1/2)\, B^2 T^2\, \bar{v}^{2/3} = (11/12)\, B^2 T^2\ \bar{v}^{2/3}.$$

The above relation suggests that the value of the chemical potential becomes larger with an increase in the temperature. A temperature gradient results in a gradient involving the chemical potential of electrons. The state equation, $P = (1/6)\, B^2 T^2/\bar{v}^{1/3}$ indicates that $\bar{v}$ increases (or the electron concentration decreases) as T increases at fixed P. Hence, the warmer portion can have a lower electron concentration.

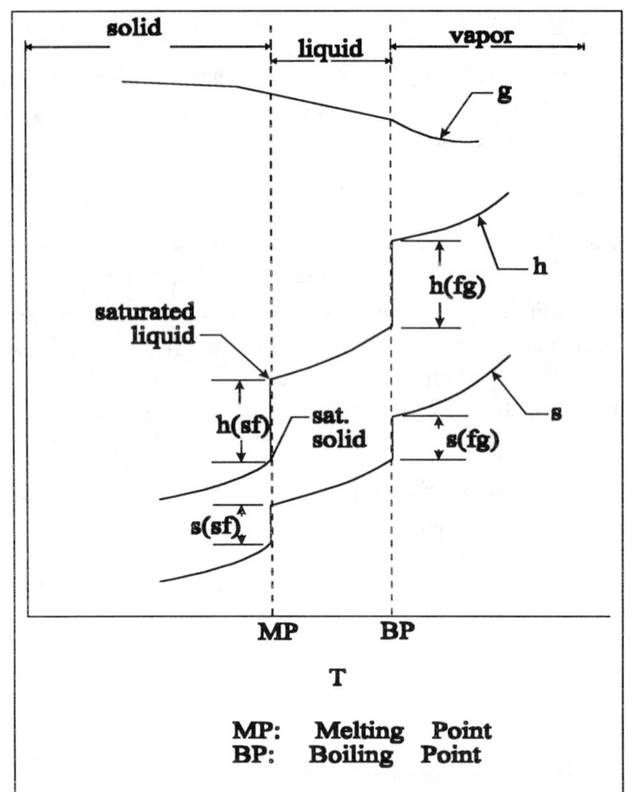

Figure 3: Illustration of the variation in some properties, e.g., h, s and g, with temperature.

Example 3 illustrates that the relation $\bar{a} = \bar{a}$ (T,v) is a fundamental equation that contains all the relevant information to construct a table of properties for P,u, h, g, s, etc., (e.g. Tables A-4 for H_2O, A-5 for R134a, etc.). One can plot the variation in h, g, and s with respect to temperature as illustrated in Figure 3.

---

d. Example 4

Obtain an expression for the entropy change in an RK gas when the gas is isothermally compressed. Determine the entropy change when superheated R–12 is isothermally compressed at 60°C from 0.0194 m^3 kg^{-1} (state 1) to 0.0126 m^3 kg^{-1} (state 2). Compare the result with the tabulated value of $s_1 = 0.7259$, $s_2 = 0.6881$.

Solution

Consider the RK state equation

$$P = RT/(v-b) - \mathbf{a}/(T^{1/2}v(v+b)) \qquad (A)$$

Note that the attractive force constant **a** is different from "a" Helmholtz function. From the third relation Eq. (22) and Eq. (A),

$$(\partial s/\partial v)_T = (\partial P/\partial T)_v = R/(v-b) + (1/2)\, a/(T^{3/2}v(v+b)). \qquad (B)$$

Integrating Eq. (B),

$$s_2(T,v_2) - s_1(T,v_1) =$$

$$R\ln((v_2-b)/(v_1-b)) + (1/2)(a/(T^{3/2}b))\ \ln(v_2(v_1+b)/(v_1(v_2+b))). \qquad (C)$$

From Table 1 for R–12, T_c = 385 K, and P_c = 41.2 bar. Therefore $\bar{a}$ =208.59 bar (m^3 $kmole^{-1}$)^2 $K^{1/2}$, and $\bar{b}$ = 0.06731 m^3 $kmole^{-1}$. The molecular weight M = 120.92 kg $kmole^{-1}$, and

$a = \bar{a}/M^2 = 208.59$ bar $(m^3\ kmole^{-1})^2 K^{1/2} \div 120.92^2 (kg\ kmole^{-1})^2$

$= 1.427$ k Pa $(m^3\ kg^{-1})^2\ K^{1/2}$, and

$b = \bar{b}/M = 0.557\times10^{-3}\ m^3\ kg^{-1}$.

Since, R = 8.314 ÷ 120.92 = 0.06876 kJ kg^{-1} K^{-1},

$s_2 - s_1 = 0.06876 \ln[(0.0126 - 0.000557) \div (0.0194 - 0.000557)]$

$+ (1/2)\{172.5 \div (333^{1.5}\ 0.000557)\} \ln [0.0126\ (0.0194 + 0.000557)$

$\div \{0.0194 \times (0.0126 + 0.000557)\}]$.

$= -0.06876 \times 0.448 - 0.211 \times 0.01495 = -0.03396$ kJ kg^{-1} K^{-1}.

---

4. Fourth Maxwell Relation

By subtracting d(Ts) from both sides of Eq. (16) and using the relation g = h – Ts, we obtain the state equation

$$dg = v\ dP - s\ dT \quad (24)$$

where g represents the Gibbs function (named after Josiah Willard Gibbs, 1839–1903). Equation (24) is another form of the fundamental equation. The intensive form g (= g(T,P)) is also known as the chemical potential μ.

a. Remarks

Since, dg = v(T,P) dP – s(T,P) dT,

$$(\partial g/\partial T)_P = -s(T,P) \text{ and } (\partial g/\partial P)_T = v(T,P) \quad (25)$$

so that

$$-ds = g_{TT}\ dT + g_{TP}\ dP \text{ and } dv = g_{PT}\ dT + g_{PP}\ dP.$$

The fourth Maxwell relation is represented by the equality

$$(\partial v/\partial T)_P = -(\partial s/\partial P)_T. \quad (26)$$

The LHS of this expression is measurable while the RHS is not. Since $s = -(\partial g/\partial T)_P$,

$$g = h - Ts = h + T\ (\partial g/\partial T)_P \text{ or } G = H + T\ (\partial G/\partial T)_P.$$

This relation is called the Gibbs–Helmholtz equation.

Furthermore,

$$(\partial(g/T)/\partial T)_P = (\partial h/\partial T)_P - h/T^2 - (\partial s/\partial T)_P = -h/T^2 \text{ or} (\partial(g/T)/\partial(1/T))_P = h \quad (27)$$

Similarly from Eq. (24),

$$(\partial g/\partial P)_T = v. \quad (28)$$

These relations are used to prove the Third law of thermodynamics and are useful in chemical equilibrium relations.

The phase change at a specified pressure (e.g., a piston containing an incompressible fluid with a weight placed on it) occurs at a fixed temperature. In that case, dP = dT = 0 and Eq. (24) implies that dg = 0, i.e., g_f (for a saturated liquid) = g_g (for saturated vapor at that pressure). Figure 3 illustrates the behavior of the properties h, s, and g when matter is heated from the solid to the liquid, and, finally, to the vapor phase.

Note the discontinuities regarding the entropy and Helmholtz function during the phase change, while g is continuous.
At constant temperature, $dg_T = vdP$. Therefore, the area under the v–P curve for an isotherm represents Gibbs function.
The Planck function is represented by the relation

$$r = r(T,P) = -g/T = s - h/T, \text{ i.e.,}$$

$$dr = -\,dg/T + g/T^2\, dT = (1/T^2)(-\, vT\, dP + h\, dT).$$

The relations for du, dh, da and dg can be easily memorized by using the phrase "Great Physicists Have Studied Under Very Articulate Teachers" (G, P, H, S, U, V, A, T) by considering the mnemonic diagram (cf. Figure 4). In that figure a square is constructed by representing the four corners by the properties P, S, V, T, and by representing the property H by the space between the corners represented by P and S, the property U by the space between S and V, etc., as illustrated in Figure 4. Diagonals are then drawn pointing away from the two bottom corners. Such a diagram is also known as a thermodynamic mnemonic diagram. If an expression for dG is desired (that is located at the middle of the line connecting the points T and P), we first form the differentials dT and dP, and then link the two with their conjugates as illustrated below
dG = – S(the conjugate of T with the minus sign due to the diagonal pointing towards T) × dT + V (that is conjugate of P with the plus sign due to the diagonal pointing away from P) × dP.

5. Summary of Relations

These are four important relations, namely

$$(\partial T/\partial v)_s = -(\partial P/\partial s)_v,\ (\partial T/\partial P)_s = (\partial v/\partial s)_P,$$

$$(\partial P/\partial T)_v = (\partial s/\partial v)_T, \text{ and } (\partial v/\partial T)_P = -(\partial s/\partial P)_T.$$

Even though the relations were derived by using thermodynamic relations for closed systems, the derivations are also valid for open systems as long as we follow a fixed mass.

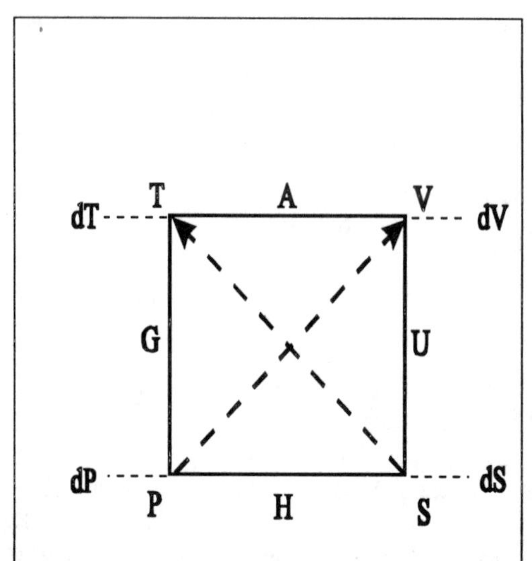

Figure 4: Thermodynamic mnemonic diagram.

For a point function, say, $P = P(T,v)$, it can be proven that

$$(\partial P/\partial T)_v\,(\partial T/\partial v)_P\,(\partial v/\partial P)_T = -1\text{, i.e., }(\partial v/\partial T)_P = -(\partial P/\partial T)_v/(\partial P/\partial v)_T \qquad (29)$$

Equation (29) is useful to obtain the derivative $(\partial v/\partial T)_P$ if state equations are available for the pressure (e.g., in the form of the VW equation of state). Likewise, if $u = u(s,v)$,

$$(\partial u/\partial s)_v\,(\partial s/\partial v)_u\,(\partial v/\partial u)_s = -1.$$

Using the relation $\partial u/\partial s = T$, $\partial s/\partial v = P/T$, and $\partial v/\partial u = -P$, we find that, as expected,

$$T\,(P/T)(-1/P) = -1.$$

---

e. *Example 5*

Show that both the isothermal expansivity $\beta_P = (1/v)(\partial v/\partial T)_P$ and the isobaric compressibility coefficient $\beta_T = -(1/v)(\partial v/\partial P)_T$ tend to zero as $T \to 0$.

Solution

Example 1 shows that $(\partial s/\partial T)_v \to 0$ and $(\partial s/\partial P)_v \to 0$ as $T \to 0$. From the fourth of the relations,

$$(\partial v/\partial T)_P = -(\partial s/\partial P)_T\text{, so that }(\partial v/\partial T)_P \to 0.$$

Similarly, using the third relation and the cyclic relations it may be shown that $(\partial P/\partial T)_v = -((\partial v/\partial T)/(\partial v/\partial P)) = (\partial s/\partial v)_T \to 0$ as $T \to 0$. Since $\partial v/\partial T \to 0$, it is apparent that $\partial v/\partial P \to 0$ as $T \to 0$.

Remark

The experimentally measured values of β_P and β_T both tend to 0 as $T \to 0$. Using the relations the reverse can be shown, i.e., $(\partial s/\partial T)_v \to 0$ and $(\partial s/\partial P)_v \to 0$.

---

f. *Example 6*

When a refrigerant is throttled from the saturated liquid phase using a short orifice, a two-phase mixture of quality x is formed. We are asked to determine the choking flow conditions for the two-phase mixture, which occurs when the mixture reaches the sound speed ($c^2 = -v^2\,(\partial P/\partial v)_s$). We must also derive an expression for the speed of sound in a two-phase mixture. Assume ideal gas behavior for the vapor phase and that the liquid phase is incompressible.

Solution

During the elemental expansion of a two phase mixture of a specified quality x from P to P + dP, and v to v + dv,

$$dh = Tds + v\,dP\text{, and} \qquad (A)$$

$$du = Tds - Pdv. \qquad (B)$$

Since,

$$dh_f = d(u_f + Pv_f)$$

For incompressible liquids,

$$dh_f = du_f + v_f\,dP.$$

For a two phase mixture of vapor and liquid,

$$dh = x\,dh_g + (1-x)\,dh_f = x\,dh_g + (1-x)(du_f + v_f\,dP).$$

Assuming ideal gas behavior for the vapor phase, and if $du_f = cdT$, then

$$dh = x\, c_{p,o}\, dT + (1 - x)(c_f dT + v_f\, dP). \quad \text{(C)}$$

Similarly,

$$du = x\, c_{vo}\, dT + (1 - x)\, c_f dT. \quad \text{(D)}$$

Considering constant entropy in Eqs. (A) and (B), using Eqs. (C) and (D), dividing by dT, we obtain the relation

$$(dP/dT)_s = (x\, c_{p,o} + (1 - x)\, c_f)/(v - (1 - x)v_f), \text{ i.e.,} \quad \text{(E)}$$

$$(dv/dT)_s = -(x\, c_{vo} + (1 - x)c_f)/P. \quad \text{(F)}$$

Dividing Eq. (E) by Eq. (F), we obtain the relation

$$-(dP/dv)_s = (x\, c_{p,o} + (1 - x)c_f)(P/(v - (1 - x)v_f))/(x\, c_{vo} + (1 - x)c_f). \quad \text{(G)}$$

Using the definition of the sound speed,

$$c^2 = -v^2(dP/dv)_s,$$

where

$$v = xv_g + (1 - x)v_f, \quad \text{(H)}$$

Eq. (G) can be written as

$$c^2 = v^2(xc_{p,o} + (1 - x)c_f)\, P/((v - (1 - x)v_f)(xc_{vo} + (1 - x)c_f. \quad \text{(I)}$$

Since $v_g = RT/P$,

$$v = (xRT/P + (1 - x)v_f), \text{ and}$$

$$c^2 = RT(x + (1 - x)(Pv_f/(RT)))^{\,2}\, (xc_{p,o} + (1 - x)c_f)/(x(xc_{vo} + (1 - x)c_f)). \quad \text{(J)}$$

If x =1, then, as expected,

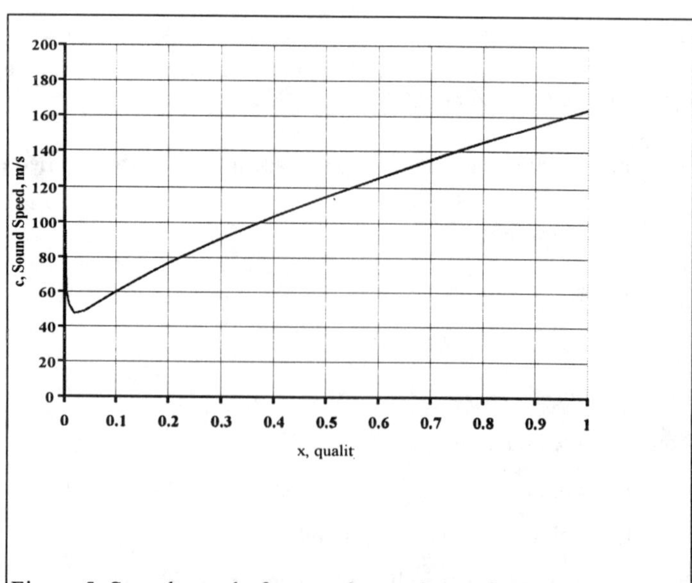

Figure 5: Sound speed of a two phase mixture for R –134a

$c^2 = c_{p,o} RT/c_{vo} = \gamma RT.$ (K)

If $x \to 0$, then

$c^2 = RT(Pv_f/RT)^2/x \to \infty.$

Using tabulated values for R–134a,

$c_f = 1.464$ kJ kg^{-1} K^{-1}, and $c_{p,o}$ at 298 K = 0.851 kJ kg^{-1} K^{-1}.

Using the values for P = 690 kPa, R = 0.08149 kJ kg^{-1} K^{-1}, c_{vo} = 0.7697 kJ kg^{-1} K^{-1}, v_f = 0.000835 m^3 kg^{-1}, and γ = 1.1. For the conditions x = 1, T = 298 K,

c = 163.4 m s^{-1}.

A plot for c with respect to x is illustrated in Figure 5. The expression for the speed of sound in a solid–vapor mixture is similar except that c_f must be replaced by c_S, specific heat of solid.

D. GENERALIZED RELATIONS

We will now derive generalized thermodynamic relations that express the nonmeasurable properties of any substance, such as s, u, and h, in terms of its measurable properties, e.g., P, T, and v. We will first obtain generalized relations for the differential quantities (e.g., ds, du, and dh,) and, then, we will integrate these relations after applying the state equations.

1. Entropy ds Relation

Any thermodynamic property for a simple compressible substance can be expressed as a function of two other independent thermodynamic properties. Therefore, for instance,

s = s (v,T), i.e.,

$$ds = \left(\frac{\partial s}{\partial v}\right)_T dv + \left(\frac{\partial s}{\partial T}\right)_v dT. \tag{30}$$

Eliminating $(\partial s/\partial T)_v$ with Eq. (14), we obtain

$$ds = \left(\frac{\partial s}{\partial v}\right)_T dv + c_v\left(\frac{dT}{T}\right), \tag{31}$$

where $c_v = f(T,v)$. Using the third relation (cf. Eq. 22) the above expression can be expressed in terms of measurable properties, namely,

$$ds = \left(\frac{\partial P}{\partial T}\right)_v dv + c_v\left(\frac{dT}{T}\right). \tag{32}$$

Likewise, considering s = s(T,P),

$$ds = (\partial s/\partial T)_P\, dT + (\partial s/\partial P)_T\, dP.$$

Substituting from Eq. (17b) for the first term on the RHS of the above equation, and using the fourth relation Eq. (26) for its second term, we obtain the expression

$$ds = c_p\, dT/T - (\partial v/\partial T)_P\, dP \tag{33}$$

Both Eqs. (32) and (33) provide relations for ds in terms of the measurable properties P, v, and T. Equation (32) is suitable with a state equation of the form P = P(T,v) (e.g., the Van der Waals equation), while Eq. (33) is more conveniently used when the state equation is expressed in the form v = v(T,P).

a. Remarks

Since $(\partial s/\partial T)_v = c_v/T$ (and $c_v > 0$ when $T > 0$), the gradient $(\partial s/\partial T)_v$ (taken along an isometric curve) has a finite and positive slope for all $T > 0$. Similarly, the gradient $(\partial s/\partial T)_P = c_P/T$ has positive values (along isobars), since $c_P > 0$ when $T > 0$.

Along an isotherm, $ds_T = -(\partial v/\partial T)_P\, dP_T$ (cf. Eq. 33). Since in the vicinity of T = 0 K, $(\partial v/\partial T)_P = 0$, this implies that s is independent of pressure at a temperature of absolute zero.

Incompressible solids and liquids undergo no volumetric change, i.e., $dv = 0$. Since, Eq. (32) states that $ds = c_v\, dT/T$, and $c_v = c_p = c$ = constant for this case,

$$s = c \ln T + \text{constant}.$$

Consider Eq. (33),

$$ds = c_p\, dT/T - (\partial v/\partial T)_P\, dP = c_p\, dT/T - \beta_P\, v\, dP, \quad (34)$$

where $\beta_P = (1/v)(\partial v/\partial T)_P$ (and, likewise, $\beta_T = -(1/v)(\partial v/\partial P)_T$). (In Chapter 6 we have learned that for liquids and solids the following relation relates the volume to the two compressibilities, namely, $v/v_{ref} = \exp(\beta_P\,(T-T_{ref}) - \beta_T\,(P-P_{ref}))$.)

For isentropic processes, $ds = 0$. and Eq. (34) yields the relation

$$dT/T = (\beta_p\, v\, dP)/c_p, \text{ so that}$$

$$(\partial T/\partial P)_s = T\,(\partial v/\partial T)_P/c_P = T\, v\, \beta_P/c_P = T\, \beta_P/c_p{}', \quad (35)$$

where $c_p{}' = c_p/v$, kJ/K m^3. If c_P, β_P and v are all approximately constant, then by integrating Eq. (34), we can obtain isentropic relations for solids and liquids, i.e.,

$$\ln T/T_{ref} = (\beta_P)\, v(P-P_{ref})/c_P. \quad (36)$$

If $\beta_P \approx 0$, the temperature remains constant as the pressure changes. In case $\beta_P > 0$ (i.e., the liquid or solid expands upon heating), the temperature increases during compression and decreases during expansion. Likewise, if $\beta_P < 0$ (e.g., for rubber, or water between 0 and 4°C at 1bar), the temperature decreases with an increase in pressure.

Alternately, using Eq. (32),

$$ds = c_v dT/T + (\partial P/\partial T)_v\, dv.$$

Since $(\partial P/\partial T)_v = -(\partial v/\partial T)_P/(\partial v/\partial P)_T = \beta_P/\beta_T$,

$$ds = c_v dT/T + (\beta_P/\beta_T)\, dv.$$

For isentropic processes, $ds = 0$, and

$$(\partial T/\partial v)_s = -\,T\,(\partial P/\partial T)_v/c_v \text{ or } (\partial T/\partial v)_s = -T\,\beta_P/(\beta_T\, c_v),$$

where both β_T and β_P are almost constant for most solids and liquids. Therefore,

$$dv_s = -dT_s\, \beta_T\, c_v/(T\, \beta_P). \quad (37)$$

The term β_T is related to the stress–strain relation in solids. From the above relation we find that when $dv_s < 0$ (i.e., during isentropic compression) $dT_s > 0$ if $\beta_P > 0$ (i.e., the material expands upon heating) which implies that the temperature rises. On the other hand, when a material is isentropically stretched ($dv_s > 0$), $dT_s < 0$. (This is similar to the phenomenon of first compressing a gas and then releasing it.).

We now examine the phenomenon of the bending of a beam that is illustrated in Figure 6. During the bending process, the upper layers of the beam are compressed and $dT_s > 0$. The bottom layers are stretched where $dT_s < 0$. Therefore, temperature gradients develop within the beam. The increase in the energy in the upper layers is stored in the form of a vibrational energy increase in the atoms (that results in the higher temperature) and partly as intermolecular potential energy. If the beam changes periodically between the expanded and compressed states, the oscillation causes the material to stretch, thereby increasing the intermolecular potential energy, but decreasing the thermal energy. If the material has a negligible thermal conductivity, there is no heat transfer between the various isothermal layers within the beam, and each layer will act as an adiabatic reversible system. Thus, if the beam is placed in a vacuum, the material will keep vibrating. However, material with a finite thermal conductivity will behave differently, since there is heat transfer within the various layers. This heat transfer is a result of the transfer of vibrational energy between the various layers in the beam, which results in a lower stretching for successive portions of the cycle. The amount of energy available for stretching is less resulting in thermo–elastic damping.

Consider a liquid at its saturation temperature at a specified pressure. As heat is supplied to it isobarically, a portion of the liquid vaporizes, but the temperature is unchanged. Since $dT = dP = 0$, Eq. (33) *seems* to suggest that $ds = 0$, which is an incorrect interpretation. In this case, c_p ($= (\partial h/\partial T)_P$) and $\partial v/\partial T$ both tend to infinity (due to the fact that h_{fg} and v_{fg} are finite, while the temperature remains constant during the boiling process). However, we can use other relations, such as $T\,ds + v\,dP = dh$ to obtain the relation $ds = dh/T$ during isobaric vaporization which on integration yields $s_{fg} = h_{fg}/T$

g. *Example 7*

Weights are gradually placed on an insulated copper bar in order to compress it to 1000 bars. If the compression is adiabatic and reversible (i.e., the material reverts to its original state once the load is removed) determine:

The change in the solid temperature.

The internal energy change.

The temperature after the load is removed.

Assume that $T_o = 250$ K, $\beta_P = 48\times10^{-6}$ K^{-1}, $\beta_T = 7.62\times10^{-7}$ bar^{-1}, $v = 1.11\times10^{-4}$ m^3 kg^{-1}, $c_p = 0.372$ kJ kg^{-1} K^{-1}, and $c_v = 0.364$ kJ kg^{-1} K^{-1}. Treat Cu as simple compressible substance

Solution

Since the process is adiabatic and reversible we will use the relation $(\partial T/\partial P)_s = Tv\beta_P/c_P$ or $(\Delta T/\Delta P)_s = T_o v\beta_P/c_P.= \{250\text{K}\times1.1\times10^{-4}\text{m}^3\text{kg}^{-1}\times48\text{x}10^{-6}\text{K}^{-1}\}/0.372\text{kJkg}^{-1}\text{K}^{-1} = 3.548\text{x}10^{-6}$ K/Kpa and $\Delta P = 100\times1000$ Kpa. Hence $dT_s = 0.36$ K and T will rise to 250.36 K.

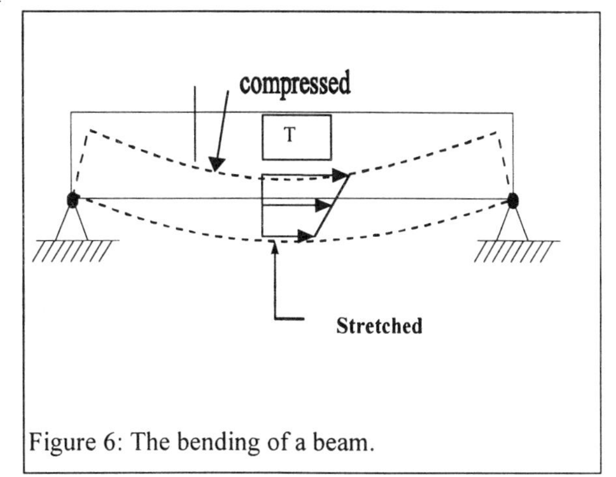

Figure 6: The bending of a beam.

Applying the First law to an adiabatic reversible process
$du_s = - Pdv_s$.
Recall from Eq. (37) that
$dv_s = -dT_s\ \beta_T c_v/(T\ \beta_P)$, i.e.,
$du_s = \{P\ \beta_T c_v/(T\ \beta_P)\}dT_s$
Integrating and assuming that P is not a function of temperature and remaining at an average value of ***500*** bar.
$du_s = \{500 \text{ bar} \times 7.62\times10^{-7} \text{ bar}^{-1} \times 0.364 \text{ kJ kg}^{-1} \text{ K}^{-1}$
$\div (250 \text{ K} \times 48\times10^{-6} \text{ K}^{-1})\}\ 0.36 \text{ K} = 0.00416 \text{ kJ kg}^{-1}$.
The temperature after the load is removed is 250 K, since the process is reversible.

h. *Example 8*

Obtain a relation for ds for an ideal gas. Using the criterion for an exact differential show that for this gas c_v is only a function of temperature.

Solution

For an ideal gas

$$P = RT/v. \qquad (A)$$

Using Eq. (A) and Eq. (32), we obtain

$$ds = R\ (dv/v) + c_v\ (dT/T). \qquad (B)$$

Comparing Eq. (B) with the relation $dZ = Mdx + Ndy$, and using the criterion for an exact differential we obtain

$$\partial\{(c_v/T)/\partial v\}_T = \partial\ \{(R/v)/\partial T\}_v = 0.$$

since at constant volume (R/v) is not a function of temperature (or pressure). Therefore, the term $\partial\{(c_v/T)/\partial v\}_T$ is not a function of v and, at most, is a function of temperature alone.

i. *Example 9*

Obtain a relation for ds for a gas that follows the RK equation of state.

Solution

For an RK gas, a state relation of the form

$$P = RT/(v - b) - a/(T^{1/2}\ v(v + b)) \qquad (A)$$

can be used. Using Eqs. (A) and (32), we obtain the expression

$$ds = c_v(dT/T) + (R/(v - b) + a/(2(T^{3/2}\ v(v + b))). \qquad (B)$$

Remark

We can show that c_v is a function of both v and T for an RK gas by using the criterion for an exact differential.

j. *Example 10*

A VW gas is used as the working fluid in an ideal power cycle. A relation between T and v is required for an isentropic process (data for $c_{vo}(T)$ is available). If $v_1 = 0.006$ $m^3 \text{ kg}^{-1}$, $T_1 = 200$ K, the compression ratio $v_1/v_2 = 3$, determine the values of T_2 and P_2 if the gas is air.

Solution

Recall that

$$ds = c_v dT/T + (\partial P/\partial T)_v dv, \text{ i.e., } ds = cv\ dT/T + R\ dv/(v-b) \qquad (A)$$

Using the criterion for an exact differential,

$$(\partial c_v/\partial v)_T = \partial\,[\{R/(v-b)\}/\partial T]_v = 0.$$

This implies that c_v is not a function of volume and is a function of temperature alone, i.e., $c_v = c_{vo}(T)$. Since ds = 0 for the ideal cycle, upon integrating Eq. (A),

$$\int c_{vo}\, dT/T = -R \ln(v-b) + C. \qquad \text{(B)}$$

Since, $s^o = \int c_{p,o} dT/T$, we define

$$(s')^o(T) = \int c_{vo} dT/T = \int (c_{p,o} - R) dT/T = s^o - R \ln T. \qquad \text{(C)}$$

We use Eqs. (B) and (C) to obtain the relation

$$(s')^o = -R \ln \{(v-b)\} + C'.$$

Therefore,

$$(s_2')^o - (s_1')^o = R \ln((v_1 - b)/(v_2 - b)). \qquad \text{(D)}$$

Simplifying,

$$\exp\,((s_2')^o/R - (s_1')^o/R) = (\exp\,(s_2{}^o/R)/T_2)/(\exp(s_1{}^o/R)/T_1) = (v_2 - b)/(v_1 - b). \qquad \text{(E)}$$

Upon defining $v_r = \exp\,(s^o/R)/T$, Eq. (E) can be written in the form

$$v_{r2}/v_{r1} = (v_2 - b)/(v_1 - b). \qquad \text{(F)}$$

Values of v_r are usually tabulated. Once the volume ratio v_2/v_1 is specified, T_2 can be determined from Eq. (F). Using the VW equation of state, we can then determine P_2. Since, $v_1 = 0.006\ m^3\ kg^{-1}$ at $T_1 = 200$ K, the VW equation yields

$P_1 = 0.08314 \times 200 \div (0.006 \times 28.97 - 0.0367) - 1.368 \div (0.006 \times 28.97)^2$

$= 121.3 - 45.3 = 76$ bar.

At T= 200 K, $v_{r1} = 1707$. We will use the relation

$$v_{r2}/v_{r1} = (v_2 - b)/(v_1 - b),$$

and the values $v_1 = 0.006\ m^3\ kg^{-1}$, $b = 0.0367 \div 28.97 = 0.00127\ m^3\ kg^{-1}$.

Therefore,

$v_2 = 0.006 \div 3 = 0.002\ m^3\ kg^{-1}$, and

$v_{r2}/v_{r1} = (0.002 - 0.00127) \div (0.006 - 0.00127) = 0.154$, so that

$v_{r2} = 1707 \times 0.154 = 262.9$.

The tabulated values indicate that at $v_{r2} = 263$, $T_2 = 423$ K.

Finally, using the VW equation of state

$P_2 = 0.08314 \times 423 \div (0.002 \times 28.97 - 0.0367) - 1.368 \div (0.002 \times 28.97)^2$

$= 1656 - 408 = 1248$ bar.

Remarks

If a = b =0, Eq. (E) represents the state relation for ideal gas. For an ideal gas, the relation yields $v_{r2} = 469$, $T_2 = 335$ K, and $P_2 = 480$ bar.

Applying Eq. (B) and assuming c_{vo} to be constant, $c_v \ln(T_2/T_1) = R \ln((v_1 - b)/(v_2 - b))$, which can be simplified and written in the form

$$T_2/T_1 = [(v_1 - b)/(v_2 - b)]^{(k-1)}, \qquad \text{(G)}$$

where $k = c_{p,o}/c_{vo}$.

Using (G) in VW equation of state

$$(P_2 - a/v_2{}^2)(v_2 - b)^k = (P_1 - a/v_1{}^2)(v_1 - b)^k$$

---

k. *Example 11*

Derive an expression for the sound speed ($c^2 = -v^2(\partial P/\partial v)_s = v/\beta_s$) in terms of the measurable properties of a simple compressible substance.

Show that $c_p/c_v = k = \beta_T/\beta_s$.

Determine a relation for the sound speed for an ideal gas.

Determine a relation for the sound speed for a VW gas.

Solution

Recall that the speed of sound

$$c^2 = -v^2(\partial P/\partial v)_s = v/\beta_s$$

$$ds = 0 = c_v\, dT/T + (\partial P/\partial T)_v\, dv, \text{ and} \quad \text{(A)}$$

$$ds = 0 = c_p dT/T - (\partial v/\partial T)_P\, dP. \quad \text{(B)}$$

We multiply Eq. (A) by (T/c_v) and Eq. (B) by (T/c_p) and then subtract one of the resulting relations from the other to obtain

$$(\partial P/\partial T)_v\, (T/c_v)\, dv_s + (\partial v/\partial T)_P\, (T/c_p)\, dP_s = 0, \text{ or} \quad \text{(C)}$$

$$(\partial P/\partial v)_s = -\, k\, (\partial P/\partial T)_v/(\partial v/\partial T)_P, \text{ where} \quad \text{(D)}$$

$$k(T,v) = c_p(T,v)/c_v(T,v). \quad \text{(E)}$$

Applying the expression for the speed of sound $c^2 = -v^2(\partial P/\partial v)_s = v/\beta_s$ in Eq. (D),

$$c^2 = v^2\, k(T,v)(\partial P/\partial T)_v/(\partial v/\partial T)_P. \quad \text{(F)}$$

Using the cyclical rule

$$(\partial P/\partial v)_T(\partial v/\partial T)_P(\partial T/\partial P)_v = -1 \quad \text{(G)}$$

we obtain

$$(\partial v/\partial T) = -\, (\partial P/\partial T)/(\partial P/\partial v) \quad \text{(H)}$$

Substituting from Eq. (H) in Eq. (F),

$$c^2 = -\, k(T,v)\, v^2\, (\partial P/\partial v)_T. = k(T,v)\, v/\beta_T \quad \text{(I)}$$

With $c^2 = v/\beta_s$, in Eq. (I)

$$v/\beta_s = k(T,v)\, v/\beta_T, \text{ or } k(T,v) = \beta_T/\beta_s.$$

In the case of ideal gases,

$$k = -\, (c^2/v^2)/(-RT/v^2) = c^2/RT \text{ or } c^2 = kRT. \quad \text{(J)}$$

Typically we denote c as c_0 for ideal gases.

For a VW gas,

$$\partial P/\partial v = -\, RT/(v-b)^2 + 2a/v^3 \quad \text{(K)}$$

Thereafter, combining Eqs. (I) and (K)

$$c^2 = k(T,v)\, v^2\, (RT/(v-b)^2 + a/v^3) \quad \text{(L)}$$

Remarks

If, in the VW state relation, a = b= 0, the expression reduces to the sound speed for an ideal gas. In that case, $k = k(T)$.

High pressures often develop within the clearance space in turbine seals, and gas leaks are governed by the resulting choked flow conditions. The value of the sound speed through a real gas is required in order to evaluate this condition.

In the case of liquids and solids, a very large pressure is required to cause a small change in the volume so that $(\partial P/\partial v)_T \rightarrow \infty$ and, consequently, $c \rightarrow \infty$. Therefore, sound travels at faster speeds in liquids and solids.

Applying Eq. (L) to the case of an ideal gas, $c_o^2 = k(T)\ RT$, and dividing Eq.(I) by c_o^2 we obtain the expression,

$$(c^2/c_o^{\ 2}) = -\,(k(T_R,\ v_R')/(k_o(T_R)\ T_R))\ v_R^{\ 2}\ (\partial P_R/\partial v_R')_{TR}.$$

---

2. Internal Energy (du) Relation

Combining the First and Second laws

$$du = T\ ds - P\ dv$$

Using Eq. (32) to eliminate ds, we obtain the expression

$$du = c_v\ dT + (T(\partial P/\partial T)_v - P)\ dv, \tag{38}$$

which implies that the change in internal energy equals the energy stored in the form of translational, vibrational, and rotational energies plus the intermolecular potential energy.

a. Remarks

In the case of an ideal gas $P = RT/v$, so that

$$T(\partial P/\partial T)_v - P = (RT/v)\ - P = 0$$

Therefore, the potential energy term in Eq. (38) equals zero for ideal gases. Recall from the discussion regarding the Lennard–Jones potential energy curve in Chapter 1 that the intermolecular potential energy term equals zero at larger intermolecular distances at which the intermolecular attraction force is zero. Consequently, energy is not stored in the form of potential energy as the volume is changed at large volumes, and a change in the gas volume (or the intermolecular distance) does not affect the intermolecular potential energy.

Equation (38) is called the "Calorific Equation of State" and indicates that the internal energy of a simple compressible substance (the solid, liquid, or gas phases) is a function of temperature and volume (cf. Figure 7). In it, the first term, $c_v\ dT$, represents the thermal portion of the energy (which varies due to changes in the translational, vibrational, and rotational energies, with each degree of freedom contributing an amount equal to $2k_BT$ per molecule, as discussed in Chapter 1).

Applying the First law to a closed system undergoing a reversible process, in the context of Eq. (38)

$$\delta Q_{rev} = dU + PdV = mc_v dT + (T\partial P/\partial T - P)dV + PdV \tag{38'}$$

Eq.(38') states that energy transfer δQ_{rev} is used to raise the thermal energy by the incremental amount $mc_v dT$ (e.g., te, re, ve), to overcome the intermolecular potential energy barrier $(T\partial P/\partial T - P)dV$ (i.e., ipe required to move the molecules against attractive forces) and to perform boundary PdV work. For an ideal gas, the intermolecular potential energy equals zero so that heat transfer is used for PdV work if the process is isothermal.

In an adiabatic system the internal energy change $c_v dT + (T\partial P/\partial T - P)dv$ equals the work performed on the system. Note that if V and T are held fixed, $\delta Q_{rev} = 0$. However, this does not imply that Q = Q(V,T) since the functional form for the relation changes, depending upon the process path between the initial and final states (i.e., Q is not a property).

Consider Example 6 in Chapter 2. As we place a large weight on the piston, the gas is adiabatically compressed and we perform Pdv work. The potential energy decrease of the weight is converted into internal energy increase of the gas. The intermolecular potential energy decreases during compression $(T\partial P/\partial T - P)dv < 0$, since the second term in Eq. (38) decreases as the intermolecular spacing becomes closer. Eq. (38) implies that the term $c_v dT$ must increase. If gas is ideal, then change in ipe =0; this implies that temperature changes to a greater extent in a real gas than in an ideal gas).

For incompressible substances, dv = 0 and, consequently from Eq. (38), du = c_v dT. Attractive forces are very strong in such substances so that the intermolecular potential energy remains virtually constant.

In the case of solids and liquids it is useful to express Eq. (38) in terms of the isobaric expansivity and isothermal compressibility. Since $(\partial P/\partial v)_T(\partial v/\partial T)_P(\partial T/\partial P)_v = -1$, and $(\partial P/\partial T)_v = \beta_P/\beta_T$, Eq. (38) can be written in the form

$$du = c_v\, dT + (T\, \beta_P/\beta_T - P)\, dv.$$

The values of β_P and β_T are approximately constant for most solid and liquid substances.

The volume of a two-phase mixture can be changed at a specified temperature and pressure (e.g., that of water at 100°C and 1 bar pressure) by altering its quality. Therefore, while changes in the mixture volume are possible during phase transition at given T, those in the pressure are not. Consequently, $(\partial v/\partial P)_T$ (= a finite value ÷ 0) $\to\infty$, i.e., $\beta_T \to \infty$. Similarly, $(\partial v/\partial T)_P \to \infty$ and hence $\beta_p \to \infty$ for a two-phase mixture. Thus as we approach the critical point, $\beta_T \to \infty$ and $\beta_P \to \infty$.

Since the relation for du in Eq. (38) is an exact differential, applying the appropriate criterion,

$$\partial c_v/\partial v = (\partial/\partial T)(T(\partial P/\partial T)_v - P) = T\partial^2P/\partial T^2 + \partial P/\partial T - \partial P/\partial T = T(\partial^2P/\partial T^2)_v. \quad (39)$$

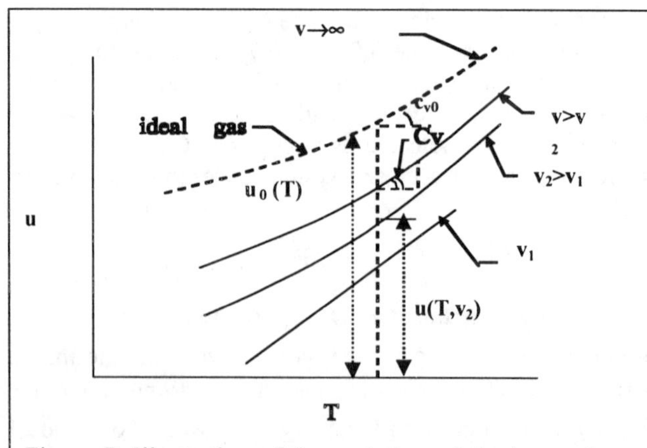

Figure 7: Illustration of the variation of the internal energy with respect to temperature.

Applying the state relations for liquids and solids (cf. Chapter 6) it can be shown that

$T(\partial^2P/\partial T^2)_v \approx 0$, i.e.,

$c_v = c_v(T)$ alone. Integrating Eq. (39) between the limits $v \to \infty$ (i.e., as $P \to 0$) and $v = v$, we obtain an expression for the deviation of the constant volume specific heat from its ideal gas value, i.e.,

$$c_v - c_{vo} = \int_{\infty}^{v} T\,(\partial^2P/\partial T^2)_v\,dv$$

If the P–v–T behavior of a gas is known from either measurements or theory, and data for c_{vo} is available, the above relation can be integrated to evaluate values of c_v.
At T = 0 K, it will be later shown that $\partial P/\partial T = 0$. At that condition, $du \to (-Pdv)$, i.e., $\partial u/\partial v \to -P$.

---

l. Example 12

The state of a copper bar is initially at a pressure of 1 bar and temperature of 250 K. It is compressed so that the exerted pressure is 1000 bar. Assume that the compression is adiabatic and reversible (i.e., the material reverts to its original state once the load is removed), and that $\beta_P = 48\times10^{-6}\ K^{-1}$, $\beta_T = 7.62\times10^{-7}\ bar^{-1}$, $v = 1.11\times10^{-4}\ m^3\ kg^{-1}$, $c_p = 0.372\ kJ\ kg^{-1}\ K^{-1}$, and $c_v = 0.364\ kJ\ kg^{-1}\ K^{-1}$. Determine the change in the internal energy and the intermolecular potential energy of the solid.

Solution

Integrating the relation $\beta_T = -\frac{1}{v}\left(\frac{\partial v}{\partial P}\right)_T$,

$$\ln(v_2/v_1) = -\beta_T(P_2 - P_1) \quad \text{or} \quad v_2 - v_1 \approx -\beta_T v_1(P_2 - P_1) \qquad \text{(A)}$$

Therefore, $v_2\text{-}v_1 = -0.845\text{x}10^{-7}\ m^3\ g^{-1}$, $v_2 = 1.1092\times10^{-4}\ m^3\ kg^{-1}$ and which are similar to the results obtained in Example 7.
Integrating Eq. (37),

$$(\ln(T_2/T_1) = -\{\beta_P/\beta_T c_v\}(v_2 - v_1) \quad \text{or} \quad (T_2 - T_1) \approx -(\beta_P/\beta_T c_v)T_1(v_2 - v_1) \qquad \text{(B)}$$

Using the results from Eq. (A),

$$T_2 - T_1 = (T_1\,\beta_P/c_v)v_1(P_2 - P_1) = 0.365\ K$$

Consequently, the temperature following compression is 250.365 K. For an adiabatic reversible process, the First law yields,

$$\Delta u = -\int P\,dv = -\int P\,(\partial v/\partial P)\,dP = \int \beta_T\,P\,v\,dP = (P_2^2 - P_1^2)\beta_T v/2$$
$$= ((1000 \times 100)^2 \div 2 - 100^2 \div 2) \times 1.11\times10^{-4} \times 7.62\times10^{-9}$$
$$= 0.00423\ kJ\ kg^{-1} \text{ or } 4.23\ J\ kg^{-1} \text{ (similar to example 7)} \qquad \text{(C)}$$

We may also use following equation:

$$du = c_v\,dT + (T\,\beta_P/\beta_T - P)\,dv. \qquad \text{(D)}$$

The thermal portion of change in "u" is given as,

$$c_v\,dT = 0.364 \times 0.365 = 0.1329 J\ kg^{-1}. \qquad \text{(E)}$$

The intermolecular potential energy of the solid

$$\Delta(\text{ipe}) = (T\beta_P/\beta_T - P)\ dv$$

where the second term is same as on the RHS of Eq. (C). Therefore

$$\Delta(\text{ipe}) = (250\ \text{K} \times 48\times10^{-6}\ \text{K}^{-1} \div 7.62\text{x}10^{-7}\ \text{bar}^{-1})$$
$$\times (-8.45\text{x}10^{-8}\ \text{m}^3\text{kg}^{-1})\times100\ \text{kPa bar}^{-1} + 0.00423, \text{ or}$$

$$\Delta\ (\text{ipe}) = -0.1331 + 0.00423 = -0.1288\ \text{kJkg}^{-1} \qquad \text{(F)}$$

The net change in the internal energy of the solid is

$$du = 0.1329 - 0.1288 = 0.0041\ \text{kJ kg}^{-1}.$$

Which is approximately the same as the answer in Eq. (C). In this example, the temperature increases by 0.365 K, increasing the thermal portion of the internal energy by 0.1329 kJ, but the IPE decreases by 0.1288 kJ (Eq.(F)) . At the minimum intermolecular potential energy (Chapter 1) , compression should cause the "ipe" to increase. Since the ipe decreases with compression, this indicates that the solid is not at that minimum value.

---

m. Example 13

A substance undergoes an adiabatic and reversible process. Obtain an expression for $(\partial T/\partial v)_s$ in terms of c_v, β_P, β_T and T. What is the value of $(\partial T/\partial v)_s$ for copper, given that $\beta_P = 5\times10^{-5}\ \text{K}^{-1}$, $\beta_T = 8.7\times10^{-7}\ \text{bar}^{-1}$, $c = c_v = 0.386\ \text{kJ kg}^{-1}\ \text{K}^{-1}$, $v = 1.36\times10^{-4}\ \text{m}^3\ \text{kg}^{-1}$, and the temperature is 25°C? What is the temperature rise if $dv = -8.106\times10^{-7}\ \text{m}^3\ \text{kg}^{-1}$?

Solution

We will use the relations

$$ds = c_v\ dT/T + (\partial P/\partial T)_v\ dv, \text{ and} \qquad \text{(A)}$$

$$(\partial P/\partial T)_v\ (\partial T/\partial v)_P\ (\partial v/\partial P)_T = -1. \qquad \text{(B)}$$

Therefore,

$$(\partial P/\partial T)_v = -(\partial v/\partial T)_P/(\partial v/\partial P)_T = \beta_P/\beta_T, \text{ i.e.,} \qquad \text{(C)}$$

$$ds = c_v\ dT/T + (\beta_P/\beta_T)\ dv \qquad \text{(D)}$$

For an adiabatic and reversible process, ds = 0, and

$$(\partial T/\partial v)_s = .-T\beta_P/(\beta_T c_v). \qquad \text{(E)}$$

For copper,

$$(\partial T/\partial v)_s = -\ 5\times10^{-5}\ \text{K}^{-1} \times 298\text{K} \div (8.7 \times 10^{-7}\ \text{bar}^{-1} \div 100\ \text{kPa bar}^{-1})$$
$$\div\ 0.386\ \text{kN m kg}^{-1}\ \text{K}^{-1} = -4.4370\times10^6\ \text{K kg m}^{-3}.$$

On compression

$$dT_s = (\partial T/\partial v)_s\ dv = -4.437\times10^6 \times (-8.106\times10^{-7}) = 3.6\ \text{K}.$$

Remarks

The compression of a solid in the elastic regime is isentropic. The First law indicates that the work done (–Pdv) during adiabatic compression and du >0. Equation (D) provides the answer for the change in T when the volume changes for any simple compressible substance. Typically $\beta_T > 0$ for any substance. Those substances that expand upon heating (e.g., gases, steel, water above 4°C at P = 1 bar) $\beta_P > 0$. From Eq. (E), $(dT/dv)_s < 0$. Thus, upon compression ($dv < 0$), and $dT > 0$. Those substances which contract upon heating (with a negative coefficient of thermal expansion, e.g., rubber, water below 4°C), $\beta_P < 0$. Eq. (E) yields, $dT > 0$, upon expansion, since $dv > 0$.

If c_v, v and β_P are constants, then Eq. (E) can be integrated to yield

$$\ln (T_2/T_1) = (v_2 - v_1)\, \beta_P/(\beta_T\, c_v). \tag{F}$$

which is the same as Eq.(B) in Example 12.

---

3. Enthalpy (dh) Relation

Obtaining the total differential of the enthalpy h = h (T, P)

$$dh = (\partial h/\partial T)_P\, dT + (\partial h/\partial P)_T\, dP = c_P\, dT + (\partial h/\partial P)_T\, dP. \tag{40}$$

Since dh = Tds + v dP, dividing the expression by dP, the following relation applies at constant T

$$(\partial h/\partial P)_T = T(\partial s/\partial P)_T + v. \tag{41}$$

Applying the fourth relation, Eq.(26) to Eq. (41), we obtain the expression

$$(\partial h/\partial T)_P = -T(\partial v/\partial P)_T + v. \tag{42}$$

Subsequently, using Eq. (42) in (40), we obtain

$$dh = c_P\, dT + \{v - T(\partial v/\partial T)_P\}\, dP. \tag{43}$$

The relation h = h(T,P) is called the calorific equation of state. If a state equation is available for v = v(T,P), Eq. (43) can be used to obtain dh.

a. Remarks

During vaporization T and P are constant, i.e., dT = dP = 0. Therefore, since $c_p \to \infty$ and $\partial v/\partial T \to \infty$, $dh = c_p\, dT - T(\partial v/\partial T)_P\, dP \neq 0$.

It can be shown from relation for Eq. (43) that

$$\partial c_p/\partial P = -T\partial^2 v/\partial T^2 \tag{44}$$

---

n. Example 14

Obtain an expression for dh and du for a liquid in terms of c_p, β_P, β_T, c_v, dT and dP. Simplify the relations for an incompressible liquid.

Solution

Rewriting Eq. (43),

$$dh = c_p\, dT + (v - T\, \beta_p\, v)\, dP, \tag{A}$$

where $\beta_p = (1/v)\,(\partial v/\partial T)_p$. Therefore,

$$du = dh - d(Pv) = c_p\, dT - Pdv - T\, \beta_p\, v\, dP$$

However, v = v(T,P), so that

$$dv = (\partial v/\partial T)_P\, dT + (\partial v/\partial P)_T\, dP = v\,(\beta_p\, dT - \beta_T\, dP). \tag{B}$$

Hence,

$$du = (c_p - Pv\beta_p)\, dT + v\,(P\beta_T - T\beta_P)\, dP \tag{C}$$

For incompressible liquids $\beta_P = \beta_T = 0$, and Eqs. (A) can be expressed in the form

$$dh = c_p\, dT + v\, dP., \text{ i.e., } h = h(T,P), \text{ and} \quad (D)$$

Further,

$$(dh/dT)_P = c_p. \quad (E)$$

For an incompressible fluid Eq. (C) assumes the form

$$du/dT = c_p, \text{ i.e., } u = u\,(T) \text{ alone.} \quad (F)$$

Upon comparing Eqs. (E) and (F),

$$(\partial h/\partial T)_p = du/dT = c_p. \quad (G)$$

Note that $du/dT = (\partial u/\partial T)_v$, since v is constant. Therefore,

$$(\partial h/\partial T)_p = du/dT = c_p = c_v = c \text{ only for incompressible liquids} \quad (H)$$

---

4. Relation for $(c_p - c_v)$

We will develop relations for $(c_p - c_v)$ in terms of measurable properties and show that $c_p > c_v$, which implies that $k > 1$. Differentiating the relation for enthalpy $h = u + Pv$,

$$(\partial h/\partial T)_P = c_p = (\partial u/\partial T)_P + P(\partial v/\partial T)_p.$$

The first term on the RHS $(\partial u/\partial T)_P$ can be simplified by dividing Eq. (38) by dT in order to obtain , $(du/dT)_P = c_v + (T(\partial P/\partial T)_v - P)\,(dv/dT)_P$; then the above equation becomes,

$$c_p = c_v + (T\partial P/\partial T - P)(\partial v/\partial T)_p + P(\partial v/\partial T)_p, \text{ or } (c_p - c_v) = (T(\partial P/\partial T)(\partial v/\partial T)_p).$$

Using the cyclic relation $(\partial P/\partial T)(\partial T/\partial v)(\partial v/\partial P) = -1$, $(\partial v/\partial T)_P = -(\partial P/\partial T)/(\partial P/\partial v))$, we obatin the expression

$$(c_p - c_v) = -\,(T(\partial P/\partial T)^2/(\partial P/\partial v)_T). \quad (45a)$$

Likewise, using the cyclic relation,

$$(\partial P/\partial T)_v = -(\partial v/\partial T)_P/(\partial v/\partial P)_T = \beta_P/\beta_T, \text{ i.e.,}$$
$$(c_p - c_v) = v\,T\,\beta_P^2/\beta_T \quad (45b)$$

a. *Remarks*

For ideal gases $\beta_P = 1/T$ and $\beta_T = -1/P$. Therefore, using Eq. (45b), $(c_p - c_v) = (c_{p,o} - c_{vo}) = R$.
We can rewrite Eq. (45b) in the form $(c_p - c_v) = Ac_p^2T$, where the relation $A = v\beta_P^2/(\beta_T c_P^2)$ is called the Nernst–Lindmann equation. For liquids and solids, v, β_P, β_T, and c_P are approximately constant and, consequently, A is a constant.
For incompressible liquids $c_p = c_v$. However for water at a temperature of 80°C, c_p = 4.19 kJ kg^{-1} K^{-1} and c_v = 3.86 kJ kg^{-1} K^{-1}.
The Gruneisen constant $\gamma_g = (\partial P/\partial T)_v v/c_v$. Its value is constant over a wide temperature range for many solid metals. Typical values of γ_g lie between 1 and 3. Using the RHS of the Eq. (45a), the difference in specific heats equals (–T (a finite value) ÷ 0) which tends to infinity at the critical point.

If $(\partial P/\partial v) > 0$ (recall from Chapter 6 that P decreases with v for real gases at temperatures lower than T_c for a certain volumetric range), then $(c_p - c_v) < 0$, i.e., $c_p < c_v$. You will learn in Chapter 10 that these states are unstable.
Using the relation $\beta_s = -1/v(\partial v/\partial P)_s$, we note that $\beta_T/\beta_s = (\partial v/\partial P)_T/(\partial v/\partial P)_s$.

---

o. *Example 15*

Obtain an expression for the enthalpy change dh in a Clausius I fluid that follows the relation

$$P = RT/(v-b), \qquad \text{(A)}$$

and show that c_p is a function of T alone.

Solution

Using Eq. (A)

$$v = b + RT/P, \text{ and} \qquad \text{(B)}$$

using Eq. (43),

$$dh = c_p\, dT + (v - T\, R/P)\, dP = c_p\, dT + b dP, \text{ i.e., } h = h(T,P). \qquad \text{(C)}$$

Using the criterion for an exact differential we can show that

$$dc_p/dP = db/dT = 0. \qquad \text{(D)}$$

Therefore, c_p is a function of temperature alone.
Integrating Eq. (C),

$$h = \int c_p\,(T)\, dT + bP + \text{constant}. \qquad \text{(E)}$$

E. EVALUATION OF THERMODYNAMIC PROPERTIES

1. Helmholtz Function

Most thermodynamic properties can be derived in terms of differentials of the Helmholtz function. Hence it is useful to derive a relation for "a" in terms of P, T, v, etc.. Consider Eq. (20)

$$da = - s\, dT - Pdv$$

Given the state equation P(T, V), Eq. (20) can be integrated to obtain "a" as a function of T and v. Since "da" represents an exact differential, we can integrate the relation keeping either T or v constant (see Chapter 1 for a discussion regarding the integration of an exact differential). We prefer to keep the temperature constant, since the P(T,v) (e.g., from the state equation) relation is known. At constant temperature, Eq. (20) assumes the following form for RK fluid.

$$da_T = - Pdv = -(RT/(v-b) - \mathbf{a}/(T^{1/2}\, v(v+b)))\, dv$$

Integrating this expression (at constant temperature),

$$a(T,v) = - RT\,\ln(v - b) + (\mathbf{a}/(bT^{1/2}))\,\ln(v/(v + b)) + f(T). \qquad (46)$$

If $b \to 0$, then $\ln(v/(v+b)) \to -b/v$ and after application of the Le Hospital rule Eq. (46) implies that

$$a(T,v) = - RT\,\ln v + (\mathbf{a}/(T^{1/2})) + f(T),\ b \to 0.$$

Since the properties of ideal gases are generally known, we will evaluate the constant f(T) in terms of ideal gas properties. For an ideal gas $b \to 0$ (since the molecules are point masses) and $\mathbf{a} \to 0$ (as there are no intermolecular attraction forces). Therefore,

$$a_o(T,v) = -RT \ln v + f(T), \text{ i.e., } f(T) = a_o(T,v) + RT \ln v.$$

The same result follows by integrating the expression $da_o = -s_o\, dT - P_o\, dv$, where $P_o = RT/v$, at constant temperature,

$$da_{o,t} = -P_{ig} dv; \text{ then } da_T - da_{o,T} = -\left(P - P_{ig}\right) \text{ or } -P(T,v)\,dv - RT\ d\ln v.$$

In general, for any state equation P(T,v),

$$a(T,v) - a_o(T,v) = \int P(T,v)\, dv - RT \ln v.$$

The latter procedure is appropriate for Martin–Hou and Kesler type state relations that do not contain specific terms (such as **a**) related to the intermolecular attraction forces and body volume b. For RK fluid

$$a(T,v) - a_o(T,v) = RT \ln \{v/(v-b)\} + \{\mathbf{a}/(bT^{1/2})\} \ln \{v/(v+b)\}, \text{ where} \quad (47)$$

$$a_o(T,v) = u_o(T) - T\, s_o(T,v)$$

Note that as $v \to \infty$, the RHS of Eq. (47) approaches 0 indicating that $a \to a_o$. We can describe a residual Helmholtz function a^{Res}, which is a correction to ideal gas behavior, as

$$a^{Res} = a(T,v) - a_o(T,v). \quad (48)$$

Thereafter, dividing Eq. (47) by RT_c, and using the equalities $a = 0.4275\ R^2T_c^{2.5}/P_c$, $b = 0.08664\ RT_c/P_c$, $v = v_R'v_c'$ (where $v_c' = RT_c/P_c$), that equation can be expressed in dimensionless form, i.e.,

$$(a(T,v_R') - a_o(T,v_R'))/(RT_c) =$$

$$-T_R \ln(1 - (0.08664/v_R')) - (4.934/T_R^2) \ln (1 + (0.08664/v_R)), \text{ or} \quad (49)$$

$$(a_o(T,v_R') - a(T,v_R'))/(RT_c) = T_R \ln (v_R'/(v_R' - 0.08664))$$

$$- (4.934/T_R^{\ 2}) \ln ((v_R' + 0.08664)/v_R')$$

---

p. *Example 16*

A mass of water exists at a temperature of 600°C. Its specific volume is given as 0.0136 $m^3\ kg^{-1}$. Determine:

The residue $\{a(T,v) - a_o(T,v)\}$ and the pressure using the RK state equation.

The pressure using ideal gas state equation.

The ideal gas volume v_o at 600°C at which the pressure equals that predicted by the RK state equation.

Solution

Using Eqs. (4) and (48)

$$a^{Res} = a\,(T,v) - a_o(T,v) = RT \ln (v/(v-b)) + (\mathbf{a}/(bT^{1/2})) \ln (v/(v+b)).$$

Substituting the values $\mathbf{a} = 142.64$ bar $m^6\ K^{1/2}\ kmole^{-2}$, $b = 0.02110\ m^3\ kmole^{-1}$, $\overline{v} = 0.0136 \times 18.02 = 0.245\ m^3\ kmole^{-1}$,

$$a^{Res}(T,v) = 0.08314 \times 100 \times 873 \ln \{0.245 \div (0.245 - 0.0211)\} +$$

$$\{142.64 \times 100 \div (0.0211 \times 873^{1/2})\} \ln \{0.2425 \div (0.245 + 0.0211)\}$$

$= 653.6 - 1890.2 = - 1236.6$ kJ kmole^{-1}
The pressure, using the RK state equation,
$P = RT/(v-b) - \mathbf{a}/(T^{1/2}v(v+b))$, i.e.,
$P = 0.08314\times873 \div (0.245 - 0.0211) - 142\div(873^{1/2}\ 0.245\times(0.245 + 0.0211))$
$= 250$ bar.
Likewise, the pressure, using the ideal gas state equation,
$P = RT/v = 0.08314\times873\div0.245 = 296$ bar.
The equivalent ideal gas volume
$\bar{v}_o = RT/P = 0.08314\times873\div250$, i.e., $v_o = 00.0161$ m^3 kg^{-1}.

Remarks

If one kmole of water occupies 0.245 m^3 at 600°C, in the presence of intermolecular attraction forces the gas pressure will be 250 bar. If the attractive forces are somehow removed (i.e., the gaseous water is made to behave ideally) while maintaining the same specific volume, the gas pressure will rise to 296 bar.
From Example (16) it is apparent that $(a(T,v) - a_o(T,v)) \neq (a(T,P) - a_o(T,P))$, since although the temperature and volume are unchanged, the ideal gas pressure differs from the real gas pressure. If the specific volume of the ideal gas state is changed so as to obtain the same pressure P as that in the real gas state, then
$$(a(T,v)-a_o(T,v_o) = a(T,P)-a_o(T,P)) = (a(T,v)-a_o(T,v)) - (a_o(T,v_o)-a(T,v)). \quad (50)$$
The first term in parentheses on the RHS can be evaluated using Eq. (50), while the second term in parentheses can be evaluated as outlined below.
$$a_o(T,v_o)-a_o(T,v) = (u_o(T)-Ts_o(T,v_o)) - (u_o(T)-Ts_o(T,v)) = T(s_o(T,v) - s_o(T,v_o)).$$
Recall that $ds_o = c_{vo}\ dT/T + Rdv/v$, at a specified temperature
$$ds_o = Rdv/v.$$
Upon integrating this expression, we obtain
$$(s_o(T,v) - s_o(T,v_o)) = R\ \ln(v/v_o) = R\ \ln(v/(RT/P)) = R\ \ln(Z).$$
Typically, $Z<1$, so that $s_o(T,v_o)>s_o(T,v)$, since a larger number of quantum states are available as the volume is increased. Finally, Eq. (50) can be written in the form
$$a(T,P) - a_o(T,P) = a^{Res}(T,P) = (a(T,v) - a_o(T,v)) - RT\ \ln Z. \quad (51)$$
In dimensionless form
$$(a(T,P) - a_o(T,P))/RT_c = a^{Res}(T_R,v_R')/RT_c + T_R \ln Z(T_R,v_R') \quad (52)$$

---

q. Example 17

A mass of water is maintained at a temperature of 600°C and a pressure of 250 bar. Determine $a(T,P) - a_o(T,P)$ according to RK equation of state.

Solution

We will use the RK state equation
$$P = \bar{R}T/(\bar{v}-\bar{b}) - \bar{a}/(T^{1/2}\bar{v}(\bar{v}+\bar{b})), \quad (A)$$
with values for $P = 250$ bars, $T = 873$ K, $\bar{R} = 0.08314$ bar m^3 kmole^{-1} K^{-1}, $\bar{a} = 142.6$ bar m^6 K$^{0.5}$ kmole^{-2} and $\bar{b} = 0.02110$ m^3 kmole^{-1}.
Therefore,
$\bar{v} = 0.245$ m^3 kmole^{-1}, i.e., $v = 0.245/18.02 = 0.0136$ m^3 kg^{-1}, and
$Z = Pv/RT = 250 \times 0.245 \div (0.08314 \times 873) = 0.844$.

From Example 16,
$a^{Res}(T,v) = a(T,v) - a_o(T,v) = - 1236.6$ kJ kmole^{-1}.

Using Eq. (51)

$a^{Res}(T,P) = a(T,P) - a_o(T,P) = -1236.6 + 8.314 \times 873 \ln(0.844) = -2467.6$ kJ kmole^{-1}.

Remark

At the same pressure, the molecules in the ideal gas move farther apart than in the RK gas. Consequently, the ideal gas volume increases from 0.245 to 0.290 m^3 kmole^{-1}. Therefore, $s_o(T,v) < s_o(T,v_o)$ so that $a_o(T,v) > a_o(T,v_o)$, i.e., $(a(T,P) - a_o(T,P)) < (a(T,v) - a_o(T,v))$.

2. Entropy

Since

$$-s = (\partial a/\partial T)_v,\ -s_o = (\partial a_o/\partial T)_v,$$

upon differentiating Eq. (47), we obtain the relation

$$s^{Res}(T,v) = s(T,v) - s_o(T,v) = R \ln\{1-(b/v)\} - (1/2)\{a/(bT^{3/2})\} \ln\{1+(b/v)\}, \quad (53)$$

where $s^{Res}(T,v)$ denotes the residual entropy. The same result can be obtained using the relation

$$ds = c_v\, dT/T + (\partial P/\partial T)_v dv$$

so that, at a specified temperature, $ds_T = (\partial P/\partial T)_v\, dv$. Using the RK equation of state,

$$ds_T = (R(v-b) + (1/2)\, a/(T^{3/2}\, v(v+b)))dv.$$

Upon integrating this relation,

$$s(T,v) = R \ln(v-b) + ((1/2)a/(bT^{3/2})) \ln(v/(v+b)) + f(T).$$

If both a and b$\rightarrow$0, we obtain ideal gas entropy $s_o(T,v)$. Using a similar procedure as for the Helmholtz function, we can obtain a relation for s^{Res}.

$$s^{Res} = s(T,v) - s_o(T,v) = R \ln\left\{1-\frac{b}{v}\right\} - \frac{1}{2}\frac{1}{bT^{3/2}} \ln\left\{1+\frac{b}{v}\right\} \quad (42a)$$

Dividing this expression by R and rearranging,

$$-s^{Res}(T_R, v_R{'})/R = (s_o(T,v) - s(T,v))/R$$

$$= \ln(1-(0.08664/v_R{'})) - (2.4670/T_R^{3/2})(\ln(1+(0.08664/v_R{'}))) \quad (54)$$

We observed from Example 16 that when the temperature and volume are maintained the same for both the real and ideal gas states (i.e., they have the same intermolecular spacing at a specified temperature), the ideal gas pressure differs from that of the real gas. To obtain identical pressures at both states, the ideal gas volume must be increased so that its intermolecular spacing is larger than that in the real gas. Clearly $s_o(T,v_o) = s_o(T,P)$, but $s_o(T,v_o) > s_o(T,v)$, since a larger number of quantum states are available at the larger intermolecular spacing. Therefore, $(s(T,P) - s_o(T,P)) < (s(T,v) - s_o(T,v))$. Using a similar procedure as that employed in case of the Helmholtz function,

$$s^{Res}(T,P) = s(T,P) - s_o(T,P) = s(T,v) - s_o(T,v_o), \text{ and}$$

$$s(T,P) - s_o(T,P) = \{s(T,v) - s_o(T,v)\} + \{s_o(T,v) - s_o(T,v_o)\} \quad (55)$$

For ideal gases at a specified temperature,

$$(s_o(T,v) - s_o(T,v_o)) = R \ln(v/v_o) = R \ln(v/(RT/P)) = R \ln(Z). \quad (56)$$

Using Eqs. (56) in (55), we obtain the relation

$$s(T,P) - s_o(T,P) = (s(T,v) - s_o(T,v)) + R \ln(Z), \text{ where} \quad (57)$$

$$Z = Pv/(RT) = P_R v_R'/T_R.$$

Dividing Eq. (57) throughout by R and employing Eq. (54), we obtain the entropy departure function

$$-s^{res}/R = (s_o(T,P) - s(T,P))/R =$$

$$(2.4670/(T_R^{(3/2)}) \ln(1+ (0.08664/v_R')) - \ln(1-(0.08664/v_R')) - \ln(T_R/(P_R v_{R'})). \quad (58)$$

Appendix Figure B-6 illustrates a plot of $(s_O(T,P)\text{-}s(T,P))/R$ vs P_R with T_R as a parameter.

---

r. *Example 18*

Determine the entropy of water assuming the relation s= 0 for the saturated liquid at the triple point, and h_{fg} = 2503 kJ kg^{-1}. If the ideal gas specific heat of water is reproduced by the relation $\bar{c}_{p,o}$ (kJ kmole^{-1} K^{-1}) = 28.85 + 0.01206 T+100,600/T^2, determine its entropy at a pressure of 250 bar and a temperature of 873 K.

Solution

Typically, properties are tabulated with respect to arbitrary reference conditions, e.g., T_{ref} and P_{ref}. For saturated liquid water, it is customary to set that reference condition at the triple point, i.e., $T_{ref} = T_{TP}$ = 273 K and P_{TP} = 0.00611 bar at which the entropy is assumed to have a value of zero. Since

$$Tds + vdP = dh,$$

during vaporization at a specified pressure,

$$ds = dh/T, \text{ i.e., } s_g - s_f = (h_g - h_f)/T = h_{fg}/T.$$

Therefore,

$s_g(T_{TP},P_{TP}) - 0 = h_{fg}/T = 2503 \div 273 = 9.17$ kJ kg^{-1} K^{-1}.

Applying the RK equation at this state,

$0.00611 = 0.08314\times273\div(\bar{v}-0.0211) - 142.64\div(273^2\times\bar{v}(\bar{v}+0.0211))$, i.e.,

$\bar{v}$ = 3715 m^3 kmole^{-1}.

The compressibility factor based on this value of the specific volume

$Z(T_{TP},P_{TP}) = P\bar{v}/(\bar{R}T) = 0.00611\times3715\div(0.08314\times273) = 1$.

Furthermore, employing Eq. (53),

$s(273, 3715) - s_o(273,3715) \approx 0$, and

using Eq. (57)

$s(273, 0.00611) - s_o(273,0.00611) \approx 0$.

This result is expected, since the pressure is low so that the vapor behavior is like that of an ideal gas. Hence,

$$s(273,0.00611) = s_o(273, 0.006 \text{ bar}) = 9.17 \text{ kJ kg}^{-1}. \quad (A)$$

Since,

$d\bar{s}_o = \bar{c}_{p,o}\, dT/T - \bar{R}\, dP/P$,

integrating this expression between the states (873 K, 250 bar) and (273 K, 0.00611 bar), we obtain the relation

$$\bar{s}_o(873\text{K}, 250\text{ bar}) - \bar{s}_o(273\text{K}, 0.006\text{ bar}) = \left(\int_{273}^{873} \bar{c}_{p,o}\, dT/T\right) - \bar{R}\ln(250 \div 0.006) \quad (B)$$

Using the given relation for $\bar{c}_{p,o}$,

$\bar{s}_o(873\text{K}, 250 \text{ bar}) - \bar{s}_o(273 \text{ K}, 0.006 \text{ bar}) =$

$(28.85\times\ln(873\div273)+0.01206\times(873-273)-(100600\div2)(873^{-2}-273^{-2}))-$

$(8.314\times\ln(250\div0.006)$

$= 41.38 - 88.30 = -46.92$ kJ kmole^{-1} K^{-1}.

On a mass basis

s_0(873 K, 250 bar) – s_0(273K,0.006 bar) = –46.92 ÷ 18.02 = –2.604 kJ kg^{-1} K^{-1}, and

s_0(873 K, 250 bar) = 9.17 – 2.604 = 6.566 kJ kg^{-1} K^{-1}.

From the results of Example (16), at P = 250 bars, and T = 873 K,

$\bar{v} = 0.245$ m^3 kmole^{-1} and Z = 0.844.

Thereafter, using Eq. (53).

$$\bar{s}(T,v) - \bar{s}_0(T,v) = \bar{R}\ \ln(1-(b/\bar{v})) - (1/2)(a/(bT^{3/2}))\ \ln(1+(b/\bar{v}))$$

$$= 8.314\times\ln(1-(0.0211\div0.245) -$$

$$(0.5\times142.64\times100\div(0.0211\times873^{1.5}))\ \ln(1+(0.0211/0.245))$$

$$= -1.826 \text{ kJ kmole}^{-1} \text{ K}^{-1}.$$

Then using Eq. (57),

$\bar{s}(T,P) - \bar{s}_0(T,P) = s(T,v) - s_0(T,v) + \bar{R}\ln Z = -1.826 - 1.409 = -3.235$ kJ kmole^{-1} K^{-1},

or

$s(T,P) - s_0(T,P) = -0.180$ kJ kg^{-1} K^{-1}, i.e.,

$s(T,P) = 6.566 - 0.180 = 6.386$ kJ kg^{-1} K^{-1}.

Remarks

Using the RK equation, we determined the values of s at specified temperatures and pressures, and at specified temperatures and specific volumes. This enables the production of T–s diagrams along with superimposed isotherms, isobars, and isometric contours.

Instead of the RK equation, we can use the entropy departure charts, at P_R (= 250÷220.9) = 1.132, and T_R (=873÷647) = 1.349,

$$(s_0(T,P) - s(T,P))/R = 0.389. \qquad \text{(C)}$$

The value of s(873 K,250 bar) can be calculated thereafter.

3. Pressure

Oftentimes the state equation for a(T,v) are provided empirically. Thereafter the pressure can be determined through the relation

$$(\partial a/\partial v)_T = -P \qquad (59)$$

s. *Example 19*

Assume that

$$a(T,v) = a_0(T,v) + RT\ \ln(v/(v-b)) + (a/(bT^{1/2}))\ \ln(v/(v+b)). \qquad \text{(A)}$$

Determine an expression for the pressure.

Solution

Consider the derivative of Eq. (A) with respect to the specific volume, i.e.,

$$\partial a/\partial v = -P(T,v) = \partial a_o/\partial v - RT\ (1/v - 1/(v-b)) - (a/(bT^{1/2}))(1/v - 1/(v+b)).$$

Since

$$\partial a_o/\partial v = -P_o(T,v) = -RT/v, \qquad \text{(B)}$$

employing Eqs. (A) and (B),

$$-P(T,v) = -RT/v - RT\ (1/v - 1/(v-b)) - (a/(bT^{1/2}))(1/v - 1/(v+b)). \qquad \text{(C)}$$

Upon simplification, Eq. (C) can be expressed in the form

$$P(T,v) = RT/(v-b) - (\mathbf{a}/(T^{1/2}\, v\,(v+b))) \qquad (D)$$

Remark

Since we have used the RK equation to obtain an expression for a(T,v), the same state equation is obtained in Eq. (D).

---

4. Internal Energy

If the volume of an ideal gas is expanded from v to v_o while maintaining the same pressure as in a corresponding real gas, its internal energy u_o is unchanged, since the intermolecular potential energy remains unaltered. Therefore, $u_o(T,v_o) = u_o(T,v) = u_o(T)$. Furthermore, since $u = a + Ts$,

$$u(T,v) - u_o(T) = a(T,v) - a_o(T) + T\,\{s(T,v) - s_o(T,v)\}. \qquad (60)$$

Using Eqs. (47) and (53) in Eq. (60), we obtain the relation

$$u(T,v) - u_o\,(T) = u(T,P) - u_o(T) = RT\,\ln\,\{v/(v-b)\} + \{\mathbf{a}/(bT^{1/2})\}\,\ln\,\{v/(v+b)\} +$$

$$RT\,\ln\,\{v/(v-b)\} + (1/2)\{\mathbf{a}/(bT^{1/2})\}\,\ln\,\{v/(v+b)\}.$$

Simplifying this expression.

$$u(T,v) - u_o\,(T) = -(3/2)(\mathbf{a}/(bT^{1/2}))\,\ln(1 + (b/v)). \qquad (61a)$$

The term $u(T,v) - u_o(T)$ or $u(T,P) - u_o(T) = u^{Res}$ is the residual internal energy, and is typically less than zero. Substituting for a and b in terms of critical properties, Eq. (60) assumes the form

$$u_{C,R} = -\frac{u^{Res}}{RT_C} = \frac{u_O(T) - u(T,v)}{RT_C} = \frac{7.401}{T_R^{0.5}}\ln\left(1 + \frac{0.08664}{v'_R}\right), \qquad (61b)$$

where the difference $u_o(T) - u\,(T,v)$ represents the departure of the internal energy from that in a corresponding ideal gas. Recall that

$$v_R' = v/v_c',\ v_c' = RT_c/P_c.$$

These expressions form the basis for charts illustrating the behavior of $-(u^{Res}/RT_c)$ with respect to the reduced pressure, temperature, and specific volume.

a. *Remarks*

We recall from Eq. (23)) that $\partial(a/T)/\partial(1/T) = u$. Thereafter, dividing Eq. (47) throughout by T and then differentiating with respect to (1/T), we obtain Eq. (61).

The internal energy is generally higher in an ideal gas than in a corresponding real gas.

Figure 8a presents a plot of u with respect to T at different specific volumes. At a specified temperature, as the volume is increased to a large value, $u \rightarrow u_o$ (the dashed line QR). The slope of line QR yields the value of c_{vo}, while the slope of line AB yields the value of c_v. Further $u_{o,R} = u_o + \int_{T_o}^{T_R} c_{vo}dT$.

The free volume per molecule is much larger in the gas or vapor phase than in the liquid phase due to the greater intermolecular distance. Therefore, since the vapor has a higher intermolecular potential energy (cf. the LJ diagram), the liquid has a lower internal energy.

The constant volume specific heat c_v is obtained by differentiating Eq. (61a) with respect to temperature at a specified specific volume, i.e.,

$$c_v(T,v) - c_{v0}(T) = 3/4(\mathbf{a}/(bT^{(3/2)}) \ln(1 + b/v). \quad (62)$$

This difference is always positive, i.e., $c_v(T,v) > c_{vo}(T)$. If a fluid is heated from the saturated liquid to the saturated vapor state at constant temperature, then Eq. (62) suggests that

$$c_v(T,v_f) - c_v(T,v_g) = 3/4\ (\mathbf{a}/(bT^{(3/2)}) \ln((1 + b/v_f)/(1+ b/v_g)). \quad (63)$$

In general, since $v_f < v_g$, $c_v(T,v_f) > c_v(T,v_g)$. Dividing Eq. (62) by R and using the relations for a and b in terms of the critical properties T_c and P_c, Eq. (62) assumes the form

$$(c_v(T,v) - c_{v0}(T))/R = 3.7007/T_R^{3/2} \ln(1 + 0.08664/v_R'). \quad (64)$$

The variation of the reduced specific heat with P_R and v_R' with respect to volume is illustrated in Figure 9. The RK based relation for c_v *vs.* v for H_2O at 593 K is plotted in Figure 10. Instead of the RK equation, if one uses the following a hypothetical RK equation with

$$P = RT/(v-b) - a/(T^n v(v+b)),$$

then we can show that

$$(c_v(T,v) - c_{v0}(T))/R = (4.9342\ n(n+1)/T_R^{n+1}) \ln(1 + 0.08664/v_R')).$$

If $-1 < n < 0$ (i.e., attractive forces increase with temperature), $c_v(T,v) < c_{v0}(T)$ and c_v could become negative! A negative specific heat implies that the temperature will in-

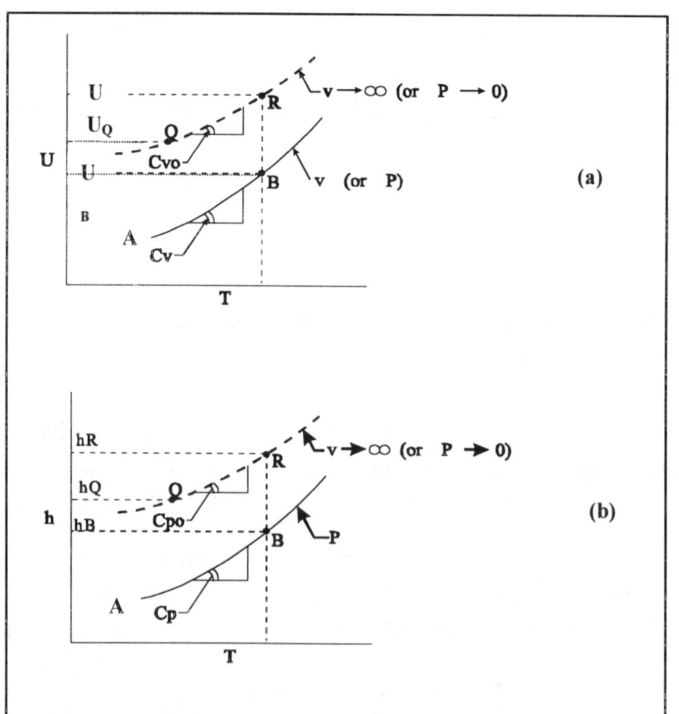

Figure 8: (a) Illustration of the determination of the value of u at state B from a known value u_o at state Q. (b Illustration of the determination of the value of h at state B from a known value h_o at state Q.

crease upon heat loss and *vice versa.* Further discussion is provided in Chapter 10.

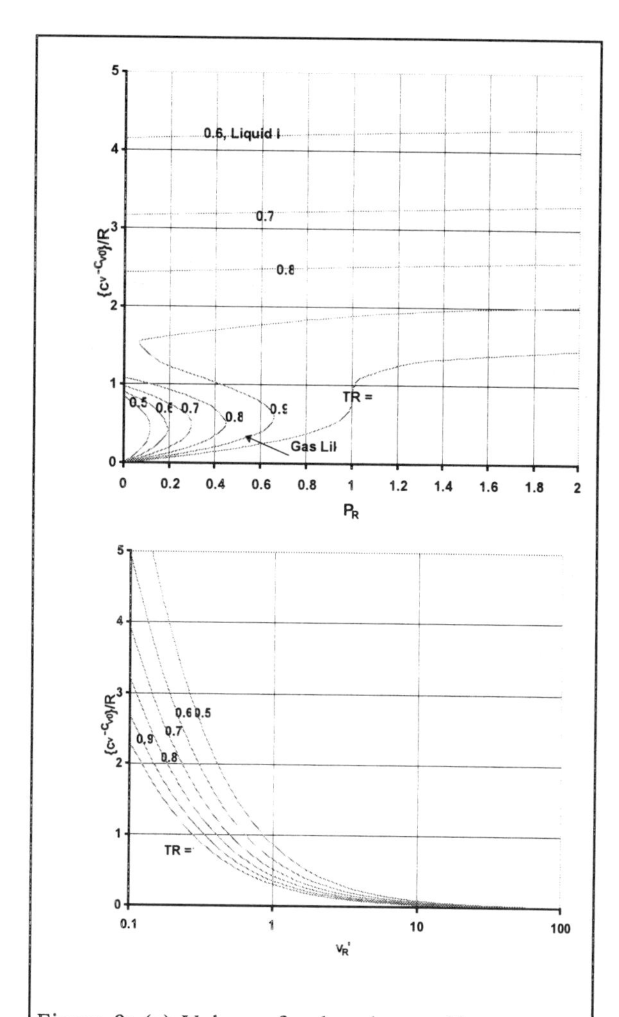

Figure 9: (a) Values of reduced c_{vR} with respect to P_R for an RK fluid. The middle values are unstable and the upper values at specified values of P_R and T_R correspond to a liquid–like solution, while the lower values correspond to a gas–like solution. The value of c_v for a liquid is always higher than that for a gas. (b) Values of reduced c_{vR} with respect to v_R'. The variation in the values is monotonic unlike those in relation to P_R.

Using Equation (38),

$$(\partial T/\partial v)_u = -(T(\partial P/\partial T)_v - P)/c_v.$$

If the fluid state is defined by the RK equation, the numerator in the above relation is always negative, i.e., when an RK substance expands at constant internal energy, the temperature decreases. Such an adiabatic throttling process for a closed system is discussed later.

Recall that $c_p - c_v = -T(\partial P/\partial T)^2/(\partial P/\partial v)$. If $(\partial P/\partial v) \approx 0$ (in Chapter 6 we discussed that for real gases the pressure decreases with a decrease in the specific volume over a range of values of v when $T < T_c$), then $c_p - c_v < 0$, i.e., $c_p < c_v$. Therefore, the value of c_p may be negative and ratio k ($=c_p/c_v$) <1 over a range of values of the specific volume.

We will show in Chapter 8 that the internal energy of q mixture or any component in a mixture can be obtained by using Eq. (61a).

---

t. Example 20

Determine the internal energy of water at 250 bar and 600°C if u_o = 3302.7 kJ kg^{-1} at that temperature. The corresponding tabulated value of the internal energy from the steam tables (Table A-4C) is 3138 kJ kg^{-1}. Employ the RK equation in your solution.

Solution

Since the critical properties for water are P_c = 220.9 bar and T_c = 647.3 K, P_R = 1.132 and T_R = 1.349, Therefore, upon applying the appropriate results from Chapter 6, v_R' = 1.007, and

$-u_{C,R} = 7.401/1.349^{0.5} \ln(1 + 0.08664/1.007) = 0.526$, i.e.,

$$u_o - u = 0.526 \times 0.461 \times 647.3 = 157 \text{ kJ kg}^{-1}, \text{ or } u = 3146 \text{ kJ kg}^{-1}.$$

The difference with respect to the steam tables is 0.25 %.

---

5. Enthalpy

The enthalpy residue can be obtained from the relation

$$h(T,P) - h_o(T) = u(T,v) + Pv - (u_o + (P_o v_o)) = u(T,v) - u_o + Pv - RT, \quad (65)$$

where h(T,P) = h(T,v).Using the appropriate state equation for the pressure, and Eqs. (61) and (65), we obtain a relation for the residual enthalpy

$$h^{Res}(T,P) = h(T,P) - h_o(T) = [(3/2)\{a/(bT^{1/2})\}] \ln\{v/(v+b)\} + \{RTv/(v-b) - a/(T^{1/2}(v+b))\} - RT \quad (66)$$

Figure 8b presents a plot of h with respect to T with P as a parameter. As $h \to h_o$, $P \to 0$. In relation to $h_{o,o}$, the ideal gas enthalpy at R is provided by the expression

$$h_{o,R} = h_{o,Q} + \int_{T_Q}^{T_R} P_o \, dT.$$

In dimensionless form,

$$h_{C,R}(T_R,P_R) = 7.401/T_R^2 \ln(1 + 0.08664/v_R') - (T_R/(1 - 0.08664/v_R')) +$$

$$0.4275/(T_R^2 (v_R' + 0.08664)) + T_R \quad (67)$$

Sometimes, h_R is denoted as $h^{(o)}{}_R$ in case a simple fluid is used or a two parameter equation of state is used. At a specified value of P_R, the dimensionless difference between the vapor and liquid state enthalpies, i.e., the dimensionless enthalpy of vaporization is $h_{fg}/RT_c = (h_g - h_f)/RT_c$.

a. Remarks

The difference $(c_P - c_{P,o})$ is

$$c_p - c_{P,o} = (\partial(h - h_o)/\partial T)_P. \quad (68)$$

In dimensionless form,

$$(c_{P,o} - c_P)/R = (\partial h_{C,R}/\partial T_R)_{P_R}.$$

Using Eq. (65),

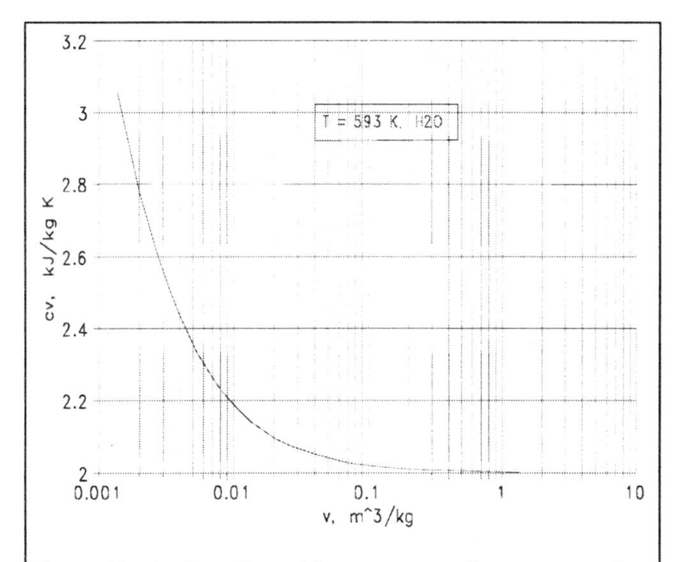

Figure 10: A plot of c_v with respect to v for water employing the RK equation at 593 K.

$$c_p - c_{P,o} = (\partial/\partial T)(u-u_o)_v + (\partial/\partial v)(u-u_o)_T\,(\partial v/\partial T)_P + P\,\partial v/\partial T - R.$$

Therefore,

$$c_p - c_{P,o} = (c_v - c_{v,o})_v + ((\partial/\partial v)(u-u_o)_T + P)(\partial v/\partial T)_P - R. \quad (69)$$

This provides a recipe to determine c_P for a real fluid. In general, the difference $(c_{P,o} - c_p) < 0$, i.e., the specific heat of a real fluid is higher than that of an ideal gas.
For a non–ideal fluid, $c_v \neq c_P - R$.
Consider the relation $(h_o - h) = u_o - u + RT - P\,v$, so that

$$\partial(h_o - h)/\partial P = \partial(u_o - u)/\partial P - v - P\,\partial v/\partial P.$$

Since $\ln(1 + b/v) \to b/v$ as v becomes large (or as $P \to 0$) and $v \approx RT/P$ under these conditions, Eq. (60) yields the result $u_o - u \approx 3\mathbf{a}P/(2R^2T^{3/2})$ so that

$$\partial(h_o - h)/\partial P = (3/2)\{\mathbf{a} / (R^2\,T^{3/2})\} - v + Pv^2/RT. \quad (70)$$

Since $Z \to 1$ under these conditions,

$$\partial(h_o - h)/\partial P = (3/2)(\mathbf{a}/RT^{3/2}).$$

In dimensionless form

$$\partial h_{C,R}/\partial P_R = 0.6413/T_R^{1.5},\ P_R \to 0. \quad (71)$$

In addition,

$$(\partial Z/\partial P_R)_{P_R \to 0} = 0.08664/T_R - 0.4275/T_R^{5/2}. \quad (72)$$

As $T \to 0$, dh $(= T\,ds + v\,dP) \to vdP$, and $\partial h/\partial P = v$. Since v has a finite value, the enthalpy must change accordingly with pressure as $T \to 0$.
For liquids and solids

$$dh = c_p dT + (v - T\,\partial v/\partial T)\,dP = c_p dT + (1 - T\beta_P)\,v dP.$$

We will use the expression developed in Chapter 6 for v of liquids

$$v/v_{ref} = \exp(\beta_P\,(T-T_{ref}) - \beta_T\,(P-P_{ref})).$$

If c_p, β_T, and β_P are constant, then along an isotherm

$$h(T,P) - h(T,P_{ref}) = -(1 - T\beta_P)\,(v - v_{ref})/\beta_T.$$

Similarly, along an isobar

$$h(T,P) - h(T_{ref},P) = c_p\,(T - T_{ref}).$$

---

u. *Example 21*

Determine the value of h_o–h at the triple point of water vapor. The conditions are T_{TP} = 273 K, P_{TP} = 0.0061 bar.

If the value of h_{fg} = 2501.3 kJ kg^{-1} at the triple point, assuming that h = 0 kJ kg^{-1} for the saturated liquid at that point, what are the real and ideal gas enthalpies of the vapor at that state?

What is the ideal gas enthalpy at 600°C if $\bar{c}_{p,o}$ = 28.85 + 0.01206 T+100,600/T^2 (in units of kJ $kmole^{-1}$ K^{-1})?

Determine h at P = 250 bar and T = 600°C.

Determine u at P = 250 bar and T = 600°C.

Solution

The line ABM in Figure 11 represents the isotherm T_{TP} = 273 K. At P = 0.00611 bar and T = 273 K, the RK equation of state yields $\bar{v}$ = 3717 m^3 $kmole^{-1}$ (or 206.3 m^3 kg^{-1}). This volume is the same as that predicted by the ideal gas state relation $\bar{v}$ = $\bar{R}$ T/P, since at low pressure the vapor behaves as an ideal gas. Recall that

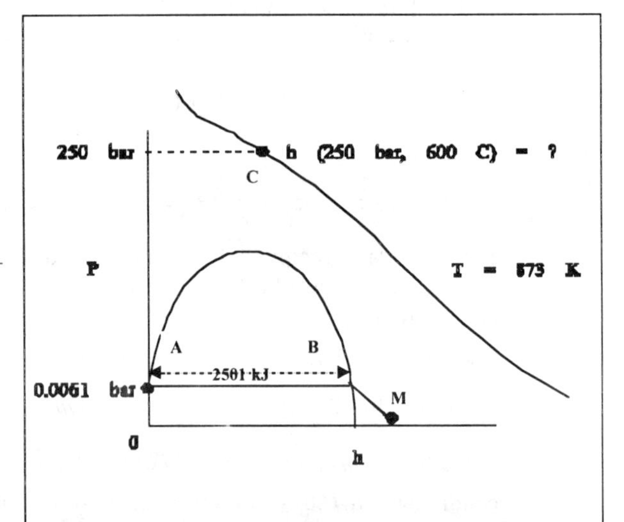

Figure 11: P–h diagram illustrating the determination of the enthalpy using the triple point as a reference.

$$u^{Res} = u(T,v) - u_o(T) = -(3/2)(\mathbf{a}/(bT^{1/2}))\,\ln(1+(b/v)), \text{ i.e.,}$$

$$\bar{u}^{Res} = -(3\div 2)\times(142.64\times 100\div(0.0211\times 273.15^{1/2}))\,\ln(1+(0.0211\div 3717))$$

$$= -\,0.348 \text{ kJ kmole}^{-1} \text{ or } 0.019 \text{ kJ kg}^{-1}.$$

$$\bar{h}^{Res} = \bar{h}(T,P) - \bar{h}_o(T) = (\bar{u}(T,P) + P\bar{v} - \bar{u}_o(T)) - \bar{R}T = \bar{u}^{Res} + P\bar{v} - \bar{R}T$$

$$= -0.348 \text{ kJ kmole}^{-1}.$$

Since the vapor behaves like an ideal gas at the low pressure of 0.00611 bar, the value of $\overline{h}^{Res}$ at this condition is small.
The enthalpy at point B can be evaluated by using the relation

$$h(273.15\text{ K}, 0.00611\text{ bar}) = h_g(273.15\text{ K}, 0.00611\text{ bar})$$

$$= h_f(273\text{ K}, 0.0061\text{ bar}) + 2501.3 = 2501.3\text{ kJ kg}^{-1}.$$

The ideal gas enthalpy at point C is obtained by using the relation

$$h_o(873\text{ K}) - h_o(273.15\text{ K}) = \int c_{p,o}dT = (1/18.02)\int(28.85+0.01206\ T+100{,}600/T^2)dT$$

$$= 1205\text{ kJ kg}^{-1},\text{ i.e.,}$$

$$h_o(873\text{K}) = 2501.3 + 1205 = 3706.3\text{ kJ kg}^{-1}.$$

At P = 250 bar, T = 873 K, $\overline{v} = 0.245\text{ m}^3\text{ kmole}^{-1}$ or $0.0136\text{ m}^3\text{ kg}^{-1}$.

$$\overline{u}^{Res} = -((1.5)\times 142.64\times 100\div(0.02110\times 1873^{0.5}))(\ln(1 + 0.0211\div 0.245))$$

$$= -2835\text{ kJ kmole}^{-1}\text{ or } -157\text{ kJ kg}^{-1}.$$

Therefore,

$$h(873\text{ K}, 250\text{ bar}) - h_o(873\text{ K}) = u(873\text{ K}, 250\text{ bar}) + Pv - \{u_o(873\text{ K}) + RT\}$$

$$= u(T,P) - u_o(T) + Pv - RT$$

$$= -157 + 250\times 100\times 0.245\div 18.02 - (8.314\div 18.02)\times 873 = -220\text{ kJ kg}^{-1},\text{ and}$$

$$h(873\text{ K}, 250\text{ bar}) = 3706.3 - 220 = 3486.3\text{ kJ kg}^{-1}.$$

$$u = h - Pv = 3486.3 - 250\times 100\times 0.0136 = 3146\text{ kJ kg}^{-1}.$$

Remarks

For determining the value of h_o for H_2O we use the triple point as the reference. Since H_2O behaves as an ideal gas at low pressures, we can select a value for h(0.35 bars, 200° C) from the steam tables for superheated vapor (Table A-4C) as 2878.4 kJ.kg. The pressure is low so that h(0.35 bar, 200 °C) ≈ h_0(200°C). In this case, h_0(873 K) can be determined. The real gas enthalpy h(250 bars, 873 K) can be determined using the corrections. Such a procedure does not require knowledge of the reference conditions h_{fg}, T_{TP}, T_{TP}, etc..

Instead of using the RK equation, one can use the enthalpy correction charts that are available. From these charts at $T_R = 1.349$, $P_R = 1.132$,

$$((h_o(T) - h(T,P)))/(RT_c) = 0.735,\text{ i.e.,}$$

$$h = 3706 - (8.314\times 647.3\div(18.02))\times 0.735 = 3486\text{ kJ kg}^{-1}.$$

Below, we compare the theoretical results with tabulated values obtained from the steam tables (A-4C) at 600°C and 250 bar.

v, $m^3\ kg^{-1}$	h, $kJ\ kg^{-1}$	u, $kJ\ kg^{-1}$	s, $kJ\ kg^{-1}\ K^{-1}$	Source
0.01361	3486	3146	6.2904	RK equation
0.01414	3491	3138	6.361	Steam Tables

The reference condition for the steam tables (Table A-4) is at the saturated liquid state of water at its triple point. Consequently, at 273 K steam has an enthalpy of 2501.3 kJ

kg^{-1}. The ideal enthalpy has almost the same value of 2503 kJ kg^{-1}. With this ideal gas enthalpy value, we can obtain values of the real gas enthalpy at any temperature and pressure.
Similar procedures can be adopted for s. using the value s(273.15 K 0.0061 bar) = 0 for saturated liquid, s (273.15 K 0.0061 bar) for saturated vapor can be determined using $s_{fg} = h_{fg}/T$ where h_{fg} is specified at triple point. Using Eqs. (53) and (57) we evaluate s^{Res} and hence s_O (273.15 K, 0.0061 bar) then obtain s_O (873 K, 250 bar) at point C, s^{Res} (873 K, 250 bar) and finally get s (873 K, 250 bar). We see from the above examples that the general formulae for h and s at any state (T,P) can be expressed in terms of reference values $h_{ref} = s_{ref} = 0$ T_{ref} or P_{ref} in the form

$$h(T,P) = h_{fg,ref} + \int_{T_{ref}}^{T} c_{p0}\, dT + h^{Res}(T,P) - h^{Res}(T_{ref},P_{ref}),$$

where

$$h^{Res}(T,P) = h(T,P) - h_o(T) = u(T,P) - u_o(T) + Pv - RT, \text{ and}$$

$$h^{Res}(T_{ref},P_{ref}) = h(T_{ref},P_{ref}) - h_o(T_{ref}) = u(T_{ref},P_{ref}) - u_o(T_{ref}) + P_{ref}v_{ref} - RT.$$

Furthermore, using the relation

$$u(T,P) - u_o(T) = -(3/2)(\mathbf{a}/(bT^{1/2}))\ \ln\{1 + (b/v)\},$$

we can invoke the RK state equation, i.e.,

$$P = RT/(v-b) - a/(T^{1/2}\, v(v+b)), \text{ where}$$

$\mathbf{a} = 0.4275\, R^2 T_c^{\ 2.5}/P_c$, and $b = 0.08667\, RT_c/P_c$. Typically $h^{Res}(T_{ref},P_{ref}) \approx 0$, if $P_{ref} \ll P_c$. Similarly,

$$s(T,P) = s_{fg,ref} + \int_{T_{ref}}^{T} (c_{po}/T)dT - R\ln(P/P_{ref}) + s^{Res}(T,P) - s^{Res}(T_{ref},P_{ref}),$$

where $s_{fg,ref} = h_{fg,ref}/T_{ref}$,

$$(s(T,P) - s_o(T,P)) = s^{Res} = R\ln(1-b/v) - (1/2)(a/(bT^{3/2}))\ \ln(1+b/v) + R\ln Z.$$

A similar relation can be obtained for the reference state. Typically $s^{Res}(T_{ref}, P_{ref}) \approx 0$.

---

v. *Example 22*

Determine h_{fg} for water at T = 593 K if the saturated liquid and vapor specific volumes are, respectively, $v_f = 0.0014988$ m^3 kg^{-1}, $v_g = 0.01549$ m^3 kg^{-1}.

Solution

Vaporization occurs at constant temperature and pressure. Integrating the expression dh = du + d(Pv), we obtain the relations

$$h_{fg} = u_g - u_f + P(v_g - v_f), \text{ and} \tag{A}$$

$$u_{fg} = u_g(T,v_g) - u_f(T,v_f) = -((3/2)(\mathbf{a}/(bT^{1/2}))\ \ln((v_g + b)v_f/((v_f + b)\, v_g)). \tag{B}$$

Using the values $\bar{a}$ = 14259 kN m^4 $kmole^{-2}$ $K^{1/2}$, i.e., $\mathbf{a}$ = 43.912 kN m^4 $K^{1/2}$ kg^{-2} = 0.43912 bar m^2 $K^{1/2}$ kg^{-2}, $\bar{b}$ = 0.0211 m^3 $kmole^{-1}$, i.e., b = 0.00117 m^3 kg^{-1}, and the provided values of v_f and v_g we obtain
$u_{fg} = -1.5 \times 1541 \times \ln(0.604) = 1165$ kJ kg^{-1}.
For sake of comparison, the steam tables provide a value of 1081 kJ kg^{-1}.
If $\mathbf{a}$ = 0, i.e., there are no attractive forces, u_{fg} = 0. We can use equation (A) to obtain the value of h_{fg}. Since the pressure is not provided, we will employ the RK state

equation using the data for v_g and T. (We will not use the v_f data, since the RK equation is inaccurate in this limit.) Thus,

$P = RT/(v_g - b) - (a/T^{1/2})(1/(v_g(v_g+b))$

$= (0.08314 \div 18.02) \times 593 \div (0.01549 - 0.00117) -$

$(0.43912 \div 593^2)(1 \div (0.01549(0.01549+0.00117))) = 121.2$ bar.

The tables (A-4A) yield a value $P^{sat} = 112.7$ bar. Likewise,

$h_{fg} = u_{fg} + P(v_g - v_f) = 1165 + 121.2 \times 100 \times (0.01549 - 0.0014988) = 1335$ kJ kg^{-1}.

From the tables A-4A $h_{fg} = 1238$ kJ kg^{-1}.

Remark

The liquid possesses lower intermolecular potential and the translational energy. The internal energy u_{fg} is the energy required to overcome the strong intermolecular attractive forces in the liquid due to a close molecular spacing and spread the molecules apart (Chapter 1). It represents the potential energy increase in the Lennard–Jones potential function at a specified temperature during the phase transition. Since each unit mass must perform a boundary work of P (v_g-v_f) and hence energy transfer via heat is used to supply "ipe" and boundary work. The vaporization enthalpy $h_{fg} = u_{fg} + P(v_g - v_f)$ includes both the intermolecular potential energy gain and the boundary work performed in increasing the volume of unit mass from liquid to vapor states as given by Eq. (B).

6. Gibbs Free Energy or Chemical Potential

We are interested in the difference $(g(T,P) - g_o(T,P))$ (rather than $(g(T,v) - g_o(T,v))$), since g(T,P) is generally applied to phase equilibrium problems at specified T and P. Recall that

$$g(T,P) = h(T,P) - Ts(T,P), \text{ so that}$$

$$g(T,P) - g_o(T,P) = (h(T,P) - h_o(T)) - T\{s(T,P) - s_o(T,P)\}.$$

Applying Eqs. (66), (57) and (61),

$$g(T,P) - g_o(T,P) = (3/2)(\mathbf{a}/(bT^{1/2}))\ln(v/(v+b)) + (RTv/(v-b) - \mathbf{a}/(T^{1/2}(v+b)) - RT$$

$$+ RT\ln(v/(v-b)) - (1/2)(\mathbf{a}/(bT^{1/2}))\ln(v/(v+b)) - RT\ln Z_{RK}.$$

$$= (RTv/(v-b) - \mathbf{a}/(T^{1/2}(v+b)) - RT + RT\ln(v/(v-b)) +$$

$$(\mathbf{a}/(bT^{1/2}))\ln(v/(v+b)) - RT\ln Z_{RK} \quad (73)$$

w. *Example 23*

Determine the relations for properties s, v, u, and h if g(T,P) is known.

Solution

Using the relations g = h – Ts, and dg = vdP – sdT,

$$(\partial g/\partial T)_P = -s \text{ and } (\partial g/\partial P)_T = v \quad (A)$$

Thereafter,

$$h = g + Ts \text{ and } u = h - Ts. \quad (B)$$

Remarks

Manipulating the relations for g (= h-Ts) and dg (= -s dT + vdP), we obtain the expression

$$dg = vdP - (h - g)dT/T, \text{ i.e., } dg - g\,dT/T = v\,dP - h\,dT/T$$

Therefore,

$T\, d(g/T) = vdP - h\, dT/T$, or $d(g/RT) = v\, dP/RT - h\, dT/(RT^2)$, and

$$(\partial(g/RT)/\partial T)_P = -h/RT^2. \quad \text{(C)}$$

Similarly,

$(\partial(g_o/RT)/\partial T)_P = -h_o/RT^2$ so that $(\partial(g_{C,R}/T_R)/\partial T_R)_{P_R} = -h_{C,R}/T_R^2$, and

$$(\partial g_{C,R}/\partial T_R)_{PR} = -s_{C,R}, \text{ and } (\partial g_{C,R}/\partial P_R)_{TR} = v_{C,R}. \quad \text{(D,E)}$$

---

x. *Example 24*

Determine the reversible work required for the steady reversible isothermal compression of methane at 230K from P_1 = 150 bar to P_2 = 250 bar. You may use the Kesler charts.

Solution

We will use the conservation equation

$\bar{q} - \bar{w}_s = \bar{h}_2 - \bar{h}_1$ and $\delta\bar{q} = T\, d\bar{s}$.

For an isothermal process,

$\bar{q} = \int T d\bar{s} = T(\bar{s}_2 - \bar{s}_1)$.

Therefore,

$-\bar{w}_s = (\bar{h}_2 - \bar{h}_1) - T(\bar{s}_2 - \bar{s}_1)$.

For methane, T_c = 191 K, and P_c = 46.4 bar. Hence,

$T_{R,1} = T_{R,2} = 230 \div 191 = 1.2$, $P_{R,1} = 150 \div 46.4 = 3.2$, $P_{R,2} = 250 \div 46.4 = 5.4$.

From the discussion in Chapter 2, using the enthalpy correction charts (Appendix Figure B-3)or the Kessler tables (Table A-24A) at $T_{R,2}$ = 1.2, $P_{R,2}$ = 5.4, Z_2 = 0.75. Thus,

$(\bar{h}_{o2} - \bar{h}_2)/\bar{R}T_c = 3.172$

Similarly, at T_{R1} = 1.2, P_R = 3.2, and

$(\bar{h}_{o1} - \bar{h}_1)/\bar{R}T_c = 2.834$.

Therefore,

$\bar{h}_1 = \bar{h}_{o1} - 2.834 \times 8.314 \times 191 = (\bar{h}_{o1} - 4500)$ kJ kmole^{-1},

$(\bar{h}_2 = \bar{h}_{o2} - 3.172 \times 8.314 \times 191 = (\bar{h}_{o2} - 5037)$ kJ kmole^{-1}, and

$\bar{h}_2 - \bar{h}_1 = (\bar{h}_{o2}(T_2) - \bar{h}_{o1}(T_1) - 537)$ kJ k mole^{-1}.

Since $T_2 = T_1$, $\bar{h}_{o2}(T_2) = \bar{h}_{o1}(T_1)$, and

$\bar{h}_2 - \bar{h}_1 = -537$ kJ kmole^{-1}.

For $T_{R,1} = 1.2$, $P_{R,1} = 3.2$, and $((\bar{s}_{01} - \bar{s}_1)/\bar{R}) = 1.737\ 9$ (Tables A-25A or Appendix Figure B-4). Therefore,

$(\bar{s}_{o1} - \bar{s}_1) = 14.4$ kJ kmole^{-1} K^{-1}.

For $T_{R,2} = 1.2$, $P_{R,2} = 5.4$, and $((\bar{s}_{02} - \bar{s}_2)/\bar{R}) = 1.819$, and

$(\bar{s}_{o2} - \bar{s}_2) = 15.1$ kJ kmole^{-1} K^{-1}.

Consequently,

$\bar{s}_2 - \bar{s}_1 = 4.95$ kJ kmole^{-1} K^{-1}, and

$\bar{w}_s = -601$ kJ kmole^{-1}.

Remark

If the ideal gas state equation is used,

$w_s = -\int \bar{v} dP = -\int(\bar{R}T/P)dP = -\bar{R}T \ln(P_2/P_1) = -8.314 \times 230 \ln(250 \div 150)$

$= -977$ kJ kmole^{-1}.

---

7. Fugacity Coefficient

The fugacity coefficient ϕ is defined as

$$\ln \phi = (g(T,P) - g_o(T,P))/(RT). \quad (74)$$

Using Eq. (73),

$$\ln \phi = (v/(v-b) - \mathbf{a}/(RT^{3/2} (v+b)) - 1$$

$$+ \ln (v/(v-b)) + (\mathbf{a}/(bRT^{3/2})) \ln(v/(v+b)) - \ln Z_{RK}. \quad (75)$$

This coefficient will be discussed later.

F. PITZER EFFECT

So far we have used two parameter state equations. Most such equations are explicit expressions for the pressure in terms of v and T. Inclusion of the Pitzer factor ω improves the accuracy of the state relations.

1. Generalized Z Relation

Applying Eqs. (43) and (38) at constant temperature, namely,

$$dh_T = (v - T (\partial v/\partial T)_P) \, dP, \quad (76a)$$

$$du_T = (T(\partial P/\partial T)_v - P) \, dv, \quad (76b)$$

and the relation

$$v = ZRT/P. \quad (77)$$

Differentiating Eq. (77) with respect to temperature, we obtain the relation

$$(\partial v/\partial T)_P = (\partial Z/\partial T)_P \, RT/P + ZR/P.$$

Using this expression in Eq. (76a) and (76b), we obtain the expression

$$dh_T = -(RT^2/P)(\partial Z/\partial T)_P \, dP \quad (78a)$$

$$du_T = (RT^2/v)(\partial Z/\partial T)_v \, dv \quad (78b)$$

Since, $T_R = T/T_c$ and $P_R = P/P_c$, and omitting subscript "T", the Eq. (78a) becomes

$$dh/RT_c = -(T_R^2/P_R)(\partial Z/\partial T_R)_{P_R} \, dP_R. \quad (79)$$

We now introduce the Pitzer factor in the form

$$Z = Z^{(o)} (T_R,P_R) + \omega \, Z^{(1)} (T_R,P_R), \quad (80)$$

and introduce Eq. (80) into Eq. (79), so that

$$dh/RT_c = -(T_R^2) \{(\partial Z^{(o)}/\partial T_R)_{P_R} + \omega \, (\partial Z^{(o)}/\partial T_R)_{P_R}\} (dP_R/P_R). \quad (81)$$

Upon integrating Eq. (81)

$$h/RT_c = -(T_R^2) \int\{(\partial Z^{(o)}/\partial T_R)_{P_R} + \omega \, (\partial Z^{(o)}/\partial T_R)_{P_R}\} (dP_R/P_R) + f(T_R). \quad (82)$$

As $P_R \to 0$, $h \to h_o$, and $Z \to 1$. Therefore, $\partial Z/\partial T_R = 0$, and Eq. (71) assumes the form

$$h_o/RT_c = -(T_R^2)(\int\{(\partial Z^{(o)}/\partial T_R)_{P_R} + \omega \, (\partial Z^{(o)}/\partial T_R)_{P_R}\}(dP_R/P_R)_{P_R \to 0} + f(T_R). \quad (83)$$

Subtracting Eq. (82) from Eq. (83), we obtain the expression

$$h_{c,R} = -\frac{h^{Res}}{RT_c} = \frac{h_0 - h}{RT_c} = (T_R^2 \int_{P_R \to 0}^{P_R} \frac{\partial Z^{(0)}}{\partial T_R} \frac{dP_R}{P_R}) + \omega(T_R^2 \int_{P_R \to 0}^{P_R} \frac{\partial Z^{(1)}}{\partial T_R} \frac{dP_R}{P_R}), \quad (84)$$

where the first term on the RHS can be represented as

$$(h_0 - h)/(RT_c)^{(0)} = T_R^2 \int_{P_R \to 0}^{P_R} \frac{\partial Z^{(0)}}{\partial T_R} \frac{dP_R}{P_R}, \quad (85)$$

and the second term is as

$$(h_0 - h)/(RT_c)^{(1)} = T_R^2 \int_{P_R \to 0}^{P_R} \frac{\partial Z^{(1)}}{\partial T_R} \frac{dP_R}{P_R}. \quad (86)$$

The superscript (0) implies that $Z^{(o)}$ is based on a simple fluid state equation, while the superscript (1) represents a correction to the simpler fluid properties by considering more complex effects. From Eqs. (84) to (86), we obtain the expression

$$(h_o - h)/(RT_c) = (h_o - h)^{(o)}/(RT_c) + \omega(h_o - h)^{(1)}/(RT_c). \quad (87)$$

Similarly,

$$(s_o(T,P) - s(T,P))/R = (s_o(T,P) - s(T,P))^{(o)}/R + \omega(s_o(T,P) - s(T,P))^{(1)}/R, \quad (88)$$

$$(g_o(T,P) - g(T,P))/R = (g_o(T,P) - g(T,P))^{(o)}/R + \omega(g_o(T,P) - g(T,P)^{(1)}/R, \text{ and} \quad (89)$$

$$\phi(T,P) = \phi^{(o)}(T,P) + \omega\phi^{(1)}(T,P) \quad (90)$$

We can obtain $Z^{(o)}$ and $Z^{(1)}$ using the Kesler equation of state. Kesler charts have been generated for $Z^{(o)}$, $Z^{(1)}$, $(h_o - h)^{(o)}/(RT_c)$, and $(h_o - h)^{(1)}/(RT_c)$. See next section.

G. KESLER EQUATION OF STATE (KES) AND KESLER TABLES

Instead of the RK equation one can use the Kesler equation of state presented in Chapter 6 to obtain values of thermodynamic properties for simple fluids (with the constants in Table A-21) in the form $((h_0\text{-}h)/RT_c)^{(0)}$, $(s_o(T,P) - s(T,P))^{(o)}$, etc. The procedure can be repeated for reference fluids (with with appropriate constants) using the Kessler equation of state and obtain relations for $((h_0\text{-}h)/RT_c)^{(ref)}$, $(s_o(T,P)- s(T,P))^{(ref)}$, etc. Defining

$$((h_0\text{-}h)/RT_c)^{(1)} = (((h_0\text{-}h)/RT_c)^{ref} - ((h_0\text{-}h)/RT_c)^{(0)})/w_{ref},$$

we can tabulate values for $((h_0\text{-}h)/RT_c)^{(0)}$ and for $((h_0\text{-}h)/RT_c)^{(1)}$ at any specified P_R and T_R. Then for any other fluid,

$$((h_0\text{-}h)/RT_c) = ((h_0\text{-}h)/RT_c)^{(0)} + w\ ((h_0\text{-}h)/RT_c)^{(1)}.$$

Similarly other properties like the entropy can be expressed, i.e.,

$$((s_0\text{-}s)/RT_c)^{(1)} = ((s_0\text{-}s)/RT_c)^{ref} - ((s_0\text{-}s)/RT_c)^{(0)},$$

See Tables A-24A to A-26B for tabulations of $((h_0\text{-}h)/RT_c)^{(0)}$, $((h_0\text{-}h)/RT_c)^{(1)}$, $(s_0\text{-}s)^{(0)}/R$, $((s_0\text{-}s)^{(1)}/R$, $\phi^{(o)}$, and $\phi^{(1)}$.

H. FUGACITY

Information regarding fugacity enables the evaluation of the chemical potential and allows the characterization of phase and chemical equilibrium, as will be discussed later.

1. Fugacity Coefficient

One can use a state relation for g to obtain an expression for ϕ, e.g.,

$$dg = -s\, dT + v\, dP.$$

Assuming that a state relation is available for $v = v(T,P)$, at constant temperature,

$$dg = v\, dP, \text{ or} \tag{91}$$

$$dg = d(Pv) - P\, dv \tag{92}$$

a. RK Equation

If the RK equation is used in the context of Eq. (92),

$$g(T,v) = RTv/(v-b) - \mathbf{a}/(T^{1/2}(v+b)) - RT\ln(v-b) + (\mathbf{a}/(T^{1/2}\, b))\ln(v/(v+b))) + f(T)$$

If $\mathbf{a} = b = 0$, for an ideal gas:

$$g_o(T,v) = RT - RT\ln v + f(T).$$

This enable us to determine f(T), and

$$g(T,v) - g_o(T,v) = RT\, b/(v-b) - \mathbf{a}/(T^{1/2}(v+b))$$

$$- RT\ln((v-b)/v) + (\mathbf{a}/(T^{1/2}b))\ln(v/(v+b))). \tag{93}$$

Dividing Eq. (93) by RT we obtain the relation

$$(g(T,v) - g_o(T,v))/RT = b/(v-b) - \mathbf{a}/(RT^{3/2}(v+b))$$

$$- \ln((v-b)/v) + (\mathbf{a}/(RT^{3/2}b))\ln(v/(v+b))). \tag{94}$$

The fugacity coefficient can be obtained from the expression

$$\ln\phi = (g(T,P) - g_o(T,P))/RT = g^{Res}(T,P)/RT.$$

Proceeding as before,

$$g(T,v) - g_o(T,v) = g(T,P) - g_o(T,v) = g(T,P) - g_o(T,v_o),$$

where v_o denotes the ideal gas volume at the state (T, P). Manipulation of this expression results in

$$g(T,P) - g_o(T,v_o) = g(T,v) - g_o(T,v) + g_o(T,v) - g_o(T,v_o).$$

Since,

$$g_o(T,v) - g_o(T,v_o) = h_o(T) - Ts_o(T,v) - (h_o(T) - T\, s_o(T,v_o))$$

$$= RT\ln v_o/v = RT\ln(RT/Pv) = -RT\ln Z,$$

$$\ln\phi = (g(T,P) - g_o(T,P))/RT = g(T,v) - g_o(T,v) - RT\ln Z. \tag{95}$$

Using Eq. (94) in Eq. (95) we obtain

$$\ln\phi = 0.08664/(v_R' - 0.08664) - 0.4275/(T_R^{\ 3/2}(v_R' + 0.08664))$$

$$- \ln(1 - 0.08664/v_R') - (4.9342/T_R^{\ 3/2})\ln(1 + 0.08664/v_R') - \ln Z, \tag{96}$$

where $v_R'(T_R, P_R)$ is known. As $a \to 0$ and $b \to 0$, $\phi \to 1$. Further as $v_R' \to \infty$ (i.e., as $P_R \to 0$), $\phi \to 1$.

y. *Example 25*

Determine the value of ϕ for water at 250 bar and 673 K (i.e., P_R = 1.132, and T_R = 1.349).

Solution

From previous examples, v_R' = 1.007 at P_R = 1.132, T_R = 1.349. Therefore,

$\ln \phi = 0.08664 \div (1.007 - 0.08664) - 0.4275 \div (1.349^{3/2} \times (1.007 - 0.08664))$

$- \ln(1 - 0.08664 \div 1.007) - 4.9342 \div 1.349^{3/2} \times \ln(1 + 0.08664 \div 1.007)$

$= 0.09414 - 0.2495 + 0.08997 - 0.2599 + 0.168 = -0.1573$, i.e.,

$\phi = 0.854$

b. *Generalized State Equation*

Consider the generalized state equation

$$Pv = ZRT \quad (97)$$

At given T, dg = v dP. Using Eq. (97),

$$dg = vdP = (ZRT/P)\, dP. \quad (98)$$

For ideal gases

$$dg_{o,T} = v_o\, dP = (RT/P)\, dP = RT\, d\ln P. \quad (99)$$

Therefore,

$$d(g - g_o) = (v - v_o)\, dP = v_o(Z-1)\, dP = RT\,(Z-1)\, dP/P.$$

Integrating between the limits of $P \to 0$ and P,

$$g^{Res}/(RT) = (g - g_0)/(RT) = \int_{P\to 0}^{P} (Z-1)\frac{dP}{P}, \text{ i.e.,} \quad (100)$$

$$\ln\phi = \int_{P_R\to 0}^{P_R} (Z-1)\frac{dP_R}{P_R}. \quad (101)$$

Differentiating Eq. (101) at a specified temperature, we obtain the relation

$$d\ln\phi = d\,g^{res}/RT = (Z-1)\, dP/P. \quad (102)$$

2. Physical Meaning

We introduce the fugacity f, which has the same units as pressure. Following Lewis (1875–1946), we define the fugacity f as

$$dg = v\, dP = RT\, d\ln f. \quad (103)$$

Noting the similitude with Eq. (99),

$$d((g - g_o)/RT) = (Z-1)\, dP/P = d\ln(f/P) = d\ln\phi \quad (104)$$

Upon comparing Eqs. (104) and (102) we realize that

$$\phi = f/P \quad (105)$$

Integrating Eq. (104) at constant temperature

$$((g - g_o)/RT) = \ln\phi + F(T).$$

As $g \to g_o$ (i.e., $P \to 0$), $f \to P$ (i.e., $\phi \to 1$). (The ideal gas equation of state can, therefore, be expressed in the form $fv = RT$ where $f = P$ for ideal gas). Hence $F(T) = 0$, and

$$((g - g_o)/RT) = \ln \phi.$$

The fugacity coefficient ϕ is a measure of the deviation of the Gibbs function from its ideal gas value. One may express real gas equation of state $Pv = ZRT$ as $fv = Z'(T_R,P_R)RT$ where $Z'(T_R,P_R) = \phi(T_R,P_R)\, Z(T_R,P_R)$. In the presence of intermolecular attraction forces, typically, for a real gas $h < h_o$ and $s < s_o$. The corresponding value of g ($= h-Ts$) is less than or greater than g_o ($= h_o-Ts_o$), when, respectively, ϕ is smaller than or larger than unity.

Accounting for the Pitzer factor $Z = Z^{(o)} + \omega Z^{(1)}$, Eq. (105) assumes the form

$$\ln \phi = \ln \phi^{(o)} + w \ln \phi^{(1)} \tag{106}$$

where

$$\ln\ \phi^{(0)} = \int_{P_R \to 0}^{P_R} (Z^{(0)} - 1)\frac{dP_R}{P_R}, \text{ and } \ln\ \phi^{(1)} = \int_{P_R \to 0}^{P_R} Z^{(1)}\frac{dP_R}{P_R}. \tag{107}$$

a. Phase Equilibrium

In general, during boiling, the pressure and temperature remain constant. Since $dg = -s\,dT + v\,dP$, for this process $dg = 0$. Consequently, as a fluid of unit mass undergoes change from the saturated liquid state to the saturated vapor state, $g^f = g^g$. Using Eq. (105), we note that the differences

$$(g^f(T,P) - g_o(T,P))/(RT) = \ln \phi^f, \text{ and } (g^g(T,P) - g_o(T,P))/(RT) = \ln \phi^g$$

are identical, since $g^f = g^g$. Therefore,

$$\phi^f = \phi^g, \text{ and } f^f = f^g = f^{sat}. \tag{108}$$

For a phase change from an α (say, the solid phase) to a β phase (say, vapor) $g^\alpha = g^\beta$ (for a single component g is also called the chemical potential μ so that $\mu^\alpha = \mu^\beta$), $\phi^\alpha = \phi^\beta$, and $f^\alpha = f^\beta$. This implies the existence of a single saturation curve along which the fugacities for both the saturated liquid and vapor states are the same at given T. The inflexion of the curve occurs at the critical point.

b. Subcooled Liquid

Integrating Eq. (103) along an isotherm from the saturated liquid state (T,P^{sat}) at which $f = f_f$ to a compressed liquid state (T, P)

$$(g(T,P) - g(T,P^{sat}))/RT = \ln(f(T,P)/f_f(T,P^{sat})) = \int_{P^{sat}}^{P} (v/RT)dP \tag{109}$$

The Poynting correction factor POY is related to the RHS of Eq. (109), namely,

$$POY(T,P) = f(T,P)/f_f(T,P^{sat}) = \exp\left(\int_{P^{sat}}^{P} (v/RT)dP\right) \tag{110}$$

In general, along an isotherm $v \approx v_f$ so that

$$POY = \exp(v_f(P - P^{sat}(T))/RT), \text{ i.e.,} \tag{111}$$

$$(g(T,P) - g(T,P^{sat}))/RT = \ln(f/f_f) = v_f(P - P^{sat})/RT. \tag{112}$$

A similar procedure can also be adopted for solids.

Treating the liquid as incompressible, i.e., $u(T,P) \approx u(T, P^{sat})$, $s(T,P) \approx s(T,P^{sat})$, and

$$h(T,P) = u(T,P) + Pv(T,P) \approx u(T,P^{sat}) + Pv_f(T,P^{sat}).$$

c. *Supercooled Vapor*

Sometimes a vapor can be cooled to a temperature below saturation temperature without causing condensation (Chapter 10). In the case of super-cooled vapor the thermodynamic properties can be related to the saturation properties, At low pressure $u(T,P) \approx u(T,P^{sat})$, $h(T,P) \approx h(T,P^{sat})$, $s(T,P) \approx s(T,P^{sat}) - R \ln(P/P^{sat})$, and $g(T,P) \approx g_o(T,P)$. Therefore,

$$g(T,P) = h - Ts \approx h(T,P^{sat}) - T(s(T,P^{sat}) - R \ln(P/P^{sat})) = g(T,P^{sat}) + RT \ln(P/P^{sat}).$$

where P^{sat} is at T.

---

z. *Example 26*

Determine the fugacity of pure water for the following cases:
Saturated vapor at 100°C,
Saturated liquid at 100°C,
Compressed liquid at 100°C, and 200 bar.
Superheated vapor at 100°C, and 0.5 bar.
Saturated vapor at 350°C.
Super-cooled vapor at 90°C, 1 bar, assume ideal gas behavior

Solution

The saturation pressure at 100°C is P^{sat} = 1 bar. Since P_c = 220.9 bars, $P_R = P^{sat}/P_c \ll 1$, at this state water vapor $H_2O(g)$ behaves as an ideal gas. Therefore,
$f = P^{sat}$ = 100 kPa or 1 bar.
Since P and T are constant, f is unchanged during the phase change. Therefore, for the saturated liquid $H_2O(\ell)$ f = 100 kPa or 1 bar.
At constant temperature,
$d(\ln f) = vdP/(RT)$.
For liquids, $v \approx$ constant. For this problem, $v = 0.001\ m^3\ kg^{-1}$. Integrating from the saturated liquid state at 100°C and 1 bar to the compressed liquid state at 100°C and 2 bar,
$\ln(f(T,P)/f^{sat}(T)) = v(P - P^{sat})/(RT)$, i.e.,
$\ln(f(100°C, 200\ bar))/(f^{sat}(100°C)) = ((0.001\times(20{,}000-100))/(8.314\times373/18.02))$
$= 0.116$.
Therefore,
$(f(100°C, 200\ bar))/(f^{sat}(100°C)) = POY = \exp(0.116) = 1.123$, and
$f(100°C, 200\ bar)) \approx 1.122\times(f^{sat}(100\ °C) = 1.123\times f^{sat}(100°C) = 1.123$ bar.
Superheated vapor behaves as an ideal gas at low pressures. Therefore, the fugacity equals the pressure, i.e., f = 0.5 bar.
At 350°C, P^{sat} = 165 bar, $P_R = 165\div220.9 = 0.75$, $T_R = 623\div641 = 0.98$, and $Z \approx 0.3$. Therefore, under these conditions water vapor behaves as a real gas and $f \neq P$. Using the fugacity coefficient charts, $\phi = 0.74$, and $f = 0.74\times165 = 124$ bar.
Since the behavior a ideal gas, $f = P^{sat}$ at 90°C = 0.7014 bars.

Remarks

The example illustrates that when the pressure is increased by a factor of 200, in the case of liquids f changes by only 12 %. At a specified temperature the changes in the values of f with respect to pressure ($d \ln f = v_f\, dP$) are small, since the liquid specific volume v_f is small. However, for the gaseous state, v is much larger than v_f (oftentimes by three orders of magnitude) so that f changes significantly as the pressure is altered.

---

aa. Example 27

Employ the RK equation of state for the following problems.
Determine the value of g for $H_2O(g)$ at its triple point of (T_{TP} = 273 K, P_{TP} = 0.0061 bar).
Determine the corresponding value of g_o.
Determine the value of g for $H_2O(g)$ at P = 250 bar, and T = 873 K if $\bar{c}_{p,o} = 28.85 + 0.01206\ T + 100{,}600/T^2$ kJ $kmole^{-1}$ K^{-1}.

Solution

For water at its triple point, $g(f, T_{TP}, P_{TP}) = h_{f,TP} - T\ s_{f,TP} = 0$.
Since the vapor behaves as an ideal gas at the triple point,
$h_o - h(T_{TP}, P_{TP}) \approx 0$, and $s_o - s(T_{TP}, P_{TP}) \approx 0$

Therefore, $h_o - Ts_o \approx h - Ts$ so that $g(T_{TP}, P_{TP}) = g_o$.

Consequently, $\phi = 1$, and $f = P_{TP} = 0.006$ bar.

We can get g(250, 873) by using the definition g = h–Ts and using correction charts to determine h_o– h and s_o–s entropy, ideal gas enthalpy and entropy at 250 bar, 873 K or we can use

$$g\ (T,P) = g_o(T,P) + RT \ln \phi$$

Referring to Example 17,
$h_o(873K) = 3706$ kJ kg^{-1}.
Referring to Example 16
$s_o(873, 250bar) = 6.566$ kJ kg^{-1} K^{-1}.
$g_o(873,250) = h_o(T) - T\ s_o(T,P) = 3706 - 873\times6.566 = -2{,}026.1$ kJ kg^{-1}.

$g(873,250) = g_o(873,250) + RT \ln \phi$

From example 26,
$\phi = 0.854$, $f = 0.854 \times 250 = 214$ bars.

$g(873,250) = -2{,}026.1 + (8.314/18.02) \times 873 \ln 0.854 = -2089.7$ kJ kg^{-1}.

---

bb. Example 28

Determine the properties u, h and f for liquid water at 120°C and 250 kPa. Assume that data for u^{sat}(120°C), and v^{sat}(120°C) are available.
Determine the properties u, h, s, g, and f for liquid water at120°C and 100 kPa.

Solution

P^{sat}(120°C) = 199 kPa. Since P = 250 kPa, the liquid is in a compressed state. For this state u(120°C, 199 kPa) = u^{sat}(120°C) = 503.5 kJ kg^{-1}. Recall that

$$du = c_v dT + (T(dP/dT)_v - P)\ dv \quad (A)$$

Assume that v = v(T) so that its value (v = 0.001063 m^3 kg^{-1}) along the 120°C isotherm does not change. Therefore, since dv = dT = 0, du = 0, and
u^{sat}(120°C, 250 kPa) ≈ u (120°C, 199 kPa) = 503.5 kJ kg^{-1}.

Furthermore,
$h = u + Pv = 503.5 + 250 \times 0.001063 = 503.5$ kJ kg^{-1}.

Using Eq. (112)
$\ln (f(T,P)/f^{sat}(T)) = v_f(P - P^{sat})/(RT)$
$= 0.001063\times(250 - 199)/((8.314/18.02)\times393) = 0.000179,$

$POY = f(T,P)/f(T, P^{sat}) = 1.000179$, i.e.,
$f(T,P) \approx f^{sat(g)}(T)$.

Since the pressure is relatively low, $f^{sat(g)} = P^{sat} = 199$ kPa, i.e., Thus $f^{sat(l)} = f^{sat(g)} = 199$ kPa.

Water boils at 100°C when P ≈ 100 kPa. Therefore, we can expect the water vapor at 120°C to be superheated. However, under some circumstances (to be discussed in Chapter 10) water can exist as a superheated liquid at 120°C and 100 kPa, instead of as a superheated vapor. Since u is a function temperature alone for liquids,

$u(120°C,100\text{ kPa}) \approx u^{sat}(120°C) = 503.5\text{ kJ kg}^{-1}$.

Therefore,

$h(120°C,100\text{ kPa}) = u^{sat}(120°C) + Pv = 503.5 + 100\times0.001063 = 503.51\text{ kJ kg}^{-1}$,

$s(120°C,100\text{ kPa}) = s^{sat}(120°C) = 1.5276\text{ kJ kg}^{-1}\text{ K}^{-1}$,

$g(120°C,100\text{ kPa}) = g^{sat}(120°C) + v_f(P - P^{sat})$

$= (503.71 - 393\times1.5276) + 0.00106\times(199-100) = -96.53\text{ kJ kg}^{-1}$, and

$\ln (f/f^{sat}) = v_f(P - P^{sat})/RT = 0.001063(100-199)\div(0.4614\times393) = -0.000578$, i.e.,

$f/f^{sat} = 0.9994$ or $f = f^{sat} \times 0.9994 = 199 \times 0.9994 = 198.4$ kPa.

Remark

Note that the fugacity of the superheated liquid is lower than that of the corresponding saturated liquid at the same temperature.

cc. Example 29

Methane is reversibly compressed at 230 K in a steady state steady flow (sssf) device from 150 bar to 1000 bar. Using the fugacity charts, determine work done in kJ kmole^{-1}.

Solution

The sssf energy balance for a reversible process has the form

$$\delta w = -\,vdP. \qquad (A)$$

For an isothermal process,

$$dg_T = v\,dP. \qquad (B)$$

From Eqs. (A) and (B) we determine that

$$-\delta w = dg_T, \text{ i.e.,} \qquad (C)$$

$$-w = g_2 - g_1. \qquad (D)$$

Since dg = RT d ln f, integrating this relation at constant temperature,

$$-w = g_2 - g_1 = RT \ln f_2/f_1 = RT \ln (\phi_2 P_2/(\phi_1 P_1)). \qquad (E)$$

From the fugacity charts, $P_{R,2} = P_2/P_c = 250\div46.4 = 5.4$, $P_{R,1} = P_1/P_c = 150\div46.4 = 3.2$, and $T_{R,1} = T_{R,2} = T_2/T_c = 230\div190.7 = 1.2$, i.e., $\log_{10}\phi_2 = -0.358$, $\log_{10}\phi_1 = -0.275$. Therefore, $\phi_2 = 0.439$, and $\phi_1 = 0.531$ so that

$-\bar{w} = \bar{R}T \ln(0.439\times250\div(0.531\times150)) = 8.314 \times 230 \times \ln 1.378 = 613\text{ kJ kmole}^{-1}$.

Remarks

If the gas behaves like an ideal gas, $w_{12} = -\int vdP = -\bar{R}T \ln(P_2/P_1) = -8.314 \times 230 \times \ln(250\div150) = -977\text{ kJ kmole}^{-1}$. The magnitude for the work done in case of an ideal gas is much larger since the ideal gas involves higher pressure for the same volume and hence requires more boundary work during compression.

If gas is compressed isothermally in a closed system $w_{12} = \int P\,dv = (P_2v_2 - P_1v_1 - \int vdP)$ where second part on the right can be determined using fugacity charts

I. EXPERIMENTS TO MEASURE $(u_O - u)$

It is possible to measure the difference $(u_o - u)$ for real gases using the Washburn experiments (see Chapter 2). A mass m of high pressure gas stored in a tank A is discharged through s narrow tube C into the atmosphere at B (Figure 12). The tank and the tube are maintained in a constant temperature bath D. Recall the following relation from Chapter 2, during a short period of time "dt".

$$m\,du + u\,dm = \delta Q - dm(h(T_B,P_o) + ke_o). \quad (a)$$

Since the discharged gas pressure is atmospheric,

$$h(T_B, P_o) = h_o(T_B).$$

Ignoring the kinetic energy ke, Eq.(a) becomes

$$d(mu) = \delta Q + dm\,h_o(T_B), \text{ i.e., } m_2u_2 - m_1u_1 = Q + (m_2 - m_1)\,h_o(T_B).$$

Where states (1) and (2) are initial and final states of tank A With $u_2 = u_0$, $m_2 = m_0$,

$$(m_0 u_0 (T_B) - m_1u_1 (T_1, P_1)) = Q + (m_0 - m_1) RT_B \text{ and } m_0 = v/v_0, \text{ with } m_1 = v/v_1$$

$$V/v_o\, u_o - V/v_1\, u_1 = Q + (V/v_o - V/v_1)(u_o + RT_B), \text{ i.e.,}$$

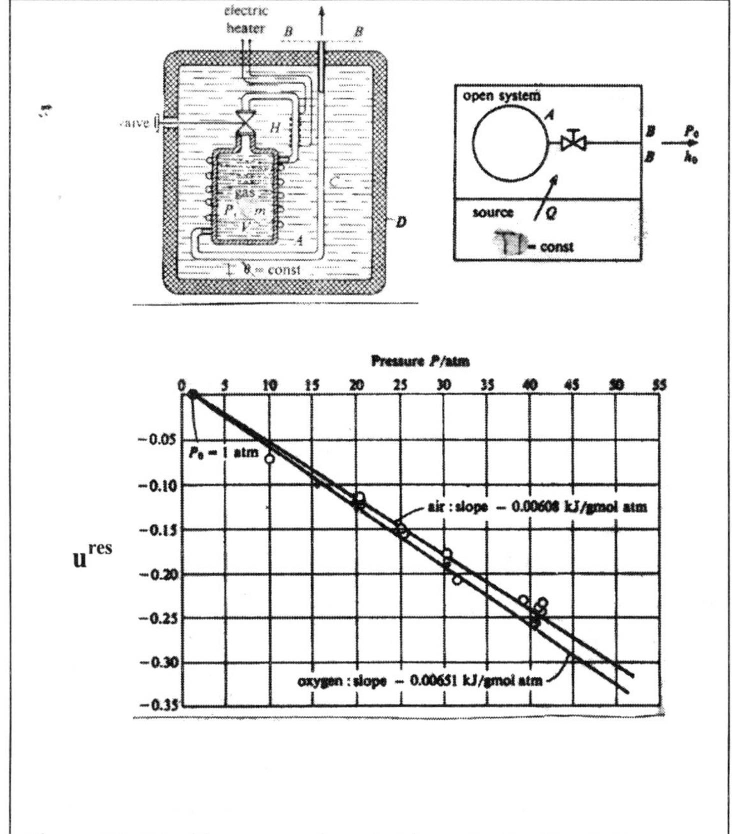

Figure 12. Washburn experiments (from A. Kestin, A Course in Thermodynamics, McGraw Hill, NY, 1979, p 262, Volume I. With permission.).

$$u_o(T_B) - u_l(T_B, P) = Q\, v_l/V + (RT_B\, v_l/v_o - RT_B) = Q\, m_l + (P_o\, V/m_l - RT_B). \quad \text{(b)}$$

With known Q, V, T_B and P_O we can determine the difference between the ideal and real gas internal energies in this manner. The "P" in tank A can be altered and corresponding $u_0 - u_l$ can be determined from Eq. (b). Likewise, using the expression h = u + Pv,

$$h_o(T_B) - h_l(T_B, P) = Q\, m_l - (P - P_o)\, V/m_l.$$

The Washburn coefficient $(\partial u/\partial P)_T$ represents the slope of the difference $(u_o(T_B) - u_l(T_B, P))$ with respect to pressure. The slopes for molecular oxygen and air, respectively, tend to approach values of 6.51 and 6.08 kJ kmole^{-1} bar^{-1} as $P \to P_o$.

Differentiating Eq. (60), we obtain the relation

$$\partial(u_o(T) - u(T,v))/\partial P = -(3/2)(\mathbf{a}/(bT^{1/2}))(\partial v/\partial P)/(v^2(1+b/v)). \quad \text{(c)}$$

Since v has a relatively large value (as $P \to P_o$), neglecting higher order terms in v,

$$\partial(u_o(T) - u(T,v))/\partial P = -(3/2)(\mathbf{a}/(bT^{1/2}))(\partial v/\partial P)_{T,v\to\infty}/v^2.$$

Using the expression for $\partial v/\partial P$ given in Chapter 6 for RK equation,

$$(\partial v/\partial P)_T = 1/((\mathbf{a}(2v+b)/T^{1/2}\, v^2\, (v+b)^2) - RT/(v-b)^2).$$

As $v \to \infty$,

$$\partial v/\partial P = -v^2/RT. \quad \text{(d)}$$

Therefore using Eq.(d) in Eq.(c),

$$\partial(u_o(T) - u(T, v))/\partial P = (3/2)(\mathbf{a}/RT^{3/2}).$$

Rewriting this expression in reduced form and using the RK state equation relation for $\mathbf{a}$ = $0.4275\, R^2\, T_c^{1.5}/P_c$,

$(\partial u_{C,R}/\partial P_R) = 0.6413/T_R^{1.5}$ Using the values for $\mathbf{a}$ = 17.39 bar K$^{1/2}$ m^6 kmole^{-2} and $\overline{R}$ = 0.08314 bar m^3 kmole^{-1} K^{-1} for molecular oxygen, at 301 K, $\partial(u_o(T) - u(T,v))/\partial P$ = 0.0601 m^3 kmole^{-1} or 6.01 kJ kmole^{-1} bar^{-1} while the experimental value is given as 0.0651 kmole^{-1} bar^{-1}.

J. VAPOR/LIQUID EQUILIBRIUM CURVE

In the previous sections we have used the real gas equations of state to determine thermodynamic properties. In this section, we will obtain saturation properties, such as P^{sat}, T^{sat}, and h_{fg}, and the Joule Thomson coefficient using these state equations.

1. Minimization of Potentials

a. *Helmholtz Free Energy A at specified T, V and m*

In Chapters 1 and 3 we have discussed the phenomena of evaporation, condensation, and phase equilibrium. Evaporation occurs as a result of the chemical potential difference between the liquid and vapor phases of a fluid. If an evacuated rigid vessel of volume V is injected with a liquid and then immersed in a constant temperature bath at conditions conducive to evaporation, μ_ℓ is initially higher, which is why evaporation occurs. As the vapor fills the space within the vessel, the pressure increases, thereby increasing μ_g. Evaporation stops at a saturation pressure P^{sat} that is characteristic of the bath temperature T at which $\mu_\ell = \mu_g$. Recall from Chapter 3 that

$$dA = -P\, dV - S\, dT - T\, \delta\sigma.$$

If we consider the evaporation to be an irreversible process that occurs in a rigid closed system, $\delta\sigma > 0$ at the specified temperature, volume, and mass, i.e.,

$$dA = -T\,\delta\sigma \text{ so that } dA < 0.$$

---

dd. Example 30

A rigid container that has a volume of 0.35 m^3 is completely evacuated and then it is filled with 0.1 kmole of liquid water (Figure 13a). It is then immersed in an isothermal bath at a temperature of 50°C. The liquid evaporates to form vapor, and the vapor pressure is measured. We will refer to the liquid water as subsystem A and the vapor–filled space above the liquid as subsystem B. Phase equilibrium is reached when the vapor reaches a saturation pressure, i.e., when there is no net change in the mass of either the liquid or the vapor. This occurs when the net evaporation ceases. Determine the change in Helmholtz function with respect to the vapor pressure P_v (= P_B) and determine the value of that pressure when Helmholtz function reaches a minimum value. Assume that at T = 273.15 K $h_f = s_f = 0$, h_{fg} = 2501.3 kJ kg^{-1} and that s (323 K, P_B) ≈ s_g (273 K, P_{TP}) + $c_{p,o,v}$ $\ln(T/T_{TP})$ – R ln P_B/P_{TP}, $c_{p,o,v}$ = 1.8 kJ kg^{-1} K^{-1}, R = 0.46 kJ kg^{-1} K^{-1}, c = 4.184 kJ kg^{-1} K^{-1}. The constant volume reactor is typically adopted to measure Reid's vapor pressure.

Solution

The Helmholtz energy of systems liquid (A) and vapor (B) (see Figure 13a)

$$A = A_A + A_B, \text{ where} \quad (A)$$

$$A_A = N_A\,\bar{a}_A, \text{ and } A_B = N_B\,\bar{a}_B \text{ (or } A_v = N_v\,\bar{a}_v). \quad (B)$$

The value of A_A changes, since the liquid mass decreases during vaporization. The value of a_v changes since P_v changes. For the liquid phase,

$u_f(323\text{ K}) \approx h_f(323\text{ K}) = c_w(T - 273) = 4.184(323 - 273) = 209.2$ kJ kg^{-1},

$s_A = c\ln(T/273.15) = 0.7033$ kJ kg^{-1} K^{-1},

$a_A = u_A - Ts_A = 209.2 - 323.15 \times 0.7033 = -18.075$ kJ kg^{-1},

so that initially

$A_A = N_A\,\bar{a}_A = 0.1 \times (-17.97 \times 18.02) = -32.57$ kJ (and $A_v = 0$), and

$A = A_A + A_v = -32.57 + 0 = -32.57$ kJ.

For an arbitrary amount of vapor accumulation, say 0.0002 kmole, since the total number of moles of water N is unchanged,

$N_A = N - N_v = 0.1 - 0.0002 = 0.0998$ kmole, i.e.,

$A_A = 0.0998 \times (-18.075 \times 18.02) = -32.506$ kJ.

The vapor pressure

$P_v = N_v\,\bar{R}T/V_B$, where

$V_B = V - V_A = V - N_A W_A/v_A = 0.35 - 0.0998 \times 18.02 \div 1000 \approx 0.35$ m^3, i.e.,

$P_v = 0.0002 \times 0.08314 \times 323.15 \div 0.35 = 0.01535$ bar.

Furthermore,

$u_v \approx u_{vo} = h_{vo} - RT = (h_{g,ref} + c_{p,o}(T - 273.15)) - RT$

$= (2501.3 + 1.8\times(323.15\text{-}273.15)) - (8.314 \div 18.02) \times 323.15 = 2442.2$ kJ kg^{-1}.

$s_v(T, P_v) = s_v(323\text{ K}, P_v) \approx s_g(273\text{ K}, P_{TP}) + c_{p,o}\ln(323/273) - R\ln p_v/P_{TP}$

$= 2501.3\text{+}273.15 + 1.8\ln(323.15\text{+}273.15) - 0.461\ln(0.015\text{+}0.0061)$

$= 9.046$ kJ kg^{-1} K^{-1}, and

$a_v = u_v(T) - Ts_v(T, P_v) = 2442.2 - 323.15 \times 9.046 = -477.36$ kJ kg^{-1}, i.e.,

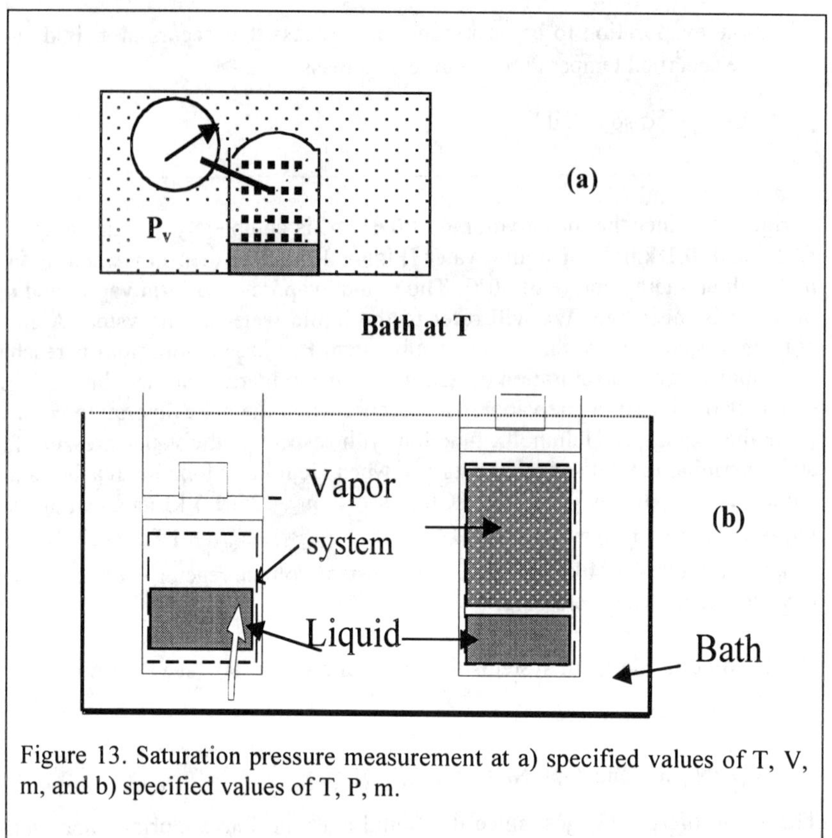

Figure 13. Saturation pressure measurement at a) specified values of T, V, m, and b) specified values of T, P, m.

$A_v = 0.0002\times(-477.36 \times 18.02) = -1.7204$ kJ. $A = A_A + A_B = -32.51 - 1.72 = -34.23$ kJ

By repeating the calculations we can obtain a chart for A with respect to P_v. The value of A reaches a minimum at P_v = 0.124 bar (Figure 14) (which approximately equals the tabulated value of 0.125 bar).

Remarks

If $c = c_{po}$ or $c = c_{po} = 0$, then the saturation pressure at A_{min} can be derived as

$$P^{sat}/P_{TP} = \exp((h_{fg,TP}/R)\,(1/T_{ref} - 1/T)),$$

which is known as the Clausius Clapeyron equation (see later sections).

The saturation pressure P^{sat} can also be obtained by minimizing the value of $A = N_A\,\bar{a}_A + N_v\,\bar{a}_v$ using the Lagrange multiplier method subject to the constraint $N_v + N_w = N$. In this manner we can prove that $\mu_w = \mu_g$.

Solids can similarly undergo a phase transition within the solid phase. These processes are unsteady, and thermodynamics–based analyses cannot provide information on the time–dependent rate behavior of phase transition.

b. G at Specified T, P and m

If the experiment described in the previous section is repeated in an appropriate piston-cylinder-weight assembly containing a mixture of liquid water and air (Figure 13b) and subsequently immersed in an isothermal bath, the temperature, pressure, and mass can be held constant while evaporation occurs. The vapors so accumulated will move the piston. The piston motion will stop when the mixture volume reaches a particular value at which phase equilibrium is reached. For this case,

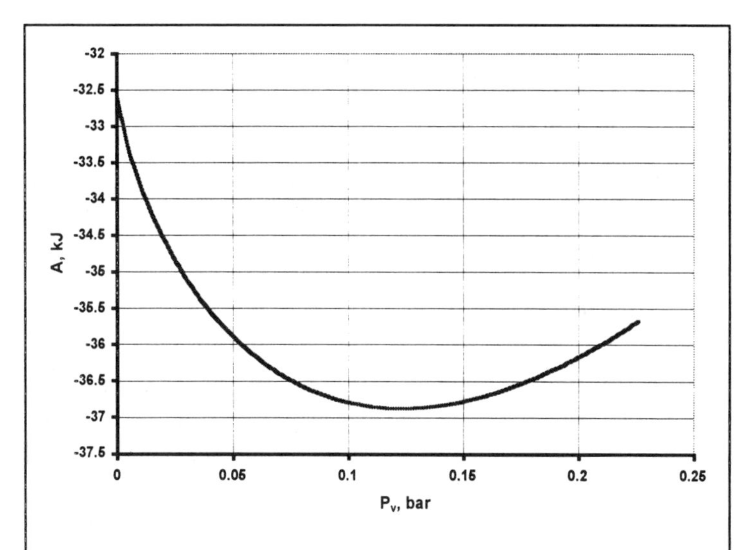

Figure 14. The Helmholtz function with respect to P_v for water at 323 K. The minima in A corresponds to the saturation pressure at 323 K

$$dG = V\,dP - S\,dT - T\,\delta\sigma.$$

And from Chapter 3 we know that G must reach a minimum at fixed T, P, m. Equilibrium is reached when the Gibbs free energy in the liquid and gas phases, $G = G_P + G_g$ reaches a minimum value, i.e., when

$$\mu_{H_2O(P)} = \mu_{H_2O(g)}.$$

This energy minimum can be obtained by following a similar procedure to that used in Example 30. If we assume that the pressure is low enough so that the gas mixture can be treated as ideal gas. At phase equilibrium,

$$P_{H_2O} = P^{sat}_{H_2O}(T) = X_{H_2O}P, \text{ where}$$

The water vapor mole fraction $X_{H2O} = N_{H_2O(g)}/(N_{H_2O(g)} + N_{air})$

2. Real Gas Equations

The previous section required a knowledge of "c", c_{p0}, h_{fg} and relations for "s". Instead, a real gas state equation can be used to determine a relation for $P^{sat}(T)$ and by applying the equality $g_P = g_v$. In order to do so, the critical conditions of the fluid must be known.

a. Graphical Solution

It is possible to characterize the variation in pressure with respect to specific volume at $T<T_c$, e.g., the curve BECGNDUHJMAFL contained in Figure 15. At constant temperature,

$$dg = v\,dP. \tag{113}$$

where "v' is given by any real gas state equation presented in Chapter 6. As the fluid is compressed from a large volume at low pressure, the pressure first increases (e.g., the curve BECGN along which $dP > 0$). Therefore, due to the work input, the Gibbs energy, which is a measure of availability, also increases (cf.. Eq. (113)). With a further decrease in volume past point N along the curve NHM, $dP < 0$, i.e., $dg < 0$ and hence "g' decreases. The pressure decrease continues until point M is reached, beyond which $dP > 0$, i.e., $dg > 0$ (along curve MFL) and hence g increases again. Rewriting Eq. (113) we obtain the relation

$$dg = d(Pv) - Pdv. \quad (114)$$

Using the RK state equation, $P = RT/(v-b) - a/(T^{1/2}\, v(v+b))$ and integrating between the resultant expression between the limits v_{ref} and v, we obtain the relation

$$g - g_{ref} = (Pv - P_{ref}v_{ref}) - RT \ln((v - b)/(v_{ref} - b)) + \quad (115)$$

$$(a/bT^{1/2}) \ln((v/(v+b))((v_{ref}+b)/v_{ref})). \quad (116)$$

We arbitrarily set g_{ref} $(v_{ref}) = 0$. The resultant plot is presented in Figure 15. The vapor–like curve BECKN and the liquid–like curve MFL are apparent. The unstable branch NDUHJM (along which the pressure decreases with decreasing specific volume) will be discussed in Chapter 10. At the points F and G, $g_f = g_g$, which, according to the phase equilibrium condition, must correspond to the saturation pressure.

We note from Eq. (113) that the slopes $\partial g/\partial P = v_f$ along the liquid–like curve and

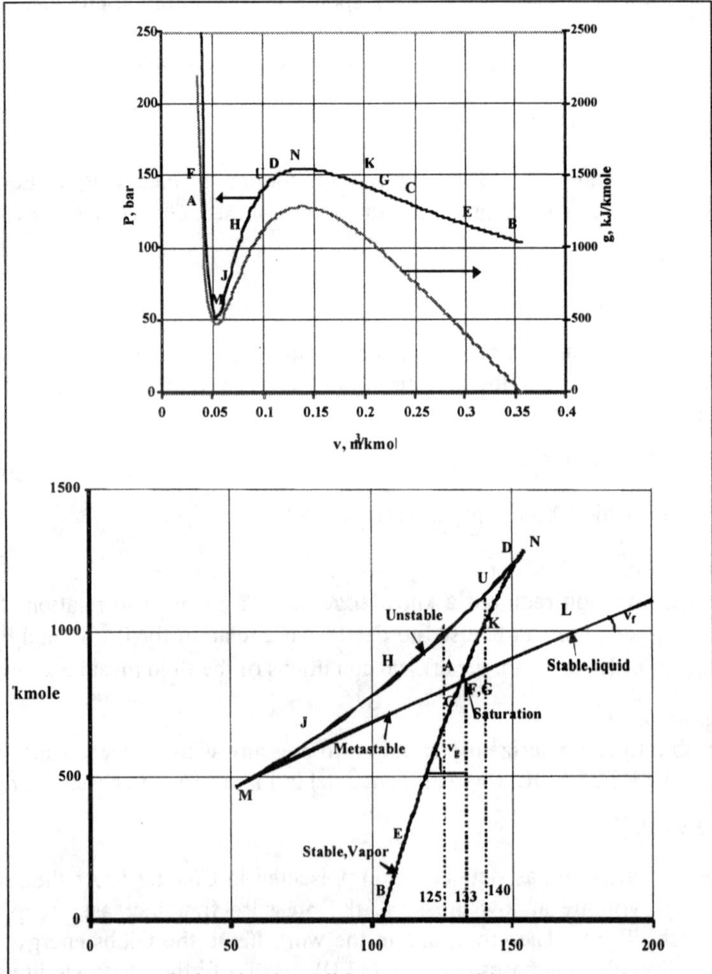

Figure 15. Variation in the Gibbs energy with respect to pressure for water vapor modeled as an RK gas at 593 K.

$\partial g/\partial P = v_g$ along the vapor–like curve. (Further discussion of this behavior is contained in Chapter 10.) A geometrical interpretation of the equality of $g_f = g_g$ follows. Since, dg = d(Pv) – Pdv,

$$\int_{g^\alpha}^{g^\beta} dg = P^{sat}(v^\alpha - v^\beta) - \int_{v^\alpha}^{v^\beta} Pdv = 0. \quad (117)$$

Since, at equilibrium, $g^\alpha(T, P^{sat}) = g^\beta(T, P^{sat})$,

$$P^{sat}(v^\alpha - v^\beta) = \int_{v^\alpha}^{v^\beta} P(T,v)\, dv = 0.$$

If $v^\alpha = v_g$, $v^\beta = v_f$, the LHS of this relation represents the area FCEDGMF in Figure 16, while the RHS represents the area FCEDGCMAF. This implies that the area FAMF = area MCGM. The equality of the two areas is called Maxwells equal area rule

Apply Eq. (115) for the two phases α and β. At the equilibrium condition,

$$(g^\alpha(T,v^\alpha) - g_{ref}(T,v_{ref})) = ((RTv^\alpha/(v^\alpha - b) - (a/T^{1/2})/(v^\alpha+b)) - (RTv_{ref}/(v_{ref} - b)$$

$$- (\mathbf{a}/T^{1/2})/(v_{ref}+b))) - RT\ln((v^\alpha - b)/(v_{ref} - b))$$

$$+ (\mathbf{a}/(T^{1/2}b))(\ln((v^\alpha(v_{ref}+b))/(v_{ref}(v^\alpha+b)))), \text{ and,} \quad (118a)$$

$$(g^\beta(T,v^\beta) - g_{ref}(T,v_{ref})) = ((RTv^\beta/(v^\beta - b) - (a/T^{1/2})/(v^\beta + b)) - (RTv_{ref}/(v_{ref} - b)$$

$$- (\mathbf{a}/T^{1/2})/(v_{ref} + b))) - RT\ln((v^\alpha - b)/(v_{ref} - b))$$

$$+ (\mathbf{a}/(T^{1/2}b))(\ln((v^\beta(v_{ref}+b))/(v_{ref}(v^\beta+b))). \quad (118b)$$

Subtracting the first of Eqs. (118a) from Eq.(118b), and using the condition $g^\alpha = g^\beta$, after simplification we obtain the expression

$$(RTv^\beta/(v^\beta - b) - (\mathbf{a}/T^{1/2})/(v^\beta + b)) - (RTv^\alpha/(v^\alpha - b) - (a/T^{1/2})/(v^\alpha + b))$$

$$- RT\ln((v^\beta - b)/(v^\alpha - b)) - (a/(T^{1/2}b))\ln((v^\beta(v^\alpha + b))/(v^\alpha(v^\beta + b))) = F = 0. \quad (119)$$

Eq. (119) is satisfied at $P = P^{sat}$.

---

ee. Example 31

Determine the saturation pressure for water at 593 K.

Solution

We will use an iteration method. First, we assume that $P^{sat} =$ 150 bar, $\bar{a}$ = 142.64 bar m^6 $kmole^{-2}$ K^{-1}, b = 0.0211 m^3 $kmole^{-1}$. Using the RK state equation, v^α= 0.0418 m^3 $kmole^{-1}$, v^β = 0.174 m^3 $kmole^{-1}$. Therefore, Eq. (119) implies that F = 3.0123.

Now, we will assume that P^{sat} = 133 bars, so that v^α= 0.0427 m^3 $kmole^{-1}$, v^β = 0.237 m^3 $kmole^{-1}$, and F = 0.238 which is almost zero.

The saturation pressure can be identified in this manner.

Upon normalizing Eq.(119) we obtain the relation

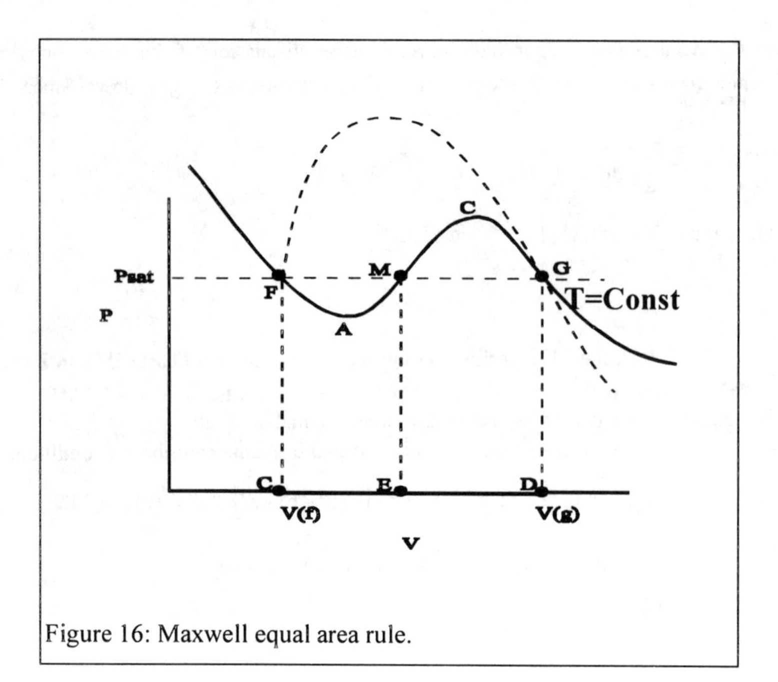

Figure 16: Maxwell equal area rule.

$$(T_R v_R'^{\beta}/(v_R'^{\beta} - 0.08664) - 0.4275/(T_R^{1/2}(v_R'^{\beta} + 0.08664)))$$

$$- (T_R v_R'^{\alpha}/(v_R'^{\alpha} - 0.08664) - 0.4275/(T_R^{1/2}(v_R'^{\alpha} + 0.08664)))$$

$$= T_R \ln((v_R'^{\beta} - 0.08664)/(v_R'^{\alpha} - 0.08664))$$

$$- (4.9342/(T_R^{1/2})) \ln((v_R'^{\beta} (v_R'^{\alpha} + 0.08664))/(v_R'^{\alpha} (v_R'^{\beta} + 0.08664))).$$

This equation can be applied to determine the behavior of P_R^{sat} with respect to T_R for RK fluids.

b. Approximate Solution

The RK relation suggests that (cf. Chapter 6)

$$(dP_R/dT_R)_{v_R} = 1/(v_R' - 0.08664) + 0.2138/(T_R^{3/2} v_R'(v_R' + 0.08664)). \quad (120)$$

Along the critical isometric curve $v'_{R,C} = 0.3333$, so that

$$(dP_R/dT_R)_{v_R = v_{R,C}} = 4.054 + 1.528/T_R^{3/2}. \quad (121)$$

At the critical point, $T_R = 1$, and

$$(dP_R/dT_R)_c = 5.582.$$

The behavior of many substances can be described by the relation

$$\ln P_R = A - B/T_R.$$

Differentiating this relation, $d \ln(P_R)/d(1/T_R) = - B$. At the critical point $P_R = T_R = 1$, so that $A = B = 5.582$. Therefore,

$$\ln P_R = 5.582(1 - 1/T_R). \quad (122)$$

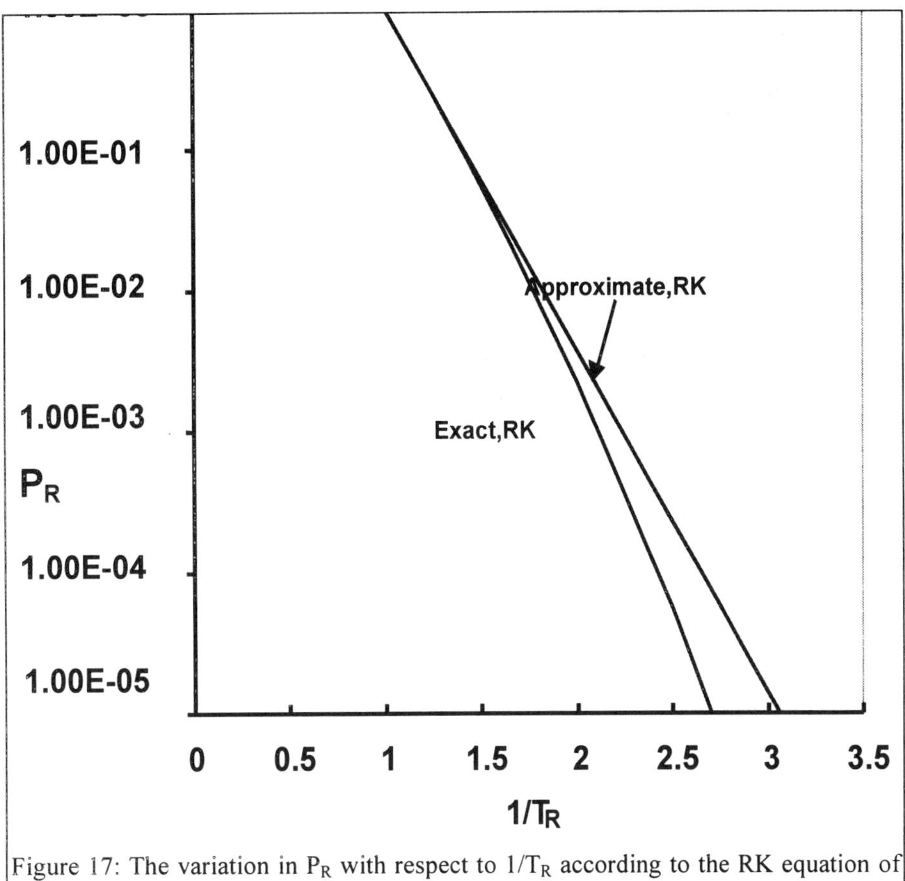

Figure 17: The variation in P_R with respect to $1/T_R$ according to the RK equation of state.

The approximate and exact solutions are compared in Figure 17, and found to be in excellent agreement in the region $T_R > 0.5$. In general, for simple fluids, the experimental curvefit yields

$$\ln P_R \approx 5.31(1 - 1/T_R). \tag{123}$$

(According to this equation, Pitzer factor P_R^{sat} at $T_R = 0.7$ is 0.651; w = - 0.814).

3. Heat of Vaporization

The heat of vaporization h_{fg} can be determined by directly applying a real gas state equation. Since

$$ds = c_v\, dT/T + (\partial P/\partial T)_v\, dv,$$

at a specified vaporization temperature,

$$ds = (\partial P/\partial T)_v\, dv, \text{ i.e.,}$$

$$s_{fg} = s_g - s_f = \int(\partial P/\partial T)_v\, dv.$$

The value of the RHS of the relation can be determined using a real gas state equation. Furthermore, since

$$dh = T\, ds + v\, dP,$$

integrating this expression from the saturated fluid state to the saturated vapor state at the vaporization temperature,

$$h_{fg} = h_g - h_f = T(s_g - s_f) + \int v\, dP. \quad (124)$$

Recall that $\int v dP = g_f - g_g = 0$. Therefore,

$$h_{fg} = T(s_g - s_f) = T s_{fg} = T\int(\partial P/\partial T)_v\, dv.$$

ff. *Example 32*

Obtain a relation for s_{fg} and h_{fg} in terms of T, v_f and v_g for a gas whose behavior can be described by the RK equation of state. Determine s_{fg} and h_{fg} for water using this state equation at 593 K. (The previous example showed that, for water at this temperature, P^{sat} = 133 bar, $v_g = 0.237\ m^3\ kmole^{-1}$, and $v_f = 0.0427\ m^3\ kmole^{-1}$.)

Solution

Recall that

$$s_{fg} = \int(\partial P/\partial T)_v\, dv = \int(R/(v-b) + (1/2)(a/T^{3/2})/(v(v+b)))\, dv. \quad (A)$$

Integrating this expression between the limits of v_f and v_g we obtain the relation

$$s_{fg}/R = \ln((v_g-b)/(v_f-b)) + (1/2)(a/(RbT^{3/2}))\ln((v_g(v_f+b)/(v_f(v_g+b))). \quad (B)$$

Applying Eqs. (112) and (B),

$$h_{fg,R} = h_{fg}/(RT_c) = T_R s_{fg}/R = T_R \ln((v_{R,g}'-0.08664)/(v_{R,f}'-0.08664))$$

$$+ (2.4671/T_R^2)\ln((v_{R,g}'(v_{R,f}'+0.08664)/(v_{R,f}'(v_{R,g}'+0.08644))). \quad (C)$$

For water at 593 K, P^{sat} = 133 bar, and v_g and v_f are, respectively, 0.237 and 0.0427 $m^3\ kmole^{-1}$. Using the values of $\bar{a}$ = 142.64 bar $m^6\ K^{0.5}\ kmole^{-2}$, b = 0.0211 $m^3\ kmole^{-1}$ in Eq. (B), we obtain

$$s_{fg}/R = \ln((0.2368 - 0.0211) \div (0.04234 - 0.0211))$$
$$+ 0.5 \times 142.64 \times 100 \div (8.314 \times 0.0211 \times 593^{1.5}))$$
$$\times \ln(0.2368 \times (0.04234 + 0.0211) \div (0.04234 \times (0.2368 + 0.0211)))$$
$$= 2.318 + 2.814 \ln(1.3758) = 3.216, \text{ i.e.,}$$

$s_{fg} = (8.314 \div 18.02) \times 3.216 = 1.4839\ kJ\ kg^{-1}\ K^{-1}$, and

$h_{fg} = T_{fg} = 593 \times 1.4839 = 880\ kJ\ kg^{-1}$.

According to the Steam Tables A-4A, the corresponding value of $h_{fg} = 1239\ kJ\ kg^{-1}$.

Remark

The real gas equation itself is approximate. The large error in the value obtained using the real gas equation is due to the derivative $(\partial P/\partial T)_v$, which produces divergent errors.

4. Vapor Pressure and the Clapeyron Equation

Determination of saturation pressure involves inherent errors due to errors in real gas state equations and critical properties. In this section we will not use real gas state equations; rather we will use the criterion of equality of differential of Gibbs function for vapor and liquid phases and knowledge of saturation pressure at a single temperature (or vice versa) to deduce saturation relations.

Inside the vapor dome $g_f = g_g$ along any isotherm (or isobar, FG in Figure 18). As the state of a fluid is changed inside the dome, the saturation temperature and pressure change in such a manner that

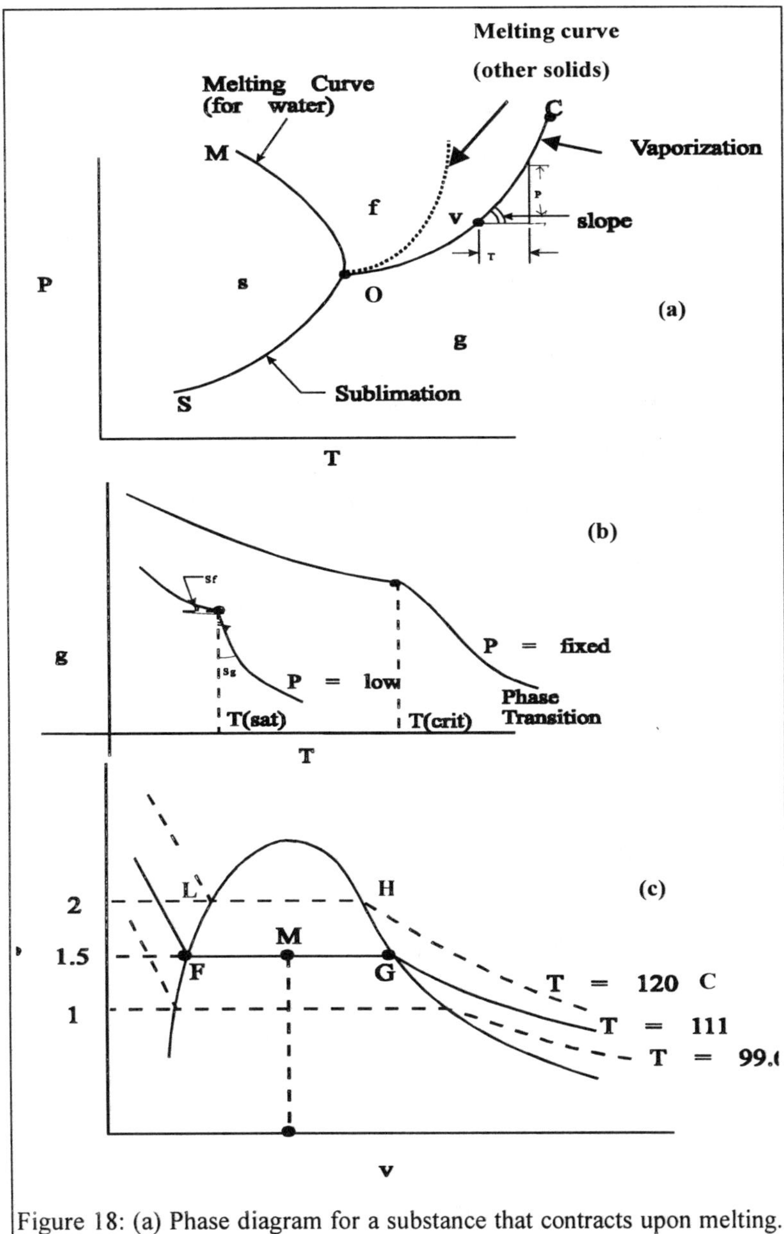

Figure 18: (a) Phase diagram for a substance that contracts upon melting. (b) the variation of g with respect to T.

dg_f along FL = dg_g along GH where (125)

For example, a change of g_f along FL= change of g_g along GH (Fig.18)

$$dg_f = -s_f\, dT + v_f\, dP, \text{ and } dg_g = -s_g\, dT + v_g\, dP. \qquad (126)$$

Using Eq. (125) in Eq. (126) we find that

$$(dP/dT) = s_{fg}/v_{fg}.$$

Since Tds + vdP = dh, at constant pressure ds = (dh/T) so that $s_{fg} = h_{fg}/T$. Therefore,

$$(dP/dT) = h_{fg}/Tv_{fg}. \quad (127)$$

Equation (127) is called the Clapeyron equation. Referring to Figure 18, P^{sat}= 15 at T= 111°C while P^{sat} = 2 bar at 120°C. This slope is related to h_{fg}, T and v_{fg} at T= 111^{0}C if ΔP and ΔT are small. Thus, in bar °C^{-1}, ΔP/ΔT = (2–1.5)/(120–111) = 0.56.

The slope (dP/dT) is represented by Eq. (127), and it varies, since h_{fg} and v_{fg} vary along OC. It predicts the variation of saturation pressure with respect to temperature along the vaporization curve OC in the phase diagram illustrated in Figure 18. The points O and C, respectively, represent the triple point and the critical point. Using the relation Pv = ZRT in Eq. (127), we obtain the relation

$$dP/dT = h_{fg}P/(T^2RZ_{fg}),$$

where $Z_{fg} = Z_g - Z_f$. We may rewrite this expression in the form

$$d\ln(P)/d(1/T) = -h_{fg}/(RZ_{fg}) \quad (128)$$

Equation (128) can be generalized to any phase transition from phase α to phase β phase, i.e.,

$$d\ln(P)/d(1/T) = -h_{\alpha\beta}/(R(Z_\beta - Z_\alpha)). \quad (129)$$

Since both h_{fg} and Z_{fg} both decrease with an increase in the temperature, the RHS of Eq. (129) is a weak function of temperature and can be generally treated as a constant. In that case,

$$\ln(P) = -(1/T)\, h_{fg}R/Z_{fg} + C. \quad (130)$$

Using the equalities P = P^{sat}_{ref} and T= T_{ref}, to determine the constant,

$$(P^{sat}/P^{sat}_{ref}) = \exp((h_{fg}/(RZ_{fg}))(1/T_{ref} - 1/T)). \quad (131)$$

a. Remarks

The slope dP/dT in Eq. (127) must be the same near the saturated liquid and the saturated vapor states.

For the VW equation of state, the slope along the isochoric curve at the critical point, $(\partial P/\partial T)_{v_c,T_c} = R/(v_c-b)$. Using the value b = (1/3) v_c, $(\partial P/\partial T)_{v_c,T_c} = (3/2)(R/v_c) = (3/2)(P_c/Z_cT_c)$. In dimensionless form, the slope $(\partial P_R/\partial T_R)_{v_c} = 4$, since $Z_c = 3/8$ in context of the VW state equation. Similarly, using the RK state equation $(\partial P/\partial T)_{v_c,T_c} = 1.861\, P_c/(Z_cT_c)$, i.e., $(\partial P_R/\partial T_R)_{v_c} = 5.583$. Using values for water, i.e., T_c = 647.3 K and P_c = 220.9 bars, in the expression obtained using the RK state equation, $(\partial P/\partial T)_{v_c}$ = 2.433 bar K^{-1}.

A first order phase transition is the one for which the property g remains continuous from a phase α to another phase β. In general, the slopes $(\partial g/\partial T)_\alpha \neq (\partial g/\partial T)_\beta$, and $(\partial g/\partial P)_\alpha \neq (\partial g/\partial P)_\beta$. The properties v, u, h, s, and a are discontinuous (cf. Figure 19). During a second order phase transition (e.g., critical point), the two phases cannot be distinguished and the values of g and the first derivatives are continuous. However, in this case the second derivatives are discontinuous, for instance, at the critical point. Since $\partial g_f/\partial T = -s_f$ and $\partial g_g/\partial T = -s_g$, $\partial g_f/\partial P = v_f$ and $\partial g_g/\partial P = v_g$, at the critical point $\partial g_f/\partial T = \partial g_g/\partial T$, $\partial g_f/\partial P = \partial g_g/\partial P$, and $g_f = g_g$. However, the value of $\partial^2 g/\partial P^2 = (\partial v/\partial P)_T$ is very large, since $\partial P/\partial v = 0$ at the critical point. Alternately, the transition from phase α to β is a first order transition if $h_{\alpha\beta} \neq 0$. The transition at the critical point is a second order transition and $h_{\alpha\beta} = 0$ there.

A fluid containing one component exhibits a single critical point.

In context of Eq. (128), we can evaluate the value of h_{fg}/RZ_{fg} at low pressures, since $Z_g \gg Z_f$, i.e., $Z_{fg} \approx Z_g$. Assuming ideal gas behavior for the vapor, $Z_g = 1$. Since $v_f \ll v_g$, then $Z_f \ll Z_g$, Eq.(131) transforms into the relation

$$(P^{sat}/P^{sat}_{ref}) = \exp\left((h_{fg}/R)(1/T_{ref} - 1/T)\right), \tag{132}$$

where T_{ref} denotes the saturation temperature at the reference pressure $P^{sat}_{ref}(T_{ref})$. Equation (132) is called the Clausius Clapeyron equation and it also has the form

$$\ln P = A - B/T, \text{ where } A = \ln P^{sat}_{ref} + h_{fg}/RT_{ref} \text{ and } B = h_{fg}/R. \tag{133}$$

For water, A = 13.082 and B = 4962 K. The slope of ln(P) with respect to T^{-1} is proportional to heat of vaporization, and its value is nearly constant except in the vicinity of the critical point

The Antoine equation is a modified form of Eq. (133), namely,

$$\ln P = A - B/(T+C). \tag{134}$$

One can use experimental saturation data at three different temperatures, evaluate A, B and C and tabulate them for computer applications. An empirical relation that describes fluid behavior is of the form:

$$\ln P^{sat} = A + B/T + C \ln T + DT.$$

(For water, $\ln P^{sat} = 12.58 + (-4692)/T + 0.0124 \ln T$, where the value of P^{sat} is in bar and that of temperature in K.)

Differentiating this empirical relation and equating it with Eq. (126) we can determine the h_{fg}, provided v_{fg} is known. (One solution is to consider $v_g \approx RT/P$ at low pressures and $v_f \approx 0$.)

Applying Eq. (127) during any phase change

$$(dP/dT) = h_{\alpha\beta}/(Tv_{\alpha\beta}) = (h_\beta - h_\alpha)/(T(v_\beta - v_\alpha)). \tag{135}$$

Let α denote the initial solid phase s and β the final liquid phase f during the melting of a solid so that $v_\beta - v_\alpha = v_{sf} = v_f - v_s$. In the case of ice, $v_f < v_s$ and $h_{sf} = (h_f - h_s) > 0$, i.e.,

$$(dP/dT) < 0.$$

As the pressure increases, the ice melting point T_{MP} decreases, as illustrated in Figure 18(a). In case of a sublimation process, Eq. (135) suggests that

$$A = \ln P^{sat}_{ref}(T_{ref}) + h_{sg}/RT_{ref}, \text{ and } B = h_{sg}/R. \tag{136}$$

gg. Example 33

The boiling point of water T_{ref} at P_{ref} = 0.1 MPa is 99.6°C. What is the saturation (or boiling) temperature at P = 0.5 MPa? Assume that h_{fg} = 2258 kJ kg^{-1}.

Solution

The value for water of R = 8.315/18.02 = 0.461 kJ kg^{-1} K^{-1}. Applying Eq. (132), $0.5 \div 0.1 = \exp((2258 \div 0.461) \times (372.6^{-1} - T^{-1}))$, i.e.,

T = 425 K (the steam tables (A-4) provide a value of 424.6 K).

hh. Example 34

Obtain a relation for h_{fg}/RT_c with respect to P_R for a substance using the RK state equation in terms of $v_{R,f}'$, $v_{R,g}'$, and T_R.

Solution

Consider the relation

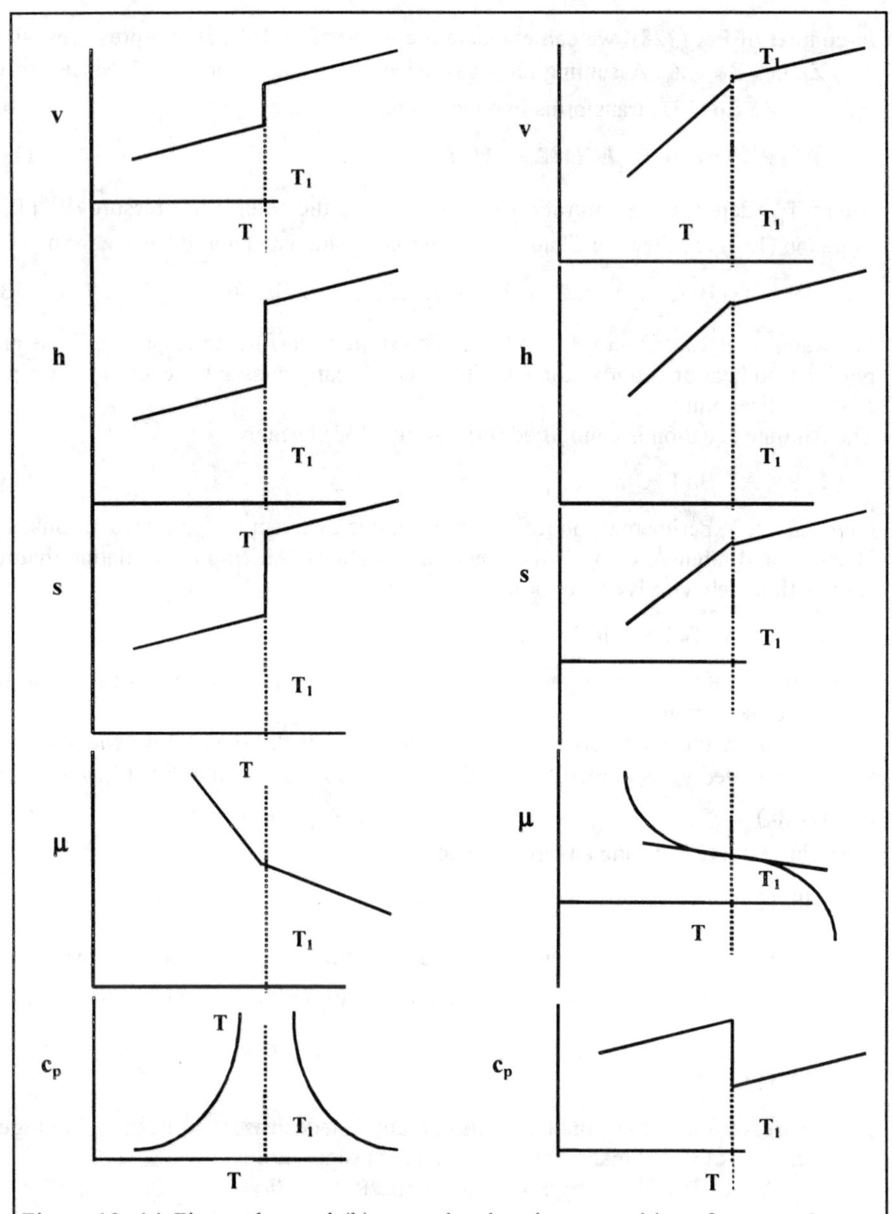

Figure 19: (a) First order, and (b) second order phase transitions for several properties.

$h_{fg}/RT_c = (d\,P^{sat}_{ref}/dT_R)T_R(v_{R,g}{'} - v_{R,f}{'}).$ (A)

During phase change $\int (P(T,v))dv = P^{sat}\,(v_g - v_f)$, where P(T,v) is given by any real gas state equation. Differentiating with temperature,

$(dP^{sat}/dT)\,(v_g - v_f) - \int(\partial P/\partial T)\,dv = 0,$ (B)

$(dP_R{}^{sat}/dT_R)(v_{R,g}{'} - v_{R,f}{'}) - \int(\partial P_R/\partial T_R)\,dv_R{'} = 0,$ and (C)

$P_R = T_R/(v_R{'} - 0.0864) - 0.4275/(T_R{}^2\,v_R{'}\,(v_R{'} + 0.0864)).$ (D)

Therefore, using Eq. (D) in Eq. (C) for $(\partial P_R/\partial T_R)$ and then applying Eq. (127) in reduced form

$$h_{fg}/(RT_c) = T_R \ln ((v_{R,f}' - 0.0864)/(v_{R,f}' - 0.0864))$$

$$+ (2.4671/T_R^{3/2}) \ln((v_{R,g}'(v_{R,f}' + 0.0864))/((v_{R,g}' + 0.0864)v_{R,f}')). \quad \text{(E)}$$

If v_g and v_f are known, h_{fg} can be determined.

5. Empirical Relations

a. Saturation Pressures

Correcting for higher order effects, it is possible to write

$$\ln P_R^{sat} = (\ln P_R)^{(o)} + \omega(\ln P_R)^{(1)}. \quad (137)$$

In general, for several hydrocarbons,

$$(\ln P_R)^{(o)} = 5.92714 - (6.09648)/T_R - 1.28862 \ln T_R + 0.169347\, T_R^6, \text{ and}$$

$$(\ln P_R)^{(1)} = 15.2518 - (15.6875)/T_R - 13.4721 \ln T_R + 0.43577\, T_R^6.$$

The following relations are also applicable, namely,

$$(\ln P_R)^{(o)} = 5.92714\,(1 - 1/T_R), \text{ and}$$

$$(\ln P_R)^{(1)} = 7.49408 - 11.18177\, T_R^3 + 3.68769\, T_R^6 + 17.92998 \ln T_R.$$

b. Enthalpy of Vaporization

The Clausius relation can be written in terms of reduced variables, i.e.,

$$h_{fg,R} = h_{fg}/RT_c = (d\, P_{ref}^{sat}/dT_R)\, T_R\, (v_{R,g}' - v_{R,f}') \quad (138)$$

An experimental correlation for $\ln P_R$ with respect to $1/T_R$ is available, i.e.,

$$(1/P_R)\, dP_R/dT_R = C/T_R^2,$$

where C = 5.31 from experiments and 5.582 for an RK gas. Since

$$dP/dT = h_{fg}\, P/(T^2 R Z_{fg}), \text{ i.e., } dP_R/dT_R = (h_{fg}/RT_c)\, P_R/(T_R^2\, Z_{fg}),$$

$$h_{fg}/RT_c = C\, Z_{fg}. \quad (139)$$

A plot of $\ln(P_R)$ with respect to $1/T_R$ is linear, and has a slope proportional to h_{fg}. The higher the critical temperature, the higher this slope. If we assume that C= 5.31, then at the normal boiling point $(T_{NB})_{P=1}$ Eq.(123) yields,

$$-\ln P_c = 5.31\,(1 - T_c/T_{NB}),$$

where T_{NB} denotes the normal boiling point in K and P_c in bars. Hence,

$$T_c/T_{NB} = 1 + (1/5.31) \ln P_c = 1 + \ln (P_c^{\,0.179}),\ P_c \text{ in bars}$$

The dependence of T_c/T_{NB} on P_c is a weak one so that $T_c/T_{NB} \approx$ constant. Typically, the ratio T_c/T_{NB} has values in the range 1.5–1.8. Manipulating the above two relations, we obtain

$$h_{fg}/RT_{NB} \approx 5.31 + \ln P_c.$$

In general, for most nonpolar organic liquids, $(h_{fg,NB}/RT_{NB}) \approx$ 8–10. Other formulas have the forms

$h_{fg}(T)/h_{fg,ref}(T_{ref}) = ((1 - T_R)/(1 - T_{R,ref}))^{0.38}$ (Watson's correlation), or

$$h_{fg}(T)/h_{fg}(T_{TP}) = b_1\Psi^{1/3} + b_2\Psi^{0.79} + b_3\Psi^{1.2o8} + b_4\Psi + b_5\Psi^2 + b_6\Psi^3,$$

where $\Psi = (T_c - T)/(T_c - T_{TP})$, $b_1 = 0.60176$, $b_2 = 3.45913$, $b_3 = 4.62671$, $b_4 = -6.89614$, $b_5 = -1.10643$, and $b_6 = 0.31522$.

6. Saturation Relations with Surface Tension Effects

We have presented a physical interpretation of surface tension in Chapter 5. We now discuss a more rigorous explanation of the phenomena based on the energy minimum principle. We first write an expression for energy change for a bubble (called the embryo phase β) that has just appeared in a single component liquid (called the mother phase α). Surface tension forces exist between the vapor and liquid phases within a thin layer γ that lies adjacent to bubble and has a thickness δ, volume V^γ, pressure P^γ, and entropy S^γ. As $\delta \to 0$, then the volume V^γ, appears like a surface. Now, consider an isolated system containing three subsystems α, β and γ. Assume that the state of the combined system ($\alpha+\beta+\gamma$) lies infinitesimally away from equilibrium. Consequently, infinitesimal changes in volume occur in the α and β phases, which also change the volume of the subsystem γ.

For the open subsystem α, the infinitesimal energy changed $U^\alpha = T^\alpha\, dS^\alpha - P^\alpha\, dV^\alpha + \mu^\alpha\, dN^\alpha$,

where dN^α denotes the moles transferred from it. Similarly, for the β phase,

$$dU^\beta = T^\beta\, dS^\beta - P^\beta\, dV^\beta + \mu^\beta\, dN^\beta.$$

We assume that the surface tension region "γ" has uniform properties that are distinct from the α and β phases. Consequently,

$$dU^\gamma = T^\gamma\, dS^\gamma - P^\gamma\, dV^\gamma - (-\sigma')\, dV^\gamma + \mu^\gamma\, dN^\gamma.$$

For sake of illustration, consider a section of a cylindrical bubble of length L that has an arbitrary arc length ds and a surface tension layer thickness δ. Its volume $dV^\gamma \approx \delta L\, ds$. Let σ' denote the tensile stress and P^σ the compressive stress exerted on the bubble. As $\delta \to 0$, $\sigma' dV^\sigma = \sigma'\, \delta L\, ds \approx \sigma\, ds\, L = \sigma\, dA$ where $\sigma = \sigma'\, \delta$, and

$$dU^\gamma = T^\gamma\, dS^\gamma - P^\gamma\, dV^\gamma + \sigma\, dA + \mu^\gamma\, dN^\gamma.$$

For the combined system

$$dU = dU^\alpha + dU^\beta + dU^\gamma = T^\alpha dS^\alpha - P^\alpha dV^\alpha + \mu^\alpha dN^\alpha + T^\beta dS^\beta - P^\beta dV^\beta + \mu^\beta dN^\beta$$

$$+ T^\gamma dS^\gamma - P^\gamma dV^\gamma + \mu^\gamma dN^\gamma + \sigma\, dA. \quad (140)$$

(If the change in the properties of the region γ are neglected, $dU = T^\alpha dS^\alpha - P^\alpha dV^\alpha + \mu^\alpha dN^\alpha + T^\beta dS^\beta - P^\beta dV^\beta + \mu^\beta dN^\beta + \sigma dA$.) The entropy of the isolated system cannot change, and for constant S, V and M,

$$dS = dS^\alpha + dS^\beta + dS^\gamma = 0, \text{ i.e., } dS^\alpha = -(dS^\beta + dS^\gamma) \quad (141)$$

Likewise,

$$dV^{\alpha} = -(dV^{\beta} + dV^{\gamma}), \text{ and } dN^{\alpha} = -(dN^{\beta} + dN^{\gamma}). \quad (142)$$

Using Eqs. (140)-(141) and the energy minimum condition dU = 0, we obtain the relation

$$dS^{\beta}(T^{\beta}-T^{\alpha}) + dS^{\gamma}(T^{\gamma}-T^{\alpha}) - dV^{\beta}(P^{\beta}-P^{\alpha}) - dV^{\gamma}(P^{\gamma}-P^{\alpha})$$

$$+ dN^{\beta}(\mu^{\beta}-\mu^{\alpha}) + dN^{\gamma}(\mu^{\gamma}-\mu^{\alpha}) + \sigma\, dA = 0. \quad (143)$$

Arbitrary changes in dS^{β} and dS^{γ} must satisfy Eq. (143) so that $T^{\beta} = T^{\alpha} = T^{\gamma} = T$, which is the thermal equilibrium condition, and the same argument regarding dN^{β} and dN^{γ} implies that $\mu^{\alpha} = \mu^{\beta} = \mu^{\gamma} = \mu$, which is the phase equilibrium condition. Therefore,

$$dV^{\beta}(P^{\beta} - P^{\alpha}) + dV^{\gamma}(P^{\gamma} - P^{\alpha}) - \sigma\, dA = 0.$$

The compressive pressure within the thin region on which surface tension forces occur varies from P^{β} to P^{α}. If we include dV^{γ} with the mother phase volume, then $P^{\alpha} = P^{\gamma} = P$ (which is the condition related to vaporization from the mother phase α, which is a liquid). Omitting the subscript for the mother phase α,

$$dV^{\beta}(P^{\beta} - P) + 0 - \sigma\, dA, = 0, \text{ i.e., } (P^{\beta} - P) = \sigma\, dA/dV^{\beta}. \quad (144)$$

where β is the embryo phase. Similarly, for a condensation process from β (which is mother now) to α (embryo) and now combining γ with β, $P^{\beta} = P$, and $dV^{\beta} = -dV^{\alpha}$, so that Eq. (144) can be written in the form

$$(P^{\alpha} - P) = \sigma\, dA/dV^{\alpha}.$$

Generalizing, if P^E is the pressure of embryo phase (say, vapor for a boiling process or a liquid drop in a condensation process) and P is the pressure of mother phase (say, liquid in a vaporization process or vapor in condensation process), then

$$(P^E - P) = \sigma\, dA^E/dV^E.$$

Pressure of embryo- Pressure of mother) = Surface tension × surface area change with volume change of the embryo phase. Considering a spherical embryo of radius a,

$$A^E/dV^E = d(4\pi a^2)/d(4/3\pi a^3) = 2/a, \text{ and}$$

$$(P^E - P) = \sigma\, dA^E/dV^E = 2\sigma/a$$

a. Remarks

The pressure P denotes the mother phase pressure in case of both evaporation and condensation. In general, for spherical drops, the difference between the pressures inside curved surface and in the bulk phase equals 2 σ/a.

The vapor bubbles may be generated sometimes at $T > T^{sat}(P)$ (cf. Chapter 10). The liquid at this state is called superheated liquid. However, tables of properties are not available at this condition. Hence, we have to generate properties of superheated liquid in terms of saturation properties. Thus, a general derivation is presented below for both phases. If the surface tension $\sigma \neq 0$, the pressure in vapor and liquid phases is not the same, even though the system is isothermal. Supposing that β is the vapor phase and α is the liquid mother phase, at phase equilibrium,

$$\mu^{\beta}(P^{\beta}, T) = \mu^{\alpha}(P^{\alpha}, T) \text{ or } f^{\beta}(P^{\beta}, T) = f^{\alpha}(P^{\alpha}, T) \quad (145)$$

Using the Poynting correction,

$$\ln \frac{f^{\beta}(P^{\beta},T)}{f(P^{sat},T)} = \int_{P^{sat}(T)}^{P^{\beta}} \frac{v^{\beta}}{RT} dP \qquad (146)$$

Since $f(P^{\beta}, T) = \phi(P^{\beta}, T)P^{\beta}$ and $f(P^{sat}(T), T) = \phi(P^{sat}(T), T)P^{sat}(T)$, the above equation assumes the form

$$f^{\beta} = \phi^{\beta}(P^{\beta},T)P^{\beta} = \phi(P^{sat},T) \quad P^{sat}(T)(\exp[\int_{P^{sat}(T)}^{P^{\beta}} \frac{v^{\beta}}{RT} dP] \qquad (147)$$

Similarly

$$f^{\alpha} = \phi^{\alpha}(P^{\alpha},T)P^{\alpha} = \phi(P^{sat},T) \quad P^{sat}(T)(\exp[\int_{P^{sat}(T)}^{P^{\alpha}} \frac{v^{\alpha}}{RT} dP] \qquad (148)$$

Therefore, at equilibrium for any fluid,

$\phi(P^{\beta}, T)P^{\beta} = \phi(P^{\alpha}, T)P^{\alpha}$.

Then if β is an ideal gas with $f^{\beta} = P^{\beta}$ and using this result in Eq. (148), and since $\phi(P^{sat}(T), T) = 1$ for ideal gas behavior, then Eq. (148) simplifies to

$$P^{\beta} = P^{sat}(T) \exp(v^{\alpha}(P^{\alpha} - P^{sat}(T))/RT). \qquad (149)$$

Omitting the superscript for the mother phase and using $\beta = E$ (e.g., vapor embryo phase)

$$P^{E} = P^{sat}(T) \exp(v(P - P^{sat}(T))/RT) \qquad (150)$$

which is also known as the Kelvin equation. Typically if P of mother phase is known, then P^E (e.g., vapor bubble) is determined from Eq. (150). Then using $P^E - P = 2\sigma/a$, then the size "a" at equilibrium can be determined.
If the embryo is a liquid drop (α) and the mother phase is vapor (β) with vapor as an ideal gas, $v^{\beta} = RT/P$, and $v^{\alpha} \approx$ constant; we obtain the Eq.(149) again.
Replacing α by E (now liquid), and omitting subscript β for mother phase

$$P = P^{sat}(T) \exp(v^{E}(P^{E} - P^{sat}(T))/RT)$$

Where $P^E - P = 2\sigma/a$.

$$P = P^{sat}(T) \exp(v^{E}(P + 2\sigma/a - P^{sat}(T))/RT) \qquad (151)$$

If $P - P^{sat}(T) \ll 2\sigma/a$, then

$$P = P^{sat}(T) \exp(v^{E} 2\sigma/(aRT)) \qquad (152)$$

If $\sigma = 0$, then $P^E = P + 2\sigma/a = P$
In a multicomponent system, the Lagrange multiplier method described in Chapter 3 can be used to show that for a component k, $\mu_k^{\alpha} = \mu_k^{\beta} = \mu_k^{\gamma} = \mu_k$.

---

ii. *Example 35*

A water bubble has a radius of 10^{-4} m, the surface tension $\sigma = 7\times10^{-5}$ kN m^{-1}. Calculate the embryo (vapor) pressure, if $P_{mother} = 100$ kN m^{-2}.

Solution

Using the surface tension and radius correction,

$P_{embryo} = P_{mother} + (2\times7\times10^{-5} \div 10^{-4}) = 101.4$ kN m^{-2}.

---

b. Pitzer Factor from Saturation Relations

The following empirical relation can be used to describe the variation in the saturation pressure with respect to temperature, namely,

$$\omega = -1 - (\log_{10}(P_R^{sat}))_{T_R=0.7}. \quad (153)$$

K. THROTTLING PROCESSES

Fluids can be cooled in heat exchangers in which the cooling is limited by the two–phase nature of most coolants. Cooling can also be achieved by a sudden adiabatic expansion, called a throttling process. The Joule Thomson coefficient provides information about the extent of cooling during such a process.

1. Joule Thomson Coefficient

The application of the Joule Thomson Coefficient is illustrated through the following example (that is schematically described in Figure 20). Assume that you can plug a thick ceramic pipe with a porous sponge and flow a high–pressure gas through it. Significant pressure losses will occur without a consequential change in the kinetic or potential energies. From energy balance analysis for an open system we can show that the sum of the enthalpies and the kinetic and potential energies is constant, i.e., $h_i + ke_i + pe_i = h_e + ke_e + pe_e =$ constant. Here, the subscripts i and e, respectively, denote the inlet and exit. When the inlet and exit areas are equal, i.e., $A_i = A_e$, then $\rho_i v_i = \rho_e v_e$, where v denotes the flow velocity. For a negligible change in the potential energy, $h_i + ke_i = h_e + ke_i(\rho_i/\rho_e)^2$. Consequently, the temperature across the sponge changes as the pressure is decreased, as illustrated in Figure 20. In general, changes in the kinetic energy are not large and we can assume that $h_i = h_e$. (Although, strictly speaking, for compressible fluids $\rho_e \neq \rho_i$ and, hence, $h_e \neq h_i$.)

Throttling is essentially an isenthalpic expansion process. The Joule Thomson coefficient (μ_{JT}) is defined as,

$$\mu_{JT} = (\partial T/\partial P)_h \quad (154)$$

Since h = h(T,P),

$$dh = (\partial h/\partial T)_P\, dT + (\partial h/\partial P)_T\, dP = c_p\, dT + (\partial h/\partial P)_T\, dP.$$

For an isenthalpic process, dh = 0, i.e.,

$$\mu_{JT} = (\partial T/\partial P)_h = -(\partial h/\partial P)_T/c_P.$$

For an ideal gas, $\partial h_o/\partial P = 0$, so that the above equation can be re-written as

$$\mu_{JT} = -\,\partial((h - h_o)/c_p)/\partial P. \quad (155)$$

where $h - h_0$ is known for any specified equation of state (Eq. (66)) or from enthalpy correction charts (Figure B-3) As illustrated by curve "iIABe" in Figure 20(b), for inlet pressures $P_i < P_I$, the temperature decreases with a decrease in pressure and $\mu_{JT} > 0$. On the other hand, for a flow with the same enthalpy, if $P_i > P_I$, initially, $\mu_{JT} < 0$ and temperature increases with decrease in pressure. Later, when $P < P_I$, the temperature will fall. The point I Figure 20(b) is called the inversion point where μ_{JT} changes sign. If inlet temperature is changed at same Pi enthalpy changes and as such inversion points change as shown by inversion curve KIL.

---

jj. Example 36

The experimentally determined value of μ_{JT} for N_2 is 0.31 K bar^{-1} at–70°C and 5 MPa. If the gas is throttled from 6 MPa and –67°C to 4 MPa, what is the final exit temperature?

Solution

$\mu_{JT} = (\Delta h/c_p)/\Delta P = \Delta T/\Delta P = (T_e - 203)/(40 - 60) = 0.31$ K bar^{-1}. Therefore, $T_e = -73.2$°C.

---

a. Evaluation of μ_{JT}

Recall that

$$dh = c_p dT + (v - T(\partial v/\partial T)_P)dP.$$

Since $dh \approx 0$ during throttling, we obtain the relation

$$\mu_{JT} = (dT/dP)_h = -(v - T(\partial v/\partial T)_P)/c_P = (T^2/c_P)\, \partial(v/T)/\partial T, \text{ or} \quad (156)$$

$$\mu_{JT} = (dT/dP)_h = -(1 - T\beta_P)/(c_P/v). \quad (157)$$

b. Remarks

For ideal gases, $v = RT/P$ so that $T(\partial v/\partial T) = v$ and, hence, $\mu_{JT} = 0$. There is no temperature change due to throttling for ideal gases.

For incompressible fluids $\partial v/\partial T = \beta_P = 0$, and $\mu_{JT} = -v/c_P$ has a negative value. Therefore, liquids generally heat up upon throttling.

If $v < T(\partial v/\partial T)_P$ or $T\beta_P < 1$, $\mu_{JT} > 0$, and vice versa.

At the inversion point, $\mu_{JT} = 0$. Inversion occurs when $T_{inv}\, \beta_P = (T/v)(\partial v/\partial T)_P = 1$. For a real gas $Pv = ZRT$, i.e., $\partial v/\partial T = ZR/P + (RT/P)\partial Z/\partial T$. If $(T/v)(\partial v/\partial T)_P = 1$, this implies that $ZRT/Pv + (RT^2/Pv)\partial Z/\partial T = 1$, or $(\partial Z/\partial T)_P = 0$.

For cooling to occur, the inlet pressure must be lower than the inversion pressure.

Cooling of a gas can also be accomplished using isentropic expansion. We define $\mu_s = (\partial T/\partial P)_s$ that is related to the temperature decrease due to the work delivered during an isentropic process. Recall that

$$ds = c_p\, dT/T - (\partial v/\partial T)_P\, dP.$$

For an isentropic process, $ds = 0$, and

$$\mu_s = T(\partial v/\partial T)_P/c_P = Tv\beta_P/c_P. \quad (158)$$

Dividing Eq.(157) by Eq.(158),

$$\mu_{JT}/\mu_s = 1 - (1/(T\beta_P)) \quad (159)$$

Equation (159) provides a relation to indicate relative degree of heating of the isentropic to isenthalpic throttling process.

For substances that expand upon heating $\beta_P > 0$ and, hence, $\mu_s > 0$. If $\beta_P T > 1$, then $\mu_{JT} < \mu_s$, i.e., the isentropic expansion results in greater cooling than isenthalpic expansion for the same pressure ratio.

The values of T_{inv}, β_P, β_T, and P^{sat} can be directly obtained from the state equations, while μ_{JT}, h, u, c_v, and other such properties depend both on the equations of state and

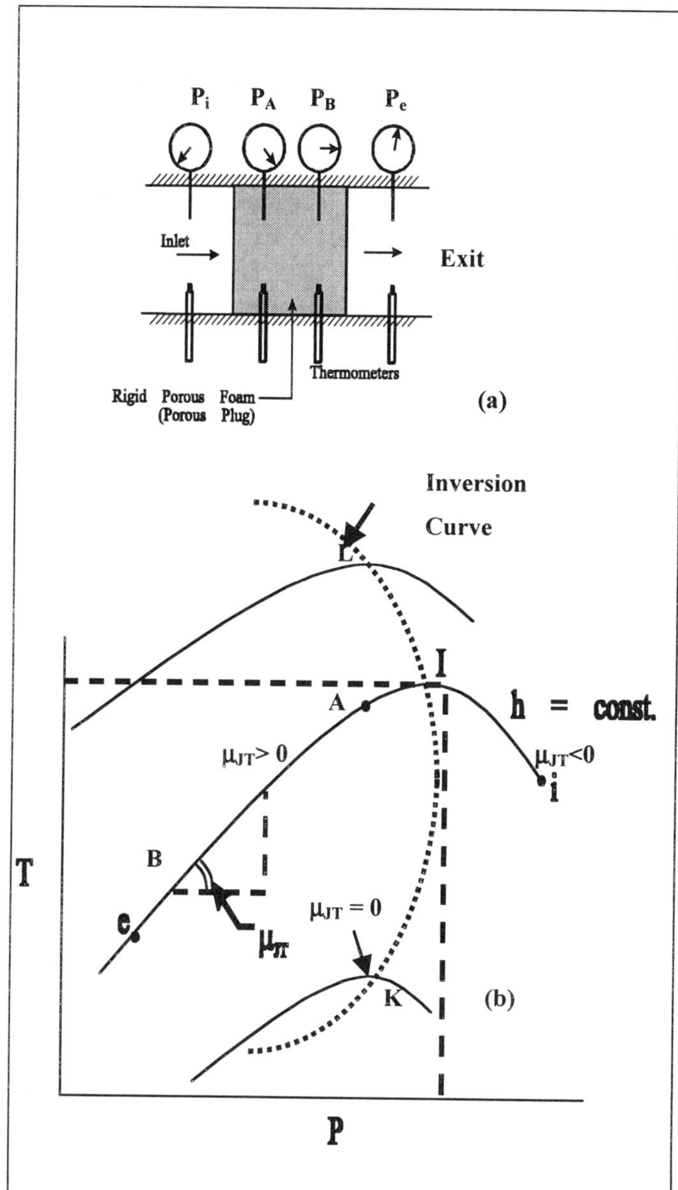

Figure 20: (a) Illustration of a throttling process. (b) The variation of temperature with respect to pressure during throttling.

the ideal gas properties. Therefore, the accuracy of the state equations of state can be directly inferred by comparing the predicted values of T_{inv}, β_P, β_T, and P^{sat} with experimental data.

The entropy change during an adiabatic throttling process equals the entropy generated. Therefore, its value must be positive, since throttling is an inherently irreversible process. We can determine ds by using the following relation:

$$dh = T\, ds + v\, dP.$$

Since $dh = 0$,

$$(\partial s/\partial P)_h = -(v/T). \tag{160}$$

During throttling, $dP_h < 0$ and $v/T > 0$; hence Eq.(160) dictates that $ds_h>0$. From second law for fixed mass adiabatic system, ds_h (= $\delta\sigma$) =$-(v/T)\ dP_h > 0$.

---

kk. Example 37

Obtain an expression for μ_{JT} for a VW gas in terms of v and T.

Solution

The Joule Thomson coefficient

$$\mu_{JT} = -(\partial h/\partial p)_T/c_p, \text{ where} \tag{A}$$

$$(\partial h/\partial p)_T = v - T(\partial v/\partial T)_p. \tag{B}$$

Using the VW equation of state

$$dP = R\,dT/(v-b) + (-RT/(v-b)^2 + 2\,\mathbf{a}/v^3)dv. \tag{C}$$

Since dP = 0,

$$(dv/dT)_p = (R/(v-b))/(RT/(v-b)^2 - 2\,\mathbf{a}/v^3) \tag{D}$$

Therefore,

$$\mu_{JT} = -(1/c_p)\,(v - ((T\,R/(v-b))/(RT/(v-b)^2 - 2\,\mathbf{a}/v^3))), \text{ i.e.,} \tag{E}$$

$$\mu_{JT} = -(v/c_p)(R\,T\,b\,v^2 - 2\,\mathbf{a}\,(v-b)^2)/(RTv^3 - 2\,\mathbf{a}\,(v-b)^2)). \tag{F}$$

Remarks

For many liquids $v \approx b$, i.e.,

$$\mu_{JT} \approx -v/c_p \text{ for liquids.}$$

Since $\mu_{JT} < 0$, incompressible liquids will heat up upon throttling.
In context of Eq.(F), if $b \ll v$, $(v-b)^2 \approx v^2$. Consequently, using Eq.(F),

$$\mu_{JT} = -(v/c_p)\,(RTbv^2 - 2\,\mathbf{a}\,v^2)/(RTv^3 - 2\,\mathbf{a}\,v^2)). \tag{G}$$

Dividing the denominator and numerator by RTv^2,

$$\mu_{JT} = -(v/c_p)(b - 2\mathbf{a}/RT)/(v - 2\mathbf{a}/RT) \tag{H}$$

If $v \gg (2\mathbf{a}/RT)$, e.g., for high temperature vapors,

$$\mu_{JT} \approx ((2\,\mathbf{a}\,/RT) - b)/c_p. \tag{I}$$

At low temperatures, $2a/RT \gg b$, and $\mu_{JT} > 0$, i.e., cooling occurs upon throttling. At higher temperatures, $\mu_{JT} < 0$ (i.e., the fluid is heated upon throttling).
In the limit $T\rightarrow\infty$ (when the fluid approaches ideal gas behavior), Eq. (F) implies that $\mu_{JT} = -b/c_p$, i.e., $\mu_{JT}c_p/v_c' = -b/v_c' = -1/8$. This limiting value is based on the real gas equation of state. However, for ideal gases, $\mu_{JT} = 0$, since $b\rightarrow 0$ for the ideal gas point mass molecules. Recall that b equals the geometrical free volume available for molecules to move around. The corresponding result using the RK equation is $\mu_{JT}\,c_P/v_c' \approx -b/v_c' \approx -0.08664$.

2. Temperature Change During Throttling

Recall from Chapter 2 that the internal energy of unit mass of any fluid can be changed by frictional process and by performing boundary deformation work (Pdv). For incompressible fluid, dv=0 and hence boundary deformation work is zero; thus "u" can change only with a frictionless process (e.g., flow of liquids through pipes) where the mechanical part of energy "vdP" is converted into internal energy using frictional processes. Thus a combination of both of the processes Pdv and vdP result in temperature change during the throttling process.

a. Incompressible Fluid

Assume that an incompressible fluid is throttled from a higher to a lower pressure under steady state conditions. Let us follow unit mass as it enters and exits the throttling device. Since dv=0, there is no deformation work and hence u cannot change due to Pdv. However the unit mass is pushed into the throttling device with a pump work of vP_i, while the same mass is pushed out with a pump work of vP_e. Thus there must have been destruction of mechanical energy from vp_i to vP_e which is converted into thermal energy (see Chapter 2) resulting in an increase of u and hence an increase of T during throttling. The energy increases by an amount $du = -vdP \approx -v(P_e - P_i)$. (Alternately dh =du+ pdv+ vdp = du +0+ vdp=0.) Also, recall that $\mu_{JT} \approx -v/c_P < 0$. Further throttling is inherently irreversible process and hence entropy always increases in adiabatic throttling process (e.g., increased T causes increased s).

b. Ideal Gas

Now consider a compressible ideal gas. Visualize the throttling process as a two step procedure. First the specific volume is maintained as though fluid is incompressible and the energy rises by the amount du = –vdP. This leads to gas heating. Now let the volume increase during the second stage (i.e expansion to low pressure). The volume increase cannot change the intermolecular potential energy, since the gas is ideal, but Pdv work is performed. This leads to decrease in the internal energy. The total energy change is du = – vdP – Pdv = – d(Pv), i.e., or du + d(Pv) = dh = 0. For ideal gases, $dh = c_p dT$, and, hence, dT = 0. In this case, the energy decrease by the Pdv deformation work equals the energy increase due to pumping (=-vdP)

c. Real Gas

In a real gas the additional work required to overcome the intermolecular attraction forces (or the increase in the molecular intermolecular potential energy, "ipe") must be accounted for. Then the temperature can decrease if increase in "ipe" is significant. Consider the relation

$$dh = du + d(Pv) = du + P\,dv + v\,dP.$$

Since $du = c_v\,dT + (T\,(\partial P/\partial T)_v - P)\,dv$,

$$dh = c_v dT + (T(\partial P/\partial T)_v - P)dv + Pdv + vdP. \quad (161)$$

The terms on the RHS of this equation, respectively, denote (1) the change in the temperature due to change in the molecular translational, vibrational, and rotational energies, (2) the intermolecular potential energy change, (3) the deformation work, and (4) the work required for pumping. During throttling, dh = 0, the intermolecular potential energy increases, the work deformation is positive, and dP < 0 so that vdP <0. In case if vdP ≈ 0, then it is apparent that the net change in the molecular translational, vibrational, and rotational energies is negative, i.e., dT < 0. Recall (15), i.e.,

$$\mu_{JT} = -(v - T\,(\partial v/\partial T)_P)/c_p = -\,v/c_p + \{T\,(\partial v/\partial T)_P\}/c_p.$$

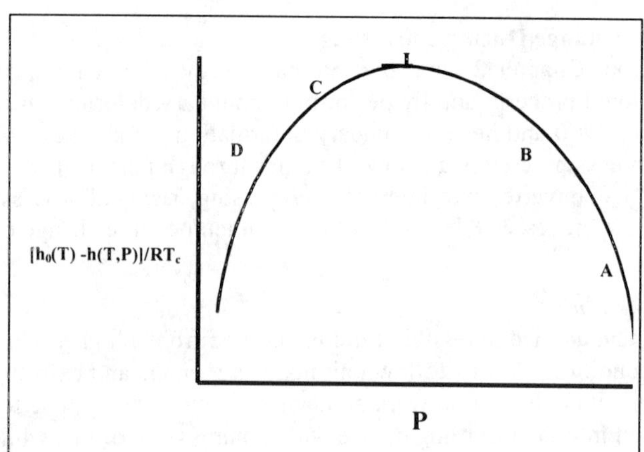

Figure 21: Use of the enthalpy correction charts to determine the inversion conditions.

The first term on the RHS of this expression represents the heating effect due to flow work, while the second term accounts for the entire Pdv work and the energy required to overcome the intermolecular forces. Generally, both the terms are important for fluids.

3. Enthalpy Correction Charts

The Joule Thomson coefficient $\mu_{JT} = -(\partial h/\partial P)_T/c_P$. However,

$$-(\partial h/\partial P)_T = -(\partial(h-h_o)/\partial P)_T - (\partial h_o/\partial P)_T = -(\partial(h-h_o)/\partial P)_T - 0 = \{\partial h_C/\partial P\}_T.$$

Therefore,

$$\mu_{JT} = \{\partial h_C/\partial P\}_T/c_P, \text{ and } \mu_{JT,R} = \mu_{JT}c_P P_c/RT_c = (\partial h_{R,C}/\partial P_R)_{T_R}. \quad (162)$$

where $h_{R,c}$ are given in enthalpy correction charts (Appendix, Figure B.3). The behavior of $h_{R,C}$ is illustrated in Figure 21. Its value increases with a decrease in the pressure at a specified value of T_R along the curve ABI. Consequently, $(\partial h_{R,c}/\partial P_R)_{T_R} < 0$, and $\mu_{JT} < 0$. On the other hand along curve ICD $(\partial h_{R,c}/\partial P_R)_{T_R} > 0$. Point I represents the inversion point. The temperature change during throttling that accompanies a pressure change can be determined using the enthalpy correction charts. Recall that $\Delta h \approx 0$, i.e., $h_2 - h_1 = 0$, i.e., $(h_{o2} - h_{C,2}) - (h_{o1} - h_{C,1}) = 0$. Assuming a constant specific heat and considering ideal gases, this leads to the relation $(T_2 - T_1) = (h_{C,1}(T_{R1}, P_{R1}) - h_{C,2}(T_{R2}, P_{R2}))/c_{P,o}$.

ll. Example 38

Determine μ_{JT} for N_2 at –70°C and 4.6 MPa using the enthalpy correction charts. Assume that $c_{P,o} = 1.039$ kJ kg^{-1} K^{-1}.

Solution

We will select pressures of 4.1 and 5.1 MPa (that lie in the vicinity of 4.6 MPa) at $T_R = 203/126.2 = 1.6$, and $P_{R,1} = 41/33.9 = 1.2$ and $P_{R,2} = 51/33.9 = 1.5$.

At these conditions $h_{R,C,1} = 0.667$, $h_{R,C,2} = 0.531$, $h_{C,1} = 8.314\times 126.2 = 700$ kJ kmole^{-1}, $h_{C,2} = 557$ kJ kmole^{-1} so that $\mu_{JT,R} = (0.667 - 0.531)/(1.5 - 1.2) = 0.453$.

Consequently, $(c_{P,o} - c_P)/R = (\partial h_{R,C}/\partial T_R)_{P_R = 1.35} \approx (\partial h_{R,C}/\partial T_R)_{P_R = 1.5} = (0.583 - 0.774)/0.2 = -0.955$.

Now $R = (8.314/28.02) = 0.297$ kJ kg^{-1} K^{-1}, $c_P = 0.955\times0.297+1.039 = 1.322$ kJ kg^{-1} K^{-1}, i.e.,

$\mu_{JT} = \mu_{JT,R}\, RT_c/(P_c\, c_P)$

$= 0.453 \times 0.297$ (kJ kg^{-1} K^{-1})126.2 K $\div$(33.9 bar$\times$1.322 kJ kg^{-1} K^{-1}) = 0.379 K bar^{-1}.

---

4. Inversion Curves

a. State Equations

The inversion conditions can be obtained by several means. For instance, either of the cubic equations of state, the enthalpy correction charts, or empirical state equations can be used. Inversion occurs when $1 = T\beta_P = (T/v)(\partial v/\partial T)_P$ (cf. Eq. (157))

$$(\partial \ln v/\partial \ln(T))_P = 1. \quad (163)$$

---

mm. Example 39

Obtain an expression that describes the inversion curve of a VW gas.

Solution

We will use Eq. (F) developed in Example 37, namely,

$$\mu_{JT} = -(1/c_p)(R\,T\,b\,v^3 - 2\,\mathbf{a}\,v\,(v-b)^2)/(RTv^3 - 2a\,(v-b)^2)). \quad (A)$$

At the inversion, $\mu_{JT} = 0$, so that

$$T_{inv} = 2\,\mathbf{a}\,(v-b)^2/(R\,b\,v^2). \quad (B)$$

In dimensionless form,

$$T_{inv,R} = (27/4)\,(v_R' - (1/8))^2/v_R'^2 \quad (C)$$

If the value of v_R' is known, $T_{inv,R}$ can be obtained using Eq.(C). The corresponding pressure $P_{inv,R}$ is obtained by applying the normalized VW equation of state, i.e.,

$$P_R = T_R/(v_R' - (1/8)) - (27/64)/v_R'^2. \quad (D)$$

The inversion curve is obtained by assigning a sequence of values for v_R' and calculating the corresponding values of T_R (cf. Eq. (C)), and then determining P_R from the state equation. Inversion curves for various state equations are presented in Figure 22.

Remarks

For a large specific volume, $(v-b)^2 \approx v^2(1 - 2b/v)$. Using Eq. (A) one can subsequently show that

$$\mu_{JT} \approx ((2\mathbf{a}/RT) - b)/c_p, \text{ and}$$

$$T_{inv}/T_c \approx 2\mathbf{a}/bR\,T_c = 2\,(27/64)\,(R^2T_c^2/P_c)/((1/8)(RT_c/P_c)RT_c) = 27/4. \quad (D)$$

---

b. Enthalpy Charts

The enthalpy correction charts (Appendix, Figure B-3) that plot $(h_o - h)/RT_c$ with respect to P_R with T_R as a parameter can be used to determine the inversion points.

The Joule Thomson coefficient $\mu_{JT} = 0$ when $(\partial T/\partial P)_h = 0$. The inversion condition can be determined using the relation $(\partial h/\partial P)_T = 0$, i.e., $[\partial\,\{(h_o - h)/RT_c\}/\partial P]_T = 0$. The peak value of $(h_o - h)/RT_c)$ with respect to P_R at a specified value of T_R yields the inversion point.

c. Empirical Relations

For several gases, such as CO_2, N_2, CO, CH_4, NH_3, C_3H_8, Ar, and C_2H_4, the inversion curve is approximately described by the expression

$$P_R = 24.21 - 18.54/T_R - 0.825\,T_R^2. \quad (164)$$

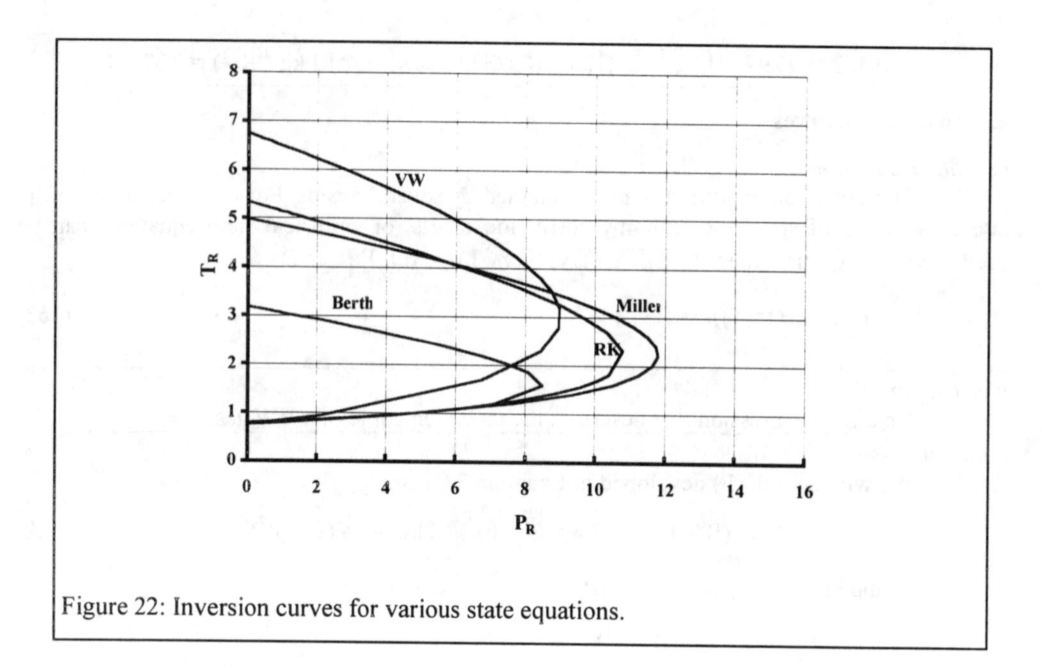

Figure 22: Inversion curves for various state equations.

5. Throttling of Saturated or Subcooled Liquids

The cooling or heating of a vapor during throttling is partly due to the destruction of the mechanical part of the energy as well as boundary deformation work that occurs. When a saturated liquid is throttled from (P_1,T_1) with enthalpy h_{f1}, the final pressure $P_2 < P_1$ and temperature $T_2 < T_1$. As the pressure decreases, T^{sat} also decreases with decreased enthalpy of saturated liquid to h_{f2}. Then the difference $h_{f1}- h_{f2}$ is used to evaporate a portion of the liquid since $h = h_{f1}$. Recall that $\ln P = A - B/T$, (eq. (153)) i.e.,

$$\ln (P_1/P_2) = B/T_1\,(T_1/T_2 - 1) \text{ or } T_2 - T_1 = -\,T_1/(1 + B/(T_1 \ln (P_1/P_2))).$$

Therefore, defining the average Joule Thomson coefficient,

$$\mu_{JT} = (T_2 - T_1)/(P_2 - P_1) = -T_1/(P_2 - P_1)\,(1 + B/(T_1 \ln (P_1/P_2))), \text{ or}$$

$$\approx (T_1/(P_2 - P_1) - BP_1/T_1) \text{ when } (P_1 - P_2) \ll P_1. \quad \text{(A)}$$

The quality x_2 at the exit can be determined as follows. Assuming state 1 is a saturated liquid,

$$h_2 = h_1 = h_{f1} = x_2 h_{g2} + (1 - x_2)h_{f2} = x_2 h_{fg,2} + h_{f1} + c_\ell(T_2 - T_1), \text{ i.e.,}$$

$$x_2 h_{fg,2} = -c_\ell\,(T_2 - T_1), \text{ or}$$

$$x_2 = c_\ell\,(T_1 - T_2)/h_{fg,2} \quad \text{(B)}$$

Using Eqs. (A) and (B)

$$x_2 = (c_\ell T_1/h_{fg,2})/(1 + B/(T_1 \ln(P_1/P_2))). \quad \text{(C)}$$

where $B = h_{fg}/R$. If we assume that $h_{fg,2} = h_{fg,1}$, then

$$x_2 \approx (c_\ell\, T_1/h_{fg,1})/(1 + B/(T_1 \ln(P_1/P_2))).$$

For water, $B \approx 529$ K, while for R134 A, $B \approx 2762$ K.

6. Throttling in Closed Systems

Consider a large insulated tank that is divided into two sections A and B. Section A consists of high pressure gases at the conditions ($P_{A,1}$, $T_{A,1}$) and section B consists of low pressure gases at the state ($P_{B,1}$, $T_{B,1}$). If the partition is ruptured, the tank will assume a new equilibrium state. The state change occurs irreversibly and the entropy reaches a maximum value. The new equilibrium state can be obtained either by differentiating $S = S_A + S_B$ with respect to T_A subject to the constraints that U,V and m are fixed, or by applying the energy balance equations.

---

nn. Example 40

Eight kmole of molecular nitrogen is stored in sections A and B of a rigid tank. Section A corresponds to a pressure $P_{A,1}$ = 119 bar and temperature $T_{A,1}$ = 177 K. In section B, $P_{B,1}$ = 51 bar and $T_{B,1}$ = 151 K. The partition is suddenly ruptured. Determine the final equilibrium temperature T_2. Assume that $c_v = c_{vo}$ = 12.5 kJ kmole^{-1} K^{-1} (c_{vo} does not depend upon the temperature), and that the gas behavior can be described by the RK equation of state $P = RT/(v-b) - \mathbf{a}/(T^{1/2}\, v(v+b))$, where **a** = 15.59 bar m^6 kmole^{-2}, b = 0.02681 m^3 kmole^{-1}, T_c = 126.2 K, and P_c = 33.9 bar. Assume that the volume $V_B = 3V_A$.

Solution

Applying the RK equation to section A

$$119 = 0.08314\times177\div(\bar{v}_{A,1}-0.02681) - 15.59\div(177^{1/2}\ \bar{v}_{A,1}(\bar{v}_{A,1}+0.02681)\ \text{bar}. \quad (A)$$

Therefore,, $\bar{v}_{A,1}$= 0.0915 m^3 kmole^{-1}. (Alternatively, we can use the values T_R = 1.4 and P_R = 3.5 to obtain Z = 0.74 from the appropriate charts. Thereafter, using the relation $P\bar{v} = ZRT$, that value of $\bar{v}$ can be obtained.)

Similarly, $\bar{v}_{B,1}$= 0.167 m^3 kmole^{-1} (for which, $P_{R,B,1}$ = 1.5 $T_{R,B,1}$ = 1.2, and Z = 0.68).

Using the relations, $V_A/\bar{v}_{A,1} + V_B/\bar{v}_{B,1} = 8$ and $V_B = 3V_A$, we obtain the expression

$V_A\,(1/v_{A,1} + 3/v_{B,1}) = 8$, i.e.,

V_A = 0.277 m^3, V_B = 0.831 m^3 so that V = 1.108 m^3.

Thereafter,

$\bar{v}_2$ = 0.139 m^3 kmole^{-1}, N_A = 3.024 kmole, and N_B = 4.976 kmole.

Recall that the internal energy $u - u_o = (3/2\ \mathbf{a}/bT^{1/2})\ \ln(v/(v+b))$. In section A,

$u - u_o$ = –1684.7 kJ kmole^{-1}, i.e., $U_{A,1} - U_{A,1,o}$ = –5094.8 kJ.

Similarly, in section B,

$u - u_o$ = –1056.8 kJ kmole^{-1}, i.e., $U_{B,1} - U_{B,1,o}$ = –5258.6 kJ.

Applying the First law to the tank, $Q - W = \Delta U = 0$, so that

$U_{A,1} + U_{B,1} = U_2$, i.e.,

$U_2 = U_{A,1,o} - 5094.8 + U_{B,1,o} - 5258.6 = U_{A,1,o} + U_{B,1,o} - 10353.4$ kJ.

Now,

$U_2 - U_{2,o} = U_{A,1,o} + U_{B,1,o} - 10353.4 - U_{2o} = N(3/2\ \mathbf{a}/bT^{1/2})\ \ln(v/(v+b))$

$= 697799/T_2^{1/2}\ \ln(0.139\div(0.139+0.02681)) = -123{,}070/T_2^{1/2}$ kJ.

If c_{vo} is a constant, then

$N_A c_{vo} T_{A1} + N_B c_{vo} T_{B1} - N c_{vo} T_2 - 10353.4 = -123{,}070/T_2^{1/2}$ kJ.

Therefore,

$12.5\times(3.024\times177 + 4.976\times151 - 8\times T_2) = 10353.4 - 123{,}070/T_2^{1/2}$ kJ, or

T_2 = 156 K.

Using the RK equation,

$P_2 = 0.08314\times156\div(0.139 - 0.02681)\times15.59\div(156^{1/2}\times0.139\times(0.139+0.0261))$

= 61.45 bars

Remarks

Likewise, for isenthalpic throttling in sssf devices we can use the relation for $(h - h_o)$.

The entropy generated during adiabatic throttling can be determined using a similar procedure.
Such calculations are useful in determining the final pressures and temperatures for shock tube experiments. These experiments involve a pressurized gas in a section A that is separated by a diaphragm from section B. During the experiment, this diaphragm is ruptured.

---

7. Euken Coefficient – Throttling at Constant Volume

During the adiabatic expansion of pressurized gases, the following relation applies.

$$du = c_v\, dT + (T\, \partial P/\partial T - P)\, dv.$$

The Euken coefficient μ_E is related to a constant volume throttling process during which du = 0, i.e.,

$$\mu_E = (\partial T/\partial v)_u = -(T\, \partial P/\partial T - P)/c_v = -T^2(\partial/\partial T(P/T))_u/c_v. \quad (165)$$

For an RK fluid,

$$\mu_E = -(3/2)\, \mathbf{a}/(T^{1/2}\, v\, (v-b))/c_v. \quad (166)$$

This coefficient is always negative, i.e., a specific volume increase is accompanied by a temperature decrease during adiabatic irreversible expansion in a rigid system. The corresponding entropy change is obtained by applying the relation

$$du = Tds - P\, dv, \text{ i.e.,}$$

$$(\partial s/\partial v)_u = P/T. \quad (167)$$

Using Eqs. (165) and (167), we obtain the expression

$$\mu_E = -T^2(\partial^2 s/\partial T \partial v)_u/c_v.$$

For an adiabatic throttling process, $ds = \delta\sigma$. Hence from Eq. (167),

$$\delta\sigma = ds_u = P/T\, dv_u.$$

Since $dv > 0$ and $P/T > 0$, $ds > 0$.

For an equation of state $P = RT/(v-b) - a/T^n v^m$, Eq.(165) transforms to $\mu_E = (\partial T/\partial v)_u = -(n+1)\, a/(c_v\, T^n\, v^m)$. If c_v has a constant value, one can integrate and obtain $T^{(n+1)} = (n+1)^2\, a/(c_v\, (m-1)\, v^{(m-1)}) + C$, $m \neq 1$. If m=2, n=1, then for constant u, $T^2 = (4\, a/(c_v\, v)) + C$.

As per this model, at constant values of u, as $v \to 0$, $T \to \infty$ and $T \to (T_{ig})^{(n+1)}$ as $v \to \infty$, since attractive forces become negligible as we approach ideal gas (ig) limit. Hence, adiabatic throttling of a closed system yields, $T^{(n+1)} - T_{ig}^{(n+1)} = (n+1)^2\, a/\{c_v\, (m-1)\, v^{(m-1)}\}$

Recall from Chapter 6 that $a = RT_c^{(n+1)}(m+1)^2\, v_c^{(m-1)}/(4\, m)$. Therefore, $(T_R^{(n+1)} - T_{ig,R}^{(n+1)})(c_v/R) = (n+1)^2\, (m+1)^2 ((m^2-1)/4m)^{(m-1)}/(4\, m\, (m-1)\, v_R'^{(m-1)})$. For m=2, n=1 (Berthelot equation), the reduced temperature change with pseudo-reduced volume change at constant energy is provided by the expression $(T_R^{(n+1)} - T_{ig,R}^{(n+1)})(c_v/R) = (27/16)\, (1/v_R')$.

a. Physical Interpretation

Consider a container with two sections A and B. Section A is filled with pressurized gases and the second part B contains a vacuum. If the partition in Section A is instantaneously removed, the gas in Section A expands into Section B. As a result of this process the overall internal energy remains constant, but the intermolecular spacing increases (hence, the term $(T\partial P/\partial T - P)dv > 0$). Consequently, thermal portion of the energy $c_v dT$ must decrease (i.e., dT

< 0) in order to compensate for the increase in the intermolecular potential energy. (In the case of an ideal gas, there is a negligible change in the intermolecular potential energy, since the specific volume is very large. Therefore, $(T\partial P/\partial T - P)\, dv = 0$ and there is no change in temperature.) Note that no net boundary work is performed for a rigid system.

L. DEVELOPMENT OF THERMODYNAMIC TABLES

It is apparent from the information contained in the Chapters 6 and 7 thus far that a set of expressions can be developed for the thermodynamic properties of a fluid that exists in any phase if a state relation is known. For example, properties for superheated vapors as H_2O and R134 (Tables A-4 and A-5) can be generated using the real gas state equations.

Table 2: Reference conditions and ideal gas properties for a few fluids.

Property	Steam	Freon 22	R134A	R152A	Ammonia	Nitrogen	Carbon dioxide	Freon 12
Chemical formula	H_2O		CF_3CH_2F		NH_3	N_2	CO_2	
W_m	18.015	86.476	102.03	66.05	17.03	28.013	44.01	120.92
P_c (bar)	220.9	49.775	40.67	45.20	112.8	33.9	73.9	41.2
T_c (K)	647.3	369.15	374.30	386.44	405.5	126.2	304.15	385.0
T_{ref} (K)	273.16	233.14		233.14	233.15	64.143	216.55	233.15
P_{ref} (bar)	0.006113	1.0495		0.512615	.7177	.1253	5.178	.6417
h_{ref} (kJ kg^{-1})	0.01	0.0			0.0	150.3	301.45	0.0
u_{ref} (kJ kg^{-1})	0.0	0.0			0.0	150.4	301.01	−0.04
s_{ref} (kJ kg^{-1} K^{-1})	0.0	0.0			0.0	2.431	3.72	0.0
h_{fg}(kJ kg^{-1})	2501.3	233.18			1389.0	215.188	524.534	169.59
A_o	30.54	22.54	16.778	17.229	29.75	28.58	29.1519	26.765†
B_o	0.01030	0.1141077	0.2865	0.4757	.025	.00377	−.001573	0.17594†
C_o	0.0	130196.35	0	0	−154808.	50208	0.292×10^{-6}	$-.27\ 10^{-7}$†
D_o	0.0	-0.329×10^{-4}	-2.276×10^{-4}	2.893×10^{-4}	0.0	0.0	0.5283×10^{-5}	-0.103×10^{-6}†
E_o			1.135×10^{-7}	6.740×10^{-8}				

1. Procedure for Determining Thermodynamic Properties

Thermodynamic properties can be determined, once the state equation, critical constants, and corresponding ideal gas properties are known. Some useful formulas are listed below and some thermodynamic data is listed in Table 2. The reference conditions should be specified. For example for water, the reference condition is generally specified as that of the saturated liquid at the triple point. The choice of reference conditions is arbitrary. Here,

$v_R' = v/v_c'$, $v_c' = RT_c/P_c$

$v_R = v/v_c$,

$Z = P_R\, v_R'/T_R$

$u_{c,R} = -u^{Res}/(RT_c) = (u_o(T) - u(T,P))/RT_c$,

$h_{c,R} = (h_o(T) - h(T,P))/RT_c$,

$s_{c,R} = (s_o(T,P) - s(T,P))/R$,

$c_{P,c\,R} = (c_P(T,P) - c_{P,o}(T))/R$,

$c_{v,c,R} = (c_v(T,P) - c_{v,o}(T))/R$,

$\mu_{JT,R} = \mu_{JT} c_p/v_c'$,

$g_{c,R} = (g_o(T,P) - g(T,P))/RT_c$,

$a_{c,R} = (a_o(T,P) - a(T,P))/RT_c$

$\phi = f/P = \exp((g(T,P) - g_o(T,P))/RT)$

T^{sat} with $g^f = g^g$ and

T_{inv} with $\mu_{JT} = 0$

The constants are used in the formula $\bar{c}_{po} = A_o + B_oT + C_oT^{-2} + D_oT^2 + E_oT^3$. In the

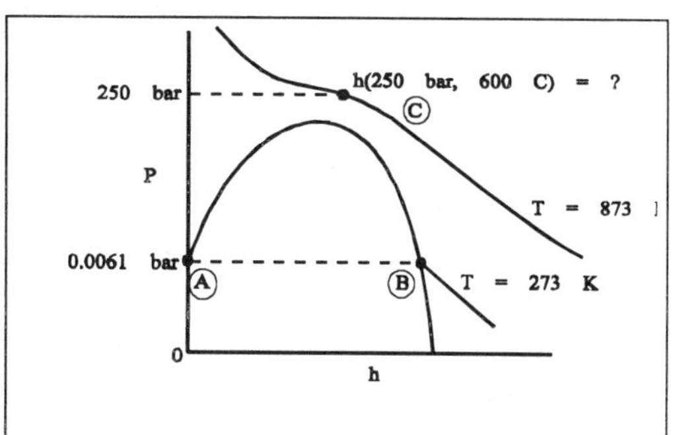

Figure 23: Schematic illustration of a method of determining the thermodynamic properties of a material using a P–h diagram.

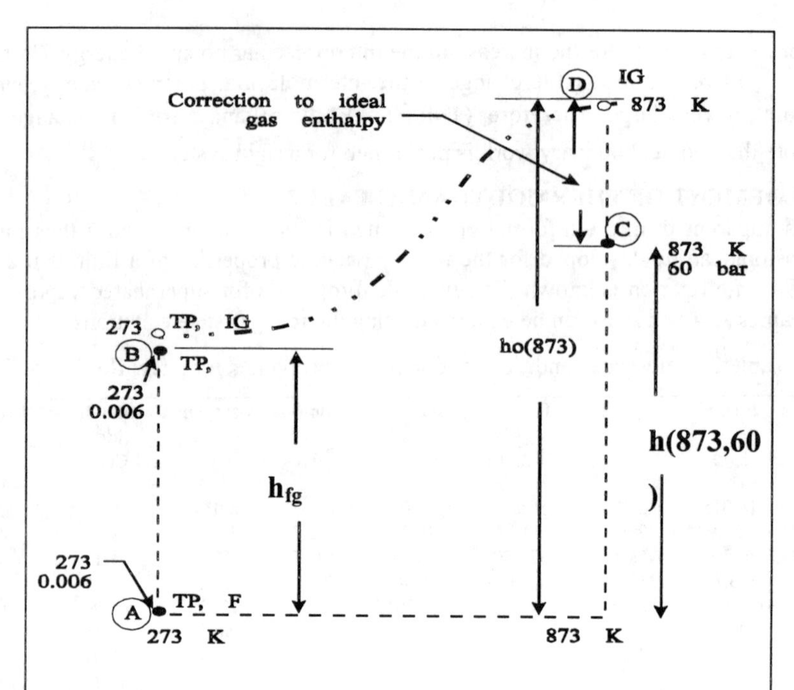

Figure 24: Schematic illustration of the determination of enthalpy of a vapor or a real gas with respect to the values at the reference condition.

range 0 K < T < 1000 K, the maximum error is less than 8%.

oo. Example 41

Determine the thermodynamic properties of water at 250 bar and 600°C using the RK equation of state. Assume that $c_{p,o} = 28.85 + 0.01206\ T + 1.002\times10^5/T^2$ kJ kmole^{-1} K^{-1}, $h_{fg,ref} = 2501.3$ kJ kg^{-1}, $P_c = 220.9$ bar, and $T_c = 647.3$ K. The reference conditions are those for the saturated liquid at its triple point, i.e., 273.15 K and 0.006113 bar. The molecular weight of water is 18.02 kg kmole^{-1}.

Solution

A schematic diagram of the procedure followed is illustrated in on a P–h diagram in Figure 24 and Figure 23.

First, the reference condition is selected at which $h_f = s_f = 0$ (e.g., the saturated liquid state at the triple point of water at point A in the figure).

At the reference condition $h_g = 0 + h_{fg} = 2501$ kJ kg^{-1} (point B in Figure 24 and Figure 23). Therefore,

$s_g = h_{fg}/T = 2501.3 \div 273 = 9.17$ kJ kg^{-1} K^{-1},

The entropy of the saturated vapor at 273.15 K and 0.0061 bar is 9.17 kJ kg^{-1} K^{-1}above the entropy of saturated liquid at same temperature and pressure.

For the vapor at 273 K and 0.0061 bar (i.e., at $P_{R,ref} = 0.000028$ and $T_{R,ref} = 0.422$) the reduced correction factor $(h_o - h)/RT_c \approx 0$, since the pressure is low and the intermolecular attraction forces are weak. Therefore,

$h_o = h = 2501$ kJ kg^{-1}

at the triple point (point B in Figure 24 and Figure 23). Similarly,

$s_o = s = 9.17$ kJ kg^{-1} K^{-1} or 165.2 kJ kmole^{-1} K^{-1}.

The values of h_o(873 K) and s_o(873 K, 250 bar) can be obtained using the specific heat relations for an ideal gas, i.e.,

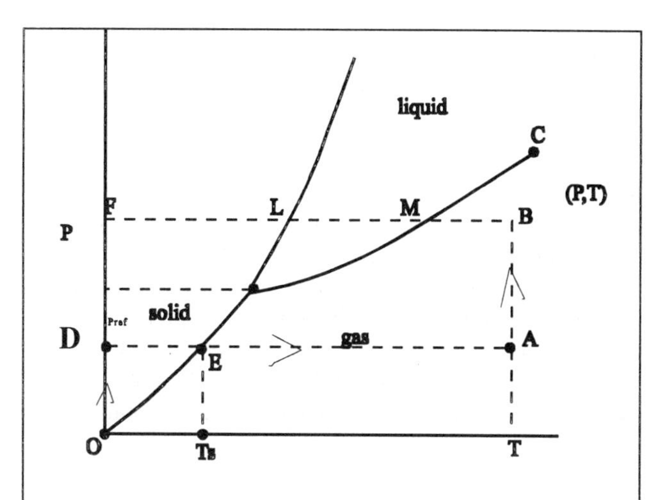

Figure 25: Schematic illustration of an entropy calculation starting from a temperature of absolute zero; C: critical point.

$h_o = 3706$ kJ kg^{-1}
$= 66782$ kJ kmole^{-1}.
This corresponds to the point D. Similarly,
$s_0(873$ K, 250 bar$) = 6.47$ kJ kg^{-1} K^{-1} or 116.6 kJ kmole^{-1} K^{-1}.
The ideal gas internal energy u_o at 873 K can be determined using the relation
$u_o = h_o - RT = 66782 - 8.314 \times 873 = 59{,}524$ kJ kmole^{-1}.

The correction or residual factors at 873 K and 250 bar can be obtained. From charts we see that at $P_R = 1.13$, $T_R = 1.35$, $Z = 0.845$. Therefore,
$(h_o - h)/RT_c = 0.735$, $(u_o - u)/RT_c = 0.526$, and $(s_o - s)/R = 0.389$.
Consequently,
$u = 3146$ kJ kg^{-1}, and $s = 6.29$ kJ kg^{-1} K^{-1}.
Thereafter, the enthalpy can be determined using the relation $h = u + Pv = u + ZRT$. Hence,
$h = 3146 + 0.845 \times (8.314 \div 18.02) \times 873 = 3486$ kJ kg^{-1},
which is represented by point C.
Other properties can be similarly obtained.

2. Entropy

Applying the relation at a temperature of absolute zero, $(\partial v/\partial T)_P = -\partial s/\partial P = 0$. Therefore, there is no entropy change with pressure at T = 0 K and $s(P_{ref}, 0\text{ K}) = 0$. Two paths reach state B from state "0": path 0 FLMB, path 0 DEAB. Consider the path 0DEAB illustrated in Figure 25. The entropy s(T,P) equals the sum of the entropy change along the paths 0D and DE), the entropy addition at E due to the change from the solid to gas phase, entropy change due to superheating EA and the entropy change along AB. Therefore, at any state (T,P)

$$s = s_o + \int_0^{P_{ref}} \left(\frac{\partial s}{\partial P}\right)_T dP + \int_0^{T_s} \left(\frac{c_s(P_{ref}, T)}{T}\right) dT + \frac{h_{sg}}{T} + \int_{T_s}^{T} \left(\frac{c_p(P_{ref}, T)}{T}\right) dT + \int_{P_{ref}}^{P} \left(\frac{\partial s}{\partial P}\right)_T dP \quad (168)$$

where $\partial s/\partial P = -(\partial v/\partial T)_P$. Given a state equation, the above equation can be integrated to determine the conditions at point B in Figure 25. If P_{ref} is a small pressure, the entropy at point A corresponds to the ideal gas value and, hence, $c_p(P_{ref}, T) = c_{p,o}$ and $s_B(T,P) - s_A(T,P_{ref}) = s(T,P) - s_o(T,P_{ref}) = s(T,P) - s_0\,(T,P) + R \ln (P/P_{ref})$.

M. SUMMARY

Ideal and real models of gas behavior are important, since they can be used to obtain the thermodynamic properties of substances, which are used in conservation and balance equations. The ideal gas models are accurate at high temperatures and low pressures. In states in the vicinity of saturated vapor and at relatively higher pressures, the real gas models provide more accurate predictions of properties. Using the real gas state equations we are able to determine differences in fluid behavior from its ideal state for properties, such as u, h, s, c_v and c_p. However, determination of μ_{JT}, inversion temperature, saturation temperature do not require ideal gas properties. Further, the corresponding values of Z, $h_{R,c}$, and $s_{R,c}$ and h_{fg} along the saturation lines can be obtained.

Chapter 8

8. THERMODYNAMIC PROPERTIES OF MIXTURES

A. PARTIAL MOLAL PROPERTY

In Chapter 7, we discussed how the thermodynamic properties of pure components of a substance could be determined and used in the four important conservation and balance equations for design of thermal systems (mass, energy, entropy and availability). Thermal systems, however, generally involve mixtures of several substances, possibly in multiple phases. In this chapter, we will discuss the state equations, generalized thermodynamic relations, and thermo-physical-chemical properties for species k in a mixture as well as non-reacting mixtures.

For pure simple substances, a pair of properties can be used to determine the state of a system. Using those two properties, we can then determine the other system properties. More than two properties are required for a corresponding situation involving a mixture to account for the mixture composition.

1. Introduction

The mixture composition can be represented by the mole fraction X_K or mass fraction Y_K, or the molality Mo.

a. *Mole Fraction*

The mole fraction is denoted by X_k for gases, and x_k for liquids. For gases,

$$X_k = X_{k,g} = N_{k,g}/N, \text{ and} \quad (1a)$$

for liquids,

$$X_{k(\ell)} = N_{k(\ell)}/N. \quad (1b)$$

The subscript k denotes the k–th component of a mixture. Generally, the subscript "g" is omitted for gas phases herein.

b. *Mass Fraction*

Likewise, the mass fraction can be defined. For gases

$$Y_{k,g} = m_{k,g}/m, \text{ and} \quad (1c)$$

for liquids

$$Y_{k(\ell)} = m_{k(\ell)}/m. \quad (1d)$$

c. *Molality*

The molality is used to describe liquid solutions, i.e.,

$$Mo = 10^{-3}\times\text{kmole of solute} \div \text{kg of solvent}. \quad (1e)$$

d. *Molecular Weight of a Mixture*

The mixture molecular weight M is the mixture per unit mole of the mixture, namely,

$$M = \Sigma X_k M_k \quad (1f)$$

A solution is dilute if the mole fraction of the solute is much smaller than that of the solvent. More generally, a mixture is dilute when the value of the mole fraction of a particular component dominates the mole fractions of the other components.

---

a. Example 1

Consider a lead–acid battery containing 3.75 Mo of an H_2SO_4 acid solution. The acid density is 1230 kg m^{-3}. Determine the solution composition and molecular weight.

Solution

Mo = 3.75 gmole of H_2SO_4/kg of water

In a mixture containing a kg of solvent and 3.75×10^{-3} kmole of H_2SO_4, the total moles are

$(1000\div18.02) + 3.75 = 59.24$ gmole.

Since M for H_2SO_4 is 98 kg $kmol^{-1}$, the total mixture mass is

$1000 \text{ g} + 3.75 \times 98 = 1367.5$ g.

Therefore,

$x_{H_2SO_4} = 3.75\div59.24 = 0.063$,

$Y_{H_2SO_4} = 367.5 \div 1367.5 = 0.27$, and

$M = 1367.5 \div 59.24 = 23.08$ kg $kmole^{-1}$.

---

2. Generalized Relations

Recall that the internal energy of a mixture containing K components is

$$U = U(S,V,N_1, N_2, ..., N_K), \tag{2a}$$

$$dU = TdS - PdV + \Sigma_K\mu_k dN_k, \text{ and} \tag{2b}$$

$$(\partial U/\partial S)_{V,N} = T,\ (\partial U/\partial V)_{S,N} = -P,\ (\partial U/\partial N_i)_{S,V,\ N_1,N_2,...N_{j\neq i},...N_K} = \mu_1, \tag{2c}$$

where T denotes the thermal potential, P the pressure potential, μ_i the chemical potential of the i–th species in the mixture. All three potentials are expressed in terms of partial derivatives of U. Likewise,

$$H = H(S, P,N_1, N_2.....N_K), \tag{3a}$$

$$dH = TdS + VdP + \Sigma_K\mu_k dN_k, \text{ and} \tag{3b}$$

$$(\partial H/\partial S)_{P,N} = T,\ (\partial H/\partial P)_{S,N} = V,\ (\partial H/\partial N_i)_{S,\ P,\ N_1,N_2,...N_{j\neq i},...N_K} = \mu_1. \tag{3c}$$

Since A = U – TS, subtracting the term (TdS + S dT) from Eq. (2b) provides the relations for A shown below.

$$A = A(T,V,N_1, N_2.....N_K), \tag{4a}$$

$$dA = -SdT - PdV + \Sigma_K\mu_k dN_k, \text{ and} \tag{4b}$$

$$(\partial A/\partial S)_{V,N} = -T,\ (\partial A/\partial V)_{S,N} = -P,\ (\partial A/\partial N_i)_{S,V,\ N_1,N_2,...N_{j\neq i},...N_K} = \mu_1. \tag{4c}$$

We now subtract (TdS + SdT) from Eq. (3b) to obtain

$$G = G(T, P,N_1, N_2.....N_K), \tag{5a}$$

$$dG = SdT + VdP + \Sigma_K\mu_k dN_k, \text{ and} \tag{5b}$$

$$(\partial G/\partial T)_{P,N} = -S,\ (\partial G/\partial P)_{S,N} = V,\ (\partial G/\partial N_i)_{S,\ P,\ N_1,N_2,...N_{j\neq i},...N_K} = \mu_1. \tag{5c}$$

a. Remarks

The nonmeasurable properties, e.g., A, G, and S, can be expressed in terms of the measurable properties P, V, T, N_1, N_2, N_3, …, etc. The chemical potential, which governs the direction of species transfer in a mixture, can be expressed in various forms. For a closed system with no chemical reactions $dN_k = 0$.

Based on the previous discussion, a thermodynamic property B (for instance, B = V) can be expressed as

$$B = B(T, P, N_1,..., N_K). \tag{6a}$$

The extensive property B is a partly homogeneous function of degree 1 with respect to N_1, N_2, N_3, …, i.e., if the temperature and pressure are held constant and the number of moles of each species in the mixture is doubled (although the corresponding mole fractions remain unchanged), the value of B is also doubled.

The corresponding partial molal property of the i–th species can be written in the form

$$\hat{b}_i = (\partial B/\partial N_i)_{T,\,P,\,N_1,N_2,...N_{j\neq i},...N_K}. \tag{6b}$$

(For instance, $\hat{b}_i = \hat{v}_i$.) On a mass basis,

$$\hat{b}_i^m = (\partial B/\partial m_i)_{T,\,P,\,N_1,N_2,...N_{j\neq i},...N_K}. \tag{6c}$$

3. Euler and Gibbs–Duhem Equations

The total differential of Eq. (6a) is

$$dB = (\partial B/\partial T)_{P,N}\, dT + (\partial B/\partial P)_{T,N}\, dP + \sum_{k=1}^{K} \hat{b}_k\, dN_k, \text{ i.e.,} \tag{7a}$$

Since $B = N\,\bar{b}$,

$$d(N\bar{b}) = Nd\bar{b} + \bar{b}dN = N(\partial\bar{b}/\partial T)_{P,N}\, dT + N(\partial b/\partial P)_{T,N}\, dP + \sum_{k=1}^{K} \hat{b}_k\, dN_k \tag{7b}$$

Furthermore, since $N_k = X_kN$, $dN_k = X_kdN + NdX_k$, further simplification of Eq. (7b) results in the expression

$$N\{\, d\bar{b} - \frac{d\bar{b}}{dT}dT - \frac{d\bar{b}}{dP}dP - \sum_{k=1}^{K}\hat{b}_k\, dX_k\,\} + dN\,\{\bar{b} - \sum_{k=1}^{K}\hat{b}_k\, X_k\,\} = 0 \tag{8}$$

Since N is arbitrary and the value of dN can vary, it is apparent from Eq. (8) that the coefficients of N and dN must vanish. Equating the coefficient of dN to zero, we obtain

$$\bar{b} = \sum_{k=1}^{K}\hat{b}_k\, X_k\,, \text{ i.e., } B = \bar{b}N = \sum_{k=1}^{K}\hat{b}_k\, N_k, \tag{9a}$$

where B= S, U, H, V, etc. Likewise,

$$d\bar{b} = \frac{d\bar{b}}{dT}dT + \frac{d\bar{b}}{dP}dP - \sum_{k=1}^{K}\hat{b}_k\, dX_k. \tag{9b}$$

The relation in Eq. (9a) is known as the Euler equation. Since all the differentials are exact, we infer from Eq. (9b) that

$$\bar{b} = \bar{b}(T, P, X_1, X_2, \ldots, X_{K-1}). \tag{10}$$

Note that $X_1 + X_2 + ... + X_K = 1$.

The number of independent variables in Eq. (10) is ((K–1)+2) = K+1, while in Eq. (6) it is K+2. At constant temperature and pressure, Eq. (9b) assumes the form

$$d\bar{b} = \sum_{k=1}^{K} \hat{b}_k \, dX_k. \tag{11}$$

Differentiating Eq. (9a) ,

$$d\bar{b} = \sum_{k=1}^{K} \hat{b}_k \, dX_k + \sum_{k=1}^{K} d\hat{b}_k \, X_k. \tag{12a}$$

Equating Eq. (12a) with Eq. (9b), we obtain the relation

$$\frac{d\bar{b}}{dT} dT + \frac{d\bar{b}}{dP} dP - \sum_{k=1}^{K} X_k d\hat{b}_k = 0. \tag{12b}$$

Upon multiplying this expression by N,

$$\left(\frac{\partial B}{\partial T}\right)_{P,N} dT + \left(\frac{\partial B}{\partial P}\right)_{T,N} dP - \sum_{k=1}^{K} N_k d\hat{b}_k = 0. \tag{12c}$$

At a specified temperature and pressure

$$\sum X_K d\hat{b}_K = 0. \tag{13}$$

The expressions in Eqs. (12a)–(12c) and (13) are various forms of the Gibbs–Duhem (GD) equations, and apply to liquid, solid, and gas mixtures. Combining Eqs. (12a) and (13) we obtain

$$d\bar{b} = \sum_{k=1}^{K} \hat{b}_k \, dX_k. \tag{14}$$

a. Characteristics of Partial Molal Properties

Since

$$\hat{b}_k = (\partial B/\partial N_k)_{T,P} = \hat{b}_k (T, P, N_1, N_2, ..., N_K),$$

applying Euler's theorem for species k =1 for a partly homogeneous function of order zero,

$$N_1 \partial \hat{b}_1/\partial N_1 + N_2 \partial \hat{b}_1/\partial N_2 + N_3 \partial \hat{b}_1/\partial N_3 + ... = 0, \text{ i.e.,}$$

or, more generally,

$$\Sigma_k(N_k \partial \hat{b}_j/\partial N_k) = 0, j=1, 2,...K. \tag{15a}$$

Dividing Eq. (15) by N

$$\Sigma_k(X_k \partial \hat{b}_j/\partial N_k) = 0, j=1, 2,...K. \tag{15b}$$

Consider partial molal property of species 1. Differentiating the partial molal property $\hat{b}_1 = \partial B/\partial N_1$ with respect to N_2, we obtain the relation

$$\partial \hat{b}_1/\partial N_2 = \partial^2 B/\partial N_2 \partial N_1 = \partial/\partial N_1(\partial B/\partial N_2) = \partial \hat{b}_2/\partial N_1, \tag{15c}$$

which is a form of Maxwell's relations. Similarly, $\partial \hat{b}_1/\partial N_3 = \partial \hat{b}_3/\partial N_1$. Using Eq. (15c) in Eq. (15a)

$$\Sigma_j(N_k\, \partial \hat{b}_k/\partial N_j) = 0, j=1, 2,...K. \quad (15d)$$

Dividing by N

$$\sum_{k=1}^{K} X_k \frac{\partial \hat{b}_k}{\partial N_j} = 0, \qquad j=1,2...K.$$

With j=1 in Eq. (15d) ,

$$N_1 \partial \hat{b}_1/\partial N_1 + N_2\, \partial \hat{b}_2/\partial N_1 + N_3 \partial \hat{b}_3/\partial N_1 + ... = 0. \quad (15e)$$

Note that the partial derivatives $\partial \hat{b}_1/\partial N_1$ imply that N_2, N_3, etc., are constant Since $N_1 = X_1$ N, then $dN_1 = dX_1\, N + X_1\, dN$. If only N_1 is altered, then the values of all $N_{j\neq 1}$ are constant. In that case

$$dN_1 = dX_1\, N + X_1\, dN_1, \text{ or } (1 - X_1)\, dN_1 = dX_1\, N, \text{ i.e., } dN_1 = dX_1\, N/(1 - X_1).$$

Using this result in Eq. (15e)

$$\partial \hat{b}_1/\partial N_1 = (1\text{-}X_1)(\partial \hat{b}_1/N\partial X_1),\ \partial \hat{b}_2/\partial N_1 = (1\text{-}X_1)(\partial \hat{b}_1/ N\partial X_1),.....$$

Hence, Eq. (15e) with j=1 simplifies to the form

$$X_1 \partial \hat{b}_1/\partial X_1 + X_2 \partial \hat{b}_2/ \partial X_1 + X_3 \partial \hat{b}_3/\partial X_1 + ... = 0, \text{ i.e., } \Sigma_k X_k(\partial \hat{b}_k/\partial X_1) = 0,$$

Generalizing for any "j"

$$\Sigma_k X_k(\partial \hat{b}_k/\partial X_j) = 0, j=1,2...K \quad (16)$$

Gibbs function is extensively used in phase and chemical equilibrium calculations. Thus it is useful to summarize the relations for B=G. i.e.,

$$= G(T, P, N_1, N_2, ..., N_K), \text{ then}$$

$$\bar{g} = \Sigma_k\, \hat{g}_k\, X_k, \text{ and } g = \Sigma_k\, \hat{g}_k{}^m Y_k. \quad (17)$$

where the second summation is on mass basis. The partial molal Gibbs function $\hat{g}_k$ is the chemical potential μ_k of a species, i.e.,

$$\hat{g}_k = (dG/dN_k)_{T, P, N_{j\neq i}}. \quad (18a)$$

Equation (12c) implies that

$$0 = -\bar{s}\, dT + \bar{v}\, dP + X_1\, d\,\hat{g}_1 + X_2\, d\,\hat{g}_2 + ..., \text{ and} \quad (18b)$$

$$0 = - S\, dT + V\, dP + N_1\, d\,\hat{g}_1 + N_2\, d_2 + ... , \quad (19a)$$

where $(\partial G/\partial T)_{P,N} = -S$ and $(\partial G/\partial P)_{T,N} = V$. At a specified temperature and pressure

$$X_1\, d\,\hat{g}_1 + X_2\, d\,\hat{g}_2 + ... = 0. \quad (19b)$$

b. Physical Interpretation

Consider the extensive property V of a k–component mixture. Each species in the mixture contributes an amount V_k towards the total mixture volume, and its partial molal volume

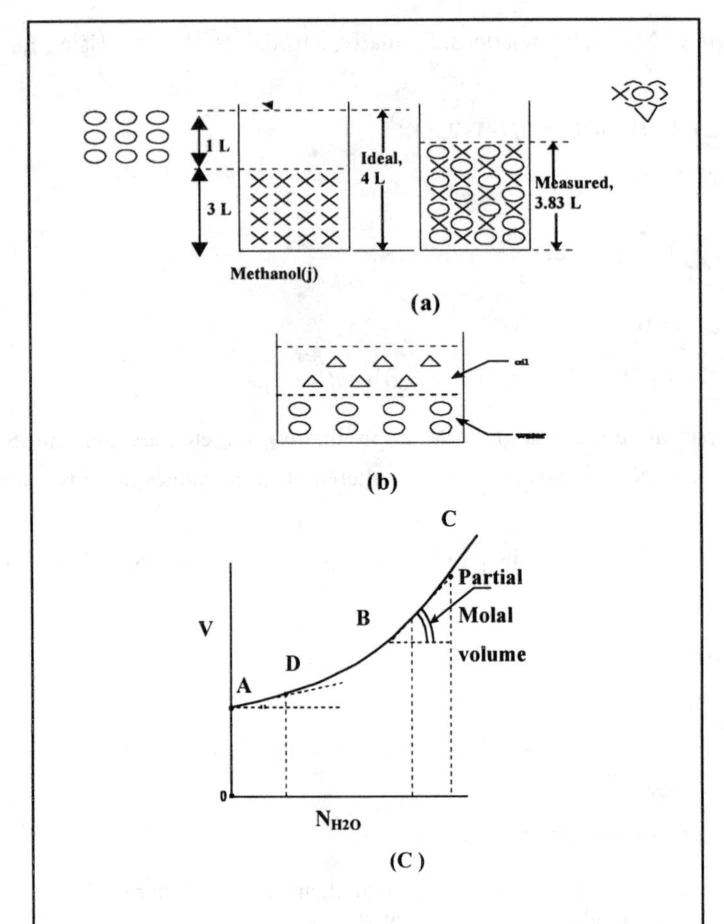

Figure 1 a. Mixing of two miscible species. b. Mixing of two immiscible species. c. Determination of the partial molal volume from a plot of total mixture volume *vs.* the number of moles of water in a water/alcohol mixture.

$$\hat{v}_k = (\partial V/\partial N_k)_{T,P} = \hat{v}_k(T, P, N_1, N_2, ..., N_K),$$

When the same component is considered in its pure state at the same temperature and pressure, its specific volume $\bar{v}_k$, and, in general, $\hat{v}_k \neq \bar{v}_k$.

If a liter of water is added at standard conditions to three liters of pure alcohol, you will find that both species completely mix at the molecular level. This is an example of a miscible mixture. Under standard conditions, the total mixture volume is not four liters, suggesting that the mixture must have contracted due to a change in the intermolecular attractive forces. The volume occupied by 1 kmole of water, i.e., 6×10^{26} water molecules in the mixture is its partial molal volume. As more water is added to the mixture the total volume increases as shown by curve ADBC in Figure 1. The slope of the mixture volume with respect to the number of moles of water provides a measure of the partial molal volume. The point A in the figure represents a condition corresponding to trace amounts of water in the mixture, while point B represents alcohol in trace quantities. An immiscible mixture is formed if two species do not mix at a molecular level and, in that case, the partial molal volume loses meaning.

i. Remarks

In Chapters 1 and 6 we have discussed the functional form for the intermolecular attraction forces given by the Lennard– Jones empirical potential $\Phi(\ell)$ between a pair of molecules. For like pairs of molecules,

$$\Phi(\ell) = 4\varepsilon((\sigma/\ell)^{12} - (\sigma/\ell)^6). \quad (20a)$$

For an unlike molecular pair consisting of species k and j,

$$\Phi(\ell) = 4\varepsilon_{kj}((\sigma_{kj}/\ell)^{12} - (\sigma_{kj}/\ell)^6), \quad (20b)$$

where $\varepsilon_{kj} = (\varepsilon_k\varepsilon_j)^{1/2}$, and the collision diameter $\sigma_{kj} = (\sigma_k+\sigma_j)/2$. Note that concentration effects on σ_{kj} and ε_{kj} are not included in this model. If F_k (= $\partial\phi/\partial\ell$) denotes the intermolecular attraction forces between the molecules of species k, and F_{kj} denotes the corresponding forces between dissimilar molecules of the two species k and j,

$$F_{kk} = 4(\varepsilon/\ell)(6(\sigma/\ell)^6 - 12(\sigma/\ell)^{12}), \text{ and} \quad (21a)$$

$$F_{kj} = 4(\varepsilon_{kj}/\ell)((\sigma_{kj}/\ell)^6 - (\sigma_{kj}/\ell)^{12}). \quad (21b)$$

Consider the following scenarios: (1) $F_{kj} = F_{kk}$. In this case an ideal solution (or homologous series) is formed, e.g., a mixture of toluene and benzene. (2) $F_{kj} > F_{kk}$. This is an example of a non ideal solution (e.g., the volume contraction upon mixing of water in alcohol). (3) $F_{kj} < F_{kk}$. Also a non ideal solution, but in this case there is a volumetric expansion upon mixing).

If $F_{kj} \gg F_{kk}$ at all concentrations, the mixture is miscible at a molecular level. Alternatively, if $F_{kj} \ll F_{kk}$, at all concentrations, the mixture is completely immiscible. In some cases, the mixture is miscible up to a certain mole fraction beyond which $F_{kj} \ll F_{kk}$. Such mixtures are called partially miscible mixtures.

In miscible mixtures, energy must be initially utilized to overcome the intermolecular attraction forces between like molecules (e.g., k–k). Inserting a molecule of j and forming the j–k pairs alters the intermolecular attraction forces. Consequently, the system may reject energy (during exothermic mixing) or require it (endothermic mixing).

In an ideal solution $F_{kk} = F_{kj}$, and $\hat{v}_k(T, P, X_1, X_2, ...) = \bar{v}_k(T, P)$. Recall that

$$V = \Sigma_k\, \hat{v}_k(T, P, X_1, X_2, ...)N_k.$$

Therefore, for an ideal mixture

$$V^{id} = \Sigma_k\, \bar{v}_k(T, P)N_k. \quad (22)$$

This relation is called law of additive volumes or the Amagat–Leduc Law for mixtures. It is particularly valid for gas mixtures at low pressures.

Similarly, for any property (other than the entropy),

$$B^{id} = \Sigma_k\, \bar{b}_k(T, P)N_k. \quad (23)$$

Mixing is always an irreversible process. For adiabatic ideal mixing, i.e., when there is no volumetric change and heat is neither absorbed nor removed. Since the Second Law states that

$$dS - \delta Q/T_b = \delta\sigma,$$

the difference in the entropy after and before mixing is given by the expression

$$S_{final} - S_{initial} - 0 = \sigma,$$

where $\sigma > 0$. Furthermore, since $S_{initial} = \Sigma_k \bar{s}_k(T, P)N_k$ and $S_{final} = \Sigma_k \hat{s}_k(T, P)N_k$,

$$\Sigma_k(\hat{s}_k(T, P) - \bar{s}_k(T, P))\, N_k > 0.$$

Even after ideal mixing, the entropy of a species k inside the mixture at T and P is larger than the entropy of the pure species at same temperature and pressure. This is due to the increase in the intermolecular spacing between the k and j molecules, which increases the number of quantum states for each species (Chapter 1).

4. Relationship Between Molal and Pure Properties

a. *Binary Mixture*

For a two component system, Eq. (14) suggests that at specified values of T and P

$$d\bar{b} = \hat{b}_1\, dX_1 + \hat{b}_2\, dX_2 \quad (24a)$$

Since $X_1 + X_2 = 1$, Eq. (24a) assumes the form

$$d\bar{b} = \hat{b}_1 \{-dX_2\} + \hat{b}_2\, dX_2, \text{ i.e.,. } d\bar{b}/dX_2 = -\hat{b}_1 + \hat{b}_2 \quad (24b)$$

Since the mixture property is provided by Eq (9a),

$$\bar{b} = \hat{b}_1 X_1 + \hat{b}_2 X_2 \quad (24c)$$

Multiplying Eq (24b) by X_2, rearranging and simplifying,

$$\hat{b}_1 X_2 = \hat{b}_2 X_2 - X_2\, d\bar{b}/dX_2$$

and using Eq. (24c) to eliminate $\hat{b}_2 X_2$

$$\hat{b}_1 = \bar{b} - X_2\, d\bar{b}/dX_2 \quad (24d)$$

Similarly

$$\hat{b}_2 = \bar{b} - X_1\, d\bar{b}/dX_1 \quad (24e)$$

Equations (24d) and (24e) express the partial molal property of the species in terms of the molal property of the mixture.

b. *Multicomponent Mixture*

Similarly we can extend the derivation to a multi-component mixture, i.e.,

$$\hat{b}_i = \bar{b} - \sum_{k=1,k\neq i}^{K} X_k (\partial b/\partial X_k)_{T,P,k\neq i}\,. \quad (25)$$

---

b. *Example 2*

A small amount of liquid water (species 2) is added to liquid methanol (species 1) in a constant temperature bath held at 25°C and 1 atm. The variation of X_1 is shown in Figure 2. Assume that $\bar{v}_1 = 0.04072$ m^3 kmole^{-1}, $\bar{v}_2 = 0.0181$ m^3 kmole^{-1}.
Determine the specific volume $\bar{v}^{id}$ under ideal mixing.
Plot $\hat{v}_1$ and $\hat{v}_2$ with respect to X_1.
$\hat{v}_{1,\, X_1\to 1}$, $\hat{v}_{1,\, X_1\to 0}$ (species 1 is in infinite dilution or in trace amounts).
$\hat{v}_{2,\, X_2\to 1}$, $\hat{v}_{2,\, X_2\to 0}$ (species 2 is in infinite dilution or in trace amounts)
Determine an approximate expression for $\bar{v}$ in terms of X_1 when $X_1\to 0$.

Solution

The specific volume of any mixture or solution

$$\bar{v} = \hat{v}_1 x_1 + \hat{v}_2 x_2. \quad (A)$$

For an ideal solution,

$$\hat{v}_1^{id} = \bar{v}_1,\ \hat{v}_2^{id} = \bar{v}_2, \text{ i.e.,} \quad (B)$$

$$\bar{v}^{id} = \bar{v}_1 X_1 + \bar{v}_2 X_2 \quad (C)$$

Similarly on mass basis $v^{id} = \hat{v}_1^{m} Y_1^{m} + \hat{v}_2^{m} Y_2^{m}$. Since $\bar{v}_1$, and $\bar{v}_2$ are fixed once the temperature and pressure are specified, and $X_2 = (1-X_1)$, using Eq. (C), we obtain the relation

$$\bar{v}^{id} = \bar{v}_1 X_1 + \bar{v}_2 (1-X_1). \quad (D)$$

Equation (D) indicates that a plot of $\bar{v}^{id}$ with respect to X_1 is linear. However, measurements indicate that this is not so.

Using the measured results for $\bar{v}$ with respect to X_1 shown in Figure 2, it is possible to obtain $\hat{v}_1$ and $\hat{v}_2$ using Eqs. (24d) and (24e) with $\bar{b} = \bar{v}$. Since $X_1 + X_2 = 1$, $dX_1 = -dX_2$, and

$$\bar{v}_1 = \bar{v} + (d\bar{v}/dX_1)(1-X_1), \quad (E)$$

$$\hat{v}_2 = \bar{v} - X_1 d\bar{v}/dX_1, \quad (F)$$

Figure 2 shows a plot of specific volume, partial molal volumes of methanol (1) and

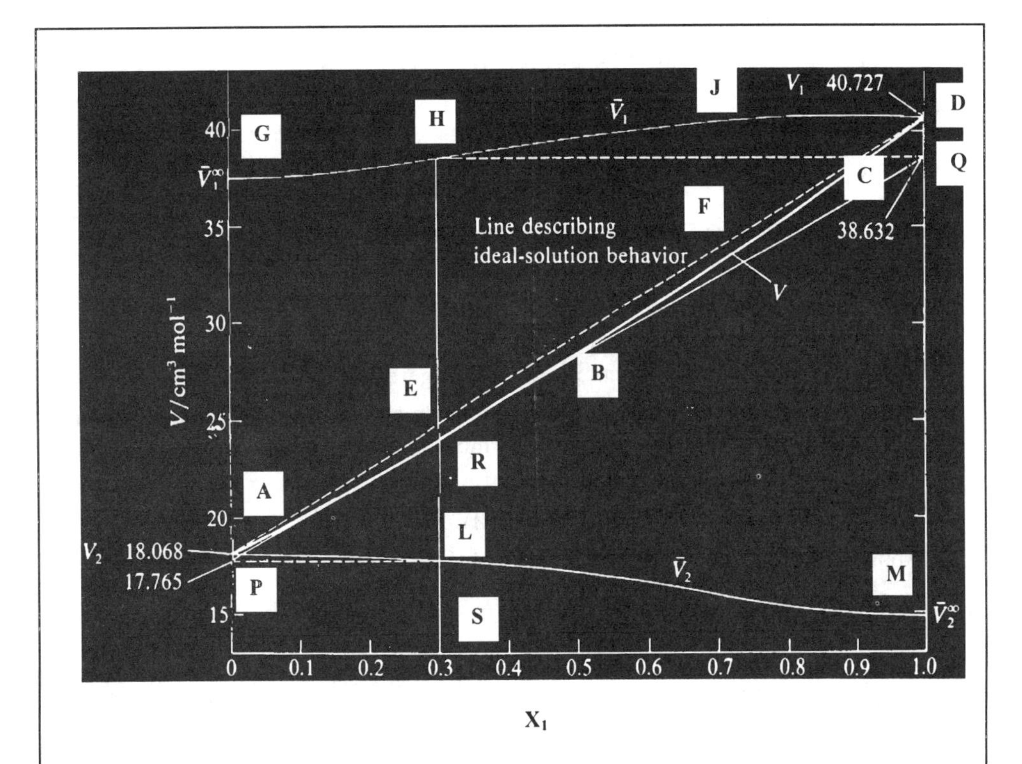

Figure 2: Partial and molal volumes of methanol (1) and water (2) at 25 C, 1 atm (From Smith and Van Ness, *Introduction to Chemical Engineering Thermodynamics*, 4th Edition, McGraw Hill Book Company, 1987, p. 428. With permission.)

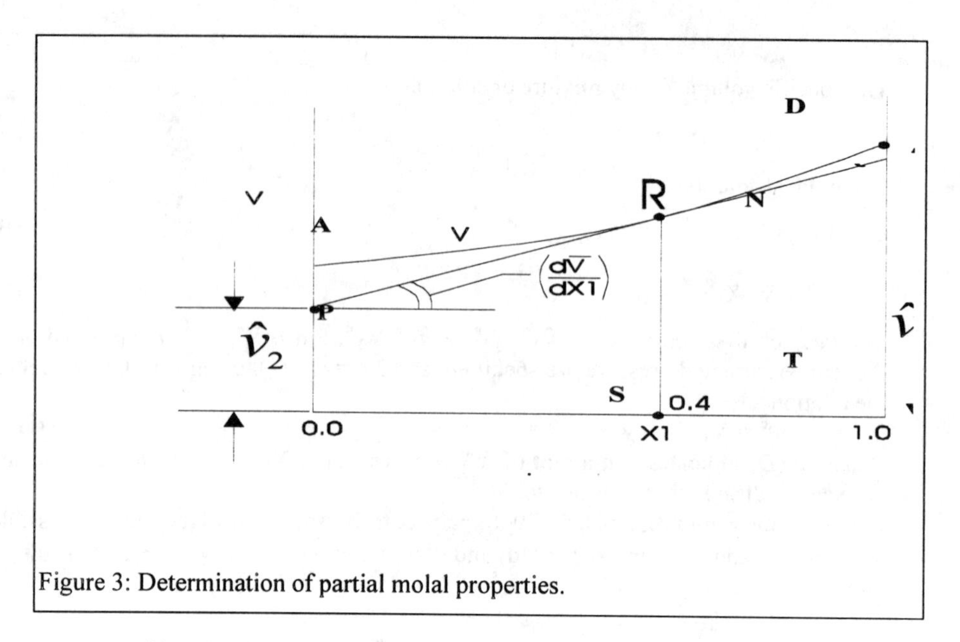

Figure 3: Determination of partial molal properties.

water (2) at 25°C, 1 atm while Figure 3 illustrates a graphical method that can be used to determine $\hat{v}_2$ by applying Eq. (F). For instance, a tangent to the $\bar{v}- X_1$ curve at point R yields the slope $d\bar{v}/dX_1$, and the intercept PS represents $\hat{v}_2$ at that value of X_1 (i.e., at R). As $X_1 \to 1$, the mixture is virtually pure so that species 1 is mainly surrounded by like molecules, and

$$\hat{v}_{1,\,X_1\to 1} = \bar{v}_1 = 0.04072 \text{ m}^3 \text{ kmole}^{-1}. \text{ (Point D, Figure 2)}.$$

As $X_1 \to 0$, Eq. (E) yields

$$\hat{v}_{1,\,X_1\to 0} = \bar{v}_{X_1\to 0} + (d\bar{v}/dX_1)_{X_1\to 0} \approx \bar{v}_2 + (d\bar{v}/dX_1)_{X_1\to 0}. \quad \text{(G)}$$

However the slope $(d\bar{v}/dX_1)_{X_1\to 0} \neq 0$. Then (G) yields

$$(d\bar{v}/dX_1)_{X_1\to 0} = 0.0194. \quad \text{(H)}$$

We must, therefore, resort to experiments from which we find that
$\hat{v}_{1,\,X_1\to 0} = \hat{v}_1 = 0.0375 \text{ m}^3 \text{ kmole}^{-1}$ (point G, Figure 2)
At this condition, species 1 is in trace amounts and is surrounded mainly by unlike molecules with force fields dominated by molecules of species 2. Likewise,
$\hat{v}_{2,X1\to 1} = \bar{v}_2 = 0.0181 \text{ m}^3 \text{ kmole}^{-1}$, (point A) and
$\hat{v}_{2,X2\to 0} = \bar{v}_2 = 0.015 \text{ m}^3 \text{ kmole}^{-1}$ (point M).
Applying Eq. (A) as. $X_1 \to 0$, and $X_2 \to 1$, using Eq.(G),

$$\bar{v} = X_1\, \hat{v}_{1,X1\to 0} + X_2\, \hat{v}_{2,X2\to 1} = X_1\, \hat{v}_{1,X1\to 0} + (1-X_1)\, \bar{v}_2$$

Simplifying,

$$v = \bar{v}_2 + X_1\,(\hat{v}_{1X_1\to 0} - \bar{v}_2) \quad \text{(I)}$$

An extension to multiple components is given in remarks. Since,
$\hat{v}_{1,\,X_1\to 0} = 0.0375 \text{ m}^3 \text{ kmole}^{-1}$, and
$\hat{v}_2 = \bar{v}_2 = 0.0181 \text{ m}^3 \text{ kmole}^{-1}$, since $X_2 \to 1$.

Then for small values of X_1, Eq. (I) yields

$\bar{v} = 0.0181 + 0.0194\ X_1$.

Thus the mixture volume increases linearly with X_1 as $X_1 \to 0$. Similarly, as $X_2 \to 0$, one can show that

$\bar{v} = 0.0407\ (1 - X_2) + 0.015\ X_2$

Remarks

The expression $\bar{v}^{id} = \bar{v}_1 X_1 + \bar{v}_2 X_2 = \bar{v}_1 X_1 + \bar{v}_2 (1-X_1) = (\bar{v}_1 - \bar{v}_2) X_1 + \bar{v}_2$ is also known as the Law of Additive Volumes (or the Lewis–Randall rule for volume). The law presumes that the intermolecular attraction forces between unlike molecules are the same as those between like molecules. This is a reasonable assumption for successive homologous series of hydrocarbons (e.g., pentane, hexane, etc).

Using a Taylor series expansion,

$$V(N_1+dN_1,N_2+dN_2,N_3+dN_3,..) = V(N_1,N_2,N_3,..) + \partial V/\partial N_1 dN_1 + \partial V/\partial N_2 dN_2 + \ldots .$$

Suppose initially there is only species 1 and hence initial amounts of N_2, N_3 .. etc. are equal to zero. Assume that small amounts of species 2, 3, 4, ... are being added to species 1. In that case, $\delta N_1 = 0$, and $\delta N_2 = N_2$, $\delta N_3 = N_3$, $\delta N_4 = N_4$, ... are small. Therefore,

$$V(N_1, 0+N_2, 0+N_3) = V(N_1) + \partial V/\partial N_1 \times (0) + \partial V/\partial N_2\, N_2 + \partial V/\partial N_3\, N_3 + \ldots, \text{ i.e.,}$$

$$V(N_1, N_2, N_3, \ldots) = V(N_1) + \hat{v}_2 N_2 + \hat{v}_3 N_3 + \ldots .$$

Dividing throughout by the total number of moles N (recall that $N \to N_1$),

$$\bar{v}\ (X_1 \to 1, X_2 \to 0) = \bar{v}_1 + \hat{v}_2 N_2/N + \hat{v}_3 N_3/N + \ldots = \bar{v}_1 + \hat{v}_2 X_2 + \hat{v}_3 X_3 + \ldots .$$

where X_2, X_3, ... denote mole fractions of trace species. In a salt (solute – species 2) and water (solvent – species 1) solution, an upper limit $X_{2,upper} \approx 0.3$ exists at standard conditions, which is called the solubility limit. In a mixing tank, the addition of salt beyond a 30% salt mole fraction usually results in the settling of solid salt. Therefore, one may not reach the limit $X_2 \to 1$ or $X_1 \to 0$ in a solution.

c. *Example 3*

The following relation describes the volume change in a water (solvent – species 1) and salt (solute – species 2) solution:

$$V = 1.001 + 16.625 N_2 + 56.092\ N_2^{3/2} + 119.4\ N_2^2, \quad (A)$$

where V is expressed in units of liters, and N_2 in kmole. Obtain an expression for $\hat{v}_2$ in terms of N_2, and determine the value of $\hat{v}_1$. Determine $\hat{v}_{2X_2 \to 0}$ or $\hat{v}_2^\infty$ in liter per kmole. Obtain an expression for $\hat{v}_1$ in terms of N_2, and one in terms of X_1.

Solution

Differentiating Eq. (A),

$$\hat{v}_2 = \partial V/\partial N_2 = 16.625 + 84.138\ N_2^{1/2} + 238.8\ N_2 \text{ in units of l kmole}^{-1}. \quad (B)$$

As $N_2 \to 0$ (or $X_2 \to 0$), the solute volume at infinite dilution

$\hat{v}_2 = 16.625$ l kmole^{-1}.

Furthermore, $V = \hat{v}_1 N_1 + \hat{v}_2 N_2$, i.e.,

$$\hat{v}_1 = (V - \hat{v}_2 N_2)/N_1. \quad (C)$$

Since a liter of water corresponds to a 1 kg mass,

$N_1 = 1$ kg/18.02 kg kmole^{-1} = 0.0555 kmole,
Therefore, applying Eq. (C),

$$\hat{v}_1 = 18.02 - 505.4\ N_2^{3/2} - 2151.6\ N_2^2 \text{ in units of l kmole}^{-1}. \quad \text{(D)}$$

The mole fraction $X_2 = (1 - X_1) = N_2/(N_2 + N_1) = N_2/(N_2 + 0.055)$, i.e.,
$N_2 = 0.055\ (1/X_1 - 1)$, and

$$\hat{v}_1 = 18.04 - 6.625(1/X_1 - 1)^{3/2} - 6.567(1/X_1 - 1)^2,\ X_1 > 0, \text{ in units of l kmole}^{-1}.$$

Example 4

A large flexible tank is divided into two sections A and B by a partition. Section A consists of 2 kmole of C_2H_2 (species 1) at 320 K and 100 bar, and section B consists of 3 kmole of CO_2 (species 2) at the same temperature and pressure. The partition is removed, but the temperature and pressure are maintained constant. Determine V_A, and V_B using the RK equation of state for each component before mixing and the total mixture volume using an ideal mixture model.

Solution

The RK equation has the form

$$P = RT/(\bar{v} - \bar{b}) - \bar{a}/(T^{1/2}\,\bar{v}(\bar{v} + \bar{b})).$$

Therefore,
$\bar{v}_1 = 0.09$ m^3 kmole^{-1}, and $V_A = 2 \times 0.09 = 0.18$ m^3.
$\bar{v}_2 = 0.10$ m^3 kmole^{-1}, and $V_B = 3 \times 0.1 = 0.3$ m^3.
$V^{id} = \bar{v}N = (\bar{v}_1 X_1 + \bar{v}_2 X_2)N = \bar{v}_1 N_1 + \bar{v}_2 N_2 = 2 \times 0.09 + 3 \times 0.1 = 0.48$ m^3.

Remarks

This is an illustration of the law of additive volumes. When the partition is removed, the total volume is the same as the combined original volume at the same temperature and pressure.
The actual volume $V = \hat{v}_1 N_1 + \hat{v}_2 N_2$ can differ from V^{id}, i.e., if the partition is removed in the rigid system, the final pressure may not be 100 bar.

---

d. *Example 5*

Determine V^{id} for a gaseous mixture of 0.3 kmole of H_2O (species 1) and 9.7 kmole of N_2 (species 2) at 160°C and 100 kPa, and 25°C and 100 kPa.

Solution

Since the 433 K temperature is high, we expect each species to behave as though it is an ideal gas in its pure state. Therefore,
$V^{id} = 0.3\,\bar{v}_1 = (433$ K,100 kPa$) + 9.7\,\bar{v}_2(433$ K,100 kPa), where
$\bar{v}_1(433$ K, 100 kPa$) = 8.314 \times 423 \div 100 = 36$ m^3 kmole^{-1} (from the tables for superheated vapor $\bar{v}_1 = 35.8$ m^3 kmole^{-1}), and
$\bar{v}_2(433$ K, 100 kPa$) = 8.314 \times 423 \div 100 = 36$ m^3 kmole^{-1}.
$V^{id} = 0.3 \times 36 + 9.7 \times 36 = 10 \times 36 = 360$ m^3.

At 298 K,
$V^{id} = 0.3\,\bar{v}_1(298$ K, 100 kPa$) + 9.7\,\bar{v}_2(298$ K, 100 kPa).
Although water exists as a liquid under these conditions, in the mixture it exists as a vapor. Therefore,
$V^{id} = 0.3\,\bar{v}_1(298$ K, 100 kPa, liquid$) + 9.7\,\bar{v}_2(298$ K, 100 kPa)
$= 0.3 \times 0.018 + 9.7 \times 24.8 = 240.5$ m^3.

If we use a hypothetical gaseous state for water at 25°C and 100 kPa, then using the ideal gas law
$V^{id} = 0.3\,\bar{v}_1(298$ K, 100 kPa$) + 9.7\,\bar{v}_2(298$ K, 100 kPa)

$= 0.3 \times 24.8 + 9.7 \times 24.8 = 248\ m^3$.

Remarks

It is more accurate to determine V^{id} using the specific volumes of pure components in the same state as they exist in the mixtures.

We have used a hypothetical state to determine the volume pure water, since the water changes phase in the mixture.

Suppose we have 0.3 kmole of $H_2O(l)$ in compartment A at P =100 kPa and 9.7 kmole of N_2 at P= 100 kPa in compartment of B of a PCW assembly, which is immersed in a bath at 25°C. If the partition is removed and allowed to equilibrate at 25°C, 100KPa, then the vapor will become $H_2O(g)$. Since the water changes phase after mixing, it must be endothermic, i.e., heat must be supplied from the thermal bath to evaporate the water molecules. Therefore, after mixing, $F_{kj} \ll F_{kk}$.

---

5. Relations between Partial Molal and Pure Properties

We have discussed the partial molal volume and now focus on other partial molal properties.

a. *Partial Molal Enthalpy and Gibbs function*

Since the enthalpy $H(T, P, N_1, N_2, \ldots) = U(T, P, N_1, N_2, \ldots) + PV(T, P, N_1, N_2, \ldots)$, the partial molal enthalpy $\hat{h}_i = (\partial H/\partial N_i)_{T,\ P,\ N_1, N_2, \ldots N_{j\neq i}, \ldots N_K}$ can be expressed as

$$\hat{h}_i = (\partial U/\partial N_i)_{T,\ P,\ N_1, N_2, \ldots N_{j\neq i}, \ldots N_K} + P(\partial V/\partial N_i)_{T,\ P,\ N_1, N_2, \ldots N_{j\neq i}, \ldots N_K}, \text{ i.e.,} \quad (26)$$

$$\hat{h}_i = \hat{u}_i + P\,\hat{v}, \text{ where} \quad (27)$$

$$\hat{u}_i = (\partial(N\,\bar{u})/\partial N_i)_{T,\ P,\ N_1, N_2, \ldots N_{j\neq i}, \ldots N_K} = N(\partial\,\bar{u}/\partial N_i)_{T,\ P,\ N_1, N_2, \ldots N_{j\neq i}, \ldots N_K} + \bar{u}.$$

Similarly, since $G = H - TS$,

$$\hat{g}_i = (\partial H/\partial N_i)_{T,\ P,\ N_1, N_2, \ldots N_{j\neq i}, \ldots N_K} - T(\partial S/\partial N_i)_{T,\ P,\ N_1, N_2, \ldots N_{j\neq i}, \ldots N_K}, \text{ i.e.,} \quad (28)$$

$$\hat{g}_i = \hat{h}_i - T\,\hat{s}_i. \quad (29)$$

Likewise,

$$\hat{a}_i = \hat{u}_i - T\,\hat{s}_i. \quad (30)$$

b. *Differentials of Partial Molal Properties*

Applying the Gibbs–Duhem equation

$$(\partial B/\partial T)_{P,N}dT + (\partial B/\partial T)_{T,N}dP - \Sigma_k d\,\hat{b}_k N_k = 0 \quad (12c)$$

in terms of the Gibbs energy, i.e., $B = G$, $\partial G/\partial T = -S$, and $\partial G/\partial P = V$, so that

$$-S\,dT + V\,dP - \Sigma_k d\,\hat{g}_k\,N_k = 0.$$

In terms of intensive properties, this relation may be written in the form

$$-\Sigma_k\,\hat{s}_k N_k dT + \Sigma_k\,\hat{v}_k N_k dP - \Sigma_k d\,\hat{g}_k N_k = -\Sigma_k N_k(d\,\hat{g}_k + \hat{s}_k dT - \hat{v}_k dP) = 0, \text{ i.e.,}$$

For arbitrary $N_k > 0$,

$$d\,\hat{g}_k = -\hat{s}_k\,dT + \hat{v}_k\,dP. \quad (31)$$

Differentiating Eq. (29) and using (31) to eliminate $d\hat{g}_k$,

$$d\hat{h}_k = Td\hat{s}_k + \hat{v}_k dP. \quad (32a)$$

Dividing Eq. (32) by dT at constant pressure, the partial molal specific heat

$$\hat{c}_{pk} = \partial \hat{h}_k/\partial T = T\,\partial \hat{s}_k/\partial T. \quad (32b)$$

Subtracting the term $d(Pv_k)$ from Eq. (32a), we obtain the relation

$$d\hat{u}_k = T\,d\hat{s}_k - P\,d\hat{v}_k. \quad (33a)$$

Similarly

$$\hat{c}_{vk} = (\partial \hat{u}_k/\partial T) = T\,(\partial \hat{s}_k/\partial T). \quad (33b)$$

These relations for partial molal properties are similar to those for pure substances. Maxwell's relations can be likewise derived. Subtracting $d(T\hat{s}_k)$ from Eq. (33a), we obtain the relation

$$d\hat{a}_k = -\hat{s}_k\,dT - P\,d\hat{v}_k. \quad (34)$$

This implies that $\hat{s}_k = -\partial \hat{a}_k/\partial T$, $P = -\partial \hat{a}_k/\partial \hat{v}_k$. These expressions are similar to those for pure properties. Maxwell's relations can be likewise derived. Furthermore, from Eq. (32a)

$$\partial \hat{h}_k/dP = T\partial \hat{s}_k/dP + \hat{v}_k. \quad (35)$$

Using the Maxwell's relations we can show that

$$\partial \hat{h}_k/dP = -T\,\partial \hat{v}_k/dT + \hat{v}_k.$$

For the entropy, the G–D relation is

$$(\partial S/\partial T)_{P,N}\,dT + (\partial S/\partial P)_{T,N}\,dP - \Sigma_k d\hat{s}_k\,N_k = 0 \quad (36)$$

Since $S = \Sigma_k \hat{s}_k N_k$ and $\partial \hat{s}_k/\partial T = \hat{c}_{pk}/T$, we may use Maxwell's relations to simplify the second term, i.e., $(\partial S/\partial P)_{T,N} = -(\partial V/\partial T)_P$, where $V = \Sigma_k N_k \hat{v}_k$ so that

$$\Sigma_k N_k(\hat{c}_{p,k}/T)dT - \Sigma_k(N_k\partial \hat{v}_k/\partial T)\,dP - \Sigma_k d\hat{s}_k N_k = 0, \text{ or}$$

$$\Sigma_k N_k((\hat{c}_{p,k}/T)dT - (\partial \hat{v}_k/\partial T)dP - d\hat{s}_k) = 0.$$

Therefore,

$$d\hat{s}_k = (\hat{c}_{p,k}/T)dT - (\partial \hat{v}_k/\partial T)\,dP. \quad (37)$$

Using Eqs. (37) in Eq. (32a),

$$d\hat{h}_k = \hat{c}_{p,k}\,dT + (\hat{v}_k - T\,(\partial \hat{v}_k/\partial T))\,dP, \quad (38)$$

which is again similar to the corresponding expression for a pure substance. Likewise,

$$d\hat{u}_k = \hat{c}_{v,k}\,dT + (T(\partial P/\partial T) - P)\,d\hat{v}_k \quad (39)$$

Equations (37) to (39) are similar to the corresponding expressions for a pure substance. Thus if state equations are available for mixtures, $\hat{u}_k$, $\hat{h}_k$ and $\hat{s}_k$ can be determined.

i. Remarks

Maxwell's relations can be obtained using Eqs. (31)–(34). These relations are similar to those for pure components.

Consider the derivative $(\partial/\partial N_i(\partial V/\partial T)_{P, N_1, N_2, \ldots N_{j\neq i}, \ldots N_K})_{T, P, N_1, N_2, \ldots N_{j\neq i}, \ldots N_K}$. Switching order of differentiation, the expression equals the term

$$(\partial/\partial T(\partial V/\partial N_i)_{T, P, N_1, N_2, \ldots N_{j\neq i}, \ldots N_K})_{P, N_1, N_2, \ldots N_{j\neq i}, \ldots N_K} = \partial \hat{v}_i/\partial T.$$

Likewise,

$$(\partial/\partial N_i(\partial S/\partial T)_{P, N_1, N_2, \ldots N_{j\neq i}, \ldots N_K})_{T, P, N_1, N_2, \ldots N_{j\neq i}, \ldots N_K} = T\, \partial \hat{s}_i/\partial T = \hat{c}_{pi}, \quad (40)$$

which is again an expression that is similar to that for a pure substance.

6. Ideal Gas Mixture

a. Volume

Since gases are ideal, there are no intermolecular forces. Hence an ideal gas $\hat{v}_k = \bar{v}_k$, and, hence,

$$V(T, P, N_1, N_2, \ldots) = \Sigma_k\, \bar{v}_k(T, P)\, N_k.$$

Using the ideal gas law for each component,

$$V(T, P, N_1, N_2, \ldots) = \Sigma_k(\bar{R}T/P)N_k = V_1 + V_2 + \ldots = N(\bar{R}T/P). \quad (41)$$

Equation (41) is a representation of the law of additive volumes. For a mixture the relation assumes the form

$$PV = N\bar{R}T = (N_1 + N_2 + \ldots)\bar{R}T. \quad (42)$$

Equation (42) suggests that

$$(\partial V/\partial N_k)_{T, P, N_1, N_2, \ldots N_{j\neq k}, \ldots N_K} = \hat{v}_k = \bar{R}T/P = \bar{v}_k. \quad (43a)$$

Therefore,

$$\bar{v}^{ig} = \Sigma_k X_k\, \hat{v}_k = \Sigma_k X_k\, \bar{v}_k \quad (43b)$$

b. Pressure

Another form of Eq. (42) is

$$P = N_1\bar{R}T/V + N_2\bar{R}T/V + \ldots = p_1(T,V,N_1) + p_2(T,V,N_2) + \ldots. \quad (44)$$

If each component alone occupies the whole volume, the pressure exerted by component k is $p_k = N_k\, \bar{R}T/V$ (which is called component pressure and is the same as the partial pressure for ideal gases). Then the pressure exerted by the mixture

$$P = \Sigma_k p_k(T,V,N_k), \quad (45)$$

which is also known as Dalton's law of additive pressure.

---

e. Example 6

Partially combusted products consist of propane (C_3H_8): 0.5, O_2 : 2.5, CO_2: 1.5, and H_2O: 2 kmoles at T = 298 K and P= 1bar. Determine the partial molal volume of CO_2 using the ideal gas mixture model at 1 bar and 298 K.

Solution

The mixture consists of 0.5, 2.5, 1.5 and 2 kmole of C_3H_8, O_2, CO_2 and H_2O, respectively.

$V = (\bar{R}T/P)\,\Sigma_k N_k$, and

$\hat{v}_{CO_2} = (\partial V/\partial N_{CO_2})_{T,\,P,\,N_1,N_2,...,N_{j\neq k},...,N_K} = \bar{R}T/P = \bar{v}$

$= 0.08314\times 298 \div 1 = 24.78\ m^3\ kmole^{-1}$, and is the same for all species.

Thus $V = \bar{v}N = 161.07\ m^3$.

If a room contains the mixture at T and P, we can hypothesize that 0.5 kmole (i.e., 3×10^{26} molecules) occupies $12.39\ m^3$ while O_2 occupies $61.95\ m^3$.

---

c. *Internal Energy*

The internal energy of a system is the combined energy contained in all of the molecules in the system. In ideal gases the internal energy of a species in a mixture equals its energy in a pure state at the temperature and total pressure of the mixture. Therefore,

$$U_0(T,N) = \Sigma_k\, \bar{u}_{k,0}(T)\, N_k. \tag{46}$$

Eq. (46) is known as Gibbs-Dalton (GD) law Dividing throughout by N

$$\bar{u}_0(T,X) = \Sigma_k X_k\, \bar{u}_0(T) \tag{47}$$

d. *Enthalpy*

Since

$$H_0 = U_0 + PV_0, \tag{48}$$

as before,

$$H_0(T,N) = \Sigma_k\, \bar{h}_{k,0}(T) N_k$$

$$\bar{h}_k(T) = \bar{u}_{k,0} + P\,\bar{v}_k = \bar{u}_{k,0} + \bar{R}T, \text{ ideal gas} \tag{49a}$$

Similarly

$$\bar{h}_0(T) = \Sigma_k X_k\, \bar{h}_{k,0}(T) \tag{49b}$$

e. *Entropy*

Two gases A and B are contained in a chamber on two sides of a rigid partition in volumes V_A and V_B at specified values of temperature and pressure. The partition is removed and the gases are allowed to mix in a volume V, but the pressure and temperature are unchanged. For an ideal gas

$$dS = N_B\, \bar{c}_{v,0} dT/T + N_B \bar{R}(dV/V).$$

A change in volume occupied by, say, species B from V_B to V (e.g., if there are two adjacent rooms with O_2 and N_2 in each room and if the partition between them is removed, the N_2 now occupies a larger volume making more quantum states available for energy storage) results in an entropy change at same temperature. Then,

$$S_{B,mix}(T,V,N_B) - S_B(T,V_B,N_B) = N_B \bar{R}\ln(V/V_B).$$

The volume $V = N\bar{R}\ T/P$ and $V_B = N_B \bar{R}\ T/P$, i.e.,

$$S - S_B = N_B \bar{R}\ln(V/V_B).$$

The volume $V = N\bar{R}\ T/P$ and $V_B = N_B \bar{R}\ T/P$, i.e.,

$$S_{B\,mix}(T,V,N_B) - S_B(T,V_B,N_B) = N_B \bar{R}\ln(N/N_B) = -N_B \bar{R}\ln(X_B).$$

Dividing by N_B,

$$\hat{s}_B(T, P, X_B) - \bar{s}_B(T, P) = -\bar{R}\ln(X_B).$$

Further,

$$S_{B,final} - S_{B,initial} = N_B\bar{R}\ln(N/N') = -N_B\bar{R}\ln(X_B), \text{ i.e.,}$$

$$\hat{s}_B(T, P, X_B) - \bar{s}_B(T, P) = -\bar{R}\ln(X_B). \quad (50)$$

Similarly $\hat{S}_A(T,P,X_A) - \bar{S}_A(T,P) = -\bar{R}\ln X_A$. Since $X_k < 1$, $\hat{s}_k > \bar{s}_k$, indicating that mixing causes the entropy to increase. The mixture entropy

$$S^{ig} = S_0 = N_B\hat{s}_B(T, P, X_B) + N_A\hat{s}_A(T, P, X_A). \quad (51)$$

Recall that

$$\bar{s}(T, P) = s^o(T) - \bar{R}\ln(P/1), \text{ i.e., } \hat{s}_k(T, p_k) = \bar{s}_k{}^o(T) - \bar{R}\ln(p_k/1), \text{ where} \quad (52)$$

$$\bar{s}_k^0 = \int_{T_{ref}}^{T}(c_{p,k,o}/T)dT. \quad (53)$$

Now we will generalize Eq. (51) for several components in a mixture and add a subscript "0" to denote that the gas is ideal, i.e.,

$$S_0(T, P, N) = \Sigma_k\hat{s}_{k0}N_k, \text{ and } \bar{s} = S_0/N = \Sigma_k\hat{s}_{k0}X_k, \text{ or} \quad (54)$$

$$S_0(T, P, N) = \Sigma_k\hat{s}_{k0}(T, p_k)N_k, \text{ and } \bar{s} = S_0/N = \Sigma_k\hat{s}_{k0}(T, P_k)X_k, \text{ and} \quad (55)$$

$$\bar{s} = \Sigma_k\hat{s}_k X_k = \Sigma_k\hat{s}_k(T, P_k)X_k = \Sigma_k(\hat{s}_k(T, P) - \bar{R}\ln X_k)X_k. \quad (56)$$

f. Gibbs Free Energy

The Gibbs free energy $G = H - TS$ so that

$$\hat{g}_k = \hat{h}_k - T\hat{s}_k. \text{ Then}$$

$$\hat{g}_{k,0} = \mu_{k,0} = \bar{h}_{k,0} - T(\bar{s}_{k,0}(T, P) - \bar{R}\ln X_k) = \bar{g}_{k,0}(T, P) + \bar{R}T\ln X_k. \quad (57)$$

In mixtures, as X_k increases, so does $\hat{g}_k$, since $RT\ln X_k < 0$. The property $\hat{g}_k$ is a measure of the free energy or availability of the k–th species in the mixture. Rewriting,

$$\bar{g}_0 = \Sigma_k X_k\hat{g}_{k,0}(T, P, X_1, Y_2, \ldots) = \Sigma_k X_k\bar{g}_{k,0}(T, P_k) = \Sigma_k X_k(\bar{g}_{k,0}(T, P) + \bar{R}T\ln X_k). \quad (58)$$

---

f. Example 7

A mixture contains 60% N_2 and 40% O_2 at 298 K and 2 bar. The mixture is placed in a cylinder A that is connected by a rigid semipermeable membrane to another cylinder B that contains pure N_2 at 298 K. Cylinder B is maintained at constant pressure through a piston with adjustable weights. When the pressure in cylinder B is relatively low, N_2 is transferred from A to B. As the pressure in cylinder B is raised (by placing weights on the piston), the transfer of N_2 from cylinder A to B ceases at 1.19 bar. What is this chemical potential N_2 species in the mixture?

Solution

Recalling that partial molal Gibbs function is same as chemical potential, and omitting the subscript "o"

$$\mu_{N_2,mix} = \mu_{N_2}(298, 2\text{ bar}) + RT\ln X_{N_2}$$

$$= 298 - 298\times(\ln(298\div273) - 0.297\times\ln(2\div1)) + 0.297\times298\ln(0.6)$$

= 333.2 – 45.2 = 288.0 kJ per kmole of N_2

7. Ideal Solution

a. Volume

A liquid mixture in which all of the components are miscible at the molecular level is called an ideal solution, provided the following condition is satisfied, i.e.,

$$\hat{v}_k = \bar{v}_k, \text{ and } \bar{v}^{id} = \Sigma_k X_k \bar{v}_k^{\ id} = \Sigma_k X_k \bar{v}_k. \quad (59)$$

In an ideal solution, the forces between the unlike molecules are assumed to be the same as those between like molecules.

b. Internal Energy and Enthalpy

In ideal mixture of liquids or real gases,

$$\bar{h}^{id} = \Sigma_k X_k \bar{h}_k, \text{ and} \quad (60)$$

$$\bar{u}^{id} = \bar{h}^{id} - P\,\bar{v}^{id} = \Sigma_k X_k(\bar{h}_k - P\,\bar{v}_k). \quad (61)$$

c. Gibbs Function

At a specified temperature, the change in the Gibbs free energy of a pure component (i.e. when it is alone at T and P) is given as (Chapter 7)

$$d\,\bar{g}_k = \bar{v}_k(T, P)\, dP, \text{ i.e.,} \quad (a)$$

If the composition and temperature in a mixture are held fixed and the pressure is altered then applying Eq.(31) for k in a mixture

$$d\,\hat{g}_k = \hat{v}_k\,(T, P, X_k)\, dP \quad (62a)$$

Then from Eqs. (a) and (62a)

$$d(\hat{g}_k - \bar{g}_k) = (\hat{v}_k(T, P, X_k) - \bar{v}_k(T, P))dP. \quad (62b)$$

If a mixture of fixed composition is subjected to an incremental pressure dP, and the pure component is also subjected to the same pressure increment, this expression provides the difference between two Gibbs function $\hat{g}_k$ and $\bar{g}_k$ due to difference in $\hat{v}_k$ and $\bar{v}_k$. For example, H_2O in the solution is compressed from 200 to 250 kPa, with $\hat{v}_{H_2O}$ = 0.015 m^3 $kmole^{-1}$, $d\hat{g}_{H_2O}$ = 0.75 kJ $kmole^{-1}$. On the other hand, $\bar{v}_{H2O}$ = 0.018 m^3 $kmole^{-1}$ and $d\,\bar{g}_{\ H2O}$ = 0.9 $kJkmole^{-1}$. Integrating

$$\hat{g}_k - \bar{g}_k = \int(\hat{v}_k - \bar{v}_k)dP + f(T, X_k).$$

In an ideal solution at any temperature and pressure, $\hat{v}_k = \bar{v}_k$ and hence

$$\hat{g}_k^{\ id} - \bar{g}_k = f(T, X_k).$$

As P→0 at a specified temperature and composition the same relation should hold good. Hence $\hat{g}_k^{\ id}$ and $\bar{g}_k$ approach their corresponding values in an ideal gas. Therefore, f(T, X_k) = $\bar{R}T \ln X_k$. consequently, in an ideal solution

$$\hat{g}_k^{\ id} = \bar{g}_k(T, P) + \bar{R}T \ln X_k \text{ or } \hat{g}_k id - \bar{g}_k\,(T, P) = \bar{R}T \ln X_k. \quad (62c)$$

d. Entropy

Since,

$\hat{g}_k^{id} = \hat{h}_k - T\hat{s}_k(T, P, X_k) = \hat{h}_k - T\bar{s}_k(T, P) + \bar{R}T \ln X_k$ i.e.,

then

$$\hat{s}_k(T, P, X_k) = \bar{s}_k(T, P) - \bar{R} \ln X_k. \quad (63)$$

The entropy of the ideal mixture is

$$S^{id}(T, P, N) = \Sigma_k \bar{s}_k(T, P, X_k)N_k, \text{ and } \bar{s}^{id} = S^{id}/N = \Sigma_k \bar{s}_k(T, P, X_k)X_k. \quad (64)$$

---

g. Example 8

Lake water at 25°C and 1 bar absorbs air from the atmosphere. If the air mole fraction in the liquid is 0.001, what is the entropy of H_2O in the lake water?

Solution

Air consists of O_2 and N_2 molecules that have weak attractive forces between themselves in the atmosphere. However once these molecules enter liquid water, they are surrounded by H_2O molecules which exert strong attractive forces and hold the gas molecules in the liquid phase. Thereby, small amounts of air become dissolved in liquid water.

$\hat{s}_{H_2O}(25°C, 1 \text{ bar}, 0.999) = \bar{s}_{H_2O}(25°C, 1 \text{ bar}) - \bar{R} \ln X_{H_2O}$.

Pure water exists at 25°C and 1 bar as a compressed liquid. We will assume that the liquid is incompressible and that $\bar{s}_{H_2O}(25°C, P^{sat} = 0.032 \text{ bar}) = \bar{s}_{H_2O}(25°C, 1 \text{ bar}) =$ 6.621 kJ kmole^{-1} K^{-1}. Therefore,

$\hat{s}_{H_2O}(25°C, 1 \text{ bar}, 0.999) = 6.621 - 8.314 \times \ln 0.999 = 6.629$ kJ kmole^{-1}.

---

8. Fugacity

a. Fugacity and Activity

As in Chapter 7 we define the fugacity for a component as

$$d\bar{g}_k = \bar{R}T\, d(\ln f_k(T, P)) \quad (65)$$

In analogy with the pure component (Eq. (65)), we can write Eq. (62a) in the form

$$d\hat{g}_k = \hat{v}_k\, dP = \bar{R}T\, d \ln(\hat{f}_k(T, P, X_k)). \quad (66)$$

The fugacity of species k in the mixture $\hat{f}_k$ is different from f_k since $\hat{v}_k \neq \bar{v}_k$. Note that $\hat{f}_k$ is not the partial molal fugacity of the k–th species in the mixture.

If $\hat{v}_k(P, T, X_k)$ is known, Eq. (66) may be integrated at a given composition to obtain $\hat{f}_k(T, P)$. Subtracting Eq. (66) from Eq. (65)

$$d(\hat{g}_k - \bar{g}_k) = \bar{R}T\, d(\ln(\hat{f}_k(T, P, X_k)/f_k(T, P))) = (\hat{v}_k(T, P, X_k) - \bar{v}_k(T, P))dP. \quad (67)$$

We introduce the activity of the k–th species in the mixture

$$\hat{\alpha}_k = (\hat{f}_k(T, P, X_k)/f_k(T, P)), \quad (68)$$

$$\hat{\alpha}_k = \text{fugacity of species k in the mixture} \div \text{fugacity of pure species k}. \quad (69)$$

Note that fugacity of k in mixture depends upon concentration. Generally, the species k is presumed to be more active in the mixture if its concentration is higher. The activity could also be different if intermolecular forces between dissimilar molecules are different from those of similar molecules. Substituting with Eq. (68) in Eq. (67), and integrating the results at a specified temperature,

$$\hat{g}_k(T, P, X_1, X_2..) - \bar{g}_k(T, P) = \bar{R}T \ln \hat{\alpha}_k + f(T). \quad (70)$$

Since the above equation is applicable even when k exists in large amounts at which $\hat{g}_k \to \bar{g}_k, \hat{\alpha}_k \to 1.0$ then f(T) =0, and

$$\hat{g}_k(T, P, X_1, X_2..) - \bar{g}_k(T, P) = \bar{R}T \ln \hat{\alpha}_k = \int (\hat{v}_k(T, P, X_k) - \bar{v}_k(T, P))dP \qquad (71)$$

Note that if $(\hat{v}_k - \bar{v}_k)$ is known as a function of P. Then $(\hat{g}_k - \bar{g}_k)$ and hence $\hat{\alpha}_k$ can be obtained as a function of T, P, X_k. The partial molal Gibbs function of a species is very important in determining the condition for its phase equilibrium (cf. Chapter 9) in a multicomponent mixture and as well as at chemical equilibrium (cf. Chapter 12). The fugacity is a useful property. For instance, since the chemical potentials $\mu_k = \hat{g}_k$ equal one another for the vapor and liquid in multiphase systems, i.e., $\hat{f}_{k,liq} = \hat{f}_{k,vapor}$. The value of the Gibbs function is normally a large negative number ($g \to (-\infty)$ when $X_k \to 0$ and $g \to 0$ as $X_k \to 1$), particularly for chemically reacting species, but the corresponding fugacity values lie between zero and unity.

b. Approximate Solutions for $\hat{g}_k$

In order to determine the value of $\hat{g}_k$, we require the values of pure properties and activities. Here, we present some approximate schemes to evaluate $\hat{g}_k$.

i. Ideal solution or the Lewis–Randall Model

For an ideal mixture,

$$\hat{g}_k^{id}(T, P, X_1, X_2, ...) - \bar{g}_k(T, P) = \bar{R}T \ln \hat{\alpha}_k^{id}. \qquad (72)$$

Using Eqs. (62c), (68) and (72),

$$\hat{\alpha}_k^{id} = \hat{f}_k^{id}(T, P, X_k)/f_k(T, P) = X_k, \text{ i.e.,} \qquad (73a)$$

$$\hat{f}_k^{id}(T, P, X_k) = X_k f_k(T, P). \qquad (73b)$$

Equation (73b) is known as the Lewis–Randall (LR) Rule, which assumes that $\hat{v}_k = \bar{v}_k$. Errors due to the model are lowest for adjacent homologous series, e.g., n–hexane and n–heptane, and methanol and ethanol. Line AC in Figure 4 illustrates the LR model for H_2O: N_2 mixture.

ii. Henry's Law

The ideal solution model for component 1 of a binary mixture oftentimes fails when $X_1 \to 0$, i.e., when species 1 is surrounded by a large amount of species 2. Consider a mixture of 3 mole percent of H_2O (component 1) and 97 % (mole %) of N_2 (component 2) at 1 bar and 300 K. Under these conditions, water (component 1) exists as a vapor in the mixture. The relation $\hat{f}_1^{id} = X_1 f_1(300\text{ K}, 1\text{ bar}) = f_1(300\text{ K}, P^{sat}) + \bar{v}_f(1 - P^{sat})/(\bar{R}T)$ yields liquid–like fugacities (see Chapter 7 for f_k). This is clearly inappropriate for the H_2O which exists as vapor in the mixture. Since there are very few water molecules in the mixture, the N_2 molecules impose the force fields in the H_2O–N_2 mixture which are negligible attractive force fields. Therefore, the water exists in vapor form in the mixture. At low values of X_1, the actual fugacity of component 1 in the mixture corresponds to that of gaseous H_2O.

We can obtain $(d\hat{f}_1/dX_1)$ at low values of X_1 (e.g., slope at A for curve AEC in Figure 4) assuming that this gradient is constant and determine the value of $\hat{f}_1$, a method known as Henry's Law (HL), i.e.,

$$\hat{f}_1^{id}(HL) = X_1(d\hat{f}_1/dX_1)_{X_1 \to 0}. \qquad (74)$$

Line ADB represents $\hat{f}_{H_2O}^{HL}$ in a H_2O–N_2 mixture, which is valid when $X_{H_2O} \to 0$. The extrapolated fugacity of H_2O at B is called a hypothetical fugacity at $X_{H_2O}=1$ and is obtained from HL. Note that when water exists at low concentrations at 25°C and 1 bar, the water in the mixture may exist as a gas while at high concentrations it may exist as liquid with N_2 dissolved in liquid. Similarly we can use the slope at point C to determine $\hat{f}_{H_2O}^{HL}$ when $X_{N_2}\rightarrow 0$. Henry's Law accurately predicts the fugacity of a component k when $X_k\rightarrow 0$, while the Lewis-Randall Rule predicts $\hat{f}_k$ reasonably well when $X_k\rightarrow 1$.

c. Standard States

Instead of expressing $\hat{g}_k$ in terms of $\bar{g}_k(T, P)$, we can express $\hat{g}_k$ in terms of $\bar{g}(T, P_o)$, where P^o denotes a reference pressure. If we add and subtract $\bar{g}_k(T, P^o)$ to the LHS of Eq. (71), then

$$\hat{g}_k(T, P, X_k) - \bar{g}_k(T, P^o) - (\bar{g}_k(T, P) - \bar{g}_k(T, P^o))$$

$$= \hat{g}_k(T, P, X_k) - \bar{g}_k(T, P^o) - \bar{R}T \ln (f_k(T, P)/f_k(T, P^o)) = \bar{R}T \ln \hat{\alpha}_k, \text{ i.e.,}$$

$$\hat{g}_k(T,P,X_k) - \bar{g}_k(T,P^o) = \bar{R}T \ln (f_k(T,P)\ \hat{\alpha}_k/f_k(T,P^o)) = \bar{R}T \ln(\hat{f}_k(T, P)/f_k(T,P^o)). \quad (75)$$

If we choose $P^o = 1$ bar, then Eq. (75) assumes the form

$$\hat{g}_k(T, P, X_k) - \bar{g}_k(T,1) = \bar{R}T \ln (\hat{f}_k(T, P)/f_k(T,1)). \quad (76)$$

Alternately, sometimes the standard state can be selected for species k in a mixture itself rather than for the pure component at T, P^o, i.e., in the context of Eq. (75) $P = P^o$ for the k–th species in the mixture. Consequently,

$$\hat{g}_k (P^o, T) - \bar{g}_k(T, P) = \bar{R}T \ln \hat{\alpha}_k (T, P^o; T, P), \text{ where} \quad (77)$$

$$\hat{\alpha}_k (T, P^o; T, P) = \hat{f}_k(T, P_o)/f_k (T, P). \quad (78)$$

Subtracting Eq. (77) from Eq. (71) ,

$$\hat{g}_k (T,P, X_k) - \hat{g}_k(T, P^o) = \bar{R}T \ln(\hat{\alpha}_k(T,P)/\hat{\alpha}_k(T,P^o; T, P) = \bar{R}T \ln (\hat{f}_k(T, P)/\hat{f}_k(P^o,T)). \quad (79)$$

The state (P^o,T) is called the standard state of component k in the mixture. Typically P^o=1bar.

We have discussed four possible ways to express $\hat{g}_k(T, P)$ in terms of the Gibbs function of (1) a pure component at specified T and P (cf. Eq. (71)); (2) for a pure component at (T, P^o) (cf. Eq. (75)); (3) for a pure component at (T, 1 bar) (cf. Eq. (76)); and (4) for the k–th species in mixture at (T, P^o) (cf. Eq. (79)).

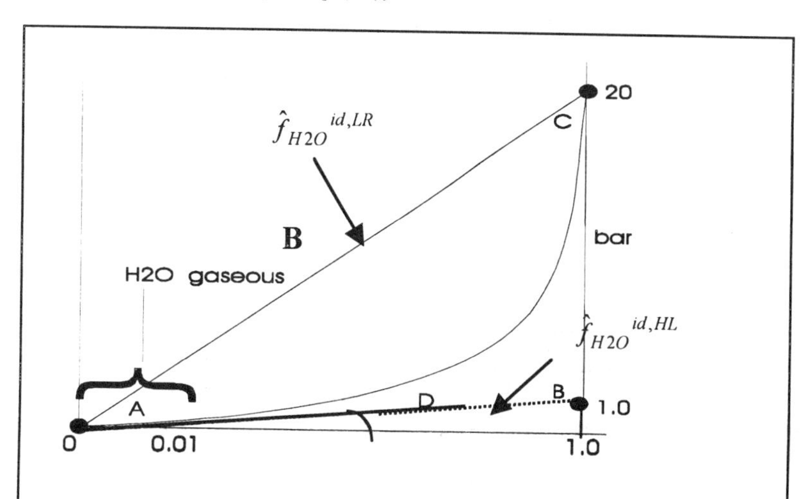

Figure 4: Illustration of Lewis Randall Rule and Henry's Law for the estimation of fugacity.

i. Gas Mixtures

In the context of Eq.(76), if the mixture consists of real gases and the component k is a gas in its pure state at any temperature and pressure,

$$\hat{g}_k(T, P, X_k) - \bar{g}_{k,o}(T, 1) = \bar{R}T \ln (f_k(T, P)\ \hat{a}_k/1).$$

For an ideal mixture of real gases $\hat{a}_k = X_k$, i.e.,

$$\hat{g}_k(T, P, X_k) - \bar{g}_{k,o}(T, 1) = \bar{R}T \ln (f_k(T, P)\ X_k/1). \quad (80)$$

For an ideal gas mixture at $f_k(T, P) = P$, i.e.,

$$\hat{g}_k(T, P, X_k) - \bar{g}_{k,o}(T, 1) = \bar{R}T \ln (P\ X_k/1) = \bar{R}T \ln(p_k/1). \quad (81)$$

ii. Liquid Mixtures

For a liquid mixture, the standard state can be chosen for the pure liquid component at any temperature, but at atmospheric pressure. Applying Eq. (76),

$$\hat{g}_k(T, P, X_{k(\ell)}) - \bar{g}_k(T, 1) = \bar{R}T \ln (f_{k(\ell)}(T, P)\ \hat{\alpha}_{k(\ell)}/f_k(T,1)). \quad (82)$$

The Poynting correction can be used thereafter to simplify this expression for $f_k(T,1)$ (cf. Chapter 7) in terms of saturation properties. The procedure for solid mixtures is similar to that for liquid mixtures.

d. Evaluation of the Activity of a Component in a Mixture.

We have discussed how to determine the value of $\hat{g}_k^{id}$ for ideal solutions (Eqs. (72) and (62c)) For non ideal mixtures, if a relation for $\hat{v}_k$ in terms of X_k, T, and P is available, then a relation for $\hat{g}_k(T, P, X_k)$ can be obtained (Eqs. (62a) and (71). Recall from Eq. (62c)

$$\hat{g}_k^{id} - \bar{g}_k = \bar{R}T \ln X_k. \quad (62c)$$

Further, from Eq. (71)

$$\bar{R}T \ln \hat{\alpha}_k = \int (\hat{v}_k - \bar{v}_k)dP, \quad (71)$$

and the activity $\hat{\alpha}_k = \hat{f}_k(T, P)/f_k(T, P)$. can be obtained as a function of T, P, X_k.

e. Activity Coefficient

Subtracting Eq. (62c) from Eq. (71), we obtain the relation

$$\hat{g}_k - \hat{g}_k^{id} = \bar{R}T \ln (\hat{\alpha}_k/\hat{\alpha}_k^{id}) = \bar{R}\ T \ln \gamma_k, \quad (83)$$

where the activity coefficient γ_k is defined as

$$\gamma_k = \hat{\alpha}_k/\hat{\alpha}_k^{id} \quad (84)$$

$$\gamma_k = (\hat{f}_k/(X_k\ \phi_k\ P)) = (\hat{\phi}_k/\ \phi_k), \text{ where} \quad (85)$$

$$\hat{\phi}_k = \hat{f}_k/(X_k\ P). \quad (86)$$

While activity includes the effects of concentration and intermolecular forces between dissimilar molecules (component k in mixtures), the activity coefficient separates the effects of concentration on the activity. Thus, the activity coefficient is strongly dependent upon the degree of intermolecular forces. Rewriting Eq. (83).

$$\hat{g}_k - \hat{g}_k^{id} = \bar{R}\ T \ln (\gamma_k) = \bar{R}\ T \ln (\hat{\phi}_k/\ \phi_k) \quad (87)$$

If the ideal state is selected as that for an ideal gas,

$$\hat{g}_k - \hat{g}_{k,o} = \hat{g}_k^{Res} = \overline{R}\, T \ln\, (\hat{\phi}_k), \qquad (88)$$

since $\phi_k = 1$ for an ideal gas.

We will now relate the fugacity coefficient $\hat{\phi}_k$ in terms of Z. Multiplying Eq. (88) by N_k,

$$G- G_0 = \Sigma \overline{R}\, T\ \ N_k \ln\, (\hat{\phi}_k)\ \text{ or }\ \overline{g} - \overline{g}_0 = \Sigma \overline{R}\, T X_k\ \ln (\hat{\phi}_k). \qquad (89)$$

Define

$$N \ln \phi = \Sigma N_k \ln\ (\hat{\phi}_k), \text{ so that} \qquad (90a)$$

$$\ln \phi = \Sigma X_k \ln\ (\hat{\phi}_k). \qquad (90b)$$

Then Eq. (89) assumes the form

$$G- G_0 = \Sigma \overline{R}\ T N_k \ln\ (\hat{\phi}_k) = \overline{R}\ T N \ln \phi, \text{ or } \overline{g} - \overline{g}_0 = \overline{R}\ T \ln \phi.$$

Differentiating with respect to N_k and using Eqs. (90a) and (90b),

$$(\partial(N \ln \phi)/\partial N_k)_{T,P,\ nj} = \ln \phi + N\ (\partial(\ln \phi)/\partial N_k)_{T,P,\ nj}, \text{ i.e.,}$$

$$(\partial\ (N \ln \phi)\ /\partial N_k)_{T,P,\ nj} = \ln\ (\hat{\phi}_k) \qquad (91)$$

Recall from Chapter 7 that

$$\ln \phi = \int_0^P (Z\text{-}1)\ dP/P. \qquad (92a)$$

Multiply Eq. (91a) by N and differentiate with respect toN_k,

$$\ln\ (\hat{\phi}_k) = (\partial\ (N \ln \phi)/\partial N_k)_{T,P,\ nj} = (\partial/\partial N_k)(\int_0^P (NZ\text{-}N)\ dP/P)(\int_0^P (\hat{Z}_k - 1)\ dP/P), \qquad (92b)$$

where $(\partial/\partial N_k)(NZ) = \hat{Z}_k$, since $\hat{Z}_k = (\partial/\partial N_k)(NZ) = (\partial/\partial N_k)(PV/\overline{R}\ T) = \hat{v}_k/\ (\overline{R}\ T/P) = \hat{v}_k/\ \overline{v}_{ko} = \hat{v}_k/\ \overline{v}_o$ = specific volume of k-th component in the mixture /ideal gas specific volume of the same component.

f. Fugacity Coefficient Relation in Terms of State Equation for P

Recall that G = A+ PV, i.e.,

$$G- G_0 = \Sigma \overline{R}\ T\ \ N_k \ln (\hat{\phi}_k) = A(T,P,N) - A_0\ (T,P,\ N) + (PV - N \overline{R}\ T)$$

$$= A(T,V,N) - A_0\ (T,V,N) - N \overline{R}\ T \ln Z + (PV - N \overline{R}\ T)$$

$$= -\int_\infty^V (P\text{-}\ NRT/V)\ dV - N \overline{R}\ T \ln Z + (PV - N \overline{R}\ T). \qquad (93)$$

Differentiating Eq. (93) with respect to N_k,

$$\overline{R}\ T \text{Ln}\ (\hat{\phi}_k) = -\int_\infty^V (\partial P/\partial N_k - \overline{R} T/V)\ dV - (P - N \overline{R} T/V)\ (\partial V/\partial N_k)$$

$$-\ \overline{R}\ T \ln Z - N \overline{R} T\ \partial/\partial N_k\ (\ln Z) + P\ (\partial V/\partial N_k) - \overline{R} T)$$

$$= \int_\infty^V (\partial P/\partial N_k - \overline{R} T/V)\ dV + (N \overline{R} T/V)(\hat{v}_k)$$

$$- \bar{R}\,T \ln Z - N\bar{R}T(\partial/\partial N_k)(\ln Z) - \bar{R}T.$$

Since $\ln(Z) = \ln(PV/N\bar{R}T)$, then, $((1/Z)((\partial Z/\partial N_k) = (\hat{v}_k/V - 1/N)$, and

$$\bar{R}T \ln(\hat{\phi}_k) = -\int_\infty^V(\partial P/\partial N_k - \bar{R}T/V)\,dV + (N\bar{R}T/V)(\hat{v}_k)$$

$$- \bar{R}T \ln Z - N\bar{R}T(\hat{v}_k/V - 1/N) - \bar{R}\,T), \text{ i.e.,}$$

$$\bar{R}\,T \ln(\hat{\phi}_k) = \int^\infty_V(\partial P/\partial N_k - (\bar{R}T/V))\,dV - \bar{R}T \ln Z \quad (94)$$

g. *Duhem– Margules Relation*

i. Multiple Components

At a specified temperature and pressure, the Gibbs–Duhem Equation (after multiplying Eq. (13) by N) yields the expression

$$N_1\,d\hat{g}_1 + N_2\,d\hat{g}_2 + \ldots = 0. \quad (95)$$

Using the definition

$$\hat{g}_k = \hat{g}_k^{id} + \bar{R}T \ln \gamma_k, \text{ where } \hat{g}_k^{id} = \bar{g}_k(T, P) + \bar{R}T \ln X_k,$$

we obtain the relation

$$\hat{g}_k = \bar{g}_k(T, P) + \bar{R}T \ln(X_k\,\gamma_k) \quad (96)$$

Using Eq. (96) in the Gibbs–Duhem Equation (95), one obtains the Duhem–Margules Relation at given T and P , i.e.,

$$\Sigma_k N_k\,d\ln(X_k\gamma_k) \text{ or } \Sigma_k X_k\,d\ln(X_k\gamma_k) = 0. \quad (97)$$

ii. Binary Components

In a mixture containing two components,

$$N_1\,d(\bar{g}_1 + \bar{R}T \ln(X_1\gamma_1)) + N_2\,d(\bar{g}_2 + \bar{R}T \ln(X_2\gamma_2)) = 0, \text{ i.e.,}$$

$$N_1\,d\ln(X_1\gamma_1)) + N_2\,d\ln(X_2\gamma_2)) = X_1(d\ln X_1 + d\ln\gamma_1) + X_2(d\ln X_2 + d\ln\gamma_2) = 0, \text{ or}$$

$$X_1\,dX_1/X_1 + X_1\,d\ln\gamma_1 + X_2\,dX_2/X_2 + X_2\,d\ln\gamma_2 = 0.$$

Since $X_1 + X_2 = 1$, $dX_1 + dX_2 = 0$. Therefore,

$$X_1\,d\ln\gamma_1 + (1-X_1)\,d\ln\gamma_2 = 0, \text{ i.e., } X_1 d/dX_1(\ln\gamma_1)\,dX_1 + X_2\,d/dX_1(\ln\gamma_2)dX_1 = 0, \text{ or}$$

$$X_1\,(d/dX_1)(\ln\gamma_1) = X_2\,(d/dX_2)(\ln\gamma_2) = 0. \quad (98)$$

Recall that $\gamma_k = \hat{f}_k/(X_k\,f_k(T, P))$, i.e.,

$$X_1\,(d/dX_1)(\ln\hat{f}_1) = X_2\,(d/dX_2)(\ln\hat{f}_2) = 0.$$

With $X_2 = 1 - X_1$,

$$X_1\,(d/dX_1)(\ln\hat{f}_1) = -(1-X_1))\,(d/dX_1)(\ln\hat{f}_2) = 0. \quad (99)$$

We will see later that for ideal gas mixtures $\hat{f}_k = p_k$. Thus if we know the experimental value of the partial pressure p_1 (or $\hat{f}_1$) and its variation with X_1, then the variation of p_2 (or $\hat{f}_2$) with respect to X_1 is provided by Eq. (99).

h. Ideal Mixture of Real Gases

In an ideal mixture of real gases,

$$(\hat{f}_k^{id}/ f_k)) = X_k = \hat{\alpha}_k^{id} \text{ i.e., } \hat{f}_k^{id} = X_k f_k = X_k \phi_k P, \text{ where} \quad (100)$$

$$\hat{\alpha}_k^{id} = X_k, \text{ and } \gamma_k^{id} = 1 \quad (101)$$

The activity of k–th species in an ideal mixture of real gases equals its mole fraction in the mixture.

i. Mixture of Ideal Gases

In an ideal gas, $\phi_k = 1$, and $f_k = P$. Therefore,

$$\hat{f}_k^{ig} = P X_k = p_k, \quad (102)$$

which is the partial pressure of the k–th species. The activity of the k–th species in a mixture of ideal gases equals its mole fraction in the mixture, and its fugacity equals its partial pressure.

---

h. Example 9

Determine f_{O_2} for pure oxygen at 100 bar and 200 K, and $\hat{f}_{O_2}^{id}$ and $\hat{\alpha}_{O_2}^{id}$ in an 80 % N_2 and 20% O_2 mixture at 300 K and 100 bar.

Solution

At 100 bar and 200 K, $P_R = 100 \div 50 = 2$, and $T_R = 1.3$. From the fugacity charts $\phi_{O_2} = 0.75$, and

$f_{O_2} = \phi_{O_2} P = 75$ bar.

Recall that

$(\bar{g}_{O_2}(T, P) - \bar{g}_{O_2,o}(T, P))/\bar{R}T = \ln(\phi_{O_2}) = -0.288$.

The value of $\bar{g}_{O2}$ of a real gas is lower than in an ideal gas.

At 300 K and 100 bar, the activity.

$\hat{\alpha}_{O_2}^{id} = \hat{f}_{O_2}^{id}/f_{O_2} = X_{O_2} f_{O_2}/f_{O_2} = X_{O_2} = 0.2$, and

$\hat{f}_{O_2}^{id} = 0.2 \times 75 = 15$ bar.

Remarks

The value of the partial Gibbs function of oxygen $\hat{g}_{O_2}$ in the mixture is lower than in its pure state.

Example 10

Determine the enthalpy H^{id} for a mixture containing 3 kmole of H_2O (species 1) and 97 kmole of N_2 (species 2) at 160°C and 100 kPa, and at 25°C and 100 kPa.

Solution

Since the pressure is low, we expect each gas to behave as an ideal gas in its pure state. Therefore, at 160°C and 100 kPa,

$H^{id} = 3\bar{h}_1$ (433 K ,100 bar) + 97 $\bar{h}_2$ (433 K ,100 bar), where

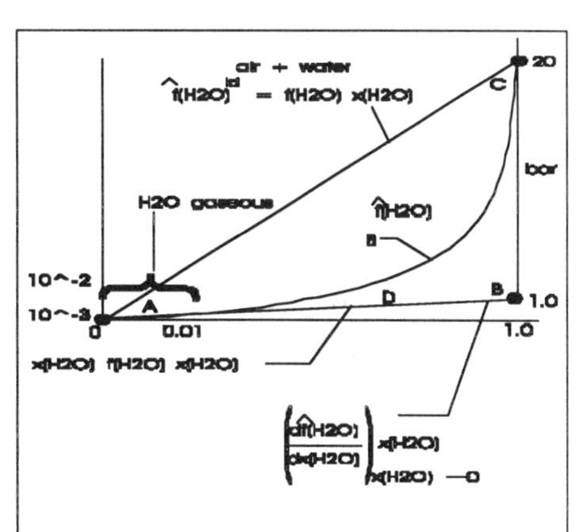

Figure 5: Schematic illustration of actual and hypothetical states.

$\bar{h}_1 = 2796.2$ kJ kg^{-1} × 18.02 kg kmole^{-1} = 50387 kJ kmole^{-1} (Steam Tables A-4 C).
$\bar{h}_2 = 433$ kJ kg^{-1} × 28.97 kg kmole^{-1} = 12544 kJ kmole^{-1}.9 (or use N2 Tables , A-16)
Therefore,
$H^{id} = 3 \times 50{,}387 + 97 \times 12{,}544 = 1{,}342{,}000$ kJ.

At 298 K and100 bar, water exists as a liquid when in its pure state, but in the mixture it exists as a vapor. We will specify
$H^{id} = 3\ \bar{h}_1(298\text{ K, 100 bar, liquid}) + 97\ \bar{h}_2(298\text{ K, 100 bar})$
$= 3 \times 105 \times 18.02 + 97 \times 298 \times 28.97 = 843000$ kJ.

If we use a hypothetical ideal gas state for pure water at 25°C and 100 kPa, then
$H^{id} = 3\ \bar{h}_1(298\text{ K, 100 bar, ideal gas}) + 97\ \bar{h}_2(298\text{ K, 100 bar})$
$= 3 \times 18.02 \times 2547 + 97 \times 28.97 \times 298 = 975000$ kJ.

Remarks

It is more accurate to determine H^{id} using enthalpies of pure components in the same state as they exist in a mixture. We have defined a hypothetical state to determine the pure substance property to account for the phase change from the "natural phase" of the pure state. This process is schematically illustrated in Figure 5. The fugacity can be likewise determined.

i. *Example 11*

Determine the specific volume of sea water (1.1% $NaCl_2$ or common salt on mole basis) based on the ideal solution model. Assume that $\bar{v}_s = 193.12$ l kmole^{-1}, $\bar{v}_w = 18$ l kmole^{-1} (cf. Figure 5), where the subscripts s and w, respectively, denote salt and water. We will use the subscript sw to denote sea water. We wish to separate the sea water into pure water and salt. What is the amount of work required to produce pure water from a kmole of sea water at 298 K? Assume an ideal solution model. If a very large amount of sea water is processed to produce a kmole of pure water (accompanied with a negligible change in composition in the remaining sea water), what is the work required to produce the pure water ?

Solution

$$\bar{v} = \hat{v}_w X_w + \hat{v}_s X_s \quad \text{(A)}$$

For an ideal solution $\hat{v}_w = \bar{v}_w$, $\hat{v}_s = \bar{v}_s$. Therefore, the specific volume of sea water $\bar{v} = 18\text{ L} \times 0.989 + 193.12 \times 0.011 = 19.93$ l kmole^{-1}.

$$\dot{W}_{opt} = (\Sigma_k \dot{N}_k\ \hat{\psi}_k)_i - (\Sigma_k \dot{N}_k\ \hat{\psi}_k)_e, \text{ i.e.,} \quad \text{(B)}$$

$$\bar{w}_{opt} = \bar{\psi}_{sw} - (X_w \hat{\psi}_w(T, P) + X_s\ \hat{\psi}_s(T, P)), \text{ and} \quad \text{(C)}$$

$$\bar{\psi}_{sw} = X_w \hat{\psi}_w(T, P, X_w) + X_s \hat{\psi}_s(T, P, X_s). \quad \text{(D)}$$

Recall that
$\hat{\psi}_k (T, P, X_k) = \hat{h}_k(T, P, X_k) - T_o\ \hat{s}_k(T, P, X_k)$.
For an ideal solution

$$\hat{\psi}_k^{id}(T, P, X_k) = \bar{h}_k(T, P) - T_o\,(\bar{s}_k(T, P) - \bar{R} \ln X_k) = \bar{\psi}_k(T, P) + \bar{R} T_o \ln X_k. \quad \text{(E)}$$

Using Eq. (E) in Eq.(D),

$$\bar{\psi}_{sw} = X_w(\bar{\psi}_w(T, P) + \bar{R} T_o \ln X_w) + X_s\,(\bar{\psi}_s(T, P) + \bar{R} T_o \ln X_s). \quad \text{(F)}$$

Substituting Eq. (F) in Eq. (C) ,we obtain the relation
$\bar{w}_{opt} = X_w\ \bar{R} T_o \ln X_w + X_s\ \bar{R} T_o \ln X_s = \bar{R} T_o(X_w \ln X_w + X_s \ln X_s)$,

Since $X_s = 0.011$, $X_w = 0.989$,

$\bar{w}_{opt} = 8.314\times298\times(0.989\times\ln 0.989\ 0.011\times\ln\ 0.011)$

$= -150$ kJ per kmole of sea water or -152 kJ/kmole of pure water.

Note that there is complete separation of water and salt here.

Suppose $d\dot{N}_{w,pure}$ is small amount of pure water flow rate leaving the unit. Then

$$\delta\dot{W}_{opt}=(\Sigma_k \dot{N}_k \hat{\psi}_k)_i-(\Sigma_k \dot{N}_k \hat{\psi}_k)_e=(\dot{N}_{sw,i}\bar{\psi}_{sw,i})-(\dot{N}_{w,e}\hat{\psi}_{w,e}+\dot{N}_{s,e}\hat{\psi}_{s,e})$$

$$-(d\dot{N}_{w,pure}\bar{\psi}_{w,pure}). \quad \text{(H)}$$

Since $\dot{N}_{w,e} = \dot{N}_{w,i} - d\dot{N}_{w,pure}$, and $\dot{N}_{s,e} = \dot{N}_{s,i}$, then,

$$\dot{W}_{opt} = (\dot{N}_{w,i}\ \hat{\psi}_{w,i} + \dot{N}_{s,i}\ \hat{\psi}_{s,i}) - (\dot{N}_{w,i}\ \hat{\psi}_{w,e} + \dot{N}_{s,i}\ \hat{\psi}_{s,e})$$

$$+ d\dot{N}_{w,,pure}\ \hat{\psi}_{w,e} - d\dot{N}_{w,pure}\ \bar{\psi}_{w,pure}. \quad \text{(I)}$$

Furthermore, since the composition change is negligible, $\hat{\psi}_{w,e} \approx \hat{\psi}_{w,i}$, $\hat{\psi}_{s,e} \approx \hat{\psi}_{s,i}$; Eq.(I) assumes the form,

$$\delta\dot{W}_{opt} = d\dot{N}_{w,pure}(\hat{\psi}_{w,e} - \bar{\psi}_{w,e}) = d\dot{N}_{w,pure}(\bar{\psi}_{w,e} + \bar{R}T\ln X_{w,e} - \bar{\psi}_{w,e})$$

$$= d\dot{N}_w\bar{R}T\ln X_{w,e}, \text{ i.e.,}$$

$\delta\dot{W}_{opt}/d\dot{N}_{w,pure} = \bar{R}T\ln X_{w,e} = 8.314 \times 298 \times \ln(0.989)$

≈ -27.40 kJ kmole^{-1} of pure water.

Note that there is no complete separation of salt from sea water. Thus, the salt concentration in sea water leaving the system is higher.

Relation between Gibbs Function and Enthalpy

Recall from Chapter 7 that

$$\bar{g}_k - \bar{g}_{k,o} = \bar{R}T\ln f_k/P, \text{ i.e., } \partial((\bar{g}_k - \bar{g}_{k,o})/\bar{R}T)/\partial T = \partial/\partial T(\ln f_k/P). \quad (103)$$

Furthermore,

$$g_k/T = h_k(T, P)/T - s_k(T, P), \text{ i.e., } \partial/\partial T\,(g_k/T) = \partial/\partial T\,(h_k/T) - \partial s_k/\partial T, \text{ and}$$

$$T\,ds_k + v_k\,dP = dh_k, \text{ i.e., } T\,\partial s_k/\partial T = \partial h_k/\partial T.$$

Following Eq. (27) of Chapter 7 and applying for k the component in a mixture,

$$\partial/\partial T(\hat{g}_k/T) = \partial/\partial T(\hat{h}_k/T) - (\partial\hat{s}_k/\partial T)$$

$$= \partial/\partial T(\hat{h}_k/T) - (1/T)(\partial\hat{h}_k/\partial T) = -\hat{h}_k/T^2, \text{ i.e.,} \quad (104)$$

$$\partial/\partial T((\hat{g}_k - \bar{g}_{k,o})/\bar{R}T) = -(\hat{h}_k - \bar{h}_{k,o})/T^2. \quad (105)$$

Since $d(\hat{g}_k - \bar{g}_{k,o}) = d\ln(\hat{f}_k/P)$,

$$\partial/\partial T(\ln(\hat{f}_k/P)) = -(\hat{h}_k - \bar{h}_{k,o})/RT^2. \quad (106)$$

k. Excess Property

The difference between the actual property of a mixture and the corresponding property if it is considered as an ideal mixture is called the excess property. For instance, the excess volume

$$V^E = V - V^{id} = \Sigma_k N_k \hat{v}_k - \Sigma_k N_k \bar{v}_k = \Sigma_k N_k(\hat{v}_k - \bar{v}_k) = N\bar{v} - N\bar{v}^{id} = N(\bar{v} - \bar{v}^{id}), \text{ and} \quad (107)$$

$$\bar{v}^E = \bar{v} - \bar{v}^{id}, \text{ i.e., } V^E = N\,\bar{v}^E. \quad (108)$$

At a specified temperature and pressure $\bar{v}_k$ remains unchanged, and the change in the excess property due to a change in the number of moles of a substance is

$$dV^E = dV - \Sigma dN_k\,\bar{v}_k, \text{ or} \quad (109)$$

$$\bar{v}^E = \Sigma_k \hat{v}_k X_k - \Sigma_k \bar{v}_k^{id} X_k, \text{ i.e.,} \quad (110)$$

$$\bar{v}^E = \Sigma_k X_k(\hat{v}_k - \bar{v}_k^{id}), \text{ so that} \quad (111)$$

$$\bar{v}^E = \Sigma_k \bar{v}_k^E X_k, \text{ where} \quad (112)$$

$$\hat{v}_k^E = (\hat{v}_k - \hat{v}_k^{id}). \quad (113)$$

Then for any extensive property B, the excess property is defined as,

$$\hat{b}_k^E = (\hat{b}_k - \hat{b}_k^{id}).$$

Likewise, if two pure components are mixed, then the corresponding Gibbs free energy change is called the free energy of mixing. The excess Gibbs function G^E can be determined in terms of the state (T, P, X_k). We can also determine the excess enthalpy and entropy of a nonideal mixture. For instance,

$$\bar{g}^E = \bar{g} - \bar{g}^{id} = (\bar{h} - \bar{h}^{id}) - T(\bar{s} - \bar{s}^{id}), \text{ where} \quad (114)$$

$$\partial(\bar{g}^E/T)/\partial(1/T) = \bar{h}^E \text{ (cf. Eq. (27), Chapter 7).} \quad (115)$$

Similarly,

$$(\partial \bar{g}^E/\partial P)_T = \bar{v}^E, \text{ and } -\bar{s}^E = (\partial \bar{g}^E/\partial T)_P, \text{ i.e.,} \quad (116)$$

$$\hat{g}_k^E = \partial G^E/\partial N_k = \partial(N\bar{g}^E)/(\partial N_k) = \hat{g}_k - \hat{g}_k^{id} = \bar{R}T \ln \gamma_k \quad (117)$$

The Margules correlation for $\bar{g}^E$ in a binary mixture is given by the relation

$$\bar{g}^E/\bar{R}T = X_1 X_2 (A_{12} X_2 + A_{21} X_1) = X_1 (1-X_1)(A_{12}(1-X_1) + A_{21} X_1),$$

where A_{12} etc are generally functions of T and P. Expressing in terms of moles

$$G^E/RT = N\bar{g}^E/\bar{R}T = N_1(N_2/(N_1+N_2))(A_{12}N_2/(N_1+N_2)+A_{21}N_1/(N_1+N_2))$$

$$= N_1N_2/(N_1+N_2)^2)(A_{12}N_2+A_{21}N_1).$$

Differentiating with respect to N_1 and using Eq. (117)

$$((N_2/(N_1+N_2)^2)-2N_1N_2/(N_1+N_2)^3)(A_{12}N_2 + A_{21} N_1)+(N_1N_2/(N_1+N_2)^2)A_{21} = \ln \gamma_1.$$

This relation can be simplified, and the activity coefficient determined, i.e.,

$\ln \gamma_1 = X_2^2(A_{12} X_2 + 2(A_{21}-A_{12}) X_1)$, and $\ln \gamma_2 = X_1^2(A_{21} X_1 + 2(A_{12}-A_{21}) X_2)$.(118)

As $X_1\to 0$, $X_2\to 1$, and $\ln \gamma_1^4 = A_{12}$. Likewise, as $X_2\to 0$, then $X_2\to 1$, and $\ln \gamma_2^4 = A_{21}$. We can solve for the constants A_{12} and A_{21} if the values of the activity coefficients are known functions of concentration. Using the Equations (118),

$$A_{12} = ((\ln \gamma_2^2)/X_1^2-(1-2X_2)(\ln \gamma_1)/X_2^2)/(4-(1-2X_1)(1-2X_2)), \text{ and} \quad (119a)$$

$$A_{21} = ((\ln \gamma_1^2)/X_2^2-(1-2X_1)(\ln \gamma_2)/X_1^2)/(4-(1-2X_1)(1-2X_2)) \quad (119b)$$

We will see later that liquid–vapor equilibrium data can be used to determine γ_k and, consequently, $\bar{g}^E$ correlation and the constants A_{12} and A_{21} can be obtained. Thereafter, the excess enthalpy can be determined. The enthalpy of mixing

$$\Delta \bar{h}_{mixing} = \bar{h}^E = \bar{h} - \bar{h}^{id} = \Sigma_k X_k(\hat{h}_k - \bar{h}_k).$$

Then for any extensive property B,

$$b_k^{\ E} = (b_k - b_k^{id}). \quad (120)$$

Similarly one may define U^E, $\bar{u}^{\ E}$, H^E, $\bar{h}^{\ E}$, G^E and $\bar{g}^E$.

Since mixing can be endothermic or exothermic, the enthalpy of a mixture can change compared to a corresponding ideal solution. Figure 6 illustrates excess enthalpy for ethanol-water solution. The enthalpy of mixing describes the amount of heat that is removed or added when a kmole of mixture is formed and maintained at the same temperature and pressure that its pure components were. If $H<H^{id}$ at that temperature and pressure, (e.g., the mixing of H_2SO_4 and H_2O), then heat must be removed during mixing; hence the mixture contains a lower enthalpy as compared to its pure components. Similarly, if $H^E>0$, heat must be added, e.g., during the mixing of methanol and benzene. One may also define the properties U^E, $\bar{u}^{\ E}$, H^E, $\bar{h}^{\ E}$, G^E S^E, $\bar{s}^E$, and $\bar{g}^E$. In an ideal solution, all excess properties except for the excess entropy and excess Gibbs function equal zero. In a regular solution, the excess entropy of mixing is equal to zero. A regular solution is a non-ideal solution in which the excess entropy increase (if any) is balanced by entropy decrease through energy removal from the mixture in order to maintain the same temperature and pressure.

l. Osmotic Pressure

Consider a U tube that is partitioned into symmetrical columns A and B by a semipermeable membrane that is permeable only to water (W), as illustrated in Figure 7. Distilled water (W) at 20°C is poured into column A up to a height $L_{A,1}$ and column B is filled with sea water (S) at the same temperature to a height $L_{B,1} = L_{A,1}$. (Ignoring density differences, the pressure on either side of the membrane is approximately equal.) Water molecules will permeate from column A to column B and, consequently, the solution in B will become diluted. At species equilibrium, the height in section A will decrease to $L_{A,2}$, and that in section B will increase to $L_{B,2}$.

Water permeation occurs, since $\bar{g}_{W,A} > \hat{g}_{W,B}$ at the initial state. For the distilled water in Section B , $\hat{g}_{W,B}^{\ id} = \bar{g}_W(T, P) + \bar{R}T \ln X_{W,B}$. The water mole fraction in column B, $X_{W,B} < 1$ while for water in portion A, $\bar{g}_W(T, P) = \bar{g}_{W,A}(T, P)$. Therefore, $\hat{g}_{W,B}^{\ id}<\bar{g}_{W,A}(T, P)$. As the flow continues, $\hat{g}_{W,B}$ increases due to increase in $X_{W,B}$ and as well as slight increase in pressure. The $\bar{g}_{W,A}$ decreases since the pressure decreases. At equilibrium, $\hat{g}_{W,B} = \bar{g}_{W,A}$, the water transport ceases. Note that volume displaced in column A may not equal the volume gain in column B, since $\hat{v}_W$ may not equal to $\bar{v}_W$ unless the solution is ideal.

The product $(L_{B,2}-L_{A,2})\rho_{B,2}g$ is referred to as the osmotic pressure. It represents the pressure required by the solution to maintain equilibrium with the solvent. In this case solution

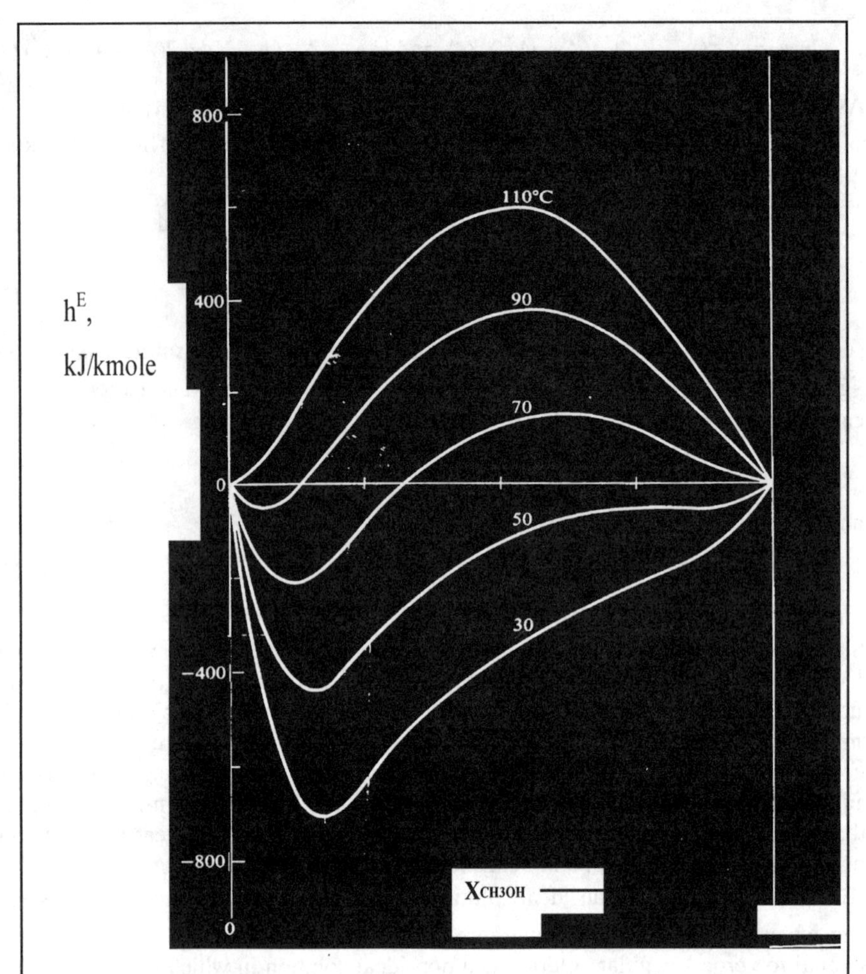

Figure 6: Enthalpy of an ethanol–water solution. (from Smith and Van Ness, Introduction to Chemical Engineering Thermodynamics, 4th Edition, McGraw Hill Book Company, 1987, p. 431. With permission.)

in column B is dilute when equilibrium is reached as compared to the initial condition so that $\rho_{B,2} \neq \rho_{B,1}$.

Instead of allowing distilled water to be transferred through the membrane, we can place a piston on column B, thereby increasing the pressure in that column. If the pressure is increased to the point when mass transfer ceases, the difference $P_B–P_A$ equals the osmotic pressure for the solution at the conditions (T, P, $X_{W,B}$). The higher the solute concentration (or the lower the solvent concentration $X_{W,B}$), the larger is the osmotic pressure required to maintain the species at equilibrium. Tap water has a lower salt concentration as compared to sea water. Therefore, tap water has a lower osmotic pressure (is hypoosmotic) compared to sea water (which may be hyperosmotic).

Rainwater, which is almost as pure as distilled water, has negligible solutes, can permeate through the surfaces of leaves, which contain biological cell–sustaining solutions. As water transport continues, the leaves swell, and, consequently, the pressure in the leaves increases. At a critical value of the internal pressure, water permeation ceases due to species equilibrium. A similar process occurs in the root of a plant where distilled water enters and permeates to other parts of the plant containing solutions.

i. Ideal Solution

For the k–th component in an ideal solution, recall that

$$\hat{\mu}_k(T, P, X_1, X_2, ...) = \mu_k(T, P) + \bar{R}T \ln X_k. \quad (121)$$

In an ideal solution of salt (s), sugar (su) and water (w),

$$\hat{\mu}_{w,B}(T, P_B, X_s, X_{su}) = \mu_w(T, P_B) + \bar{R}T \ln X_w.$$

For incompressible liquids,

$$\hat{\mu}_{w,B}(T, P_B) = \mu_w(T, P_A) + v_w(P_B - P_A). \quad (122)$$

Using Eq. (122) in Eq. (121),

$$\hat{\mu}_{w,B} (T, P_B, X_s, X_{su}) = \mu_w(T, P_A) + v_w(P_B - P_A) + \bar{R}T \ln X_w. \quad (123)$$

In the context of Figure 7, at species equilibrium,

$$\hat{\mu}_{w,B}(T, P_B, X_s, X_{su}) = \mu_w(T, P_A). \quad (124)$$

Therefore, Eq. (123) assumes the form,

$$\bar{R}T \ln X_w = -v_w(P_B - P_A), \quad (125)$$

where $(P_B - P_A)$ is the osmotic pressure, which prevents the mass transport across the membrane.

If water is the dominant component in the solution, in that case

$$\bar{R}T \ln X_w = \bar{R}T \ln (1 - (X_s + X_{su})) \approx - \bar{R}T (X_s + X_{su}) = - \bar{R}T (1-X_w)$$

Using Eq. (125),

$$(P_B - P_A) = \bar{R}T(1 - X_w)/v_w, \quad (126)$$

which is known as the van't Hoff relation for osmotic pressure. We note from Eq. (126) that the higher the solute concentration, the larger the osmotic pressure.

Rheological fluids involving solutes and solvents are used to minimize friction effects in bearings. The oil bearings generally involve a pressure gradient, which causes the local chemical potentials to differ. Therefore, a solute may move from a region of higher potential to one of lower potential. This results in irreversibility and loss of work as energy is lost due to the chemical damping of the oscillations in the local conditions. Another example pertains to a diesel or gasoline liquid droplet, which consists of a solution of several pure liquid hydrocarbons. A species gradient within the liquid

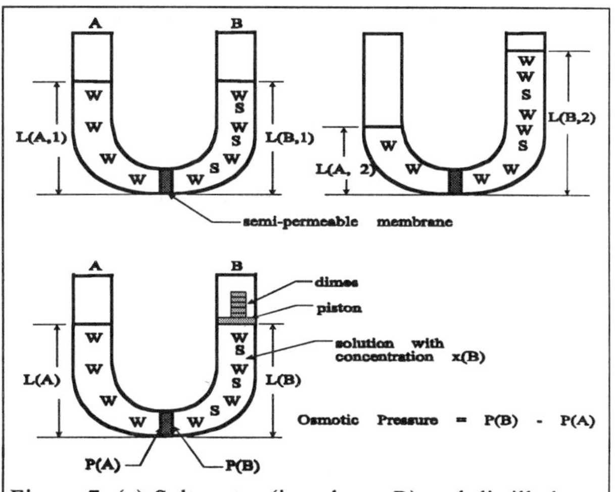

Figure 7: (a) Salt water (in column B) and distilled water (in column A) are separated by a semipermeable membrane, (b) The flow of distilled water occurs from column A into the salt water solution in column B, (c) a possible method to prevent the transport of water from column A to B by the application of pressure in column B.

droplet can cause differences in the local chemical potentials causing diffusive mass transport but resulting in irreversibility.

---

j. Example 12

A rigid semipermeable membrane connects the mixture container B at 320 K, 100 bar with another piston–cylinder assembly A, which contains pure CO_2 also at a temperature of 320 K but unknown pressure P. The section B consists of a mixture of 40% CO_2 and 60% acetylene at 320 K. Determine μ_{CO_2} in Section B. The weight in Section A is adjusted that equilibrium is reached between Sections A and B (i.e, ther is no flow of CO_2). Determine the required pressure P in Section A.
Assume c_{p0} = 10.08 kJ/kmole K (or see Table A-6C) and select the reference temperature as 273 K in estimating enthalpy and entropy of ideal gases.

Solution

For an ideal mixture of real gases in Section B

$$\hat{\mu}^{id}_{CO2}(T,P\, X_{CO_2}) = \bar{\mu}_{CO2}(T, P) + \bar{R}T \ln X_{CO_2}, \text{ where} \quad (A)$$

$$\bar{\mu}_{CO2}(T, P) = \bar{\mu}_{CO2,\,o}(T,P) + R\,T \ln \phi. \quad (B)$$

We will now determine the value of ϕ.

$$T_R = 320 \div 304 = 1.05\,,\ P_R = 100 \div 73.9 = 1.35\,, \text{ and } \phi = 0.62. \quad (C)$$

For the ideal gas state

$$\bar{\mu}_{CO2,\,o} = \bar{h}_{CO2,\,o} - T\,\bar{s}_{CO2,\,o}, \text{ where} \quad (D)$$

$\bar{h}_{CO2,\,o} = \bar{c}_{p,o}(T-273) = 10.08 \times (320-273) = 474$ kJ kmole^{-1},
(or use Tables A-9) and
$\bar{s}_{CO2,\,o} = 10.08 \times \ln(320 \div 273) - 8.314 \times \ln(100 \div 1) = -36.686$ kJ kmole^{-1} K^{-1}
(or use Tables A-9)
Using these results in Eqs. (B) and (D),

$$\bar{\mu}_{CO2,\,o} = 474 - 320 \times (-36.69) = 12214 \text{ kJ kmole}^{-1}, \quad (E)$$

$$\bar{\mu}_{CO2}(T, P) = 12{,}214 + 8.314 \times 320 \times \ln(0.62) == 10942 \text{ kJ kmole}^{-1}, \text{ and} \quad (F)$$

$$\hat{\mu}^{id}_{CO2}(320 \text{ K}, 100 \text{ bar}, X_{CO_2}=0.4) = 10{,}929 + 8.314 \times 320 \ln(0.4)$$
$$= 8505 \text{ kJ kmole}^{-1}.$$

At equilibrium

$$\hat{\mu}^{id}_{CO2}(320 \text{ K}, 100 \text{ bar}, X_{CO_2}=0.4) \text{ in B} = \bar{\mu}_{CO_2}(320 \text{ K}, P) \text{ in A} \quad (G)$$
$$= 8505 \text{ kJ kmole}^{-1}, \text{ where}$$

$$\bar{\mu}_{CO_2}(320 \text{ K}, P) = \bar{\mu}_{CO2,\,o}(320, P) + R\,T \ln \phi\,(320, P). \quad (H)$$

Here, ϕ is evaluated using fugacity charts at 320 K and specified pressure. The procedure is to assume a value for P (say, 30 bar), obtain the value of ϕ from the tables or charts and assesss whether Eq. (G) is satisfied, and finally calculate $\bar{\mu}_{CO_2}(320 \text{ K}, P)$ from Eq. (H). For P =30 bar
$T_R = 320 \div 304 = 1.05$, $P_R = 30 \div 73.9 = 0.41$, and $\phi = 0.89$, i.e.,
$\bar{s}_{CO2,\,o}(320 \text{ K}, 30 \text{ bar}) = -26.7$ kJ kmole^{-1} K^{-1}.

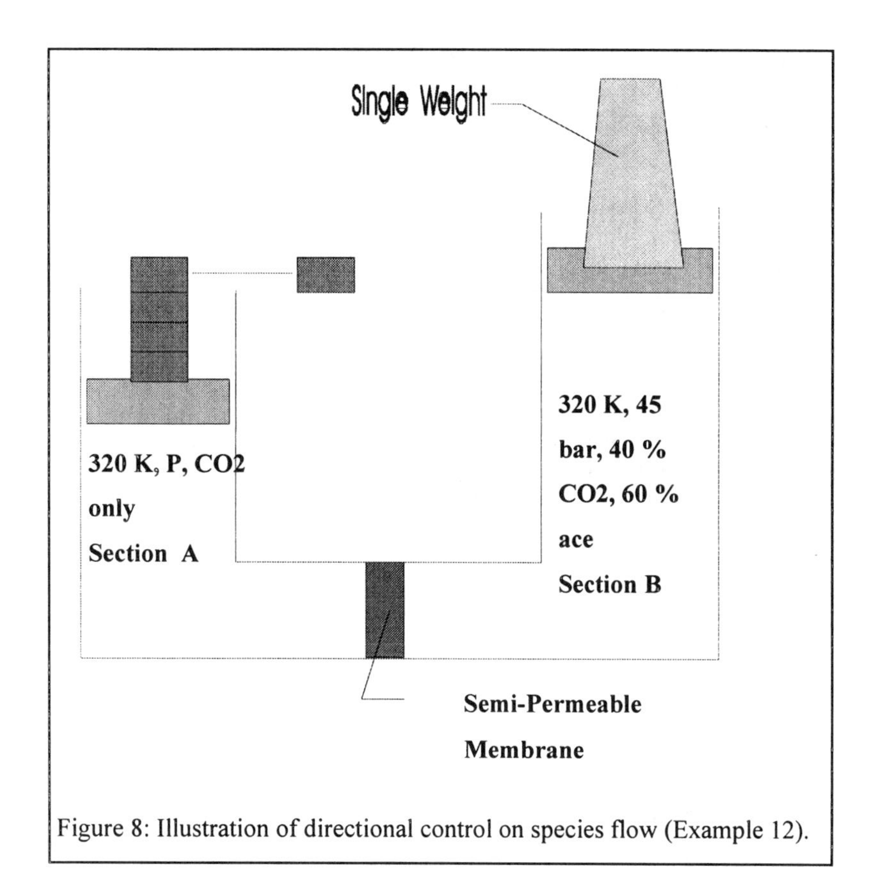

Figure 8: Illustration of directional control on species flow (Example 12).

$\bar{\mu}_{CO2,\,o}$ (320, 30) = 474 – 320 × (–26.7) = 9008kJ kmole^{-1}, and
$\bar{\mu}_{CO_2}$ (320, 30) = 9008+ 8.314 × 320 × ln 0.89 = 8698 kJ kmole^{-1}.

---

k. *Example 13*

Determine $\bar{g}_{H_2O}$ for pure liquid water at 100°C and 100 kPa. Determine the partial molal Gibbs function for liquid water $\hat{g}_{H_2O}$ at 100°C and 100 kPa in sea water that has molal salt concentration of 8 %. Use the ideal solution model. What is the osmotic pressure between the salt water and pure water at 373 K?

Solution

$\bar{g}_{H_2O}$(373 K, 100 kPa) = $\bar{h} - T\bar{s}$ = 419.04 – 373 × 1.3069 = –68.4 kJ kg^{-1}.
= –1233.2 kJ kmole^{-1}.

Using the ideal solution model,
$\hat{g}_{H_2O}$(373 K, 100 kPa, X_{H_2O}=0.92) = $\bar{g}_{H_2O}$(373 K, 100 kPa) + RT ln(0.92)
= –68.4 + (0.4614×373 ln(0.92)
= –1491.2 kJ kmole^{-1}.

Using Eq.(126)
$P_{salt\ water} - P_{water}$ = - (8.314× 373 ln (0.92))/(18.02× 0.001) = 14349 kPa.

Remark

$\bar{g}_{H_2O}$ decreases as salt is added so that the availability of liquid water in the salt solution is lower compared to in the pure component, since mixing causes irreversibility.

B. MOLAL PROPERTIES USING THE EQUATIONS OF STATE

We have seen before that the partial molal and hence mixture properties can be determined if the partial molal volume ($\hat{v}_k$) is known. The mixture equations of state are useful in obtaining relations for $\hat{v}_k$ in terms of $\bar{v}_k$.

1. Mixing Rules for Equations of State

The simplest model is the ideal solution model. The specific volume of species k within the mixture is assumed to be same as the specific volume of pure species k, i.e.,

$$\hat{v}_k = \bar{v}_k.$$

a. *General Rule*

The equation of state for a mixture can be obtained by implementing mixing rules (e.g., in order to the VW equation of state to a mixture, the constants a and can be determined by applying these rules). It is possible to formulate various mixing rules, i.e.,

$$\beta = \Sigma_k X_k \beta_k, \tag{127a}$$

$$\beta = (\Sigma_k X_k \beta_k^{1/2})^2, \tag{127b}$$

$$\beta = (\Sigma_k X_k \beta_k^{1/3})^3, \tag{127c}$$

$$\beta = (1/4)\, \Sigma_k X_k \beta_k + (3/4)\, (\Sigma_k X_k \alpha \beta_k^{1/3})(\Sigma_k X_k \beta_k^{2/3}), \text{ or} \tag{127d}$$

$$\beta = \Sigma_k X_j X_k \beta_{kj}, \tag{127e}$$

where β_k represents appropriate constant in a state equation. In Kay's rule to be discussed later, β_k could represent either $T_{c,k}$ or $P_{c,k}$ in Eq. (127a). Consider the Clausius–I Equation of State for a pure substance (Chapter 6). The same equation can be applied by replacing "b" with "b_m", and

$$P = \bar{R}T/(\bar{v} - \bar{b}_m),$$

where "b_m" can be defined using one of the mixing rules (Eqs. (127a) to (127e)) with $\bar{b}_m = \bar{\beta}$, and $\bar{b}_k = \bar{\beta}_k$.

---

l. Example 14

A piston–cylinder assembly contains two sections. Section A contains 2 kmole of acetylene, while section B contains 1 kmole of CO_2. The body (or collisional) volume of CO_2 and acetylene are, respectively, 0.043 and 0.0326 m^3 $kmole^{-1}$. Determine the volumes of both gases in their pure states at 100 bar and 320 K using the state equation

$$P = \bar{R}T/(\bar{v} - \bar{b}), \tag{A}$$

where $\bar{b}$ denotes the body volume. What is the total volume $V_A + V_B$ under those conditions? If the partition between the two sections is removed and the gases allowed to mix, what is the partial molal volume of the two gases if they are maintained at 100 bar and 320 K? What is the total volume at this state?

Solution

For a pure fluid,

$$\bar{v} = \bar{R}T/P + \bar{b}.$$

Therefore, for CO_2,
$\bar{v}_{CO_2} = 0.08314 \times 320 \div 100 + 0.043 = 0.306 \text{ m}^3 \text{ kmole}^{-1}$, and
for acetylene,
$\bar{v}_{C_2H_2} = 0.08314 \times 320 \div 100 + 0.036 = 0.302 \text{ m}^3 \text{ kmole}^{-1}$.
The total volume
$V_A + V_B = 0.302 \times 2 + 0.306 = 0.910 \text{ m}^3$.
For the mixture
$\bar{v} = \bar{R}T/P + \bar{b}_m$, i.e., $V = N\bar{R}T/P + N\bar{b}_m = (N_1+N_2+...)\bar{R}T/P + (N_1\bar{b}_1 + N_2\bar{b}_2 + ...)$, and
$\hat{v}_i = (dV/dN_i)_{T,\,P,\,N_1,N_2,...,N_{j\neq i},...N_K} = \bar{R}T/P + \bar{b}_i$.
The partial volumes of the gases in the mixture are
$\hat{v}_{CO_2} = (0.08314 \times 320 \div 100) + 0.043 = 0.306 \text{ m}^3 \text{ kmole}^{-1}$, and
$\hat{v}_{C_2H_2} = 0.198 \text{ m}^3 \text{ kmole}^{-1}$.
We find that the partial molal volumes are the same as the pure volumes, since the mixture is an ideal mixture in context of the given state equation (B).
Since
$V = \hat{v}_{CO_2} N_{CO_2} + \hat{v}_{C_2H_2} N_{C_2H_2}$, then
$V = 2 \times 0.302 + 1 \times 0.306 = 0.91 \text{ m}^3 = V^{id}$.

---

b. Kay's Rule

Kay's Rule is based on a pseudo–critical temperature T_{cm} and a like pressure P_{cm} for a gas mixture, while adopting the same equation of state as that used for the pure components of the mixture. The pseudo–critical temperature for a mixture is obtained by applying a linear mixing rule (cf. Eq. (127a)). The pseudo–critical properties

$$T_{cm} = X_1 T_{c1} + X_2 T_{c2} + ..., \text{ and } P_{cm} = X_1 T_{c1} + X_2 T_{c2} + \dots . \quad (128)$$

The assumption is that the real gas state equations for pure components are also valid for mixtures, but with these pseudo–critical properties. In the context of the RK state equation

$$P = \bar{R}T/(\bar{v} - \bar{b}_m) - \bar{a}_m/(T^{1/2}\,\bar{v}(\bar{v} + \bar{b})), \quad (129)$$

The constants $\bar{a}_m$ and $\bar{b}_m$ are evaluated by using the pseudo–critical properties, i.e.,

$$\bar{a}_m = (0.4275\,\bar{R}^2 T_{cm}^{2.5}/P_{cm}, \text{ and } \bar{b}_m = 0.08664\,\bar{R}T_c/P_c. \quad (130)$$

Other state equations can be similarly applied.

---

m. Example 15

A 2 m^3 rigid tank contains a mixture of 40% carbon dioxide and 60% acetylene by volume at 320 K and 100 bar. Determine N_{CO_2}, and P_{CO_2} using Kay's rule and applying both the RK equation and the compressibility charts.

Solution

The pseudo–critical temperature and pressure are, respectively,

$$T_{cm} = X_{C_2H_2} T_{c,\,C_2H_2} + X_{CO_2} T_{c,\,CO_2} = 0.60 \times 308.3 + 0.4 \times 304 = 307 \text{ K, and} \quad (A)$$

$$P_c = X_{C_2H_2} P_{c,\,C_2H_2} + X_{CO_2} P_{c,\,CO_2} = 0.6 \times 61.4 + 0.4 \times 73.9 = 67 \text{ bar.} \quad (B)$$

Therefore, in the context of the RK state equation

$$P = \bar{R}T/(\bar{v} - \bar{b}_m) - \bar{a}_m/(T^{1/2}\,\bar{v}(\bar{v} + \bar{b})), \quad (C)$$

$$\bar{b}_m = 0.08664\,\bar{R}T_{cm}/P_{cm} = 0.08664 \times 0.08314 \times 307 \div 67 = 0.033 \text{ m}^3 \text{ kmole}^{-1}, \text{ and} \quad (D)$$

$$\bar{a}_m = 0.4275\,\bar{R}^2 T_{cm}^{2.5}/P_c = 0.4275\times 0.08314^2\times 307^{2.5}\div 67 = 73.83 \text{ bar m}^6\text{K}^{0.5}\text{kmole}^{-2}. \quad (E)$$

At 100 bar and T = 320 K, we can solve for the three roots of $\bar{v}$. Choosing the root with the highest value,
$\bar{v}$= 0.08655 m^3 $kmole^{-1}$, i.e., N = V/ $\bar{v}$ = 2 ÷ 0.08655 = 23.2 kmole.
Therefore,
$N_{CO_2} = X_{CO_2} \times 23.2 = 0.4 \times 23.2 = 9.28$ kmole.
Now, consider the state equation

$$PV = NZ\bar{R}T, \quad (F)$$

where Z is obtained from the compressibility charts using the reduced properties $P_R = P/P_{cm}$ = 100 ÷ 67 = 1.49, and $T_R = T/T_{cm}$ = 320 ÷ 308 = 1.04. At these conditions, from the charts, Z = 0.32. Using this result in Eq. (F),
100.0 × 2 = N × 0.32 × 0.08314 × 320, i.e., N = 23.6 kmole, and
N_{CO_2} = 0.4× 23.6 = 9.4 kmole.

---

n. Example 16

A 2 m^3 rigid tank contains a mixture of 40% carbon dioxide and 60% acetylene by volume at 320 K and 100 bar. Use Kay's rule and the RK equation to determine the internal energy and enthalpy of the mixture. Assume that $c_{p,o,\,CO_2}$ and $c_{p,o,\,C_2H_2}$ remain constant and select values at 300 K. In order to determine u and h of real gas mixture, assume that each component exists in the ideal gas state and that the enthalpy is zero–valued at 0 K. Furthermore, $c_{v,o,\,C_2H_2}$ = 36.12 kJ $kmole^{-1}$ K^{-1} (or see Tables A-6C)

Solution

Treating the mixture as an ideal gas, its internal energy

$$\bar{u}_{o}(T) = X_{CO_2}\,\bar{u}_{o,\,CO_2}(T) + X_{C_2H_2}\,\bar{u}_{o,\,C_2H_2}(T), \quad (A)$$

where $\bar{u}_{ko} = \bar{c}_{vk,o}T$. Therefore,
$\bar{u}_{o,\,C_2H_2} = \bar{c}_{v,o,\,C_2H_2}T$ = 36.12 × 320 = 11559 kJ $kmole^{-1}$.
$\bar{u}_{o,\,CO_2} = \bar{c}_{v,o,\,CO_2}T$ = 0.657 × 44.01× 320= 28.91 × 320 = 9253 kJ $kmole^{-1}$.
$\bar{u}_o$ = 0.6×11559 + 0.4× 9253 = 10,637 kJ $kmole^{-1}$ of mixture.
The mixture behaves like a pure component with a pseudo-critical temperature and pressure. We will assume that the relations derived for pure components are valid for mixtures, i.e.,
$(\bar{u}_o - \bar{u})/\bar{R}T_{cm} = 7.4013/T_R^{1/2} \ln(1 + 0.08664/v_R')$.
Recall that
$\bar{v}$= 0.08655 m^3 $kmole^{-1}$, and $\bar{v}_c'$ = 0.08314 × 307 ÷ 67 = 0.381 m^3 $kmole^{-1}$.
Therefore,
v_R' = 0.08655 ÷ 0.381 = 0.227, and T_R = 320 ÷ 307 = 1.042, and
$(\bar{u}_o - \bar{u})/\bar{R}T_{cm}$ = 7.4013 ÷ $(1.042)^{1/2}$ ln(1 + 0.08664 ÷ 0.227) = 2.344, or
$(\bar{u}_o - \bar{u}) = 2.344\,\bar{R}\,T_{cm}$ = 2.344 × 8.314 × 307 = 5983 kJ $kmole^{-1}$, i.e.,
$\bar{u}$ = 10637 – 5983 = 4654 kJ $kmole^{-1}$.
Similarly,
$\bar{h} = \bar{u} + P\bar{v}$ = 4654 + 67 × 100 × 0.08655 = 5234 kJ $kmole^{-1}$.

---

c. *Empirical Mixing Rules*

Various other mixing rules are available. For instance, we can use a square rule (cf. Eq. (127b)) to determine $\bar{a}$, and a linear rule (cf. Eq. (127a)) for $\bar{b}$.

$$\bar{a}_m = (\Sigma_k X_k\ \bar{a}_k^{1/2})^2 \quad (131)$$

$$\bar{b}_m = \Sigma_k X_k \bar{b}_k \quad (132)$$

Therefore, the RK equation assumes the form

$$Z^3 - Z^2 + (\bar{a}_m^* - \bar{b}_m^{*2} - \bar{b}_m^*)\,Z - \bar{a}_m^*\ \bar{b}_m^* = 0, \text{ where} \quad (133)$$

$$\bar{a}_m^* = (\Sigma_k X_k\, \bar{a}_k^{*1/2})^2,\ \bar{b}_m^* = \Sigma_k X_k\, \bar{b}_k^*, \quad (134a, b)$$

$$\bar{a}_k^* = 0.4275 P_{R,k}/T_{R,k}^{2.5},\ \bar{b}_k^* = 0.08664 P_{R,k}/T_{R,k}, \quad (134c,d)$$

$P_{R,k} = P/P_{c,k}$, and $T_{R,k} = T/T_{c,k}$.

Once $\bar{a}_m^*$ and $\bar{a}_m^*$ are defined, one can derive pseudo–critical temperature T_{cm} and pressure P_{cm} at which Eqs. (134a) and (134b) are satisfied, i.e.,

$$\bar{a}_m^* = 0.4275\, \bar{R}^2\, T_{cm}^{2.5}/P_{cm}, \text{ and } \bar{b}_m^* = 0.08664\, \bar{R}\, T_{cm}/P_{cm}.$$

Using Eqs (134c) and (134d) in Eqs. (134a) and (134b), we obtain the relations

$$T_{cm} = \Sigma\, X_k\, (T_{ck}^{\ 5/2}/P_{ck}^{\ 1/2})^{\ 2} / \Sigma\, X_k\, (T_{ck}/P_{ck}), \text{ and}$$

$$P_{cm} = T_{cm} / \Sigma\, X_k\, (T_{ck}/P_{ck}).$$

Another mixing rule for the RK equation involves using Eq. (127e) for $\bar{a}$and Eq. (127a) for $\bar{b}$, i.e., .

$$\bar{a}_m = \Sigma_i \Sigma_j X_i X_j\ \bar{a}_{ij},\ \bar{b}_m = \Sigma_i X_i\, \bar{b}_i, \text{ where} \quad (135)$$

$$\bar{a}_{ij} = 0.42748\, \bar{R}^2 T_{c,ij}^{2.5}/P_{c,ij},\ P_{c,ij} = Z_{c,ij}\, \bar{R}\, T_{c,ij}/\, \bar{v}_{c,ij}, \quad (136)$$

$$Z_{c,ij} = (Z_{c,i} + Z_{c,i})/2, \text{ and } \bar{v}_{c,ij} = ((\bar{v}_{c,i}^{\ 1/3} + \bar{v}_{c,j}^{\ 1/3})/2)^3. \quad (137)$$

d. *Peng Robinson Equation of State*

The Peng Robinson state equation can be written in the form

$$P = \bar{R}T/(\bar{v} - \bar{b}_m) - \bar{a}_m/(\bar{v}^2 + 2\,\bar{b}_m\,\bar{v} - \bar{b}_m^{\ 2}), \text{ where} \quad (138)$$

$$\bar{a}_{ij} = (1-\delta_{ij})(\bar{a}_i\, \bar{a}_j)^{0.5},\ i \neq j,\ \bar{a}_{ii} = \bar{a}_i,$$

and the mixing rule of Eqs (127a) and (127e) apply. i.e., $\bar{a}_m = \Sigma_k X_j X_k\, \bar{a}_{\ kj}$ and $\bar{b}_m = \Sigma_k X_k\, \bar{b}_k$.

e. *Martin Hou Equation of State*

The Martin Hou state equation has the form

$$P = \bar{R}T/(\bar{v} - \bar{b}) - \sum_{i=2}^{5} F_i/(\bar{v} - \bar{b})^i + F_6(T)/e^{av}, \text{ where} \quad (139)$$

$$F_i(T) = A_i + B_i T + C_i e^{(-T/\ T_c)}. \quad (140)$$

f. Virial Equation of State for Mixtures

Recall that

$$Z = 1 + BP/\overline{R}T, \text{ where} \quad (141)$$

$$B = \Sigma_i\Sigma_j X_i X_j B_{ij},\ B_{ij} = (\overline{R}T_{c,ij}/P_{c,ij})(B^o + \omega_{ij}B^1), \quad (142)$$

$$\omega_{ij} = (\omega_i + \omega_j)/2, \text{ and } T_{c,ij} = (T_{c,i}\ T_{c,i})^{1/2}(1 - k_{ij}). \quad (143)$$

The parameter $k_{ij} \approx 0$ for most cases and is zero for a pure component.

---

o. Example 17

The compressibility factor Z can be determined using the following relation

$$Z = 1 + (BP/(\overline{R}T)), \text{ where} \quad (A)$$

$$B = \Sigma_i\Sigma_j X_i X_j B_{ij}, \quad (B)$$

$$B_{ij} = (RT_{c,ij}/P_{c,ij})B^o, \quad (C)$$

$$B^o = 0.083 - 0.422/T_R^{1.6}, \quad (D)$$

$$T_{c,ij} = (T_{c,i}T_{c,j})^{(1/2)}(1 - k_{ij}), \quad (E)$$

$$P_{cij} = Z_{c,ij}\overline{R}T_{cij}/v_{c,ij}, \quad (F)$$

$$Z_{c,ij} = (Z_{c,i} + Z_{c,j})/2, \text{ and} \quad (G)$$

$$v_{c,ij} = (v_{c,i}^{1/3} + v_{c,j}^{1/3})/2. \quad (H)$$

In a binary mixture, show that the partial molal volume of component 1 is

$$\hat{v}_1 = (\partial V/\partial N_1)_{T,P,N_2} = \overline{R}T/P + b', \text{ where} \quad (I)$$

$$b' = d(NB)/dN_1 = B_{11}(1 - X_1)\ X_2(2B_{12} - B_{11} - B_{22}). \quad (J)$$

Determine the partial molal volume of component 1 in a binary mixture.

Solution

The total volume

$V = NZ\overline{R}T/P = (1 + BP/\overline{R}T)\ N\overline{R}T/P = N\overline{R}T/P + NB.$

The partial molal volume

$\hat{v}_1 = \partial V/\partial N_1 = \overline{R}T/P + b'$, where

$b' = \partial(NB)/\partial N_1.$

Using Eq. (B),

$b' = (\partial/\partial N_1)\ (N(X_1X_1B_{11} + X_1X_2B_{12} + X_2X_1B_{21} + X_2X_2B_{22})).$

We assume that $B_{21} = B_{12}$ so that

$\partial(NB)/\partial N_1 = B_{11}(1 - X_1)X_2(2B_{12} - B_{11} - B_{22})$, i.e.,

$\hat{v}_1 = (dV/dN_1)_{T,P,N_2} = \overline{R}T/P + b'.$

The specific volume of the pure component is

$\overline{v}_1 = \overline{R}T/P + B_{11}$, i.e.,

$\hat{v}_1 - \overline{v}_1 = (1 - X_1)\ X_2\ (2B_{12} - B_{11} - B_{22}).$

---

2. Dalton's Law of Additive Pressures (LAP)

According to Dalton's Law of Additive Pressures (LAP),

$$P(T,V, N_1, N_2, ...) = p_1(T,V,N_1) + p_2(T,V,N_2) + ... + p_k(T,V,N_k) + ... , \text{ i.e.,} \quad (144)$$

$$P = \Sigma_k p_k(T,V,N_k) \quad (145)$$

where p_k denotes the component pressure of the k–th species. For a real gas,

$$PV = N_m Z_m \bar{R} T, \quad (146)$$

and Dalton's law yields the relation

$$p_k V = N_k Z_k \bar{R} T. \quad (147)$$

Applying Eqs. (26) and (130)

$$Z_m = \Sigma_k X_k Z_k(p_k,T). \quad (148)$$

Then using Eqs. (147) and (146),

$$p_k(V,T,N_k) = Z_k N_k \bar{R} T/V = P (X_k Z_k (p_k, T)/ [\Sigma_k X_k Z_k(p_k,T)))$$

Note that in this case

$$p_k(V,T,N_k) \neq X_k P \text{ unless } Z_k\text{'s are equal or } Z_k = 1. \quad (149)$$

---

p. Example 18

A 2 m^3 rigid tank contains a mixture of 40% carbon dioxide and 60% acetylene by volume at 320 K and 100 bar. Determine N_{CO_2} and P_{CO_2} using Dalton's Law of Additive Pressures and the VW equation of state.

Solution

The CO_2 mole fraction $X_{CO_2} = 0.4$, and the ratio

$$\bar{v}_{CO_2}/\bar{v}_{C_2H_2} = 40 \div 60 = 0.67. \quad (A)$$

The pressure

$$P_{CO_2} = \bar{R}T/(\bar{v}_{CO_2} - \bar{b}_{CO_2}) - (\bar{a}_{CO_2}/\bar{v}^2_{CO_2}), \text{ where} \quad (B)$$

$\bar{a}_{CO_2} = 3.643$ m^6 kmole^{-2}, and $\bar{b}_{CO_2} = 0.0427$ m^3 kmole^{-1}. Likewise,

$$P_{C_2H_2} = \bar{R}T/(\bar{v}_{C_2H_2} - \bar{b}_{C_2H_2}) - (\bar{a}_{C_2H_2}/\bar{v}^2_{C_2H_2}), \text{ where} \quad (C)$$

$\bar{a}_{C_2H_2} = 4.41$ m^6 kmole^{-2}, and $\bar{b}_{C_2H_2} = 0.051$ m^3 kmole^{-1}.

The total pressure

$$P = P_{CO_2} + P_{C_2H_2} = 100 \text{ bar}. \quad (D)$$

Assume a value for $\bar{v}_{C_2H_2}$ and use Eq. (A) to solve for $\bar{v}_{CO_2}$. Use these values in Eqs. (B) and (C) to determine p_{C2H2}, p_{CO2} and finally use Eq. (D) to check if the total pressure is indeed 100 bar. If not, iterate again.

The converged values are

$\bar{v}_{C_2H_2} = 0.324$ kmole m^{-3} and $\bar{v}_{CO_2} = 0.486$ kmole m^{-3}.

Using Eqs. (B) and (C),

$P_{CO2} = 44.6$ bar and $P_{C_2H_2} = 55.6$ bar, and

Therefore,

$N_{CO_2} = V/\bar{v}_{CO_2} = 2 \div 0.486 = 4.12$ kmoles, $N_{C2H2} = 6.15$ kmoles.

The total number of moles , N= 10.27 kmole.

Remarks

The pressure p_{CO2} = 44.6 bar, while the CO_2 partial pressure is 0.4 × 100 = 40 bar. Therefore, $p_{CO2} \neq X_{CO2}$ P.

3. Law of Additive Volumes (LAV)

When a mixture is separated into its components, but the volume is held fixed, the intermolecular separation distance increases, and, consequently, the intermolecular forces decrease in value. The same molecules in a mixture have a closer intermolecular spacing, thereby exerting stronger intermolecular forces. Then LAP does not model the force fields particularly at high pressures. If the pressure is held constant instead of the volume, in that case the intermolecular forces are appropriately simulated. Therefore,

$$V(T,P,N_1,N_2,...) = N_1\ \bar{v}_1(T,P) + N_2(T,P)\ \bar{v}_2(T,P) + ... -= \Sigma_k N_k\ \bar{v}_k. \quad (150)$$

However the LAV may yield erroneous values when a component is in trace amounts. For instance, if humid air is at high pressure and low temperature, separating air and water into the two components results in liquid water with a volume corresponding to the liquid state while the actual state in the mixture is gaseous.

4. Pitzer Factor for a Mixture

We can apply Kay's rule to determine the compressibility factor for a mixture $Z_m^{(0)}$ from the compressibility charts. This value can be corrected by including the Pitzer factor by applying Eq. (127a) with $\beta_k = \omega_k$ and $\beta = \omega$. Therefore,

$$\omega = \Sigma_k X_k\ \omega_k. \quad (151)$$

However the method will have problems for a component in trace amounts (e.g., wet air at 25°C, 1 bar in which H_2O exists in trace amounts; Dalton's law may yield better results.)

5. Partial Molal Properties Using Mixture State Equations

Any partial molal property can be determined if a mixture equation of state can be developed.

a. Kay's rule

Since $\bar{v}$ = V/N, for a mixture, VW equation of state assumes the form

$$P = N\bar{R}T/(V - N\,\bar{b}_m) - N^2\ \bar{a}_m/V^2, \text{ where} \quad (152)$$

$$\bar{a}_m = (27/64)\,\bar{R}^2\ T_{cm}^{\ 2}/P_{cm} \text{ and } \bar{b}_m = (1/8)\,\bar{R}\,T_{cm}/P_{cm}. \quad (153)$$

The pseudo–critical temperature and pressure are functions of composition. The partial molal volume

$$\hat{v}_i = (\partial V/\partial N_i)_{T,\,P,\,N_1,N_2,...,N_{j\neq i},...,N_K} \quad (154)$$

can be obtained by differentiating Eq (152). It is noted that the pseudo–critical temperature $T_{cm}$and pressure P_{cm} are also functions of composition. See example below.

q. Example 19

A two–component mixture consisting of 40% carbon dioxide (1) and 60% acetylene (2) on molal basis at 320 K and 100 bar is well–described by Kay's rule and the VW equation of state. Obtain an expression for the partial molal volume $\hat{v}_{CO2}$ and determine $\hat{v}_{CO_2}$.

Solution

Using Kay's rule, for a 2 component mixture

$T_{cm} = X_1 T_{c1} + X_2 T_{c2}$, and $P_{cm} = X_1 P_{c1} + X_2 P_{c2}$, where
$X_1 = N_1/N$, $X_2 = N_2/N$, and $N = N_1 + N_2$.
The total differential of Eq. (152)

$$dP = N\bar{R}dT/(V - \bar{b}_m N) + dN\bar{R}T/(V - \bar{b}_m N)$$

$$-(N\bar{R}T/(V - \bar{b}_m N)^2)(dV - d\bar{b}_m N - \bar{b}_m dN)$$

$$-2NdN\,\bar{a}_m/V^2 - N^2 d\bar{a}_m/V^2 + (2N^2\bar{a}_m/V^3)dV. \quad \text{(A)}$$

If we keep T, P, and N_2 are held constant, then $dN = dN_1$. Consequently, Eq. (A) assumes the form,

$$0 = 0 + dN_1\bar{R}T/(V - \bar{b}_m N) - (N\bar{R}T/(V - \bar{b}_m N)^2)(dV - d\bar{b}_m N - \bar{b}_m dN_1)$$

$$-2N\,dN_1\,\bar{a}_m/V^2 - N^2 d\bar{a}_m/V^2 + (2N^2\bar{a}_m/V^3)\,dV, \text{ i.e.,} \quad \text{(B)}$$

$$0 = \bar{R}T/(V - \bar{b}_m N) - (N\bar{R}T/(V - \bar{b}_m N)^2)\,(dV/dN_1 - (d\bar{b}_m/dN_1)N - \bar{b}_m)$$

$$-2N\bar{a}_m/V^2 - N^2(d\bar{a}_m/dN_1)/V^2 + (2N^2\bar{a}_m/V^3)dV/dN_1. \quad \text{(C)}$$

The partial molal volume

$$\hat{v}_1 = dV/dN_1 = (2N\bar{a}_m/V^2 - \bar{R}T/(V - \bar{b}_m N) - (N\bar{R}T/(V - \bar{b}_m N)^2)$$

$$((d\bar{b}_m/dN_1)N + \bar{b}_m) + N^2(d\bar{a}_m/dN_1)/V^2)/((2N^2\bar{a}_m/V^3) - (N\bar{R}T/(V - \bar{b}_m N)^2). \quad \text{(D)}$$

Differentiating Eq. (153) with respect to N_1,

$$(d\bar{a}_m/dN_1)_{N_2} = (54/64)\bar{R}^2T_{cm}(dT_{cm}/dN_1)/P_{cm} - (27/64)\bar{R}^2T_{cm}^2(dP_{cm}/dN_1)/P_{cm}^2, \quad \text{(E)}$$

$$(d\bar{b}_m/dN_1)_{N_2} = (1/8)\bar{R}(dT_{cm}/dN_1)/P_{cm} - (1/8)\bar{R}T_{cm}(dP_{cm}/dN_1)/P_{cm}^2, \quad \text{(F)}$$

$$(dT_{cm}/dN_1)_{N_2} = (1/N)T_{c,1} - (N_1/N^2)T_{c,1} - (N_2/N^2)T_{c,2} = (1/N)(T_{c,1} - X_1T_{c,1} - X_2T_{c,2}), \text{ and} \quad \text{(G)}$$

$$(dP_{cm}/dN_1)_{N_2} = (1/N)P_{c,1} - (N_1/N^2)P_{c,1} - (N_2/N^2)P_{c,2} = (1/N)(P_{c,1} - X_1P_{c,1} - X_2P_{c,2}). \quad \text{(H)}$$

In case of infinite dilution, as $N_1 \to 0$, $T_{cm} \to T_{c2}$ and $P_{cm} \to P_{c2}$, so that

$$(dT_{cm}/dN_1)_{N_2} = \to (T_{c1} - T_{c,2})/N, \text{ and } (dP_{cm}/dN_1)_{N_2} \to (P_{c1} - P_{c,2})/N, \text{ i.e.,}$$

$$\hat{v}_1 = ((2\bar{a}_m/\bar{v}^2) - (\bar{R}T/(\bar{v} - \bar{b}_m)) - (\bar{R}T/(\bar{v} - \bar{b}_m)^2)(N(d\bar{b}_m/dN_1) + \bar{b}_m)$$

$$+ (Nd\bar{a}_m/dN_1)/V^2))/((2\bar{a}_m/\bar{v}^3) - (\bar{R}T/(\bar{v} - \bar{b}_m)^2), \text{ (cf. Eq. (D)).} \quad \text{(I)}$$

As $X_1 \to 0$, $N_1 \to 0$, $b_m \to b_2$ and $\bar{v} \to \bar{v}_2$. Similarly Eqs. (E) and (F) can be rewritten as

$$N(d\bar{a}_m/dN_1)_{N_2} = (27/64)2\bar{R}^2T_{cm}(NdT_{cm}/dN_1)/P_{cm}$$

$$-(27/64)\bar{R}^2T_{cm}^2(N(dP_{cm}/dN_1))/P_{cm}^2, \text{ and} \quad \text{(J)}$$

$$N(d\bar{b}_m/dN_1)_{N_2} = (1/8)\bar{R}(N(dT_{cm}/dN_1))/P_{cm} - (1/8)\bar{R}T_{cm}(N(dP_{cm}/dN_1))/P_{cm}^2. \quad \text{(K)}$$

Using these relations, for the acetylene and CO_2 mixture

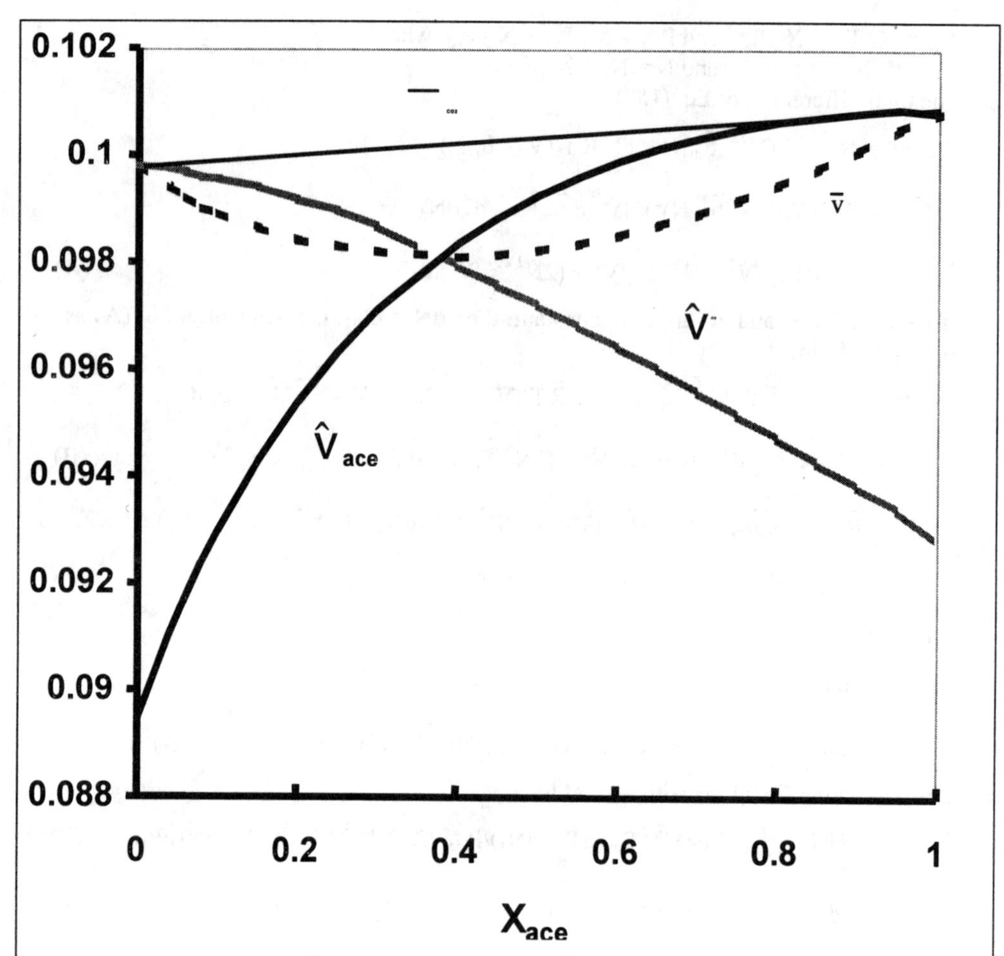

Figure 9: The variation of $\hat{v}_{C_2H_2}$, $\hat{v}_{CO_2}$, $\bar{v}$, and $\bar{v}^{id}$ with respect to the composition for an acetylene and CO_2 mixture modeled using Kay's mixing rule and the VW equation of state at 320 K and 100 bar.

$T_{c,m} = 0.6\times309 + 0.4\times304.2 = 307.08$ K, and

$P_{cm} = 0.6\times62.4 + 0.4\times73.9 = 67$ bar, i.e.,

$\bar{v} = 0.0985$ m^3 kmole^{-1}, $\bar{v}_{C_2H_2} = 0.09989$ m^3 kmole^{-1}, and $\hat{v}_{C_2H_2} = 0.10085$ m^3 kmole^{-1}.

$\bar{v} = \hat{v}_{C_2H_2} X_{C_2H_2} + \hat{v}_{CO_2} X_{CO_2} = 0.10085 \times 0.6 + 0.4 \times \hat{v}_{CO_2}$, i.e.,

$\hat{v}_{CO_2} = 0.0964$ m^3 kmole^{-1}.

The variation of $\bar{v}^{id}$, $\hat{v}_{ace}$, and $\hat{v}_{CO2}$ are illustrated in Figure 9.

r. *Example 20*

A two–component mixture consisting of 40% carbon dioxide (1) and 60% acetylene (2) by volume at 320 K and 100 bar is well–described by Kay's rule and the VW equation of state. Determine $(\bar{u}_o - \bar{u})$, $(\bar{u}_{1,o} - \hat{u}_1)$, $(\bar{u}_{2,o} - \hat{u}_2)$, $(\bar{h}_o - \bar{h})$, $(\bar{h}_{1,o} - \hat{h}_1)$, $(\bar{h}_{2,o} - \hat{h}_2)$.

Solution

Recall the expression for a mixture at given T

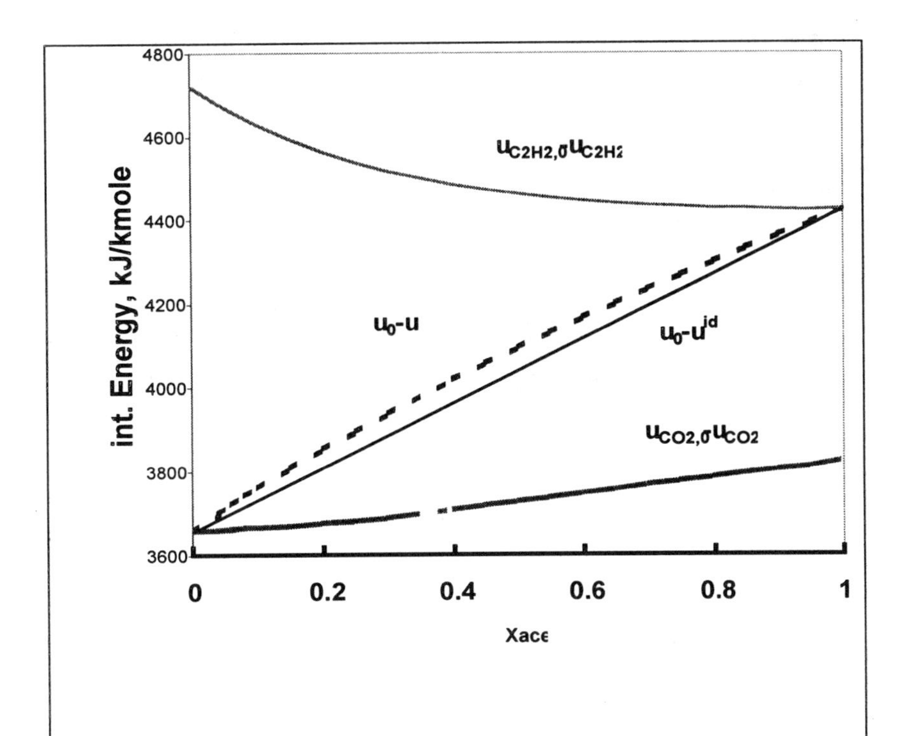

Figure 10. The variation of $(\bar{u}_o - \bar{u})$, $(\bar{u}_{CO2,o} - \hat{u}_{CO2})$, $(\bar{u}_{ace,o} - \hat{u}_{ace})$, and $(\bar{u}_o - \bar{u}^{id})$ with respect to the composition for an acetylene and CO_2 mixture modeled using Kay's mixing rule and the VW equation of state at 320 K and 100 bar.

$$d\bar{u}_T = (T\partial P/\partial T - P)d\bar{v} = (T\bar{R}/(\bar{v} - \bar{b}_m) - RT/(\bar{v} - \bar{b}_m) + \bar{a}_m/\bar{v}^2)dv.$$

Integrating (see Chapter 7)

$$\bar{u}_o(T) - \bar{u}(T,v) = a_m/\bar{v}, \text{ i.e.,} \quad \text{(A)}$$

$$U_o - U = N\bar{a}_m/\bar{v} = N^2\bar{a}_m/V, \; N = N_1 + N_2 \quad \text{(B)}$$

From the previous example, $\bar{v} = 0.0985$ m3 kmole^{-1}, so that

$\bar{u}_o - \bar{u} = \bar{a}_m/\bar{v} = 4.1042\times100\div0.0985 = 4{,}166.7$ kJ kmole^{-1}, and

$\bar{h}_o - \bar{h} = \bar{u}_o + \bar{R}T - (\bar{u} + P\bar{v}) = (\bar{u}_o - \bar{u}) + \bar{R}T - P\bar{v}$

$= 4166.7 + 8.314\times320 - 100\times100\times0.0985 = 5842.2$ kJ kmole^{-1}.

Differentiating Eq. (B) with respect to N_1, and using definition of a partial molal property,

$\hat{u}_{1,o} - \hat{u}_1 = 2N\bar{a}_m/V + N^2 d\bar{a}_m/dN_1/V - N^2\bar{a}_m\hat{v}_1/V^2$

$= 2\bar{a}_m/\bar{v} + Nd\bar{a}_m/dN_1/\bar{v} - \bar{a}_m\hat{v}_1/\bar{v}^2$.

Since $\hat{u}_{1,o} = \bar{u}_{1,o}$; assuming $\bar{u}_{1,o} = \bar{c}_{vo,1}T$,

$\bar{u}_{C_2H_2,o} = 36.12\times320 = 11558$ kJ kmole^{-1}.

Recall that $\hat{v}_{C_2H_2} = 0.09989$ m^3 kmole^{-1}, i.e.,

$\hat{u}_{1,o} - \hat{u}_1 = 2\times4.1042\times100\div0.0985 + 0.333\times100\div0.0985 - 4.1042\times100\times0.09989\div0.0985^2 = 4446.1$ kJ kmole^{-1}, and

$\bar{h}_{1,o} - \hat{h}_1 = 4446.1 + 8.314 \times 320 - 100 \times 100 \times 0.09984 = 6107.7$ kJ kmole^{-1}.
Since $\bar{u} = X_1 \hat{u}_1 + X_2 \hat{u}_2$,
$\bar{u}_{2o} - \hat{u}_2 = (\bar{u}_o - \bar{u} - (\bar{u}_{1o} - \hat{u}_1) X_1)/(1 - X_1)$
$= (4166.7 - 4446.1 \times 0.6) \div (1 - 0.6) = 3747.7$ kJ kmole^{-1}.
Similarly,
$\bar{h}_{2o} - \hat{h}_2 = 5444.0$ kJ kmole^{-1}.

---

b. *RK Equation of State*

We will now develop relations based on the RK state equation. Consider RK equation of state. For component 1 in a k–component mixture, at given P and T as N_1 is varied

$$\partial P/\partial N_1 = 0 = (\partial P/\partial \bar{a}_m)(\partial \bar{a}_m/\partial N_1) + (\partial P/\partial \bar{b}_m)(\partial \bar{b}_m/\partial N_1) + (\partial P/\partial \bar{v})(\partial \bar{v}/\partial N_1) \text{ i.e.,} \quad (155)$$

Multiplying Eq.(155) by N

$$N\partial P/\partial N_1 = 0 = (\partial P/\partial \bar{a}_m)(N\partial \bar{a}_m/\partial N_1) + (\partial P/\partial \bar{b}_m)(N\partial \bar{b}_m/\partial N_1) + (\partial P/\partial \bar{v})(N\partial \bar{v}/\partial N_1). \quad (156)$$

Furthermore,

$$N\partial v/\partial N_1 = N\partial/\partial N_1(V/N) = \hat{v}_1 - \bar{v}, \text{ i.e.,} \quad (157a)$$

Generalizing, $N\partial \bar{\tau}/\partial N_1 = \hat{\tau}_1 - \bar{\tau}$ where τ = s, u, h etc (157b)

$$\hat{v}_1 - \bar{v} = -(\partial P/\partial \bar{a}_m N\partial \bar{a}_m/\partial N_1 + \partial P/\partial \bar{b}_m N\partial \bar{b}_m/\partial N_1)/(\partial P/\partial \bar{v})), \text{ or} \quad (157c)$$

$$\hat{v}_1 - \bar{v}_1 = (\bar{v} - \bar{v}_1) - (\partial P/\partial \bar{a}_m N\partial \bar{a}_m/\partial N_1 + \partial P/\partial \bar{b}_m N\partial \bar{b}_m/\partial N_1)/(\partial P/\partial \bar{v})). \quad (158)$$

(Note that the sum $\Sigma_k X_k \hat{v}_k = \bar{v}$.) Applying the RK equation of state,

$$\partial P/\partial \bar{a}_m = -1/(T^{1/2}\,\bar{v}(\bar{v} + \bar{b}_m)), \quad (159a)$$

$$\partial P/\partial \bar{b}_m = \bar{R}T/(\bar{v} - \bar{b}_m)^2 + \bar{a}_m/(T^{1/2}\,\bar{v}(\bar{v} + \bar{b}_m)^2) \quad (159b)$$

$$\partial P/\partial \bar{v} = -\bar{R}T/(\bar{v} - \bar{b}_m)^2 + (\bar{a}_m/(T^{1/2})(1/(\bar{v}^2(\bar{v} + \bar{b}_m)) + 1/(\bar{v}(\bar{v} + \bar{b}_m)^2))$$

$$= -\bar{R}T/(\bar{v} - b_m)^2 + (\bar{a}_m/(T^{1/2}\,\bar{v}^2(\bar{v} + \bar{b}_m)^2))(2\bar{v} + \bar{b}_m). \quad (159c)$$

Applying the mixing rules,

$$\bar{a}_m = (\Sigma_k X_k \bar{a}_k^{1/2})^2, \text{ i.e., } N^2\bar{a}_m = A_m = (\Sigma_k N_k\, \bar{a}_k^{1/2})^2, \text{ and} \quad (159d)$$

$$\partial A_m/\partial N_1 = 2(\Sigma_k N_k \bar{a}_k^{1/2})\bar{a}_1^{1/2} = 2N(\Sigma_k X_k \bar{a}_k^{1/2})\bar{a}_1^{1/2} = 2N\bar{a}_m^{1/2}\bar{a}_1^{1/2}. \quad (159e)$$

$$N\bar{b}_m = B_m = \Sigma_k N_k \bar{b}_k, \text{ so that} \quad (159f)$$

$$\partial B_m/\partial N_1 = \bar{b}_1, \text{ and} \quad (159g)$$

$$\partial \bar{a}_m/\partial N_1 = \partial(A_m/N^2)/\partial N_1 = (\partial A_m/\partial N_1)/N^2 - 2A_m/N^3 = 2\bar{a}_m^{1/2}\bar{a}_1^{1/2}/N - 2\bar{a}_m/N. \quad (159h)$$

Therefore,

$$N\partial \bar{a}_m/\partial N_1 = 2\,\bar{a}_m^{1/2}(\bar{a}_1^{1/2} - \bar{a}_m^{1/2}), \text{ and} \quad (160)$$

$$N\partial \bar{b}_m/\partial N_1 = (\bar{b}_1 - \bar{b}_m). \quad (161)$$

As $\bar{a}$ and $\bar{b}$ both tend to zero, $\partial P/\partial \bar{a}_m \rightarrow (-1/T^{1/2})\bar{v}^2$, $\partial P/\partial \bar{b}_m \rightarrow -\bar{R}T/\bar{v}^2$, $\partial P/\partial \bar{v} \rightarrow -\bar{R}T/\bar{v}^2$, $N\partial \bar{a}_m/\partial N_1 \rightarrow 0$, and $N\partial b_m/\partial N_1 \rightarrow 0$. Therefore, $(\hat{v}_1 - \bar{v}) \rightarrow (\bar{v}_{1,o} - \bar{v}_o) \rightarrow 0$.

If $\bar{a}_1 = \bar{a}_2 = ...$, and $\bar{b}_1 = \bar{b}_2 = ...$, then $N\,\partial \bar{a}_m/\partial N_1 \rightarrow 0$ and $N\,\partial \bar{b}_m/\partial N_1 \rightarrow 0$ so that $\hat{v}_1 \rightarrow \bar{v}_1 - \bar{v}$ and $\hat{v}_2 = \bar{v}_2 - \bar{v}$, although $\hat{v}_1 \neq v_{1,o}$. Since $\bar{a}_m = \bar{a}_1$, $\bar{v} = \bar{v}_1$, i.e., the mixture specific volume equals that of the pure component at the same temperature and pressure, in a manner similar to the ideal solution model (Law of additive volumes).

Other partial molal properties can be similarly obtained. For instance, differentiating $(U_o - U)$ with respect to N_1 at a specified temperature and pressure and keeping the number of moles of all species other than species 1 constant, one obtains the following:

$$(\partial/\partial N_1)(U_o - U) = \bar{u}_{1,o} - \hat{u}_1, \text{ where} \quad (162)$$

$$U_o - U = N\,\bar{u}_o - N\,\bar{u}. \quad (163)$$

Differentiating Eq. (163) with respect to N_1, and definition of partial molal property

$$\bar{u}_{1,o} - \hat{u}_1 = (\bar{u}_o - \bar{u}) + N\partial(\bar{u}_o - \bar{u})/\partial N_1, \text{ and} \quad (164a)$$

$$\hat{u}_1 - \bar{u}_1 = (\bar{u}_{1,o} - \bar{u}_1) - ((\bar{u}_o - \bar{u}) + N\partial(\bar{u}_o - \bar{u})/\partial N_1). \quad (164b)$$

Similarly

$$\hat{s}_1 - \bar{s}_1 = (\bar{s}_{1,o} - \bar{s}_1) - ((\bar{s}_o - \bar{s}) + N\partial(\bar{s}_o - \bar{s})/\partial N_1). \quad (164c)$$

Dividing this relation throughout by $\bar{R}T_{c1}$

$$(\hat{u}_1 - \bar{u}_1)/(\bar{R}T_{c1}) = f(T_{R1},P_{R1}) - (f(T_{Rm},P_{Rm},T_{c1}/T_{cm}) + f(T_{Rm},P_{Rm},T_{c1}/T_{cm},P_{c1}/P_{cm})).$$

Furthermore, using the mixture RK equation of state,

$$\bar{u}_o - \bar{u} = (3/2)(\bar{a}_m/(b_m T^{1/2})\ln(1 + \bar{b}_m/\bar{v}) = F(\bar{a}_m, \bar{b}_m, \bar{v}, T).$$

At specified pressure and temperature,

$$\partial(\bar{u} - \bar{u}_o)/\partial N_1 = (\partial F/\partial \bar{a}_m)(\partial \bar{a}_m/\partial N_1) + (\partial F/\partial \bar{b}_m)(\partial \bar{b}_m/\partial N_1) + (\partial F/\partial \bar{v})(\partial \bar{v}/\partial N_1), \text{ i.e.,}$$

$$N\partial(\bar{u} - {}_o)/\partial N_1 = (\partial F/\partial a_m)(N\partial \bar{a}_m/\partial N_1) + (\partial F/\partial \bar{b}_m)(N\partial \bar{b}_m/\partial N_1)$$

$$+ (\partial F/\partial \bar{v})(N\partial \bar{v}/\partial N_1), \text{where} \quad (165)$$

$$F = \bar{u} - \bar{u}_o = (3/2)(\bar{a}_m/(\bar{b}_m T^{1/2})\ln(1 + \bar{b}_m/\bar{v}) = F(\bar{v}, T, \bar{a}_m, \bar{b}_m), \text{ i.e.,} \quad (166a)$$

$$\partial F/\partial \bar{a}_m = (3/2)(1/(\bar{b}_m T^{1/2})\ln(1 + \bar{b}_m/\bar{v}),$$

$$\partial F/\partial \bar{b}_m = (3/2)(\bar{a}_m/(\bar{b}_m^2 T^{1/2})\ln(1 + \bar{b}_m/\bar{v})$$

$$+ (3/2)(\bar{a}_m/(\bar{b}_m T^{1/2})(1/(1 + \bar{b}_m/\bar{v}))(1/\bar{v}), \text{ and} \quad (166b)$$

$$\partial F/\partial \bar{v} = (3/2)(\bar{a}_m/(\bar{b}_m T^{1/2})(1/(1 + \bar{b}_m/\bar{v}))(-\bar{b}_m/\bar{v}^2). \quad (166c)$$

Using Eqs. (166) in Eq. (165), we can obtain an expression for $N\partial(\bar{u} - \bar{u}_o)/\partial N_1$. Using the result in Eq. (164b) we can, thereafter, obtain $(\bar{u}_{1,o} - \hat{u}_1)$. Likewise,

$$\bar{h}_{1,o} - \hat{h}_1 = \bar{u}_{1,o} + \bar{R}T - (\hat{u}_1 + P\hat{v}_1) = \bar{u}_{1,o} - \hat{u}_1 + (\bar{R}T - P(\hat{v}_1 - \bar{v} + \bar{v}))$$

$$= \bar{u}_{1,o} - \hat{u}_1 - P\bar{v} + \bar{R}T - P(\hat{v}_1 - \bar{v}) = (\bar{u}_{1,o} - \hat{u}_1) + \bar{R}T(1-Z) - P(\hat{v}_1 - \bar{v}), \text{ i.e.,} \quad (167)$$

$$\hat{h}_1 - \bar{h}_1 = (\bar{h}_{1,o} - \bar{h}_1) - (\bar{u}_{1,o} - \hat{u}_1) - \bar{R}T(1 - Z) + P(\hat{v}_1 - \bar{v}).$$

In case of the entropy, the evaluation of Eq. (164c) requires the following term. Following Chapter 7 we can obtain expression $(\bar{s}_o - \bar{s})$ of a mixture. Let

$$F_1 = ((\bar{s}_o - \bar{s}) = -\bar{R}\ln(1-(\bar{b}_m/v)) + (1/2)\bar{a}_m/(\bar{b}_m T^{3/2}))\ln(1+(\bar{b}_m/\bar{v})) - \bar{R}\ln Z_{RK}, \quad (168)$$

$$(\partial(\bar{s}_o - \bar{s})/\partial N_1) = (\partial F_1/\partial a_m)(N\partial\bar{a}_m/\partial N_1) + (\partial F_1/\partial\bar{b}_m)(N\partial\bar{b}_m/\partial N_1)$$

$$+ (\partial F_1/\partial\bar{v})(N\partial\bar{v}/\partial N_1) + (\partial F_1/\partial Z)(N\partial Z/\partial N_1), \text{ where}$$

$$\partial F_1/\partial\bar{a}_m = (1/2)(1/(\bar{b}_m T^{3/2}))\ln(1+(\bar{b}_m/\bar{v})). \quad (169)$$

$$\partial F_1/\partial\bar{b}_m = \bar{R}(1/(1-(\bar{b}_m/\bar{v}))) - (1/2)(\bar{a}_m/(\bar{b}_m^2 T^{3/2}))\ln(1+(\bar{b}_m/\bar{v}))$$

$$+ (1/2)(\bar{a}_m/(\bar{b}_m T^{3/2}))(1/(1+(\bar{b}_m/\bar{v})))(1/\bar{v}), \quad (170)$$

$$\partial F_1/\partial\bar{v} = \bar{R}(\bar{b}_m/\bar{v}^2)(1/(1-(\bar{b}_m/\bar{v})^2) - (1/2)$$

$$(\bar{a}_m/(\bar{b}_m T^{3/2}))(1/(1+(\bar{b}_m/\bar{v})))(1/\bar{v}^2), \quad (171)$$

$$\partial F_1/\partial Z = \bar{R}/Z, \text{ and} \quad (172)$$

$$N\partial Z/\partial N_1 = N\partial/\partial N_1(V/V_o) = N(\hat{v}_1/V_o - \bar{v}_o V/V_o^2) = (\hat{v}_1/\bar{v}_o - Z)$$
$$= (\hat{v}_1 - \bar{v} + \bar{v})/\bar{v}_o - Z) = (\hat{v}_1 - \bar{v})/\bar{v}_o + Z - Z) = (\hat{v}_1 - \bar{v})/\bar{v}_o. \quad (173)$$

The results from Eqs. (168) to (173) can be used in Eq. (164c) to determine $\hat{s}_{1o} - \bar{s}_1$.

---

s. *Example 21*

A two–component mixture consists of 40% carbon dioxide (component 2) and 60% acetylene (component 1) by volume at 320 K and 100 bar. Assume $c_{p,o,1}$ = 44.43 kJ kmole^{-1} K, and $c_{p,o,2}$ = 37.4 kJ kmole^{-1} K. Determine its thermodynamic properties. Use the RK mixing rule.

Solution

The critical properties of the components are T_{c1} = 308.3 K, P_{c1} = 61.4 bar, T_{c2} = 304.2, and P_{c2} = 73.8 bar.

Furthermore,

$\bar{a}_1 = 80.316$ m^6 kmole^{-2}, $\bar{b}_1 = 0.0362$ m^3 kmole^{-1}, and

$\bar{a}_2 = 64.622$ m^6 kmole^{-2}, $\bar{b}_2 = 0.0297$ m^3 kmole^{-1}, i.e.,

$\bar{a}_m = (0.6 \times 80.316^{1/2} + 0.4 \times 64.622^{1/2})^2 = 73.834$ m^6 kmole^{-2}, and

$\bar{b}_m = 0.6 \times 0.0362 + 0.4 \times 0.0297 = 0.0336$ m^3 kmole^{-1}.

Thereafter (Eqs. (160) and (161)),

$N\, d\bar{a}_m/dN_1 = 2 \times 73.834^{1/2}\,(80.316^{1/2} - 73.834^{1/2}) = 6.346\ m^6\ kmole^{-2}$.

$N\, d\bar{b}_m/dN_1 = (0.0362 - 0.0336) = 0.0026\ m^3\ kmole^{-1}$.

The pressure,

$P = \bar{R}T(\bar{v} - \bar{b}_m) - \bar{a}_m/(T^{1/2}\ \bar{v}\,(\bar{v} + \bar{b}_m)$,

At 100 bar and 320 K,

$\bar{v} = 0.0894\ m^3\ kmole^{-1}$, and $Z = 0.336$.

Therefore, using Eq. (157c)

$\hat{v}_1 - \bar{v} = -(\partial P/\partial\bar{a}_m\, N\partial\bar{a}_m/\partial N1 + \partial P/\partial\bar{b}_m N\partial\bar{b}_m/\partial N_1)/(\partial P/\partial\bar{v}_m))$, where Eqs (159) yield

$\partial P/\partial\bar{a}_m = -1/(320^{1/2}\ 0.0894 \times (0.0894 + 0.0336)) = -5.0837\ bar\ kmole^2\ m^{-6}$, and

$\partial P/\partial\bar{b}_m = 0.08314 \times 320 \div (0.0894 - 0.0336)^2 + 73.834 \div (320^{1/2}\ 0.0894(0.0894 + 0.0336)^2)$

$= 8545 + 3052 = 11597\ bar\ kmole^1\ m^{-3}$.

$\partial P/\partial\bar{v} = -\ 8545 + 7250 = -1295\ bar\ kmole^1\ m^{-3}$.

Hence,

$\hat{v}_1 - \bar{v} = -(-5.0837 \times 6.346 + 11597 \times 0.0026) \div (-1295) = -0.00163\ m^3\ kmole^{-1}$.

This value compares favorably with that obtained using Kay's Rule.

Furthermore,

$F = \bar{u}_o - \bar{u} = (3/2)\,(\bar{a}_m/(\bar{b}_m T^{1/2})\ \ln(1 + \bar{b}_m/\bar{v}) = 5879\ kJ\ kmole^{-1}$.

$\partial F/\partial\bar{a}_m = (3/2)\,(1/(\bar{b}_m T^{1/2})\ \ln(1 + \bar{b}_m/\bar{v}) = 0.796\ kJ\ kmole^1\ m^{-6}$

$\partial F/\partial\bar{b}_m = -(3/2)(\bar{a}_m/(\bar{b}_m^2 T^{1/2})\ \ln(1+\bar{b}_m/\bar{v}) + (3/2)(\bar{a}_m/(\bar{b}_m\ T^{1/2})(1/(1+\bar{b}_m/\bar{v}))(1/\bar{v})$

$= -1749 + 1498 = -251\ kJ\ m^{-3}$.

$\partial F/\partial\bar{v} = (3/2)\,(\bar{a}_m/(\bar{b}_m\ T^{1/2})\,(1/(1 + \bar{b}_m/\bar{v}))(-\bar{b}_m/\bar{v}^2) = -563\ kJ\ m^{-3}$.

$N\partial(\bar{u} - \bar{u}_o)/\partial N_1 = 0.796 \times 6.346 - 51 \times 0.0026 + (-563) \times (-0.00163)$

$= 531.65\ kJ\ kmole^{-1}$.

$\bar{u}_{1,o} - \hat{u}_1 = (\bar{u}_o - \bar{u}) + N\partial(\bar{u}_o - \bar{u})/\partial N_1 = 5879 + 532 = 6411\ kJ\ kmole^{-1}$.

$\bar{h}_{1,o} - \hat{h}_1 = (u_{1,o} - \hat{u}_1) + \bar{R}T\,(1 - Z) - P(\hat{v}_1 - \bar{v})$

$= 6409 + 8.314 \times 320 \times (1 - 0.336) - 100 \times 100 \times (-0.00163)$

$= 8192\ kJ\ kmole^{-1}$.

$F_1 = s_o(T, P, \bar{a}_m, \bar{b}_m) - s(T, P, \bar{a}_m, \bar{b}_m)$

$= -\bar{R}\ \ln(1 - (\bar{b}_m/\bar{v})) + (1/2)(\bar{a}_m/(\bar{b}_m T^{3/2}))\ \ln(1 + (\bar{b}_m/\bar{v})) - \bar{R}\ \ln Z_{RK}$

$= -8.314 \times \ln(1 - 0.0336 \div 0.0894) + 0.5 \times (73.834 \times 100/0.0336 \times 320^{1.5})$

$\times \ln(1 + 0.0336 \div 0.0894) - 8.314 \times \ln(0.336) = 19.11\ kJ\ kmole^{-1}\ K^{-1}$.

$\partial F_1/\partial\bar{a}_m = (1/2)\,(1/(\bar{b}_m T^{3/2}))\ \ln(1 + (\bar{b}_m/\bar{v}))$

$= 0.5 \times \ln(1 + 0.0336 \div 0.0894) \div (0.0336 \times 320^{1.5})$

$= 0.000829\ kJ\ kmole^1\ m^{-6}\ K^{-1}$.

$\partial F_1/\partial\bar{b}_m = (\bar{R}/\bar{v})(1/(1 - (\bar{b}_m/\bar{v}))) - (1/2)(\bar{a}_m/(\bar{b}_m^2 T^{3/2}))\ \ln(1 + (\bar{b}_m/v)) +$

$+ (1/2)(\bar{a}_m/(\bar{b}_m T^{3/2}))\,(1/(1 + (\bar{b}_m/\bar{v})))\,(1/\bar{v})$.

$= (8.314 \div 0.0894) \div (1 - 0.0336 \div 0.0894) - 0.5 \times 73.834 \times 100$

$\div (0.0336^2 \times 320^{1.5})\ \ln(1 + 0.0336 \div 0.0894) + 0.5 \times 73.834 \times 100$

$\div (0.0336 \times 320^{1.5}) \div (1 + (0.0336 \div 0.0894))\,(1 \div 0.0894)$

$= 149 - 182.3 + 156.1 = 122.8\ kJ\ m^{-3}\ K^{-1}$.

$\partial F_1/\partial\bar{v} = \bar{R}\,(\bar{b}_m/\bar{v}^2)\,(1/(1 - (\bar{b}_m/\bar{v})^2) - (1/2)(\bar{a}_m/(T^{3/2}))\,(1/(1 + (\bar{b}_m/\bar{v})))(1/\bar{v}^2)$

$= 8.314 \times 0.0336 \div 0.0894^2 \div (1 - (0.0336 \div 0.0894)^2)$

$- 0.5 \times 73.834 \times 100 \div (320^{1.5})(1 \div 0.0894^2) \div (1 + 0.0336 \div 0.0894)$

$= -17.95\ kJ\ m^{-3}\ K^{-1}$.

$\partial F_1/\partial Z = -\bar{R}/Z = -8.314/0.336 = -24.74\ kJ\ kmole^{-1}\ K^{-1}$.

$N\,\partial Z/\partial N_1 = N\partial/\partial N_1(V/V_o) = N(\hat{v}_1/V_o - \bar{v}_o V/V_o^2) = (\hat{v}_1/\bar{v}_o - Z)$
$= (\hat{v}_1 - \bar{v} + \bar{v})/\bar{v}_o - Z = (\hat{v}_1 - \bar{v})/\bar{v}_o$.
$\bar{v}_o = 0.08314 \times 320/100 = 0.267\ m^3\ kmole^{-1}$, i.e.,
$(\hat{v}_1 - \bar{v})/\bar{v}_o = -0.00163 \div 0.267 = -0.0610$.
Recall that,
$\hat{s}_{1,o}(T,P,N_1,N_2,,..) - \hat{s}_1(T,P,N_1,N_2,,..) = (\bar{s}_o(T,P) - \bar{s}(T,P)$
$+ N\partial/\partial N_1(\bar{s}_o(T,P,\bar{a}_m,\bar{b}_m) - \bar{s}(T,P,\bar{a}_m,\bar{b}_m)$.
$N\partial/\partial N_1(\bar{s}_o(T,P,\bar{a}_m,\bar{b}_m) - \bar{s}(T,P,\bar{a}_m,\bar{b}_m)) = 0.000829 \times 6.346 + 122.8 \times 0.0026$
$+ 34.99 \times (-0.00163) - 24.74 \times (-0.061) = 1.78\ kJ\ kmole^{-1}\ K^{-1}$, i.e.,
$\hat{s}_{1,o}(T,P,N_1,N_2,,..) - \hat{s}_1(T,P,N_1,N_2,,..) = 19.11 + 1.78 = 20.89\ kJ\ kmole^{-1}\ K^{-1}$.
$\hat{s}_1 = (\bar{s}_{1,o} - \bar{R}\ln X_1) - 20.89\ kJ\ kmole^{-1}\ K^{-1}$.
$\hat{g}_{1,o} - \hat{g}_1 = (\hat{h}_{1,o} - T\hat{s}_{1,o}) - (\hat{h}_1 - T\hat{s}_1) = (\hat{h}_{1,o} - \hat{h}_1) - T(\hat{s}_{1,o} - \hat{s}_1)$
$= 8192 - 320 \times 20.89 = 1507.2\ kJ\ kmole^{-1}$.
$\ln \hat{f}_1/P = -(\hat{g}_{1,o} - \hat{g}_1)/\bar{R}T = -0.5665$, and
$\phi_1 = \hat{f}_1/P = 0.5675$.

Remarks

We have shown before that as $\bar{a}_m \to 0$, $\bar{b}_m \to 0$, $\hat{v}_1 \to \bar{v} = \bar{v}_1$. Similarly, it can be shown that the relation $N\partial(\bar{u}_o - \bar{u})/\partial N_1$ tends to negligible values as $\bar{a}_m \to 0$ and $\bar{b}_m \to 0$. Therefore, $\bar{u}_{1,o} - \hat{u}_1 = \bar{u}_o - \bar{u}$. However, $\bar{u}_o = \bar{u}$, i.e., $\bar{u}_{1o} = \hat{u}_1$. Similarly, $\partial F_1/\partial\bar{a}_m \to 1/(2\ T^{3/2}\ \bar{v})$, $\partial F_1/\partial\bar{b}_m \to \bar{R}/\bar{v}$, $\partial F_1/\partial\bar{v} \to 0$, and $\partial F_1/\partial Z \to \bar{R}$ as $\bar{a}_m \to 0$ and $\bar{b}_m \to 0$. Consequently, $\partial/\partial N_1(N(s_o(T,P,N_1,N_2,,..) - s(T,P,N_1,N_2,,..)) \to 0$ so that $(\hat{s}_{1,o}(T,P,N_1,N_2,,..) - \hat{s}_1(T,P,N_1,N_2,,..)) = \bar{s}_o(T,P) - \bar{s}(T,P) \to 0$, and $\hat{s}_1 - \hat{s}_{1,o} = \bar{s}_1(T,P) - \bar{R}\ln X_1$. According to the ideal solution model, $\bar{a}_1 = \bar{a}_2 = ... = \bar{a}_m$ and $\bar{b}_1 = \bar{b}_2 = ... = \bar{b}_m$. Therefore, $\partial\bar{a}_m/\partial N_1 = 0$, $\partial\bar{b}_m/\partial N_1 = 0$, $\partial\bar{v}/\partial N_1 = 0$, and $\partial Z/\partial N_1 = 0$ with the result that $(\hat{s}_{1,o} - \hat{s}_1) = s_o(T,P) - s(T,P)$. However, the difference $s_o(T,P) - s(T,P)$ is not necessarily negligible, and $(\hat{s}_1 - \hat{s}_{1,o}) = (\hat{s}_2 - \hat{s}_{2,o}) = ... =$ a constant value at a specified temperature irrespective of composition. In a like manner we can determine the value of $\hat{v}_1^E = \hat{v}_1 - \hat{v}_1^{id} = \hat{v}_1 - \bar{v}_1$, and values of other thermodynamic properties.

t. *Example 22*

Determine the mixture properties of the acetylene– CO_2 mixture discussed in the previous examples. For acetylene, $T_{ref} = 188.7$ K, $P_{ref} = 1$ bar, $c_{p,o} = 50.98 + 0.01623\ T - 1079988/T^2\ kJ\ kmole^{-1}$, T in K and $h_{fg} = 653.61\ kJ\ kg^{-1}$. In case of CO_2, $T_{ref} = 216.55$ K, $P_{ref} = 5.178$ bar, $c_{p,0} = 29.159 - 0.001573\ T - 2.92\times10^{-7}/T^2 + 5.2813\times10^{-6}\ T^2\ kJ\ kmole^{-1}$, and $h_{fg} = 524.53\ kJ\ kg^{-1}$. (You can also use Tables A-6C for specific heats.)

Solution

We can determine that $(h_{C_2H_2} - h_{C_2H_2,o}) = 653.6\ kJ\ kg^{-1}$ at T_{ref} and P_{ref}, i.e., $h_{C_2H_2,o}(320\ K) = 21907\ kJ\ kmole^{-1}$ using the ideal gas specific heat data. Similarly, $h_{CO_2,o}(320\ K) = 40172.5\ kJ\ kmole^{-1}$, and

$$h_o(320\ K) = 0.6 \times 21907 + 0.4 \times 40172.5 = 29213\ kJ\ kmole^{-1}.$$

The mixture properties $\bar{a}_m$ and $\bar{b}_m$ are obtained using the mixing rule, i.e.,

$$\bar{a}_m = 73.78\ m^6\ kmole^{-2},\ \text{and}\ \bar{b}_m = 0.03356\ m^3\ kmole^{-1}.$$

Consequently, using the RK state equation, the specific volume of mixture at 100 bar and 320 K is 0.0904 m^3 $kmole^{-1}$ or 0.0027 m^3 kg^{-1}.
Utilizing the relation,

$$\bar{h}_o - \bar{h} = (\bar{u}_o - \bar{u}) + \bar{R}T - P\bar{v} = 7577 \text{ kJ kmole}^{-1}, \text{ i.e.,}$$

$$\bar{h} = 21636 \text{ kJ kmole}^{-1}.$$

The mixture molecular weight
M_{mix} = 33.228 kg $kmole^{-1}$, i.e.,

$$h = 651.13 \text{ kJ kg}^{-1}, \text{ and } u = h - Pv = 623.925 \text{ kJ kg}^{-1}.$$

Since, $s_{1,o} = h_{fg}/T_{ref}$ at T_{ref}, $s_{1,o}$(100×6 bar, 320 K) and $s_{2,o}$(100×0.4 bar, 320 K) can be determined. The mixture entropy is

$$s_{o,mix} = 0.6 \times s_{1o}(100\times0.6 \text{ bar}, 320 \text{ K}) + 0.4\times s_{2o}(100\times0.4 \text{ bar}, 320 \text{ K}).$$

Applying the entropy correction equations,

$$s_{o,mix}(100 \text{ bar}, 320 \text{ K}, X_1=0.4) - s(100 \text{ bar}, 320 \text{ K}, X_2=0.6)$$
$$= 18.89 \text{ kJ kmole}^{-1} \text{ K}^{-1}, \text{ and}$$
$$s_{mix}(100 \text{ bar}, 320 \text{ K}, X_1=0.4, X_2=0.6)$$
$$= 132.15 \text{ kJ kmole}^{-1} \text{ K}^{-1} \text{ or } 3.98 \text{ kJ kg}^{-1} \text{ K}^{-1}.$$

C. SUMMARY

This chapter presents relations for determining the partial properties of a component and of mixtures of gases (real or ideal), liquids and solids. The simplest scheme is to assume an ideal mixing model. The concepts of excess property and activity coefficients are introduced to describe the deviation of real behavior from the ideal model. Mixing rules are introduced for real gas mixtures and a methodology is presented for deriving partial molal properties from state equations.

Chapter 9

9. PHASE EQUILIBRIUM FOR A MIXTURE

A. INTRODUCTION

In this chapter we will consider the phase equilibrium of a multicomponent mixture, e.g., gasoline, diesel and kerosene fuels. We will discuss variations in the boiling (or condensation) temperatures and the vapor (or liquid) phase composition with respect to changes in the liquid (or vapor) mixture composition at a specified pressure.

1. Miscible, Immiscible and Partially Miscible Mixture

In a miscible mixture its components are mixed at a molecular level, e.g., molecular oxygen and nitrogen in the atmosphere, and salt and water in the oceans. Typically, most gas phase mixtures are miscible, since the intermolecular distances are far apart and $X_k < 1$. In immiscible mixtures the intermolecular forces between like molecules are too strong to allow a dissimilar molecule sandwiched between similar molecules. In that case, $X_k = 1$ for the k–th species, e.g., in case of an oil–water mixture. Partially miscible fluids are miscible in a mixture until a critical mole fraction beyond which they become immiscible. For instance, salt can be dissolved in water up to a certain concentration beyond which it settles, and it is possible to have a mixture consisting of a salt–water solution and pure salt. Slurries are mixtures that contain additional species called surfactants, which act as a bridge between the (usually two) immiscible species. For instance, the introduction of surfactants into an oil–water mixture forms a slurry that is physically mixed, but is unmixed at the molecular level.

2. Phase Equilibrium

Water exists as a liquid at 20°C at pressures in excess of 2.34 kPa. This is the saturation pressure or bubble point of water at 20°C. During an isothermal process, a vapor bubble forms as the pressure equals the saturation pressure. As the pressure is further lowered, all of the water will exist as superheated vapor. Therefore, at a specified temperature, water can exist as a compressed liquid ($P>P^{sat}$), as a two–phase mixture ($P = P^{sat}$), or as superheated vapor ($P<P^{sat}$). In the compressed liquid or superheated states, the water exists in a single phase and two independent properties, say (T, P), are required to specify the state, i.e., there are two degrees of freedom present. In the saturated liquid or vapor region, one independent property, e.g., temperature, specifies P^{sat} (conversely, the pressure specifies T^{sat}). In a mixture consisting of two components (say, methanol and water) an additional parameter related to the component concentration, e.g., the mole fraction is required to describe the state of the system. Methanol is highly volatile and has a higher P^{sat} as compared to water at the same temperature. Therefore, a mixture may have different compositions in the corresponding liquid and vapor phases. The phase change phenomenon is schematically illustrated in Figure 1.

a. Two Phase System

Consider a two component mixture consisting of water (normal boiling point 100°C) and methanol (normal boiling point 65°C) in a piston–cylinder–weight assembly immersed in an isothermal bath. Suppose the molal concentration of water is 60 % and, consequently, 40 % for methanol and the mixture temperature is 20°C. At a pressure of 200 kPa there is no phase change for this mixture (cf. Figure 2). The same holds true for any methanol–water mixture of arbitrary composition at the same temperature and pressure. If the pressure is decreased to 6.2 kPa (while the mixture temperature is still 20°C), a vapor bubble appears, but inside the vapor bubble the mole fraction of water vapor at steady state is 0.23 (so that the mole fraction of methanol is 0.77) even though the liquid water content is 60 % . This is an example of phase equilibrium between a liquid mixture and a vapor bubble. Upon decreasing the pressure to 5.0 kPa additional vapor is formed, the composition of which can be determined by developing phase equilibrium criteria, as will be discussed below. During evaporation, it is possible to

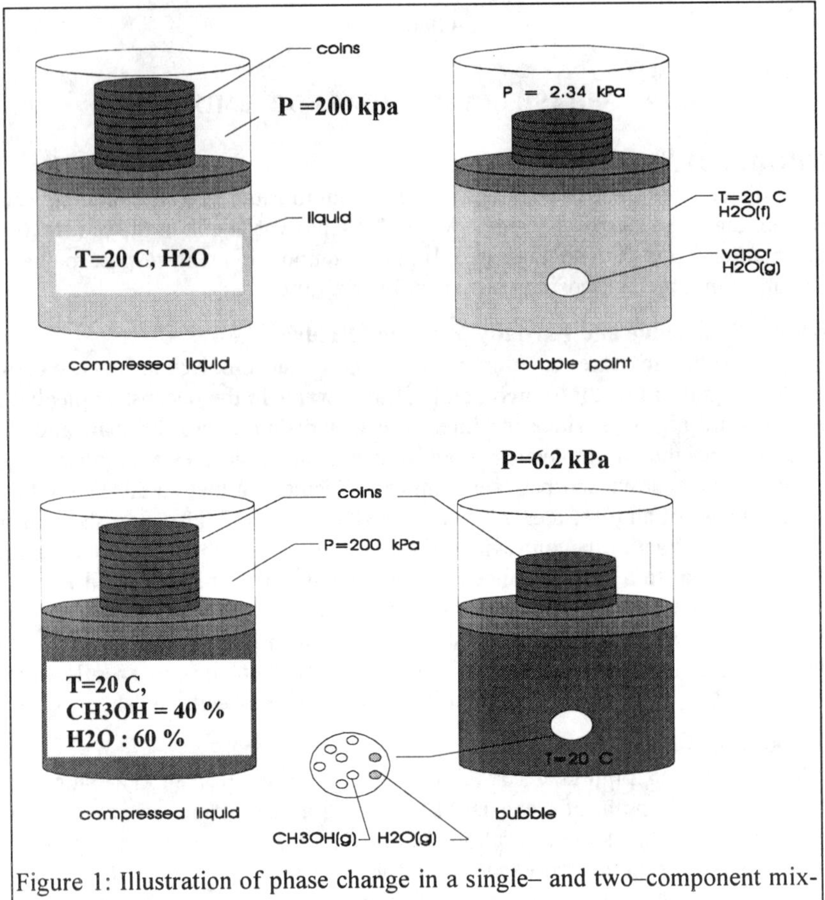

Figure 1: Illustration of phase change in a single– and two–component mixture.

maintain the liquid at constant composition by introducing appropriate amounts of the component(s) into the piston–cylinder–weight assembly.

Recall from Chapter 3 that $\hat{g}_k{}^\alpha = \hat{g}_k{}^\beta$ at phase equilibrium. This can also be inferred by using the results from Chapter 8 for a fixed mass system, i.e.,

$$d\,\hat{g}_k = -\hat{s}_k\,dT + \hat{v}_k\,dP. \tag{1}$$

At constant temperature and pressure, $d\,\hat{g}_k = 0$ so that during phase transition $\hat{g}_k$ remains unchanged. (This result is true also for pure components.) The partial molal Gibbs function is related to the fugacity (since, $d\,\hat{g}_k = d\ln\hat{f}_k$) so that at equilibrium $\hat{g}_k{}^\alpha = \hat{g}_k{}^\beta$ or $\hat{f}_k{}^\beta = \hat{f}_k{}^\alpha$.

b. Multiphase Systems

Consider a k–component mixture consisting of π phases. At a specified temperature and pressure (cf. Appendix A)

$$\hat{g}_{1(1)} = \hat{g}_{1(2)} = \hat{g}_{1(3)} = \ldots = \hat{g}_{1(\pi)},\ \hat{g}_{2(1)} = \hat{g}_{2(2)} = \hat{g}_{2(3)} = \ldots = \hat{g}_{1(\pi)},$$

$$\ldots,\text{ and } \hat{g}_{k(1)} = \hat{g}_{k(2)} = \hat{g}_{k(3)} = \ldots = \hat{g}_{1(\pi)}, \tag{2}$$

where the subscripts 1(2) refer to the component 1 in phase 2 .

$$\hat{g}_{k(\alpha)} = \bar{g}_{k(\alpha)}(T,P) + \bar{R}T\ln(\hat{f}_{k(\alpha)}/f_{k(\alpha)}) = (\bar{g}_{k(\alpha),0}(T,P) + \bar{R}T\ln f_{k(\alpha)}) + \bar{R}T\ln\hat{f}_{k(\alpha)}/f_{k(\alpha)}$$

$$= \bar{g}_{k(\alpha),0}(T,P) + \bar{R}T \ln \hat{f}_k^{\alpha}. \quad (3)$$

Where $\bar{g}_{k(\alpha),0}(T,P)$ refers to ideal gas Gibbs free energy.
Similarly,

$$\hat{g}_{k(\beta)} = \bar{g}_{k(\beta),0}(T,P) + \bar{R}T \ln \hat{f}_{k(\beta)}, \text{ and} \quad (4)$$

$$\hat{f}_{k(\alpha)} = \hat{f}_{k(\beta)} \quad (5)$$

Expanding the last equation

$$\hat{f}_{1(1)} = \hat{f}_{1(2)} \ldots \hat{f}_{1(\pi)}, \hat{f}_{2(1)} = \hat{f}_{2(2)} \ldots \hat{f}_{2(\pi)} \hat{f}_{k(1)} = \hat{f}_{k(2)} \ldots \hat{f}_{k(\pi)}.$$

c. *Gibbs Phase Rule*

In a single phase consisting of a pure component, the temperature and pressure can be independently varied, i.e., it possesses two degrees of freedom (F = 2). In order for two phases of a single component to coexist, a single degree of freedom (F = 1) must exist, i.e., either the temperature or pressure. For example, one can maintain vapor and liquid phases for H_2O at say T= 20 C (P^{sat} = 2.34kPa), T = 100 C (P^{sat} = 100 kPa) etc. In this case "T" is an independent variable and P^{sat} is dependent on T within the wet region. If a single component simultaneously exists in three phases of (e.g., at the triple point of the substance), there are no degrees of

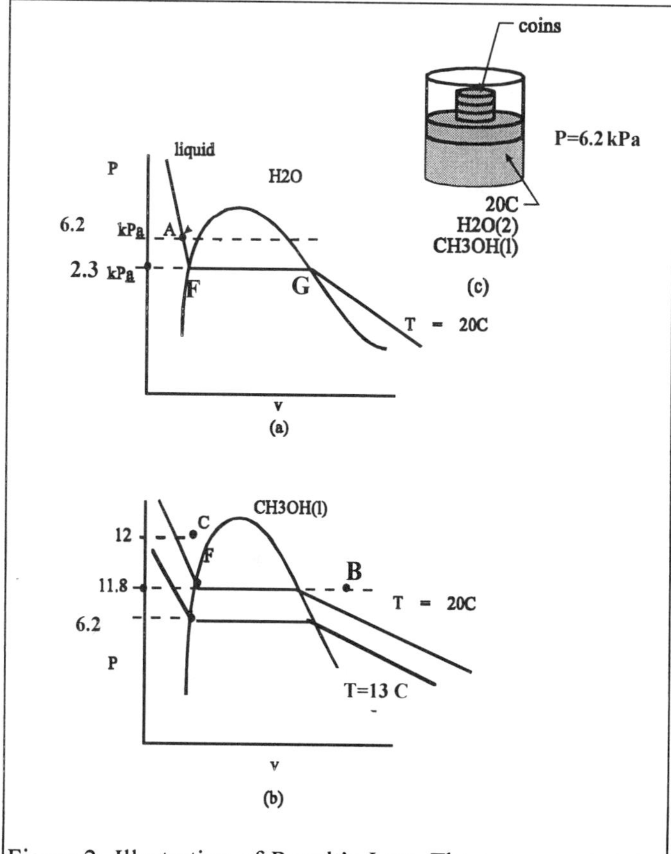

Figure 2: Illustration of Raoult's Law. The component states are described in terms of a P–v diagram for the pure components.

freedom, and F = 0 (for instance, at the triple point of ice, liquid water, and water vapor at 273.15 K and 0.0061 bar). If all three phases are desired, then neither the temperature nor the pressure can be altered.

For a single phase consisting of two components, there are three degrees of freedom (the temperature, pressure, and the mole fraction of either species). Two phases of two components must possess two degrees of freedom in order to exist. For three phases of two components to exist, there must be a single degree of freedom. In general, the number of degrees of freedom of an N–component mixture containing π phases is (cf. Appendix A)

$$F = K + 2 - \pi. \tag{6}$$

B. SIMPLIFIED CRITERIA FOR PHASE EQUILIBRIUM

1. General Criteria for Any Solution

The higher the intermolecular attractive force in a substance, in general, the higher is its boiling temperature. The normal boiling points of water and methanol in their pure states are, respectively, 100°C and 80°C, indicating that the attractive forces between water molecules are higher than between methanol molecules.

Consider a liquid mixture of 40% water and 60% alcohol at 20°C and 8.03 kPa. If the water existed by itself at that temperature and pressure, it would be a compressed liquid (cf. point A in Figure 2), since its saturation pressure is 2.3 kPa. Likewise, if methanol existed alone at that temperature and pressure, it would exist as superheated vapor (point B in Figure 2), since the saturation pressure for methanol at 20°C is 11.8 kPa. At 20°C and 8.03 kPa, the mixture exists in a liquid state due to the relatively small number of methanol molecules that can vaporize per unit surface area of the liquid mixture. (Consequently, the partial pressure of the methanol vapor in the gas phase is lower and corresponds to the methanol evaporation rate. The methanol molecules, which have lower intermolecular attraction forces among themselves than do water molecules, and, consequently, a lower boiling point than water, are held in a liquid state due to the surrounding water molecules). In summary, intermolecular forces between dissimilar molecules and the reduced surface area for evaporation/condensation of each species in the mixture, alter the evaporation and condensation characteristics.

Oftentimes, we must determine the "bubble point" or the pressure at which an initial bubble appears in the mixture at a specified temperature and the corresponding vapor phase compostion using a phase equilibrium condition for a multicomponent mixture (Eqs.(2) and (6))

2. Ideal Solution and Raoult's Law

a. Vapor as Real Gas Mixture

If α represents the liquid and β the vapor phase, then for an ideal solution in the liquid phase and an ideal mixture of real gases,

$$X_{k(\alpha)}\, f_{k(\alpha)}(T,P) = X_{k(\beta)}\, f_{k(\beta)}, \tag{7}$$

where $f_{k(\alpha)}$ denotes the fugacity of component k in its pure state in the α phase (i.e., the same phase as the mixture) and at the same temperature and pressure and X_k mole fraction. The fugacity of the liquid phase at any temperature and pressure

$$f_{k(\alpha)}(T,P) = f_{k(\alpha)}(T,P^{sat})\, POY_{(\alpha)},\ \text{where } POY = \exp\left(\int_{P^{sat}}^{P} (v_{(\alpha)}(T,P)/RT)\, dP\right). \tag{8}$$

where POY is the Poynting correction factor (Chapter 7). Likewise, in the vapor state

$$f_{k(\beta)}\ (T,P) = f_{k(\beta)}(T,P^{sat})\ POY_{(\beta)},\ \text{where } POY = \exp\left(\int_{P^{sat}}^{P} (v_{(\beta)}/RT)dP\right). \quad (9)$$

where $v_{(\beta)}$ can be obtained as a function of T and P using any real gas equation of state (Appendix B).

b. Vapor as Ideal Gas Mixture

In general, the RHS of Eq. (7) is related to the partial pressure of the k–th component if β is the gas phase and Eq. (7) assumes the form

$$X_{k(\alpha)}\ f_{k(\alpha)}\ (T,P) = X_k\ P \quad (10)$$

Typically, v_f has a small value and we can neglect the term POY so that

$$f_{k(\alpha)}\ (T,P) = f_{k(\alpha)}\ (T,P^{sat}) = f_{k(\beta)}(T,P^{sat}) = P_k^{sat},\ \text{and} \quad (11)$$

$$p_k = X_{k(\alpha)}\ P_k^{sat} \quad (12)$$

Note that the capital lettered P_k^{sat} represents the saturation pressure of pure species k at given T. The relation in Eq. (12) is known as Raoult's law, which can be used to solve problems pertaining to multicomponent phase equilibrium

Consider two components (namely, 1 and 2) at a specified temperature and pressure in two phases α (liquid) and β (vapor). In that case

$$f_{1(\alpha)}X_{1(\alpha)} = X_{1\beta}f_{1(\beta)},\ f_{2(\alpha)}X_{2(\alpha)} = X_{2,\beta}f_{2(\beta)},\ X_{1(\alpha)} + X_{2(\alpha)} = 1,\ \text{and } X_{1(\beta)} + X_{2(\beta)} = 1 \quad (13)$$

Unknowns, $X_{1(\alpha)}$, $X_{2(\alpha)}$, $X_{1(\beta)}$, and $X_{2(\beta)}$. Knowing $f_{1(\alpha)}$, $f_{1(\beta)}$, $f_{2(\alpha)}$, $f_{2(\beta)}$, and four equations (cf. Eq. 13), we can solve for the four unknowns at specified T and P.

i. Remarks

The ideal solution model allows us to express partial molal properties in terms of pure properties and molal concentrations.

For a miscible liquid mixture under phase equilibrium

$$\hat{f}_{1(f)} = \hat{f}_{1(g)}. \quad (14)$$

In the future, subscript (g) for gas phase will be omitted

For an ideal liquid mixture, Eq.(14) becomes

$$\hat{f}_{1(f)}(T,P,X_1) = X_{1(\ell)}\ f_{1(f)}(T,P). \quad (15)$$

For a mixture immiscible in the liquid phase, but which is miscible in the gas phase, the phase equilibrium condition implies that $\hat{f}_1(g)$ e., fugacity of component 1 in the vapor mixture = f_1 (T, P), i.e., the fugacity of the pure component in the liquid phase since the component 1 is immiscble.

---

a. Example 1

Sea water consists of salt (the solute) and water (the solvent). Determine the boiling temperature of solution with an 8% salt concentration at 1 bar. Use both Raoult's Law, and the equality $\hat{g}_{H_2O(\ell)} = \hat{g}_{H_2O(g)}$.

Solution

The partial pressure of water

$$p_{H_2O} = X_{H_2O\,l(\ell)}\ P_{H_2O}^{sat}(T).$$

At 100 kPa, since $X_{H_2O(\ell)} = 0.92$,
$100 = 0.92\ P^{sat}_{H_2O}(T)$, i.e., $P^{sat}_{H_2O}(T) = 108.7$ kPa.
At this pressure, using the tables
T = 102.1°C,
which is the boiling point of the salt water.
At phase equilibrium

$$\hat{g}_{H_2O(\ell)}(T,P,X_{H_2O(\ell)}) = \hat{g}_{H_2O(g)}(T,P,X_{H_2O}) \quad (A)$$

Using the ideal solution model for the liquid phase in context of Eq. (A),

$$\hat{g}_{H_2O(\ell)}(T,P,X_{H_2O}) = \bar{g}_{H_2O(\ell)} + \bar{R}T \ln X_{H_2O}. \quad (B)$$

Similarly, for the gas phase,

$$\hat{g}_{H_2O(g)}(T,P,X_{H_2O}) = \bar{g}_{H_2O(g)}(T,P) + \bar{R}T \ln X_{H_2O}. \quad (C)$$

Assuming the gas phase to consist of only water vapor, i.e., $X_{H_2O,\ell} = 1$, using Eqs. (A)–(C),

$$\bar{g}_{H_2O(\ell)} + \bar{R}T \ln X_{H_2O(\ell)} = \hat{g}_{H_2O(g)}(T,P,1). \quad (D)$$

If we assume the boiling point to be 100°C,

$$\hat{g}_{H_2O(g)}(373\text{ K}, 100\text{ kPa}, 1) = \bar{g}_{H_2O(g)}(373\text{ K}, 100\text{ kPa})$$

$$= \bar{h}_{H_2O(g)}(373\text{ K}, 100\text{ kPa}) - 373 \times \bar{s}_{H_2O(g)}(373\text{ K}, 100\text{ kPa}) + 0$$

$$= 2676.1 \times 18.02 - 373 \times 7.3549 \times 18.02 = -1212\text{ kJ kmole}^{-1}. \quad (E)$$

In the liquid state, $X_{H_2O,\ell} = 0.92$, and using Eq. (B),

$$(373\text{ K},100\text{ kPa},0.92) = \bar{g}_{H_2O(\ell)}(373\text{ K},100\text{ kPa}) + 8.314 \times 373 \times \ln 0.92, \text{ and} \quad (F)$$

$$\bar{g}_{H_2O(\ell)}(373\text{ K},100\text{ kPa}) = \bar{h}_{H_2O(\ell)} - T\,\bar{s}_{H_2O(\ell)}$$

$$419.04 \times 18.02 - 373 \times 1.3069 \times 18.02 = -1212\text{ kJ kmole}^{-1}. \quad (G)$$

Using Eqs. (E)–(G),
$\hat{g}_{H_2O(\ell)}(373\text{ K}, 100\text{ kPa}, 0.92) = -1212 - 8.314 \times 373 \ln 0.92 = -1407.6\text{ kJ kmole}^{-1}$, i.e.,
$\hat{g}_{H_2O(\ell)} - \hat{g}_{H_2O(g)} = -1470.6 - (-1212) = 258.6\text{ kJ kmole}^{-1}$.
The partial molal Gibbs free energy of the liquid phase water is lower than in the gas phase, since the water mole fraction in the liquid phase is lower than unity. The liquid temperature must be increased in order to raise the value of $\hat{g}_{H_2O(\ell)}$ so that it equals that of the vapor phase water.
In this context assume that the boiling point T_{BP} equals 110°C or 383 K. In that case, for the liquid phase water, Eq. (E) yields

$$\hat{g}_{H_2O(\ell)}(383\text{ K},100\text{ kPa},0.92) = \bar{g}_{H_2O(\ell)}(383\text{ K},100\text{ kPa}) + 8.314*383*\ln 0.92. \quad (H)$$

The Gibbs energy of liquid water at 383 K and 100 kPa is unavailable from the tables, since this is a superheated vapor state for pure water. The Gibbs energy of saturated liquid water at 383 K is available at a pressure of 143 kPa, i.e.,

$$\bar{g}_{H_2O(\ell)}\ (483\ \text{K},\ 143\ \text{kPa}) = \bar{h}_{H_2O(\ell)} - T\,\bar{s}_{H_2O(\ell)}$$

$$= 461.3 - 383 \times 1.4185 = -81.99\ \text{kJ kg}^{-1} = -\ 1477.5\ \text{kJ kmole}^{-1}. \quad (J)$$

We will use the relation

$$dg_T = v\ dP. \quad (K)$$

Assuming the liquid to be incompressible, and integrating Eq. (K) between the limits (T,P^{sat}) and (T,P), for the hypothetical liquid water state,

$$\bar{g}_{H_2O(\ell)}\ (T,P) - g_f(T,P^{sat}) = \bar{v}_f(P - P^{sat}).$$

where for water $\bar{v}_f = 0.018\ m^3\ kmole^{-1}$ at 373 K. Thus

$$\bar{g}_{H_2O(\ell)}\ (383\ \text{K},\ 100\ \text{kPa}) = \bar{g}_f(383\ \text{K},\ 143\ \text{kPa}) + 0.018\times(100 - 143)$$

$$= -1477.5 + 0.018\times(100 - 143) = -1478.2\ \text{kJ kmole}^{-1},\ \text{i.e.,} \quad (L)$$

It is seen that $v_f\ (P-P^{sat})$ term is negligible. Hence

$$\bar{g}_{H_2O(\ell)}\ (T,P) \approx \bar{g}_f\ (T,Psat) = -1477.5\ \text{kJ kmole}^{-1}. \quad (K)$$

Using Eq. (B), the partial molal Gibbs function of liquid water can be determined, i.e.,

$$\hat{g}_{H_2O(\ell)}\ (383\ \text{K},100\ \text{kPa},\ 0.92) = -1476.7 + \bar{R}T\ \ln\ (0.92) = -1742\ \text{kJ kmole}^{-1}. \quad (M)$$

For the gaseous water,

$$\hat{g}_{H_2O(g)}(383\ \text{K},\ 100\ \text{kPa}) = \bar{g}_{H_2O(g)}(383\ \text{K},\ 100\ \text{kPa}) + \bar{R}T\ \ln\ (1).$$

In the gaseous state, water is superheated, and from the tables

$$\hat{g}_{H_2O(g)}(383\ \text{K},100\ \text{kPa}) = 2696.4\times18.02 - 383\times7.41\times18.02 = -2580.5\ \text{kJ kmole}^{-1}. \quad (N)$$

Therefore,

$$\hat{g}_{H_2O(\ell)}\ (383\ \text{K},\ 100\ \text{kPa},\ 0.92) - \hat{g}_{H_2O(g)}(383\ \text{K},\ 100\ \text{kPa}) = 838\ \text{kJ kmole}^{-1}.$$

Using Eqs. (H) and (N) we can interpolate for the temperature at which $(\hat{g}_{H_2O(\ell)} - \hat{g}_{H_2O(g)}) = 0$. That temperature is 102.2°C.

Remarks

We have assumed the ideal solution model applies in the liquid phase. This model assumes that the attractive forces between water–water molecules equal those between water–salt molecules. Therefore, the change in the boiling temperature of water in the mixture compared to that of pure water is uninfluenced by attractive forces.

When additional salt is introduced into the solution the mole fraction of liquid water is reduced. At phase equilibrium, the random condensation rate of a component k of a mixture must equal the random evaporation of that component from the liquid phase. The surface of a mass of pure water exposed to the gas phase contains only liquid water molecules, and the condensation and evaporation rates equal one another at a gas–phase pressure of 100 kPa at a temperature of 373 K. In the case of salt water the same area will accommodate a smaller number of liquid water molecules (due to the presence of salt molecules), which reduces the evaporation rate. Phase equilibrium considerations imply that a smaller number of vapor molecules consequently condense, and the vapor pressure at 373 K must be lower than 100 kPa. In order to achieve the same vapor pressure as is generated by the pure component, the liquid

temperature must be raised as salt is added to water in order for the same number of molecules to evaporate as for the pure component. (Similar arguments apply for the depression of the freezing point of a salt–water solution with salt addition).
We will now obtain a simple relation for change in boiling point when solute is added in a solvent. The application of Raoult's law yields,

$$P^{sat}_{H_2O} = (P/X_{k(\ell)}) \tag{O}$$

Recall from Chapter 7 (Clausius Clapeyron relation) that we can approximately express $P^{sat}_{H_2O}$ in the form

$$\ln P^{sat}_{H_2O} = A - B/T. \tag{P}$$

Employing Eq. (O) in (P) and solving for T,

$$T = B/(A - \ln(P/X_{k(\ell)})). \tag{Q}$$

For the pure component $X_{k,\ (\ell)}=1$, and

$$T_{pure} = B/(A - \ln (P)). \tag{R}$$

Using Eqs. (Q) and (R),

$$(1/T_{pure} - 1/T) = (1/B) \ln X_{k(\ell)}. \tag{S}$$

Since $T = T_{pure} + \delta T$, and $\delta T/T_{pure} \ll 1$, the LHS of the above relation can be expanded in a binomial series. Retaining only the first order terms, Eq. (S) yields

$$\delta T = -(T^2_{pure}/B) \ln X_{k(\ell)} \tag{T}$$

Recall from the Clausius Clapeyron relation, Chapter 7 that $B = h_{fg}/R$. Therefore,

$$\delta T = -(T^2_{pure} R/h_{fg}) \ln X_{k(\ell)}. \tag{U}$$

In case of water B = 5205.2 K and T_{pure} = 373 K. For the case $X_{k(\ell)} = 0.92$, $\delta T = 2.23$ K, i.e., the boiling temperature $T_{BP} = 102.3$ C.
At a specified pressure, the boiling temperature is increased by the amount δT. In the current example, the salt concentration is much less compared to water. Thus $X_{H2O(\ell)}= 1- X_{salt}, X_{salt} << 1$, then $\ln X_{H2O(\ell)} = \ln (1- X_{salt})\ .\ -X_{salt}$. Using this result in Eq. (U), the boiling point elevation is given as

$$\delta T = (T_{pure}{}^2 R/h_{fg}) X_{salt}, \tag{V}$$

For H_2O, one can show that $\delta T = 28.441 X_{salt}$. Sometimes we write $\delta T = k_b Mo_{solute}$, where Mo_{solute} is the molality of the solution (Eq.(1e), Chapter 8). Comparing this empirical expression with Eq. (V) and replacing the molal fraction with molality (see Example 1 in Chapter 8), we can show that

$$k_b = R\, T_{pure}{}^2/(1000 \times h_{fg}) = (8.314/18.02)\times 373^2/(1000\times 2257) = 0.51 \tag{W}$$

Since gas phase consists of only H_2O, the salt water is purified by distillation process. The anti–freeze in your car consists of glycol and water solution. The glycol is almost non–volatile while water is volatile. Thus we can use Eq. (U) to determine the rise in boiling temperature with addition of glycol. The freezing point depression with addition of salt can also be determined using similar derivation.

b. Example 2

In general, when hydrocarbon droplets burn in air, the droplets first vaporize, following which the vapor is transported away from the droplet surface before burning. Phase equilibrium is often assumed between the drop and the surrounding air. In this problem we will determine the partial pressure of a vapor at the surface of a droplet. Consider a water droplet at a uniform temperature of 90°C that is vaporizing in air at a pressure of 1 bar. The air consists of a mixture of oxygen and nitrogen, while the droplet consists of a pure water. Assume ideal gas behavior to apply in the gaseous state. Determine:

The water vapor pressure.

The mole fraction of the vapor at the droplet surface.

The mole fraction of water vapor in case the pressure is raised to 10 bar, but the pressure is held constant (e.g., injection of a liquid fuel spray in gas turbines).

The Gibbs energy of water in the liquid and vapor phases at the elevated pressure.

Solution

The pressure in the liquid droplet equals that of the ambient, while the water vapor pressure p_v is lower. In the droplet $X_{H_2O(\ell)} = 1$, and applying Raoult's Law at the droplet surface

$$p_v = X_{H_2O(\ell)} P^{sat}_{H_2O}(T).$$

The saturation pressure of water $P^{sat}_{H_2O}$ = 0.7014 bar at 90°C. Hence p_{H2O} = 0.7014bar

The gas phase consists of multiple components so that

$p_v = X_{H_2O(\ell)}$ P, i.e., $X_{H_2O(\ell)}$ = 0.7014.

If total pressure is 10 bar, the at T = 90 C,

$X_{H_2O(\ell)} = p_v/P = 0.7014 \div 10 = 0.07014$, i.e., $X_{Air} = 0.92986$.

Note that increased P at same T reduces the mole % .

The Gibbs free energy

$$g_{H_2O\,(P)}(90°C, 10\text{ bar}) = h_{H_2O(P)}(90°C, 10\text{ bar}) - Ts_{H_2O\,(P)}(90°C, 10\text{ bar})$$

$$\approx h^{sat}_{H_2O(\ell)}(90°C) - T\,s^{sat}_{H_2O(\ell)}(90°C)$$

$$= 376.92 - 363 \times 1.1925 = -55.96 \text{ kJ kg}^{-1}.$$

The vapor adjusts its mole fraction in the gas phase such that

$$g_{H_2O\,(P)}(90°C, 10\text{ bar}) = g_{H_2O\,(g)}(90°C, 10\text{ bar}, X_{H2O}).$$

In an ideal gas mixture, pH2O(g) = 0.07014 * 10 = 0.7014 bar

$$\bar{g}_{H_2O(1)}\ (90°C, 10\text{ bar}) = \hat{g}_{H_2O(g)}\ (90°C, 0.7014\text{ bar}).$$

C. PRESSURE AND TEMPERATURE DIAGRAMS

1. Completely Miscible Mixtures

a. Liquid–Vapor Mixtures

This case is illustrated through the following example.

c. Example 3

A solution consists of 40% methanol (species 1) and 60% water (species 2). Assume that methanol is completely miscible in water and that the solution behaves according to Raoult's Law. The saturation pressure correlations for the pure components are:

$$\ln p_1^{sat}\ (\text{mm of Hg}) = 20.61 - 4719.2/T\text{ K, and} \quad (A)$$

$$\ln p_2^{sat}\ (\text{mm of Hg}) = 20.60 - 5205.2/T\text{ K}. \quad (B)$$

If the temperature is maintained at 20°C, at what pressure will a vapor bubble begin to form?
At this pressure and at 20°C, determine the quality W, i.e., the ratio of moles of vapor in the vapor–liquid mixture, and the vapor–phase composition.
At 20°C, what is the pressure at which virtually the entire liquid has vaporized? Determine the liquid composition and the quality at this pressure.

Solution

Water is a low volatility substance and methanol is highly volatile. At 20°C a vapor bubble first appears in pure water at P_2^{sat} = 17.03 mm of (0.0224 bar), while for methanol it appears at a higher pressure P_1^{sat} = 90.33 mm of mercury (0.119 bar). However we are interested in determining the pressure at which bubble is formed from a liquid mixture. A two-phase mixture forms when a bubble embryo first appears in the continuous mother phase of liquid, and Raoult's Law can be applied at this point. Applying Raoult's Law for each component in the mixture,

$$p_1 = X_{1(\ell)}\, P_1^{sat}, \quad \text{(C)}$$

$$p_2 = X_{2(\ell)}\, P_2^{sat} \quad \text{(D)}$$

Using Eq. (A), at T = 20°C, P_1^{sat} = 90.33 mm of Hg, (0.119 bar) i.e., from Eq. (C)

$$p_1 = 0.4 \times 90.33 = 36.13 \text{ mm of Hg}(0.048 \text{ bar}). \quad \text{(E)}$$

Similarly from Eq.(D)

$$p_2 = 0.6 \times 17.03 = 10.22 \text{ mm of mercury } (0.013 \text{ bar}). \quad \text{(F)}$$

Adding Eqs. (E) and (F), the pressure at which a vapor bubble forms

$$P = p_1 + p_2 = 46.35 \text{ mm of mercury } (0.061 \text{ bar}). \quad \text{(G)}$$

Generalizing the result for a multicomponent mixture

$$P = \Sigma X_{k(\ell)}\, P_k^{sat}(T)$$

Since this state (0.0612 bars ,20°C) lies on the saturated liquid line, at this pressure, the quality W = 0.

$$p_1 = X_1\, P, \quad \text{(H)}$$

where X_1 is the vapor phase mole fraction of species 1. From Eqs, (E), (G), and (H),

$$X_1 = p_1/P = 36.13 \div 46.35 = 0.78. \quad \text{(I)}$$

Similarly,

$$X_2 = 10.22 \div 46.35 = 0.22. \quad \text{(J)}$$

The vapor phase has a different composition as compared to the liquid composition. Generalizing the result for a multicomponent mixture

$$X_k = p_k / \Sigma X_{k(\ell)}\, P_k^{sat} = X_{k(\ell)}\, P_k^{sat} / \Sigma X_{k(\ell)}\, P_k^{sat}. \quad \text{(K)}$$

Applying Raoult's law,

$$X_1 = (X_{1(\ell)}\, P_1^{sat}/(X_{1(\ell)}\, P_1^{sat} + X_{2(\ell)}\, P_2^{sat}), \text{ and} \quad \text{(L)}$$

$$X_2 = (X_2\, P_2^{sat}/(X_1\, P_1^{sat} + X_2\, P_2^{sat}), \text{ i.e.,} \quad \text{(M)}$$

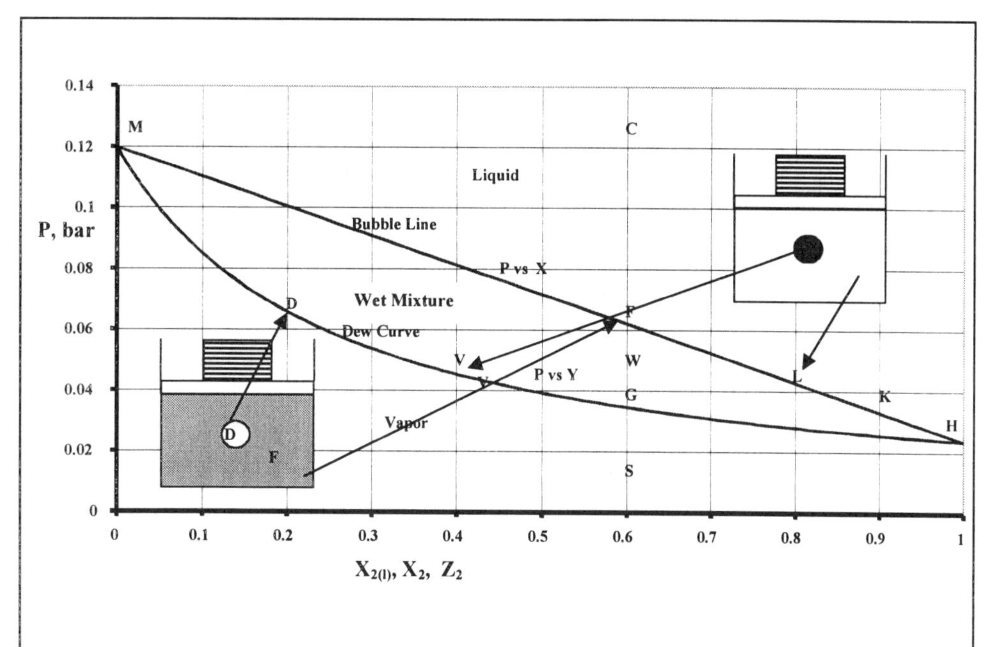

Figure 3: Illustration of $P\text{–}X_{k(\ell)}$ or $P\text{-}X_{k(\ell)}\text{-}X_k\text{-}Z$ diagram at 20°C for a methanol (1)–water (2) solution.

When almost the entire liquid is vaporized $X_1 = 0.4$ and $X_2 = 0.6$, and from Eqs. (L) and (M) we have

$X_{1(\ell)} = 0.112$, $X_{2(\ell)} = 0.888$, and

$P = (X_{1(\ell)}\, P_1^{sat} + X_{2(\ell)}\, P_2^{sat}) = 0.112 \times 90.33 + 0.888 \times 17.03 = 25.24$ mm of Hg.

The last liquid drop contains 11% methanol and 89% water. One can alter the liquid composition and obtain plots of P vs $X_{H_2O(\ell)}$ as shown by curve MFLKH in Figure 3.

Generalizing for K components, one can solve for $X_{k(P)}$ using the K linear relations (Eqs. (K)) and the known vapor phase mole fractions, X_k. At the saturated vapor state the quality W= 1.

Remarks

Consider a 40% methanol and 60% water mixture at 20 °C in a PCW assembly so that pressure is at 0.132 bar (State C). As we start slowly removing smaller weights one after another, the first vapor bubble (embryo phase) appears in the liquid (mother phase) at a pressure of 46.35 mm of mercury (or 0.061 bar, Figure 3 point F). This pressure is the vapor pressure at 20°C and the pressure and temperature specify the saturated liquid state of the mixture. At phase equilibrium, inside that bubble, the mole fraction of water vapor is 0.22 and mole fraction of methanol is 0.78 (point D). The vapor in the embryo is still at 46.35 mm of Hg. Although water constitutes 60 % of the liquid phase (point F), it constitutes only 22 % of the vapor phase (Point D), since it is less volatile due to a lower vapor pressure. By altering the composition, the $P\text{-}X_{k(P)}$ diagram can be obtained as M DVGH.

Now consider an 88.8% water and 11.2% methanol solution at 20°C. The first vapor bubble appears at a pressure of 24.36 mm of mercury (point K, 0.0325 bar) and the vapor composition is 60% water and 40% methanol (point G). Thus, for $X_{2(P)} = 0.6$, P = 0.061 bar, and $X_2 = 0.22$ while at $X_{2(l)} = 0.888$, P = 0.0322 bar, and $X_2 = 0.6$. As $X_{2(l)} \to 1.0$, $P \to P_2^{satt}$ (i.e., 17.03 mm of mercury or 0.0225 bar, point H). As $X_{2(l)} \to$

0 (and, consequently, $X_{1(l)} \rightarrow 1$) $P = P_2^{sat}$ (90.33 mm of mercury or 0.119 bar, point M). A line connecting the pressures through the bubble points is called the bubble line (saturated liquid line, MFKH) at which the first bubble appears.

Consider a methanol–water mixture with $X_{2(l)} = 0.6$ at 20ºC. The mixture is in compressed liquid state at a pressure of 100 mm of mercury (point C, 0.132 bar). As the pressure is reduced below 0.061 bar (46.4 mm of Hg or the bubble pressure), say, to 0.0422 bar (32 mm Hg, point W), phase equilibrium requires that $X_{2(l)} = 0.8$ and $X_2 =$ 0.43, Representing the initial molal fraction of species 2 as Z_2, then at "W" (cf. Figure 3),

$$Z_2 = (N_2(\ell) + N_2\,(g))/N, \text{ i.e.,} \quad \text{(K)}$$

Rewriting,

$$Z_2 = (X_{2(\ell)}N_\ell + X_2\,N_g)/N = X_{2(\ell)}(1-W) + X_2W, \quad \text{(L)}$$

where $W = N_g/N$ denotes the molal quality. Consequently,

$$W = (X_{2(\ell)} - Z_2)/(X_{2(\ell)} - X_2). \quad \text{(M)}$$

If $Z_2 = 0.6$, $P = 32$ mm of mercury (i.e., 0.042 bar), $X_{2(l)} = 0.8$, $X_2 = 0.43$, and, hence, $W = 0.54$. Therefore, 54 % of the original mixture exists as vapor, while 46% is in the liquid state. Water (species 2) constitutes 43% of the vapor. We can apply the ideal gas equation for gas phase to compute the vapor volume. In order to specify the state at 20ºC for a two-phase mixture, we require additional information, such the overall

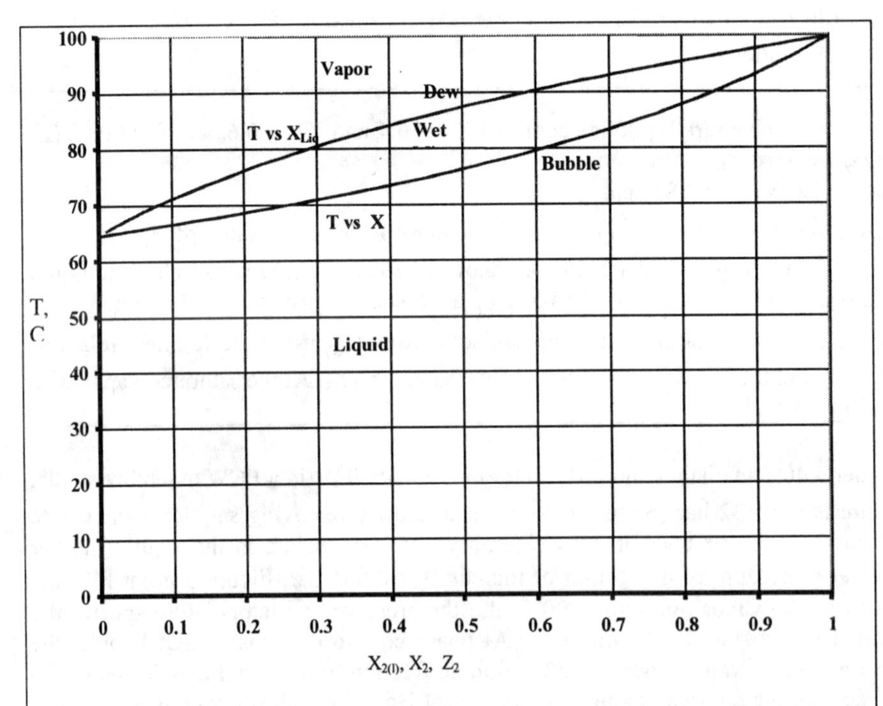

Figure 4: Illustration of a T–X–Y diagram at a pressure of 760 mm of mercury (i.e., 100 kPa) for a methanol (species 1)–water (species 2) solution.

mixture composition Z. Different pressures then lead to different vapor phase compositions. At same T, for a two component mixture, we can have two phase mixture within MFLKH (saturated liquid line for a mixture or bubble line) and MDVGH (saturated vapor or dew line) with different pressures. For example, at 20°C, Z_2 or $X_{2(l)}$ = 0.6, the pressure can be varied from 46.5 mm of mercury (0.0614 bar) at saturated liquid line to 24.36 mm of mercury (0.0322 bar) at saturated vapor line (see the vertical line FWG) but still maintain two phase mixture. Thus, we can have two degrees of freedom (say T and P) for existence of two phases at given composition. On the other hand, for a pure component such as water if T = 100 C, two phase mixture (vapor and liquid) exists only at a single pressure of P =760 mm of mercury or 1 bar. If the pressure is raised, liquid is formed. Decreasing the pressure produces superheated vapor.

If Z_2 = 0.6 then at X_2 = 0.6, Eq. (M) suggests that W = 1 regardless of the value of $X_{2(l)}$. Since W = 1, all of the liquid has evaporated at this condition, which is called the saturated vapor state (that starts along (line MDVGH – Figure 4). At this state, the vapor concentrations equal the original liquid molal concentrations. Further reduction of pressure causes the vapor mixture to become superheated (state S).

Consider point S, which is at the superheated state, say at X_2 = 0.6 If the pressure of the superheated vapor mixture (the mother phase) is increased, at some pressure a liquid drop (the embryo phase) appears (Point G). In this manner, varying the mole fraction of a component X_2, one can identify the dew point (Figure 3) curve or the saturated vapor curve. In context of Figure 3, the region below the dew point curve MDVGH (cf. Figure 3) is the superheated region, while the region within it and the saturated liquid line MFLKH is called the wet region. The region above the saturated liquid line MFLKH is called the compressed (or subcooled) liquid region.

$P–X_{k(l)}–T$ Diagram: The methodology of Example 3 can be repeated at various temperatures to obtain a three–dimensional space diagram. The maximum temperature at which a species exists in the liquid phase is its critical temperature (647 K in the case of water and 313 K for methanol. At this temperature the intermolecular separation distances are the same for the saturated liquid and the saturated vapor. When $T_{c,1} < T < T_{c,2}$, in that case the diagram does not extend to $X_{1(l)}$ =1, since $T > T_{c,1}$.

T–Xk Diagram: This case is illustrated through Example 4 and Figure 4.

---

d. *Example 4*

A solution consists of 40% of methanol (species 1) and 60% water (species 2). Assume that it follows Raoult's Law and the following relations apply

$$\ln P_1^{sat} \text{ (in mm of Hg)} = 20.61 - 4719.2/T, \text{ T in K and} \quad \text{(A)}$$

$$\ln P_2^{sat} \text{ (in mm of Hg)} = 20.60 - 5205.2/T, \text{ Tin K .} \quad \text{(B)}$$

The normal boiling point is defined when P^{sat} = 100 kPa. From the above relations, the normal boiling point of species 1 is 64.7°C and that of species 2 is 100°C. Draw a graph of T with respect to $X_{2(\ell)}$, X_2, and Z_2.

Solution

Applying Raoult's Law to each component,

$$p_1 = X_{1(\ell)}\, P_1^{sat}, \text{ and} \quad \text{(C)}$$

$$p_2 = X_{2(\ell)}\, P_2^{sat}. \quad \text{(D)}$$

Using Eq. (A) at 20°C, P_1^{sat} = 90.33 mm of Hg. Then

$$P = p_1 + p_2 = (1 - X_{2(\ell)})\ P_1^{sat}(T) + X_{2(\ell)}\ P_2^{sat}(T) = 760 \text{ mm of Hg}. \tag{E}$$

From Eqs. (A), (B), and (E),

$$P = 760 = (1 - X_{2(\ell)}) \exp(20.61 - 4719.2/T) + X_{2(\ell)} \exp(20.60 - 5205.2/T). \tag{F}$$

Eq.(F) is non-linear in T at specified $X_{2(\ell)}$; however it is linear in $X_{2(\ell)}$ at specified T. The graphs can be generated as follows. If T = 75°C, P = 760 mm, then one solves for $X_{2(\ell)}$ from Eq.(F): $X_{2(\ell)} = 0.451$, $X_{1(\ell)} = 1 - X_{2(\ell)} = 0.549$, $P_2^{sat} = 128.41$ mm, then using Eq. (D)

$$p_2 = X_{2(\ell)}\ P_2^{sat}(T) = 0.451 \times 128.41 = 57.9 \text{ mm} = X_2 P = X_2 \times 760, \text{ i.e.,}$$

$X_2 = 0.169$ so that $X_1 = 0.831$. Figure 4 illustrates T-$X_{2(\ell)}$-X_2 diagram at specified pressure while Figure 5 qualitatively illustrates T-$X_{2(\ell)}$-X_2 at various pressures.

b. Relative Volatility

The relative volatility is defined as

$$R_{v,21} = (X_2/X_{2(\ell)})/(X_1/X_{1(\ell)}) \tag{16}$$

If $X_2 = X_{2(\ell)}$ and $X_1 = X_{1(\ell)}$, then $R_{v,21} = 1$. If $R_{v,21} > 1$, this implies that component 2 is highly volatile compared to 1 and *vice versa.*

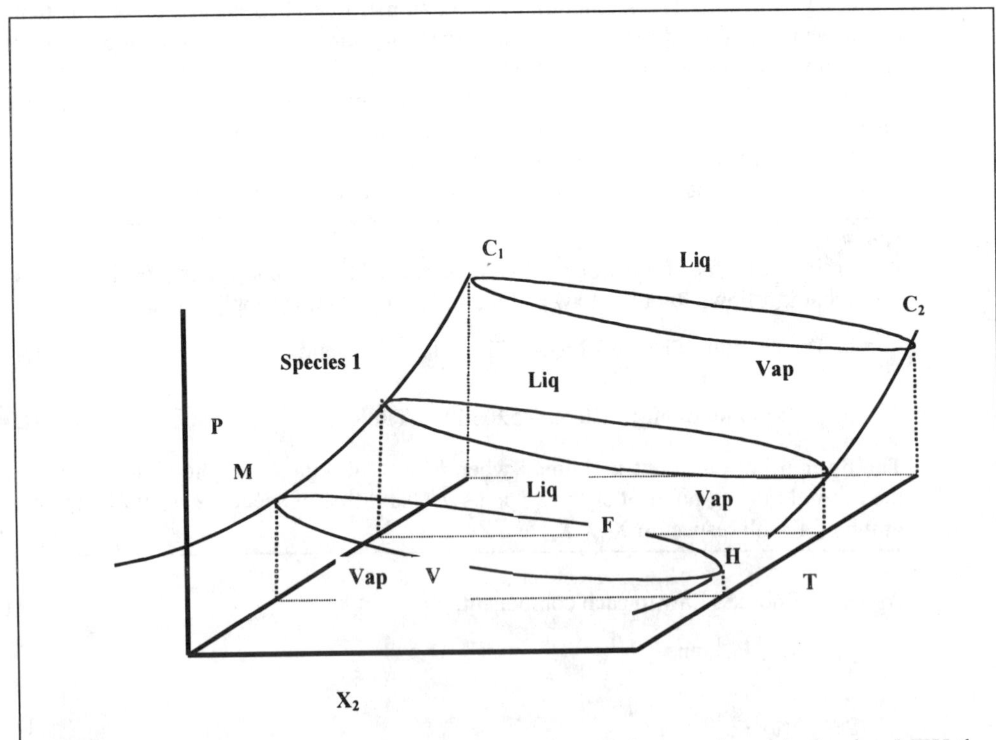

Figure 5: P–X–T diagram for a binary mixture, where C denotes the critical point; MFH the bubble line, and MVH the vapor line (from G. J. Wiley, and R. Sonntag, *Fundamentals of Classical Thermodynamics*, 3 rd Ed., John Wiley & Sons, 1986. With permission.).

c. *P–T Diagram for a Binary Mixture*

The generation of this graph is illustrated through the following example.

---

e. *Example 5*

A solution consists of 40% of methanol (species 1) and 60% water (species 2). Assume that it follows Raoult's Law and the following relations apply

$$\ln P_1^{sat} \text{ (in mm of Hg)} = 20.61 - 4719.2/T, \text{ T in K, and} \quad (A)$$

$$\ln P_2^{sat} \text{ (in mm of Hg)} = 20.60 - 5205.2/T, \text{ T in K}. \quad (B)$$

The normal boiling point is defined when P^{sat} = 100 kPa. From above relations, the normal boiling point of species 1 is 64.7°C and that of species 2 is 100°C. Draw a graph of P with respect to T with $X_{2(\ell)}$ as a parameter.

Solution

Applying Raoult's Law to each component,

$$p_1 = X_{1(\ell)} P_1^{sat}, \text{ and} \quad (C)$$

$$p_2 = X_{2(\ell)} P_2^{sat}, \text{ so that} \quad (D)$$

$$P = p_1 + p_2 = (1 - X_{2(\ell)}) P_1^{sat}(T) + X_{2(\ell)} P_2^{sat}(T). \quad (E)$$

Fix a value for $X_{2(l)}$. Vary T, obtain P from Eq. (E). Thus P vs T at given $X_{2(l)}$ can be plotted.

d. *P–$X_{k(l)}$–T diagram*

We can repeat Example 3 for various temperatures and qualitatively obtain 3 D plot as shown in Figure 5. For example, if species 2 is water at 20°C, the curve MFH represents the bubble line while MVH represents the dew line. Repeating the calculation at T_2^{sat} = 40°C, we can obtain similar curves. The line HC_2 descrbes the saturation pressure of pure species 2 while the line MC_1 shows the corresponding curve for species 1. We know that maximum temperature for species 2 is the critical temperature (647 K) while the corresponding temperature is 313 K for methanol. When $T_{c,1} < T < T_{c,2}$, then the P-T-$X_{k(l)}$-X_k diagram will not extend to $X_{1(l)}$ =1 since $T > T_{c,1}$ and hence it is always vapor at x_1 =1. Similarly if $T > T_{c,2}$ then it will not extend near $X_{2(l)}$ =1. But it can have curves for some intermediate values of $X_{1(l)}$ and $X_{2(l)}$.

e. *Azeotropic Behavior*

The term azeotrope is used for situations that have no composition change. During the boiling of a binary mixture, at a certain liquid composition, the vapor composition can be the same as the liquid composition i.e., X_2 = $X_{2(\ell)}$. Therefore, the liquid components cannot be separated. This state is an azeotrope. At a specified total pressure, as the mixture composition is changed, an azeotropic mixture reaches a minimum (Q in Figure 6) or maximum in the boiling temperature (e.g., water–antifreeze mixtures). In a nonazeotropic mixture the boiling temperature varies monotonically with composition (e.g., water–NH_3 mixtures, Figure 4 for H2O:methanol mixtures). Figure 6 presents a T–$X_{k(\ell)}$ diagram that illustrates the azeotropic behavior of a binary mixture of species A and B. At point Q the azeotropic composition has a corresponding temperature T_Q.

---

f. *Example 6*

Lake water at 25°C is exposed to ambient air at 1 bar. Phase equilibrium occurs and Raoult's Law can be assumed to be applicable. Assume that air is dissolved in the liquid water so that $X_{air(l)}$ = 0.019. What is the mole fraction of the water vapor in the gas phase? What will this vapor mole fraction be if no air is dissolved in the liquid water?

Include the POY correction factor to answer the question. If the partial pressure of the water vapor at phase equilibrium is 3.1 kPa, determine the mole fraction of air in the liquid solution.

Solution

We will use the expressions

$$\mu_k(T,P,X_{k,}) = \mu_k(T,P) + RT \ln X_k = \mu_k(T,P^{sat}) + RT \ln (p_k/P^{sat}(T)), \text{ and} \quad (A)$$

$$\mu_k (T,P,X_{k(\ell)}) = \mu_k (T,P^{sat}) + v_f(P - P^{sat}) + RT \ln X_{k(\ell)}. \quad (B)$$

Equating Eqs. (A) and (B),

$$v_f(P - P^{sat}) + RT \ln X_{k(\ell)} = RT \ln (p_k/P^{sat}(T)), \text{ where,}$$

$$p_k/(X_{k(\ell)} P^{sat}(T)) = \exp (v_f(P - P^{sat})/RT) = POY. \quad (C)$$

Ignoring the Poynting correction, and applying Raoult's Law for water,
$p_{H_2O} = X_{H_2O,\ell} P^{sat}_{H_2O}(25C) = 0.981 \times 3.16.$
When $p_{H_2O} = 3.1$ kPa and P = 100 kPa,
$X_{H_2O} = 3.1 \div 100 = 0.031.$
In the absence of any dissolved air, $X_{H_2O,\ell} = 1$ so that $P_{H_2O} = P^{sat} = 3.16$ kPa, and
$X_{H_2O} = 3.16 \div 100 = 0.0316.$
The water vapor mole fraction at 25°C is reduced if air is dissolved in liquid water.

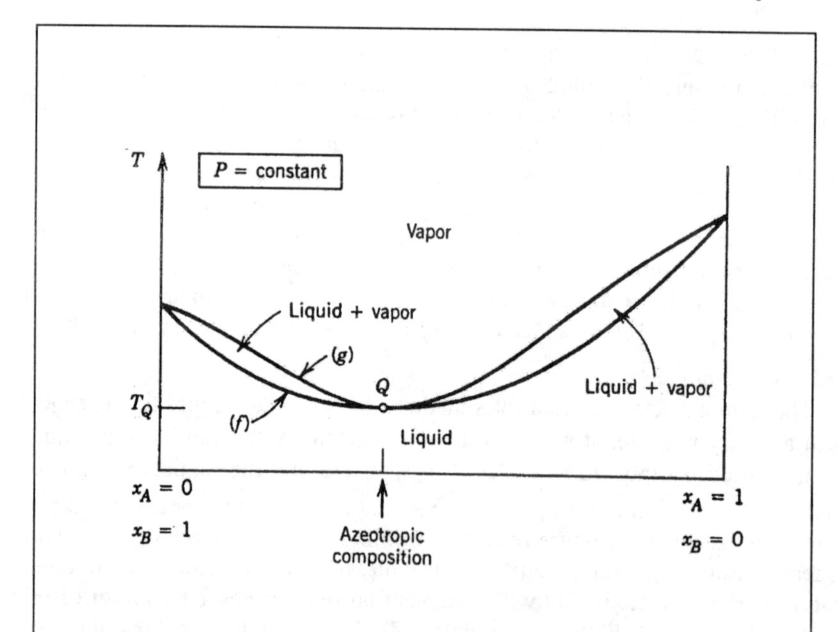

Figure 6: The T–X_k, –X_k diagram for a binary azeotropic mixture containing two species A and B, where g denotes the dew line (saturated vapor), f the bubble line (saturated liquid), and Q the azeotropic point. (From A. Bejan, *Advanced Engineering Thermodynamics*, John Wiley and Sons., 1988, p. 271. With permission)

If the RHS of Eq. (C) has a significant value, then

$p_{H_2O}/(0.981 \times 3.16) = \exp(0.001(100-3.16) \div (0.461 \times 298)) = 1.000705$, i.e.,

$p_{H_2O} = 3.10215$, and $X_{H_2O} = 0.0310215$.

---

g. *Example 7*

An ideal solution contains alcohol (species 1 present on a 60% mole basis in the liquid phase and 87% in the vapor phase) and water (species 2) at a temperature and pressure of 60°C and 433 mm of Hg. At 60°C, $P_1^{sat} = 625$ mm of Hg, and $P_2^{sat} = 144$ mm of Hg. Determine $\hat{f}_{1(g)}$, $\hat{f}_{2(g)}$, $\hat{f}_{1(\ell)}$, and $\hat{f}_{2(\ell)}$.

Solution

The specified pressure is greater than P_2^{sat} but less than P_1^{sat}. Therefore, at the specified temperature and pressure, pure water exists as compressed liquid while pure alcohol exists as superheated vapor.

$\hat{f}_{1(g)} = X_{1(\ell)}\ f_{1(\ell)}$ (60°C, 433 mm of hg) = $X_{1(\ell)} \times 433$ mm of mercury = $0.87 \times 433 = 377$,

since the fugacity of the vapor equals the vapor pressure, assuming ideal gas behavior.

Similarly,

$\hat{f}_{2(\ell)} = X_{2(\ell)}\ f_{2(\ell)}(60°C, 433 \text{ mm of Hg}), \approx X_{2(\ell)}\ f_{2(g)}(60°C, 144 \text{ mm of Hg})$, where

$f_{2(g)} \approx 144$ mm of Hg,

since the liquid fugacities do not change significantly with pressure. Hence,

$\hat{f}_{2(\ell)} = 0.4 \times 144 = 57.6$ mm of Hg.

Likewise.

$\hat{f}_{1(\ell)} = X_{1(\ell)}\ f_{1(\ell)}(60°C, 433 \text{ mm of Hg})$.

In its pure state, the alcohol exists in the form of a vapor. We must determine the fugacity at the hypothetical liquid state (T,P = 60°C and 433 mm of Hg). We will assume that

$f_{1(\ell)}$ (60°C, P_1^{sat} at 60 C) $\approx f_{1(\ell)}(60°C, 625 \text{ mm of Hg})$,

since the term

$v_{(\ell)}(433 - 625) \ll f^{\ell}$ (60 C, 625 mm of Hg).

However,

$f_{1(\ell)}(60°C, 625 \text{ mm of Hg}) = f_{1(g)}(60°C, 625 \text{ mm of Hg})$

= 625 mm of Hg, since the vapor is assumed to be ideal.

Therefore,

$\hat{f}_{1(\ell)} = 0.6 \times 625 = 375$ mm of Hg.

$\hat{f}_{2(\ell)} = 0.13 \times f_2$ (60°C, 433 mm of Hg).

We will use a hypothetical ideal gas state for f_2(60°C, 433 mm of Hg), i.e.,

$\hat{f}_{2(g)} = 0.13 \times 433$ mm of mercury = 56 mm of Hg.

Remarks

We see that $\hat{f}_{2(g)} \approx \hat{f}_{2(\ell)}$, and $\hat{f}_{1(g)} \approx \hat{f}_{1(\ell)}$, thereby satisfying the phase equilibrium criterion.

We have assumed ideal gas behavior for the hypothetical vapor state. Real gas behavior can also be accounted for, as shown below. Recall that f_2(60°C, 433 mm of Hg) must be calculated for a hypothetical liquid state. To do so, we can first determine the fugacity of saturated vapor or liquid at the state (60°C, 144 mm of Hg). Then we can employ the relation

$d(\ln(f)) = \int v dP/(RT) = \int (Pv/(RT)) d(\ln P) = \int Z d(\ln P)$.

Integrating between the limits P^{sat} and P, and assuming that $Z = Z^{sat}$, we obtain the relation

$\ln(f/f^{sat}) \approx Z^{sat}\ln(P/P^{sat})$, or $(\ln(\phi))/(\ln(f^{sat}/P^{sat})) = Z^{sat}$ (60°C, 144 mm of Hg).

A plot of $\ln(\phi)$ with respect to P_R at specified T_R is approximately linear.

h. Example 8

A fuel droplet contains a binary mixture of 60% n–heptane (species 1) and 40% hexadecane (species 2). Assume that air (species 3) is insoluble in the liquid phase. The Cox–Antoine relation can be assumed to apply, i.e.,

$$\ln P_k^{sat} = E_k + F_k/(T + G_k),\ k = 1,2 \quad (A)$$

where E_k, F_k, and G_k are constant for a specified fuel component. If the pressure is expressed in mm of Hg, $E_1 = 15.89$, $F_1 = -2911.32$, and $G_1 = -56.4$, and $E_2 = 24.66$, $F_2 = -10660.2$, and $G_2 = 54.1$ when $1 < P < 40$ mm of Hg. Determine the partial pressures of the two species 297 K in air if the ambient pressure is 100 kPa and the gas–phase composition.

Solution

Applying Raoult's Law for miscible mixtures $p_k = x_k\ P_k^{sat}$, k=1,2, since

$$p_1 + p_2 + p_3 = P,\ X_{1,\ell} P_1^{sat} + X_{2,\ell} P_2^{sat} + p_3 = P, \text{ i.e.,} \quad (B)$$

$$p_3 = p_{air} = P - (X_{1,\ell}\ P_1^{sat} + X_{2,\ell}\ P_2^{sat}). \quad (C)$$

At 20°C, $P_1^{sat} = 44.245$ mm of Hg, and $P_2^{sat} = 0.00338$ (i.e., species 2 is almost nonvaporizable). Therefore,

$p_1 = X_{1,\ell} P_1^{sat} = 0.6 \times 44.245 = 26.55$ mm of Hg, and

$p_2 = (1-X_{1,\ell}) P_2^{sat} = 0.4 \times 0.00338 = 0.001352$ mm of Hg.

Using Eq. (C),

$p_{air} = 760 - (26.55 + 0.001335) = 733.45$ mm of Hg, i.e.,

$X_1 = 26.55 \div 760 = 0.0349$, and

$X_{air} = 733.45 \div 760 = 0.965$,

(Since hexadecane is virtually nonvaporizing, we neglect its mole fraction.) The mass fraction of n–heptane

$Y_1 = 0.0349 \times (7 \times 12 + 16 \times 1) \div (0.0349 \times 100 + 0.965 \times 28.97) = 0.11$, i.e.,

$Y_3 = 0.89$

The variation of the temperature of a droplet containing a binary mixture consisting of n–heptane (60 %) and hexadecane with respect to the n–heptane mole fraction is illustrated in Figure 7 at P=100 kPa.

i. Example 9

Dissolved air in water provides the oxygen and nitrogen that are necessary to sustain marine life. Obtain approximate relations that can be used to determine the trace mole fractions of oxygen and nitrogen in liquid water. Recall that in air $(X_{O_2}/X_{N_2}) = 3.76$.

Solution

We will assume that Raoult's Law applies, i.e.,

$$X_{O_2(\ell)} = X_{O_2}\ P/P_{O_2}^{sat}(T), \text{ and} \quad (A)$$

$$X_{N_2(\ell)} = X_{N_2}\ P/P_{N_2}^{sat}(T). \quad (B)$$

In air $X_{N_2}/X_{O_2} = 3.76$. Furthermore,

$$(X_{O_{2(\ell)}}/X_{N_{2(\ell)}}) = (X_{O_2}/X_{N_2})/(P^{sat}_{N_2}(T)/P^{sat}_{O_2}(T)) \quad (C)$$

Since $P^{sat}_{N_2}(T) \approx P^{sat}_{O_2}(T)$,
$(X_{O_{2(\ell)}}/X_{N_{2(\ell)}}) = (X_{O_2}/X_{N_2}) = 3.76$.

Remarks

As the temperature rises, the value of P^{sat}_k for a substance increases. Hence $X_{k,\ell}$ decreases. The warming of river water decreases the O_2 and N_2 concentrations in it.

j. *Example 10*

A 20-liter rigid volume consists of 80% liquid and 20% vapor by mass at 111.4°C and 1.5 bar. A pin is placed on piston to prevent its motion. Gaseous nitrogen is isothermally injected into the volume until the pressure reaches 2 bar. What is the nitrogen mole fraction in the gas phase? Assume that N_2 does not dissolve in the liquid.
What happens if there is no pin during the injection of N_2. Instead of adding N_2, discuss the effects with salt addition at 111.4°C.

Solution

$P^{sat}(111.4°C) = 1.5$ bar. The total volume

$$V = m_g\, v_g + m_f\, v_f = m\,(x\, v_g + (1-x)\, v_f) \quad (A)$$

$$= m\,(0.2\times 1.159 + 0.8\times 0.001053) = m\,(0.233).$$

Therefore,
$m = 20\times0.001/0.233 = 0.0858$ kg
Using the ideal gas law for the vapor phase

$$m = V_f/v_f + (V - V_f)/(RT/P_o), \quad (B)$$

$$V_f = (m - P_o\, V/RT)/(1/v_f - P_o/RT) \approx v_f\,(m - P_o\, V/RT). \quad (C)$$

The pressure increases as additional gas is injected, thereby increasing the Gibbs energy of the liquid and vapor phases.
In case of liquid water,

$$g_\ell(T,P) = g_\ell(T,P^{sat}) + v_\ell(P - P_{sat}).$$

For an ideal gas mixture in the vapor phase,

$$\hat{g}_{H_2O}(T,P,X_{H2O}) = \bar{g}_{H_2O}(T,p_{H_2O}) = \bar{g}_{H2O}(T,P^{sat}) + \int \bar{v}_{H2O(g)}dP$$

$$= \bar{g}_g\,(T,P^{sat}) + \bar{R}T\,\ln(p_{H_2O}/P^{sat}). \quad (D)$$

Equating Eq. (C) with Eq. (D)

$$\bar{v}\,(P - P^{sat}) = RT\,\ln\,(p_{H_2O}/P^{sat}), \text{ or } \ln(p_{H_2O}/P^{sat}) = (v_\ell(P - P^{sat}))/(RT) \quad (E)$$

This relation is known as the Kelvin–Helmholtz formula which shows the effect of total pressure on partial pressure of vapor. Note that the partial pressure of H_2O in the vapor phase is not the same as saturation pressure at T.
For water, $v_\ell = 0.001053\ m^3\ kmole^{-1}$, $P^{sat} = 1.5$ bar, and for this case P = 2 bar, and T = 384.56 K. Therefore, the partial pressure of H_2O in vapor phase,

$$p_{H_2O} = 1.500445 \text{ bar}.$$

a value close to saturation pressure at T= 384.6K since v_f is small. Further

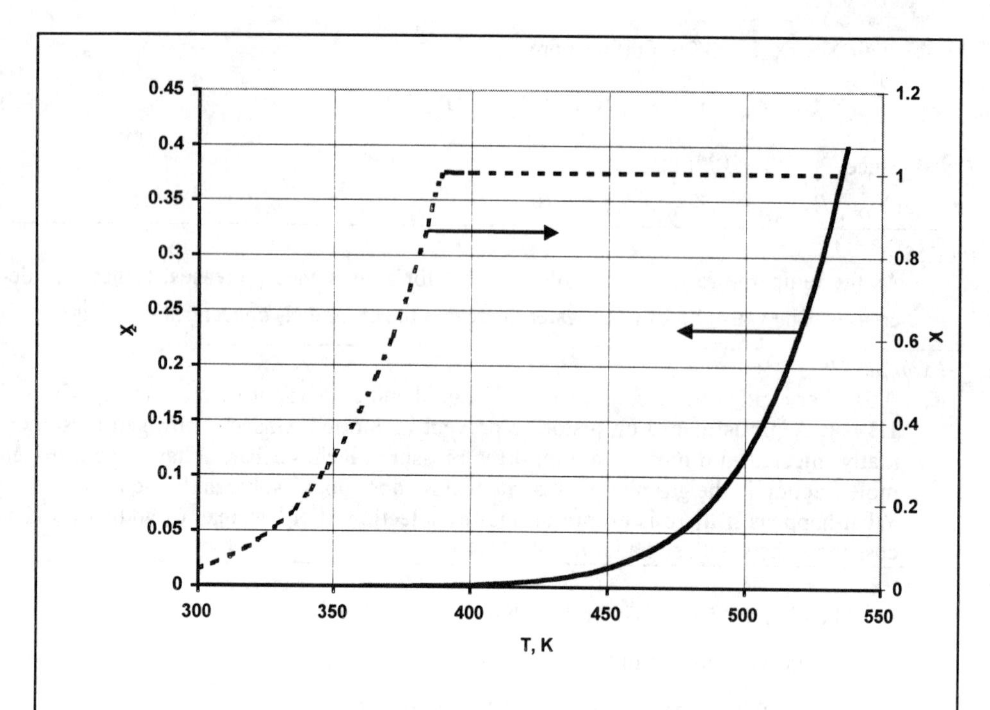

Figure 7: The variation of the mole fraction of heptane vapor with droplet temperature for a mixture containing 60 % n–heptane and 40 % hexadecane at 100 kPa.

$X_{H_2O} = 0.75022$, and $X_{N2} = 0.24798$.

The vapor mass

$$m_v = p_vV_v/RT = p_v(V - V_f)/RT, \text{ and}$$

the liquid mass

$$m_f = V_f/v_f.$$

Adding the two masses,

$$m = p_v(V - V_f)/RT + V_f/v_f.$$

Therefore,

$$V_f = (m - p_vV/RT)/(1/v_f - p_v/RT) \approx v_f(m - p_vV/RT). \quad \text{(F)}$$

Since P_v after N_2 injection is slightly higher than P_v before N_2 injection, there should be more vapor; thus the volume of liquid decreases. According to Le Chatelier, the system counteracts the pressure increase by increasing the volume of the vapor phase. If we ignore the term $(v_f(P - P_{sat}))/(RT)$, in Eq. (A) this implies that $p_{H2O} = P^{sat}$ and $X_v = 0.75$.

The injection of nitrogen implies that $X_{H2O}<1$. Pressure remains constant. Therefore, $\hat{g}_{H_2O} = \bar{g}_{H_2O}(T,P) + \bar{R}T \ln X_{H2O}$. Since $X_{H2O} <1$, $\hat{g}_{H_2O} < \hat{g}_{H_2O(\ell)}(T,P)$, as long as the temperature and pressure are maintained, vaporization continues until all of the liquid vaporizes.

Similarly when we add salt in water(or an impurity), the Gibbs function of the liquid H_2O decreases which causes the vapor molecules to cross over from the vapor into the liquid phase.

Remarks

At a specified temperature, an increase in pressure causes the "g" of liquid to increase slightly. The Gibbs free energy of the vapor equals that of the liquid. If the vapor is an ideal gas, the enthalpy of the vapor will remain unchanged. The slight Gibbs energy increase must then cause the entropy of vapor to decrease which corresponds to an increase in the partial pressure of vapor.

Consider a component k of a liquid mixture that exists in equilibrium with a vapor phase that also contains a mixture of insoluble inert gases. In this case,

$$\mu_{k(l)}(T,P) = \mu_{k(g)}(T, P).$$

If the vapor phase is isothermally pressurized, then

$$\mu_{k(l)}(T,P) + d\mu_{k(l)} = \mu_{k(g)}(T,P) + d\mu_{k(g(}, v_{k(l)}dP_l = v_{k(g)}\, dP_g \text{ and } dP_l/dP_g = v_k^g/v_k^\ell.$$

An increase in the pressure in the vapor phase requires a large change in the liquid phase pressure to ensure that liquid–vapor equilibrium is maintained.

2. Immiscible Mixture

a. *Immiscible Liquids and Miscible Gas Phase*

This case is illustrated through the following example.

---

k. *Example 11*

Consider binary vapor mixture of methanol (species 1) and water (species 2) that are assumed to be immiscible in the liquid phase. Illustrate their behavior with respect to pressure and temperature. You may assume that

$$\ln P_2^{sat} = 13.97 - 5205.2/T\ (K), \text{ and} \tag{A}$$

$$\ln P_1^{sat} = 13.98 - 4719.2/T\ (K). \tag{B}$$

Solution

Employing Eqs. (A) and (B), the normal boiling points of species 1 (methanol) and 2 (water) are, respectively, 64.4 and 100°C.

We will employ Raoult's Law, in which the liquid mole fractions for water and methanol must be set to unity, since they are immiscible. Therefore,

$$P_2^{sat}(T) = X_2\, P, \text{ and} \tag{C}$$

$$P_1^{sat}(T) = X_1\, P. \tag{D}$$

Upon adding Eqs. (C) and (D), we obtain the expression

$$P_2^{sat}(T) + P_1^{sat}(T) = P. \tag{E}$$

Figure 8 shows the T- $X_{k(\ell)}$-X_k diagram.

Remarks

In case of immiscible mixtures, partial pressures are only a function of temperature alone. Irrespective of the liquid phase composition, at a specified temperature, P_2^{sat} can be obtained from Eq. (A), while P_1^{sat} can be, likewise, obtained using Eq. (B). Using Eqs. (C) and (D), we obtain the values of X_1 and X_2 for a specified pressure, and plots of temperature can be plotted with respect to composition, as shown in Figure 8. The lines BME and EJGA in that figure are called the dew lines for species 1 and 2, respectively. The region above the curve BMEJGA is the superheated vapor mixture region while that below the curve CELD is the compressed liquid region.

Consider the following scenario. A vapor mixture is contained in a piston–cylinder–weight assembly, such that P = 1 bar, X_2 = 0.6, and T = 100°C (cf. point S). Species 2 exists in the form of superheated vapor, since p_2 = 0.6 bar at T = 100°C. The cylinder is now cooled. The saturation temperatures T_2^{sat} = 86.5°C at p_2 = 0.6 bar, and T_1^{sat} = 43.87°C at p_1 = 0.4 bar. The assembly contains a vapor mixture only, as long as T>86.5°C. As the vapor mixture is cooled, a liquid drop appears at T = 86.5°C

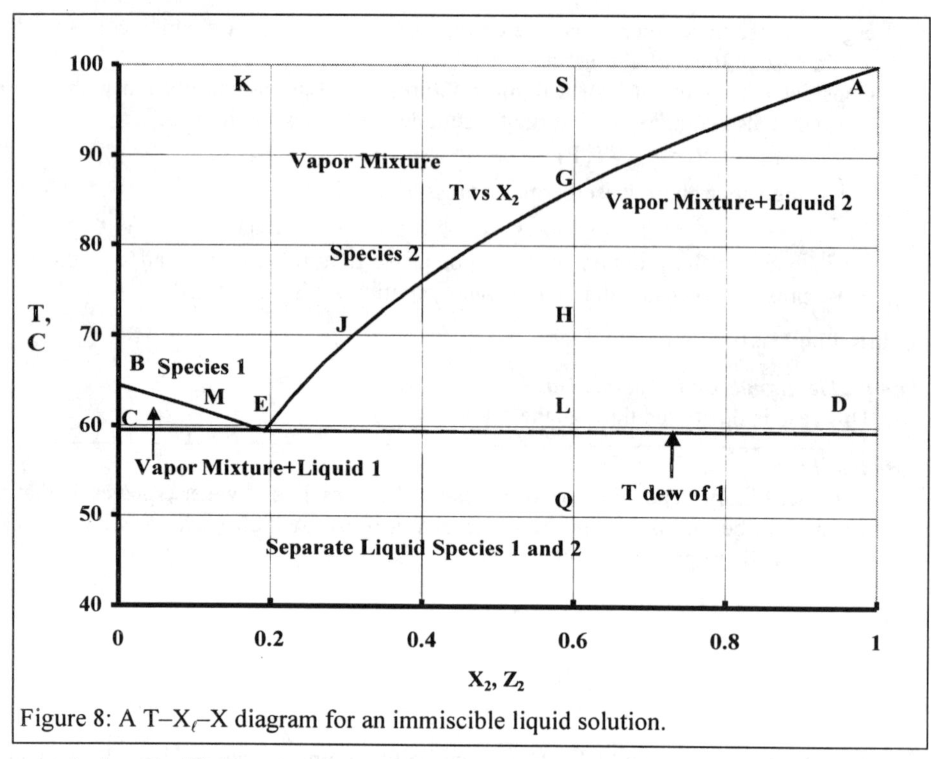

Figure 8: A T–X_ℓ–X diagram for an immiscible liquid solution.

(point G). (If the gas phase composition is changed to $X_2 = 0.2$, in that case the first liquid drop appears at 61°C (point E)). If the mixture is cooled to 70°C (cf. point H), phase equilibrium – that is manifested in the form of Eq. (C) – implies that vapor phase mole fraction must reduce to $X_2 = 0.3$ (cf. point J), i.e., more of species 2 must condense. It also implies that X_1 must increase to 0.7 from the initial mole fraction of 0.4. Eqs (D) and (B) dictate that $T_1^{sat} = 52°C$, so that species 1 at 70°C exists in the form of a superheated vapor. Upon further cooling to 60°C, phase equilibrium requires that $X_2 = 0.19$ (cf. point E), and T_1^{sat} increases to 60°C. Any further cooling causes both species 1 and 2 to condense, where the condensate phase is an immiscible binary mixture. Within the region EJGADE (i.e., for $X_2 > 0.19$, $60°C < T < 100°C$), liquid species 2 and vapor mixture must coexist. In region BMEC ($X_2 < 0.19$, $60°C < T < 64.7°C$), liquid species 1 and vapor mixture must coexist. At 60°C, $X_2 = 0.19$, and both the liquid and vapor mixture coexist.

At point E there are two liquid phases and one vapor phase. According to Gibbs phase rule, $F = K + 2 - \pi = 2 + 2 - 3 = 1$. Therefore, there is one independent variable in the set (P, T, X_2). In case the pressure is fixed, then the temperature and X_2 are fixed (i.e., 60°C, and $X_2 = 0.19$) for coexistence the three phases to coexist. If the mixture is cooled from 100°C (cf. point K), species 1 condenses, increasing the mole fraction of species 2 until $X_2 = 0.19$.

Now assume that the liquid mixture is heated at the condition $X_{2(\ell)} = 0.6$ and P = 1 bar in a piston–cylinder–weight assembly. At low temperatures, the sum of the saturation pressures (cf. Eq. (E)) is insufficient to create the imposed 1 bar pressure. Therefore, at T < 60°C (point Q), the fluid exists as a compressed liquid. At ≈60°C (cf. point L), the sum of the saturation pressures is roughly 1 bar. The temperature at this condition can be predicted using Eqs. (A), (B), and (E) as 60°C. Consequently, the values of X_1 and X_2 can be determined as 0.81 and 0.19 using Eqs. (C) and (D). Thus first vapor bubble at a 1 bar pressure appears at 60°C. At this point there are three phases (two immiscible liquid phases, since they are immiscible, and a vapor phase). As more heat is isobarically added, the temperature cannot rise according to Eq. (E), but the vapor bubble can grow. If the heating process begins with 0.4 kmole of species 1 and 0.6 kmole of species 2 and vaporization occurs until the vapor phase is at state E (i.e., T_1^{sat} = 60°C), since the vapor phase mole fraction of species 1 is 0.81, the ratio of the moles of species 2 that are vaporized to those of species 1 is 0.19÷0.81. Therefore, for every 0.4 kmole of species 1 that are vaporized, the moles of species 2 that are vaporized equal 0.4×0.19÷0.81 = 0.094 kmole. Hence, the vapor mixture contains 0.4 kmole of species 1, 0.094 kmole of species 2, and 0.6–0.094 = 0.506 kmole of species 2 remain in the liquid phase. Now the species 1 from liquid have been completely vaporized. Once T>60°C, further vaporization of species 2 occurs, thereby increasing the mole fraction of species 2 in the vapor state, and it is possible to determine the value of X_2 along the curve EJGA. As the temperature reaches 86.5°C (cf. point G), all of the initial 0.6 kmole of species 2 in the liquid phase vaporize so that X_2= 0.6.

b. Miscible Liquids and Immiscible Solid Phase

Oftentimes two species 1 and 2 are miscible in the liquid phase, but are immiscible in the solid phase and each species forms its own aggregate in the solid phase (i.e., upon cooling of the liquid mixture, the two species form two separate solid phases). In this case, at phase equilibrium,

$$\hat{f}_{1(\ell)} = \hat{f}_{1(s)}, \text{ and } \hat{f}_{2(\ell)} = \hat{f}_{2(s)} \quad (17a)$$

Under the ideal solution assumption and since $X_{1(s)} = 1$ due to immiscibility

$$X_{1,\ell}\, f_{1(\ell)}(T,P) = f_{1(s)}(T,P). \quad (17b)$$

For example, pure H_2O at a temperature of –5°C and a pressure of 1 bar should exist as ice. However, if the water is a component of a binary solution (e.g. salt addition), then $\hat{f}_{H_2O(\ell)}$ = $X_{H_2O(\ell)}\, f_{H2O(\ell)}$ (–5°C, 1 bar) = $X_{H_2O(\ell)}\, f_{H2O(\ell)}$ (–5°C, P^{sat}) $POY_{(\ell)}$, where POY = exp ($v_{(\ell)}$ (P – P^{sat})/RT). Generalizing,

$$X_{1(\ell)}\, f_{1(\ell)}(T,P) = X_{1(\ell)}\, f_{1(\ell)}\, (T,P_1^{sat})\, POY_{1(\ell)} = f_{1,s}(T,P_1^{sat})\, POY_{1(s)}. \quad (18)$$

Since $f_{1(\ell)}\, (T,P_1^{sat}) = f_{1(s)}\, (T,P_1^{sat})$,

$$X_{1(\ell)}\, POY_{1(\ell)} = POY_{1(s)}, \text{ and } X_{2(\ell)}\, POY_{2(\ell)} = POY_{2(s)}. \quad (19)$$

However, $X_{1(\ell)} + X_{2(\ell)} = 1$, so that

$$POY_{1(s)}/POY_{1(\ell)} + POY_{2(s)}/POY_{2(\ell)} = 1 \quad (20)$$

Following example 11, the pressure can be determined from Eq. (20) at a specified temperature. The mole fractions $X_{1(\ell)}$ or $X_{2(\ell)}$ at that pressure can be obtained using Eq. (19).

3. Partially Miscible Liquids

a. Liquid and Gas Mixtures

Many liquids are miscible within a certain range of concentrations. The solubility of liquids with one another generally increases with the temperature. The corresponding pressure–temperature relationships are a combination of the corresponding relationships for miscible and immiscible liquids. Figure 9 illustrates the T-$X_{k(\ell)}$ -X_k diagram for a partially miscible liquid.

In the context of Figure 9 assume that methanol and water are partially miscible. Let $X_{1(\ell)}$ denote the methanol mole fraction and $X_{2,\ell}$ the water mole fraction in the liquid. Suppose that water (species 2) is soluble in methanol (species 1) up to 10% by mole fraction at 40°C. As the temperature is increased, the increased "ve" can overcome the attractive forces between methanol molecules and hence its solubility increases as the temperature approaches 60°C. Line FLC represents the boundary between miscibility and immiscibility. When solubility remains constant, the line is vertically oriented (cf. line VF). Water is insoluble from $X_{2(\ell)} = 0.1$ to $X_{2(\ell)} = 0.8$ say at temperatures less than 40°C, but at 60°C it is immiscible for values of $X_{2(\ell)} < 0.7$. Region I is a miscible liquid mixture but rich in species 2, while region II is a miscible liquid mixture but rich in species 1. The boundary DQG represents the variation of miscibility with temperature in the region richer in $X_{2,\ell}$. The region above line CED is similar to the immiscible case we have just discussed.

Consider the vapor mixture at a 90% water vapor concentration (point K). As we cool the vapor from state K to M, first a liquid drop appears containing both species that has a composition corresponding to point R, while the vapor has a composition corresponding to point M as discussed for miscible liquids. As the temperature is decreased to point N, the last liquid will have composition at N (at the bubble line) while the vapor is at state T. If temperature is further decreased, a liquid mixture fixed at a composition N forms.

If we start at point S, then we obtain the first drop at point T with drop composition corresponding to N, which is in the miscible region. As we cool further to point U, the liquid composition is at D (miscible limit at 60°C) while the vapor is at E. However there is still water and methanol vapor left in the mixture. Condensation will occur at a constant vapor composition with the liquid-I composition at D (rich in species 2) and liquid-II composition C (rich in species 1). If the temperature drops below 60°C, there are two separate liquid phases I (composition rich in species 2 along DQG) and II (composition lean in species 2 along FLC). However, the fraction of species 2 in liquid–II will increase since the solubility of species 2 increases (DQG) while the fraction of species 2 in liquid–II will decrease (FLC).

b. Liquid and Solid Mixtures

When a solid (a solute, such as salt) is dissolved in a liquid (a solvent, e.g., water), the dissolved solid can be considered as a liquid in the liquid solution. It is pertinent to know the maximum amount of solute that can be dissolved in a solvent. We will denote the salt in solid phase as s(s) and that in the liquid as s(ℓ). At the equilibrium state of a saturated liquid solution with a solid salt,

$$\hat{f}_{s(\ell)} = X_{s\ell}\, f_{s(\ell)}(T,P) = f_{s(s)}(T,P), \text{ where} \tag{21}$$

$$f_{s(s)}(T,P) = f_{s(s)}(T,P^{sub})\, POY_{s(s)}, \text{ and} \tag{22}$$

$$POY_{s(s)} = \exp\,[v_{s(s)}(P-P^{sub})/RT]. \tag{23}$$

The P^{sub} denotes the saturation pressure for the sublimation of a salt at a specified temperature. Since $f_{s(s)}(T,P) = f_{s(g)}(T,P^{sub})$,

$$f_{s(s)}(T,P) = f_{s(g)}(T,\, P^{sub})\, POY_{s(s)}. \tag{24}$$

If the vapor phase behaves as an ideal gas,

$$f_{s(s)}(T,P) = P^{sub}\, POY_{s(s)}. \tag{25}$$

Similarly,

$$f_{s(\ell)}(T,P) = \phi_{s(\ell)}(T,P)\, P. \tag{26}$$

Employing Eqs. (21) and (26)

$$X_{s(\ell)}\, \phi_{s(\ell)}(T,P)\, P = P^{sub}(T)\, POY_{s(s)}, \text{ and} \tag{27}$$

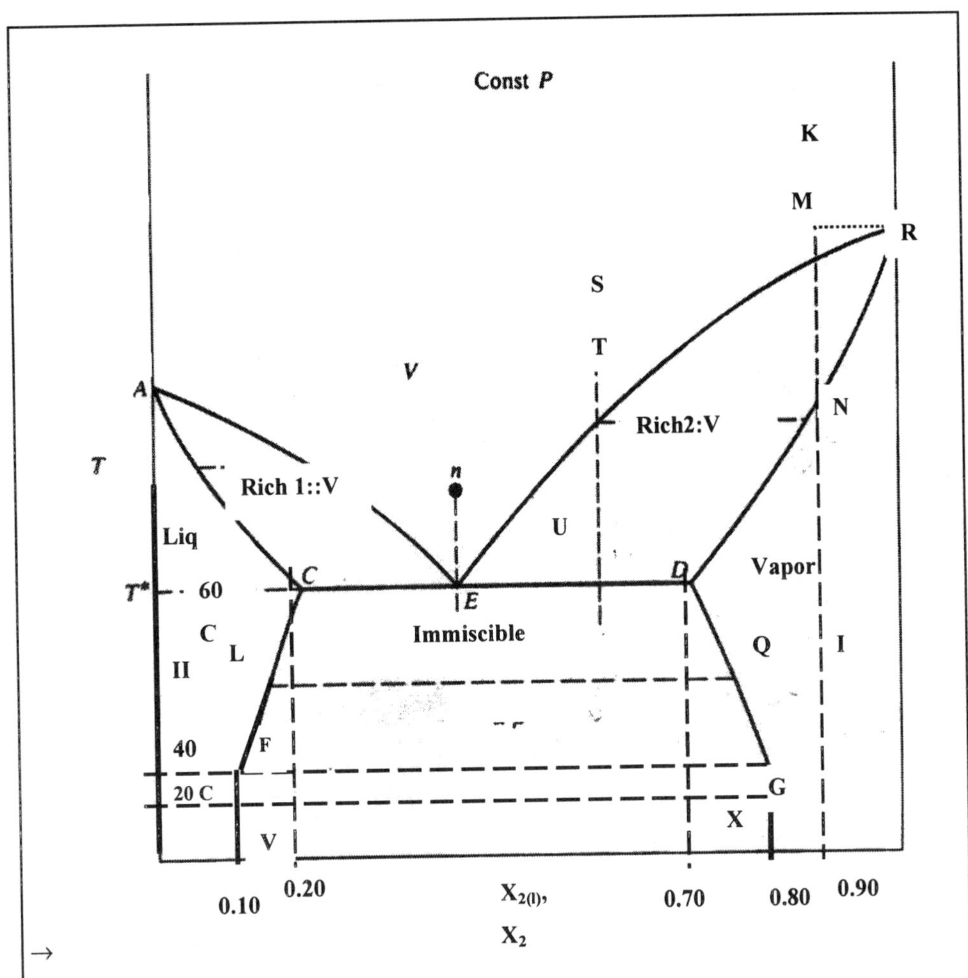

Figure 9: $T–X_{k,,\ell}–X_k$ diagram for partially immiscible liquid mixture, where I denotes the miscible vapor, II the miscible liquid, AE and BMTE the dew lines; AC and BRND the bubble lines, and CLF and DQG the boundaries separating the miscible and immiscible regions. (From Smith and Van Ness, *Introduction to Chemical Engineering Thermodynamics*, 4th Edition, McGraw Hill Book Company, 1987, p. 455. With permission.)

$$X_{s(\ell)} = (P^{sub}(T)\ POY_{s(s)})/(P\ \phi_{s(\ell)}). \tag{28}$$

At low pressures, an increase in the pressure causes the solubility to decrease while at higher pressures the value of $\phi_{s(\ell)}$ may decrease and, consequently, the solubility may also increase.

D. DISSOLVED GASES IN LIQUIDS

Gases dissolve in liquid solutions through a process called absorption. (This should not be confused with adsorption, which is a process during which molecules are attached to a the surface of a solid material due to strong intermolecular forces.) The solubility of a component in a mixture is expressed as a ratio of the maximum amount of solute that can be present in a specified amount of solvent. In case of gases, the solubility is typically expressed in units of ppm. We will treat dissolved gaseous species within a liquid as though they behave like liquids.

1. Single Component Gas

As carbon dioxide is dissolved in water, at some concentration a vapor bubble containing pure CO_2 will start to form. At that saturated condition.

$$\hat{f}_{CO_2(\ell)}(T,P,X_{CO_2,\ell}) = f_{CO_2}(T,P). \tag{29}$$

Employing the ideal solution model,

$$X_{CO_2(\ell)}\ f_{CO_2(\ell)}(T,P) = f_{CO_2}(T,P).$$

If the gas phase behaves as an ideal gas $X_{CO_2,\ell}\ f_{CO_2(\ell)}(T,P) = P$, else

$$f_{CO_2(\ell)}(T,P) = f_{CO_2(\ell)}(T,P^{sat})POY_{CO_2(\ell)},\ \text{where}$$

$$POY_{CO_2(\ell)} = \exp\{v_{CO_2(\ell)}\ (P - P^{sat})/RT\},\ \text{and}\ f_{CO_2(\ell)}(T,P^{sat}) = f_{CO2(g)}(T,P^{sat}) = P^{sat}.$$

Therefore,

$$X_{CO_2(\ell)} = P/\{P_{CO2}{}^{sat}\ POY_{CO_2\ (\ell)}\}. \tag{30}$$

In general, $POY_{CO_2\ (\ell)} \approx 1$, and

$$X_{CO_2(\ell)} = P/P_{CO2}{}^{sat}. \tag{31}$$

Generalizing for solute k dissolved in a solvent,

$$X_{k\ (\ell)} = P/P_k{}^{sat}.$$

This methodology works for values of $P^{sat}(T) > P$. In the case of carbon dioxide dissolved in soda water at 25°C, P^{sat} = 66.7 bar. Consequently, at P=1 bar, $X_{CO_2(\ell)} = 1\div 67 = 0.015$, implying a solubility of 1.5%. As the temperature increases, P^{sat} increases and the solubility of gases decreases. This is an opposite trend to the solubility of liquid components in liquids or of solids in liquid. Recall that chemical potentials of solute in vapor and liquid phases determine whether k is absorbed in or distilled from the solvent. For example, when the pure distilled water is exposed to pure carbon dioxide, if $\mu_{CO_2,g} > \mu_{CO_2(\ell)}$, the carbon dioxide is transferred (absorbed) from the gas into the liquid phase. If $\mu_{CO_2,g} < \mu_{CO_2(\ell)}$, carbon dioxide is transferred (distilled) from the liquid to the gas phase.

The relation shown in Eq. (31) presumes Raoult's law or ideal solution behavior. At low carbon dioxide mole fractions, since a relatively large number of water molecules surround the molecules of carbon dioxide, the liquid water molecules dominate the intermolecular attraction forces. If the attractive forces between water molecules significantly differ from those between the CO_2 molecules, the ideal solution model breaks down. The ideal solution model is also not applicable at higher pressures, since the dioxide no longer behaves as an ideal gas.

2. Mixture of Gases

Consider a gaseous mixture (e.g., of carbon dioxide and oxygen) above a liquid surface. In that case $\hat{f}_{CO_2(\ell)} = \hat{f}_{CO_2(g)}$, and using the ideal solution model

$$X_{CO_2(\ell)}\ f_{CO_2(\ell)}(T,P) = X_{CO_2}\ f_{CO_2(g)}(T,P)$$

Treating the gases as ideal,

$$X_{CO_2(\ell)}\ f_{CO_2(\ell)}\ (T,P) = X_{CO_2}\ P = p_{CO_2},$$

and proceeding as before

$$X_{CO_2(\ell)} \, POY_{CO_2(\ell)} \, P^{sat}_{CO_2} = X_{CO_2} \, P = p_{CO_2}, \text{ i.e.,}$$

$$X_{CO_2} = p_{CO_2}/(P^{sat}_{CO_2} \, POY_{CO_2(\ell)}), \text{ or} \quad (32)$$

$$p_{CO_2} = X_{CO_2(\ell)} \, P^{sat}_{CO_2} \, POY_{CO_2(\ell)}. \quad (33)$$

In this case, the total pressure that appears in Eq. (31) is replaced by the partial pressure. In power plants, water exists under large pressures and hence air may be dissolved in it in the boiler drums. Since solubility decreases at low pressures, the air is released in the condenser sections (Eq. (31)). Oxygen is corrosive to metals, and it, therefore, becomes necessary to remove the dissolved air or oxygen from water prior to sending water to the boiler. Deaerators are used to remove the dissolved gases from water. They work by heating the water with steam (P_{sat} increases, Eq. (31)), and then allowing it to fall over a series of trays in order to expose the water film so that the gases are removed from the liquid phase as much as possible.

Another example pertains to diving in deep water. The human body contains air cavities (e.g., the sinuses and lungs). As a diver proceeds to greater depths, the surrounding pressure increases. In order to prevent the air cavities from collapsing at greater depths, the divers must adjust the air pressure they breathe in. They do so by manipulating their diving equipment to equalize the cavity pressures with the surrounding water pressure. Consequently, the pressurized air gets dissolved in the blood (Eq. (31)). Upon rapid depressurization, in the process of reaching phase equilibrium, the dissolved air is released into the blood stream in the form of bubbles that can be very harmful to human health. Raoult's Law may be applied to estimate the concentration of air in blood. Similarly when a person develops high blood pressure, the amount of soluble O_2 and CO_2 may increase.

If we assume blood to have the same properties as water, we can determine the solubility of oxygen at a 310 K temperature and 1 atm pressure as follows. The vapor pressure data of oxygen can be extrapolated from a known or reference condition to 310 K using Clausius–Clayperon equation (which is valid if (h_{fg}/Z_{fg}) is constant), namely,

$$(P^{sat}_k/P_{ref}) = \exp((h_{fg,k}/(R_k Z_{fg,k}))(1/T_{ref} - 1/T)). \quad (34)$$

The saturation pressure at 310 K can be determined using the relation $\ln(P^{sat}) = 9.102 - 821/T$ (K) bar, i.e., $P^{sat}(310 \text{ K}) = 635$ bar. In air, at 1 atm $p_{O_2} = 0.21$ bar, and the resulting solubility of O_2 in water is 300 ppm.

Another example pertains to hydrocarbon liquid fuels (e.g., fuel injected engines) that are injected into a combustion chamber at high pressures (≈ 30 bar). The gaseous carbon dioxide concentration in these chambers is of the order of 10%. At 25°C, the solubility of the dioxide in the fuels is ≈0.1×3 MPa÷61MPa = 0.005. This solubility increases as the pressure is increased.

3. Approximate Solution–Henry's Law

Rewrite Eq. (33) as,

$$p_k = X_{k,\ell} \, H_k(T,P), \text{ where} \quad (35)$$

$$H_k(T,P) = P^{sat}_k \, (POY)_{k(\ell)}. \quad (36)$$

Where "k is the solute dissolved in a liquid solvent. The symbol H_k denotes Henry's constant for the k–th gaseous species dissolved in the liquid solution. The units used for H_k are typically those of pressure. Since v_f has a relatively small value, POY ≈ 1. Therefore,

$$H_k(T,P) \approx H_k(T) = p^{sat}_k(T), \text{ i.e.,} \quad (37)$$

Hence Eq. (35) is written as,

$$p_k = H_k(T)\, p_k^{sat}(T). \tag{38}$$

At 25°C, for molecular oxygen and nitrogen, respectively, H(25°C) = 4.01×10^4 and 8.65×10^4 bar when $p_{O_2}<1$ bar, and $p_{N_2}<1$ bar. The ideal solubility of oxygen in water is $X_{k,\ell} = 0.21\times10^6/40100$ = 5.2 ppm. This result is only approximate. The solubility of O_2 in water is found to be as high as 170 ppm.

Rewriting Eq. (35)

$$X_{k,\ell} = p_k/H_k(T,P). \tag{39}$$

Multiplying Eq. (39) by the number of moles per unit volume n_ℓ,

$$X_{k(\ell)} n_\ell = p_k\, n_\ell/H_k(T,P), \text{ so that} \tag{40}$$

$$n_{k(\ell)} = (X_k\, n_\ell/H_k(T,P))\, P, \text{ i.e., } n_{v,k} = H_k'\, P, \text{ where} \tag{41}$$

$$H_k' = X_{k(\ell)}/H_k(T,P) \approx X_k\, n_\ell/H_k(T) = X_k\, n_\ell/\, p_k^{sat}(T). \tag{42}$$

Eq. (41) states that moles of gas dissolved in a unit volume liquid is proportional to total pressure. For carbon dioxide that exists in a liquid solution with water, H_k' = 0.0312 k mole of CO_2 m^{-3} bar^{-1} at 25°C. Since the volume at STP for the gas is 24.5 m^3, H_k' = 0.764 m^3 of CO_2 (STP) per m^3 of liquid per bar. The solubility of oxygen in blood is 0.03 mL (STP) per liter of blood per mm of Hg. If partial pressure of oxygen in the lungs is 100 mm of Hg, the dissolved oxygen is 3 mL per liter of blood or 4 mg per liter of blood. Figure 10 presents the variation of H(T) for various dissolved gases in liquid water.

E. DEVIATIONS FROM RAOULT'S LAW

Consider two species k and j that form a binary mixture. The attraction force between similar molecules of species k is denoted as F_{kk} and between dissimilar molecules as F_{kj}. The following scenarios ensue: (1) $F_{kj} = F_{kk}$ so that the ideal solution model and Raoult's Law apply, e.g., toluene–benzene mixtures and mixtures of adjacent homologous series; (2) $F_{kj} > F_{kk}$ implying a nonideal solution in which contraction occurs upon mixing, e.g., acetone–water mixtures and other examples of hydrogen bonding; and (3) $F_{kj} < F_{kk}$, which corresponds to a nonideal solution in which the volume expands upon mixing, e.g., ethanol–hexane and other polar–non polar liquids. In case of the second scenario, since the intermolecular attraction forces are stronger between k–j pairs than between k–k molecular pairs, the vapor pressure of species k can be lower than that predicted using Raoult's Law, which is referred to as a negative deviation from the Law. In case (3) the attraction forces are lower, and a larger amount of vapor may be produced as compared with the Raoult's Law prediction, i.e., both the second and third scenarios suggest that we must involve activity coefficients, $\gamma_{k(\ell)}$. It will now be shown that

$$p_k = \gamma_{k(\ell)}\, X_{k,\ell}\, p_k^{sat}. \tag{43}$$

where $\gamma_{k(\ell)} = \hat{f}_{k(\ell)}(T,P)/\hat{f}_{k(\ell)}^{id}(T,P) = \hat{f}_{k(\ell)}(T,P)/(X_{k(\ell)}\, f_{k(\ell)}(T,P))$.

1. Evaluation of the Activity Coefficient

We have previously employed the ideal solution model to predict the vapor pressure of a component k in an ideal solution. If the measured component vapor pressure differs from that prediction, then it is apparent that the ideal solution model is not valid. We can determine the activity coefficient (that represents the degree of non-ideality from the measured vapor pressure data) as follows.

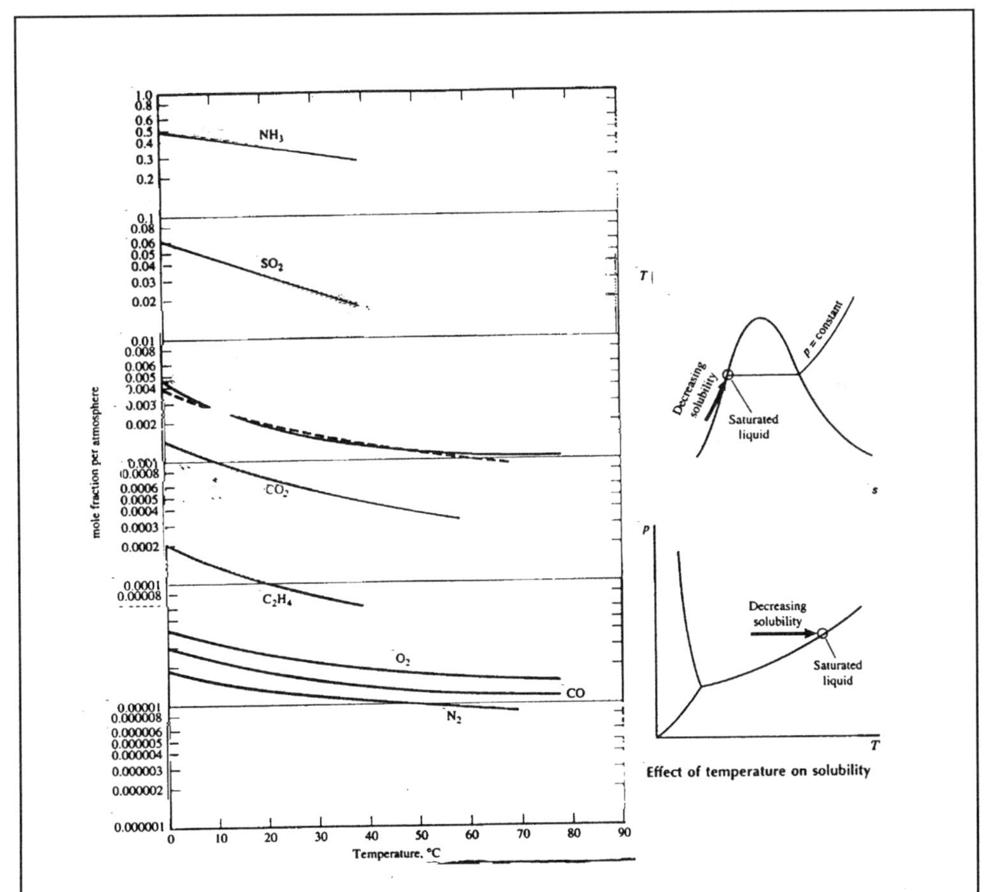

Figure 10: Henry's constant H(T) for the solubility of gases in water. (From S. S. Zumdahl, *Chemistry*, DC Heath and Company, Lexington, Mass, 1986. With permission.)

$$\gamma_k = \hat{f}_k/\hat{f}_{k,id} = \hat{f}_k/X_k\, f_k(T,P). \tag{44}$$

At phase equilibrium,

$$\hat{f}_{k(g)} = \gamma_{k(g)}\, X_k\, f_{k(g)}(T,P) = \hat{f}_{k(\ell)} \text{ and} \tag{45}$$

$$\gamma_{k(g)}\, X_k\, f_{k(g)}(T,P) = \hat{f}_{k(\ell)} = \gamma_{k(\ell)}\, X_{k,\ell} f_{k(\ell)}\,(T,P). \tag{46}$$

Since the vapor is assume to be an ideal gas mixture $\gamma_{k(g)} = 1$, and $\bar{f}_{k(g)}(T,P) = P$. Therefore,

$$p_k = \gamma_{k(\ell)}\, X_{k,\ell}\, f_{k(\ell)}(T,P), \text{ where} \tag{47}$$

$$f_{k,\ell}(T,P) = f_{k(\ell)}(T,P^{sat})\ POY \approx f_{k(\ell)}(T,P^{sat}) = P^{sat}, \text{ i.e.,}$$

$$p_k = \gamma_{k(\ell)}\, X_{k,\ell}\, f_{k(\ell)}(T,P) = \gamma_{k(\ell)}\, X_{k,\ell}\, P^{sat} = \gamma_{k(\ell)}\, p_{k,Raoult}. \tag{48}$$

Thus, $\gamma_{k(\ell)}$ is a measure of the deviation from Raoult's law. With respect to the measured vapor pressure at a specified value of X_k, namely,

$$\gamma_{k(\ell)} = p_k/(X_{k,\ell}\, p_k^{sat}), \text{ or } \gamma_{k,\ell} = p_k/p_{k,Raoult}. \tag{49}$$

Recall that the γ_k's for any phase are related to $(\bar{g}^E/(\bar{R}T))$ (see Eqs. (114) to (117), Chapter 8). If,

$$X_{1(\ell)}\, X_{2(\ell)}\, \bar{R}T/\bar{g}^E = B' + C'\,(X_{1(\ell)} - X_{2(\ell)}), \tag{50}$$

then we obtain the Van Laar Equations:

$$\ln \gamma_1 = A(T)\, X_{2(\ell)}^2/((A(T)/B(T))\, X_{1(\ell)} + X_{2(\ell)})^2,\ \ln \gamma_2 = B(T)\, X_{1(\ell)}^2/((X_{1(\ell)} + (B(T)/A(T))\, X_{2(\ell)})^2, \text{ where} \tag{51}$$

$$A(T) = 1/(B' - C'), \text{ and} \tag{52}$$

$$B(T) = 1/(B' + C'). \tag{53}$$

The constants A and B are generally weak functions of temperature, and for the binary mixture, one can solve for A and B as

$$A = \ln \gamma_{1(\ell)}\,(1 + (X_{2(\ell)} \ln \gamma_{2(\ell)})/(X_{1(\ell)} \ln \gamma_{1(\ell)}))^2, \tag{54}$$

$$B = \ln \gamma_{2(\ell)}\,(1 + (X_{1(\ell)} \ln \gamma_{1(\ell)})/(X_{2(\ell)} \ln \gamma_{2(\ell)}))^2. \tag{55}$$

At the azeotropic condition, at which $X_{k(\ell)} = X_k$, Eq. (47) assumes the form

$$p_k = X_{k(\ell)}\, P = \gamma_{k(\ell)}\, X_{k(\ell)}\, p_k^{sat}. \tag{56}$$

Thus,

$$\gamma_{k(\ell)}(X_{k,azeotropic}) = P/p_k^{sat}(T). \tag{57}$$

The Wilson equation for activity coefficient has the form

$$\ln \gamma_{1(\ell)} = A X_{2,\ell}^2/((A/B)\, X_{1(\ell)} + X_{2(\ell)})^2,\ \ln \gamma_2 \tag{58}$$

$$= B\, X_{1(\ell)}^2/(X_{1(\ell)} + (B/A) X_{2(\ell)})^2, \tag{59}$$

$$\ln \gamma_{2(\ell)} = -\ln(X_{2(\ell)} + X_{1(\ell)} A_{21}) + X_{1(\ell)}(A_{12}/(X_{1(\ell)} + X_{2(\ell)}\, A_{12}) - A_{21}/(X_{2(\ell)} + X_{1(\ell)}\, A_{21})).$$

As $X_{1(\ell)} \rightarrow 0$,

$$\ln \gamma_{1(\ell)} = -\ln A_{12} + 1 - A_{21}, \text{and } \ln \gamma_{2(\ell)} = -\ln A_{21} + 1 - A_{12}, \text{ where} \tag{60}$$

$$A_{ij} = (v_j/v_i)\, \exp(-a_{ij}/R\,T),$$

and v_j denotes the molal volume of species j, and a_{ij} is a known constant that is independent of composition and temperature, where $a_{ij} \approx a_{ji}$.

F. SUMMARY

This chapter summarizes relations for obtaining saturation properties for miscible and immiscible mixtures. Using the phase equilibrium criteria of equal fugacities of any given component in all phases, the composition in any phase can be determined at specified values of T and P. Relations for dew point, bubble point, vapor/liquid composition, solubilities, etc., are obtained. The methodology is extended to nonideal solutions.

G. APPENDIX

1. Phase Rule for Single Component

a. Single Phase

Both pressure and temperature can be varied independently for a single phase for a single component. The system is bivariant with two degrees of freedom, i.e., F = 2.

b. Two Phases

If water is maintained at 180°C with the condition that two phases must coexist, the pressure P must be held at 1 MPa. Likewise, at 170°C, P = 0.8 Mpa. Therefore, only one of the two properties (P,T) is an independent variable in a two phase system. This system is monovariant with a single degree of freedom.

c. Three Phases

The three phases of water coexist only at T = 0.01°C and P = 0.006 bar. It is not possibly to vary either T or P if the three phase condition is desired. This system is invariant with a zero degree of freedom.

d. Theory

We have seen that the phase rule

$$F = 3-\pi \quad \text{(A)}$$

applies, where π denotes the number of phases. At phase equilibrium between three phases, say, the α, β, and γ phases,

$$\mu^{\alpha}(P,T) = \mu^{\beta}(P,T) = \mu^{\gamma}(P,T). \quad \text{(B)}$$

Eqs. (B) represent two equations $\mu^{\alpha} = \mu^{\beta}$ and $\mu^{\beta} = \mu^{\gamma}$ and, since there are two unknowns P and T, the solutions are uniquely fixed so that F = 0.

2. General Phase Rule for Multicomponent Fluids

In a K component system,

$$\Sigma_K X_k = 1. \quad \text{(B)}$$

In this case, the number of variables specifying the chemical potential of each component in each phase is ((K–1) +2). Thus for any phase j

$$\mu_{k(j)}(X_{1(j)}, X_{2(j)}, \ldots, X_{K-1\,(j)}, P, T),\ j = 1,2,\ \ldots,\ \pi, \text{ and } k = 1,2,\ \ldots,\ K-1. \quad \text{(C)}$$

The mole fractions of any component in the different phases are, in general, different.
From the phase equilibrium condition for any species k

$$\mu_{k(1)}(X_{1(1)}, \ldots, X_{K-1\,(1)}, P, T) = \mu_{k(2)}(X_{1(2)}, \ldots, X_{K-1(2)}, P, T) = \ldots = \mu_{K(\pi)}(X_{1(\pi)}, \ldots, X_{K-1\,(\pi)}, P, T), \quad \text{(D)}$$

which represents a set of (π–1) equations. Overall, there are (π–1)K equations available that relate the composition variables $(X_1^1, \ldots, X_{k-1(1)})$, $(X_{1(2)}, \ldots, X_{K-1(2)})$, ..., $(X_{1(\pi)}, \ldots, X_{K-1(\pi)})$ and the two intensive variables P and T. The total number of variables then equals (π(K–1)+2). Since these variables are related by (π–1)K equations, the number of degrees of freedom is

$$F = \pi\,(K-1) + 2 - (\pi-1)K = K - \pi + 2, \quad \text{(E)}$$

which is known as Gibbs phase rule. The phase rule is usually applied to a system of K components and π phases at specified values of P and T.

---

l. *Example 12*

How many independent intensive variables are required to fix the state of superheated steam.?

Solution

$K = 1$ (for a single component), $\pi = 1$ (for a single phase). Therefore, $F = 2$.

Remarks

For a mixture of liquid water and steam, $K = 1$, $\pi = 2$, and $F = 1$. It is a monovariant system.

Likewise, for a mixture of ice, liquid water, and steam, $F = 0$.

---

m. *Example 13*

A single liquid phase is desired for a mixture of water and alcohol. Determine the degrees of freedom.

$$F = 2 + 2 - 1 = 3.$$

Solution

This is a trivariant system, e.g., we can independently assign the variables T, P, X_1.

$$F = 2 + 2 - 2 = 2$$

---

n. *Example 14*

Consider a nonreacting system containing $H_2SO_4(\ell)$, Ca(s), and H_2(g) in two phases, namely, liquid and gaseous. Determine the number of degrees of freedom.

Solution

If the system is nonreacting,

$$F = 3 + 2 - 2 = 3$$

If the system is reacting according to the chemical reaction

$$H_2SO_4(\ell) + Ca(s) \rightarrow H_2\,(g) + CaSO_4(s),$$

the chemical equilibrium condition requires that

$$\hat{g}_{H_2SO_4} + \hat{g}_{Ca(s)} = \hat{g}_{H_2} + \hat{g}_{CaSO_4(s)}.$$

This restriction reduces the number of freedom by 1, and

$$F = 3 - 1 = 2.$$

If the number of equilibrium reactions are R (Chapter 12), the number of degrees of freedom is modified to be

$$F = K + 2 - \pi - R.$$

Generalizing for all work modes,

$$F = K + 1 + \text{work modes} - \pi - R$$

---

o. *Example 15*

Consider a binary mixture of species 1 and 2 in the liquid phase, which on cooling forms two separate pure solids, one for each component. These solids are in equilibrium with the liquids. What is the phase rule for this case?

Solution

$F = K + 2 - \pi$, where $\pi = 3$ (the two pure solid phases and the one liquid phase) and $K = 2$, so that

$$F = 2 + 2 - 3 = 1.$$

---

3. Raoult's Law for the Vapor Phase of a Real Gas

If a liquid mixture exists at a high pressure and low temperature, its vapor phase must be treated as a real gas mixture, i.e.,

$$\hat{f}_k^{l}(T,P,X_{k,\ell}) = \hat{f}_k^{g}(T,P,X_k).$$

Applying an ideal solution and ideal mixture model,

$$X_{k,\ell}\, \hat{f}_k^{l}(T,P) = X_k\, \hat{f}_k^{g}(T,P).$$

A species may exist as a gas at a specified temperature and pressure when alone (e.g., water at 110°C, 100 kPa), but as a liquid in a liquid mixture with another higher boiling temperature component. In that case,

$$\ln (f_k(T,P)/f_k(T,P^{sat})) = \int_{P^{sat}}^{P} (v_k/(RT))\, dP.$$

If the hypothetical state is a liquid, then

$$\ln (f_k^{\ell}(T,P)/ f_k^{\ell}(T,P^{sat})) = \int_{P^{sat}}^{P} (v_k^{\ell}/(RT))\, dP.$$

Since v_k^{ℓ} is a small quantity,

$$f_k^{\ell}(T,P) = f_k^{\ell}(T,P^{sat}).$$

Similarly, for the gaseous state,

$$\ln (f_k^{g}(T,P)/ f_k^{g}(T,P^{sat})) = \int_{P^{sat}}^{P} (v_k^{g}/(RT))\, dP.$$

Since $\int Z\, d(\ln P) = \int (v/RT)\, dP = (Pv/RT) - \int((P/RT)\, dv)$, applying the RK equation either for the liquid or gaseous state, we can apply the expression

$$\int (v/RT)dP = (v/(v-b) - a/(RT^{3/2}(v+b))) - \ln (v-b) + (a/(RT^{3/2}b)) \ln (v/(v+b)).$$

Chapter 10

10. STABILITY

A. INTRODUCTION

The entropy maximum and energy minimum principles will be used to derive the stability criteria for a fluid that exists at a specified state. This will allow us to stipulate the phase change conditions (e.g., evaporation and condensation) for single and multicomponent fluids. Applications will also be presented.

The various equilibrium states of mechanical systems are illustrated in Figure 1. States B, C, D, and F represent mechanical equilibrium positions (or states) while A is a non-equilibrium position (and, hence, not a state). The equilibrium states B, C, and D can be classified according to their stability behavior by conducting perturbation tests as follows. An equilibrium state is disturbed from its initial state by a small amplitude perturbation that changes the potential energy. If a ball returns to its original state, that state is stable. The mechanical stability behavior can be characterized by several states. Figure 1 illustrates a stable state (e.g., state D corresponding to a minimum potential energy), an unstable state (e.g., state C that has a locally maximum potential energy), a metastable state (e.g., state B that can be perturbed to potential energy levels, but which have finite constraints that require a relatively large disturbance to overcome), and a neutrally stable state (e.g., state F that has an invariant potential energy). A disturbance at state C that is a point of maximum potential energy will cause a ball placed there to move to either of positions B or D that are more stable. Therefore, stability can also be defined with respect to a disturbance from an equilibrium state. State C is unstable since any disturbance causes the ball to move towards either State B or State D which are the next equilibrium states in the immediate vicinity of State C. Thus State C is impossible to achieve in a practical system. The state C is equivalent to a nickel standing on its edge in the absence of a disturbance. A slight disturbance, however, can cause the nickel to fall flat on the surface.

If the ball at B is disturbed (e.g., by a potential energy disturbance) by a finite amount, then ball may move to a more stable state D which has the lowest energy of all states illustrated in Figure 1. The rate at which the ball returns to its original state depends upon the friction between the ball and surface. If the potential energy constraint mgZ is removed, the ball will eventually roll to position D. State D is the most stable equilibrium state, state C is unstable and state B is a metastable equilibrium state. As long as constraints exist in the mechanical system of Figure 1 and the disturbances are minor, a metastable system is also a stable system.

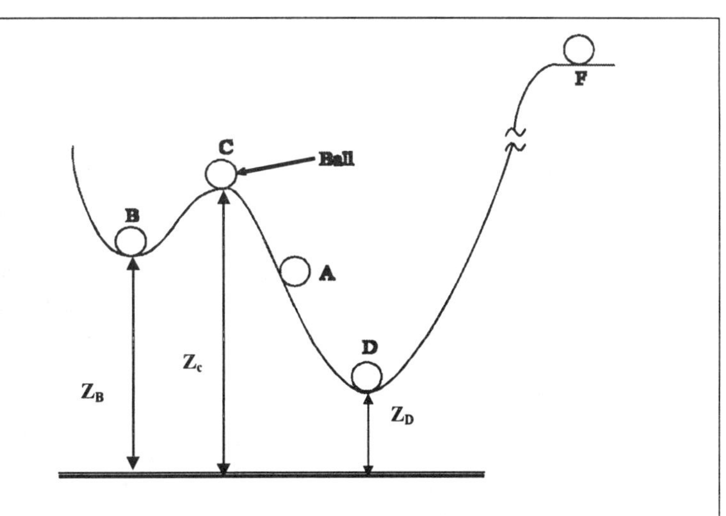

Figure 1: Illustration of various mechanical states.

During a mechanical disturbance test the potential energy is changed from its initial level in order to determine if its value increases or decreases. In that context, state C is unstable since d(PE) < 0, but state D is stable since d(PE) > 0 with disturbance. A disturbance to State D (cf. Figure 1) creates a nonequilibrium situation (i.e., towards higher potential energy locations as d(PE) > 0), which induces the ball to roll back. This process brings the ball back to state D and decreases the potential energy by converting it into kinetic energy. Any perturbation of a stable equilibrium state causes a process that tends to attenuate the disturbance. This is also known as the Le Chatelier principle and is a consequence of the Second Law.

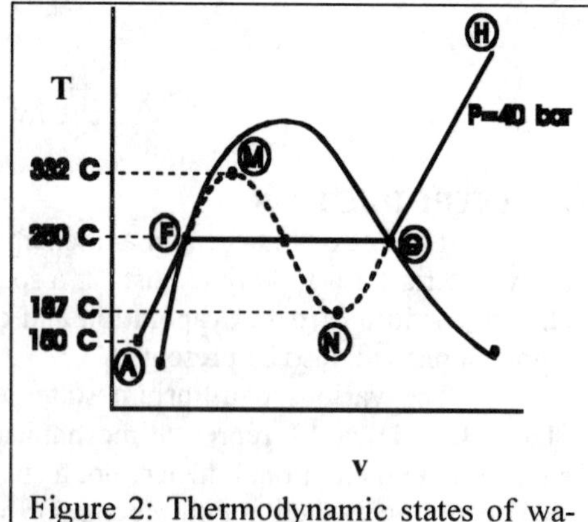

Figure 2: Thermodynamic states of water.

Consider a thermodynamic system with water as working fluid (cf. Figure 2). The path AFGH is a 40 bar isobar and all states along AFGH are stable analogous to the state D in Figure 1. However, it is possible to reach superheated liquid states along FM (except the point M) which are analogous to state B in Figure 1. Similarly, if vapor at state H is cooled. Similarly if vapor at state G is cooled, it can be cooled to a vapor state to along GN in the form of subcooled vapor. Again, states along path GN (except at point N) are analogous to state B. Likewise, the states along the path MN are analogous to state C in Figure 1. In a manner similar to mechanical stability tests, we can test the stability of fluids along the paths HGN or AFM by perturbing the system (e.g., by disturbing the volume from a value V to V + dV) and determining if the system returns to its original state. The rate of return to the stable equilibrium state for any fluid depends upon transport rate processes, e.g., heat and mass transfer, which is beyond the scope of classical thermodynamics.

If one strikes a match in air, the match simply burns and extinguishes. This implies that the constituents of air do not react at a significant rate. On the other hand, if a match is ignited in air in the presence of a significant amount of gasoline vapors (i.e., if the reactive mixture is in metastable equilibrium), small temperature disturbances can ignite the mixture. In general, if a disturbance decreases the entropy of an isolated system at specified values of U, V and m, then the system must initially have been at a stable equilibrium state (SES). Note that what is known as "equilibrium" in the context of classical thermodynamics yields only an average state (e.g., an average system temperature or pressure) while any real system is incessantly dynamic at its microscopic level. Thus, a system should be stable with inherent microscopic and small natural disturbances.

B. STABILITY CRITERIA

1. Isolated System

An example of an isolated system is one constrained by rigid adiabatic and impermeable walls, i.e., with specified values of U, V, and m. An isolated system is at equilibrium when its entropy reaches a maximum value so that $\delta S = 0$. At this state, if the system is perturbed such that the values of U, V, and m of the system are unchanged (e.g., by increasing the temperature or internal energy by an infinitesimal amount), the perturbations are dampened since the system is stable. The perturbations at fixed values of U, V, and m actually decrease the entropy, i.e., $\delta S < 0$ at stable equilibrium. (e.g., consider adiabatic chemical reactions, Chapter 11).

a. Single Component

In Chapters 6 and 7 we employed the real gas state equation and evaluated various thermodynamic properties. Let us consider the compression of water in the context of the RK equation of state is, say, at 593 K and 1 bar. The volume of water decreases (or its intermolecular spacing decreases) and the pressure first increases, then decreases with a further decrease in the volume, and again increases. Since the intermolecular spacing continuously decreases, stability analysis can help determine the state at which the fluid becomes a liquid.

First, let us consider an isolated system with specified values of U, V and m and express S = S(U, V, m). From Chapter 7,

$$s = s(T,v) \text{ and } u = u(T,v), \text{ i.e., } s = s(u,v), \quad (1)$$

which is the entropy fundamental equation.

---

a. Example 1

Consider the Van der Waals equation of state. Obtain an expression for s = s(u,v) for water assuming the specific heat to be constant and equal to the ideal gas value c_{v0} = 28 kJ kmole^{-1}. Select the reference condition such that

$$s = c_{v0} \ln(u/c_{v0} + a/(v\, c_{v0})) + R \ln(v-b).$$

Plot the entropy *vs.* volume for an internal energy value of 7000 kJ kmole^{-1}.

Solution

Since

$$du = c_v\, dT + (T(\partial P/\partial T)_v - P)\, dv,$$

$$du_T = (T(\partial P/\partial T)_v - P)\, dv. \quad (A)$$

Using

$$P = RT/(v-b) - \mathbf{a}/v^2 \quad (B)$$

in Eq. (A),

$$du_T = T(R/(v-b)) - (RT/(v-b) - \mathbf{a}/v^2) = (\mathbf{a}/v^2)\, dv$$

Integrating at constant temperature,

$$u(T,v) = -\mathbf{a}/v + f(T). \quad (C)$$

In case $\mathbf{a} = 0$,

$$u_0(T) = f(T). \quad (D)$$

Eliminating f(T) between Eqs. (C) and (D),

$$u(T,v) = u_0(T) - \mathbf{a}/v. \quad (E)$$

Assuming constant (ideal gas) specific heats and $u = u_{ref,0}$ at $T = T_{ref}$,

$$u_0(T) - u_{ref,0} = c_{v0}(T - T_{ref}), \quad (F)$$

Eq. (E) assumes the form

$$u = c_{v0}(T - T_{ref}) + u_{ref,0} - \mathbf{a}/v, \text{ i.e.,} \quad (G)$$

$$T = ((u - u_{ref,0}) + \mathbf{a}/v)/c_{v0} + T_{ref}. \quad (H)$$

Similarly, we can integrate the expression

$$ds = c_v dT/T + \partial P/\partial T\, dv$$

at constant temperature to obtain the relation

$$s(T,v) = s_0(T,v) - R \ln(v/(v-b)). \quad (I)$$

Using the ideal gas relation for $s_0(T,v)$,

$$s_0(T,v) - s_{ref,0}(T_{Ref}, v_{ref}) = c_{v0} \ln(T/T_{Ref}) + R \ln(v/v_{ref}),$$

$$s(T,v) = c_{v0} \ln (T/T_{Ref}) + R \ln (v/v_{ref}) + s_{ref,0}(T,v_{ref}) - R \ln (v/(v-b)). \quad (J)$$

Further, using Eqs. (H) and (J) to eliminate the temperature,

$$s = c_{v0} \ln((u-u_{ref,0})/(c_{v0}T_{ref}) + (\mathbf{a}/(vc_{v0}T_{ref}) + 1) + R \ln((v-b)/v_{ref} + s_{ref,0}(T_{Ref}, v_{ref}). \quad (K)$$

Setting ,

$$s_{ref,0}(T_{ref}, v_{ref}) = c_{v0} \ln(c_{v0}\, T_{ref}) + R \ln v_{ref}, \qquad u_{ref,0} = c_{v0}\, T_{ref}$$

we obtain the expression

$$s = c_{v0} \ln (u + \mathbf{a}/v) + R \ln ((v-b)). \quad (L)$$

For ideal gases, $\mathbf{a} = b = 0$, and Eq. (K) leads to the relation

$$s_0 = c_{v0} \ln \ (u/(c_{v0}\, T_{Ref})) + R \ln (v/v_{Ref}) \quad (M)$$

This expression leads to a plot of s *vs.* both u and v for a real or ideal gas.
Using the values $\bar{\mathbf{a}}$ = 5.3 bar m^6 $kmole^{-2}$, $\bar{b}$ = 0.0305 m^3 $kmole^{-1}$, $\bar{u}$ = 7000 kJ $kmole^{-1}$, T_{ref} = 1 K, $\bar{v}_{ref}$ = 1 m^3 $kmole^{-1}$, $\bar{c}_{v0}$ = 28 kJ $kmole^{-1}$ K^{-1}, $\bar{u}_{ref,0} = \bar{c}_{v0}\, T_{ref}$ = 28 kJ $kmole^{-1}$ in Eq. (K), a plot of s *vs.* v at u = 7500 kJ $kmole^{-1}$ is presented in Figure 3.

Remarks

The relation s =s (u,v) is the entropy fundamental equation. It is somewhat more difficult to manipulate the RK equation and obtain an explicit expression for s = s(u,v) using Eq. (L).

b. Example 2

Consider a rigid and insulated 0.4 m^3 volume tank filled with 4 kmole (24×10^{26} molecules) of water, and an internal energy of 7500 kJ $kmole^{-1}$. Divide the tank into two equal 0.2 m^3 parts. Assume that water follows the VW state equation. What is the entropy of each section? Assume that section 1 is slightly compressed to 0.14 m^3 while the section 2 is expanded to 0.26 m^3, but maintaining the same total volume and total internal energy. What is the entropy change during the process? (Use Figure 3.)

Solution

The fluid molecules are distributed uniformly in the tank and initially, $\bar{v}_1 = \bar{v}_2 = \bar{v}$ = 0.4÷4 = 0.1 m^3 $kmole^{-1}$. Using Figure 3 or Eq. (K) of Example 1, we note that $\bar{s}$ = 149.83 kJ $kmole^{-1}$. The state is represented by point D in the figure, and the extensive

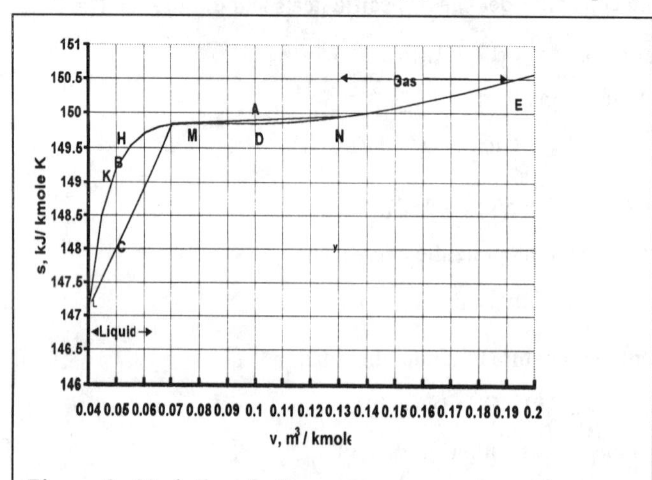

Figure 3: Variation in the entropy *vs.* volume for water modeled as a VW fluid at u = 7500 kJ $kmole^{-1}$.

entropy of each section is initially

$$S_D = (S_1 = N_1\ \bar{s}_1) = (S_2 = N_2\ \bar{s}_2) = 2\times149.83 = 299.7 \text{ kJ}. \quad \text{(A)}$$

The total entropy of the tank is $S = 2S_D = 299.7 + 299.7 = 599.4$ kJ. After the perturbation,

$\bar{v}_1 = 0.14\div2 = 0.07$ m^3 kmole^{-1}, $\bar{u}_1 = 7500$ kJ kmole^{-1}, and

$\bar{v}_2 = 0.26\div2 = 0.13$ m^3 kmole^{-1}, $\bar{u}_2 = 7500$ kJ kmole^{-1}.

The corresponding states are represented by points M and N in Figure 3, i.e.,

$$S_M = S_1 = N_1\ \bar{s}_1 = 2\times149.8 = 299.6 \text{ kJ, and}$$

$$S_N = S_2 = N_2\ \bar{s}_2 = 2\times149.9 = 299.9 \text{ kJ}.$$

The total entropy after the disturbance

$$S = S_1 + S_2 = S_M + S_N = 599.5 \text{ kJ}.$$

Remarks

We note that the entropy (at specified values of U, V and m) increases after the disturbance, i.e., $S_M + S_N > 2\ S_D$. Therefore, the initial state is unstable and changes in the direction of increasing entropy. At states B and E disturbances no longer cause a further increase in the entropy.

---

2. Mathematical Criterion for Stability

a. Perturbation of Volume

i. Geometrical Criterion

Consider the state B illustrated in Figure 3 which undergoes a small disturbance ΔV at a specified value of U. Due to the disturbance, Section 1 of the system in Example 2 reaches state L while Section 2 reaches state M (Figure 3) . With respect to stability

$$\delta S = \delta S_1 + \delta S_2 < 0, \text{ i.e.,} \quad (2)$$

$$\delta S = (S_L(U,V-\Delta V,N) - S_B(U,V,N)) + (S_M(U,V+\Delta V,N) - S_B(U,V,N)) < 0.$$

Since

$$\delta S = S_L(U,V-\Delta V,N) + S_M(U,V-\Delta V,N) - 2S_B(U,V,N) < 0, \text{ i.e.,}$$

$$(S_L(U,V-\Delta V,N) + S_M(U,V-\Delta V,N))/2 < (S_B(U,V,N). \quad (3)$$

The entropy after a disturbance decreases in order that the initial state of the system is stable. In the context of Figure 3, the ordinate of the midpoint C of the chord LCM that connects the points L and M is represented by the LHS of Eq. (3) while the RHS represents the ordinate of the point B. Therefore, the chord LCM must lie below the curve LBM for the system to be stable. The curve LKBHM satisfying the criteria given by Eq. (3), is a concave curve with respect to the chord LCM. On the other hand, the midpoint of the chord MAN lies above the convex curve MDN, thereby violating this stability criterion. We find that for a system to be stable the fundamental relation for S = S (U, V, N) one must satisfy the concave condition, which is established by Eq. (3).

ii. Differential Criterion

The discussion so far pertains to disturbances, which are of large magnitude ΔV. Consider a disturbance is in the neighborhood of state B (cf. Figure 3) that extends from state K at (U, V–ΔV, N) to state H at (U, V + ΔV, N). In that case

$$(\delta S = (S_K(U,V-\Delta V,N)-S_B(U,V,N)) + (S_H(U,V+\Delta V,N)-S_B(U,V,N))) < 0, \quad (4)$$

where δS denotes the entropy change due to the disturbance. The entropy should decrease following disturbance at the stable points A, B and E illustrated in Figure 4. Expanding S_K and S_H in a Taylor series

$$S_K(U, V-\Delta V, N) = S_B + (\partial S/\partial V)_B(-dV) + (1/2!)(\partial^2 S/\partial V^2)_B(-dV)^2 + \ldots\ldots, \text{ and}$$

$$S_H(U, V+dV, N) = S_B + (\partial S/\partial V)_B(dV) + (1/2!)(\partial^2 S/\partial V^2)_B(dV^2) + \ldots, \text{ i.e.,}$$

$$\delta S = S_B + (\partial S/\partial V)_B(-dV) + (1/2!)(\partial^2 S/\partial V^2)_B(-dV)^2 + \ldots +$$

$$S_B + (\partial S/\partial V)_B(dV) + (1/2!)(\partial^2 S/\partial V^2)_B(dV)^2 + \ldots - 2\, S_B(U, V, N) < 0.$$

We will represent the contribution to the disturbance by the first derivatives in the form

$$dS_B = (\partial S/\partial V)_B(-dV) + ((\partial S/\partial V)_B(dV) = 0,$$

and by the higher–order derivatives as

$$d^2S_B = (1/2!)(\partial^2 S/\partial V^2)_B(-dV)^2 + (1/2!)(\partial^2 S/\partial V^2)_B(dV)^2 + \ldots.$$

So that Eq. (4) assumes the form

$$(\delta S = (dS)_B + (d^2S)_B) < 0.$$

Since $(dS)_B = 0$,

$$(\delta S = (d^2S)_B = (\partial^2 S/\partial V^2)_B(dV)^2) < 0.$$

Omitting the subscript B, the general criteria for stability at any given state that

$$(dS) = 0, \text{ and } (d^2S) < 0, \quad (5)$$

which are the same as the conditions for entropy being maximized at specified values of U, V and N (stable points A, B and E and unstable point D in Figure 4). Since $(dV)^2 > 0$, the relation

$$(\partial^2 S/\partial V^2) < 0 \quad (6)$$

must be satisfied in the neighborhood of an equilibrium state for it to be stable. If case the second derivatives are zero, the third derivatives in Taylor series are included to obtain the stability condition $d^3S < 0$, and so on.

---

c. Example 3

Assume that water follows the ideal gas state equation and show that for a unit mass, $s_{vv} = \partial^2 s/\partial v^2 < 0$ at all volumes.

Solution

We will employ the relations

$$T ds - P\, dv = du \text{ and } ds = du/T + P/T\, dv.$$

At constant internal energy

$(\partial s/\partial v)_u = P/T$, and

since for ideal gases, $P/T = R/v$,

$s_v = (\partial s/\partial v)_u = R/v = f(v)$ alone, and

$s_{vv} = (\partial^2 s/\partial v^2)_u = -R/v^2 < 0$

Remark

This illustrates that ideal gases are stable at all states.

Figure 5 contains plots of $\bar{s}$ *vs.* $\bar{v}$ at various values of $\bar{u}$. At very high values of $\bar{u}$ (e.g., at temperatures larger than the critical temperature), the fluid is stable for any volume, i.e., there is no increase in entropy in the presence of a disturbance at any state. For a solid there are certain regimes in which stability criteria are not satisfied and disturbances result in phase transitions. e.g., in case of water from ice–phase I, to ice–phase II.

Since, in our illustrations, U, V and N (or m) were specified during a disturbance, the

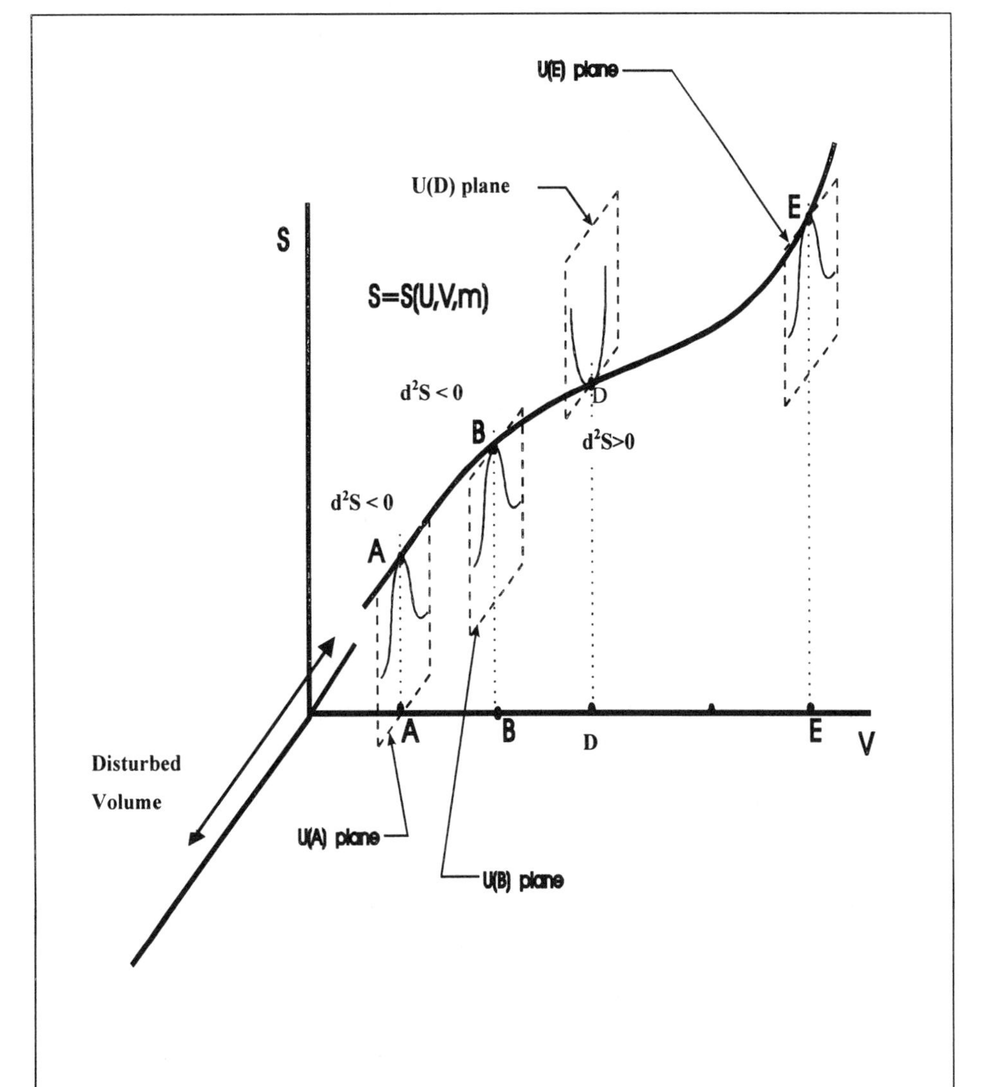

Figure 4: Illustration of the entropy perturbation following a disturbance in the volume of sections 1 and 2 of a rigid container.

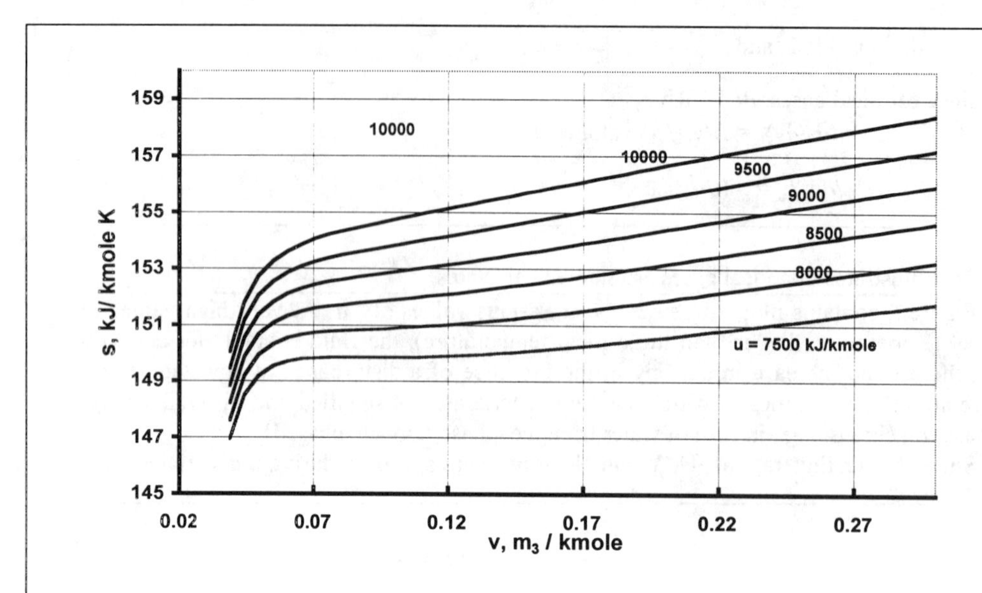

Figure 5: Variation of s *vs.* v in case of water according to the VW state equation.

relation

$$\delta W_{opt} = -\, dU + T_0\, \delta S,$$

with disturbance at a stable state yields

$$\delta W_{opt} = T_0\, \delta S < 0. \tag{7}$$

If the system is initially unstable (cf. State D in Figure 1), then $\delta W_{opt} > 0$, i.e., the system could have performed work during the entropy increase in the isolated system.

b. Perturbation of Energy

Repeating the procedure outlined in the previous section, but with the disturbance parameter in terms of U, it is again possible to show that for stability

$$S_{UU} = (\partial^2 S/\partial U^2) < 0, \tag{8}$$

which is identical to the "concavity condition" described previously.

Consider the plot of entropy *vs.* internal energy at specified values of N (or m) illustrated by the curve ABCDEFGH in Figure 6 for a fluid following a real gas equation of state. At some equilibrium states $d^2S > 0$ (i.e., these states are unstable) and at others, $d^2S < 0$ (stable states). In the context of Figure 6 consider the fluid at state B at which the curve ABC is concave with respect to U. Here, the stability criterion is satisfied. On the other hand, curve DEF is convex, the criterion is violated.

A plot of $(\partial S/\partial U)_V = 1/T$ *vs.* U is presented in Figure 6b and that of T *vs.* U is illustrated in Figure 6c. It is apparent that for certain ranges of U (U_D to U_F), T decreases with increasing U. Mathematically, the stability condition $\partial^2 S/\partial U^2 < 0$ is equivalent to the condition

$$(\partial/\partial U)(\partial S/\partial U) < 0,\ \text{i.e.,}\ \partial/\partial U(1/T) < 0\ \text{or}\ -1/T^2(\partial T/\partial U) < 0. \tag{9}$$

In the context of Eq. (9), since $(\partial U/\partial T) = mc_v$, then $mc_v > 0$, and at a specified value of either m or N,

$$c_v > 0 \tag{10}$$

for a state to be stable. In Figure 6c, c_v has positive values along the branch ABCD but negative values along DEF, which violates the *thermal stability* criterion provided by Eq. (10). The negative c_v values imply that temperature decreases when energy is added and increase when heat is removed. Consider water at 25°C and 1 bar in an insulated cup and assume that it has negative specific heat. Divide the water into two equal sections A and B. We now discuss the meaning of thermal stability. If the water is thermally disturbed with a small amount of heat transfer δQ from A to B, then A will continue to become warmer than B (due to larger storage of energy as intermolecular potential energy and less as vibrational energy) with the disturbance. The entropy generation is still positive for the adiabatic cup since $\delta\sigma = \delta Q\ (1/T_A - 1//T_B) > 0$ as $T_A > T_B$. It still satisfies Second Law with entropy generation indicating that the initial state is thermally unstable.

The specific heat $c_v > c_{vo}$ for fluids following the RK equation, where $c_{vo} > 0$. This implies that a curve representing variations in S *vs.* U must be concave (i.e., it must have a decreasing radius of curvature) with respect to U at a specified value of V. In other words, thermal stability is always satisfied for satisfying most of the real gas state equations. Read Example 17 for instances when this may be violated.

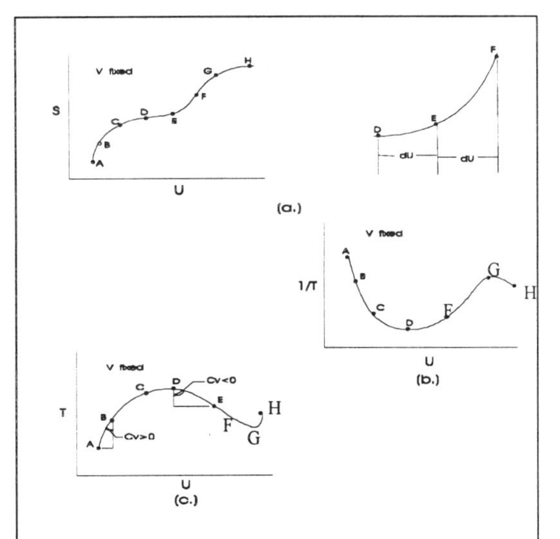

Figure 6: Qualitative illustration of the variation of a) entropy, b) $\partial S/\partial U = 1/T$ and c) temperature with respect to the internal energy.

d. Example 4

Consider a VW gas and show that $(s_{uu} = \partial^2 s/\partial u^2) < 0$ at all volumes.

Solution

Consider the relation $T\,ds - P\,dv = du$. At fixed volume

$$\partial s/\partial u = 1/T.$$

Differentiating this expression,

$$\partial^2 s/\partial u^2 = -1/T^2 (\partial T/\partial u)_v.$$

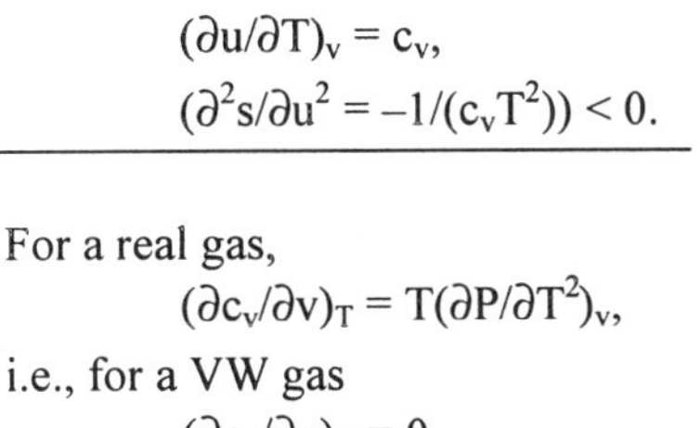

Since

$$(\partial u/\partial T)_v = c_v,$$

$$(\partial^2 s/\partial u^2 = -1/(c_v T^2)) < 0.$$

Remarks

For a real gas,

$$(\partial c_v/\partial v)_T = T(\partial P/\partial T^2)_v,$$

i.e., for a VW gas

$$(\partial c_v/\partial v)_T = 0,$$

which implies that $c_v = f(T) = c_{v0}(T) > 0$

A VW gas is thermally stable at all points.

c. Perturbation with Energy and Volume

The discussion so far has pertained to disturbance in the internal energy at specified volume or in the volume at specified values of U. If the volume and energy are both perturbed, the following conditions must be satisfied, i.e.,

$$\delta S = S(U+dU, V+dV, N) + S(U-dU, V-dV, N) - 2\,S(U, V, N) < 0, \text{ i.e.,}$$

$$S(U+dU, V+dV, N) + S(U-dU, V-dV, N) < 2\,S(U, V, N).$$

Expanding this expression in a Taylor series, and retaining terms up to the second derivative,

$$\delta S = (\partial S/\partial U\ dU + \partial S/\partial V\ dV + (\partial/\partial U + \partial/\partial V)^2 S) +$$

$$(-\partial S/\partial U\ dU - \partial S/\partial V dV + (\partial/\partial U + \partial/\partial V)^2 S) < 0$$

Since,

$$dS = (\partial S/\partial U\ dU + \partial S/\partial V\ (-dU)) + (\partial S/\partial V\ dV + \partial S/\partial V\ (-dV)) = 0, \text{ then}$$

$$(\partial/\partial U + \partial/\partial V)^2\ S < 0$$

Expanding this relation, we obtain the expression,

$$(d^2S = \partial^2 S/\partial U^2\ dU^2 + 2(\partial^2 S/\partial U\partial V)dU\ dV + \partial^2 S/\partial V^2\ dV^2) < 0, \text{ i.e.,}$$

$$(d^2S = S_{UU}dU^2 + 2\ S_{UV}dU\ dV + S_{VV}dV^2) < 0. \quad (11)$$

Multiplying Eq. (11) by S_{UU}, since $S_{UU} < 0$, Eq. (11) assumes the form

$$(d^2S = S_{UU}^2\ dU^2 + 2\ S_{UU}\ S_{UV}\ dU\ dV + S_{UU}\ S_{VV}\ dV^2) > 0, \text{ i.e.,} \quad (12)$$

$$(S_{UU}\ dU + S_{UV}\ dV)^2 + (S_{UU}\ S_{VV} - S_{UV}^2)\ dV^2 > 0 \quad (13)$$

Now, $(S_{UU}\ dU + S_{UV}\ dV)^2 > 0$, and since $dV^2 > 0$, it is apparent from the perspective of stability that

$$(S_{UU}\ S_{VV} - S_{UV}^2) > 0, \quad (14)$$

which is the stability condition in the presence of volumetric and energetic fluctuations within a system. The stability criterion at a given state can be summarized as

$$D_{1,U} = S_{UU} = \partial^2 S/\partial U^2 < 0, \text{ and } D_{1,V} = S_{VV} = \partial^2 S/\partial V^2 < 0, \quad (15)$$

In determinant form, the stability criterion is

$$D_2 = \begin{vmatrix} S_{UU} S_{UV} \\ S_{VU} S_{VV} \end{vmatrix} > 0 \quad (16)$$

where D_2 is determinant of second order. If $D_{1,U} < 0$, and $D_2 > 0$, then $D_{1,V} < 0$. It is noted that since $\partial s/\partial u = 1/T$, then $s_{vu} = (\partial^2 s/\partial v\ \partial u) = -(\partial T/\partial v)_u/T^2$.

---

e. *Example 5*

For an ideal gas show that Eq. (16) applies for all volumes.

Solution

From Example 1 we note that $s_{vv} < 0$, and from Example 2 that $s_{uu} < 0$.

Since $\partial s/\partial u = 1/T$, then $s_{vu} = \partial^2 s/\partial v\ \partial u = -(\partial T/\partial v)_u/T^2$.

If the energy of an ideal gas is specified, its temperature cannot change with respect to changes in the volume. For this reason $(\partial T/\partial v)_u = 0$. Therefore,

$$s_{vu} = 0, \text{ and}$$

using Eq. (16)

$$s_{uu}\ s_{uu} - s_{vu}\ s_{vu} > 0,$$

which satisfies the stability criterion.

---

f. *Example 6*

Apply the RK equation of state

$$P = RT/(v-b) - \mathbf{a}/(T^{1/2}\, v(v+b)) \quad \text{(A)}$$

to water at 593 K. (a) Plot the pressure with respect to volume; (b) determine s_{VV}, s_{UU} and s_{UV} at $\bar{v}$ = 0.1 and 0.4 m^3 $kmole^{-1}$; and (c) check whether the stability criteria are satisfied at these states.

Solution

The plot of pressure *vs.* volume is contained in Figure 7. Since,

$$\partial s/\partial v = P/T, \text{ and} \quad \text{(B)}$$

$$\partial s/\partial u = 1/T, \text{ then} \quad \text{(C)}$$

$$\partial s/\partial v = R/(v-b) - \mathbf{a}/(T^{3/2}\, v(v+b)). \quad \text{(D)}$$

Differentiating this expression with respect to v at constant u, we obtain the relation

$$s_{vv} = (\partial^2 s/\partial v^2)_u = -R/(v-b)^2 + \mathbf{a}(2v+b)/(T^{3/2}\, v^2(v+b)^2) + (3/2)\mathbf{a}(\partial T/\partial v)_u/(T^{5/2}\, v(v+b)). \quad \text{(E)}$$

Recall that

$$du = c_v dT + (T\partial P/\partial T - P)\, dv, \text{ hence,}$$

$$(dT/dv)_u = -((T\partial P/\partial T - P))/c_v.$$

In the context of the RK equation of state

$$(\partial T/\partial v)_u = -((3/2)\, \mathbf{a}/(c_v\, T^{1/2}\, v(v+b))) \quad \text{(F)}$$

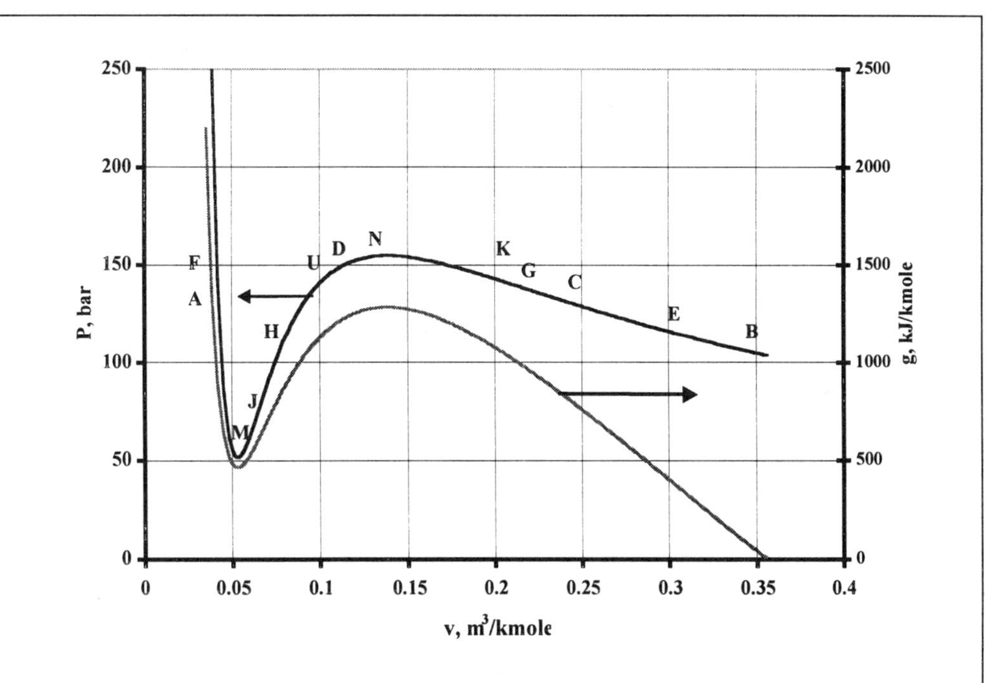

Figure 7: Plot of pressure and Gibbs energy with respect to volume for water using the RK equation of state.

Similarly differentiating Eq. (C) with respect to u, we obtain the relation

$$s_{uu} = \partial^2 s/\partial u^2 = -(\partial T/\partial u)_v/T^2 = -1/(c_v T^2), \tag{G}$$

and differentiating Eq. (C) with respect to v, we have

$$s_{vu} = \partial^2 s/\partial v\partial u = -(\partial T/\partial v)_u/T^2. \tag{H}$$

Thus, all of the differentials involved with the stability criterion represented by Eq. (16) can be evaluated.
For v = 0.4 m^3 $kmole^{-1}$; and T = 593 K, P = 95.4 bar, c_v = 30.09 kJ $kmole^{-1}$ K, **a** = 142.64 bar m^3 kJ $kmole^{-2}$, b = 0.02110 m^3 $kmole^{-1}$, $(\partial T/\partial v)u = -173.04$ kmole K m^{-3}, $\partial^2 s/\partial v^2 = -31.92$ kJ kmole m^{-6} K^{-1}, $\partial^2 s/\partial u^2 = -9.4\times10^{-8}$ kmole kJ^{-1} K^{-1}, and $\partial^2 s/(\partial v\partial u) = 0.000492$ kmole m^{-3} K^{-1}. For stability,

$$D_2 = \begin{vmatrix} S_{UU} S_{UV} \\ S_{VU} S_{VV} \end{vmatrix} > 0, \text{ so that} \tag{16}$$

$$(s_{uu}s_{vv} - s_{vu}{}^2 = = 2.77\times10^{-6} \text{ kmole}^2 \text{ m}^{-6} \text{ K}^{-2}) > 0.$$

When calculations are repeated at $\overline{v}$= 0.1 m^3 $kmole^{-1}$, D_2 < 0

Remarks

The fluid is unstable at (0.1 m^3 $kmole^{-1}$, 593 K), while at (0.4 m^3 $kmole^{-1}$, 593 K) it is stable.
Even though $s_{uu} < 0$, $s_{vv} < 0$, $s_{uu}\, s_{vv} - s_{vu}{}^2$ can change sign at selected range of volumes at 593 K.

d. *Multicomponent Mixture*

We will now extend the procedure to multiple components using Kestin's approach. Recall that

$$dS = (1/T)\, dU + (P/T)\, dV.$$

Since $\partial S/\partial U = S_U = 1/T$ and $\partial S/\partial V = S_V = P/T$, then

$$dS = S_U(U,V)\, dU + S_V(U,V)\, dV.$$

Furthermore,

$$d^2S = d(1/T)\, dU + d(P/T)\, dV = d(S_U(U,V))\, dU + d(S_V(U,V))\, dV, \text{ or}$$

$$d^2S = (S_{UU}\, dU + S_{UV}\, dV)\, dU + (S_{VV}\, dV + S_{VU}\, dU)\, dV$$

$$= S_{UU}\, dU^2 + 2\, S_{UV}\, dU\, dV + S_{VV}\, dV^2. \tag{17}$$

For a system to be stable, dS = 0 and $d^2S < 0$. (This is somewhat analogous to the criteria describing the maximum value of a function y = y (x) where dy/dx = 0 at the maxima, but $d^2y/dx^2 < 0$.)

Consider the expression for the entropy change in a multicomponent mixture (cf. Chapter 3),

$$dS = dU/T + P/T\, dV - \Sigma_k(\mu_k/T)\, dN_k,$$

where S = S $(U,V,N_1,N_2,...)$. this expression leads to the relation

$$d^2S = d(1/T)\, dU + d(P/T)\, dV - \Sigma_k\, d(\mu_k/T)\, dN_k$$

$$= -dT/T^2\, dU + dP/T\, dV - P\, dT/T^2\, dV - \Sigma_k d\mu_k/T\, dN_k + \Sigma_k \mu_k\, dT/T^2\, dN_k$$

$$= dT/T\, (-dU/T - P/T\, dV + \Sigma_k\, \mu_k/T\, dN_k) + (dP/T)\, dV - \Sigma_k (d\mu_k/T)\, dN_k.$$

$$= dT/T\, (dS) + (dP/T)\, dV - \Sigma_k (d\mu_k/T)\, dN_k.$$

The stability criterion is represented by the expression

$$(d^2S = -(dT/T)(dS) + (dP/T)\,(dV) - \Sigma_k (d\mu_k/T)(dN_k)) < 0,\ U,V,\ m \text{ specified, or} \quad (18)$$

$$-\, d(T)\, dS + d(P)\, dV - \Sigma_k d(\mu_k)(dN_k) < 0,$$

where the terms within the parentheses () are all intensive properties. Rewriting this expression, we obtain

$$d^2S_{\,U,V,N} = -\, d(T\, dS - P\, dV + \Sigma_k \mu_k\, dN_k) < 0 \quad (19)$$

The term within the parentheses in Eq. (19) is dU and it is apparent that $U = U(S,V,N)$, i.e.,

$$d^2S_{U,V,N} = -\, d\,(dU) < 0$$

Thus the criterion $(d^2S)_{U,V,N} < 0$ at specified values of U, V, and N implies that

$$(d^2U)_{S,V,N} > 0, \quad (20)$$

which indicates that for a stable system U reaches a minimum value at specified values of S, V and N. Similar expressions can be obtained in terms of A, H, and G. In developing expressions analogous to Eq. (20) we will use the relations

$$dU = T\, dS - PdV + \Sigma_k \mu_k\, dN_k,$$

$$dH = T\, dS + V\, dP + \Sigma_k \mu_k\, dN_k, \quad (21a)$$

$$dA = -\, SdT - P\, dV + \Sigma_k \mu_k\, dN_k, \text{ and} \quad (21b)$$

$$dG = -S\, dT + V\, dP + \Sigma_k \mu_k\, dN_k. \quad (21c)$$

For instance, if we consider that S(H,P,N), then

$$(d^2S_{H,P,N} = -d(dH)) < 0, \text{ and } (d^2H)_{S,P,N} > 0 \quad (22a)$$

for the enthalpy to be a minimum at specified values of S, P, and N. Similarly,

$$(-d(dA) = -d^2A_{T,V,N}) < 0, \text{ or } d^2A_{T,V,N} > 0 \quad (22b)$$

which is the criterion for A having a minimum value at specified values of T, V and N. Likewise.

$$-d(dG) = -\, d^2G_{T,P,N} < 0, \text{ or } d^2G_{T,P,N} > 0 \quad (23)$$

is a criterion for G being minimum at specified values of T, P and N.

i. Remark
 The differential

$$d(\mu_k)dN_k = (\partial\mu_k/\partial N_1\, dN_1 + \partial\mu_k/\partial N_2\, dN_2 + \ldots)\, dN_k = (\Sigma_j(\partial\mu_k/\partial N_j)\, dN_j)\, dN_k.$$

Hence,

$$\Sigma_k d(\mu_k)\, dN_k = \Sigma_k(\Sigma_j(\partial\mu_k/\partial N_j)\, dN_j)\, dN_k. \tag{24}$$

ii. Criterion for Binary Mixture

Consider a binary mixture at specified values of U, V, and N (= $N_1 + N_2$) that undergoes disturbances in U, V, N (i.e., of the third order). We can expand Eq. (16) as a third order determinant to obtain the stability criterion, i.e.,

$$D_3 = \begin{vmatrix} S_{UU} & S_{UV} & S_{UN} \\ S_{VU} & S_{VV} & S_{VN} \\ S_{NU} & S_{NV} & S_{NN} \end{vmatrix} > 0. \tag{25}$$

For a k–component mixture the generalized criterion is

$$D_{K+1} = \begin{vmatrix} S_{UU} & S_{UV} & S_{UN_1} & S_{UN_2} \ldots\ldots\ldots\ldots S_{UN_{k-1}} \\ S_{VU} & S_{VV} & S_{VN_1} & S_{VN_2} \ldots\ldots\ldots\ldots S_{VN_{k-1}} \\ S_{N_1U} & S_{N_1V} & S_{N_1N_1} & S_{N_1N_2} \ldots\ldots\ldots\ldots S_{N_1N_{k-1}} \\ & \vdots & & \\ S_{N_{K-1}U} & S_{N_{K-1}V} & S_{N_{K-1}N_1} & S_{N_{K-1}N_2} \ldots\ldots\ldots S_{N_{K-1}N_{K-1}} \end{vmatrix} > 0. \tag{26}$$

For disturbance of order K+1, the stability criterion $D_1 < 0$, $D_2 > 0$, $D_3 > 0$, ... $D_{K+1} > 0$ results in k+1 inequalities.

e. System With Specified Values of S, V, and m

In this case the criterion for a system to be stable is that the internal energy U must reach a minimum value, i.e., $d^2U > 0$. For a perturbation in volume at a specified entropy, since

$$dU = T(S,V)\, dS - P(S,V)\, dV \tag{27}$$

$$dU_S = -P(S,V)\, dV_S, \text{ and } d^2U_S = -dP\, dV_S = -(\partial P/\partial V)_S\, dV_S\, dV_S.$$

Since $d^2U_S > 0$, this implies that

$$-(\partial P/\partial V)_S > 0 \text{ or } (\partial P/\partial V)_S < 0. \tag{28}$$

Recall from Chapter 3 that the sound speed $c^2 = -v^2(\partial P/\partial v)$. The fact that $(\partial P/\partial v)_s < 0$ implies that the $c^2 > 0$ and is a real quantity. (The sound speed is of the order of the average molecular velocity, which cannot be negative, Chapter 1.) On a plot of U *vs.* V the slope $\partial U/\partial V$ represents (–P) provided the entropy is held fixed (isentropic expansion or compression).

Recall from Chapter 7 that

$$((\partial P/\partial V)_S = k(\partial P/\partial V)_T) < 0, \text{ and } c_p = c_v + T\, v\, \beta_P^2/\beta_T, \tag{29}$$

where $\beta_T = -(1/v)(\partial v/\partial P)_T$ and $\beta_P = (1/v)(\partial v/\partial T)_P$. Using the two relations in Eqs. (29) we obtain the expression

$$((\partial P/\partial v)_s = (\partial P/\partial v)_T\, T\, v^2\, \beta_P^2/c_v) < 0.$$

Since $c_v > 0$ and $\beta_P^2 > 0$, stability is ensured for all possible variations in case

$$(\partial P/\partial v)_T < 0 \text{ or } \beta_T > 0.$$

Then from Eq. (29), it is apparent that, from the perspective of stability, $c_p > 0$ and $k > 1$, i.e.,

$c_p > c_v$.

The criterion given by Eq. (29) is called the mechanical stability condition (i.e., related to a mechanical or volumetric disturbance) in the context of which $(\partial P/\partial V)_T < 0$. Due to this condition a fluid does not generally break down into two phases upon mechanical perturbation (one a dense liquid and the other a dilute vapor phase). Even a minute disturbance will cause unstable behavior in case $(\partial P/\partial V)_T < 0$. Therefore, if the volume decreases during an isothermal process, the fluid pressure must increase in order for the fluid to be stable.

---

g. Example 7

Is an ideal gas stable at all pressures and volumes at a specified temperature?

Solution

Considering the ideal gas equation of state $P = m\,RT/V$,

$$((\partial P/\partial v)_T = -mRT/V^2) < 0.$$

An ideal gas is stable at all states.

---

h. Example 8

Is an RK gas stable at all pressures and volumes at a specified temperature?

Solution

An RK gas behaves according to the equation

$$P = RT/(v-b) - a/(T^{1/2}v(v+b)), \text{ i.e.,} \quad (A)$$

$$(\partial P/\partial v)_T = -RT/(v-b)^2 + \mathbf{a}(2v + b)/((T^{1/2}v^2(v+b)^2). \quad (B)$$

For stability,

$$(\partial P/\partial v)_T < 0 \text{ if } \mathbf{a}(2v + b)/((T^{1/2}\, v^2(v+b)^2) < RT/(v-b)^2, \text{ or} \quad (C)$$

$$\mathbf{a}(v-b)^2\,(2v+b)/(v^2(v+b)^2) < RT^{3/2}. \quad (D)$$

and for critical isotherm

$$\mathbf{a}(v-b)^2\,(2v+b)/(v^2(v+b)^2) < RT_c^{3/2}. \quad (E)$$

Remark

At $T>T_c$, the pressure decreases monotonically with increasing volume. Therefore, since, $(\partial P/\partial v)_T < 0$ and the stability criterion is always satisfied, it is not possible to form two phases in the supercritical region.

---

f. Perturbation in Entropy at Specified Volume

For a system to be stable, the internal energy must be at a local minimum when the entropy is disturbed but the volume is held fixed. Recall that

$$(\partial U/\partial S)_V = T(S,V),$$

so that for a local minimum,

$$D_1 = U_{SS} = (\partial^2 U/\partial S^2) > 0, \text{ where } (\partial^2 U/\partial S^2) = (\partial T/\partial S)_V = T/C_v, \quad (30)$$

which implies that C_v or $c_v > 0$, since the temperature is a positive quantity. The condition $c_v > 0$ implies that the internal energy and entropy must be a monotonic function of temperature with positive slope at a specified volume. The temperature will increase if the entropy is increased for a specified volume of a fluid.

g. *Perturbation in Entropy and Volume*

Again, we will employ the relation $dU = TdS - PdV$ so that

$$d(dU) = d^2U = d(T(S,V))\, dS - d(P(S,V))\, dV.$$

Since $T = U_S$ and $P = U_V$,

$$dT = U_{SS}\, dS + U_{SV}\, dV, \text{ and } -dP = U_{VV}\, dV + U_{VS}\, dS, \text{ so that}$$

$$(d^2U = U_{SS}\, dS^2 + 2U_{SV}\, dS\, dV + U_{VV}\, dV^2) > 0.$$

Since $U_{SS} > 0$, multiplication by U_{SS} yields

$$(U_{SS}\, dS + U_{SV}\, dV)^2 + (U_{VV}\, U_{SS} - U_{SV}{}^2)dV^2 > 0 \tag{31}$$

Hence, the condition for stability is

$$(U_{VV}U_{SS} - U_{SV}{}^2) > 0. \tag{32}$$

If $(U_{VV}U_{SS} - U_{SV}{}^2) < 0$, Eq. (31) still holds if $|(U_{SS}\, dS + U_{SV}\, dV)^2| > (U_{VV}U_{SS} - U_{SV}{}^2)dV^2$, i.e., depending upon strength of the disturbances dS and dV, Eq. (31) may or may not be satisfied when Eq. (32) is violated.

Writing Eq. (32) in determinant form

$$D_2 = \begin{vmatrix} U_{SS} U_{SV} \\ U_{VS} U_{VV} \end{vmatrix} > 0. \tag{33}$$

In case $U_{SS} > 0$, then dividing Eq. (32) by U_{SS}, we obtain the inequality $(U_{VV} - U_{SV}{}^2/U_{SS}) > 0$ or $U_{VV} > U_{SV}{}^2/U_{SS}$. Since $U_{SV}{}^2 > 0$, if $U_{SS} > 0$, this implies that $U_{VV} > 0$. Therefore, if $D_1 > 0$ and $D_2 > 0$, we satisfy the condition that $U_{VV} > 0$. At the limit of intrinsic stability, $D_1 = D_2 = 0$.

i. Binary and Multicomponent Mixtures

For a multicomponent mixture,

$$d^2U = d(T)\, dS - d(P)\, dV + \Sigma_k d(\mu_k) dN_k, \text{ i.e.,}$$

in case of a binary mixture

$$d^2U = d\,(T)\, dS - d(P)\, dV + d(\mu_1)\, dN_1 + d\mu_2\, dN_2.$$

In addition to the conditions $D_1 > 0$, and $D_2 > 0$ the following condition applies to a binary mixture, namely,

$$D_3 = \begin{vmatrix} U_{SS} U_{SV} U_{SN_1} \\ U_{VS} U_{VV} U_{VN_1} \\ U_{NS} U_{NV} U_{NN_1} \end{vmatrix} > 0. \tag{34}$$

At the limits of intrinsic stability, $D_1 = D_2 = D_3 = 0$ for a binary mixture. Extending the result to a k–component mixture, it can be shown that the determinant of (k+1)th order must be positive, i.e.,

$$D_{k+1} = \begin{vmatrix} U_{SS} U_{SV} U_{SN_1} U_{SN_2} \cdots U_{SN_{k-1}} \\ U_{VS} U_{VV} U_{VN_1} U_{VN_2} \cdots U_{VN_{k-1}} \\ U_{NS} U_{NV} U_{NN_1} U_{N_1N_2} \cdots U_{NN_{k-1}} \end{vmatrix} > 0. \tag{35}$$

h. System With Specified Values of S, P, and m

Consider the relation

$$dH = T\,dS + V\,dP.$$

Since intensive properties are not additive, we cannot obtain enthalpy variations by perturbing P to P±ΔP at a specified entropy and mass, but keeping the overall pressure constant. If the entropy is specified

$$(\partial H/\partial P)_S = V, \text{ and } (\partial^2 H/\partial P^2)_S = (\partial V/\partial P)_S.$$

We have previously shown that $(\partial P/\partial V)_S < 0$, i.e., $(\partial^2 H/\partial P^2)_S < 0$. In the case of entropy perturbations

$$(\partial^2 H/\partial S^2)_P = (\partial T/\partial S)_P = T/C_P > 0, \quad (36)$$

which implies that C_p or $c_p > 0$. Thus the enthalpy and entropy increase monotonically with increasing temperatures at a specified pressure. A plot of H *vs.* S at specified P shows convexity with respect to S for stable states. The derivation for a multicomponent system is left as an exercise for the reader.

i. System With Specified Values of T, V, and m

In this case A is to be minimized with respect to the disturbance at specified values of T, V, and m. Figure 8 illustrates plots of A with respect to V and T. For a single component,

$$dA = -S\,dT - P\,dV.$$

The system can undergo disturbances in V while keeping the overall volume constant, i.e.,

$$dA_T = -P\,dV, \text{ or} \quad (36)$$

$$A_V = (\partial A/\partial V)_T = -P, \text{ and } dA_T = -P\,dV_T. \quad (37)$$

Consequently,

$$(d^2A_T = d(dA_T) = -d(P(V,T))\,dV = -(\partial P/\partial V\,dV)\,dV = -(\partial P/\partial V)_T\,dV^{\;2}) > 0.$$

The relation $d^2A > 0$ implies that

$$A_{VV} = -(\partial P/\partial V)_T > 0, \text{ or } (\partial P/\partial V)_T < 0.$$

The resultant curve of A *vs.* V will be convex with respect to V for a system to be stable.

Figure 8b and Figure 8d illustrate the stable and unstable branches. At the critical point $\partial P/\partial V = 0$, $\partial^2 P/\partial V^2 = 0$ and, hence, $A_{VV} = 0$. Here, $\partial^3 P/\partial V^3$ must have a negative value in order for stability conditions to be satisfied.

---

i. Example 9

Show that at the critical point $\partial^3 P/\partial V^3 < 0$ for a fluid following the VW equation.

Solution

For a VW fluid,

$$\partial P/\partial v = -RT/(v-b)^2 + 2a/v^3, \text{ and } \partial^2 P/\partial v^2 = 2RT/(v-b)^3 - 6\,\mathbf{a}/v^4, \text{ i.e.,} \quad (38)$$

$$\partial^3 P/\partial v^3 = -6RT/(v-b)^4 + 24\mathbf{a}/v^5.$$

At $T = T_c$, and $v = v_c$, $\partial P/\partial v = 0$, $\partial^2P/\partial v^2 = 0$. The third derivative w.r.t. v,

$$(\partial^3P/\partial v^3 = -6RT_c/(v_c-b)^4 + 24\,a/v_c^5 = -6RT_c/((2/3)v_c)^4 + 24(9/8)(RT_cv_c)/v_c^5$$

$$= -(243/8)\,RT_c/v_c^4 + 27\,RT_c/v_c^4 = -(27/8)\,RT_c/v_c^4) < 0. \quad (39)$$

This stability condition is satisfied at the critical point.

Remark

The specific heat c_v is finite at the critical point, but $(c_p-c_v) = -T(\partial P/\partial T)^2/(\partial P/\partial v) \to \infty$ at that state.

i. Perturbations With Respect to Temperature

Since the temperature is an intensive property, if two sections of a specified volume are disturbed, $\delta T \neq \delta T_1 + \delta T_2$. In this case,

$$A_T = (\partial A/\partial T)_V = -S.$$

Recall that, from the energy minimum principle, $((\partial S/\partial T)_V = c_v/T) > 0$. Consequently,

$$(A_{TT} = (\partial^2A/\partial T^2)_V = -(\partial S/\partial T)_V = -c_v/T) < 0.$$

ii. Binary and Multicomponent Mixtures

For a binary mixture,

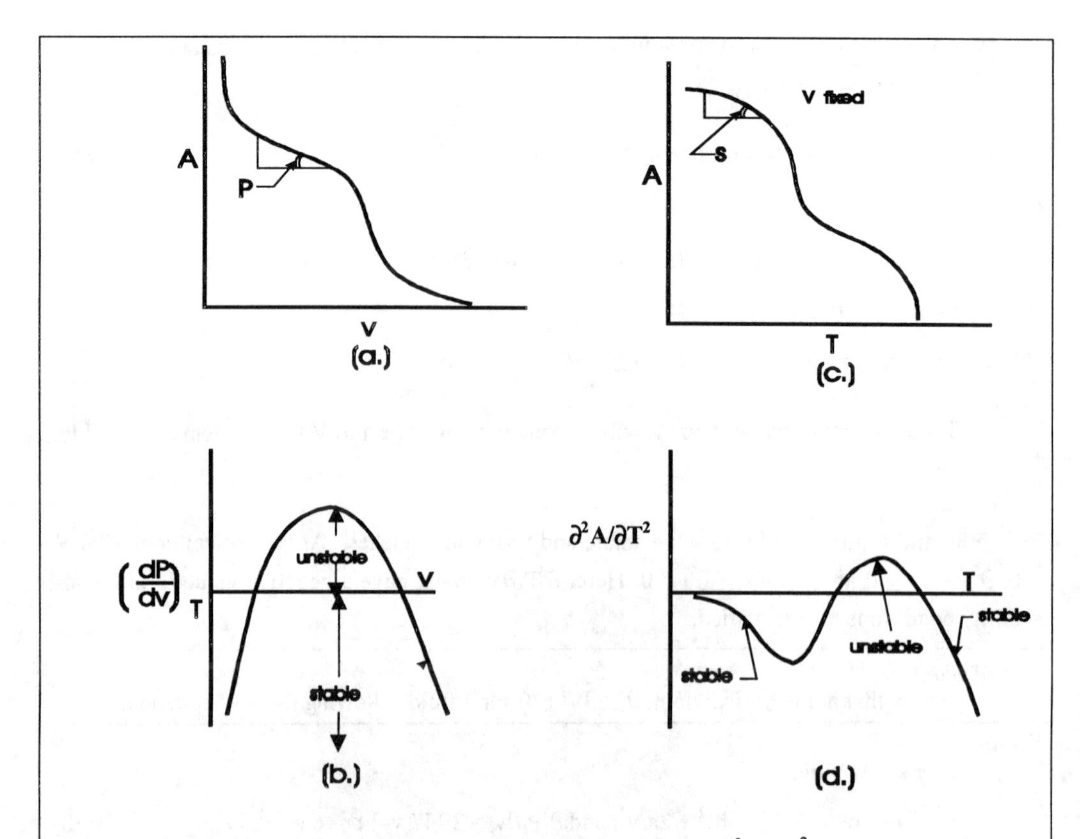

Figure 8: Qualitative variation of a) A with respect to T, b) $-(\partial^2A/\partial V^2)_T = (\partial P/\partial V)_T$ with respect to V, c) A with respect to T, and d) $(\partial^2A/\partial T^2)_V$ with respect to T.

$$dA = - S\, dT - P\, dV + \Sigma(\mu_1\, dN_1 + \mu_2\, dN_2), \text{ i.e.,} \quad (40)$$

$$D_{1,V} = A_{VV} > 0,\ D_{1,T} = A_{TT} < 0, \text{ and } D_2 = \begin{vmatrix} A_{VV} A_{VN} \\ A_{N1V} A_{N1N1} \end{vmatrix} > 0. \quad (41)$$

These relations are readily extended for multicomponent fluids.

j. System With Specified Values of T, P, and m

In this case, for a system to be stable, the Gibbs energy must be minimized in the presence of disturbances in P and T. Consider the relation,

$$dG = - S\, dT + V\, dP. \quad (42)$$

i. Perturbations With Respect to Pressure

Considering perturbations with respect to the temperature,

$$G_P = (\partial G/\partial P)_T = V, \text{ and } (G_{PP} = (\partial^2 G/\partial P^2)_T = (\partial V/\partial P)_T) < 0 \quad (43a)$$

At the critical point,

$$(\partial P/\partial v)_T = 0, \text{ and } (\partial^2 P/\partial v^2)_T = 0, \text{ i.e.,}$$

both $(\partial v/\partial P)_T$ and $(\partial^2 g/\partial P^2)_P$ tend to infinity.

ii. Perturbation With Respect to temperature

Similarly,

$$G_T = (\partial G/\partial T)_P = -S,\ G_{TT} = (\partial^2 G/\partial T^2)_P = - (\partial S/\partial T)P = - C_P/T. \quad (43b)$$

Since C_p or $c_P > 0$, $(\partial^2 G/\partial T^2)_P < 0$.

iii. Perturbations With Respect to P and T

Since G is to be minimized at stable conditions, one can show that

$$(G_{PP}\, G_{TT} - G_{TP}^2) > 0 \quad (44)$$

Recall that $G_{PP} = \partial V/ \partial P$, $G_{TT} = - C_p/T$, $G_{TP} = (\partial V/ \partial T)_P$

k. Multicomponent Systems

For a mixture, $\Delta G^E = G - \Sigma_\kappa(\bar{g}_k(T, P)N_k)$. A plot of ΔG vs. Z_1 (= $N_1/(N_1 + N_2)$) is presented in Figure 10 for a binary mixture. Upon mixing two pure components, the Gibbs energy of the mixture is lower than the sum of Gibbs energies of the two pure components, $\Sigma_\kappa(\bar{g}_k(T, P)N_k)$, i.e., $\Delta G_{T,P} < 0$ for $1 > Z_1 > 0$.

Disturbances in the mixture state can place the local properties along either curve (a) or (b). If there is local disturbance at state C on curve (a), a negative increment in Z_1 causes a larger Gibbs energy drop dG^- (= G_A–G_C) to state A as compared to a positive increment to state D, i.e., dG^+ (= G_D–G_C). Therefore, in this hypothetical case, the net disturbance $(dG^- + dG^+) > 0$, causing the mixture to return to its original state C. Curve (b) also satisfies the condition $\Delta G < 0$. However, we have introduced a hypothetical kink in the curve at states M and N. If there is a local disturbance in N_1 (or Z_1) at state E, then $|dG^+| < |dG^-|$ so that $(dG^- + dG^+) < 0$, which implies that at fixed temperature and pressure the disturbance is undamped. This can result in the formation of two phases or two components, depending upon the local state.

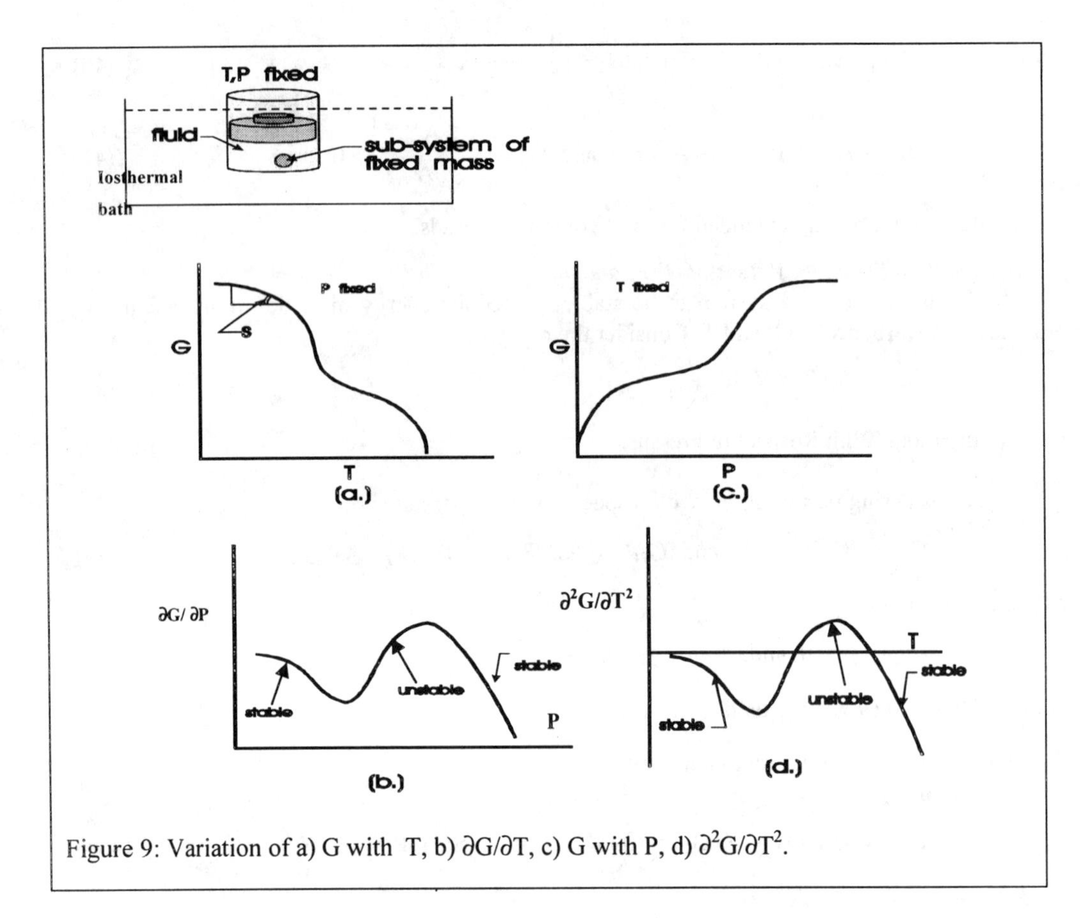

Figure 9: Variation of a) G with T, b) $\partial G/\partial T$, c) G with P, d) $\partial^2 G/\partial T^2$.

In the context of this discussion, it is apparent that

$$G_A + G_D > 2\,G_C.$$

Expanding G_A and G_D in terms of G_C we obtain the expression

$$(G_C + (\partial G/\partial N_1)_C(dN_1)_A + (\partial^2 G/\partial N_1^2)_C(dN_1^2)_A + \ldots + G_C + (\partial G/\partial N_1)_C(dN_1)_D$$

$$+(\partial^2 G/\partial N_1^2)_C(dN_1^2)_D + \ldots) < 2\,G_C. \quad (45)$$

However,

$$(dN_1)_D + (dN_1)_A = 0, \text{ i.e.,}$$

$$(\partial^2 G/\partial N_1^2)_C\,((dN_1^2)_D + (\partial^2 G/\partial N_1^2)_C(dN_1^2)_A) > 0. \quad (46)$$

C. APPLICATION TO BOILING AND CONDENSATION

We will illustrate an application of stability criteria to boiling and condensation (i.e, the formation of two phases) through the following example.

j. Example 10

Water is contained in a piston–cylinder–weight assembly that is immersed in a constant–temperature bath at 593 K. Assume that the fluid obeys the RK equation of state irrespective of its phase, i.e.,

$$P = RT/(\bar{v} - \bar{b}) - \bar{a}/(T^{1/2}\,\bar{v}\,(\bar{v} + \bar{b})) \quad (A)$$

Obtain a plot for P *vs.* $\bar{v}$; Using the relation $dg_T = v\,dP$, obtain a plot for $\bar{g}$ *vs.* $\bar{v}$.

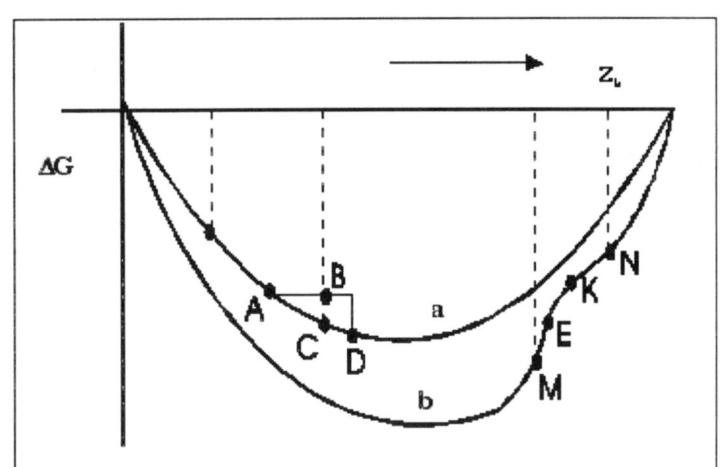

Figure 10: Variation of ΔG with mole fraction in a binary mixture.

Solution

This problem was solved in Chapter 7. A brief summary is presented. Using the values $\bar{a}$ = 142.64 bar m^3 $K^{0.5}$ $kmole^{-1}$, $\bar{b}$ = 0.02110 m^3 k $mole^{-2}$, and T = 593 K in Eq. (A), a plot of P *vs.* $\bar{v}$ is readily obtained as shown in Figure 11. The first term in Eq. (A) occurs due to the collisions of high velocity molecules, while the second term appears due to attractive forces that result in a pressure reduction. As the fluid is compressed from state B, the pressure increases along the path BECGKN. If the fluid at state N is compressed further, it instantaneously condenses into a liquid state L. Similarly, the fluid that is initially at state L can be expanded to a low pressure along the path LFARM. If the fluid at M is expanded further, it vaporizes instantaneously.

Since $d\bar{g}_T = \bar{v}\,dP$, it is possible to integrate Eq. (A) between the limits v and v_{ref}, i.e.,

$$\int d\bar{g}_T = \bar{g}(T,\bar{v}) - \bar{g}(T,\bar{v}_{ref}) = \int \bar{v}\,dP, \text{ to obtain}$$

$$\bar{g}(T,\bar{v}) - \bar{g}(T,\bar{v}_{ref}) = P\bar{v} - P^{ref}\bar{v}^{ref} - (\bar{R}T\ln((\bar{v}-\bar{b})/(\bar{v}^{ref}-\bar{b})) -$$

$$(\bar{a}/(\bar{b}T^{1/2}))\ln((\bar{v}/\bar{v}_{ref})(\bar{v}_{ref}-\bar{b})/(\bar{v}-\bar{b})).$$

We will now arbitrarily set $\bar{v}_{ref}$ = 4.83 m^3 $kmole^{-1}$ (so that P_{ref} = 10 bar at 593 K) and $\bar{g}_{ref}$ = 0. This enables us to produce a plot of $\bar{g}$ *vs.* $\bar{v}$ (cf. Figure 11). Using the same values of v, one can obtain g *vs.* P as shown in Figure 12.

Remarks

The fluid is in a saturation state at states G and F at which the vapor and liquid coexist. The Gibbs free energy for both phases is equal. (The saturation pressure according to the RK equation is 133 bar at 593 K.)

Path QBECGN is a stable vapor branch, since $\partial P/\partial\bar{v}<0$ (or $\partial\bar{v}/\partial P<0$). The lowest value of $\bar{g}$ at specified values of T and P (cf. Figure 12) indicates that $P < P^{sat}$ = 133 bar, i.e., the vapor has a lower free energy when it is compared to the liquid curve QRAF.

Path FL is a stable liquid branch since $\partial P/\partial\bar{v}<0$ (or $\partial\bar{v}/\partial P<0$).

Path GKN represents metastable vapor (i.e., an equilibrium condition with a finite constraint).

The state N represents an intrinsic stability limit for the vapor at which $\partial P/\partial v = 0$.

The state M is an intrinsic stability limit for the liquid at which $\partial P/\partial v = 0$.

Path MRAF corresponds to metastable liquid with intermediate values of $\bar{g}$.
Path NDHJM is an unstable branch since $\partial P/\partial \bar{v} > 0$ and the highest values of $\bar{g}$ are to be found here.

---

1. Physical Processes and Stability

Consider the isothermal compression of water at 593 K. (cf. Figure 11). As the fluid is compressed from state B towards states E, C, G, etc., the volume decreases with the increase in pressure, since the intermolecular spacing decreases. The intermolecular attraction forces slowly increase as the states E, C, G, etc., are approached. At larger volumes (i.e. lower pressures) the first term in RK equation dominates, i.e., $b \ll v$, $P \approx RT/v$ so that $P \propto 1/v$ at a specified temperature. The Gibbs energy value is lower at larger volumes, and gradually increases as the pressure is raised (BECG in Figure 11), indicating that the fluid accumulates a larger potential to perform work. The rate of pressure increase with decreasing volume is lower at smaller volumes due to the larger intermolecular attraction forces and the second term in the RK equation $a/(T^{1/2}v(v+b))$ becomes significant. Beyond a maximum pressure at state N, the intermolecular attraction forces are so large that this second term dominates and tends to lower the pressure. Consequently, the pressure starts to decrease with compression at smaller volumes due to the very small intermolecular spacing. (The pressure can sometimes be negative indicating that the fluid is under tension. In case of water the tension can be as high as –40 bar without evaporation occurring.) Consequently the "g" decreases along NDUHJM. Upon compression beyond states M, R, A, etc., the body volume effect (which reduces the space available for the movement of molecules) becomes dominant and results in a higher number density so that the first term in the RK equation again dominates. Thereupon, the pressure again increases rapidly and $P \approx RT/(v-b)$, i.e., $P \propto 1/(v-b)$ and the Gibbs energy again increases (MRAFL).

All fluid states along the path BECGKNDHJMRFL are equilibrium states. The stability criterion $\partial P/\partial v < 0$ suggests that the path NDUHJM is an unstable branch along which inherent disturbances exist and hence a uniform intermolecular spacing or stable state cannot be maintained.

a. *Physical Explanation*

At T = 593 K, $g_F = g_G$ at P = 133 bars. Hence P^{sat} = 133 bar; the fluid in the context of Example 10 typically changes its state from vapor to liquid (from state G to F along line GHF (cf. Figure 11)). However the Gibbs energy changes $\Delta g_{GH} = g_H - g_G$ and $\Delta g_{FH} = g_H - g_F$ represent the adverse potentials at the saturation condition that the fluid has to overcome to either form a vapor embryo at state G from the liquid mother phase at state L, or a liquid embryo at state F from the vapor mother phase at

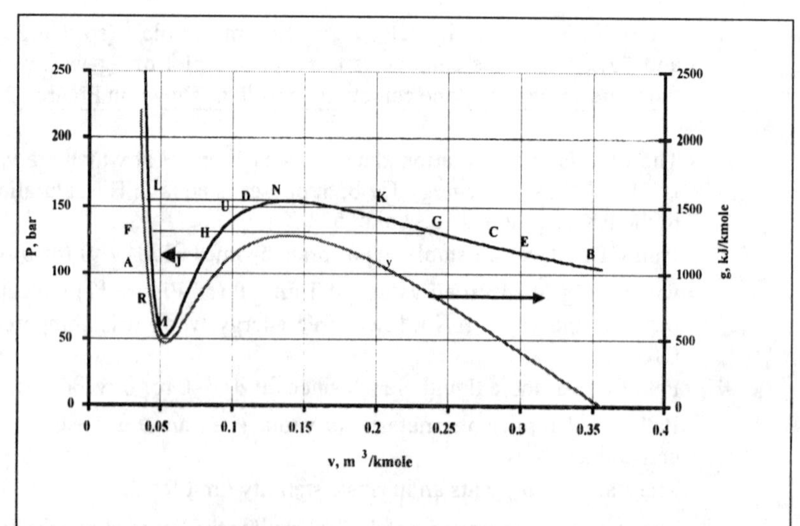

Figure 11: Plot of P (and $\bar{g}$) *vs.* volume using the RK equation.

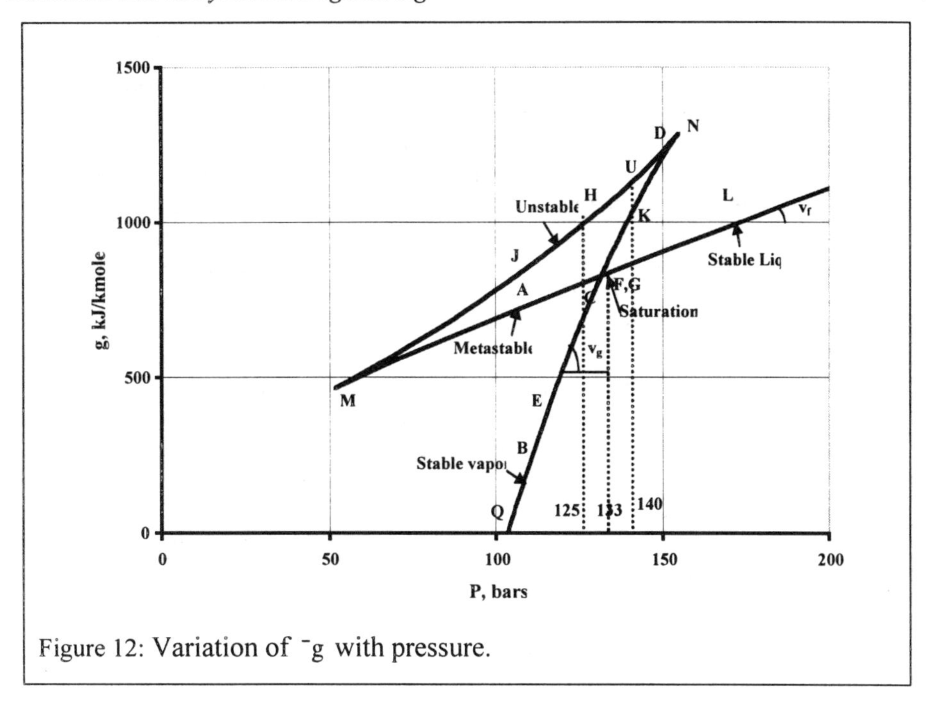

Figure 12: Variation of $\bar{g}$ with pressure.

state G (Figure 13a). We note from Figure 11 or Figure 12 that when P < 50 bar, a single state is possible (i.e., superheated vapor). If 50 < P < 155 bar, there are three possible states for the same Gibbs energy value, and when P > 155 bar there is a single liquid solution (i.e., compressed liquid).

Consider a constant T, P, and m system at an arbitrary state J. A reduction in the fluid volume from a value v_J reduces the internal pressure exerted by the fluid further compressing the fluid. Assume that equilibrium is achieved at a *liquid* volume v_A at which $P_A = P_J$ (Figure 12). Likewise a corresponding vapor state C exists at which $P_C = P_J$ with a volume v_C. The implication is that at the pressure P_J there are three plausible solutions for the Gibbs energy. The questions are as follows. Which are stable states? Which are nonstable?

Since $dG_T = VdP$, $dG_{embryo} = (V(\partial P/\partial V))\, dV$. An increased volume (e.g., during evaporation) implies that $\partial P/\partial V < 0$ so that $\partial G/\partial V < 0$. Suppose a disturbance at J causes the embryo phase to expand to a volume slightly higher than v_J from state J, the volume increase tends to increase the embryo phase pressure. Since the embryo phase pressure is higher than the mother phase pressure, which is held fixed, the embryo expands to larger and larger volumes, eventually to the vapor state B. The first bubble during boiling is formed through this process. The embryo phase bubble is associated with a lower Gibb's free energy (Figure 12) as compared to the mother phase that is still at state J. In Chapter 3 we discussed that a Gibbs energy gradient produces a species flow from a system at a higher Gibbs energy to that at a lower Gibbs energy. In that context, the molecules from the mother phase migrate to the vapor phase during vaporization as long as the pressure and temperature are maintained constant. If the disturbance at J results in reduction of volume, the embryo pressure decreases; since the mother phase is at higher pressure, embryo is compressed further until it forms a liquid droplet at A. In the case of flow processes, a sudden condensation or vaporization produces a severe pressure disturbance or a sudden acceleration of the flow, leading to local turbulence (e.g., in boiler tubes and in clouds).

In the context of the above discussion, we now consider the H_2O at 140 bar and 593 K for which the specific volume v_D is at an unstable state. An embryo may form, but a decrease in the embryo volume causes the pressure to decrease below 140 bar, which results in compression of the embryo by the mother phase that is still at 140 bar. This accelerates the formation of a drop at state L. Since the Gibbs energy of the compressed liquid at state L is lower than that of the mother phase at state K, the fluid molecules will tend to migrate to the liquid state. State H is also unstable and only a microscopic disturbance is required to drive the state to either of states F or G. At saturation, the two minima are equal, i.e., $g_F = g_G = g^{sat}$. This process is called a first order phase transition during which both the liquid and vapor states are probable, which is a consequence of boiling at a specified temperature and pressure. In this case, the liquid and vapor molecules can exchange phase if a disturbance is strong enough to overcome the potential $(g_H - g_G)$, which results in a wet mixture. Note that thermodynamics do not specify the time scales (called relaxation time scale) required to effect the change from meta-stable or unstable state to stable state. Constitute equations for the transport processes are required to determine those time scales.

These examples pertain to phase equilibrium. An analogous situation exists during chemical equilibrium where, at a specified temperature and pressure, the Gibbs energy of products reaches a minimum value.

2. Constant Temperature and Volume

Consider a fluid of mass m within a rigid tank of volume V that is immersed in a bath at a temperature T for which the real gas state equation yields the P-v diagram presented in Figure 11. The states F and G represent the saturated states at a specified temperature. At state K (Figure 14) the fluid is in the form of slightly sub–cooled vapor. We will divide the system into two parts 1 and 2, and disturb the volume such that portion 1 expands slightly to state K″, while portion 2 shrinks to K′, but the total volume is fixed. Then, $P_{K'} > P_{K''}$ and portion 2 will expand back so that the system eventually reverts to its original state K.

Consider the fluid at State D (cf. Figure 14). If the fluid is disturbed so that the mother phase volume (say, portion 1) slightly expands to $v_{D''}$ while the embryo volume (portion 2) contracts to $v_{D'}$ with the total volume held fixed, since $v_{D''} \approx v_D$ and $v_{D'} \ll v_D$, $P_{D''} \approx P_D$, while $P_D > P_{D'}$. The mother phase is at a higher pressure than the embryo phase, which undergoes increasing compression. The pressure in the embryo first decreases along path HJM (cf. Figure 11) but starts increasing again along path MRL (the liquid path) until state L is reached at which the pressure in the embryo equals P_D. A disturbance can occur at several locations inside the system creating a large number of drops that can combine to form a continuous liquid phase. However, since liquid drops generally occupy a smaller volume than the gas phase in a fixed volume sys-

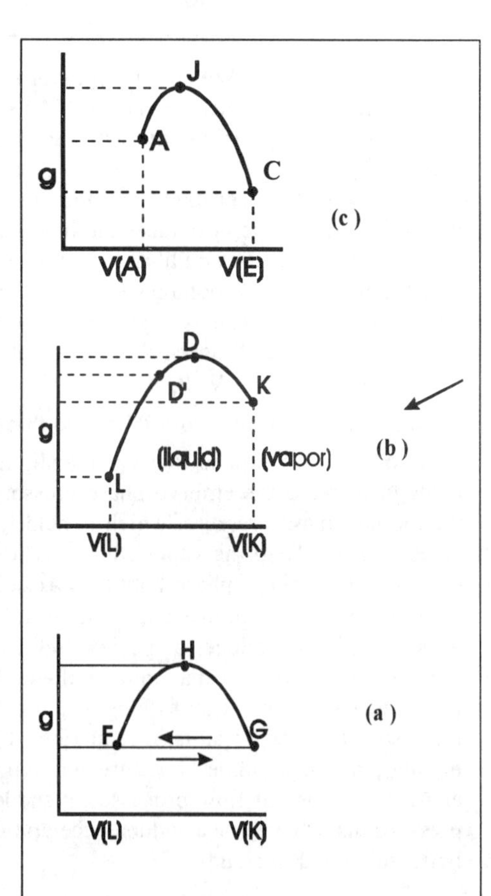

Figure 13: Variation in the value of g at a specified value of T, but for different pressures. (a) $P = P^{sat}$; (b) $P>P^{sat}$; (c) $P < P^{sat}$.

tem, the entire vapor body may not condense, and a wet mixture may form.

At equilibrium, there cannot be any potential gradients between the liquid and vapor phases so that $g_{liquid} = g_{vapor}$. At a specified temperature, this condition occurs at a particular value of the saturation pressure P^{sat}, which allows us to determine the volume or quality x, i.e.,

$$v = xv_g + (1-x)\, v_f, \text{ or } x = (v-v_f)/(v_g - v_f). \quad (47)$$

The volume at a stable state $v > v_G$. For the condition $v_N < v < v_G$, a metastable vapor state exists and a mixture of vapor and liquid is formed, but at a higher quality. If $v_F < v < v_M$, then a metastable liquid state exists and a vapor-liquid mixture of lower quality is formed. The condition $v_M < v < v_N$ is unstable and a mixture of vapor and liquid of medium quality is formed.

When phase transformation from a metastable state occurs at a specified temperature and volume, the Helmholtz energy is minimized. Figure 15 presents a plot of $\bar{a}$ and P *vs.* $\bar{v}$. State D on the P-$\bar{v}$ diagram (P = 140 bar, v_D = 0.1 m^3 kmole^{-1}, T = 593) corresponds to the point Q on the $\bar{a}$-$\bar{v}$ curve. A disturbance increases $\bar{a}$ within the mother phase if the volume decreases and *vice versa*. However, the value of the Helmholtz energy increase is smaller than its decrease with the result that $\bar{a}$ decreases (at fixed T and V). Consequently, the fluid eventually reaches a vapor state and $\bar{a}$ is minimized. At a specified temperature and volume the pressure (i.e., P = 133 bar, v = 0.1 m^3 kmole^{-1}, T = 593 K) and quality adjust such that the Helmholtz energy is at a minimum value (e.g., with vapor at State G where $\bar{a}$ = 20.6 MJ and liquid at State F with a higher value of $\bar{a}$ = 23.2 MJ).

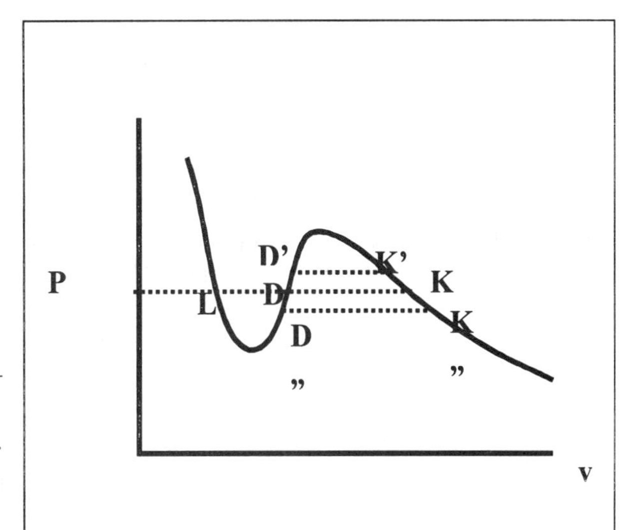

Figure 14. A Pressure-volume diagram for illustrating fluid stability along an isotherm.

k. Example 11

2.662 kmole of water are contained in a rigid tank of volume 27 m^3 at 320°C and 140 bar. What are the values of $\bar{s}$, $\bar{u}$, and $\bar{a}$ at this state? If the temperature and volume are maintained constant, what is the most stable equilibrium state, and what are the values of $\bar{s}$, $\bar{u}$, and $\bar{a}$ at this state? Assume that c_P = 5.96 kJ kg^{-1} K^{-1}.

Solution

$\bar{v}$ = 2.662/27 = 0.0986 m^3 kmole^{-1}.

Applying the RK equation at 593 K and 140 bar (State D in Figure 11),

P^{sat} = 133 bar.

Since $P > P^{sat}$ at 593 K, and $P < P_N$ = 155 bars the state is metastable.

We will use the method described in Chapter 7 to determine the fluid properties. For instance,

$\bar{h}$(320°C, 0.0986 m^3 kmole^{-1}) = 44633 kJ kmole^{-1} or 2477 kJ kg^{-1},

$\bar{u}$(320C, 0.0986 m^3 kmole^{-1}) = h – P$\bar{v}$

= 43252 kJ kmole^{-1} or 2400 kJ kg^{-1}, and

$\bar{s}$ (320C, 0.0986 m^3 kmole^{-1}) = 91.34 kJ kmole^{-1} or 5.07 kJ kg^{-1}.

Thereafter,

$\bar{a} = \bar{u} - T\bar{s} = 43252 - 593 \times 91.34 = -10{,}912$ kJ kmole^{-1} or -606 kJ kg^{-1},

and

$\bar{g} = \bar{h} - T\bar{s} = -9{,}532$ kJ kmole^{-1}.

At a specified value of T and V, the value of $\bar{a}$ decreases until it reaches a minimum. The fluid at state D transforms into a wet mixture, but $g_{liquid} = g_{vapor}$. Hence, P must equal 133 bar so that $\bar{a}$ is minimized, i.e.,

$\bar{v} = (1-x)\, v_f(320°C, P_{new}) + x\, v_g(320°C, P_{new}) = 0.0986$ m^3 kmole^{-1}

The pressure $P_{new} = P^{sat}$, which is the saturation pressure at phase equilibrium (that equals 133 bar).

Applying the RK equation at 593 K and 133 bar,

v_f (320°C, 133 bar) = 0.04275, m^3 kmole^{-1}or 0.00237 m^3 kg^{-1}, and

v_g (320°C, 133 bar) = 0.236 m^3 kmole^{-1}or 0.0131 m^3 kg^{-1}, i.e.,

$x = (0.0986 - 0.04275)/(0.236 - 0.04275) = 0.289$, and

$\bar{a} = \bar{a}_f(1-x) + \bar{a}_g x = -11{,}114$ kJ kmole^{-1}.

Remarks

At equilibrium $\bar{a} = -11{,}114$ kJ kmole^{-1}, which is lower than the $-10{,}912$ kJ kmole^{-1} value at the metastable point at the same temperature and volume. The disturbance creates a wet mixture. State K (593 K, 0.211 m^3 kmole^{-1}) is a metastable state which, when disturbed, produces a wet mixture with a higher quality at the same temperature and volume, since the initial volume is larger.

At specified values of T and P, g reaches two possible minimums. Such multiple states do not exist for the Helmholtz energy at specified values of T and v.

---

3. Specified Values of S, P, and m

Consider water at 140 bar and 0.211 m^3 kmole^{-1} (cf. state K in (cf. state K in Figure 11). When S and P are specified, a disturbance creates two phases. At a fixed pressure, $g_f = g_g$ and $T = T^{sat}$. The mixture quality is adjusted such that

$$x = (s_{initial} - s_f)/s_{fg}.$$

4. Specified Values of S (or U), V, and m

In this case, we can assume a temperature T_1, and determine $P^{sat}(T_1)$, $v_f(T_1)$, and $v_g(T_1)$ so that

$$x(T_1) = (v - v_f(T_1))/(v_g(T_1) - v_f(T_1)).$$

The quality must satisfy the relation

$$x(T_1) = (s - s_f(T_1))/(s_g(T_1) - s_f(T_1)),$$

otherwise another temperature T_2 must be assumed and the procedure repeated. In this case, the internal energy is minimized.

In case U is specified, then the relation

$$x(T_1) = (u - u_f(T_1))/(u_g(T_1) - u_f(T_1))$$

must be satisfied.

D. ENTROPY GENERATION DURING IRREVERSIBLE TRANSFORMATION

The change from an unstable or a metastable state to a stable state is irreversible and, hence, entropy is generated. Recall that

$$\delta q/T - ds = \delta\sigma, \tag{48}$$

where δq denotes the heat transferred from a reservoir at a temperature T. At a specified volume,

$$\delta q_v = du, \text{ i.e., } \delta\sigma = du/T - ds = (du - Tds)/T = da/T. \quad (49)$$

Therefore, in the context of the example # 11,

$$\sigma=(a_{metastable/unstable}-a_{stable})/T=(-10{,}912-(-11{,}114))/593=0.3406 \text{ kJ kmole}^{-1} \text{ K}^{-1}. \quad (50)$$

For a unit mass

$$\delta\sigma = ds - \delta q/T, \quad (51)$$

and if the transformation occurs at specified values of T and P, applying the First Law $\delta q_P = dh$ so that Eq. (51) assumes the form

$$\delta\sigma = ds - dh/T = (Tds-dh)/T.$$

Since T is constant,

$$\delta\sigma = -\, dg/T.$$

Integrating this expression, we obtain

$$\sigma = (g_{metastable/unstable} - g_{stable})/T. \quad (52)$$

For instance, during the change from state D (metastable) to L (stable liquid) (cf. Figure 11), $\sigma = (-9530.5 - (-9776.2)) \div 593 = 0.414$ kJ kmole^{-1} K^{-1}, while for a change from D to the metastable vapor state K, $\sigma = (-9530.5 - (-9638.8)) \div 593 = 0.183$ k kmole^{-1} K^{-}.

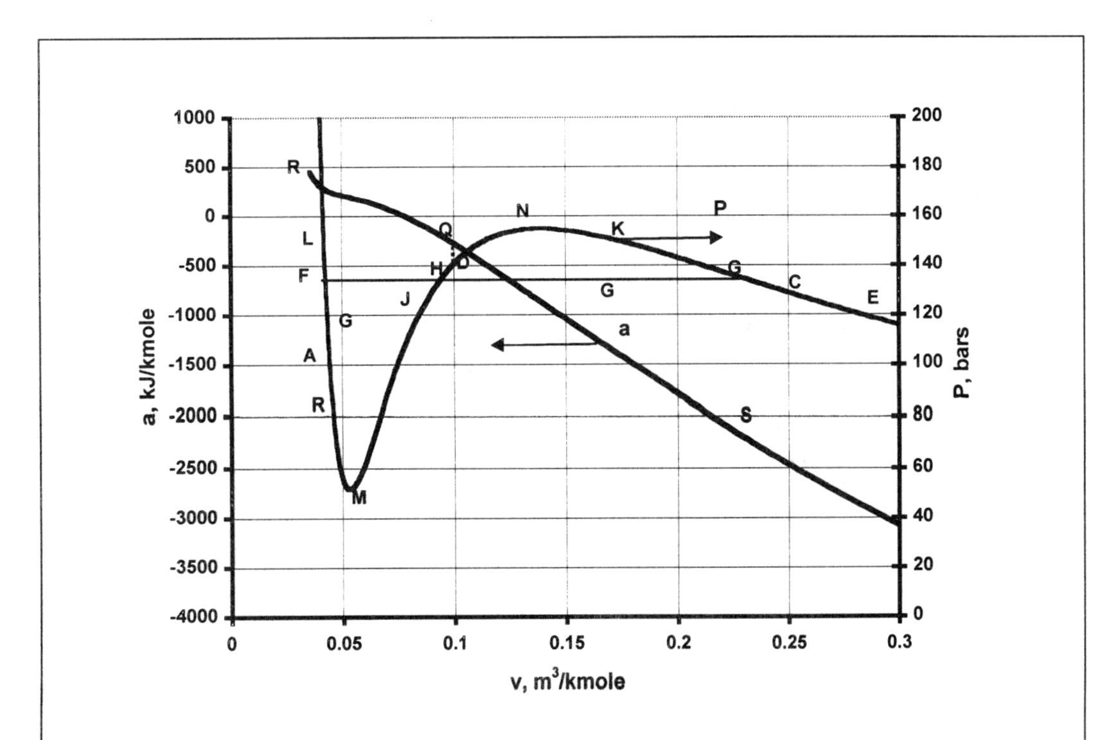

Figure 15: The variation in the Helmholtz function and pressure with respect to the volume for water modeled using the RK equation (P^{sat} =133 bar, T =593 K).

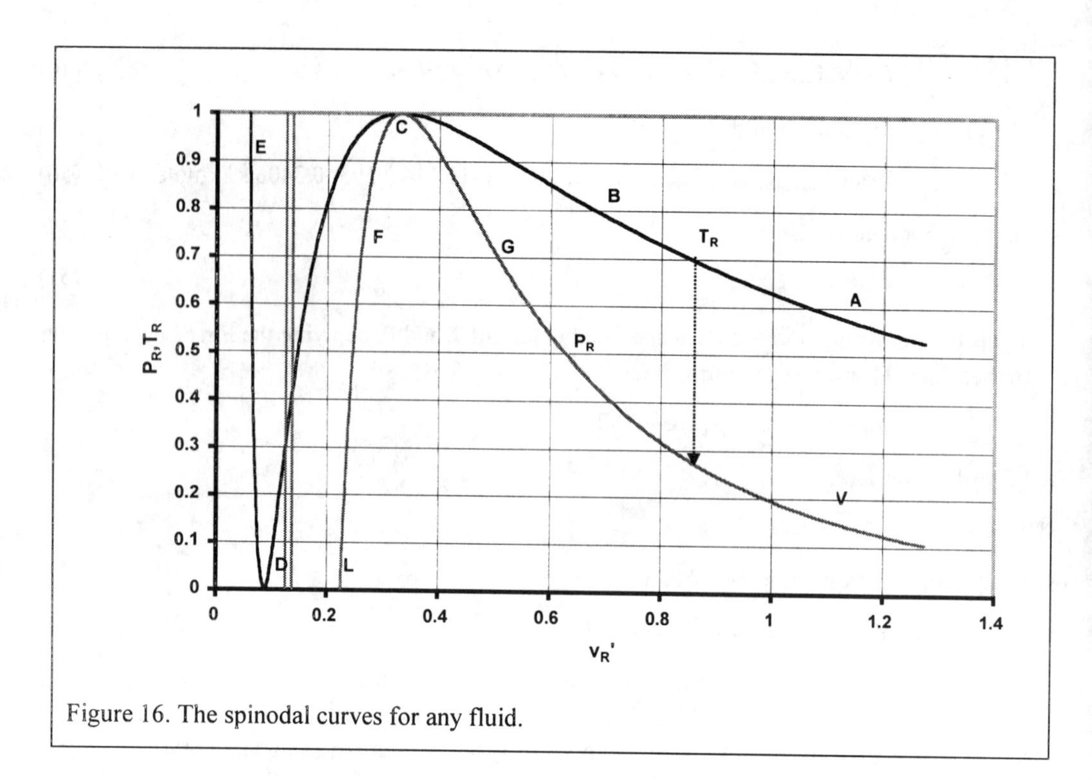

Figure 16. The spinodal curves for any fluid.

E. SPINODAL CURVES

1. Single Component

For the branches ECGKN and MRAFL in Figure 15, $\partial P/\partial v < 0$ and $(d^2A)_{T,V,m} > 0$. Along the branch NDHJM, $\partial P/\partial v > 0$ and $(d^2A)_{T,V,m} < 0$. The points M and N at which $(\partial P/\partial v)_T = 0$, are the limits of intrinsic stability and are called "spinodal points". At the spinodal points M and N, $\partial P/\partial v = 0$ and $(d^2A)_{T,V,m} = 0$. At a specified temperature, the spinodal points yield the maximum pressure at which the fluid exists as vapor (cf. state N) and the minimum pressure at which it exists as liquid (state M). The spinodal points can be predicted by applying the state equations as illustrated through the following example.

l. Example 12

Use the relation

$$P = RT/(v-b) - a/(T^{1/2}\, v(v+b)) \tag{A}$$

To obtain an expression for P *vs.* v and T *vs.* v for water along the spinodal points plot P *vs.* v at 567, 593 and 615 K for water. Obtain an expression for the reduced values P_R and T_R *vs.* v_R' along the spinodal points, and plot P_R and T_R *vs.* v_R' and P_R *vs.* T_R along these points.

Solution

At the spinodal points $\partial P/\partial v = 0$. Differentiating Eq. (A),

$$(\partial P/\partial v = -\, RT/(v-b)^2 + (\mathbf{a}/(T^{1/2}\, v(v+b)))\,(1/v + 1/(v+b))) = 0.$$

Thereafter, solving for the temperature,

$$T = ((\mathbf{a}(v-b)^2/(R\, v(v+b)))\,(1/v + 1/(v+b)))^{2/3.} \tag{B}$$

Using Eqs. (A) and (B) we obtain the relation

$$P = (R(v-b))^{1/3} ((\mathbf{a}/(v(v+b))) (1/v + 1/(v+b)))^{2/3}$$
$$- (\mathbf{a}/(v(v+b)))^{2/3} (R/((v-b)^2 (1/v + 1/(v+b))))^{1/3} \quad \text{(C)}$$

With the values $\bar{a}$ = 142.59 bar m^6 $K^{1/2}$ $kmole^{-2}$ and $\bar{b}$ = 0.0211 m^3 $kmole^{-1}$ in the context of Eq. (B), we obtain a plot of T *vs.* $\bar{v}$, and using Eq. (C) we can obtain a plot of P *vs.* $\bar{v}$.

Using Eq. (C), one can also obtain the spinodal pressure *vs.* volume for water at the temperatures 567, 593 and 615 K (cf.).

We can normalize Equations (B) and (C) so that

$$T_R=((0.4275(v_R'-0.08664)^2/(v_R'(v_R'+0.08664))) (1/v_R'+1/(v_R'+0.08664)))^{2/3}, \text{ and (D)}$$

$$P_R = T_R/(v_R' - 0.08664) - 0.4275/(T_R^{1/2} v_R' (v_R' + 0.08664)) \quad \text{(E)}$$

Figure 16 presents a plot of P_R *vs.* v_R' and T_R *vs.* v_R' and Figure 17 contains a plot of P_R *vs.* T_R along the spinodal points. The curves CGV and CFL denote the vapor and liquid spinodal curves. Figure 18 presents plots of Z vs P_R with T_R as a parameter for nonpolar fluids for the spinodal and saturation conditions at any given state.

Remarks

Along the spinodal points on the curves CGV and CFL, c_v > 0. Therefore, $(\partial^2U/\partial S^2)$ = $(T/mc_v) \neq 0$, but the condition $(\partial^2U/\partial^2V)_{S,m} = (\partial P/\partial V)_T = 0$ is satisfied.

As the critical temperature T_c is approached, the vapor and liquid spinodal volumes merge into a single value v_c so that $d^2A = 0$ without exhibiting a maxima or a minima.

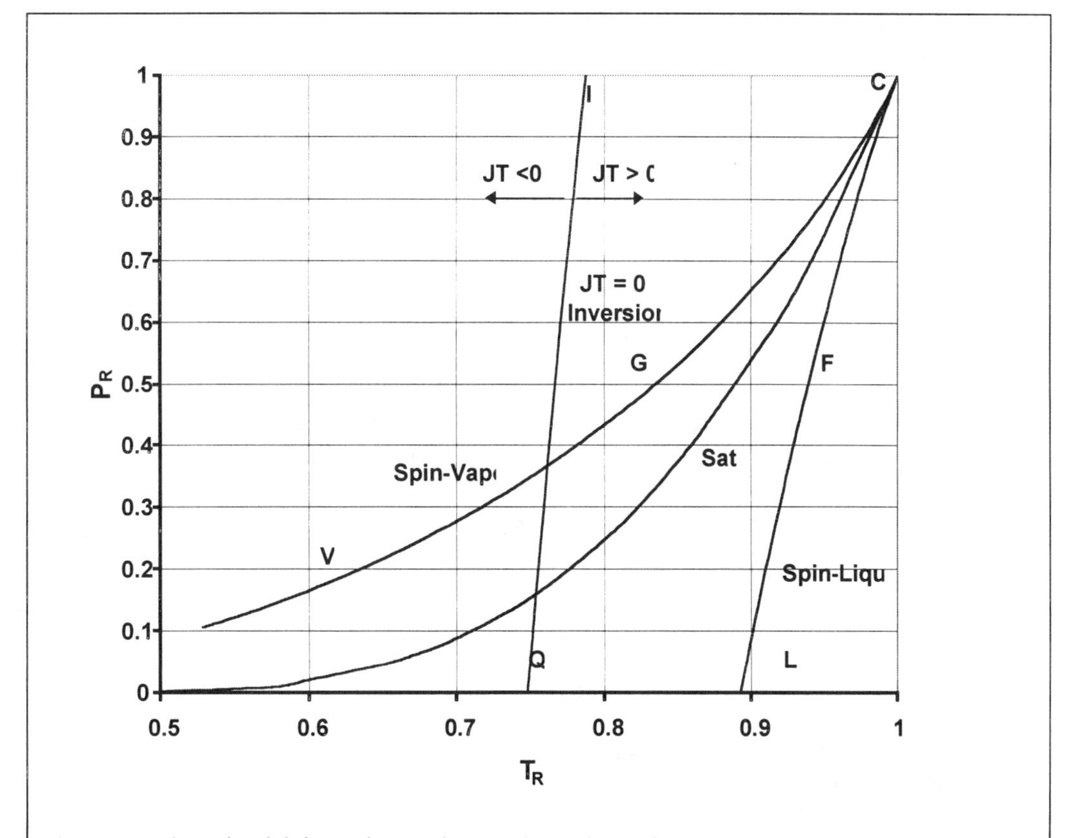

Figure 17: The spinodal, inversion, and saturation values of P_R *vs* .T_R for an RK fluid.

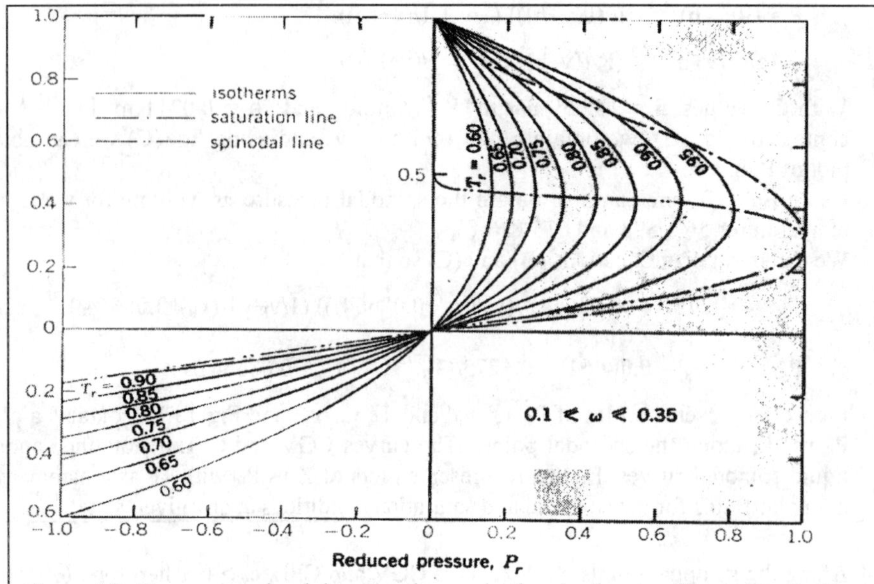

Figure 18. Z curves: spinodal, saturation and other isotherms. (From W. G. Dongand, J. H. Lienhard, *Can. J. Chem.Eng.*, 64, pp. 158-161, 1986. With permission.)

Since $\partial P/\partial v = -(\partial T/\partial v)_P\,(\partial P/\partial T)_v$, if $\partial P/\partial v = 0$ (i.e., $\beta_T \to \infty$), then $(\partial T/\partial v)_P = 0$. Therefore, the coefficient of thermal expansion $\beta_P \to \infty$ at the spinodal point (since, typically, $\partial P/\partial T \neq 0$).

Applying the results from Chapter 7, we can plot the inversion pressure with respect to the temperature (curve IQ in Figure 17). The regimes where $\mu_{JT} > 0$, $\mu_{JT} < 0$, and $\mu_{JT} = 0$ are indicated in the figure.

The spinodal curves are useful to predict the degree of liquid superheat and vapor subcooling of vapors at a specified pressure. Consider a fluid which has saturation temperature of $T_1 = T^{sat}$ at P_1 (cf. Figure 19, 593 K at P =133 bar). Figure 19 plots three curves of pressure with respect to volume for the temperatures T_1, T_2, and T_3. The temperatures T_2 and T_3 are selected such that the vapor spinodal pressure at N corresponding to T_2 coincides with the liquid spinodal pressure at M corresponding to T_3. The points F and G represent the saturated liquid and vapor states at the condition (P_1, $T_1 = T^{sat}$). The liquid at a pressure P_1 can be superheated to the temperature T_3 (i.e., from state F to M) without boiling and the vapor can be likewise subcooled to T_2 (i.e., from state G to N) without condensation. This phenomenon is illustrated through the following example.

m. Example 13

Consider the RK equation of state

$$P = RT/(v-b) - a/(T^{1/2}\, v(v+b)). \qquad (A)$$

Determine the maximum temperature to which water can be superheated at 133 bar without boiling and the temperature to which water vapor can be subcooled at the same pressure without condensation occurring. Assume that $T^{sat} = 593$ K at P = 133 bar for water when it is modeled by the RK equation of state. What is the fluid state at 615.001°C and 133 bar?

Solution

From Example 12, the temperatures at which $\partial P/\partial v = 0$ are represented by the relation

$$T = ((a\,(v-b)^2/(R\,v(v+b)))\,(1/v + 1/(v+b)))^{2/3}. \qquad (B)$$

There are two spinodal vapor and liquid volumes at particular temperature, e.g., the states M and N at 133 bar in Figure 19.
Using (B) and the RK equation to eliminate the temperature, we obtain the expression

$$P = (R\,(v-b))^{1/3}((a/(v(v+b)))(1/v + 1/(v+b)))^{2/3}$$

$$-\,(a/(v(v+b)))^{2/3}(R/((v-b)^2(1/v + 1/(v+b))))^{1/3}. \qquad (C)$$

Eq. (C) is useful for obtaining the vapor spinodal curves AN and liquid spinodal ON (cf. Figure 19).
Using the values $\bar{R}$ = 0.08314 bar m^3 $kmole^{-1}$ K^{-1}, $\bar{a}$ = 142.59 bar m^6 $K^{1/2}$ $kmole^{-2}$, $\bar{b}$ = 0.0211 m^3 $kmole^{-1}$, P = 133 bar in Eq. (C), v_M = 0.0597 m^3 $kmole^{-1}$ (cf. Figure 19, liquid–like spinodal point) and v_N = 0.158 m^3 $kmole^{-1}$ (vapor-like spinodal point). and v_N = 0.158 m^3 $kmole^{-1}$ (vapor-like spinodal point). Thereafter, applying these results in Eq. (B), for the liquid $T_{Sp,\ell}$ = 615 K and for the vapor $T_{Sp,v}$ = 567 K. Therefore, water can be superheated at 133 bar by 615–593 = 22 K and the vapor subcooled by 567–593 = –26 K.

One can use Charts in Fig. 17 to predict the degree of superheating and subcooling at any given pressure. With P = 133 bars, P_R = 0.602, then T_R along spinodal liquid is 0.95 while along spinodal vapor is 0.88 while T_R along saturated line is 0.92. The corresponding T's are 614.5, 569, and 593 K. Sometimes the RK equation may be crude to determine the degree of superheating. Recall from Chapter 6 that $\bar{a} = c_3\,\bar{R}^2 T_c^{2.5}/P_c$, and $\bar{b} = c_4\,\bar{R}T_c/P_c$ where from theory c_3 = 0.4275, c_4 = 0.08664. If

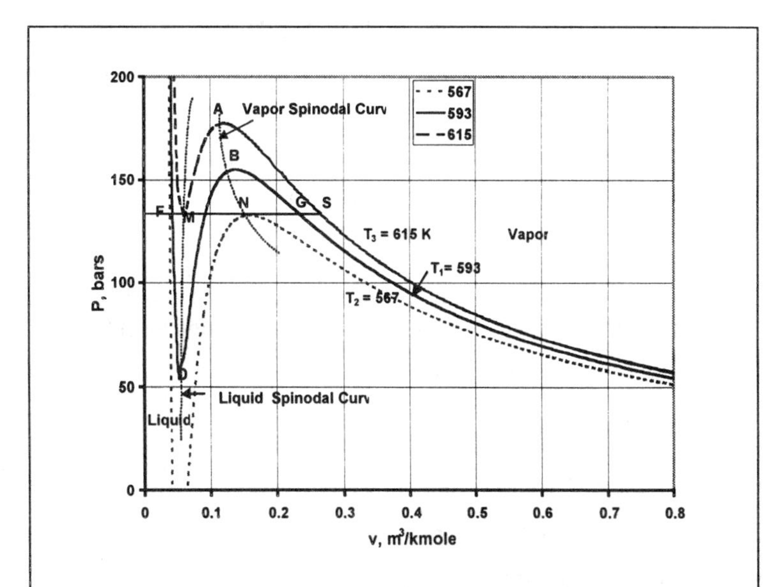

Figure 19: Spinodal curves at 567, 593, and 615 K for water modeled according to the RK equation of state. If the pressure is maintained at 133 bar, T^{sat} = 593 K according to the relation with vapor at state G and liquid at state F. The vapor can be subcooled to 567 K (cf. the spinodal point N) and the liquid superheated to 615 K (cf. spinodal point M) at the same value of P^{sat} = 133 bar.

c_3 and c_4 are selected to be different to fit the experimental data for P,v,and T, it is equivalent to modifying T_c and P_c's (Chapter 6) . The charts in Figure 17 can still be used with modified critical properties (Chapter 6).

Subcooling also occurs during ice formation and in sublimation processes. Pure distilled water can be cooled to –8°C without forming ice.

2. Multicomponent Mixtures

The spinodal analyses can also be applied to mixtures.

n. Example 14

Consider a mixture that contains 60% water (species 1) and 40% methyl alcohol (species 2). Determine the spinodal curves for the mixture as a function of pressure and temperature. Assume that the mixture follows the RK equation of state.

Solution

The stability of two components require that

$$D_1 = \mathbf{A}_{VV} \geq 0, \text{ and} \quad (A)$$

$$D_2 = \begin{vmatrix} A_{VV} A_{VN1} \\ A_{N1V} A_{N1N1} \end{vmatrix} \geq 0 \quad (B)$$

Selecting the mixing rule for the RK equation of state

$$a_m = (\Sigma\, Y_k\, \bar{a}_k^{1/2})^2, \text{ so that} \quad (C)$$

$$A_m = a_m N^2,\ \bar{b}_m = \Sigma\, Y_k\, \bar{b}_k, \text{ and } B_m = b_m N. \quad (D)$$

The pseudo critical temperature and pressure can be expressed as

$$T_c' = (0.08664/(0.4275\ R)(a_m/b_m))^{2/3}, \text{ and } P_c' = 0.08664\ RT_c'/b_m, \text{ i.e.,}$$

$$a_m = (0.6 \times 142.6^{0.5} + 0.4 \times 220^{1/2})^2 = 171.6 \text{ bar k}^{1/2}\text{m}^6 \text{ kmole}^{-2},$$

$$b_m = (0.6 \times 0.0211 + 0.4 \times 0.0462) = 0.03115 \text{ m}^3 \text{ kmole}^{-1}, \text{ and}$$

$$T_c' = 564.9 \text{ K and } P_c' = 130.6 \text{ bar.}$$

Using the condition $A_{VV} = 0$ we determine that $\partial P/\partial V = 0$. In Figure 20 the vapor spinodal points are represented by the curve VGC and the liquid spinodal points by the curve CFL. However, this condition alone does not satisfy the stability criteria for a mixture. The following additional spinodal condition must be satisfied, i.e.,

$$\mathbf{A}_{V N_1} = \partial/\partial N_1(\mathbf{A}_V) = -\partial P/\partial N_1. \quad (E)$$

Since

$$P = NRT/(V-B_m) - A_m/(T^{1/2}\, V(V+B_m)), \quad (F)$$

where $B_m = Nb_m$, $A_m = N^2 a_m$, $\partial B_m/\partial N_1 = b_1$, $\partial A_m/\partial N_1 = 2\,(a_1\, a)^{1/2}$ (cf. Chapter 8),

$$\partial P/\partial N_1 = RT/(V-B_m) + NRT\,(\partial B_m/\partial N_1)/(V-B_m)^2 - (\partial A_m/\partial N_1)/(T^{1/2}\, V(V+B_m))$$

$$+ A_m\,(\partial B_m/\partial N_1)/(T^{1/2}\, V(V+B_m)^2) \quad (G)$$

Substituting Eq. (G) in Eq. (E), and multiplying the resultant expression by N, we obtain the relation

$$N\,\mathbf{A}_{V N_1} = -RT/(v - b_m) - RTb_1/(v - b_m)^2$$

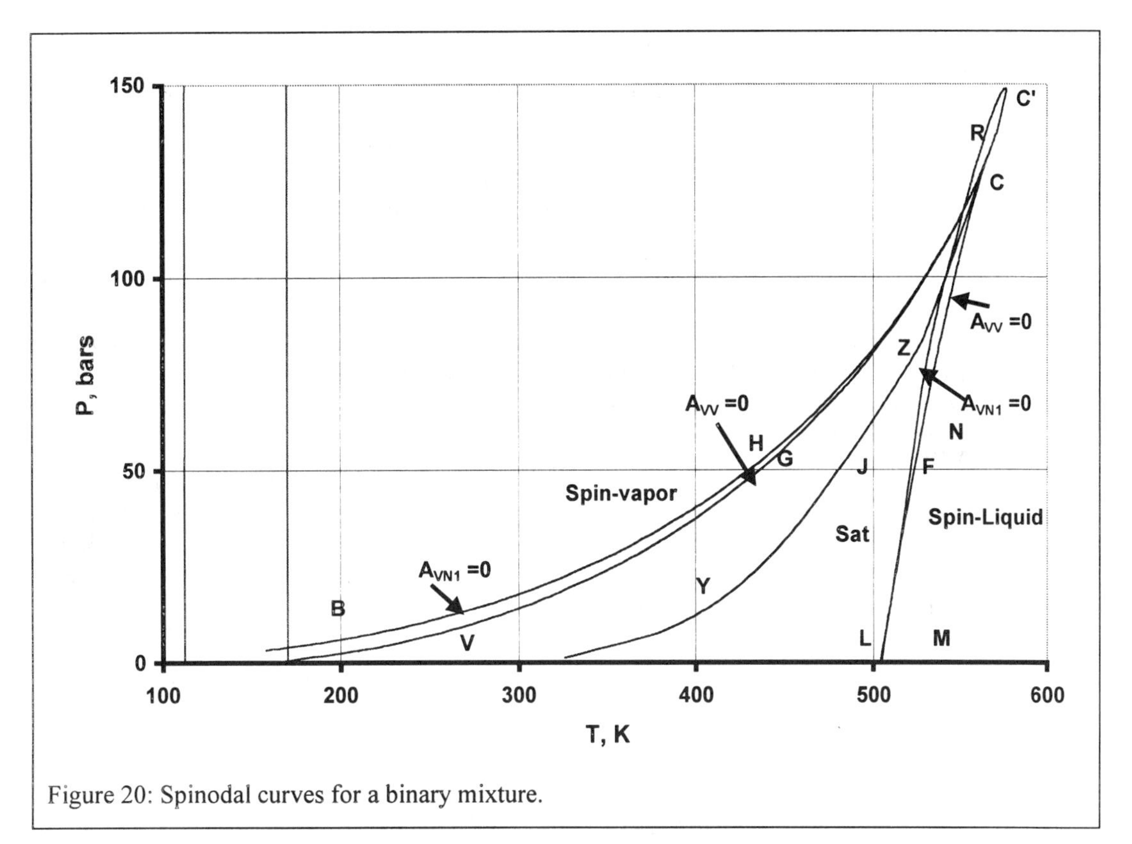

Figure 20: Spinodal curves for a binary mixture.

$$+ 2(a_1a_m)^{1/2}/(T^{1/2}v(v + b)) - ab_1/(T^{1/2}v(v + b_m)^2). \quad \text{(H)}$$

An additional spinodal condition for a mixture is obtained by using the following equality, i.e.,

$$\mathbf{A}_{V_{N_1}} = 0, \text{ since } N > 0. \quad \text{(I)}$$

Substituting Eq. (I) in Eq. (H), we obtain a relation for the temperature, i.e.,

$$T^{3/2} = (((v - b_m)^2/(R(v - b_m + b_1)\, v(v + b_m)^2))\, (2\, (a_1a_m)^{1/2}\, (v + b_m) - ab_1))^{2/3}. \quad \text{(J)}$$

Thereafter, using Eq. (F) the pressure at which $A_{V_{N_1}} = 0$ is obtained (cf. Figure 20 - curve MNC′ for the liquid and C′HB for vapor). The pressure along which $A_{VV} = 0$ is represented by the curves LFC (for the liquid) and CGV (for the vapor). Figure 20 also illustrates the bubble point J along the saturation curve YJC at 50 bar for T^{sat} = 475 K.

We now discuss the criteria. When $A_{VV} = 0$, the mixture can be superheated to 520 K (point F) without a bubble forming, and a vapor mixture can be subcooled to 430 K (point G) without condensation. The presumption is that any minor disturbance within the system occurs due to volumetric changes alone (i.e., there is a uniform composition within the disturbed space). However, if the composition is also locally nonuniform due to a disturbance (i.e., due to fluctuations in N_1) then the spinodal condition corresponds to $A_{V_{N_1}} = 0$. Accordingly, the curve BHC′RNM is the spinodal vapor curve.

Within HG, $\partial^2A/\partial N_1{}^2 > 0$ but $\partial^2A/\partial V^2 < 0$. If a disturbance occurs due to changes in both V and N_1, then the conditions $\partial^2A/\partial V^2 > 0$ (i.e., $\partial P/\partial V < 0$) and $\partial^2A/\partial N_1{}^2 > 0$ require that the system lie within GF at 50 bar.

F. DETERMINATION OF VAPOR BUBBLE AND DROP SIZES

From Chapters 5 and 7 we note that for a vapor bubble (embryo phase, E) in a liquid (mother phase, M), the pressure difference is given as

$$(P_E - P_M) = 2\,\sigma/r_v, \tag{53}$$

which is called the Laplace equation. This relation allows us to determine the minimum bubble size during evaporation and the maximum drop size at a metastable state using the spinodal pressure relations at a specified temperature.

o. *Example 15*

Water boils with the mother liquid phase at 593 K and 133 bar. The vapor embryo is at the same temperature as the liquid, but at a different pressure. Assume that the RK equation applies and that the vapor pressure is 155 bar. (According to the RK equation of state, for water P^{sat} = 133 bar at 593 K.) Determine the bubble size. Assume that the surface tension $\sigma = 10.5\times10^{-6}$ kN m^{-1}.

Solution

The bubble pressure is higher than the liquid pressure due to the effect of surface tension. The vapor can exist at a metastable state. The bubble radius

$$r_B = 2\,\sigma/(P_E - P_M),\text{ i.e.,} \tag{A}$$

$$r_B = (2\times10.5\times10^{-6})/((155\text{–}133)\times100) = 9.549\times10^{-9}\text{ m.}$$

Eq. (A) indicates that higher the value of P_E, the smaller the bubble size.

Remarks

Various bubbles have different diameters depending upon the various metastable states of the bubbles. For instance, at 133.1 bar, $r_B = 2.1$ μm.

If the mother phase is a vapor at 593 K and 133 bar, the liquid drops exist at a higher pressure in the form of compressed liquid.

Consider superheated vapor at 593 K and 50 bar. The lowest pressure in the liquid state at 593 K is 55 bar, which leads to the formation of metastable liquid drops. The associated drop size is $r_d = 2\times10.5\times10^{-6}/((55\text{–}50)\times100) = 0.042$ μm

p. *Example 16*

Superheated liquid water (α phase) exists at 593 K and 1 bar along with a vapor embryo (β phase) at the same temperature. The saturation pressure at 593 K is 133 bar. Determine the size of the embryo bubble assuming that the surface tension $\sigma = 10.5\times10^{-6}$ kN m and the liquid molar volume to be 0.018 m^3 kmole^{-1}.

Solution

The pressure

$$P^\beta = P^{sat}\exp\left(v^\alpha(P^\alpha - P^{sat})/RT\right)$$

$$= 133\exp\left(0.018\times(1-133)/(0.08314\times593)\right) = 126.7\text{ bar.}$$

Since $P^\beta = P^\alpha + 2\,\sigma/r$, $r = 1.67\times10^{-9}$ m.

G. UNIVERSE AND STABILITY

The size of the universe is determined to be 10 billion light years. There may be 100 billion galaxies in the universe. The dimension of our galaxy, the Milky Way, is of the order of 100,000 light years (1 light year is equivalent to roughly 6 trillion miles and 1 megaparsec equals the distance light travels in 3.26 million years). Galaxies tend to group together in clusters and the clusters in super clusters.

Since the gravitational forces within the universe exceed those determined just with the observable matter, the imbalance is attributed to dark matter (black holes, white stars, etc.,) unseen with existing technology. The amount of dark matter controls the gravitational force and, hence, balances the forces due to the motion of the galaxies. When balanced, it is a flat universe, but when galaxies expand, it is an open universe. When it collapses upon itself, it is called a closed universe. It is thought that there is 90% dark matter in the universe. The current evidence seems to suggest that the universe is expanding with increasing velocity, which appears to contradict the Big Bang theory that suggests a slowing velocity. The expanding universe is believed to be open so that a light beam will never return to its point of origination. The density of the open universe is believed to roughly equal 1.3 times the mass of a single H atom. If that density were instead equal to 130 times the mass of an H atom, it is believed that the universe would be closed.

New theories suggest a pouring of energy from "virgin" vacuum into the bubble of the universe containing dark matter. Since this violates the First law, the existence of an anti-energy in virgin vacuum has been proposed. Thus, energy input to a bubble produces an outward pressure that causes the universe to expand with increasing velocity. While the Big Bang problem reduces to a fixed mass with dark matter and fixed energy, the new approach suggests adiabatic throttling into "virgin" vacuum (no dark matter). The new theory also implies a non-adiabatic closed system (i.e., fixed dark matter and other observable mass) universe.

A recent image obtained by Chandra X-ray Observatory shows a cosmic "cloud system" with pressure fronts in the system and they show a colder 50 million degree central region embedded in an elongated cloud of 70 million degree gas all of which is muddied in an "atmosphere" of 100 million degree gas. The size of the cosmic cloud system containing hundreds of galaxies is six million light years across. There is enough gaseous matter to make more galaxies. The galaxies may collide and merge over billions of years and release a large amount of energy, which can heat the cluster gas.

The universe is also presumed to have a "nonzero" substance called "aether" with non-zero levels of energy, pressure, and density; it is also the medium of transmission of light and electromagnetic radiation. From the thermodynamics perspective, a question arises if we can we apply the stability criteria developed so far to the "cosmic clouds" or to the universe with a modified real gas state equation with a generalized power law for attractive forces. The following is a hypothetical analysis of the authors. First, we will assume that the universe obeys the state equation

$$P = R'T/(v-b) - a/T^n v^m, \tag{54}$$

where P denotes the repulsive pressure. It is proportional to energy content per unit volume of free space minus the reduction in pressure due to attractive forces between matter. The variable b refers to the volume of the matter per unit mass, **a** is the attractive force constant, R′ is a proportionality constant that is not necessarily equal to universal gas constant, and the "pseudo-temperature" T is proportional to the sum of the average translational and rotational energies. The attractive force model between clouds is assumed be similar to the model derived for real gases (Appendix of Chapter 6) except that the power law is modified. Since the boundary of the universe is at zero pressure, any small positive pressure will expand the universe and vice versa. Thus if P= 0, forces are balanced or the universe is flat. From Eq. (54),

$$T^{(n+1)} = (a\,(v-b)/\,(R'v^m)) \approx = (a\,/\,(R'v^{m-1}))\ \text{if } v \gg b.$$

For the Berthelot type of equation, n=1, m=2,

$$T = \pm(a / (R'v))^{1/2}.$$

The flat universe solutions seem to suggest both positive and negative temperatures. While we rule out negative temperatures for planetary matters, we are not sure about the dark matter or "aether". Remember that if $T<0$, the first term in Eq. (54) causes attraction while the second term causes repulsion. Since $T = T(v)$ for a flat universe, $u= u(T)$ and $u_0(a=0,T)-u(T) = (2 (aR'/v))^{1/2}$ or $2 R'T$. Further, $h = u + Pv = u$, since $P = 0$.

We now discuss what happens during expansion or contraction if the universe follows Eq. (54) at fixed values of u. One consequence is that an increase in volume will form liquid like or condensed subsections (i.e., regions in which the matter/dark matter/clouds is closely packed) within the universe. Alternately, is may be possible to form vapor like (or expanded) subsections in which the matter is loosely packed. An example of the stability of the universe follows. Note that a true stability model of the universe must include all planets in all the galaxies, comets, asteroids, space dust, cosmic clouds, etc.

---

q. *Example 17*

We will address the stability of mater in the universe through a crude "elementary" thermodynamic analysis. First, we will assume that the universe obeys the state equation

$$P = R'T/(v-b) - a/T^n v^m. \quad \text{(A)}$$

Where the attractive force model is assumed to be similar to the model derived for real gases (Appendix of Chapter 6) except that the power law is modified.. Due to increased pressure in the beginning, a cloud of mass starts expanding. We will derive the criterion for the stability of the system at a specified internal energy and mass in terms of the temperature and volume. What should be the range of values of n for which thermal stability maybe affected? What will be the change in average temperature with change in volume as the universe expands or contracts? Next, for the purpose of illustration, what will be the relation for T if the universe/cosmic clouds behaves as a Berthelot gas (m=2, n=1)?

Solution

Using Eq. (A)

$$(\partial P/\partial T)_v = R'/(v-b) + n\, \mathbf{a}/T^{n+1}\, v^m. \quad \text{(B)}$$

Recall that

$$du = c_v\, dT + (T\, (\partial P/\partial T)_v - P)\, dv. \quad \text{(C)}$$

Note that this expression yields only change in the value of u for the universe due to a change in the random energy (the dT term) and ipe (the dv term)
At a specified value of u,

$$(\partial T/\partial v)_u = -((T\, (\partial P/\partial T)_v - P)/c_v\, (u,v)). \quad \text{(D)}$$

Substituting Eq. (B) in Eq. (D) we obtain the relation

$$(\partial T/\partial v)_u = -\, \mathbf{a}\, (n+1)/(T^n\, v^m\, c_v\, (u,v)). \quad \text{(E)}$$

Similarly,

$$(\partial s/\partial v)_u = P/T = R'/(v-b) - \mathbf{a}/(T^{n+1}\, v^m). \quad \text{(F)}$$

Differentiating Eq. (F) with respect to v

$$\partial^2 s/\partial v^2 = -R'/(v-b)^2 + m\mathbf{a}/(T^{n+1}v^{m+1}) + (\mathbf{a}(n+1)/(T^{n+2}\,v^m))\,(\partial T/\partial v)_u. \quad \text{(G)}$$

Substituting Eq. (D) in Eq. (G) we have

$$\partial^2 s/\partial v^2 = -R'/(v-b)^2 + m\,\mathbf{a}/(T^n v^{m+1}) - (\mathbf{a}\,(n+1)/(T^{n+1}\,v^m))^2/c_v(u,v)). \quad \text{(H)}$$

Mechanical Stability:
For a system to be stable, $\partial^2 s/\partial v^2 < 0$. Multiplying Eq. (H) by $T^{(2n+2)}v^{(2n+1)}$ and applying this criterion,

$$m\,\mathbf{a}\,T^{(n+1)}\,v^m < (\mathbf{a}^2(n+1)^2\,v/(c_v(u,v))) + R'\,v^{(2m+1)}\,T^{(2n+2)}/(v-b)^2.$$

Assuming that m =2, n =0 (i.e., assuming that the universe behaves as a VW gas,

$$2\,\mathbf{a}\,T\,v^2 < (\mathbf{a}^2\,v/(c_v(u,v))) + R'\,v^5\,T^2/(v-b)^2, \quad \text{(I)}$$

which is a quadratic equation in terms of T. For a VW gas $c_v = c_{v0}(T)$, and

$$2\,\mathbf{a}\,T\,v^2 < (\mathbf{a}^2\,v/(c_{v0}(T))) + R'\,v^5\,T^2/(v-b)^2. \quad \text{(J)}$$

A quantitative criterion for stability can be provided by Eq. (J). (The current temperature in the universe is about 3 deg. K.) For a VW gas

$$u - u_{ref} = c_{v0}(T - T_{ref}) - \mathbf{a}((1/v) - (1/v_{ref})). \quad \text{(K)}$$

Assume that $u_{ref} = -c_{v0}\,T_{ref} + \mathbf{a}/v_{ref}$, i.e.,

$$u = c_{v0}\,T - \mathbf{a}\,/v, \text{ so that} \quad \text{(L)}$$

$$T = (u + \mathbf{a}\,/v)/c_{v0}. \quad \text{(M)}$$

Note that the temperature as defined in real gas state equations (e.g., Eq. (M)) is proportional average energy per molecule. Amo Penzios and Robert Wilson of Bell Labs found that the universe emits radiation with a pattern similar to a blackbody radiation of a container at temperature 3 K (more precisely 2.726 K). The universe has cooled to the current temperature from 10^{32} K due to expansion from the Big Bang which occurred about 12 billion years back. Thus $\sigma_{SB}T_{univ}^4 = u_{rad}$, where σ_{SB} is the Stefan-Boltzmann constant (5.67×10^{-11} kWm^{-2}K^{-4}). Once the temperature is known, u can be determined using Eq. (M). The temperature decreases with increasing volume and vice versa.

Thermal Stability:
Consider the relations

$$(\partial c_v/\partial v)_T = T(\partial P/\partial T^2)_v,$$

$$(\partial P/\partial T)_v = R'/(v-b) + na/T^{n+1}\,v^m, \; m>0, \text{ and}$$

$$(\partial^2 P/\partial T^2)_v = -\,n(n+1)a/T^{n+2}\,v^m, \; m>0.$$

Note that **a** can be evaluated in terms of force field constants as described in Chapter 6 in context of the VW equation of state. Hence, $\partial c_v/\partial v = -\,n(n+1)a\,/T^{n+2}v^m$. In order that $c_v > c_{v0}$, this slope must be negative. Otherwise, $c_v < c_{v0}$ as the volume is decreased and the slope may become negative causing thermal instability, thereby creating a sun and a freezing system with a positive entropy generation. Therefore, $n(n+1) > 0$. If $n > 0$, $n(n+1)$ is always positive. If $(n+1) < 0$ i.e, $n < -1$, then $n(n+1) > 0$. Thermal stability is affected in the universe if $-1 < n < 0$. It is possible to create a warming planet and a freezing planet simultaneously as the volume is decreased if $-1 < n < 0$ below certain volumes.

Adiabatic Throttling of the Universe:
As the volume of the adiabatic universe changes (in the context of the Big Bang theory) in the empty space, energy must remain constant. Recall from Chapter 7, Eqs. (165) and (166), that for adiabatic throttling at constant internal energy $(\partial T/\partial v)_u = -(T \partial P/\partial T - P)/c_v = -T^2(\partial/\partial T(P/T))_u/c_v = -(n+1)\, a/(c_v\, T^n\, v^m)$. If $c_v > 0$. Therefore, the average temperature decreases with an increase in volume and *vice versa* for any value of n (the inverse happens if $c_v < 0$). If c_v is constant, one can integrate to obtain $T^{(n+1)} = (n+1)^2\, a/(c_v\, (m-1)\, v^{(m-1)}) + C^{(n+1)}$ where $m \neq 1$ and C is a constant. Then $(T^{(n+1)} - C^{(n+1)}) = ((n+1)^2\, a/(c_v\, (m-1)\, v^{(m-1)}))$. Per this model, at constant values of u, as $v \to 0$, $T \to \infty$ and $T \to$ constant as $v \to \infty$, since attractive forces become negligible. If m=2, n=1 (Berthelot), then at constant values of u, $T^2 = (4\, a/(c_v\, v)) + C^2$. Then $(T^2 - C^2) = (4\, a/(c_v\, v))$ for adiabatic throttling of the universe. For continuum $C = T^{ig}$, $(T^2 - (T^{ig})^2) = (4\, a/(c_v\, v))$.

Remarks

Recall Eq. (C). If a fixed mass system undergoes a reversible adiabatic process, $du = -\delta w = -P\, dv = c_v\, dT + (T\, (\partial P/\partial T)_v - P)\, dv$ and, hence, $dT = -T\, (\partial P/\partial T)_v\, dv/c_v$. Using Eq. (B), $dT = -(RT/(v-b) + n\, \mathbf{a}/T^n\, v^m)dv\, /c$. While the numerator is always positive, a negative specific heat implies that $dT>0$ if $dv>0$ (expansion process). If the process is isometric, one can show that $dT = \delta q/c_v$ so that heat rejection will cause $dT>0$ implying negative specific heats.

For a continuum, the RMS of the microscopic thermal part of the energy (te+re+ve) is proportional to the temperature. Similarly, if one makes an approximation that the RMS otational energy for a dense cluster of objects is also proportional to the temperature, then it is possible to use the Maxwell–Boltzmann type of distribution for the various speeds of the objects within a cluster.

We quote "Because universal gravitation is, well, **universal**, *every* body in the universe attracts *every* other body. A true model of the system would not only include the solar system but also the thousands of asteroids, billions of comets, and every spec of dust in the solar system, every space probe that has ever been launched, and every passing atom of hydrogen, but it would also have to take into account things like the gravitational attraction of Alpha Centauri, M31, and the most distant quasar," http://math.bu.edu/INDIVIDUAL/jeffs/stability.html.

H. SUMMARY

Criteria for thermodynamic stability are derived at specified values of (U,V,m), (S,V,m), (H,P,m), (S,P,m), (T,V,m), and (T,P,m). The real gas state equations fail the stability criteria within certain ranges of volumes for specified values of the temperature, which is the result of phase change. Spinodal conditions are provided for fluids. Physical explanations of boiling and condensation, superheating and supercooling are provided. Finally, the state equations are extrapolated to explain the stability of the universe.

Chapter 11

11. CHEMICALLY REACTING SYSTEMS

A. INTRODUCTION

Chemical reactions result in a rearrangement of atoms among molecules. Examples of chemical reactions include photosynthesis, metabolism, combustion, petroleum refining, and plastics manufacture. Combustion occurs due to exothermic oxidation reactions.

Energy cannot be extracted from a system that is constrained unless the thermodynamic constraint is removed. fossil fuels such as coal, oil, and natural gas are stored below the earth's surface and are in thermal, mechanical and chemical equilibrium. If the fuels are allowed physical contact with air, i.e., if the chemical constraint is removed, it is possible to release chemical energy. The direction of heat flow is determined by a temperature gradient or potential, that of fluid flow by a pressure gradient or a mechanical potential. We will learn in this chapter that the gradient of the Gibbs energy or chemical potential determines the direction in which a chemical reaction proceeds.

B. CHEMICAL REACTIONS AND COMBUSTION

1. Stoichiometric or Theoretical Reaction

A stoichiometric or a theoretical reaction results in the complete combustion of fuel as illustrated below. Consider the reaction

$$CH_4 + a\ O_2 \rightarrow b\ CO_2 + c\ H_2O \tag{1}$$

involving one kmole of fuel and "a" kmoles of molecular oxygen. The species on the left hand side of Eq. (1) are usually called reactants (which react and are consumed during the overall chemical reaction) and those on the right hand side are termed products (which are produced as a result of the chemical reaction). In the case of a steady flow reactor, the input and output streams contain the following quantity in kmole of the elements (C, H, and O).

	Input	=	Output
C	1	=	b
H	4	=	2 c
O	2a	=	2b + c

We deduce from the carbon atom balance that b = 1, from the H atom balance that c = 2, and from the O atom balance that a = (2b+c)/2. Consequently, Eq. (1) can be written in the form of the stoichiometric reaction

$$CH_4 + 2\ O_2 \rightarrow CO_2 + 2\ H_2O. \tag{2}$$

Consider, for the sake of illustration, the combustion of 10 kmole of CH_4, 25 kmole of O_2, 5 kmole of CO_2 and 3 kmole of H_2O during which the fuel completely burns. Thus only 10 kmole of CH_4 are consumed in the reactor. Applying Eq. (2), we readily determine that 20 kmole of O_2 are utilized during the process, and hence 5 kmole of O_2, 15 (=10 + 5) kmole of CO_2 and 23 (= 20+3) kmole of H_2O leave the combustor. The stoichiometric coefficients do not necessarily represent the amounts of species entering or leaving a reactor.

Note that when air is to supply the oxidizer, each kmole of O_2 is associated with 3.76 kmole of molecular nitrogen. During the combustion of methane in air, the stoichiometric air–to–fuel (A:F) ratio is

$$(A:F)_{mole} = (2 + 2\times3.76)\div1 = 9.52 \text{ kmole of air to 1 kmole of fuel.} \tag{3}$$

On mass basis, the A:F ratio is

$$(A{:}F)_{mass} = (2\times32 + 2\times3.76\times28)\div(1\times16) = 17.16 \text{ kg of air/kg of fuel.} \quad (4)$$

More accurately, air is a mixture containing 78% N_2, 1% Ar and 21% O_2. If the argon is included,

$$CH_4 + 2\,(O_2 + 3.71\,N_2 + 0.05\,Ar) \rightarrow CO_2 + 2\,H_2O + (7.42\,N_2 + 0.1\,Ar). \quad (5a)$$

In this case, $(A{:}F)_{mole} = 9.52$, but on a mass basis it changes to

$$(A{:}F)_{mass} = (2\times32 + 7.42\times28 + 0.1\times40) \div 1617.24 = 17.24.$$

2. Reaction with Excess Air (Lean Combustion)

Fuel and air are often introduced separately (i.e., without premixing) into a combustor, e.g., in a boiler. Due to the large flowrates and short residence times there is no assurance that each molecule of fuel is surrounded by the appropriate number of oxygen molecules required for stoichiometric combustion. Therefore, it is customary to supply excess air in order to facilitate better mixing and thereby ensure complete combustion. The excess oxygen remains unburned and appears in the products, e.g.,

$$CH_4 + 3\,(O_2 + 3.76\,N_2) \rightarrow CO_2 + 2\,H_2O + O_2 + 11.28\,N_2. \quad (5b)$$

In this case, the A:F ratios are

$$(A{:}F)_{mole} = (3\times3.76+3)\div1=14.28, \text{ and } (A{:}F)_{mass} = (3\times3.76\times28+3\times32)\div16=25.74.$$

The excess air percentage (mole % basis which is equally valid for mass % basis) is

$$((A{:}F) - (A{:}F)_{stoich})\times100\div(A{:}F)_{stoich} = (3-2)\times100\div2 = 50\,\%.$$

3. Reaction with Excess Fuel (Rich Combustion)

Incomplete combustion occurs when the air supplied is less than the stoichiometric amount required. For this condition, the products of incomplete oxidation may contain a mixture of CO, CO_2, H_2, and H_2O.

4. Equivalence Ratio, Stoichiometric Ratio

The equivalence ratio

$$\phi = (F{:}A)/(F{:}A)_{stoich} = (A{:}F)_{stoich}/(A{:}F) = (O_2{:}F)_{stoich}/(O_2{:}F) \quad (6)$$

As an example, if $\phi = 0.5$ for methane–air combustion, this implies that the excess is air supplied for every kmole of fuel that is burned. In general, for methane–air combustion

$$CH_4 + (2/\phi)(O_2 + 3.76\,N_2) \rightarrow CO_2 + 2H_2O + 2((1/\phi)-1)\,O_2 + (2/\phi)\times3.76N_2, \text{ or} \quad (7)$$

Another term used to represent the fuel–air mixture composition is the stoichiometric ratio

$$SR = (\text{actual air supplied})/(\text{stoichiometric air demand of fuel}) = 1/\phi.$$

For instance, if $\phi = 0.5$, then $SR = 2$, i.e., the air supplied is two times as large as the stoichiometric or theoretical air demand of the fuel.

---

a. Example 1

Consider the metabolism of glucose in the human body. As we breathe in air, we transfer oxygen from our lungs into our bloodstream. That oxygen is transported to the cells of our tissues where it oxidizes glucose. Write down the stoichiometric reaction for the consumption of glucose (s) $C_6H_{12}O_6$ by pure oxygen and by air. Determine the amount of air required if 400% excess air is involved, express (A:F) in terms of the percentage of theoretical air and the equivalence ratio, and write the associated reaction equation. If the human breathing rate ≈360 L(STP) hr^{-1}, how much glucose is consumed per minute?

Solution

The stoichiometric or theoretical reaction equation for this case is

$$C_6H_{12}O_6 + a\ O_2 \rightarrow b\ CO_2 + c\ H_2O \quad (A)$$

Applying an atom balance,

Carbon C: 6 = b, (B)

Hydrogen H: 12 = 2 c, i.e., c = 6, and (C)

Oxygen O: 6 + 2a = 2 b + c, i.e., a = 6. (D)

Therefore, the stoichiometric relation assumes the form

$$C_6H_{12}O_6 + 6\ O_2 \rightarrow 6\ CO_2 + 6\ H_2O. \quad (E)$$

The corresponding stoichiometric or theoretical reaction equation in the case of air is

$$C_6H_{12}O_6 + 6\ (O_2 + 3.76\ N_2) \rightarrow 6\ CO_2 + 6\ H_2O + 22.56\ N_2. \quad (F)$$

When excess air is supplied,

400% excess air = (supplied air–stoichiometric air)×100/stoichiometric air

= (supplied air – (6 + 22.56)) × 100 ÷ (6+22.56), i.e.,

the supplied air = 142.8 kmole of air/kmole of glucose. (G)

Therefore, the excess air as a percentage of theoretical air equals

$$(A{:}F/(A{:}F)_{stoich}) \times 100 = 142.8 \div 28.56 = 500\%. \quad (H)$$

The equivalence ratio

$$\phi = ((A{:}F)_{stoich}/(A{:}F)) = 28.56 \div 142.8 = 0.2. \quad (I)$$

The actual reaction equation with 400% excess air or five times the theoretical air is

$$C_6H_{12}O_6 + 5\times6(O_2 + 3.76\ N_2) \rightarrow 6CO_2 + 6\ H_2O + a\ O_2 + b\ N_2. \quad (J)$$

Applying the atom balance for

Oxygen O: 5×6×2 = 6×2 + 6×1 + 2a, i.e., a = 24, and (K)

Nitrogen N: b = 30×2×3.76÷2 = 112.8, i.e., (L)

$$C_6H_{12}O_6 + 5\times6\ (O_2 + 3.76\ N_2) \rightarrow 6CO_2 + 6\ H_2O + 24\ O_2 + 112.8\ N_2. \quad (M)$$

The air consumption is specified as 360 L hr^{-1} or 0.0147 kmole hr^{-1}. Therefore, the glucose consumption per hour is

$$\omega = 0.0147 \div 142.8 = 0.0001029 \text{ kmole hr}^{-1}. \quad (N)$$

The mass of one kmole of glucose is 180.2 kg, so that the glucose consumption per hour in terms of mass is

$$m = 00.0001029 \text{ kmole hr}^{-1} \times 180.2 \text{ kg kmole}^{-1} = 18.6 \text{ g hr}^{-1} \text{ or } 0.31 \text{ g min}^{-1}. \quad (O)$$

Remarks

The ratio of CO_2 to O_2 is called respiratory quotient (RQ) in the medical literature. For glucose, the stoichiometric reaction indicates that RQ =1. Fats (e.g., palmitic acid) are also used by the body for metabolism. The stoichiometric reaction for fat is given as

$$C_{15}H_{31}COOH + 23\, O_2 \rightarrow 16CO_2 + 16\, H_2O.$$

The RQ is given as 0.7. Since older persons have problems with excreting CO_2, then fats are preferable compared to glucose due to lower RQ values.

---

5. Dry Gas Analysis

The products of combustion are analyzed on a volumetric or molar basis in order to determine if combustion is complete. In the context of Eq. (5b) the percentage of CO_2 in the products equals $(1\times100\div(1+2+1+11.28) = 7\%$. This is an example of a wet analysis, since the products also include water vapor. If the combustion products are analyzed with water removed as a constituent, this is called a dry gas analysis. Again, applying Eq. (5b), the CO_2 percentage on a dry basis is $1\times100/(1+1+11.28) = 7.53\%$. The volume percentages of the constituents of an ideal gas mixture are identical to their molar percentages. The measured dry or wet gas compositions can be used to determine (A:F) as illustrated in the following example.

---

b. Example 2

A dry gas analysis of the gas exhaled by a human lung is as follows– O_2:16.5% and CO_2:3.1%. Assume the "fuel" burned by humans is characterized by the chemical formula CH_x and is completely burned. Determine the values of "x" and (A:F).

Solution:

If the reactant CH_x and is completely consumed, the exhaled products consist of only CO_2 H_2O, N_2 and O_2. The stoichiometric relation for this chemical activity is

$$CH_x + a\, O_2 + b\, N_2 \rightarrow c\, CO_2 + d\, H_2O + e\, O_2 + f\, N_2. \quad (A)$$

There are seven unknowns, namely x, a, b, c, d, e, f. From an atom balance of the elements C, H, O, and N we obtain the following four equations, i.e.,

$$C: 1 = c, \quad (B)$$

$$H: x = 2\, d, \quad (C)$$

$$O: 2\, a = 2\, c + d + 2\, e, \text{ and} \quad (D)$$

$$N: 2\, b = 2f, \text{ where} \quad (E)$$

$$b/a = 3.76. \quad (F)$$

From the dry gas analysis:

$$\text{Percentage of } CO_2 = c \times 100 \div (c + e + f) \quad (G)$$

$$\text{Percentage of } O_2 = e \times 100 \div (c + e + f) \quad (H)$$

The seven equations are obtained using the seven equations Eqs. (B)–(G), i.e.,

$c = 1$, (I)

$d = x/2$, (J)

$e = (2\ a - 2\ c - d)/2 = a - 1 - x/4$, (K)

$f = b = 3.76\ a$, (L)

(Percentage of CO_2)/(Percentage of O_2) = $c/e = 3.1 \div 16.5 = 0.19$, (M)

$c + e + f = c \div 0.031 = 32.26$, (N)

Hence,

$e = 1 \div 0.19 = 5.26$,
$f = b = 32.26 - 1 - 5.26 = 26$,
$a = 6.91$,
$5.26 = 6.91 - 1 - x/4$, i.e., $x = 2.6$, and
$d = 1.3$.

Consequently, the chemical relation assumes the form

$$CH_{2.6} + 6.91\ O_2 + 26\ N_2 \rightarrow CO_2 + 1.3\ H_2O + 5.26\ O_2 + 26\ N_2. \quad \text{(A)}$$

The air fuel ratio

$A{:}F = (6.91 + 26) \div 1 = 32.91$, and $(A{:}F)_{mass} = 65$.

For a stoichiometric reaction, the corresponding relation is

$$CH_{2.6} + 1.65\ O_2 + 1.65 \times 3.76\ \ N_2 \rightarrow 1\ CO_2 + 1.3\ H_2O + O_2 + 1.65 \times 3.76\ N_2, \text{ and}$$

$(A{:}F)_{mass} = 16$.

Remarks

Once the exhaust gas composition is known, the fuel used and the A:F ratio for the combustion process can be determined. Modern gas analyzers that incorporate the appropriate software for determining (A:F) are widely used.

For the sake of illustration, consider the oxidation of the following fuel

$CH_mO_n\ N_pS_q$+ air $\rightarrow$ (1-e) CO_2, e CO, H_2O, SO_2, O_2, N_2, Ar.

For stoichiometric oxidation, $a_{stoich} = (1+q+m/4-n/2)$.

If the percentages CO_2 and CO are measured, then

N_{dry}, number of dry moles in products = $1/(x_{CO2} + x_{CO})$.

$e = x_{CO} \times N_{dry}$.

$a = (N_{dry} - (e/2 - m/4 + n/2 + p/2)) \times x_{O2,a}$, $\phi = a_{stoich}/a = (A{:}F)_{stoich}/(A{:}F)$.

Here, $x_{O2,a}$ denotes the ambient oxygen concentration (mole fraction), and a the actual oxygen content supplied.

If the percentages O_2 and CO are measured, then

$B = (x_{O2} - x_{CO}/2)/(1 - x_{CO}/2)$.

$a = (1+q+m/4-n/2 + B(n/2+p/2-m/4))/(1-B/x_{O2,a})$,

$\phi = a_{stoich}/a == (A{:}F)_{stoich}/A{:}F$.

$N_{dry} = (a/x_{O2,a} - m/4 + n/2 + p/2)/(1 - x_{CO}/2)$.

A generic wet gas analysis of human exhalation is as follows – N_2: 78%, O_2: 16%, CO_2: 3%, and H_2O: 3%.

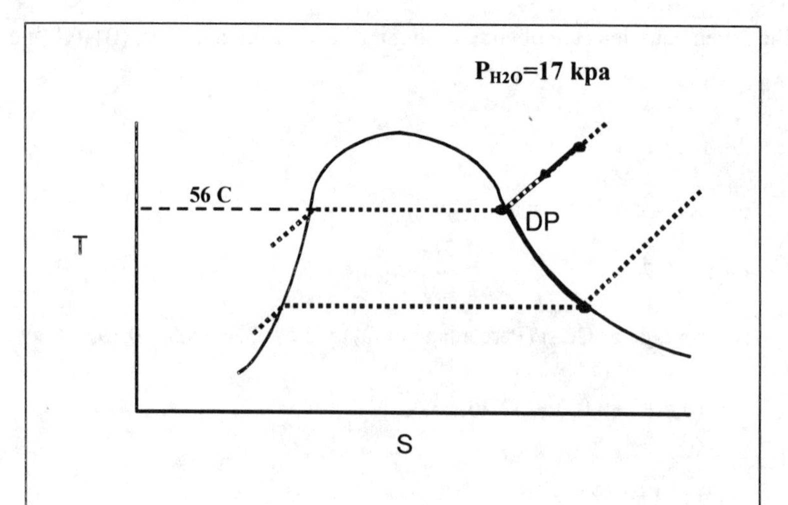

Figure 1: Schematic illustration of the determination of the dew point.

The wet exhaust from a diesel engine generally comprises N_2: 77%, O_2: 13.54%, CO_2: 5%, and H_2O: 4%.

---

c. *Example 3*

Consider the combustion of natural gas (which is assumed to have the same properties as CH_4). Determine (A:F) if the products contain 3% O_2 (on a dry basis), and the composition on both a wet and dry basis.

Solution:

The reaction equation is

$$CH_4 + a\,(O_2 + 3.76\,N_2) \rightarrow CO_2 + 2H_2O + b\,N_2 + d\,O_2 \quad (A)$$

From an O atom balance,

$$2\,a = 2 + 2 + 2\,d, \text{ or } a = 2 + d. \quad (B)$$

The exhaust contains 3% O_2 on a dry basis, i.e.,

$$d/(1 + 3.76\,a + d) = 0.03. \quad (C)$$

Therefore,

$$d = 0.2982, \text{ and} \quad (D)$$

$$a = 2.2982. \quad (E)$$

Also, from an N atom balance

$$b = 3.76\,a. \quad (F)$$

Therefore,

$$(A{:}F) = 2.2982 \times 4.76 \div 1 = 10.9, \text{ and} \quad (G)$$

Eq. (A) assumes the form

$$CH_4 + 2.3\,(O_2 + 3.76\,N_2) \rightarrow CO_2 + 2\,H_2O + 8.6\,N_2 + 0.3\,O_2. \quad (H)$$

The composition is as follows:

CO_2 % in exhaust (wet) = 100×(1÷(1+2+3.76×2.2982+0.2982)) = 8.4%,

CO_2 % in exhaust (dry) = 100×(1÷(1+3.76×2.2982+0.2982)) = 10.1%,

N_2 % in exhaust (wet) = 100×(3.76×2.2982÷(1+2+3.76×2.2982+0.2982)) = 72.4%,

N_2 % in exhaust (dry) = 100×(3.76×2.2982÷(1+3.76×2.2982+0.2982)) = 86.9%.

Remark

Using Eq. (H), the mole fraction of water on wet basis has a value equal to 2÷(1+2+8.6+0.3) = 0.17. Consequently, the partial pressure of water is 0.17 bar if the mixture pressure is atmospheric. At this pressure T^{sat} = T_{DP} = 56°C, where T_{DP} denotes the dew point temperature. Inversely, if T_{DP} is known, the water percentage and (A:F) can be determined (cf. Figure 1).

C. THERMOCHEMISTRY

1. Enthalpy of Formation (Chemical Enthalpy)

Elemental C(s) can be burned in O_2 at 25°C and 1 bar to form CO_2 The combustion reaction process is exothermic, but the product temperature can be reduced through heat transfer. The enthalpy of formation of CO_2 , h^o_{f,CO_2} in this context equals the amount of heat added or removed to form the compound at 25°C and 1 atm from its elemental constituents (which themselves exist in their natural state at that temperature and pressure). The subscript f denotes formation and the superscript o denotes that the value corresponds to a pressure of 1 atm.

Another example involves the formation of NO from elemental nitrogen and oxygen in their natural form at STP (at that condition nitrogen exists as N_2 and oxygen in the form of O_2). In this case.

$$1/2\ N_2 + 1/2\ O_2 \rightarrow NO + 90{,}592 \text{ kJ kmole}^{-1} \text{ of NO}$$

This chemical reaction requires a heat input of 90,592 kJ kmole^{-1}. Using the First law,

$$Q = \bar{h}^0_{f,NO} = 90{,}512 \text{ kJ kmole}^{-1} = \bar{h}_{NO} - (1/2\ \bar{h}_{N_2} + 1/2\ \bar{h}_{O_2}). \tag{8}$$

It is apparent that $\bar{h}_{NO} > (1/2\ \bar{h}_{N_2} + 1/2\ \bar{h}_{O_2})$. If we arbitrarily set $\bar{h}_{N_2} = \bar{h}_{O_2} = 0$, then $\bar{h}_{NO} = \bar{h}^o_{f,NO}$ = 90,592 kJ kmole^{-1}, which implies that the energy contained in NO at 298 K and 1 atm is 90,592 kJ more than the energy associated with N_2 and O_2 . This energy can be interpreted as that absorbed or released when the atoms form the compound. (cf. Figure 2). See Tables A-27A and A-8 to A-19 for a tabulation of enthalpy of formation of various substances.

2. Thermal or Sensible Enthalpy

Once a compound is formed at 298 K and 1 atm, its temperature can be raised by adding heat isobarically, and, consequently, the enthalpy of that compound increases. The difference between the enthalpy at a temperature T and at 298 K is the thermal or sensible enthalpy

$$h^o_t = \int_{298}^{T} c^o_p dT, \tag{9}$$

where c^o_p refers to the specific heat of the species at 1 bar. (For an ideal gas the superscript ° is redundant, since $c_p \neq c_p(P)$.) As an example, NO requires 22,230 kJ kmole^{-1} of energy to raise its temperature from 298 K to 1000 K (cf. Figure 2). In this context

$$h_{t,T} - h_{t,298} = \Delta h_{t,1000K} = \int_{298}^{1000} c^o_{p,k}\, dT = h^o_{t,1000K} - h^o_{t,298K}. \tag{10}$$

See Tables A-8 to A-19 for tabulation of $\Delta h_{t,T} = h_{t,T} - h_{t,298}$ of several ideal gases.

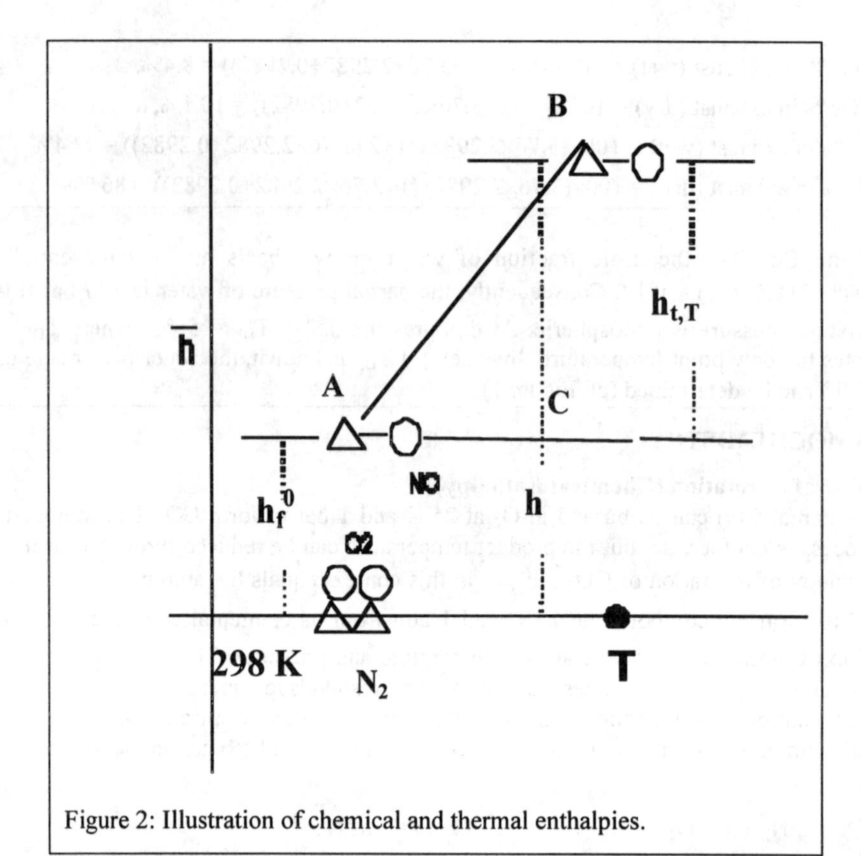

Figure 2: Illustration of chemical and thermal enthalpies.

3. Total Enthalpy

The total enthalpy of a species at any state (T,P) equals the value of its chemical enthalpy at 298 K at 1 atm (when species are formed from elements) plus the additional enthalpy required to raise the temperature of the same species from 298 K to T and pressure from 1 bar to P. At P = 1 bar (≈ 1 atm),

$$\bar{h}_k^{\,0}(T) = \bar{h}_{k,f}^{\,0} + \bar{h}_{t,\,298\to T} = \bar{h}_k^{\,0}(T) = \bar{h}_{k,f}^{\,0} + \int_{298}^{T} \bar{c}_{p0}\, dT \qquad (11)$$

In the case of NO at 1000 K and 1 atm, the total enthalpy is

$$\bar{h}^{o}_{f,NO}(1000\text{ K}) = \bar{h}^{o}_{f,NO} + (\bar{h}^{o}_{f,NO,1000K} - \bar{h}^{o}_{f,NO,298K}) = 90592 + 22230 = 112822 \text{ kJ kmole}^{-1}.$$

Figure 2 schematically illustrates how the total enthalpy can be determined. This concept is useful when considering chemical reactions.

Now consider the pressure effect. The enthalpy of a species at an arbitrary pressure and temperature equals its enthalpy at that temperature and a 1 bar pressure and the additional enthalpy required (or extracted) to change the pressure to P, i.e.,

$$h(T,P) = h^o(T) + \Delta h_{T,1\to P}, \text{ where} \qquad (12)$$

$$h^o(T) = h_f^{\,o} + h^o_{t,\,298\to T}, \text{ and} \qquad (13)$$

Recall from Eq. (43) of Chapter 7 that for any pure substance we had defined the change in enthalpy $dh = c_P\, dT + (v - T(\partial v/\partial T)_P)\, dP$ = change in thermal part of enthalpy + change in intermolecular potential part of enthalpy. Note that this expression does not contain $\bar{h}_{k,f}^0$ since Eq.(43) of Chapter 7 expresses only the change in enthalpy of given species k but not the absolute value of enthalpy of species k. Thus the term $\Delta h_{T,1\to P}$ in Eq. (12) represents the change in intermolecular potential part of enthalpy as pressure is raised from 1 to P bars for the specified species. The term $\Delta h_{T,1\to P}$ represents a correction for real gas behavior. For ideal gases $\Delta h_{T,1\to P} = 0$.

4. Enthalpy of Reaction

The amount of heat added ($\Delta H_R < 0$) or removed ($\Delta H_R > 0$) when the reactants enter and the products leave an isothermal system is called the enthalpy of reaction, i.e.,

$$\Delta H^o_{R,T} = H^o_{Products,T} - H^o_{Reactants,T}. \quad (14)$$

For instance, consider the oxidation of CO to CO_2 at 298 K and 1 atm, i.e.,

$$CO + 1/2\ O_2 \rightarrow CO_2,$$

for which $H_{Products} = 1 \times h_{CO2} = -\ 393{,}520$ kJ, and $H_{Reactants} = 1 \times h_{CO} + 1/2\ O_2 = -110{,}530$ kJ. Therefore,

$$\Delta H^o_{R,T} = -282{,}990 \text{ kJ (kmole of CO)}^{-1}, \text{ i.e.,}$$

when a kmole of CO is burned 282,990 kJ of heat must be removed in order for the products to leave at the same temperature as the reactants (298 K). since the values of the enthalpy of reaction are normally tabulated at a temperature of 298 K, the subscript T=298 K is omitted. If a reaction involves either oxidation or combustion, then the enthalpy of reaction ΔH^o_R is termed as the enthalpy of combustion ΔH^o_C.

5. Heating Value

The heating value of a fuel

$$HV = -\Delta H^o_C = H_{Reac} - H_{prod}. \quad (15)$$

The lower heating value

$$LHV = H_{Reac} - H_{prod,\ H2O(g)}$$

when the products are assumed contain water in gaseous form, and the higher or gross heating value

$$HHV = H_{Reac} - H_{prod,\ H2O(\ell)}$$

when the products are assumed to contain water in liquid form. Generally, HV, ΔH^o_R, and ΔH^o_C are tabulated at 298 K and 1 atm.

For gaseous fuels the heating values are stated on a volumetric basis. If the HV on a mole basis is known, one can obtain HV′ based on a volumetric basis using the relation HV′ = HV in kJ/24.5 m^3 or HV' = HV in BTU/lb mole /392 ft^3.

---

d. Example 4

In the context of the combustion of methane determine the values of HHV, LHV, ΔH^o_R, and ΔH^o_C at 298 K. Further, if biogas is assumed to consist of 60% methane and 40% carbon dioxide (volume %), what is its higher heating value at 298 K?

Solution

The stoichiometric combustion of methane can be represented by the chemical reaction

$CH_4 + 2\ O_2 \rightarrow CO_2 + 2\ H_2O$, i.e., (A)

$$- HHV = \Delta H_C^o = \Delta H_R^o = H^o_{Products} - H^o_{Reactants} = h^o_{f,CO_2} + 2h^o_{f,H_2O} - h^o_{f,CH_4}, \text{ i.e.,} \quad (B)$$

$- HHV_{CH_4} = \Delta H_C^o = \Delta H_R^o$

$= (-393{,}520$ kJ kmole^{-1}. of CO_2)+2×(−285,830 kJ kmole^{-1}. of H_2O) –
(−74,850 kJ kmole^{-1}. of CH_4) =−890,330 kJ kmole^{-1} of fuel

HHV_{CH_4} = 890,330 kJ kmole^{-1} of CH_4,

HHV'= 890330/24.5 = 36340 kJ/m^3 of CH_4

$LHV_{CH_4} = -\Delta H^o_{R,H2O(g)}$ = =[(−74,850 kJ kmole^{-1}. of fuel) – (−393,520 kJ kmole^{-1} of CO_2) - (2 kmole of H_2O(g)/kmole of fuel)×(−241,820 kJ kmole^{-1}. of H_2O(g))] = 802310 kJ kmole^{-1} of fuel

LHV' = 802310/24.5 = 32747 kJ m^{-3}.

Since HHV of N_2 =0, HHV_{Biogas} = 0.60 × 36340 + 0.4 × 0 = 21804 kJ m^{-3}.

Remarks

For methane, ΔH_R^o = – 890330 kJ kmole^{-1}. of fuel represents the amount of heat that is to be removed for the products to exit a combustor at 298 K. If this amount of heat is not removed, the temperature of the products rises, which is characteristic of combustion.

At 25°C, for any fuel the LHV = (HHV – moles of H_2O produced per mole of fuel × h_{fg} × 18.02 ÷ moles of fuel). For instance, in the case of methane at 298 K the LHV = (890330 – 2×2442.3×18.02) = 802,310 kJ kmole^{-1} of fuel.

The HHV per unit mass of O_2 for most hydrocarbon fuels is approximately the same. For methane, this value equals (55472÷4) = 13870 kJ kg^{-1} of O_2, in the case of gasoline (i.e., $CH_{2.46}$), the value is (48304÷(1.615×32÷14.5) = 13550 kJ kg^{-1} of O_2, and for n–octane it is (47880÷(8+4.5)×32÷114 = 13640 kJ kg^{-1} of O_2.

6. Entropy, Gibbs Function, and Gibbs Function of Formation

The discussion about the enthalpies of reacting species is useful for applying the First law of thermodynamics Now we will introduce methodologies for determining the entropy, Gibbs function and other species properties, which are useful in the application of the Second law.

Since for an ideal gas,

$$ds = c_{p0}\ dT/T - R\ dP/P,$$

then for a pure component

$$s_k(T,P) - s_k(T_{ref},1) = s_k^0(T) - R \ln (P/1), \text{ where} \quad (16)$$

$$s_k^0 = \int_{T_{ref}}^{T} (c_{p.k}/T)dT$$

It is usual to select the conditions T_{ref} = 0 K, P = 1 bar, and $s_k(T_{ref},1) = 0$ for an ideal gas. The Gibbs function for a pure component k is

$$\bar{g}_k = \bar{h}_k - T\ \bar{s}_k. \quad (17)$$

For instance, the Gibbs function under these conditions for molecular hydrogen is

$$\bar{g} = \bar{h} - T\ \bar{s} = 0 - 298\times130.57 = -38910 \text{ kJ kmole}^{-1}. \text{ of } H_2,$$

and in the case of water it is

$$\bar{g} = \bar{h} - T\bar{s} = -285830 - 298\times69.95 = -306675 \text{ kJ kmole}^{-1}. \text{ of } H_2O.$$

Since combustion problems typically involve mixtures, the entropy of the k–th component in a mixture must be first determined, e.g., following the relation

$$\hat{s}_k(T,P,X_k) = s_k^o(T) - R \ln (p_k/1), \text{ where } p_k = X_k P.$$

The Gibbs function of the k–th mixture component $\hat{g}_k$ can be obtained by applying the expression

$$\hat{g}_k = \hat{h}_k - T\hat{s}_k.$$

For ideal gases.

$$\hat{h}_k = \bar{h}_k, \text{ and } \bar{h}_k = \bar{h}_{f,k} + \Delta\bar{h}_k.$$

Alternately, the values of $\bar{g}_k$(298K, 1 bar) for any compound can also be determined by ascertaining the Gibbs function of formation $\bar{g}_{k,f}$ under those conditions. The value of $\bar{g}_{k,f}$ is identical to the Gibbs energy of formation ΔG for a reaction that forms the compound from its elements that exist in a natural form. For instance, in the case of water,

$$\bar{g}_{f,H_2O} = \Delta G^o_{R,H_2O} = \bar{g}_{H_2O}(298\text{K},1 \text{ bar}) - \bar{g}_{H_2}(298 \text{ K},1 \text{ bar}) - \bar{g}_{O_2}(298 \text{ K},1 \text{ bar}). \quad (18)$$

The value of $\bar{g}_{k,f}$(298K, 1 bar) is assigned as zero for elements that exist in their natural form. Table A-27A contains values of $\bar{g}_{k,f}$(298K, 1 bar) of many substances.

---

e. Example 5

Determine the Gibbs energy for water at 25°C and 1 bar. Under those conditions, also determine the change in the Gibbs energy when $H_2O(\ell)$ is formed from its elements, and the Gibbs energies of formation for $H_2O(\ell)$ and of $H_2O(g)$.

Solution

$$\bar{g}_{H_2O(\ell)} = \bar{h}_{H_2O(\ell)} - T\bar{s}_{H_2O(\ell)} = -285830 - 298 \times 69.95 = -306675 \text{ kJ kmole}^{-1}.\text{(A)}$$

For the chemical reaction

$$H_2 + 1/2\ O_2 \rightarrow H_2O(\ell), \quad \text{(B)}$$

$$\Delta G = \bar{g}_{H_2O(\ell)} - \bar{g}_{H_2} - 1/2\ \bar{g}_{O_2}, \text{ where,} \quad \text{(C)}$$

$$\bar{g}_{H_2} = \bar{h}_{H_2} - T\bar{s}_{H_2} = 0 - 298\times130.57 = -38910 \text{ kJ kmole}^{-1}, \text{ and} \quad \text{(D)}$$

$$\bar{g}_{O_2} = 0 - 298\times205.04 = -61102 \text{ kJ kmole}^{-1}. \quad \text{(E)}$$

Applying Eqs. (A), (D), (E) in Eq. (C),

$$\Delta G = -306675 - (-38910) - 0.5\times(-61102) = -237{,}214 \text{ kJ kmole}^{-1}. \quad \text{(F)}$$

Therefore, the Gibbs function of a kmole of $H_2O(\ell)$ is 237214 kJ lower than that of its elements (when they exist in a natural form at 25°C and 1 bar). As in the case of the enthalpy of formation, the Gibbs function of all elements in their natural form can be arbitrarily set to zero (i.e., $\bar{g}_{f,H_2} = 0$, $\bar{g}_{f,O2} = 0$). Thereafter, the Gibbs energy of formation is

$$\bar{g}_{f,\,H_2O(\ell)} = \Delta G = -237214 \text{ kJ kmole}^{-1}. \quad (G)$$

For water vapor at an arbitrary temperature and pressure,

$$\bar{g}_{f,\,H_2O(g)}(T,P) = \bar{h}_{H_2O(g)}(T,P) - T\,\bar{s}_{H_2O(g)}(T,P). \quad (H)$$

If the vapor is assumed to be an ideal gas, then

$$\bar{h}_{H_2O(g)}(T,P) \approx \bar{h}_{H_2O(g)}(T,P^{sat}) = \bar{h}_{H_2O(\ell)}(T,P^{sat}) + \bar{h}_{fg}(T,P^{sat}). \quad (I)$$

Similarly,

$$\bar{s}_{H_2O(g)}(T,P) = \bar{s}_{H_2O(g)}(T,P^{sat}) - R \ln\ (P/P^{sat})$$

$$= \bar{s}_{H_2O(\ell)}(T,P^{sat}) + h_{fg}(T,P^{sat})/T - R \ln (P/P^{sat}). \quad (J)$$

Applying Eqs. (I) and (J) in Eq. (H), we obtain

$$\bar{g}_{H_2O(g)}(T,P) = \bar{h}_{H_2O(\ell)}(T,P^{sat}) - T\ \bar{s}_{H_2O(\ell)}(T,P^{sat}) + RT \ln (P/P^{sat}). \text{ i.e.,}$$

$$\bar{g}_{H_2O(g)}(T,P) = \bar{g}_{H_2O(\ell)}(T,P^{sat}) + \bar{R}T \ln (P/P^{sat}). \quad (K)$$

If water is treated as an incompressible fluid, then,

$$dg_{H_2O(g)} = v_f\, dP.$$

Integrating this expression between the limits P^{sat} and P

$$g_{H_2O(\ell)}(T,P) - g_{H_2O(\ell)}(T,P^{sat}) = -v_f(P-P^{sat}), \text{ i.e.,}$$

$$g_{H_2O(g)}(T,P) = (g_{H_2O(\ell)}(T,P) - v_f(P-P^{sat})) + RT \ln (P/P^{sat}). \quad (L)$$

Typically, $v_f\,(P-P^{sat}) \approx 0$ so that Eq. (L) assumes the form

$$g_{H_2O(g)}(T,P) \approx g_{H_2O(\ell)}(T,P) + RT \ln(P/P^{sat}). \quad (M)$$

for water at 298 K, P^{sat} = 0.03169 bar. Therefore, at 1 bar water does not exist as gas. However, we can define a hypothetical state for water as an ideal gas at 25°C and 1 bar and thereafter use Eq. (K) to determine $g_{H_2O(g)}$, i.e.,

$$\bar{g}_{H_2O(g)}(298\text{K},1 \text{ bar}) \approx \bar{g}_{H_2O(\ell)}(298\text{K}, 1 \text{ bar}) + RT \ln (1/0.03169), \text{ so that}$$

$$\bar{g}_{H_2O(g)} - \bar{g}_{H_2O(\ell)} = 8.314\times 298\times \ln (1/0.03169) = 8552 \text{ kJ kmole}^{-1}.$$

Therefore,

$$\bar{g}_{f,\,H_2O(g)} = \bar{g}_{f,\,H_2O(\ell)} + 8552 = 228662 \text{ kJ kmole}^{-1}.$$

Remark

Generalizing Eq. (L) for any ideal gas species,

$$g_{k(g)}(T,P) = (g_{k(\ell)}(T,P) - v_{f,k}(P-P^{sat})) - RT \ln (P/P^{sat}) \quad (N)$$

D. FIRST LAW ANALYSES FOR CHEMICALLY REACTING SYSTEMS

1. First Law

Consider a control volume (illustrated in Figure 3) whose boundary is penetrated by k material streams that bring in various species at the rates $\dot{N}_1$, $\dot{N}_2$, $\dot{N}_3$... into the volume. The chemical processes within the CV are governed by the First law of thermodynamics, i.e.,

$$dE_{cv}/dt = \dot{Q}_{cv} - \dot{W}_{cv} + \Sigma_{k,i}\dot{N}_k\,\bar{e}_{T,k} - \Sigma_{k,e}\dot{N}_k\,\bar{e}_{T,k} \quad (19)$$

The methalpy $\bar{e}_{T,k} = (\bar{h} + \bar{ke} + \bar{pe})_k$ includes the enthalpy of formation, thermal enthalpy, as well as the kinetic and potential energies wherever appropriate. Methalpy, $\bar{e}_{T,k} = (\bar{h} + \bar{ke} + \bar{pe})_k$ The values of $\bar{ke}_k = M_kV_k^2/2000$, $\bar{pe}_k = M_k g_k Z_k/1000$ (in SI units), and $M_kV_k^2/(2\,g_c\,J)$, $pe_k = M_k g_k Z_k/(g_c\,J)$ (in English units). If the latter two energies in the methalpy are neglected (as they often can be), then

$$dE_{cv}/dt = \dot{Q}_{cv} - \dot{W}_{cv} + \Sigma_{k,i}\dot{N}_k\,\bar{e}_{T,k} - \Sigma_{k,e}\dot{N}_k\,\bar{e}_{T,k}$$

Let $\dot{N}_F$ denote the molal flow rate of fuel through one of the inlet streams that flow into a combustion chamber. Dividing Eq. (19) by $\dot{N}_F$ and simplifying the resultant expression we obtain the relation

$$(1/\dot{N}_F)\,dE_{cv}/dt = (\dot{Q}_{cv}/\dot{N}_F) - (\dot{W}_{cv}/\dot{N}_F) + \Sigma_{k,I}(\dot{N}_k/\dot{N}_F)\,\bar{e}_{T,k} - \Sigma_{k,e}(\dot{N}_k/\dot{N}_F)\,\bar{e}_{T,k}$$

For SSSF processes d( )/dt = 0, and

$$(\dot{Q}_{cv}/\dot{N}_F) - (\dot{W}_{cv}/\dot{N}_F) + \Sigma_{k,i}(\dot{N}_k/\dot{N}_F)\,\bar{e}_{T,k} - \Sigma_{k,e}(\dot{N}_k/\dot{N}_F)\,\bar{e}_{T,k} = 0.$$

Replacing the subscript "i" by R and "e" by P to indicate the reactant and product streams, respectively,

$$\bar{e}_{T,R} = \Sigma_{k,i}(\dot{N}_k/\dot{N}_F)\,\bar{e}_{T,k} \text{ and } \bar{e}_{T,P} = \Sigma_{k,e}(\dot{N}_k/\dot{N}_F)\,\bar{e}_{T,k}$$

and we obtain the following more convenient form

$$\bar{q} - \bar{w} = \bar{e}_{T,P} - \bar{e}_{T,R}. \quad (20)$$

Neglecting the potential and kinetic energies, we obtain the relation

$$\bar{q} - \bar{w} = \bar{h}_P - \bar{h}_R,$$

where $\bar{h}_R$ denotes the enthalpy of the reactant stream $\Sigma_{k,i}(\dot{N}_k/\dot{N}_F)\,\bar{h}_{,k}$, and $\bar{h}_P$ the enthalpy of the product stream $\Sigma_{k,e}(\dot{N}_k/\dot{N}_F)\,\bar{h}_k$.

---

f. Example 6

Glucose, i.e., $C_6H_{12}O_6$, is oxidized in the human body in its cells. Air is inhaled at 298 K, and a mixture of air and metabolism products (CO_2, H_2O, O_2 and N_2) is exhaled at 310 K. Assume that glucose is supplied steadily to the cells at 25°C and that 400% excess air is utilized during its consumption. If the breathing rate of humans is 360 L/hr, determine the amount of heat loss from the human body, $\bar{h}_{f,glucose} = -1.26\times10^6$ kJ kmole^{-1}, and the higher heating value of glucose is 15628, kJ kg^{-1}.

Solution

The reaction equation with 400 % of excess air (or 5× theoretical air) is

$$C_6H_{12}O_6 + 5\times 6\,(O_2 + 3.76\,N_2) \rightarrow 6CO_2 + 6\,H_2O + 24\,O_2 + 112.8\,N_2.$$

We now conduct an energy balance, i.e.,

$$dE_{cv}/dt = \dot{Q}_{cv} - \dot{W}_{cv} + \Sigma_{k,i}\dot{N}_k\,\bar{h}_{,k} - \Sigma_{k,e}\dot{N}_k\,\bar{h}_k$$

Ignoring the breathing (PdV) and other forms of work

$$0 = \dot{Q}_{cv} + 1\times(-1.26\times10^6) - 6\times(-393520 + 9807 - 9364)$$
$$- 6\times(-241820 + (10302 - 9904)) - 24 \times (0 + 9030 - 8682)$$
$$-112.8\times(09014 - 8669).$$

Hence,

$$\bar{q}_{cv} = -2.5\times10^6 \text{ kJ kmole}^{-1} \text{ of glucose.}$$

$M_F = 6\times12.01 + 12\times1.01 + 6\times16 = 180.2$ kg kmole^{-1}. Therefore,

$$q_{cv} = 13872 \text{ kJ kg}^{-1} \text{ of glucose.}$$

The air consumption is 360 L hr^{-1} or 0.360 m^3 hr^{-1}×(24.45 m^3 kmole^{-1}) = 0.0147 kmole hr^{-1} so that the glucose consumption is

$$0.0147\div142.8 = 0.0001031 \text{ kmole hr}^{-1}, \text{ or}$$

$$0.0001031 \text{ kmole} \times 180.2 \text{ kg kmole}^{-1} \times 1000 \text{ g kg}^{-1} \text{ hr}^{-1} \div 60 \text{ min hr}^{-1}$$
$$= 0.31 \text{ g min}^{-1}.$$

The heat loss rate

$$\dot{Q}_{cv} = \dot{m}_F\, q_{cv} = 0.31 \text{ g min}^{-1} \times 13872 \text{ J g}^{-1} \div 60 \text{ s min}^{-1} = 72 \text{ W}.$$

We can interpret this result that the hypothetical human body considered in this example has the same energy consumption as a 70 W light bulb. In reality, the body burns a mixture of glucose (which is a carbohydrate with a H/C ratio of two) and fats (e.g., palmitic acid). Palmitic acid has a higher heating value HHV = 39125.5 kJ kg^{-1} and participates in the energy consumption through the stoichiometric reaction

$$C_{15}H_{31}COOH + 23\ O_2 \rightarrow 16CO_2 + 16\ H_2O.$$

Our typical food consists of carbohydrates (HHV = 18000 kJ dry kg^{-1}), fatty acids (HHV = 40000 kJ dry kg^{-1}), and proteins (HHV = 22000 kJ dry kg^{-1}) of which almost 96 %, 98 % and 78 % are metabolized, respectively. In general, the larger the amount of moisture in food, the lower is its heating value. Fats have a lower moisture content and, therefore, have a larger heating value while the carbohydrates having relatively more moisture have a smaller value.

---

g. *Example 7*

Determine the amount of carbon dioxide (which is a greenhouse gas) emitted if coal is used as a fuel. Assume that the chemical formula for coal is $CH_{0.8}O_{0.3}$, its gross heating value is 30,000 kJ kg^{-1}, and the power plant efficiency is 30%. Compare your answer with the result for natural gas (CH_4) that has a gross heating value of 50,000 kJ kg^{-1}.

Solution

The chemical reaction governing the burning of coal is

$$CH_{0.8}O_{0.3} + 1.05\ O_2 \rightarrow CO_2 + 0.4\ H_2O.$$

The amount of CO_2 produced per kg of coal consumed is

$$44.01/(12.01 + 0.8\times1.01 + 16\times0.3) = 2.498 \text{ kg per kg of coal.}$$

Note that 1 kWh = 1 kJ s^{-1} × 3600 s hr^{-1} × 1 hr^{-1} = 3600 kJ.

The overall efficiency = Work output/Gross heat value released by combustion, i.e., per kWh of power produced

$$0.3 = 3600 \text{ kJ} \div (\text{gross heat value released by combustion}).$$

Therefore, the

$$\text{gross heat value released by combustion per kWh} = 3600\div0.3 = 12000 \text{ kJ}.$$

Consequently, the

amount of fuel burned per kWh $= 12000 \div 30000 = 0.4$ kg, and the

amount of CO_2 released $= 2.498 \times 0.4 = 0.999$ kg per kWh of power.

In the case of methane, the

CO_2 produced per kg of fuel consumed $= 44.01 \div 16.05 = 2.74$, and the

fuel consumed per kWh $= 12000 \div 50000 = 0.24$ kg.

Therefore, the

CO_2 produced $= 0.24 \times 2.74 = 0.658$ kg kWh^{-1} or 0.183 kg MJ^{-1}.

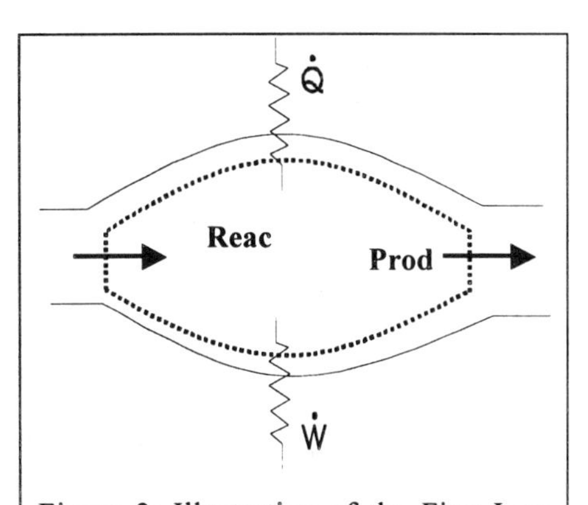

Figure 3: Illustration of the First Law for a chemically reacting system.

Remarks

The CO_2 emitted when natural gas is used as a fuel is reduced due to the higher heating value of methane.

The Boie equation is an empirical relation that can be used to determine the HHV of many C-H-N-O fuels including solid and liquid fuels. The relation is

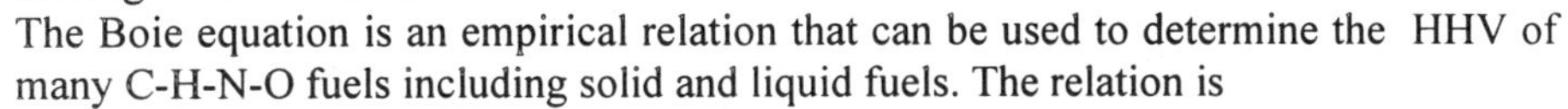

$$HHV, kJ\ kg^{-1} = 35{,}160 \times C + 116{,}225 \times H - 11{,}090 \times O + 6{,}280 \times N + 10{,}465 \times S$$

where C, H, O, N, and S denote the mass fractions of carbon, hydrogen, oxygen, nitrogen, and sulfur in the fuel. Using this relation, the formula for the mass of CO_2 emitted per MJ of heat input is

$$\text{kg of } CO_2 \text{ per MJ of heat input} = C \times 44.01 \times 1000 \div$$

$$(35{,}160 \times C + 116{,}225 \times H - 11{,}090 \times O + 6{,}280 \times N + 10{,}465 \times S), \text{ or}$$

$$\text{kg of } CO_2 \text{ per MJ of heat input} = 1000 \div (798.9 + 2640.9\ (H/C) - 252\ (O/C) +$$

$$142.7\ (N/C) + 237.8\ (S/C)).$$

Therefore, the higher the H/C ratio and the lower the O/C ratio, the lower the CO_2 emission to the atmosphere.

h. Example 8

Consider the metabolism of glucose (i.e., $C_6H_{12}O_6$) in the human body that breathes in air at 298 and exhales a mixture of CO_2, H_2O, O_2, and N_2 at 37°C. Assume that there is a steady supply of glucose to the human body at 25°C, that 400% excess air is inhaled, and $\bar{h}_{f,glucose} = -1.26 \times 10^6$ kJ $kmole^{-1}$. If the inhalation occurs at a rate of 360 l hr^{-1}, determine the amount of heat loss from the human body.

Solution

400% excess air is equivalent to 5 times theoretical air, i.e.,

$$C_6H_{12}O_6 + 5 \times 6(O_2 + 3.76\ N_2) \rightarrow 6CO_2 + 6H_2O + 24O_2 + 112.8N_2.$$

The energy balance is

$$dE_{cv}/dt = \dot{Q}_{cv} - \dot{W}_{cv} + \Sigma_{k,i} \dot{N}_k\, \bar{e}_{T,k} - \Sigma_{k,e} \dot{N}_k\, \bar{e}_{T,k} \quad (19)$$

Neglecting the potential and kinetic energies,

$$dE_{cv}/dt = \dot{Q}_{cv} - \dot{W}_{cv} + \Sigma_{k,i} \dot{N}_k \bar{h}_k - \Sigma_{k,e} \dot{N}_k \bar{h}_k \text{ i.e.,}$$

$$0 = \dot{Q}_{cv} - \dot{W}_{cv} + \Sigma_{k,i} \dot{N}_k \bar{h}_{,k} - \Sigma_{k,e} \dot{N}_k \bar{h}_k$$

Ignoring the breathing (PdV) and other forms of work,

$$0 = \dot{Q}_{cv} + 1 \times (-1.26\times10^6) + 0 - 6\times(-393520+443) -$$

$$6\times(-241820+398) - 24\times(0+348) - 112.8\times345, \text{ i.e.,}$$

$$\bar{q}_{c.v.} = (\dot{Q}/\dot{N}_F) = -2.5\times10^6 \text{ kJ per kmole of glucose.}$$

Since the molecular weight of glucose is

$$6\times12.01 + 12\times1.01 + 6\times16 = 180.2 \text{ kg kmole}^{-1},$$

$$q_{cv} = -13872 \text{ kJ per kg of glucose.}$$

The air consumption is 360 l hr^{-1} or

$$0.360 \text{ m}^3 \text{ hr}^{-1} \div (24.45 \text{ m}^3 \text{ kmole}^{-1}) = 0.0147 \text{ kmole hr}^{-1}.$$

Therefore, the

glucose consumption per hr = $0.0147 \div 142.8 = 0.0001031$ kmole hr^{-1}, or

$$0.0001031 \text{ kmole} \times 180.2 \text{ kg kmole}^{-1} \times 1000 \text{ g kg}^{-1} \text{ hr}^{-1}$$

$$= 18.6 \text{ g hr}^{-1} \text{or } 0.31 \text{ g min}^{-1}.$$

The heat loss rate

$$= \dot{Q}_{cv} = \dot{m}_F q_{cv} = 0.31 \text{ g min}^{-1} \times 13872 \text{ J g}^{-1} \div (60 \text{ s min}^{-1}) = 72 \text{ W}.$$

Remarks

The human body is a complex system. For a detailed analysis of this problem we must determine the work required for a body to function and a knowledge of the chemical enthalpies of the constituents of the metabolic process. Hence, due to change in metabolic activity, increased amount of heat may be liberated which will be exhibited through composition of nasal gases. If glucose is not transported to cells or is not metabolized, its concentration will increase in the bloodstream, which is known as the diabetic condition.

The body burns 0.7 kJ min^{-1} kg^{-1} of energy during aerobic dancing, 0.8 kJ min^{-1} kg^{-1} while running a mile in 9 minutes, 1.2 kJ min^{-1} kg^{-1} for a 6 minute mile, and 0.3 kJ min^{-1} kg^{-1} while walking. Due to a change in metabolic activity, an increased amount of heat may sometimes be liberated (e.g., fever), which will also be manifested through the composition of the human exhalation gases.

2. Adiabatic Flame Temperature

Consider a flammable premixed mixture of methane and air jet that emerges from a burner. The contour of the hottest location is often called the flame (cf. Figure 4). The maximum possible flame temperature is called the adiabatic flame temperature which is attained in the absence of heat losses and work transfer.

a. *Steady State Steady Flow Processes in Open Systems*

We will apply the energy balance equation in the context of Figure 4 at steady state, in the absence of work transfer, and neglecting the kinetic and potential energies so that

$$dE_{cv}/dt = \dot{Q}_{cv} - \dot{W}_{cv} + \Sigma_{k,i} \dot{N}_k \bar{e}_{T,k} - \Sigma_{k,e} \dot{N}_k \bar{e}_{T,k}$$

Dividing by the molar flow of fuel, for an adiabatic combustor,

$$0 = \bar{h}_P - \bar{h}_R, \quad (21)$$

where $\Sigma_{k,e}\,(\dot{N}_k / \dot{N}_F)\,\bar{h}_k = \bar{h}_P$ and $\Sigma_{k,i}\,(\dot{N}_k / \dot{N}_F)\,\bar{h}_k = \bar{h}_R$. This implies that the total enthalpy of the products leaving the combustor equals that of the entering reactants, which allows us to calculate the adiabatic flame temperature.

b. Closed Systems

The energy balance for a closed system results in the relation

$$Q - W = U_P - U_R = \Sigma_{k,P}\,(N_{k,P} / N_F)\,\bar{h}_{k,p} - \Sigma_{k,R}\,(N_{k,R} / N_F)\,\bar{h}_{k\,R}, \text{ where} \qquad (22)$$

$$\bar{u}_k = \bar{h}_k - (P\,\bar{v})_k$$

For ideal gases,

$$\bar{u} = \bar{h} - RT, \text{ i.e.,}$$

$$\bar{q} - \bar{w} = \bar{u}_P - \bar{u}_R.$$

Here, $\bar{u}_R$ denotes the internal energy of the reactants per kmole of fuel $\Sigma_{k,R}\,(N_{k,R} / N_F)\,\bar{u}_{k\,R}$, and $\bar{u}_P$ the enthalpy of the products per kmole of fuel, $\Sigma_{k,P}\,(N_{k,P} / N_F)\,\bar{u}_{k,p}$

---

i. Example 9

Liquid octane enters a reactor with 40% excess air, with both the fuel and air being at 298 K and 1 atm. Determine the adiabatic flame temperature in the case of (a) steady state steady flow device, (b) a rigid closed system. Also determine the final pressure for case (b). Assume ideal gas behavior and that $\rho_{octane(\ell)} = 703$ kg m^{-3}.

Solution

Normally, the adiabatic flame temperature is calculated assuming complete combustion. Therefore, if a stoichiometric amount of oxygen is supplied

$$C_8H_{18}(\ell) + 12.5\,O_2 \rightarrow 8CO_2 + 9\,H_2O.$$

In this context, when 40% excess air is supplied,

Supplied $O_2 = 12.5 \times 1.4 = 17.5$ moles per mole of fuel, and

Supplied $N_2 = 17.5 \times (79 \div 21) = 65.83$ moles per mole of fuel.

Therefore, the reaction equation assumes the form

$$C_8H_{18}(\ell) + 17.5\,O_2 + 65.83\,N_2 \rightarrow 8CO_2 + 9\,H_2O + 5\,O_2 + 65.83\,N_2.$$

a) The energy balance equation for a steady state steady flow device is

$$0 = (\Sigma\,\dot{N}_k\,\bar{h}_k)_i - (\Sigma\,\dot{N}_k\,\bar{h}_k)_e.$$

This equation can be solved by iteration. The adiabatic flame temperature in this case is 1900 K.

b) The corresponding energy balance equation for a rigid closed system is

$$Q - W = U_P - U_R.$$

The work $W = \int PdV = 0$, since $dV = 0$, and $Q = 0$ since we assume that the system is adiabatic. Therefore,

$$U_P = U_R.$$

In this context,

$$\bar{u}_{octane(\ell)} = \bar{h}_{octane(\ell)} - P\,\bar{v} = -249950 \text{ kJ} - 1 \text{ bar} \times 100 \text{ kN m}^{-2} \text{ bar}^{-1} \div (703 \text{ m}^3 \text{ kg}^{-1}) \times 114 \text{ kg kmole}^{-1} = -249950 - 16.2 \approx -249950 \text{ kJ kmole}^{-1}.$$

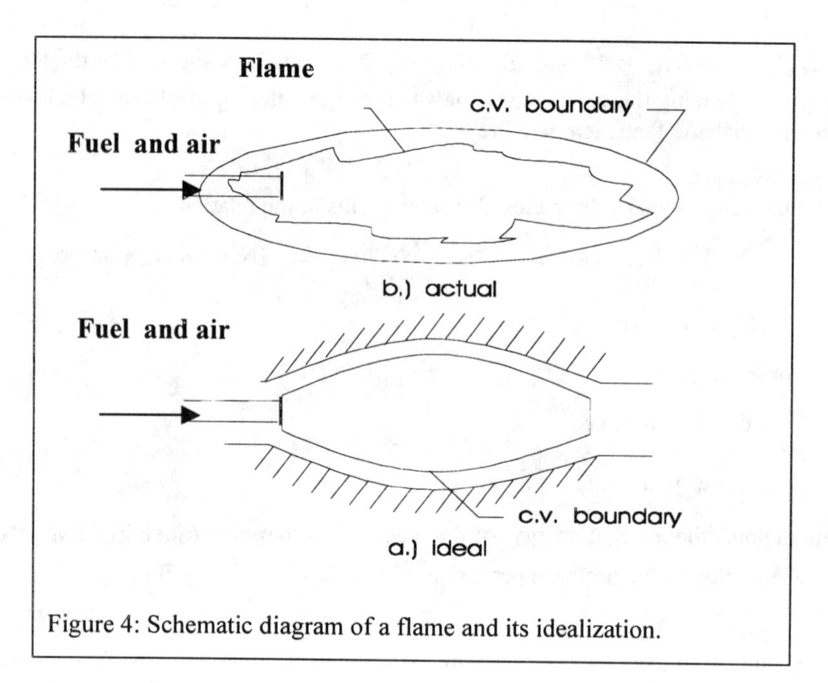

Figure 4: Schematic diagram of a flame and its idealization.

For an ideal gas $P\bar{v} = \bar{R}T$, i.e.,

$$\bar{u}_{O_2,298\,K} = h_{O_2,298\,K} - 8.314\times 298 = -2478 \text{ kJ kmole}^{-1}.$$

$$\bar{u}_{N_2,298\,K} = h_{N_2,298\,K} - 8.314\times 298 = -2478 \text{ kJ kmole}^{-1}.$$

Therefore,

$$\bar{u}_R = (1\times(-249950)+17.5\times(-2478)+65.83\times(-2478)) = -456406 \text{ kJ kmole}^{-1}.$$

At 2400 K,

$$\bar{u}_{CO_2} = h_{CO_2} - 8.314 \times 2400 = -393520 + (125152 - 9364) - 8.314 \times 2400$$
$$= -297685 \text{ kJ kmole}^{-1}.$$

Hence,

$$\bar{u}_P - \bar{u}_R = (8(-393520+115798)-8.314\times2400) + 9\times(-241820+93744)-8.314\times2400) + 5\times(0+74467)-8.314\times2400) + 65.83\times(0+70645)-8.314\times2400)) - (-456406)$$
$$\approx 174000 \text{ kJ kmole}^{-1}.$$

Assume that T = 2200 K,

$$\bar{u}_P - \bar{u}_R \approx -389{,}000 \text{ kJ kmole}^{-1}.$$

Interpolating, we obtain that the adiabatic flame temperature

$$T \approx 2340 \text{ K}.$$

The pressure rise,

$$P_2/P_1 = N_2\bar{R}T_2/(N_1\bar{R}T_1)$$
$$= 87.83\times2300\div(83.33\times298) = 8.13 \text{ bar}.$$

Remarks

You can repeat these calculations for air that is preheated to 700 K. The open system adiabatic flame temperature is roughly 2200 K, which is about 300 K higher than 1910 K when 298 K air is used. If the preheat required for heating the secondary air is supplied by the combustion products at 2200 K, then the combustion products temperature drops to 1910 K. In other words, if one tracks the temperature in such a system, it will rise gradually from 298 K to 2200 K (with an enthalpy above the conventional value without preheating) during an adiabatic process and decrease 1910 K once heat is removed for preheating the combustion air. Such a scheme is called Excess Enthalpy Combustion (EEC), which is useful for fuels particularly of low heating values or combustion involving large amount of excess air in order to stabilize the flame. On the other hand, if one wants to maintain a temperature of 1910 K without preheating, then the oxygen concentration must be reduced as is done in High Temperature Air Combustion (HiTAC) systems.

E. COMBUSTION ANALYSES IN THE CASE OF NONIDEAL BEHAVIOR

Nonideal gas behavior can manifest itself at high pressures (e.g., in gas turbines) or at low temperatures (e.g., in the context of gas mixtures stored in containers). We address that in this section.

1. Pure Component

If the reference condition pertaining to the enthalpy of any species k is set in the context of the natural state of its elements at 298 K and 1 bar, then

$$h_k(298\text{ K, 1 bar}) = h^o_{f,k}. \qquad (23)$$

We define an enthalpy correction $\Delta h_{k,corr}$ to account for departures from ideal gas behavior so that

$$\Delta h_{k,corr}(298\text{ K, 1 bar}) = h^o_k(298\text{ K}) - h_k(298\text{ K, 1 bar}).$$

2. Mixture

The component k exists in a mixture. Thus enthalpy $\hat{h}_k$ of the k-th species in a mixture is different from the enthalpy of the component in its pure state $\bar{h}_k$. Defining the enthalpy difference in terms of an excess function $\hat{h}^E_k$,

$$\hat{h}^E_k = \hat{h}_k(T,P) - \bar{h}_k(T,P). \qquad (24)$$

For an ideal gas $\hat{h}^E_k = 0$. In the case of entropy, the usual reference condition is set at 0 K and 1 bar, and the excess function is similarly defined, i.e., $\hat{s}^E_k = \hat{s}_k(T,P) - \bar{s}_k(T,P)$. Assuming an ideal gas mixture.

$$\hat{s}^E_{k,id} = \hat{s}_{k,id}(T,P) - \bar{s}_k(T,P) = -\bar{R}\ln X_k. \qquad (25)$$

In the case of liquid mixtures,

$$\hat{s}^E_{k,id} = \hat{s}_{k,id}(T,P) - \bar{s}_k(T,P) = -\bar{R}\ln X_{k,\ell}. \qquad (26)$$

j. Example 10

Determine the enthalpy, internal energy, entropy, and Gibbs energy of NO at 200 K and 195 bar in an ideal mixture of real gases, when $X_{NO} = 0.05$.

Solution

The enthalpy of NO at 298 K and 1 bar is 90592 kJ kmole^{-1}. If it were considered as an ideal gas under those conditions, $T_R = 1.66$ and $P_R = 1\div 64.8 = 0.0154$. Therefore, $\Delta h_{corr} \approx 0.001$ kJ kmole^{-1}, and

$$h^o_{NO}(298\text{ K, 1 bar}) = 90592 + 0.001 \approx 90592\text{ kJ kmole}^{-1}.$$

Since

$$h^o_{NO}(200\text{ K}) - h^o_{NO}(298\text{ K}) = -2950\text{ kJ kmole}^{-1},$$
$$h^o_{NO}(200\text{ K}) = 90592 - 2950 = 87642\text{ kJ kmole}^{-1}.$$

In order to determine the real gas enthalpy at 200 K and 195 bar, we use the relations

$P_R = 195 \div 64.8 = 3$, $T_R = 200 \div 180 = 1.11$, and $\Delta h_{corr}/RT_c = 3.353$,

so that

$\Delta h^{corr} = 8.314 \times 180 \times 3.353 = 5018$ kJ kmole^{-1}, and

$h^o_{NO}(200, 195) = 87642 - 5018 = 82624$ kJ kmole^{-1}.

Since we assume an ideal solution model,

$\hat{h}_k(T,P) = \bar{h}_k(T,P)$.

Recall that

$\hat{v}^E_k = \hat{v}_k - \bar{v}_k$.

If the ideal solution model is adopted $\hat{v}^E_k = 0$, i.e., $\hat{v}_k = \bar{v}_k$.

Using the relation,

$\hat{u}_k = \hat{h}_k - P\,\hat{v}_k$

at the conditions $T_R = 1.11$ and $P_R = 3$, $Z = 0.475$. Therefore,

$\hat{u}_k = \hat{h}_k - \hat{Z}_k\,\bar{R}T = 86624 - 0.475 \times 8.314 \times 200 = 85834$ kJ kmole^{-1}.

At 298 K,

$\bar{s}^o_k(298\text{ K}) = 210.7$ kJ kmole^{-1} K^{-1}.

At $T_R = 1.66$, $P_R = 0.0154$,

$\Delta s_{NO,\,corr} = \bar{s}_{k,0}(298\text{ K}) - \bar{s}_k(298\text{ K, 1bar}) = 0.001$ kJ kmole^{-1} K^{-1} i.e.,

$\bar{s}^o_{NO}$ (298 K, 1 bar) = 210.761 -0.001 ≈ 210.8 kJ kmole^{-1} K^{-1}.

Likewise,

$\bar{s}^o_{NO}$ (200 K) = 198.8 kJ kmole^{-1} K^{-1}.

From the entropy charts

$(\bar{s}_{NO,o}(T,P) - (\bar{s}_{NO}(T,P)) = 2.202 \times 8.314 = 18.31$ kJ kmole^{-1} K^{-1}, i.e.,

$\bar{s}_{NO}(T, P) = -18.3 + 198.8 = 180.5$ kJ kmole^{-1} K.

Finally, using the ideal solution model

$\hat{s}_{NO} = \bar{s}\,(T,P) - R \ln X_{NO}$, and

$\hat{s}_{NO}(200\text{ K},195\text{ bar}) = 180.5 - 8.314 \times \ln 0.05 = 205.4$ kJ kmole^{-1} K.

Therefore,

$\hat{g}_{NO} = \hat{h}_{NO} - T\,\hat{s}_{NO} = 82{,}624 - 200 \times 205.4 = 41544$ kJ kmole^{-1}.

Remark

The partial molal Gibbs function or chemical potential plays an important role in determining the direction of a chemical reaction in a system and the composition of the system at chemical equilibrium.

F. SECOND LAW ANALYSIS OF CHEMICALLY REACTING SYSTEMS

1. Entropy Generated During an Adiabatic Chemical Reaction

If a chemical reaction is irreversible under adiabatic conditions, then, according to Second law, $\delta\sigma > 0$. The question is "why?". Combustion is usually a process of oxidation of fossil fuels to CO_2 and H_2O. Fuel and oxygen are called reactants in this case, while CO_2 an H_2O are called products. The chemical bonds between the atoms in fuel are broken dring combustion, energy is converted into thermal energy which is eventually stored as te, ve, re (random energy) in a large number of quantum states. Hence, the entropy of the products is higher than that of the reactants. When combustion occurs in an adiabatic vessel, the energy is conserved. In such a process, the entropy difference between the products and reactants is due to the entropy generated during the process.

k. *Example 11*

One kmole of CO, 0.5 kmole of O_2, and 1.88 kmole of N_2 enter an adiabatic reactor and produce CO_2 due to the exothermic reaction of the monoxide and oxygen. Determine the temperature and entropy of the products assuming that CO reacts in varying amounts, i.e., 0.1, 0.2, ..., 0.9, 1 kmole.

Solution

For a steady adiabatic process that involves no work,

$$dE_{cv}/dt = \dot{Q}_{cv} - \dot{W}_{cv} + \Sigma_{k,i} \dot{N}_k \bar{h}_{,k} - \Sigma_{k,e} \dot{N}_k \bar{h}_{k\ e}. \quad \text{(A)}$$

Denoting $\bar{h}_R = \Sigma_{k,i} \dot{N}_k \bar{h}_{,k}$, $\bar{h}_P = \Sigma_{k,e} \dot{N}_k \bar{h}_k$, Eq. (A) may be expressed in the form

$$\bar{h}_P - \bar{h}_R = 0. \quad \text{(B)}$$

For every kmole of CO consumed, a half kmole of oxygen is consumed and a kmole of carbon dioxide is produced. For instance, in the case 0.2 moles of CO react,

$$\bar{h}_R = 1\times \bar{h}_{CO}(298\text{ K}) + 0.5\times \bar{h}_{O_2}(298\text{ K}) + 1.88\ \bar{h}_{N_2}(298\text{ K})$$

$$= 1\times(-110530) + 0 + 0\times(-393520)+1.88\times 0 = -110530\text{ kJ, and} \quad \text{(C)}$$

$$\bar{h}_P = 0.8\times \bar{h}_{CO}(T) + 0.4\times \bar{h}_{O_2}(T) + 0.2\times \bar{h}_{CO_2}(T) + 1.88\times \bar{h}_{N_2}(T)$$

$$=0.8\times(-110530+(h_T-h_{298})) + 0.4\times(0+(h_T- h_{298}))$$

$$+ 0.200\times(-393520+(h_T-h_{298})) + 1.88 \times (0+(h_T-h_{298})). \quad \text{(D)}$$

Equating Eqs. (C) and (D) one can iteratively solve for T, which in this case is 846.6 K. Figure 5 presents the variation of T and entropy generated *versus* the moles of CO that are burned. The corresponding entropy balance equation is

$$dS_{cv}/dt = \dot{Q}_{cv}/T_b + \Sigma_{k,i} \dot{N}_k \bar{s}_k - \Sigma_{k,e} \dot{N}_k\ \bar{s}_k + \dot{\sigma} \quad \text{(E)}$$

which for steady state and adiabatic conditions assumes the form

$$0 = \dot{Q}_{cv}/T_b + \Sigma_{k,i} \dot{N}_k \bar{s}_k - \Sigma_{k,e} \dot{N}_k\ \bar{s}_k + \dot{\sigma}\text{, or } \dot{\sigma} = + \Sigma_{k,e} \dot{N}_k \bar{s}_k - \Sigma_{k,i} \dot{N}_k\ \bar{s}_k \quad \text{(F)}$$

Denoting $S_R = (\Sigma N_k \bar{s}_k)_i$, $S_P = (\Sigma\ N_k\ \bar{s}_k)_e$, and $\hat{s}_k\ (T,P,X_k) = \bar{s}_k\ (T,p_k) = \bar{s}_k{}^o - \bar{R}\ \ln (p_k/1)$, we obtain

$$\bar{\sigma} = S_P - S_R. \quad \text{(G)}$$

Since S_R is specified and $\sigma > 0$, S_P must increase with the burned fraction. For instance, if 0.2 moles of CO react, the product stream contains 0.8 moles of CO, 0.4 moles of oxygen, 0.2 moles of the dioxide, and 1.88 moles of nitrogen which is inert. In that case,

$$p_{CO} = N_{CO}P/N = 0.8\times 1/(0.8+0.4+0.2+1.88) = 0.244\text{ bar.}$$

Similarly, $p_{O_2} = 0.122$, $p_{CO_2} = 0.0.061$, and $p_{N_2} = 0.573$. Hence,

$$S_P = 0.8\times \bar{s}_{CO}(T, p_{co}) + 0.40\times \bar{s}_{O_2}(T, p_{O_2}) + 0.20\times \bar{s}_{CO_2}(T, p_{CO_2}) +$$

$$1.88\times \bar{s}_{N_2}(T, p_{N_2}) = 0.8\times(\bar{s}^o_{CO}(846.6\text{ K}) - 8.314 \times \ln 0.57/1) +$$

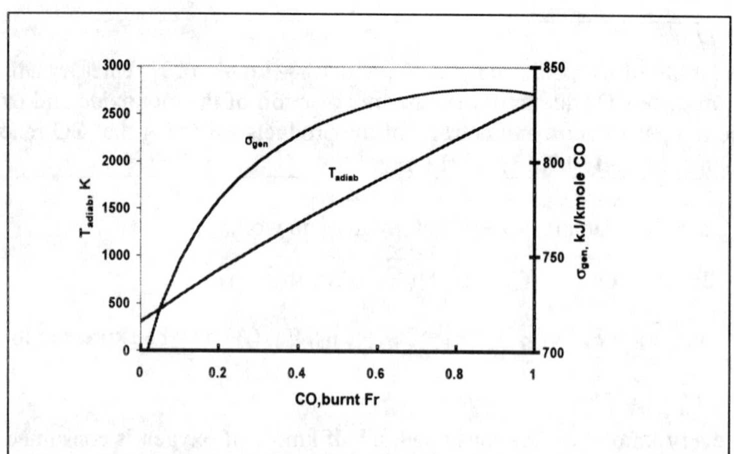

Figure 5: Variation of the adiabatic temperature and entropy with the CO burned fraction.

$$0.4\times(\bar{s}^{0}_{O_2}(846.6\text{ K}) - 8.314\times\ln 0.122/1) + 0.2\times(\bar{s}^{0}_{CO_2}(846.6\text{ K}) -$$

$$8.314\times \ln 0.061/1)+1.88\times(\bar{s}^{0}_{N_2}(846.6\text{ K})-8.314\times\ln\ 0.573/1) = 779.3\text{ kJ K}^{-1}.$$

Figure 5 plots the variation in S_p with respect to the CO burned fraction.

---

l. Example 12

Determine the entropy generated during the oxidation of glucose within the cells of the human body assuming that the reaction takes place at 310 K under steady state conditions using 400 % excess air that is at 300 K and 1 bar. The entropy of glucose at 310 K is 288.96 kJ kmole^{-1} K. Determine the entropy generation per kmole of glucose burned assuming that no work is done during the purely biochemical process. The typical consumption rate of glucose is 0.31 g min^{-1} for a 65 kg human. If the accumulated entropy generation over the lifetime of any biological species is limited to 10000 kJ kg^{-1} K^{-1}, obtain the life expectancy for a human. The value of "s" for glucose at 298 K, 1 bar is 212 kJ kmole^{-1} K^{-1}.

Solution

The chemical reaction can be represented by the equation

$$C_6H_{12}O_6 + 5\times6(O_2 + 3.76\ N_2) \rightarrow 6\ CO_2 + 6\ H_2O + 24\ O_2 + 112.8\ N_2.$$

We will apply the entropy balance equation

$$dS_{cv}/dt = \dot{Q}_{cv}/T_b + \Sigma_{k,i}\dot{N}_k\bar{s}_k - \Sigma_{k,e}\dot{N}_k\bar{s}_k + \dot{\sigma} \qquad (A)$$

at steady state. At the inlet p_{O_2} = 0.21 bar, and p_{N_2} = 0.79 bar, and since $\bar{s}^{o}_{k}(300\text{ K})$ has values of 205.03 and 191.5 kJ kg^{-1} K^{-1} for O_2 and N_2, respectively, $\bar{s}_k(T, p_k)$ for these two species is, respectively, 218.01 and 193.46 kJ kg^{-1} K^{-1}. Therefore, the inlet entropy is given as

$$S_i = \Sigma_{k,i}\dot{N}_k\bar{s}_k = 1\times212+30\times218.01+30\times3.76\times193.4662$$

$$= 28663\text{ kJ K}^{-1}\text{ per kmole per second of glucose consumed.} \qquad (B)$$

At the exit (at 310K in units of kJ kg^{-1} K^{-1}),

$\bar{s}^o_{O_2} = 205$, $\bar{s}^o_{N_2} = 191.54$, $\bar{s}^o_{CO_2} = 213.7$, and $\bar{s}^o_{H_2O} = 188.8$.

The total amount of products for a kmole of glucose consumed is (24+112.8+6+6) = 148.8 kmole. Therefore,

$X_{O_2} = 24 \div 148.8 = 0.161$, Similarly $X_{N_2} = 0.758$, $X_{CO_2} = 0.0403$, and $Y_{H_2O} = 0.0403$.

$\bar{s}_{O_2}(310\text{ K}, p_{O_2}) = 205.066 - 8.314 \times \ln(1 \times 0.16) = 220.2$ kJ K^{-1} per kmole of O_2.

The corresponding values of the other entropies at the exit (at 310K in units of kJ kmole^{-1} K^{-1}) are

$\bar{s}_{N_2} = 193.84$, $\bar{s}_{CO_2} = 240.42$, and $\bar{s}_{H_2O} = 215.49$.

$$S_e = \Sigma_{k,e} N_k\ \bar{s}_k$$

$$= 6 \times 24.42 + 6 \times 215.49 + 24 \times 220.2 + 112.8 \times 193.64$$

$$= 29900 \text{ kJ K}^{-1} \text{ per kmole per second of glucose consumed.}$$

Applying this value in the context of Eqs. (A) and (B) and assuming the heat loss from an average human body to be 2.5×10^6 kJ per kmole of glucose consumed (example 6) ,

$$0 = -2.5\times10^6 \div 310 + 28633 - 29900 + \dot{\sigma}, \text{ i.e.,}$$

$\dot{\sigma} = 9300$ kJ K^{-1} per kmole per second of glucose consumed, or

$\dot{\sigma} = 9300 \div 180.2 = 52$ kW K^{-1} per kg per second of glucose consumed.

The typical consumption rate of glucose is 1.031×10^{-4} kmole per hr (i.e., 0.31 g min^{-1}) for a person weighing 65 kg. Therefore, the entropy generated per second due to the irreversible metabolic activity is $52\times0.31\div(1000\times60) = 2.67\times10^{-4}$ kW K^{-1}. The entropy generated per unit mass of that person is $2.67\times10^{-4}\div65 = 4.1\times10^{-6}$ kW kg^{-1} K^{-1}. Since the total entropy that can be generated $\sigma_m = 10{,}000$ kJ kg^{-1} K^{-1}, the lifetime of that person is

10000 kJ kg^{-1} K^{-1}$\div$(4.1$\times10^{-6}$ kJ s^{-1} kg^{-1} K^{-1}$\times$3600 s hr^{-1}$\times$24 hr day^{-1} $\times$ 365 days year^{-1}) = 77 years.

Remarks

The entropy generated per unit mass of a human is 4.1×10^{-6} kW kg^{-1} K^{-1}. The human body has more than 60 trillion cells, which are constantly repaired and restored to an "original" state, but with some imperfections (e.g., aging). The cells require energy from metabolism to do so. The imperfections accumulate over time affecting the cell performance. If the size of each is of the order of 1 mm, there are 10^6 cells m^{-3}. Assume that the density of humans is roughly 1000 kg m^{-3} and $\dot{\sigma}_{cell} = 4.1\times10^{-9}$ kW K^{-1} per cell. Other organisms may have to burn more energy per cell (see Chapter 2) and hence generate more entropy. Consequently, they may have a shorter life span. If T_i = 37°C, the only irreversibility is due to the chemical reaction.

Typically, the entropy advection terms are small. Therefore,

$$\dot{\sigma} = -\dot{Q}_{cv}/T_b = \dot{m}_{F,b}\,HV/T_b$$

where the fuel burn rate $\dot{m}_{F,b}$ is a function of body temperature and generally increases with an increase in body temperature (e.g., thorough fever, jogging, etc.). Under normal conditions, the term $\dot{m}_{F,b}$ HV is the basal metabolic rate (BMR). The BMR seems to decrease due to aging. However it cannot decrease below a critical BMR (1 W kg^{-1}) which supports life functions. The corresponding critical entropy

generation is 0.00322 W kg^{-1} K^{-1}. The above relations neglect entropy advection through perspiration and excretion.
In general the body burns a mixture of glucose, palmitic acid and proteins. The entropy of palmitic acid under standard conditions is 452.37 kJ kg^{-1} K^{-1}. Most metabolic thermodynamic calculations assume a mixture of glucose and palmitic acid. Figure 6 shows the entropy generated in kg^{-1} K^{-1} per kmole of fuel and in kg^{-1} K^{-1} per kJ of metabolism when a mixture of fat and glucose is burned. Since the lifetime = $\sigma_m/\dot{\sigma}_m$ $\approx \sigma_m T_b/\dot{q}_{cv}$, the higher the specific metabolic rate, the lower is the life expectancy. It can be shown from scaling laws that $\dot{q}_{cv}$, $\approx C\, m_B^{-n}$ (n = 1/3 from the theory in Chapter 2 and from experiments n= 0.26 and C = 0.003552 for analyses SI units). There fore, the life span, t_{life}(years) $\approx$ (3.17×10^{-8} σ_m $m_B^{\,n}T_b/C$) from theory. From experiments t_{life}(years) = 11.8 $m_B^{\,k}$ where k =0.20. Assuming n $\approx$ k, T_b = 310 K, C= 0.003552, σ_m = 4265 kJ kg^{-1} K^{-1} rather than 10,000 as we had assumed. The earlier empirical law seems to predict a lower life span. Similarly, the warmer the body temperature, the longer the life span assuming that biological/metabolical reactions are unaffected (this is not generally true). Experiments on water fleas have revealed an increase in their life spans as the ambient temperature is increased from 5 to 15°C, but this reduces thereafter, since their metabolism is affected.
If the human body is unhealthy, it requires a higher energy or metabolic rate to cure itself, and, consequently, the entropy generated is larger for a sick person, which shortens the anticipated life span.

2. Entropy Generated During an Isothermal Chemical Reaction

---

m. Example 13

The following species are introduced into a reactor at 3000 K and 101 kPa: 5 kmole s^{-1} of CO, 3 kmole s^{-1} of O_2, and 4 kmole s^{-1} of CO_2. One percent of the inlet CO oxidizes to form CO_2. Determine $\delta\sigma$ for a fixed mass of the reactants. Assume instantaneous mixing at entry and ideal gas behavior.

Solution

In general, mixing and reaction occur on finite time scales. However, we will assume that the mixing time is so short compared to the reaction time that an instantaneous mixing assumption applies. Consider a reactant mixture consisting of 5 kmole of CO, 3 kmole of O_2, and 4 kmole of CO that has a total mass of 5×28 + 3×32 + 4×44 = 412 kg. We will follow this mass as the reaction occurs. Assume that CO is oxidized according to the chemical reaction

$$CO + 1/2\, O_2 \rightarrow CO_2.$$

Since only 1% of the inlet CO undergoes this reaction, then at the exit the number of moles of CO

$$N_{CO,e} = 5 - 0.01\times5 = 4.95 \text{ kmole.}$$

Hence, the depletion in the CO $dN_{CO} = -0.05$ kmole. Likewise,

$$N_{CO_2,e} = 4 + 0.05 = 4.05 \text{ kmole.}$$

$$N_{O_2,e} = 3 - 0.05 \times 0.5 = 2.975 \text{ kmole.}$$

At constant pressure, the First law for a closed system yields

$$\delta Q_P = \Delta H.$$

The enthalpy change per kmole of the mixture

$$\Delta\bar{h} = \bar{h}_P - \bar{h}_R.$$

The product enthalpy,

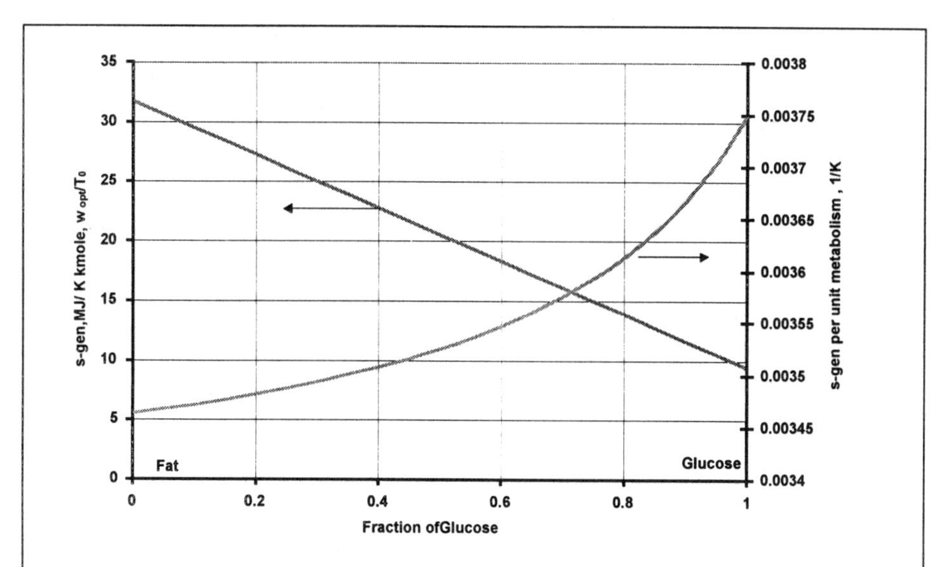

Figure 6: Variation of entropy generation per kmole of fuel and per unit metabolism.

$$\bar{h}_P = 4.95 \times \bar{h}_{CO,e} + 4.05 \times \bar{h}_{CO_2,e} + 2.975 \times \bar{h}_{O_2,e}.$$

Now,

$$\bar{h}_{CO}(T,P) = \bar{h}_{CO,o}(T) = (\bar{h}_f^o + \bar{h}_{t,3000} - \bar{h}_{t,298})_{CO}, \text{ i.e.,}$$

$$\bar{h}_{CO} = -110{,}530 + 93562 = -16979 \text{ kJ kmole}^{-1}.$$

Similarly,

$$\bar{h}_{CO_2} = -393520 + 152891 = -240629 \text{ kJ kmole}^{-1}, \text{ and}$$

$$\bar{h}_{O_2} = 0 + 98036 = 98036 \text{ kJ kmole}^{-1}.$$

Therefore,

$$\Delta H_R^o(3000 \text{ K}) = -272720 \text{ kJ kmole}^{-1} \text{ of CO}.$$

Since the gases are assumed to be ideal, $\hat{h}_k(T) = \bar{h}_k(T)$, and

$$H_e = 4.95 \times (-16979) + 2.975 \times (98036) + 4.05 \times (-240{,}629)$$

$$= -766936 \text{ kJ (or -1862 kJ/kg of mixture)}$$

The reactant enthalpy

$$H_R = 5 \times (-16989) + 3.0 \times (98098) + 4 \times (-240{,}658)$$

$$= -753283 \text{ kJ kmole}^{-1} \text{ (or -1828 kJ/kg of mixture)}$$

Therefore,

$$\delta Q = H_P - H_R = -13653 \text{ kJ (or -33.1 kJ/kg of mixture)}$$

This is the amount of heat that has to be removed from a kmole of the product mixture so that the products can be maintained at 3000 K.

The entropies for ideal gases are

$$\bar{s}_{CO}(T,P) = (\bar{s}_{CO}^o(3000\text{K}) - 8.314 \times \ln(P/1))$$

$$= 273.508 - 8.314 \times \ln 1 = 273.58 \text{ kJ K}^{-1} \text{ kmole}^{-1} \text{ of CO},$$

$$\bar{s}_{CO_2}(T,P) = 334.08 \text{ kJ K}^{-1} \text{ kmole}^{-1}, \text{ and}$$

$$\bar{s}_{O_2}(T,P) = 284.4 \text{ kJ K}^{-1} \text{ kmole}^{-1}.$$

At the inlet,

$$X_{CO} = 5/(5+4+3) = 0.417,\ X_{O_2} = 0.25, \text{ and } X_{CO_2} = 0.3323, \text{ i.e.,}$$

$$\hat{s}_{CO}(T,P,X_{CO}) = \bar{s}_{CO}(T,P) - \bar{R} \ln X_{CO}$$

$$= 273.58 - 8.314 \times \ln 0.417 = 280.85 \text{ kJ K}^{-1} \text{ kmole}^{-1},$$

$$\hat{s}_{O_2}(T,P,X_{O_2}) = 284.4 - 8.314 \times \ln 0.25 = 295.9 \text{ kJ K}^{-1} \text{ kmole}^{-1}, \text{ and}$$

$$\hat{s}_{CO_2}(T,P,X_{CO_2}) = 334.08 - 8.314 \ln 0.33 = 343.3 \text{ kJ K}^{-1} \text{ kmole}^{-1}.$$

At the exit,

$X_{CO} = 0.413$, $X_{O_2} = 0.248$, and $X_{CO_2} = 0.338$, so that

$$\hat{s}_{CO}(T,P,X_{CO,e}) = 273.58 - 8.314 \times \ln (0.413) = 280.93 \text{ kJ K}^{-1} \text{ kmole}^{-1},$$

$$\hat{s}_{O_2} = 284.4 - 8.4314 \times \ln 0.248 = 296.0 \text{ kJ K}^{-1} \text{ kmole}^{-1}, \text{ and}$$

$$\hat{s}_{CO_2} = 334.08 - 8.314 \times \ln 0.338 = 343.1 \text{ kJ K}^{-1} \text{ kmole}^{-1}.$$

The exit entropy,

$$S_e = 4.95 \times 280.93 + 2.975 \times 296 + 4.05 \times 343.1 = 3660.8 \text{ kJ}.$$

The reactant entropy

$$S_R = 5.00 \times 280.85 + 3 \times 295.9 + 4 \times 343.3 = 3665.1 \text{ kJ K}^{-1}, \text{ i.e.,}$$

$$d S = 3665.1 - 3660.8 = -4.3 \text{ kJ K}^{-1} \text{ kmole}^{-1}.$$

Therefore,

$\delta\bar{\sigma} = dS - \delta Q/T = -4.3 - (-13{,}636) \div 3000 = 0.245 \text{ kJ K}^{-1}$ or 0.0006 kJ/kg of mixture

---

Remarks

For an elemental reaction $\delta\sigma > 0$ (cf. Figure 7).

Since no entropy is generated due to nonuniform temperature or pressure gradients within the fixed mass, the finite value of $\delta\sigma$ arises due to the irreversible chemical reaction.

In such a system there are no thermal or mechanical irreversibilities. For this system

$$dS - \delta Q/T = \delta\sigma, \text{ i.e.,}$$

$$T\,\delta\sigma = T\,dS - \delta Q = T(S_e - S_i) - (H_e - H_i) = G_i - G_e = T\,\delta\sigma$$

$$= 3000 \times 0.935 = (2806 \text{ kJ K}^{-1} \text{ kmole}^{-1} > 0), \text{ and}$$

$$dG = G_e - G_i = -\ \delta\sigma < 0.$$

Here,

$$G = H - TS = \Sigma \hat{h}_k N_k - T\,\Sigma N_k \hat{s}_k = \Sigma N_k(\hat{h}_k - T\hat{s}_k) = \Sigma N_k \hat{g}_k.$$

The Gibbs energy of a fixed mass at a specified temperature and pressure decreases as the chemical reaction proceeds. Since G is a measure of availability, the availability decreases during this irreversible process (Fig. 7).

In the next section we find that the direction of reaction (i.e., from the reactant CO to the product CO_2) must be such that $\delta\sigma > 0$ or $dG_{T,P} < 0$. If this is untrue, then the assumed reaction is impossible.

G. MASS CONSERVATION AND MOLE BALANCE EQUATIONS

In Chapter 2 we derived the conservation equations for mass and energy. In multicomponent reacting systems, species are both consumed and generated due to chemical reactions. For instance, if a mole of CO and 2 moles of O_2 are admitted into a combustor, and if we will find that 1 kmole of CO_2 is produced and 1.5 kmoles of O_2 remain, then we can assume that complete combustion has occured. Even though the mass leaving ($1\times 44 + 1.5 \times 32 = 92$ kg) is same as the mass entering ($1\times28+2\times32 = 92$ kg) the combustor, the total moles exiting the reactor ($1+1.5 = 2.5$) are different from those entering it ($1+2 = 3$), implying that moles (or species) have been generated. The reactor produced 1 kmole of CO_2 (or 44 kg) but consumed 1 kmole of CO (= 28 kg) and 1/2 kmole of O_2 (= 16 kg). In chemically reacting systems, the mass production rate of species k is $\dot{m}_{k,gen}$ and mass conservation the species is written in the for

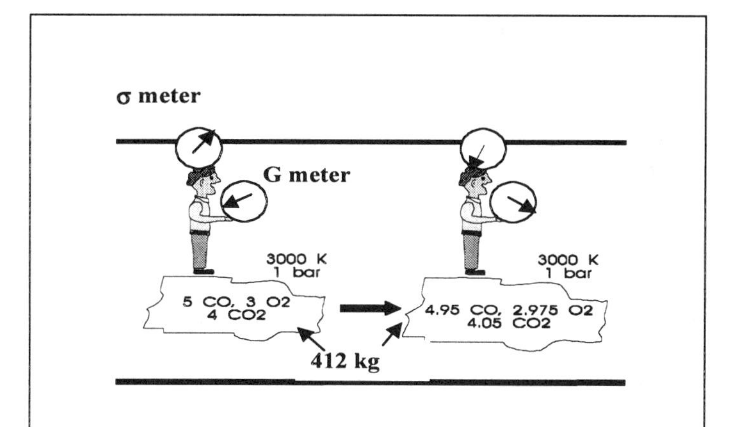

Figure 7: Illustration of entropy generation and decrease in G of a fixed mass with reaction. The value of G keeps decreasing as a reaction proceeds.

$$dm_k/dt = (\dot{m}_{k,i} + \dot{m}_{k,gen} - \dot{m}_{k,e}), \tag{27}$$

where dm_k/dt denotes the mass accumulation rate within the reactor volume V and $\dot{m}_{k,gen}$ the generation rate of species k due to chemical reaction. Summing over all species,

$$dm_{cv}/dt = \Sigma dm_k/dt = (\Sigma \dot{m}_{k,i} + \Sigma \dot{m}_{k,gen} - \Sigma \dot{m}_{k,e}) = \dot{m}_i - \dot{m}_e, \tag{28}$$

where $\dot{m}_i = \Sigma \dot{m}_{k,i}$, $\dot{m}_e = \Sigma \dot{m}_{k,e}$, $\Sigma \dot{m}_{k,gen} = 0$, since mass is conserved.

Writing Eq. (27) in vectorial form,

$$(d/dt) \int \rho_k\, dV = \int \rho_k \cdot \vec{V}_k d\vec{A} + \int \dot{m}_{k,gen}'''dV \tag{29}$$

where $\int \dot{m}_{k,gen}'''$ denotes the mass generated per unit volume of reactor and ρ_k the density of species k. Using the Gauss divergence theorem, we can write the species conservation equation in differential form as

$$\partial \rho_k/\partial t + \nabla \cdot \rho_k \vec{V}_k = \dot{m}_{k,gen}'''. \tag{30}$$

Summing over all species

$$\partial \rho/\partial t + \nabla \cdot \rho \vec{V} = 0. \tag{31}$$

Since $m_k = N_k M_k$, Eq. (27) has the following form in terms of mole balance

$$dN_k/dt = \dot{N}_{k,i} + \dot{N}_{k,gen} - \dot{N}_{k,e} \tag{32}$$

where $\dot{N}_{k,gen}$ denotes the number of moles of species k produced by the chemical reaction. As mentioned before $\Sigma \dot{N}_{k,gen} = \dot{N}_{gen} \neq 0$. Summing over all species

$$\Sigma dN_k/dt = dN/dt = \Sigma \dot{N}_{k,i} + \Sigma \dot{N}_{k,gen} - \Sigma \dot{N}_{k,e} = \dot{N}_i + \dot{N}_{gen} - \dot{N}_e \tag{33}$$

Similarly proceeding as in Chapters 2 to 4 for the energy conservation, entropy and availability balance equations, we write the mole balance equation in integral form as

$$(d/dt) \int n_k\, dV = \int n_k \cdot \vec{V} d\vec{A} + \int \dot{N}_{k,gen}''' dV, \tag{34}$$

where n_k = moles of species k per unit volume. Further using the Gauss divergence theorem,

$$\partial n_k/\partial t + \nabla . n_k \vec{V} = \dot{N}'''_{k,gen}. \quad (35)$$

1. Steady State System

Under steady state conditions $dN_k/dt = 0$, $dN/dt = 0$, $dm_k/dt = 0$, $dm/dt = 0$ and from Eq. (28), $\dot{m}_i = \dot{m}_e$. Likewise, from Eq.(33), $\dot{N}_i + \dot{N}_{gen} = \dot{N}_e$.

---

n. Example 13

A combustor is fired with 2 kmole of CO and 3 k mole of O_2. When the combustor is just started, very little CO burns. As it warms up, more and more CO are burnt. Assume that mass does not accumulate within the reactor. At the point when the combustor achieves 40% efficiency, write the mass conservation, mole balance and energy conservation equations. If the combustor reaches a steady state with combustion efficiency of 90%, write the mass conservation and mole balance equations.

Solution

The reaction equation is written as follows:

$$2\ CO + 3\ O_2 \rightarrow N_{CO,e}\ CO + N_{CO2,e}\ CO_2 + N_{O2,e}\ O_2. \quad (A)$$

Since mass does not accumulate within the reactor, the atom balance for C and O atoms yields

$$2 = N_{CO,e} + N_{CO2,e}, \text{ and} \quad (B)$$

$$2 + 6 = N_{CO,e} + 2\ N_{CO2,e} + 2\ N_{O2,e}, \text{ i.e.,} \quad (C)$$

$$0.4 = (2 - N_{CO,e})/2. \quad (D)$$

With the three equations Eqs. (B) through (D), we can solve for the three unknowns $N_{CO,e}$, $N_{CO2,e}$ and $N_{O2,e}$, i.e.,

$$N_{CO,e} = 1.6,\ N_{CO2,e} = 0.4, \text{ and } N_{O2,e} = 2.8.$$

The mass conservation implies

$$dm_{CO}/dt = 2\times28 + \dot{m}_{CO,gen} - 1.6\times28 = 11.2 + \dot{m}_{CO,gen}.$$

Similarly,

$$dN_{CO}/dt = 2 + \dot{N}_{CO,\ gen} - 1.6 = 0.4 + \dot{N}_{CO\ gen}$$

Normally $\dot{N}_{CO,\ gen}$ is a negative quantity since CO is consumed. The term dN_{CO}/dt represents the accumulation (destruction, since negative) rate of CO in the reactor. Similarly for CO_2,

$$dN_{CO2}/dt = 0 + \dot{N}_{CO2\ gen} - 0.8,$$

where $\dot{N}_{CO2,\ gen} > 0$, since CO_2 is a product that is generated. In the initial periods when the combustor is being fired, the CO_2 concentration gradually increases due to the term dN_{CO2}/dt.

Similarly for 90% efficiency,

$$N_{CO,e} = 0.2\ N_{CO2,e} = 1.8, \text{ and } N_{O2,e} = 2.1$$

$$dm_{CO}/dt = 2\times28 + \dot{m}_{CO,gen} - 0.2\times28\times28 = 50.4 + \dot{m}_{CO,gen}.$$

The combustor is operating steadily so that $dm_{CO}/dt = 0$,

$$\dot{m}_{CO,gen} = -50.4 \text{ kg/s}$$

Similarly ,

$$dN_{CO}/dt = 2 - 0.2 + \dot{N}_{CO,\ gen} = 1.8 + \dot{N}_{CO,\ gen}.$$

Since $dN_{CO}/dt = 0$,

$$\dot{N}_{CO,\ gen} = -1.8 \text{ k mole s}^{-1}, \text{ and}$$

$$dN_{CO2}/dt = 0 + \dot{N}_{CO2,gen} - 1.8,$$

If the combustor is operating steadily then $dN_{CO2}/dt = 0$, and

$$\dot{N}_{CO2,gen} = 1.8 \text{ k mole s}^{-1}, \quad \dot{N}_{O2,gen} = -0.9 \text{ kmole s}^{-1}.$$

H. SUMMARY

Chemical reactions occur when species rearrange their atoms and different compounds with different bond energies are produced. Dry and wet gas analyses are presented in this chapter, which are an analytical tool to measure species transformations. Examples are presented for determining (A:F) from dry gas analyses. The enthalpy of formation or chemical enthalpy, thermal enthalpy and the total enthalpy are defined. Energy conservation (First law) and entropy balance (Second law) of reacting systems are introduced and illustrative examples are provided. Finally mass conservation and mole balance equations for reacting systems are presented.

Chapter 12

12. REACTION DIRECTION AND CHEMICAL EQUILIBRIUM

A. INTRODUCTION

Nature is inherently heterogeneous and, consequently, natural processes occur in such a direction so as to create homogeneity and equilibrium (which is a restatement of the Second Law). In the previous sections, we assumed that hydrocarbon fuels react with oxygen to produce CO_2, H_2O, and other products. We now ask the question whether these products, e.g., CO_2, and H_2O, can react among themselves to produce the fuel and molecular oxygen. If not, then why not? What governs the direction of reaction? Now we will characterize the parameters that govern the predominant direction of a chemical reaction. We will also discuss the composition of reaction products under equilibrium conditions.

B. REACTION DIRECTION AND CHEMICAL EQUILIBRIUM

1. Direction of Heat Transfer

Prior to discussing the direction of a chemical reaction, we will consider the direction of heat transfer. Heat transfer occurs due to a thermal potential, from a higher to a lower temperature. Thermal equilibrium is reached when the temperatures of the two systems (one that is transferring and the other that is receiving heat) become equal. Due to the irreversible heat transfer from a warmer to a cooler system, $\delta\sigma > 0$.

2. Direction of Reaction

The direction of heat transfer is governed by a thermal potential. For any infinitesimal irreversible process $\delta\sigma > 0$. (For heat transfer to take place from a lower to a higher temperature $\delta\sigma < 0$, which is impossible.) Likewise, the direction of a chemical reaction under specified conditions is also irreversible and occurs in such a direction such that $\delta\sigma > 0$.The direction of a chemical reaction within a fixed mass is also such that $\delta\sigma > 0$ due to chemical irreversibility. For instance, consider the combustion of gaseous CO at low temperatures and high pressures, i.e.,

$$CO + 1/2\ O_2 \rightarrow CO_2 \tag{1a}$$

At high temperatures and at relatively low pressures

$$CO_2 \rightarrow CO + 1/2\ O_2. \tag{1b}$$

Reaction (1a) is called an oxidation or combustion reaction, and Reaction (1b) is termed a dissociation reaction. The direction in which a reaction proceeds varies depending upon the temperature and pressure. We will show that the chemical force potentials $F_R(= \hat{g}_{CO}+(1/2)\,\hat{g}_{O2})$ for the reactants and F_P $(=\hat{g}_{CO2})$ for the products govern the direction of chemical reaction at specified values of T and P. If $F_R>F_P$, Reaction 1a dominates and *vice versa* if $F_R<F_P$, just as the thermal potential T governs the direction of heat transfer.

Consider a premixed gaseous mixture that contains 5 kmole of CO, 3 kmole of O_2 and 4 kmole of CO_2, placed in a piston–cylinder–weight assembly (PCW) at a specified constant temperature and pressure. It is possible that the oxidation of CO within the cylinder releases heat, in which case heat must be transferred from the system to an ambient thermal reservoir. An observer will notice that after some time the oxidation reaction (Reaction (1a)) ceases when chemical equilibrium is reached (at the specified temperature and pressure). Assume that the observer keeps an experimental log that is reproduced in Table 1.

In the context of Reaction 1a, if 0.002 kmole of CO (dN_{CO}) are consumed, then $1/2\times(0.002) = 0.001$ kmole of O_2 (dN_{O_2}) are also consumed and 0.002 kmole of CO_2 (dN_{CO2}) are produced. Assigning a negative sign to the species that are consumed and using the associated stoichiometric coefficients,

Table 1: Experimental log regarding the oxidation of CO at specified conditions.

Time, sec	CO, kmole	CO_2, kmole	O_2, kmole
0	5	3	4
t_A	4.998	2.999	4.002
t_B	4.5	2.75	4. 5
t_C	4.25	2.625	4.75
t_D	3.75	2.375	5.25
t_E	3.75	2.375	5.25

$$\frac{dN_{CO}}{-1}=\frac{-0.002}{-1}=+0.002;\ \frac{dN_{O2}}{-1/2}=+0.002;\ \frac{dN_{CO2}}{+1}=+0.002,$$

yielding a constant number of 0.002. One can now define the extent of the progress of reaction ξ by the relation

$$d\xi = dN_{CO_2}/\nu_{CO_2}, \tag{2}$$

where dN_{CO_2} denotes the increase in the number of moles of CO_2 as a result of the reaction and ν_{CO_2} the stoichiometric coefficient of CO_2 in the reaction equation. During times $0 < t < t_A$,

$$d\xi = (4.002–4.0)/1 = 0.002.$$

Generalizing, we obtain the relation

$$d\xi = dN_k/\xi_k.$$

The production of CO_2 ceases at a certain mixture composition when chemical equilibrium is attained (time t_E in Table 1).

3. Mathematical Criteria for a Closed System

Specified Values of U, V, and m

For a closed, fixed mass system (operating at specified values of U, V and m) undergoing an irreversible process (cf. Chapter 3)

$$(dS - \delta Q/T_b = \delta\sigma) > 0.$$

We wish to determine irreversibility due to reaction alone and eliminate other irreversibilities due to temperature and pressure gradients within the system, we set $T_b = T$. Thus

$$(dS - \delta Q/T = \delta\sigma) \ge 0 \text{ with ">0" for irreversible, "=0" for reversible process} \tag{3}$$

For an adiabatic reactor,

$$(dS_{U,V} = \delta\sigma) \ge 0 \tag{4}$$

Thus for adiabatic reactions within a rigid vessel, the entropy S reaches a maximum.

ii. Specified Values of S, V, and m

Recall from Chapter 3 that

$$dU = TdS - P\,dV - T\,\delta\sigma. \tag{5a}$$

Note that Eq. (5a) is valid for a process where irreversible process ($\delta\sigma > 0$) or reversible process ($\delta\sigma = 0$) occurs. For a system operating at specified values of S, V, and m,

$$(dU_{S,V} = -T\,\delta\sigma) \le 0, \text{ with “<0” for irreversible, “=0” for reversible process} \tag{5b}$$

iii. Specified Values of S, P, and m
Likewise, since

$$dH = T\,dS + V\,dP - T\delta\sigma,$$

for specified values of S, P, and m

$$(dH = -T\,\delta\sigma) \le 0. \tag{6a}$$

iv. Specified values of H, P, and m

$$(dS_{H,P} = \delta\sigma) \ge 0. \tag{6b}$$

which is similar to Eq. (4). Note that adiabatic reactions in a constant pressure closed system involves constant enthalpy and the entropy reaches a maximum value.

v. Specified Values of T, V, and m
Recall that

$$dA = -S\,dT - P\,dV - T\,\delta\sigma, \text{ i.e., } (dA_{T,V} = -T\,\delta\sigma) \le 0. \tag{7}$$

For isothermal reactions within a rigid vessel, the Helmholtz function A reaches a minimum.

vi. Specified Values of T, P, and m
Similarly,

$$dG = -S\,dT + V\,dP - T\,\delta\sigma, \text{ i.e.,} \tag{8a}$$

$$(dG_{T,P} = -T\,\delta\sigma) \le 0. \tag{8b}$$

For isothermal and isobaric reactions within a fixed mass, the Gibbs' function G reaches a minimum. Note that Eq. (8a) is valid for a process where irreversible process ($\delta\sigma > 0$)or reversible process ($\delta\sigma = 0$) occurs. However if state change occurs reversibly between two equilibrium states G and G+dG, then

$$dG = -S\,dT + V\,dP.$$

In order to validate equation (8b) for an irreversible process of a fixed mass, we must determine the value of either $\delta\sigma$ or dG during an irreversible process (cf. Example 8). As was shown in Chapter 3 when we considered an irreversible mixing process and in Chapter 7 when we considered an irreversible evaporation process at specified (T,P), we must determine the value of G for the reacting system as the reaction proceeds.

4. Evaluation of Properties During an Irreversible Chemical Reaction

The change in entropy between two *equilibrium states* for an open system is represented by the relation (cf. Chapter 3 and Chapter 8)

$$dS = dU/T + P\,dV/T - \Sigma\mu_k\,dN_k/T. \tag{9a}$$

Similarly,

$$dU = T\,dS - P\,dV + \Sigma\mu_k\,dN_k. \tag{9b}$$

Equation (9b) can be briefly explained as follows. For a fixed mass closed system, $dN_k = 0$. Thus, $dU = TdS - PdV$, and the change in internal energy ≈ heat added - work performed. However, if mass crosses the system boundary and the system is no longer closed (e.g., pumping of air into a tire), chemical work is performed for the species crossing the boundary ($=\Sigma\mu_k\,dN_k$). Likewise,

$$dH = T\,dS + V\,dP + \Sigma\mu_k\,dN_k, \tag{9c}$$

$$dA = -S\,dT - P\,dV + \Sigma\mu_k\,dN_k, \tag{9d}$$

$$dG = -S\,dT + V\,dP + \Sigma\mu_k\,dN_k \tag{9e}$$

It is apparent from Eqs. (9a) to (9e) that

$$-T\,(\partial S/\partial N_k)_{U,V} = -T\,(\partial S/\partial N_k)_{H,P} = (\partial U/\partial N_k)_{S,V} = (\partial A/\partial N_k)_{T,V} = (\partial G/\partial N_k)_{T,P} = \hat{g}_k = \mu_k, \text{ and} \tag{9f}$$

$$-TdS_{U,V,,m} = -\,TdS_{H,,P,m} = dU_{S,V,m} = dH_{S,P,m} = dA_{T,V,,m} = dG_{T,P,,m} = \Sigma\mu_k\,dN_k. \tag{9g}$$

a. Nonreacting Closed System

In a closed nonreacting system in which no mass crosses the system boundary $dN_k = 0$. Therefore for a change in state along a reversible path,

$$dS = dU/T + P\,dV/T, \tag{10}$$

$$dU = T\,dS - P\,dV,\ dH = T\,dS + V\,dP$$

$$dA = -S\,dT - P\,dV, \text{ and } dG = -\,S\,dT + V\,dP.$$

It is apparent that for a closed system

$$\Sigma\mu_k\,dN_k = 0. \tag{11}$$

b. Reacting Closed System

Assume that 5 kmole CO, 3 kmole of O_2 and 4 kmole of CO_2 (with a total mass equal to 5×28+3×32+4×44= 412 kg) are introduced into two identical piston–cylinder–weight assemblies A and B. We will assume that system A contains anti-catalysts or inhibitors which suppress any reaction while system B can engage in chemical reactions which result in the final presence of 4.998 kmole of CO, 2.999 kmole of O_2 and 4.002 kmole of CO_2. The species changes in system B are $dN_{CO} = -0.002$, $dN_{O2} = -0.001$ and $dN_{CO2} = 0.002$ kmole, respectively. The Gibbs energy change dG of system B is now determined by hypothetically injecting 0.002 kmole of CO_2 into system A and withdrawing 0.002 kmole of CO and 0.001 kmole of O_2 from it (so that total mass is still 412 kg) so as to simulate the final conditions in system B. The Gibbs energy $G_A = G + dG_{T,P}$. System A is open even though its mass has been fixed. The change $dG_{T,P}$, during this process is provided by Eq. (9e). Thus,

$$dG_{T,P;A} = (-0.002)\mu_{CO} + (-0.001)\,\mu_{O_2} + (+0.002)\,\mu_{CO_2}. \tag{12}$$

Since the final states are identical in both systems A and B, the Gibbs energy change $dG_{T,P;B}$ during this process must then equal $dG_{T,P;A}$. Therefore,

$$dG_{T,P;B} = -T\delta\sigma = dG_{T,P;A}, \text{ i.e., } (\Sigma\mu_k\, dN_k)_A < 0. \tag{13}$$

The sum of the changes in the Gibbs energy associated with the three species CO, CO_2, O_2 are

$$dG_{T,P;B} = -T\delta\sigma = (\mu_{CO}\, dN_{CO} + \mu_{O2}\, dN_{O2} + \mu_{CO2}\, dN_{CO2}) < 0, \text{ i.e.,} \tag{14a}$$

$$dG_{T,P,B} = (-0.002)\mu_{CO} + (-0.001)\mu_{O_2} + (+0.002)\mu_{CO_2} = -T\delta\sigma < 0 \tag{14b}$$

Note that system B is a chemically reacting closed system of fixed mass.

Recall that for irreversible processes involving adiabatic rigid closed systems $dS_{U,V,m} > 0$ and from Eq. (9g), $\Sigma\mu_k\, dN_k < 0$. This inequality involves constant (T,V) processes with A being minimized or constant (T,P) processes with G being minimized. If (S,V) are maintained constant for a reacting system (e.g., by removing heat as a reaction occurs), then $dU_{S,V,m} = \Sigma\mu_k dN_k < 0$ In this case S is maximized at fixed values of U, V and m, while U is minimized at specified values of S, V, and m. The inequality represented by Eq. (13) is a powerful tool for determining the reaction direction for any process.

c. Reacting Open System

If in one second, a mixture of 5 kmole of CO, 3 kmole of O_2, and 4 kmole of CO_2 flows into a chemical reactor and undergoes chemical reactions that oxidize CO to CO_2, the same criteria that are listed in Eqs. (4) through (8) can be used as long we follow a fixed mass. As reaction proceeds inside the fixed mass, the value of G should decrease at specified (T, P) so that $dG_{T,P} \le 0$.

5. Criteria in Terms of Chemical Force Potential

The reaction $CO + 1/2O_2 \rightarrow CO_2$, must satisfy the criterion provided by Eq. (14b) to proceed. In the context of the above discussion, dividing Eq. (15) by 0.002 or the degree of reaction,

$$dG_{T,P}/0.002 = (-1)\,\mu_{CO} + (-1/2)\,\mu_{O_2} + \mu_{CO_2} < 0. \tag{14c}$$

Recall that $\mu_k = \hat{g}_k$, i.e.,

$$dG_{T,P} = \Sigma\nu_k\,\hat{g}_k = \Sigma\nu_k\mu_k \le 0, \tag{14d}$$

where k and ν_k represent the reacting species and its stoichiometric coefficient for a reaction. In case of the reaction $CO + 1/2O_2 \rightarrow CO_2$, $\nu_{CO} = -1$, $\nu_{O2} = -1/2$, $\nu_{CO2} = 1$. Equation (14c) can be alternately expressed in the form

$$\mu_{CO} + \nu_{O2}\mu_{O_2} > \mu_{CO_2}.$$

Defining the chemical force for the reactants and products as

$$F_R = \mu_{CO} + (1/2)\,\mu_{O_2}, \text{ and } F_P = \mu_{CO_2} \text{ for reaction CO+ 1/2O2} \rightarrow \text{CO2} \tag{15}$$

The criterion $dG_{T,P} < 0$ leads to the relation $F_R > F_P$. This criterion is equally valid for an adiabatic closed rigid system (U, V, m specified), and adiabatic and isobaric system (H, P, m specified), isentropic rigid closed system (S, V, m specified), isentropic and isobaric systems (S, P, m specified), and, finally, isothermal and isovolume systems (T, V, m specified).

From Eq. (2),

$$d\xi = dN_k/\nu_k = dN_{CO}/(-1) = dN_{CO2}/(+1) = 0.002.$$

Replacing 0.002 in Eq. (14c) by $d\xi$, the stoichiometric coefficients by ν_k, and generalizing for any reaction

$$(\partial G/\partial \xi)_{T,P} = \Sigma \mu_k \nu_k < 0. \quad (16)$$

The Gibbs energy decreases as the reaction progresses and eventually reaches a minimum value at equilibrium. Defining the chemical affinity as

$$F = -(\partial G/\partial \xi)_{T,P}, \quad (17)$$

Equation (16) assumes the form

$$(-(\partial G/\partial \xi)_{T,P} = F = -\Sigma \mu_k \nu_k) > 0. \quad (18a)$$

Similarly, following the relations for dA, dU and dS, we can show that

$$(-(\partial A/\partial \xi)_{T,V} = F = -\Sigma \mu_k \nu_k) > 0, \quad (18b)$$

$$(-(\partial U/\partial \xi)_{S,V} = F = -\Sigma \mu_k \nu_k) > 0, \quad (19a)$$

$$(-(\partial H/\partial \xi)_{S,P} = F = -\Sigma \mu_k \nu_k) > 0, \text{ and} \quad (19b)$$

$$(T(\partial S/\partial \xi)_{U,V} = T(\partial S/\partial \xi)_{H,P} = F = -\Sigma \mu_k \nu_k) > 0. \quad (20)$$

The last expression shows that the entropy increases in an isolated system as chemical reaction proceeds. For a reaction to proceed under any of these constraints, the affinity $F > 0$.

In the CO oxidation example, the values of F for the reactants and products are

$$F_R = \mu_{CO} + (1/2)\,\mu_{O_2}, \text{ and } F_P = \mu_{CO_2}. \quad (21)$$

Since $(F_R - F_P) > 0$ for oxidation to proceed,

$$F_R > F_P, \quad (22)$$

which is similar to the inequality $T_{hot} > T_{cold}$ that allows heat transfer to occur from a hotter to a colder body. In a manner similar to the temperature (thermal potential), F_R and F_P are analogous intensive properties called chemical force potentials. The chemical potential μ_k is the same as partial molal Gibb's function $\hat{g}_k$, $(= \hat{h}_k - T\,\hat{s}_k)$, which is a species property. Each species has a unique way of distributing its energy and, thus, fixing the entropy. A species distributing energy to a larger number of states has a low chemical potential and is relatively more stable. During chemical reactions, the reacting species proceed in a direction to form more stable products (i.e., towards lower chemical potentials). The physical meaning of the reaction potential is as follows: For a specified temperature, if the population of the reacting species (e.g., CO and O_2) is higher (i.e., higher value of F_R) than the product molecules (i.e., CO_2 at lower F_P), then there is a high probability of collisions amongst CO and O_2 resulting in a reaction that produces CO_2. On the other hand, if the population of the product molecules (e.g., CO_2) is higher (larger F_P value) as compared to the reactant molecules CO and O_2 (i.,e., lower F_R), there is a higher probability of collisions amongst CO_2 molecules which will break into CO and O_2. If the temperature is lowered, the molecular velocities are reduced and the translational energy may be insufficient to overcome bond energy among the atoms in the molecules that is required to the potential $F(T, P\ X_i)$.

---

a. *Example 1*

Five kmole of CO, three of O_2, and four of CO_2 are instantaneously mixed at 3000 K and 101 kPa at the entrance to a reactor. Determine the reaction direction and the values of F_R, F_P, and G. What is the equilibrium composition of the gas leaving the reactor? How is the process altered if seven kmole of inert N_2 is injected into the reactor?

Solution

We assume that if the following reaction occurs in the reactor:

$$CO + 1/2\ O_2 \rightarrow CO_2, \text{ then} \quad (A)$$

$$F_R > F_P \quad (B)$$

so that the criterion $dG_{T,P} < 0$ is satisfied. The reaction potential for this reaction is

$$F_R = (1)\ \mu_{CO} + (1/2)\ \mu_{O_2}, \text{ and} \quad (C)$$

$$F_P = (1)\ \mu_{CO_2}. \quad (D)$$

For ideal gas mixtures,

$$\mu_{CO} = \hat{g}_{CO} = \bar{g}_{CO}(T,P) + \bar{R}T \ln X_{CO} = \bar{g}_{CO}(T,p_{CO}). \quad (E)$$

The larger the CO mole fraction, the higher the value of μ_{CO} and, hence, F.

$$\bar{g}_{CO}(T,P) = \bar{h}_{CO}(T,P) - T\ \bar{s}_{CO}(T,P)$$

$$= (\bar{h}_{f,CO}^{0} + (\bar{h}_{t,3000K} - \bar{h}_{t,298K})_{CO}) - 3000\times(\bar{s}^{o}_{CO}(3000) - 8.314(\ln\times P/1))$$

$$= (-110530+93541) - 3000\times273.508 - 8.314\times\ln 1)$$

$$\bar{g}_{CO} = -837513 \text{ kJ per kmole of CO.} \quad (F)$$

Similarly, at 3000K and 1 bar,

$$\bar{g}_{O_2} = -755099 \text{ kJ kmole}^{-1}, \text{ and } \bar{g}_{CO_2} = -1242910 \text{ kJ kmole}^{-1}. \quad (G)$$

The species mole fractions

$$X_{CO} = 5\div(5+3+4) = 0.417,\ X_{O_2} = 3\div(5+4+3) = 0.25, \text{ and } X_{CO_2} = 0.333. \quad (H)$$

Further,

$$\mu_{CO} = \hat{g}_{CO}\ (3000K, 1 \text{ bar}, X_{CO} = 0.417)$$

$$= \bar{g}_{CO}(3000K, 1 \text{ bar}) + 8.314\times3000\times\ln(0.417)$$

$$= -837513 + 8.314 \times 3000 \times \ln 0.467$$

$$= -856504 \text{ kJ kmole}^{-1} \text{ of CO in the mixture.} \quad (I)$$

Similarly,

$$\mu_{O_2} = (3000K, 1 \text{ bar}, X_{O_2}=0.25) = -789675 \text{ kJ per kmole of } O_2. \quad (J)$$

$$\mu_{CO_2} = (3000K, 1 \text{ bar}, X_{CO_2}=0.333) = -1270312 \text{ kJ per kmole of } CO_2. \quad (K)$$

Therefore, based on the oxidation of 1 kmole of CO,

$$F_R = -856504 + 1/2(-789675) = -1254190 \text{ kJ, and} \quad (L)$$

$$F_P = -1270312 \text{ kJ, i.e.,} \quad (M)$$

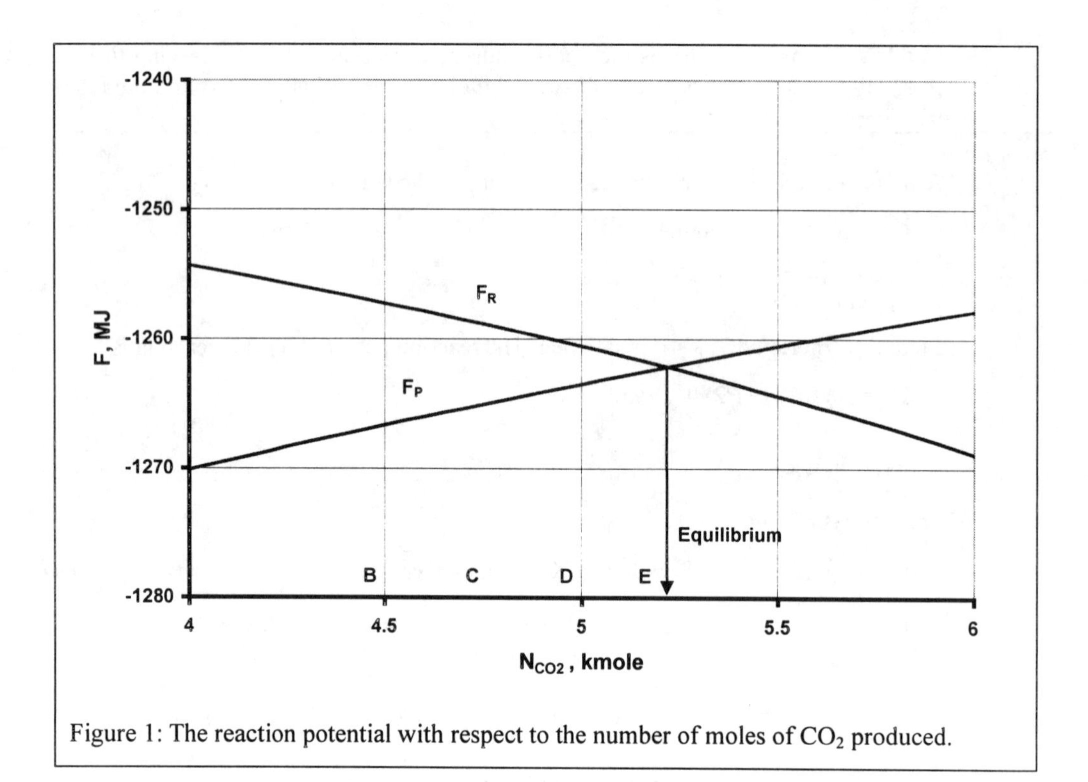

Figure 1: The reaction potential with respect to the number of moles of CO_2 produced.

$$F_R > F_P, \qquad (N)$$

which implies that assumed direction is correct and hence CO will oxidize to CO_2.
The oxidation of CO occurs gradually. As more and more moles of CO_2 are produced, its molecular population increases, increasing the potential F_P. Simultaneously, the CO and O_2 populations decrease, thereby decreasing the reaction potential F_R until the reaction ceases when chemical equilibrium is attained. Thus chemical equilibrium is achieved when $F_R = F_P$, i.e., $dG_{T,P}=0$. This is illustrated in Figure 1. The corresponding species concentrations are

$$N_{CO_2} = 5.25 \text{ kmole}, N_{CO} = 3.75 \text{ kmole, and } N_{O_2} = 2.375 \text{ kmole.}$$

(Recall the evaporation example discussed in Chapter 7 where A reaches a min imum value at specified values of T, V and G. From a thermodynamic perspective, this problem is similar to placing a cup of cold water in bone dry air. Evaporation will occur when $dG_{T,P} < 0$, but after a finite amount of water is transformed into the vapor, evaporation will cease at which $g_{H2O(\ell)} = g_{H2O(g)}$ and $dG_{T,P} = 0$.)

The Gibbs energy at any section

$$G = \Sigma\mu_k N_k = \mu_{CO} N_{CO} + \mu_{O_2} N_{O_2} + \mu_{CO_2} N_{CO_2}, \text{ i.e.,}$$

$$G = -856504 \times 5 - 789675 \times 3 - 1270312 \times 4 = -11,732,793 \text{ kJ.}$$

Figure 2 plots values of G vs N_{CO2}. The plot in Figure 2 shows that G reaches a minimum value when $F_R=F_P$.

Nitrogen does not participate in the reaction. Therefore, $dN_{N_2}= 0$ and, so, the expressions for F_R and F_P are unaffected. However, the mole fractions of the reactants change so that the values of F_R and F_P are different, as is the equilibrium composition. The G expression for this case is

$$G = \Sigma\mu_k N_k = \mu_{CO} N_{CO} + \mu_{O_2} N_{O_2} + \mu_{CO_2} N_{CO_2} + \mu_{CO_2} N_{CO_2} + \mu_{N2} N_{N2}$$

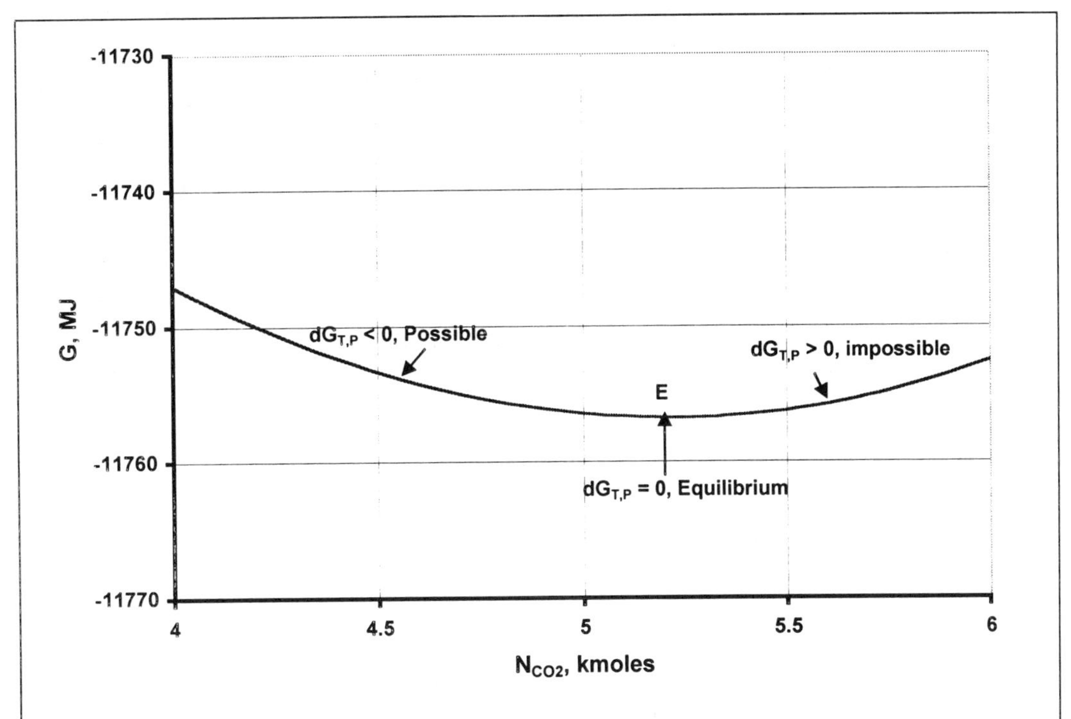

Figure 2: Illustration of the minimization of the Gibbs energy at equilibrium with respect to the number of moles of carbon dioxide produced.

Remarks

The overall reaction has the form

$$5\ CO + 3\ O_2 + 4\ CO_2 \rightarrow 3.75\ CO + 2.375\ O_2 + 5.25\ CO_2.$$

The assumed direction (i.e., $CO + 1/2\ O_2 \rightarrow CO_2$) is possible if $dG_{T,P} < 0$ or $F_R > F_P$. The mixture is at equilibrium if $dG_{T,P} = 0$ (as illustrated in Figure 2) or $F_R = F_P$. If $F_R < F_P$, the reverse reaction $CO_2 \rightarrow CO + 1/2\ O_2$ becomes possible.

---

6. Generalized Relation for the Chemical Potential

Recall from Chapter 8 that

$$\mu_k = \hat{g}_k = \bar{g}_k(T,P) + \bar{R}T \ln \hat{\alpha}_k, \text{ where} \tag{23}$$

$\hat{a}_k = \hat{f}_k/f_k$ is the ratio of the fugacity of species k in a mixture to the fugacity of the same species in its pure state. Equation (15) can be generalized for any reaction in the form

$$\Sigma\, \hat{g}_k\, dN_k = \Sigma\, \bar{g}_k(T,P) + \bar{R}T \ln \hat{\alpha}_k)\, dN_k \le 0, \tag{24}$$

where the activity coefficient $\hat{a}_k$ equals the species mole fraction for ideal mixtures and the equality applies to the equilibrium state.

---

b. Example 2

Consider the reactions

$$C\ (s) + 1/2\ O_2 \rightarrow CO, \text{ and} \tag{I}$$

$$C(s) + O_2 \rightarrow CO_2 \tag{II}$$

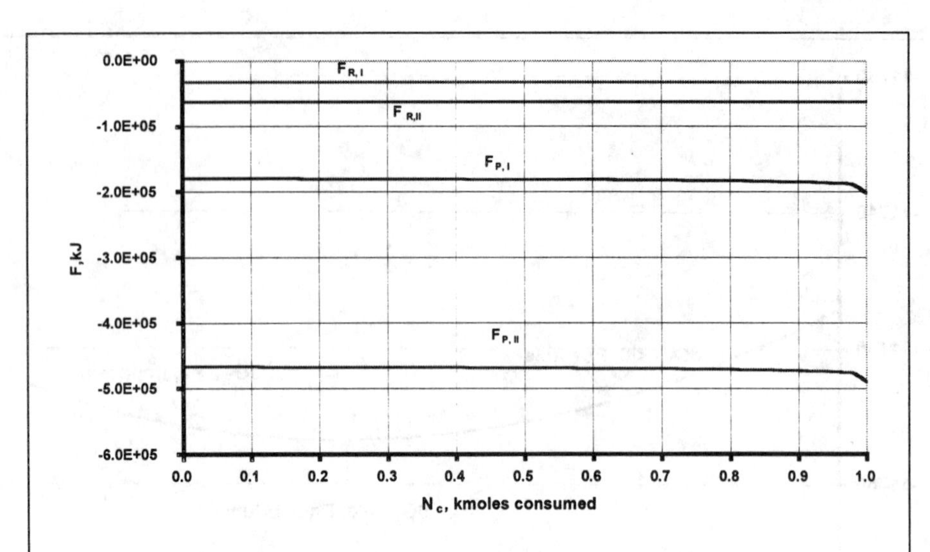

Figure 3: The reaction potentials for reactions I and II with respect to the number of moles of carbon that are consumed.

Which of the two reactions is more likely when 1 kmole of C reacts with 50 kmole of O_2 in a reactor at 1 bar and 298 K. Assume that $\bar{c}_{p,C}/\bar{R}$ = 1.771+0.000877 T–86700/T^2 in SI units and T is in K. Assume ideal mixture.

Solution

If $|(F_R - F_P)|_I > |(F_R - F_P)|_{II}$, then the first reaction dominates and vice versa. Note that the reaction potentials are functions of the species populations and hence vary as a reaction proceeds. Using Eq. (23),

$$F_R = \bar{g}_C(T,P) + \bar{R}T \ln \hat{\alpha}_k. \quad \text{(A)}$$

Since solid carbon (C(s)) is a pure component and hence the activity $\hat{a}_{C(s)} = 1$. Further,

$$\bar{h}_C = \bar{h}^o_{f,C} + \int_{298K}^{T} \bar{c}_{p,C} dT.$$

where $\bar{h}^o_{f,C}$= 0 kJ kmole^{-1}, and (B)

$$\bar{s}_C = \bar{s}^o_C(298K) + \int_{298K}^{T} (\bar{c}_{p,C}/T) dT.$$

Now,

$$\bar{s}^o_C\,(298K) = 5.74 \text{ kJ kmole}^{-1}\text{ K}^{-1}. \quad \text{(C)}$$

Hence, using Eqs. (B) and (C), $\bar{g}^o_C = \bar{g}_C(298K, 1 \text{ bar}) = \bar{h}_{298K} - 298 \times \bar{s}_C\,(298K)$, i.e.,

$$\bar{g}^o_C = 0 - 298 \times 5.74 = -1711 \text{ kJ kmole}^{-1}. \quad \text{(D)}$$

For solids and liquids, $\bar{g}_k(T,P) \approx \bar{g}^o_k(T)$. Assume that 0.001 moles of C(s) react with 0.0005 moles of O_2 to produce 0.001 moles of CO. Hence,

$$p_{O_2} = X_{O_2} P = (50 - 0.0005) \div (0.001 + (50 - 0.0005)) = 0.9999 \text{ P} = 0.9999 \text{ bar}.$$

Therefore,

$$\bar{s}_{O_2} = 205.03 - 8.314 \times \ln 0.9999 = 205.03 \text{ kJ K}^{-1}\text{ kmole}^{-1}, \text{ i.e.,}$$

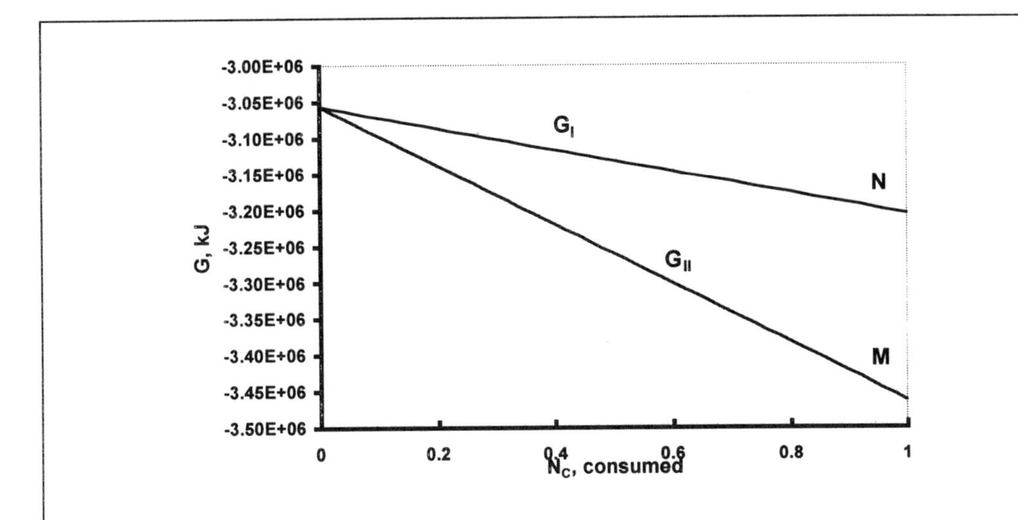

Figure 4: Variation in G_I and G_{II} with respect to the number of moles of carbon consumed for reactions I and II at 298 K.

$$\bar{g}_{O_2}(298K,\ 1\ bar) = 0 - 298 \times 205.03 = -61099\ kJ\ kmole^{-1}. \tag{E}$$

Similarly,

$X_{CO} = 0.001 \div (0.001 + 49.9995) \approx 0.00002$, and

$\bar{s}_{CO}(T, p_{CO}) = 197.54 - 8.314 \times \ln(0.00002) = 287.5\ kJ\ K^{-1}\ kmole^{-1}$, so that

$$\bar{g}_{CO}(298K,\ 1\ bar) = -110530 - 298 \times 287.5 = -196205\ kJ\ kmole^{-1}. \tag{F}$$

Employing Eqs. (D) and (E),

$F_R = \bar{g}_C + 1/2\ \bar{g}_{O_2} = -1710 + 0.5 \times (61099) = -32260$ kJ, and

$F_P = \bar{g}_{CO} = -196205 kJ\ kmole^{-1}$, i.e.,

$F_R - F_P = -32260 + 196205 = 163945$ kJ.

For reaction I,

$(dG/|dN_C|)_I = (dG/|d\xi|)_I = -(F_R - F_P)_I = -163945$ kJ.

For reaction (II), the corresponding amount of O_2 consumed is 0.0001 kmole while 0.0001 kmole of CO_2 is produced. Therefore,

$N_{O_2} = 50 - 0.001 = 49.999$,

$X_{O_2} = 0.999 \times (0.0001 + 49.999) = 0.999$,

$X_{CO_2} = 0.001 \times (0.001 + 49.999) \approx 0.00002$, and

Consequently,

$\bar{s}_{O_2} = 205.03 - 8.314 \times \ln(0.999) \approx 205.03\ kJ\ K^{-1}\ kmole^{-1}$,

$\bar{s}_{CO2} = 213.74 - 8.314 \times \ln(0.0002) \approx 303.70\ kJ\ K^{-1}\ kmole^{-1}$,

$\bar{g}_{O_2} = -61099\ kJ\ kmole^{-1}$, and

$\bar{g}_{CO_2} = -393546 - 298 \times 303.70 = -484048\ kJ\ kmole^{-1}$.

For this reaction

$$F_R = -1710 + (-61099) = -62810\ kJ, \text{ and } F_P = = -484048\ kJ\ kmole^{-1}, \text{ i.e.,} \tag{G}$$

$F_R - F_P = -62810 + 484048 = 421238$ kJ.

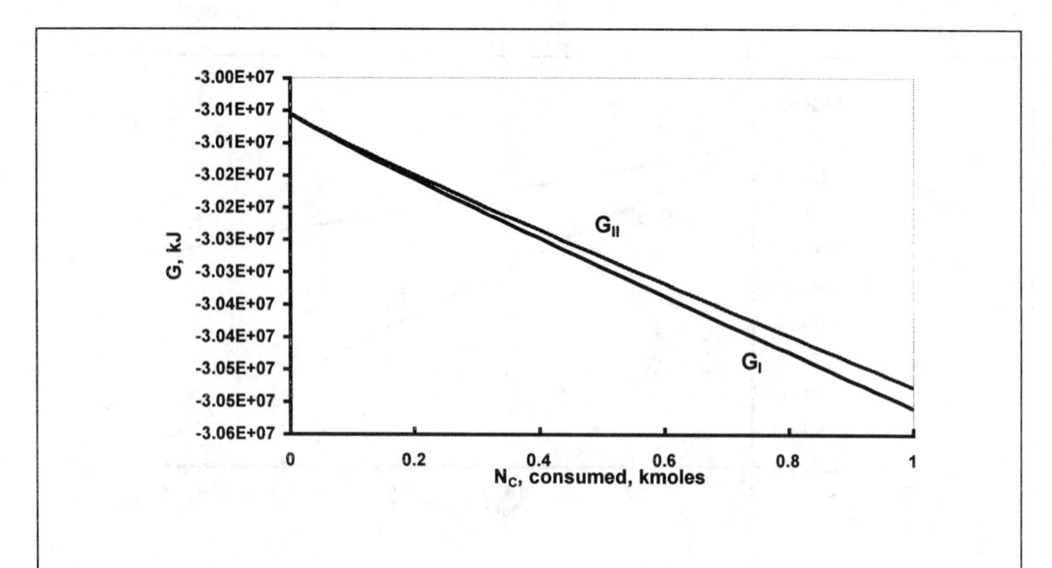

Figure 5: Variation in G_I and G_{II} with respect to the number of moles of carbon consumed for reactions I and II at 3500 K.

Hence,

$$(dG/|dN_C|)_{II} = (dG/d\xi)_{II} = -(F_R - F_P)_{II} = -394390 \text{ kJ}.$$

The variations in the reaction potentials for reactions I and II with respect to the number of moles of carbon that are consumed at a reactant temperature of 298 K are presented in Figure 3, and the corresponding variation in G_I and G_{II} in Figure 4. At 298 K CO_2 production dominates. The analogous variations in G_I and G_{II} at 3500 K are presented in Figure 5. At the higher temperature CO formation is favored.

Remarks

Since,

$$\bar{g}_k(T, P, X_k) = \bar{h}_k - T\bar{s}_k = \bar{h}_k - T(\bar{s}_k^o - \bar{R}\ln P X_k/1)$$
$$= \bar{h}_k - T\{\bar{s}_k^o - \bar{R}\ln(P/1)\} + \bar{R}T\ln X_k$$
$$= \bar{g}_k(T,P) + \bar{R}T\ln X_k,$$

in general, the values of $\bar{g}_k(T, P, X_k)$ are a function of the species mole fractions. If we assume that $|\bar{g}_k(T,P)| \gg |\bar{R}T\ln X_k|$, then $\bar{g}_k(T, P, X_k) \approx \bar{g}_k(T,P)$.

This offers an approximate method of determining whether reaction I or II is favored. For instance, if the reactions are assumed to go to completion, $\Delta G_I = \bar{g}_{CO} - (\bar{g}_C + 1/2\,\bar{g}_{O_2})$. Likewise, we can evaluate ΔG_{II} to determine whether $|\Delta G_{II}| > |\Delta G_I|$;if so, the CO_2 production reaction is favored. Values of $\Delta G(T,P)$ at 1 bar, i.e., $\Delta G^o(T)$ are tabulated. (Tables 27A and 27B at T= 298 K)

In addition to reactions I and II, consider the following reactions:

$$C(s) + CO_2 \rightarrow 2\,CO, \quad \text{(III)}$$

$$CO + 1/2\,O_2 \rightarrow CO_2, \text{ and} \quad \text{(IV)}$$

$$H_2 + 1/2\,O_2 \rightarrow H_2O. \quad \text{(V)}$$

Figure 6 plots value of ΔG^o with respect to the temperature for these five reactions. For instance, for reaction III, $\Delta G^o\,(298\text{ K}) = 2\bar{g}_{CO} - (\bar{g}_C + \bar{g}_{CO_2}) = 120080$ kJ, which a positive number or $F_R = \bar{g}_{CO2} + \bar{g}_C < F_P = 2\,\bar{g}_{CO}$. This implies that the reaction cannot proceed in the indicated direction. In reaction III, the reaction potential of the products (F_P) is initially low and the value of F_R is higher. However, the equilibrium state is reached at a very low CO concentration when $dG_{T,P} = 0$, i.e., $F_R = F_P$. Thereafter, $F_P > F_R$ or $dG_{T,P} > 0$, and the reaction does not proceed. In other words, $\Delta G^o > 0$ implies that

$$C + CO_2 \rightarrow \text{large amounts of leftover C and } CO_2 + \text{small amounts of } CO_2. \quad \text{(H)}$$

On the other hand for reaction II, $\Delta G^o < 0$ implies that

$$C + O_2 \rightarrow \text{small amounts of leftover C and } O_2 + \text{large amounts of } CO_2. \quad \text{(I)}$$

Generally, the value of ΔG^o for a reaction indicates the extent of completion of that reaction. A relatively large negative value of ΔG^o implies that $F_R \gg F_P$, and this requires the largest decrease in the reactant population (or extent of completion of reaction) before chemical equilibrium is reached. Normally, a positive value for ΔG implies that the reaction will produce an insignificant amount of products (reaction III).

We will now show that the value of ΔG^o for reaction IV can be obtained in terms of the corresponding values for reactions I and II. For reactions I, II, and IV, respectively

$$\Delta G^o_I(298\text{ K}) = \bar{g}_{CO} - (\bar{g}_C + 1/2\,\bar{g}_{O_2}), \quad \text{(J)}$$

$$\Delta G^o_{II}(298\text{ K}) = \bar{g}_{CO_2} - (\bar{g}_C + \bar{g}_{O_2}), \text{and} \quad \text{(K)}$$

$$\Delta G^o_{IV}(298\text{ K}) = \bar{g}_{CO_2} - (\bar{g}_{CO} + 1/2\,\bar{g}_{O_2}), \quad \text{(L)}$$

where the $\bar{g}_k$'s are evaluated at 298 K, i.e., $\bar{g}_k = \bar{g}^o_k$. Equation (L) assumes the form

$$\Delta G^o_{IV}(298\text{ K}) = \bar{g}_{CO_2} - (\bar{g}_C + \bar{g}_{O_2}) - \{\bar{g}_{CO} - (\bar{g}_C + 1/2\,\bar{g}_{O_2})\} = \Delta G^o_{II} - \Delta G^o_I, \text{ i.e.,} \quad \text{(M)}$$

$$\Delta G^o_{IV}(298\text{ K}) = \Delta G^o_{II} - \Delta G^o_I = \bar{g}_{CO_2}(298\text{ K}) - \bar{g}_{CO}(298\text{ K})$$
$$= -394390 + 137137 = -257253\text{ kJ}.$$

We can arbitrarily set $\bar{g}^o_k = 0$ for the elemental species C and O_2 at T=298 K so that

$$\Delta G^o_I(298\text{ K}) = \bar{g}^0_{f,CO}(298\text{ K}),\ \Delta G^o_{II}(298\text{ K}) = \bar{g}^0_{f,CO2}(298\text{ K}).$$

where $\bar{g}^0_{f,k}$ is called Gibbs' function of formation of species k from elements in natural form.

C. CHEMICAL EQUILIBRIUM RELATIONS

For the reaction $CO_2 \rightarrow CO + 1/2\, O_2$ to occur, $F_R (= \hat{g}_{CO_2}) > F_P (= \hat{g}_{CO} + 1/2\,\hat{g}_{O_2})$. In general,

$$dG_{T,P} = \Sigma\, \hat{g}_k\, dN_k \le 0. \quad (25)$$

Since the change in the mole numbers is related to the stoichiometric coefficients, then at specified values of T and P,

$$\Sigma \nu_k \{\bar{g}_k\, dN_k(T,P) + \bar{R}T \ln \hat{\alpha}_k\} \le 0. \quad (26)$$

e.g., for the reaction $CO_2 \rightarrow CO + \_\ O_2$, $\nu_{CO2} = -1$, $\nu_{CO} = +1$, $\nu_{O2} = +1/2$.

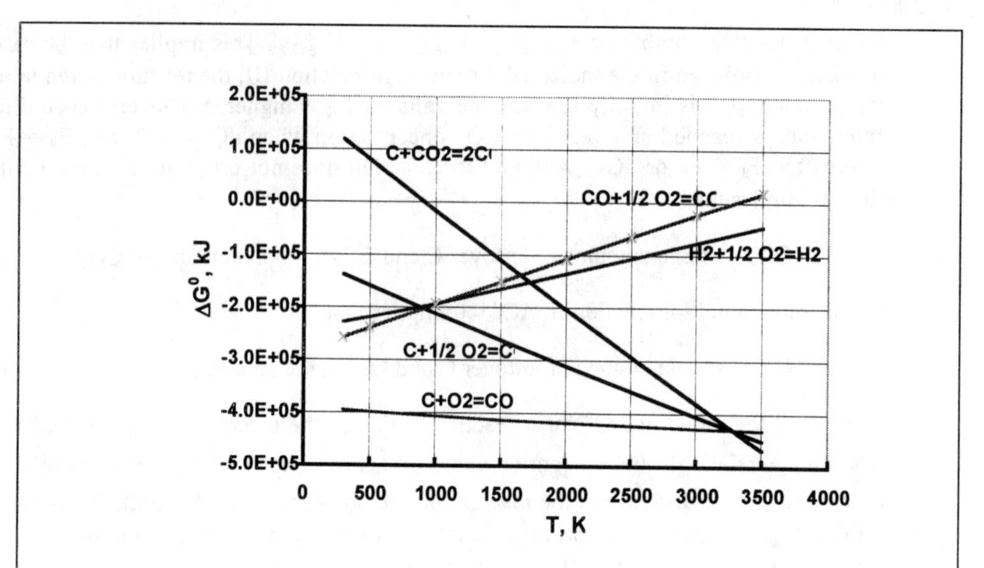

Figure 6: The variation in the value of ΔG^o with respect to the temperature for several reactions.

1. Nonideal Mixtures and Solutions

Rewriting Eq. (26),

$$\Sigma \ln \hat{\alpha}_k^{\nu_k} \le -\Sigma \nu_k \, \bar{g}_k (T,P)/(\bar{R}T). \quad (27)$$

We define the term

$$K(T,P) = \Pi \, \hat{\alpha}_k^{\nu_k} = \exp(-\Sigma \nu_k \, \bar{g}_k (T,P)/(\bar{R}T)) \quad (28)$$

which is constant at specified values of T and P and Eq. (27) assumes the form

$$\Pi \, \hat{\alpha}_k^{\nu_k} \le K(T,P), \text{ or} \quad (28a)$$

$$K(T,P) \ge \Pi \, \hat{\alpha}_k^{\nu_k} \quad (28b)$$

The physical meaning of this relation is as follows. Consider an ideal gas mixture of 5 kmole of CO, 3 kmole of O_2 and 4 kmole of CO_2 in a PCW assembly in which a reaction proceeds at fixed (T,P). Here, $\hat{a}_k = X_k$ and $K(T,P)$ remains constant while X_k changes. Equation (28) must be satisfied as the reaction proceeds and the equality holds good at chemical equilibrium.

Unlike the superheated steam tables in which properties are tabulated as functions of (T,P), K(T,P) is tabulated typically only at P_0, the standard pressure, since simple relations are available (particularly for ideal gases) to relate K(T,P) to $K(T,P_0)$. Such a relation is provided below. Recall from Chapter 8 that for a species k at a state characterized by specified (T,P)

$$\bar{g}_k (T,P) = \bar{g}_k (T,P_o) + (\bar{R}T \ln (f_k(T,P)/f_k(T,P_o)), \quad (29)$$

where f is the fugacity of species k. If that species is an ideal gas, $f_k(T,P) = P$ and $f_k(T,P_0) = P_0$. The second term on the RHS represents the deviation from this behavior at P_o. Selecting $P_o = 1$ bar,

$$\bar{g}_k(T,P) = \bar{g}_k^o(T) + (\bar{R}T \ln (f_k(T,P)/f_k(T, 1\text{bar})), \text{ so that} \quad (30)$$

using this relation in Eq. (28),

$$K(T,P) = \exp(-\Sigma\nu_k \bar{g}_k^o(T)/(\bar{R}T)) \exp(-\Sigma\nu_k \ln(f_k(T,P)/f_k(T, 1bar))). \quad (31)$$

Now let us define

$$\ln K^o(T) = -\Sigma\nu_k \bar{g}_k^o(T)/(\bar{R}T), \text{ i.e.,} \quad (32)$$

$$K^o(T) = \exp(-\Sigma\nu_k \bar{g}_k^o(T)/(\bar{R}T)) = \exp(-\Delta G^o/(\bar{R}T)), \text{ where} \quad (33)$$

the term $K^o(T)$ is *conventionally* called the equilibrium constant (a misnomer since it is a function of temperature and is constant only if the temperature is also held constant) and is tabulated for many standard reactions in Table A-28B. Using Eqs. (31) and (32), the relation between $K(T,P)$ and $K^o(T)$ is

$$K(T,P) = K^o(T)\, \Pi\, (f_k(T,P)/f_k(T, 1bar))^{(-\nu k)}. \quad (34)$$

Subsequently, using Eq. (34) in Eq. (28b), as the reaction proceeds, the following inequality must be satisfied:

$$K^o(T) \geq \Pi\, (f_k(T,P)\, \hat{\alpha}_k/f_k(T, 1bar))^{\nu_k}. \quad (35')$$

At chemical equilibrium

$$K^o(T) = \Pi\, (f_k(T,P)\, \hat{\alpha}_k/f_k(T, 1bar))^{\nu_k}. \quad (35a)$$

In order to evaluate $f_k(T, 1bar)$ and the corresponding $K^o(T)$ it is convenient to define the state of each species k at a standard state corresponding to a pressure of 1 bar.

a. Standard State of an Ideal Gas at 1 Bar

If one considers the state of species k to be an ideal gas at 1 bar, then $f_k(T, 1bar) = 1$. Hence, using Eq. (30),

$$\bar{g}_k(T,P) = \bar{g}_k^o(T) + \bar{R}T \ln(f_k(T,P)/1) \quad (35b)$$

where $\bar{g}_k^o(T)$ is evaluated at the temperature T assuming the species k to be an ideal gas. Since the substance is non-ideal at the state (T,P), the second term in Eq. (35b) accounts for the correction due to non-ideal behavior.

The values of $\bar{g}_k^o(T)$ can be determined using the expression

$$\bar{g}_k^0(T) = \bar{h}_k^0(T) - T\, \bar{s}_k^0(T), \quad (35c)$$

where the term $\bar{h}_k^0(T)$ includes chemical and thermal enthalpies. The $\bar{g}_k^0(T)$ can also be estimated using Gibbs function of formation (Chapter 11). Unless otherwise stated, Eq. (35c) will be for determining the values of $\bar{g}_k^0(T)$.

b. Standard State of a Nonideal Gas at 1 Bar

Consider the following chemical reaction

$H_2O(\ell) + CO(g) \rightarrow CO_2(g) + H_2(g)$.

The reaction involves both liquid $H_2O(\ell)$ and gaseous species. With Eq. (30), we obtain:

$$\bar{g}_{k(\ell)}(T, P) = \bar{g}_{k(\ell)}^0(T) + \bar{R}T \ln \frac{f_{k(\ell)}(T, P)}{f_{k(\ell)}(T, 1)}, \text{ where} \quad (36a)$$

$$f_{k(\ell)}(T, P) = f_{k(\ell)}(T, P^{sat})\, POY_k \approx f_{k(\ell)}(T, P^{sat}),$$

and the Poynting correction factor is (cf. Chapter 8)

$$POY_k = \exp\left\{\int_{P^{sat}}^{P} v_{k(\ell)} dP/(RT)\right\} \approx 1$$

Since POY ≈1 for most liquids and solids, Eq. (36a) assumes the form

$$\bar{g}_{k(\ell)}(T, P) \approx \bar{g}^0_{k(\ell)}(T) \quad (36b)$$

---

c. *Example 3*

Determine the value of $\bar{g}_{H_2O(\ell)}$(40°C, 10 bar) and $\bar{g}_{H_2O(g)}$(40°C, 10 bar) for water vapor. Select the standard state to be at 1 bar at 40°C. Assume ideal gas behavior at the standard state. Also determine the value of $\hat{g}^{id}_{H_2O(\ell)}$ at 40°C and 10 bar for a salt water mixture in which water constitutes 85% on mole basis.

Solution

For water,

$$\bar{g}_{H2O(\ell)}(313\text{ K},10\text{ bar}) =$$

$$\bar{g}^0_{H_2O(g)}(313\text{ K}) + 8.314\times313\times\ln(f_{H_2O(\ell)}(313\text{ K, 10 bar})/1)$$

$$= \bar{g}_{H_2O(g)}(313\text{ K, 1 bar}) + 8.314\times313\times\ln(f_{H_2O(\ell)}(313\text{ K, 10 bar})/1), \text{ where} \quad (A)$$

$$f_{H_2O(\ell)}(T,P)/\ f_{H_2O(\ell)}(T,P^{sat}) = POY. \quad (B)$$

In the case of water, P^{sat} = 0.07384 bar at 40°C, and

$$POY = \exp\left(\int_{0.074}^{10} \bar{v}dP/(\bar{R}T)\right) = \exp(0.00101\times(10-0.074)\times100\div(0.461\times313))$$

$$= \exp(0.0000695) = 1.007 \approx 1. \quad (C)$$

Therefore,

$$f_{H_2O(\ell)}(T,P)/\ f_{H_2O(\ell)}(T,P^{sat}) \approx 1. \quad (D)$$

Since $f_{H2O(\ell)}(T,P^{sat}) = f_{H2O(g)}(T, P^{sat})$ and the vapor is an ideal gas, then $f_{H2O(g)}(T, P^{sat}) = P^{sat}$ = 0.07384 bar and

$$f_{H_2O(\ell)}(313\text{ K, 1 bar}) = 0.074\text{ bar}. \quad (E)$$

Using Eqs. (A) and (E),

$$\bar{g}^0_{H_2O(\ell)}(313\text{ K},10\text{ bar}) \approx \bar{g}^0_{H_2O(g)}(313\text{ K, 1 bar}) + 8.314\times313\times\ln(0.07384/1). \quad (F)$$

Now,

$$\bar{g}^0_{H_2O(g)}(313\text{ K, 1 bar}) = (\bar{h}^0_{H_2O(g)} - T\bar{s}^0_{H_2O(g)})_{313\text{ K, 1 bar}}$$

$$= -241321 - 313\times190.33 = -300894\text{ kJ kmole}^{-1}. \quad (G)$$

Using Eqs. (A), (E), and (F),

$$\bar{g}^0_{H_2O(\ell)}(313\text{ K},10\text{ bar}) = -300894 + 8.314\times313\times\ln(0.07384/1)$$

$$= -307675\text{ kJ kmole}^{-1}. \quad (H)$$

If the liquid state is selected at 1 bar instead of an ideal gas state, Eq. (A) becomes

$\bar{g}_{H_2O(\ell)}(T,P) = \bar{g}_{H_2O(\ell)}(T,1\text{ bar}) + 8.314 \ln(f_{H_2O(\ell)}(T,P)/ f_{H_2O(\ell)}(T,1\text{ bar}))$, where (I)

At given T, RT d ln (f) = v dP (Chapter 8) and hence

$$\ln(f_{H_2O(\ell)}(T,P)/ f_{H_2O(\ell)}(T,1\text{ bar})) = \int_1^{10} \bar{v}\, dP/\bar{R}T$$

$$= 0.00101\times(10-1)\times100\div(0.461\times313) = 0.0063, \text{ so that} \quad \text{(J)}$$

$$f_{H_2O(\ell)}(313\text{ K}, 10\text{ bar}) \approx (f_{H_2O(\ell)}(313\text{ K}, 1\text{ bar}) \quad \text{(K)}$$

Using Eqs. (I) and (K),

$\bar{g}^{o}_{H_2O(\ell)}(313\text{ K},10\text{ bar}) = \bar{g}^{o}_{H_2O(\ell)}(313\text{ K},1\text{ bar})$, where ($\ell$)

$$\bar{g}^{o}_{H_2O(\ell)}(313\text{ K}, 1\text{ bar}) = (\bar{h}^{o}_{H_2O(\ell)} - T\bar{s}^{o}_{H_2O(\ell)})_{313\text{ K}, 1\text{ bar}}$$

$$= -285830+4.184\times(40-25)\times18.02-313\times(69.95+4.184\times18.02\times\ln(313\div298))$$

$$= -307752\text{ kJ kmole}^{-1} = \bar{g}^{o}_{H_2O(\ell)}(313\text{ K}, 10\text{ bar}), \quad \text{(M)}$$

which almost equals the previous answer (Eq. (H)).

$\hat{g}^{id}_{H_2O(\ell)}(T,P) = \bar{g}_{H_2O(\ell)}(T,P) + \bar{R}T \ln X_{H_2O}$, i.e.,

$$\hat{g}^{id}_{H_2O(\ell)}(313\text{ K}, 10\text{ bar}) = \bar{g}_{H_2O(\ell)}(313\text{ K}, 10\text{ bar}) +$$

$$8.314 \times 313 \times \ln 0.85 = -308159\text{ kJ kmole}^{-1}.$$

Remarks

We have selected the standard state with regard to both the liquid and gaseous states. For liquids or solids, $\bar{g}(T,P) \approx \bar{g}^{o}(T)$ $f_k(T,P) \approx f_k(T,1)$ and hence $K(T,P) \approx K^{o}(T)$.

---

2. Reactions Involving Ideal Mixtures of Liquids and Solids

For ideal mixtures of liquids and solids $\hat{\alpha}_k = X_k$, and Eq. (28b) assumes the form

$$K(T,P) \geq \Pi X_k^{\nu_k}, \quad k: \text{liquid and solid phases}, \quad (37)$$

where X_k denotes the mole fraction of solid or liquid species k (e.g., $CaSO_4(s)$ in a mixture of $Fe_2O_3(s)$, $CaSO_4(s)$, and $CaO(s)$). For a solid,

$$\bar{g}_{k(s)}(T,P) = \bar{g}_{k(s)}(T, 1\text{ bar}) + \bar{R}T \ln (f_{k(s)}(T,P)/f_k(T,1\text{ bar})).$$

Since,

$$(f_{k(s)}(T, P)/f_k(T, 1\text{ bar})) = POY = \exp(-v_s(P-1)/(\bar{R}T)), \text{ and } POY \approx 1,$$

$$\bar{g}_{k(s)}(T, P) \approx \bar{g}_{k(s)}(T, 1\text{ bar}).$$

If K(T, P) is evaluated with respect to $\bar{g}_{k(s)}(T, P)$ ($\approx \bar{g}_{k(s)}$ (T, 1 bar)), then for solids and liquids

$$K(T, P) \approx K^{o}(T) \geq \Pi X_k^{\nu_k}, \text{ liquid or solid mixtures} \quad (38)$$

3. Ideal Mixture of Real Gases

For an ideal mixture of real gases

$$\hat{\alpha}_k = \frac{\hat{f}_k^{id}(T,P,X_k)}{f_k(T,P)} = X_k, \tag{39}$$

where $\hat{f}_k$ denotes the fugacity of component k inside the mixture and f_k the pure species fugacity (cf. Chapter 8). Selecting the standard state for all species to be that for an ideal gas at 1 bar, using Eq. (35a)

$$K^o(T) \geq \Pi(X_k f_k(T,P)/1), \text{ ideal mix of real gas mixtures,} \tag{40}$$

where $f_k(T,P)$ are in units of pressure (bar). Recall that $f_k(T,P) = \phi_k P$ where ϕ_k is the fugacity coefficient

4. Ideal Gases

For ideal gases, $f_k(T, P) = P$ and $\hat{\alpha}_k(T,P,X_k) = X_k$. Therefore, $f_k(T, 1 \text{ bar}) = 1$, and Eq. (35a) assumes the form

$$K^o(T) \geq \Pi\{X_k (P/1)^{\nu_k}\}, \text{ or} \tag{41a}$$

$$\Pi\{X_k (P/1)^{\nu_k}\} \leq K^o(T). \tag{41b}$$

Alternately, one may use the relation $\hat{g}_k(T,P) = \bar{g}_k^o(T) + \bar{R}T \ln(p_k/1)$ where $p_k = X_k P$ and group all terms involving $\bar{g}_k^o(T)$ on one side and the partial pressure terms on the other side. For the reaction $CO + 1/2O_2 \rightarrow CO_2$, it can be shown that

$$-\Delta G^0(T)/\bar{R}T \geq \ln(p_{CO2}/1)^1 - \ln(p_{CO}/1)^1 - \ln(p_{O2}/1)^{1/2}), \text{ where}$$

$\Delta G^0(T) = \bar{g}^0_{CO2}(T) - \bar{g}^0_{CO}(T) - 1/2\ \bar{g}^0_{O2}(T)$. For convenience the $\bar{g}^0_k$ values are tabulated in Tables A-8 to A-19. Equation (41b) stipulates that at any specified temperature, $K^o(T)$ is constant and while the reaction occurs, the partial pressure terms on the left hand side of Eq. (41b) keep increasing to approach the value of $K^o(T)$.

Since

$$p_k = X_k, \tag{42}$$

$$\Pi(p_k/1)^{\nu_k} \leq K^o(T). \tag{43}$$

Some texts use the nomenclature K_p instead of K^o to indicate that partial pressures are involved on the left hand side of Eq. (43).

a. Partial Pressure

Consider the following reaction

$$CO_2 \rightarrow CO + 1/2\ O_2. \tag{44}$$

The relation for the equilibrium constant is

$$K^o(T) = \exp(\Delta G^o(T)/\bar{R}T)), \text{ i.e.,} \tag{45}$$

Rewriting Eq. (43),

$$K^o(T) \geq (p_{CO}/1)^1 (p_{O_2}/1)^{1/2}/(p_{CO_2}/1)^1. \tag{46}$$

b. Mole Fraction

Replacing the partial pressures with mole fractions (e.g., $p_{CO} = X_{CO}\ P$),

$$K^o(T) \geq \{ (P/1)^{1/2}\ (X_{CO})^1 (X_{O_2})^{1/2}/(X_{CO_2})^1 \}. \tag{47}$$

Furthermore, since

$$X_k = N_k/N, \tag{48}$$

Equation (47) assumes the form,

$$K^o(T) \geq\ [\{P/(1\times N)\}^{1/2} (N_{CO})^1 (N_{O_2})^{1/2}]/(N_{CO_2})^1. \tag{49}$$

We have retained the 1 bar term in Eqs. (47)–(49) to indicate that the equilibrium constant is a dimensionless quantity.

---

d. Example 4

Consider a mixture with the following composition at 1800 K and 2 MPa, i.e., N_{CO_2} = 1.2 kmole, N_{O_2} = 0.6 kmole, N_{CO} = 3.6 kmole, and N_{N_2} = 6.6 kmole. In which direction will the following reaction proceed: $CO + 1/2\ O_2 \rightarrow CO_2$, or $CO_2 \rightarrow CO + 1/2\ O_2$ if we maintain T and P?

Solution

Consider one of the directions for the reaction, say,

$$CO_2 \rightarrow CO + 1/2\ O_2,$$

$$\bar{g}_k^{\,0}(T) = \bar{h}_k^{\,0}(T)\ -\ T\ \bar{s}_k^{\,0}(T)$$

Using Table A-8,

$$\bar{g}^o_{CO} = -110530 + 49517 - 1800 \times 254.8 = -519650\ \text{kJ kmole}^{-1}.$$

Similarly, using Table A-9,

$$\bar{g}^o_{O_2} = 51660\ -\ 1800\ \times 264.701 = -424800\ \text{kJ kmole}^{-1},\ \text{and}$$

from Table A-19

$$\bar{g}^o_{CO_2} = -\ 393546 + 79399 - 1800 \times 302.892 = -859355\ \text{kJ kmole}^{-1},\ \text{and}$$

$$\Delta G^o = (1)\times(-519650)+(1/2)\times(-424800)+(-1)\times(-859355)= 127305\ \text{kJ}.$$

(If one uses the "g' values in Tables A-8, A-9, and A-19, then ΔG^o = (1)×(–269164)+(1/2)×(0)+(–1)×(–396425)= 127261 kJ, which is the almost the same answer that we have obtained above.)

$$K^o(T) = \exp\ (-127305 \div (8.314\times 1800)) = 0.00020.$$

From Tables A-28 B log 10 ($K^0(T)$) = -3.696, i.e., $K^0(T)$ = 0.0002.

For the reaction $CO_2 \rightarrow CO + 1/2\ O_2$ to occur, Eq. (46) must be satisfied. For the specified composition,

$$X_{CO} = 3.6/12 = 0.3,\ p_{CO} = 0.3 \times 20 = 6\ \text{bar}.$$

Similarly p_{O2}= 1 bar, p_{CO2} = 2 bar. The ratios of the partial pressures

$$(p_{CO}/1)^1\ (p_{O_2}/1)^{1/2}/(p_{CO_2}/1)^1 = (6/1)^1\ (1/1)^{1/2}/(2/1)^1 = 3.$$

The criterion

$$K^o(T) = 0.0002 \geq (p_{CO}/1)^1\ (p_{O_2}/1)^{1/2}/(p_{CO_2}/1)^1,\ \text{or}$$

$$p_{CO}/1)^1\ (p_{O_2}/1)^{1/2}/(p_{CO_2}/1)^1 \leq K^o(T) = 0.0002$$

is violated.

Therefore, CO will oxidize to CO_2, i.e., the reverse path is favored.

---

e. *Example 5*

A piston–cylinder assembly contains 2 kmole of O_2. A weight is placed on the top of piston such that the pressure is 1 bar. The gas is then instantaneously heated to 3000 K and maintained at this temperature. We find that O atoms are formed and the concentration of O_2 molecules decreases. Determine the equilibrium composition.

Solution

If x denotes the moles of O_2 and y the moles of O, then

$$2x + y = 4. \qquad (A)$$

Chemical equilibrium is attained after the following reaction

$$O_2 \rightarrow 2\,O \qquad (B)$$

ceases, i.e., according to equality sign in Eq. (46)

$K^o(T) = (p_O/1)^2/(p_{O_2}/1)$, where

$$p_{O_2} = X_{O_2}\,P = (x/(x+y))P, \text{ and} \qquad (C)$$

$$p_O = (y/(x+y))P \qquad (D)$$

Therefore, equilibrium is attained when

$K^o(T) = (y^2/(x(x+y)))(P/1)$.

Since P = 1 bar, $K^o(T) = 0.0127$, and

$$0.0127 = (y^2/(x(x+y))) \qquad (E)$$

Using Eqs. (A) and (E), we obtain a quadratic equation in terms of x, i.e.,

$(4-2x)^2 = 0.0127 \times (4-x)$.

We can solve for x and select the root, such that x>0, and y>0, i.e.,

x = 1.8875 kmole, and y = 4 – 2x = 4 – 2 × 1.8875 = 0.225 kmole.

Remarks

This problem can also be solved by minimizing G $(= x\,\hat{g}_{O2}(T,P,X_{O2}) + y\,\hat{g}_{N2}(T,P,X_{N2}))$ at the specified values of T and P, subject to restriction given by Eq. (A). When there are a large number of species, say O, O_2, O_3, etc., we resort to the minimization of the Gibbs energy and the LaGrange multiplier method can be used. See later parts of this chapter.

---

f. *Example 6*

One kmole of air is in a closed piston–cylinder–weight assembly placed at 298 K and 1 bar. Trace amounts of NO and NO_2 are generated according to the overall reaction

$$0.79\,N_2 + 0.21\,O_2 \rightarrow a\,NO + b\,NO_2 + c\,N_2 + d\,O_2. \qquad (A)$$

Determine the NO and NO_2 concentrations at chemical equilibrium, assuming the following decomposition reactions

$$N_2 + O_2 \rightarrow 2\,NO, \text{ and} \qquad (B)$$

$$N_2 + 2\,O_2 \rightarrow 2\,NO_2. \qquad (C)$$

The values of Gibbs function of formation $\bar{g}^o(298\text{ K})$ for NO and NO_2 are, respectively 86550 and 51310 kJ $kmole^{-1}$, and for the elements "j" in their natural forms $\bar{g}_j^o(298\text{ K}) = 0$ kJ $kmole^{-1}$. Assume that the gases are ideal.

Solution

For reactions (B) and (C)

$$K^o_{NO} = (p_{NO}/1)^2/((p_{N_2}/1)(p_{O_2}/1)), \text{ and} \quad \text{(D)}$$

$$K_{NO_2} = (p_{NO_2}/1)^2/((p_{N_2}/1)(p_{O_2}/1)^2). \quad \text{(E)}$$

(We will use the expression for $K^o_{NO,\,NO_2} = \exp(-\Delta G^o_{NO,NO_2}/\bar{R}T)$, where

$\Delta G^o_{NO} = 2\ \bar{g}^o_{NO} - \bar{g}^o_{N_2} - \bar{g}^o_{O_2} = 2 \times 86550 = 173100$ kJ per kmole of N_2.)

Therefore,

$K^o_{NO} = 4.54\times10^{-31}$,

Since NO exists in trace quantities the partial pressures of N_2 and O_2 in Eq. (D) are virtually unaffected by reactions (B) and (C). Hence,

$(p_{NO}/1) = (4.54 \times 10^{-31}\times(0.79/1)\times(0.21/1))^{1/2} = 2.75\times10^{-18}$, i.e.,

$X_{NO} = 2.75\times10^{-18}$ or NO = 2.75×10^{-12} ppm.

Similarly,

$K^o_{NO_2} = \exp(-\Delta G^o_{NO_2}/RT) = (p_{NO_2}/1)^2/((p_{N_2}/1)(p_{O_2}/1)^2)$, where

$\Delta G^o_{N_2O} = 2\ \bar{g}^o_{NO_2} - \bar{g}^o_{N_2} - 2\ \bar{g}^o_{O_2} = 2\times51310 = 102620$ kJ per kmole of N_2.

Thus,

$K^o_{NO_2} = 1.03\times10^{-18}$,

$(p_{NO_2}/1) = (4.54\times10^{-31}\times(0.79/1)\times(0.21/1)^2)^{1/2} = 1.69\times10^{-10}$, and

$X_{NO_2} = 1.69\times10^{-4}$ ppm.

g. *Example 7*

5 kmole of CO, 3 of O_2, 4 of CO_2, and 7 of N_2 are introduced into a reactor at 3000 K and 2000 kPa. Determine the equilibrium composition of gas leaving reactor, assuming that the outlet (product) stream contains CO, O_2, N_2, and CO_2. Will the equilibrium composition change if the feed is altered to 6 kmole of CO, 3 kmole of CO_2, 3.5 kmole of O_2, and 7 kmole of N_2 enter the reactor? Assume that the outlet stream contains the same species. Will the CO concentration at the outlet change if the pressure changes, say to 101 kPa?

Solution

Assume that the chemical reaction proceeds according to the reaction

$$CO_2 \rightarrow CO + 1/2\ O_2$$

so that

$K^o(T) = ((p_{CO}/1)(p_{O_2}/1)^{1/2})/(p_{CO_2}/1)$, where

$p_{CO} = X_{CO}\ P = (N_{CO}/N)\ P$, and

$N = N_{CO} + N_{O_2} + N_{CO_2} + N_{N2}$

Therefore,

$$K^o(T) = (N_{CO}N^{1/2}_{O_2})\{P/(1\times N)\}^{1/2}/N_{CO_2}. \quad \text{(A)}$$

The conservation of C and O atoms provide two additional equations. The overall balance equation in terms of the three unknown concentrations is

$$5CO + 3O_2 + 4CO_2 + 7N_2 \rightarrow N_{CO}CO + N_{O_2}O_2 + N_{CO_2}CO_2 + N_{N2}\ N_2 \quad \text{(B)}$$

There are four unknowns N_{CO}, NCO_2 and NO_2) and we have three atom balance equations.

Carbon atoms: $N_C = 5 + 4 = N_{CO} + N_{CO2}$ (C)

Oxygen atoms: $N_O = 5 \times 1 + 3\times 2 + 4 \times 2 = N_{CO} \times 1 + N_{CO_2} \times 2 + N_{O_2} \times 2$ (D)

Nitrogen atoms : $N_N = 7\times 2 = N_{N2}\times 2$ (D')

The fourth equation is given by equilibrium condition at 3000 K: $CO_2 \Leftrightarrow CO + 1/2\ O_2$ reaction, $\log_{10}K = -0.48$. Using this value in Eq. (A)

$$0.327 = (N_{CO}N_{O_2}^{1/2})(P/(1\times N))^{1/2}/N_{CO_2}. \quad \text{(E)}$$

Using Eq. (C),

$$N_{CO} = N_C - N_{CO_2},\ \text{i.e.,}\ N_{CO} = 9 - N_{CO_2}. \quad \text{(F)}$$

Further, using Eqs. (D) and (F),

$$N_{O_2} = (N_O - N_C - N_{CO_2})/2,\ \text{i.e.,} \quad \text{(G)}$$

$$N_{O_2} = (19 - 9 - N_{CO_2})/2. \quad \text{(H)}$$

Therefore, the number of moles at the exit

$$N = N_{CO} + N_{O_2} + N_{CO_2} + N_{N2} = (N_C - N_{CO_2}) + (N_O - N_C)/2 + N_{CO_2}$$

$$= N_{N2} + (N_O + N_C - N_{CO_2})/2 = 21 - N_{CO_2}/2 \quad \text{(I)}$$

Applying Eqs. (A) and (G)–(I), at 20 bar, at the exit

$N_{CO_2} = 6.96$ kmole, and

$N_{CO} = 2.04$ kmole, $N_{O_2} = 1.52$ kmole, and $N = 17.52$ kmole.

When the feed stream is altered to react 6 kmole of CO, 3 kmole of CO_2, 3.5 kmole of O_2, and 7 kmole of N_2, the respective inputs of C, O and N atoms remain unaltered at 9, 19 and 14 respectively. Therefore, the equilibrium composition is unchanged. This indicates that it does not matter in which form the atoms of the reacting species enter the system. The same composition, for instance, could be achieved by reacting a feed stream containing 9 kmole of C(s) (solid carbon, such as charcoal), 9.5 kmole of O_2 and 7 kmole of N_2 (which is treated as an inert in this problem).

From Eq. (A) we note that for a specified temperature, the value of $K^o(T)$ is unique. Therefore, if the pressure changes, the temperature does not. Eq. (E) dictates that the composition is altered and more CO_2 is produced as the pressure is increased.

---

h. Example 8

Consider a PCW assembly that is immersed in an isothermal bath at 3000 K. It initially consists of 9 kmole of C atoms and 19 kmole of O atoms (total mass = $9\times 12.01 + 19\times 16 = 412$ kg) is allowed to reach chemical equilibrium at 3000 K and 1 bar. What is the equilibrium composition? What is the value of the Gibbs energy? If we keep placing sand particles one at a time on the piston to a final pressure of 4 bar, i.e., we have allowed sufficient time for chemical equilibrium to be reached at that pressure. What is the resulting equilibrium composition and Gibbs energy?

Solution

We leave it to the reader to show that at equilibrium

N_{CO_2} = 5.25 kmole, N_{CO} = 3.7 kmole, and N_{O_2} = 2.37 kmole. (A)

Therefore,

$N = \Sigma N_k$ = 11.37 kmole. (B)

The Gibbs energy,

$G = N_{CO_2}\, \hat{g}_{CO_2} + N_{CO}\, \hat{g}_{CO} + N_{O_2}\, \hat{g}_{O_2}$, where (C)

$\hat{g}_{CO_2} = \bar{g}^0_{CO_2}(T) + \bar{R}T \ln(p_{CO_2}/1) = \bar{h}^0_{f,CO_2} + (\bar{h}_{t,T} - \bar{h}_{t,298K}) - [\bar{s}^0 - \bar{R}T \ln(X_{CO_2}P/1)]$, (D)

$X_{CO_2} = N_{CO_2}/N = 0.462$.

At 3000 K and 1 bar, $\hat{g}_{CO2} = -1262000$ kJ kmole^{-1}, $\hat{g}_{CO} = -865200$ kJ kmole^{-1} and $\hat{g}_{O2} = -794400$ kJ kmole^{-1}. Hence,

$G = -11,753,000$ kJ,

which at this equilibrium state must be at a minimum value.

At a temperature of 3000 K and a pressure of 4 bar, the equilibrium composition changes to

N_{CO_2} = 6.4 kmole, N_{CO} = 2.6 kmole, and N_{O_2} = 1.8 kmole, and (E)

$N = \Sigma N_k$ = 10 kmole, and $G = -11374000$ kJ. (F)

Remarks

There is no entropy generated since there is no irreversibility. The difference in the minimum Gibbs free energies (i.e., at the equilibrium states) between the two states (3000 K,10 bar) to (3000 K,4 bar) is

$dG = -SdT + VdP$, $dG_T = VdP = (N\bar{R}T/P)\, dP$, or

$G_2(3000,1) - G_1(3000, 4) = (-11374000) - (-11757000) = 383000$ kJ.

If $N \approx$ constant $\approx (11.37+10)/2 = 10.69$ kmole, then $dG_T = (N\bar{R}T/P)\, dP$. Integrating,

$G_2 - G_1 \approx N\bar{R}T \ln(P_2/P_1) = 369455$ kJ.

The relations $dU = T\,dS - P\,dV$, $dH = T\,dS + V\,dP$, $dG = -S\,dT + V\,dP$, etc., for closed systems can be applied even for chemical reactions as long as we connect a reversible path between the two equilibrium states. However these equations can not be applied during irreversible chemical reactions. Such a statement is also true for non-reacting systems.

If we compress the products very slowly from 1 to 4 bar isothermally at 3000 K, the reaction tends to produce more CO_2, i.e., N_{CO2} increases from 5.25 to 6.4 kmole. Instead, rapid compression to 4 bar produces an insignificant change from the composition at 1 bar, i.e., the products will be almost frozen at $N_{CO2} = 5.25$, $N_{CO} = 3.75$, $N_{O2} = 2.37$ even though the state is now at 3000 K and 4 bar. The products during this initial time are in a nonequilibrium state. The value of G at this state is G_{Frozen} = $5.25\times \hat{g}_{CO2}(3000, 4\text{ bar}, X_{CO2}=5.25/11.37)$ + $\hat{g}_{CO}(3000, 4\text{ bar}, X_{CO}=3.75/11.37)$ + $\hat{g}_{O2}(3000, 4\text{ bar}, X_{O2}=3.75/11.37) = 5.25 \times (-1,227,400) + 3.75 \times (-830,600) + 2.37 \times (-759,800) = -11,360,000$ kJ, which is higher than $G = -11,374,000$ kJ at the equilibrium composition corresponding to 3000 K, 4 bar. If we allow more time at the state (3000 K, 4 bar) and examine the composition after a long while, then chemical equilibrium will have been reached, and G will have approached its minimum value of $-11,360,000$ kJ.

A similar phenomenon occurs when these reacting gases flow at the slowest possible velocity through a diffuser where the pressure at the diffuser exit is 4 bar. If we follow the 412 kg mass when it flows through the diffuser, it will reach its equilibrium composition given by Eq. (D). However, if the same mass flows at high velocity, the composition at the exit of the diffuser can be almost the same as at the inlet.

5. Gas, Liquid and Solid Mixtures

Consider the following chemical reactions

$$CaCO_3(s) \rightarrow CaO(s) + CO_2(g), \quad (50)$$

$$H_2O(\ell) + CO(g) \rightarrow CO_2(g) + H_2(g), \text{ and} \quad (51)$$

$$CaO(s) + SO_2(g) + 1/2\ O_2(g) \rightarrow CaSO_4(s). \quad (52)$$

All three reactions consider species in two separate phases called heterogeneous reactions. For reaction given by Eq. (52) the equilibrium relation follows from Eq. (35a), i.e.,

$$K^0(T) = \frac{f_{CaSO4(s)}(T,P)\hat{\alpha}_{CaSO4(s)} / f_{CaSO4(s)}(T,1)}{\dfrac{f_{SO2(g)}(T,P)\hat{\alpha}_{SO2(g)}}{f_{SO2(g)}(T,1)} \dfrac{f_{O2(g)}(T,P)\ \hat{\alpha}_{O2(g)}}{f_{O2(g)}(T,1)} \dfrac{f_{CaO(s)}(T,P)\hat{\alpha}_{CaO(s)}}{f_{CaO(s)}(T,1)}} \quad (53)$$

If $CaSO_4$ and CaO are fully mixed at the molecular level, assuming that the solid mixture is an ideal solution,

$$\hat{\alpha}_{CaSO_4(s)} = X_{CaSO_4(s)}, \text{ and } \hat{\alpha}_{CaO(s)} = X_{CaO(s)}, \text{ where } X_{CaSO_4(s)} + X_{CaO(s)} = 1. \quad (54)$$

For the ideal solution $f_{k(s)}(T, P) \approx f_{k(s)}(T, 1)$. Assuming ideal gas behavior for SO_2,

$$\hat{\alpha}_{SO_2} = X_{SO_2}, \ \hat{\alpha}_{O_2} = X_{O_2}, f_{SO_2}(T, P) = P, \text{ and } f_{O_2}(T, P) = P. \quad (55)$$

Therefore, Eq. (53) assumes the form

$$K^o(T) = (X_{CaSO_4(s)}/(X_{SO_2} X_{O_2}^{1/2} X_{CaO(s)}) (P/1)^{-3/2}. \quad (56)$$

---

i. Example 9

Determine the relations between the partial pressures and temperature for the following scenarios: pure $H_2SO_4(\ell)$ dissociating upon evaporation, $H_2SO_4(\ell) \rightarrow H_2O(g) + SO_3(g)$, and an ideal mixture of 40% volatile $H_2SO_4(\ell)$ and 60% nonvolatile liquid or solid participating in the same reaction. Assume that $\bar{g}_{H_2SO_4} = -690013$ kJ kmole^{-1}, $\bar{g}_{H_2O(g)} = -228572$ kJ kmole^{-1}, $\bar{g}_{SO_3(g)} = -371060$ kJ kmole^{-1} at 298 K and 1 bar. Determine $(p_{H_2O(g)})(p_{SO_3(g)})$ at 298 K for pure $H_2SO_4(\ell)$ and for an ideal mixture of 40% volatile $H_2SO_4(\ell)$ and 60% nonvolatile liquid.

Solution

Pure $H_2SO_4(\ell)$

The problem involves a mixture of phases. We will select the standard state to be the liquid state for $H_2SO_4(\ell)$. Then from Eq. (35a),

$$K^o(T) = \Pi(f_k(T,P)\ \hat{\alpha}_k/f_k(T, 1 \text{ bar}))^{\nu_k}.$$

The gaseous species are assumed to be ideal so that

$$f_{k(g)}(T, P) = P, f_{k(g)}(T, 1 \text{ bar}) = 1, \text{ and } \hat{\alpha}_{k(g)}(T, P, X_k) = X_k.$$

In the liquid phase at 1 bar

$$f_{H_2SO_4}(T, P)/f_{H_2SO_4}(T, 1 \text{ bar}) \approx 1.$$

Further, since the liquid is pure,

$\hat{\alpha}_{H_2SO_4} = 1$, and

$\hat{\alpha}\ K^o(T) = (p_{H_2O(g)}/1)^1 (p_{SO_3(g)}/1)^1$.

At 298 K, when $X_{H_2SO_4} = 1$

$(p_{H_2O(g)})(p_{SO_3(g)}) = 1.44\times10^{-16}$,

which indicates that the partial pressures are very low, i.e., there is negligible dissociation.

Ideal Liquid Mixture:

Since the liquid phase is in an ideal mixture, the activity of $H_2SO_4(\ell) = X_{H_2SO_4}$, and

$K^o(T) = (P\,X_{H_2O(g)}/1)^1\ (P\,X_{SO_3(g)}/1)^1/X_{H_2SO_4}$

$= (p_{H_2O(g)}/1)(p_{SO_3(g)}/1)/X_{H_2SO_4}$, i.e.,

$(p_{H_2O(g)})(p_{SO_3(g)}) = X_{H_2SO_4}\,K^o(T)$.

The Gibbs energy change

$\Delta G^o(T) = \bar{g}_{H_2O(g)} + \bar{g}_{SO_3(g)} - \bar{g}_{H_2SO_4}$

$= -228572 - 371060 + 690013 = 90381$ kJ kmole^{-1}, and

$\ln K^o(T) = -\Delta G^o(T)/(\bar{R}T) = -36.48$, i.e.,

$K^o(T) = 1.44\times10^{-16}$.

At 298 K, when X = 0.4,

$(p_{H_2O(g)})(p_{SO_3(g)}) = 0.4\times1.44\times10^{-16} = 0.576\times10^{-16}$.

Remarks

During the vaporization of $H_2SO_4(\ell)$ it is possible to produce H_2O, SO_2, SO_3, and O_2, rather than $H_2SO_4(g)$. The pertinent reactions are

$H_2SO_4(\ell) \rightarrow H_2O(g) + SO_2(g) + 1/2\ O_2(g)$, and

$H_2SO_4(\ell) \rightarrow H_2O(g) + SO_3(g)$

At equilibrium, can you determine the SO_2 and SO_3 concentrations?

j. *Example 10*

Consider the water gas shift reaction $H_2O(\ell) + CO \rightarrow H_2(g) + CO_2(g)$. Determine the equilibrium constant for the reaction at 298 K, treating the gaseous species as ideal.

Solution

Since CO, H_2 and CO_2 are treated as ideal gases, $f_k = P$, $\hat{a}_{H2O} = 1$, $f_{H2O(\ell)}(T,P) \approx f_{H2O(\ell)}(T,1)$ and for others $\hat{\alpha}_k = X_k$ so that Eq. (35a) transforms to

$$K^o(T) = (p_{H_2}/1)(p_{CO_2}/1)/((p_{CO}/1)\,X_{H_2O}) = (p_{H_2}/1)(p_{CO_2}/1)/(p_{CO}/1)$$

$$= \exp(-(\bar{g}^o_{H_2} + \bar{g}^o_{CO_2} - \bar{g}^o_{CO} - \bar{g}^o_{H_2O})/\bar{R}T). \quad (A)$$

Using tabulated values,

$\bar{g}^o_{H_2O(\ell)} = \bar{h}_{H_2O} - T\bar{s}_{H_2O}$

$= -285830 - 298\times69.95 = -306675$ kJ kmole^{-1} of $H_2O(\ell)$,

$\bar{g}^o_{CO} = -110530 - 298\times197.56 = -169403$ kJ kmole^{-1} of CO,

$\bar{g}^o_{CO_2} = -393520 - 298\times213.7 = -457203$ kJ kmole^{-1} of CO_2, and

$\bar{g}^o_{H_2} = 0 - 298\times130.57 = -38910$ kJ kmole^{-1} of H_2.

Therefore,

$\Delta G^o = -38910 - 457203 - (-169403 - 306675)$

$= -20035$ kJ kmole^{-1} of CO, and

$K^{\circ}(298 \text{ K}) = \exp(20{,}035 \div (8.314 \times 298)) = 3250.4$.

Remarks

Since K°(298 K) is extremely large, $X_{CO_3} X_{H_2}/X_{CO}$ is also large, and, consequently, the value of X_{CO} at chemical equilibrium is extremely small. Therefore, if CO gas is bubbled through a vast reservoir of $H_2O(\ell)$ at 298 K, very little unreacted CO is left over. Note that the results pertain only to an equilibrium condition. However, the time scale required to reach it may be inordinately large.

k. *Example 11*

Consider the reaction of SO_3(g) with CaO(s), a process that is used to capture the SO_3 released during the combustion of coal, i.e., CaO(s) + SO_3(g) $\rightarrow$ $CaSO_4$(s). Determine the equilibrium relation assuming that the sulfates and CaO are mixed at the molecular level (i.e., they are mutually soluble) and are unmixed. What is the partial pressure of SO_3 at 1200 K if $K^{\circ}(1200 \text{ K}) = 2.93 \times 10^7$.

Solution

Assume the standard for solids to be solid and for gases to be ideal gases at P = 1 bar and use the approximation that $f_k(s)(T,P) \approx f_k(s)(T, 1\,\text{bar})$.

Since gas SO_3 behaves like an ideal gas,

$$\hat{\alpha}_{SO_3} = X_{SO_3}, \; f_{SO_3}(T, P) = P.$$

The solid phase contains both $CaSO_4$ and CaO. Assuming the ideal solution model for the solid phase,

$$\hat{\alpha}_{CaSO_4} = X_{CaSO_4}, \text{ and } \hat{\alpha}_{CaO} = X_{CaO} = 1 - X_{CaSO_4}.$$

The equilibrium relation for the reaction is,

$$K^{\circ}(T) = X_{CaSO_4}/(X_{CaO}\,(P\,X_{SO_3})/1).$$

If the solids are not mixed at the molecular level, they exist separately. Therefore, $X_{CaSO_4} = X_{CaO} = 1$, and

$$K^{\circ}(T) = 1/((P\,X_{SO_3})/1).$$

In the unmixed case,

$$2.93 \times 10^7 = 1/(p_{SO_3}/1), \text{ i.e., } p_{SO_3} = 0.41 \times 10^{-8} \text{ bar}.$$

Remarks

If the pressure P = 1 bar, then for the unmixed case $X_{SO_3} = p_{SO_3}/P = 0.41 \times 10^{-8}$ or 0.0041 ppm (parts per million). If the pressure is isothermally increased, the value of p_{SO_3} remains unchanged, but X_{SO_3} decreases, i.e., more of the sulfate will be formed.

In many instances, in power plants SO_2 released due to coal combustion is allowed to react with lime in order to produce sulfates according to the following reaction

$$CaO(s) + SO_2 + 1/2O_2 \rightarrow CaSO_4\,(s).$$

The equilibrium relation for this reaction is

$$K^{\circ}(T) = 1/((p_{SO_2}/1)(p_{O_2}/1)^{1/2})) = (1/(X_{SO_2} X_{O_2}^{1/2}))\,(P/1)^{-1/2}.$$

Increasing the pressure at constant T, causes X_{SO_2} to decrease so that a lesser amount of SO_2 will be emitted, i.e., more SO_2 is captured from the combustion gases.

l. *Example 12*

One kmole each of C(s) and O_2 enter a reactor at 298 K. The species CO, CO_2, and O_2 leave the reactor at 3000 K and 1 bar at equilibrium. Find the value of the equilibrium composition at the exit. What is the heat transfer across the boundary? What will happen if the inlet stream is altered to contain 1/2 kmole of oxygen and one kmole of CO? Also explain what happens if the outlet contains C(s), CO, and O_2.

Solution

The overall chemical reaction is

$$C(s) + O_2 \rightarrow a\ CO_2 + b\ CO + c\ O_2 \quad (A)$$

The species leaving the reactor are in an equilibrium state so that the following reaction must be in equilibrium, namely,

$$CO_2 \rightarrow CO + 1/2\ O_2. \quad (B)$$

From an atom balance,

$$\text{C atoms } 1 = a + b \quad (C)$$

$$\text{O atoms: } 2 = 2a + b + 2c. \quad (D)$$

Therefore,

$$b = 1 - a, \text{ and } c = (1 - a)/2. \quad (E,F)$$

The total moles leaving the reactor are

$$N = a + b + c = (3 - a)/2. \quad (G)$$

The exit equilibrium condition requires that

$$K^o(T) = p_{CO}\ p_{O2}\ \_ /p_{CO2} \quad (H)$$

For the carbon dioxide dissociation reaction at 3000 K,

$K^o(3000\text{ K}) = 0.327$.

Since,

$X_{CO} = b/N$, $X_{CO_2} = a/N$, $X_{O_2} = c/N$, and $p_k = X_k \times 1$ bar,

Solving the three unknowns a,b and c from Eqs. (B), (C) and (H)

a = 0.563, b = 0.437, and c = 0.219.

Applying the First Law

$$dE_{cv}/dt = \dot{Q}_{cv} - \dot{W}_{cv} + \Sigma_{,k}\ \dot{N}_{ik}\ \hat{h}_{i,k} - \Sigma_k\ \dot{N}_{e,k}\ \hat{h}_{e,k}, \quad (I)$$

Under steady state, $dE_{cv}/dt = 0$, and there is no work transfer. Thus, for every kmole of C(s),

$$\bar{q} = \frac{\dot{Q}}{\dot{N}_{c(s)}} == 0.563\ \bar{h}_{CO_2}(3000\text{ K}) + 0.437\ \bar{h}_{CO}(3000\text{ K}) + 0.219\ \bar{h}_{O_2}(3000\text{ K}) -$$

$$\bar{h}_{C(s)}(298\text{ K}) - 1/2\ \bar{h}_{O_2}(298\text{ K}) = 121426 \text{ kJ per kmole of C(s) consumed.} \quad (J)$$

If the inlet stream is altered to contain 1/2 kmole of oxygen and one kmole of CO, the atom balance remains unchanged. Therefore, the outlet composition will remain unaltered. However, the heat transfer will change, since the inlet stream containing CO and O_2 has a lower enthalpy as compared to the mixture of C(s) and O_2. Hence, the value of Q will be lower.

If CO,O2 and C(s) are present at the outlet, the overall chemical reaction is

$$C(s) + 1\ O_2 \rightarrow b\ CO + c\ O_2 + d\ C(s) \quad (K)$$

There are two atom balance equations, and we will also consider the following reaction to be in equilibrium, i.e.,

$$C(s) + 1/2\ O_2 \rightarrow CO, \text{ so that} \quad (L)$$

$$K^o(T) = p_{CO}/1/((p_{O_2}/1)^{1/2} (f_{C(s)}(T,P)/f_{C(s)}(T,1)))$$

Since $(f_{C(s)}(T,P) \approx f_{C(s)}(T,1))$,

$$K^o(T) = p_{CO}/1/((p_{O_2}/1)^{1/2}) = (X_{CO}/X_{O_2}) (P/1)^{1/2}. \quad (M)$$

Hence, from O atom and C atom balances

$2= b+2c$, $b + d = 1$; thus $c = (2-b)/2$, $d = (1-b)$, and $N = b+c = (b + 2)/2$, so that

$$(b = 2/\{(P/2)^{1/2}/K^o(T) + 1\}) < 1, \text{ and } d = 1 - b.$$

For the $C(s) + 1/2\ O_2 \rightarrow CO$,

$$K^o(3000\ K) = 10^{6.4}.$$

Since $K^o(3000\ K)$ is relatively large, Eq. (M) suggest that $X_{O_2} \approx 0$, i.e., almost all of the oxygen is consumed and converted into product.

---

m. Example 13

Let the number of kmole of O_2 entering a reactor equal the value a, while the number of moles of O_2, CO, CO_2, and C(s) leaving that reactor equal N_{O_2}, N_{CO}, N_{CO_2}, and $N_{C(s)}$. Assume the following reactions (respectively, A and B) to be in equilibrium: $C(s)+ 1/2\ O_2 \rightarrow CO$, and $CO_2 \rightarrow CO + 1/2\ O_2$. Determine $N_{CO}(T)$ and minimum amount of carbon(s) that should enter the reactor so as to maintain equilibrium at the exit.

Solution

O atom balance: $2a = 2 N_{O_2} + N_{CO} + 2 N_{CO_2}$, and (C)

$$K_A = N_{CO}/N_{O_2}^{1/2} (P/(1\times N))^{1/2}. \quad (D)$$

Likewise,

$$K_B = (N_{CO} N_{O_2}^{1/2}/N_{CO_2} (P'/N)^{1/2}, \quad (E)$$

where $P' = P/1$. Hence,

$$K_B/K_A = N_{O_2}/N_{CO_2}, \text{ and} \quad (F)$$

Since $2a = 2 N_{O_2} + N_{CO} + 2 N_{O_2}(K_A/K_B)$,

$$N_{O_2} = (2a - N_{CO})/(2(K_A/K_B +1)), \text{ and} \quad (H)$$

$$N = N_{CO} + (2a - N_{CO})/(2(K_A/K_B + 1)) + (K_A/K_B) (2a - N_{CO})/(2(K_A/K_B + 1))$$

$$= (2\ a + N_{CO})/2. \quad (I)$$

Therefore, Eq. (D) becomes

$$K_A = (N_{CO}/((2a - N_{CO})/(2(K_A/K_B + 1)))^{1/2}) (2\ P'/(2\ a + N_{CO}))^{1/2}, \text{ or}$$

$$K_A/(K_A/K_B + 1)^{1/2} = N_{CO} (4\ P'/(4a^2 - N_{CO}^{\ 2}))^{1/2}, \text{ i.e.,}$$

$$K_A^{\ 2} K_B/(K_B + K_A) = 4\ N_{CO}^{\ 2} P'/(4a^2 - N_{CO}^{\ 2}), \quad (J)$$

$$N_{CO}(T) = 2\ a/(1 + 4\ (P'/K_A)\ (1/K_A+1/K_B))^{1/2}, \text{ and} \quad (K)$$

$$X_{CO}(T) = N_{CO}/N = 2/(1 + (1 + 4\ (P'/K_A)\ (1/K_A + 1/K_B))^{1/2})). \quad (L)$$

X_{CO_2} can be similarly expressed. The required carbon atom input is

$$N_{C,in} \geq N_{CO} + N_{CO_2}, \quad \text{(M)}$$

where the equality applies to the minimum carbon input required for achieving chemical equilibrium. A mathematical expression for the minimum carbon input is

$$N_{C,in,min}/a = [\{(K_A/K_B)+1\}/\{1+4(1/K_B+1/K_A)P'/K_A\}^{/2} + 2(K_A/K_B)]/(2+K_B/K_A).$$

---

6. van't Hoff Equation

The van't Hoff Equation for $K^o(T)$ is due to Jacobus Henricus van't Hoff (1852–1911). It presents a relation between the equilibrium constant $K^o(T)$ and the enthalpy of reaction ΔH_R.

a. Effect of Temperature on $K^o(T)$

Recall from Chapter 7 that

$$(\partial(\bar{g}_k/T)/\partial(1/T))_P = \bar{h}_k. \quad (57)$$

Consider the reaction: $CO_2 \rightarrow CO + 1/2\ O_2$, for which

$$(\partial(\bar{g}_k/T)/\partial(1/T))_P = \bar{h}_k, \text{ and} \quad (58)$$

$$\Delta G(T, P) = \bar{g}_{CO}(T, P) + 1/2\ \bar{g}_{O_2}(T, P) - \bar{g}_{CO_2}(T, P), \text{ i.e.,} \quad (59)$$

$$\partial(\Delta G/T)/\partial(1/T) = \Delta H_R(T,P). \quad (60)$$

The enthalpy of reaction

$$\Delta H_R\ (T, P) = \bar{h}_{CO}(T, P) + 1/2\ \bar{h}_{O_2}(T, P) - \bar{h}_{CO_2}(T, P). \quad (61)$$

At 1 bar,

$$d(\Delta G^o/T)/d(1/T) = \Delta H_R^o(T). \quad (62)$$

Since $\ln K^o(T) = -\Delta G^o/\bar{R}T$,

$$d \ln K^o(T)/dT = \Delta H_R^o(T)/\ \bar{R}T^2, \quad (63)$$

which is known as the van't Hoff equation. If $\Delta H_R^0 \approx$ constant,

$$\ln K^o(T) = -\ \Delta H_R^o/\bar{R}T + \text{constant}, \quad (64)$$

which is a linear relationship. Figure 7 presents plots of $\ln K^o(T)$ vs $1/T$ for various reactions and the approximate relation of Eq. (64) appears to be valid. If $K^o = K_{ref}^o$ at $T = T_{ref}$, the constant in Eq. (64) can be eliminated, i.e.,

$$\ln (K^o/K_{ref}^o) = -(\Delta H_R^o/\bar{R})\ (1/T - 1/T_{ref}). \quad (65)$$

This relation can be written in the form

$$\ln K^o(T) = A - B/T, \text{ where} \quad (66)$$

$$A = \ln K_{ref}^o + (\Delta H_R^o/\bar{R})\ (1/T_{ref}), \text{ and} \quad (67)$$

$$B = (\Delta H_R^o/\bar{R}). \quad (68)$$

Eq. (66) is a linear relationship. Figure 7 presents plots of $\ln K^o(T)$ *vs.* $1/T$. If $T_{ref} = T_o$, we can simplify the constant A so that

$$A = (-\Delta G^o(T_o)/\overline{R}\,T_o) + (\Delta H^{\,o}_R(T_o)/\overline{R}\,T_o).$$

Since $\Delta G^o = \Delta H^{\,o}_R - T_o\Delta S^{\,o}_R$

$$A = \Delta S^{\,o}_R(T_o)/\overline{R}, \text{ and} \tag{69}$$

$$B = \Delta H^{\,o}_R(T_o)/\overline{R}. \tag{70}$$

If $\Delta H^{\,o}_R > 0$ (e.g., endothermic decomposition reactions), $B > 0$ and $K^o(T)$ increases with tem-

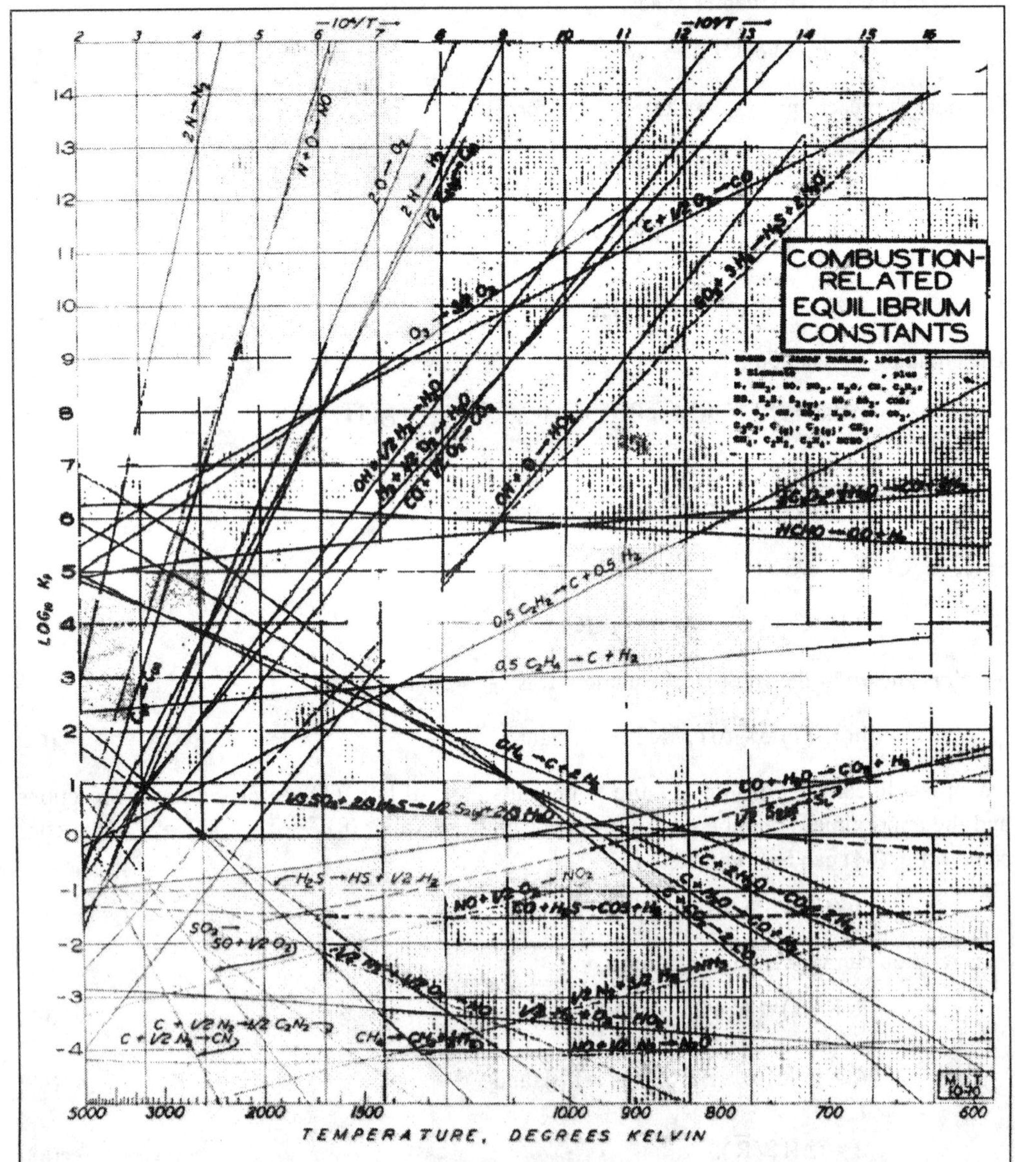

Figure 7: Plots of $\ln K^o(T)$ vs $1/T$ for various reactions. (Adapted from M. Modell and R. C. Reid, *Thermodynamics and its Applications*, Second Edition, Prentice Hall, 1983.)

perature. Conversely, if $\Delta H_R^0 < 0$ (e.g., exothermic combustion reactions), $B < 0$. At the transition temperature T_{trans} the value of K equals unity. Applying Eqs. (66), (69), and (70),

$$K^o(T) = 1 = \exp(\Delta S_R^o/\bar{R}) \exp(-\Delta H_R^o/\bar{R} T_{trans})$$

$$= \exp(-(\Delta H_R^o - T_{trans}\Delta S_R^o)/\bar{R} T_{trans})), \text{ i.e.,}$$

$$T_{trans} = (\Delta H_R^o/\Delta S_R^o). \quad (71)$$

The transition temperature is the temperature at which significant amount of products start to be formed.

---

n. *Example 14*

Determine the equilibrium constant for the reaction $NH_4HSO_4(\ell) \rightarrow NH_3(g) + H_2O(g) + SO_3(g)$. At 298 K, $\Delta H^o = 336500$ kJ kmole^{-1} and $\Delta S^o = 455.8$ kJ kmole^{-1} K. Determine the transition temperature (i.e. at which $K^0(T) = 1$).

Solution

Recall from Eq. (65) that

$$K^o(T) = K^o(T_o) \exp\{(-\Delta H^o/\bar{R})(1/T - 1/T_o)\}, \quad (A)$$

where $K(T_0) = \exp(-\Delta G^o/\bar{R} T_0)$ and

$$\Delta G^o = \Delta H^o - T\Delta S^o = 336500 - 298\times 455.8 = 200672 \text{ kJ kmole}^{-1}, \text{ i.e.,}$$

$$K^o(T_o) = \exp(-\Delta G^o/\bar{R} T_o) = 6.67\times 10^{-36}, \text{ or} \quad (B)$$

$$K^o(T) = 6.67\times 10^{-36} \exp\{-(336500/8.314)(1/T - 1/298)\} \quad (C)$$

By setting $K^o(T) = 1$, the transition temperature $T_{trans} = 336500 \div 456 = 738$ K.

Remarks

Since,

$$K^o(T) = (p_{NH_3}/1 (p_{H_2O}/1)(p_{SO_3}/1), \quad (D)$$

and K^o increases with temperature, decomposition is also favored at higher temperatures.

---

o. *Example 15*

Derive the Clausius–Clapeyron Relation from the van't Hoff equation by considering the equilibrium of liquid and vapor.

Solution

Consider an isothermal and isobarically maintained air duct into which water droplets are injected. The chemical potentials of the liquid drops and the vapor are different, which cause a transfer of species from one phase into the other (from liquid to vapor during evaporation). The liquid droplets will eventually reach equilibrium with the vapor. At the equilibrium condition

$$H_2O(\ell) \rightarrow H_2O(g), \text{ and} \quad (A)$$

$$K^o(T) = p_{H_2O(g)}/1 \quad (B)$$

If the equilibrium constant is known at a reference temperature T_{ref},

$$K^o_{ref}(T) = p_{H_2O,ref(g)}/1, \text{ and} \quad (C)$$

from the van't Hoff equation

$$\ln (K^o/K^o_{ref}) = -(\Delta H^o_R/\bar{R})(1/T - 1/T_{ref}). \quad (D)$$

For the vaporization process,

$$\Delta H^o_R = \bar{h}_g - \bar{h}_f, \text{ or } \Delta H^o_R = \bar{h}_{fg}, \text{ i.e.,} \quad (E)$$

$$\ln (p_{H_2O(g)}/p_{H_2O(g),ref}) = -(\bar{h}_{fg}/\bar{R})(1/T - 1/T_{ref}), \quad (F)$$

which is a relation for the change in the partial pressure of the vapor as the temperature changes. This is almost same as the Clausius–Clapeyron Relation

$$\ln (P/P_{ref}) = -((\bar{h}_g - \bar{h}_f)/T/\bar{R})(1/T - 1/T_{ref}). \quad (72)$$

b. Effect of Pressure

Since,

$$d\,\hat{g}_{k,T} = \hat{v}_k dP \text{ and } d\,\hat{g}/\partial P = \hat{v}_k,$$

$$(\partial(\Delta G)/\partial P)_T = \Sigma \upsilon_k\ \hat{v}_k = \Delta V_R, \quad (73)$$

where ΔV_R denotes the volume change between the products and the reactants. In this context, since $\ln K(T,P) = -\Delta G(T,P)/\bar{R}T$ and using Eq. (73),

$$\partial(\ln K(T,P))/\partial P = -\partial(\Delta G/\bar{R}T)/\partial P = -(1/\bar{R}T)\,\Delta V_R. \quad (74)$$

For an incompressible species (e.g, during the reaction $Na(s) + Cl(s) \rightarrow NaCl(s)$), $\Delta V_R \approx$ constant, and

$$\ln (K(T,P)) = -(1/\bar{R}T)\,P\,\Delta V^o_R + \text{constant}.$$

If the value of $K(T,P_{ref})$ is known, then,

$$\ln (K(T,P)/K_{ref}(T,P_{ref})) = -(1/\bar{R}T)\,\Delta V^o_R(P - P_{ref}). \quad (75)$$

For ideal gases, $\Delta V_R = \Sigma\,\upsilon_k(\bar{R}T/P)$, i.e.,

$$\ln (K(T,P)) = -(\Sigma \upsilon_k)\ \ln P + \text{Constant}. \quad (76)$$

p. Example 16

Consider the reaction

$$CO_2 \rightarrow CO + 1/2\ O_2. \quad (A)$$

That occurs in an isobaric and isothermal reactor. Discuss the effect on the equilibrium composition when the temperature is increased at a specified pressure and vice versa.

Solution

Recall that

$$d(\ln K^o)/dT = \Delta H^o_R/\bar{R}T^2, \quad (B)$$

where $\Delta H_R^0 > 0$ for the reaction. Hence, $d(\ln K^o)/dT > 0$, and the value of K increases with an increase in the pressure. Consequently, since

$$K^o(T) = (p_{CO}/1)\,(p_{O_2}/1)^{0.5}/(p_{CO_2}/1), \quad \text{(C)}$$

The value of p_{CO_2} decreases (as does that of X_{CO_2}). The effect of increasing the temperature is to dissociate more CO_2.
Simplifying Eq. (C),

$$K^o(T) = X_{CO}(X_{O_2})^{0.5}\,(P/1)^{0.5}/X_{CO_2}. \quad \text{(D)}$$

The value of K^o is a function of temperature alone. Therefore, increasing the pressure should cause the value of X_{CO_2} to increase, i.e., a relatively lower amount of dissociation will occur. Since each mole of CO_2 that dissociates produces 1.5 moles of the other two species, a lower dissociation results in a smaller amount of product (in terms of moles), thereby lowering the pressure (which counteracts the pressure increase). This is an example of the Le Chatelier principle which states that any inhomogeneity or disturbance that is introduced into a system must result in a process which counteracts that inhomogeneity or disturbance.

---

q. Example 17

Consider the combustion of CH_4 in air at 1 bar. Determine the number of moles of products produced (and, in particular, the CO concentration) per kmole of fuel consumed if the oxygen mole fraction in the products is 3% on a dry basis. Assume that for the reaction $CO_2 \rightarrow CO + 1/2\ O_2$,

$$\ln K^o = A - B/T \quad \text{(A)}$$

where A = 9.868 and B = 33742.4 in the appropriate units. Assume that CO is in trace amounts

---

Solution

Assume that the overall combustion reaction is represented by the equation

$$CH_4 + a\,(O_2 + 3.76\,N_2) \rightarrow CO_2 + 2H_2O + b\,N_2 + d\,O_2. \quad \text{(B)}$$

It is by now straightforward to determine that a = 2.2982, b = 8.641, and d = 0.2982.
The carbon dioxide concentration in the exhaust on a wet basis is

$$100 \times (1 \div (1 + 2 + 3.76 \times 2.2982 + 0.2982))\% = 8.4\%.$$

Likewise the nitrogen concentration is 72.4%. Note that CO is not produced during combustion according to our model, but is instead formed due to dissociation of CO_2 through the reaction $CO_2 \rightarrow CO + 1/2\ O_2$. The equilibrium constant for that reaction

$$K^o(T) = p_{CO}(p_{O_2})^{0.5}/(p_{CO_2}),\ \text{i.e.,}\ p_{CO} = K^o(T)\,p_{CO_2}/(p_{O_2})^{0.5}. \quad \text{(C)}$$

Since CO exists in trace amounts, $p_{CO_2} = 0.084$ bar and $p_{O_2} = 0.025$ bar (in proportion to their concentrations).
Applying Eq. (A) at 1500 K, $K^o(T) = 3.28\times10^{-6}$, i.e.,

$$p_{CO}/1 = 3.28\times10^{-6}\,(0.084/1)/(0.025/1)^{0.5} = 1.743\times10^{-6}.$$

At one bar, $X_{CO} = p_{CO}/P = p_{CO}/1\ \text{bar} = 1.743\times10^{-6}$.

Remark

Figure 8 presents the CO concentration in ppm for methane and solid carbon combustion with 20% excess air. The CO concentration does not significantly differ when the fuel changes from methane to carbon. Similarly, one can determine the equilibrium concentrations of other trace species, such as O, H_2.

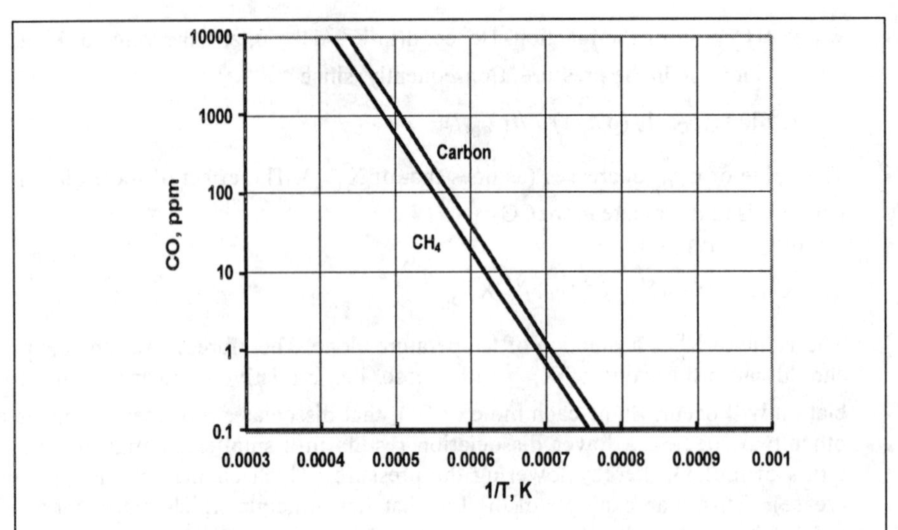

Figure 8: The CO emission due to the combustion of carbon and CH_4 at a 20% excess air level.

7. Equilibrium for Multiple Reactions

This concept will be illustrated through the following example.

---

r. *Example 18*

Consider the stoichiometric combustion of one kmole of CH_4 with air. The products are at 2250 K and 1 bar stream and contain CO_2, CO, H_2O, H_2, O_2, N_2, and OH. Determine the equilibrium composition.

Solution

The overall chemical reaction is

$$CH_4 + 2(O_2 + 3.76\ N_2) \rightarrow$$

$$N_{CO_2} CO_2 + N_{CO}\ CO + N_{H_2O} H_2O + N_{H_2} H_2 + N_{O_2} O_2 + N_{N_2} N_2 + N_{OH}\ OH. \quad (A)$$

There are seven species of unknown composition. The four atom conservation equations for C, H, N, and O atoms are:

$$\text{C atoms: } 1 = N_{CO_2} + N_{CO}, \quad (B)$$

$$\text{H atoms: } 4 = 2\times N_{H_2O} + 2\times N_{H_2} + N_{OH}, \quad (C)$$

$$\text{N atoms: } 7.52\times 2 = 2\times N_{N_2}, \text{ and} \quad (D)$$

$$\text{O atoms: } 2\times 2 = 2\times N_{CO_2} + N_{CO} + N_{H_2O} + 2\times N_{O_2} + N_{OH}. \quad (E)$$

We, therefore, require three additional relations. At equilibrium, for the reactions

$$CO_2 \rightarrow CO + 1/2\ O_2,\ K^{o}_{CO_2} = p_{CO}(p_{O_2})^{0.5}/(p_{CO_2}), \quad (F)$$

$$H_2O \rightarrow H_2 + 1/2\ O2,\ K_{H_2O} = (p_{H_2})(p_{O_2})^{0.5}/(p_{H_2O}), \text{ and} \quad (G)$$

$$OH \rightarrow 1/2H_2 + 1/2\ O_2,\ K_{OH} = (p_{H_2})^{0.5}(p_{O_2})^{0.5}/(p_{OH}). \quad (H)$$

Assume for example N_{CO}, N_{O2}, N_{CO2}. Solve for other species from Eqs. (B) to (E). Then check whether Eqs. (F) to (H) are satisfied. If not iterate. We have developed a spreadsheet based program which presents solution for all species at any given T and P, and any given air composition. The results (in terms of one kmole of methane consumed) are

$N_{CO_2} = 0.910$, $N_{CO} = 0.09$, $N_{H_2O} = 1.96$, $N_{H_2} = 0.04$, $N_{O_2} = 0.064$, and $N_{N_2} = 7.52$

Remark

If it exists in trace amounts, the NO concentration at equilibrium can be determined during combustion by considering the following reactions, i.e.,

$$NO \to 1/2\ N_2 + 1/2\ O_2, \text{ for which} \quad \text{(I)}$$

$$K_{NO} = (p_{H_2}/1)^{0.5}(p_{O_2}/1)^{0.5}/(p_{NO}/1). \quad \text{(J)}$$

8. Adiabatic Flame Temperature with Chemical Equilibrium

The energy balance equation applicable during combustion in a reactor is

$$dE_{cv}/dt = \dot{Q}_{cv} - \dot{W}_{cv} + \Sigma_{,k}\ \dot{N}_{ik}\ \hat{h}_{i,k} - \Sigma_k\ \dot{N}_{e,k}\ \hat{h}_{e,k},. \quad (77)$$

a. Steady State Steady Flow Process

At steady state $dE_{cv}/dt = 0$. For an adiabatic reactor, $\dot{Q}_{cv} = 0$. Since $\dot{W}_{cv} = 0$,

$$\Sigma_k\ \dot{N}_{ik}\ \hat{h}_{i,k} - \Sigma_k\ \dot{N}_{e,k}\ \hat{h}_{e,k} = 0 \text{ or } \dot{H}_I = \dot{H}_e = 0 \text{ where } = \dot{H} = \Sigma_k\ \dot{N}_k\ \hat{h}_{ik} \quad (78)$$

For ideal gas mixtures or ideal solutions, $\hat{h}_k = \bar{h}_k = \bar{h}^0_{f,k} + (\bar{h}_{t,T} - \bar{h}_{t,298})_k$

b. Closed Systems

The energy balance for an adiabatic closed system is

$$dE_{cv}/dt = \dot{W}_{cv}.$$

If the boundary work is neglected (e.g., for a fixed volume), $dE_{cv}/dt = 0$. If other forms of the energy are neglected and we assume that $E = U$,

$$(\Sigma\ N_k\ \hat{u}_k)_{products} - (\Sigma\ N_k\ \hat{u}_k)_{reactants} = 0 \text{ or } U_{reactants} = U_{products} \quad (79)$$

where $\hat{u}_k = \hat{h}_k - P\ \hat{v}_k$ and for ideal gas mixtures $\hat{u}_k = \bar{u}_k = \bar{h}_k - \bar{R}T$.

s. Example 19

Consider the stoichiometric combustion of 1 kmole of CH_4 with air at 1 bar. The species enter an adiabatic reactor at 298 K (state 1) and the products leaving it are determined to be the species CO_2, CO, H_2O, H_2, O_2, and N_2 at equilibrium. Determine the product temperature and composition.

Solution

First let use assume T. The problem of solving for the composition becomes similar to previous examples. The overall chemical reaction is

$$CH_4 + 2(O_2 + 3.76\ N_2) \to$$

$$N_{CO_2}CO_2 + N_{CO}\ CO + N_{H_2O}H_2O + N_{H_2}H_2 + N_{O_2}O_2 + N_{N_2}N_2. \quad \text{(A)}$$

There are six species of unknown composition. The atom conservation equations for C, H, N, and O atoms are:

C atoms: $1 = N_{CO_2} + N_{CO}$, (B)

H atoms: $4 = 2\times N_{H_2O} + 2\times N_{H_2}$, (C)

N atoms: $7.52\times 2 = 2\times N_{N_2}$, and (D)

O atoms: $2\times 2 = 2\times N_{CO_2} + N_{CO} + N_{H_2O} + 2\times N_{O_2}$. (E)

We, therefore, require two additional relations to solve for six unknowns. At equilibrium, for the reactions

$CO_2 \rightarrow CO + 1/2\ O_2$, $K^o_{CO_2} = p_{CO}(p_{O_2})^{0.5}/(p_{CO_2})$, (F)

$H_2O \rightarrow H_2 + 1/2\ O_2$, $K_{H_2O} = (p_{H_2})(p_{O_2})^{0.5}/(p_{H_2O})$. (G)

The energy balance is used to solve for T. For ideal gas mixture

$(\Sigma \overline{N}_k \bar{h}_k)_i - (\Sigma \overline{N}_k \bar{h}_k)_e = 0$. (H)

This problem is solved iteratively. Solving, the adiabatic temperature is 2249 K. The equilibrium composition is

$N_{CO_2} = 0.91$, $N_{CO} = 0.09$, $N_{H_2O} = 1.96$, $N_{H_2} = 0.04$, $N_{O_2} = 0.064$,
$N_{N_2} = 7.52$, and Q = 29 kJ.per kmole of CH_4

Table A-27C tabulates the equilibrium flame temperature and entropy generated at 298 K in an adiabatic reactor for selected fuels.

9. Gibbs Minimization Method

a. General Criteria for Equilibrium

At a specified temperature and pressure, for any composite system, $(dG_{T,P} = \Sigma\mu_k dN_k) \le 0$, where the equality holds at equilibrium. The Gibbs energy for a chemical reaction decreases as the reaction progresses (cf. $dG_{T,P} < 0$) until it reaches a minimum value. Since, G = G(T,P,N_1, N_2...), the criteria for the G minima are

$$dG_{T,P} = 0 \text{ and } d^2G_{T,P} > 0. \quad (80a,b)$$

Consider methane–air combustion at specified temperature and pressure according to the overall reaction

$$CH_4 + a\ O_2 + b\ N_2 \rightarrow c\ CO + d\ CO_2 + e\ O + f\ O_2 + g\ NO + h\ H_2O + i\ H_2 + j\ N_2 + k\ C(s),$$

where a and b are known There are nine unknown species concentrations and four atom balance equations for C, H, N, and O atoms. Therefore, five equilibrium relations are required, which may include the reactions

$CH_4 \rightarrow C(s) + 2\ H_2$, (A)

$CO_2 \rightarrow CO + 1/2\ O_2$, (B)

$H_2O \rightarrow 1/2\ O_2 + H_2$, (C)

$O_2 \rightarrow 2\ O$, and (D)

$$N_2 + O_2 \rightarrow 2\ NO. \tag{E}$$

We may also select a linear combination of reactions. For instance, subtracting reaction (B) from reaction (C) we obtain the reaction

$$H_2O - CO_2 \rightarrow 1/2\ O_2 + H_2 - CO - 1/2\ O_2, \text{ i.e.,}$$

$$H_2O + CO \rightarrow CO_2 + H_2, \tag{F}$$

which is the familiar water gas shift reaction. Note that reaction (F) does not provide an independent equilibrium relation that can be selected in addition to reactions (A)–(E). The solution procedure becomes far more complex as we encounter literally hundreds of species in a realistic applications. Therefore, it is useful to adopt a more general procedure during which the species concentrations are adjusted until the Gibbs energy reaches a minimum value at equilibrium, i.e., $dG_{T,P} = 0$ and $d^2G_{T,P} > 0$ subject to the atom balance constraints and the specified temperature and pressure. The LaGrange multiplier method is useful in this regard.

---

t. *Example 20*

A piston–cylinder assembly contains two kmole of O_2 at 1 bar and 3000 K. The pressure and temperature are maintained constant. Chemical reaction proceeds and O atoms are formed at the expense of O_2. Determine the equilibrium composition starting with $G= G(T, P, N_{O_2}, N_O)$ and minimizing G subject to atom conservation.

Solution

The O atom conservation equation is

$$2\ N_{O_2} + N_O = 4 \tag{A}$$

The Gibbs energy of the mixture must keep decreasing as the reaction proceeds and equilibrium is achieved when it is at the minimum.

$$G = \Sigma\mu_k N_k, \text{ i.e.,} \tag{B}$$

$$G = G(T, P, N_{O_2}, N_O) = \mu_{O_2} N_{O_2} + \mu_O N_O. \tag{C}$$

We will minimize Eq. (C) at the specified pressure and temperature subject to the atom balance constraint Eq. (A). Using the LaGrange multiplier scheme

$$F = G(T,P,N_{O_2},N_O) + \lambda\ (2N_{O_2} + N_O - 4) = 0, \text{ and } \partial F/\partial N_{O_2} = 0,\ \partial F/\partial N_O = 0. \tag{D}$$

From Eq. (D),

$$\partial F/\partial N_{O_2} = \partial G/\partial N_{O_2} + 2\lambda = 0, \text{ i.e., } \mu_{O_2} + 2\lambda = 0 \text{ (and } \mu_O + \lambda = 0). \tag{E}$$

Assuming the ideal mixture model to apply,

$$\mu_{O_2} = \hat{g}_{O_2} = \bar{g}_{O_2}(T,P) + \bar{R}T \ln X_{O_2}. \tag{F}$$

Further, assuming ideal gas behavior,

$$\bar{g}_{O_2}(T,P) = \bar{g}^0_{O_2} + \bar{R}T \ln (P/1). \tag{G}$$

Therefore,

$$\mu_{O_2} = \bar{g}^0_{O_2} + \bar{R}T \ln (p_{O_2}/1), \text{ and} \tag{H}$$

$$\mu_O = \bar{g}^0_O + \bar{R}T \ln (p_O/1), \text{ so that} \tag{I}$$

$$(\bar{g}^{o}_{O_2} + \bar{R}T \ln (p_{O_2}/1)) + 2\lambda = 0, \text{ and} \quad (J)$$

$$(\bar{g}^{o}_{O} + \bar{R}T \ln (p_O/1)) + \lambda = 0. \quad (K)$$

Multiplying Eq. (K) by 2 and subtracting it from Eq. (J),

$$\bar{g}^{o}_{O_2} - 2\,\bar{g}^{o}_{O} + \bar{R}T \ln (p_{O_2}/1) - 2\bar{R}T \ln (p_O/1) = 0, \text{ i.e.,}$$

$$(p_O/1)^2/(p_{O_2}/1) = \exp(-(2\,\bar{g}^{o}_{O} - \bar{g}^{o}_{O_2})/(\bar{R}T)) \text{ or } (N_O)^2(P/N)^{(2-1)}/N_{O_2} = K^o, \text{ where} \quad (L)$$

$$N = N_{O_2} + N_O. \quad (M)$$

The equilibrium constant

$$K = \exp(-\Delta G^o/\bar{R}T), \text{ where } \Delta G^o = 2\,\bar{g}^{o}_{O} - \bar{g}^{o}_{O_2}. \quad (N)$$

With the values $N_O = 4 - 2N_{O_2}$ and $N = 4 - 2N_{O_2} + N_{O2} = 4 - N_{O_2}$,

$$(4 - 2N_{O_2})^2(P/(1(4 - N_{O_2})))/N_{O_2} = K^o. \quad (O)$$

the pressure P = 1 bar,

$$(N_{O_2})^2 - 4\,N_{O_2} + 4^2/(4+K^o) = 0. \quad (P)$$

Now,

$$\bar{g}^{o}_{O_2} = -755102 \text{ kJ kmole}^{-1}, \text{ and } \bar{g}^{o}_{O} = -323359 \text{ kJ kmole}^{-1}.$$

We can solve for N_{O_2} and selecting the root, such that $N_{O_2} > 0$ and $N_O > 0$, i.e.,

$$N_{O_2} = 1.8875, N_O = 0.225, \text{ and } G_{min} = -1.52\times10^6.$$

---

b. Multiple Components

The solution can be explicitly obtained for two components. We will now generalize the methodology for multicomponent systems. The procedure to minimize $G = G(T,P,N_1, N_2, \ldots, N_K)$ subject to the atom balance equations is as follows.

Formulate the atom balance equations for each element "j"

$$\Sigma_k d_{jk} N_k = A_j, j = 1, \ldots, J, k = 1, \ldots, K, \quad (81)$$

where d_{jk} denotes the number of atoms of an element "j" in species k (e.g., for the element j=O in species k $=CO_2$, d = 2) and A_j is the number of atoms of type j entering the reactor. This relation can be expressed using the La Grange multiplier method, i.e.,

$$\lambda_j(\Sigma_k d_{jk} N_k - A_j),\ j = 1, \ldots, J, k = 1, \ldots, K. \quad (82)$$

The Gibbs energy at equilibrium must be minimized subject to this condition.

We create a function

$$F = G + \Sigma_j\lambda_j(\Sigma_k d_{jk} N_k - A_j),\ j = 1, \ldots, J, k = 1, \ldots, K, \quad (83)$$

such that

$$\partial F/\partial N_k = (\partial G/\partial N_k)_{T,P} + (\Sigma_j\lambda_j d_{jk})_{T,P} = 0, j = 1, \ldots, J, k = 1, \ldots, K, \text{ i.e.,} \quad (84)$$

Next, we minimize F with respect to N_k, k = 1, …, K, and for each species,

$$\mu_k + \Sigma_j\lambda_j d_{jk} = 0,\ j = 1, \ldots, J, k = 1, \ldots, K. \quad (85)$$

the term $\mu_{E,k} = \Sigma_j \lambda_j d_{jk}$ is the combined element potential of species k. We can interpret λ_j as the element potential of element j in species k, λ_j d_{jk} as contribution by element j to the k-th species potential and the summation of the potential of all elements j in species k as the combined elemental potentials in species k.

Example 21

A steady flow reactor is fired with 1 kmole of $C_{10}H_{20}$, with 5 % excess air. The species (1 to 5) leaving are CO, CO_2, H_2 H_2O, OH, O_2, NO, and N_2 at T= 2500 K and 1 bar. Determine equilibrium composition of species leaving the reactor. Assume ideal gas behavior.,

---

Solution

The stoichiometric amount of O_2 can be determined from the stoichiometric relation

$$(C_{10}H_{20}) + 15\ (O_2 + 3.76\ N_2) = 10\ CO_2 + 10\ H_2O + 56.4\ N_2$$

With 5% excess air, the O_2 supplied = 16.5 kmole, the N_2 supplied is 62.04 kmole.

$$(C_{10}H_{20}) + 15.\ 8\ (O_2 + 3.76\ N_2) \rightarrow \text{Products.}$$

This system is an open system. Now we follow a fixed mass (140+528+ 1737= 2405 kg) as it travels the reactor. We will assume that this mixture is instantaneously heated to 2500 K at 1 bar and then calculate assuming various values for the moles of the 8 species subject to the atom conservation relations for C, H, N, and O. We will then select the composition at which G has a minimum value at this T and P using the LaGrange multiplier method to arrive at the composition.

The four elements C, H, N and O are denoted by the subscript j and the eight species denoted by subscript k. The coefficients d_{jk}, i.e., d_{11} = 1, d_{12} = 1, ... are provided in the following table:

Coefficients d_{jk}

Element j→	C	H	N	O
Species k ↓				
CO	1			1
CO_2	1			2
H_2		2		
H2O		2		1
NO			1	1
N_2			2	
OH		1		1
O_2				2

The atom conservation equations (Σ_k d_{jk} $N_k - A_j$) (see Eq. (81)) yield the relations:

j = 1 (C atoms):

$$1N_{CO} + 1N_{CO_2} + 0N_{H_2} + 0N_{H_2O} + 0N_{NO} + 0N_{N2} + 0N_{OH} + 0N_{O2} - 10 = 0, \quad \text{(A)}$$

j = 2 (H atoms):

$$0\ N_{CO} + 0\ N_{CO_2} + 2\ N_{H_2} + 2\ N_{H_2O} + 0N_{NO} + 0N_{N2} + 1N_{OH} + 0N_{O2} - 20 = 0, \quad \text{(B)}$$

j = 3 (O atoms):

$$1N_{CO} + 2\ N_{CO_2} + 0\ N_{H_2} + 1\ N_{H_2O} + 1N_{NO} + 0N_{N2} + 1N_{OH} + 2N_{O2}\ - 31.6 = 0, \quad \text{(C)}$$

j = 4 (N atoms):

$$0\ N_{CO} + 0\ N_{CO_2} + 0N_{H_2} + 0\ N_{H_2O} + 1N_{NO} + 2N_{N2} + 0N_{OH} + 0N_{O2}\ - 118.6 = 0, \quad \text{(D)}$$

Dividing these equations by the total moles (N= ΣN_k), we obtain the relations

$$1X_{CO} + 1X_{CO_2} + 0X_{H_2} + 0X_{H_2O} + 0X_{XO} + 0X_{N2} + 0X_{OH} + 0X_{O2} - 10/N = 0, \quad (E)$$

$$0\,X_{CO} + 0\,X_{CO_2} + 2\,X_{H_2} + 2\,X_{H_2O} + 0X_{NO} + 0X_{N2} + 1X_{OH} + 0X_{O2} - 20/N = 0, \quad (F)$$

$$1X_{CO} + 2\,X_{CO_2} + 0\,X_{H_2} + 1\,X_{H_2O} + 1X_{NO} + 0X_{N2} + 1X_{OH} + 2X_{O2} - 31.6/N = 0, \quad (G)$$

$$0X_{CO} + 0\,X_{CO_2} + 0X_{H_2} + 0\,X_{H_2O} + 1X_{NO} + 2X_{N2} + 0X_{OH} + 0X_{O2} - 118.6/N = 0, \quad (H)$$

and N is solved from the identity

$$\Sigma X_k = 1. \quad (I)$$

$$G = G(T,P,N_1, N_2, \ldots, N_K). \quad (J)$$

Multiplying Eqs. (A)–(D), respectively by λ_C, λ_H, λ_N and λ_O, and adding with Eq. (J) we form a function

$$F = G + \lambda_C\,(N_{CO} + N_{CO_2} - 10) + \lambda_H(2N_{H_2} + 2N_{H_2O} + N_{OH} - 20) +$$

$$\lambda_O(N_{CO} + 2N_{CO_2} + N_{H_2O} + N_{OH} + N_{NO} + 2\,N_{O2} - 31.6) +$$

$$\lambda_N\,(N_{NO} + 2\,N_{N2} - 118.6) \quad (K)$$

that is minimized at equilibrium, i.e.,

$$\partial F/\partial N_{CO} = (\partial G/\partial N_{CO})_{T,P} + \lambda_C + \lambda_O = 0, \quad (L)$$

$$\partial F/\partial N_{CO2} = (\partial G/\partial N_{CO2})_{T,P} + \lambda_C + 2\lambda_O = 0, \text{ and} \quad (M)$$

$$\partial F/\partial N_{H_2} = (\partial G/\partial N_{H_2})_{T,P} + 2\lambda_H = 0. \quad (N)$$

Rewrite Eqs. (L) to (N) in the form

$$\mu_{CO} + \lambda_C + \lambda_O = 0, \quad (O)$$

$$\mu_{CO2} + \lambda_C + 2\,\lambda_H = 0, \text{ and} \quad (P)$$

$$\mu_{H_2} + 2\lambda_H = 0, \text{where } \mu_k = (\partial G/\partial N_k)_{T,P}. \quad (Q)$$

Recall that for an ideal mixture of gases or ideal mix of liquids or solids

$$\mu_k = \hat{g}_k = \bar{g}_k(T,P) + \bar{R}T \ln (X_k). \quad (R)$$

Divide by $\bar{R}T$, namely,

$$\mu_k / \bar{R}T = (\bar{g}_k(T,P)/\bar{R}T) + \ln (X_k 1). \quad (R')$$

Divide Eqs. (O) through (Q) by $\bar{R}T$ and using Eq. (R′)

$$(\bar{g}_{CO}(T,P)/\bar{R}T) + \ln (X_{CO}).+ \lambda_C' + \lambda_O' = 0, \quad (S)$$

$$(\bar{g}_{CO2}(T,P)/\bar{R}T) + \ln (X_{CO2}).+ \lambda_C' + 2\,\lambda_H' = 0, \text{ and} \quad (T)$$

$$(\bar{g}_{H2}(T,P)/\bar{R}T) + \ln(X_{H2}) + 2\lambda_H' = 0, \qquad (U)$$

where $\lambda_j' = \lambda_j/\bar{R}T$.

The values of $\ln(X_k)$ can be determined from the linear equations Eqs. (S) to (U) for assumed values of modified LaGrange multipliers of λ_C', λ_H', λ_N', and λ_O'. Then we can check whether these four multipliers satisfy the four-element conservation equations Eqs. (E) to (H). Thus, knowing the values of μ_k, those of X_k can be determined from linear equations such as Eq. (R). Irrespective of the number of unknown species, the assumptions for the LaGrange multipliers are made only equal to number of atom balance equations. Good starting values can be obtained by assuming complete combustion, determining those mole fractions to obtain initial guesses for employing Eqs. (S) to (U). For $T = 2500$ K and $P = 1$ bar, $\bar{g}_k = \bar{h}_k - T\bar{s}_k$, and

$$\bar{g}_{CO} = -701455 \text{ kJ kmole}^{-1}, \bar{g}_{H_2} = -419840 \text{ kJ kmole}^{-1},$$

$$\bar{g}_{CO_2} = -1078464 \text{ kJ kmole}^{-1}, \text{ and so on} \qquad (S)$$

Good starting values could be λ_C', λ_H', λ_N', and λ_O' = 19.7, 12.5, 14.0, 17.1, respectively. The converged solutions are provided below.

$N_{CO} = 2.04$ kmole, $N_{CO_2} = 7.96$, $N_{H_2} = 0.42$, $N_{H_2O} = 9.24$,

$N_{NO} = 0.57$, $N_{N2} = 59$, $N_{OH} = 0.69$, $N_{O2} = 1.52$,

$N = \Sigma N_k = 81.4$ kmole.

The NASA equilibrium code uses a descent Newton-Raphson method with a typical number of iterations equal to 8-12.

D. SUMMARY

The chemical force potentials of reacting and product species can be used to determine the direction of reaction just as the temperature (thermal potential) can be used to determine the direction of heat transfer. While the First Law is used to determine the equilibrium temperature, the Second Law is used to maximize the entropy at specified values of U, V, and m or minimize G at given values for T, P, and m for determining the chemical equilibrium composition.

E. APPENDIX

Combustion reactions customarily involve compounds consisting of elemental C, H, N, and O. Consider the reaction

$$H_2O \rightarrow H + OH, \qquad (A)$$

For which we are asked to determine the value of K^o. If the K^o value is available for the reactions involving elements in their natural forsm,

$$H_2 + 1/2\ O_2 \rightarrow H_2O, \qquad (B)$$

$$1/2 H_2 \rightarrow H, \text{ and} \qquad (C)$$

$$1/2\ H_2 + 1/2\ O_2 \rightarrow OH, \qquad (D)$$

Then for reactions (B)–(D)

$$-\bar{R}T \ln K^o_{H_2O} = \bar{g}^o_{H_2O} - \bar{g}^o_{H_2} - 1/2\,\bar{g}^o_{O_2}, \qquad (E)$$

$$-\bar{R}T \ln K^o_H = \bar{g}^o_H - 1/2\,\bar{g}^o_{H_2}, \text{ and} \qquad (F)$$

$$-\overline{R}T \ln K^o_{OH} = \overline{g}^o_{OH} - 1/2\,\overline{g}^o_{H_2} - 1/2\,\overline{g}^o_{O_2}. \quad \text{(G)}$$

Likewise, for reaction Eq. (A),

$$-\overline{R}T \ln K^o = = \overline{g}^o_{H} - \overline{g}^o_{OH} - 1/2\,\overline{g}^o_{H_2O}, \text{ i.e.,} \quad \text{(H)}$$

Substituting $\overline{g}^o_H$ from Eq. (F), $\overline{g}^o_{OH}$ from Eq. (G) and $\overline{g}^o_{H_2O}$ from Eq. (E), we obtain

$$K^o = K^o_{OH}\, K^o_H / K^o_{H_2O}. \quad \text{(I)}$$

Similarly, for the equilibrium constant for the reaction $CO_2 \rightarrow CO + 1/2\ O_2$, one can write by observation, $K^0 = K^o_{CO}/K^o_{CO_2}$. Note that K^o (T) will not appear for elements in natural form.

Chapter 13

13. AVAILABILITY ANALYSIS FOR REACTING SYSTEMS

A. INTRODUCTION

Combustion is a process during which chemical energy is converted into thermal energy. The ultimate objective is to convert chemical energy into useful work. The extent to which this conversion is possible can be determined through an availability analysis. We have discussed availability concepts for nonreacting systems in Chapter 4. This chapter presents availability analyses for reacting systems. We will also discuss a methodology of determining the maximum possible work potential of various fuels. For instance, in this context, it is often required to determine the maximum possible work per gallon of gasoline or other liquid fuel produced from an automobile engine operating in the ambient.

B. ENTROPY GENERATION THROUGH CHEMICAL REACTIONS

Availability (or exergy) analyses employ the Second Law to predict system irreversibilities. The combustion–related chemical reactions generate irreversibilities that decrease the availability and lower the conversion efficiency of chemical energy into useful work. The general entropy balance equation is

$$dS_{cv}/dt = \dot{Q}_{cv}/T_b + \Sigma_{k,i}\dot{N}_k\bar{s}_k - \Sigma_{k,e}\dot{N}_k\ \bar{s}_k + \dot{\sigma} \qquad (1)$$

$$(1/\dot{N}_F)\ dS_{cv}/dt = (\dot{Q}_{cv}/T_b)(1/\dot{N}_F) + \Sigma_{k,I}(\dot{N}_k/\dot{N}_F)\bar{s}_k - \Sigma_{k,e}(\dot{N}_k/\dot{N}_F)\ \bar{s}_k + \bar{\sigma}, \qquad (2)$$

where $\dot{N}_F$ denotes the rate of change in the moles of fuel, and $\bar{\nu}_k = (\dot{N}_k/\dot{N}_F)$ is the reaction coefficient (which is not necessarily equal to the stoichiometric coefficient). It follows that

$$\bar{\sigma} = (1/\dot{N}_F)\ dS_{cv}/dt - (\dot{Q}_{cv}/T_b)(1/\dot{N}_F) - \Sigma_{k,I}\ \bar{\nu}_k\ \bar{s}_k + \Sigma_{k,e}\ \bar{\nu}_k\ \bar{s}_k, \qquad (3)$$

and under steady state steady flow conditions

$$\bar{\sigma} = (\Sigma\,\bar{\nu}_k\,\hat{s}_k)_e - \Sigma(\bar{q}_j/T_{b,j}) - (\Sigma\,\bar{\nu}_k\,\hat{s}_k)_i. \qquad (4)$$

---

a. Example 1

Determine the entropy generated during the oxidation of glucose within the cells of the human body assuming that the reaction occurs at 310 K. Assume that steady state conditions apply, the entropy of glucose at 310 K is 288.96 kJ kmole^{-1} K^{-1}, the inlet conditions are 300 K and 1bar, and 400% excess air is involved.

Solution

The overall reaction with 400 % excess air can be expressed as

$$C_6H_{12}O_6 + 5\times 6\,(O_2 + 3.76\,N_2) \rightarrow 6CO_2 + 6\,H_2O + 24\,O_2 + 112.8\,N_2, \qquad (A')$$

where $\bar{q}$ = -2.5×10^6 kJ kmole^{-1} of glucose (from examples 1 and 6 in Chapter 11). At steady state

$$\bar{\sigma} = (\Sigma\,\bar{\nu}_k\,\hat{s}_k)_i - \Sigma(\bar{q}_j/T_{b,j}) - (\Sigma\,\bar{\nu}_k\,\hat{s}_k)_e. \qquad (A)$$

Recall that $\bar{s}_k^o(T) = \hat{s}_k = (T,p_k)$. Therefore, for the species on the LHS of Eq. (A) $(\Sigma\,\bar{\nu}_k\,\hat{s}_k)_i$ = 1×288.96+30×218.01+30×3.76×193.4662, i.e.,

$$(\Sigma\,\bar{\nu}_k\,\hat{s}_k)_i = 28652\ \text{kJ K}^{-1}\ \text{k}^{-1}\ \text{glucose}. \qquad (B)$$

At the exit (involving product species on the RHS of Eq. (A′)),

X_{O_2} = 24÷148.8= 0. 161, X_{N_2}= 0.758, X_{CO_2}= 0.0403, and X_{H_2O}= 0.0403.
Hence,

$\bar{s}_{O_2}(310K, p_{O_2})$ = 205.066 – 8.314×ln (1×0.16) = 220.2 kJ kmole^{-1} K^{-1},
$\bar{s}_{N_2}$= 193.8 kJ kmole^{-1} K^{-1}, $\bar{s}_{CO_2}$= 240.4 kJ kmole^{-1} K^{-1}, and
$\bar{s}$= 215.5 kJ kmole^{-1} K^{-1}, i.e.,
$(\Sigma \bar{\nu}_k \hat{s}_k)_e$ = 6×213.7+ 6×238.6 + 24×220 + 112.8×193.9
= 29866 kJ kmole^{-1} of glucose.

Using Eq. (A) and specifying the control volume boundary just inside the human skin,
$\bar{\sigma}$ = 29866- (- 2.5×10^6÷310) – 28652 = 9279 kJ K^{-1} kmole^{-1} of glucose.

Remarks

Entropy is generated since the inhalation temperature is different from the exhalation temperature, and due to the irreversible chemical reaction.

If T_b = T_e = T_I = 37 C, the entropy generation is due to chemical irreversibilities alone.

The typical consumption of glucose is 0.31 g min^{-1} for a 65 kg person. Therefore, the entropy generated per second due to that person's irreversible metabolic rate is (9279/180.2)×0.31/(1000×60) = 2.66×10^{-4} kW K^{-1} or 4.1 ×10^{-6} kW kg^{-1} K^{-1}. This number is typically higher for smaller–sized animal species.

C. AVAILABILITY

1. Availability Balance Equation

The general availability balance equation is

$$d(E_{cv}- T_0 S_{cv})/dt = \Sigma \dot{Q}_{R,j}(1 - \frac{T_0}{T_{R,j}}) + (\Sigma \dot{N}_k \hat{\psi}_k)_i - (\Sigma \dot{N}_k \hat{\psi}_k)_e - \dot{m} - \dot{I}, \quad (5)$$

where $\dot{I} = T_o \dot{\sigma}_{cv} \geq 0$ since combustion is typically an irreversible process and the absolute stream availability $\hat{\psi}_k = (\hat{e}_T - T_o \hat{s})_k = (\hat{h} + \overline{ke} + \overline{pe} - T_o \hat{s})_k$. Recall that species stream availability (or species stream exergy or relative availability) is defined as

$$\hat{\psi}_k' = \hat{\psi}_k - \hat{\psi}_{k,0}, \quad (6a)$$

where $\hat{\psi}_{k,0}$ ($p_{k,e}$, T_0) is the thermo-mechanical absolute stream availability of each species and $p_{k,e}$ is the partial pressure of species k, $p_{k,e} = X_{k,e} P_0$. One may write Eq. (5) in terms of species stream exergy also.

For ideal gases, $s_k = s_k^o - R \ln (p_k/P_{ref})$, where the reference state is generally assumed to be at P_{ref} = 1 bar. Therefore, neglecting "ke" and "pe",

$$\psi_k = \psi_k^o + R\, T_o \ln(p_k/1) \text{ where } p_k \text{ is expressed in units of bar.} \quad (6b)$$

Dividing Eq. (5) by the molal flow rate of the fuel,

$$(1/\dot{N}_F)\, d(E_{cv}-T_oS_{cv})/dt =$$

$$(\Sigma \bar{\nu}_k \hat{\psi}_k)_i + \Sigma \bar{q}_{R,j}(1-T_o/T_{R,j}) - (\Sigma \bar{\nu}_k \hat{\psi}_k)_e - \bar{w}_{cv} - \bar{i}, \quad (7a)$$

where $\bar{w}_{cv} = (\dot{W}_{cv}/\dot{N}_F)$ and $\bar{i} = (\dot{I}/\dot{N}_F) = T_o \bar{\sigma}$. In terms of "i",

$$\bar{i} = -(1/\dot{N}_F)d(E_{cv}-T_oS_{cv})/dt + (\Sigma \bar{\nu}_k \hat{\psi}_k)_i +$$

$$\Sigma \bar{q}_{R,j}(1-T_o/T_{R,j}) - (\Sigma \bar{\nu}_k \hat{\psi}_k)_e - \bar{w}. \quad (7b)$$

If a gas turbine combustor is started from cold conditions, the energy within it (dE_{cv}/dt) starts accumulating and the entropy increases (dS_{cv}/dt) due to a temperature rise within the combustor. Consequently, the stream availability at the exit will be zero initially and starts raising over time so that the work output from turbine increases until a steady flow steady state is achieved, when

$$\bar{i} = (\Sigma \bar{v}_k \hat{\psi}_k)_i + \Sigma \bar{q}_{R,j}(1-T_o/T_{R,j}) - (\Sigma \bar{v}_k \hat{\psi}_k)_e - \bar{w}. \quad (8)$$

The optimum work is obtained when $\bar{i} = 0$, i.e.,

$$\bar{w}_{opt} = (\Sigma \bar{v}_k \hat{\psi}_k)_i + \Sigma \bar{q}_{R,j}(1-T_o/T_{R,j}) - (\Sigma \bar{v}_k \hat{\psi}_k)_e. \quad (9)$$

In the absence of work and thermal reservoirs (that are typical of an adiabatic combustor),

$$\bar{i} = (\Sigma \bar{v}_k \hat{\psi}_k)_i - (\Sigma \bar{v}_k \hat{\psi}_k)_e. \quad (10)$$

In general, we are interested in the change in the availability of streams entering or exiting the reactor. Let

$$\Psi_R = \Psi_1 = (\Sigma \bar{v}_k \hat{\psi}_k)_I, \text{ and } \Psi_P = \Psi_2 = (\Sigma \bar{v}_k \hat{\psi}_k)_e, \text{ then} \quad (11)$$

$$\bar{i} = \Psi_R - \Psi_P = \Psi_1 - \Psi_2 = \text{change in absolute availability, where} \quad (12)$$

$$\psi_{1,2}(T,P,T_o) = e_{T,1,2}(T,P) - T_o s_{1,2}(T,P) = (h_{1,2}(T,P) + ke_{1,2} + pe_{1,2}) - T_o s_{1,2}(T,P). \quad (13)$$

For a steady flow process from an entry (state 1) to an exit (state 2), under negligible kinetic and potential energy changes,

$$\psi_1 - \psi_2 = ((H_1 - H_o) - T_o(S_1 - S_0)) - ((H_2 - H_o) - T_o(S_2 - S_0)) = H_1 - H_2 - T_o(S_1 - S_2), \quad (14a)$$

where on a unit mass basis

$$\psi_1 - \psi_2 = h_1 - h_2 - T_o(s_1 - s_2). \quad (14b)$$

If $T_o = T_1 = T_2 = T$, Eq. (14) assumes the form

$$\psi_1 - \psi_2 = G_1 - G_2. \quad (15)$$

For ideal gases, $s = s^o - R \ln(P/P_{ref})$, where the reference state is generally assumed to be at P_{ref} = 1 bar. Therefore,

$$\psi_{1,2} = \psi^o + R\, T_o \ln(P_{1,2}/P_{ref}).$$

---

b. *Example 2*

1 kmole of butane enters a steady state steady flow reactor at 298 K and 250 kPa with 50% excess air. Combustion is assumed to be complete, and the products leave the reactor at 1000 K and 250 kPa. Determine the heat transfer, reactant and product entropies, the absolute availability of the reactants and products, the entropy change between the exit and inlet, the entropy generation, the optimum work and the irreversibility.

Solution

The overall reaction can be expressed as

$$C_4H_{10} + 9.75O_2 + 36.66N_2 \rightarrow 4CO_2 + 5H_2O + 3.25O_2 + 36.66N_2 \quad (A)$$

The reaction coefficients are

$\nu_{CO_2} = (N_{CO_2}/N_F)_e = 4$, $\nu_{H_2O} = 5$, $\nu_{O_2} = 3.25$ and $\nu_{N_2} = 36.66$, for the exit stream and $\nu_{N_2} = 36.66$ and $\nu_{O_2} = 9.75$ for the inlet stream.

The energy equation is

$$dE_{cv}/dt = \dot{Q}_{cv} - \dot{W}_{cv} + \Sigma_{k,i} \dot{N}_k \bar{e}_{T,k} - \Sigma_{k,e} \dot{N}_k \bar{e}_{T,k} \quad \text{(B)}$$

neglecting the kinetic and potential energies, we obtain the expression

$$(1/\dot{N}_F)\, dU_{cv}/dt = \bar{q} - \bar{w} + (\Sigma\, \bar{\nu}_k\, \hat{h}_k)_i - (\Sigma\, \bar{\nu}_k\, \hat{h}_k)_e.$$

At steady state and ideal gas conditions, $\hat{h}_k = \bar{h}_k$, and since $\bar{w} = 0$, we have the simplified relation

$$\bar{q} + (\Sigma\, \bar{\nu}_k\, \hat{h}_k)_i - (\Sigma\, \bar{\nu}_k\, \hat{h}_k)_e = 0, \text{ i.e.} \quad \text{(C)}$$

$$\bar{q} = H_p - H_R, \text{ where} \quad \text{(D)}$$

$$H_p = (\Sigma\, \bar{\nu}_k\, \hat{h}_k)_e \text{ and } H_R = (\Sigma\, \bar{\nu}_k\, \hat{h}_k)_i. \quad \text{(E)}$$

$H_R = -126148$ kJ kmole^{-1} of C_4H_{10}, and
$H_P = 4\times(-393{,}520 + 33{,}425) + 5\times(-241{,}827 + 25{,}978) + 3.25\times(22{,}707) + 36.66\times(21{,}460) = -1{,}659{,}104$ i.e.,
$\bar{q} = H_P - H_R = -1533044$ kJ kmole^{-1} of C_4H_{10}.

Likewise, we may show that
$S_R = (\Sigma\,\bar{\nu}_k\,\hat{s}_k)_i = 9212.3$ kJ kmole^{-1} of fuel K^{-1},
$\Psi_R = (\Sigma\,\bar{\nu}_k\,\bar{\psi}_k)_i = -2869892$ kJ kmole^{-1} of fuel,
$S_P = (\Sigma\,\bar{\nu}_k\,\hat{s}_k)_e = 11363.7$ kJ kmole^{-1} of fuel K^{-1},
$\Psi_P = (\Sigma\,\bar{\nu}_k\,\bar{\psi}_k)_i = -5045283$ kJ kmole^{-1} of fuel K^{-1}, and
$S_P - S_R = 11363.7 - 9212.3 = 2154.5$ kJ kmole^{-1} of fuel K^{-1}.

Note that $\hat{s}_k = \bar{s}_k^o - \bar{R}\ \ln (p_k/1)$, $p_k = X_k P$

Table 1. Units: p_k are in bar, $\bar{h}$ in kJ kmole^{-1}, and the entropies in kJ kmole^{-1} K^{-1}.

	$\bar{\nu}_k$	$\bar{h}$	X_k	$p_k=X_kP$	$\hat{s}$	$\bar{s}_k^o$	$(\bar{h}-T_o\,\hat{s})_k$
Reactants							
C_4H_{10}	1	−126148	0.021	0.0527	310.2	334.7	−225887
O_2	9.75	0	0.2057	0.5141	205.1	210.7	−62781
N_2	36.66	0	0.7733	1.933	191.6	186.1	−55428
Products							
CO_2	4	−360470	0.0818	0.205	269.3	282.52	−444302
$H_2O(g)$	5	−215849	0.1022	0.256	232.7	244.05	−288561
O_2	3.25	22707	0.0665	0.166	243.6	258.51	−54326
N_2	36.66	21460	0.7495	1.874	228.2	222.99	−44973

Applying the entropy balance equation for a steady state steady flow process,

$$(1/\dot{N}_F)\, dS_{cv}/dt = (\Sigma\,\bar{\nu}_k\,\hat{s}_k)_i + \Sigma(\bar{q}_j/T_{b,j}) - (\Sigma\,\bar{\nu}_k\,\hat{s}_k)_e + \bar{\sigma},$$

$$\bar{\sigma} = S_P - S_R - \Sigma\,(\bar{q}_j/T_{b,j}). \quad \text{(F)}$$

Using the values $T_b = 298$ K and $\bar{q} = -1533044$ kJ,

$$\bar{\sigma} = 11363.7 - 9212.3 - (-1533044) \div 298 = 7294 \text{ kJ kmole}^{-1} \text{of fuel K}^{-1}.$$

The optimum work relation is

$$(1/\dot{N}_F) d(E_{cv} - T_o S_{cv})/dt = (\Sigma\ \bar{\nu}_k \bar{\psi}_k)_i + \Sigma\ \bar{q}_{R,j}(1 - T_o/T_{R,j}) - (\Sigma\ \bar{\nu}_k \bar{\psi}_k)_e - \bar{w} - \bar{i}.$$

Under steady state conditions if there is no thermal reservoir, then

$$\bar{w}_{opt} = \bar{\Psi}_R - \bar{\Psi}_P, \text{ where } \bar{\Psi}_R = (\Sigma\ \bar{\nu}_k \bar{\psi}_k)_I, \text{ and } \bar{\Psi}_P = (\Sigma\ \bar{\nu}_k \bar{\psi}_k)_e, \text{ i.e.,} \quad \text{(G)}$$

$$\bar{w}_{opt} = 2175291 \text{ kJ kmole}^{-1} \text{ of } C_4H_{10}.$$

Therefore,

$$\bar{i} = \bar{w}_{opt} - \bar{w}_{cv} = 2175291 \text{ kJ kmole}^{-1} \text{ of } C_4H_{10}, \text{ since } \bar{w}_{cv} = 0.$$

---

c. *Example 3*

Consider that fuel and air enter an automobile engine at 298 K and 1 bar (state 1), and the car exhaust exits at the same pressure, but at 400 K (state 2). Assume complete combustion with 20% excess air. Show that

$$\bar{w} = \bar{\Psi}_R - \bar{\Psi}_P - \bar{i} - P\,\bar{v}_F, \text{ kJ kmole}^{-1} \text{of burnt fuel} \quad \text{(A)}$$

Determine the optimum work per kmole of burned liquid octane (C_8H_{18}).

Solution

The mass of the automobile is variable, since the liquid fuel weight in the fuel tank decreases over time due to fuel consumption. Hence, this is an unsteady problem. Assume that the species in the exhaust are CO_2, H_2O, O_2, and N_2. From the availability balance relation

$$d(E_{cv} - T_0 S_{cv})/dt = \Sigma \dot{Q}_{R,j}\left(1 - \frac{T_0}{T_{R,j}}\right) + (\Sigma \dot{N}_k \hat{\psi}_k)_i - (\Sigma \dot{N}_k \hat{\psi}_k)_e - \dot{W}_{cv} - \dot{I} \quad \text{(B)}$$

Consider the system control surface to exist around the automobile where $T_R = T_o$. Neglecting the kinetic and potential energies and considering an adiabatic car, Eq. (B) simplifies to the form

$$d(U_{cv} - T_o S_{cv})/dt = (\Sigma \dot{N}_k \hat{\psi}_k)_i - (\Sigma \dot{N}_k \hat{\psi}_k)_e - \dot{W}_{cv} - \dot{I} \quad \text{(C)}$$

Since $U_{cv} = N_F \bar{u}_F$ and $S_{cv} = N_F \bar{s}_F$ and $\bar{u}_F$ and $\bar{s}_F$ are constants, then

$$d(N_F(\bar{u}_F - T_o \bar{s}_F))/dt = +(\Sigma \dot{N}_k \hat{\psi}_k)_i - (\Sigma \dot{N}_k \hat{\psi}_k)_e - \dot{W}_{cv} - \dot{I}$$

$$-\dot{N}_{F,b}(\bar{u}_F - T_o \bar{s}_F) = (\Sigma \dot{N}_k \hat{\psi}_k)_i - (\Sigma \dot{N}_k \hat{\psi}_k)_e - \dot{W}_{cv} - \dot{I} \quad \text{(D)}$$

where the $_{\text{fuel}}$ burn rate $(dN_F/dt) = -\ \dot{N}_{F,b}$. Now

$$(\bar{u}_F - T_o \bar{s}_F) \approx \bar{h}_F - P\,\bar{v}_F - T_o \bar{s}_F = \bar{\psi}_F - P\,\bar{v} \quad \text{(E)}$$

Dividing Eq. (D) by the fuel burn rate,

$$\bar{\psi}_F + (\Sigma(\dot{N}_k/\dot{N}_{F,b})\hat{\psi}_k)_i - (\Sigma(\dot{N}_k/\dot{N}_{F,b})\hat{\psi}_k)_e - (\dot{W}/\dot{N}_{F,b}) - (\dot{I}/\dot{N}_{F,b}) = 0, \text{ or}$$

$$\bar{\psi}_F + (\Sigma\ \bar{\nu}_k \hat{\psi}_k)_i - (\Sigma\ \bar{\nu}_k \hat{\psi}_k)_e - \bar{w} - \bar{i} - P\,\bar{v}_F = 0, \text{ i.e.,} \quad \text{(F)}$$

$$\bar{w} = \bar{\Psi}_R - \bar{\Psi}_P - \bar{i} - P\,\bar{v}_F, \text{ where } \bar{\Psi}_R = \bar{\psi}_F + (\Sigma\ \bar{\nu}_k \hat{\psi}_k)_i, \ \bar{\Psi}_P = (\Sigma\ \bar{\nu}_k \hat{\psi}_k)_e.$$

Note that $\bar{\Psi}_R = (\Sigma \dot{N}_k \hat{\psi}_k)_i$ denotes the availability of air crossing the boundary of car. The optimum work

$\overline{w}_{opt} = \overline{\Psi}_R - \overline{\Psi}_P - P\,\overline{v}_F$.

The reaction equation that represents the complete combustion of a kmole of C_8H_{18} with 20% excess air is

$C_8H_{18(liq)} + 15\ O_2 + 15\times3.76\ N_2 = 8CO_2 + 9H_2O + 2.5O_2 + 15\times3.76\ N_2$.

At the inlet

$\overline{\Psi}_R = (\overline{\psi}_F + \overline{\nu}_{O_2}\hat{\psi}_{O_2} + \overline{\nu}_{N_2}\hat{\psi}_{N_2})_i$, where

$$\overline{\psi}_F = \overline{h}_F - T_o\,\overline{s}_F = -249910 - 298\times360.79 = -357425 \text{ kJ per kmole of fuel.} \quad (G1)$$

In the case of air

$X_{O_2} = 15\div(15 + 15\times3.76) = 0.21$, and

$\hat{\psi}_{O_2} = \overline{h}_{O_2} - T_o\,\overline{s}_{O_2}(T_o, p_{O_2}) = 0-298\times(205.04-8.314\times\ln(0.21\times1\div1))$, i.e., (G2)

$\hat{\psi}_{O_2} = -64969$ kJ per kmole of O_2.

Likewise,

$X_{N_2} = 0.79$, and $\hat{\psi}_{N_2} = -57651$ kJ per kmole of N_2.

Using Eqs. (E)–(G2),

$\overline{\Psi}_R = \overline{\psi}_F + (\Sigma\ \overline{\nu}_k\hat{\psi}_k)_i = (-357{,}425) + 15\times(-64969) + 56.43\times(-57{,}651) =$
-4585118 per kmole of fuel burned.

$$\overline{\Psi}_P = (\overline{\nu}_{CO_2}\hat{\psi}_{CO_2} + \overline{\nu}_{H_2O}\hat{\psi}_{H_2O} + \overline{\nu}_{O_2}\hat{\psi}_{O_2} + \overline{\nu}_{N_2}\hat{\psi}N_{N_2})_e, \text{ where} \quad (H)$$

$X_{CO_2} = 8\div(8+9+2.5+56.43) = 0.1053$ and $p_{CO_2} = 0.1053\times1 = 0.1053$ bar,

Similarly, $X_{H_2O} = 0.1185$, $X_{O_2} = 0.0329$, $X_{N_2} = 0.7432$.

We can now determine the values of $\hat{\psi}_{k,e}$, e.g.,

$$\hat{\psi}_{CO_2} = (-393520+13372-9364)-298\times225.225 = -456{,}629 \text{ kJ (kmole } CO_2)^{-1}. \quad (I)$$

Hence,

$$\overline{\Psi}_P = (\Sigma\ \overline{\nu}_k\hat{\psi}_k)_e = -9835835 \text{ kJ per kmole of fuel burnt, and} \quad (J)$$

$\overline{w}_{opt} = (-4585118 - (-9835835)) - 100\times114\div703 = 5250700$ kJ per kmole of burned fuel, or 122569 kJ gallon^{-1} assuming the liquid density to be 703 kg m^{-3}.

Remarks

The work that can be developed by a 20 gallon (1 gallon = 3.7854 l) tank of octane is 2451374 kJ. Since 1 kW hr = 3600 kJ, in the case of this example W_{opt} = 681 kW hr, and an automobile with a power output of 100 kW (a 6 cylinder engine) can be ideally driven for 6.81 hrs and at 50 kW can be driven for 13.62 hrs. At a speed of 60 mph for a 6 cylinder car this allows the vehicle to cover a distance of 409 miles at a maximum of 20.5 miles per gallon for 100 kW engine and 41 MPH for 50 kW engine. The typical work output of an automobile is roughly 42500 kJ gallon^{-1} due to nonideal conditions.

d. Example 4

Five kmole of CO, three kmole of O_2, and two kmole of CO_2 are fully mixed when they enter a combustor at 3000 K and 1 bar. The products leave the combustor at the same temperature and pressure at equilibrium. Determine the optimum work.

Solution

The equilibrium composition for this problem is readily determined as 2.888 CO, 1.944 O_2 and 4.112 CO_2 using a similar procedure given in Example 7 of Chapter 12. The overall reaction can thereafter be represented by the equation

$$5CO + 3O_2 + 2CO_2 \rightarrow 2.888CO + 1.944O_2 + 4.112CO_2. \quad (A)$$

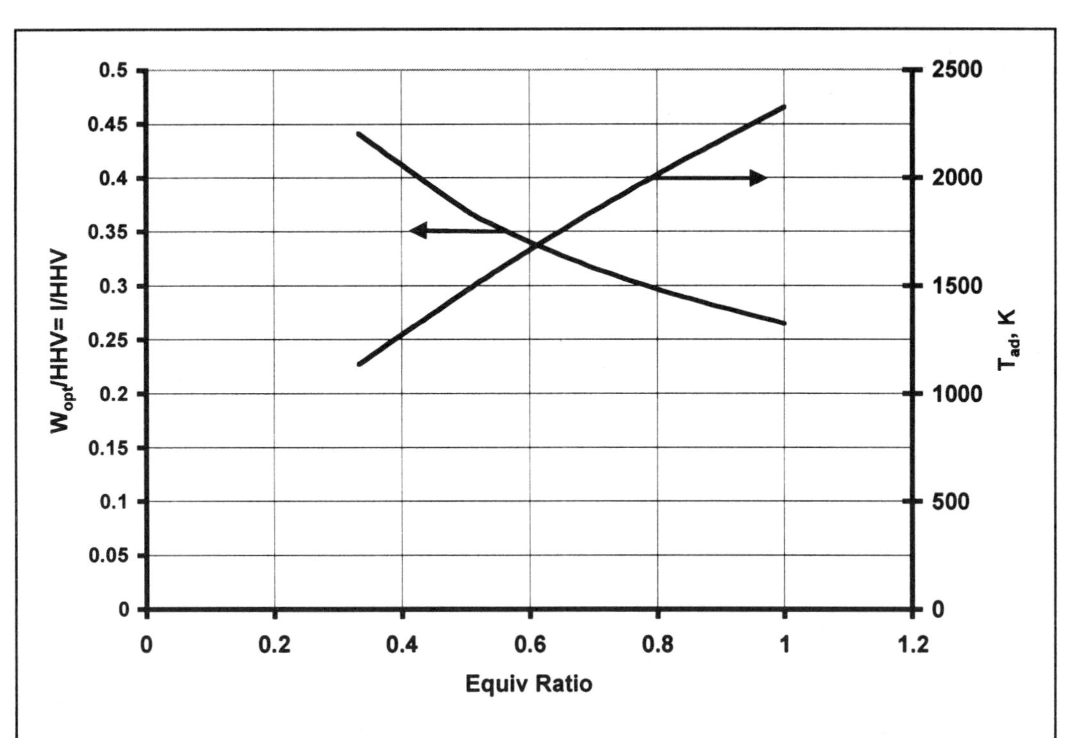

Figure 1: The irreversibility produced by burning methane and air, with both reactant streams at 298.15 K.

W_{opt} = availability in – availability out = $\Psi_R - \Psi_P$, where (B)

$\Psi_R = (5\,\hat{\psi}_{CO} + 3\,\hat{\psi}_{O_2} + 2\,\hat{\psi}_{CO_2})_i$, and (C)

$\Psi_P = (2.888\,\hat{\psi}_{CO} + 1.944\,\hat{\psi}_{O_2} + 4.112\,\hat{\psi}_{CO_2})_e$. (D)

Now,

$\hat{\psi}_{CO,i} = \bar{h}_{CO}(T,p_{CO}) - T_o\,\bar{s}_{CO} = \bar{g}_{CO} + (T - T_o)\,\bar{s}_{CO} =$
$(-110530) + 96395.7 - 298\times(274.6 - 8.314\times\ln((5\div(5+3+2))\times1)) =$
(-0.9772×10^5) kJ per kmole of CO,

$\hat{\psi}_{O_2,i} = 0 + 98152.9 - 298\times(284.4 - 8.314\times\ln((3\div10)\times1)) =$
0.1042×10^5 kJ per kmole of O_2, and

$\hat{\psi}_{CO_2,i} = -393520 + 152347.9 - 298\times(332.9 - 8.314\times\ln 0.2) =$
$(=-0.3444\times10^6)$ kJ per kmole of CO_2.

Therefore,

$\Psi_R = -1.146\times10^6$ kJ (E)

At the exit (in units of kJ),

$\hat{\psi}_{CO,e} = -0.9866\times10^5$, $\hat{\psi}_{O_2,e} = 9699.5$, and $\hat{\psi}_{CO_2,e} = -0.3424\times10^6$ kJ, i.e.,

$\Psi_P = -1.6215\times10^6$ kJ (F)

Hence,

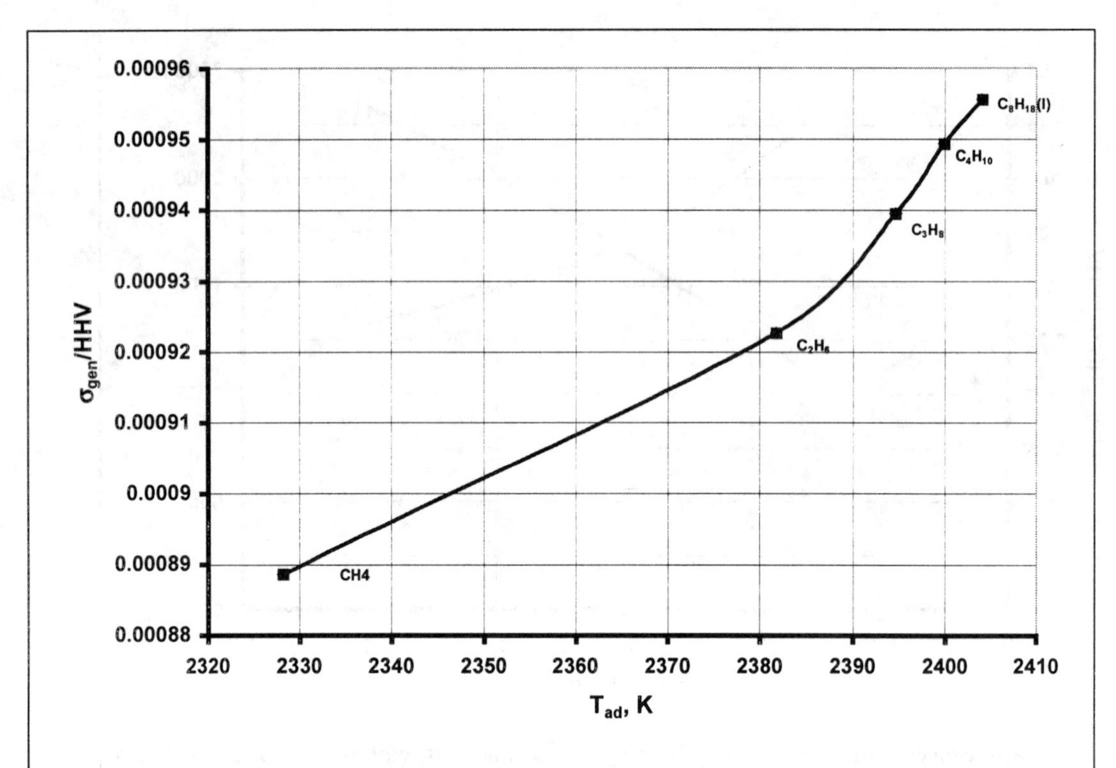

Figure 2: Entropy generation per unit amount of heat released (based on the higher heating value of the fuel) during adiabatic combustion of various hydrocarbon fuels under stoichiometric conditions.

$W_{opt} = (-1.146 + 1.6215) \times 10^6 = 475500$ kJ per 5 kmole of CO consumed, or (G)

95100 kJ per kmole of CO entering the reactor.
This is an example of isothermal availability since $T_{reac} = T_{prod}$.

2. Adiabatic Combustion

In several applications, such as gas turbines, boilers, and residential gas burners, combustion occurs at constant pressure. The higher the combustion temperature, the larger is the availability. Therefore, when excess air is used (which induces smaller equivalence ratios ϕ), the combustion product temperatures decrease, which increases irreversibility and lowers the availability. This is illustrated in Figure 1 in terms of the irreversibility (= $T_o\,\sigma$). (Recall that the overall combustion reaction, say, of methane, in terms of the equivalence ratio can be represented by the reaction equation $CH_4 + (2/\phi)(O_2 + 3.76N_2) \rightarrow CO_2 + 2H_2O + 2(1/\phi-1)\,O_2 + (2/\phi)\,3.76N_2$.) Figure 2 presents the entropy generated per unit amount of heat released for hydrocarbon fuels under adiabatic combustion. Hence even though enthalpy is conserved, availability is lost during irreversible chemical reactions.

e. *Example 5*

Consider the adiabatic and stoichiometric combustion of molecular hydrogen with air. The inlet conditions to the burner are at 298 K and 1 bar (state 1). The products leave the combustor at state 2. Determine the irreversibility. If the products are cooled back to 298 K (state 3), determine the optimum work. Assume that combustion of H_2 is complete and H_2O exists as a gas at 298 K, 1bar.

Solution

The overall chemical reaction is

$$H_2 + 1/2\ O_2 + 1.88\ N_2 \rightarrow H_2O + 1.88\ N_2.$$

Recall that $\sigma = W_{opt}/T_o$, where

$$W_{opt} = \Psi_1 - \Psi_2.$$

With the values

$X_{H_2} = 1 \div (1+0.5+1.88) = 0.296$, $X_{O_2} = 0.148$, $X_{N_2} = 0.556$,

$\Psi_1 = N_{H_2} \hat{\psi}_{H_2} + N_{O_2} \hat{\psi}_{O_2} + N_{N_2} \hat{\psi}_{N_2}$, where, for instance

$\hat{\psi}_{H_2} = \bar{h}_{H_2}(T, p_{H_2}) - T_o \bar{s}_{H_2}$ and $p_{H_2} = X_{H_2} P$,

$\Psi_1 = (0 - 298 \times (130.57 - 8.314 \times \ln(1 \times 0.296 \div 1))) +$

$1/2 \times (0 - 298 \times (205.03 - 8.314 \times \ln(1 \times 0.148 \div 1)))$

$1.88 \times (0 - 209 \times (191.5 - 8.314 \times \ln(1 \times 0.556 \div 1))) = -184920$ kJ

The adiabatic flame temperature is obtained by considering the energy balance

$$dE_{cv}/dt = \dot{Q}_{cv} - \dot{W}_{cv} + \Sigma_{k,i} \dot{N}_k \bar{e}_{T,k} - \Sigma_{k,e} \dot{N}_k \bar{e}_{T,k}.$$

Under ideal gas conditions $\hat{h}_k = \bar{h}_k$, and assuming negligible contributions from the potential and kinetic energies, steady state, and a single step overall chemical reaction, and in the absence of work

$(\Sigma \bar{\nu}_k \bar{h}_{T,k})_i = (\Sigma \bar{\nu}_k \bar{h}_{T,k})_e$, or $H_1 = H_2$, where

$H_1 = \bar{h}_{H_2,i} + \bar{\nu}_{O_2,i} \bar{h}_{O_2,i} + \bar{\nu}_{N_2} \bar{h}_{N_2,i} = 0$ (elemental species), and

$H_1 = \bar{\nu}_{N_2,e} \bar{h}_{N_2,e} + \bar{\nu}_{H_2O,e} \bar{h}_{H_2O,e} = 0.$

Solving iteratively, $T_2 = T_{ad} = 2528.7$ K.

$\Psi_2 = N_{H_2O,e} \hat{\psi}_{H_2O,e} + N_{N_2,e} \hat{\psi}_{N_2,e}$, i.e.,

$\Psi_2 = 1 \times (-241820 + 99704 - 298 \times (276.4 - 8.314 \times \ln(1 \times 0.347 \div 1)))$

$+ 1.88\ (0 + 75597 - 298 \times (260.7 - 8.314 \times \ln(1 \times 0.653 \div 1))) =$

(-233079) kJ.

$W_{opt,12} = -184920 - (-233079) = 48159$ kJ per kmole of fuel burnt.

$\bar{i}_{12} = 48159$ kJ per kmole of fuel.

$\bar{\sigma}_{12} = 48159 \div 298 = 162$ kJ K^{-1} per kmole of fuel.

If the gases are cooled from $T_2 = 2528.7$ to $T_3 = T_o = 298$ K, but with the same composition,

$W_{opt,23} = \Psi_2 - \Psi_3$, where

$\Psi_3 = 1 \times (-241820 + 298 \times (188.72 - 8.314 \times \ln(1 \times 0.347 \div 1)))$

$+ 1.88 \times (0 - 298 \times (191.5 - 8.314 \times \ln(1 \times 0.653 \div 1))) =$

(-408259) kJ per kmole of fuel, i.e.,

$W_{opt,23} = -233079 - (-408259) = 175180$ kJ per kmole of fuel.

Remarks

The stream exergy at state 1 is $\Psi_1 - \Psi_0 = \Psi_1 - \Psi_3 = W_{opt\ 13} = -184{,}920 - (-408259) = 223{,}339$ kJ; Even though state 2 has the same enthalpy as state 1, the stream exergy at state 2 is $\Psi_2 - \Psi_0 = \Psi_2 - \Psi_3 = W_{opt,23} = 175{,}180$. Defining the availability fraction at state 2 as $\xi_{opt,23} = W_{opt,23} / W_{opt,13} = 0.78$, which indicates that approximately 22% of maximum possible optimum work that was otherwise available at state 1 is lost during the irreversible adiabatic combustion.

The calculations can be repeated for a stoichiometric mixture of methane and air entering the burner at 298 K and 1 atm. In this case, the flame temperature, optimum work and availability fraction at the flame temperature are $T_{ad} = 2325$ K, $W_{loss} = 236$

MJ/kmol, and $\xi_{opt,23}$= 0.72 (i.e., the availability loss fraction is 0.28). The combustion temperature of 2325 K is achieved when the combustion proceeds without any loss of heat to the surroundings ($H_{products}$-$H_{reactants}$). The only irreversibility is due to the chemical reaction. We see, therefore, that a combustion reaction is highly irreversible and wasteful of availability, even when it is conducted so that enthalpy of combustion is retained in the combustion products. An availability analysis reveals that chemical processes may be degrading the energy quality while conserving its quantity.

3. Maximum Work Using Heat Exchanger and Adiabatic Combustor

In a typical boiler, the hot combustion gases from a nearly adiabatic combustor are provided to the heat exchanger section where they are cooled from a temperature T_{ad} to T_{exch}. During this process the gases generally transfer heat to a fluid, e.g., liquid water in the boiler section of a Rankine cycle steam engine. The hot fluid then transfers heat to a heat engine producing work with a certain η. In case of a Carnot cycle $\eta = (1-T_o/T_{HE})$, where T_{HE} denotes the characteristic heat engine temperature, where, $T_{exch} \geq T_{HE}$. Ideally, $T_{exch} = T_{HE}$.

The work obtained

$$W = Q\,\eta. \tag{16}$$

If the combustor heat loss equals

$$Q_c,\ H_{products} = H_{reactants} - Q_c, \tag{17}$$

for a Carnot engine

$$W = (H_{reactants} - Q_c - H_{exch})\,(1 - T_o/T_{HE}). \tag{18}$$

For an adiabatic combustor

$$W = (H_{reactants} - H_{exch})\,(1 - T_o/T_{HE}).$$

The Carnot efficiency increases by raising T_{HE}, but the heat transfer from the combustion products to the heat exchanger is lowered. Due to this tradeoff, there is an optimum value of T_{HE} at which W is maximized. Figure 3 presents the variation of W with respect to $T_{HE} = T_{exch}$ for stoichiometric methane–air combustion with variable $c_{po}(T)$. The optimum temperature value T_{opt} = 858 K at which W = 397 MJ/kmole.

The assumption of constant specific heats simplifies the calculation. Let $(H_{reac} - H_{exch}) = m\,c_p(T_{ad} - T_{exch})$, where $T_{exch} = T_{HE}$. In this case,

$$W = mc_p(T_{ad} - T_{exch})(1 - T_o/T_{HE}) = mc_p(T_{ad} - T_{HE})(1 - T_o/T_{HE}). \tag{19}$$

Differentiating Eq. (19) with respect to T_{HE}, and equating the resultant expression to zero,

$$T_{HE} = (T_{ad}\,T_o)^{1/2}. \tag{20}$$

Using this result, the expression for the maximum work assumes the form

$$W_{max} = mc_pT_{ad}(1 - (T_o/T_{ad})^{1/2})^2. \tag{21}$$

In addition to the chemical irreversibility during adiabatic combustion, irreversibilities exist during the heat transfer from hot products to the heat exchanger fluid. (For stoichiometric methane–air T_{ad} = 2328 K and T_o =298 K. Eq. (19) yields the temperature optimum $T_{exch} = T_{HE}$ = 833 K assuming constant specific heat values.)

4. Isothermal Combustion

A fuel cell and biological reactions, such as photosynthesis, are examples of situations involving isobaric and isothermal chemical reactions. Here,

$$d(E_{cv} - T_0 S_{cv})/dt = \Sigma \dot{Q}_{R,j}(1 - \frac{T_0}{T_{R,j}}) + (\Sigma \dot{N}_k \hat{\psi}_k)_i - (\Sigma \dot{N}_k \hat{\psi}_k)_e - \dot{W}_{cv} - \dot{I} \quad (22)$$

Assuming steady state conditions in the absence of work, and heat exchange only with the ambient, i.e., $\dot{Q}_{R,j} = \dot{Q}_0$ and $T_{R,j} = T_o$,

$$\dot{I} = (\Sigma \dot{N}_k \hat{\psi}_k)_i - (\Sigma \dot{N}_k \hat{\psi}_k)_e. \quad (23)$$

In this case the irreversibility is caused by entropy generated during the combustion process $\dot{\sigma}_c$, and during the irreversible heat transfer due to the temperature gradient within the system $\dot{\sigma}_{HE}$ (i.e., the combustion chamber temperature T and the ambient temperature T_o). The total entropy generated

$$\dot{\sigma}_{cv} = \dot{\sigma}_c + \dot{\sigma}_{HE}. \quad (24)$$

We can evaluate these entropies by using the entropy balance equation, i.e.,

$$dS_{cv}/dt = \Sigma \frac{\dot{Q}_j}{T_{b,j}} + (\Sigma \dot{N}_k \hat{s}_k)_i - (\Sigma \dot{N}_k \hat{s}_k)_e - \dot{\sigma}_{cv}. \quad (25)$$

If the system boundary lies just within the reactor walls, leaving out the thin thermal boundary layer, then $T_{b,j} = T$

$$dS_{cv}/dt = \Sigma \frac{\dot{Q}_0}{T} + (\Sigma \dot{N}_k \hat{s}_k)_i - (\Sigma \dot{N}_k \hat{s}_k)_e - \dot{\sigma}_c \quad (26)$$

Employing the First Law, i.e.,

$$dE_{cv}/dt = \dot{Q}_0 - \dot{W}_{cv} + \Sigma_{k,i} \dot{N}_k e_{T,k} - \Sigma_{k,e} \dot{N}_k e_{T,k}, \quad (27)$$

And eliminating $\dot{Q}_0$ between Eqs. (26) and (27),

$$\dot{\sigma}_c = -(dE_{cv}/dt)(1/T) + \dot{W}_{cv}/T + \Sigma_{k,i} \dot{N}_k e_{T,k}/T - \Sigma_{k,e} \dot{N}_k e_{T,k}/T + dS_{cv}/dt + (\Sigma \dot{N}_k \hat{s}_k)_e - (\Sigma \dot{N}_k \hat{s}_k)_I. \quad (28)$$

At steady state conditions, for pure combustion process with negligible ke and pe,

$$T\dot{\sigma}_c = (\Sigma \dot{N} \hat{g}_k)_e - (\Sigma \dot{N} \hat{g}_k)_I \text{ , reaction at temperature T} \quad (29)$$

The irreversibility for an isothermal and isobaric process due to chemical reaction alone can be expressed as

$$T_0 \dot{\sigma}_c = (T_o/T)((\Sigma N_k \hat{g}_k)_i - (\Sigma N_k \hat{g}_k)_e), \text{ reaction at temperature T}$$

For ambient temperature reactions (e.g., in a fuel cell or in plant leaves),

$$T_0 \dot{\sigma}_c = (\Sigma_k N_k \hat{g}_k)_e - (\Sigma N_k \hat{g}_k)_I \text{ , reaction at ambient temperature } T_0 \quad (30)$$

The irreversibility due to heat transfer alone is

$$\dot{\sigma}_{HE} = -\dot{Q}_0 (1/T_o - 1/T).$$

---

f. *Example 6*

The stoichiometric combustion of molecular hydrogen in air proceeds in a premixed state. The reactants enter a combustor at 298 K and 1 bar and the products leave at the same temperature and pressure. Determine the values of σ and Δh_c if the water formed exists in a liquid or in a gaseous state.

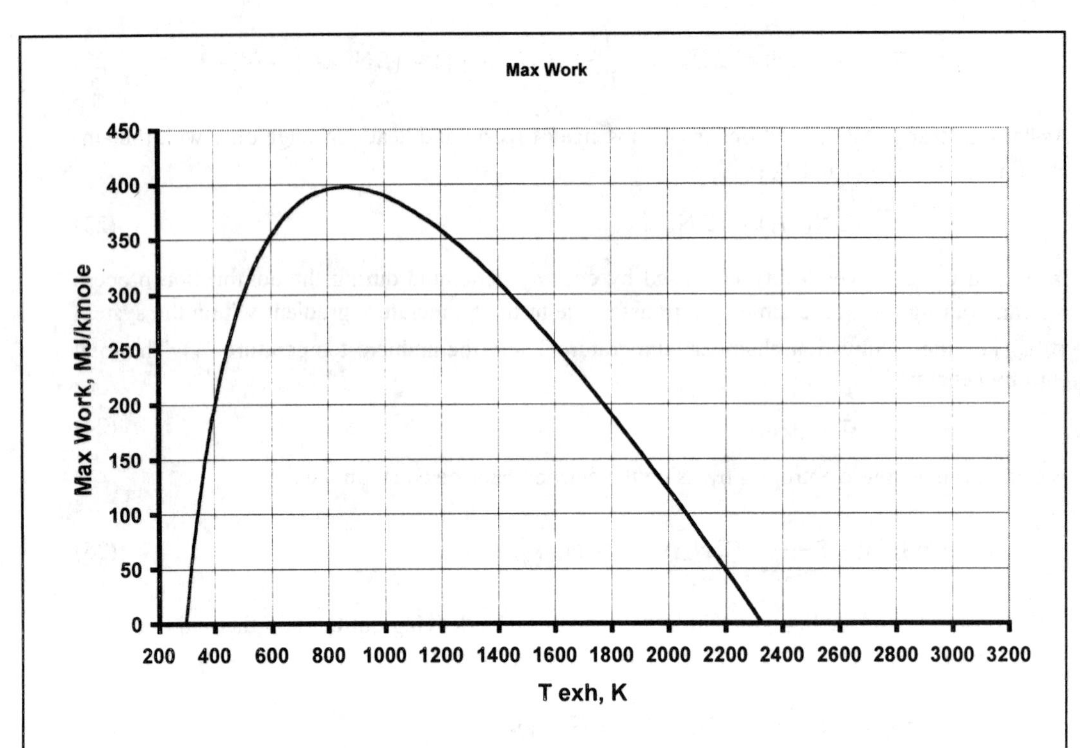

Figure 3: Work output with respect to $T_{HE.}$ for stoichiometric methane–air combustion considering variable specific heat $c_{po}(T)$.

Solution

The overall chemical reaction is

$$H_2 + 1/2\ O_2 + 1.88\ N_2 \rightarrow H_2O + 1.88\ N_2.$$

Recall that $\sigma = W_{opt}/T_o$, where

$$\overline{w}_{opt} = \Psi_1 - \Psi_2.$$

With the values

$X_{H_2} = 1\div(1+0.5+1.88) = 0.296$, $X_{O_2} = 0.148$, $X_{N_2} = 0.556$,

$\hat{g}_{H_2} = (0–298\times(130.57 – 8.314 \times \ln(1\times0.296\div1)) = –41926$ kJ kmole^{-1},

$\hat{g}_{O_2} = 0–298\times(205.03 – 8.314 \times \ln(1\times0.148\div1)) = – 65832$ kJ kmole^{-1},

$\hat{g}_{N_2} = –298\times(191.5 – 8.314 \times \ln(1\times0.556\div1)) = –58521$ kJ kmole^{-1},

$\Psi_1 = G_1 = (–41,926) + 1/2\times(–65,832) + 1.88\times(–58,521) =$
$–184861$ kJ per kmole of H_2.

Treating H_2O as a gas,

$X_{H_2O} = 1\div(1+1.88) = 0.347$, $X_{N_2} = 0.653$,

$\hat{g}_{H_2O} = –241820–298\times(188.71–8.314\times\ln(1\times0.347/1)) = –300678$ kJ kmole^{-1},

$\hat{g}_{N_2} = –298\times(191.5–8.314\times\ln(1\times0.653/1)) = –58123$ kJ kmole^{-1},

$\Psi_2 = G_2 = (–300768) + 1.88\times(0 –58\ 123) = –4100392$ kJ per kmole of H_2,

$\overline{w}_{opt} = –184861 –(–4100392) = 22178$ kJ per kmole of H_2.

All of this work is lost. Therefore,

$\overline{\sigma} = 225090\div298 = 755.34$ kJ kmole^{-1} K, and

$\Delta h_c = 0\ –(–241820) = 241820$ kJ per kmole of H_2.

Treating H_2O as a liquid,

$\hat{g}_{H_2O} = \bar{g}_{H_2O} = -285830 - 298 \times 69.95 = -306675$ kJ per kmole of H_2,

$\Psi_2 = 1\times 1\,(-285830-298 \times (69.95)) +$

$1.88\,(0 - 298 \times (191.5 - 8.314 \times \ln(1\times1 \div 1))) =$

(-413961) kJ per kmole of H_2, and

$\bar{w}_{opt} = 229098$ kJ per kmole of H_2.

Consequently,

$\bar{\sigma} = 229098 \div 298 = 768.8$ kJ kmole^{-1}, and

$\Delta h_c = 285830$ kJ per kmole of H_2.

---

5. Fuel Cells

Combustion is a process during which irreversible exothermic oxidation reactions convert the chemical energy of fuels into thermal energy and then, oftentimes, into electrical energy through an elaborate conversion process that results in low conversion efficiencies. Typically, only about 35% of the chemical energy of a fuel is eventually converted into electrical energy in a steam power plant. Fuel cells offer a more efficient alternative. In a fuel cell chemical energy is converted into electricity through the release of electrons by chemical reaction. The process is almost reversible and hence the irreversibility is greatly reduced in these cells in comparison with combustion. Prior to description of the fuel cell, let us illustrate the basics of oxidation states and the electron transfer during chemical reaction.

a. Oxidation States and electrons

Consider the reaction of butane C_5H_{12} with O_2 producing the products CO_2 and H_2O. The oxidation states for this reaction are provided by the number of electrons gained or lost. The mass of a C atom is twelve times larger than of H atom so that the electrons surrounding H atoms are pulled more strongly towards the carbon atoms (Chapter 1). The five C atoms in the fuel gain y=12 electrons. The oxidation state of each C atom in the fuel is -12/5 (= y/x, where x denotes the number of carbon atoms in the fuel) while that of each H atom is +1. The oxygen molecule is symmetric and its oxidation state is zero. Consider each of the five CO_2 molecules in the products. Atomic O atom is heavier than C atoms. The four outer electrons of each of the carbon atoms are pulled towards the O atom. In this context, the oxidation state of each C atom is +4 while the oxidation state of each O atom is -2. There is a net loss of 5×4 = 20 electrons from the 5 carbon atoms in 5 CO_2. Thus as the carbon in C_5H_{12} is burned, the oxidation state of C changes from -12/5 to +4 in CO_2. There is a net loss of (12/5) + 4 electrons (=y/x + 4) per C atom in the fuel. For this example, the 5 C atoms in the fuel lose 32 (= xx ((y/x) +4)= y+4 x) electrons. The oxidation state of H atom is unchanged. The oxidation states of O atoms in CO_2 and H_2O are -2 and -2. Hence, the O atoms in the five CO_2 molecules and six H_2O molecules gain 5×2×(-2) +6×1×(-2) = 32 electrons, which are transferred from the five C atoms in C_5H_{12} to the oxygen atoms in the products during the chemical reaction. The fuel cell achieves such a transfer through an external load. Generalizing for any fuel denotes by C_xH_y, the number of electrons transferred is 4x + y. In the case of pure H_2 (i.e., x=0), the analogous number of electrons is two.

b. H_2-O_2 Fuel Cell

The operational principle of a H_2-O_2 fuel cell is illustrated in Figure 4. At the anode

$$H_2 \rightarrow 2e^- + 2H^+,$$

and at the cathode

$$2H^+ + 2e^- + 1/2O_2 \rightarrow H_2O\,(l).$$

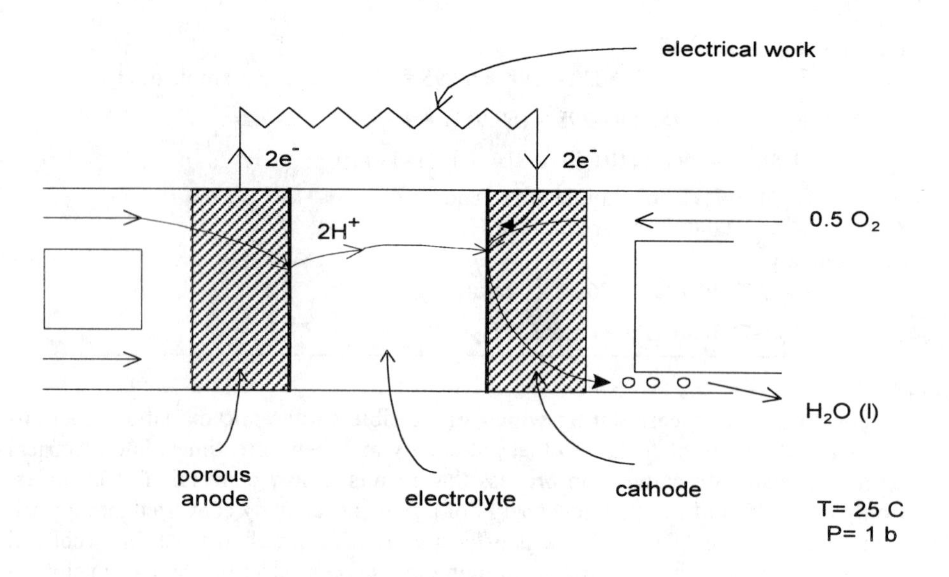

Figure 4: Principle of operation of a fuel cell.

In a fuel cell chemical reactions occur at the ambient temperature T_o. In a hydrogen–powered fuel cell, separate streams of H_2 and O_2 are converted into liquid water at T_o. Consequently, the absolute availability for the H_2 at the inlet is $\bar{g}_{H_2}$. The maximum possible work under the steady state steady flow characteristic of a fuel cell can be determined using the familiar relation:

$$d(E_{cv} - T_0 S_{cv})/dt = \Sigma \dot{Q}_{R,j}(1 - \frac{T_0}{T_{R,j}}) + (\Sigma \dot{N}_k \hat{\psi}_k)_i - (\Sigma \dot{N}_k \hat{\psi}_k)_e - \dot{W}_{cv} - \dot{I} \quad (31)$$

$$\dot{W}_{opt} = (\Sigma \dot{N} \hat{g}_k)_i - (\Sigma \dot{N} \hat{g}_k)_e \quad (32)$$

The fuel cell efficiency is

$$\eta_{fc} = w/\Delta h_c. \quad (33)$$

The maximum possible efficiency is

$$(\eta_{fc})_{opt} = w_{opt}/\Delta h_c = (-\Delta g)/\Delta h_c = -(\Delta h_R - T\Delta s_R)/\Delta h_c. \quad (34)$$

Assuming the higher heating value applies, $\Delta h_c = HHV = -\Delta h_R$, then

$$(\eta_{fc})_{opt} = 1 + T\Delta s_R/\Delta h_c. \quad (35)$$

If the value of Δs_R (= s_{prod}- s_{reac}) >0 at the fuel cell operating temperature, (e.g., by adding heat to maintain isothermal conditions for an endothermic reaction), then $(\eta_{fc})_{opt}>1$. The availability or exegetic efficiency for a hydrogen–oxygen fuel cell can be defined as

$$\eta_{avail} = W/(\text{exergy of } H_2). \quad (36)$$

g. *Example 7*

Find the maximum work deliverable in a fuel cell by 1 kmole of H_2 with O_2 if it is isothermally reacted at 25°C and 1 bar to produce liquid water. Both reactants enter the cell separately. Determine the maximum voltage developed by the fuel cell. Consider also the scenario for the reaction of a stoichiometric amount of H_2 with O_2. What is the maximum possible fuel cell efficiency. Assume that Δh_c = 285830 kJ $kmole^{-1}$.

Solution

The fuel cell temperature $T = T_o$ = 25°C, and the two reactant streams enter separately. Consequently,

$$\dot{W}_{opt} = (\Sigma_k \dot{N} \bar{g}_k)_i - (\Sigma \dot{N} \bar{g}_k)_e.$$

Assume that $\dot{N}_{H_2}$= 1 kmole s^{-1}. In that case, $\dot{N}_{O_2}$= 1/2 kmole s^{-1}, and $\dot{N}_{H_2O}$= 1 kmole s^{-1}, and

$$(\Sigma \dot{N}_k \bar{g}_k)_i = \bar{g}_{H_2} + (1/2)\bar{g}_{O_2}, \text{ where}$$

$$\bar{g}_{H_2} = 0 - 298 \times 130.574 = -38911 \text{ kJ kmole}^{-1}, \text{ and}$$

$$\bar{g}_{H_2} = 0 - 298 \times 205.033 = -61100 \text{ kJ kmole}^{-1}, \text{ i.e.,}$$

$$(\Sigma \dot{N}_k \bar{g}_k)_i = -69461 \text{ kJ per kmole of } H_2.$$

Likewise,

$$(\Sigma \dot{N}_k \bar{g}_k)_e = -306675 \text{ kJ per kmole of } H_2, \text{ so that}$$

$$\bar{w}_{opt} = G_{reactants} - G_{products} = -69461 + 306675 = 237214 \text{ kJ per kmole of } H_2.$$

At the cathode of a fuel cell positive potential is applied; the reaction is given as

$$2H^+ + 2e^- + 1/2O_2 \rightarrow H_2O(\ell),$$

and at the anode the pertinent reaction is

$$H_2 \rightarrow 2e^- + 2H^+.$$

Overall, a molecule of H_2 generates 2 electrons, i.e., a kmole of H_2 or 6.023×10^{26} molecules generates $2{\times}6.023{\times}10^{26}$ electrons. An electron carries a charge of $1.602{\times}10^{-19}$ Coulomb so that a kmole of H_2 generates a charge of $1.602{\times}10^{-19} \times 2 \times 6.023 \times 10^{26}$ Coulomb.

The electrical work

$$W_{elec} = \text{Voltage in volts} \times \text{charge in coulombs} =$$

$$237214 \text{ kJ per kmole of } H_2 \times 1000 \text{ J kJ}^{-1} \div$$

$$(1.602{\times}10^{-19} \times 2 \times 6.023 \times 10^{26} \text{ Coulomb}) =$$

$$1.229 \text{ V}.$$

The optimum fuel cell efficiency

$$(\eta_{fc})_{opt} = \bar{w}_{opt}/\Delta h_c = 237214/285830 = 0.83.$$

Remarks

A short formula for determining the voltage of a fuel cell is Volts = (ΔG per kmole of fuel in kJ) $\times$ $1.036{\times}10^{-5}$÷(Number of electrons generated per molecule of the fuel).

Fuel cells may be connected in series to obtain a higher voltage than an individual cell provides.

For fuel cells using hydrocarbon fuels, the anodic reaction is

$$C_xH_y + 2x\, H_2O \rightarrow x\, CO_2 + (4x{+}y)H^+ + (4x{+}y)e^-,$$

and the cathodic reaction is

$$(x{+}y/4)O_2 + (4x{+}y)H^+ + (4x{+}y)e^- \rightarrow (2x{+}y/2)H_2O.$$

The overall reaction can be represented as

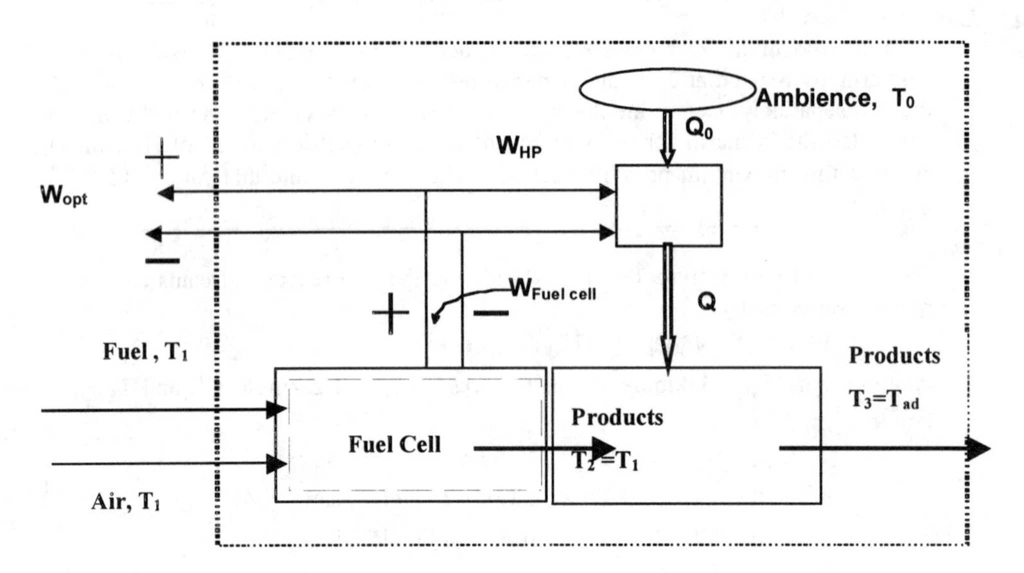

Figure 5: A schematic illustration for achieving a temperature equal to the adiabatic flame temperature but without irreversibility. The fuel is H_2 at 298K, oxidizer is air at 298K. Here, $W_{Fuel\ Cell}$ = 229,100 kJ, W_{HP} = 180,940, W_{opt} = 48,160 kJ, T_{ad} = 2529 K, Q_0 =108,900 kJ, and Q = 285830 kJ.

$$C_xH_y + (x + y/4)\ O_2 \rightarrow x\ CO_2 + y/2\ H_2O.$$

Therefore, for a hydrocarbon fuel C_xH_y, the electrons generated per molecule are represented by the relation (4x+y).

In the context of the H_2 fuel cell, had we supplied work or applied an emf of 1.23 V, the reaction could have been reversed to produce H_2 and O_2 from H_2O. In this case, water that is formed from the hydrogen and oxygen could be converted back into the respective fuels through a reversible process so that $\sigma = 0$.

In Example 5, we considered the adiabatic combustion of H_2 in air that produced hot gaseous products at 2529 K. That process produced an irreversibility of 48160 kJ $kmole^{-1}$. If our objective is to obtain gases at 2529K but with no entropy generation, i.e., $\sigma = 0$ it can be achieved, through a thought experiment illustrated in Figure 5. We can supply H_2 and stoichiometric air first to a fuel cell to produce 229,100 kJ of work and water(g) at 298 K as illustrated in this example. Then, we can heat a mixture of H_2O and N_2 to 2529 K using a Carnot type heat pump to supply necessary heat to raise the temperature from 298 K to 2529 K. You can readily calculate that the heating requires 180940 kJ of work input. In the ideal fuel cell and heating combination case, we obtain a net electrical work of 48160kJ (= 229100 – 180940) kJ, while still producing hot gases at 2529 K; therefore this work capability of 48,160 kJ is lost when we use adiabatic combustion.

D. FUEL AVAILABILITY

At steady state, the maximum work under thermo-mechanical equilibrium conditions is expressed by the relation

$$\dot{W}_{opt} = (\Sigma \dot{N}_k \hat{\psi}_k(T_i, p_{k,i}))_i - (\Sigma \dot{N}_k \hat{\psi}_k(T_e, p_{k,0}))_e.$$

where exit partial pressures are not necessarily same as the ambient. Recall that in case of combustion this equation can be divided by the number of moles of fuel $\dot{N}_F$ flowing into the combustor to yield the availability under thermomechanical equilibrium conditions.

$$(\bar{w}_{opt} = \Psi_R - \Psi_P)_{TM}, \text{ where} \tag{37}$$

$$\Psi_{R \text{ or } P, TM} = (\Sigma\ \bar{\nu}_k \hat{\psi}_k)_{i \text{ or } e, TM} = (\Sigma\ \bar{\nu}_k \hat{\psi}_k(T_o, p_{k,0})). \tag{38}$$

Here, the exit is considered to be in thermomechanical equilibrium at a total pressure of P_0, p_{k0} denotes the partial pressure of a species k in the thermomechanical equilibrium state with the ambient (or standard atmosphere), but each component is not necessarily in chemical equilibrium with the corresponding components in the ambient.

Consider a special case in which fuel and air, although not premixed, both enter a combustor separately at the state (T_o, P_o). The reaction products are discharged at partial pressures $p_{k,\infty}$ so that products are in thermo-mechanical-chemical equilibrium with the ambient. The optimum work under such a condition where each species is discharged at the partial pressure corresponding to the ambient is called the fuel availability. In this case,

$$\Psi_R = (\Sigma\ \bar{\nu}_k \hat{\psi}_k)_i = (\bar{\psi}_F(T_o,P_o) + \bar{\nu}_{O_2} \bar{\psi}_{O_2}(T_o, p_{O_2,o})) =$$

$$(\bar{g}_F(T_o, P_o) + \bar{\nu}_{O_2} \bar{g}_{O_2}(T_o, p_{O_2,o})), \text{ and} \tag{39a}$$

$$\Psi_{P,TMC} = (\Sigma\ \bar{\nu}_k \hat{\psi}_k)_{e,TM} = (\Sigma\ \bar{\nu}_k \bar{g}_k(T_o, p_{k,\infty}))_e. \tag{39b}$$

Then,

$$\bar{w}_{opt,TMC} = \text{Avail}_F = \Psi_R - \Psi_{P,TMC}$$

$$= (\bar{g}_F(T_o, P_o) + \bar{\nu}_{O_2} \bar{g}_{O_2}(T_o, p_{O_2,o})) - (\Sigma\ \bar{\nu}_k \bar{g}_k(T_o, p_{k,\infty}))_e.$$

Nitrogen need not be considered, since the partial pressures and moles of N_2 at the inlet and exit are equal. Table A-27B tabulates the fuel availability assuming $X_{O_2,\infty} = 0.2035$, $X_{CO_2,\infty} = 0.0003$, $X_{H_2O,\infty} = 0.0303$, $X_{N2,\infty} = 0.7659$

---

h. Example 8

Determine the fuel availability for methane. Assume that $T_o = 298$ K, $p_{O_2,o} = 0.2055$, $p_{CO_2,o} = 0.003$, and $p_{H_2O,o} = 0.0188$. If the lower heating value LHV of methane is 802330 kJ kmole^{-1}, determine the ratio of the fuel availability to LHV.

Solution

Consider the overall reaction

$$CH_4 + 2O_2 \rightarrow CO_2 + 2H_2O.$$

The CO_2 and H_2O so produced joins the atmosphere in gaseous form at partial pressures corresponding to the environment. The fuel availability

$$\text{Avail}_F = \bar{w}_{opt} = \Psi_R - \Psi_{P,TMC}.$$

$$\Psi_R = (\bar{\psi}_{CH_4}(T_o,P_o) + 2\bar{\psi}_{O_2}(T_o, p_{O_2,o})), \text{ where}$$

$$\bar{\psi}_{CH_4}(T_o,P_o) = \bar{h}_{CH_4} - T_o \bar{s}_{CH_4}(T_o,P_o) = -74850 - 298 \times 186.16 = (-1.303\times10^5) \text{ kJ kmole}^{-1},$$

$$\bar{\psi}_{O_2}(T_o, p_{O_2,o})) = \bar{h}_{O_2} - T_o \bar{s}_{O_2}(T_o, p_{O_2,o}) = 0 - 298 \times (205.04 - 8.314 \times \ln(0.2055 \div 1)) = (-65022) \text{ kJ kmole}^{-1}, \text{ i.e.,}$$

$\Psi_R = 1\times(-130300) + 2\times(-65022) = -260300$ kJ per kmole of CH_4.

Likewise,

$\overline{\psi}_{CO_2} = \overline{h}_{CO_2} - T_o\,\overline{s}_{CO_2}(T_o, p_{CO_2,\infty}) =$

$(-393520 - 298 \times (213.7 - 8.314 \times \ln(0.003 \div 1)) =$

(-472000) kJ per kmole of CO_2,

$\overline{\psi}_{H_2O} = \overline{h}_{H_2O} - T_o\,\overline{s}_{H_2O}(T_o, p_{H_2O,\infty}) =$

$(-241820) - 298 \times (188.7 - 8.314 \times \ln(0.0188 \div 1)) =$

-307900 kJ per kmole of H_2O, i.e.,

$\Psi_{P,TMC} = 1\times(-472000) + 2\times(-307000) = -1088000$ kJ per kmole of fuel.

Therefore,

$Avail_F = -260300 - (-1088000) = 827500$ kJ per kmole of fuel.

The ratio

$Avail_F/LHV = 827500 \div 802330 = 1.031$.

Remarks

This procedure can be repeated for butane, for which $Avail_F = 2767296$ kJ per kmole of fuel and the ratio $Avail_F/LHV = 2767296 \div 2708330 = 1.0218$. For most hydrocarbon fuels, the ratio of fuel availability to the lower heating value is in the range 1.02–1.07. An empirical relation for (Moran)

$Avail_F/LHV = 1.033 + 0.0169\ (H/C) - 0.0698/C$, gaseous hydrocarbon, C: carbon atom, H: hydrogen atom and

$Avail_F/LHV = 1.0422 + 0.0119\ (H/C) - 0.042/C$, liquid HC

The fuel availability $Avail_F = \overline{w}_{opt,TMC}$, where

$$\overline{w}_{opt,TMC} = (\overline{g}_F(T_o, P_o) + \overline{\nu}_{O_2}\,\overline{g}_{O_2}(T_o, p_{O_2,o})) - (\Sigma\,\overline{\nu}_k\,\overline{g}_k(T_o, p_{k,\infty}))_e.$$

For ideal gases

$$\overline{g}(T,p_k) = \overline{h}_k(T_o) - T_o\,\overline{s}_k(T,p_k) = \overline{h}_k(T_o) - T_o(\overline{s}_k(T,1) - \overline{R}\ln(p_k/1)) =$$

$$= (\overline{h}_k(T_o) - T_o\,\overline{s}_k(T,1)) + \overline{R}T_o\ln(p_k/1) =$$

$$\overline{g}^o_k(T_o) + \overline{R}T_o\ln(p_k(\text{bar})/1), \text{ i.e.,}$$

$$\overline{w}_{opt,TMC} = (\overline{g}^o_F(T_o) + \overline{\nu}_{O_2}\,\overline{g}^o_{O_2}(T_o) + \overline{\nu}_{O_2}\,\overline{R}T_o\ln(p_{O_2}/1)) -$$

$$(\Sigma\,\overline{\nu}_k\,\overline{g}^o_k(T_o))_e - (\Sigma\,\overline{\nu}_k\,\overline{R}T_o\ln(p_{k,\infty}/1))_e, \text{ i.e.,}$$

$$\overline{w}_{opt,TMC}/(\overline{R}T_o) = -\Delta G^o/(\overline{R}T_o) + \ln((p_{O_2,\infty}/1)^{\nu_{O_2}}/(\Pi(p_{k,\infty}/1)^{\nu_k})_e). \quad (40)$$

Typically, $\Delta G^o/(\overline{R}T_o) \gg \ln((p_{O_2,\infty}/1)^{\nu_{O_2}}/(\Pi(p_{k,\infty}/1)^{\nu_k})_e)$, so that

$$\overline{w}_{opt,TMC} \approx -\Delta G^o.$$

Further, since $\Delta G^o = \Delta H^o - T_o\,\Delta S^o$ and $\Delta H^o \gg T_o\,\Delta S^o$

$$\overline{w}_{opt,TMC} \approx -\Delta H^o = -(LHV).$$

Therefore, the ratio of fuel availability to LHV is roughly unity.

The chemical availability of a fuel $Avail_F$ is a measure of the maximum possible work. The actual work output W is generally lower than $Avail_F$, and the exergetic efficiency

$$\eta_{avail} = W/Avail_F.$$

E. SUMMARY

This chapter illustrates the irreversibility during adiabatic and isothermal reactions. For specified inlet and exit states in a combustor, the optimum work relations are derived and illustrated for a car burning gasoline and fuel cells. Finally, a methodology is provided for calculating the maximum possible work from a given fuel and a schematic is presented for achieving the maximum work.

14. PROBLEMS

A. CHAPTER 1 PROBLEMS

Problem A1

Must a mixture be necessarily homogeneous?

Problem A2

What is irreversible thermodynamics?

Problem A3

If a differential is exact, does this mean that it is related to a point function?

Problem A4

Can the electron mass be ignored when it crosses the system boundary, but still treat a system as closed during electric heating?

Problem A5

Is a composite system a homogeneous system?

Problem A6

Consider the sum δz or $dZ = 6xy^3\,dx + 9x^2y^2\,dy$. Is Z a point function of x and y (i.e., a property), or a path function (non–property)?

Problem A7

Compute the partial derivative $(\partial v/\partial T)_P$ for the relation $P(T, \bar{v}) = \bar{R}T/(\bar{v} - \bar{b}) - \bar{a}/(T^{1/2}\,\bar{v}(\bar{v} + \bar{b}))$ when T = 873 K and $\bar{v}$ = 0.245 m^3 $kmol^{-1}$, assuming that $\bar{a}$ = 142.64 bar m^6 $K^{1/2}$ $kmol^{-2}$ and $\bar{b}$ = 0.0305 m^3 $kmol^{-1}$.

Problem A8

Perform line integration of the following differentials, first along constant values of x and then along constant values of y from (2, 5) to (4, 7): $xy^3\,dx + 3\,x^2y^2dy$, and $2e^xydy + e^xy^2dx$ and determine which one is path independent. Verify your results employing the mathematical criteria for exact differentials. Obtain an expression for z(x, y) for any one of the above differentials that can be expressed as dz.

Problem A9

Perform cyclical integration along constant values of x, i.e., (2,5) to (2,7), and along constant values of y, i.e., (2,7) to (4,7), and along constant values of x, i.e., (4,7) to (4,5), and along constant values of y, i.e., (4,5) to (2,5) for the following expressions: $xy^3\,dx + 3\,x^2y^2dy$, and $2e^xydy + e^xy^2dx$ and show which one is path independent.

Problem A10

Using the LaGrange multiplier method solve the following problem. Find the maximum volume V of a tent for a fixed cloth surface area S. Assume that the tent is of triangular cross section of equal side x and y units long. Note floor is also laid with cloth. If S = 2 m^2 determine x in m, y in m and V in m^3. Show that at the optimum condition.

Problem A11

Using the LaGrange multiplier method solve the following problem. Find the maximum volume V of closed tent for a fixed cloth surface area of S m^2. Assume that (1) the tent is of triangular cross section of equal side x and y units long. Note that the floor is also to be laid with cloth, and (2) the tent is a rectangular parallelepiped of dimensions x, y and z. If S = 20 m^2, determine x, and y in m and V in m^3 at the optimum condition for both cases.

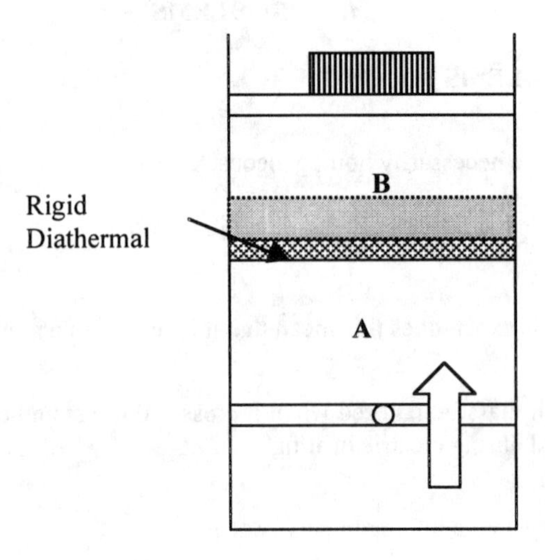

Figure Problem A.19

Problem A12

Consider the function $\phi = x^3 y/t + x^2y^2/t^3 + x y^3/t^7$. Is this a fully homogeneous function? Is this function partly homogeneous and, if so, with respect to what variables? Show that the Euler equation applies if this is a partly homogeneous function.

Problem A13

Using the LaGrange multiplier method find the maximum volume V of tent for a fixed cloth surface area of 2 m^2. Note that the floor is also laid with cloth. (Hint: $z \geq 0$, $x \geq 0$, $x^2 > 2(1 - xz)$).

Problem A14

Determine $(\partial u/\partial x)_y$ and $(\partial u/\partial y)_x$ for the following equations: $u - x^2y + y^3u + yu^2 + 8x + 3$, and $u^2xy + ux^2 + xy^2 + u^3 + uxyz - 0$.

Problem A15

Show whether the following equations are exact or inexact: (a) $du = 3\,x^2\,dy + 2\,y^2\,dx$. (b) $du = y\,dx + x\,dy$. (c). $du = 2xy\,dx + (x^2+1)\,dy$. (d) $du = (2x+y)\,dx + (x-2y)\,dy$. (e) $du = (xy \cos(xy) + \sin(xy))dx + (x^2 \cos(xy) + e^y)\,dy$.

Problem A16

Obtain the value of the line integrals using the equations in Problem 10a–c and the path described by moving clockwise along the sides of a square whose vertices are (1, 1), (1, –1), (–1, –1), (–1, 1).

Problem A17

Minimize the distance between the point (1,0) to a parabola (choose a parabolic equation) without using the LaGrange Multiplier method, and using the LaGrange multiplier method.

Problem A18

Consider the exact differential $dS = dU/T + (P/T)\,dV$. Since $S = S(U,V)$, let $M = 1/T = M(U,V)$ and $N = P/T = N(U,V)$. Write the criteria for the exact differential of dS. Similarly, write down the criteria for the following exact differentials: $dS = dH/T -$

(V/T) dP; dT = –dA/S –P/S dV; dT = –dG/S + V/S dP; dP = dG/V + S/V dT; dV = –S dT/P –dA/P.

Problem A19

1 kg of Ar is contained in Section A at P = 1 bar, T = 100ºC. This gas is in contact through a diathermal wall with another piston–cylinder section B) assembly containing 1 kg partly liquid water (quality x = 0.5) and vapor at 100ºC with a weight at top. As we compress the gas in Section A, the temperature tries to increase, but because of the contact with Section B, T remains at 100ºC. Answer the following True or False questions:

a) The composite system consists of a pure substance.
b) The composite system has two phases for H_2O and one single phase for Ar gas.
c) The composite system is homogeneous.
d) The total volume cannot be calculated for the composite system.
e) There is no heat transfer between Sections A and B.
f) There is no work transfer between Sections A and B.
g) There is work transfer from Section A to B.
h) The quality in Section B decreases.

Problem A20

If the number of molecules per unit volume (n') ≈ $1/l^3$ where l is the average distance (or mean free path between molecules), determine the value of l for the gases in your classroom at 25°C, 1 bar (assume the ideal gas law is applicable). Express your answer in μm and Angstrom (1 A = 10^{-10} m) units. Assume that your classroom is filled with pure oxygen.

Problem A21

In the context of the above problem, do you believe that the ideal gas law is applicable at this intermolecular distance (i.e., that the attractive force between adjacent molecules is negligible)? Assume that ℓ_0 corresponds to the liquid state of oxygen. The molal liquid volume of oxygen is given as 0.02804 m^3/kmol.

Problem A22

Natural gas has the following composition based on molal percentage: CH_4: 91.27, ethane: 3.78, N_2: 2.81, propane: 0.74, CO_2: 0.68, n–Butane: 0.15, i–Butane: 0.1, He: 0.08, i–pentane: 0.05, n–pentane: 0.04, H_2: 0.02, C–6 and heavier (assume the species molecular weight to be 72 kg $kmol^{-1}$): 0.26, Ar: 0.02. Determine the molecular weight, the methane composition based on weight percent, and the specific gravity of the gas at 25°C and 1 bar.

Problem A23

Prove Eq. (7a) beginning your analysis from Eq. (7).

Problem A24

A bottle of 54.06 kg of distilled water is purchased from a grocery store. a) How many kmols of water does it contain? b) If the bottle volume is 54.05 L, what is the specific volume of water in m^3/kg and m^3/kmol? c) How many molecules of H_2O are there in the bottle? d) What is the mass of each molecule? e) Determine the approximate distance and force between adjacent molecules.

B. CHAPTER 2 PROBLEMS

Problem B1

Is the relation h = u + Pv valid only for a constant pressure process?

Problem B2

Is the earth a closed or an open system?

Problem B3

If you type this entire text on a computer, will the mass of the computer increase?

Problem B4

Is $\int Pdv$ work boundary work or flow work?

Problem B5

What is physical interpretation of c_v and c_p?

Problem B6

What is the Poincare Scheme?

Problem B7

Is it true that in a closed or an open system, work and heat transfer can occur across the system?

Problem B8

Is there a difference between a quasiequilibrium and an internally reversible process?

Problem B9

An incompressible liquid (v = constant) undergoes adiabatic internally reversible compression in a open system. If you follow a unit mass, then is the change in internal energy a) zero, or b) non-zero?

Problem B10

An incompressible liquid (v = constant) undergoes internally reversible compression in a closed system. Then is the work input per unit mass zero?

Problem B11

What does the term quasiequilibrium mean?

Problem B12

When can a nonquasiequilibrium process not be represented on a P–v diagram or a T–s diagram?

Problem B13

Is it generally true that we can use the equality $\delta W = -PdV$ in the relation $\delta Q - \delta W = dU$?

Problem B14

When is the relation $\delta Q - \delta W = dU$ equivalent to the expression $\delta Q - \delta W = dH$?

Problem B15

What is the physical meaning of a characteristic time for a process involving the heating of a house?

Problem B16

For a steady state process involving an open system $dm_{cv}/dt = 0$, i.e., m_{cv} is constant. Is this always true for a closed system?

Problem B17

For a steady state open system $dU_{cv}/dt = 0$. Is it true that the reverse statement, $dU_{cv}/dt = 0$, implies that the open system must be steady. The latter statement is a) true b) false. (Hint: consider the heating of ideal gas in an oven where P is constant.)

Problem B18

Gas cylinders normally use pressure regulators to control the downstream pressure at P_R. At any set regulator pressure P_{Rrg}, the mass flow leaving the regulator is given by $\dot{m} = 4040\ A_{reg}\ P_{Reg}/T_{Reg}{}^{1/2}$, kg/s, P in bar, T_{reg} in K and A in m^2 (choked flow). Thus, the pressure downstream of the regulator is fixed at P_{reg} and the flow from the cylinder leaves at conditions P and T. As the flow leaves, the cylinder pressure (P) decreases with time (t). Assume $T_{reg} = T$, i.e., the cylinder temperature. Assume that the cylinder is adiabatic. If the initial cylinder pressure and temperature are 100 bar and 300 K, and regulator pressure is 3 bar, determine the time for pressure in the cylinder to decrease to 50 bar.

Problem B19

The space shuttle is powered by booster rockets and the main shuttle engines are fired with H_2 and O_2. The empty weight of the system is 76000 lb and after charging the booster and shuttle tank, the weight increases to 1500,000 lbs. Given the following conditions, calculate the altitude reached by the shuttle 2 minutes after firing: Thrust (F) 3000,000 lb_f, where $F = mv_{gas}/g_c$, $v_{gas} = 0.4\ (kg_c\ RT)^{1/2}$, T = 6000 R, Firing Rate = constant. Note that the mass of the system varies during lift off due to discharge of propellants.

Problem B20

A closed adiabatic system with a weightless piston at the top contains air at 100 kPa, 227°C. The ambient pressure is also 100 kPa (Area of piston: 100 cm^2; volume at state 1: 0.1 m^3). Suddenly a weight of 1 kN is placed and system reaches state 2. a) Sketch the process on a PV diagram, and b) determine P_2, T_2 and V_2.

Problem B21

Air at 100 psia, and 40 F, is held in a tank of 20 ft^3 volume. Heat is added until the remaining air in the tank is at 240°F, while some air is bled from the tank to hold the pressure constant at 100 psia. Determine the heat transfer, assuming the air to be an ideal gas with constant specific heats.

Problem B22

A cooker "A" of 30 cm diameter and volume 30 L is initially filled with liquid water of 4 kg. It is then heated until the pressure in the cooker rises to 5 bar at which pressure it contains a mixture of pure water vapor and liquid water. Then, assume that we insulate the cooker and attach a metal tube of cross sectional area A to it that is placed slightly away from the bottom surface of the cooker. Assume also that we provide a valve at the top of this metal tube. When the valve is opened, the water left in the cooker can be injected into another open adiabatic cooker B of equal dimension in order to conserve energy. We will neglect evaporation from cooker B. However, we would like to monitor the pressure in the first cooker. As an expert in thermodynamics you are asked to predict the pressure *vs.* time until no liquid water remains. Assume that vapor behaves like an ideal gas with a specific heat of 1.59 kJ kg^{-1} K^{-1}. Water is incompressible with a specific heat of 4.184 kJ kg^{-1} K^{-1} and v = 0.001 m^3/kg. The cross sectional area of the metal tube is 10 mm^2. Assume the power is off when we open the valve and the cooker free space is occupied with vapor only. Neglect the potential energy change.

a) What is the quality when the valve in the metal tube is opened?

b) Write down the mass and energy conservation equations for the vapor phase in the system (assume no condensation of vapor or vaporization of water) and obtain a relation for P *vs.* v for the vapor phase.

c) What is the quality when all of the water has been expelled from cooker A?

d) Sketch the process for the cooker A on a P–v diagram.

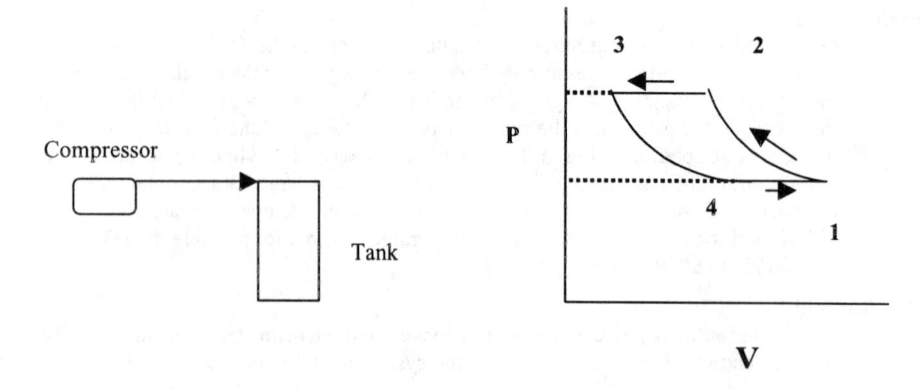

P-v diagram of a compressor in context of Problem B.23.

e) Write down the energy balance equation for the metal tube and obtain an expression for velocity through the metal tube assuming that steady state exists for the c.v. (metal tube)
f) Obtain an expression for mass flow through the tube.
g) Derive the expression for P (t) in terms of vapor volume in the cooker?
h) What is the pressure in the cooker when all water is gone?
i) If $\ln P^{sat}$ (bar) = 13.09 –4879/T, plot T^{sat} *vs.* t and compare with T *vs.* t. Check your assumption in (b)

Problem B23

A tank must be charged with gases from a reciprocating compressor that is running at a speed of N. The displacement volume is V_{disp} and the compresion ratio is r. The outlet valve for the compressor opens only when the pressure in the cylinder exceeds tank pressure. The compression process in the piston–cylinder follows the relation Pv^n = Constant. Determine the P(t) *vs.* time if the tank is assumed to be adiabatic.

Problem B24

When propane is burned in a gas turbine, one produces a gaseous mixture of 6.48% CO_2, 10.12% O_2, 7.28% H2O and 76.12%% N_2. Ten kmol of mixture /min. enter an adiabatic gas turbine at 1200 K, and 10 bar and leave at 700 K, 1 bar . There are no chemical reactions

a) Write down the species balance in terms of k mol of flow in and out,
b) Write down the energy balance equation, and
c) Simplify the equations in parts (a) and (b) for a steady state and steady flow process.

Calculate the work produced by the turbine in kW

Problem B25

Consider an insulated rigid tank containing 2 m^3 of air. The tank is divided into two parts A and B by a partition mounted with a thick insulation (a constraint which prevents heat transfer) and locked in place by a pin (another constraint which prevents work transfer). The portion A consists of 1 kg of hot air at 320 K and 1.48 bar while portion B consists of cold air at 290 K and 0.42 bar. Determine the final conditions in both chambers for the following cases.

a) If the pin is removed but insulation stays with the partition
   i) If QE expansion occurs for chamber A but the process in B may not be QE,

ii) If QE compression occurs for chamber B but the process in A may not be QE.

b) If the pin is removed and the insulation is also removed from the partition

Problem B26

You have been hired by PRESSCOOKER, Inc. to analyze the thermodynamic processes in cookers. A pressure cooker of mass m_C contains water (mass m_w, volume V_w). There is some air m_a left in the cooker. The cooker is covered and tightly sealed. A small weight is placed at its top and the cooker is heated. You must analyze the problem of water heat up on an electric range that supplies an electrical power W_{elec}. Water starts evaporating and releases vapor (mass = m_v) to the air space above the water. After time t_L, the pressure in the cooker reaches P_L, the weight is lifted and steam is released. We can assume that the cooker is adiabatic. Neglect the kinetic and potential energy contributions. Assume uniform temperature and pressure inside the cooker and the temperatures of the contents are the same as that of the steel. The specific heats of air($c_{P0,a}$), vapor($c_{p0,v}$), water(c_w) and steel cooker (c_S) are assumed to be constant.

a) Write down the mass and energy conservation equations for $t < t_L$. Simplify. Explain the various terms (not exceeding two lines for each term).

b) Write down the mass and energy conservation equations for $t > t_L$. Simplify. Explain the various terms (not exceeding two lines for each t the entropy generation term).

c) Mention how will you solve the problem for T *vs.* t, m *vs.* t for $t < t_L$ and $t > t_L$

Note: m denotes the total mass of cooker including the contents.

Problem B27

Assume that Mt. St. Helen erupts again and releases a rock of 2000 kg mass. The surface of the rock is at T = 2000 K while the interior is a liquid at 2000 K. The entire rock moves at a speed of 300 m/s and rises in altitude. As the rock moves in ambient air, there is also heat loss. Analyze the problem by applying the First Law.

Problem B28

Assume that scientists have measured the following with "inert" (nonfusion–causing) electrodes in heavy water in a nuclear reactor: Work input through a stirrer: 100 mW; Electric input through a heater: 200 mW (controllable); Power input through electrodes: 350 mW. What should be the rate of heat loss if the heavy water is maintained at 35°C? Assume that scientists change the electrodes that may cause a fusion reaction to occur. Under identical conditions as before, they have found that they will have to reduce the heater input to 150 mW. The heavy water is still at 35°C. What is the rate of "unaccountable energy" or "excess energy" or the so called "fusion energy" in mW?

Problem B29

Using Eqs (J) and (K) and ideal gas law prove Eq. (L) in Example (17).

C. CHAPTER 3 PROBLEMS

Problem C1

An Otto cycle involves reversible processes. Otto cycle A uses Ar while Otto cycle B involves N_2. Then using Clausius theorem and T-s diagram for Otto cycle one of the following must be true for the same initial (T_1) and peak temperatures (T_3) a) $\eta_A = \eta_B$, b) $\eta_A > \eta_B$, c) $\eta_A < \eta_B$. Which one is it?

Problem C2

Innovations, Inc. claims that they have developed a heart pump driven by a heat engine which uses the warm reservoir of a human body 98.6°F and the cold reservoir of ambience 60°F. It is claimed that for every 100 BTU of heat absorbed from the body, the engine could deliver work of 10.5 BTU. Can you verify the claim?

Problem C3

The S *vs.* T_a during the cooling of a coffee cup in room air (T_a) shows a maximum at equilibrium. Can the same curve yield the time scale required for equilibrium? What is a spontaneous process?

Problem C4

How can changing heat transfer into work reduce entropy production?

Problem C5

Can we define equilibrium in terms of E, V, m?

Problem C6

Will the equilibrium condition within a tank containing N_2 change if the tank walls are porous?

Problem C7

Will the work added to a system always change entropy? Specifically, consider QE adiabatic work and NQE adiabatic work.

Problem C8

If the concept related to σ is important, why do not engineers use it in common practice?

Problem C9

Consider a cup full of coffee placed in room air. If the pressure and entropy are maintained constant within the rigid room, in practice how can there be a heat loss?

Problem C10

Why is there a negative sign associated with the equality $\partial S/\partial N_1 = - \mu_1/T$?

Problem C11

Consider problem B.10. The entropy change after compression process is a) zero, or b) positive?

Problem C12

Explain entropy using physical principles. What is an "endoreversible" engine?

Problem C13

If the entropy increases, do you believe S=0 at the beginning of the universe?

Problem C14

What prevents O_2 being on one side within a room?

Problem C15

Is entropy generated as a result of electrical and gravitational fields?

Problem C16

Can you predict time to reach equilibrium be using thermodynamics?

Problem C17

The temperature of air in a rigid container initially at 300K (T1) and 100Kpa (P1) has to be increased to a final temperature of 600K. The student found that the gas temperature was not uniform inside the tank during the heating process with a gas burner even though the temperatures were uniform at 300K (State 1) and 600K (State 2). Assume constant specific heats and c_{vo}=0.7 kJ/kg-K. Then which of the following statement is true?

a) The specific entropy change (s_2-s_1) in kJ/kg-K is given as 0.49 kJ/kg-K.
b) The specific entropy change cannot be determined since the process is internally irreversible.
c) The entropy change is zero.

Problem C18

What is the physical significance of A and G reaching a minimum value?

Problem C19

How does the entropy change if a partition between two fluids is porous?

Problem C20

What is the physical meaning of the entropy flux? (Hint: heat transfer through fins.)

Problem C21

Consider the cooling of coffee and air for which $S_1 = S_{coffee + air}$ = 2 kJ/K, $S_2 = S_{max} = S_M$ = 5 kJ/K after reaching a maximum (at state M). Can this entropy be lowered to $S_3 < S_2$ through an adiabatic process?

Problem C22

Why should σ be different if energy is supplied via heat or via electrical work crossing a system boundary?

Problem C23

If microscopic fluctuations occur at equilibrium, why does it not violate the Second Law?

Problem C24

What are the ways of increasing the temperature of a closed system?

Problem C25

Are there ways of determining irreversibility without evaluating σ?

Problem C26

If the entropy never decreases we obtain greater disorder. If a room is messy we can clean it up . Why can't we do the same for a thermodynamic system?

Problem C27

What is the difference between compressible and incompressible materials?

Problem C28

When can you assume constant specific heats and when can you assume variable specific heats?

Problem C29

What is the difference between s_2-s_1 and σ_{12}?

Problem C30

Physically, what is the difference between chemical and deformation work ?

Problem C31

Recall that we can use the expression $dS = (\delta Q/T)_{rev} = dU/T + P\ dV/T$ for a fixed mass system to develop tables for s = s(T,P). Can we use this expression to determine the entropy of a fluid which enters a turbine or for any other open system?

Problem C32

If the human body is assumed to have the same properties as liquid water, then during fever, the entropy of the human body should be higher, since a) the volume of the human body slightly decreases, and b) its energy increases during fever. State which of the two is the correct answer.

Problem C33

When a violinist performs, only 2% of the work done is used to produce musical sound, and the remainder is dissipated in the form of heat. Assume that violin strings are made of steel. As a string is heated, the damping coefficient changes, which alters the sound. How can you employ the Second Law to relate the damping and heating processes in order to produce a better violin design?

Problem C34

Consider an insulated piston cylinder assembly. The piston mass is negligible and is fabricated of good insulation material. A ambient pressure P_0 = 100 kPa acts on the piston. The temperature of the gas (which is air) is 300K (T_a). A mass of 100 kg (m_E) is suddenly placed on the piston and the surface area of the piston is 1 cm^2 (area A). The initial volume is 0.01m^3 (V_1). The local gravitational acceleration is 10m/s^2 (g). Make any reasonable approximation and indicate the formulations required to obtain answers to the following questions.

a) The final pressure (P_2),
b) The final temperature (T_2) and volume (V_2), and
c) Entropy change between states 1 and 2.

Problem C35

Consider the reversible polytropic process pv^n = constant between the states i (initial) and f (final). We must devise an equivalent process that consists of an adiabatic process ig, an isothermal process gh, and an adiabatic process hf, such that work performed by the equivalent processes is the same as the work done by the reversible polytropic process.

a) Show that for the isothermal process,
$T_g = (T_i - T_f)\,(((k-1)/(n-1)) - 1)/(\ln((P_f v_f^{\,k})\,(P_i v_i^{\,k})))$.
b) What is the value of n for constant P process? Using the answer to part (a) obtain an expression for a constant pressure process.
c) What is the value of n for a constant volume process? Using the answer to part (a) obtain an expression for a constant volume process.
d) Plot the processes within the range P_i = 60 bar to P_f = 48.2 bar, with v_i = 0.1m^3/kg and v_f = 0.12 $m^3\,kg^{-1}$. Assume the medium to be air and treat it as an ideal gas.

Problem C36

Derive an expression for the efficiency of the reversible cycles illustrated in the figure in terms of T_L and T_H, and for a Carnot cycle operating between the same temperatures.

Problem C37

Consider an adiabatic reversible process for an incompressible liquid flowing through an expanding duct.

a) What is the entropy change?
b) Can the process be considered isothermal?

Problem C38

Consider an adiabatic reversible process for an incompressible liquid flowing through an expanding duct. Using the basic energy and mass conservation relations for an

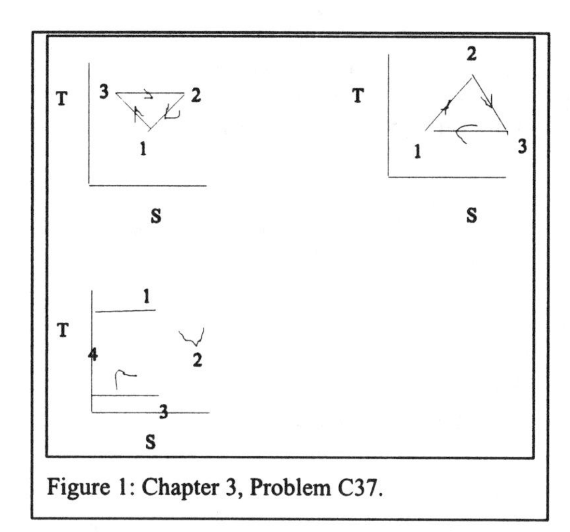

Figure 1: Chapter 3, Problem C37.

open system obtain a relation between the pressure and the area of the duct. Show that this reduces to Bernoulli energy equation. (Recall that for liquids du = cdT.)

Problem C39

Heat (Q_H) is rejected from the condenser to the ambient in a refrigeration cycle for which the temperature T_H (ambience) is 10°C below the condenser temperature. Similarly, heat is added to the evaporator from a cold space at a temperature T_L. The evaporator coil is at a temperature that is 10°C below T_L. Is it possible to use the heat transfer Q_H to reduce the work input to the compressor?

Problem C40

a) Determine the entropy generation rate in a bar of cross sectional area A which is maintained at T_o (x = 0) and at T_L (at x = L). The peripheral area is insulated. Derive expressions for the: i) entropy generation per unit volume in terms of local temperature, ii) total entropy generation rate per unit cross sectional area in the bar, and iii) entropy generation rate per unit heat loss rate.

b) What is the: i) entropy generation per unit volume at x = 0.1 m for aluminum if T_o = 500 K, T_L = 300K, L = 0.3m, ii) total entropy generated within the whole bar, iii) entropy generated per unit heat loss, iv) lost work rate per unit mass loss rate, and v) lost work rate per unit mass loss rate?

c) Determine the difference between the specific entropies at x = L and x = 0.

d) The heat flux Q″ enters at x= 0 and leaves at x=L. Calculate the entropy flux in and out? Why is the exit entropy flux different?

Problem C41

As a patent officer you receive a patent application for a cyclic device which consists of irreversible adiabatic compression from (P_1,V_1) to (P_2,V_2), quasi-static isothermal heat addition from (P_2,V_2) to (P_3,V_3) (V_2<V_3<V_1) and then finally adiabatic and quasi-static expansion from (P_3,V_3) to (P_1,V_1). Will you issue the patent? Justify.

Problem C42

One kmol of CO is contained in a piston–cylinder assembly at 3000 K and 1 bar. What is the system entropy? Now, one kmol of O_2 at 1 bar and 3000 K is introduced, but constant pressure is maintained in the system. What is the entropy change for the inert mixture?

Problem C43

An ideal gas can be heated in a closed system using a) an isobaric process or b) an isometric process through a similar temperature rise. What is the ratio of isobaric entropy change to isometric entropy change? Discuss the results briefly.

Problem C44

An ideal Otto cycle which uses the same gas for all four processes (adiabatic compression, isometric heat addition, adiabatic expansion, isometric heat rejection) has the following expression for efficiency

$$\eta = 1 - 1/(r_v{}^{(k-1)})$$

where r_v denotes the compression ratio v_1/v_2, v_1 the initial volume, and v_2 the volume after compression.

a) Determine η for with $k = 1.4$, $r_v = 8$.

b) Determine η for a monatomic gas for which $k = 1.6667$, and $r_v = 8$.

c) Discuss the results for a) and b).

Problem C45

Consider the transient three–dimensional heat conduction equation and the entropy balance equations for solids. Show that the entropy generation rate per unit volume is given by the expression $\dot{\sigma} = -(q''/T^2)\nabla T$.

a) What is the result for $\dot{\sigma}$ if $q'' = -\lambda\,\nabla T$,

b) What is the result for $\dot{\sigma}$ if $q'' = \lambda\,\nabla T$,

c) What is your conclusion?

Problem C46

Ten kg of hot air is stored at 1000K and 1 bar in compartment A of a rigid insulated tank, and 5 kg of cold air at 500 K is stored in compartment B of the same rigid insulated tank at 1 bar.

a) If you open the partition between compartments A and B
   i) What is the final temperature T_2?
   ii) What is the final pressure P_2?

b) What is the entropy generated?

c) If we connect a Carnot engine between reservoirs A and B (i.e., by extracting heat from compartment A and rejecting it to compartment B to produce work),
   i) What is the final temperature in A and B?
   ii) What is the maximum possible work?
   iii) What is the entropy generated?

Problem C47

Assume that you have a "finned head" on your body. "Finned heads" argue that they are the "coolest" people on the earth, since each strand of hair behaves as a fin to keep their heads cool. However "bald heads" argue that the radiation heat loss from their heads is a more important form of cooling. Looking at the problem from a thermodynamic point of view, which "head" generates more entropy for the same ambient temperature?. The temperature of a "head" can be assumed to be 98.6°F for both groups. State your assumptions.

Problem C48

Consider the piston–cylinder assemblies A and B, where the head of cylinder A is in contact with head of cylinder B. Equal masses of ideal gases exist initially at P =10 bar and T = 500 K in both cylinders. Cylinder A is quasi-statically compressed to 15 bar while the pressure in B is reduced to 5 bar. Each cylinder is insulated except at the head. Combining systems A and B, answer the following questions: Is the process isentropic? What is the final temperature? If we expand the gas in cylinder A back to 10 bar and compress the gas in B back to 10 bar, what are the final temperatures?

Problem C49

The following problem will illustrate the Clausius inequality. A professor asked his research assistant (student A) to compress 0.002 kg of air in an insulated cylinder which was initially at 100 kPa and 300 K (state 1). He wanted the student to do this as slowly as possible so that when the volume was reduced to $1/10^{th}$ of the original volume, he obtained a pressure of 25120 kPa and a temperature of 753 K. He advised the student that it might take many hours to do the job. As soon as the professor left the room, the student got anxious and in a few milliseconds he/she moved the piston to reach $1/10^{th}$ of the original volume and left. The professor returned after a few hours. He found to his dismay that the temperature reading was 900 K and correspondingly the pressure was also high (state 2). He figured out what had happened and he immediately fired the student. He hired a new research assistant (student B) who he thought would act more responsibly. He told the research assistant to move the piston back as slowly as possible to the original volume to the initial state. The cylinder was still insulated. After a few hours the student moved the piston back to the original position. But the temperature and pressure after expansion (state 3) were not the same as the values at state 1. He reported the results to the professor. The professor told the student to remove the insulation and to isometrically cool the cylinder to a temperature of 300 K. The student did that and found the pressure to be almost the same as the pressure at state 1.

a) If the compression process was adiabatic and reversible what would the temperature (T_{2s}) and pressure (P_{2s}) have been after compression? Assume constant specific heats (evaluated at 300 K). Why are these values different from the values measured by student A? What would be a best determine for the actual pressure at state 2 for the measured temperature?
b) Why is state 3 different from state 1?
c) How much heat must be removed between states 3 and 1?
d) Do you believe the pressure measurement after the cooling process?
e) Determine the cyclic integral of $\delta Q/T$. What is the sign of this quantity? Explain the significance of the sign.
f) Assuming that the entropy at 300 K, 100 kPa is 2.515 kJ kg^{-1} K^{-1}, evaluate the entropies at states (2) and (3).
g) What is the cyclic integral of dS? Why is this integral different from the answer to part (e)?
h) If the atmospheric temperature is 300 K, what is the entropy change of an isolated system during a single cyclical process?

Problem C50

Consider a system A of mass 2 kg at 8 bar at 500 K, a system B of mass 1 kg at 10 bar at 300 K.

a) The two systems are adiabatic and divided by an insulated partition and by a pin. The pin is released. What is the final pressure? What are T_A and T_B? Assume that the process is quasi-steady. Assume an adiabatic expansion process for one cylinder.
b) The two systems are divided by a diathermal wall. Except at the partition, there is no heat transfer. What is the final pressure? What are the final temperatures?

Problem C51

Obtain an expression for the entropy generated over a time period t when a pressurized gas at the state (T_i, P_i) enters an adiabatic piston–cylinder–weight assembly of cross sectional area A and the weight W is just lifted. The ambient temperature is T_o.

Problem C52

The generalized entropy relation for any simple compressible substance following the state equation $v = v(T,P)$ is $ds = c_v\, dT/T + (\partial P/\partial T)_v\, dv$. Assume that a solid substance undergoes adiabatic reversible compression or expansion.

a) Obtain an expression for $(\partial T/\partial v)_s$ in terms of c_v, β_P, β_T and T where $\beta_P = (1/v)\,(\partial v/\partial T)_P$ and $\beta_T = -(1/v)\,(\partial v/\partial P)_T$.

b) Discuss the results qualitatively for a substance that expands or contracts upon heating.

c) Determine $(\partial T/\partial v)_s$ for copper assuming its properties at 25°C.

d) Irreversibilities exist in systems (e.g., temperature gradients involved in the bending of the copper beam) and as a result entropy $\delta\sigma$ is generated. Since $ds - \delta q/T = \delta\sigma$., qualitatively compare $(\partial T/\partial v)$ obtained under adiabatic irreversible conditions with those obtained for part (a).

Problem C53

A family returns from vacation to find their house at a temperature of 15°C, while the outside temperature is 5°C. The house has a volume of 2000 cubic meters and the effective heat capacity of the house, furniture, and fixtures (exclusive of the air) is 3×10^5 kcal K^{-1}, while the heat capacity of air (c_p) may be assumed to be constant and its value can be fixed with respect to some average temperature. As the house is heated, air is expelled (through the sides of windows, the chimney, etc.) to maintain the pressure at one atmosphere.

a) How much heat is required to raise the temperature of the house and its contents to 25°C (assuming negligible heat losses, except through the expelled air)?

b) How much electrical energy would it take to run a heat pump to achieve the same objective? Assume that the heat pump and motor combined run at 35% of the theoretical efficiency (independent of temperature) and that the expansion coils are outside the house and are maintained at 0°C, while the compressor and condenser and motor are inside the house at a temperature equal to that of the house.

Problem C54

Suppose that saturated liquid data for h, v and s are available for water from 10 to 180°C. Produce a compressed liquid table for h, u (= h–Pv) and s at P = 10 bar for temperatures from 10°C to 180°C using the saturated liquid data. Assume that the liquid specific volume does not change with pressure at a given temperature.

Problem C55

Consider a high intensity discharge lamp in which the electric discharge occurs with an energy U in a narrow ultra violet wavelength range band $\Delta\lambda$ around λ. The energy intensity of the photons is expressed by Planck's law. The entropy of the photons S = 4(U/3)T. Mercury vapor under high pressure absorbs this energy but emits in the visible range of wavelength. assuming that the vapor absorbs all of the energy in the visible wave length range.

a) determine the temperature of vapor (T_v) assuming it to be a black body ($U = 4\sigma T_v^4$), and

b) determine the entropy generated.

Problem C56

Consider a fin of arbitrary cross sectional area with its base maintained at the temperature T_w (x = 0) that loses heat to its ambient at the temperature T_∞. Show that the expression for the work loss rate per unit heat loss rate is the same as the Carnot efficiency. (Hint: use a control volume which includes the base area with other boundaries extending far away from the fin.)

Problem C57

Hot water (W) at the temperature $T_{W,0}$ is kept in an adiabatic classroom which contains air (A) initially at a temperature $T_{A,0}$. Initially, there is a constraint on the cup in the form of an insulation around the cup. Once we remove the insulation, there is irreversible heat transfer that leads to an increase in the entropy of the combined system. Concerned with this, a graduate student connects a Carnot engine between the water and room air (A) and delivers work to the outside of the classroom to run an elevator.

a) What is the energy of the combined system (A+W)?

b) What are the energy of the subsystems, A and W?

c) What is the entropy of the combined system?

d) Compare the final temperatures of the subsystems, A and W.

e) Determine the work done.

f) Discuss the internal energy variation of A+W with time as work is delivered.

Problem C58

Consider an Otto cycle (which is also a reversible cycle) operating with compression ratio of 5 ($=V_1$-V_2). It can be shown that the temperature ratio $(T_4/T_1) = (T_3/T_2)$. If T_1=300K, what is T_2? Assume T_3 = 800K. What is the Carnot efficiency for the same maximum and minimum temperatures? Do you believe that the Otto efficiency is greater, smaller or equal to the Carnot efficiency? Provide reasons for your answer in five or six lines.

Problem C59

Consider a reversible Otto cycle operating with a compression ratio ($=v_1/v_2$) of 5. The temperature ratio $(T_4/T_1) = (T_3/T_2)$. If T_1 = 300 K and T_3 = 800 K, determine T_2? What is the Carnot efficiency for the same maximum and minimum temperatures? What is the Otto efficiency and the corresponding Carnot efficiencies?

Problem C60

A cooker A of 30 cm diameter and a 30 L volume is filled with 4 kg of water. The cooker operates at a pressure of 5 bar. A metal tube of 10 mm^2 cross sectional area is contained inside the cooker from a position slightly removed from its bottom surface and attached to a valve at the top of the cooker. When the valve is opened, the remaining water in the cooker is injected into another open adiabatic cooker B of equal dimension in order to conserve energy. However we would like to monitor the pressure in cooker A with respect to time until there is no liquid left in it. Assume that water vapor behaves as an ideal gas with a specific heat of 1.65 kJ kg^{-1} K^{-1}. Liquid water is incompressible with a specific heat of 4.184 kJ kg^{-1} K^{-1} and a specific volume v = 0.001 m^3 kg^{-1}. The area of the metal tube is 10 mm^2. Assume that the vessels are insulated, and that there is no heat transfer when the valve is opened and that the cooker free space is occupied by vapor alone.

a) What is the water quality when the valve in the metal tube is opened?

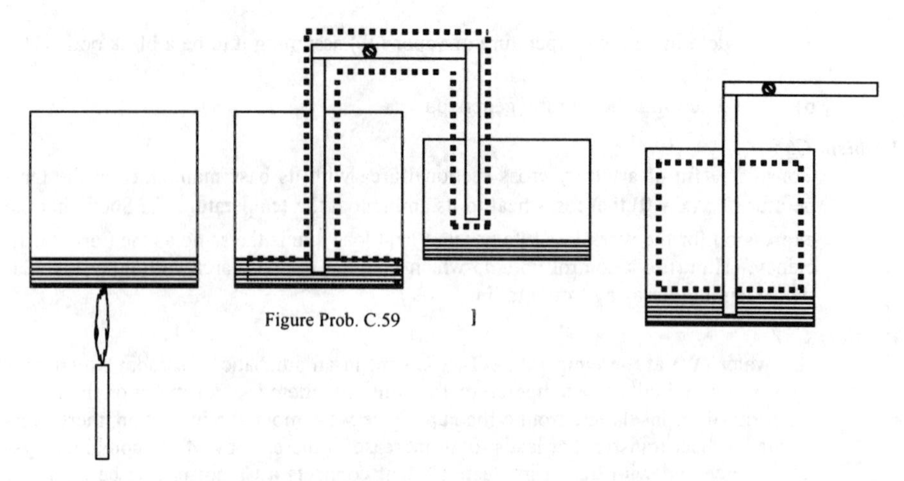

Figure Prob. C.59

b) What is the water quality when all water has exited from cooker A?
c) Illustrate the process for cooker A on a P–v diagram.
d) Write the energy balance equation for the process occurring in the metal tube and obtain an expression for the gas velocity through the tube for a steady state process.
e) Obtain an expression for mass flow through the tube.
f) Write the mass conservation equation for liquid phase (assume phase equilibrium between vapor and liquid).
g) Write the entropy balance equation for the combined system.
h) Derive the expression for P(t) in terms of the vapor volume in the cooker?
i) What is the pressure in cooker A when all of the water has evaporated?
h) If P^{sat} (bar) = 1.8^6 exp ($5199/T^{sat}$ (^{0}K)) compare the $T^{sat}(t)$ behavior with T(t). Check the assumption.

Problem C61

Consider a Carnot cycle in which the air is adiabatically and reversibly compressed say from $V_1 = 0.1$ m^3, $P_1 = 100$ kPa, $T_1 = 300$ K to $V_2 = 0.06$ m$_3$, and $P_2 = 205$ kPa. Heat is then isothermally added (i.e., T_H during heat addition) where $Q_{in} = Q_H = 14.75$ kJ, and the air expanded to state 3. The gases are adiabatically and quasistatically expanded to a temperature $T_4 = 300$ K. Finally heat is isothermally rejected so that the volume returns to its original value. Assume ideal gas behavior, $c_{vo} = 0.714$ kJ kg^{-1} K^{-1}, and constant specific heats.

a) Determine Q_{out} (= Q_L).
b) Determine Q_L/Q_H.
c) What is W_{cycle}?
d) Is $Q_L/Q_H = T_H/T_L$?

Problem C62

Determine the entropy of N_2(g) at 373 K and 1 bar. If N_2 is a solid at 0 K, and $h_{sf} = h_{fg} = 0$, what is the entropy s of N_2(g) at 0 K? Assume ideal gas behavior between 0 and 373 K (undoubtedly, a drastic assumption).

Problem C63

Consider a gas turbine that is 2 m long, with a net power output of 100 kW, and operating with a monatomic gas for which $c_{po} = 20.79$ kJ kmol^{-1} K^{-1}. Let $P_i = 10$ bar and $P_e = 1$ bar. Since the gas is monatomic, the specific heat is not a function of T. Assume steady state steady flow and that the turbine walls are well insulated. The inlet and exit velocities are very low. In order to obtain additional work from the turbine it is necessary to raise the inlet temperature to $T_i = 1500$ K. However, the turbine blades cannot withstand a temperature greater than 900 K (T_{bl}), and blade cooling is adopted. The temperature of the gas within the turbine varies according to the relation

$$T = T_i - (T_i - T_e)\, x/L,$$

where L denotes the length of turbine, and x the distance from inlet. The turbine walls are insulated, and the blades are cooled as long as the gas temperature exceeds the turbine blade temperature. We also assume that when the gas temperature falls below 1100 K, there is no appreciable heat loss to the blades. The heat loss rate per unit length and per unit mass flow is given by the relation $h'(T - T_{bl})$ where h' denotes the heat loss per unit length of turbine and is 0.5 kW/m K. The exit temperature is 600 K.

a) Start from generalized mass and energy conservation and entropy balance equations
$dN_{cv}/dt = N_i - N_e$,
$dE_{cv}/dt = Q_{cv} - W_{cv} + N_i\,(h + ke + pe)_i - N_e(h + ke + pe)_e$, and
$dS_{cv}/dt = \int \delta Q/T_b + N_i\, s_i - N_e\, s_e + \sigma$.
Using assumptions stated in the problem and any additional assumption, present the mass and energy conservation and entropy balance equations in a simplified form.

b) What is the work under steady state operation?

c) Determine the entropy generated per unit mass flow in the turbine.

d) If the turbine runs at an inlet temperature of 1100 K, but with no cooling and no heat loss, what is the work done for the same exit conditions?

Problem C64

Assume that there is a secondary system, which is a reservoir at a fixed pressure, inside a spacecraft at the state $T_{s,o}$ and $P_{s,o}$,. Determine the optimum work done if an ideal gas initially at $T_{p,o}$, $P_{p,o}$ in a primary system that undergoes change of state to T_{pf}, P_{pf} due to interaction with the secondary system. Assume that $T_{po} = 1000$ K, $P_{po} =$ 20 bar, $m_p = 4$ kg, the gas in primary system is Ar, $P_{pf} = 10$ bar, $T_{pf} = 600$ K, $T_{so} =$ 300 K, $P_{so} = 1$ bar, and $m_s = 8$ kg. Assume that the gas in the secondary system is He.

Problem C65

In a conventional Carnot cycle $T_L = T_4 = T_o$. In an unconventional cycle an ideal gas is expanded to the temperature T_4' ($< T_o$) and volume v_4'. Show that the additional work during the expansion process $w_{add} < P_o(v_4' - v_4)$. Assume constant specific heats.

Problem C66

Consider a cycle consisting of reversible adiabatic compression from state 1 to 2, isothermal heat addition at T_H, and reversible adiabatic expansion to state 1, which is at 0 K. Draw a T–P diagram for the cycle. Is this cycle possible?

Problem C67

Is it possible to obtain an efficiency $\eta = 1$ if compressed gas at room temperature is available at the state (P_1, T_o), then adiabatically and reversibly compressed to state 2 (P_2, T_2) heat is added at a constant high temperature T_H to state 3, and the gas is finally expanded to state 4 (P_0, T_o) and then discharged to the atmosphere?

Problem C68

Consider saturated water in an insulated blender at 100° C, P =–101kPa (state 1). A weightless piston is kept above the water. The ambient pressure is 101kPa. As the motor is turned on, the water just starts evaporating and reaches saturated vapor state (state 2). Sketch the process on P-v and T-s diagrams.

a) What is the boundary work?

b) What is the work input through the blender shaft?

c) Is there any entropy generated during the process? If so, how much for unit mass?
d) Comment on the areas under process 1-2 in the P-v and T-s diagrams.

Problem C69

10 kg of Ar is contained in the piston–cylinder section A of a system at the state (1.0135 bar, 100ºC). The gas is in contact through a rigidly fixed diathermal wall with a piston–cylinder section B of the system that contains a wet mixture of water with a quality x = 0.5 that is constrained by a weight. As the gas in section A is compressed the temperature in A remains at 100ºC using QE process due to contact with section B. Assume that the quality in section B increases to 90%. Both systems are well insulated except at the diathermal wall. Determine:

a) the initial pressure in Chamber B,
b) the heat transfer $Q_{12,B}$ in kJ to Chamber B during compression of Ar in Chamber A,
c) the work for sections A and B in kJ,
d) the change in the entropies of Ar and H_2O (both liquid and vapor), and
e) the volume V_2 in Chamber A
f) Is the process for the composite system (A+B combined together) isothermal and isentropic?

Problem C70

A piston–cylinder assembly contains Ar(g) at 60 bar and 1543 K (state 1).

a) Determine the work done if the gas undergoes isothermal expansion to 1 bar (state 2). What is the heat transfer? Does this work process violate the Second Law?
b) Determine the work done if the gas undergoes quasistatic adiabatic expansion to 1 bar (state 3). Can we continue the expansion to $v_3 \rightarrow \infty$ by removing the insulation and adding heat?

Problem C71

A rigid container of volume V is divided into two rigid subsystems A and B by a rigid partition covered with insulation. Both subsystems are at the same initial pressure P_o. Subsystem B contains 4 kg of air at 350 K, while subsystem A contains 0.4 kg of air at 290 K. The insulation is suddenly removed and A and B are allowed to reach thermal equilibrium.

a) What is the behavior of the overall entropy with respect to the temperature in subsystem A. What is the equilibrium temperature?
b) As heat is transferred, the entropy of subsystem A increases while that of subsytem B decreases. The entropy in the combined system A and B is held constant by removing heat from subsystem A. Plot the behavior of the overall internal energy with respect to the temperature in subsystem A. What is the equilibrium temperature?
c) Both subsystems are allowed to move mechanically in order to maintain the same pressure as the initial pressure P_o. The entropy is held constant by allowing for heat transfer. Plot the behavior of the overall enthalpy with respect to the temperature in subsystem A. What is the equilibrium temperature?

Problem C72

A piston–cylinder–weight assembly is divided into two insulated subsystems A and B separated by a copper plate. The plate is initially locked and covered with insulation. The subsystem A contains 0.4 kg of N_2 while subsystem B contains 0.2 kg of N_2.

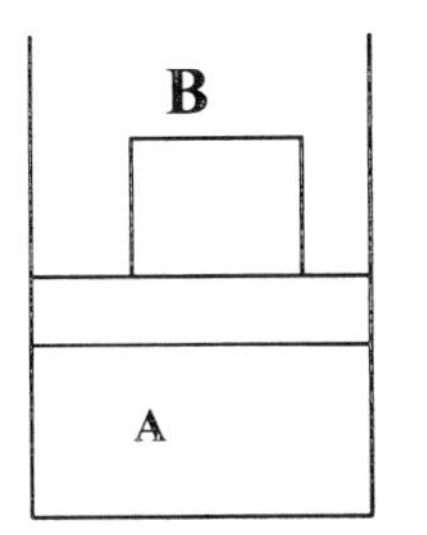

Figure Problem C.72

a) The insulation is removed, but the plate is kept locked in locked positions. Both subsystems are at the same initial pressure $P_{1A} = P_{1B}$ = 1.5 bar with temperatures T_{1A} = 350 K, and T_{1B} = 290 K. Both A and B reach thermal equilibrium slowly. Assuming that internal equilibrium exists within each subsystem, plot ($S = S_A + S_B$) with respect to T_B for specified values of U, V, and m. What is the value of T_B at equilibrium?

b) The plate insulation is maintained, but the lock is removed. Assume P_{1B} = 2.48 bar and P_{1A} = 1.29 bar and equal temperatures $T_{A,1} = T_{B,1}$ = 335 K. Assume quasiequilibrium expansion in subsystem B and plot S with respect to P_A for specified values of U, V, and m

b) The insulation is removed, but heat transfer to outside ambience is allowed with the restraint that the entropy of the combined system A+B is constant. Plot U with respect to T_B. What is the value of T_B at equilibrium?

Problem C73

An adiabatic rigid tank is divided into two sections A (one part by volume) and B (two parts by volume) by an insulated movable piston. Section B contains air at 400 K and 1 bar, while section A contains air at 300 K and 3 bar. Assume ideal gas behavior. The insulation is suddenly removed. Determine:

a) The final system temperatures.
b) The final volumes in sections A and B.
c) The final pressures in sections A and B.
d) The entropy generated per unit volume.

Problem C74

Steam enters a turbine at 40 bar and 400°C, at a velocity of 200 m s^{-1} and exits at 36.2°C as saturated vapor, at a velocity of 100 m/s. If the turbine work output is 600 kJ kg^{-1}, determine:

a) The heat loss.
b) The entropy generation assuming that the control surface temperature T_b is the average temperature of the steam considering both inlet and exit.
c) The entropy generation if the control surface temperature $T_b = T_o$= 298 K, which is the ambient temperature

Problem C75

Determine entropy generated during the process of adding ice to tap water. A 5 kg glass jar (c = 0.84 kJ kg^{-1} K^{-1}) contains 15 kg of liquid water (c = 4.184 kJ kg^{-1} K^{-1}) at 24°C. Two kg of ice (c = 2 kJ kg^{-1} K^{-1}) at –25°C wrapped in a thin insulating foil of negligible mass is added to water. The ambient temperature T_o = 25°C. The insulation is suddenly removed. What is the equilibrium temperature assuming that no ice is left (the heat of fusion is 335 kJ kg^{-1}), and what is the entropy generated?

Problem C76

Consider the isentropic compression process in an automobile engine. The compression ratio $r_v = (V_1/V_2)$ = 8 and T_1 = 300 K. Assuming constant specific heats, determine the final temperature and T_2 and the work done if the fluid is air and Ar respectively. Explain your answers.

Problem C77

The fuel element of a pool–type nuclear reactor is composed of a core which is a vertical plate of thickness 2L and a cladding material of thickness t on both sides of the plate. It generates uniform energy q''' , and there is heat loss $h_H(T_s - T_\infty)$ from the plate surface, where T_s denotes the surface temperature of the cladding material. The temperature profiles are as follows:

In the core,

$$(T - T_\infty)/(q''' L_{core}^2/2k_{core}) = 1 - (x/L)^2 - B, \text{ where}$$

$$B = 2(k_{core}/k_{clad}) + 2 (L_{clad}/L_{core}) (k_{core}/k_{clad}) (1 + k_{clad}/(h_H L_{clad})).$$

For the cladding material

$$(T - T_\infty)/(q''' L_{core}^2/2k_{clad}) = -(x/L)^2 + c, \text{ where}$$

$$C = (L_{clad}/L_{core})(1 + k_{clad}/(h_H L_{clad})) \text{ and } L_{clad} = L_{core} + t.$$

Here L denotes length, k the thermal conductivity, h_H the convective heat transfer coefficient, and t thickness.

a) Obtain expressions for the entropy generated per unit volume for the core and clad.

b) Simplify the expression for the entropy generated per unit volume at the center of core?

c) Determine the entropy generated per unit surface area for the core and clad.

Problem C78

The energy form of the fundamental equation for photon gas is $U = (3/4)^{4/3} (c/(4\ \sigma))^{1/3} S^{4/3} V^{-1/3}$ where c denotes speed of light, σ Stefan Boltzmann constant, and V volume.

a) Obtain an expression for T(S,V).

b) Obtain an expression for (P/T) in terms of S and V.

c) Using the results for parts (a) and (b) determine P(T,V).

Problem C79

A heat engine cycle involves a closed system containing an unknown fluid (that is not an ideal gas). The cycle involves heat addition at constant volume from state 1, which is saturated liquid, to state 2, adiabatic reversible expansion from state 2 to state 3 which is a saturated vapor, and isobaric and isothermal heat rejection from state 3 to state 1 (that involves condensation from saturated vapor to saturated liquid). The cycle data are contained in the table below. The heat addition takes place from a thermal energy reservoir at 113°C to the system. Heat rejection occurs from the system to the ambient at 5°C. Determine the heat added and rejected, the cycle efficiency, the associated Carnot efficiency, and the entropy generated during the cyclical process

State	*P, bar*	*T, °C*	*v, $m^3\ kg^{-1}$*	*h, $kJ\ kg^{-1}$*
1	50	5	0.003	720
2	310	113	0.003	965
3	50	5	0.004	860

Problem C80

An ideal gas available at state (P_1,T_1) is to be isentropically expanded to a pressure P_2. Given the choice that you can either use a turbine or a piston–cylinder assembly, which one do you recommend? Are the isentropic efficiencies the same for both devices if the final states are the same?

Problem C81

Show that the reversible work for an isothermal process undergoing expansion from a pressure of P_1 to P_2 in a closed system is same as the work in an open system (neglect kinetic and potential energies in the open system) for the same pressure change with an ideal gas as the medium of fluid. Is this statement valid for an adiabatic reversible process for the same pressure changes in both the open and closed systems and with the same initial/inlet conditions? Justify.

Problem C82

Show that the expression

$$dU = T\,dS - P\,dV + \mu dN \qquad (A)$$

reduces to the expression du = Tds – Pdv.

Problem C83

Assume that we have 2 kmol of N_2 at 400 K and 1 bar in a rigid tank, and S_1 = 200.1×2 = 400.2 kJ/K. We add 0.1 kmols of N_2 and transfer heat from the system such that $S_2 = S_1$.

a) Determine U at states 1 and 2.
b) Determine the temperature at state 2.
b) Determine the chemical potential $\mu(= \partial U/\partial N)_{S,V}$

Problem C84

Consider a counter-flow heat exchanger in which two streams H and C of specific heats c_{pH} and c_{pC} flow counter to each other. The inlet is denoted as i and the exit as e. If $T_{H,i}$ and $T_{H,e}$ are the inlet and exit temperatures of stream H, and $T_{C,i}$ is the inlet of stream C., then obtain an expression for the maximum most temperature $T_{C,e}$. Assume that $C_{p,H}m_H < C_{pC}m_C$ and $T_{H,e} = T_{C,i}$. Determine the entropy generated per kg of smaller heat capacity fluid

Problem C85

Consider an adiabatic reversible compression from 1 to 2 via path A from volume v_1 to v_2 followed by irreversible adiabatic expansion from 2-3 and cooling from 3-1 (path B: 2-3 and 3-1). Apply Clausius in-equality for such a cycle and discuss the result.

D. CHAPTER 4 PROBLEMS

(Unless otherwise stated assume T_0 = 25°C and P_0 = 1 bar)

Problem D1

Is the relation $s(T_o, p_{H_2O,o}) = s(T_o, p^{sat}_{H_2O,})$ R ln $(p_{H_2O,o}/p_{H_2O,sat})$ equivalent to $s(T_o, p_{H_2O,o}) = s^o(T_o) - R \ln (X_{H_2O}P/P_o)$?

Problem D2

In the condenser part of a power plant, is there an irreversibility due to Q_o?

Problem D3

Is it more practical to design for w_{opt} than w_s?

Problem D4

Is the notion of availability based on an isentropic concept?

Problem D5

Is optimum work the same as reversible work?

Problem D6

When is $g \equiv \psi$?

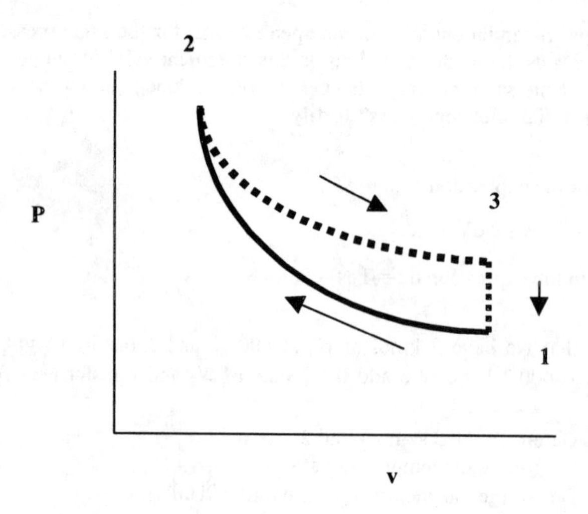

Figure C. 84

Problem D7

Are ke and pe included in the definition of ψ?

Problem D8

Describe the concept of chemical availability.

Problem D9

Use an example to describe the availability for gasoline.

Problem D10

Differentiate between the absolute (availability-Europe) and the relative availability (exergy).

Problem D11

Explain the physical implications of the expression $\psi = RT \ln X_k$. Does this mean that $\psi_{chem} < 0$?

Problem D12

Is chemical equilibrium satisfied when $\mu = \mu_o$?

Problem D13

What is the typical range of COP?

Problem D14

What is the difference between isentropic and optimum work?

Problem D15

What is the absolute stream availability? Can it have negative values? Does the value depend upon the reference condition used for the properties, such as h, s, etc.?

Problem D16

What is the (relative) stream availability or exergy? Can it have negative values? Does the value depend upon the reference condition used for the properties, such as h, s, etc.?

Problem D17

What is the difference between closed system availability and open system availability ?

Problem D18

Can we assume that $P_o\Delta v \approx 0$ for liquids?

Problem D19

What do we mean by useful and actual work?

Problem D20

Consider the universe. As $S \to \infty$, does $\phi \to 0$?

Problem D21

What does a dead state imply?

Problem D22

How are irreversibilities avoided in practice?

Problem D23

For G to have a minimum value in a multicomponent system at specified values of T and P, what is the partial pressure of the species?

Problem D24

Can the availability be completely destroyed?

Problem D25

What are your thoughts regarding current oil consumption and availability?

Problem D26

What is the implication of $W_{u,opt}$ for compression work?

Problem D27

An irreversible expansion occurs in a piston–cylinder assembly with air as the medium. The initial and final specific volumes and temperatures are, respectively, 0.394 $m^3\ kg^{-1}$ and 1373 K, and 2.049 $m^3\ kg^{-1}$ and 813 K. Assume constant specific heats, $c_{v0} = 0.717$ kJ kg^{-1} K^{-1} and $c_{p0} = 1.0035$ kJ kg^{-1} K^{-1}.

a) Determine the actual work delivered if the process is adiabatic and the adiabatic efficiency.

b) Assume that this is a reversible process between the two given states (not necessarily adiabatic for which Pv^n = constant). What is the value of n? Determine the reversible work delivered.

c) What is the maximum possible work if the only interactions are with the environment, T_{amb} = 300 K, and P_{amb} = 100 kPa. What is the availability efficiency of this process? Is this the same as the adiabatic efficiency?

d) What is the total entropy generated and the irreversibility?

Problem D28

Water flows through a 30 m long insulated hose at the rate of 2 kg min^{-1} at a pressure of 7 bar at its inlet (which is a faucet). The water hose is well insulated. Determine the entropy generation rate. What could have been the optimum work?

Problem D29

Steam enters a turbine at 5 bar and 240°C (state 1).

a) Determine the absolute availability at state 1? What is the absolute availability at the dead state (considering thermomechanical equilibrium)?

b) What is the optimum work if the dead state is in mechanical and thermal equilibrium?

c) What is the chemical availability?

d) What is the optimum work if the steam eventually discharges at the dead state? The environmental conditions are 298 K, 1 bar, and air with a water vapor mole fraction of 0.0303.

Problem D30

Saturated liquid water (the mother phase) is contained in a piston–cylinder assembly at a pressure of 100 kPa. An infinitesimal amount of heat is added to form a single vapor bubble (the embryo phase).

a) If the embryo phase is assumed to be at the same temperature and pressure as the mother phase, determine the absolute availabilities $\psi = h - T_o s$ and Gibbs functions of the mother and embryo phases.

b) If the pressure of the embryo (vapor) phase is 20 bar at 100°C, while the mother phase is at 1 bar, what are the values of the availability and Gibbs function of the vapor embryo? (Assume the properties for saturated vapor at 100°C and that the vapor phase behaves as an ideal gas from its saturated vapor state at 1 bar and 100°C to 20 bar and 100°C to determine the properties.)

Problem D31

You've been engaged as a consultant for a manufacturing facility that uses steam. Their steam generator supplies high pressure steam at 800 psia, but they use the steam at 300 psia. How would you advise them to decrease the pressure such that they minimize irreversibilities? Be sure to explain your answer. If so, explain what and the mechanism responsible for the destruction. Show both the process and the throttling process on an h-s diagram and refer to it to illustrate your answer.

Problem D32

Consider the energy from the sun at $T_{R,1}$ and the ocean water at T_0. Derive expressions for W_{opt}. Look at Figure Problem D.32 and interpret your results in terms of the figure.

Problem D33

Ice is to be heated at the North Pole where the ambient temperature is –30°C to temperature of –25°C, –20°C, ..., 90°C. Determine the minimum work required. The heat of melting of ice is 334.7 kJ kg^{-1}, and c_{ice} is 1.925 kJ kg^{-1} K^{-1}.

Problem D34

A gas tank contains argon at T and P.

a) Obtain an expression for the maximum possible work if an open system is used when tank pressure is T and P. Assume that there is negligible change in T and P of the tank and constant specific heats for the ideal gas. The ambient temperature is T_o and the ambient pressure is P_o.

b) Suppose the gas is slowly transferred from the tank to a large piston–cylinder (PC) assembly in which the pressure and temperature decrease to the ambient values. Treat the tank and PC assembly as one closed system. What is the behavior of $\phi/(RT_o)$ with respect to T/T_o with P/P_o as a parameter? Consider the case when the gas state is at 350 K and 150 bar, and T_o = 298 K and P_o = 100 kPa.

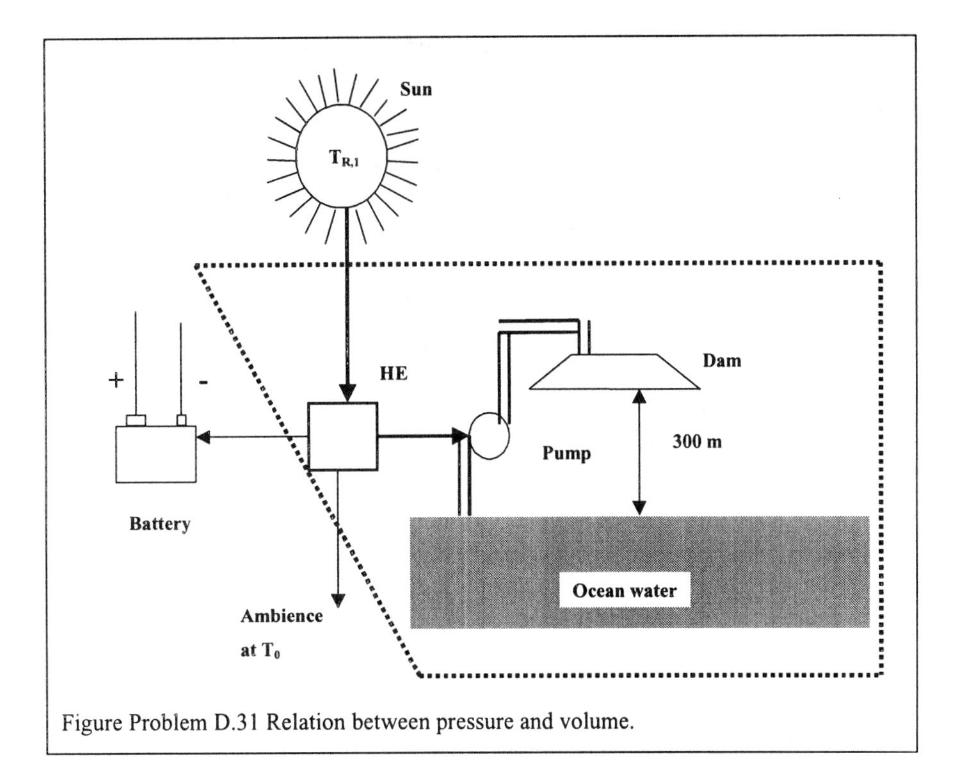

Figure Problem D.31 Relation between pressure and volume.

Problem D35

Natural gas (that can be assumed to be methane) is sometimes transported over thousands of miles in pipelines. The flow is normally turbulent with almost uniform velocity across the pipe cross sectional area. There is a large pressure loss in the pipe due to friction. The friction also generates heat that raises the gas temperature, which can result in an explosion hazard. Assume that the pipes are well insulated and the specific heats are constant. Assume that initially P_1 = 10 bar and T_1 = 300 K, and finally P_2 = 8 bar for a mass flow rate of 90 kg s^{-1} m^{-2}. What is the entropy change per unit mass? What is the corresponding result if the velocity changes due to the pressure changes?

Problem D36

The adiabatic expansion of air takes place in a piston–cylinder assembly. The initial and final volume and temperature are, respectively, 0.394 kg m^{-3} and 1100°C, and 2.049 kg m^{-3} and 813 K. Assume constant specific heats c_{v0} = 0.717 kJ kg^{-1} K^{-1} and c_{p0} = 1.0035 kJ kg^{-1} K^{-1}.

a) What is the actual work?

b) What is the adiabatic efficiency of the process?

c) Assuming that a reversible path is followed between the same initial and final states according to the relation Pv^n = constant, what is the work delivered? Why is this different from the actual work?

d) Now assume isentropic expansion from the initial state 1 to a volume of 2.333 kg m^{-3} and isometric reversible heat addition until the final temperature is achieved. What is the heat added in this case?

e) If the heat is first added isometrically and reversibly, and then isentropically expanded to achieve the final state, what is the value of the reversible work?

f) What is the maximum possible work for a closed system if the ambient temperature is 300 K? What is the value of the irreversibility?

Problem D37

Consider an ideal Rankine cycle nuclear power plant. The temperature of the heat source is 1400 K. The turbine inlet conditions are 6 MPa and 600°C. The condenser pressure is 10 kPa. The ambient temperature is 25°C. What is the irreversibility in KJ/kg and the maximum possible cycle work in KJ/kg?

Problem D38

Steam enters a ***non-adiabatic*** steady state steady flow turbine at 100 bar as saturated vapor and undergoes irreversible expansion to a quality of 0.9 at 1 bar. The heat loss from the turbine to the ambience is known to be 50 kJ/kg. Determine the

a) actual work,
b) optimum work, and
c) availability or exergetic or Second law efficiency for the turbine.

Problem D39

Consider the generalized equation for work from a open system in terms of entropy generation. Using the Gauss divergence theorem, derive an expression for the work done per unit volume w'' by a device undergoing only heat interaction with its environment and show that $w'' = -d/dt(e - T_o s) - \nabla(\rho v(e_T - T_o s)) - T_o\sigma$. Obtain an expression for the steady state maximum work.

Problem D40

Water is heated from the compressed liquid state of 40°C and 60 bar (state 1) to saturated vapor at a pressure P_2. Heat is supplied from a large reservoir of burnt gases at 1200 K. If the final pressure $P_2 = 60$ bar, calculate $s_2 - s_1$ and the value of the reversible heat transfer q_{12} to the water. If $P_2 = 58$ bar due to frictional losses (state 2′) but $h_2' = h_1$, calculate $s_2' - s_1$. Is this process internally reversible? Is there any entropy generated and, if so, how much? If the value of Q_H is identical for both cases (without and with frictional losses), what is the net entropy generated due to the irreversible heat transfer? Determine the changes in the availabilities $(\psi_2 - \psi_1)$ and $(\psi_{2'} - \psi_1)$.

Problem D41

A water drop of radius a at a temperature T_ℓ is immersed in ambient air at a temperature T_∞ and it vaporizes. The temperature and water vapor mole fraction profile can in terms of the radial spatial coordinate r be expressed through the following expression under "slow evaporation" conditions

$$X_v/X_{v,s} = (T - T_\infty)/(T_\ell - T_\infty) = a/r, \text{ where } r \ge a$$

where X_v denotes the mole fraction of the vapor and $X_{v,s}$ that at surface. Determine the difference between absolute availabilities at two locations r = a, and r = b. Plot the variation of availability in kJ/kg of mix with a/r where r is the radius.

Problem D42

Electrical work is employed to heat 2 kg of water from 25°C to 100°C. The specific heat of water is 4.184 kJ kg^{-1} K^{-1}. Determine the electrical work required, and the minimum work required (e.g., by using a heat pump instead).

Problem D43

Six pounds of air at 400°F and 14.7 psia in a cylinder is placed in a piston-cylinder assembly and cooled isobarically until the temperature reaches 100°F. Determine the optimum useful work, actual useful work, irreversibility and the availability or exergetic or so called 2nd law efficiency.

Problem D44

An adiabatic turbine receives 95,000 lbm of steam per hour at location 1. Steam is bled off (for processing use) at an intermediate location 2 at the rate of 18,000 lbm per hour. The balance of the steam leaves the turbine at location 3. The surroundings are at a pressure and temperature of 14.7 psia and 77°F, respectively. Neglecting the changes in the kinetic and potential energies and with the following information: P_1 = 400 psia, T_1 = 600°F, P_2 = 50 psia, T_2 = 290°F, P_3 = 2 psia, T_3 = 127 °F, v_3 = 156.4 $ft^3\ lbm^{-1}$, determine the maximum sssf work per hour, the actual work per hour, and the irreversibility.

Problem D45

In HiTAC (High temperature Air Combustion systems), preheating of air to 1000°C is achieved using either a recuperator or a regenerator. The recuperator is a counterflow heat exchanger while the regenerator is based on a ceramic matrix mounted in a tank through which hot gases and cold air are alternately passed. The hot gas temperature or this particular application is 1000 K. Assume c_p to be constant for the hot gas, and for it to be the same as that for the cold air. If the recuperator is used, cold air enters it at 25°C and the flowrate ratio of the hot to cold gases $\dot{m}_H/\dot{m}_C = 0.5$. The temperature differential between the air leaving the recuperator and the hot gases entering it is 50 K. Determine the availability efficiency for the recuperator. Will you recommend a regenerator instead? Why?

Problem D46

Large and uniformly sized rocks are to be lifted in a quarry from the ground to a higher level. The weight of a standard rock is such that the pressure exerted by it alone on the surrounding air is 2 bar. The rocks are moved by a piston–cylinder assembly that contains three pounds of air at 300°F when it is at ground level. Heat is transferred from a reservoir at 1000°F until the temperature of the air in the cylinder reaches 600°F so that piston moves up, thereby lifting a rock. Assume that air is an ideal gas with a constant specific heat. If the surrounding temperature and pressure are 60°F and 14.7 psia, determine:

a) The gas pressure.
b) The work performed by the gas.
c) The useful work (i.e., during the lifting of rocks) delivered by the gas.
d) The optimum work.
e) The optimum useful work.
f) The irreversibility and the availability efficiency (based on the useful work).

Problem D47

A jar contains 1 kg of pure water at 25°C. It is covered with a nonporous lid and placed in a rigid room which contains 0.4 kg of dry air at a temperature and pressure of 25°C and 1 bar. The lid is suddenly removed. The specific heat of water is 4.184 kJ $kg^{-1}\ K^{-1}$, and that of air is 0.713 kJ $kg^{-1}\ K^{-1}$.

a) Determine the temperature and composition of the room, the atmosphere of which contains water vapor and dry air at equilibrium. Ignore the pressure change.
b) The change in the availability.

Problem D48

Hot combustion products enter a boiler at 1 bar and 1500 K (state 1). The gases transfer heat to water and leave the stack at 1 bar and 450 K (state 2). Water enters the boiler at 100 bar and 20°C (state 3) and leaves as saturated vapor at 100 bar (state 4). The saturated vapor enters a non-adiabatic turbine at 100 bar and undergoes irreversible expansion to a quality of 0.9 at 1 bar (state 5). The combustion gases may be approximated as air. And the total gas flow is 20 kg s^{-1}. Determine the:

a) Absolute availabilities at all states.
b) Absolute availability at the dead state for gas and water.
c) Relative availabilities at all states.
d) Optimum power for the gas loop, i.e., with the same inlet and exit conditions of the gas.
e) Optimum work for the entire plant including gas and water loops.
f) Irreversibilities in the heat exchanger and turbine.

Problem D49

A nuclear reactor transfers heat at a 1727°C temperature to water and produces steam at 60 bar and 1040°C. The vapor enters the turbine at 60 bar and 1040°C and expands isentropically to 0.1 bar. The vapor subsequently enters the condenser where it is condensed to a saturated liquid at 0.1 bar and then pumped to the boiler using an isentropic pump. What are the values of η_{cyc}, the optimum work and the availability efficiency, the overall cycle irreversibility, and the irreversibility in the boiler and condenser? Perform an availability balance for the various states.

Problem D50

A house contains an air equivalent mass of 150 kg at 0°C. It must be warmed to 25°C. The only allowed interaction is with environment that is at a temperature $T_o = 273$ K. What is the minimum work input? Assume that air leaves the house at a constant temperature of 12.5°C and that the pressure in the house is near ambient. What is the minimum work input if outside air is circulated at the rate of 0.335 kg s^{-1} and the house must be warmed within 15 min?

Problem D51

Two efficiencies can be defined for heat exchangers. In a closed system $Q_s = Q_{used} + Q_{loss}$, and $\eta_h = Q_{used}/Q_{source}$ = (end use)÷(source energy). Since the end use and source availabilities are respectively, $Q_{used}(1-T_o/T_{used})$, and $Q_{source}(1-T_o/T_{source})$, show that $\eta_{avail} = \eta_h(1-T_o/T_{used})/(1-T_o/T_{source})$. Discuss the two efficiencies.

Problem D52

During a cold wave the ambient air temperature is –20°C. The temperature of a lake in the area is initially a uniform 25°C, but, gradually, a thick layer of ice is formed. Under the ice layer there is water at 25°C. The surface temperature of the ice layer is –10°C, and the heat transfer from the warm water to the ice is 100 kJ kg^{-1} of ice. Determine the optimum work. The heat of melting for ice is 335 kJ kg^{-1}, and the specific heats of ice and water, respectively, are 1.925 kJ kg^{-1} K^{-1} and 4.184 kJ kg^{-1} K^{-1}.

Problem D53

Consider a non-adiabatic fire tube boiler. Hot gases at a temperature of 400°C flow into the fire tube at a rate of 20 kg s^{-1}. The gas is used to heat water from a saturated liquid state to a saturated vapor condition at 150°C. The heat loss from fire tube boiler is 50 kJ kg^{-1} of gas. If the gases exit the heat exchanger at 200°C, determine the water flow required, the entropy generation if the control volume boundary is selected to be just inside the heat exchanger, entropy generation if control volume boundary is selected to be just outside the heat exchanger. and the optimum work. Assume that gases have the same properties as air (with $c_p = 1$ kJ kg^{-1} K^{-1}), and where $T_o = 298$ K and $P_0 = 1$ bar.

Problem D54

A 10 m^3 tank contains air at 1 bar, 300 K. A compressor is used to evacuate the tank completely. The compressor exhausts to the ambience at 1 bar and 300 K. Assume that the tank temperature remains constant through heat transfer from ambience at 300 K. You are asked to determine the minimum (optimum) work required. Select the

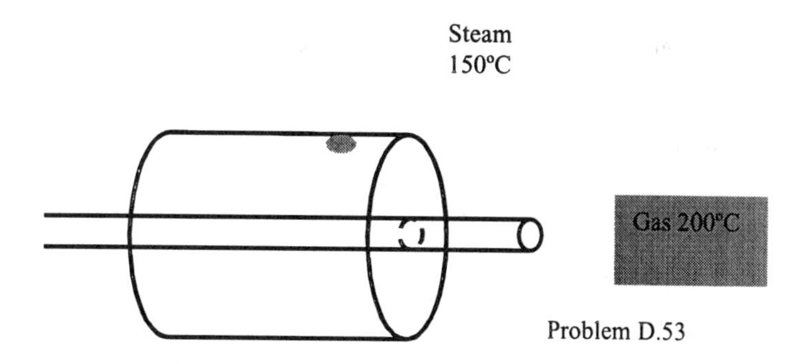

Problem D.53

control volume which includes the tank, compressor and the outlet from the compressor.

a) Does the tank mass remain constant?

b) Does the internal energy of unit mass within the tank remain constant if gas is assumed to be an ideal gas?

c) Does the absolute availability at the exit of the compressor change with time

d) Starting from mass conservation and generalized availability balance, then simplify the equation for the current problem., Indicate all the steps clearly and integrate over a period of time within which the tank is emptied.

e) Assuming that $h = c_{p0}\,T$, $u = c_{v0}\,T$, $s = c_{p0} \ln(T/T_{ref}) - R \ln(P/P_{ref})$, $T_{ref} = T_0$, $P_{ref} = 1$ bar, determine the work in kJ.

E. CHAPTER 5 PROBLEMS

Problem E1

Consider the state equation $S = C\,N^{1/6}\,V^{1/3}\,U^{1/2}$. Obtain a state equation for S(T,V,N), P(T,V,N), and A(T,V,N).

Hint: $T = (dU/dS)_{V,N}$. *Also use the first Legendre transform of S with respect to U*

Problem E2

Consider the state equation $U^o = U^o(S,V,N_1,N_2, \ldots, N_n) = U^o(x_1,x_2, \ldots, x_{n+2})$. Show that the second Legendre transform with respect to S and V is G. Obtain the $(n+2)^{th}$ Legendre transform of the expression, and show that it is zero. By using the total differential of the Legendre transform, derive the Gibbs–Duhem equation.

Problem E3

Consider an electron gas in a metal. For instance, about 3 trillion electrons flow per second in a 50 W lamp. An electron has the weight of 1/1836 of an H atom. These mobile electrons are responsible for the large thermal and electrical conductivity of metals. In theory, these electrons can be treated as a gas that obeys Fermi–Dirac statistics. Because certain integrals are approximately evaluated, the theory is restricted to low or moderate temperatures. This limitation is not significant, however, since the approximation is actually accurate up to the melting point of metals. We obtain the following entropy equation from the theory, i.e., $S = C_1 N^{1/6} V^{1/3} (U-U_o)^{1/2}$, where $C_1 = (2^{3/2}\pi^{4/3}/3^{1/3})(k/h)m^{1/2}$, k denotes the Boltzmann constant (1.3804×10^{-23} J K^{-1}), h the Planck constant (6.62517×10^{-34} J s), and m the electron mass (9.1086×10^{-31} kg), N denotes the number of free electrons in the metal, $U_o = (3/5)N\mu_o$ the internal energy of the electron gas at 0 K, $\mu_o = C_2(N/V)^{2/3}$, and $C_2 = 3^{2/3}h^2/(8\pi^{2/3}m)$. Show that (a) S =

$C_1N^{1/6}(V^{2/3}U - (3/5)C_2N^{5/3})^{1/2,}$ and that the entropy is a homogeneous function of degree 1, obtain an expression for the electron gas (b) temperature, and (c) pressure, and (d) assume that when $U \gg U_o$ whether the conditions of the fundamental equation are satisfied.

Problem E4

Consider the n–th Legendre transform of a homogeneous function of degree m $y^{(0)}(x_1,x_2, ..., x_n)$. Using the Euler equation and Legendre transform method, show that $y^{(n)} = y^{(0)}(m-1)$.

F. CHAPTER 6 PROBLEMS

Problem F1

Can the combustion gases emerging from a boiler be considered to have the same properties as air or should we employ the real gas equation of state?

Problem F2

What is v_c'?

Problem F3

Which two–parameter equation of state is best to use?

Problem F4

Which two–parameter equation of state does not yield negative pressures?

Problem F5

What are the important differences between the Dietrici and VW equations of state?

Problem F6

Why do we obtain two solutions when we neglect the parameter b in the VW equation of state?

Problem F7

Is there an analytical method for determining the stability of solutions?

Problem F8

Are there generalized equations of state for complex fluids that do more than just Pdv (i.e., compressible) work?

Problem F9

Is the real gas equation of state valid for high speed flows as long as they are in continuum?

Problem F10

Does $v \rightarrow 0$ as $P \rightarrow \infty$, $T \rightarrow 0$?

Problem F11

Is it true that at specified values of T_R and Z, P_R is single valued?

Problem F12

Is it possible to develop a real gas equation of state for a subcooled liquid?

Problem F13

Is the Pitzer factor constant for any given fluid?

Problem F14

Why do equations of state sometimes fail in the compressed liquid and vapor domes?

Problem F15

Why is the vapor dome region difficult to predict with a two–parameter equation of state?

Problem F16

Can we extend the real gas equation of state to liquids?

Problem F17

Can you determine the value of Z with just the values of T_R and P_R for the Clausius II equation of state?

Problem F18

What are the values of Z_c for the RK, VW, Berthelot, and Dietrici equations of state?

Problem F19

Is it true that real gas equations of state are applicable only for the vapor state?

Problem F20

How are real gas state equations derived?

Problem F21

Why do some state equations predict saturated properties well, while others do not?

Problem F22

Why is the "middle" solution for v at specified values of T and P meaningless in context of a cubic equation?

Problem F23

How many distinct real solutions exist in context of the RK equation at a specified temperature if $T > T_c$, and $T < T_c$?

Problem F24

The fundamental equation for an electron gas is $S = C_1N^{1/6}(V^{2/3}U - (3/5)C_2N^{5/3})^{1/2}$. Obtain an equation of state in terms of P, V and T. Does this electron gas behave as an ideal gas? What is the compressibility factor at 200 bar and 100 K?

Problem F25

Consider the VW equation $P = RT/(v-b) - a/v^2$. Plot the P(v) behavior of water. Show that $P_R = T_R/(v'_R-1/8) - (27/64)/v'^2_R$ and plot P_R with respect to v'_R for $T_R = 0.6$ and 1.2. Prove that the $Z > 1$ when the body volume effect dominates attractive forces (i.e., $a \approx 0$ at very high pressures) and vice versa (i.e., $b/v \ll 1$, $(b/v)^2 \approx 0$, but $b \neq 0$). Using the relation, $Z = P_R\ v'_R/T_R$ plot $Z(P_R)$ for $T_R = 0.6$ and 1.2 and $Z(P_R)$ for $v'_R=0.3$ and 0.4, and discuss your results.

Problem F26

Derive an expression for a and b in terms of T_c, and P_c for the Dietrici equation of state $P = (RT/(v-b)) \exp(-a/(RTv))$ and show that $a = (4/P_c)(RT_c/e)^2$ and $b = RT_c/(P_c\ e^2)$ where $e = 2.7182818$. Note that one cannot obtain negative pressures with the Dietrici equation as opposed to the RK equation unless $v \ll b$, which is physically impossible. Plot P(v) for water at various temperatures and obtain gas like solutions for volume *vs.* T (if they exist) at 113 bar.

Problem F27

For the Clausius II equation, obtain the relations for a, b, and c in terms of critical properties and critical compressibility factor. (Hint: Solve for a and b in terms of c and v_c using the inflection condition. Then, use the tabulated value of Z_c to determine that of c.) Determine the corresponding values for H_2O and CH_4.

Problem F28

Calculate the specific volume of H_2O(g) at 20 MPa. and 673 K by employing the (a) compressibility chart, (b) Van der Waals equation, (c) ideal gas law, (d) tables, (e)

Pitzer correction factor and Kessler tables. What is the mass required to fill a 0.5 m^3 cylinder as per the five methods?

Problem F29

Determine the specific volume and mass of CH_4 contained in a 0.5 m^3 cylinder at 10 MPa and 450 K using the following methods:

a) Ideal gas law.
b) Compressibility charts.
b) van der Waals equation.
c) Approximate virial equation of state.
d) Compressibility factor tables including the Pitzer factor.
e) Approximate equation for v(P,T) given by expanding the Berthelot equation $v = (1/2)(b + (RT/P))(1 \pm (1-(4a/(PT(b+RT/P))))^{1/2})$, $b/v \ll 1$.

Problem F30

Consider the virial equation of state $(Pv/RT) = Z = 1 + B(T)/v + C(T)/v^2$.

a) Determine B(T) and C(T) if $P = RT/(v-b)$ and $b/v \ll 1$.
b) Determine B(T) and C(T) if $P = RT/(v-b) - a/v^2$ and $b/v \ll 1$.
   i) Obtain an expression for the two solutions for v(T,P) from the quadratic equation. Are these solutions for the liquid and vapor states? Discuss.
   ii) Discuss the two solutions for steam at 373 K and 100 kPa. Explain the significance of these solutions.
   iii) Show that the expression for the Boyle temperature (at which Z = 1) is provided by the following relation if second order effects are ignored, namely, $T_{Boyle} = a/(Rb)$.
   iv) What is the Boyle temperature for water?

Problem F31

CF_3CH_2F (R134A) is a refrigerant. Determine the properties (v, u, h, s, etc.) of its vapor and liquid states. The critical properties of the substance are $T_c = 374.2$ K, $P_c =$ 4067 kPa, $\rho_c = 512.2$ kg m^{-3}, M = 102.03 kg $kmol^{-1}$, $h_{fg} = 217.8$ kJ kg^{-1}, $T_{freeze} = 172$ K, $T_{NB} = 246.5$ K (this is the normal boiling point, i.e., the saturation temperature at 100 kPa).

a) Determine the value of vs^{at}(liquid) at 247 K. Compare your answer with tabulated values (e.g., in the ASHRAE handbook).
b) Determine the density of the compressed liquid at 247 K and 10 bar.
c) Use the RK equation to determine the liquid and vapor like densities at 247 K and 1 bar. Compare the liquid density with the answer to part (b).

Problem F32

If $c^2 = -kv^2 (dP/dv)_T$, deduce the relation for the sound speed of a RK gas in terms of v_R', T_R, and k.

Problem F33

Using steam tables, determine β_P and β_T for liquid water at (25°C, 0.1 MPa)., (70°C, 0.1 MPa), and (70°C, 10 MPa). What is your conclusion?

Problem F34

Show that if $(b/v)^2 \ll (b/v)$, the explicit solutions for v(P,T) and a in the context of the state equation $P = RT/(v-b) - a/T^n v^2$ are provided by the relations $v = \alpha(1+(1-\beta/\alpha^2)^{1/2})$, $\beta/\alpha^2 < 1$, where $\alpha(T,P) = RT^{n+1}/(2PT^n)$, $\beta(T,P) = (a - bRT^{n+1})/(PT^n)$. (Hint: expand the term $1/(v-b)$ as a polynomial in terms of (b/v).) Using the explicit solutions with n = 0 (i.e., the VW state equation), determine the solution(s) for v(593 K, 113 bar) in the case of water. If $bRT^{n+1} \ll a$, simplify the solution for v. Is solution for (593 K, 113 bar) possible? Show that if $v \gg b$, $Z < 1$ and if $RT/(v-b) \gg a/T^n v^2$ (i.e. $v \approx b$ when the molecules are closely packed), $Z > 1$.

Problem F35

A diesel engine has a low compression ratio of 6. Fuel is injected after the adiabatic reversible compression of air from 1 bar and 300 K (state 1) to the engine pressure (state 2). Assume that for diesel fuel $P_c = 17.9$ bar, $T_c = 659$ K, $\rho_l = 750$ kg m^{-3}, $C_{pl} = 2.1$ kJ kg^{-1} K^{-1}, $\Delta h_c = 44500$ kJ kg^{-1}, $L_{298} = 360$ kJ kg^{-1}, $L(T) = L_{298}((T_c - T)/(T_c - 298))^{0.38}$, and $\log_{10} P^{sat} = a - b/(T^{sat} - c)$, where a = 4.12, b = 1626 K, c = 93 K. Determine the specific volume of the liquid at 1 bar and 300 K. Assume that the value of Z_c can be provided by the RK equation. Since the liquid volume does not significantly change with pressure, using the value of the specific volume and ρ_l determine the fuel molecular weight. Determine the liquid specific volume at state 2. What are the specific volumes of the liquid fuel and its vapor at the state (P^{sat}, T_2)?

Problem F36

Derive an expression for f/P for VW gas using the definition $dg = RT\, d\, \ln(f) = v\, dP$ and $dg^{ig} = RT\, d\ln(P) = v^{ig}\, dP$; determine f/P at critical point using the expressions of "a" and "b" for VW gas.

Problem F37

Determine the values of v_ℓ and v_g for refrigerant R–12 at 353 K and 16 bar by applying the following models: a) ideal gas, b) RK equation, c) PR equation, d) Rackett equation, e) PR equation with w = 0. Discuss the results.

Problem F38

Experimental data for a new refrigerant are given as follows:

$P_1 = 111$ bar, $T_1 = 365$ K, $v_1 = 0.1734$ m^3/kmol

$P_2 = 81.29$ bar, $T_2 = T_1 = 365$, $v_2 = 0.2805$

a) If VW equation of state is valid, determine "a" and "b"

b) If critical properties P_c, T_c of the fluid are not known, how will you determine T_c, P_c? Complete solution is not required.

Problem F39

The VW equation of state can be expressed in the form $Z^3 - (P_R/(8T_R)+1)Z^2 + (27 P_R/(64T_R))\, Z - (27\, P_R^2/(512\, T_R^{\,3})) = 0$. Obtain an expression for $\partial Z/\partial P_R$ and its value as $P_R \to 0$. At what value of T_R is $\partial Z/\partial P_R = 0$. Obtain an expression for an approximate virial equation for Z at low pressures.

Problem F40

For the Peng–Robinson equation of state: $a = 0.4572\, R^2T_c^2/P_c$ and $b = 0.07780\, R\, T_c/P_c$. Determine the value of Z_c, and Z(673 K, 140 bar) for H_2O.

Problem F41

Consider the state equation: $P_R = T_R/(v'_R - b^*) - a^*(1+\kappa(1-T_R^{1/2}))^2/(T_R^n((v'_R+c^*)+(v_R'+d^*)))$, where n = 0 or 0.5, and κ is a function of w only. If $P_R((v'_R+c^*) + (v'_R+d^*))/a^* = A$, and $((v'_R+c^*) + (v'_R+d^*))/(a^*(v'_R-b^*)) = B$, show that for n = 0, $P_R = T_R/(v'_R-b^*) - a^*(1+\kappa(1-T_R^{1/2}))^2/(T_R^n((v'_R+c^*) + (v'_R+d^*)))$, and $T_R^{1/2} = -(\kappa+\kappa^2)/(B-\kappa^2) \pm ((1+2\kappa+\kappa^2+A)/(B-\kappa^2) + (\kappa+\kappa^2)^2/(B-\kappa^2)^2)^{1/2}$.

Problem F42

Consider the state equation $P_R = T_R/(v'_R - b^*) - a^*/(T_R^n((v'_R + c^*)\, v'_R)))$. Show that for the Berthelot and Clausius II equations ($n = 1$), $T_R = P_R(v'_R - b^*)/2(1+(4a^*(v'_R - b^*)/(v'_R + c^*)^2+1)^{1/2})$. Show that for the VW equation of state, both n and c^* equal zero, that $c^* = d^* = 0$ for the Berthelot equation, and $d^* = 0$ for the Clausius II equation.

Problem F43

Plot the pressure with respect to the specific volume of H_2O by employing the RK state equation at 600 K and determine the liquid– and vapor–like solutions at 113 bar.

Problem F44

Plot the product $P\bar{v}$ with respect to the pressure for water (you may use tabulated values). Does low pressure $P\bar{v}$ provide any insight into the temperature. Can you construct a constant volume ideal gas thermometer which measures the pressure in a glass bulb containing a known gas and then infer the temperature?

Problem F45

Using the inflection conditions for the Redlich–Kwong equation $P = (RT/(v-b)) - a/(T^{1/2}\, v(v+b))$, derive expressions for a and b in terms of T_c, and P_c. and show that (a) $(b/v_c)^3 - 3(b/v_c)^2 - 3(b/v_c) + 2 = 0$, or $b/v_c = 0.25992$, (b) $a/(RT_c^{3/2}v_c) = (1+(b/v_c)^2)/((1-(b/v_c)^2)\,(2+(b/v_c))$, or $a/(RT_c^{3/2}v_c) = 1.28244$, and (c) $Z_c = 1/3$.

Problem F46

Determine explicit solutions for v(P,T) if $(b/v)^2 \ll (b/v)$ for the state equation $P = RT/(v-b) - a/(T^n\, v(v+b))$. Show that $v = \alpha + (-\beta+\alpha^2)^{1/2} = \alpha(1 \pm (1-\beta/\alpha^2)^{1/2})$, $\beta/\alpha^2<1$, where $\alpha\,(T,P) = RT^{n+1}/(2PT^n)$, $\beta(T,P) = (a - bRT^{n+1})/(PT^n)$. (Hint: expand $1/(v-b)$ and $1/(v+b)$ in terms of polynomials of (b/v).) Using the explicit solutions and $n = 1/2$ (RK equation), determine solutions for v(593 K, 113 bar) for H_2O. Show that if $v \gg b$ then $Z < 1$, and if $RT/(v-b) \gg a/T^n v^2$ (i.e., $v \approx b$, or that the molecules are closely packed) then $Z > 1$.

Problem F47

Using the RK equation obtain an approximate expression for v by neglecting terms of the order of $(b/v)^3$.

Problem F48

Convert the Berthelot, VW, and Dietrici state equations to their reduced forms using the relations $P_R = P/P_c$, $T_R = T/T_c$, and $v'_R = v/v'_c$, $v'_c = RT_c/P_c$.

Problem F49

For the state equation $P = RT/(v-b) - a/(T^n\, v^m)$ show that (a) $a = ((m+1)^2/4m)\,(RT_c^{n+1}v_c^{m-1})$, $b = v_c\,(1-(2/(m+1)))$, and $Z_c = ((m^2-1)/(4m))$. Obtain a reduced form of this real gas equation, i.e., $P_R = f(v'_R, T_R)$.

Problem F50

For the state equation $P = RT/(v-b) - a/v^2$ determine the values of a and b without using the inflection conditions, but using the facts that at critical point there are three equal real roots (at $T<T_c$ there are three roots, and for $T > T_c$ only one real root exists).

Problem F51

Determine the Boyle curves for T_R *vs.* P_R for gases following the VW equation of state. Also obtain a relationship for $P_R(v'_R)$.

Problem F52

If number of molecules per unit volume $n' = 1/\ell^3$ where ℓ denotes the average distance (or mean free path between molecules). determine the value of ℓ for N_2 contained in a cylinder at –50°C and 150 bar by applying the (a) ideal gas law and (b) the RK equation. Compare the answer from part (b) with the molecular diameter determined from the value of $\bar{b}$. Apply the LJ potential function concept (Chapter 1) in order to determine the ratio of the attractive force to the maximum attractive force possible.

Problem F53

In the case of the previous problem determine the value of ℓ for the H_2O at 360°C and 120 bar.

Problem F54

Using the RK equation plot the pressure with respect to specific volume at the critical temperature for the range $0.25v_c<v<2v_c$. Here, v_c has its value based on the RK equation at specified values of P_c and T_c. From the tables plot the function P(v) for the same conditions and discuss your results.

Problem F55

Apply the RK equation for H_2O at 473 K, 573 K, and 593 K and obtain gas–like solutions (if they exist) at 113 bar. Compare these values with the liquid/vapor volumes obtained from the corresponding tables.

Problem F56

A person thinks that the higher the intermolecular attractive forces, the larger the amount of energy or the higher the temperature required to boil a fluid at a specified pressure. Consequently, since the term a in the real gas equation of state is a measure of the intermolecular attractive forces, you are asked to plot T^{sat} with respect to a. Use the normal boiling points (i.e., T^{sat} at 1 bar) for monatomic gases such as Ar, Kr, Xe, He, and Ne, and diatomic gases such as O_2, N_2, Cl_2, Br_2, H_2, CO, and CH_4. Also determine T^{sat} using the correlation $\ln(P_R) = 5.3(1-(1/T_R^{sat}))$ where $P_R = P/P_c$ and P = 1 bar. Use the RK and VW state equations. Do you believe the hypothesis?

Problem F57

A fixed mass of fluid performs reversible work $\delta W = Pdv$ according to the processes 1–2 isometric compression, 2–3 isothermal heating at T_H, 3–3 isometric expansion, and 4–1 isothermal cooling at T_L. The cycle can be represented by a rectangle on a T–v diagram. Determine the value of $\int\delta W/T$ if the medium follows the VW and ideal gas equations of state.

Problem F58

Flammable methane is used to fill a gas cylinder of volume V from a high–pressure compressed line. Assume that the initial pressure P_1 in the gas tank is low and that the temperature T_1 is room temperature. The line pressure and temperature are P_i and T_i. Typically, $P_i \gg P_2$, the final pressure. There is concern regarding the rise in temperature during the filling process. We require a relation for T_2 and the final mass at a specified value of P_2. Assume two models: (a) the ideal gas equation of state $P = RT/v$ for which $du_0 = c_{v0}dT$, and (b) the real gas state equation $P = RT/(v-b) - a/v^2$ with $c_v = c_{vo}$ and $du = c_vdT + (T\partial P/\partial T - P)dv$.

Problem F59

Determine v for water at P =133 bar, T= 593 K using VW, RK, Berthelot, Clausius II, SRK and PR equations.

Problem F60

Consider generalized equation of state $P = RT/(v-b) - a\,\alpha\,(w, TR) / (T^n\,(v+c)\,(v+d))$. Using the results in text, determine Z and v for H2O atT1 = 473K, P1 = 150 bar, T2= 873K, P2 = 250 bar using VW, RK, Berthelot, Clausius II, SRK and PR. Compare results with steam tables.

G. CHAPTER 7 PROBLEMS

Problem G1

For an ideal gas $c_{vo} = c_{vo}(T)$ and, hence, $u_o = u_o(T)$. Is this true for a VW gas?

Problem G2

How will you analyze transient flow processes discussed in example 14 of Chapter 2 for real gases?

Problem G3

Recall that $du_T = (a/T\,v^2)\,dv$ for a Berthelot gas. The integration constant F(T) can be evaluated at the condition $a \to 0$. Is the expression for F(T) identical to that for an ideal gas?

Problem G4

If, for a gas, $du = c_v dT + f(T,v)dv$ and $c_v = c_v(T,v)$, which is unknown, can we determine the value of u by integrating the expression at constant values of v?

Problem G5

Is it possible to predict the properties s_{fg}, and h_{fg} using "real gas" state equations?

Problem G6

An insulated metal bar of cross sectional area A is stretched through a length dx by applying a pressure P. Does the bar always cool or heat during this process?

Problem G7

The residual internal energy of a Berthelot fluid $u(T,v) - u_o(T) = -2a/(Tv)$. Determine an expression for the residual specific heat at constant volume $c_v(T,v) - c_{vo}(T)$.

Problem G8

A rubber product contracts upon heating in the atmosphere. Does the entropy increase or decrease if the product is isothermally compressed? (Hint: Use the Maxwell's relations.)

Problem G9

a) Using the generalized thermodynamic relation for du, derive an expression for u^R/RT_c for a Clausius II fluid. b) What is the relation for $c_{vo}(T) - c_v(T,v)$ for the fluid? c)Determine the values of u^R/RT_c and h^R/RT_c for CO_2 at 425 K and 350 bar.

Problem G10

Determine an expression for $\partial c_v/\partial v$ for a Clausius II fluid in terms of v and T.

Problem G11

Consider the isothermal reversible compression of Ar gas at 180 K from 29 bar to 98 bar in a steady state steady flow device. Using the fugacity charts determine the work in kJ per kmol of Ar.

Problem G12

Assume that air is a single component fluid. Air is throttled in order to cool it to a temperature at which oxygen condenses out as a liquid.

a) In order to determine the inlet conditions for the throttling process you are asked to determine the inversion point. Looking at the charts presented in text for RK equation, determine the inversion pressure at 145.38 K.

b) Using the $\bar{c}_v$ relations, determine $\bar{c}_v$ of air at the inversion point.

c) Determine $\bar{c}_p$ at this inversion condition. Assume that $\bar{c}_{po} = 29$ kJ $kmol^{-1}$. Is the value of $\bar{c}_p - \bar{c}_v = \bar{R}$?

d) What is the value of the Joule Thomson Coefficient at 1.2 times the inversion pressure at 145.38 K. Assume that the value of $\bar{c}_p$ at this pressure equals that at the inversion point. Do you believe air will be cooled at this point?

Problem G13

Near 1 atm, the Berthelot equation has been shown to have the approximate form $Pv = RT\,(1 + (9PT_c/(128P_cT))\ 1{-}6(T_c^2/T^2)))$. Obtain an expression for s(T,P).

Problem G14

In a photon gas the radiation energy is carried by photons, which are particles without mass but carry energy. The gas behaves according to the state equation $P = 4\,\sigma\,T^4/(3\,c_0)$, where σ denotes the Stefan Boltzmann constant and c_o the light speed in vacuum. Obtain an expression for the internal energy by applying the relation $du = c_v\,dT + (T(\partial P/\partial T)_v - P)\,dv$.

Problem G15

Oxygen enters an adiabatic turbine operating at steady state at 152 bar and 309 K and exits at 76 bar and 278 K. Determine the work done using the Kessler charts. Ignore Pitzer effects. What will be the work for the same conditions if a Piston-cylinder system is used?

Problem G16

The Joule Thomson effect can be depicted through a porous plug experiment that illustrates that the enthalpy remains constant during a throttling process. In the experiment a cylinder is divided into two adiabatic variable volume chambers A and B by a rigid porous material placed between them. The chamber pressures are maintained constant by adjusting the volume. Freon vapor with an initial volume $V_{A,1}$, pressure $P_{A,1}$ and energy $U_{A,1}$ is present in chamber A. The vapors penetrate through the porous wall to reach chamber B. The final volume of chamber A is zero. Determine the work done by the gas in chamber B, and the work done on chamber A. Apply the First Law for the combined system A and B and show that the enthalpy in the combined system is constant.

Problem G17

Obtain a relation for the Joule Thomson coefficients for a VW gas and an RK gas in terms of a, b, c_p, R, and T. Determine the inversion temperature.

Problem G18

Obtain an expression for f/P for a VW gas and write down the expression at the critical point. Assume that the gas behaves like an ideal gas at a low pressure P_o and large volume v_o. (Hint: $\int vdP = Pv - \int Pdv$, and $P_0v_0 = RT$.)

Problem G19

The Cox–Antoine equation is $\ln P = A - B/(T+C)$. Determine A, B and C for H_2O and R134A using tabulated data for T^{sat} *vs.* P. Compare T^{sat} at $P = 0.25P_c$ and $0.7P_c$ obtained from the relation with the tabulated values.

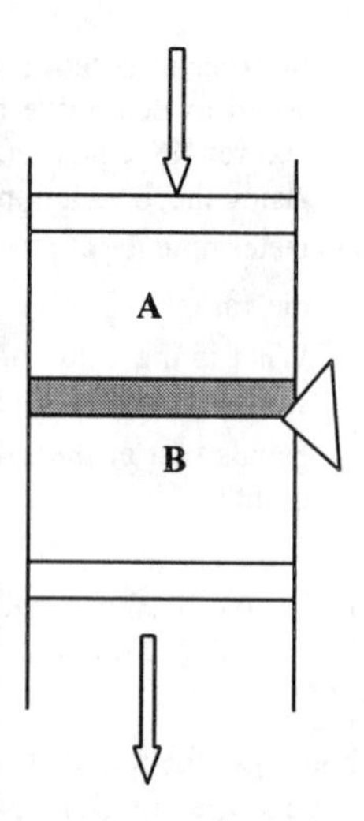

Problem Figure G.15

Problem G20

Determine the chemical potential of liquid CO_2 at 25°C and 60 bar. The chemical potential of CO_2, if treated as an ideal gas, at those conditions is –451,798 kJ $kmol^{-1}$.

Problem G21

Plot P(v) in case of H_2O at 373 K in the range $v_{min}= 0.8*v_f$ and $v_{max}= 1.5*v_g$ assuming that the fluid follows the RK state equation. The values of v_f and v_g are (for 523 K, P^{sat}) exp(.582(1-T_c/T)). What are the values for v_f and v_g for P^{sat} ?. Assume that h = 0 kJ $kmol^{-1}$ and s = 0 kJ $kmol^{-1}$ K^{-1} at $v = 0.8v_f$ and 523 K. From the g(P) plot, determine the RK saturation pressure at 523 K.

Problem G22

The properties of refrigerant R–134A (CF_3CH_2F) are required. The critical properties of the fluid are T_c = 374.2 K, P_c = 4067 kPa, ρ_c = 512.2 kg/m^3, M = 102.03, h_{fg} = 217.8 kJ kg^{-1}, T_{freeze} = 172 K, and T_{NB} = 246.5 K (the normal boiling point is the saturation temperature at 100 kPa). Plot the values of ln (P_{sat}) with respect to 1/T using Clausius–Clapeyron equation. Use the RK equation of state and plot P_R with respect to V_R with T_R as a parameter. Use the relation $dg_T = vdP = (\int d(pv) - \int Pdv)$ to plot the values of g/RT_c with respect to v_R' at specified values of T_R. Assume that $g/RT_c = 0$ at 373 K when $v_R' = 0.1$.

Problem G23

You are asked to analyze the internal energy of photons which carry the radiation energy leaving the sun. Derive an expression for change in the internal energy of the photons if they undergo isothermal compression from a negligible volume to a volume v. The photons behave according to the state equation $P = (4\,\sigma/3\,c_0)\,T^4$, where $\sigma = 5.67\times10^{-11}$ kW m^{-2} K^{-4}denotes the Stefan Boltzmann constant, $c_0 = 3\times10^{10}$ m s^{-1} the speed of light in vacuum, and T the temperature of the radiating sun.

a) Show that $c_v = c_v(T, v)$ for the photons.

b) Obtain a relation for μ.

Problem G24

From the relation s = s(T,P), obtain a relation for $(\partial T/\partial P)_s$ in terms of c_p, β_P, v and T. If $Z = 1 + (\alpha T_R + \beta T_R^{\,m})P_R$, where $\alpha = 0.083$, $\beta = -0.422$, and m = 0.6, obtain an expression for $(s_o - s)/R$.

Problem G25

How much liquid can you form by throttling CO_2 gas that is at 200 bar and 400 K to 1 bar? The property tables are not available. How much liquid can you form if you use an isentropic turbine to expand the gas to 1 bar? Make reasonable assumptions.

Problem G26

Recall that $du = c_v dT + (T(\partial P/\partial T)_v - P)\, dv$. A) Obtain an expression for du for a VW gas. Is c_v a function of volume? (Hint: use the Maxwell's relations.) B) If c_{vo} is independent of temperature, obtain an expression for the internal energy change when the temperature and volume change from T_1 to T_2 and from v_1 to v_2. Assume c_v is constant.

Problem G27

Gaseous N_2 is stored at high pressure (115 bar and 300 K) in compartment A (that has a volume V_A) of a rigid adiabatic container. The other compartment B (of volume $V_B = 3V_A$) contains a vacuum. The partition between them is suddenly ruptured. If $c_v = c_{vo} = 12.5$ kJ kmol^{-1} K^{-1}, determine the temperature after the rupture. Assume VW gas.

Problem G28

Gas from a compressed line is used to refill a gas cylinder from the state (P_1, T_1) to a pressure P_2. The line pressure and temperature are P_i and T_i. Determine the final pressure and temperature if (a) the cylinder is rapidly filled (i.e. adiabatic) and (b) slowly filled (i.e. isothermal cylinder). Use the real gas state equation $P = RT/(v-b) - a/v^2$.

Problem G29

Using the relation $\ln P^{sat} = (A - B/T)$, show that $\Delta h^{vapor} = \Delta v^{vapor} (BP^{sat}/T)$.

Problem G30

Derive an expression for $(a(T,v)-a_o(T,v_o))$ using the Peng–Robinson equation $P = (RT/(v-b)) - a(w,T)/(v(v+b) + b(v-b))$. Derive expressions for $(s-s_o)$ and $(h-h_o)$.

Problem G31

a) Calculate the fugacity of $H_2O(\ell)$ at 400 psia and 300°F (assume that $\bar{v} = c$ for the liquid state). b) If the condition A denotes compressed liquid, then is $f_A(P,T) \approx f(T,P^{sat})$ (i.e., the fugacity of the saturated liquid at the same temperature)? c) At phase equilibrium the fugacity of the saturated liquid equals that of saturated vapor, i.e., $f(T, \bar{v}_f) = f(T, \bar{v}_g)$ and $P(T, \bar{v}_f) = P(T, \bar{v}_g)$. Predict P^{sat} at 200°C using these relations and the RK state equation.

Problem G32

For a Clausius gas, $P(V - Nb) = NRT$, and for a Van der Waals gas $P = NRT/(V - Nb) - N^2a/V^2$. For either gas obtain expressions for $(\partial P/\partial T)_V$, $(\partial P/\partial v)_T$, and $(\partial v/\partial T)_P$. If for a pure substance, $ds = (\partial P/\partial T)_v\, dv + (c_v/T)dT$, show that for both gases c_v is independent of the volume.

Problem G33

The differential entropy change for a gas obeying the molar equation of state $p = RT/v - aT^2/v$ is $ds = (A/T - 2a \ln v)\, dT + (R/v - 2aT/v)\, dv$. Perform the line integration from state (v_1, T_1) to (v_2, T_2) along the paths $(v_1, T_1) \rightarrow (v_2, T_1) \rightarrow (v_2, T_2)$, and $(v_1, T_1) \rightarrow (v_1, T_2) \rightarrow (v_2, T_2)$ and show that (TdS – Pdv) is exact.

Problem G34

In the section of the liquid–vapor equilibrium region well below the critical point $v_\ell \ll v_g$ and the ideal gas law is applicable for the vapour. Derive a simplified Clapeyron equation using these assumptions and show how the mean heat of vaporization can be determined if the vapor pressures of the liquid at two specified adjacent temperatures are known.

Problem G35

For ice and water $c_p = 9.0$ and 1.008cal K^{-1} $mole^{-1}$, respectively, and the heat of fusion is 79.8 cal g^{-1} at 0°C. Determine the entropy change accompanying the spontaneous solidification of supercooled water at –10°C and 1 atm.

Problem G36

For water at 110°C, dP/dT = 36.14 (mm hg) K^{-1} and the orthobaric specific volumes are 1209 (for vapor) and 1.05 (for liquid) cc g^{-1}. Calculate the heat of vaporization of water at this temperature.

Problem G37

The specific heat of water vapor in the temperature range 100°–120°C is 0.479 cal g^{-1} K^{-1}, and for liquid water it is 1.009 cal g^{-1} K^{-1}. The heat vaporization of water is 539 cal g^{-1} at 100°C. Determine an approximate value for h_{fg} at 110°C, and compare this result with that obtained in the previous problem.

Problem G38

Recall that $dg_T = v\,dP$, and plot g_R' (= (g/RT_c)) and P_R with respect to v_R' at 593 K for H_2O and determine the liquid like and vapor like solutions at 113 bar. Determine saturation pressure at T = 593 K for RK fluid. Assume that g = 0 at $v_R' = 200$.

Problem G39

Use the expression $du = c_v dT + (T(\partial P/\partial T)_v - P)\,dv$ to determine c_v for N_2 at 300 K and 1 bar. Integrate the relation along constant pressure from 0 to 300 K at 1 bar, and then from 1 to 100 bar at 300 K in the context of the RK equation. What is the value of u at 300 K and 100 bar if u(0 K, 1 bar) = 0?

Problem G40

Since T = T(S,V,N) is an intensive property, it is a homogeneous function of degree zero. Use the Euler equation and a suitable Maxwell relation to show that $(\partial T/\partial v)_s = -sT/c_v v$, and $(\partial P/\partial s)_v = sT/(c_v v)$. For a substance that follows an isentropic process with constant specific heats, show that $T/v^{(s/cv)}$ = constant

Problem G41

Show that generally real gases deliver a smaller amount of work as compared to an ideal gas during isothermal expansion for a (a) closed system from volume v_1 to v_2 (Hint: use the VW equation ignoring body volume), and (b) an open system from pressure P_1 to P_2 (Hint: use the fugacity charts in the lower pressure range).

Problem G42

Plot the values of $(c_v - c_{vo})$ with respect to volume at the critical temperature using the RK state equation. What is the value at the critical point?

Problem G43

Assume that the Clausius Clapeyron relation for vapor–liquid equilibrium is valid up to the critical point. Show that the Pitzer factor w =0.1861 (h_{fg}/RTc)-1. Determine the Pitzer factor of H_2O if h_{fg} = 2500 kJ kg^{-1}.

Problem G44

An electron gas follows the relation $S = C\,N^{1/6}V^{1/3}U^{1/2}$. Obtain an expression for c_v and show that $c_v = c_v(T,v)$. Also obtain expressions for u(T,v) and h(T,v).

Problem G45

Determine the values of u, h and s at 444 K and 1000 kPa for Freon 22, (Chlorodifluromethane) if $\bar{s}_o$ = 105.05 kJ $kmol^{-1}$ K^{-1}, $\bar{h}_o$ = 32667 kJ $kmol^{-1}$, and M = 86.47 kg $kmol^{-1}$. Use the RK equation.

Problem G46

Upon the application of a force F a solid stretches adiabatically and its volume increases by an amount dV. The state equation for the solid is $P = BT^m(V/V_o - 1)^n$. Show that the solid can be either cooled or heated depending upon the value of m.

Problem G47

Use the Peng-Robinson equation to determine values of $P^{sat}(T)$ for H_2O.

Problem G48

Apply the Clausius Clapeyron equation in case of refrigerant R–134A. Assume that $h_{fg,} = 214.73$ kJ kg^{-1}, at $T_{ref} = 247.2$ K, and $P_{ref} = 1$ bar. Discuss your results, and the impact of varying h_{fg}.

Problem G49

A superheated vapor undergoes isentropic expansion from state (P_1,T_1) to (P_2,T_2) in a turbine. It is important to determine when condensation begins. Assume that vapor behaves as an ideal gas with constant specific heats. Assume that ln P^{sat} (in units of bar) = A – B/T(in units of K) where for water A = 13.09, B = 4879, and $c_{vo} = 1.67$ kJ kg^{-1}.

a) Obtain an expression for the pressure ratio P_1/P_2 that will cause the vapor to condense at P_2.

b) Qualitatively sketch the processes on a P-T diagram.

Problem G50

Determine the chemical potential of CO_2 at 34 bar and 320 K assuming real gas behavior, $h_o = c_{po}(T - 273)$, $s_o = c_{po} \ln (T/273) - R \ln (P/1)$, and $c_{po} = 10.08$ kJ kmol^{-1}.

Problem G51

Does $H_2O(g)$ (for which $P_c = 221$ bar and $T_c = 647$ K) behave as an ideal gas at 373 K and 1.014 bar? Determine the value of v_g.

Problem G52

What is the enthalpy of vaporization h_{fg} of water at 373 K if $P^{sat} = 1.014$ bar? Assume RK equation and $v_f = 0.001$ m^3/ kg.

Problem G53

R134A is stored in a 200 ml adiabatic container at 5 bar and 300 K. It is released over a period of 23 ms during which the mass decreased by 0.32 g. Assume that R134A behaves according to the RK state equation and its ideal gas specific heats are not functions of temperature. Obtain a relation between temperature and volume for the isentropic process in the tank. Using this relation, determine the pressure and temperature in the tank after the R134A release. If the process were isentropic with constant specific heats, what would be the pressure and temperature in the tank after the release of R134A?

Problem G54

Determine the closed system absolute availability ϕ of a fluid that behaves according to the RK equation of state as it is compressed from a large volume v_0 at a specified temperature. Assume that u = 0, s =0, and $\phi = 0$ at the initial condition. Obtain an expression for f(v, T, a, b). (Hint: first obtain expressions for u and s.) Determine ϕ for H_2O at 593 K and a specific volume of 0.1 m^3 kmol^{-1}. Use v_1 at 1 bar and 593 K.

Problem G55

Using the result $(c_p - c_v) = T(\partial v/\partial T)_P(\partial P/\partial T)_v$ show that if Pv = ZRT, then$(c_p - c_v/R) = Z + T_R((\partial Z/\partial T_R)_{v_R} + (\partial Z/\partial T_R)_{P_R}) + (T_R^2/Z)(\partial Z/\partial T_R)_{v_R}(\partial Z/\partial T_R)_{P_R}$. Can you use the

"Z charts" for determining values of $(c_p - c_v)$ for any real gas at specified temperatures and pressures?

Problem G56

It is possible to show that $(c_p - c_v) = v\,T\,\beta_P^2/\beta_T$, and, for VW gases, $c_v = c_{vo}$. For a VW gas show that $(c_p - c_v) = c_p(T,v) - c_{vo}(T) = R/(1 - (2a(v-b)^2)/(RTv^3))$. Determine the value of c_p at 250 bar and 873 K for H_2O if it is known that $c_{vo}(873\ K) = 1.734\ kJ\ kg^{-1}$ K. Compare your results with the steam tables.

Problem G57

If $(b/v)^2 \ll (b/v)$ in context of the state equation $P = RT/(v-b) - a/\,T^n\,v^2$, an approximate explicit solution for $v(P,T,a)$ is $v = \alpha + (-\beta + \alpha^2)^{1/2} = \alpha\,(1\,(1-\beta/\alpha^2)^{1/2})$, $\beta/\alpha^2 < 1$, where $\alpha(T,P) = RT^{n+1}/(2PT^n)$, and $\beta(T,P) = (a-bRT^{n+1})/(PT^n)$. If $h = u_o - a/v + Pv$, obtain an expression for c_p.

Problem G58

Use the RK state equation to plot $(g-g_{ref})/RT_c$ with respect to P_R for values of T_R = 0.1,0.2, ..., 1.0. Also plot P_R^{sat} *vs.* T_R.

Problem G59

Develop a computer program that calculates P_R^{sat} with respect to T_R using the RK equation of state and the criterion that $g_f = g_g$.

Problem G60

Obtain values of $T_{inv,R}$ with respect to v_R', and $T_{inv,R}$ and Z_{inv} with respect to $P_{inv,R}$ using the RK equation of state.

Problem G61

In the context of throttling, cooling occurs only if the temperature T<100 K for H_2 and T<20 K for He. Check this assertion with the expression for μ_{JT} based on VW state equation $\mu_{JT} = -(1/c_p)\,(v - ((RT/(v-b))/(RT/(v-b)^2 - 2a/v^3))$ for both fluids.

Problem G62

A rigid adiabatic container of volume V is divided into two sections A and B. Section A consists of a fluid at the state $(P_{A,0}, T_{A,0})$ while section B contains a vacuum. The partition separating the two sections is suddenly ruptured. Obtain a relation for the change in fluid temperature with respect to volume (dT/dv) after partition is removed in terms of β_P, β_T, P, and c_v. What is the temperature change if the fluid is incompressible? What is the temperature change in case of water if $V_A = 0.99$ V, P = 60 bar, and T = 30°C, $\beta_P = 2.6\times10^{-4}\ K^{-1}$, $\beta_T = 44.8\times10^{-6}\ bar^{-1}$, $v_A = 0.00101\ m^3\ kg^{-1}$, and $c_p = c_v = c = 4.178\ kJ\ kg^{-1}$ K?

Problem G63

Trouton's empirical rule suggests that $\Delta s_{fg} \approx 88\ kJ\ kmol^{-1}\ K^{-1}$ at 1 bar for many liquids liquids (another form is $h_{fg} = 9\ RT_{NB}$). Obtain a general expression from the Clausius Clapeyron equation for the variation of saturation temperature with pressure.

Problem G64

Using the state equation $P = RT/(v-b) - a/(T^n v^m)$ and the equality $g_f = g_g$, show that $P^{sat} = (1/(v_g - v_f))\,(RT\ \ln((v_g-b)/(v_f-b)) + (a/(m-1)T^n)\,(1/v_g^{(m-1)} - 1/v_f^{(m-1)}))$. Simplify the result for the VW and Berthelot equations of state.

Problem G65

Show that the inversion temperature for an RK gas $T_{R,inv}^{3/2}$ = (4.9342(1 – 008664/ v_R')^2/(1+ 0.08664/ v_R')) (1.5 + 1/(1+0.08664/ v_R')).

Problem G66

Using the relations ds = c_v dT/T + ∂P/∂T dv and ds = c_p dT/T – ∂v/∂T dP, show that (c_P – c_v) = T v $\beta_P{}^2/\beta_T$.

Problem G67

The Helmholtz function A for a Debye solid A = –N R π^4 $T^4/(5\ \theta_D{}^3)$, where θ_D is Debye temperature and N denotes the number of moles. Obtain expressions for u, s, and c_v, in terms of the temperature.

Problem G68

Determine the relation between the temperature and volume during an isentropic process for a VW gas. If at the initial state 1, T_1 = 200 K, v_1 = 0.006 m^3 kg^{-1}, and if v_1/v_2 = 3, determine the final state 2 (P_2, T_2) if the gas is air.

Problem G69

In the context of the relation s = s(u,T) show that P/T is only a function of volume as v→ ∞ for any simple compressible substance.

Problem G70

About 0.1 kmol of liquid methanol at 50°C in system A is separated by a thin foil in thermal and mechanical equilibrium from dry N_2 occupying 1 % of liquid volume at 2 bar and 50°C in system B. The foil is removed and the liquid temperature falls. Heat must be consequently added to maintain the state at 50°C and 2 bar in both subsystems. Determine the partial pressure of vapor at which the vaporization stops. Assume that $\bar{h}_{fg}$ = 37920 kJ $kmol^{-1}$. If $\mu_{methanol}(\ell)$ = $g_{methanol}$ = h(ℓ) – T s(ℓ), $\mu_{methanol}$(g) = $g_{methanol}$(g) = h(g) – T s(T,$p_{methanol}$), and $p_{methanol}$ = $X_{methanol}$ P. Neglect the volume change in the liquid methanol. Determine G = G_A + G_B with respect to $p_{methanol}$.

Problem G71

Show that the chemical potential of a pure VW gas is μ(T,v) = μ(T,P) = μ^o(T) + RTv/(v–b) – 2a/v – RT – RT ln (pv/RT) + RT ln (v/(v–b)).

Problem G72

Apply the Martin–Hou state equation P = RT/(v–b) + $\Sigma_{i=2,5}$ $F_i(T)/(v-b)^i$ + $F_6(T)/e^{(B\ v)}$, for which b and B are constants to obtain expressions for a(T,v)–a_0(T,v), s(T,v) –s_0(T,v), and u(T,v)–u_0(T). Let dF(T)/dT = F′(T). What are the expressions for the case if $F_i = A_i + B_i\ T + C_i\ e^{-KT/T_c}$? (ASHRAE tabulates these constants for various refrigerants.)

Problem G73

Determine the temperature after C_3H_8 is throttled from 20 bar and 400 K 1 bar with $\bar{c}_{p.o}$ = 94.074 kJ $kmol^{-1}$ K^{-1} . Use a) RK equation and b) Kessler charts for h^R/RT_c.

Problem G74

Consider du = c_v dT + (T(∂P/∂T)$_v$ – P) dv. Obtain a relation for u_0-u and c_{v0}-c_v in terms of a,b, n,T and v for generalized RK equation of state P = RT/(v-b)- a/(T^n v (v+b)).

Problem G75

Using RK equation of state and appropriate reference conditions determine the following for steam at 180 bar and 400°C and compare the values with steam tables for v, h,u and s and fugacity charts for f/P: v, h, u, s, c_v, c_p-c_v, c_p, f/P and μ_{JT} in K /bar.

Problem G76

The following expression for the Helmholtz function has been used to determine the properties of water

$$a(\rho,T) = a_0(T) + RT[1n\rho + \rho Q(\rho,\tau)],$$

where ρ denotes density, T denotes temperature on the Kelvin scale, τ denotes 1000/T. The functions a_0 and Q are sums involving the indicated independent variables and a number of adjustable constants, i.e.,

$$a_0 = \sum_{i-1}^{6} C_i/\tau^{i-1} + C_7 \ln T \text{ €}_8 \ln T/\tau, \text{ and}$$

$$Q = (\tau - \tau_c) \sum_{j}^{7} (\tau - \tau_{aj})^{j-2} \left[\sum_{i-1}^{8} A_{ij}(\rho - \rho_{aj})^{i-1} + e^{-E\rho} \sum_{i-9}^{10} A_{ij}\rho^{i-9} \right].$$

Here, R = 4.6151 bar cm^3/g K or 0.46151 J/g K, τ_c/1000/T_c = 1.544912, E = 4.8, and $\tau_{aj} = \tau_c$ if j=1, τ_{aj} = 2.5 if j>i, ρ_{aj} = 0.634 if j=1, ρ_{aj} =1.0 if j>i.

The coefficients for a_0 in joules per gram are given as follows;

C_1 = 1857.065 C_4 = 36.6649 C_7 = 46.

C_2 = 3229.12 C_5 = -20.5516 C_8 = -1011.249

C_3 = -419.465 C_6 = 4.85233

Values for the coefficients A_{aj} are listed in the original source. Obtain expressions for (a) pressure, (b) specific entropy, (c) specific internal energy and specific enthalpy resulting from this fundamental function.

See also J.H. Keenan, F.G. Keyes, P.G. Hill, and J.G. Moore, *Steam Tables*, Wiley, New York, 1969; L. Haar, J.S. Gallagher, and G.S. Kell, *NBS/NRC Steam Tables*, Hemisphere, Washington, D.C., 1984. The properties of water are determined in this reference using a different functional form for the Helmholtz function than given by Eqs. (1)-(3).

Problem G77

Ammonia is throttled from P_1=169 bar and T_1= 214 C to a very low pressure P_2 (<< critical pressure). Determine

a) T_2 in C and

b) Change in internal energy u_2 - u_1 in kJ/kg

Use Kessler tables and ignore Pitzer factor. The ideal gas specific heat can be assumed to be a constant and equal to c_{p0} = 2.130 kJ/kg K, M= 17.03 kg/kmol.

H. CHAPTER 8 PROBLEMS

Problem H1

Helmholtz function A is generally a function of A = A (T, V, N_1 .... N_n) ; a) Write down the Euler equation for A. Then obtain $\bar{a}$, b) Find the differential $d\bar{a}$, c) Write down the Gibbs-Duhem equation for A. Express it on a unit kmol basis, d) Use (c) in (b) to obtain simplified expression for $d\bar{a}$, e) What is $(d\bar{a}/dx_2)$ at constant $\bar{v}$, x_3, $x_4 \ldots x_K$.

Problem H2

One wishes to prepare a mix of 60% acetylene and 40% CO_2 (mole basis) at a pressure of 100 bar and at a temperature of 47°C. Your boss asks you to determine the number of kmol of acetylene and CO_2 required to form the mixture. Assume tank volume to be 1 m^3. Determine the kmol using the following method: a) Ideal gas law, b) Kay's rule and compressibility charts, c) Law of additive pressures. and RK equation for pure component, d) Law of additive volumes and RK equation for pure component, e) Empirical equation for $\bar{a}_m$ and $\bar{b}_m$ and RK equation for the mixtures By looking at the answers you must report to your boss regarding the expected minimum and maximum requirements.

Problem H3

Consider a mixture of methane species (1) and propane species (2) (40: 60, Kmol basis). Assume that Kay's rule is applicable for the mixture and the mixture follows VW eq. of state. Assume P = 30 bar and T = 300 K. a) Are $\bar{a}$ and $\bar{b}$ functions of mole fractions?, b) Determine $\bar{v}$ for the mixture (m^3/kmol), c) Determine the partial molal volume of species (1) ($\hat{v}_1$), d) Determine the molal volumes in the pure state for species (1) and (2). e) Comment on the molal volumes of species (1) in the pure states ($\bar{v}_1$), f) If you assume a hypothetical ideal gas state for species (2) ($\bar{v}_2$) in the pure state, what is the ideal gas volume of species (2)?, g) Determine the ideal molal volumes of mixture ($\bar{v}_{id}$) based on answers for i) part (d), and ii) part (e).

Problem H4

Consider the VW equation: P = RT/(v-b) - a/v^2. Neglect body volume "b". Solve for v. Suppose this equation is valid for two component mixtures (say H2O vapor- species 1 and air-species 2) at T = 300 K, P = 200 bar. a) Plot $\hat{v}_1$, $\hat{v}_2$ *vs.* x_1 using Kay's rule and a spreadsheet program. Compare the solution for ($\bar{v}$) with ideal solution model following LR rule and HL.

Problem H5

Consider the equation of state for a mixture: P V = N Z $\bar{R}$ T where N = N_1 + N_2 + ...N_K. Test whether Z (T, P, N_1, N_2...) is an extensive property ? Hint: Use the definition of partial molal property b_1 = $(\partial B/\partial T)_{T,P,N2,..}$, and show that $N_1\ \partial Z/\partial N_1 + N_2\ \partial Z/\partial N_2 + ...$ = 0 using the Euler equation.

Problem H6

Consider the approximate virial equation of state valid at low to moderate pressures: Z = 1 + BP/RT. This equation can be used for mixtures with n components

$B = \Sigma\Sigma Y_i Y_j\ B_{ij}$, i=1,..n, j=1... n,

B_{ii},B_{jj}: virial coefficient of pure species i,

$B_{ij} = (R\ T_{c,ij}\ /P_{c,ij}) = (B^o + w_{ij}\ B^1)$

$w_{ij} = (w_i + w_j)/2$

$T_{c,ij} = (T_{c,i}\ T_{c,j})\ 1/2\ (1 - k_{ij})$

(kij = 0 when i =j, kij >0 when i is not equal to j; assume kij =0)

$P_{c,ij} = Z_{c,ij}\ R\ T_{c,ij}/\ v_{c,ij}$

$Z_{c,ij} = (Z_{c,i} + Z_{c,j})/2$

$v_{c,ij} = ((v_{c,i}^{\ 1/3} + v_{c,j}^{\ 1/3})\ /2)^3$

$B^o = 0.083 - 0.422/Tr^{\ 1.6}$

$B^1 = 0.139 - 0.172/Tr^{\ 4.2}$

Obtain an expression for partial molal volume of species 1.

Problem H7

Consider a 60:40 NH_3-H_2O mixture at 10 bar, 400 K. a) Obtain the partial molal volume of H_2O at 10 bar and 400 K. Use the VW relation and Kay's rule. Since $\ln(\phi) = \int_0^P (Z-1)\, dP/P$, treating ln ($\phi$) as an intensive property and (N ln(ϕ)) as an extensive property, $\ln(\hat{\phi}_1) = (\partial/\partial N_1)[N \ln\phi)]$. Show that for any real gas, ln $\hat{\phi}_1$ = ln $\phi + \int_0^P (\hat{Z}_1 - Z)\, dP/P$, where $\hat{\phi}_1$ is the partial molal fugacity coefficient of species 1.

Problem H8

Determine u,h and f of H_2O(P) at T = 90 C and P =100 kPa., b) Determine u,h and f of H_2O(P) at T = 90 C and P = 50 kPa. Assume that u^{sat} (90 C), v^{sat} (90 C) are available.

Problem H9

Determine the chemical potential of CO_2 at P = 34 bar, 320 K. Assume real gas behavior. For ideal enthalpy use $h_0 = c_{p0}$ (T- 273), s_0 c_{p0} ln (T/273) - R ln (P/1), c_{p0} = 10.08 kJ/ k mole. Use a) charts, b) RK equation.

Problem H10

Using the relations for s_{fg} for RK equation of state (Chapter 07) for pure component, obtain the relations for a) $\hat{s}_{fg,1}$ and b) $\hat{h}_{fg,1}$ using RK mixing rule. Note that $\hat{h}_{fg,1}$, enthalpy of vaporization when component 1 is inside the mixture.

Problem H11

Obtain the relations for a) $\hat{u}_k - \bar{u}_{k,0}$, and b) $\hat{h}_k - \hat{h}_{k,0}$ for a gas mixture following Berthelot equation and Kay's rule for critical constants.

Problem H12

A mixture of of 60% acetylene and 40% CO_2 (mole basis) is compressed isothermally from 1 bar, 47°C to a pressure of 100 bar. Determine the amount of work in kJ/kmol of mixture if a) a closed system is used, b) an open system is used assuming ideal gas law and Kay's rule and RK equation of state.

Problem H13

A piston-cylinder assembly with a weight at the top consists of a wet H_2O mixture of 20% quality at 1.5 bar. The initial total volume is 20 L. The whole system is immersed in a bath maintained at 111.4°C. Through a hole in the vapor phase section of cylinder we inject inert gas say N_2 until mole fraction of N_2 in vapor phase is 25%. The N_2 does not dissolve in liquid. Do you believe there will be more liquid or more vapor, or will the mixture remain as before? You can use either arguments or calculations.

Problem H14

Ammonia is manufactured using hydrogen and nitrogen. A mixture having a molar ratio of H_2 to N_2 equal to 3 is compressed to 400 atm and heated to 573 K. Determine the specific volume at this condition using the following methods for RK mixture: a) Ideal gas. b). Law of additive pressures and generalized Z charts. c). Law of Additive Volumes and generalized Z charts. d). Kay's rule.

Problem H15

Obtain an expression for partial molal volume of component 1 in a mixture following RK equation of state and the mixing rule $\bar{a}_m = (\Sigma_k X_k\, \bar{a}_k^{1/2})^2$, $\bar{b}_m = \Sigma_k X_k\, \bar{b}_k$.

Problem H16

A real gaseous mixture of acetylene (species 2) and CO_2 (species 1) is considered. The mole fraction of (1) is x_1. Assume that Kay's rule applies for the critical pressure and temperature of the mixture. The Redlich-Kwong equation of state (EOS) for the mixture is

$\bar{a} = 0.4275\, \bar{R}^2 T_c^{2.5}/P_c$, and $\bar{b} = 0.08664\, \bar{R} T_c/P_c$, where

T_{cm} denotes the critical temperature and P_{cm} the critical pressure of the mixture.

a) Obtain an expression for $\hat{v}_2$ and then reduce the expression for $\hat{v}_2$ when x_2 goes to a very small value (say 0.01).
b) Determine $\hat{v}_2$ when x_2 is small (say, 0.01) at 320 K and 100 bar.
c) If $x_1 = 0.6$, what is the value of $\hat{v}_2$ at T = 320 K and P = 100 bar? Compare with the answer from part b.

Problem H17

Obtain the relations for $\hat{u}_k - \bar{u}_{k,0}$, $\hat{s}_{10}(T,P) - \hat{s}_1(T,P)$ and $\hat{h}_k - \hat{h}_{k,0}$ for a gas mixture following RK equation and RK mixing rule $\bar{a}_m = (\Sigma\, Y_k\, \bar{a}_k^{\,1/2})^2$, $\bar{b}_m = \Sigma\, Y_k\, \bar{b}_k$.

Problem H18

A gas mixture containing CO_2 and acetylene exists at 100 bar and 0.0938 K. If the mass fraction of CO_2 is 0.4, determine the temperature. Use LAV, LAP and RK.

I. CHAPTER 9 PROBLEMS

Problem I1

Consider a mixture of $O_2(1)$ and $N_2(2)$ at low temperatures in the form of a liquid mixture. You are asked to draw the T (K) *vs.* X and T *vs.* $X_{k,l}$ diagrams. Assume the following vapor pressure relations: ln (P^{sat} bar) = A - B/(T in K +C) where A, B and C are as follows: for O_2: 8.273075661, 666.0593179, -9.69072568, respectively, and for N_2: 6.394732229, 369.1680573, and -19.61997409, respectively. Use a spreadsheet program. Determine (a) $X_{1,e}$ and X_l for the equilibrium phases at 100 K and 100 kPa. b) T and X_l at 100 kPa and $X_{1,e} = 0.4$. (c) P and X_l for T = - 170 C and $X_{1,e} = 0.4$. d) T and $X_{1,e}$ for 100 kPa and $X_l = 0.4$. e) P and $X_{1,e}$ for –160°C and $X_l = 0.4$. f) The fraction of the system that is liquid, $X_{1,e}$, and X_l at –160 °C and 100 kPa, when the overall composition of the system is 21 mole percent of oxygen.

Problem I2

Consider water in the atmosphere. Normally air is dissolved in liquid water. The normal boiling point of water is 100°C. Plot the mole fraction of X_{N2} and X_{O2} (given that $X_{N2}/X_{O2} = 3.76$) *vs.* T. Assume that the mole fraction of H_2O in liquid is close to unity so that mole fraction of water vapor in the gas mixture could be immediately determined. The value of p_{O2} = 1 mm at –219°C, 10 mm at -210.6, 40 mm at -204.1°C, -198.8°C at 100mm, -188.8 °Cat 400 mm and -182.96°C at 760 mm; p_{N2} = 1 mm at -226.1°C, 10 mm at -219.1°C, 40 mm at -214.0°C, -209.7°C at 100mm, -200.9°C at 400 mm and -195.8°C at 760 mm. (First evaluate the constants A, B and C for the Cox Antoine relation for N_2 and O_2 and then use spreadsheet.)

Problem I3

During the evaporative desalinization of sea water, salt water is first heated to its boiling point and then partially vaporized. The vapor, which is essentially pure water, is then condensed and collected to obtain water. If the entering sea water is initially 1.5 mol percent NaCl, determine the boiling temperature of the solution at 1 atm. ΔH_v = 970 Btu/lb_m and may be assumed constant. The vapor may be assumed to be ideal and to have negligible density with respect to the liquid. List any assumptions you make.

Problem I4

The addition of glycol to water can lower the freezing point and, hence, more thermal energy can be stored in the mixture. Further the enthalpy of melting (h_{sm}) increases with a lowered freezing point. Plot a) the enthalpy of melting *vs.* glycol%, b) the freezing point *vs.* glycol concentration (upto 50% by weight) if the following data is available: h_{sm} = 334 kJ/kg at 273 K, $c_{p,H2O(P)}$ =4.184 kJ/kg K, $c_{p,H2O(s)}$ = 2 kJ/kg K. Assume the ideal solution model.

Problem I5

Find the partial pressure of benzene vapor for a mixture containing 30% benzene (species 1) and 70% toluene (species 2) solution at 92°C, if the saturation vapor pressures of benzene and toluene at 92°C are 1078 torr and 432 torr.

Problem I6

Consider a mixture of water (species 1) and ammonia (species 2). The vapor pressure relations are given as follows: ln P (bar) = 12.867 - 3063 /T (K) for ammonia; ln P (bar) = 13.967 - 5205.2/T (K) for water. Plot P *vs.* $X_{1,e}$, X_1 at 0°C, 25°C, 50°C and T *vs.* $X_{1,e}$ and X_1 at 0.5, 1, 10 bar.

Problem I7

The following is the composition of an acid which is vaporized and burnt in a hazardous waste plant: H_2SO_4: 92% by mass, Hydrocarbons: 4%, H_2O: 4%. Lump hydrocarbons with water. The vapor pressure relations are as follows: ln (p) = A - B/(T(K)+C), with p expressed in units of bar. The values of A, B and C are as follows: water: 11.9559, 3984.849, -39.4856, respectively; H_2SO_4: 8.346772, 4240.275, -119.155, respectively. Determine the vapor phase mole fraction of each component at a pressure of 1 bar at 100°C. Assume an ideal solution. Will the vapor phase composition change if N_2 is present in the vapor phase at 1 bar and 100°C? If so, determine the value of this change. If p_{H2O} and p_{H2SO4} at 270°C for a strong acid are 0.335 bar and 0.0525 bar, respectively, determine the activity coefficients for the two species.

Problem I8

Justify if the ideal solution model is valid for a H_2SO_4 and water liquid mixture.

Problem I9

Phase equilibrium is reached when the Gibbs energy has a minimum value at specified values of temperature, pressure, and mass. One kg of water at 343 K water is poured into a cylinder of piston-cylinder-weight assembly. The space above water initially contains 0.4 kg of dry air at 343 K and 100 kPa (c_{pw} = 4.184 kJ kg^{-1}). As water vaporizes the temperature of water drops and heat is added to maintain a constant water temperature. Consider values of p_{vapor} in the range 0–0.9 bar in increments of 0.01 bar and determine G(p_{vapor}). Assume that water vapor and air behave as ideal gases.

Problem I10

a) Obtain an expression for vapor pressure in air and vapor mixture just above the liquid surface of a lake which is at T. Assume that liquid is pure distilled water and pressure is P bar.

b) Derive the expression for mole fraction of vapor in the gas phase if gas phase is assumed to be an ideal gas mixture.

c) Determine p_v and Y_v at 30°C and 0.9 bar.

Problem I11

A methanol (component 1, 60% mole basis) and water (2, 40%) mixture at 60°C exists at a pressure P. Assume an ideal solution with p_1^{sat} = 625 mm of Hg and p_2^{sat} = 144 mm of Hg and determine the values of P, Y_1, and Y_2 at equilibrium.

Problem I12

Seven gmole of methanol (species 1 in both the liquid and vapor phases) and 3 gmole of water (species 2 as liquid and vapor) coexist in a piston cylinder assembly at 60°C, and 433 kPa. With the values p_1^{sat} = 625 mm of Hg, p_2^{sat} = 144 mm of Hg, determine x_1, x_2, Y_1, Y_2, the vapor fraction or quality $\bar{w}$, and the moles of vapor of species 1 and 2.

Problem I13

Plot P *vs.* x_{H2O} and P *vs.* Y_{H2O} at T = 65 C for methanol and water solution. Assume ideal solution behavior.

J. CHAPTER 10 PROBLEMS

Problem J1

If you strike a match in a class room, it may not result in explosion; but if you strike (never try this) a match near a gas station, there may be an explosion. Relate this to stability.

Problem J2

If there can be phase change (i.e., the formation of two regions with two different densities) at given T and P, why cannot there be different thermal layers at specified values of ρ and P?

Problem J3

Derive an expression for the spinodal condition for a fluid following the Peng Robinson equation of state. Obtain the spinodal curve for both liquid and vapor n–hexane and plot P(V), P(T), T(v) at specified T, v and P respectively.

Problem J4

Plot P(v) for H_2O at 373 K using the RK state equation for volumes in the range 0.0867 $\bar{v}_c$' to 200 $\bar{v}_c$', $\bar{v}_c$'= $\bar{R}T_c/P_c$. If g (373K, 200 $\bar{v}_c$') = 0 and dg = vdP, plot g(v) at 373 K. (Hint: use integration by parts.) Determine the states at which $g_{liquid} = g_{vapor}$, and at which dP/dv = 0.

Problem J5

Consider the relation G–A = PV. Since $d^2G < 0$ at constant values of T and P for a single component fluid, show that $d^2A < -P\, d^2V$ at these values of T and P.

Problem J6

Refrigerant R–134A enters a short 1 mm diameter and 12 mm long capillary tube at 6 bar and with 30°C subcooling (i.e., it is 30°C below its saturation temperature at 6 bar). The fluid is discharged into a 1 bar atmosphere. The pressure first decreases rapidly due to vena contracta (expansion in a converging) and then rises due to the increased diameter of the flow (as in a diffuser). At a certain pressure R–12 starts flashing (or vaporizing). Since its density is low, the flow velocity rapidly increases with a subsequent decrease in pressure and may reach the sound speed (or choked flow condition). Data is needed on spinodal conditions. Determine the spinodal pressures at 30°C, 20°C, and 10°C subcooled conditions. Use Dietrici equation of state. How does this compare with values determined from the charts for the RK equation.

Problem J7

Obtain an expression for vapor spinodal curve for both P and T with respect to v assuming that b«v. Use the Berthelot equation.

Problem J8

Prove that if S = S (U, V, N_1, N_2, ..., N_n) at equilibrium, and if dS < 0 due to a perturbation, then for the perturbed state $dH_{T,S} > 0$, $dA_{T,V} > 0$, and $dG_{T,P} > 0$.

Problem J9

If u = u(s,v), show that the stability criteria for u_{min} are $c_v>0$ and $(\partial P/\partial v)_s<0$.

Problem J10

Starting with $S-S^e = ((U-U^e)/T) + (P/T)(V-V^e) + C$ and $C < 0$, show that (a) at specified values of S, V and m, (b) $U > U^e$, at specified values of T, V, and m, $A>A^e$, and c) at specified values of T, P and m, $G > G^e$.

Problem J11

Obtain the spinodal condition for the state equation $P = RT/(v-b) - a/(T^n v^2)$.

Problem J12

Obtain the spinodal curve for a RK gas in terms of P_R *vs.* T_R and P_R *vs.* v'_R.

Problem J13

Determine the maximum temperatures to which liquid water can be superheated and vapor can be subcooled at 133 bar.

Problem J14

Show that $d^2S = \partial^2S/\partial U^2\, dU^2 + \partial^2S/\partial V^2\, dV^2 + 2\, \partial^2S/\partial U\partial V\, dU\, dV = -(C_v/T^2)\, dT^2 + (\partial P/\partial V)_T(1/T)\, dV^2$.

Problem J15

For a VW gas $s = c_{vo} \ln((u+a/v)/c_{vo}) + R \ln(v-b)$. Using the criterion $\partial^2s/\partial v^2 < 0$, obtain the following expression for values of v which satisfy above criteria at specified u, namely, $(k-1)(1+x)^2/(1-b^*x)^2 > x(2+x)$, where $x = a/(vu)$, $b^* = (bu)/a$, and $k = c_{po}/c_{vo}$.

Problem J16

Using the Berthelot equation of state $P = RT/(v-b) - a/(Tv^2)$ for water plot T(v) for P = 1, 10, 20, 40, 60, 80, 100, 200 bar. At P = 60 bar determine the maximum temperature to which water can be superheated without forming vapor and the minimum temperature to which water can be cooled without causing condensation. Assume that $\ln P_R^{,sat} \approx 7(1-1/T_R)$. Plot the saturated liquid and vapor curves and determine the degree of superheat and subcooling at 60 bar.

Problem J17

Consider n-hexane (C_6H_{14}) at P=4 bar, v= 0.3m3/kmol following Van der Waals equation of state. Verify whether n-hexane is mechanically stable at this condition.

Problem J18

For the state equation $P = RT/(v-b) - a/(T^n v^m)$ for what values of m is it possible to obtain the spinodal conditions for $T<T_c$? Can we relate this to the universe, treating galaxies as point masses?

Problem J19

In the context of the Peng Robinson state equation solve for T(P,v). Plot T with respect to P at 1 bar and 60 bar. Determine the degree of superheat and subcooling.

Problem J20

Use the VW equation of state $P = RT/(v-b) - a/v^2$ for H_2O. Plot P(v) at 593 K. Recall the expression $s = c_{vo} \ln((u+a/v)/c_{vo}) + R \ln(v-b)$, where $T = (u+a/v)/c_{vo}$. Plot s(v, 593 K), assuming that $c_{vo} = 54$ kJ kmol^{-1} K^{-1}. Determine d^2s along the 593 K isotherm. Discuss your results in the context of stability criteria.

Problem J21

Saturated liquid water (the mother phase) is kept in a piston cylinder assembly at a pressure of 100 kPa. A minute amount of heat is added to form a single vapor bubble (the embryo phase). a) If the embryo phase is assumed to be at the same temperature and pressure as the mother phase, determine the absolute stream availabilities $\psi = h - T_0\ s$ and Gibbs functions of the mother and embryo phases. b) If the embryo vapor phase is at the spinodal pressure corresponding to 100°C while the liquid mother phase is still at 1 bar, what are the absolute stream availability and Gibbs function of the vapor embryo? Compare the answers from parts (a) and (b). For the spinodal pressure assume that the RK equation applies. (In order to use the values from saturation tables assume that the vapor phase behaves as an ideal gas to calculate the enthalpies and entropies between the saturated and spinodal states.) c) If the embryo at the spinodal pressure condenses back to the mother liquid phase at 100°C and 1 bar what is the change in the Gibbs function?

Problem J22

Obtain the stability criteria for an ideal gas using the criteria related to h_{PP}, h_{ss}, and h_{sP}. Apply the relations $dh = c_{po}\ dT$ and $s = c_{po}\ dT/T - R\ dP/P$.

Problem J23

Show that $u_{ss}\ u_{vv} - u_{sv}^{\ 2} = -a_{vv}/a_{TT}$.

K. CHAPTER 11 PROBLEMS

Problem K1

Is the reaction expression written on a molecular basis the same as the mole basis?

Problem K2

Under what condition is the A: F ratio based on a mole basis the same as the volume basis?

Problem K3

Consider propane gas C_3H_8, which can be empirically written as $CH_{2.6667}$. The stoichiometric air: fuel ratio per kmol of C_3H_8 is exactly equal to the stoichiometric air per kmol of $CH_{2.6667}$. True or False?

Problem K4

For an adiabatic reaction, one can reasonably assume that the entropy of products leaving is the same as entropy of the reactants. True or False?

Problem K5

For an adiabatic reaction involving PCW, one can reasonably assume that the enthalpy of products is the same as the enthalpy of the reactants. True or False?

Problem K6

For an adiabatic reaction within a closed system, the entropy increases at fixed values U,V, m. True or False?

Problem K7

In an adiabatic reaction within a PCW, the entropy increases at fixed H, P, m. True or False?

Problem K8

During a CO_2 sequestration process, the reaction $CaO(s) + CO2\ (g) \rightarrow CaCO3(s)$ occurs. How much heat in kJ/kmol of CO_2 is required for the reaction at 298 K?

Problem K9

Currently at a 370 ppm level of CO_2 in the ambient, about 7 billion metric tons of carbon is emitted every year. which is expected to rise to 1 gigatons by 2015 and 4 gigatons by 2025. If the dominant fuel used is coal $CH_{0.7589}O_{0.1816}N_{0.0128}S_{0.00267}$, how much fuel can be burned each year to reach these levels?

Problem K10

About 90% of the CO_2 emitted dissolves into the oceans and forms a methane hydrate sediment through the reaction $CO_2 + bH_2O \rightarrow CH_4$: $7H_2O(s) + CO_2$. How much CO_2 is captured by 1 kmol of H_2O?

Problem K11

A glass jar that contains N moles of dry air is placed in a pool of water. A combustible solid, CH_mO_n is ignited and dropped into the jar. A lid is snugly fitted at the neck of the jar. Sometimes the lid is seen to fall into the jar. (a) Determine the appropriate conditions for this to occur. Discuss the problem for (i) m = 1, n = 0, (ii) m = 0, n = 0 (pure carbon). (b) What are the results if CO is formed instead of CO_2? Will the lid be expelled away from the jar?

Problem K12

At 25°C and 1 bar is the reaction $4Fe(s) + 3O_2 \rightarrow 2Fe_2O_3(s)$ exothermic?

Problem K13

The dry ash free cattle waste can be represented as $CH_{1.253}N_{0.0745}O_{0.516}S_{0.00813}$. The heating value of dry waste with 53% ash is known to be as 9215 kJ/kg . Determine a) stoichiometric air fuel ratio for a dry ash free fuel, b) enthalpy of formation in kJ/kmol for the empirical fuel, and c) adiabatic flame temperature in K for a dry fuel with 53% ash. Assume that c_p of ash = 0.800 kJ/kg K.

Problem K14

In a HiTAC (high temperature air combustion) process the air is mixed with flue gases in order to reduce O_2 concentration to 2-5%. If methane is used with a stoichiometric ratio of 2 with 2% oxygen concentration in the oxidant stream (air + flue gas mixture), and air temperatures are a) 298 K, b) 1000 K, then determine the adiabatic flame temperatures. Assume constant specific heats for all species.

Problem K15

As opposed to burning glucose ($\bar{s}_f$ = 212 kJ/kmol), the body burns a mixture of fat (palmitic acid, $C_{16}H_{32}O_2$, $\bar{h}_f$= - 834694.4 kJ/kmol, $\bar{s}_f$ = 452.37 kJ/kmol) and glucose. Let the heat loss rate be specified at 110 W for 70 kg person and breathing rate at 0.1 liter per second. Determine the entropy generation per kmol of the mixture *vs.* the fraction of glucose in the fuel metabolized and entropy generation per unit amount metabolism. Comment on the results. What happens to the results if the fat is replaced by cholesterol $C_{27}H_{45}OH$? Assume that cholesterol has the same properties as fat.

Problem K16

The human body is an open system and some arbitrary person, on average, loses body heat at the rate of 110 W. Assume that person's body temperature remains constant at 37°C, the ambient temperature is 25°C, the specific heat of air is 1 kJ kg^{-1} K^{-1}, the inhalation (and exhalation) mass flow rates are both 6 g min^{-1} and properties of exhaled gas are the same as that of air. Determine the entropy generation rate:

a) If the control volume is assumed just inside the human body.
b) If the control volume is assumed just outside of the human body. Explain the difference between answers in (a) or (b)

Problem K17

Normally for closed system: $dS = dH/T - VdP/T$ or $dU/T + PdV/T$. Consider a closed system which is suddenly loaded with 1 K mole of CO and 1 K mole of O_2 at 3000 K. a) What are the entropy and enthalpy at this (meta) state (1). b) If you leave the system for a sufficiently long period of time and at the same time maintain the pressure at 1 atm and temperature at 3000 K, you find that there are 0.34 CO, 0.66 CO_2 and 0.67 oxygen moles (state 2). What is the entropy and enthalpy at state (2). c) What is the entropy change (S_2 - S_1)? d) What is the enthalpy change? e) Is the entropy change equal to $\Delta H/T$ where $\Delta H = H_2 - H_1$? Comment.

Problem K18

Consider the growth of leaves on a tree. Consider a single leaf as it is growing. The gaseous CO_2 and liquid water are used to produce a solid leaf which is assumed to be cellulose $C_6 H_{10} O_5$. a) Develop an overall reaction scheme. The sunlight is used as an energy source for such a reaction. b) Write down the mass, energy and entropy balance equations. Assume that reactions occur at 25°C, 1 bar ? Determine a) sunlight required in kJ/kg of cellulose, b) entropy change for the reaction in kJ/kg K, c) repeat parts (a) and (b) if the solid is lignin ($C_{40} H_{44} O_6$), and d) If wood consists of 40-45% cellulose, 15-30% lignin and the rest is hemi-cellulose, how will you determine the answers for (a) and (b)?

Problem K19

The body burns a mixture of glucose ($C_6H_{12}O_6$, h_{f0} = -1260268 kJ/kmol, s(298,s)=212 kJ/kmol K, HHV 2815832 kJ/kmol,10034905.6 kJ/kmol and fat ($C_{16}H_{32}O_2$, h_f= -834694.4, s(298,s) =452 kJ/kmol K, HHV = 10034905.6 kJ/kmol K. If inhaled air temperature is 25°C, and exhaled air temperature is 37°C . Plot entropy generation in kJ per kmol of mixture K and in kJ per kJ of heat released per K *vs.* glucose fraction in the mixture. Assume 400% excess air.

Problem K20

Natural gas has the following composition based on molal%: CH_4: 91.27, Ethane 3.78, N_2 = 2.81, Propane 0.74, CO_2: 0.68, n-Butane: 0.15, i-Butane 0.1, He 0.08, i pentane 0.05, n-pentane 0.04, H2: 0.02, C-6 and heavier (assume the species to be of mole wt: 72): 0.26, Ar: 0.02. Determine a) the molecular weight, b) gross heating value in BTU/SCF, kJ/m^3, c) LHV.

Problem K21

a) In a constant volume combustion chamber one kmol of CH_4 and 3 kmol of O_2 are burned at 298 K and 1 bar. Heat Q_v is removed so that the products are at 298 K a) What is the final pressure? Assume that H_2O does not condense. b) If the same reaction involving the same molar content occurs in a sssf reactor at 298 K and 1 bar and the products leave at 298 K and 1 bar, the heat removed is Q_p. c) Determine the difference (if any) between Q_p and Q_v. d) If H_2O partially condenses, what is the value of Q_P for case (b), e) If water partially condenses, what is the value of Q_V and the final pressure?

Problem K22

CH_4 was supplied to a reactor along with air. The dry gas analysis yields the following composition CO_2: 4%, O_2: 7%. a) Determine the CO content in the products, b) A: F, c) equivalence ratio, and d) air required in m^3/hr for combustion of 15 m^3/hr of fuel at STP.

Problem K23

Gaseous CO_2 and liquid water are used to produce a hydrocarbon during photosynthesis that leads to leaf growth. Sunlight is used as the energy source for the reaction.

Describe the mass, energy and entropy balance equations for this process. It is argued that the leaf is formed by groups of organized molecules while CO_2 is disorganized and as such order increases and hence the entropy may decrease. Is this a violation of Second Law?

Problem K24

Octane C_8H_{18} is burned with dry air at P = 14.7 psia. a) Calculate stoichiometric A: F ratio. If volumetric analyses of dry products are CO_2: 7%, O_2: 10.90%, N_2: 82.10%, then determine b) equivalence ratio for actual combustion and c) dew point temperature of H_2O in the products.

L. CHAPTER 12 PROBLEMS

Problem L1

Can there be plant which can make C_3H_8 by the reaction 3CO2 + 4H2O → C_3H_8+ $5O_2$? Is this endothermic or exothermic?

Problem L2

During coal liquefaction, the coal is gasified in the presence of oxygen and steam to produce a gas mixture of carbon monoxide and hydrogen called synthesis gas and then converted to liquid hydrocarbons in the presence of iron catalysts. This is called the Fischer-Tropsch gas synthesis process, i.e., a CO + b H_2 → c C_nH_{2n+2} + d O_2. If liquid octane is produced determine heat required for the reaction to proceed at 25°C?

Problem L3

Consider the mixing of 3.76 kmol of N_2 with 1 mole of O_2. does the following reaction to go to completion at 25°C and 1 bar, namely, 3.76 N_2 + O_2 → 2 NO + 2.76 N_2?

Problem L4

Plot $X_{NO}(T)$ for NO in air at chemical equilibrium at 100 kPa. Apply the reaction 1/2 N_2 + 1/2 O_2 = NO. Assume NO to exist in trace amounts.

Problem L5

Determine the trace amounts of SO_2 and NO exhausted from a smoke stack with respect to the temperature under chemical equilibrium if Illinois No. 6 coal is combusted with 20% excess air. Empirical formulae of coal: $C_{0.6671}$ $H_{0.5610}$ $N_{0.011001}$ $O_{0.06738}$ $S_{0.01322}$. Assume complete combustion of C and H to CO_2 and H_2O.

Problem L6

For the reaction H_2S→H_2+ S determine the equilibrium relation if sulfur exists as gas at 1000 K and as solid at 298 K. Will the amount of S be affected by a change in pressure at either 298 K or 1000 K?

Problem L7

Consider the dissociation (dimerization) of N_2O_4, i.e., N_2O_4 → $2NO_2$ for which ΔG^0 = 57330–176.7 T kJ $kmol^{-1}$. Plot the degree of dissociation as a function of pressure at 298 K, and as a function of temperature at 1 bar.

Problem L8

For the reaction C(s) + CO_2 (g) → 2CO(g) determine the equilibrium composition as a function of pressure at 2000 K. Assume ideal gas behavior for CO_2 and CO and 1 kmol of carbon initially.

Problem L9

The equilibrium constant for the reaction $H_2 + 1/2O_2 \rightarrow H_2O(liq)$ has been determined to be 1.6×10^4 at 25°C. The standard states are defined with respect to the pure components at 1 atm and 25°C. Calculate the minimum work required to dissociate 1 kmol of water at 25°C and 1 atm if (a) pure and separated hydrogen and oxygen are to be produced, and (b) a stoichiometric mixture of oxygen and hydrogen is to be produced.

Problem L10

Both hydrogen and air enter a welding torch at 25°C and burn according to the reaction $H_2 + 1/2\ O_2 \rightarrow H_2O\ (g)$. If the torch is adjusted to give 200 percent more air than the stoichiometric amount and combustion is adiabatic, what is the flame temperature? The values of c_p for O_2, N_2, H_2, and H_2O in units of cal $gmol^{-1}\ K^{-1}$ are, respectively, $6.14 + 3.102\times10^{-3}T$, $6.524 + 1.250\times10^{-3}T$, $6.947 + 0.120\times10^{-3}T$, and $7.256 + 2.290\times10^{-3}T$ with T in units of K, and $(\Delta H_{react})_{25°C} = -57.8$ kcal per g–mol of H_2, and $K^0 = 1.0\times10^{40}$ at 25°C. The standard states for all components is 1 bar.

Problem L11

In order to increase the operating temperature of the hydrogen torch of the previous problem, pure oxygen at the stoichiometric rate replaces air as the oxidizer. Neglecting all reactions other than the combination of hydrogen and oxygen, determine the adiabatic flame temperature for these operating conditions.

Problem L12

The following reactions are believed to occur during the catalytic oxidation of ammonia to nitric oxide:

$4NH_3 + 5O_2 \rightarrow 4NO + 6H_2O,$ (A)

$4NH_3 + 3O_2 \rightarrow 2N_2 + 6H_2O,$ (B)

$4NH_3 + 6NO \rightarrow 5N_2 + 6H_2O,$ (C)

$2NO + O_2 \rightarrow 2NO_2,$ (D)

$2NO \rightarrow N_2 + O_2,$ and (E)

$N_2 + 2O_2 \rightarrow 2NO_2.$ (F)

It is essential to determine the equilibrium composition when air is used to oxidize ammonia. Determine the minimum number of independent chemical equilibrium relations necessary to completely solve for the composition. Just outline the procedure in solving for the composition using equilibrium constants.

Problem L13

An electric generating station burns anthracite (essentially, pure carbon) in air to provide heat for its main boilers. Determine the equilibrium composition of the gases leaving the combustion chamber at 900 K and 1.0 bar. The following reactions are known to occur:

$C + 1/2\ O_2 \rightarrow CO,$ (1)

$CO + 1/2\ O_2 \rightarrow CO_2,$ and (2)

$CO_2 + C \rightarrow 2CO,$ (3)

where the standard states are pure gaseous O_2 and CO_2, and pure solid carbon at 1 atm. (As long as any unreacted carbon remains, it is always in its standard state. Thus the activity of carbon is equal to unity and independent of the amount of carbon left.)

Problem L14

The JANAF tables list values of K^0 for reactions involving natural forms of elements. Determine the value of K^0 for the reaction $CO + H_2O \rightarrow CO_2 + H_2$ at 2000 K and 1 bar using tabulated g^0 values. A chemicals company suddenly charges a tank with a mixture of 2.85 CO, 0.15 CO_2, 0.15 H_2, and 3.85 H_2O (all in kmol) at a total pressure of 2 bar and 900 K. The tank is maintained at 900 K and 2 bar. There is concern by engineers that $CO + H_2O$ (g) $\rightarrow CO_2 + H_2$ which is exothermic and as such tank may explode; since H_2O dominates the mixture, the management argued that $CO_2 + H_2 \rightarrow CO + H_2O$ (g) which is endothermic may be happening.

a) Determine the chemical forces of reactants (F_R) and products (F_P) for any of the assumed direction.

b) Settle the issue of direction of reaction.

Answer whether changing the pressure will affect the direction of reaction? Do not calculate

Problem L15

Recall that $g_k = (h - Ts)_k$ and $g'_k = (g_f^o + \Delta g)_k$ where g_f denotes the specific Gibbs energey of formation. Show that for CO, $g'_{CO} - g_{CO} = T_o(s_C^o + 1/2\, s_{O_2}^o)$. Similarly show that $g'_O - g_O = T_o(1/2)\, s_{O_2}^o$ and $g'_{O_2} - g_{O_2} = T_o\, s_{O_2}^o$. Also, show that $\Sigma\upsilon_k g_k = \Sigma\upsilon_k g'_k = 0$ when the reaction $C + 1/2\, O_2 = CO$ is at equilibrium.

Problem L16

For the steam reforming reaction CH_3OH (liq) + H_2O (liq) $\rightarrow 3\, H_2 + CO_2$ both liquid methanol and liquid H_2O are supplied 298 K and 1 bar to a reactor which should produce a mixture of H_2 and CO_2 also at 298 K and 1 bar. Is the reaction possible for this case?

Problem L17

Many power plants in U.S. fire either coal or natural gas to produce electrical power. Coal can be represented by C(s) and natural gas by CH_4. The excess air for a particular application is such that the oxygen content in the exhaust on dry basis is 3%. Assume complete combustion and the pressure of the products to be 1 bar. For both fuels determine the (a) A: F ratio, (b) CO_2 and N_2 percent in the exhaust, (c) the CO and NO present in the exhaust if it is at 1500 K assuming the following reaction: $N_2 + O_2 \rightarrow 2\, NO$, $N_2 + 2\, O_2 \rightarrow 2\, NO_2$, and $CO_2 \rightarrow CO + 1/2\, O_2$. Assume that NO and CO are in trace amounts, d) CO, NO and CO_2 in g/GJ for both the fuels.

Problem L18

Which of the two reactions $C(s) + 1/2\, O_2 \rightarrow CO$ or $C(s) + O_2 \rightarrow CO_2$ is favored at (a) 2000 K and (b) 3000 K?

Problem L19

One kmol of C(s) at 2 bar, and premixed 2 kmol of O_2 and 0.001 kmol of CO_2 at 1000 K and 2 bar are introduced into a steady flow reactor. Will the CO_2 concentration increase or decrease in the product stream due to the reaction $C(s) + O_2 \rightarrow CO_2$?

Problem L20

Methanol(ℓ) can be produced from syngas ($CO + H_2$) according to the reaction $CO(g) + 2H_2(g) \rightarrow CH_3OH(\ell)$. Determine the suitable conditions for the feasibility of its production at 25°C and 1 bar.

Problem L21

Consider the reaction $SO_2(g)$ + $CaO(s)$ + 1/2 $O_2(g) \rightarrow CaSO_4(s)$, which is used to capture the SO_2 released due to combustion of coal. What is the equilibrium relation, assuming that the $SO_2$and CaO are fully mixed at molecular level? How much SO_2 and O_2 is left over at 1200 K?$c_{p,CaO(s)}$ = 42.8 ,$c_{p,CaSO4(s)}$ = 100 kJ/kmol K?

Problem L22

Show that $\hat{g}_1(T,P,X_1) = \bar{h}_1(T) - \bar{s}_1(T,P) + \bar{R}T \ln X_1 = \bar{g}_1(T,p_1)$.

Problem L23

A reactor is supplied with elements 9 kmol of C and 19 kmol of O and allowed to reach chemical equilibrium at 3000 K and 1 bar. What is the equilibrium composition? What is the value of the Gibbs energy at equilibrium? If the products are isobarically cooled to 2800 K and allowed to reach chemical equilibrium, what is the new equilibrium composition and the new value of the Gibbs energy?

Problem L24

Determine the value of ΔG(298 K, 2 bar) for the water gas shift reaction H_2O + CO(g) $\rightarrow H_2(g) + CO_2(g)$ considering the water to be (a) liquid and (b) gas.

Problem L25

It is necessary to determine $\hat{g}_{CO_2}$ in a mixture containing 20% CO, 10% CO_2, 10% O_2 with the remainder being N_2. Assume that the mixture is an ideal mix of real gases at 66 bar and 370 K (you may use the fugacity charts).

Problem L26

In an application there are two possible reactions for the oxidation of carbon C(s) + $1/2O_2 \rightarrow$ CO, and C(s) + $O_2 \rightarrow CO_2$. Determine the affinity at the point when 40% of the carbon is consumed separately by the first and second reactions at 1 bar and 3500 K. Assume that $c_{p,C}/R = 1.771 + 0.000877\ T - 86700/T^2$ with T in units of K.

Problem L27

The steam reforming reaction is $CH_4 + H_2O \rightarrow CO + 3H_2$. Is this reaction possible at 298 K if equal molal mixture of CH_4, $H_2O(g)$, CO, H_2 are sent to the reactor ? Is heat absorbed or released at 298 K ? Is 50% conversion possible at 298 K, and if it were to be obtained, what would be the molal ratio of H_2 to CH_4 in the products?

Problem L28

A combustor is fired with coal having atomic composition $CH_{0.755}N_{0.0128}O_{0.182}S_{0.00267}$. For every kmol of coal fired, 0.234 kmol of moisture enters the combustor. If 20% excess air is used and combustion is complete, a trace amount of NO is formed (according to the reaction $1/2N_2 + 1/2O_2 \rightarrow$ NO), the sulfur is burned to SO_2, and the products leave at 2800 K, determine the equilibrium composition.

Problem L29

Gaseous propane is burned with 60% of theoretical air in a steady flow process at 1 atm. Both the fuel and air are supplied at 298 K. The products, which consist of CO_2, CO, H_2O, H_2, and N_2 in equilibrium, leave the combustion chamber at 1500 K. Determine the composition of the products and the amount of heat transfer in the process per kg of propane burned. The standard enthalpy of formation for propane is –103,847 J $gmol^{-1}$.

Problem L30

A mixture contains 20% CO, 10% O_2 and the remainder CO_2 at a temperature of 1500 K and 10 bar. Obtain the values of G, $\partial G/\partial N_{CO_2}$, $\partial^2 G/\partial N^2_{CO_2}$. $g_{CO}(1500,10)$=-415434 kJ/kmol, g_{O2}= -317622 kJ/kmol, g_{CO2}=-769977 kJ/kmol.

Problem L31

Air is supplied to a compressor in a gas turbine at 298 K and 1 bar, and is adiabatically and reversibly compressed to 10 bar. The air then proceeds to that combustor that is fired with iso–octane fuel at 298 K. Combustion occurs adiabatically with 100% excess air. Determine the adiabatic flame temperature assuming (a) complete combustion, and (b) complete chemical equilibrium with CO, NO, and OH present in the products. (c)What is the equilibrium composition for part (b)?

Problem L32

Evaporative cooling of inlet air is suggested for a gas-turbine power plant since it is expected to provide denser air and hence more mass flow for the same velocity to the compressor. The evaporative cooling results in decrease of temperature of air to 20°C from saturated wet air at 40°C. Assume fuel to be CH_4 at 298K burning with 100% excess air. a) What will be the decrease in adiabatic flame temperature compared to a dry air case? b) What will be decrease in NO if any compared to dry air case. Assume that air is saturated with vapor.

Problem L33

If $F = G + \Sigma_k \lambda_k \Sigma_j (d_{kj} N_j - A_k)$, then prove that $G_{min} = - \Sigma_k \lambda_k A_k$, $G = \Sigma_k \mu_j N_j$ or $\Sigma_k g_j N_j$.

Problem L34

Obtain a set of relations for determining the equilibrium composition of gases at given T for a fuel $C_C H_H O_O N_N$ burning in air with the following set of reactions.

$H_2O \Leftrightarrow H_2 + 1/2\, O_2$,

$CO_2 \Leftrightarrow CO + 1/2\, O_2$,

$H_2O \Leftrightarrow 1/2\, H_2 + OH$,

$NO \Leftrightarrow 1/2\, N_2 + 1/2\, O_2$, and

$OH \Leftrightarrow 1/2\, H2 + 1/2\, O_2$.

Describe a procedure for solving the composition using a spread sheet.

Problem L35

One kmol each of carbon monoxide and water vapor enter an adiabatic reactor at 298 K and 1 bar and produce CO_2 and H_2. (a) Plot the temperature and entropy of the products, the sums $(g_{CO}+g_{H_2O})$, $(g_{CO_2}+g_{H_2})$, and σ with respect to the degree of reaction of CO (i.e., assume that 0, 0.1, ..., 1.0 kmol of CO react). (b) If the conversion of reactants to products is adiabatic, what is the value of σ when one kmol of CO is consumed? (c) What is the degree of reaction when σ reaches a maximum value? (d) What should be the criterion for the direction of reaction to occur? $\sigma > 0$ or $d\sigma > 0$. Which criterion is the more appropriate to use?

M. CHAPTER 13 PROBLEMS

Unless otherwise stated assume the ambient temperature and pressure to be 298 K and 1 bar.

Problem M1

In an adiabatic reaction, one can reasonably assume that the availability of products leaving an sssf reactor is the same as availability of the reactants. True or False?

Problem M2

In an adiabatic reaction within a PCW, one can reasonably assume that the availability of products is the same as availability of the reactants. True or False?

Problem M3

The fuel availability is independent of environmental conditions. True or False?

Problem M4

If you know the heating value of cattle manure as 5000 Btu/lb, can you determine the fuel availability?

Problem M5

Methane is burned with 40% of excess air isobarically at 1 atm. Methane and air enter the combustor fully mixed at 77°F. What is the absolute stream availability unmixed reactants? What is the irreversibility due to adiabatic mixing, and adiabatic combustion? What is the absolute stream availability if the product temperature is 800°R?

Problem M6

Octane at 298 K and 4 bar and air at 710 K and 4 bar enter an adiabatic reactor. Reaction products leave the reactor at 1100 K and 4 bar. Determine the A: F ratio, the maximum possible sssf work, assuming the ambient temperature to be 298 K, and the entropy generated.

Problem M7

In an open system determine the adiabatic irreversibilities of (a) methanol, and (b) gasoline (i.e., $CH_{2.6}$ with a lower heating value of 47720 kJ kg^{-1}) under stoichiometric conditions. Express your answers in terms of fractional heating values. (c) Which of the two fuels displays a larger irreversibility? (d) Do the values change if a closed system is considered? (e) Which fuel will you recommend for an automobile engine? Assume that entropy of $CH_{2.6}$ is same as entropy of octane on unit mass basis

Problem M8

One kmol of CO and 1/2 kmol of O_2 react at 0 K in an isothermal reactor. (a) What is the entropy of the reactants and products? . (b) Since $\sigma \geq 0$, what is the heat transfer Q at 0 K ? (c) Since $Q = \Delta H = H_P - H_R$, what are the value of H_P and H_R, at 0 K?

Problem M9

A four stroke diesel engine has a 80.26 mm bore, and 88.9 mm stroke, and runs at 2400 RPM with 20% excess air. The fuel is approximated by the chemical formula $C_{14}H_{24.9}$ with a heating value of 42940 kJ kg^{-1}. a) How much power can be developed if the cycle efficiency is 40%? b) If instead of drawing air at 100 kPa and 300 K, turbocharged air is provided during the suction stroke at 300 kPa and 330 K, how much additional power is developed? Assume the same cycle efficiency as case (a). c) If the radiator takes in half of the heat rejected for case (a), what will be the exhaust temperature and hence what would have been the maximum possible power for case (a)?

Problem M10

In an open system determine the adiabatic irreversibilities of (a) methanol, and (b) gasoline (i.e., $CH_{2.6}$ with a lower heating value of 47720 kJ kg^{-1}). Express your answers in terms of fractional heating values. (c) Which of the two fuels displays a larger irreversibility? (d) Do the values change if a closed system is considered? (e) Which fuel will you recommend for an automobile engine?

Problem M11

In a water gas shift reactor ($CO + H_2O \rightarrow CO_2 + H2$) each of the reactant species enter at 298 K and 1 bar. The products leave at 1000 K and 1 bar. Calculate the irreversibility in kJ/kmol of CO. Is it possible to produce a work output of 2×10^6 kJ/kmol of CO for the specified conditions?

Problem M12

The equilibrium constant for the reaction $H_2 + 1/2O_2 \rightarrow H_2O(\ell)$ is 1.6×10^4 at 298 K, and the standard states are assumed to be the pure components at 1 bar pressure and 298 K. Determine the minimum work required to dissociate 1 kmol of water at 298 K and 1 bar if (a) pure and separated hydrogen and oxygen are to be produced (each at 1 bar, 298 K), and (b) a stoichiometric mixture of oxygen and hydrogen is to be produced.

Problem M13

Determine the maximum work deliverable by a fuel cell consuming 1 kmol of H_2 that reacts at 25°C and 1 atm to form H_2O(liq).

Problem M14

Determine the availability of iso-octane fuel and verify the result with the value tabulated in Table A-27B

Problem M15

A kmol of CH_4 and a stoichiometric amount of O_2 enter a fuel cell at 1 bar, and 298K and produce CO_2 and H_2O(liq) at 298K, and 1 bar. Calculate the change in availabilities between the inlet and exit and the maximum possible work.

Problem M16

For fuel cells using hydrocarbon fuels, the anodic reaction is

$$C_xH_y + 2x\ H_2O \rightarrow x\ CO_2 + (4x+y)H^+ + (4x+y)e^-,$$

and the cathodic reaction is

$$(x+y/4)O_2 + (4x+y)H^+ + (4x+y)e^- \rightarrow (2x+y/2)H_2O.$$

The overall reaction can be represented as

$$C_xH_y + (x + y/4)\ O_2 \rightarrow x\ CO_2 + y/2\ H_2O.$$

Therefore, for a hydrocarbon fuel C_xH_y, the electrons generated per molecule are represented by the relation (4x+y). a) If methane fuel is used in a fuel cell determine the maximum possible voltage. b) What will be the answers for octane fuel?

Problem M17

Determine the fuel availability for octane. Assume that T_o = 298 K, $p_{O_2,o}$ = 0.2055, $p_{CO_2,o}$ =0.003, and $p_{H_2O,o}$ = 0.0188. Determine the fuel availability in kJ/Kmol and the ratio of fuel availability to LHV.

Problem M18

Hydrocarbon fuels are used to power submarine diesel engines. The exhaust passes through a cooling system so that almost all the H_2O in the products condenses and provides drinking water. The CO_2 left in the products is mixed with pure oxygen and fired back into the diesel engine. The intake gas is not air but a mixture of CO_2 and O_2, such that the mole fraction of O_2 is 21%. Is the optimum work affected? Determine the optimum work if a) air alone is used for burning the fuel, or b) the CO_2 and O_2 mixture is used for HC fuel. Assume that the reactants enter at 25°C and 1 bar and the products leave at 25°C and 1 bar with H_2O in the liquid state. If Δw_{opt} = $(w_{opt,a}-w_{opt,CO2})$ plot $\Delta w_{opt}/(nRT)$ *vs.* m/n

APPENDIX A

A. TABLES

TABLE 0: UNITS, CONVERSIONS

Other Conversions
$1 \text{ atm} = 14.696 \ \frac{\text{lbf}}{\text{in.}^2} = 1.0133 \text{ bar}$
$1 \text{ atm} = 1.013 \text{ bar}$
$1 \text{ Bbl} = 42 \text{ gal} = 5.615 \ \text{ft}^3$
$1 \text{ Bbl} = 158.98 \text{L}$
$1 \text{ BTU} = 778.17 \ \text{lbf} \cdot \text{ft} = 1.0551 \text{ kJ}$
$1 \text{ BTU} = 25{,}037 \frac{\text{lbm} \cdot \text{ft}^2}{\text{s}^2}$
$1 \text{ ft} = 12 \text{ in} = 0.3048 \text{ m}$
$1 \text{ gal} = 231 \text{ in}^3 = 0.1337 \ \text{ft}^3$
$1 \text{ gal} = 3.785 \text{L}$
$1 \text{ hp} = 0.7064 \ \text{Btu/s} = 0.746 \text{ kW}$
$1 \text{ lbf} = 32.174 \ \frac{\text{lbm} \cdot \text{ft}}{\text{s}^2}$
$1 \text{ mile} = 5280 \text{ ft} = 1609 \text{ m}$
$1 \text{ psi} = 1 \ \frac{\text{lbf}}{\text{in.}^2} = 144 \frac{\text{lbf}}{\text{ft}^2}$
$1 \text{ psi} = 6.894 \text{ kPa}$

Unit	SI units	
acre	4046.9	m^2
Angstrom	$1.0\text{x}10^{-10}$	m
atmosphere	101325	Pa
bar	1×10^5	Pa
BTU	1055.1	J
day	86400	s
°R	0.555556	K
ft	0.3048	m
gal	3.7854×10^{-3}	m^3
hour	3600	s
hp	745.70	W
inch	0.0254	m
lbf	4.4482	N
lbm	0.45359	kg
liter	0.001	m^3
mile	1609.3	m
minute	60	s
mm hg	133.32	Pa
psi	6894.8	Pa

Multiplier	Prefix	Symbol
10^{12}	tera	T
10^9	giga	G
10^6	mega	M
10^3	kilo	k
10^2	hecta	h
10	deca	da
10^{-1}	deci	d
10^{-2}	centi	c
10^{-3}	milli	m
10^{-6}	micro	μ
10^{-9}	nano	n
10^{-12}	pico	p
10^{-15}	femto	f
10^{-18}	atto	a

Others

1 short ton = 2000 lb
1 metric ton = 1000 kg
1 long ton = 2240 lb
1 lb = 7000 grains
$g_c = 32.174\ lb_m\ ft/s^2\ lb_f$
$J = 778.14\ ft\ lb_f/\ BTU$
One (food) calorie = 1000calories or 1 kcal
$T(K) = T(^oC) + 273.15$
$T(^oR) = T(^oF) + 459.67$
$N_{avag} = 6.023 \times 10^{26}$ molecules/kmol for a molecular substance (e.g., oxygen)
$= 6.023 \times 10^{26}$ atoms/atom mole for an atomic substance (e.g., He)
k_B, Boltzmann Constant $=1.38\text{x}10^{-26}$ kJ/molecule K
h_P , Planck's constant = 6626x10-37kJ-s/molecule
c, Speed of light in vacuum = $2.998\text{x}10^8$ m/s.

Ideal Gas Law

$Pv = RT$; $PV = m\ R\ T$; $PV = n\bar{R}T$, $P\bar{v} = \bar{R}\ T$,

$$\bar{R} = 8.314 \frac{kPa\ m^3}{k\ mole\ K} = 0.08314 \frac{bar\ m^3}{k\ mole\ K}$$

$$= 1.986 \frac{Btu}{lbmole^oR} = 1545 \frac{ftlbf}{lbmole^oR} = 0.7299\ atm\ ft^3/lb\ mole\ R$$

Volume of 1 kmol (SI) and 1 lb mole (English) of an ideal gas at STP conditions as defined below:

STP:25°C (77°F), 101 kPa (14.7 psia)	STP: 15°C (60°F), 101kPa (14.7) psia	STP: 0°C (32°F), 101 kPa (14.7 psia)
24.5 m^3/kmol, 392 ft^3/lb mole	23.7 m^3/kmol, 375.6 ft^3/lb mole	22.4 m^3/kmol, 359.2 ft^3/lb mole

TABLE 1: Molecular weights, critical and other properties of selected elements and compounds

TMP: Temperature of melting or freezing point, TBP: temperature of normal boiling points, h_{sf}: heat of melting, h_{fg}: heat of evaporation, w: Pitzer factor

Substance	Formulae	M	T_c	P_c	z_c	$\bar{v}_c$	w	TMP	TBP	h_{sf}	h_{fg}	$\bar{a}$, RK	$\bar{b}$, RK	$\bar{a}$, VW	$\bar{b}$, VW
			K	bar		m^3/kmol		K	K	kJ/kg	kJ/kg	(bar $m^6K^{0.5}$)/ $kmol^2$	m^3/kmol	(bar m^6)/ $kmol^2$	m^3/kmol
Acetic acid	$C_2H_4O_2$	60.06	594.4	57.9	0.200	0.171	0.462	289.8	391.1	195.4	394.6	439.62	0.0739	17.7944	0.10669
Acetone	CH_3COCH_3	58.08	508.1	47	0.233	0.209	0.306	178.2	329.4	98.0	501.7	365.87	0.0779	16.0179	0.11235
Acetylene	C_2H_2	26.04	308.8	61.4	0.270	0.113	0.190	189.2	189.2	144.6	653.6	80.65	0.0362	4.52887	0.05227
Air		28.97	133	37.7	0.283	0.0829			78.7			15.99	0.0254	1.36825	0.03666
Ammonia	NH_3	17.03	405.6	112.8	0.243	0.0725	0.250	195.4	239.7	332.4	1371.8	86.79	0.0259	4.25295	0.03737
Aniline	C_6H_7N	93.14	699	540				266.9	457.6			70.69	0.0093	2.63854	0.01345
Argon	Ar	39.95	150.8	48.7	0.291	0.0749	0.001	83.8	87.3	30.4	163.5	16.94	0.0223	1.36169	0.03218
Benzene	C_6H_6	78.11	562.1	48.9	0.271	0.259	0.212	278.7	353.3	126.0	394.1	452.67	0.0828	18.8418	0.11946
Bromine	Br_2	159.81	584	103	0.269	0.127	0.108	265.9	332			236.46	0.0408	9.65589	0.05892
Butane–n	C_4H_{10}	58.12	425.2	38	0.274	0.255	0.199	134.8	272.6	80.2	387.1	289.90	0.0806	13.8742	0.11629
Butanol	C_3H_9OH	62.13	562.95	44.18					390.9			502.93	0.0918	20.9179	0.13242
Carbon dioxide	CO_2	44.01	304.2	73.8	0.274	0.094	0.239	194.7	194.7	196.6	347.6	64.62	0.0297	3.6565	0.04284
Carbon disulfide	CS_2	76.13	552	79	0.293	0.17	0.115	162.0	319.4			267.78	0.0503	11.2475	0.07262
Carbon monoxide	CO	44.01	132.9	35	0.295	0.0931	0.066	68.1	81.7	29.9	215.8	17.19	0.0274	1.47159	0.03946
Carbon tetrachloride	CCl_4	153.8	556.4	45.6	0.272	0.276	0.193	250.3	349.9	16.5	194.0	473.22	0.0879	19.7976	0.12681
Chlorine	Cl_2	70.9	417	77.11	0.276	0.124	0.073	171.5	238.7	90.4	288.2	136.08	0.0390	6.57605	0.0562
Chloroform	$CHCl_3$	119.37	536.4	55	0.295	0.239	0.216	209.6	334.3	79.9	249.0	358.03	0.0703	15.2552	0.10136
Cyanogen	C_2N_2	52.04	400	59.8				238.75	251.98			158.13	0.0482	7.8023	0.06952
Cyclopentane	C_5H_{10}	70.13	511.8	45.02	0.273	0.2583	0.194	179.3	322.4	8.7	388.7	388.96	0.0819	16.9668	0.11814
Decane–n	$C_{10}H_{22}$	142.3	619	21.23	0.249	0.6031	0.484	243.5	447.0	201.8	281.2	1326.88	0.2100	52.6302	0.30301
Ethane	C_2H_6	30.54	305.4	48.8	0.285	0.1483	0.099	89.9	184.5	95.1	489.4	98.70	0.0451	5.57342	0.06504
Ethanol	C_2H_5OH	46.07	516.2	63.8	0.248	0.1671	0.644	159.1	351.5	107.9	841.6	280.40	0.0583	12.1792	0.08408
Ether(diethyl)	$C_4H_{10}O$	74.14	466.7	36.4	0.263	0.28	0.281	155.4	307.6	98.5	365.4	381.99	0.0924	17.4493	0.13325
Ethyl benzene		106.2	617.2	36.09	0.263	0.3738	0.304	178.2	409.4	86.5	338.1	774.88	0.1232	30.78	0.17773
Ethyl chloride	C_2H_5Cl	64.51	460.4	52.69	0.275	0.2	0.191	136.8	285.4	69.0	383.7	255.07	0.0629	11.7313	0.09081
Ethylene	C_2H_4	28.05	282.4	50.4	0.280	0.1304	0.089	104.0	169.4	119.5	483.1	78.58	0.0404	4.61427	0.05823
Ethylene Glycol		62.07	645.1	75.3	0.268	0.191	1.136	260.2	469.1	187.3	845.7	414.79	0.0617	16.1162	0.08903
Fluorine	F_2	38	144	56.8	0.314	0.0662	0.054	53.5	85.0		171.9	12.95	0.0183	1.06459	0.02635

Substance	Formulae	M	T_c	P_c	z_c	$\bar{v}_c$	w	TMP	TBP	h_{sf}	h_{fg}	$\bar{a}$, RK	$\bar{b}$, RK	$\bar{a}$, VW	$\bar{b}$, VW
			K	bar		m^3/kmol		K	K	kJ/kg	kJ/kg	(bar $m^6K^{0.5}$)/ $kmol^2$	m^3/kmol	(bar m^6)/ $kmol^2$	m^3/kmol
Freon 114	$C_2Cl_2F_4$	170.92	418.9	32.6	0.274	0.293	0.256	179.3	276.9		136.2	325.55	0.0926	15.6966	0.13354
Freon 12	CCl_2F_2	120.91	385	41.2	0.279	0.217	0.176	115.4	243.4	34.3	165.2	208.60	0.0673	10.4913	0.09711
Freon 13	$CClF_3$	104.46	302	39.2	0.281	0.18		92.0	191.7		148.5	119.48	0.0555	6.78472	0.08006
Freon 152a	CHF_2CH_3	66.06	386.4	45.2	0.253	0.1795	0.275					191.87	0.0616	11.5033	0.09078
Freon 21	$CHCl_2F$	102.92	451.6	51.7	0.271	0.197		138.0	282.0		242.4	247.71	0.0629	7.98175	0.07705
Freon 22	$CHClF_2$	86.47	369.2	49.8	0.268	0.165		113.0	232.4		233.7	155.41	0.0534	16.5875	0.12938
Freon123	$CHCl_2CF_3$	152.93	456.9	36.7	0.269	0.2781	0.282					359.29	0.0897	10.0574	0.09579
Freon134a	CF_3CH_2F	102.04	374.2	40.6	0.258	0.198	0.327					197.15	0.0664	14.8916	0.11022
Freon141	CH_3CCl_2F	116.95	481.5	45.4	0.286	0.252	0.215					331.12	0.0764	11.551	0.10033
Freon142b	CH_3CClF_2	100.5	410.3	42.5	0.288	0.231	0.250					237.09	0.0695	9.63251	0.08884
Glycol–1,2 Propylene	$C_3H_8O_2$	76.1	626	61	0.280	0.239	1.106	213.2	460.8		715.9	474.97	0.0739	18.7337	0.10665
Glycol–Ethylene	$C_2H_6O_2$	62.07	645.1	75.3	0.268	0.191	1.136	260.2	470.5	187.3	845.7	414.79	0.0617	16.1162	0.08903
Helium	He	4	5.2	2.27	0.301	0.0573	–0.365	1.8	4.3	125.0	20.8	0.08	0.0165	0.03474	0.02381
Heptane–n	C_7H_{16}	100.2	540.2	27.4	0.264	0.432	0.349	182	371.7	140.3	316.7	731.46	0.1420	31.0572	0.20489
Hexadecane (cetane)	$C_{16}H_{34}$	226.4	720.6	14.19	0.220	0.93	0.747	291	559.8	235.7	232.0	2902.75	0.3658	106.711	0.52775
Hexane–n	C_6H_{14}	86.18	507.4	29.7	0.260	0.37	0.299	178	341.0	151.8	337.9	577.00	0.1231	25.2783	0.17755
HFC–125	CHF_2CF_3	120.03	339.2	35.95			0.302		225.1			174.18	0.0680	9.3329	0.09806
HFC–32	CH_2F_2	52.03	351.6	58.3			0.276		222.2			117.49	0.0434	6.18348	0.06268
Hydrogen	H_2	2.02	33.2	13	0.307	0.0651	–0.218	14.0	20.4	58.2	448.6	1.44	0.0184	0.24725	0.02654
Hydrogen chloride	HCl	36.47	324.6	83	0.249	0.081	0.132	159.0	188.1	54.7	443.2	67.58	0.0282	3.70188	0.04064
Hydrogen sulfide	H_2S	34.08	373.2	89.4	0.284	0.0985		187.7	213.7	69.7	554.3	88.94	0.0301	4.54307	0.04338
Iodine	I_2	253.8	785.15	117.5				386.5	457.5			434.25	0.0481	15.2937	0.06942
Isobutane	C_4H_{10}	58.12	408.1	36.5	0.283	0.2627	0.177		260.9	78.1	368.2	272.38	0.0805	13.3059	0.1162
Isohexane	C_6H_{14}	86.18	497.45	30.4					330.9			536.50	0.1179	23.738	0.17006
Isooctane	C_8H_{18}	114.2	543.9	25.6	0.265	0.468		165.8	372.4	79.2	271.6	796.37	0.1530	33.6978	0.2208
Isopentane	C_5H_{12}	72.15	460.4	33.8	0.270	0.306	0.227		300.9			397.63	0.0981	18.2876	0.14156
Isopropanol	C_3H_7OH	60.1	508.3	47.64	0.248	0.2201	0.669	184.7	355.4	90.0	663.4	361.31	0.0769	15.8151	0.11088
Krypton	Kr	83.8	209.4	55	0.292	0.0924	0.005	104.3	120.9			34.09	0.0274	2.32485	0.03957
Mercury	Hg	200.59	1733	1077				233.8	629.0	11.4		343.03	0.0116	8.13177	0.01672
Methane	CH_4	16.04	190.6	46	0.288	0.0992	0.011	90.7	111.7	58.7	510.2	32.22	0.0298	2.30299	0.04306
Methanol	CH_3OH	32.04	512.6	81	0.224	0.118	0.556	175.5	337.8	99.2	1101.0	217.03	0.0456	9.45967	0.06577

Substance	Formulae	M	T_c	P_c	z_c	$\bar{v}_c$	w	TMP	TBP	h_{sf}	h_{fg}	$\bar{a}$, RK	$\bar{b}$, RK	$\bar{a}$, VW	$\bar{b}$, VW
			K	bar		m^3/kmol		K	K	kJ/kg	kJ/kg	(bar $m^6K^{0.5}$)/ $kmol^2$	m^3/kmol	(bar m^6)/ $kmol^2$	m^3/kmol
Methyl Chloride	CH_3Cl	50.49	416.3	66.8	0.268	0.1389	0.153	175.4	249.3	129.7	426.8	156.42	0.0449	7.56554	0.06477
Methylene chloride	CH_2Cl_2	84.93	416.3	66.8	0.372	0.193	0.192	178.1	313.0	54.2	329.8	156.42	0.0449	7.56554	0.06477
Naphthlene	$C_{10}H_8$	128.2	748.4	40.5	0.267	0.41		353.5	491.1	44.7	337.8	1117.98	0.1331	40.3289	0.19204
Neon	Ne	20.18	44.4	27.6	0.312	0.0417	–0.029	24.5	27.0	16.0	91.3	1.41	0.0116	0.20829	0.01672
Neopentane	C_5H_{12}	72.17	433.8	32	0.269	0.303	0.197	253.2	282.6			361.93	0.0976	17.1488	0.14088
Nitric Oxide	NO	30.01	180	65	0.252	0.058	0.588	109.5	121.4	76.7	460.5	19.76	0.0199	1.45357	0.02878
Nitrogen	N_2	28.02	126.2	33.9	0.289	0.0895	0.039	63.3	77.4	25.7	199.2	15.60	0.0268	1.37001	0.03869
Nitrogen dioxide	NO_2	46.01	431.4	101.3	0.480	0.17	0.834	261.9	294.3		414.5	112.76	0.0307	5.35741	0.04426
Nirous oxide	N_2O	44.01	309.6	72.4	0.274	0.0974	0.165	182.3	184.7	148.7	376.2	68.84	0.0308	3.86071	0.04444
Nonane	C_9H_{20}	128.3	595	22.73	0.252	0.5477	0.437	219.7	423.7	120.6	293.8	1122.66	0.1886	45.4191	0.27204
Octane–n	C_8H_{18}	114.2	568.8	24.8	0.258	0.492	0.398	216.4	398.8	181.6	301.5	919.40	0.1652	38.0427	0.23836
Octene	C_8H_{16}	112.24	578.15	25.84				171.5	394.2			919.13	0.1612	37.723	0.23253
Oxygen	O_2	32	154.6	50.5	0.288	0.0734		54.4	90.2	13.9	213.3	17.39	0.0221	1.38017	0.03182
Ozone	O_3	48	261	53.7	0.220	0.0889		80.5	161.3		232.9	60.56	0.0350	3.69922	0.05051
Pentane–n	C_5H_{12}	72.15	469.6	33.7	0.262	0.304	0.251	143.4	309.0	1163.3	360.2	419.03	0.1004	19.0823	0.14482
Propane	C_3H_8	44.1	369.8	42.5	0.281	0.203	0.153	85.5	231.1	79.9	426.0	182.85	0.0627	9.38315	0.09043
Propanol	C_3H_7OH	60.1	536.7	51.7	0.253	0.2185	0.628	147.0	370.4	86.5	688.5	381.41	0.0748	16.2471	0.10789
Propylene	C_3H_6	42.08	365	46.2	0.276	0.181	0.148	87.9	225.4	71.4	437.8	162.80	0.0569	8.40906	0.08211
Propyne	C_3H_4	40.065	402.4	56.3	0.276	0.164	0.215	170.5	250			170.49	0.0515	8.38709	0.07428
Sodium hydroxide	NaOH	40	2815	253.3	0.216	0.2	1.015	596.0	1830.0	165.3		4904.56	0.0800	912.238	1.1549
Styrene	C_8H_8	104.2	647.6	39.99	0.261	0.3518	0.234	242.5	418.3	105.1	351.4	788.63	0.1166	30.582	0.1683
Sulfur dioxide	SO_2	64.06	430.8	78.8	0.268	0.122	0.256	290.0	263.2	115.5	385.4	144.45	0.0394	6.86798	0.05682
Sulfur trioxide	SO_3	80.06	490.9	82.07	0.256	0.1271	0.422	290.0	317.9	24.6	502.5	192.24	0.0431	8.5626	0.06216
Sulfuric acid	H_2SO_4	98.08	925	40.66	0.159	0.3		197.0	610.0	109.2	510.9	1891.22	0.1639	61.3649	0.23643
Toluene	C_7H_8	92.14	591.7	41.1	0.264	0.3158		178.2	383.8	72.0	364.7	612.31	0.1037	24.8408	0.14962
Water	H_2O	18.02	647.3	220.9	0.230	0.056	0.344	273.2	373.2	333.7	2258.3	142.60	0.0211	5.5312	0.03045
Xenon	Xe	131.3	289.7	58.4	0.288	0.1186	0.008	133.2				72.28	0.0357	4.19071	0.05155
Xylene–p	C_8H_{10}	106.2	616.3	35.11	0.260	0.3791		286.5	411.5	161.1	337.3	793.61	0.1264	31.5469	0.18242

TABLE 2: Triple Points of several substances (solid, liquid and vapor phase).

Substance	T, K	P, mm Hg
Acetylene	192.4	900
Ammonia	195.40	45.57
Argon	83.81	516.8
Carbon	3900	75710
Carbon dioxide	216.55	3885.1
Carbon Monoxide	68.10	115.3
Deuterium	18.63	128.0
Ethane	89.89	0.006
Ethylene	104.00	0.9
Heavy water	272.04	4.528
Helium	none	None
Hydrogen	13.84	52.8
Hydrogen bromide	186.29	244
Hydrogen chloride	158.96	104
Hydrogen cyanide	259.91	140.4
Hydrogen sulfide	187.66	173.9
Krypton	115.6	53
Mercury	234.28	2.3×10^{-6}
Methane	90.68	87.7
Neon	24.57	324
Nitric oxide	109.50	164.4
Nitrogen	63.18	94.3
Nitrous oxide	182.34	658.9
Oxygen	54.363	1.14
Palladium	1825	0.0262
Phosphorous	139.38	27.33
Platinum	2045	1.5×10^{-4}
Silicon tetrafluoride	182.9	1320
Sulfur dioxide	197.69	1.256
Sulfur hexafluoride	222.5	1700
Titanium	1941	0.0397
Uranium Hexafluoride	337.17	1137
Water	273.16	4.587
Xenon	161.3	611
Zinc	692.65	0.487

Source: NBS (US) Cir. 500 (1952) and Kestin, J., *A Course in Thermodynamics*, Hemisphere Publishing Corporation, Washington, 1979.

TABLE 3: Lennard–Jones Parameters

If data are not found, use the approximate formulae: ε/k_B= 0.77 T_c, 1.15 T_{BP}, 1.9 T_{MP}, T_{BP} boiling point in K, T_{MP} melting point in K. σ (nm) = 0.841 $\bar{v}_c^{1/3}$ (m^3/kmol), 1.166 $\bar{v}_b^{1/3}$, 1.122 $\bar{v}_m^{1/3}$. $\bar{v}_c$, $\bar{v}_b$, $\bar{v}_m$ specific volumes – respectively, critical, boiling, and melting. The first set is from viscosity data and the second from second virial coefficient data.

Species	Molecular weight	ε/k_B, K	σ, nm	ε/k_B, K	σ, nm	b_o^*, $m^3/kmol \times 10^3$
Air	28.964	78.6	0.3711	99.2	0.3522	55.11
Ar	0.039948	93.3	0.3542	119.8	0.3405	49.80
Br	0.079916	236.6	0.3672	–	–	–
Br_2	0.159832	507.9	0.4296	–	–	–
Cl_2	0.070906	316.0	0.4217	–	–	–
C_2H_2	26.038	231.8	0.4033	–	–	–
C_2H_4	28.054	224.7	0.4163	199.2	0.4523	116.7
C_2H_6	30.07	230	0.4418	–	–	–
C_2N_2	52.04	339	0.438	–	–	–
C_3H_8	44.09	254	0.5061	–	–	–
C_4H_{10}	58.12	313	0.5341	–	–	–
C_5H_{12}	72.15	345	0.5769	–	–	–
C_6H_{14}	86.17	413	0.5909	–	–	–
C_6H_6	78.11	440	0.5270	–	–	–
C_8H_{18}	114.22	320	0.7451	–	–	–
CCl_4	153.84	327	0.5881	–	–	–
CH	13.009	68.6	0.3370	–	–	–
CH_2Cl_2	84.94	406	0.4759	–	–	–
CH_3Cl	50.49	855	0.3375	–	–	–
CH_4	16.043	148.6	0.3758	148.2	0.3817	70.16
$CHCl_3$	119.39	327	0.543	–	–	–
Cl	0.035453	130.8	0.3613	–	–	–
CN	0.026018	75.0	0.3856	–	–	–
CO	28.011	91.7	0.3690	100.2	0.3763	67.22
CO_2	44.010	195.2	0.3941	189.0	0.4486	113.9
COS	60.08	335	0.413	–	–	–
CS_2	76.14	488	0.4438	–	–	–
C_6H_{12} (cyclo)	84.16	324	0.6093	–	–	–
F	0.018999	112.6	0.2986	–	–	–
F_2	0.037998	112.6	0.3357	–	–	–
H	1.008	37.0	0.2708	–	–	–
H_2	2.016	59.7	0.2827	29.2	0.287	29.76
H_2O	18.016	809.1	0.2641	–	–	–
HBr	0.080924	449.0	0.3353	–	–	–
HCl	0.036465	344.7	0.3339	–	–	–
HCN	0.027026	569.1	0.3630	–	–	–
He	0.004003	10.22	0.2551	10.22	0.2556	21.07
HF	0.020006	330.0	0.3148	–	–	–
I_2	0.025382	550	0.4982	–	–	–
N	14.007	71.4	0.3298	–	–	–
N_2	28.013	71.4	0.3798	95.05	0.3698	63.78

Species	Molecular weight	ε/k_B, K	σ, nm	ε/k_B, K	σ, nm	b_o^*, $m^3/kmol \times 10^3$
N_2O	44.016	232.4	0.3828	189.0	0.459	122.0
Ne	0.020179	32.8	0.2820	35.60	0.2749	26.21
NH	15.015	65.3	0.3312	–	–	–
NH_3	17.031	558.3	0.2900	–	–	–
NO	30.008	116.7	0.3492	131.0	0.317	40.0
O	16.000	106.7	0.3050	–	–	–
O_2	31.999	106.7	0.3467	117.5	0.358	57.75
OH	17.008	79.8	0.3147	–	–	–
SO_2	0.06407	252	0.4290	–	–	–
Xe	0.01313	229	0.4055	–	–	–

$b_0 = (2/3)\,\pi\sigma^{3}\,N_{avag}{}^{*}\times 10^{-27}$ $m^3/kmol$. It is suggested that the viscosity data be used exclusively for transport coefficient calculations and that the second virial coefficient be used exclusively for equation of state calculations.

§ Data taken from R. A. Svehla, *NASA Tech. Report R–132*, Lewis Research Center, Cleveland, Ohio (1962). Part of data and tables taken from J. 0. Hirshfelder, C. F. Curtis, and R. B. Bird. *Molecular Theory of Gases and Liquids*, Wiley, New York (1954). (With permission.)

TABLE 4A: Properties of saturated water (liquid–vapor): temperature table

Temp	Sat. press.	Specific volume, m^3/kg		Internal energy, kJ/kg			Enthalpy, kJ/kg			Entropy, kJ/(kg·K)		
		Sat. liquid	Sat. vapor	Sat. liquid	Evap.	Sat. vapor	Sat. liquid	Evap.	Sat. vapor	Sat. liquid	Evap.	Sat. vapor
T,°C	P_{sat}, MPa	v_f	v_g	u_f	u_{fg}	u_g	h_f	h_{fg}	h_g	s_f	s_{fg}	s_g
0.01	0.0006113	0.001000	206.14	0.0	2375.3	2375.3	0.01	2501.3	2501.4	0.000	9.1562	9.1562
5	0.0008721	0.001000	147.12	20.97	2361.3	2382.3	20.98	2489.6	2510.6	0.0761	8.9496	9.0257
10	0.0012276	0.001000	106.38	42.00	2347.2	2389.2	42.01	2477.7	2519.8	0.1510	8.7498	8.9008
15	0.0017051	0.001001	77.93	62.99	2333.1	2396.1	62.99	2465.9	2528.9	0.2245	8.5569	8.7814
20	0.002339	0.001002	57.79	83.95	2319.0	2402.9	83.96	2454.1	2538.1	0.2966	8.3706	8.6672
25	0.003169	0.001003	43.36	104.88	2304.9	2409.8	104.89	2442.3	2547.2	0.3674	8.1905	8.5580
30	0.004246	0.001004	32.89	125.78	2290.8	2416.6	125.79	2430.5	2556.3	0.4369	8.0164	8.4533
35	0.005628	0.001006	25.22	146.67	2276.7	2423.4	146.68	2418.6	2565.3	0.5053	7.8478	8.3531
40	0.007384	0.001008	19.52	167.56	2262.6	2430.1	167.57	2406.7	2574.3	0.5725	7.6845	8.2570
45	0.009593	0.001010	15.26	188.44	2248.4	2436.8	188.45	2394.8	2583.2	0.6387	7.5261	8.1648
50	0.012349	0.001012	12.03	209.32	2234.2	2443.5	209.33	2382.7	2592.1	0.7038	7.3725	8.0763
55	0.015758	0.001015	9.568	230.21	2219.9	2450.1	230.23	2370.7	2600.9	0.7679	7.2234	7.9913
60	0.019940	0.001017	7.671	251.11	2205.5	2456.6	251.13	2358.5	2609.6	0.8312	7.0784	7.9096
65	0.02503	0.001020	6.197	272.02	2191.1	2463.1	272.06	2346.2	2618.3	0.8935	6.9375	7.8310
70	0.03119	0.001023	5.042	292.95	2176.6	2469.6	292.98	2333.8	2626.8	0.9549	6.8004	7.7553
75	0.03858	0.001026	4.131	313.90	2162.0	2475.9	313.93	2321.4	2635.3	1.0155	6.6669	7.6824
80	0.04739	0.001029	3.407	334.86	2147.4	2482.2	334.91	2308.8	2643.7	1.0753	6.5369	7.6122
85	0.05783	0.001033	2.828	355.84	2132.6	2488.4	355.90	2296.0	2651.9	1.1343	6.4102	7.5445
90	0.07014	0.001036	2.361	376.85	2117.7	2494.5	376.92	2283.2	2660.1	1.1925	6.2866	7.4791
95	0.08455	0.001040	1.982	397.88	2102.7	2500.6	397.96	2270.2	2668.1	1.2500	6.1659	7.4159
100	0.10135	0.001044	1.6729	418.94	2087.6	2506.5	419.04	2257.0	2676.1	1.3069	6.0480	7.3549
105	0.12082	0.001048	1.4194	440.02	2072.3	2512.4	440.15	2243.7	2683.8	1.3630	5.9328	7.2958
110	0.14327	0.001052	1.2102	461.14	2057.0	2518.1	461.30	2230.2	2691.5	1.4185	5.8202	7.2387

Temp	Sat. press.	Specific volume, m^3/kg		Internal energy, kJ/kg			Enthalpy, kJ/kg			Entropy, kJ/(kg·K)		
		Sat. liquid	Sat. vapor	Sat. liquid	Evap.	Sat. vapor	Sat. liquid	Evap.	Sat. vapor	Sat. liquid	Evap.	Sat. vapor
T,°C	P_{sat}, MPa	v_f	v_g	u_f	u_{fg}	u_g	h_f	h_{fg}	h_g	s_f	s_{fg}	s_g
115	0.16906	0.001056	1.0366	482.30	2041.4	2523.7	482.48	2216.5	2699.0	1.4734	5.7100	7.1833
120	0.19853	0.001060	0.8919	503.50	2025.8	2529.3	503.71	2202.6	2706.3	1.5276	5.6020	7.1296
125	0.2321	0.001065	0.7706	524.74	2009.9	2534.6	524.99	2188.5	2713.5	1.5813	5.4962	7.0775
130	0.2701	0.001070	0.6685	546.02	1993.9	2539.9	546.31	2174.2	2720.5	1.6344	5.3925	7.0269
135	0.3130	0.001075	0.5822	567.35	1977.7	2545.0	567.69	2159.6	2727.3	1.6870	5.2907	6.9777
140	0.3613	0.001080	0.5089	588.74	1961.3	2550.0	589.13	2144.7	2733.9	1.7391	5.1908	6.9299
145	0.4154	0.001085	0.4463	610.18	1944.7	2554.9	610.63	2129.6	2740.3	1.7907	5.0926	6.8833
150	0.4758	0.001091	0.3928	631.68	1927.9	2559.5	632.20	2114.3	2746.5	1.8418	4.9960	6.8379
155	0.5431	0.001096	0.3468	653.24	1910.8	2564.1	653.84	2098.6	2752.4	1.8925	4.9010	6.7935
160	0.6178	0.001102	0.3071	674.87	1893.5	2568.4	675.55	2082.6	2758.1	1.9427	4.8075	6.7502
165	0.7005	0.001108	0.2727	696.56	1876.0	2572.5	697.34	2066.2	2763.5	1.9925	4.7153	6.7078
170	0.7917	0.001114	0.2428	718.33	1858.1	2576.5	719.21	2049.5	2768.7	2.0419	4.6244	6.6663
175	0.8920	0.001121	0.2168	740.17	1840.0	2580.2	741.17	2032.4	2773.6	2.0909	4.5347	6.6256
180	1.0021	0.001127	0.19405	762.09	1821.6	2583.7	763.22	2015.0	2778.2	2.1396	4.4461	6.5857
185	1.1227	0.001134	0.17409	784.10	1802.9	2587.0	785.37	1997.1	2782.4	2.1879	4.3586	6.5465
190	1.2544	0.001141	0.15654	806.19	1783.8	2590.0	807.62	1978.8	2786.4	2.2359	4.2720	6.5079
195	1.3978	0.001149	0.14105	828.37	1764.4	2592.8	829.98	1960.0	2790.0	2.2835	4.1863	6.4698
200	1.5538	0.001157	0.12736	850.65	1744.7	2595.3	852.45	1940.7	2793.2	2.3309	4.1014	6.432
205	1.7230	0.001164	0.11521	873.04	1724.5	2597.5	875.04	1921.0	2796.0	2.3780	4.0172	6.3952
210	1.9062	0.001173	0.10441	895.53	1703.9	2599.5	897.76	1900.7	2798.5	2.4248	3.9337	6.3585
215	2.104	0.001181	0.09479	918.14	1682.9	2601.1	920.62	1879.9	2800.5	2.4714	3.8507	6.3221
220	2.318	0.001190	0.08619	940.87	1661.5	2602.4	943.62	1858.5	2802.1	2.5178	3.7683	6.2861
225	2.548	0.001199	0.07849	963.73	1639.6	2603.3	966.78	1836.5	2803.3	2.5639	3.6863	6.2503
230	2.795	0.001209	0.07158	986.74	1617.2	2603.9	990.12	1813.8	2804.0	2.6099	3.6047	6.2146

Temp	Sat. press.	Specific volume, m^3/kg		Internal energy, kJ/kg			Enthalpy, kJ/kg			Entropy, kJ/(kg·K)		
		Sat. liquid	Sat. vapor	Sat. liquid	Evap.	Sat. vapor	Sat. liquid	Evap.	Sat. vapor	Sat. liquid	Evap.	Sat. vapor
T,ºC	P_{sat}, MPa	v_f	v_g	u_f	u_{fg}	u_g	h_f	h_{fg}	h_g	s_f	s_{fg}	s_g
235	3.060	0.001219	0.06537	1009.89	1594.2	2604.1	1013.62	1790.5	2804.2	2.6558	3.5233	6.1791
240	3.344	0.001229	0.05976	1033.21	1570.8	2604.0	1037.32	1766.5	2803.8	2.7015	3.4422	6.1437
245	3.648	0.001240	0.05471	1056.71	1546.7	2603.4	1061.23	1741.7	2803.0	2.7472	3.3612	6.1083
250	3.973	0.001251	0.05013	1080.39	1522.0	2602.4	1085.36	1716.2	2801.5	2.7927	3.2802	6.0730

TABLE4A: Saturated water – temperature table (continued)

T°C	P_{sat}, MPa	v_f	v_g	u_f	u_{fg}	u_g	h_f	h_{fg}	h_g	s_f	s_{fg}	s_g
255	4.319	0.001263	0.04598	1104.28	1596.7	2600.9	1109.73	1689.8	2799.5	2.8383	3.1992	6.0375
260	4.688	0.001276	0.04221	1128.39	1470.6	2599.0	1134.37	1662.5	2796.9	2.8838	3.1181	6.0019
265	5.081	0.001289	0.03877	1152.74	1443.9	2596.6	1159.28	1634.4	2793.6	2.9294	3.0368	5.9662
270	5.499	0.001302	0.03564	1177.36	1416.3	2593.7	1184.51	1605.2	2789.7	2.9751	2.9551	5.9301
275	5.942	0.001317	0.03279	1202.25	1387.9	2590.2	1210.07	1574.9	2785.0	3.0208	2.8730	5.8938
280	6.412	0.001332	0.03017	1227.46	1358.7	2586.1	1235.99	1543.6	2779.6	3.0668	2.7903	5.8571
285	6.909	0.001348	0.02777	1253.00	1328.4	2581.4	1262.31	1511.0	2773.3	3.1130	2.7070	5.8199
290	7.436	0.001366	0.02557	1278.92	1297.1	2576.0	1289.07	1477.1	2766.2	3.1594	2.6227	5.7821
295	7.993	0.001384	0.02354	1305.2	1264.7	2569.9	1316.3	1441.8	2758.1	3.2062	2.5375	5.7437
300	8.581	0.001404	0.02167	1332.0.	1231.0	2563.0	1344.0	1404.9	2749.0	3.2534	2.4511	5.7045
305	9.202	0.001425	0.019948	1359.3	1195.9	2555.2	1372.4	1366.4	2738.7	3.3010	2.3633	5.6643
310	9.856	0.001447	0.018350	1387.1	1159.4	2546.4	1401.3	1326.0	2727.3	3.3493	2.2737	5.6230
315	10.547	0.001472	0.016867	1415.5	1121.1	2536.6	1431.0	1283.5	2714.5	3.3982	2.1821	5.5804
320	11.274	0.001499	0.015488	1444.6	1080.9	2525.5	1461.5	1238.6	2700.1	3.4480	2.0882	5.5362
330	12.845	0.001561	0.012996	1505.3	993.7	2498.9	1525.3	1140.6	2665.9	3.5507	1.8909	5.4417
340	14.586	0.001638	0.010797	1570.3	894.3	2464.6	1594.2	1027.9	2622.0	3.6594	1.6763	5.3357
350	16.513	0.001740	0.008813	1641.9	776.6	2418.4	1670.6	893.4	2563.9	3.7777	1.4335	5.2112
360	18.651	0.001893	0.006945	1725.2	626.3	2351.5	1760.5	720.3	2481.0	3.9147	1.1379	5.0526
370	21.03	0.002213	0.004925	1844.0	384.5	2228.5	1890.5	441.6	2332.1	4.1106	0.6865	4.7971
374.14	22.09	0.003155	0.003155	2029.6	0	2029.6	2099.3	0	2099.3	4.4298	0	4.4298

Tables A–4A to A–4 C adapted from G. J. Wylen, and R. Sonntag, *Fundamentals of Classical Thermdynamics*, 3rd Ed. John Wiley & Sons, 1986; originally published in J. H. Keenan and F. G. Keyes, P. G. Hill and J. G. Moore, *Steam Tables*, John Wiley & Sons, 1978.

TABLE 4B: Properties of saturated water (liquid–vapor): pressure table

Press.	Sat. Temp.	Specific volume, m^3/kg Sat. liquid	Sat. vapor	Internal energy, kJ/kg Sat. liquid	Evap.	Sat. vapor	Enthalpy, kJ/kg Sat. liquid	Evap.	Sat. vapor	Entropy, kJ/(kg·K) Sat. liquid	Evap.	Sat. vapor
P, MPa	T,°C	v_f	v_g	u_f	u_{fg}	u_g	h_f	h_{fg}	h_g	s_f	s_{fg}	s_g
0.00061	0.01	0.001000	206.14	0.00	2375.3	2375.3	0.01	2501.3	2501.4	0.0000	9.1562	9.1562
0.0010	6.98	0.001000	129.21	29.30	2355.7	2385.0	29.30	2484.9	2514.2	0.1059	8.8697	8.9756
0.0015	13.03	0.001001	87.98	54.71	2338.6	2393.3	54.71	2470.6	2525.3	0,1957	8.6322	8.8279
0.0020	17.50	0.001001	67.00	73.48	2326.0	2399.5	73.48	2460.0	2533.5	0.2607	8.4629	8.7237
0.0025	21.08	0.001002	54.25	88.48	2315.9	2404.4	88.49	2451.6	2540.0	0.3120	8.3311	8.6432
0.0030	24.08	0.001003	45.67	101.04	2307.5	2408.5	101.05	2444.5	2545.5	0.3545	8.2231	8.5776
0.0040	28.96	0.001004	34.80	121.45	2293.7	2415.2	121.46	2432.9	2554.4	0.4226	8.0520	8.4746
0.0050	32.88	0.001005	28.19	137.81	2282.7	2420.5	137.82	2423.7	2561.5	0.4764	7.9187	8.3951
0.0075	40.29	0.001008	19.24	168.78	2261.7	2430.5	168.79	2406.0	2574.8	0.5764	7.6750	8.2515
0.010	45.84	0.001010	14.67	191.82	2246.1	2437.9	191.83	2392.8	2584.7	0.6493	7.5009	8.1502
0.015	53.97	0.001014	10.02	225.92	2222.8	2448.7	225.94	2373.1	2599.1	0.7549	7.2536	8.0085
0.020	60.06	0.001017	7.649	251.38	2205.4	2456.7	251.40	2358.3	2609.7	0.8320	7.0766	7.9085
0.025	64.97	0.001020	6.204	271.90	2191.2	2463.1	271.93	2346.3	2618.2	0.8931	6.9383	7.8314
0.030	69.10	0.001022	5.229	289.20	2179.2	2468.4	289.23	2336.1	2625.3	0.9439	6.8247	7.7686
0.040	75.87	0.001027	3.993	317.53	2159.5	2477.0	317.58	2319.2	2636.8	1.0259	6.6441	7.6700
0.050	81.33	0.001030	3.240	340.44	2143.4	2483.9	340.49	2305.4	2645.9	1.0910	6.5029	7.5939
0.075	91.78	0.001037	2.217	384.31	2112.4	2496.7	384.39	2278.6	2663.0	1.2130	6.2434	7,4564
0.100	99.63	0.001043	1.6940	417.36	2088.7	2506.1	417.46	2258.0	2675.5	1.3026	6.0568	7.3594
0.125	105.99	0.001048	1.3749	444.19	2069.3	2513.5	444.32	2241.0	2685.4	1.3740	5.9104	7.2844
0.150	111.37	0.001053	1.1593	466.94	2052.7	2519.7	467.11	2226.5	2693.6	1.4336	5.7897	7.2233
0.175	116.06	0.001057	1.0036	486.80	2038.1	2524.9	486.99	2213.6	2700.6	1.4849	5.6868	7.1717
0.200	120.23	0.001061	0.8857	504.49	2025.0	2529.5	504.70	2201.9	2706.7	1.5301	5.5970	7.1271
0.225	124.00	0.001064	0.7933	520.47	2013.1	2533.6	520.72	2191.3	2712.1	1.5706	5.5173	7.0878
0.250	127.44	0.001067	0.7187	535.10	2002.1	2537.2	535.37	2181.5	2716.9	1.6072	5.4455	7.0527
0.275	130.60	0.001070	0.6573	548.59	1991.9	2540.5	548.89	2172.4	2721.3	1.6408	5.3801	7.0209
0.300	133.55	0.001073	0.6058	561.15	1982.4	2543.6	561.47	2163.8	2725.3	1.6718	5.3201	6.9919
0.325	136.30	0.001076	0.5620	572.90	1973.5	2546.4	573.25	2155.8	2729.0	1.7006	5.2646	6.9652

Press.	Sat. Temp.	Specific volume, m^3/kg		Internal energy, kJ/kg			Enthalpy, kJ/kg			Entropy, kJ/(kg·K)		
		Sat. liquid	Sat. vapor	Sat. liquid	Evap.	Sat. vapor	Sat. liquid	Evap.	Sat. vapor	Sat. liquid	Evap.	Sat. vapor
P, MPa	T,°C	v_f	v_g	u_f	u_{fg}	u_g	h_f	h_{fg}	h_g	s_f	s_{fg}	s_g
0.350	138.88	0.001079	0.5243	583.95	1965.0	2548.9	584.33	2148.1	2732.4	1.7275	5.2130	6.9405
0.375	141.32	0.001081	0.4914	594.40	1956.9	2551.3	594.81	2140.8	2735.6	1.7528	5.1647	6.9175
0.40	143.63	0.001084	0.4625	604.31	1949.3	2553.6	604.74	2133.8	2738.6	1.7766	5.1193	6.8959
0.45	147.93	0.001088	0.4140	622.77	1934.9	2557.6	623.25	2120.7	2743.9	1.8207	5.0359	6.8565
0.50	151.86	0.001093	0.3749	639.68	1921.6	2561.2	640.23	2108.5	2748.7	1.8607	4.9606	6.8213
0.55	155.48	0.001097	0.3427	655.32	1909.2	2564.5	665.93	2097.0	2753.0	1.8973	4.8920	6.7893
0.60	158.85	0.001101	0.3157	669.90	1897.5	2567.4	670.56	2086.3	2756.8	1.9312	4.8288	6.7600
0.65	162.01	0.001104	0.2927	683.56	1886.5	2570.1	684.28	2076.0	2760.3	1.9627	4.7703	6.7331
0.70	164.97	0.001108	0.2729	696.44	1876.1	2572.5	697.22	2066.3	2763.5	1.9922	4.7158	6.7080
0.75	167.78	0.001112	0.2556	708.64	1866.1	2574.7	709.47	2057.0	2766.4	2.0200	4.6647	6.6847
0.80	170.43	0.001115	0.2404	720.22	1856.6	2576.8	721.11	2048.0	2769.1	2.0462	4.6166	6.6628
0.85	172.96	0.001118	0.2270	731.27	1847.4	2578.7	732.22	2039.4	2771.6	2.0710	4.5711	6.6421
0.90	175.38	0.001121	0.2150	741.83	1838.6	2580.5	742.83	2031.1	2773.9	2.0946	4.5280	6.6226
0.95	177.69	0.001124	0.2402	751.95	1830.2	2582.1	753.02	2023.1	2776.1	2.1172	4.4869	6.6041
1.00	179.91	0.001127	0.19444	761.68	1822.0	2583.6	762.81	2015.3	2778.1	2.1387	4.4478	6.5865
1.10	184.09	0.001133	0.17753	780.09	1806.3	2586.4	781.34	2000.4	2871.7	2.1792	4.3744	6.5536
1.20	187.99	0.001139	0.16333	797.29	1791.5	2588.8	798.65	1986.2	2784.8	2.2166	4.3067	6.5233
1.30	191.64	0.001144	0.15125	813.44	1777.5	2591.0	814.93	1972.7	2787.6	2.2515	4.2438	6.4953

TABLE 4C: Properties of superheated water vapor

T	v	u	h	s	v	u	h	s	v	u	h	s
°C	m^3/kg	kJ/kg	kJ/kg	kJ/(kg·K)	m^3/kg	kJ/kg	kJ/kg	kJ/(kg·K)	m3/kg	kJ/kg	kJ/kg	kJ/(kg·K)
	P = 0.01 MPa (45.81°C)				P = 0.05 MPa (81.33°C)				P = 0. 10 **MPa** (99.63°C)			
Sat.	14.674	2437.9	2584.7	8.1502	3.240	2483.9	2645.9	7.5939	1.6940	2506.1	2675.5	7.3594
50	14.869	2443.9	2592.6	8.1749								
100	17.196	2515.5	2687.5	8.4479	3.418	2511.6	2682.5	7.6947	1.6958	2506.7	2676.2	7.3614
150	19.512	2587.9	2783.0	8.6882	3.889	2585.6	2780.1	7.9401	1.9364	2582.8	2776.4	7.6134
200	21.825	2661.3	2879.5	8.9038	4.356	2659.9	2877.7	8.1580	2.172	2658.1	2875.3	7.8343
250	24.136	2736.0	2977.3	9.1002	4.820	2735.0	2976.0	8.3556	2.406	2 3. 7	2974.3	8.0333
300	26.445	2812.1	3076.5	9,2813	5.284	2811.3	3075.5	8.5373	2.639	2810.4	3074.3	8.2158
400	31.063	2968.9	3279.6	9.6077	6.209	2968.5	3278.9	8.8642	3.103	2967.9	3278.2	8.5435
500	35.679	3132.3	3489.1	9.8978	7.134	3132.0	3488.7	9.1546	3.565	3131.6	3488.1	8.8342
600	40.295	3302.5	3705.4	10.1608	8.057	3302.2	3705.1	9.4178	4.028	3301.9	3704.4	9.0976
700	44.911	3479.6	3928.7	10.4028	8.981	3479.4	3928.5	9.6599	4.490	3479.2	3928.2	9.3398
800	49.526	3663.8	4159.0	10.6281	9.904	3663.6	4158.9	9.8852	4.952	3663.5	4158.6	9.5652
900	54.141	3855.0	4396.4	10.8396	10.828	3854.9	4396.3	10.0967	5.414	3854.8	4396.1	9.7767
1000	58.757	4053.0	4640.6	11.0393	11.751	4052.9	4640.5	10.2964	5.875	4052.8	4640.3	9.9764
1100	63.372	4257.5	4891.2	11.2287	12.674	4257.4	4891.1	10.4859	6.337	4257.3	4891.0	10.1659
1200	67.987	4467.9	5147.8	11.4091	13.597	4467.8	5147.7	10.6662	6.799	4467.7	5147.6	10.3463
1300	72.602	4683.7	5409.7	11.5811	14.521	4683.6	5409.6	10.8382	7.260	4683.5	5409.5	10.5183
	P =0.20 MPa (120.23°C)				P = 0.30 MPa (133.55°C)				P = 0.40 MPa (143.63°C)			
Sat.	0.8857	2529.5	2706.7	7.1272	0.6058	2543.6	2725.3	6.9919	0.4625	2553.6	2738.6	6.8959
150	0.9596	2576.9	2768.8	7.2795	0.6339	2570.8	2761.0	7.0778	0.4708	2564.5	2752.8	6.9299
200	1.0803	2654.4	2870.5	7.5066	0.7163	2650.7	2865.6	7.3115	0.5342	2646.8	2860.5	7.1706
250	1.1988	2731.2	2971.0	7.7086	0.7964	2728.7	2967.6	7.5166	0.5951	2726.1	2964.2	7.3789
300	1.3162	2808.6	3071.8	7.8926	0.8753	2806.7	3069.3	7.7022	0.6548	2804.8	3066.8	7.5662
400	1.5493	2966.7	3276.6	8.2218	1.0315	2965.6	3275.0	8.0330	0.7726	2964.4	3273.4	7.8985
500	1.7814	3130.8	3487.1	8.5133	1.1867	3130.0	3486.0	8.3251	0.8893	3129.2	3484.9	8.1913
600	2.013	3301.4	3704.0	8.7770	1.3414	3300.8	3703.2	8.5892	1.0055	3300.2	3702.4	8.4558
700	2.244	3478.8	3927.6	9.0194	1.4957	3478.4	3927.1	8.8319	1.1215	3477.9	3926.5	8.6987
800	2.475	3663.1	4158.2	9.2449	1.6499	3662.9	4157.8	9.0576	1.2372	3662.4	4157.3	8.9244

T °C	v m^3/kg	u kJ/kg	h kJ/kg	s kJ/(kg·K)	v m^3/kg	u kJ/kg	h kJ/kg	s kJ/(kg·K)	v m3/kg	u kJ/kg	h kJ/kg	s kJ/(kg·K)
900	2.705	3854.5	4395.8	9.4566	1.8041	3854.2	4395.4	9.2692	1.3529	3853.9	4395.1	9.1362
1000	2.937	4052.5	4640.0	9.6563	1.9581	4052.3	4639.7	9.4690	1.4685	4052.0	4639.4	9.3360
1100	3.168	4257.0	4890.7	9.8458	2.1121	4256.8	4890.4	9.6585	1.5840	4256.5	4890.2	9.5256
1200	3.399	4467.5	5147.5	10.0262	2.2661	4467.2	5147.1	9.8389	1.6996	4467.0	5146.8	9.7060
1300	3.630	4683.2	5409.3	10.1982	2.4201	4683.0	5409.0	10.0110	1.8151	4682.8	5408.8	9.8780
	P =0.50 MPa (151.86°C)				P =0.60 **MPa** (158.85°C)				P = 0.80 MPa (170.43°C)			
Sat.	0.3749	2561.2	2748.7	6.8213	0.3157	2567.4	2756.8	6,7600	0.2404	2576.8	2769.1	6.6628
200	0.4249	2642.9	2855.4	7.0592	0.3520	2638.9	2850.1	6.9665	0.2608	2630.6	2839.3	6.8158
250	0.4744	2723.5	2960.7	7.2709	0:3938	2720.9	2957.2	7.1816	0.2931	2715.5	2950.0	7.0384
300	0.5226	2802.9	3064.2	7.4599	0.4344	2801.0	3061.6	7.3724	0.3241	2797.2	3056.5	7.2328
350	0.5701	2882.6	3167.7	7.6329	0.4742	2881.2	3165.7	7.5464	0.3544	2878.2	3161.7	7.4089
400	0.6173	2963.2	3271.9	7.7938	0.5137	2962.1	3270.3	7.7079	0.3843	2959.7	3267.1	7.5716
500	0.7109	3128.4	3483.9	8.0873	0.5920	3127.6	3482.8	8.0021	0.4433	3126.0	3480.6	7.8673
600	0.8041	3299.6	3701.7	7.3522	0.6697	3299.1	3700.9	8.2674	0.5018	3297.9	3699.4	8.1333
700	0.8969	3477.5	3925.9	8.5952	0.7472	3477.0	3925.3	8.5107	0.5601	3476.2	3924.2	8.3770
800	0.9896	3662.1	4156.9	8.8211	0.8245	3661.8	4156.5	8.7367	0.6181	3661.1	4155.6	8.6033
900	1.0822	3853.6	4394.7	9.0329	0.9017	3853.4	4394.4	8.9486	0.6761	3852.8	4393.7	8.8153
1000	1.1747	4051.8	4639.1	9.2328	0.9788	4051.5	4638.8	9.1485	0.7340	4051.0	4638.2	9.0153
1100	1.2672	4256.3	4889.9	9.4224	1.0559	4256.1	4889.6	9.3381	0.7919	4255.6	4889.1	9.2050
1200	1.3596	4466.8	5146.6	9.6029	1.1330	4466.5	5146.3	9.5185	0.8497	4466.1	5145.9	9.3855
1300	1.4521	4682.5	5408.6	9.7749	1.2101	4682.3	5408.3	9.6906	0.9076	4681.8	5407.9	9.5575

TABLE 4C: Properties of superheated water vapor (continued)

T °C	v m³/kg	u kJ/kg	h kJ/kg	s kJ/(kg·K)	v m³/kg	u kJ/kg	h kJ/kg	s kJ/(kg·K)	v m³/kg	u kJ/kg	h kJ/kg	s kJ/(kg·K)
	P = 1 MPa (179.91°C)				P = 1.2 MPa (187.99°C)				P = 1.4 **MPa** (195.07°C)			
Sat.	0.19444	2583.6	2778.1	6.5865	0.16333	2588.8	2784.8	6.5233	0.14084	2592.8	2790.0	6.4693
200	0.2060	2621.9	2827.9	6.6940	0.16930	2612.8	2815.9	6.5898	0.14302	2603.1	2803.3	6.4975
250	0.2327	2709.9	2942.6	6.9247	0.19234	2704.2	2935.0	6.8294	0.16350	2698.3	2927.2	6.7467
300	0.2579	2793.2	3051.2	7.1229	0.2138	2789.2	3045.8	7.0317	0.18228	2785.2	3040.4	6.9534
350	0.2825	2875.2	3157.7	7.3011	0.2345	2872.2	3153.6	7.2121	0.2003	2869.2	3149.5	7.1360
400	0.3066	2957.3	3263.9	7.4651	0.2548	2954.9	3260.7	7.3774	0.2178	2952.5	3257.5	7.3026
500	0.3541	3124.4	3478.5	7.7622	0.2946	3122.8	3476.3	7.6759	0.2521	3121.1	3474.1	7.6027
600	0.4011	3296.8	3697.9	8.0290	0.3339	3295.6	3696.3	7.9435	0.2860	3294.4	3694.8	7.8710
700	0.4478	3475.3	3923.1	8.2731	0.3729	3474.4	3922.0	8.1881	0.3195	3473.6	3920.8	8.1160
800	0.4943	3660.4	4154.7	8.4996	0.4118	3659.7	4153.8	8.4148	0.3528	3659.0	4153.0	8.3431
900	0,5407	3852.2	4392.9	8.7118	0.4505	3851.6	4392.2	8.6272	0.3861	3851.1	4391.5	8.5556
1000	0.5871	4050.5	4637.6	8.9119	0.4892	4050.0	4637.0	8.8274	0.4192	4049.5	4636.4	8.7559
1100	0.6335	4255.1	4888.6	9.1017	0.5278	4254.6	4888.0	9.0172	0.4524	4254.1	4887.5	8.9457
1200	0.6798	4465.6	5145.4	9.2822	0.5665	4465.1	5144.9	9.1977	0,4855	4464.7	5144.4	9.1262
1300	0.7261	4681.3	5407.4	9.4543	0.6051	4680.9	5407.0	9.3698	0.5186	4680.4	5406.5	9.2984
	P = 1.6 MPa (201.41°C)				P = 1.8 MPa (207.15°C)				P = 2.0 MPa (212.42°C)			
Sat.	0.12380	2596.0	2794.0	6.4218	0.11042	2598.4	2797.1	6.3794	0.09963	2600.3	2799.5	6.3409
225	0.13287	2644.7	2857.3	6.5518	0.11673	2636.6	2846.7	6.4808	0.10377	2628.3	2835.8	6.4147
250	0.14184	2692.3	2919.2	6.6732	0.12497	2686.0	2911.0	6.6066	0.11144	2679.6	2902.5	6.5453
300	0.15862	2781.1	3034.8	6.8844	0.14021	2776.9	3029.2	6.8226	0.12547	2772.6	3023.5	6.7664
350	0.17456	2866.1	3145.4	7.0694	0.15457	2863.0	3141.2	7.0100	0.13857	2859.8	3137.0	6.9563
400	0.19005	2950.1	3254.2	7.2374	0.16847	2947.7	3250.9	7.1794	0.15120	2945.2	3247.6	7.1271
500	0.2203	3119.5	3472.0	7.5390	0.19550	3117.9	3469.8	7.4825	0.17568	3116.2	3467.6	7.4317
600	0.2500	3293.3	3693.2	7.8080	0.2220	3292.1	3691.7	7.7523	0.19960	3290.9	3690.1	7.7024
700	0.2794	3472.7	3919.7	8.0535	0.2482	3471.8	3918.5	7.9983	0.2232	3470.9	3917.4	7.9487
800	0.3086	3658.3	4152.1	8.2808	0.2742	3657.6	4151.2	8.2258	0.2467	3657.0	4150.3	8.1765
900	0.3377	3850.5	4390.8	8.4935	0.3001	3849.9	4390.1	8.4386	0.2700	3849.3	4389.4	8.3895
1000	0.3668	4049.0	4635.8	8.6938	0.3260	4048.5	4635.2	8.6391	0.2933	4048.0	4634.6	8.5901

T °C	v m³/kg	u kJ/kg	h kJ/kg	s kJ/(kg·K)	v m³/kg	u kJ/kg	h kJ/kg	s kJ/(kg·K)	v m³/kg	u kJ/kg	h kJ/kg	s kJ/(kg·K)
1100	0.3958	4253.7	4887.0	8.8837	0.3518	4253.2	4886.4	8.8290	0.3166	4252.7	4885.9	8.7800
1200	0.4248	4464.2	5143.9	9.0643	0.3776	4463.7	5143.4	9.0096	0.3398	4463.3	5142.9	8.9607
1300	0.4538	4679.9	5406.0	9.2364	0.4034	4679.5	5405.6	9.1818	0.3631	4679.0	5405.1	9.1329

	P = 2.5 MPa (223.99°C)				P = 3.0 MPa (233.90°C)				P = 3.5 MPa (242.60°C)			
Sat.	0.07998	2603.1	2803.1	6.2575	0.06668	2604.1	2804.2	6.1869	0.05707	2603.7	2803.4	6.1253
225	0.08027	2605.6	2806.3	6.2639								
250	0.08700	2662.6	2880.1	6.4085	0.07058	2644.0	2855.8	6.2872	0.05872	2623.7	2829.2	6.1749
300	0.09890	2761.6	3008.8	6.6438	0.08114	2750.1	2993.5	6.5390	0.06842	2738.0	2977.5	6.4461
350	0.10976	2851.9	3126.3	6.8403	0.09053	2843.7	3115.3	6.7428	0.07678	2835.3	3104.0	6.6579
400	0.12010	2939.1	3239.3	7.0148	0.09936	2932.8	3230.9	6.9212	0.08453	2926A	3222.3	6.8405
450	0.13014	3025.5	3350.8	7.1746	0.10787	3020.4	3344.0	7.0834	0.09196	3015.3	3337.2	7.0052
500	0.13993	3112.1	3462.1	7.3234	0.11619	3108.0	3456.5	7.2338	0.09918	3103.0	3450.9	7.1572
600	0.15930	3288.0	3686.3	7.5960	0.13243	3285.0	3682.3	7.5085	0.11324	3282.1	3678.4	7.4339
700	0.17832	3468.7	3914.5	7.8435	0.14838	3466.5	3911.7	7.7571	0.12699	3464.3	3908.8	7.6837
800	0.19716	3655.3	4148.2	8.0720	0.16414	3653.5	4145.9	7.9862	0.14056	3651.8	4143.7	7.9134
900	0.21590	3847.9	4387.6	8.2853	0.17980	3846.5	4385.9	8.1999	0.15402	3845.0	4384.1	8.1276
1000	0.2346	4046.7	4633.1	8.4861	0.19541	4045.4	4631.6	8.4009	0.16743	4044.1	4630.1	8.3288
1100	0.2532	4251.5	4884.6	8.6762	0.21098	4250.3	4883.3	8.5912	0.18080	4249.2	4881.9	8.5192
1200	0.2718	4462.1	5141.7	8.8569	0.22652	4460.9	5140.5	8.7720	0.19415	4459.8	5139.3	8.7000
1300	0.2905	4677.8	5404.0	9.0291	0.24206	4676.6	5402.8	8.9442	0.20749	4675.5	5401.7	8.8723

TABLE 4C: Properties of superheated water vapor (continued)

T °C	v m^3/kg	u kJ/kg	h kJ/kg	s kJ/(kg·K)	v m^3/kg	u kJ/kg	h kJ/kg	s kJ/(kg·K)	v m^3/kg	u kJ/kg	h kJ/kg	s kJ/(kg·K)
	P = 4 MPa (250.40°C)				P = 4.5 MPa (257.49°C)				P = 5.0 MPa (263.99°C)			
Sat.	0.04978	2602.3	2801.4	6.0701	0.04406	2600.1	2798.3	6.0198	0.03944	2597.1	2794.3	5.9734
275	0.05457	2667.9	2886.2	6.2285	0.04730	2650.3	2863.2	6.1401	0.04141	2631.3	2838.3	6.0544
300	0.05884	2725.3	2960.7	6.3615	0.05135	2712.0	2643.1	6.2828	0.04532	2698.0	2924.5	6.2084
350	0.06645	2826.7	3092.5	6.5821	0.05840	2817.8	3080.6	6.5131	0.05194	2808.7	3068.4	6.4493
400	0.07341	2919.9	3213.6	6.7690	0.06475	2913.3	3204.7	6.7047	0.05781	2906.6	3195.7	6.6459
450	0.08002	3010.2	3330.3	6.9363	0.07074	3005.0	3323.3	6.8746	0.06330	2999.7	3316.2	6.8186
500	0.08643	3099.5	3445.3	7.0901	0.07651	3095.3	3439.6	7.0301	0.06857	3091.0	3433.8	6.9759
600	0,09885	3279.1	3674.4	7.3688	0.08765	3276.0	3670.5	7.3110	0.07869	3273.0	3666.5	7.2589
700	0.11095	3462.1	3905.9	7.6198	0.09847	3459.9	3903.0	7.5631	0.08849	3457.6	3900.1	7.5122
800	0.12287	3650.0	4141.5	7.8502	0.10911	3648.3	4139.3	7.7942	0.09811	3646.6	4137.1	7.7440
900	0.13469	3843.6	4382.3	8.0647	0.11965	3842.2	4380.6	8.0091	0.10762	3840.7	4378.8	7.9593
1000	0.14645	4042.9	4628.7	8.2662	0.13013	4041.6	4627.2	8.2108	0.11707	4040.4	4625.7	8.1612
1100	0.15817	4248.0	4880.6	8.4567	0.14056	4246.8	4879.3	8.4015	0.12648	4245.6	4878.0	8.3520
1200	0.16987	4458.6	5138.1	8.6376	0.15098	4457.5	5136.9	8.5825	0.13587	4456.3	5135.7	8.5331
1300	0.18156	4674.3	5400.5	8.8100	0.16139	4673.1	5399.4	8.7549	0.14526	4672.0	5398.2	8.7055

T °C	v m^3/kg	u kJ/kg	h kJ/kg	s kJ/(kg·K)	v m^3/kg	u kJ/kg	h kJ/kg	s kJ/(kg·K)	v m^3/kg	u kJ/kg	h kJ/kg	s kJ/(kg·K)
	P = 6 MPa (275.64°C)				P = 7 MPa (285.88°C)				P = 8 MPa (295.06°C)			
Sat.	0.03244	2589.7	2784.3	5.8892	0.02737	2580.5	2772.1	5.8133	0.02352	2569.8	2758.0	5.7432
300	0.03616	2667.2	2884.2	6.0674	0.02947	2632.2	2838.4	5.9305	0.02426	2590.9	2785.0	5.7906
350	0.04223	2789.6	3043.0	6.3335	0.03524	2769.4	3016.0	6.2283	0.02995	2747.7	2987.3	6.1301
400	0.04739	2892.9	3177.2	6.5408	0.03993	2878.6	3158.1	6.4478	0.03432	2863.8	3138.3	6.3634
450	0.05214	2988.9	3301.8	6.7193	0.04416	2978.0	3287.1	6.6327	0.03817	2966.7	3272.0	6.5551
500	0.05665	3082.2	3422.2	6.8803	0.04814	3073.4	3410.3	6.7975	0.04175	3064.3	3398.3	6.7240
550	0.06101	3174.6	3540.6	7.0288	0.05195	3167.2	3530.9	6.9486	0.04516	3159.8	3521.0	6.8778
600	0.06525	3266.9	3658.4	7.1677	0.05565	3260.7	3650.3	7.0894	0.04845	3254.4	3642.0	7.0206
700	0.07352	3453.1	3894.2	7.4234	0.06283	3448.5	3888.3	7.3476	0.05481	3443.9	3882.4	7.2812
800	0.08160	3643.1	4132.7	7.6566	0.06981	3639.5	4128.2	7.5822	0.06097	3636.0	4123.8	7.5173
900	0.08958	3837.8	4375.3	7.8727	0.07669	3835.0	4371.8	7.7991	0.06702	3832.1	4368.3	7.7351
1000	0.09749	4037.8	4622.7	8.0751	0.08350	4035.3	4619.8	8.0020	0.07301	4032.8	4616.9	7.9384

T °C	v m^3/kg	u kJ/kg	h kJ/kg	s kJ/(kg·K)	v m^3/kg	u kJ/kg	h kJ/kg	s kJ/(kg·K)	v m^3/kg	u kJ/kg	h kJ/kg	s kJ/(kg·K)
1100	0.10536	4243.3	4875.4	8.2661	0.09027	4240.9	4872.8	8.1933	0.07896	4238.6	4870.3	8.1300
1200	0.11321	4454.0	5133.3	8.4474	0.09703	4451.7	5130.9	8.3747	0.08489	4449.5	5128.5	8.3115
1300	0.12106	4669.6	5396.0	8.6199	0.10377	4667.3	5393.7	8.5475	0.90980	4665.0	5391.5	8.4842

T °C	P = 9 MPa (303.40°C)				P = 10.0 MPa (311.06°C)				P = 12.5 MPa (327.89°C)			
	v m^3/kg	u kJ/kg	h kJ/kg	s kJ/(kg·K)	v m^3/kg	u kJ/kg	h kJ/kg	s kJ/(kg·K)	v m^3/kg	u kJ/kg	h kJ/kg	s kJ/(kg·K)
Sat.	0.02048	2557.8	2742.1	5.6772	0.018026	2544.4	2724.7	5.6141	0.013495	2505.1	2673.8	5.4624
325	0.02327	2646.6	2856.0	5.8712	0.019861	2610.4	2809.1	5.7568				
350	0.02580	2724.4	2956.6	6.0361	0.02242	2699.2	2923.4	5.9443	0.016126	2624.6	2826.2	5.7118
400	0.02993	2848.4	3117.8	6.2854	0.02641	2832.4	3096.5	6.2120	0.02000	2789.3	3039.3	6.0417
450	0.03350	2955.2	3256.6	6.4844	0.02975	2943.4	3240.9	6.4190	0.02299	2912.5	3199.8	6.2719
500	0.03677	3055.2	3386.1	6.6576	0.03279	3045.8	3373.7	6.5966	0.02560	3021.7	3341.8	6.4618
550	0.03987	3152.2	3511.0	6.8142	0.03564	3144.6	3500.9	6.7561	0.02801	3125.0	3475.2	6.6290
600	0.04285	3248.1	3633.7	6.9589	0.03837	3241.7	3625.3	6.9029	0.03029	3225.4	~604.0'	6.7810
650	0.04574	3343.6	3755.3	7.0943	0.04101	3338.2	3748.2	7.0398	0.03248	3324.4	3730.4	6.9218
700	0.04857	3439.3	3876.5	7.2221	0.04358	3434.7	3870.5	7.1687	0.03460	3422.9	3855.3	7.0536
800	0.05409	3632.5	4119.3	7.4596	0.04859	3628.9	4114.8	7.4077	0.03869	3620.0	4103.6	7.2965
900	0.05950	3829.2	4364.8	7.6783	0.05349	3826.3	4361.2	7.6272	0.04267	3819.1	4352.5	7.5182
1000	0.06485	4030.3	4614.0	7.8821	0.05832	4027.8	4611.0	7.8315	0.04658	4021.6	4603.8	7.7237
1100	0.07016	4236.3	4867.7	8.0740	0.06312	4234.0	4865.1	8.0237	0.05045	4228.2	4858.8	7.9165
1200	0.07544	4447.2	5126.2	8.2556	0.06789	4444.9	5123.8	8.2055	0.05430	4439.3	5118.0	8.0937
1300	0.08072	4662.7	5389.2	8.4284	0.07265	4460.5	5387.0	8.3783	0.05813	4654.8	5381.4	8.2717

TABLE 4C: Properties of superheated water vapor (continued)

T °C	v m^3/kg	u kJ/kg	h kJ/kg	s kJ/(kg·K)	v m^3/kg	u kJ/kg	h kJ/kg	s kJ/(kg·K)	v m^3/kg	u kJ/kg	h kJ/kg	s kJ/(kg·K)
	P = 15 MPa (342.24°C)				P = 17.5 MPa (354.75°C)				P = 20.0 MPa (365.81°C)			
Sat.	0.010337	2455.5	2610.5	5.3098	0.007920	2390.2	2528.8	5.1419	0.005834	2293.0	2409.7	4.9269
350	0.011470	2520.4	2692.4	5.4421								
400	0.015649	2740.7	2975.5	5.8811	0.012447	2685.0	2902.9	5.7213	0.009942	2619.3	2818.1	5.5540
450	0.018445	2879.5	3156.2	6.1404	0.015174	2844.2	3109.7	6.0184	0.012695	2806.2	3060.1	5.9017
500	0.02080	2996.6	3308.fi	6.3443	0.017358	2970.3	3274.1	6.2383	0.014768	2942.9	3238.2	6.1401
550	0.02293	3104.7	3448.6	6.5199	0.019288	3083.9	3421.4	6.4230	0.016555	3062.4	3393.5	6.3348
600	0.02491	3208.6	3582.3	6.6776	0.02106	3191.5	3560.1	6.5866	0.018178	3174.0	3537.6	6.5048
650	0.02680	3310.3	3712.3	6.8224	0.02274	3296.0	3693.9	6.7357	0.019693	3281.4	3675.3	6.6582
700	0.02861	3410.9	3840.1	6.9572	0.02434	3398.7	3824.6	6.8736	0.02113	3386.4	3809.0	6.7993
800	0.03210	3610.9	4092.4	7.2040	0.02738	3601.8	4081.1	7.1244	0.02385	3592.7	4069.7	7.0544
900	0.03546	3811.9	4343.8	7.4279	0.03031	3804.7	4335.1	7.3507	0.02645	3797.5	4326.4	7.2830
1000	0.03875	4015.4	4596.6	7.6348	0.03316	4009.3	4589.5	7.5589	0.02897	4003.1	4582.5	7.4925
1100	0.04200	4222.6	4852.6	7.8283	0.03597	4216.9	4846.4	7.7531	0.03145	4211.3	4840.2	7.6874
1200	0.04523	4433.8	5112.3	8.0108	0.03876	4428.3	5106.6	7.9360	0.03391	4422.8	5101.0	7.8707
1300	0.04845	4649.1	5376.0	8.1840	0.04154	4643.5	5370.5	8.1093	0.03636	4638.0	5365.1	8.0442
	P = 25 MPa				P = 30 MPa				P = 35 MPa			
375	0.0019731	1798.7	1848.0	4.0320	0.0017892	1737.8	1791.5	3.9305	0.0017003	1702.9	1762.4	3.8722
400	0.006004	2430.1	2580.2	5.1418	0.002790	2067.4	2151.1	4.4728	0.002100	1914.1	1987.6	4.2126
425	0.007881	2609.2	2806.3	5.4723	0.005303	2455.1	2614.2	5.1504	0.003428	2253.4	2373.4	4.7747
450	0.009162	2720.7	2949.7	5.6744	0.006735	2619.3	2821.4	5.4424	0.004961	2498.7	2672.4	5.1962
500	0.011123	2884.3	3162.4	5.9592	0.008678	2820.7	3081.1	5.7905	0.006927	2751.9	2994.4	5.6282
550	0.012724	3017.5	3335.6	6.1765	0.010168	2970.3	3275.4	6.0342	0.008345	2921.0	3213.0	5.9026
600	0.014137	3137.9	3491.4	6.3602	0.011446	3100.5	3443.9	6.2331	0.009527	3062.0	3395.5	6.1179
650	0.015433	3251.6	3637.4	6.5229	0.012596	3221.0	3598.9	6.4058	0.010575	3189.8	3559.9	6.3010
700	0.016646	3361.3	3777.5	6.6707	0.013661	3335.8	3745.6	6.5606	0.011533	3309.8	3713.5	6.4631
800	0.018912	3574.3	4047.1	6.9345	0.015623	3555.5	4024.2	6.8332	0.013278	3536.7	4001.5	6.7450
900	0.021045	3783.0	4309.1	7.1680	0,017448	3768.5	4291.9	7.0718	0.014883	3754.0	4274.9	6.9386
1000	0.02310	3990.9	4568.5	7.3802	0.019196	3978.8	4554.7	7.2867	0.016410	3966.7	4541.1	7.2064

1100	0.02512	4200.2	4828.2	7.5765	0.020903	4189.2	4816.3	7.4845	0.017895	4178.3	4804.6	7.4037
1200	0.02711	4412.0	5089.9	7.7605	0.022589	4401.3	5079.0	7.6692	0.019360	4390.7	5068.3	7.5910
1300	0.02910	4626.9	5354.4	7.9342	0.024266	4616.0	5344.0	7.8432	0.020815	4605.1	5333.6	7.7653
	P = 40 MPa				P = 50 MPa				P = 60 MPa			
375	0.0016407	1677.1	1742.8	3.8290	0.0015594	1638.6	1716.6	3.7639	0.0015028	1609.4	1699.5	3.7141
400	0.0019077	1854.6	1930.9	4.1135	0.0017309	1788.1	1874.6	4.0031	0.0016335	1745.4	1843.4	3.9318
425	0.002532	2096.9	2198.1	4.5029	0.002007	1959.7	2060.0	4.2734	0.0018165	1892.7	2001.7	4.1626
450	0.003693	2365.1	2512.8	4.9459	0.002486	2159.6	2284.0	4.5884	0.002085	2053.9	2179.0	4.4121
500	0.005622	2678.4	2903.3	5.4700	0.003892	2525.5	2720.1	5.1726	0.002956	2390.6	2567.9	4.9321
550	0.006984	2869.7	3149.1	5.7785	0.005118	2763.6	3019.5	5.5485	0.003956	2658.8	2896.2	5.3441
600	0.008094	3022.6	3346.4	6.0144	0.006112	2942.0	3247.6	5.8178	0.004834	2861.1	3151.2	5.6452
650	0.009063	3158.0	3520.6	6.2054	0.006966	3093.5	3441.8	6.0342	0.005595	3028.8	3364.5	5.8829
700	0.009941	3283.6	3681.2	6.3750	0.007727	3230.5	3616.8	6.2189	0.006272	3177.2	3553.5	6.0824
800	0.011523	3517.8	3978.7	6.6662	0.009076	3479.8	3933.6	6.5290	0.007459	3441.5	3889.1	6.4109
900	0.012962	3739.4	4257.9	6.9150	0.010283	3710.3	4224.4	6.7882	0.008508	3681.0	4191.5	6.6805
1000	0.014324	3954.6	4527.6	7.1356	0.011411	3930.5	4501.1	7.0146	0.009480	3906.4	4475.2	6.9127
1100	0.015642	4167.4	4793.1	7.3364	0.012496	4145.7	4770.5	7.2184	0.010409	4124.1	4748.6	7.1195
1200	0.016940	4380.1	5057.7	7.5224	0.013561	4359.1	5037.2	7.4058	0.011317	4338.2	5017.2	7.3083
1300	0.018229	4594.3	5323.5	7.6969	0.014616	4572.8	5303.6	7.5808	0.012215	4551.4	5284.3	7.4837

TABLE 5A: Properties of saturated refrigerant 134a (liquid–vapor): temperature table

Temp.	Sat. press.	Specific volume, m^3/kg		Internal energy, kJ/kg		Enthalpy, kJ/kg			Entropy, kJ/(kg·K)	
		Sat. liquid	Sat. vapor	Sat. liquid	Sat. vapor	Sat. liquid	Evap.	Sat. vapor	Sat. liquid	Sat. vapor
T,°C	P_{sat}, MPa	v_f	v_g	u_f	u_g	h_f	h_{fg}	h_g	s_f	s_g
–40	0.05164	0.0007055	0.3569	–0.04	204.45	0.00	222.88	222.88	0.0000	0.9560
– 36	0.06332	0.0007113	0.2947	4.68	206.73	4.73	220.67	225.40	0.0201	0.9506
–32	0.07704	0.0007172	0.2451	9.47	209.01	9.52	218.37	227.90	0.0401	0.9456
–28	0.09305	0.0007233	0.2052	14.31	211.29	14.37	216.01	230.38	0.0600	0.9411
–26	0.10199	0.0007265	0.1882	16.75	212.43	16.82	214.80	231.62	0.0699	0.9390
–24	0.11160	0.0007296	0.1728	19.21	213.57	19.29	213.57	232.85	0.0798	0.9370
–22	0.12192	0.0007328	0.1590	21.68	214.70	21.77	212.32	234.08	0.0897	0.9351
–20	0.13299	0.0007361	0.1464	24.17	215.84	24.26	211.05	235.31	0.0996	0.9332
– 18	0.14483	0.0007395	0.1350	26.67	216.97	26.77	209.76	236.53	0.1094	0.9315
–16	0.15748	0.0007428	0.1247	29.18	218.10	29.30	208.45	237.74	0.1192	0.9298
–12	0.18540	0.0007498	0.1068	34.25	220.36	34.39	205.77	240.15	0.1388	0.9267
–8	0.21704	0.0007569	0.0919	39.38	222.60	39.54	203.00	242.54	0.1583	0.9239
–4	0.25274	0.0007644	0.0794	44.56	224.84	44.75	200.15	244.90	0.1777	0.9213
0	0.29282	0.0007721	0.0689	49.79	227.06	50.02	197.21	247.23	0.1970	0.9190
4	0.33765	0.0007801	0.0600	55.08	229.27	55.35	194.19	249.53	0.2162	0.9169
8	0.38756	0.0007884	0.0525	60.43	231.46	60.73	191.07	251.80	0.2354	0.9150
12	0.44294	0.0007971	0.0460	65.83	233.63	66.18	187.85	254.03	0.2545	0.9132
16	0.50416	0.0008062	0.0405	71.29	235.78	71.69	184.52	256.22	0.2735	0.9116
20	0.57160	0.0008157	0.0358	76.80	237.91	77.26	181.09	258.35	0.2924	0.9102
24	0.64566	0.0008257	0.0317	82.37	240.01	82.90	177.55	260.45	0.3113	0.9089
26	0.68530	0.0008309	0.0298	85.18	241.05	85.75	175.73	261.48	0.3208	0.9082
28	0.72675	0.0008362	0.0281	88.00	242.08	88.61	173.89	262.50	0.3302	0.9076
30	0.77006	0.0008417	0.0265	90.84	243.10	91.49	172.00	263.50	0.3396	0.9070
32	0.81528	0.0008473	0.0250	93.70	244.12	94.39	170.09	264.48	0.3490	0.9064
34	0.86247	0.0008530	0.0236	96.58	245.12	97.31	168.14	265.45	0.3584	0.9058
36	0.91168	0.0008590	0.0223	99.47	246.11	100.25	166.15	266.40	0.3678	0.9053
38	0.96298	0.0008651	0.0210	102.38	247.09	103.21	164.12	267.33	0.3772	0.9047

Temp.	Sat. press.	Specific volume, m^3/kg		Internal energy, kJ/kg		Enthalpy, kJ/kg			Entropy, kJ/(kg·K)	
		Sat. liquid	Sat. vapor	Sat. liquid	Sat. vapor	Sat. liquid	Evap.	Sat. vapor	Sat. liquid	Sat. vapor
T,°C	P_{sat}, MPa	v_f	v_g	u_f	u_g	h_f	h_{fg}	h_g	s_f	s_g
40	1.0164	0.0008714	0.0199	105.30	248.06	106.19	162.05	268.24	0.3866	0.9041
42	1.0720	0.0008780	0.0188	108.25	249.02	109.19	159.94	269.14	0.3960	0.9035
44	1.1299	0.0008847	0.0177	111.22	249.96	112.22	157.79	270.01	0.4054	0.9030
48	1.2526	0.0008989	0.0159	117.22	251.79	118.35	153.33	271.68	0.4243	0.9017
52	1.3851	0.0009142	0.0142	123.31	253.55	124.58	148.66	273.24	0.4432	0.9004
56	1.5278	0.0009308	0.0127	129.51	255.23	130.93	143.75	274.68	0.4622	0.8990
60	1.6813	0.0009488	0.0114	135.82	256.81	137.42	138.57	275.99	0.4814	0.8973
70	2.1162	0.0010027	0.0086	152.22	260.15	154.34	124.08	278.43	0.5302	0.8918
80	2.6324	0.0010766	0.0064	169.88	262.14	172.71	106.41	279.12	0.5814	0.8827
90	3.2435	0.0011949	0.0046	189.82	261.34	193.69	82.63	276.32	0.6380	0.8655
100	3.9742	0.0015443	0.0027	218.60	248.49	224.74	34.40	259.13	0.7196	0.8117

Tables A–5A to A–5 C adapted from M. J. Moran, and H.N. Shapiro, *Fundamentals of Engineering Thermodynamics*, 2nd Ed. John Wiley & Sons, 1992; originally based on equations from D.P. Wilson, and R.S. Basu, *ASHRAE Transactions*, 94, Pt 2., 1988, pp. 2095–2118. (With permission.)

TABLE 5B: Properties of saturated refrigerant 134a (liquid–vapor): pressure table

Press.	Sat. Temp.	Specific volume, m^3/kg		Internal energy, kJ/kg		Enthalpy, kJ/kg			Entropy, kJ/(kg·K)	
		Sat. liquid	Sat. vapor	Sat. liquid	Sat. vapor	Sat. liquid	Evap.	Sat. vapor	Sat. liquid	Sat. vapor
MPa	T,°C	v_f	v_g	u_f	u_g	h_f	h_{fg}	h_g	s_f	s_g
0.06	–37.07	0.0007097	0.3100	3.41	206.12	3.46	221.27	224.72	0.0147	0.9520
0.08	–31.21	0.0007184	0.2366	10.41	209.46	10.47	217.92	228.39	0.0440	0.9447
0.10	–26.43	0.0007258	0.1917	16.22	212.18	16.29	215.06	231.35	0.0678	0.9395
0.12	–22.36	0.0007323	0.1614	21.23	214.50	21.32	212.54	233.86	0.0879	0.9354
0.14	–18.80	0.0007381	0.1395	25.66	216–52	25.77	210.27	236.04	0.1055	0.9322
0.16	–15.62	0.0007435	0.1229	29.66	218.32	29.78	208.18	237.97	0.1211	0.9295
0.18	–12.73	0.0007485	0.1098	33.31	219.94	33.45	206.26	239.71	0.1352	0.9273
0.20	–10.09	0.0007532	0.0993	36.69	221.43	36.84	204.46	241.30	0.1481	0.9253
0.24	–5.37	0.0007618	0.0834	42.77	224.07	42.95	201.14	244.09	0.1710	0.9222
0.28	–1.23	0.0007697	0.0719	48.18	226.38	48.39	198.13	246.52	0.1911	0.9197
0.32	2.48	0.0007770	0.0632	53.06	228.43	53.31	195.35	248.66	0.2089	0.9177
0.36	5.84	0.0007839	0.0564	57.54	230.28	57.82	192.76	250.58	0.2251	0.9160
0.4	8.93	0.0007904	0.0509	61.69	231.97	62.00	190.32	252.32	0.2399	0.9145
0.5	15.74	0.0008056	0.0409	70.93	235.64	71.33	184.74	256.07	0.2723	0.9117
0.6	21.58	0.0008196	0.0341	78.99	238.74	79.48	179.71	259.19	0.2999	0.9097
0.7	26.72	0.0008328	0.0292	86.19	241.42	86.78	175.07	261.85	0.3242	0.9080
0.8	31.33	0.0008454	0.0255	92.75	243.78	93.42	170.73	264.15	0.3459	0.9066
0.9	35.53	0.0008576	0.0226	98.79	245.88	99.56	166.62	266.18	0.3656	0.9054
1.0	39.39	0.0008695	0.0202	104.42	247.77	105.29	162.68	267.97	0.3838	0.9043
1.2	46.32	0.0008928	0.0166	114.69	251.03	115.76	155.23	270.99	0.4164	0.9023
1.4	52.43	0.0009159	0.0140	123.98	253.74	125.26	148.14	273.40	0.4453	0.9003
1.6	57.92	0.0009392	0.0121	132.52	256.00	134.02	141.31	275.33	0.4714	0.8982
1.8	62.91	0.0009631	0.0105	140.49	257.88	142.22	134.60	276.83	0.4954	0.8959
2.0	67.49	0.0009878	0.0093	148.02	259.41	149.99	127.95	277.94	0.5178	0.8934

Press.	Sat. Temp.	Specific volume, m^3/kg		Internal energy, kJ/kg		Enthalpy, kJ/kg			Entropy, kJ/(kg·K)	
		Sat. liquid	Sat. vapor	Sat. liquid	Sat. vapor	Sat. liquid	Evap.	Sat. vapor	Sat. liquid	Sat. vapor
MPa	T,°C	v_f	v_g	u_f	u_g	h_f	h_{fg}	h_g	s_f	s_g
2.5	77.59	0.0010562	0.0069	165.48	261.84	168.12	111.06	279.17	0.5687	0.8854
3.0	86.22	0.0011416	0.0053	181.88	262.16	185.30	92.71	278.01	0.6156	0.8735

TABLE 5C: Properties of superheated refrigerant 134a vapor

T °C	v m^3/kg	u kJ/kg	h kJ/kg	s kJ/(kg·K)	v m^3/kg	u kJ/kg	h kJ/kg	s kJ/(kg·K)	v m3/kg	u kJ/kg	h kJ/kg	s kJ/(kg·K)
	P = 0.06 MPa (−37.07°C)				P = 0.1 MPa (−26.43°C)				P = 0.14 MPa (−18.80°C)			
Sat.	0.31003	206.12	224.72	0.9520	0.19170	212.18	231.35	0.9395	0.13945	216.52	236.04	0.9322
−20	0.33536	217.86	237.98	1.0062	0.19770	216.77	236.54	0.9602				
−10	0.34992	224.97	245.96	1.0371	0.20686	224.01	244.70	0.9918	0.14549	223.03	243.40	0.9606
0	0.36433	232.24	254.10	1.0675	0.21587	231.41	252.99	1.0227	0.15219	230.55	251.86	0.9922
10	0.37861	239.69	262.41	1.0973	0.22473	238.96	261.43	1.0531	0.15875	238.21	260.43	1.0230
20	0.39279	247.32	270.89	1.1267	0.23349	246.67	270.02	1.0829	0.16520	246.01	269.13	1.0532
30	0.40688	255.12	279.53	1.1557	0.24216	254.54	278.76	1.1122	0.17155	253.96	277.97	1.0828
40	0.42091	263.10	288.35	1.1844	0.25076	262.58	287.66	1.1411	0.17783	262.06	286.96	1.1120
50	0.43487	271.25	297.34	1.2126	0.25930	270.79	296.72	1.1696	0.18404	270.32	296.09	1.1407
60	0.44879	279.58	306.51	1.2405	0.26779	279.16	305.94	1.1977	0.19020	278.74	305.37	1.1690
70	0.46266	288.08	315.84	1.2681	0.27623	287.70	315.32	1.2254	0.19633	287.32	314.80	1.1969
80	0.47650	296.75	325.34	1.2954	0.28464	296.40	324.87	1.2528	0.20241	296.06	324.39	1.2244
90	0.49031	305.58	335.00	1.3224	0.29302	305.27	334.57	1.2799	0.20846	304.95	334.14	1.2516
100									0.21449	314.01	344.04	1.2785
	P = 0.18 MPa (−12.73°C)				P = 0.20 MPa (−10.09°C)				P = 0.24 MPa (−5.37°C)			
Sat.	0.10983	219.94	239.71	0.9273	0.09933	221.43	241.30	0.9253	0.08343	224.07	244.09	0.9222
−10	0.11135	222.02	242.06	0.9362	0.09938	221.50	241.38	0.9256				
0	0.11678	229.67	250.69	0.9684	0.10438	229.23	250.10	0.9582	0.08574	228.31	248.89	0.9399
10	0.12207	237.44	259.41	0.9998	0.10922	237.05	258.89	0.9898	0.08993	236.26	257.84	0.9721
20	0.12723	245.33	268.23	1.0304	0.11394	244.99	267.78	1.0206	0.09339	244.30	266.85	1.0034
30	0.13230	253.36	277.17	1.0604	0.11856	253.06	276.77	1.0508	0.09794	252.45	275.95	1.0339
40	0.13730	261.53	286.24	1.0898	0.12311	261.26	285.88	1.0804	0.10181	260.72	285.16	1.0637
50	0.14222	269.85	295.45	1.1187	0.12758	269.61	295.12	1.1094	0.10562	269.12	294.47	1.0930
60	0.14710	278.31	304.79	1.1472	0.13201	278.10	304.50	1.1380	0.10937	277.67	303.91	1.1218
70	0.15193	286.93	314.28	1.1753	0.13639	286.74	314.02	1.1661	0.11307	286.35	313.49	1.1501
80	0.15672	295.71	323.92	1.2030	0.14073	295.53	323.68	1.1939	0.11674	295.18	323.19	1.1780
90	0.16148	304.63	333.70	1.2303	0.14504	304.47	333.48	1.2212	0.12037	304.15	333.04	1.2055

T	v	u	h	s	v	u	h	s	v	u	h	s
°C	m^3/kg	kJ/kg	kJ/kg	kJ/(kg·K)	m^3/kg	kJ/kg	kJ/kg	kJ/(kg·K)	m3/kg	kJ/kg	kJ/kg	kJ/(kg·K)
100	0.16622	313.72	343.63	1.2573	0.14932	313.57	343.43	1.2483	0.12398	313.27	343.03	1.2326
	P = 0.28 MPa (–1.23°C)				P = 0.32 MPa (2.48°C)				P = 0.40 MPa (8.93°C)			
Sat.	0.07193	226.38	246.52	0.9197	0.06322	228.43	248.66	0.9177	0.05089	231.97	252.32	0.9145
0	0.07240	227.37	247.64	0.9238								
10	0.07613	235.44	256.76	0.9566	0.06576	234.61	255.65	0.9427	0.05119	232.87	253.35	0.9182
20	0.07972	243.59	265.91	0.9883	0.06901	242.87	264.95	0.9749	0.05397	241.37	262.96	0.9515
30	0.08320	251.83	275.12	1.0192	0.07214	251.19	274.28	1.0062	0.05662	249.89	272.54	0.9837
40	0.08660	260.17	284.42	1.0494	0.07518	259.61	283.67	1.0367	0.05917	258.47	282.14	1.0148
50	008992	268.64	293.81	1.0789	0.07815	268.14	293.15	1.0665	0.06164	267.13	291.79	1.0452
60	0.09319	277.23	303.32	1.1079	0.08106	276.79	302.72	1.0957	0.06405	275.89	301.51	1.0748
70	0.09641	285.96	312.95	1.1364	0.08392	285.56	312.41	1.1243	0.06641	284.75	311.32	1.1038
80	0.09960	294.82	322.71	1.1644	0.08674	294.46	322.22	1.1525	0.06873	293.73	321.23	1.1322
90	0.10275	303.83	332.60	1.1920	0.08953	303.50	332.15	1.1802	0.07102	302.84	331.25	1.1602
100	0.10587	312.98	342.62	1.2193	0.09229	312.68	342.21	1.1076	0.07327	312.07	341.38	1.1878
110	0.10897	322.27	352.78	1.2461	0.09503	322.00	352.40	1.2345	0.07550	321.44	351.64	1.2149
120	0.11205	331.71	363.08	1.2727	0.09774	331.45	362.73	1.2611	0.07771	330.94	362.03	1.2417
130									0.07991	340.58	372.54	1.2681
140									0.08208	350.35	383.18	1.2941

TABLE 5C: Properties of superheated refrigerant 134a vapor (continued)

T °C	v m^3/kg	u kJ/kg	h kJ/kg	s kJ/(kg·K)	v m^3/kg	u kJ/kg	h kJ/kg	s kJ/(kg·K)	v m3/kg	u kJ/kg	h kJ/kg	s kJ/(kg·K)
	P = 0.5 MPa (15.4°C)				P = 0.6 MPa (21.58°C)				P = 0.7 MPa (26.72°C)			
Sat.	0.04086	253.64	256.07	0.9117	0.03408	238.74	259.19	0.9097	0.02918	241.42	261.85	0.9080
20	0.04188	239.40	260.34	0.9264								
30	0.04416	248.20	270.28	0.9597	0.03581	246.41	267.89	0.9388	0.02979	244.51	265.37	0.9197
40	0.04633	256.99	280.16	0.9918	0.03774	255.45	278.09	0.9719	0.03157	253.83	275.93	0.9539
50	0.04842	265.83	290.04	1.0229	0.03958	264.48	288.23	1.0037	0.03324	263.08	286.35	0.9867
60	0.05043	274.73	299.95	1.0531	0.04134	273.54	298.35	1.0346	0.03482	272.31	296.69	1.0182
70	0.05240	283.72	309.92	1.0825	0.04304	282.66	308.48	1.0645	0.03634	281.57	307.01	1.0487
80	0.05432	292.80	319.96	1.1114	0.04469	291.86	318.67	1.0938	0.03781	290.88	317.35	1.0784
90	0.05620	302.00	330.10	1.1397	0.04631	301.14	328.93	1.1225	0.03924	300.27	327.74	1.1074
100	0.05805	311.31	340.33	1.1675	0.04790	310.53	339.27	1.1505	0.04064	309.74	338.19	1.1358
110	0.05988	320.74	350.68	1.1949	0.04946	320.03	349.70	1.1781	0.04201	319.31	348.71	1.1637
120	0.06168	330.30	361.14	1.2218	0.05099	329.64	360.24	1.2053	0.04335	328.98	359.33	1.1910
130	0.06347	339.98	371.72	1.2484	0.05251	339.38	370.88	1.2320	0.04468	338.76	370.04	1.2179
140	0.06524	349.79	382.42	1.2746	0.05402	349.23	381.64	1.2584	0.04599	348.66	380.86	1,2444
150					0.05550	359.21	392.52	1.2844	0.04729	358.68	391.79	1.2706
160					0.05698	369.32	403.51	1.3100	0.04857	368.82	402.82	1.2963

T °C	v m^3/kg	u kJ/kg	h kJ/kg	s kJ/(kg·K)	v m^3/kg	u kJ/kg	h kJ/kg	s kJ/(kg·K)	v m3/kg	u kJ/kg	h kJ/kg	s kJ/(kg·K)
	P = 0.8 MPa (31.33°C)				P = 0.90 MPa (35.53°C)				P = 1.00 MPa (39.39°C)			
Sat.	0.02547	243.78	264.15	0.9066	0.02255	245.88	266.18	0.9054	0.02020	247.77	267.97	0.9043
40	0.02691	252.13	273.66	0.9374	0.02325	250.32	271.25	0.9217	0.02029	248.39	268.68	0.9066
50	0.02846	261.62	284.39	0.9711	0.02472	260.09	282.34	0.9566	0.02171	258.48	280.19	0.9428
60	0.02992	27 1 .04	294.98	1.0034	0.02609	269.72	293.21	0.9897	0.0230 1	268.35	291.36	0.9768
70	0.03131	280.45	305.50	1.0345	0.02738	279.30	303.94	1.0214	0.02423	278.11	302.34	1.0093
80	0.03264	289.89	316.00	1.0647	0.02861	288.87	314.62	1.0521	0.02538	287.82	313.20	1.0405
90	0.03393	299.37	326.52	1.0940	0.02980	298.46	325.28	1.0819	0.02649	297.53	324.01	1.0707
100	0.03519	308.93	337.08	1.1227	0.03095	308.11	335.96	1.1109	0.02755	307.27	334.82	1.1000
110	0.03642	318.57	347.71	1.1508	0.03207	317.82	346.68	1.1392	0.02858	317.06	345.65	1.1286
120	0.03762	328.31	358.40	1.1784	0.03316	327.62	357.47	1.1670	0.02959	326.93	356.52	1.1567

T °C	v m^3/kg	u kJ/kg	h kJ/kg	s kJ/(kg·K)	v m^3/kg	u kJ/kg	h kJ/kg	s kJ/(kg·K)	v m3/kg	u kJ/kg	h kJ/kg	s kJ/(kg·K)
130	0.03881	338.14	369.19	1.2055	0.03423	337.52	368.33	1.1943	0.03058	336.88	367.46	1.1841
140	0.03997	348.09	380.07	1.2321	0.03529	347.51	379.27	1.2211	0.03154	346.92	378.46	1.2111
150	0.04113	358.15	391.05	1.2584	0.03633	357.61	390.31	1.2475	0.03250	357.06	389.56	1.2376
160	0.04227	368.32	402.14	1.2843	0.03736	367.82	401.44	1.2735	0.03344	367.31	400.74	1.2638
170	0.04340	378.61	413.33	1.3098	0.03838	378.14	412.68	1.2992	0.03436	377.66	412.02	1.2895
180	0.04452	389.02	424.63	1.3351	0.03939	388.57	424.02	1.3245	0.03528	388.12	423.40	1.3149
	P = 1.2 MPa (46.32°C)				P = 1.40 MPa (52.43°C)				P = 1.60 MPa (57.92°C)			
Sat.	0.01663	251.03	270.99	0.9023	0.01405	253.74	273.40	0.9003	0.01208	256.00	275.33	0.8982
50	0.01712	254.98	275.52	0.9164								
60	0.01835	265.42	287.44	0.9527	0.01495	262.17	283.10	0.9297	0.01233	258.48	278.20	0.9069
70	0.01947	275.59	298.96	0.9868	0.01603	272.87	295.31	0.9658	0.01340	269.89	291.33	0.9457
80	0.02051	285.62	310.24	1.0192	0.01701	283.29	307.10	0.9997	0.01435	280.78	303.74	0.9813
90	0.02150	295.59	321.39	1.0503	0.01792	293.55	318.63	1.0319	0.01521	291.39	315.72	1.0148
100	0.02244	305.54	332.47	1.0804	0.01878	303.73	330.02	1.0628	0.01601	301.84	327.46	1.0467
110	0.02335	315.50	343.52	1.1096	0.01960	313.88	341.32	1.0927	0.01677	312.20	339.04	1.0773
120	0.02423	325.51	354.58	1.1381	0.02039	324.05	352.59	1.1218	0.01750	322.53	350.53	1.1069
1 30	0.02508	335.58	365.68	1.1 660	0.02115	334.25	363.86	1.1501	0.01820	332.87	361.99	1.1357
140	0.02592	345.73	376.83	1.1933	0.02189	344.50	375.15	1.1 777	0.0 1 887	343.24	373.44	1.1 638
150	0.02674	355.95	388.04	1.2201	0.02262	354.82	386.49	1.2048	0.01953	353.66	384.91	1.1912
160	0.02754	366.27	399.33	1.2465	0.02333	365.22	397.89	1.2315	0.02017	364.15	396.43	1.2181
170	0.02834	376.69	410.70	1.2724	0.02403	375.71	409.36	1.2576	0.02080	374.71	407.99	1.2445
180	0.02912	387.21	422.16	1.2980	0.02472	386.29	420.90	1.2834	0.02142	385.35	419.62	1.2704
190					0.02541	396.96	432.53	1.3088	0.02203	396.08	431.33	1.2960
200					0.02608	407.73	444.24	1.3338	0.02263	406.90	443.11	1.3212

TABLE 6A: Properties of some solids at 25°C			TABLE 6B: Properties of some liquids at 25°C		
SOLIDS			LIQUIDS		
Substance	ρ kg/m^3	c_P kJ/kg K	Substance	ρ kg/ m^3	c_P kJ/kg K
Asphalt	2120	0.92	Ammonia	604	4.84
Brick, common	1800	0.84	Benzene	879	1.72
Carbon, diamond	3250	0.51	Butane	556	2.47
Carbon, graphite	2000–2500	0.61	CCL,	1584	0.83
Coal	1200–1500	1.26	C02	680	2.9
Concrete	2200	0.88	Ethanol	783	2.46
Glass, plate	2500	0.80	Gasoline	750	2.08
Glass, wool	200	0.66	Glycerin	1260	2.42
Granite	2750	0.89	Kerosene	815	2.0
Ice (0 C)	917	2.04	1 Methanol	787	2.55
Paper	700	1.2	n–octane	692	2.23
Plexiglas	1180	1.44	Oil engine	885	1.9
Polystyrene	920	2.3	Oil light	910	1.8
Polyvinyl chloride	1380	0.96	Propane	510	2.54
Rubber, soft	1100	1.67	R–12	1310	0.97
Salt, rock	2100–2500	0.92	R–22	1190	1.26
Sand, dry	1500	0.8	R–134a	1206	1.43
Silicon	2330	0.70	Water	997	4.18
Snow, firm	560	21	Liquid metals		
Wood, hard (oak)	720	1.26	Bismuth, Bi	10040	0.14
Wood, soft (pine)	510	1.38	Lead, Pb	10660	0.16
Wool	100	1.72	Mercury, Hg	13580	0.14
Metals			Potassium, K	828	0.81
Aluminum	2700	0.90	Sodium, Na	929	1.38
Brass, 60–40	8400	0.38	Tin, Sn	6950	0.24
Copper, commercial	8300	0.42	NaK (56/44)	887	1.13
Gold	19300	0.13	Zinc, Zn	6570	0.50
Iron, cast	7272	0.42			
Iron. 304 St Steel	7820	0.46			
Lead	11310	0.128			

Tables 6A to 6C adapted from R. Sonntag, C. Borgnakke and G. J. Wylen, *Fundamentals of Classical Thermodynamics*, 5th Ed. John Wiley & Sons, 1998. (With permission.)

TABLE 6C: Properties of various ideal gases at 25°C, 100 kPa

Gas	Chemical Formula	Molecular Mass	R kJ/kg K	ρ kg/m^3	c_{p0} kJ/kg K	c_{v0} kJ/kg K	k
Acetylene	C_2H_2	26.038	0.3193	1.05	1.699	1.380	1.231
Air		28.97	0.287	1.169	1.004	0.717	1.400
Ammonia	NH_3	17.031	0.4882	0.694	2,130	1.642	1.297
Argon	Ar	39.948	0.2081	1.613	0.520	0.312	1.667
Benzene	C_6H_6	78.11	0.1064	3.151	0775	0.67	1.157
Butane	C_4H_{10}	58.124	0.1430	2.407	1.716	1.573	1.091
Carbon monoxide	CO	28.01	0.2968	1.13	1.041	0.744	1.399
Carbon dioxide	CO,	44.01	0.1889	1.775	0.842	0.653	1.289
Ethane	C_2H_6	30.07	0.2765	1.222	1.766	1.490	1.186
Ethanol	CH_3OH	46.069	0.1805	1.883	1.427	1.246	1.145
Ethylene	C_2H_4	28.054	0.2964	1.138	1,548	1.252	1.237
Helium	He	4.003	2.0771	0.1615	5.193	3.116	1.667
Hydrogen	H_2	2.016	4.1243	0.0813	14.209	1.008	1.409
Methane	CH_4	16.043	0.5183	0.648	2.254	1.736	1.299
Methanol	CH_3OH	32.042	0.2595	1.31	1.405	1.146	1.227
Neon	Ne	20.183	0.4120	0.814	1.03	0.618	1.667
Nitric oxide	NO	30.006	0.2771	1.21	0.993	0,716	1.387
Nitrogen .	N	28.013	0.2968	1.13	1.042	0.745	1.400
Nitrous oxide	N_2O	44.013	0.1889	1.775	0.879	0.690	1.274
n–octane	C_8H_{18}	114.23	0.07279	0.092	1.711	1.638	1.044
Oxygen	O_2	31.999	0.2598	1.292	0.922	0.662	1.393
Propane	C_3H_8	44.094	0.1886	1.808	1.679	1.490	1.126
R–12	CCl_2F_2	120.914	0.06876	4.98	0.616	0.547	1.126
R–22	$CHClF_2$,	86.469	0.09616	3.54	0.658	0.562	1.171
R–134a	$C_2F_4H_2$	102.03	0.08149	4.20	0.852	0.771	1.106
Sulfur dioxide	SO_2	64.059	0.1298	2.618	0.624	0.494	1.263
Sulfur trioxide	SO_3	80.053	0.10386	3.272	0.635	0.531	1.196
Water/steam	H_2O	18.02	0.4614	0.727	1.86	1.40	1.329

TABLE 6D: Curve fits for thermodynamic properties of liquids

Constants for the equation: $c_{p0}/R = A + BT + CT^2$, T from 273.15 to 373.15 K
$(h_0 - h_{0,ref})/R = A\,(T - T_{ref}) + (B/2)(T^2 - T_{ref}^2) + (C/3)\,(T^3 - T_{ref}^{\ 3})$
$(s_0 - s_{0,ref})/R = A \ln (T/T_{ref}) + B(T - T_{ref}) + (C/2)(T^2 - T_{ref}^{\ 2})$
$h_{0,ref}$ and $s_{0,ref}$ are typically set to zero at T_{ref}.

Chemical Species	A	10^3 B	10^6 C
Ammonia	22.626	−100.75	192.71
Aniline	15.819	29.03	−15.80
Benzene	−0.747	67.96	−37.78
1,3–Butadiene	22.711	−87.96	205.79
Carbon tetrachloride	21.155	−48.28	101.14
Chlorobenzene	11.278	32.86	−31.90
Chloroform	19.215	−42.89	83.01
Cyclohexane	−9.048	141.38	−161.62
Ethanol	33.866	−172.60	349.17
Ethylene oxide	21.039	−86.41	172.28
Methanol	13.431	−51.28	131.13
n–Propanol	41.653	−210.32	427.20
Sulfur trioxide	−2.930	137.08	−84.73
Toluene	15.133	6.79	16.35
Water	8.712	1.25	−0.18

Tables 6D and 6F adapted from Smith and Van Ness, *Introduction to Chemical Engineering Thermodynamics*, 4th Edition, McGraw Hill Book Company, 1987. Originally, the liquid correlations were presented by J. W. Miller, Jr., G. R. Schorr, and C. L. Yaws, *Chem. Engng.*, 83(23): 129, 1976 and the solid correlations by K.K. Kelley, *U.S. Bur. Mines. Bull.* 584, 1960; L. B. Pankratz, *U.S. Bur. Mines Bull.*, 672, 1982.

Table 6E: Curve fits for thermodynamic properties of solids

Constants for the equation $c_{p0}/R = A + BT + DT^{-2}$
T (kelvins) from 298 K to T_{max}
$(h_0 - h_{0,ref})/R = A(T - Tref) + (B/2)(T^2 - T_{ref}^2) - D(1/T - 1/T_{ref})$
$(s_0 - s_{0,ref})/R = A \ln(T/Tref) + B(T - T_{ref}) - (D/2)(1/T^{-2} - 1/T_{ref}^{-2})$
$h_{0,ref}$ and $s_{0,ref}$ are typically set to zero at T_{ref}.

Chemical Species	T_{max}	A	10^3 B	10^{-5} D
CaO	2,000	6.104	0.443	−1.047
$CaCO_3$	1,200	12.572	2.637	-3.120
$Ca(OH)_2$	700	9.597	5.435	
CaC_2	720	8.254	1.429	−1.042
$CaCl_2$	1,055	8.646	1.530	−0.302
C(Graphite)	2000	1.771	0.771	−0.867
Cu	1,357	2.677	0.815	0.035
CuO	1,400	5.780	0.973	−0.874
Fe(α)	1.043	−0.111	6.111	1.150
Fe_2O_3	960	11.812	9.697	−1.976
Fe_3O_4	850	9.594	27.t12	0.409
FeS	411	2.612	13.286	
I_2	386.8	6.481	1.502	
NH_4Cl	458	5.939	16.105	
Na	371	1.988	4.688	
NaCl	1,073	5.526	1.963	
NaOH	566	0.121	16.316	1.948
$NaHCO_3$	400	5.128	18.148	
S(Rhombic)	368.3	4.114	−1.728	−0.783
SiO_2 (quartz)	847	4.871	5.365	−1.001

From K. K. Kelley, *U.S. Bur. Mines Bull.* 584, 1960; L. B. Pankratz, *U.S. Bur. Mines Bull.* 672, 1982.

TABLE 6F1: Curve fit for thermodynamic properties of gases

$$\overline{c_{P0}}/\overline{R} = a_1 + a_2 T + a_3 T^2 + a_4 T^3 + a_5 T^4$$

$$\bar{s}_0/\overline{R} = a_1 \mathrm{Ln} T + a_2 T + \frac{a_3}{2} T^2 + \frac{a_4}{3} T^3 + \frac{a_5}{4} T^4 + a_7$$

$$\frac{\bar{h}_o}{\overline{R}T} = a_1 + \frac{a_2}{2} T + \frac{a_3}{3} T^2 + \frac{a_4}{4} T^3 + \frac{a_5}{5} T^4 + a_6$$

Species	T(K)	a_1	a_2	a_3	a_4	a_5	a_6	a_7
CO	1,000–5,000	0.03025078E+02	0.14426885E–02	–0.05630827E–05	0.10185813E–09	–0.06910951E–13	–0.14268350E+05	0.06108217E+02
	300–1,000	0.03262451E+02	0.15119409E–02	–0.03881755E–04	0.05581944E–07	–0.02474951E–10	–0.14310539E+05	0.04848897E+02
CO_2	1,000–5,000	0.04453623E+02	0.03140168E–01	–0.12784105E–05	0.02393996E–08	–0.16690333E–13	–0.04896696E+06	–0.09553959E+01
	300–1,000	0.02275724E+02	0.09922072E–01	–0.10409113E–04	0.06866686E–07	–0.02117280E–10	–0.04837314E+06	0.10188488E+02
H_2	1,000–5,000	0.02991423E+02	0.07000644E–02	–0.05633828E–06	–0.09231578E–10	0.15827519E–14	–0.08350340E+04	–0.13551101E+01
	300–1,000	0.03298124E+02	0.08249441E–02	–0.08143015E–05	0.09475434E–09	0.04134872E–11	–0.10125209E+04	–0.03294094E+02
H	1,000–5,000	0.02500000E+02	0.00000000E+00	0.00000000E+00	0.00000000E+00	0.00000000E+00	0.02547162E+06	–0.04601176E+01
	300–1,000	0.02500000E+02	0.00000000E+00	0.00000000E+00	0.00000000E+00	0.00000000E+00	0.02547162E+06	–0.04601176E+01
OH	1,000–5,000	0.02882730E+02	0.10139743E–02	–0.02276877E–05	0.02174683E–09	–0.05126305E–14	0.03886888E+05	0.05595712E+02
	300–1,000	0.03637266E+02	0.01850910E–02	–0.16761646E–05	0.02387202E–07	–0.08431442E–11	0.03606781E+05	0.13588605E+01
H_2O	1,000–5,000	0.02672145E+02	0.03056293E–01	–0.08730260E–05	0.12009964E–09	–0.06391618E–13	–0.02989921E+06	0.06862817E+02
	300–1,000	0,03386842E+02	0.03474982E–01	–0.06354696E–04	0.06968581E–07	–0.02506588E–10	–0.03020811E+06	0.02590232E+02
N_2	1,000–5,000	0.02926640E+02	0.14879768E–02	–0.05684760E–05	0.10097038E–09	–0.06753351E–13	–0.09227977E+04	0.05980528E+02
	300–1,000	0.03298677E+02	0.14082404E–02	–0.03963222E–04	0.05641515E–07	–0.02444854E–10	–0.10209999E+04	0.03950372E+02
N	1,000–5,000	0.02450268E+02	0.10661458E–03	–0.07465337E–06	0.01879652E–09	–0.10259839E–14	0.05611604E+06	0.04448758E+02
	300–1,000	0.02503071E+02	–0.02180018E–03	0.05420529E–06	–0.05647560E–09	0.02099904E–12	0.05609890E+06	0.04167566E+02
NO	1,000–5,000	0.03245435E+02	0.12691383E–02	–0.05015890E–05	0.09169283E–09	–0.06275419E–13	0.09800840E+05	0.06417293E+02
	300–1,000	0.03376541E+02	0.12530634E–02	–0.033027SOE–04	0.05217810E–07	–0.02446262E–10	0.09817961E+05	0.05829590E+02
NO_2	1,000–5,000	0.04682859E+02	0.02462429E–01	–0.10422585E–05	0.01976902E–08	–0.13917168E–13	0.02261292E+05	0.09885995E+01
	300–1,000	0.02670600E+02	0.07838500E–01	–0.08063864E–04	0.06161714E–07	–0.02320150E–10	0.02896290E+05	0.11612071E+02
O_2	1,000–5,000	0.03697578E+02	0.06135197E–02	–0.12588420E–06	0.01775281E–09	–0.11364354E–14	–0.12339301E+04	0.03189165E+02
	300–1,000	0.03212936E+02	0.11274864E–02	–0.05756150E–05	0.13138773E–08	–0.08768554E–11	–0.10052490E+04	0.06034737E+02
O	1,000–5,000	0.02542059E+02	–0.02755061E–03	–0.03102803E–07	0.04551067E–10	–0.04368051E–14	0.02923080E+06	0.04920308E+02
	300–1,000	0.02946428E+02	–0.16381665E–02	0.02421031E–04	–0.16028431E–08	0.03890696E–11	0.02914764E+06	0.02963995E+02

Table 7F1 adapted from S. R. Turns, *An Introduction to Combustion*, 2nd Edition, McGraw Hill Book Co., 2000. Originally from Kee, R. J., Rupley, F. K, and Miller, J. A., *The Chemkin Thermodynamic Data Base*, Sandia Report, SAND87–8215B, reprinted March 1991.

TABLE6F2: Curve fit for thermodynamic properties of gases (continued)

$c_{P,0}/R = A + BT + CT^2 + D\,T^{-2}$ from 298 K to T_{max}
$(h_0 - h_{0,ref})/R = A\,(T - Tref) + (B/2)(T^2 - T_{ref}^2) + (C/3)(T^3 - T_{ref}^3) - D\,(1/T - 1/T_{ref})$
$(s_0 - s_{0,ref})/R = A \ln(T/T_{ref}) + B(T - T_{ref}) + (C/2)(T^2 - T_{ref}^2) - (D/2)(1/T^{-2} - 1/T_{ref}^{-2})$
where h_{ref} and s_{ref} are typically set to zero at T_{ref}.

Chemical Species		T_{max}	A	10^3B	10^6C	$10^{-5}D$
Paraffins:						
Methane	CH_4	1,500	1.702	9.081	–2.164	
Ethane	C_2H_6	1,500	1.131	19.225	–5.561	
Propane	C_3H_8	1,500	1.213	28.785	–8.824	
n–Butane	C_4H_{10}	1,500	1.935	36.915	–11.402	
iso–Butane	C_4H_{10}	1,500	1.677	37.853	–11.945	
n–Pentane	C_5H_{12}	1,500	2.464	45.351	–14.111	
n–Hexane	C_6H_{14}	1,500	3.025	53.722	–16.791	
n–Heptane	C_7H_{16}	1,500	3.570	62.127	–19.486	
n–Octane	C_8H_{18}	1,500	8.163	70.567	–22.208	
1–Alkenes:						
Ethylene	C_2H_4	1,500	1.424	14.394	–4.392	
Propylene	C_3H_6	1,500	1.637	22.706	–6.915	
1–Butene	C_4H_8	1,500	1.967	31.630	–9.873	
1–Pentene	C_5H_{10}	1,500	2.691	39.753	–12.447	
1–Hexene	C_6H_{12}	1,500	3.220	48.189	–15.157	
1–Heptene	C_7H_{14}	1,500	3.768	56.588	–17.847	
1–Octene	C_8H_{16}	1,500	4.324	64.960	–20.521	
Miscellaneous organics:						
Acetaldehyde	C_2H_4O	1,000	1.693	17.978	–6.158	
Acetylene	C_2H_2	1,500	6.132	1.952		–1.299
Benzene	C_6H_6	1,500	–0.206	39.064	–13.301	
1,3–Butadiene	C_4H_6	1,500	2.734	26.786	–8.882	
Cyclohexane	C_6H_{12}	1,500	–3.876	63.249	–20.928	
Ethanol	C_2H_6O	1,500	3.518	20.001	–6.002	
Ethylbenzene	C_8H_{10}	1,500	1.124	55.380	–18.476	
Ethylene oxide	C_2H_4O	1,000	–0.385	23.463	–9.296	
Formaldehyde	CH_2O	1,500	2.264	7.022	–1.877	
Methanol	CH_4O	1,500	2.211	12.216	–3.450	
Toluene	C_7H_8	1,500	b.290	47.052	–15.716	
Styrene	C_8H_8	1,500	2.050	50.192	–16.662	
Miscellaneous inorganics						
Air		2,000	3.355	0.575		–0.016
Ammonia	NH_3	1,800	3.578	3.020		–0.186
Bromine	Br_2	3,000	4.493	0.056		–0.154
Carbon disulfide	CS_2	1,800	6.311	0.805		–0.906
Chlorine	CI_2	3,000	4.442	0.089		–0.344
Hydrogen sulfide	H_2S	2,300	3.931	1.490		–0.232
Hydrogen chloride	HCI	2,000	3.156	0.623		0.151
Hydrogen cyanide	HCN	2,500	4.736	1.359		–0.725
Dinitrogen oxide	N_2O	2,000	5.328	1.214		–0.928
Dinitrogen tetroxide	N_2O_4	2,000	11.660	2.257		–2.787

Chemical Species		T_{max}	A	10^3B	10^6C	$10^{-5}D$
Sulfur dioxide	SO_2	2,000	5.699	0.801		−1.015
Sulfur trioxide	SO_3	2,000	8.060	1.056		−2.028

Table 6F2 adapted from Smith and Van Ness, *Introduction to Chemical Engineering Thermodynamics*, 4th Edition, McGraw Hill Book Company, 1987 (Selected from H.M. Spencer, *Ind. Eng. Chem.*, 40: 2152, 1948; K.K. Kelley, *U.S. Bur. Mines Bull.,* 584, 1960; L.B. Pankratz, *U.S. Bur. Mines Bull.*, 672, 1982).

TABLE 7: Ideal gas properties of air

T, K	h, kJ/kg	P_r	u, kJ/kg	v_r	s^0, kJ/(kg·K)	g^0, kJ/(kg·K)
200	199.97	0.3363	142.56	1707.0	1 .29559	-59.148
210	209.97	0.3987	149.69	1512.0	1.34444	-72.3624
220	219.97	0.4690	156.82	1346.0	1.39105	-86.061
230	230.02	0.5477	164.00	1205.0	1.43557	-100.1611
240	240.02	0.6355	171.13	1084.0	1.47824	-114.7576
250	250.05	0.7329	178.28	979.0	1.51917	-129.7425
260	260.09	0.8405	185.45	887.8	1.55848	-145.1148
270	270.11	0.9590	192.60	808.0	1.59634	-160.9018
280	280.13	1.0889	199.75	738.0	1.63279	-177.0512
285	285.14	1.1584	203.33	706.1	1.65055	-185.26675
290	290.16	1.2311	206.91	676.1	1.66802	-193.5658
295	295.17	1.3068	210.49	647.9	1.68515	-201.94925
300	300.19	1.3860	214.07	621.2	1.70203	-210.419
305	305.22	1.4686	217.67	596.0	1.71865	-218.96825
310	310.24	1.5546	221.25	572.3	1.73498	-227.6038
315	315.27	1.6442	224.85	549.8	1.75106	-236.3139
320	320.29	1.7375	228.42	528.6	1.76690	-245.118
325	325.31	1.8345	232.02	508.4	1.78249	-253.99925
330	330.34	1.9352	235.61	489.4	1.79783	-262.9439
340	340.42	2.149	242.82	454.1	1.82790	-281.066
350	350.49	2.379	250.02	422.2	1.85708	-299.488
360	360.58	2.626	257.24	393.4	1.88543	-318.1748
370	370.67	2.892	264.46	367.2	1.91313	-337.1881
380	380.77	3.176	271.69	343.4	1.94001	-356.4338
390	390.88	3.481	278.93	321.5	1.96633	-375.9887
400	400.98	3.806	286.16	301.6	1.99194	-395.796
410	411.12	4.153	293.43	283.3	2.01699	-415.8459
420	421.26	4.522	300.69	266.6	2.04142	-436.1364
430	431.43	4.915	307.99	251.1	2.06533	-456.6619
440	441.61	5.332	315.30	236.8	2.08870	-477.418
450	451.80	5.775	322.62	223.6	2.11161	-498.4245
460	462.02	6.245	329.97	211.4	2.13407	-519.6522
470	472.24	6.742	337.32	200.1	2.15604	-541.0988
480	482.49	7.268	344.70	189.5	2.17760	-562.758
490	492.74	7.824	352.08	179.7	2.19876	-584.6524
500	503.02	8.411	359.49	170.6	2.21952	-606.74
510	513.32	9.031	366.92	162.1	2.23993	-629.0443
520	523.63	9.684	374.36	154.1	2.25997	-651.5544
530	533.98	10.37	381.84	146.7	2.27967	-674.2451
540	544.35	11.10	389.34	139.7	2.29906	-697.1424
550	555.74	11.86	396.86	133.1	2.31809	-719.2095
560	565.17	12.66	404.42	127.0	2.33685	-743.466
570	575.59	13.50	411.97	121.2	2.35531	-766.9367
580	586.04	14.38	419.55	115.7	2.37348	-790.5784
590	596.52	15.31	427.15	110.6	2.39140	-814.406

T, K	h, kJ/kg	P_r	u, kJ/kg	v_r	s^0, kJ/(kg·K)	g^0, kJ/(kg·K)
600	607.02	16.28	434.78	105.8	2.40902	-838.392
610	617.53	17.30	442.42	101.2	2.42644	-862.5984
620	628.07	18.36	450.09	96.92	2.44356	-886.9372
630	683.63	19.84	457.78	92.84	2.46048	-866.4724
640	649.22	20.64	465.50	88.99	2.47716	-936.1624
650	659.84	21.86	473.25	85.34	2.49364	-961.026
660	670.47	23.13	481.01	81.89	2.50985	-986.031
670	681.14	24.46	488.81	78.61	2.52589	-1011.2063
680	691.82	25.85	496.62	75.50	2.54175	-1036.57
690	702.52	27.29	504.45	72.56	2.55731	-1062.0239
700	713.27	28.80	512.33	69.76	2.57277	-1087.669
710	724.04	30.38	520.23	67.07	2.588 0	-1113.44
720	734.82	32.02	528.14	64.53	2.60319	-1139.4768
730	745.62	33.72	536.07	62.13	2.61803	-1165.5419
740	756.44	35.50	544.02	59.82	2.63280	-1191.832
750	767.29	37.35	551.99	57.63	2.64737	-1218.2375
760	778.18	39.27	560.01	55.54	2.66176	-1244.7576
780	800.03	43.35	576.12	51.64	2.69013	-1298.2714
800	821.95	47.75	592.30	48.08	2.71787	-1352.346
820	843.98	52.59	608.59	44.84	2.74504	-1406.9528
840	866.08	57.60	624.95	41.85	2.77170	-1462.148
860	888.27	63.09	641.40	39.12	2.79783	-1517.8638
880	910.56	68.98	657.95	36.61	2.82344	-1574.0672
900	932.93	75.29	674.58	34.31	2.84856	-1630.774
920	955.38	82.05	691.28	32.18	2.87324	-1688.0008
940	977.92	89.28	708.08	30.22	2.89748	-1745.7112
960	1000.55	97.00	725.02	28.40	2.92128	-1803.8788
980	1023.25	105.2	741.98	26.73	2.94468	-1862.5364
1000	1046.04	114.0	758.94	25.17	2.96770	-1921.66
1020	1068.89	123.4	776.10	23.72	2.99034	-1981.2568
1040	1091.85	133.3	793.36	23.29	3.01260	-2041.254
1060	1114.86	143.9	810.62	21.14	3.03449	-2101.6994
1080	1137.89	155.2	827.88	19.98	3.05608	-2162.6764
1100	1161.07	167.1	845.33	18.896	3.07732	-2223.982
1120	1184.28	179.7	862.79	17.886	3.09825	-2285.76
1140	1207.57	193.1	880.35	16.946	3.11883	-2347.8962
1160	1230.92	207.2	897.91	16.064	3.13916	-2410.5056
1180	1254.34	222.2	915.57	15.241	3.15916	-2473.4688
1200	1277.79	238.0	933.33	14.470	3.17888	-2536.866
1220	1301.31	254.7	951.09	13.747	3.19834	-2600.6648
1240	1324.93	272.3	968.95	13.069	3.21751	-2664.7824
1260	1348.55	290.8	986.90	12.435	3.23638	-2729.2888
1280	1372.24	310.4	1004.76	11.835	3.25510	-2794.288
1300	1395.97	330.9	1022.82	11.275	3.27345	-2859.515
1320	1419.76	352.5	1040.88	10.747	3.29160	-2925.152
1340	1443.60	375.3	1058.94	10.247	3.30959	-2991.2506
1360	1467.49	399.1	1077.10	9.780	3.32724	-3057.5564
1380	1491.44	424.2	1095.26	9.337	3.34474	-3124.3012

T, K	h, kJ/kg	P_r	u, kJ/kg	v_r	s^0, kJ/(kg·K)	g^0, kJ/(kg·K)
1400	1515.42	450.5	1113.52	8.919	3.36200	-3191.38
1420	1539.44	478.0	1131.77	8.526	3.37901	-3258.7542
1440	1563.51	506.9	1150.13	8.153	3.39586	-3326.5284
1460	1587.63	537.1	1168.49	7.801	3.41247	-3394.5762
1480	1611.79	568.8	1186.95	7.468	3.42892	-3463.0116
1500	1635.97	601.9	1205.41	7.152	3.44516	-3531.77
1520	1660.23	636.5	1223.87	6.854	3.46120	-3600.794
1540	1684.51	672.8	1242.43	6.569	3.47712	-3670.2548
1560	1708.82	710.5	1260.99	6.301	3.49276	-3739.8856
1580	1733.17	750.0	1279.65	6.046	3.50829	-3809.9282
1600	1757.57	791.2	1298.30	5.804	3.52364	-3880.254
1620	1782.00	834.1	1316.96	5.574	3.53879	-3950.8398
1640	1806.46	878.9	1335.72	5.355	3.55381	-4021.7884
1660	1830.96	925.6	1354.48	5.147	3.56867	-4093.0322
1680	1855.50	974.2	1373.24	4.949	3.58335	-4164.528
1700	1880.1	1025	1392.7	4.761	3.5979	-4236.33
1750	1941.6	1161	1439.8	4.328	3.6336	-4417.2
1800	2003.3	1310	1487.2	3.994	3.6684	-4599.82
1850	2065.3	1475	1534.9	3.601	3.7023	-4783.955
1900	2127.4	1655	1582.6	3.295	3.7354	-4969.86
1950	2189.7	1852	1630.6	3.022	3.7677	-5157.315
2000	2252.1	2068	1678.7	2.776	3.7994	-5346.7
2050	2314.6	2303	1726.8	2.555	3.8303	-5537.515
2100	2377.7	2559	1775.3	2.356	3.8605	-5729.35
2150	2440.3	2837	1823.8	2.175	3.8901	-5923.415
2200	2503.2	3138	1872.4	2.012	3.9191	-6118.82
2250	2566.4	3464	1921.3	1.864	'3.9474	-6315.25

P_r, Relative pressure, v_r: relative volume. Source: adopted from K. Wark, *Thermodynamics*, 4th Ed., McGraw Hill Book Co., 1983, pp 785-786. Originally from J.H . Keenan and J. Keye, *Gas Tables*, John Wiley & Sons, NY, 1948. $P_r = 0.00368 \exp(-s^0/R)$, $v_r = 2.87\, T/p_r$, $g^0 = h - Ts^0$

TABLE 8: Ideal gas properties of carbon monoxide, CO

MW = 28.010, $h_{f,298}$ ^{0}(kJ/kmol) = –110, 541

T (K)	$\bar{c}_{p0}$ (kJ/kmol–K)	$\bar{h}_{t,T} - \bar{h}_{t,298}$ (kJ/kmol)	$\bar{h}_f^{\,0}$ (T) (kJ/kmol)	$\bar{s}^{\,0}$(T) (kJ/kmol–K)	$\bar{g}^0$ (T) (kJ/kmol)
200	28.687	–2,835	–111,308	186.018	-150,580
298	29.072	0	–110,541	197.548	-169,410
300	29.078	54	–110,530	197.728	-169,805
400	29.433	2,979	–110,121	206.141	-190,018
500	29.857	5,943	–110,017	212.752	-210,974
600	30.407	8,955	–110,156	218.242	-232,531
700	31.089	12,029	–110,477	222.979	-254,597
800	31.860	15,176	–110,924	227.180	-277,109
900	32.629	18,401	–111,450	230.978	-300,020
1,000	33.255	21,697	–112,022	234.450	-323,294
1,100	33.725	25,046	–112,619	237.642	-346,901
1,200	34.148	28,440	–113,240	240.595	-370,815
1,300	34.530	31,874	–113,881	243.344	-395,014
1,400	34.872	35,345	–114,543	245.915	-419,477
1,500	35.178	38,847	–115,225	248,332	-444,192
1,600	35.451	42,379	–115,925	250.611	-469,140
1,700	35.694	45,937	–116,644	252.768	-494,310
1,800	35.910	49,517	–117,380	254.814	-519,689
1,900	36.101	53,118	–118,132	256.761	-545,269
2,000	36.271	56,737	–118,902	258.617	-571,038
2,100	36.421	60,371	–119,687	260.391	-596,991
2,200	36.553	64,020	–120,488	262.088	-623,115
2,300	36.670	67,682	–121,305	263.715	-649,404
2,400	36.774	71,354	–122,137	265.278	-675,854
2,500	36,867	75,036	–122,984	266.781	-702,458
2,600	36.950	78,727	–123,847	268.229	-729,209
2,700	37.025	82,426	–124,724	269.625	-756,103
2,800	37.093	86,132	–125,616	270.973	-783,133
2,900	37.155	89,844	–126,523	272.275	-810,295
3,000	37.213	93,562	–127,446	273.536	-837,587
3,100	37.268	97,287	–128,383	274'.757	-865,001
3,200	37.321	101,016	–129,335	275.941	-892,536
3,300	37.372	104,751	–130,303	277.090	-920,187
3,400	37.422	108,490	–131,285	278.207	-947,955
3,500	37.471	112,235	–132,283	279.292	-975,828
3,600	37.521	115,985	–133,295	280.349	-1,003,812
3,700	37.570	119,739	–134,323	281.377	-1,031,897
3,800	37.619	123,499	–135,366	282.380	-1,060,086
3,900	37.667	127,263	–136,424	283.358	-1,088,374
4,000	37.716	131,032	–137,497	284.312	-1,116,757
4,100	37.764	134,806	–138,585	285.244	-1,145,235
4,200	37.810	138,585	–139,687	286.154	-1,173,803
4,300	37.855	142,368	–140,804	287.045	-1,202,467
4,400	37.897	146,156	–141,935	287.915	-1,231,211

4,500	37.936	149,948	–143,079	288.768	-1,260,049
4,600	37.970	153,743	–144,236	289.602	-1,288,967
4,700	37.998	157,541	–145,407	290.419	-1,317,969
4,800	38.019	161,342	–146,589	291.219	-1,347,050
4,900	38.031	165,145	–147,783	292.003	-1,376,211
5,000	38.033	168,948	–148,987	292.771	-1,405,448

Tables 8 to 19 except for $\overline{g}^0$ (T) adapted from S. R. Turns, *An Introduction to Combustion*, 2nd Edition, McGraw Hill Book Co., 2000.

For Tbles A-8 to A-19,

$\overline{g}^0 = \overline{h}_{f,298}^{\ 0} + (\overline{h}_{t,T} - \overline{h}_{t,298}) - T\,\overline{s}^0$, $\overline{g}^0 \neq \overline{g}_f^{\ 0}$, $\overline{s}_k(T,p_k) = \overline{s}^0 - \overline{R}\ \ln\ (p_k/1)$, $\overline{g}_k(T.p_k) = \overline{g}^0 + \overline{R}T\ \ln\ (p_k/1)$

TABLE 9: Ideal gas properties of carbon dioxide, CO_2

MW = 44.011, $h_{f,298}$ ^{0}(kJ/kmol) = –393,546

T (K)	$\bar{c}_{p0}$ (kJ/kmol–K)	$\bar{h}_{t,T} - \bar{h}_{t,298}$ (kJ/kmol)	$\bar{h}_f^0$ (T) (kJ/kmol)	$\bar{s}^0$(T) (kJ/kmol–K)	$\bar{g}^0$ (T) (kJ/kmol)
200	32.387	–3,423	–393,483	199.876	-436,944
298	37.198	0	–393,546	213.736	-457,239
300	37.280	69	–393,547	213.966	-457,667
400	41.276	4,003	–393,617	225.257	-479,646
500	44.569	8,301	–393,712	234.833	-502,662
600	47.313	12,899	–393,844	243.209	-526,572
700	49.617	17,749	–394,013	250.680	-551,273
800	51.550	22,810	–394,213	257.436	-576,685
900	53.136	28,047	–394,433	263.603	-602,742
1,000	54.360	33,425	–394,659	269.268	-629,389
1,100	55.333	38,911	–394,875	274.495	-656,580
1,200	56.205	44,488	–395,083	279.348	-684,276
1,300	56.984	50,149	–395,287	283.878	-712,438
1,400	57.677	55,882	–395,488	288.127	-741,042
1,500	58.292	61,681	–395,691	292.128	-770,057
1,600	58.836	67,538	–395,897	295.908	-799,461
1,700	59.316	73,446	–396,110	299.489	-829,231
1,800	59.738	79,399	–396,332	302.892	-859,353
1,900	60.108	85,392	–396,564	306.132	-889,805
2,000	60.433	91,420	–396,808	309.223	-920,572
2,100	60.717	97,477	–397,065	312.179	-951,645
2,200	60.966	103,562	–397,338	315.009	-983,004
2,300	61.185	109,670	–397,626	317.724	-1,014,641
2,400	61.378	115,798	–397,931	320.333	-1,046,547
2,500	61.548	121,944	–398,253	322.842	-1,078,707
2,600	61.701	128,107	–398,594	325.259	-1,111,112
2,700	61.839	134,284	–398,952	327.590	-1,143,755
2,800	61.965	140,474	–399,329	329.841	-1,176,627
2,900	62.083	146,677	–399,725	332.018	-1,209,721
3,000	62.194	152,891	–400,140	334.124	-1,243,027
3,100	62.301	159,116	–400,573	336.165	-1,276,542
3,200	62.406	165,351	–401,025	338.145	-1,310,259
3,300	62.510	171,597	–401,495	340.067	-1,344,170
3,400	62.614	177,853	–401,983	341.935	-1,378,272
3,500	62,718	184,120	–402,489	343.751	-1,412,555
3,600	62.825	190,397	–403,013	345.519	-1,447,017
3,700	62.932	196,685	–403,553	347.242	-1,481,656
3,800	63,041	202,983	–404,110	348.922	-1,516,467
3,900	63.151	209,293	–404,684	350.561	-1,551,441
4,000	63.261	215,613	–405,273	353.161	-1,590,577
4,100	63.369	221,945	–405,878	353.725	-1,621,874
4,200	63.474	228,287	–406,499	355.253	-1,657,322
4,300	63.575	234,640	–407,135	356.748	-1,692,922
4,400	63.669	241,002	–407,785	358.210	-1,728,668

T (K)	$\bar{c}_{p0}$ (kJ/kmol–K)	$\bar{h}_{t,T} - \bar{h}_{t,298}$ (kJ/kmol)	$\bar{h}_f^0$ (T) (kJ/kmol)	$\bar{s}^0$(T) (kJ/kmol–K)	$\bar{g}^0$ (T) (kJ/kmol)
4,500	63.753	247,373	–408,451	359.642	-1,764,562
4,600	63.825	253,752	–409,132	361.044	-1,800,596
4,700	63.881	260,138	–409,828	362.417	-1,836,768
4,800	63.918	266,528	–410,539	363.763	-1,873,080
4,900	63.932	272,920	–411,267	365.081	-1,909,523
5,000	63.919	279,313	–412,010	366.372	-1,946,093

TABLE 10: Ideal gas properties of hydrogen atom, H

MW = 1.01, $h_{f,298}{}^{0}$(kJ/kmol) = 217,977

T (K)	$\bar{c}_{p0}$ (kJ/kmol–K)	$\bar{h}_{t,T} - \bar{h}_{t,298}$ (kJ/kmol)	$\bar{h}_f{}^0$ (T) (kJ/kmol)	$\bar{s}^{\,0}$(T) (kJ/kmol–K)	$\bar{g}^0$ (T) (kJ/kmol)
200	20.786	–2,040	217,346	106.305	194,676
298	20.786	0	217,977	114.605	183,825
300	20.786	38	217,989	114.733	183,595
400	20.786	2,117	218,617	120.713	171,809
500	20.786	4,196	219,236	125.351	159,498
600	20.786	6,274	219,848	129.351	146,640
700	20.786	8,353	220,456	132.345	133,689
800	20.786	10,431	221,059	135.121	120,311
900	20.786	12,510	221,653	137.569	106,675
1,000	20.786	14,589	222,234	139.759	92,807
1,100	20.786	16,667	222,793	141.740	78,730
1,200	20.786	18,746	223,329	143.549	64,464
1,300	20.786	20,824	223,843	145.213	50,024
1,400	20.786	22,903	224,335	146.753	35,426
1,500	20.786	24,982	224,806	148.187	20,679
1,600	20.786	27,060	225,256	149.528	5,792
1,700	20.786	29,139	225,687	150.789	-9,225
1,800	20.786	31,217	226,099	151.977	-24,365
1,900	20.786	33,296	226,493	153.101	-39,619
2,000	20.786	35,375	226,868	154.167	-54,982
2,100	20.786	37,453	227,226	155.181	-70,450
2,200	20.786	39,532	227,568	156.148	-86,017
2,300	20.786	41,610	227,894	157.072	-101,679
2,400	20.786	43,689	228,204	157.956	-117,428
2,500	20.786	45,768	228,499	158.805	-133,268
2,600	20.786	47,846	228,780	159.620	-149,189
2,700	20.786	49,925	229,047	160.405	-165,192
2,800	20.786	52,003	229,301	161.161	-181,271
2,900	20.786	54,082	229,543	161.890	-197,422
3,000	20.786	56,161	229,772	162,595	-213,647
3,100	20.786	58,239	229,989	163.276	-229,940
3,200	20.786	60,318	230,195	163.936	-246,300
3,300	20.786	62,396	230,390	164.576	-262,728
3,400	20.786	64,475	230,574	165.196	-279,214
3,500	20.786	66,554	230,748	165.799	-295,766
3,600	20.786	68,632	230,912	166.954	-314,425
3,700	20.786	70,711	231,067	166.954	-329,042
3,800	20.786	72,789	231,212	167.508	-345,764
3,900	20.786	74,868	231,348	168.048	-362,542
4,000	20.786	76,947	231,475	168.575	-379,376
4,100	20,786	79,025	231,594	169.088	-396,259
4,200	20,786	81,104	231,704	169.589	-413,193
4,300	M786	83,182	231,805	170.078	-430,176
4,400	20.786	85,261	231,897	170.556	-447,208

T (K)	$\bar{c}_{p0}$ (kJ/kmol–K)	$\bar{h}_{t,T} - \bar{h}_{t,298}$ (kJ/kmol)	$\bar{h}_f^0(T)$ (kJ/kmol)	$\bar{s}^0(T)$ (kJ/kmol–K)	$\bar{g}^0(T)$ (kJ/kmol)
4,500	20.786	87,340	231,981	171.023	-464,287
4,600	20.786	89,418	232,056	171.480	-481,413
4,700	20.786	91,497	232,123	171.927	-498,583
4,800	20.786	93,575	232,180	172.364	-515,795
4,900	20.786	95,654	232,228	172.793	-533,055
5,000	20.786	97,733	232,267	173.213	-550,355

TABLE 11: Ideal gas properties of hydrogen, H_2

MW = 2.02, $h_{f,298}$ ^{0}(kJ/kmol) =0

T (K)	$\bar{c}_{p0}$ (kJ/kmol–K)	$\bar{h}_{t,T} - \bar{h}_{t,298}$ (kJ/kmol)	$\bar{h}_f^{\,0}$ (T) (kJ/kmol)	$\bar{s}^{\,0}$(T) (kJ/kmol–K)	$\bar{g}^0$ (T) (kJ/kmol)
200	28.522	–2,818	0	119.137	-26,645
298	28.871	0	0	130.595	-38,917
300	28.877	53	0	130.773	-39,179
400	29.120	2,954	0	139.116	-52,692
500	29.275	5,04	0	145.632	-67,776
600	29.375	8,807	0	150.979	-81,780
700	29.461	11,749	0	155.514	-97,111
800	29.581	14,701	0	159.455	-112,863
906	29.792	17,668	0	162.950	-129,965
1,000	30.160	20,664	0	166.106	-145,442
1,100	30.625	23,704	0	169.003	-162,199
1,200	31.077	26,789	0	171.687	-179,235
1,300	31.516	29,919	0	174.192	-196,531
1,400	31.943	33,092	0	176.543	-214,068
1,500	32.356	36,307	0	178.761	-231,835
1,600	32.758	39,562	0	180.862	-249,817
1,700	33.146	42,858	0	182.860	-268,004
1,800	33.522	46,191	0	184.765	-286,386
1,900	33.885	49,562	0	186.587	-304,953
2,000	34.236	52,968	0	188.334	-323,700
2,100	34.575	56,408	0	190.013	-342,619
2,200	34.901	59,882	0	191.629	-361,702
2,300	35.216	63,388	0	193.187	-380,942
2,400	35.50	66,925	0	194.692	-400,336
2,500	35.811	70,492	0	196.148	-419,878
2,600	36.091	74,087	0	197.558	-439,564
2,700	36.361	77,710	0	198.926	-459,390
2,800	36.621	81,359	0	200.253	-479,349
2,900	36.871	85,033	0	201.542	-499,439
3,000	37.112	88,733	0	202.796	-519,655
3,100	37.343	92,455	0	204.017	-539,998
3,200	37.566	96,201	0	205.206	-560,458
3,300	37.781	99,968	0	206.365	-581,037
3,400	37.989	103,757	0	207.496	-601,729
3,500	38.190	107,566	0	208.600	-622,534
3,600	38.385	111,395	0	209.679	-643,449
3,700	38.574	115,243	0	210.733	-664,469
3,800	38.759	119,109	0	211.764	-685,594
3,900	38.939	122,994	0	212.774	-706,825
4,000	39.116	126,897	0	213.762	-728,151
4,100	39.291	130,817	0	214.730	-749,576
4,200	39.464	134,755	0	215.679	-771,097
4,300	39.636	138,710	0	216.609	-792,709

T (K)	$\bar{c}_{p0}$ (kJ/kmol–K)	$\bar{h}_{t,T} - \bar{h}_{t,298}$ (kJ/kmol)	$\bar{h}_f^0$ (T) (kJ/kmol)	$\bar{s}^0$(T) (kJ/kmol–K)	$\bar{g}^0$ (T) (kJ/kmol)
4,406	39.808	142,682	0	217.522	-815,720
4,500	39.981	146,672	0	218.419	-836,214
4,600	40.156	150,679	0	219,300	-858,101
4,700	40.334	154,703	0	220.165	-880,073
4,800	40.516	158,746	0	221.016	-902,131
4,900	40.702	162,806	0	221.853	-924,274
5,000	40.895	166,886	0	222.678	-946,504

TABLE12: Ideal gas Properties of water (g), H_2O

MW = 18.02, $h_{f,298}{}^0$(kJ/kmol) = –241,845, enthalpy of vaporization (kJ/kmol) = 44,010

T (K)	$\bar{c}_{p0}$ (kJ/kmol–K)	$\bar{h}_{t,T} - \bar{h}_{t,298}$ (kJ/kmol)	$\bar{h}_f{}^0$ (T) (kJ/kmol)	$\bar{s}\,^0$(T) (kJ/kmol–K)	$\bar{g}^0$ (T) (kJ/kmol)
200	32.255	–3,227	–240,838	175.602	-280,192
298	33.448	0	–241,845	188.715	-298,082
300	33.468	62	–241,865	188.922	-298,460
400	34.437	3,458	–242,858	198.686	-317,861
500	35.337	6,947	–243,822	206.467	-338,132
600	36.288	10,528	–244,753	212.992	-359,112
700	37.364	14,209	–245,638	218.665	-380,702
800	38.587	18,005	–246,461	223.733	-402,826
900	39.930	21,930	–247,209	228.354	-425,434
1,000	41.315	25,993	–247,879	232.633	-448,485
1,100	42.638	30,191	–248,475	236.634	-471,951
1,200	43.874	34,518	–249,005	240.397	-495,803
1,300	45.027	38,963	–249,477	243.955	-520,024
1,400	46.102	43,520	–249,895	247.332	-544,590
1,500	47.103	48,181	–250,267	250.547	-569,485
1,600	48.035	52,939	–250,597	253.617	-594,693
1,700	48.901	57,786	–250,890	256.556	-620,204
1,800	49.705	62,717	–251,151	259.374	-646,001
1,900	50.451	67,725	–251,384	262.081	-672,074
2,000	51.143	72,805	–251,594	264.687	-698,414
2,100	51.784	77,952	–251,783	267.198	-725,009
2,200	52.378	83,160	–251,955	269.621	-751,851
2,300	52.927	88,426	–252,113	271.961	-778,929
2,400	53.435	93,744	–252,261	274.225	-806,241
2,500	53.905	99,112	–252,399	276.416	-833,773
2,600	54.340	104,524	–252,532	278.539	-861,522
2,700	54.742	109,979	–252,659	280.597	-889,478
2,800	55.115	115,472	–252,785	282.595	-917,639
2,900	55.459	121,001	–252,909	284.535	-945,996
3,000	55.779	126,563	–253,034	286.420	-974,542
3,100	56.076	132,156	–253,161	288.254	-1,003,276
3,200	56.353	137,777	–253,290	290.039	-1,032,193
3,300	56.610	143,426	–253,423	291.777	-1,061,283
3,400	56.851	149,099	–253,561	293.471	-1,090,547
3,500	57.076	154,795	–253,704	295.122	-1,119,977
3,600	57.288	160,514	–253,852	296.733	-1,149,570
3,700	57.488	166,252	–254,007	298.305	-1,179,322
3,800	57.676	172,011	–254,169	299.841	-1,209,230
3,900	57.856	177,787	–254,338	301.341	-1,239,288
4,000	58.026	183,582	–254,515	302.808	-1,269,495
4,100	58.190	189,392	–254,699	304.243	-1,299,849
4,200	58.346	195,219	–254,892	305.647	-1,330,343
4,300	58.496	201,061	–255,093	307.022	-1,360,979
4,400	58.641	206,918	–255,303	308.368	-1,391,746

4,500	58.781	212,790	–255,522	309.688	-1,422,651
4,600	58.916	218,674	–255,751	310.981	-1,453,684
4,700	59.047	224,573	–255,990	312.250	-1,484,847
4,800	59.173	230,484	–256,239	313.494	-1,516,132
4,900	59.295	236,407	–256,501	314.716	-1,547,546
5,000	59.412	242,343	–256,774	315–915	-1,579,077

TABLE 13: Ideal gas properties of nitrogen atom, N

MW = 14.01, $h_{f,298}{}^{0}$(kJ/kmol) = 472,629

T (K)	$\bar{c}_{p0}$ (kJ/kmol–K)	$\bar{h}_{t,T} - \bar{h}_{t,298}$ (kJ/kmol)	$\bar{h}_f^{\,0}$ (T) (kJ/kmol)	$\bar{s}^{\,0}$(T) (kJ/kmol–K)	$\bar{g}^{0}$ (T) (kJ/kmol)
200	20.790	–2,040	472,008	144.889	441,611
298	20.786	0	472,629	153.189	426,979
300	20.786	38	472,640	153.317	426,672
400	20.786	2,117	473,259	159.297	411,027
500	20.786	4,196	473,864	163.935	394,858
600	20.786	6,274	474,450	167.725	378,268
700	20.786	8,353	475,010	170.929	361,332
800	20.786	10,431	475,537	173.705	344,096
900	20.786	12,510	476,027	176.153	326,601
1,000	20.786	14,589	476,483	178.343	308,875
1,100	20.792	16,668	476,911	180.325	290,940
1,200	20.795	18,747	477,316	182.134	272,815
1,300	20.795	20,826	477,700	183.798	254,518
1,400	20.793	22,906	478,064	185.339	236,060
1,500	20.790	24,985	478,411	186.774	217,453
1,600	20.786	27,064	478,742	188.115	198,709
1,700	20.782	29,142	479,059	189.375	179,834
1,800	20.779	31,220	479,363	190.563	160,836
1,900	20.777	33,298	479,656	191.687	141,722
2,000	20.776	35,376	479,939	192.752	122,501
2,100	20.778	37,453	480,213	193.766	103,173
2,200	20.783	39,531	480,479	194.733	83,747
2,300	20.791	41,610	480,740	195.657	64,228
2,400	20.802	43,690	480,995	196.542	44,618
2,500	20.818	45,771	481,246	197.391	24,923
2,600	20.838	47,853	481,494	198.208	5,141
2,700	20.864	49,938	481,740	198.995	-14,720
2,800	20.895	52,026	481,985	199,754	-34,656
2,900	20.931	54,118	482,230	200.488	-54,668
3,000	20,974	56,213	482,476	201.199	-74,755
3,100	21.024	58,313	482,723	201.887	-94,908
3,200	21.080	60,418	482,972	202.555	-115,129
3,300	21.143	62,529	483,224	203.205	-135,419
3,400	21.214	64,647	483,481	203.837	-155,770
3,500	21.292	66,772	483,742	204.453	-176,185
3,600	21.378	68,905	484,009	205.054	-196,660
3,700	21.472	71,048	484,283	205.641	-217,195
3,800	21.575	73,200	484,564	206.215	-237,788
3,900	21.686	75,363	484,853	206.777	-258,438
4,000	21.905	77,537	485,151	207.328	-279,146
4,100	21–934	79,724	485,459	207.868	-299,906
4,200	22.071	81,924	485,779	208.398	-320,719
4,300	22.217	84,139	486,110	208.919	-341,584
4,400	22.372	86,368	486,453	209.431	-362,499

4,500	22.536	88,613	486,811	209.936	-383,470
4,600	22.709	90,875	487,184	210.433	-404,488
4,700	22.891	93,155	487,573	210.923	-425,554
4,800	23.082	95,454	487,979	211.407	-446,671
4,900	23.282	97,772	488,405	211.885	-467,836
5,000	23.491	100,111	488,850	212.358	-489,050

TABLE 14: Ideal gas properties of nitric oxide, NO

MW = 30.01, $h_{f,298}{}^{0}$(kJ/kmol) = 90,297

T (K)	$\bar{c}_{p0}$ (kJ/kmol–K)	$\bar{h}_{t,T} - \bar{h}_{t,298}$ (kJ/kmol)	$\bar{h}_f{}^0$ (T) (kJ/kmol)	$\bar{s}{}^0$(T) (kJ/kmol–K)	$\bar{g}^0$ (T) (kJ/kmol)
200	29.374	–2,901	90,234	198.856	47,625
298	29.728	0	90,297	210.652	27,523
300	29.735	55	90,298	210,836	27,101
400	30.103	3,046	90,341	219.439	5,567
500	30.570	6,079	90,367	226.204	-16,726
600	31.174	9,165	90,382	231.829	-39,635
700	31.908	12,318	90,393	236.688	-63,067
800	32.715	15,549	90,405	241.001	-86,955
900	33.489	18,860	90,421	244.900	-111,253
1,000	34.076	22,241	90,443	248.462	-135,924
1,100	34.483	25,669	90,465	251.729	-160,936
1,200	34.850	29,136	90,486	254.745	-186,261
1,300	35.180	32,638	90,505	257.548	-211,877
1,400	35.474	36,171	90,520	260.166	-237,764
1,500	35.737	39,732	90,532	262.623	-263,906
1,600	35.972	43,317	90,538	264.937	-290,285
1,700	36.180	46,925	90,539	267.124	-316,889
1,800	36.364	50,552	90,534	269.197	-343,706
1,900	36.527	54,197	90,523	271.168	-370,725
2,000	36.671	57,857	90,505	273.045	-397,936
2,100	36.797	61,531	90,479	274.838	-425,332
2,200	36.909	65,216	90,447	276.552	-452,901
2,300	37.008	68,912	90,406	278.195	-480,640
2,400	37.095	72,617	90,358	279.772	-508,539
2,500	37.173	76,331	90,303	281.288	-536,592
2,600	37.242	80,052	90,239	282.747	-564,793
2,700	37.305	83,779	90,168	284.154	-593,140
2,800	37.362	87,513	90,089	285.512	-621,624
2,900	37.415	91,251	90,003	286.824	-650,242
3,000	37.464	94,995	89,909	288.093	-678,987
3,100	37.511	98,744	89,809	289.322	-707,857
3,200	37.556	102,498	89,701	290.514	-736,850
3,300	37.600	106,255	89,586	291.670	-765,959
3,400	37.643	110,018	89,465	292.793	-795,181
3,500	37.686	113,784	89,337	293.885	-824,517
3,600	37.729	117,555	89,203	294.947	-853,957
3,700	37.771	121,330	89,063	295.981	-883,503
3,800	37.815	125,109	88,918	296.989	-913,152
3,900	37.858	128,893	88,767	297.972	-942,901
4,000	37.900	132,680	88,611	298.931	-972,747
4,100	37.943	136,473	88,449	299.867	-1,002,685
4,200	37.984	140,269	88,283	300.782	-1,032,718
4,300	38.023	144,069	88,112	301.677	-1,062,845
4,400	38.060	147,873	87,936	302.551	-1,093,054

4,500	38.093	151,681	87,755	303.407	-1,123,354
4,600	38.122	155,492	87,569	304.244	-1,153,733
4,700	38.146	159,305	87,379	305.064	-1,184,199
4,800	38.162	163,121	87 184	305.868	-1,214,748
4,900	38.171	166,938	86:984	306.655	-1,245,375
5,000	38.170	170,755	86,779	307.426	-1,276,078

TABLE 15: Ideal gas properties of nitrogen dioxide, NO_2

MW = 46.01, $h_{f,298}{}^{0}$(kJ/kmol) = 33,098

T (K)	$\bar{c}_{p0}$ (kJ/kmol–K)	$\bar{h}_{t,T} - \bar{h}_{t,298}$ (kJ/kmol)	$\bar{h}_f{}^0$ (T) (kJ/kmol)	$\bar{s}^{\,0}$(T) (kJ/kmol–K)	$\bar{g}^0$ (T) (kJ/kmol)
200	32.936	–3,432	33,961	226.016	-15,537
298	36.881	0	33,098	239.925	-38,400
300	36.949	68	33,085	240.153	-38,880
400	40.331	3,937	32,521	251.259	-63,469
500	43.227	8,118	32,173	260.578	-89,073
600	45.737	12,569	31,974	268.686	-115,545
700	47.913	17,255	31,885	275.904	-142,780
800	49,762	22,141	31,880	282.427	-170,703
900	51.243	27,195	31,938	288.377	-199,246
1,000	52.271	32,375	32,035	293.834	-228,361
1,100	52.989	37,638	32,146	298.850	-257,999
1,200	53.625	42,970	32,267	303.489	-288,119
1,300	54.186	48,361	32,392	307.804	-318,686
1,400	54.679	53,805	32,519	311.838	-349,670
1,500	55.109	59,295	32,643	315.625	-381,045
1,600	55.483	64,825	32,762	319.194	-412,787
1,700	55.805	70,390	32,873	322.568	-444,878
1,800	56.082	75,984	32,973	325.765	-477,295
1,900	56.318	81,605	33,061	328.804	-510,025
2,000	56.517	87,247	33,134	331.698	-543,051
2,100	56.685	92,907	33,192	334.460	-576,361
2,200	56.826	98,583	32,233	337.100	-609,939
2,300	56.943	104,271	33,256	339.629	-643,778
2,400	57.040	109,971	33,262	342.054	-677,861
2,560	57.121	115,679	33,248	344.384	-732,846
2,600	57.188	121,394	33,216	346.626	-746,736
2,700	57.244	127,116	33,165	348.785	-781,506
2,800	57.291	132,843	33,095	350.868	-816,489
2,900	57.333	138,574	33,007	352.879	-851,677
3,000	57.371	144,309	32,900	354.824	-887,065
3,100	57.406	150,049	32,776	356.705	-922,639
3,200	57.440	155,791	32,634	358.529	-958,404
3,300	57.474	161,536	32,476	360.297	-994,346
3,400	57.509	167,285	32,302	362.013	-1,030,461
3,500	57.546	173,038	32,113	363.680	-1,066,744
3,600	57.584	178,795	31,908	365.302	-1,103,194
3,700	57.624	184,555	31,689	366.880	-1,139,803
3,800	57.665	190,319	31,456	368.418	-1,176,571
3,900	57.708	196,088	31,210	369.916	-1,213,486
4,000	57.750	201,861	30,951	371.378	-1,250,553
4,100	57.792	207,638	30,678	372.804	-1,287,760
4,200	57.831	213,419	30,393	374.197	-1,325,110
4,300	57.866	20,204	30,095	375.559	-1,561,602
4,400	57.895	224,992	29,783	376.889	-1,400,222

T (K)	$\bar{c}_{p0}$ (kJ/kmol–K)	$\bar{h}_{t,T} - \bar{h}_{t,298}$ (kJ/kmol)	$\bar{h}_f^0$ (T) (kJ/kmol)	$\bar{s}^0$(T) (kJ/kmol–K)	$\bar{g}^0$ (T) (kJ/kmol)
4,500	57.915	230,783	29,457	378.190	-1,437,974
4,600	57.925	236,575	29,117	379.464	-1,475,861
4,700	57.922	242,367	28,761	380.709	-1,513,867
4,800	57.902	~48,159	28,389	381.929	-1,552,002
4,900	57.862	253,947	27,998	383.122	-1,590,253
5,000	57,798	259,730	27,586	384.290	-1,628,622

TABLE 16: Ideal gas properties of nitrogen, N_2

MW = 28.01, $\bar{h}_{f,298}{}^0$(kJ/kmol) = 0

T (K)	$\bar{c}_{p0}$ (kJ/kmol–K)	$\bar{h}_{t,T} - \bar{h}_{t,298}$ (kJ/kmol)	$\bar{h}_f{}^0$ (T) (kJ/kmol)	$\bar{s}^0$(T) (kJ/kmol–K)	$\bar{g}^0$ (T) (kJ/kmol)
200	28.793	–2,841	0	179.959	-38,833
298	29.071	0	0	191.511	-57,070
300	29.075	54	0	191.691	-57,453
400	29.319	2,973	0	200.088	-77,062
500	29.636	5,920	0	206.662	-97,411
600	30.086	8,905	0	212.103	-118,357
700	30.684	11,942	0	216.784	-139,807
800	31.394	15,046	0	220.927	-161,696
900	32.131	18,222	0	224.667	-183,978
1,000	32.762	21,468	0	228.087	-206,619
1,100	33.258	24,770	0	231.233	-229,586
1,200	33.707	M,118	0	234.146	-252,857
1,300	34.113	31,510	0	236.861	-276,409
1,400	34.477	34,939	0	239.402	-300,224
1,500	34.805	38,404	0	241.792	-324,284
1,600	35.099	41,899	0	244.048	-348,578
1,700	35.361	45,423	0	246.184	-373,090
1,800	35.595	48,971	0	248.212	-397,811
1,900	35.803	52,541	0	250.142	-422,729
2,000	35.988	56,130	0	251.983	-447,836
2,100	36.152	59,738	0	253.743	-473,122
2,200	36.298	63,360	0	255.429	-498,584
2,300	36.428	66,997	0	257.045	-524,207
2,400	36–543	70,645	0	259.598	-552,390
2,500	36–645	74,305	0	260.092	-575,925
2,600	36–737	77,974	0	261.531	-602,007
2,700	36.820	81,652	0	262.919	-628,229
2,800	36.895	85,338	0	264.259	-654,587
2,900	36.964	89,031	0	265.555	-681,079
3,000	37.028	92,730	0	266.810	-707,700
3,100	37.088	96,436	0	268.025	-734,442
3,200	37.144	100,148	0	269.203	-761,302
3,300	37.198	103,865	0	270.347	-788,280
3,400	37.251	107,587	0	271.458	-815,370
3,506	37.302	111,315	0	272.539	-844,207
3,600	37.352	115,048	0	273.590	-869,876
3,700	37.402	118,786	0	274.614	-897,286
3,800	37.452	122,528	0	275.612	-924,798
3,900	37.501	126,276	0	276.586	-952,409
4,000	37.549	130,028	0	277.536	-980,116
4,100	37.597	133,786	0	278.464	-1,007,916
4,200	37.643	137,548	0	279.370	-1,035,806
4,300	37.688	141,314	0	280.257	-1,063,791
4400	37.730	145,085	0	281.123	-1,091,856

T (K)	$\bar{c}_{p0}$ (kJ/kmol–K)	$\bar{h}_{t,T} - \bar{h}_{t,298}$ (kJ/kmol)	$\bar{h}_f^0$ (T) (kJ/kmol)	$\bar{s}^0$(T) (kJ/kmol–K)	$\bar{g}^0$ (T) (kJ/kmol)
4,500	37.768	148,860	0	281.972	-1,120,014
4,600	37.803	152,639	0	282.802	-1,148,250
4,700	37.832	156,420	0	283.616	-1,176,575
4,800	37.854	160,205	0	284.412	-1,204,973
4,900	37.868	163,991	0	285.193	-1,233,455
5,000	37.873	167,778	0	285.958	-1,262,012

TABLE 17: Ideal gas properties of oxygen atom, O

MW = 16.00, $h_{f,298}{}^0$(kJ/kmol) = 249,197

T (K)	$\bar{c}_{p0}$ (kJ/kmol–K)	$\bar{h}_{t,T} - \bar{h}_{t,298}$ (kJ/kmol)	$\bar{h}_f^0$ (T) (kJ/kmol)	$\bar{s}^0$ (T) (kJ/kmol–K)	$\bar{g}^0$ (T) (kJ/kmol)
200	22.477	–2,176	248,439	152.085	216,604
298	21.899	0	249,197	160.945	201,235
300	21.890	41	249,211	161.080	200,914
400	21.500	2,209	249,890	167.320	184,478
500	21.256	4,345	250,494	172.089	167,498
600	21.113	6,463	251,033	175.951	150,089
700	21.033	8,570	251,516	179.199	132,328
800	20.986	10,671	251,949	182.004	114,265
900	20.952	12,768	252,340	184.474	95,938
1,000	20.915	14,861	252,698	186,679	77,379
1,100	20.898	16,952	253,033	188,672	58,610
1,200	20.882	19,041	253,350	190.490	39,650
1,300	20–867	21,128	253,650	192.160	20,517
1,400	20.854	23,214	253,934	193.706	1,223
1,500	20.843	25,299	254,201	195,145	-18,222
1,600	20.834	27,383	254,454	196.490	-37,804
1,700	20.827	29,466	254,692	197.753	-57,517
1,800	20.822	31,548	254,916	198.943	-77,352
1,900	20.820	33,630	255,127	200.069	-97,304
2,000	20.819	35,712	255,325	201.136	-117,363
2,100	20.821	37,794	255,512	202.152	-137,528
2,200	20.925	39,877	255,687	203.121	-157,792
2,300	20.831	41,959	255,852	204.047	-178,152
2,400	20.840	44,043	256,007	204.933	-198,599
2,500	20.851	46,127	256,152	205.784	-219,136
2,600	20–865	48,213	256,288	206.602	-239,755
2,700	20.881	50,300	256,416	207.390	-260,456
2,800	20–899	52,389	256,535	208.150	-281,234
2,900	20.920	54,480	256,648	208.884	-302,087
3,000	20.944	56,574	256,753	209.593	-323,008
3,100	20.970	58,669	256,852	210.280	-344,002
3,200	20.998	60,768	256,945	210.947	-365,065
3,300	21.028	62,869	257,032	211.593	-386,191
3,400	21.061	64,973	257,114	212.221	-407,381
3,500	21.095	67,081	257,192	212.832	-428,634
3,600	21.132	69,192	257,265	213.427	-449,948
3,700	21.171	71,308	257,334	214.007	-471,321
3,800	21.212	73,427	257,400	214.572	-492,750
3,900	21.254	75,550	257,462	215.123	-514,233
4,000	21.299	77,678	257,522	215.662	-535,773
4,100	21.345	79,810	257,579	216.189	-557,368
4,200	21.392	81,947	257,635	216.703	-579,009
4,300	21,441	84,088	257,688	217.207	-600,705
4,400	21.490	86,235	257,740	217.701	-622,452

T (K)	$\bar{c}_{p0}$ (kJ/kmol–K)	$\bar{h}_{t,T} - \bar{h}_{t,298}$ (kJ/kmol)	$\bar{h}_f^{\,0}(T)$ (kJ/kmol)	$\bar{s}^{\,0}(T)$ (kJ/kmol–K)	$\bar{g}^0(T)$ (kJ/kmol)
4,500	21.541	88,386	257,790	218.184	-644,245
4,600	2L593	90,543	257,840	218.658	-666,087
4,700	21.646	92,705	257,889	219.123	-687,976
4,800	21.699	94,872	257,938	219.580	-709,915
4,900	21.752	97,045	257,987	220.028	-731,895
5,000	21.805	99,223	258,036	220.468	-753,920

TABLE 18: Ideal gas properties of hydroxyl, OH

MW = 17.01, $h_{f,298}$ ^{0}(kJ/kmol) = 38,985

T (K)	$\bar{c}_{p0}$ (kJ/kmol–K)	$\bar{h}_{t,T} - \bar{h}_{t,298}$ (kJ/kmol)	$\bar{h}_f{}^0$ (T) (kJ/kmol)	$\bar{s}\,^0$(T) (kJ/kmol–K)	$\bar{g}^0$ (T) (kJ/kmol)
200	30.140	–2,948	38,864	171.607	1,716
298	29.932	0	38,985	183.604	-15,729
300	29.928	55	38,987	183.789	-16,097
400	29.718	3,037	39,030	192.369	-34,926
500	29.570	6,001	39,000	198.983	-54,506
600	29.527	8,955	38,909	204.369	-74,681
700	29.615	11,911	38,770	208.925	-95,352
800	29.844	14,883	38,599	212.893	-116,446
900	30.208	17,884	38,410	216.428	-137,916
1,000	30.682	20,928	38,220	219.635	-159,722
1,100	31.186	24,022	38,039	222.583	-181,834
1,200	31.662	27,164	37,867	225.317	-204,231
1,300	32.114	30,353	37,704	227.869	-226,892
1,400	32.540	33,586	37,548	230.265	-249,800
1,500	32.943	36,860	37,397	232.524	-272,941
1,600	33.323	40,174	37,252	234.662	-296,300
1,700	33.682	43,524	37,109	236.693	-319,869
1,800	34.019	46,910	36,969	238.628	-343,635
1,900	34.337	50,328	36,831	240.476	-367,591
2,000	34.635	53,776	36,693	242.245	-391,729
2,100	34.915	57,254	36,555	243.942	-416,039
2,200	35.178	60,759	36,416	245.572	-440,514
2,300	35.425	64,289	36,276	247.141	-465,150
2,400	35.656	67,843	36,133	248.654	-489,942
2,500	35.872	71,420	35,986	250.114	-514,880
2,600	36.074	75,017	35,836	251.525	-539,963
2,700	36.263	78,634	35,682	252.890	-565,184
2,800	36.439	82,269	35,524	254.212	-590,540
2,900	36.604	85,922	35,360	255.493	-616,023
3,000	36.759	89,590	35,191	256.737	-641,636
3,100	36.903	93,273	35,016	257.945	-667,372
3,200	37.039	96,970	34,835	259.118	-693,223
3,300	37.166	100,681	34,648	260.260	-719,192
3,400	37.285	104,403	34,454	261.371	-745,273
3,500	37.398	108,137	34,253	262.454	-771,467
3,600	37.504	111,882	34,046	263.509	-797,765
3,700	37.605	115,638	33,831	264.538	-824,168
3,800	37.701	119,403	33,610	265.542	-850,672
3,900	37.793	123,178	33,381	266.522	-877,273
4,000	37.882	126,962	33,146	267,480	60,027
4,100	37.968	130,754	32,903	268.417	-930,771
4,200	38.052	134,555	32,654	269.333	-957,659
4,300	38.135	138,365	32,397	270.229	-984,635
4,400	38.217	142,182	32,134	271.107	-1,011,704

T (K)	$\bar{c}_{p0}$ (kJ/kmol–K)	$\bar{h}_{t,T} - \bar{h}_{t,298}$ (kJ/kmol)	$\bar{h}_f^{\,0}$ (T) (kJ/kmol)	$\bar{s}^{\,0}$(T) (kJ/kmol–K)	$\bar{g}^0$ (T) (kJ/kmol)
4,500	38.300	146,008	31,864	271.967	-1,038,859
4,600	38.382	149,842	31,588	272.809	-1,066,094
4,700	38.466	153,685	31,305	273.636	-1,093,419
4,800	38.552	157,536	31,017	274.446	-1,120,820
4,900	38.640	161,395	30,722	275.242	-1,148,306
5,000	38.732	165,264	30,422	276.024	-1,175,871

TABLE 19: Ideal gas properties of oxygen, O_2

MW = 32.0, $h_{f,298}{}^0$(kJ/kmol) = 0

T (K)	$\bar{c}_{p0}$ (kJ/kmol–K)	$\bar{h}_{t,T} - \bar{h}_{t,298}$ (kJ/kmol)	$\bar{h}_f{}^0$ (T) (kJ/kmol)	$\bar{s}{}^0$(T) (kJ/kmol–K)	$\bar{g}^0$ (T) (kJ/kmol)
200	28.473	–2,836	0	193.518	-41,540
298	29.315	0	0	205.043	-61,103
300	29.331	54	0	205.224	-61,513
4010	30.210	3,031	0	213.782	-854,235
500	31.114	6,097	0	220.620	-104,213
600	32.030	9,254	0	226.374	-126,570
700	32.927	12,503	0	231.379	-149,462
800	33.757	15,838	0	235.831	-172,827
900	34.454	19,250	0	239.849	-196,614
1,000	34.936	22,721	0	243.507	-220,786
1,100	35.270	26,232	0	246.852	-245,305
1,200	35.593	29,775	0	249–935	-270,147
1,300	35.903	33,350	0	252.796	-295,285
1,400	36.202	36,955	0	255.468	-320,700
1,500	36.490	40,590	0	257.976	-346,374
1,600	36.768	44,253	0	260.339	-372,289
1,700	37.036	47,943	0	262.577	-398,438
1,800	37.296	51,660	0	264.701	-424,802
1,900	37.546	55,402	0	266.724	-451,374
2,000	37.788	59,169	0	268.656	-478,143
2,100	38.023	62,959	0	270.506	-505,104
2,200	38,250	66,773	0	272.280	-532,243
2,300	38.470	70,609	0	273.985	-559,557
2,400	38.684	74,467	0	275.627	-587,038
2,500	38.891	78,346	0	277.210	-614,679
2,600	39.093	82,245	0	278.739	-642,476
2,700	39.289	86,164	0	280.218	-670,425
2,800	39.480	90,103	0	281.651	-698,520
2,900	39.665	94,060	0	283.039	-726,753
3,000	39.846	98,036	0	284.387	-755,125
3,100	40.023	102,029	0	285.697	-783,632
3,200	40.195	106,040	0	286.970	-812,264
3,300	40.362	110,W	0	288.209	-841,022
3,400	40.526	114,112	0	289.417	-869,906
3,500	40.686	118,173	0	290.594	-898,906
3,600	40.842	122,249	0	291.742	-928,022
3,700	40.1994	126,341	0	292.863	-957,252
3,800	41.143	130,448	0	293.959	-986,596
3,900	41.287	134,570	0	295.029	-1,016,043
4,000	41.429	138,705	0	296.076	-1,045,599
4,100	41.566	142,855	0	297,101	-1,075,259
4,200	41.700	147,019	0	298–104	-1,105,018
4,300	41.830	151,195	0	299.087	-1,134,879
4,400	41,957	155,384	0	300.050	-1,164,836

T (K)	$\overline{c}_{p0}$ (kJ/kmol–K)	$\overline{h}_{t,T} - \overline{h}_{t,298}$ (kJ/kmol)	$\overline{h}_f^0$ (T) (kJ/kmol)	$\overline{s}^0$(T) (kJ/kmol–K)	$\overline{g}^0$ (T) (kJ/kmol)
4,500	42.079	159,586	0	300.994	-1,194,887
4,600	42.197	163,800	0	301.921	-1,225,037
4,700	42.312	168,026	0	302.829	-1,255,270
4,800	42.421	172,262	0	303.721	-1,285,599
4,900	42.527	176,510	0	304.597	-1,316,015
5,000	42.627	180,767	0	305.457	-1,346,518

TABLE 20A Constants for the Benedict–Webb–Rubin

Units are P in bar(s), $\bar{v}$ in m^3/kmol, $\bar{R}$=0.08314 bar m^3/kmole K and T in K

$$P = \bar{R}T/\ \bar{v} + (B_2\bar{R}T - A_2 - C_2/T^2)/\ \bar{v}^2 + (B_3\bar{R}T - A_3)/\ \bar{v}^3 + A_3C_6/\bar{v}^6 + \{D_3/(\bar{v}^3T^2)\}(1+E_2/\bar{v}^2)\ \exp(-E_2/\bar{v}^2)$$

	A_2	B_2	C_2x10^{-6}	A_3	B_3	D_3x10^{-6}	C_6x1000	E_2x100
Ammonia	3.839578	0.051646	0.180933	0.104914	0.00072	0.00016	0.004652	1.98
Argon	0.83412	0.022283	0.013312	0.029211	0.0021529	0.000809	0.0356	0.2338
Benzene	6.5944	0.0503	3.4746	5.6424	0.07633	1.1917	0.7001	2.930
Carbon dioxide	2.708886	0.045628	0.114834	0.052375	0.003082	0.007161	0.11271	0.494
Carbon Monoxide	1.3587	0.0426	0.008673	0.0371	0.002632	0.001054	0.135	0.6
Ethane	4.21072	0.062772	0.181976	0.349742	0.011122	0.033202	0.243389	1.18
Ethylene	3.383909	0.055683	0.132881	0.262438	0.0086	0.0214	0.178	0.923
Helium	0.04149	0.023661	1.64x10^{-7}	-5.81x10^{-4}	−1.97x10^{-7}	−5.59x10^{-9}	−0.007263	0.00779
i–Butane	10.36847	0.137544	0.861225	1.96335	0.039998	0.289806	1.07408	3.4
i–Butylene	9.072094	0.116025	0.939589	1.715169	0.034816	0.278569	0.910889	2.95945
i–Pentane	12.96575	0.160053	1.7695	3.806059	0.066812	0.704225	1.7	4.63
Methane	1.879623	0.0426	0.02287	0.500557	0.0338	0.002579	0.124359	0.6
n–Butane	10.21856	0.124361	1.006009	1.907296	0.039983	0.3206	1.10132	3.4
n–Heptane	17.75317	0.199005	4.808734	10.50233	0.151954	2.502787	4.35611	9
n–Hexane	14.62894	0.177813	3.363411	7.211176	0.109131	1.53284	2.81086	6.66849
Nitrogen	1.208329	0.0458	0.005967	0.015098	0.001982	0.000555	0.291545	0.75
n–Pentane	12.34107	0.156751	2.149367	4.128888	0.066812	0.83511	1.81	4.75
Oxygen	1.518695	0.046524	0.003913	−0.04104	-2.80E–05	−0.00021	0.008641	0.359
Propane	6.963471	0.097313	0.515003	0.96028	0.0225	0.130712	0.607175	2.2
Propylene	6.193333	0.085065	0.445012	0.784331	0.018706	0.103973	0.455696	1.829
Sulfurdioxide	2.148	0.026182	0.80146	0.8557	0.014653	0.1148	0.071955	0.59236

Sources: H. W. Cooper and J. Goldfrank, *Hydrocarbon Processing*, 46(12), 141 (1967); E. P. Gyftopoulos, and G. P. Beretta, *Thermodynamics, Foundations and Application*, Macmillan Publishing Co., NY, 1991.

TABLE 20B: Beatie Bridgemann Equation

$P\,\overline{v}^2 = \overline{R}\,T\,(\overline{v} + B_0\,(1-(\overline{b}/\overline{v}))\,(1-c/(\overline{v}T^3)) - A_0\,(1-(a/\overline{v}^2))$

May be used when $\rho < 0.8\,\rho_c$. Units: atm, m^3/kmol, K

Gas	A_o	a	B_o	b	$c\times10^{-4}$
Air	1.3012	0.01931	0.04611	−0.01101	4.34
Ar	1.2907	0.02328	0.03931	0.0	5.99
CH_4	2.2769	0.01855	0.05587	−0.01587	12.83
C_2H_4	6.1520	0.04964	0.12156	0.03597	22.68
C_2H_6	5.8800	0.05861	0.09400	0.01915	90.00
C_3H_8	11.9200	0.07321	0.18100	0.04293	120.00
$1-C_4H_8$	16.6979	0.11988	0.24046	0.10690	300.00
Iso–C_4H_8	16.9600	0.10860	0.24200	0.08750	250.00
n–C_4H_{10}	17.7940	0.12161	0.24620	0.09423	350.00
Iso–C_4H_{12}	16.6037	0.11171	0.23540	0.07697	300.00
n–C_5H_{12}	28.2600	0.15099	0.39400	0.13960	400.00
Neo–C_5H_{12}	23.3300	0.15174	0.33560	0.13358	400.00
n–C_7H_{16}	54.520	0.20066	0.70816	0.19179	400.00
CH_3OH	33.309	0.09246	0.60362	0.09929	32.03
$(C_2H_5)_2O$	31.278	0.12426	0.45446	0.11954	33.33
CO	1.3445	0.02617	0.05046	−0.00691	4.20
CO_2	5.0065	0.07132	0.10476	0.07235	66.00
H_2	0.1975	−0.00506	0.02096	−0.04359	0.0504
He	0.0216	0.05984	0.01400	0.0	0.0040
I_2	17.0	0.0	0.325	0.0	4000.
Kr	2.4230	0.02865	0.05261	0.0	14.89
N_2	1.3445	0.02617	0.05046	−0.00691	4.20
Ne	0.2125	0.02196	0.02060	0.0	0.101
NH_3	2.3930	0.17031	0.03415	0.19112	476.87
N_2O	5.0065	0.07132	0.10476	0.07235	66.0
O_2	1.4911	0.02562	0.04624	0.004208	4.80
Xe	4.6715	0.03311	0.07503	0.0	30.02

Equation from J. A. Beatie and O.C. Bridgemann, *Pro. Roy Am. Acad. Sc.*, 63,229 (1928). Source: adapted from G. J. Wiley, and R. Sonntag, *Fundamentals of Classical Thermdynamics*, John Wiley & Sons, 1965.

TABLE 21: Lee Kesler Constants

Lee Kesler constants for simple (e.g.: Ar, He etc) and reference (octane) fluids. This is a modified form of BWR equation of state applicable for any substance. The equation has 12 constants.

$P_R = (T_R/v_R')(1 + A/v_R + B/v_R^2 + C/v_R^5 + \{b_4/(T_R^3 v_R^2)\}\{\beta + \gamma/v_r^2\} \exp\{-\gamma/v_R^2\})$

where $A = a_1 - a_2/T_R - a_3/T_R^2 - a_4/T_R^3$

$B = b_1 - b_2/T_R + b_3/T_R^3$

$C = c_1 + c_2/T_R$

$Z = P_R v_R'/T_R = P_R v_R'/T_R = 1 + A/v_R + B/v_R^2 + C/V_r^5 + (b_4/(T_R^3 v_r^2))(\beta + \gamma/v_r^2) \exp(-\gamma/v_R^2)$,

$w_{ref} = w_{octane} = 0.398$, $z(1) = \{z^{(ref)}(P_R,T_R) - z^{(0)}(P_R,T_R)\}/w_{ref}$, $z(P_R,T_R) = z^{(0)}(P_R,T_R) + w\, z^{(1)}(P_R,T_R)$

	Simple	**Reference (Octane)**
a_1	0.1181193	0.2026579
a_2	0.265728	0.331511
a_3	0.154790	0.027655
a_4	0.030323	0.203488
b_1	0.0236744	0.0313885
b_2	0.0186984	0.0503618
b_3	0.0	0.016901
b_4	0.042724	0.041577
c_1	0.155488×10^{-4}	0.48736×10^{-4}
c_2	0.623689×10^{-4}	0.740336×10^{-5}
β	0.65392	1.226
γ	0.060167	0.03754

Adapted from R. Sonntag, C. Borgnakke and G. J. Wylen, *Fundamentals of Classical Thermodynamics*, 5th Ed. John Wiley & Sons, 1998.

TABLE 22: Pitzer generalized saturation data

T_R	$-(\log P_R)^{(0)}$	$(\partial \log P_R/\partial\omega)$	Vaporization		Vapor		Liquid	
			$\Delta s_{fg}^{(0)}/R$	$\Delta s_{fg}^{(1)}/R$	$Z^{(0)}$	$Z^{(1)}$	$Z^{(0)}$	$Z^{(1)}$
1.00	0.000	0.000	0.00	0.00	0.291	–0.090	0.291	–0.080
0.99	0.025	0.021	1.29	1.42	0.43	–0.030	0.202	–0.090
0.98	0.050	0.042	1.70	1.97	0.47	0.000	0.179	–0.093
0.97	0.076	0.064	2.01	2.38	0.51	+0.020	0.162	–0.095
0.96	0.102	0.086	2.28	2.71	0.54	0.035	0.148	–0.095
0.95	0.129	0.109	2.52	3.00	0.565	0.045	0.136	–0.095
0.94	0.156	0.133	2.74	3.28	0.59	0.055	0.125	–0.094
0.92	0.212	0.180	3.14	3.80	0.63	0.075	0.108	–0.092
0.90	0.270	0.230	3.50	4.29	0.67	0.095	0.0925	–0.087
0.89	0.330	0.285	3.82	4.73	0.70	0.110	0.0790	–0.080
0.86	0.391	0.345	4.12	5.18	0.73	0.125	0.0680	–0.075
0.84	0.455	0.405	4.42	5.64	0.756	0.135	0.0585	–0.068
0.82	0.522	0.475	4.72	6.09	0.781	0.140	0.0498	–0.062
0.80	0.592	0.545	5.02	6.54	0.804	0.144	0.0422	–0.057
0.78	0.665	0.620	5.32	7.00	0.826	0.144	0.0360	–0.053
0.76	0.742	0.705	5.64	7.50	0.846	0.142	0.0300	–0.048
0.74	0.823	0.800	5.96	9.05	0.864	0.137	0.0250	–0.043
0.72	0.909	0.895	6.29	8.56	0.881	0.131	0.0210	–0.037
0.70	1.000	1.00	6.64	9.11	0.897	0.122	0.0172	–0.032
0.68	1.096	1.12	6.99	9.72	0.911	0.113	0.0138	–0.027
0.66	1.198	1.25	7.36	10.3	0.922	0.104	0.0111	–0.022
0.64	1.308	1.39	7.73	11.0	0.932	0.097	0.0088	–0.018
0.62	1.426	1.54	8.11	11.7	0.940	0.090	0.0068	–0.015
0.60	1.552	1.70	8.52	12.4	0.947	0.083	0.0052	–0.012
0.58	1.688	1.88	8.93	13.2	0.953	0.077		

From G. N. Lewis, and M.Randall, *Thermodynamics*, 2nd Ed., McGraw Hill Inc., NY 1961. Original source K.S. Pitzer et al., *J. Am. Chem.Soc.*, 77, 3439, 1955.

TABLE 23A: Lee–Kesler values for $Z^{(0)}$ (T_R,P_R)

				P_R			
T_R	**0.01**	**0.05**	**0.1**	**0.2**	**0.4**	**0.6**	**0.8**
0.3	0.0029	0.0145	0.029	0.0579	0.1158	0.1737	0.2315
0.35	0.0026	0.013	0.0261	0.0522	0.1043	0,1564	0.2084
0.4	0.0024	0.0119	0.0239	0.0477	0.0953	0.1429	0.1904
0.45	0.0022	0.011	0.0221	0.0442	0.0882	0.1322	0.1762
0.5	0.0021	0.0103	0.0207	0.0412	0.0825	0.1236	0.1647
0.55	0.9804	0.0098	0.0195	0.039	0.0778	0.1166	0.1553
0.6	0.9849	0.0093	0.0186	0.0371	0.0741	0.1109	0.1476
0.65	0.9881	0.9377	0.0178	0.0356	0.071	0.1063	0.1415
0.7	0.9904	0.9504	0.8958	0.0344	0.0687	0.1027	0.1366
0.75	0.9922	0.9598	0.9165	0.0336	0.067	0.1001	0.133
0.8	0.9935	0.9669	0.9319	0.8539	0.0661	0.0985	0.1307
0.85	0.9946	0.9725	0.9436	0.881	0.0661	0.0983	0.1301
0.9	0.9954	0.9768	0.9528	0.9015	0.78	0.1006	0.1321
0.93	0.9959	0.979	0.9573	0.9115	0.8059	0.6635	0.1359
0.95	0.9961	0.9803	0.96	0.9174	0.8206	0.6967	0.141
0.97	0.9963	0.9815	0.9625	0.9227	0.8338	0.724	0.558
0.98	0.9965	0.9821	0.9637	0.9253	0.8398	0.736	0.5881
0.99	0.9966	0.9826	0.9648	0.9277	0.8455	0.7471	0.6138
1	0.9967	0.9832	0.9659	0.93	0.9509	0.7574	0.6353
1.01	0.9969	0.9837	0.9669	0.9322	0.8561	0.7671	0.6542
1.02	0.9969	0.9842	0.9679	0.9343	0.861	0.7761	0.671
1.05	0.9971	0.9855	0.9707	0.9401	0.8743	0.8002	0.713
1.1	0.9975	0.9874	0.9747	0.9485	0.893	0.8323	0.7649
1.15	0.9978	0.9891	0.978	0.9554	0.9081	0.8576	0.8032
1.2	0.9981	0.9904	0.9808	0.9611	0.9205	0.8779	0.833
1.3	0.9985	0.9926	0.9852	0.9702	0.9396	0.9083	0.8764
1.4	0.9988	0.9942	0.9884	0.9768	0.9534	0.9298	0.9062
1.5	0.9991	0.9954	0.9909	0.9818	0.9636	0.9456	0.9278
1.6	0.9993	0.9964	0.9928	0.9856	0.9714	0.9575	0.9439
1.7	0.9994	0.9971	0.9943	0.9886	0.9775	0.9667	0.9563
1.8	0.9995	0.9977	0.9955	0,9910	0.9823	0.9739	0.9659
1.9	0.9996	0.9982	0.9964	0.9929	0.9861	0.9796	0.9735
2	0.9997	0.9986	0.9972	0.9944	0.9892	0.9842	0.9796
2.2	0.9998	0.9992	0.9983	0.9967	0.9937	0.991	0.9886
2.4	0.9999	0.9996	0.9991	0.9983	0.9969	0.9957	0.9948
2.6	1	0.9998	0.9997	0.9994	0.9991	0.999	0.999
2.8	1	1	1.0001	1.0002	1.0007	1.0013	1.0021
3	1	1.0002	1.0004	1.0008	1.0018	1.003	1.0043
3.5	1.0001	1.0004	1.0008	1.0017	1.0035	1.0055	1.0075
4	1.0001	1.0005	1.001	1.0021	1.0043	1.0066	1.009

TABLE23A: Lee–Kesler values for $Z^{(0)}$ (T_R,P_R) (continued)

T_R	P_R							
	1.000	1.200	1.500	2.000	3.000	5.000	7.000	10.000
0.30	0.2892	0.3470	0.4335	0.5775	0.8648	1.4366	2.0048	2.8507
0.35	0.2604	0.3123	0.3901	0.5195	0.7775	1.2902	1.7987	2.5539
0.40	0.2379	0.2853	0.3563	0.4744	0.7095	1.1758	1.6373	2.3211
0.45	0.2200	0.2638	0.3294	0.4384	0.6551	1.0841	1.5077	2.1338
0.50	0.2056	0.2465	0.3077	0.4092	0.6110	1.0094	1.4017	1.9801
0.55	0.1939	0.2323	0.2899	0.3853	0.5747	0.9475	1.3137	1.8520
0.60	0.1842	0.2207	0.2753	0.3657	0.5446	0.8959	1.2398	1.7440
0.65	0.1765	0.2113	0.2634	0.3495	0.5197	0.8526	1.1773	1.6519
0.70	0.1703	0.2038	0.2538	0.3364	0.4991	0.8161	1.1241	1.5729
0.75	0.1656	0.1981	0.2464	0.3260	0.4823	0.7854	1,0787	1.5047
0.80	0.1626	0.1942	0.2411	0.3182	0.4690	0.7598	1.0400	1.4456
0.85	0.1614	0.1924	0.2382	0.3132	0.4591	0.7388	1.0071	1.3943
0.90	0.1630	0.1935	0.2383	0.3114	0.4527	0.7220	0.9793	1.3496
0.93	0.1664	0.1963	0.2405	0.3122	0.4507	0.7138	0.9648	1.3257
0.95	0.1705	0.1998	0.2432	0.3138	0.4501	0.7092	0.9561	1.3108
0.97	0.1779	0.2055	0.2474	0.3164	0.4504	0.7052	0.9480	1.2968
0.98	0.1844	0.2097	0.2503	0.3182	0.4508	0.7035	0.9442	1.2901
0.99	0.1959	0.2154	0.2538	0.3204	0.4514	0.7018	0.9406	1.2835
1.00	0.2901	0.2237	0.2583	0.3229	0.4522	0.7004	0.9372	1.2772
1.01	0.4648	0.2370	0.2640	0.3260	0.4533	0.6991	0.9339	1.2710
1.02	0.5146	0.2629	0.2715	0,3297	0.4547	0.6980	0.9307	1.2650
1.05	0.6026	0.4437	0.3131	0.3452	0.4604	0.6956	0.9222	1.2481
1.10	0.6880	0.5984	0.4580	0.3953	0.4770	0.6950	0.9110	1.2232
1.15	0.7443	0.6803	0.5798	0.4760	0.5042	0.6987	0.9033	1.2021
1.20	0.7858	0.7363	0.6605	0,5605	0.5425	0.7069	0.8990	1.1844
1.30	0.8438	0.8111	0.7624	0.6908	0.6344	0.7358	0.8998	1.1580
1.40	0.8827	0.8595	0.8256	0.7753	0.7202	0.7761	0.9112	1.1419
1.50	0.9103	0.8933	0,8689	0.8328	0.7887	0.8200	0.9297	1.1339
1.60	0.9308	0.9180	0.9000	0.8738	0.8410	0.8617	0.9518	1.1320
1.70	0.9463	0.9367	0.9234	0.9043	0.8809	0.8984	0.9745	1.1343
1.80	0.9583	0.9511	0.9413	0.9275	0.9118	0.9297	0.9961	1.1391
1.90	0.9678	0.9624	0.9552	0.9456	0.9359	0.9557	1.0157	1.1452
2.00	0.9754	0.9715	0.9664	0.9599	0.9550	0.9772	1.0328	1.1516
2.20	0.9865	0.9847	0.9826	0.9806	0.9827	1.0094	1.0600	1.1635
2.40	0.9941	0.9936	0.9935	0.9945	1.0011	1.0313	1.0793	1.1728
2.60	0.9993	0.9998	1.0010	1.0040	1.0137	1.0463	1.0926	1.1792
2.80	1.0031	1.0042	1.0063	1.0106	1.0223	1.0565	1.1016	1.1830
3.00	1.0057	1.0074	1.0101	1.0153	1.0284	1.0635	1.1075	1.1848
3.50	1.0097	1.0120	0.0156	1.0221	1.0368	1.0723	1.1138	1.1834
4.00	1.0115	1.0140	1.0179	1.0249	1.0401	1.0747	1.1136	1.1773

Tables 23A to 26 B from B. I. Lee, and M. G. Kesler, *A Generalized Thermodynamic Correlation Based on Three Parameter Corresponding States*, *AIChE Journal*, 21(3): 510–527, 1975. Reproduced by permission of the American Institute of Chemical Engineers.

TABLE 23B: Lee–Kesler values for $Z^{(1)}$ (T_R, P_R)

P_R: 0.01 0.05 0.1 0.2 0.3 0.4 0.6 0.8

T_R							
0.30	–0.0008	–0.0040	–0.0081	–0.0161	–0.0323	–0.0484	–0.064
0.35	–0.0009	–0.0046	–0.0093	–0.0185	–0.0370	–0.0554	–0.073
0.40	–0.0010	–0.0048	–0.0095	–0.0190	–0.0380	–0.0570	–0.075
0.45	–0.0009	–0.0047	–0.0094	–0.0187	–0.0374	–0.0560	–0.074
0.50	–0.0009	–0.0045	–0.0090	–0.0181	–0.0360	–0.0539	–0.071
0.55	–0.0314	–0.0043	–0.0086	–0.0172	–0.0343	–0.0513	–0.068
0.60	–0.0205	–0.0041	–0.0082	–0.0164	–0.0326	–0.0487	–0.064
0.65	–0.0137	–0.0772	–0,0078	–0.0156	–0.0309	–0.0461	–0.061
0.70	–0.0093	–0.0507	–0.1161	–0.0148	–0.0294	–0.0438	–0.057
0.75	–0.0064	–0.0339	–0.0744	–0.0143	–0.0282	–0.0417	–0.055
0.80	–0.0044	–0.0228	–0.0487	–0.1160	–0.0272	–0.0401	–0.052
0.85	–0.0029	–0.0152	–0.0319	–0.0715	–0.0268	–0.0391	–0.050
0.90	–0.0019	–0.0099	–0.0205	–0.0442	0.1118	–0.0396	–0.050
0.93	–0.0015	–0.0075	–0.0154	–0.0326	–0.0763	–0.1662	–0.051
0.95	–0.0012	–0.0062	–0.0126	–0.0262	–0.0589	–0.1110	–0.054
0.97	–0.0010	–0.0050	–0.0101	–0.0208	–0.0450	–0.0770	–0.164
0.98	–0.0009	–0.0044	–0.0090	–0.0184	–0.0390	–0.0641	–0.110
0.99	–0.0008	–0.0039	–0.0079	–0.0161	–0.0335	–0.0531	–0.079
1.00	–0.0007	–0.0034	–0.0069	–0.0140	–0.0285	–0.0435	–0.058
1.01	–0.0006	–0.0030	–0.0060	–0.0120	–0.0240	–0.0351	–0.042
1.02	–0.0005	–0.0026	–0.0051	–0.0102	–0.0198	–0.0277	–0.030
1.05	–0.0003	–0.0015	–0.0029	–0.0054	–0.0092	–0.0097	–0.003
1.10	–0.0000	0.0000	0.0001	0.0007	0.0038	0.0106	0.023
1.15	0.0002	0.0011	0.0023	0.0052	0.0127	0.0237	0.039
1.20	0.0004	0.0019	0.0039	0.0084	0.0190	0.0326	0.049
1.30	0.0006	0.0030	0.0061	0.0125	0.0267	0.0429	0.061
1.40	0.0001	0.0036	0.0072	0.0147	0.0306	0.0477	0.066
1.50	0.0008	0.0039	0.0078	0.0158	0.0323	0.0497	0.06'7
1.60	0.0008	0.0040	0.0080	0.0162	0.0330	0.0501	0.067
1.70	0.0008	0.0040	0.0081	0.0163	0.0329	0.0497	0.066
1.80	0.0008	0.0040	0.0081	0.0162	0.0325	0.0488	0.065
1.90	0.0008	0.0040	0.0079	0.0159	0.0318	0.0477	0.063
2.00	0.0008	0.0039	0.0078	0.0155	0.0310	0.0464	0.061
2.20	0.0007	0.0037	0.0074	0,0147	0.0293	0.0437	0.057
2.40	0.0007	0.0035	0.0070	0.0139	0.0276	0.0411	0.054
2.60	0.0007	0.0033	0.0066	0.0131	0.0260	0.0387	0.051
2.80	0.0006	0.0031	0.0062	0.0124	0.0245	0.0365	0.048
3.00	0.0006	0.0029	0.0059	0.0117	0.0232	0.0345	0.045
3.50	0.0005	0.0026	0.0052	0.0103	0.0204	0.0303	0.040
4.00	0.0005	0.0023	0.0046	0.0091	0.0182	0.0270	0.035

TABLE 23B: Lee–Kesler values for $Z^{(1)}$ (T_R, P_R) (continued)

T_R	P_R 1	1.2	1.5	2	3	5	7	10
0.30	–0.0806	–0.0966	–0.1207	–0.1608	–0.2407	–0.3996	–0–5572	–0.7915
0.35	–0.0921	–0.1105	–0.1379	–0.1834	–0.2738	–0.4523	–0.6279	–0.8863
0.40	–0.0946	–0.1134	–0.1414	–0.1879	–0.2799	–0.4603	–0.6365	–0.8936
0.45	–0.0929	–0.1113	–0.1387	–0.1840	–0.2734	–0.4475	–0.6162	–0.8606
0.50	–0.0893	–0.1069	–0.1330	–0.1762	–0.2611	–0.4253	–0.5831	–0.8099
0.55	–0.0849	–0.1015	–0.1263	–0.1669	–0.2465	–0.3991	–0.5446	–0.7521
0.60	–0.0803	–0.0960	–0.1192	–0.1572	–0.2312	–0.3718	–0.5047	–0.6928
0.65	–0.0759	–0.0906	–0.1122	–0.1476	–0.2160	–0.3447	–0.4653	–0.6346
0.70	–0.0718	–0.0855	–0.1057	–0.1385	–0.2013	–0.3184	–0.4270	–0.5785
0.75	–0.0681	–0.0808	–0.0996	–0.1298	–0.1872	–0.2929	–0.3901	–0.5250
0.80	–0.0648	–0.0767	–0.0940	–0.1217	–0.1736	–0.2682	–0.3545	–0.4740
0.85	–0.0622	–0.0731	–0.0888	–0.1138	–0.1602	–0.2439	–0.3201	–0.4254
0.90	–0.0604	–0.0701	–0.0840	–0.1059	–0.1463	–0.2195	–0.2862	–0.3788
0.93	–0.0602	–0.0687	–0.0810	–0.1007	–0.1374	–0.2045	–0.2661	–0.3516
0.95	–0.0607	–0.0678	–0.0788	–0.0967	–0.1310	–0.1943	–0.2526	–0.3339
0.97	–0.0623	–0.0669	–0.0759	–0.0921	–0.1240	–0.1837	–0.2391	–0.3163
0.98	–0.0641	–0.0661	–0.0740	–0.0893	–0.1202	–0.1783	–0.2322	–0.3075
0.99	–0.0680	–0.0646	–0.0715	–0.0861	–0.1162	–0.1728	–0.2254	–0.2989
1.00	–0.0879	–0.0609	–0.0678	–0.0824	–0.1118	–0.1672	–0.2185	–0.2902
1,01	–0.0223	–0.0473	–0.0621	–0.0778	–0.1072	–0.1615	–0.2116	–0.2816
1.02	–0.0062	0.0227	–0.0524	–0.0722	–0.1021	–0.1556	–0.2047	–0.2731
1.05	0.0220	0.1059	0.0451	–0.0432	–0.0838	–0.1370	–0.1835	–0.2476
1.10	0.0476	0.0897	0.1630	0.0698	–0.0373	–0.1021	–0.1469	–0.2056
1.15	0.0625	0.0943	0.1548	0.1667	0.0332	0.0611	–0.1084	–0.1642
1.20	0.0719	0.0991	0.1477	0.1990	0.1095	–0.0141	–0.0678	–0.1231
1.30	0.0819	0.1048	0.1420	0.1991	0.2079	0.0875	0.0176	–0.0423
1.40	0.0857	0.1063	0.1383	0.1894	0.2397	0.1737	0.1008	0.0350
1.50	0.0864	0.1055	0.1345	0.1806	0.2433	0.2309	0.1717	0.1058
1.60	0.0855	0.1035	0.1303	0.1729	0.2381	0.2631	0.2255	0.1673
1.70	0.0838	0.1008	0.1259	0.1658	0.2305	0.2788	0.2628	0.2179
1.80	0.0816	0.0978	0.1216	0.1593	0.2224	0.2946	0.2971	0.2576
1.90	0.0792	0.0947	0.1173	0.1532	0.2144	0.2848	0.3017	0.2876
2.00	0.0767	0.0916	0.1133	0.1476	0.2069	0.2819	0.3097	0.3096
2.20	0.0719	0.0857	0.1057	0.1374	0.1932	0.2720	0.3135	0.3355
2.40	0.0675	0.0803	0.0989	0.1285	0.1812	0.2602	0.3089	0.3459
2.60	0.0634	0.0754	0.0929	0.1207	0.1706	0.2484	0.3009	0.3475
2.80	0.0598	0.0711	0.0876	0.1138	0,1613	0.2372	0.2915	0.3443
3.00	0.0565	0.0672	0.0828	0.1076	0.1529	0.2268	0.2817	0.3385
3.50	0.0497	0.0591	0.0728	0.0949	0.1356	0.2042	0.2584	0.3194
4.00	0.0443	0.0527	0.0651	0.0849	0.1219	0.1857	0.2378	0.2994

TABLE 24A: Lee–Kesler residual enthalpy values for $((h_0 - h)^{(0)}/(R\ T_c))$

T_R	P_R 0.01	0.05	0.1	0.2	0.4	0.6	0.8
0.3	6.045	6.043	6.04	6.034	6.022	6.011	5.999
0.35	5.906	5.904	5.901	5.895	5.882	5.87	5.858
0.4	5.763	5.761	5.757	5.751	5.738	5.726	5.713
0.45	5.615	5.612	5.609	5.603	5.59	5.577	5.564
0.5	5.465	5.463	4.459	5.453	5.44	5.427	5.414
0.55	0.032	5.312	5.309	5.303	5.29	5.278	5.265
0.6	0.027	5.162	5.159	5.153	5.141	5.129	5.116
0.65	0.023	0.118	5.008	5.002	4.991	4.98	4.968
0.7	0.02	0.101	0.213	4.848	4.838	4.828	4.818
0.75	0.017	0.088	0.183	4.687	4.679	4.672	4.664
0.8	0.015	0.078	0.16	0.345	4.507	4.504	4.499
0.85	0.014	0.069	0.141	0.3	4.309	4.313	4.316
0.9	0.012	0.062	0.126	0.264	0.596	4.074	4.094
0.93	0.011	0.058	0.118	0.246	0.545	0.96	3.92
0.95	0.011	0.056	0.113	0.235	0.516	0.885	3.763
0.97	0.011	0.054	0.109	0.225	0.49	0.824	1.356
0.98	0.01	0.053	0.107	0.221	0.478	0.797	1.273
0.99	0.01	0.052	0.105	0.216	0.466	0.773	1.206
1	0.01	0.051	0.103	0.212	0.455	0.75	1.151
1.01	0.01	0.05	0.101	0.208	0.445	0.728	0.102
1.02	0.01	0.049	0.099	0.203	0.434	0.708	1.06
1.05	0.009	0.046	0.094	0.192	0.407	0.654	0.955
1.1	0.008	0.042	0.086	0.175	0.367	0.581	0.827
1.15	0.008	0.039	0.079	0.16	0.334	0.523	0.732
1.2	0.007	0.036	0.073	0.148	0.305	0.474	0.657
1.3	0.006	0.031	0.063	0.127	0.259	0.399	0.545
1.4	0.005	0.027	0.055	0.11	0.224	0.341	0.463
1.5	0.005	0.024	0.048	0.097	0.196	0.297	0.4
1.6	0.004	0.021	0.043	0.086	0.173	0.261	0.35
1.7	0.004	0.019	0.038	0.076	0.153	0.231	0.309
1.8	0.003	0,017	0.034	0.068	0.137	0.206	0.275
1.9	0.003	0.015	0.031	0.062	0.123	0.185	0.246
2	0.003	0.014	0.028	0.056	0.111	0.167	0.222
2.2	0.002	0.012	0.023	0.046	0.092	0.137	0.182
2.4	0.002	0.01	0.019	0.038	0.076	0.114	0.15
2.6	0.002	0.008	0.016	0.032	0.064	0.095	0.125
2.8	0.001	0.007	0.014	0.027	0.054	0.08	0.105
3	0.001	0.006	0.011	0.023	0.045	0.067	0.088
3.5	0.001	0.004	0.007	0.015	0.029	0.043	0.056
4	0	0.002	0.005	0.009	0.017	0.026	0.033

TABLE 24A: Lee–Kesler residual enthalpy values for $((h_0 - h)^{(0)} / (R\ T_c))$;(continued)

T_R	P_R = 1	1.2	1.5	2	3	5	7	10
0.3	5.987	5.975	5.957	5.927	5.868	5.748	5.628	5.446
0.35	5.845	5.833	5.814	5.783	5.721	5.595	5.469	5.278
0.4	5.7	5.687	5.668	5.636	5.572	5.442	5.311	5.113
0.45	5.551	5.538	5.519	5.486	5.421	5.288	5.154	4.95
0.5	5.401	5.388	5.369	5.336	5.27	5.135	4.999	4.791
0.55	5.252	5.239	5.22	5.187	5.121	4.986	4.849	4.638
0.6	5.104	5.091	5.073	5.041	4.976	4.842	4.704	4.492
0.65	4.956	4.945	4.927	4.896	4.833	4.702	4.565	4.353
0.7	4.808	4.797	4.781	4.752	4.693	4.566	4.432	4.221
0.75	4.655	4.646	4.632	4.607	4.554	4.434	4.303	4.095
0.8	4.494	4.488	4.478	4.459	4.413	4.303	4.178	3.974
0.85	4.316	4.316	4.312	4.302	4.269	4.173	4.056	3.857
0.9	4.108	4.118	4.127	4.132	4.119	4.043	3.935	3.744
0.93	3.953	3.976	4	4.02	4.024	3.963	3.863	3.678
0.95	3.825	3.865	3.904	3.94	3.958	3.91	3.815	3.634
0.97	3.658	3.732	3.796	3.853	3.89	3.856	3.767	3.591
0.98	3.544	3.652	3.736	3.806	3.854	3.829	3.743	3.569
0.99	3.376	3.558	3.67	3.758	3.818	3.801	3.719	3.548
1	2.584	3.441	3.598	3.706	3.782	3.774	3.695	3.526
1.01	1.796	3.283	3.516	3.652	3.744	3.746	3.671	3.505
1.02	1.627	3.039	3.442	3.595	3.705	3.718	3.647	3.484
1.05	1.359	2.034	3.03	3.398	3.583	3.632	3.575	3.42
1.1	1.12	1.487	2.203	2.965	3.353	3.484	3.453	3.315
1.15	0.968	1.239	1.719	2.479	3.091	3.329	3.329	3.211
1.2	0.857	1.076	1.443	2.079	2.807	3.166	3.202	3.107
1.3	0.698	0.86	1.116	1.56	2.274	2.825	2.942	2.899
1.4	0.588	0.716	0.915	1.253	1.857	2.486	2.679	2.692
1.5	0.505	0.611	0.774	1.046	1.549	2.175	2.421	2.486
1.6	0.44	0.531	0.667	0.894	1.318	1.904	2.177	2.285
1.7	0.387	0.466	0.583	0.777	1.139	1.672	1.953	2.091
1.8	0.344	0.413	0.515	0.683	0.996	1.476	1.751	1.908
1.9	0.307	0.368	0.458	0.606	0.88	1.309	1.571	1.736
2	0.276	0.33	0.411	0.541	0.782	1.167	1.411	1.577
2.2	0.226	0.269	0.334	0.437	0.629	0.937	1.143	1.295
2.4	0.187	0.222	0.275	0.359	0.513	0.761	0.929	1.058
2.6	0.155	0.185	0.228	0.297	0.422	0.621	0.756	0.858
2.8	0.13	0.154	0.19	0.246	0.348	0.508	0.614	0.689
3	0.109	0.129	0.159	0.205	0.288	0.415	0.495	0.545
3.5	0.069	0.081	0.099	0.127	0.174	0.239	0.27	0.264
4	0.041	0.048	0.058	0.072	0.095	0.116	0.11	0.061

TABLE 24B: Lee–Kesler residual enthalpy values for $((h_0 - h)^{(1)}/(RT_c))$

T_R	P_R 0.01	0.05	0.1	0.2	0.4	0.6	0.8
0.3	11.098	11.096	11.095	11.091	11.083	11.076	11.069
0.35	10.656	10.655	10.654	10.653	10.65	10.646	10.643
0.4	10.121	10.121	10.121	10.12	10.121	10.121	10.121
0.45	9.515	9.515	9.516	9.517	9.519	9.521	9.523
0.5	8.868	8.869	8.87	8.872	8.876	8.88	8.884
0.55	0.08	8.211	8.212	8.215	8.221	8.226	8.232
0.6	0.059	7.568	7.57	7.573	7.579	7.585	7.591
0.65	0.045	0.247	6.949	6.952	6.959	6.966	6.973
0.7	0.034	0.185	0.415	6.36	6.367	6.373	6.381
0.75	0.027	0.142	0.306	5.796	5.802	5.809	5.816
0.8	0.021	0.11	0.234	0.542	5.266	5.271	5.278
0.85	0.017	0.087	0.182	0.401	4.753	4.754	4.758
0.9	0.014	0.07	0.144	0.308	0.751	4.254	4.248
0.93	0.012	0.061	0.126	0.265	0.612	1.236	3.942
0.95	0.011	0.056	0.115	0.241	0.542	0.994	3.737
0.97	0.01	0.052	0.105	0.219	0.483	0.837	1.616
0.98	0.01	0.05	0.101	0.209	0.457	0.776	1.324
0.99	0.009	0.048	0.097	0.2	0.433	0.722	1.154
1	0.009	0.046	0.093	0.191	0.41	0.675	1.034
1.01	0.009	0.044	0.089	0.183	0.389	0.632	0.94
1.02	0.008	0.042	0.085	0.175	0.37	0.594	0.863
1.05	0.007	0.037	0.075	0.153	0.318	0.498	0.691
1.1	0.006	0.03	0.061	0.123	0.251	0.381	0.507
1.15	0.005	0.025	0.05	0.099	0.199	0.296	0.385
1.2	0.004	0.02	0.04	0.08	0.158	0.232	0.297
1.3	0.003	0.013	0.026	0.052	0.1	0.142	0.177
1.4	0.002	0.008	0.016	0.032	0.06	0.083	0.1
1.5	0.001	0.005	0.009	0.018	0.032	0.042	0.048
1.6	0	0.002	0.004	0.007	0.012	0.013	0.011
1.7	0	0	0	0	–0.003	–0.009	–0.017
1.8	0	–0.001	–0.003	–0.006	–0.015	–0.025	–0.037
1.9	–0.001	–0.003	–0.005	–0.011	–0.023	–0.037	–0.053
2	–0.001	–0.003	–0.007	–0.015	–0.03	–0.047	–0.065
2.2	–0.001	–0.005	–0.01	–0.02	–0.04	–0.062	–0.083
2.4	–0.001	–0.006	–0.012	–0.023	–0.047	–0.071	–0.095
2.6	–0.001	–0.006	–0.013	–0.026	–0.052	–0.078	–0.104
2.8	–0.001	–0.007	–0.014	–0.028	–0.055	–0.082	–0.11
3	–0.001	–0.007	–0.014	–0.029	–0.058	–0.086	–0.114
3.5	–0.002	–0.008	–0.016	–0.031	–0.062	–0.092	–0.122
4	–0.002	–0.008	–0.016	–0.032	–0.064	–0.096	–0.127

TABLE 24B: Lee–Kesler residual enthalpy values for $((h_0 - h)^{(1)}/(RT_c))$ (continued)

T_R	P_R = 1	1.2	1.5	2	3	5	7	10
0.3	11.062	11.055	11.044	11.027	10.992	10.935	10.872	10.781
0.35	10.64	10.637	10.632	10.624	10.609	10.581	10.554	10.529
0.4	10.121	10.121	10.121	10.122	10.123	10.128	10.135	10.15
0.45	9.525	9.527	9.531	9.537	9.549	9.576	9.611	9.663
0.5	8.888	8.892	8.999	8.909	9.932	8.978	9.03	9.111
0.55	8.238	8.243	8.252	8.267	8.298	8.36	8.425	8.531
0.6	7.596	7.603	7.614	7.632	7.669	7.745	7.824	7.95
0.65	6.98	6.987	6.997	7.017	7.059	7.147	7.239	7.381
0.7	6.388	6.395	6.407	6.429	6.475	6.574	6.677	6.837
0.75	5.824	5.832	5.845	5.868	5.918	6.027	6.142	6.318
0.8	5.285	5.293	5.306	5.33	5.385	5.506	5.632	5.824
0.85	4.763	4.771	4.784	4.81	4.872	5.008	5.149	5.358
0.9	4.249	4.255	4.268	4.298	4.371	4.53	4.688	4.916
0.93	3.934	3.937	3.951	3.987	4.073	4.251	4.422	4.662
0.95	3.712	3.713	3.73	3.773	3.873	4.068	4.248	4.497
0.97	3.47	3.467	3.492	3.551	3.67	3.885	4.077	4.336
0.98	3.332	3.327	3.363	3.434	3.568	3.795	3.992	4.257
0.99	3.164	3.164	3.223	3.313	3.464	3.705	3.909	4.178
1	2.471	2.952	3.065	3.186	3.358	3.615	3.825	4.1
1.01	1.375	2.595	2.88	3.051	3.251	3.525	3.742	4.023
1.02	1.18	1.723	2.65	2.906	3.142	3.435	3.661	3.947
1.05	0.877	0.878	1.496	2.381	2.8	3.167	3.418	3.722
1.1	0.617	0.673	0.617	1.261	2.167	2.72	3.023	3.362
1.15	0.459	0.503	0.487	0.604	1.497	2.275	2.641	3.019
1.2	0.349	0.381	0.381	0.361	0.934	1.84	2.273	2.692
1.3	0.203	0.218	0.218	0.178	0.3	1.066	1.592	2.086
1.4	0.111	0.115	0.108	0.07	0.044	0.504	1.012	1.547
1.5	0.049	0.046	0.032	−0.009	−0.078	0.142	0.556	1.08
1.6	0.005	−0.004	−0.023	−0.065	−0.151	−0.082	0.217	0.689
1.7	−0.027	−0.04	−0.063	−0.109	−0.202	−0.223	−0.028	0.369
1.8	−0.051	−0.067	−0.094	−0.143	−0.241	−0.317	−0.203	0.112
1.9	−0.07	−0.088	−0.117	−0.169	−0.271	−0.381	−0.33	−0.092
2	−0.085	−0.105	−0.136	−0.19	−0.295	−0.428	−0.424	−0.255
2.2	−0.106	−0.128	−0.163	−0.221	−0.331	−0.493	−0.551	−0.489
2.4	−0.12	−0.144	−0.181	−0.242	−0.356	−0.535	−0.631	−0.645
2.6	−0.13	−0.156	−0.194	−0.257	−0.376	−0.567	−0.687	−0.754
2.8	−0.137	−0.164	−0.204	−0.269	−0.391	−0.591	−0.729	−0.836
3	−0.142	−0.17	−0.211	−0.278	−0.403	−0.611	−0.763	−0.899
3.5	−0.152	−0.181	−0.224	−0.294	−0.425	−0.65	−0.827	−1.015
4	−0.158	−0.188	−0.233	−0.306	−0.442	−0.68	−0.874	−1.097

TABLE 25A: Lee–Kesler residual entropy values for $((s_0 - s)^{(0)}/R)$

T_R \ P_R	0.01	0.05	0.1	0.2	0.4	0.6	0.8
0.3	16.782	16.774	16.764	16.744	16.705	16.665	16.626
0.35	15.413	15.408	15.401	15.387	15.359	15.333	15.305
0.4	13.99	13.986	13.981	13.972	13.953	13.934	13.915
0.45	12.564	12.561	12.558	12.551	12.537	12.523	12.509
0.5	11.202	11.2	11.197	11.192	11.182	11.172	11.162
0.55	0.115	9.948	9.946	9.942	9.935	9.928	9.921
0.6	0.078	8.828	8.826	8.823	8.817	8.811	8.806
0.65	0.055	0.309	7.832	7.829	7.824	7.819	7.815
0.7	0.04	0.216	0.491	6.951	6.945	6.941	6.937
0.75	0.029	0.156	0.34	6.173	6.167	6.162	6.158
0.8	0.022	0.116	0.246	0.578	5.475	5.468	5.462
0.85	0.017	0.088	0.183	0.408	4.853	4.841	4.832
0.9	0.013	0.068	0.14	0.301	0.744	4.269	4.249
0.93	0.011	0.058	0.12	0.254	0.593	1.219	3.914
0.95	0.01	0.053	0.109	0.228	0.517	0.961	3.697
0.97	0.01	0.048	0.099	0.206	0.456	0.797	1.57
0.98	0.009	0.046	0.094	0.196	0.429	0.734	1.27
0.99	0.009	0.044	0.09	0.186	0.405	0.68	1.098
1	0.008	0.042	0.086	0.177	0.382	0.632	0.977
1.01	0.008	0.04	0.082	0.169	0.361	0.59	0.883
1.02	0.008	0.039	0.078	0.161	0.342	0.552	0.807
1.05	0.007	0.034	0.069	0.14	0.292	0.46	0.642
1.1	0.005	0.028	0.055	0.112	0.229	0.35	0.47
1.15	0.005	0.023	0.045	0.091	0.183	0.275	0.361
1.2	0.004	0.019	0.037	0.075	0.149	0.22	0.286
1.3	0.003	0.013	0.026	0.052	0.102	0.148	0.19
1.4	0.002	0.01	0.019	0.037	0.072	0.104	0.133
1.5	0.001	0.007	0.014	0.027	0.053	0.076	0.097
1.6	0.001	0.005	0.011	0.021	0.04	0.057	0.073
1.7	0.001	0.004	0.008	0.016	0.031	0.044	0.056
1.8	0.001	0.003	0.006	0.013	0.024	0.035	0.044
1.9	0.001	0.003	0.005	0.01	0.019	0.028	0.036
2	0	0.002	0.004	0.008	0.016	0.023	0.029
2.2	0	0.001	0.003	0.006	0.011	0.016	0.021
2.4	0	0.001	0.002	0.004	0.008	0.012	0.015
2.6	0	0.001	0.002	0.003	0.006	0.009	0.012
2.8	0	0.001	0.001	0.003	0.005	0.008	0.01
3	0	0.001	0.001	0.002	0.004	0.006	0.008
3.5	0	0	0.001	0.001	0.003	0.004	0.006
4	0	0	0.001	0.001	0.002	0.003	0.005

TABLE 25A: Lee–Kesler residual entropy values for $((s_0 - s)^{(0)}/R)$ (continued)

T_R	P_R 1	1.2	1.5	2	3	5	7	10
0.3	16.586	16.547	16.488	16.39	16.195	15.837	15.468	14.925
0.35	15.278	15.251	15.211	15.144	15.011	14.751	14.496	14.153
0.4	13.896	13.877	13.849	13.803	13.714	13.541	13.376	13.144
0.45	12.496	12.482	12.462	12.43	12.367	12.248	12.145	11.999
0.5	11.153	11.143	11.129	11.107	11.063	10.985	10.92	10.836
0.55	9.914	9.907	9.897	9.882	9.853	9.806	9.769	9.732
0.6	8.799	8.794	8.787	8.777	8.76	8.736	8.723	8.72
0.65	7.81	7.807	7.801	7.794	7.784	7.779	7.785	7.811
0.7	6.933	6.93	6.926	6.922	6.919	6.929	6.952	7.002
0.75	6.155	6.152	6.149	6.147	6.149	6.174	6.213	6.285
0.8	5.458	5.455	5.453	5.452	5.461	5.501	5.555	5.648
0.85	4.826	4.822	4.82	4.822	4.839	4.898	4.969	5.082
0.9	4.238	4.232	4.23	4.236	4.267	4.351	4.442	4.578
0.93	3.894.	3.885	3.884	3.896	3.941	4.046	4.151	4.3
0.95	3.658	3.647	3.648	3.669	3.728	3.851	3.966	4.125
0.97	3.406	3.391	3.401	3.437	3.517	3.661	3.788	3.957
0.98	3.264	3.247	3.268	3.318	3.412	3.569	3.701	3.875
0.99	3.093	3.082	3.126	3.195	3.306	3.477	3.616	3.796
1	2.399	2.868	2.967	3.067	3.2	3.387	3.532	3.717
1.01	1.306	2.513	2.784	2.933	3.094	3.297	3.45	3.64
1.02	1.113	1.655	2.557	2.79	2.986	3.209	3.369	3.565
1.05	0.82	0.931	1.443	2.283	2.655	2.949	3.134	3.348
1.1	0.577	0.64	0.618	1.241	2.067	2.534	2.767	3,013
1.15	0.437	0.489	0.502	0.654	1.471	2.138	2.428	2.708
1.2	0.343	0.385	0.412	0.447	0.991	1.767	2.115	2.43
1.3	0.226	0.254	0.282	0.3	0.481	1.147	1.569	1.944
1.4	0.158	0.178	0.2	0.22	0.29	0.73	1.138	1.544
1.5	0.115	0.13	0.147	0.166	0.206	0.479	0.823	1.222
1.6	0.086	0.098	0.112	0.129	0.159	0.334	0.604	0.969
1.7	0.067	0.076	0,087	0.102	0.127	0.248	0.456	0.775
1.8	0.053	0.06	0.07	0.083	0.105	0.195	0.355	0.628
1.9	0.043	0.049	0.057	0.069	0.089	0.16	0.286	0.518
2	0.035	0.04	0.048	0.058	0.077	0.136	0.238	0.434
2.2	0.025	0.029	0.035	0.043	0.06	0.105	0.178	0.322
2.4	0.019	0.022	0.027	0.034	0.048	0.096	0.143	0.254
2.6	0.015	0.018	0.021	0.028	0.041	0.074	0.12	0.21
2.8	0.012	0.014	0.018	0.023	0.035	0.065	0.104	0.188
3	0.01	0.012	0.015	0.02	0.031	0.058	0.093	0.158
3.5	0.007	0.009	0.011	0.015	0.024	0.046	0.073	0.122
4	0.006	0.007	0.009	0.012	0.02	0.038	0.06	0.1

TABLE 25B: Lee–Kesler residual entropy values for $((s_0 - s)^{(1)}/R)$

T_R	P_R 0.01	0.05	0.1	0.2	0.4	0.6	0.8
0.3	16.782	16.774	16.764	16.744	16.705	16.665	16.626
0.35	15.413	15.408	15.401	15.387	15.359	15.333	15.305
0.4	13.99	13.986	13.981	13.972	13.953	13.934	13.915
0.45	12.564	12.561	12.558	12.551	12.537	12.523	12.509
0.5	11.202	11.2	11.197	11.192	11.182	11.172	11.162
0.55	0.115	9.948	9.946	9.942	9.935	9.928	9.921
0.6	0.078	8.828	8.826	8.823	8.817	8.811	8.806
0.65	0.055	0.309	7.832	7.829	7.824	7.819	7.815
0.7	0.04	0.216	0.491	6.951	6.945	6.941	6.937
0.75	0.029	0.156	0.34	6.173	6.167	6.162	6.158
0.8	0.022	0.116	0.246	0.578	5.475	5.468	5.462
0.85	0.017	0.088	0.183	0.408	4.853	4.841	4.832
0.9	0.013	0.068	0.14	0.301	0.744	4.269	4.249
0.93	0.011	0.058	0.12	0.254	0.593	1.219	3.914
0.95	0.01	0.053	0.109	0.228	0.517	0.961	3.697
0.97	0.01	0.048	0.099	0.206	0.456	0.797	1.57
0.98	0.009	0.046	0.094	0.196	0.429	0.734	1.27
0.99	0.009	0.044	0.09	0.186	0.405	0.68	1.098
1	0.008	0.042	0.086	0.177	0.382	0.632	0.977
1.01	0.008	0.04	0.082	0.169	0.361	0.59	0.883
1.02	0.008	0.039	0.078	0.161	0.342	0.552	0.807
1.05	0.007	0.034	0.069	0.14	0.292	0.46	0.642
1.1	0.005	0.028	0.055	0.112	0.229	0.35	0.47
1.15	0.005	0.023	0.045	0.091	0.183	0.275	0.361
1.2	0.004	0.019	0.037	0.075	0.149	0.22	0.286
1.3	0.003	0.013	0.026	0.052	0.102	0.148	0.19
1.4	0.002	0.01	0.019	0.037	0.072	0.104	0.133
1.5	0.001	0.007	0.014	0.027	0.053	0.076	0.097
1.6	0.001	0.005	0.011	0.021	0.04	0.057	0.073
1.7	0.001	0.004	0.008	0.016	0.031	0.044	0.056
1.8	0.001	0.003	0.006	0.013	0.024	0.035	0.044
1.9	0.001	0.003	0.005	0.01	0.019	0.028	0.036
2	0	0.002	0.004	0.008	0.016	0.023	0.029
2.2	0	0.001	0.003	0.006	0.011	0.016	0.021
2.4	0	0.001	0.002	0.004	0.008	0.012	0.015
2.6	0	0.001	0.002	0.003	0.006	0.009	0.012
2.8	0	0.001	0.001	0.003	0.005	0.008	0.01
3	0	0.001	0.001	0.002	0.004	0.006	0.008
3.5	0	0	0.001	0.001	0.003	0.004	0.006
4	0	0	0.001	0.001	0.002	0.003	0.005

TABLE 25B: Lee–Kesler residual entropy values for $((s_0 - s)^{(1)}/R)$ (continued)

T_R	P_R 1	1.2	1.5	2	3	5	7	10
0.3	16.586	16.547	16.488	16.39	16.195	15.837	15.468	14.925
0.35	15.278	15.251	15.211	15.144	15.011	14.751	14.496	14.153
0.4	13.896	13.877	13.849	13.803	13.714	13.541	13.376	13.144
0.45	12.496	12.482	12.462	12.43	12.367	12.248	12.145	11.999
0.5	11.153	11.143	11.129	11.107	11.063	10.985	10.92	10.836
0.55	9.914	9.907	9.897	9.882	9.853	9.806	9.769	9.732
0.6	8.799	8.794	8.787	8.777	8.76	8.736	8.723	8.72
0.65	7.81	7.807	7.801	7.794	7.784	7.779	7.785	7.811
0.7	6.933	6.93	6.926	6.922	6.919	6.929	6.952	7.002
0.75	6.155	6.152	6.149	6.147	6.149	6.174	6.213	6.285
0.8	5.458	5.455	5.453	5.452	5.461	5.501	5.555	5.648
0.85	4.826	4.822	4.82	4.822	4.839	4.898	4.969	5.082
0.9	4.238	4.232	4.23	4.236	4.267	4.351	4.442	4.578
0.93	3.894.	3.885	3.884	3.896	3.941	4.046	4.151	4.3
0.95	3.658	3.647	3.648	3.669	3.728	3.851	3.966	4.125
0.97	3.406	3.391	3.401	3.437	3.517	3.661	3.788	3.957
0.98	3.264	3.247	3.268	3.318	3.412	3.569	3.701	3.875
0.99	3.093	3.082	3.126	3.195	3.306	3.477	3.616	3.796
1	2.399	2.868	2.967	3.067	3.2	3.387	3.532	3.717
1.01	1.306	2.513	2.784	2.933	3.094	3.297	3.45	3.64
1.02	1.113	1.655	2.557	2.79	2.986	3.209	3.369	3.565
1.05	0.82	0.931	1.443	2.283	2.655	2.949	3.134	3.348
1.1	0.577	0.64	0.618	1.241	2.067	2.534	2.767	3,013
1.15	0.437	0.489	0.502	0.654	1.471	2.138	2.428	2.708
1.2	0.343	0.385	0.412	0.447	0.991	1.767	2.115	2.43
1.3	0.226	0.254	0.282	0.3	0.481	1.147	1.569	1.944
1.4	0.158	0.178	0.2	0.22	0.29	0.73	1.138	1.544
1.5	0.115	0.13	0.147	0.166	0.206	0.479	0.823	1.222
1.6	0.086	0.098	0.112	0.129	0.159	0.334	0.604	0.969
1.7	0.067	0.076	0,087	0.102	0.127	0.248	0.456	0.775
1.8	0.053	0.06	0.07	0.083	0.105	0.195	0.355	0.628
1.9	0.043	0.049	0.057	0.069	0.089	0.16	0.286	0.518
2	0.035	0.04	0.048	0.058	0.077	0.136	0.238	0.434
2.2	0.025	0.029	0.035	0.043	0.06	0.105	0.178	0.322
2.4	0.019	0.022	0.027	0.034	0.048	0.096	0.143	0.254
2.6	0.015	0.018	0.021	0.028	0.041	0.074	0.12	0.21
2.8	0.012	0.014	0.018	0.023	0.035	0.065	0.104	0.188
3	0.01	0.012	0.015	0.02	0.031	0.058	0.093	0.158
3.5	0.007	0.009	0.011	0.015	0.024	0.046	0.073	0.122
4	0.006	0.007	0.009	0.012	0.02	0.038	0.06	0.1

TABLE 26A: Lee–Kesler fugacity coefficient values for $(\log_{10} f/P)^{(0)}$

T_R	P_R 0.01	0.05	0.1	0.2	0.4	0.6	0.8
0.3	–3.708	–4.402	–4.696	–4.985	–5.261	–5.412	–5.512
0.35	–2.471	–3.166	–3.461	–3.751	–4.029	–4.183	–4.285
0.4	–1.566	–2.261	–2.557	–2.848	–3.128	–3.283	–3.387
0.45	–0.879	–1.575	–1.871	–2.162	–2.444	–2.601	–2.707
0.5	–0.344	–1.04	–1.336	–1.628	–1.912	–2.07	–2.177
0.55	–0.008	–0.614	–0.911	–1.204	–1.488	–1.647	–1.755
0.6	–0.007	–0.269	–0.566	–0.859	–1.144	–1.304	–1.413
0.65	–0.005	–0.026	–0.283	–0.576	–0.862	–1.023	–1.132
0.7	–0.004	–0.021	–0.043	–0.4	–0.627	–0.789	–0.899
0.75	–0.003	–0.017	–0.035	–0.144	–0.43	–0.592	–0.703
0.8	–0.003	–0.014	–0.029	–0.059	–0.264	–0.426	–0.537
0.85	–0,002	–0.012	–0.024	–0.049	–0.123	–0.285	–0.396
0.9	–0.002	–0.01	–0.02	–0.041	–0.086	–0.166	–0.276
0.93	–0.002	–0.009	–0.018	–0.037	–0.077	–0.122	–0.214
0.95	–0.002	–0.008	–0.017	–0.035	–0.072	–0.113	–0.176
0.97	–0.002	–0.008	–0.016	–0.033	–0.067	–0.105	–0.148
0.98	–0.002	–0.008	–0.016	–0.032	–0.065	–0.101	–0.142
0.99	–0.001	–0.007	–0.015	–0.031	–0.063	–0.098	–0.137
1	–0.001	–0.007	–0.015	–0.03	–0.061	–0.095	–0.132
1.01	–0.001	–0.007	–0.014	–0.029	–0.059	–0.091	–0.127
1.02	–0.001	–0.007	–0.014	–0.028	–0.057	–0.088	–0.122
1.05	–0,001	–0.006	–0.013	–0.025	–0.052	–0.08	–0.11
1.1	–0.001	–0.005	–0.011	–0.022	–0.045	–0.069	–0.093
1.15	–0.001	–0.005	–0.009	–0.019	–0.039	–0.059	–0.08
1.2	–0.001	–0.004	–0.008	–0.017	–0.034	–0.051	–0.069
1.3	–0.001	–0.003	–0.006	–0.013	–0.026	–0.039	–0.052
1.4	–0.001	–0.003	–0.005	–0.01	–0.02	–0.03	–0.04
1.5	0	–0.002	–0.004	–0.008	–0.016	–0.024	–0.032
1.6	0	–0.002	–0.003	–0.006	–0.012	–0.019	–0.025
1.7	0	–0.001	–0.002	–0.005	–0.01	–0.015	–0.02
1.8	0	–0.001	–0.002	–0.004	–0.008	–0.012	–0.015
1.9	0	–0.001	–0.002	–0.003	–0.006	–0.009	–0.012
2	0	–0.001	–0.001	–0.002	–0.005	–0.007	–0.009
2.2	0	0	–0.001	–0.001	–0.003	–0.004	–0.005
2.4	0	0	0	–0.001	–0.001	–0.002	–0.003
2.6	0	0	0	0	0	–0.001	–0.001
2.8	0	0	0	0	0	0	0.001
3	0	0	0	0	0.001	0.001	0.002
3.5	0	0	0	0.001	0.001	0.002	0.003
4	0	0	0	0.001	0.002	0.003	0.004

TABLE 26A: Lee–Kesler fugacity coefficient values for $(\log_{10} f/P)^{(0)}$ (continued)

T_R \ P_R	1	1.2	1.5	2	3	5	7	10
0.3	−5.584	−5.638	−5.697	−5.759	−5.81	−5.782	−5.679	−5.461
0.35	−4.359	−4.416	−4.479	−4.547	−4.611	−4.608	−4.53	−4.352
0.4	−3.463	−3.522	−3.588	−3.661	−3.735	−3.752	−3.694	−3.545
0.45	−2.795	−2.845	−2.913	−2.99	−3.071	−3.104	−3.063	−2.938
0.5	−2.256	−2.317	−2.387	−2.468	−2.555	−2.601	−2.572	−2.468
0.55	−1.835	−1.897	−1.969	−2.052	−2.145	−2.201	−2.183	−2.096
0.6	−1.494	−1.557	−1.63	−1.715	−1.812	−1.878	−1.869	−1.795
0.65	−1.214	−1.278	−1.352	−1.439	−1.539	−1.612	−1.611	−1.549
0.7	−0.981	−1.045	−1.12	−1.208	−1.312	−1.391	−1.396	−1.344
0.75	−0.785	−0.85	−0.925	−1.015	−1.121	−1.204	−1.215	−1.172
0.8	−0.619	−0.685	−0.76	−0.851	−0.958	−1.046	−1.062	−1.026
0.85	−0.479	−0.544	−0.62	−0.711	−0.819	−0.911	−0.93	−0.901
0.9	−0.359	−0.424	−0.5	−0.591	−0.7	−0.794	−0.917	−0.793
0.93	−0.296	−0.361	−0.437	−0.527	−0.637	−0.732	−0.756	−0.735
0.95	−0.258	−0.322	−0.398	−0.488	−0.598	−0.693	−0.719	−0.699
0.97	−0.223	−0.287	−0.362	−0.452	−0.561	−0.657	−0.683	−0.665
0.98	−0.206	−0.27	−0.344	−0.434	−0.543	−0.639	−0.666	−0.649
0.99	−0.191	−0.254	−0.328	−0.417	−0.526	−0.622	−0.649	−0.633
1	−0.176	−0.238	−0.312	−0.401	−0.509	−0.605	−0.633	−0.617
1.01	−0.168	−0.224	−0.297	−0.385	−0.493	−0.589	−0.617	−0.602
1.02	−0.161	−0.21	−0.282	−0.37	−0.477	−0.573	−0.601	−0.588
1.05	−0.143	−0.18	−0.242	−0.327	−0.433	−0.529	−0.557	−0.546
1.1	−0.12	−0.148	−0.193	−0.267	−0.368	−0.462	−0.491	−0.482
1.15	−0.102	−0.125	−0.16	−0.22	−0.312	−0.403	−0.433	−0.426
1.2	−0.088	−0.106	−0.135	−0.184	−0.266	−0.352	−0.382	−0.377
1.3	−0.066	−0.08	−0.1	−0.134	−0.195	−0.269	−0.296	−0.293
1.4	−0.051	−0.061	−0.076	−0.101	−0.146	−0.205	−0.229	−0.226
1.5	−0.039	−0.047	−0.059	−0.077	−0.111	−0.157	−0.176	−0.173
1.6	−0.031	−0.037	−0.046	−0.06	−0.085	−0.12	−0.135	−0.129
1.7	−0.024	−0.029	−0.036	−0.046	−0.065	−0.092	−0.102	−0.094
1.8	−0.019	−0.023	−0.028	−0.036	−0.05	−0.069	−0.075	−0.066
1.9	−0.015	−0.018	−0.022	−0.028	−0.038	−0.052	−0.054	−0.043
2	−0.012	−0.014	−0.017	−0.021	−0.029	−0.037	−0.037	−0.024
2.2	−0.007	−0.008	−0.009	−0.012	−0.015	−0.017	−0.012	0.004
2.4	−0.003	−0.004	−0.004	−0.005	−0.006	−0.003	0.005	0.024
2.6	−0.001	−0.001	−0.001	−0.001	0.001	0.007	0.017	0.037
2.8	0.001	0.001	0.002	0.003	0.005	0.014	0.025	0.046
3	0.002	0.003	0.003	0.005	0.009	0.018	0.031	0.053
3.5	0.004	0.005	0.006	0.008	0.013	0.025	0.038	0.061
4	0.005	0.006	0.007	0.01	−0.016	0.028	0.041	0.064

TABLE 26B: Lee–Kesler fugacity coefficient values for ($\log_{10}$ f/P)[1]

T_R \ P_R	0.01	0.05	0.1	0.2	0.4	0.6	0.8
0.3	–8.778	–8.779	–8.781	–8.785	–8.79	–8.797	–8.804
0.35	–6.528	–6.53	–6.532	–6.536	–6.544	–6.551	–6.559
0.4	–4.912	–4.914	–4.916	–4.919	–4.929	–4.937	–4.945
0.45	–3.726	–3.728	–3.73	–3.734	–3.742	–3.75	–3.758
0.5	–2.838	–2.839	–2.841	–2.845	–2.853	–2.861	–2.869
0.55	–0.013	–2.163	–2.165	–2.169	–2.177	–2.184	–2.192
0.6	–0.009	–1.644	–1.646	–1.65	–1.657	–1.664	–1.671
0.65	–0.006	–0.031	–1.242	–1.245	–1.252	–1.258	–1.265
0.7	–0.004	–0.021	–0.044	–0.927	–0.934	–0.94	–0.946
0.75	–0.003	–0.014	–0.03	–0.675	–0.682	–0.688	–0.694
0.8	–0.002	–0.01	–0.02	–0.043	–0.481	–0.487	–0.493
0.85	–0.001	–0.006	–0.013	–0.028	–0.321	–0.327	–0.332
0.9	–0.001	–0.004	–0.009	–0.018	–0.039	–0.199	–0.204
0.93	–0.001	–0.003	–0.007	–0.013	–0.029	–0.048	–0.141
0.95	–0.001	–0.003	–0.005	–0.011	–0.023	–0.037	–0.103
0.97	0	–0.002	–0.004	–0.009	–0.018	–0.029	–0.042
0.98	0	–0.002	–0.004	–0.008	–0.016	–0.025	–0.035
0.99	0	–0.002	–0.003	–0.007	–0.014	–0.021	–0.03
1	0	–0.001	–0.003	–0.006	–0.012	–0.018	–0.025
1.01	0	–0.001	–0.003	–0.005	–0.01	–0.016	–0.021
1.02	0	–0.001	–0.002	–0.004	–0.009	–0.013	–0.017
1.05	0	–0.001	–0.001	–0.002	–0.005	–0.006	–0.007
1.1	0	0	0	0	0.001	0.002	0.004
1.15	0	0	0.001	0.002	0.005	0.008	0.011
1.2	0	0.001	0.002	0.003	0.007	0.012	0.017
1.3	0	0.001	0.003	0.005	0.011	0.017	0.023
1.4	0	0.002	0.003	0.006	0.013	0.02	0.027
1.5	0	0.002	0.003	0.007	0.014	0.021	0.028
1.6	0	0.002	0.003	0.007	0.014	0.021	0.029
1.7	0	0.002	0.004	0.007	0.014	0.021	0.029
1.8	0	0.002	0.003	0.007	0.014	0.021	0.028
1.9	0	0.002	0.003	0.007	0.014	0.021	0.028
2	0	0.002	0.003	0.007	0.013	0.02	0.027
2.2	0	0.002	0.003	0.006	0.013	0.019	0.025
2.4	0	0.002	0.003	0.006	0.012	0.018	0.024
2.6	0	0.001	0.003	0.006	0.011	0.017	0.023
2.8	0	0.001	0.003	0.005	0.011	0.016	0.021
3	0	0.001	0.003	0.005	0.01	0.015	0.02
3.5	0	0.001	0.002	0.004	0.009	0.013	0.018
4	0	0.001	0.002	0.004	0.008	0.012	0.016

TABLE 26B: Lee–Kesler fugacity coefficient values for ($\log_{10}$ f/P)$^{(1)}$ (continued)

T_R \ P_R	1	1.2	1.5	2	3	5	7	10
0.3	–8.811	–8.818	–8.828	–8.845	–8.88	–8.953	–9.022	–9.126
0.35	–6.567	–6.575	–6.587	–6.606	–6.645	–6.723	–6.8	–6.919
0.4	–4.954	–4.962	–4.974	–4.995	–5.035	–5.115	–5.195	–5.312
0.45	–3.766	–3.774	–3.786	–3.806	–3.845	–3.923	–4.001	–4.114
0.5	–2.877	–2.884	–2.896	–2.915	–2.953	–3.027	–3.101	–3.208
0.55	–2.199	–2.207	–2.218	–2.236	–2.273	–2.342	–2.41	–2.51
0.6	–1.677	–1.684	–1.695	–1.712	–1.747	–1.812	–1.875	–1.967
0.65	–1.271	–1.278	–1.287	–1.304	–1.336	–1.397	–1.456	–1.539
0.7	–0.952	–0.958	–0.967	–0.983	–1.013	–1.07	–1.124	–1.201
0.75	–0.7	–0.705	–0.714	–0.728	–0.756	–0.809	–0.858	–0.929
0.8	–0.499	–0.504	–0.512	–0.526	–0.551	–0.6	–0.645	–0.709
0.85	–0.338	–0.343	–0.351	–0.364	–0.388	–0.432	–0.473	–0.53
0.9	–0.21	–0.215	–0.222	–0.234	–0.256	–0.296	–0.333	–0.384
0.93	–0.146	–0.151	–0.158	–0.17	–0.19	–0.228	–0.262	–0.31
0.95	–0.108	–0.114	–0.121	–0.132	–0.151	–0.187	–0.22	–0.265
0.97	–0.075	–0.08	–0.087	–0.097	–0.116	–0.149	–0.18	–0.223
0.98	–0.059	–0.064	–0.071	–0.081	–0.099	–0.132	–0.162	–0.203
0.99	–0.044	–0.05	–0.056	–0.066	–0.084	–0.115	–0.144	–0.184
1	–0.031	–0.036	–0.042	–0.052	–0.068	–0.099	–0.127	–0.166
1.01	–0.024	–0.024	–0.03	–0.038	–0.054	–0.084	–0.111	–0.149
1.02	–0.019	–0.015	–0.018	–0.026	–0.041	–0.069	–0.095	–0.132
1.05	–0.007	–0.002	0.008	0.007	–0.005	–0.029	–0.052	–0.085
1.1	0.007	0.012	0.025	0.041	0.042	0.026	0.008	–0.019
1.15	0.016	0.022	0.034	0.056	0.074	0.069	0.057	0.036
1.2	0.023	0.029	0.041	0.064	0.093	0.102	0.096	0.081
1.3	0.03	0.038	0.049	0.071	0.109	0.142	0.15	0.148
1.4	0.034	0.041	0.053	0.074	0.112	0.161	0.181	0.191
1.5	0.036	0.043	0.055	0.074	0.112	0.167	0.197	0.218
1.6	0.036	0.043	0.055	0.074	0.11	0.167	0.204	0.234
1.7	0.036	0.043	0.054	0.072	0.107	0.165	0.205	0.242
1.8	0.035	0.042	0.053	0.07	0.104	0.161	0.203	0.246
1.9	0.034	0.041	0.052	0.068	0.101	0.157	0.2	0.246
2	0.034	0.04	0.05	0.066	0.097	0.152	0.196	0.244
2.2	0.032	0.038	0.047	0.062	0.091	0.143	0.186	0.236
2.4	0.03	0.036	0.044	0.058	0.086	0.134	0.176	0.227
2.6	0.028	0.034	0.042	0.055	0.08	0.127	0.167	0.217
2.8	0.027	0.032	0.039	0.052	0.076	0.12	0.158	0.208
3	0.025	0.03	0.037	0.049	0.072	0.114	0.151	0.199
3.5	0.022	0.026	0.033	0.043	0.063	0.101	0.134	0.179
4	0.02	0.023	0.029	0.038	0.057	0.09	0.121	0.163

Table 27A: Enthalpy of formation, Gibbs energy of formation, entropy, and enthalpy of vaporization at 25°C and 1 atm.

Substance	Formula	h_f^0, kJ/kmole	g_f^0,kJ/kmoe	$\bar{s}^0$,kJ/kmole K	$\bar{h}_{fg}^0$kJ/kmole	*T_{Flame}, K
Acetylene (Ethyne)	$C_2H_2(g)$	226,736	209,170	200.85		2559.4
Ammonia	$NH_3(g)$	–46,190	–16,590	192.33		
Benzene	$C_6H_6(g)$	82,930	129,660	269.20	33,830	2354.2
Butane–n	$C_4H_{10}(g)$	–126,150	–15,170	310.03	21,060	2279.6
Carbon	C(s)	0	0	5.74		2315.9
Carbon dioxide	$CO_2(g)$	–393,520	–394,380	213.67		971.4
Carbon monoxide	CO(g)	–110,530	–137,156	197.56		2404.2
Decane–n	$C_{10}H_{22}$	–249530	32970	545.7	40020	2287.0
Diesel(light)	$CH_{1.8}(l)$	-21,506			3706	2286.6
Ethane	$C_2H_6(g)$	–84,680	–32,890	229.49		2269.0
Ethyl alcohol	$C_2H_5OH(g)$	–235,310	–168,570	282.59	42,340	2246.2
Ethyl alcohol	$C_2H_5OH(l)$	–277,690	–174,890	160.70		2203.2
Ethylene (Ethene)	$C_2H_4(g)$	52,280	68,120	219.83		2383.2
Ethylene Glycol	$(CH_2OH)_2$	–389,320	–304470	323.55	52490	2214.9
Gasoline	$CH_{1.87}(l)$	-3363			4239	2325.3
Glucose	$C_6H_{12}O_6$	–1260,268		212		2130.3
Hydrogen	$H_2(g)$	0	0	130.57		2060
Hydrogen peroxide	$H_2O_2(g)$	–136,310	–105,600	232.63	61,090	
Hydrogen–monatomic	H(g)	218,000	203,290	114.61		
Hydroxyl	OH(g)	39,040	34,280	183.75		
Lead	Pb(c)	0	0	64.81		
Lead oxide	$PbO_2(c)$	–277,400	–217,360	68.6		
Lead sulfate	$PbSO_4(c)$	–919,940	–813,200	148.57		
Manganese	Mn(c)	0	0	31.8		
Manganese dioxide	$MnO_2(c)$	–520,030	–465,180	53.14		
Manganese trioxide	$Mn_2O_3(c)$	–958,970	–881,150	110.15		
Mercuric oxide	HgO(c)	–90,210	–58,400	70.45		
Mercury	Hg(l)	0	0	77.24		
Methane	$CH_4(g)$	–74,850	–50,790	186.16		2450.7
Methyl alcohol	$CH_3OH(g)$	–200,890	–162,140	239.70	37,900	2231.3
Methyl alcohol	$CH_3OH(1)$	–238,810	–166,290	126.80		2157.7
Nitric Oxide	NO	90,290	86,595	210.65		
Nitrogen	$N_2(9)$	0	0	191.50		
Nitrogen Dioxide	NO2	33,100	51,240	239.91		
Nitrogen–monatomic	N(g)	472,680	455,510	153.19		
Nonane–n	C_9H_{20}	–228,870	24730	506.4	37,690	2286.4
Octane–iso(g)	$C_8H_{18}(g)$	–232,191				2279.8
Octane–iso(l)	$C_8H_{18}(l)$	–263,111	–25948	425.2	31095	2272.1
Octane–n(g)	$C_8H_{18}(g)$	–208,450	17,320	463.67	41,460	2285.7
Octane–n(l)	$C_8H_{18}(l)$	–249,910	6,610	360.79		2275.4
Oxygen	$O_2(g)$	0	0	205.04		
Oxygen–monatomic	O(g)	249,170	231,770	160.95		
Palmitic Acid(fat)	$C_{16}H_{32}O_2$	–834,694		452.37		2259.3
Pentane–n	$C_5H_{12}(9)$	–146,440	–8,200	348.40	31,410	2282.1
Propane	$C_3H_8(g)$	–103,850	–23,490	269.91	15,060	2276.9
Propylene (Propene)	$C_3H_6(g)$	20,410	62,720	266.94	18,490	2346.4

Substance	Formula	h_f^0, kJ/kmole	g_f^0,kJ/kmoe	$\bar{s}^0$,kJ/kmole K	$\bar{h}_{fg}^0$kJ/kmole	*T_{Flame}, K
Silver oxide	Ag_2O(c)	–31,050	–11,200	121.7		
Sulfur	S	0	0	32.06		
Sulfur dioxide	SO2(g)	–296,842	–300,194	248.12		
Sulfur trioxide	SO3(g)	–395,765	–371,060	256.77		
Sulfuric acid	H_2SO_4(l)	–813,990	–690,100	156.90		
Sulfuric acid	(aq, m = 1)	–909,270	–744,630	20.1		
Water(g)	H_2O(g)	–241,820	–228,590	188.72		
Water(l)	H_2O(1)	–285,830	–237,180	69.95	44,010	
Zinc	Zn(c)	0	0	41.63		
Zinc oxide	ZnO(c)	–343,280	–318,320	43.64		

Some of the data are from K. Wark, *Advanced Thermodynamics for Engineers*, McGraw Hill Book Co., 1995. Originally from the *JANAF Thermochemical Tables*, Dow Chemical Co., 1971; *Chemical Thermodynamic Properties*, NBS Technical Note 270–3, 1968; and *API Research Project 44*, Carnegie Press, 1953.

*T_{Flame} denotes the adiabatic flame temperature for the fuel burning in in air under chemical equilibrium. The species that are considered are NO, OH, CO, CO_2, H_2O, H_2, N_2, O_2. Computed using THERMOLAB-1 software (available on the CRC website at http://www.crcpress.com).

Table 27 B: Values of enthalpy of combustion, Gibbs free energy change, entropy change and chemical availability in dry air during combustion of fuels at standard temperature, T_o = 25°C, and pressure, p_o = 1 atm*

Fuel Formula	M kg/kmol	Δh_c^0 MJ/kg	Δg^0 MJ/kg	Δs^0 kJ/kgK	$(\Delta h^0 - \Delta g^0)/\Delta g^0$ %	$Avail_F$ MJ/kmol
Acetylene C_2H_2	26.038	–48.3	–47.1	–3.7	+2.4	1265.6
Benzene C_6H_6	78.114	–40.6	–40.8	0.5	–0.4	3298.5
Carbon (graphite) C	12.011	–32.8	–32.9	0.2	–0.2	410.26
Carbon monoxide CO	28.01	–10.1	–9.2	–3.1	+10.1	275.10
Ethane C_2H_6	30.07	–47.5	–48.0	1.5	–1.0	1493.9
Ethanol CH_5OH	46.069	–27.8	–28.4	2.1	–2.2	1359.6
Ethylene C_2H_4	28.054	–47.2	–46.9	–1. 1	+0.7	1359.6
Ethylene gly col$(CH_2OH)_2$	62.07	–17.1	–18.6	5.1	–8.1	1226.4
Hydrogen H_2	2.016	–120.0	–113.5	–22.0	+5.8	235.2
Isooctane C_8H_{18}	114.23	–44.7	–45.8	3.7	–2.4	5375.8
Methane CH_4	16.043	–50.0	–49.9	–0.3	+0.2	830.2
MethanolCH_3OH	32.042	–21.1	–21.5	1.4	–1.9	722.3
n–Butane C_4H_{10}	58.12	–45.8	–46.6	2.7	–1.7	2802.5
n–Decane $C_{10}H_{22}$	142.29	–44.6	–45.7	3.5	–2.3	6726.4
n–HeptaneC_7H_{16}	100.21	–45.0	–45.9	3.2	–2.1	4764.3
n–Hexane C_6H_{14}	86.18	–45.1	–46.1	3.1	–2.0	4110
n–Nonane C_9H_{20}	128.26	–44.7	–45.7	3.4	–2.2	6072.3
n–Octane C_8H_{18}	114.23	–44.8	–45.8	3.3	–2.2	5418.6
n–Pentane C_5H_{12}	72.15	–45.4	–46.3	2.9	–1.9	3455.8
Propane C_3H_8	44.097	–46.4	–47.1	2.3	–1.5	2149
Propylene C_3H_6	42.081	–45.8	–45.9	0.4	–0.3	1999.9
Sulfur S	32.064	–9.2	–9.3	0.3	–0.9	609.6
Sulfur monoxide SO	48.063	–6.3	–5.8	–1.6	+8.5	–

Assumed ambient mole fractions: CO_2: 0.0003, H_2O: 0.0303, N_2: 0.7659; O_2, 0.2035.
Source: E. P. Gyftopoulos, and G. P. Beretta, *Thermodynamics, Foundations and Application*, Macmillan Publishing Co., NY, 1991. Originally from the data presented by R. C. Weast, Ed., *CRC Handbook of Chemistry and Physics*, 66th Ed., CRC Press, Boca Raton, FL, 1985; Chemical Availability were calculated by THERMOLAB-1 software (available on the CRC website at http://www.crcpress.com). Other data from A. Bejan, *Advanced Engineering Thermodynamics*, John Wiley & Sons, 1988.
Each constituent before and after combustion is assumed to be in its ideal gas state at T_0 and P_0.

TABLE 27C: Values of adiabatic flame temperature, entropy generation at 298 K, and composition of some of the product gases for the combustion of various hydrocarbons in a perfectly insulated steady state burner.

Fuel	Formula	T_{adiab} K	$(T_o\sigma)/\Delta g^o$ %	CO_2 Kmol/MJ	CO mol/MJ	H mol/MJ	NO mol/MJ	NO_2 mmol/MJ	N_2O mmol/MJ
Acetylene	C_2H_2	2598.0	22.6	1.17	457	51.4	87.3	19.8	4.6
Benzene	C_6H_6	2382.6	27.5	1.61	270	30.1	49.6	10.4	2.7
Carbon	C	2326.0	26.0	2.25	280	0	44.6	9.5	2.5
Ethane	C_2H_6	2300.5	29.0	1.23	155	53.2	35.6	6.6	1.9
Ethylene	C_2H_4	2416.6	26.2	1.27	254	56.6	52.9	10.8	2.8
Hydrogen	H_2	2448.5	20.9	0	0	231	46.3	7.4	2.3
Isooctane	C_8H_{18}	2312.1	30.2	1.35	175	45.0	37.2	7.1	2.0
Methane	CH_4	2266.0	28.3	1.12	124	57.5	31.3	5.6	1.7
n–Butane	C_4H_{10}	2311.8	29.6	1.31	170	48.6	37.1	7.01	2.0
n–Decane	$C_{10}H_{22}$	2316.4	30.1	1.36	179	44.9	37.8	7.3	2.0
n–Heptane	C_7H_{16}	2316.7	29.9	1.34	178	46.2	37.8	7.2	2.0
n–Hexane	C_6H_{14}	2313.2	29.8	1.34	175	46.4	37.4	7.1	2.0
n–Nonane	C_9H_{19}	30.0	1.36	179	45.2	37.8	7.3	2.0	2.0
n–Octane	C_8H_{18}	2316.1	30.0	1.35	178	45.6	37.8	7.2	2.0
n–Pentane	C_5H_{12}	2313.2	29.7	1.32	173	47.4	37.4	7.1	2.0
Propane	C_8H_8	2307.9	29.4	1.28	165	50.3	36.6	6.9	2.0
Propylene	C_3H_6	2378.5	27.6	1.33	227	50.8	47.0	9.4	2.5

Source: Tables 27C and 28A from E. P. Gyftopoulos, and G. P. Beretta, *Thermodynamics, Foundations and Application*, Macmillan Publishing Co., NY, 1991.
Combustion for a mixture of each hydrocarbon with the stoichiometric amount of dry air at 298 K and 1 atm. Values are determined assuming chemical equilibrium for the gases in the outlet stream with respect to the reaction mechanisms $CO_2 \rightarrow CO + (1/2)O_2$, $H_2O \rightarrow H_2 + (1/2)O_2$, $N_2 + O_2 \rightarrow 2NO$, $(1/2)N_2 + O_2 \rightarrow NO_2$ and $N_2 + (1/2)O_2 \rightarrow N_2O$. The entropy generation σ and change in Gibbs free energy during combustion Δg^o are per unit amount of fuel.

TABLE 28A: Values of the constants A_k and B_k for the reaction mechanisms of formation of various substances in ideal–gas states at standard pressure, $p_0 = 1$ atm, for use in the approximate expression

$K(T) = \exp.(\Delta A - (\Delta B/T))$, T in K, $298 < T < 5000$; $\Delta A = \Sigma\, \nu_k A_k$, $\Delta B = \Sigma\, \nu_k B_k$

(e.g.: $CO_2 \Leftrightarrow CO + 1/\, O_2$, $\Delta A = 1\times A_{CO} + (1/2)\times A_{O_2} - 1\times A_{CO_2}$)

Substance	Formula	A_k (K)	B_k (K)
Acetylene	C_2H_2	6.325	26.818
Ammonia	NH_3	−13.951	−6.462
Carbon	C	18.871	86.173
Carbon (diatomic)	C_2	22.870	100.582
Carbon dioxide	CO_2	−0.010	−47.575
Carbon monoxide	CO	10.098	−13.808
Carbon tetrafluoride	CF_4	−18.143	−112.213
Chlorine (atomic)	Cl	7.244	14.965
Chloroform	$CHCl_3$	−13.284	−12.327
Ethylene	C_2H_4	−9.827	4.635
Fluorine (atomic)	F	7.690	9.906
Freon 12	$CC1_2F_2$	−14.830	−58.585
Freon 21	$CHC1_2F$	−12.731	−34.190
Hydrogen (atomic)	H	7.104	26.885
Hydronium ion	H_3O^+	−8.312	71.295
Hydroxyl	OH	1.666	4.585
Hydroxyl ion	OH^-	−6.753	−20.168
Methane	CH_4	−13.213	−10.732
Nitric oxide	NO	1.504	10.863
Nitrogen (atomic)	N	7.966	57.442
Nitrogen dioxide	NO_2	−7.630	3.870
Nitrogen oxide	N_2O	−8.438	10.249
Oxygen (atomic)	O	7.963	30.471
Oxygen ion	O^-	0.528	10.048
Ozone	O_3	−8.107	17.307
Proton	H^+	13.437	188.141
Water	H_2O	−6.866	−29.911

Source: Regression of data from the *JANAF Thermochemical Tables*, 2nd ed., D. R. Stull and H. Prophet, Project Directors, NSRDS–NBS37. U.S. Department of Commerce National Bureau of Standards, Washington, D.C., 1971.

For any elemental species in natural form, $A_k = 0$ and $B_k = 0$.

TABLE 28B: Logarithms to the base 10 of the equilibrium constant K^0.

(1)* $H_2 \Leftrightarrow 2H$; (2) $O_2 \Leftrightarrow 2O$; (3) $N_2 \Leftrightarrow 2N$; (4) $1/2\ O_2 + 1/2\ N_2 \Leftrightarrow NO$; (5) $H_2O \Leftrightarrow H_2 + 1/2\ O_2$; (6) $H_2O \Leftrightarrow OH + 1/2\ H_2$; (7) $CO_2 \Leftrightarrow CO + 1/2\ O_2$; (8) $CO_2 + H_2 \Leftrightarrow CO + H_2O$; (9) $N_2 + 2O_2 \Leftrightarrow 2\ NO_2$.

T, K	(1)	(2)	(3)	(4)	(5)	(6)	(7)	(8)	(9)
298	-71.224	-81.208	-159.600	-15.171	-40.048	-46.054	-45.066	-5.018	-41.355
500	-40.316	-45.880	-92.672	-8.783	-22.886	-26.130	-25.025	-2.139	-30.725
1000	-17.292	-19.614	-43.0516	-4.062	- 10.062	-11.280	-10.221	-0.159	-23.039
1200	-13.414	-15.208	-34.754	-3.275	-7.899	-8.789	-7.764	+0.135	-21.752
1400	-10.630	-12.054	-28.812	-2.712	-6.347	-7.003	-6.014	+0.333	-20.826
1600	-8.532	-9.684	-24.350	-2.290	-5.180	-5.662	-4.706	+0.474	-20.126
1700	-7.666	-8.706	-22.512	-2.116	-4.699	-5.109	-4.169	+0.530	-19.835
1800	-6.896	-7.836	-20.874	-1.962	-4.276	-4.617	-3.693	+0.577	-19.577
1900	-6.204	-7.058	-19.410	-1.823	-3.886	-4.177	-3.267	+0.619	-19.345
2000	-5.580	-6.356	-18.092	-1.699	-3.540	-3.780	-2.884	+0.656	-19.136
2100	-5.016	-5.720	-16.898	1.586	-3.227	-3.422	-2.539	+0.688	-18.946
2200	-4.502	-5.142	-15.810	-1.484	-2.942	-3.095	-2.226	+0.716	-18.773
2300	-4.032	-4.614	-14.818	-1.391	-2.682	-2.798	-1.940	+0.742	-18.614
2400	-3.600	-4.130	-13.908	-1.305	-2.443	-2.525	-1.679	+0.764	-18.47
2500	-3.202	-3.684	-13.070	-1.227	-2.224	-2.274	-1.440	+0.784	-18.337
2600	-2.836	-3.272	-12.298	-1.154	-2.021	-2.042	-1.219	+0.802	-18.214
2700	-2.494	-2.892	-11.580	-1.087	-1.833	-1.828	-1.015	+0.818	-18.1
2800	-2.178	-2.536	-10.914	-1.025	-1.658	-1.628	-0.825	+0.833	-17.994
2900	-1.882	-2.206	-10.294	-0.967	-1.495	-1.442	-0.649	+0.846	-17.896
3000	-1.606	-1.898	-9.716	-0.913	-1.343	-1.269	-0.485	+0.858	-17.805
3100	-1.348	-1.610	-9.174	-0.863	-1.201	-1.107	-0.332	+0.869	-17.72
3200	-1.106	-1.340	-8.664	-0.815	-1.067	-0.955	-0.189	+0.878	-17.64
3300	-0.878	-1.086	-8.186	-0.771	-0.942	-0.813	-0.054	+0.888	-17.566
3400	-0.664	-0.846	-7.736	-0.729	-0.824	-0.679	+6.071	+0.895	-17.496
3500	-0.462	-0.620	-7.312	-0.690	-0.712	-0.552	+0.190	+0.902	-17.431

Source: Based on data from the *JANAF Tables*, NSRDS–NBS–37, 1971, and revisions published in *Journal of Physical and Chemical Reference Data* through 1982.

* If reactions are reversed (e.g. from $H_2 \Leftrightarrow 2H$ to $2H \Leftrightarrow H_2$), then reverse the sign for the numbers in table (e.g. at 2000 K, change from -5.580 for $H_2 \Leftrightarrow 2H$ to + 5.580 for $2H \Leftrightarrow H_2$).

TABLE 28B: Logarithms to the base 10 of the equilibrium constant K (continued – reactions involving solid carbon)

(10) $C + 1/2\ O_2 \Leftrightarrow CO$; (11) $C + O_2 \Leftrightarrow CO_2$; (12) $C + 2H_2 \Leftrightarrow CH_4$; (13) $C + CO_2 \Leftrightarrow 2CO$; (14) $C + H_2O \Leftrightarrow CO + H_2$

T, K	(10)	(11)	(12)	(13)	(14)
298	24.0479	69.0915	11	–20.52	–16.02
300	23.9285	68.668	–	–	–
400	19.1267	51.5365	6.65	–13.02	–10.12
500	16.2528	41.2582	4.08	–8.64	–6.62
600	14.3362	34.4011	2.36	–5.69	–4.26
700	12.9648	29.5031	1.12	–3.59	–2.59
800	11.9319	25.8266	0.2	–1.98	–1.31
900	11.1256	22.9665	–0.53	–0.74	–0.33
1000	10.4777	20.6768	–1.05	0.26	0.48
1100	9.445	18.8026	–1.49	1.08	1.15
1200	9.4983	17.24	–1.91	1.74	1.68
1300	9.1176	15.9165	–2.24	2.3	2.13
1400	8.79	14.7816	–2.54	2.77	2.5
1500	8.5045	13.7986	–2.79	3.18	2.83
1600	–	–	–3.01	3.56	3.14
1700	–	–	–3.2	3.89	3.41
1750	7.9182	12.0392	–	–	–
1800	–	–	–3.36	4.18	3.64
1900	–	–	–3.51	4.45	6.86
2000	7.4623	10.3258	–3.64	4.69	4.05
2100	–	–	–3.75	4.91	4.22
2200	–	–	–3.86	5.1	4.37
2300	–	–	–3.96	5.27	4.51
2400	–	–	–4.06	5.43	4.64
2500	6.8008	8.2212	–4.15	5.58	4.76
2600	–	–	–4.23	5.72	4.87
2700	–	–	–4.3	5.84	4.97
2800	–	–	–4.37	5.95	5.06
2900	–	–	–4.43	6.05	5.14
3000	6.3372	6.8437	–4.49	6.16	5.23
3100	–	–	–4.55	6.25	5.3
3200	–	–	–4.61	6.33	5.37
3300	–	–	–4.66	6.41	5.44
3400	–	–	–4.71	6.49	5.5
3500	5.9968	5.7954	–4.75	6.56	5.56

APPENDIX B

B. CHARTS

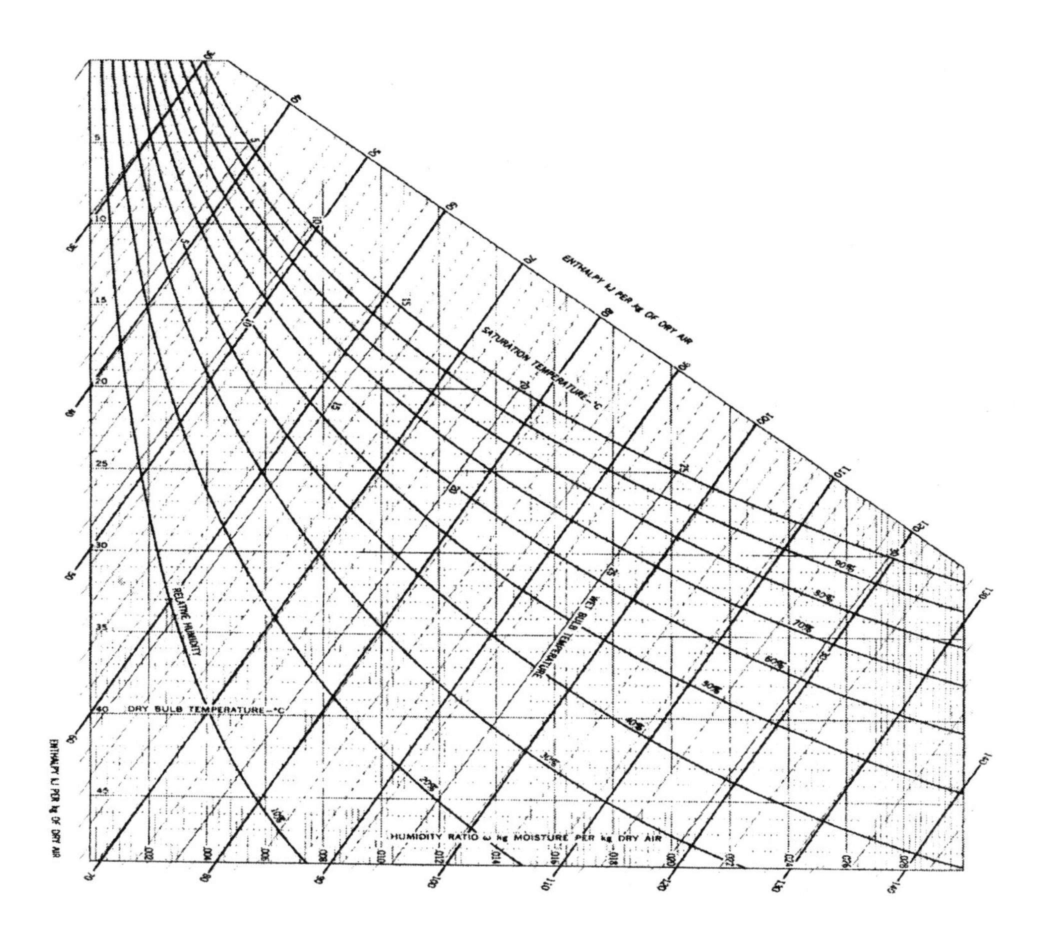

Figure B-1: Psychometric chart.

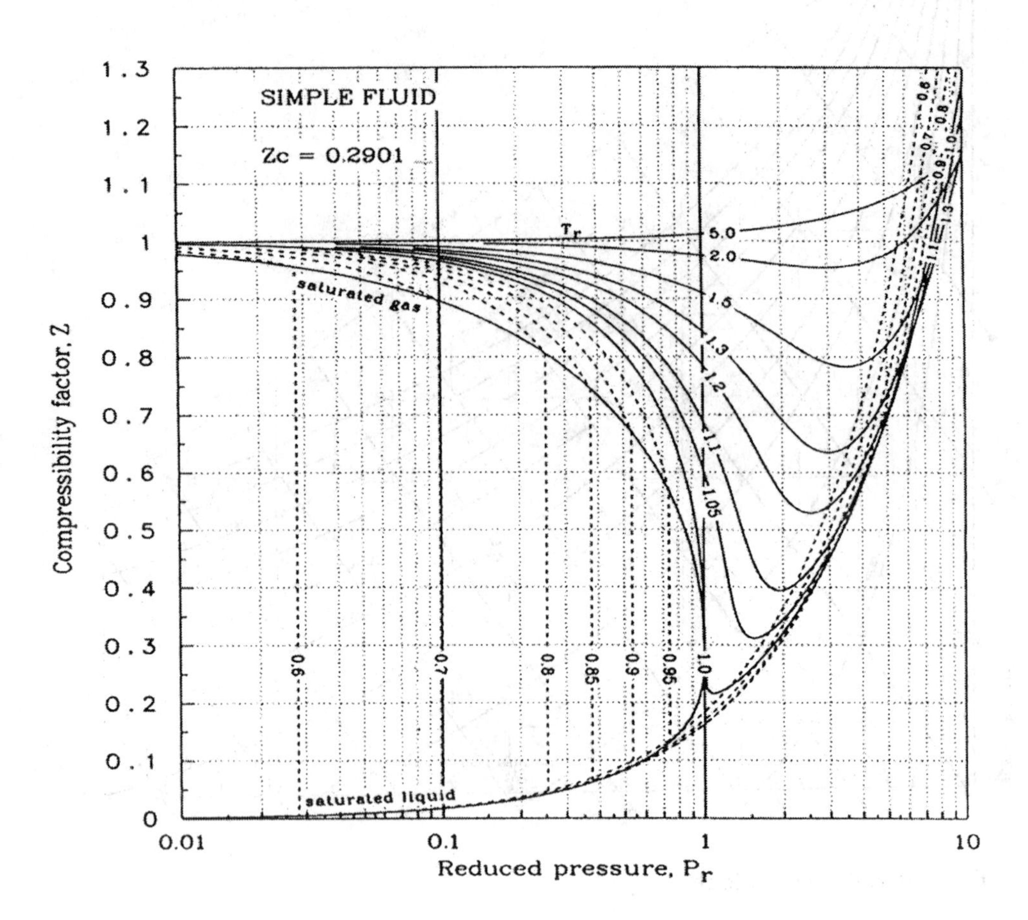

Figure B-2a: Lee Kessler simple fluid compressibility factor.

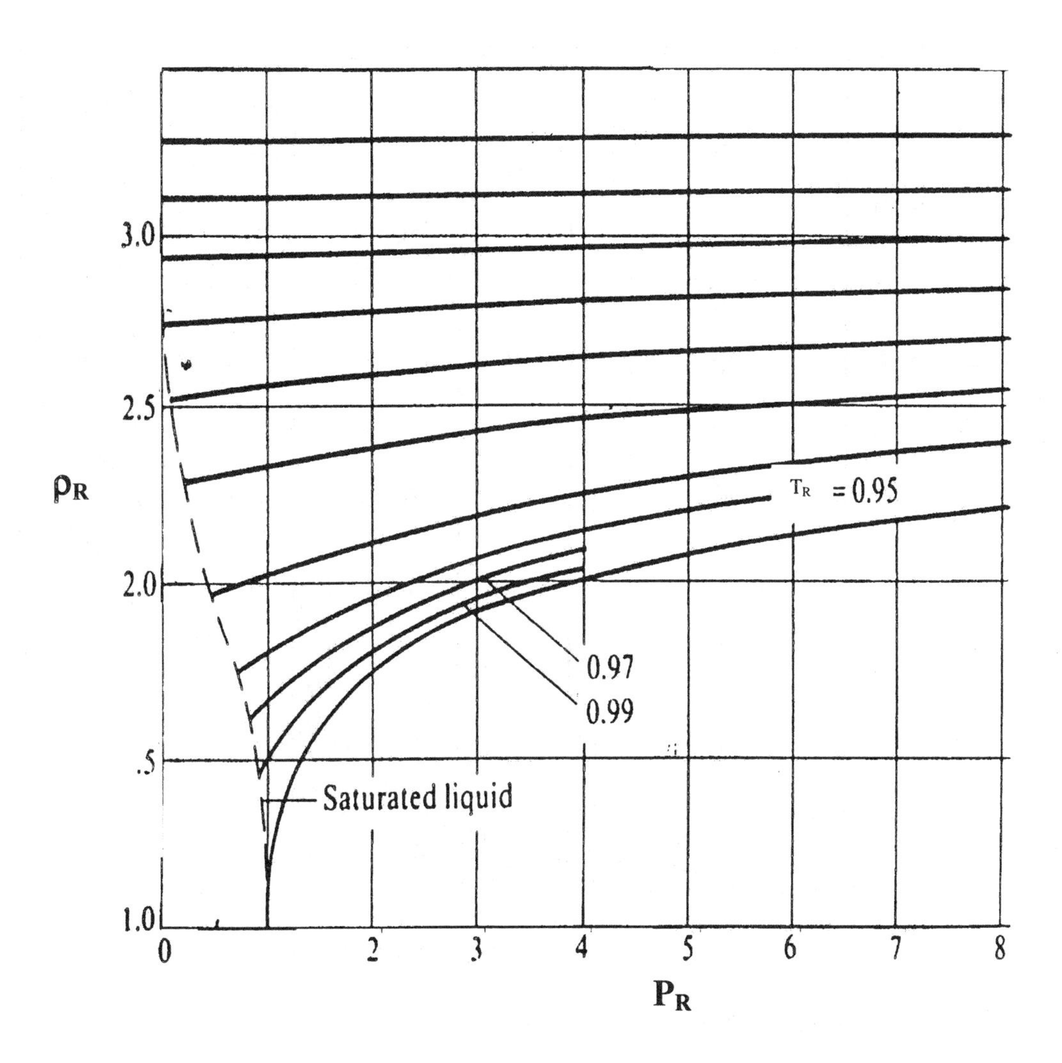

Figure B-2b: Generalized correlation for liquids.

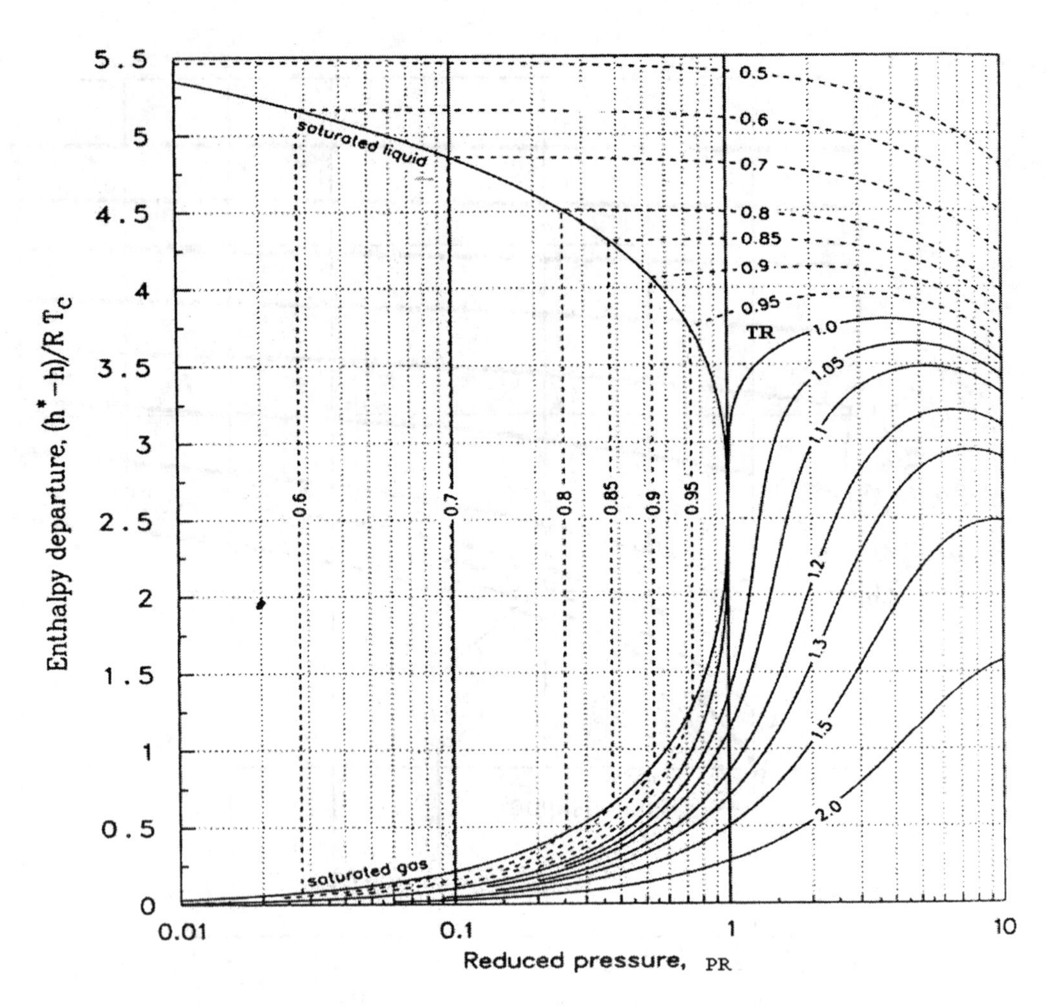

Figure B-3: Enthalpy correction for a simple fluid, $h^* = h_0$.

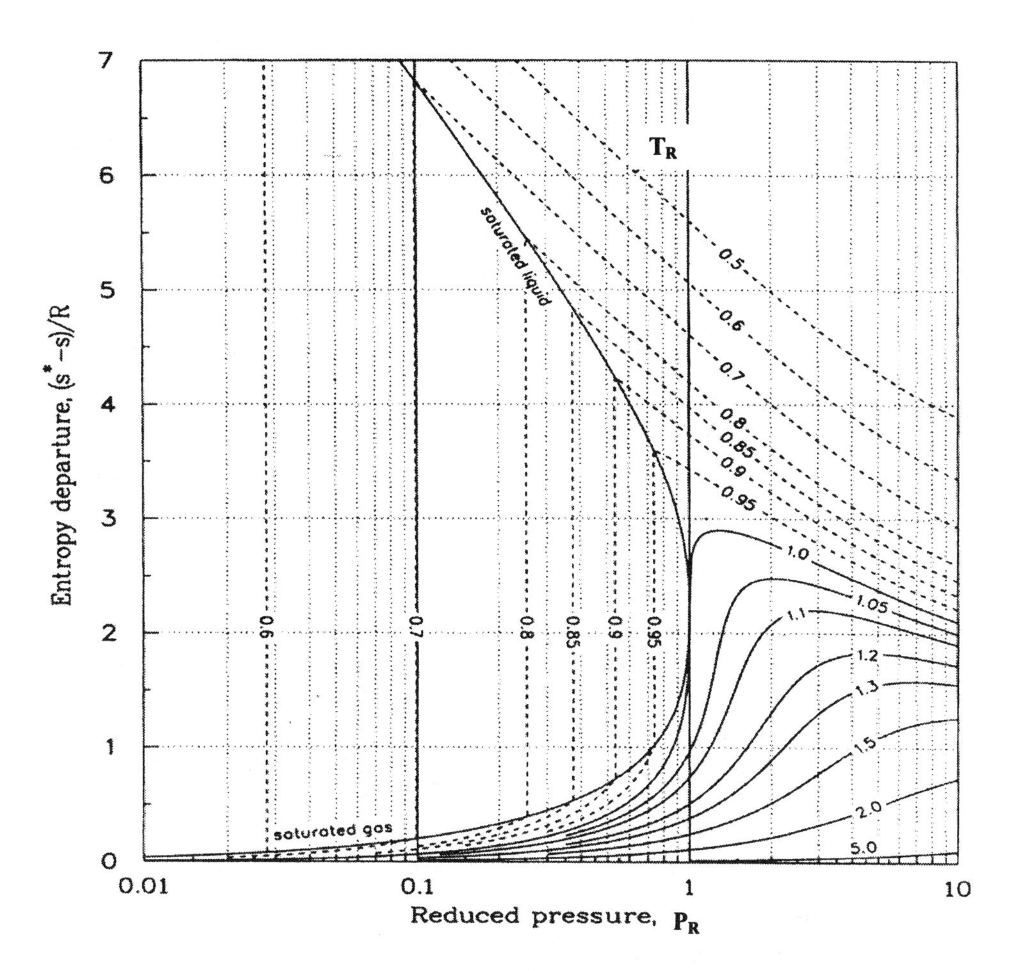

Figure B-4: Entropy correction of Simple Fluid, $s^* = s_0$.

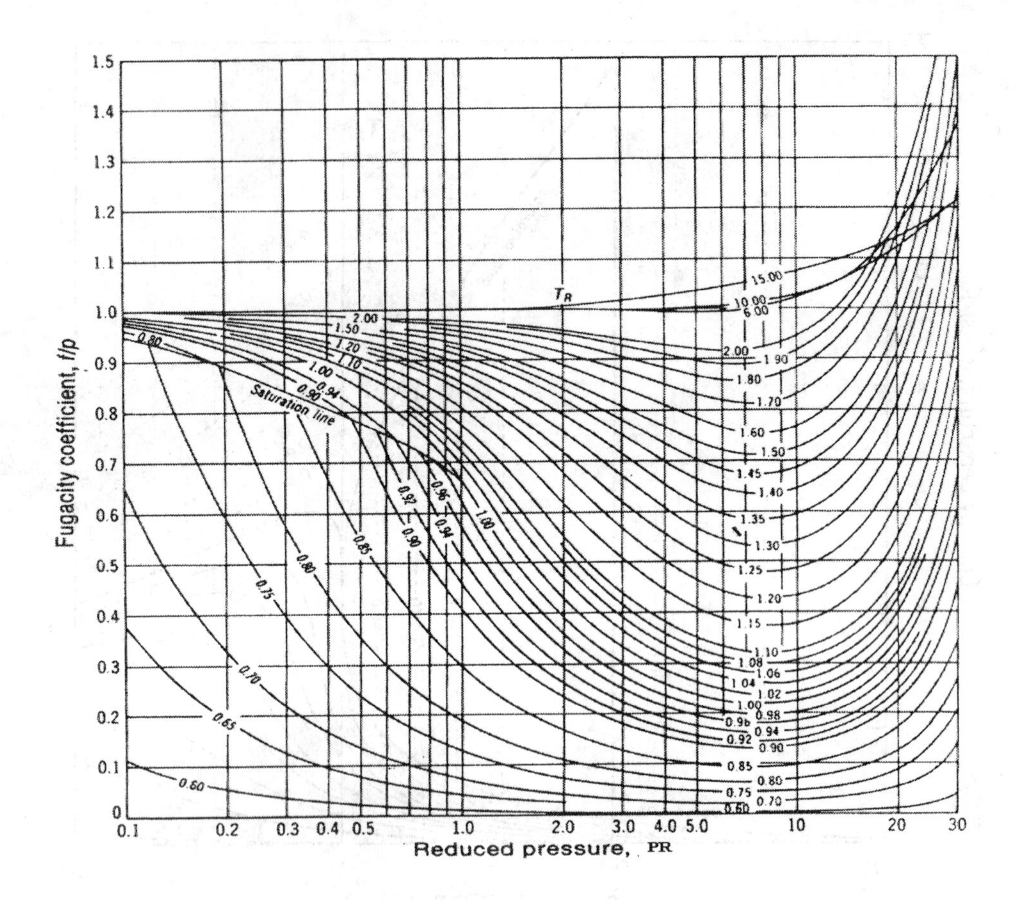

Figure B-5: Fugacity coefficient for a simple fluid.

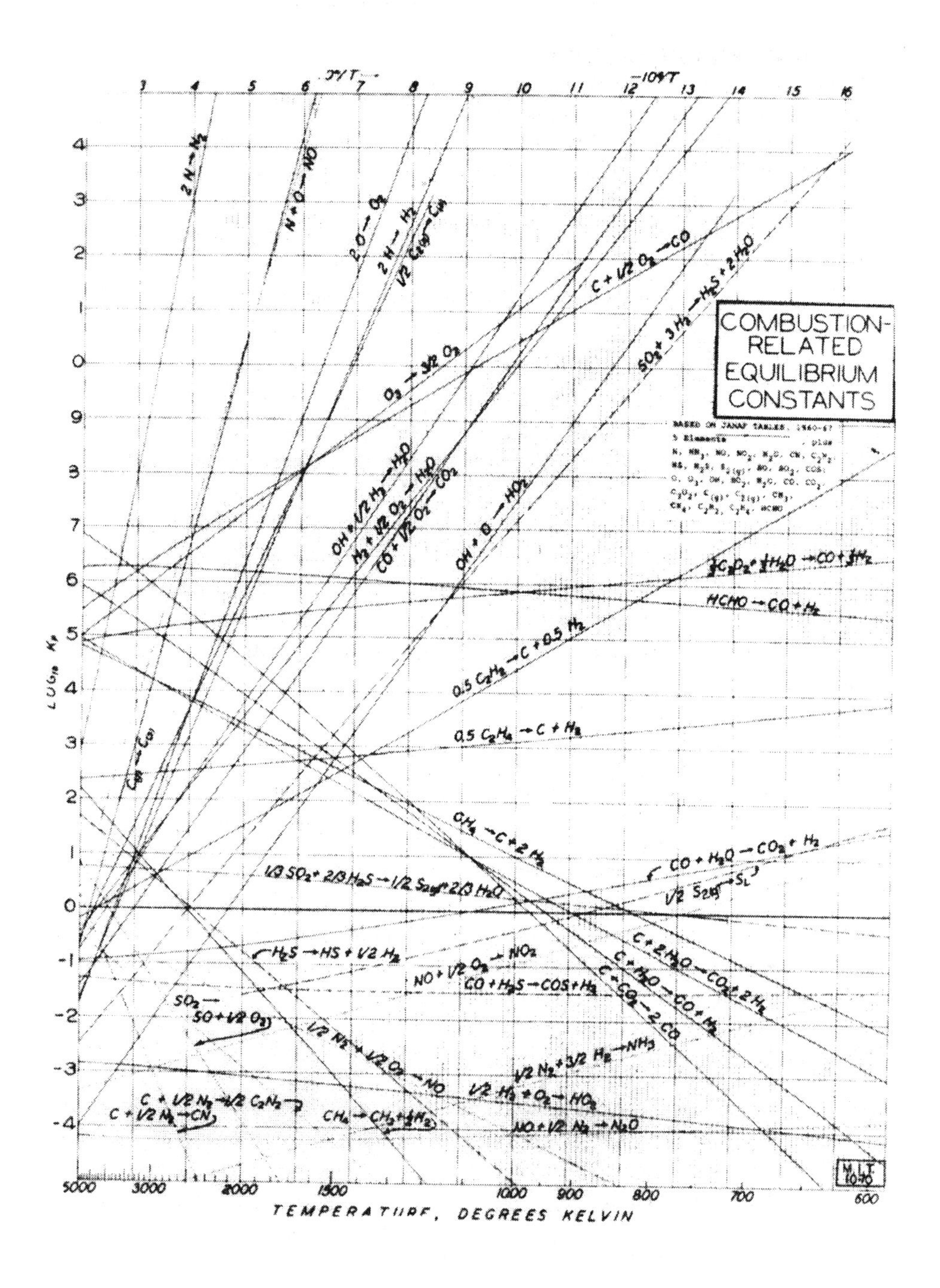

Figure B-6: Plot of ln K vs 1/T for several reactions (adopted from Modell and Reid, *Thermodynamics and Its Applications*, Second Edition, Prentice Hall, 198, p 335).

APPENDIX C

C. FORMULAE

A. CHAPTER 1 RELATIONS

Gravitational acceleration $g(r) = G\, m_E/r^2$, r in m
$G = 6.67\times10^{-14}\ m^3/kg\ s^2$ (or $kN\ m^2/kg^2$)
m_E, mass of earth = 5.97×10^{24} kg
r_E, radius of earth = 6.37×10^{6} m
At earth's surface $g(r_E) = 9.807\ m/s^2$
$g_c = 32.174\ lb_m\ ft/s^2\ lb_f$
$N_{Avog} = 6.023\times10^{26}$ molecules/kmole for a molecular substance

$= 6.023\times10^{26}$ atoms/k atom mole for an atomic substance

T(K) = T(C) + 273.15
T (R) = T(F) + 459.67
Joules equivalent of work J = 778.14 ft lb_f/BTU
1 short ton (usually referred to as a ton in the US)= 2000 lb
1 metric ton = 1000 kg
1 long ton (English): 2240 lb
Planck constant $h_P = 6.625\times10^{-37}$ kJ s/molecule
Boltzmann's constant $k_B = 1.380\times10^{-26}$ kJ/K molecule = $\bar{R}/N_{Avog}$
Speed of light in vacuum $c_0 = 2.998\times10^{8}$ m/s, 9.836×10^{8} ft/s
One light year = 5.875×10^{12} miles
Stefan–Boltzmann constant $\sigma_R = 5.670\times10^{-11}\ kW/m^2\ K^4$, $0.1714\times10^{-8}\ Btu/h\ ft^2\ R^4$
Ideal gas law

$$P\,v = RT,\quad PV = m\,R\,T$$

$$PV = n\bar{R}T,\quad P\bar{v} = \bar{R}\,T$$

$$\bar{R} = 8.314\,\frac{kPa\ m^3}{k\ mole\ K},\ 0.08314\,\frac{bar\ m^3}{k\ mole\ K},\ 1545\,\frac{ft\ lbf}{lbmole^\circ R}$$

$$\bar{R} = 0.730\,\frac{atm\ ft^3}{lbmole\ R},\ 10.7\,\frac{psia\ ft^3}{lbmole\ R}$$

$p_k/P = N_k/N = X_k$.
STP: 25°C (60°F), 101.325 kPa (14.696 psia), 760 mm of Hg (29.213 in of Hg at 32°F, 10.3323 m of water at 4°C)
1 kmole of ideal gas at STP occupies 24.8 m^3
1 lbmole of ideal gas at STP occupies 392 ft^3
Properties of pure substances

$x = m_g/(m_f + m_g)$
$v = x\,v_g + (1 - x)v_f$ or $v_f + x(v_g - v_f)$
u, h, and s relations are similar to the v expression

Mathematical background

Point function or property

For two variables $dz = Mdx + Ndy$, $\left(\frac{\partial M}{\partial y}\right)_x = \left(\frac{\partial N}{\partial x}\right)_y$

Cyclic relation $\left(\frac{\partial Z}{\partial x}\right)_y \left(\frac{\partial x}{\partial y}\right)_z \left(\frac{\partial Y}{\partial Z}\right)_x = -1$

Other useful relations $\left(\frac{\partial u}{\partial x}\right)_y = \left(\frac{(\partial u/\partial z)_y}{(\partial x/\partial z)_y}\right), \left(\frac{\partial z}{\partial x}\right)_y = \left(\frac{1}{(\partial x/\partial z)_y}\right)$

Homogeneous function

$$\phi(\lambda x_1 \;....\; \lambda x_s) = \lambda^m \phi(x_1 \;....\; x_s)$$

$$\sum_i x_i \frac{\partial \phi}{\partial x} = m\phi, \text{ Euler equation}$$

$$Z = Z(a,b,x,y) = x^m Z(a,b,y/x)$$

Lagrange multiplier

Optimize f(x,y,z) subject to g (x,y,z) = 0 and h(x,y,z) = 0. Then $F = f + \lambda_1 g + \lambda_2 h$, $\partial F/\partial x=0$, $\partial F/\partial y=0$, $\partial F/\partial z=0$.

Stokes theorem $\oint \vec{F}\cdot d\vec{s} = \oint_{cs} (\vec{\nabla}\times\vec{F})\cdot d\vec{A}$

Gauss divergence theorem $\oint_{cs} \vec{F}\cdot d\vec{A} = \oint_{cv} (\vec{\nabla}\cdot\vec{F})\cdot dV$

Microscopic thermodynamics

LJ Potential and Force

$$\phi(\ell) = 4\varepsilon\,((\ell_0/\ell)^{12} - (\ell_0/\ell)^6)$$

$$F(\ell) = -d\phi/d\ell,$$

ε/k_B in K $\approx$ 0.,77 T_c, 1.15 T_{BP}, 1.9 T_{MP}

σ (nm) $\approx 0.84\ \bar{v}_c^{1/3}$, $1.66\ \bar{v}_b^{1/3}$, $1.122\ \bar{v}_m^{1/3}$

c:critical, b: boiling, m: melting

$$l_{mean} = 1/(2^{1/2}\pi n'\sigma^2)$$

$$V_{rms} = (3k_BT/m))^{1/2} = (3\bar{R}T/M))^{1/2}$$

$$V_{avg} = \{8/(3\pi)\}^{1/2} V_{rms},\ V_{mps} = (2k_BT/m))^{1/2} = (2\bar{R}T/M))^{1/2}$$

where $V_{rms}^2 = V_x^2 + V_y^2 + V_z^2$ is based on the three velocity components

Sound Speed $c = \sqrt{k\bar{R}T/M}$

$$Kn = \ell_{mean}/d$$

B. CHAPTER 2 RELATIONS

Closed systems

$$\oint \delta Q = \oint \delta W,\ \delta Q - \delta W = dE,\ Q_{12} - W_{12} = E_2 - E_1$$

W_{out} positive, Qin positive

$$E = U + KE + PE$$

$$\delta W_{rev} = P\,dV$$

$$H = U + PV,\ h = u + Pv$$

$c_v = (\partial u/\partial T)_v$, $c_P = (\partial h/\partial T)_P$, (kJ/(kg K)), $(c_p/c_v) = k$

Ideal gases

$c_{v,0} = du/dT$, $c_{P,0} = dh/dT$, $(c_{po} - c_{vo}) = R$

(Note: Suffixes p and v are unnecessary for ideal gases. Instead let these be c_{10} and c_{20})

Monatomic gases

$$c_{P,0} = (5/2)\,R$$

$$(u_{0,2} - u_{0,1}) = \int_1^2 c_{vo}\,dT,\ (h_{0,2} - h_{0,1}) = \int_1^2 c_{po}\,dT$$

Open systems

Mass accounting or conservation

$$dm_{cv}/dt = \Sigma\,\dot{m}_i - \Sigma\,\dot{m}_e$$

$dm_{cv}/dt = \Sigma_k\, m_{k,i} - \Sigma_k\, m_{k,e}$, multiple components, single inlet and exit

Energy accounting or conservation

$$dE_{cv}/dt = \dot{Q}_{cv} - \dot{W}_{cv} + \Sigma\,\dot{m}_i\,(h+ke+pe)_i - \Sigma\,\dot{m}_e\,(h + ke + pe)_e$$

$$m_{cv} = \int dV\ /\ v\ ,\quad E_{cv} = \int (u + ke + pe)\,dV\ /\ v$$

where, ke (kJ/kg) $=V^2/2000$, pe(kJ/kg) = g Z/1000 (SI)
ke (BTU/lb) $V^2/(2\ g_c\ J)$, pe (BTU/lb) =g Z/(g_c J)
methalpy, e_T = = h + ke + pe
$\delta w_{rev} = -v\ dP$

Bernoulli type of equation applied in fluid mechanics, (f_M: Moody friction factor)

$$\frac{P_i}{\rho} + \frac{V_i^2}{2} + gz_i - (\frac{P_e}{\rho} + \frac{V_e^2}{2} + gz_e) = (u_i - u_e),$$

$$\text{Head Loss} = (u_i - u_e)/g = f_M L V^2/(2gD)$$

Uniform system and uniform flow processes
$Q_{cv} - W_{cv} = (m_2u_2 - m_1u_1)_{cv} + (m_eh_e - m_ih_i)$, assumed, ke and pe =0
where $m_i - m_e = (m_2 - m_1)_{cv}$
$(\delta\ w_{cv}\ )_{rev} = (\delta\ w_{shaft})_{rev}\ =\ -\ vdP$

Differential forms

$$\partial\rho/\partial t + \vec{\nabla}\cdot\rho\vec{V} = 0,\ \partial(\rho e)/\partial t + \vec{\nabla}\cdot(\rho e_T\vec{V}) = -\vec{\nabla}\cdot\vec{\dot{Q}}'' - w_{cv}'''$$

C. CHAPTER 3 RELATIONS

Performance of heat engines/heat pumps

Thermal efficiency (= sought/bought)
η = sought/bought= w_{cyc}/q_{in}, $w_{cyc} = q_{in} - q_{out}$
Carnot cycle $q_{out}/q_{in} = T_L/T_H$ for heat engines, $\eta = 1-(T_L/T_H)$
Coefficient of performance (= sought/bought)
$(COP)_{cooling}$ = sought/bought = |q absorbed from cooler body|/|w_{cyc}|
$(COP)_{heating}$ = sought/bought = |q rejected to hotter body|/|w_{cyc}|
Carnot heat pumps, $q_L/q_H = T_L/T_H$,
$COP_{cool} = T_L/(T_H - T_L)$, $cop_{heat\ pump} = T_H/(T_H - T_L)$,
HP/ton = 4.715/COP, 1 ton of refrigeration =211 kJ/min or 200 BTU/min

Entropy balance equation

Closed system $ds = (\delta q/T)_{rev}$

$ds = \delta q/T_b + \delta\sigma$, $s_2 - s_1 = \int \delta q/T_b + \sigma_{12}$, $\sigma_{12} \geq 0$

Open system $$\frac{dS_{cv}}{dt} = \Sigma\frac{\dot{Q}_{cv}}{T} + \Sigma\dot{m}_i s_i - \Sigma\dot{m}_e s_e + \dot{\sigma}_{cv}$$

Other relations Tds = du + pdv or ds = du/T + (p/T)dv,
Tds = dh – vdp or ds = dh/T – (v/T) dp

Entropy relations

Ideal gases
$ds = (c_{p0}/T)\ dT - (R/p)\ dp$, $ds = c_{v0}\ dT/T + (R/v)\ dv$

Summary of processes involving ideal gases with the various specific heat assumptions

	Constant specific heat	*Variable specific heat*
Entropy	$s = c_{p0} \ln (T/T_{ref}) - R \ln (P/Pref)$ P_{ref} =1 bar, P in bar, T_{ref} = 273 K or $s = c_{v0} \ln (T/T_{ref}) + R \ln (v/v_{ref})$ where $v_{ref} = R\ T_{ref}/T_{ref}$	$s = s^0 - R \ln (P/Pref)$ P_{ref} = 1 bar, $s^0 = \int c_{p0}\ (T)\ dT/T$
Isentropic Process	$Pv^k = C$, $Tv^{(k-1)} = C$, $T/P^{(k-1)/k} = C$ $P_2/P_1 = (T_2/T_1)^{k/(k-1)}$ $v_2/v_1 = (T_1/T_2)^{1/(k-1)}$	$s_2 = s_2^{\ 0} - R \ln (P_2/1) = s_1 = s_1^{\ 0} - R \ln (P_1/1)$ $P_2/P_1 = p_{r2}(T_2)/p_{r1}(T_1)$ $v_2/v_1 = v_{r2}(T_2)/v_{r1}(T_1)$

	Constant specific heat	***Variable specific heat***
Internal Energy	$u = c_{v0}\,T$	$u = \int_{T_{ref}}^{T} c_{v0}(T)\,dT$ or from tables
Enthalpy	$h = c_{p0}\,T$	$h = \int_{T_{ref}}^{T} c_{p0}(T)\,dT$ or from tables

Solids, liquids $(ds)_{incomp} = c\,(dT/T)$

Mixtures

Dalton law $P(N,V,T) = p_1(N_1,V,T) + p_2(N_2,V,T) + \ldots,$

Gibbs Dalton law $U = N_1\,\bar{u}_1(T,p_1) + N_2\,\bar{u}_2(T,p_2) + \ldots$

$S = N_1\,\bar{s}_1(T,p_1) + N_2\,\bar{s}_2(T,p_2) + \ldots$

Efficiencies: $\eta = w/w_s$,adiab.expansion, $= w_s/w$, adiab.compress; $\eta = w/w_T$, w_T/w, isothermal

Maximum entropy and minimum energy

$dU = T_b\,dS - P\,dV - (\delta W_{other}) - T_b\,\delta\sigma$

$dH = T_b dS + VdP - \delta W_{other} - T_b\,\delta\sigma$

$dA = -SdT - P\,dV - \delta W_{other} - T_b\,\delta\sigma$

$dG = -S\,dT + V\,dP - \delta W_{other} - T_b\,\delta\sigma$

$dS_{U,V,m} \geq 0$, $dS_{H,P,m} \geq 0$, $dU_{S,V,m} \leq 0$, $dH_{S,P,m} \leq 0$, $dA_{T,V,m} \leq 0$, $dG_{T,P,m} \leq 0$

$\delta\sigma/dt = (1/T_A - 1/T_B)\,dU_A/dt + (P_A/T_A - P_B/T_B)\,dV_A/dt + (\mu_{B1}/T_B - \mu_{A1}/T_A)\,dN_{A1}/dt +$

$(\mu_{B2}/T_B - \mu_{A2}/T_A)\,dN_{A2}/dt \geq 0$

D. CHAPTER 4 RELATIONS

Availability balance equation

Open system $\frac{d}{dt}(E_{c.v} - T_0\,S_{c.v.}) = \Sigma_{j=1}^{N}\,\dot{Q}_{R,j}\left(1 - \frac{T_0}{T_{R,j}}\right) + \dot{m}_i\psi_i - \dot{m}_e\psi_e - \dot{W}_{cv} - T_0\,\dot{\sigma}_{cv}$,

where the absolute stream availability (i.e open system) ψ is defined as

$\psi(T,P,T_0) = e_T(T,P) - T_0\,s(T,P)$; $e_T = h + ke + pe$

Stream exergy/availability or relative stream availability, $\psi' = \psi - \psi_0$

For sssf, $w_{opt} = \psi_i - \psi_e$

Loss in stream availability/irreversibility $i = w_{opt} - w = T_0\,\sigma$

Closed systems

$\phi = u - T_0\,s + P_0\,v$, absolute closed system availability

$\phi' = \phi - \phi_0$, closed system exergy or availability

$w_{u,opt} = \phi_1 - \phi_2$

Loss in availability or irreversibility $i = w_{u,opt} - w_u = w_{opt} - w = T_0\,\sigma$, $w_u = w - P_0\,\Delta v$

Availability or Exergetic(Work Potential) Efficiency

Heat engines $\eta_{Avail} = W_{cyc}/W_{max,cyc}$, $W_{max,cyc} = W_{cyc} + T_0\,\sigma_{cyc}$

Heat pumps: COP_{avail} or $\eta_{Avail} = W_{min,cyc}/W_{cyc}$, $W_{min,cyc} = W_{cyc} + T_0\,\sigma_{cyc}$

Work devices: $\eta_{Avail} = W/W_{max}$, $W_u/W_{u,max}$, W_{min}/W, $W_{u,min}/W_u$

Non–work systems η_{Avail} = (Exergy leaving the system) ÷ (Exergy entering the system).

Thermo–mechanical (TM) and chemical (C) equilibrium

Exit stream in TM equilibrium

$$\dot W_{opt,TM} = \Sigma_k \hat\psi_{k,i}(T_i, P_i, X_{1,i}, X_{2,i}...)\dot N_{k,i} - \Sigma_{k,e}\hat\psi_{k,0}(T_0, P_0, X_{1,e}, X_{2,e},...)\dot N_{k,e},$$

where $\hat\psi_1 = \hat h_k - T_0 \hat s_k$ (neglecting ke and pe)

Exit stream in TMC equilibrium

$$\dot W_{opt,TMC} = \Sigma_k \hat\psi_{k,i}(T_i, P_i, X_{1,i}, X_{2,i}...)\dot N_{k,i} - \Sigma_{k,e}\hat\psi_{k,0}(T_0, P_0, X_{1,\infty}, X_{2,\infty},...)\dot N_{k,e}$$

or $\dot W_{Opt,TMC} = \dot W_{Opt,TM} + \dot W_{Chem}$, where

$$\dot W_{Chem} = [\Sigma_k \hat\psi_{k,0}(T_0, P_0, x_{1,e}, x_{2,e},...) - \Sigma_k \hat\psi_{k,0}(T_\infty, P_\infty, x_{1,\infty}, x_{2,\infty},...)]\dot N_{k,e},$$

and where $T_\infty = T_0$, $P_\infty = P_0$, $X_{1,e} \neq X_{1,\infty}$,

$\Psi' = \Sigma N_k^*$ (exergy relative when exit is at thermal and mechanical equilibrium only)$_k$+ (ΣN_k^* (chemical availability)$_k$)

Chemical availability of component k = $\hat\psi_{k,e} - \hat\psi_{k,\infty} = \hat g_k(T_0, P_0, X_{1,e},...) - \hat g_k(T_0, P_0, X_{1,0},...)$

Psychrometry

Specific Humidity, $w = (M_v/M_a)(P_v/P_a) = 0.622\,(p_v/p_a)$

The degree of saturation $\mu = m_v(T,P)/m_v^{sat}(T,P) = N_v(T,P)/N_v^{sat}(T,P)$

The relative humidity $RH = X_v(T,P)/X_v^{sat}(T,P) = (N_v(T,P)/N(T,P))/(N_v^{sat}(T,P)/N^{sat}(T,P)) = P_v(T)/P_v^{sat}(T)$

$\ln P_v^{sat}(T) = A - B/(T+C)$, T in K, For H2O (0<T<50 C), 0,25 C,50 C correlation

$A = 12.21505207$, $B = 4119.460581$, $C = -35.208049$

$RH = \mu(1 - X_v) + X_v = \mu/(1 - X_v^{sat}(1-\mu))$.

Differential Forms:

$$\rho(\partial(e - T_0 s)/\partial t) + \rho\vec v \cdot \vec\nabla\psi = \vec\nabla \cdot ((\lambda\vec\nabla T)(1 - T_0/T)) - \dot w''' - \dot i'''$$

E. CHAPTER 5 RELATIONS

Fundamental equation $S = S(U, V, N_1, ..., N_s)$

$$dS = \left(\frac{\partial S}{\partial U}\right)_V dU + \left(\frac{\partial S}{\partial V}\right)_U dV + \sum_K \left(\frac{\partial S}{\partial N_k}\right)_{U,V} dN_k$$

$(\partial S/\partial U)_V = 1/T$, $(\partial S/\partial V)_U = P/T$, $(\partial S/\partial N_k)_{U,V} = -\mu_k/T$

Integrated form of U $U = TS - PV + \Sigma\mu_k N_k$

Equation of state $\frac{1}{T} = \left(\frac{\partial S}{\partial U}\right)_{V,N}, \frac{P}{T} = \left(\frac{\partial S}{\partial V}\right)_N, \frac{\mu_k}{T} = \left(\frac{\partial S}{\partial N_k}\right)_{U,V,N}$

Legendre Transform $\phi = \phi^{(0)} = \phi(x_1, x_2, ...x_n)$, $y_i = (\partial\phi/\partial x_i)$

where $\phi^{(o)}$ is the basis function . The m–th Legendre Transform $\phi^{(m)} = \phi^{(0)} - \sum_{i=1}^{m} x_i y_i$

F. CHAPTER 6 RELATIONS

Summary of P–V–T equations for real gases

$P_R = P/P_c$, $T_R = T/T_C$, $v_R' = v/v_c'$, $v_c' = RT_c/P_c$, $Z(T,P) = v(T,P)/v_0(T,P)$

Inflection conditions: $(\partial P/\partial v)_T = 0$, $(\partial^2 P/\partial v^2)_T = 0$

$\bar{b} \approx (2/3) N_{Avog} \pi \sigma^3$.

Pitzer Factor $w = -1.0 - \log_{10} (P_R^{sat})_{TR=0.7} = -1 - 0.4343 \ln (P_R^{sat})_{TR=0.7}$. $Z = Z^{(0)} (T_R, P_R) + w Z^{(1)} (T_R, P_R)$

If R is a universal gas constant, then a, v, b are based upon mole basis. If R is simply a gas constant, then a, v, b are based upon mass basis

Summary of Equations of State

	Name	*State Equations*	*Constants*	*Remarks*
1.	Virial Eq.	$Pv = RT + B_1'P + C_1'P^2 + D_1'P^3 + \ldots$, where B_1', C_1', … are functions of T or $Pv = RT + B/v + C/v^2 + \ldots$, and B, C, ... are functions of T.	B_1', C_1', D_1' are called 2nd, 3rd, and 4th virial coefficients that represent corrections to ideal gas behavior	$P/_uT = Z$
2.	Approximate Virial Eq.	$Pv/RT = Z = 1 + B(T)/P$ $Z = 1 + (P_R/T_R)(0.083 - (0.422/T_r^{1.6}))$ where $T_R = T/T_c$, $P_R = P/P_c$		$B^*(T^*) = B(T)/\bar{b}_o$, $\bar{b}_0$ in m^3/kmole, $T^* = T/(\varepsilon/k)$ For b_o, ε/k. See R.E. Sonntag and G. Van Wylen
3.	Clausius I	$P = RT/(v-b)$	b = body volume	Cannot satisfy inflection conditions.
4.	Van der Waals (VW)	$P = RT/(v-b) - a/v^2$, b is a correction for volume occupied by molecular and repulsive forces, a/v^2 is a correction for attractive forces $Z^3 - (B^* + 1) Z^2 + A^* Z - A^*B^* = 0$	$a = (27/64) v_c'^{\,2} P_c = (27/64) R^2 T_c^2/P_c \approx 2.667 \pi \varepsilon \sigma^3 N_{Avog}$, $b = v_c'/8 = (1/8) RT_c/P_c$, $Z_c = 3/8$, $A^* = (27/64) P_R/T_R^2$, $B^* = (1/8) P_R/T_R$	Does not agree with $(P_c v_c/RT_c)_{exp} = 0.2$ to 0.3 for most gases. Another form:of the VW relation is $\bar{v}^3 + \bar{v}^2 (-\bar{b}P - \bar{R}T)/P + \bar{v} (\bar{a}/P)(-\bar{a}\,\bar{b}/P) = 0$
5.	Berthelot	$P = RT/(v-b) - (a/T)(1/v^2)$	$a = (27/64)(v_c'^2 T_c P_c) = = (27/64) R^2 T_c^3/P_c$, $b = v_c'/8 = RT_c/8P_c$	

	Name	State Equations	Constants	Remarks
6.	Dieterici	$P = (RT/(v-b)) \exp\{-a/(RTv)\}$	$a = (4/e^2)\, v_c'^2\, P_c = (4/e^2)\, R^2\, T_c^2/P_c$ $b = v_c'/e^2 = RT_c/(e^2 P_c)$, $Z_c = 0.271$, $e = \exp(1) = 2.3026$	Developed to provide better agreement with experiments.
7.	Redlich–Kwong (RK)	$P = RT/(v-b) - a/(T^{1/2} v\,(v+b))$ or $Z^3 - Z^2 + (A^* - B^{*2} - B^*)\,Z - A^* B^* = 0$	$a = 0.4275\, v_c'^2\, T_c^{0.5}\, P_c = 0.4275 R^2\, T_c^{2.5}/P_c$, $b = 0.08664\, v_c' = 0.08664\, RT_c/P_c$, $Z_c = 1/3$, $A^* = 0.4275\, P_R/T_R^{2.5}$, $B^* = 0.08664\, P_R/T_R$	Good accuracy over wide range and at high pressure.
8.	Clausius II	$P = RT(v-b) - a/(T(v+c)^2)$	$a = 27/64\, v_c'^2\, T_c\, P_c = (27/64)\, R^2\, T_c^3/P_c$ $b = v_c'(Z_c - 1/4) = (RT_c/P_c)\,(Z_c - 1/4)$ $c = v_c'(3/8 - Z_c) = (RT_c/P_c)\,(3/8 - Z_c)$	
9.	Peng Robinson	$P = (RT/(v-b)) - (a\, \alpha(w,TR)/((v+b(1+\sqrt{2}))\,(v + b\,(1-\sqrt{2})))$	$a = 0.45724\, v_c'^2\, Pc = 0.45724 R^2\, T_c^2/P_c$ $\alpha(w,T_R) = (1 + f(w)\,(1 - T_R^{(1/2)}))^2$ $f(w) = 0.37464 + 1.54226\, w - 0.26992\, w^2$ $b = 0.07780\, v_c' = 0.0778\, (RT_c/P_c)$, $Z_c = 0.26$	
10.	SRK equation	$P = RT/(v-b) - a\, \alpha(w,T_R)/(v(v+b))$	$A = 0.4275 v_c'^2\, Pc = 0.4275 R^2\, T_c^2/P_c$, $b = 0.08664\, v_c' = 0.08664\, (RT_c/P_c)$, $Z_c = 1/3$ $\alpha(w,T_R) = (1 + f(w)\,(1 - T_R^{(1/2)}))^2$ $f(w) = (0.480 + 1.574\, w - 0.176\, w^2)$	
11.	Generalized Eq.	$P_R = T_R/(v_R' - b^*) - a^* \alpha\,(w,T_R)/(T_R^n\,(v_R' + c^*)$	$a^* = a/(P_c\, v_c'^2\, T_c^n)$, $b^* = b/v_c'$, $c^* =$	

	Name	State Equations	Constants	Remarks
	of state	$(v'_R + d^*)$,	c/v'_c, and $d^* = d/v'_c$. See table below	
12.	Compressibility factor	$Pv = ZRT$ or $P_R v_R' = Z T_R$ $v_R' = v/v_c'$, $v_c' = RT_c/P_c$ $P_R = P/P_c$, $T_R = T/T_c$ Also $fv = (\phi(T_R, P_R)\ Z(T_R, P_R))\ RT$ (Chapter 07)		$Z = v/v_{ideal}$ for $T_R > 2.5$, $Z > 1$, for $T_R < 2.5$ $Z < 1$ and has a minimum value. At $T_R = 1$, $P_R = 1$, Z can vary widely. For $P_R > 10$ always use real gas relations.
13.	Benedict Webb Rubin	$P = RT/v + (B_2RT - A_2 - C_2/T^2)/v^2 + (B_3R\ T - A_3)/v^3 + A_3C_6/v^6 + (D_3/(v^3T^2))(1+E_2/v^2)\exp(-E_2/v^2)$	8 constants. See Table 20A	Good accuracy over wide P–V–T condition.
14.	Martin–Hou	$P = RT/(v-b) + (A_2 + B_2 + C_2\ e^{-KT})/(v-b)^2 + (A_3 + B_3T + C_3\ e^{-KT})/(v-b)^3 + A_4/(v-b)^4 + (A_5 + B_5T + C_5\ e^{-KT})/(v-b)^5$	12 constants evaluated from P–v–T data of fluids	Mainly developed for refrigerants, 1 % accuracy for $v > 0.67\ v_c$ and $T < 1.5\ T_c$
15.	Lee Kessler	$P_R = (T_R/v_{R'})(1 + A/v'_R + B/{v'_R}^2 + C/{v'_R}^5 + (D/v'_R)(\beta + \gamma/{v'_R}^2)\exp(-\gamma/{v'_R}^2))$, $Z = P_R\ v'_R/T_R$	See Table A–21	
16.	Beattie Bridgeman equation	$P\ \bar{v}^2 = \bar{R}T\ (\bar{v} + B_0\ (1 - (\bar{b}/\bar{v}))\ (1 - c/(\bar{v}T^3)) - (A_0/\bar{v}^2)(1 - (a/\bar{v}))$	See Table A–20B	Accurate for $v > 1.25\ v_c$

Generalized cubic equation of state $P_R = (T_R/(v_R' - b^*)) - a^* \alpha(w,T_R)/(T_R^n (v_R' + c^*)(v_R^* + d^*))$, where $a^* = a/(P_c v_c'^2 T_c^n)$, $b^* = b/v_c'$, $c^* = c/v_c'$, $d^* = d/v_c'$

Equation	$c^* = c/v_c'$	$d^* = c/v_{c'}$	n	$\alpha(w,T_R)$	$b^* = b/v_c$	$a^* = a/(P_c T_c^n v_c'^2)$
Clausius–I	0	0	0	0	–	0
VW	0	0	0	1	1/8	27/64
Berthlot	0	0	1	1	1/8	(27/64)
Clausius–II	$(3/8 - Z_c)$	$(3/8 - Z_c)$	1	1	$Z_c - 1/4$	27/64
Horvath–Lin	$\gamma\beta$ (note 3)	0	1	1	β (note 3)	α (note 3)
RK	0.08664	0	1/2	1	0.08664	0.4275
Lorentz	0	0	0	note 1	0	
Martin	–	–	0	note 2		
SRK	0	0	0		0.08664	0.42748
PR	$(1+\sqrt{2})0.07780$	$(1-\sqrt{2})0.0778$	0		0.07780	0.45724

Note 1: $T_R\, b'/v_c'^2 - a'/v_c'^2$
Note 2: $c'/v_c'^2 - T_R$
Note 3: $\gamma = Z_c^{-4.72}/360$, $(1+(\gamma+1)^{1/3} + (\gamma+1)^{2/3})^{-1}$, $\alpha = ((1+\gamma f)^2 (1-2f-\gamma f^2))/((1-f)^4(2+\gamma f)^2)$, $\beta = (1-2f-\gamma f^2)/((2+\gamma)+(2+4\gamma)f + (\gamma+2\gamma^2)f^2)$

Liquids and Solids $dv = v\beta_P\, dT - v\beta_T\, dP$, $\beta_P = (1/v)(\partial v/\partial T)_P$, $\beta_T = -(1/v)(\partial v/\partial P)_T$, $\kappa_T = 1/(\beta_T P) = (-v/P)(\partial P/\partial T)_T$

Rackett equation for saturated liquid $v^{sat} / v_c = Z_c^{(1-T_R{}^{0.2857})}$

G. CHAPTER 7 RELATIONS

Differentials

Exact are denoted as d()

Criterion for exactness

$$\text{If } dZ = M(x,y)\,dx + N(x,y)\,dy, \text{ then } \left(\frac{\partial M}{\partial y}\right)_x = \left(\frac{\partial N}{\partial x}\right)_y$$

Inexact are denoted by $\delta()$ and are path dependent functions

Thermodynamic relations

$du = Tds - Pdv$

$dh = Tds + vdP$

$da = -Pdv - sdT$

$dg = vdP - sdT$

Gibbs function $g = h - Ts$

Helmholtz function $a = u - Ts$

$(\partial((a/T)/\partial T))_v = u$, $(\partial a/\partial v)_T = -P$, and $(\partial a/\partial T)_v = -s$

$(\partial (g/T)/\partial (1/T))_P = h$, $(\partial g/\partial P)_T = v$, and $(\partial g/\partial T)_P = -s$

Maxwell's relations (or criteria for exact differential of thermodynamic relations)

$$\left(\frac{\partial T}{\partial v}\right)_s = -\left(\frac{\partial P}{\partial s}\right)_v$$

$$\left(\frac{\partial T}{\partial P}\right)_s = \left(\frac{\partial v}{\partial s}\right)_P$$

$$\left(\frac{\partial P}{\partial T}\right)_v = \left(\frac{\partial s}{\partial v}\right)_T$$

$$\left(\frac{\partial v}{\partial s}\right)_P = -\left(\frac{\partial s}{\partial P}\right)_T$$

Thermodynamic properties

Given the state equation $P = P(T,V)$, properties can be determined using the relations

$$d\bar{s} = \bar{c}_v\left(\frac{dT}{T}\right) + \left(\frac{\partial P}{\partial T}\right)_{\bar{v}} d\bar{v}$$

$$= \bar{c}_p\left(\frac{dT}{T}\right) - \left(\frac{\partial \bar{v}}{\partial T}\right)_p dP$$

$$d\bar{u} = \bar{c}_v\, dT + \left[T\left(\frac{\partial P}{\partial T}\right)_{\bar{v}} - P\right] d\bar{v}$$

$$d\bar{h} = \bar{c}_p\, dT + \left[\bar{v} - T\left(\frac{\partial \bar{v}}{\partial T}\right)_p\right] dP$$

$$\left(\frac{\partial \bar{c}_v}{\partial \bar{v}}\right)_T = T\,(\partial^2 P / \partial T^2)_{\bar{v}}$$

$$\left(\frac{\partial \bar{c}_p}{\partial P}\right)_T = -T\,(\partial^2 \bar{v} / \partial T^2)_p$$

$$(\bar{c}_p - \bar{c}_v) = -T\left[\left(\frac{\partial \bar{v}}{\partial T}\right)_p\right]^2 \left(\frac{\partial P}{\partial \bar{v}}\right)_T = T\,\bar{v}\,\beta_p^2 / \beta_T$$

Fugacity Coefficient

$RT\, d\ln(\phi) = v dP$, $\phi = f/P$

$f_k(T,P) = f_k(T,P^{sat})\, POY_k$

$POY_k = \exp\left(\int_{P^{sat}}^{P} (v_k/RT)dP\right)$ for any given phase .

Other properties

Isobaric (volume) expansivity, $\beta_p = (1/v)(\partial v/\partial T)_p$, 1/K or 1/R isothermal compressibility, $\beta_T = -(1/v)\,(\partial v/\partial P)_T$, 1/bar, 1/atm

Isothermal bulk modulus, $B_T = 1/\beta_T$, bar, atm

Isentropic compressibility, $\beta_s = -(1/v)\,\partial v/\partial P$, 1/bar, 1/atm

$\partial v/\partial P = -v^2/RT$, as $P \to 0$

$\partial(u_0(T) - u(T,v))/\partial P = +(3/2)\,(a/RT^{3/2})$, as $P \to 0$

$\partial(h_0 - h)/\partial P = (3/2)\,(a/RT^{3/2}) - v + v = (3/2)\,(a/RT^{3/2})$, as $P \to 0$

Saturation properties and Joule Thomson coefficient

Correlations for h_{fg}

Empirical correlations for $h_{fg,n}$ $\dfrac{h_{fg.n}}{RT_n} = 1.092\,\dfrac{(\ln P_c - 1)}{(0.930 - T_{rn})}$

$h_{fg,n} = 13.52$, where T_n, normal boiling point, EK

$h_{fg.\,n}$, heat of vaporization at T_n

P_c, critical pressure, bar

$T_{rn} = T_n/T_c$

Correlation for any other T $\dfrac{h_{fg,2}}{h_{fg,1}} = \left\{\dfrac{(1 - T_{R,2})}{(1 - T_{R,1})}\right\}^{0.38}$

$T_{R,2} = T_2/T_c$

Approximate relations

Saturation pressures based on real gas equations

$\ln P_R = (dP_R/dT_R)_C\,(1 - 1/T_R)$

where dP_R/dT_R for various state equations are tabulated below.

	Z_c	$(dP_R/dT_R)_C$
VW	3/8	4
Berthe	3/8	7
Claus–II	Z_c	7
RK	1/3	5.582
SRK	1/3	$4.0536 + 3.05362\,f_{SRK}(w)$
PR	0.3214	$4.1051 + 3.1051\,f_{PR}(w)$

Clapeyron Equation

$(dP/dT)^{sat} = h_{\alpha\beta}/(T\,v_{\alpha\beta})$

$h_{\alpha\beta} = h_\alpha - h_\beta$, $v_{\alpha\beta} = v_\alpha - v_\beta$, β – phase, α – phase

Clausius–Clayperon (vapor/ideal gas) – an approximate relation

$$\frac{P}{P_{ref}} = \exp\left(\frac{h_{fg}}{R}\left[\frac{1}{T_{ref}} - \frac{1}{T}\right]\right) \text{ or}$$

$\ln P = A - B/T$, $A = \ln P^{sat}_{ref} + h_{fg}/RT_{ref}$ and $B = h_{fg}/R$.

Throttling coefficient

$$\mu_{JT} = (\partial T/\partial P)_h$$

Joule Thomson effect (open system):

$$\mu_{JT} = -\frac{\left[v - T\left(\dfrac{\partial v}{\partial T}\right)_P\right]}{c_p} = \frac{v[\beta_P\,T - 1]}{c_p}$$

Asymptotic limits: As $p_R \to 0$, $\mu_{JT}\,c_P/v_c' \to b/v_c' \to 0.08664$ for RK equation of state

Inversion curve $\mu_{JT} = 0$, $(\partial T / \partial P)_h = 0$, $v = T\left(\frac{\partial v}{\partial T}\right)_P$

Euken coefficient (closed system, throttling at constant volume)

$$\mu_E = (\partial T/\partial v)_u = -(T(\partial P/\partial T)_v - P)/c_v$$

H. CHAPTER 8 RELATIONS

Mole and mass fractions

Mole fraction (number fraction) $X_i = N_i/N$

Mass fraction $Y_i = m_i/m$

Molecular mass of mixture $M_m = \Sigma X_i M_i$

Conversion from X_i to Y_i: $Y_i = X_iM_i/M_m$,

Conversion from Y_i to X_i: $X_i = Y_iM_m/M_i$

Molality, Mo = $10^{-3}\times$kmole of solute ÷ kg of solvent.

Generalized relations

$$U = U(S, V, N_1, N_2 \ldots N_n)$$
$$H = H(S, P, N_1, N_2 \ldots N_n)$$
$$A = A(T, V, N_1, N_2 \ldots N_n)$$
$$G = G(T, P, N_1 \ldots\ldots N_n)$$

Differentials

$$dU = Tds - PdV + \Sigma\mu_j dN_j$$

$$dH = Tds + VdP + \Sigma\mu_j dN_j$$

$$dA = -SdT - PdV + \Sigma\mu_j dN_j$$

$$dG = -SdT - VdP + \Sigma\mu_j dN_j$$

Thermodynamc potentials

$$T = \left(\frac{\partial U}{\partial S}\right)_{V,\ N_1\ \ldots\ N_n} = \left(\frac{\partial H}{\partial S}\right)_{P,\ N_1,\ \ldots\ N_n}$$

$$P = -\left(\frac{\partial U}{\partial V}\right)_{S,\ N_1\ \ldots\ N_n},\ V = \left(\frac{\partial H}{\partial P}\right)_{S_1\ N_1\ \ldots\ N_n}$$

$$\mu_1 = \left(\frac{\partial U}{\partial N_1}\right)_{S,\ V,\ N_2\ \ldots\ N_n} = \left(\frac{\partial H}{\partial N_1}\right)_{S,\ P,\ N_2\ \ldots\ N_n}$$

$$= \left(\frac{\partial A}{\partial N_1}\right)_{T,\ V,\ N_2\ \ldots\ N_n} = \left(\frac{\partial G}{\partial N_1}\right)_{T,\ P,\ N_2\ \ldots\ N_n}$$

Partial molal property (B= U, A, H, G, etc.)

$$\left(\frac{\partial B}{\partial N_1}\right)_{T,P_1N_2,N_3,..} = \hat{b}_1,\ \text{e.g.},\ \hat{v}_1 = \left(\frac{\partial V}{\partial N_2}\right)_{T,P_1N_1,N_2,..},\ \hat{g}_1 = (\partial G/\partial N_1)_{T,P,N_2,N_3,..}$$

where $\hat{g}_1$, partial molal Gibbs' function of species 1=$\mu_1 = (\partial G/\partial N_1)_{T,P,\ N2,\ N3..}$

Mixture Property

$$B = \Sigma N_k \hat{b}_k,\ \hat{b}_1 = \bar{b} - X_2\, d\bar{b}/dX_2$$

Gibbs Duhem equation

$$\frac{d\bar{b}}{dT}dT + \frac{d\bar{b}}{dP}dP - \sum_{k=1}^{K} X_k d\hat{b}_k = 0 \text{ or } \left(\frac{\partial B}{\partial T}\right)_{P,N} dT + \left(\frac{\partial B}{\partial P}\right)_{T,N} dP - \sum_{k=1}^{K} N_k d\hat{b}_k = 0$$

Differentials of partial molal properties:

$$d\hat{g}_k = -\hat{s}_k\, dT + \hat{v}_k\, dP,\ d\hat{h}_k = Td\hat{s}_k + \hat{v}_k dP,$$

$d\hat{u}_k = T\,d\hat{s}_k - P\,d\hat{v}_k$, $d\hat{h}_k = T d\hat{s}_k + \hat{v}_k dP$.

$\hat{c}_{pk} = \partial \hat{h}_k/\partial T = T(\partial \hat{s}_k/\partial T)_P$, $\hat{c}_{vk} = T(\partial \hat{s}_k/\partial T)\,\hat{v}_k$

Generalized Thermodynamic Relations

$d\hat{u}_k = \hat{c}_{v,k}\,dT + (T(\partial P/\partial T) - P)\,d\hat{v}_k$, $d\hat{h}_k = \hat{c}_{p,k}\,dT + (\hat{v}_k - T(\partial \hat{v}_k/\partial T))\,dP$,

$d\hat{s}_k = (\hat{c}_{pvk}/T)dT + (\partial P/\partial T)\,\hat{v}_k\,dP$, $d\hat{s}_k = (\hat{c}_{p,k}/T)dT - (\partial \hat{v}_k/\partial T)\,dP$.

P–V–T relations for ideal or real gas mixtures

Dalton's law (LAP) $P = \Sigma\, p_k\,(T, V, N_k)$

Amagat Leduc law (LAV) $V = \Sigma\, V_k\,(T, P, N_k)$

For ideal gases, volume fraction vf_k / X_k

Partial pressure for ideal gases $p_k = X_k\,P$

Gibbs Dalton law $U = N_1\,\bar{u}_1\,(T,p_1) + N_2\,\bar{u}_2\,(T,p_2) + \ldots$, $H = U+PV$

$S = N_1\,\bar{s}_1\,(T,p_1) + N_2\,\bar{s}_2\,(T,p_2) + \ldots$, $\hat{s}_k = \bar{s}_k\,(T,P,X_k)$

$= \bar{s}_k^{\,0} - \bar{R}\,\ln(p_k/1) = = \bar{s}_k(T,P) - \bar{R}\,\ln X_k$

Ideal solution/ideal mixture

Any property other than g, a, or s

If $b_k = h_k, u_k, v_k$ then $\hat{b}_k = \bar{b}_k(T,P)$

For g_k, a_k, s_k

$\hat{b}_k = \bar{b}_k\,(T,p_k)$ for ideal mix of real gases

$\hat{b}_k = \bar{b}_k\,(T,P, X_k)$ for ideal mix of liquids and solids

$\hat{b}_k = \bar{b}_k(T,P) - \ln X_k$,

$\bar{b}_k = {}_k(T,P) - \ln X_k$ ideal or real gases

$\hat{s}_k(T, P, X_k) - \bar{s}_k(T, P) = \bar{R}\,\ln X_k$, $\hat{g}_k id - \bar{g}_k\,(T, P) = \bar{R}T\,\ln X_k$.

Fugacity of k

$d\hat{g}_k = \hat{v}_k\,dP = \bar{R}T\,d\ln(\hat{f}_k(T, P, X_k))$

Lewis Randall rule $\hat{f}_k^{\,id}(T, P, X_k) = X_k\,f_k(T, P)$

Henry's law $\hat{f}_1^{\,id}(HL) = X_1(d\hat{f}_1/dX_1)_{x_1 \to 0}$

Fugacity coefficient

$\hat{\phi}_k = \hat{f}_k/(X_k P)$

$(\hat{f}_k^{\,id}/f_k)) = X_k = \hat{\alpha}_k^{\,id}$, i.e., $\hat{f}_k^{\,id} = X_k\,f_k = X_k\,\phi_k\,P$, ideal mix of real gases

For ideal gas mixtures, $\hat{f}_k^{\,ig} = P\,X_k = p_k$

Activity $\hat{\alpha}_k$

$\hat{\alpha}_k = (\hat{f}_k(T, P, X_k)/f_k(T, P))$, $\hat{\alpha}_k^{\,id} = X_k$

$\hat{g}_k(T, P,X_k) - \bar{g}_k\,(T, P) = \bar{R}T\,\ln\hat{\alpha}_k = \int(\hat{v}_k\,(T, P,X_k) - \bar{v}_k(T, P))dP$

$\hat{g}_k\,(T,P, X_k) - \hat{g}_k(T, P^o) = \bar{R}T\,\ln((\hat{\alpha}_k(T,P)/\hat{\alpha}_k(T,P^o; T,P))$

$= \bar{R}T\,\ln(\hat{f}_k(T, P)/\hat{f}_k(P^o,T))$

Activity coefficient, γ_k

$\gamma_k = \hat{\alpha}_k/\hat{\alpha}_k^{\,id}$

$\hat{g}_k - \hat{g}_k^{\,id} = \bar{R}\,T\ln(\gamma_k) = \bar{R}\,T\ln(\hat{\phi}_k/\phi_k)$

Duhem–Margules relation $\Sigma_k N_k\,d\ln(X_k\gamma_k)$ or $\Sigma_k X_k\,d\ln(X_k\gamma_k) = 0$

Excess Property $\hat{b}_k^{\,E} = (\hat{b}_k - \hat{b}_k^{\,id})$, $B^E = B - B^{id} = \Sigma_k N_k(\hat{b}_k - \bar{b}_k)$

$\partial(\bar{g}^E/T)/\partial(1/T) = \bar{h}^E$, $(\partial\bar{g}^E/\partial P)_T = \bar{v}^E$, and $-\bar{s}^E = (\partial\bar{g}^E/\partial T)$

Mixing rules

$\beta = \Sigma_k X_k \beta_k$, $\beta = (\Sigma_k X_k \beta_k^{1/2})^2$, $\beta = (\Sigma_k X_k \beta_k^{1/3})^3$

$\beta = (1/4)\ \Sigma_k X_k \beta_k + (3/4)\ (\Sigma_k X_k \alpha \beta_k^{1/3})(\Sigma_k X_k \beta_k^{2/3})$, $\beta = \Sigma_k X_j X_k \beta_{kj}$,

Kay's rule $T_{cm} = X_1\ T_{c1} + X_2\ T_{c2} + \ldots$, and $P_{cm} = X_1 T_{c1} + X_2 T_{c2} + \ldots$.

RK and Other rules: $\bar{a}_m = (\Sigma_k X_k\ \bar{a}_k^{1/2})^2$, $\bar{b}_m = \Sigma_k X_k \bar{b}_k$

$\bar{a}_m = \Sigma_i \Sigma_j X_i X_j\ \bar{a}_{ij}$, $\bar{b}_m = \Sigma_i X_i\ \bar{b}_i$,

I. CHAPTER 9 RELATIONS

Phase rule

$F = K + 2 - \pi$, F: degrees of freedom, K: components, π: phases

$\hat{g}_{1(1)} = \hat{g}_{1(2)} = \hat{g}_{1(3)} = \ldots = \hat{g}_{1(\pi)}$, $\hat{g}_{2(1)} = \hat{g}_{2(2)} = \hat{g}_{2(3)} = \ldots = \hat{g}_{1(\pi)}$,

…, and $\hat{g}_{k(1)} = \hat{g}_{k(2)} = \hat{g}_{k(3)} = \ldots = \hat{g}_{1(\pi)}$,

or $\hat{f}_{k(f)} = \hat{f}_{k(g)}$.

Ideal solution model $X_{k(\alpha)}\ f_{k(\alpha)}\ (T,P) = X_{k(\beta)}\ f_{k(\beta))}$

$f_k\ (T,P) = f_k\ (T,P^{sat})$ POY,

$$POY = \exp\left(\int_{P^{sat}}^{P} (v/RT)dP\right.$$ for any given phase .

Ideal liquid solution but vapor is an ideal gas (Rauolt;s law: $p_k = X_{k(\alpha)}\ P_k^{sat}$)

Boiling point elevation $\delta T = -(T^2_{pure} R/h_{fg})\ \ln X_{k(\ell)}$ or $\delta T \approx k_b\ Mo_{solute}$,

Azeotropic $X_{k(g)} = X_{k(\ell)}$

Dissolved gases $= X_{k\ (\ell)} = p_k/(P_k^{sat})$

Henry's law $p_k = X_{k,\ell}\ H_k(T,P)$, $H_k(T,P) = P_k^{sat}(POY)_{k(l)}$.

Deviation from Rauolt's law: $p_k = \gamma_{k(\ell)}\ X_{k,\ell}\ p_k^{sat}$ or $\gamma_{k,\ell} = p_k/p_{k,Raoult}$

J. CHAPTER 10 RELATIONS

Criteria

$dS_{U,V,m} = 0$, $(d^2S)_{U,V,m} < 0$, $dS_{H,P,,m} = 0$, $d^2S_{HPV,m} < 0$,

$dU_{S,V,m} = 0$, $d^2U_{S,V,m} > 0$, $dH_{S,P,m} = 0$, $d^2H_{S,P,m} > 0$

$dA_{T,V,m} = 0$, $d^2\ A_{T,V,m} > 0$, $dG_{T,P,m} = 0$, $d^2G_{T,P,m} > 0$

$$D_2 = \begin{vmatrix} S_{UU} S_{UV} \\ S_{VU} S_{VV} \end{vmatrix} = S_{UU}\ \ S_{VV}\ -\ S_{VU}\ ^2 > 0$$

Thermal stability $\partial^2 S/\partial U^2 < 0$ or $c_v > 0$

Mechanical stability $(\partial P/\partial V)_T < 0$

Kestin formulae

If $dS = (1/T)\ dU + (P/T)\ dV = S_U\ (U,V)\ dU + S_V\ (U,V)\ dV$, then

$d^2S = d(1/T)\ dU + d(P/T)\ dV = d(S_U(U,V))\ dU + d(S_V(U,V))\ dV$,

$\sigma = (g_{metastable/unstable} - g_{stable})/T$

Spinodal Points : $(\partial P/\partial V)_T = 0$

K. CHAPTER 11 RELATIONS

Thermochemistry

Air composition mole%: $N_2 = 79\%$, $O_2 = 21\ \%$

mass %: $N_2 = 77\%$, $O_2 = 23\%$

Molecular weight of air: 28.97 kg/kmol or 28.97 lb/lbmol

Process/Variable	Formulae
Stoichiometric combustion	Complete combustion and no O_2 in products
A:F (mass basis)	$\left(\frac{\text{air required in kg}}{\text{kg of fuel}}\right)$
A:F (mole basis)	$\left(\frac{\text{air required in k moles}}{\text{k mole of fuel}}\right)$
Excess air %	$\left\{\frac{A{:}F - (A{:}F)_{stoich}}{(A{:}F)_{stoich}}\right\} \times 100$
Air supplied as % theoretical air or stoichiometric Ratio (SR)	$\frac{\text{air supplied}}{\text{theoretical stoich. air}} \times 100$
Equivalence ratio $(\phi) = 1/(SR)$	$\frac{(A{:}F)_{stoich}}{(A{:}F)} = \frac{(F{:}A)}{(F{:}A)_{stoich}}$
Lean mixture	$\phi < 1$
Rich mixture	$\phi > 1$
Partial pressure of water vapor in products	, where η denotes number of gaseous moles in products.
Total enthalpy, $(\bar{h})$	Enthalpy of formation $(\bar{h}_f^0)$ + thermal enthalpy $(\Delta\, \bar{h}_{298\ to\ T})$
Enthalpy of reaction, (ΔH°_R)	$H_{PROD,T} - H_{REACT,T}$
Enthalpy of combustion, (ΔH_c)	$\Delta H^\circ_C \equiv \Delta H^\circ_R$
Heating value (HV)	$HV = -\ \Delta H^\circ_{R,298} = H_{REACT298} - H_{PROD298}$
Higher heating value, HHV	= HV with H_2O in liquid form
Lower heating value, LHV	= HV with H_2O in gaseous state

Internal energy for use $\bar{u}_i = \bar{h}_i - P\,\bar{v}_i\,,\ = \bar{h}_i - \bar{R}T$ if species i is ideal gas

$$\text{Combustion efficiency } \eta_{comb} = \frac{\text{Actual heat release}}{\text{Theoretical heat release}}$$

$$\text{Boiler efficiency } \eta_{boiler} = \frac{\text{Heat transferred to water / Kg of fuel}}{\text{Higher heating value / Kg of fuel}}$$

Overall thermal efficiency $\eta_{thermal} = \dfrac{\text{Elec. work / kg of fuel}}{\text{Higher heating value / kg of fuel}}$

Energy and entropy balance for reacting systems

$$dE_{cv}/dt = \dot{Q}_{cv} - \dot{W}_{cv} + \Sigma_{k,i}\dot{N}_k\bar{e}_{T,k} - \Sigma_{k,e}\dot{N}_k\bar{e}_{T,k},\ \bar{e}_{T,k} = (\bar{h} + \bar{ke} + \bar{pe})_k$$

$$dS_{cv}/dt = \dot{Q}_{cv}/T_b + \Sigma_{k,i}\dot{N}_k\hat{s}_k - \Sigma_{k,e}\dot{N}_k\hat{s}_k + \dot{\sigma}$$

Mole balance

$$dN_k/dt = \dot{N}_{k,i} + \dot{N}_{k,gen} - \dot{N}_{k,e}$$

L. CHAPTER 12 RELATIONS

$dU = TdS - P\,dV - T\,\delta\sigma$

$dH = T\,dS + V\,dP - T\delta\sigma$

$dA = -S\,dT - P\,dV - T\,\delta\sigma$

$dG = -S\,dT + V\,dP - T\,\delta\sigma$

$-T\,(\partial S/\partial N_k)_{U,V} = -T\,(\partial S/\partial N_k)_{H,P} = (\partial U/\partial N_k)_{S,V} = (\partial A/\partial N_k)_{T,V} = (\partial G/\partial N_k)_{T,P} = \hat{g}_k = \mu_k$

Process/Variable/Condition	***Formula***
Direction of reaction	$\delta\sigma \geq 0$, $dG_{T,P} \leq 0$ or $(\Sigma\mu_k\,dN_k) \leq 0$, $dG_{T,P} = (\Sigma\mu_{k\,k}\,dN_k) \leq 0$, $G = \Sigma N_k\,\bar{g}_k\,(T,p_k)$, $\hat{g}_k\,(T,p_k) \equiv \mu_k$
Equilibrium condition	$\delta\sigma = 0$, $\Sigma\hat{g}_k\,dN_k = dG_{T,P} = 0$, $\Sigma\hat{g}_k\,dN_k = dA_{T,V} = 0$
Chemical potential	$\mu_j = (\partial G/\partial N_j)_{T,P,N_2\ldots}$ $\mu_j = (\partial A/\partial N_j)_{T,V,n_2\ldots}$
Chemical force potential (e.g. $H_2 + 1/2\,O_2 \to H_2O$)	$F_{chem,\,react} = \bar{g}_{H_2} - \frac{1}{2}\bar{g}_{O_2}$, $F_{chem,\,prod} = \bar{g}_{H_2O}$ $F_{chem,\,react} > F_{chem\,Prod}$
Gibbs function useful for chemical reactions, P in bars	$\hat{g}_i = \hat{h}_i - T\,\hat{s}_i = \{\bar{h}_i^o - T\,\bar{s}_i^{\,o}\} - \bar{R}\,T\,\ln\,(PX_i/1)]$ $\hat{g}_i = \bar{g}_i^{\,o}(T) + \bar{R}T\,\ln\,(PX_i/1)$, see TableA8 to A19 for $\bar{g}_i^{\,o}$ $\hat{g}_i = \bar{g}_i(T,P) + \bar{R}\,T\,\ln\,X_i$
Equilibrium constant, $K^0(T)$	$K^o(T) = \exp\,(-\Delta G^o/\bar{R}\,T)$, $\Delta G^0 = G^0_{RHS} - G^0_{LHS}$
Equilibrium constant, $K^0(T)$ from "elementary" reactions	$K^0(T)\exp\,(\Delta A - (\Delta B/T))$, T in K, $\Delta A = \Sigma\,\nu_k\,A_k$, $\Delta B = \Sigma\,\nu_k\,B_k$, (e.g.: $CO_2 \Leftrightarrow CO + 1/O_2$, $\Delta A = 1\times A_{CO} + (1/2)\times A_{O_2} - 1\times A_{CO_2}$), use Table A–28A
Van't Hoff relation	$\ln K^0(T) = A - B/T$, $A = A = \ln K^{\,o}_{ref} + (\Delta H^{\,o}_R/\bar{R})\,(1/T_{ref})$, $B = (\Delta H^{\,o}_R/\bar{R})$.

M. CHAPTER 13 RELATIONS

$g = h - Ts$

$a = u - Ts$

$$d(E_{cv} - T_0 S_{cv})/dt = \Sigma \dot{Q}_{R,j}(1 - \frac{T_0}{T_{R,j}}) + (\Sigma \dot{N}_k \hat{\psi}_k)_i - (\Sigma \dot{N}_k \hat{\psi}_k)_e - \dot{W}_{cv} - \dot{I}$$

Variable	Formula
Entropy	$\bar{s}(T,P) = \bar{s}^o(T) - \bar{R} \ln \frac{P}{1}$, P in bar
Absolute availability of species j (not availability)	$\hat{\psi}_j = \bar{h}_j - T_0 \hat{s}_j$
Exergy or availability	$\psi_j - \psi_{j,0}$
Chemical availability	$\hat{g}_k(T_0, P_0, X_{1,e}, ..) - \hat{g}_k(T_0, P_0, X_{1,\infty}, ..)$
Fuel availability, $avail_f$ with fuel at 1 bar, air at 1 bar	$\Sigma N_{k,i} \hat{g}_{k,i}(T_0, P_0, X_{1,0}, ..) - \Sigma N_{k,e} \hat{g}_{k,e}(T_0, P_0, Y_{1,\infty}, ..)$
Irreversibility/lost availability, I	$I = W_{opt} - W_{act}$
$W_{opt, TM}$, W_{chem} but not of fuel only	$w_{opt, TM} = \psi_i - \psi_e(T_0, P_0, X_{1e}, X_{2e} ..)$ $w_{chem} = \psi_e(T_0, P_0, X_{1e}, X_{2e} ..) - \psi_\infty(T_\infty, P_\infty, X_{1\infty}, X_{2\infty} ..)$
Availability efficiency	$\eta_{avail} = W/Avail_F$.
Voltage in fuel cell, number of electrons from $C_x H_y$ is (4x+y).	$V_{(volts)} = \frac{(\Delta G \text{ in kJ / K mole}) \times 1.036 \times 10^{-5}}{(\text{No of electrons per molecule of fuel})}$

APPENDIX D

D. REFERENCES

A. GENERAL REFERENCES

A. Bejan, *Advanced Engineering Thermodynamics*, John Wiley & Sons, NY, 1988.

R. E. Balzhiser, M. R. Samuels and J. D. Elissen, *Chemical Engineering Thermodynamics*, Prentice Hall, 1972.

H. B. Callen, *Thermodynamics and an Introduction to Thermostatistics*, 2nd edition, John Wiley & Sons, NY, 1985.

G. Carrington, *Basic Thermodynamics*, Oxford Science Publications, NY, 1994.

Y.A. Cengel and M.A. Boles, *Thermodynamics, an Engineering Approach,* 3rd Edition, McGraw Hill Book Co., NY, 1998.

T. E. Daubert, *Chemical Engineering Thermodynamics*, McGraw Hill Book Co., NY, 1985.

G. Emanuel, *Advanced Classical Thermodynamics*, AIAA Education Series, Ohio 1987.

E. P. Gyftopoulos, and G. P. Beretta, *Thermodynamics*, Foundations and Application, Macmillan Publishing Co., N.Y., 1991.

J. Kestin, *A Course in Thermodynamics*, Volume II and McGraw Hill, NY, 1979.

G. N. Lewis, and M. Randall, *Thermodynamics*, 2nd Edition, McGraw Hill Inc., NY, 1961.

M. L. McGlashan, *Chemical Thermodynamics*, Academic Press, NY, 1979.

M. Modell and R. C. Reid, *Thermodynamics and its Applications*, 2nd Edition, Prentice Hall, 1983.

M. Moran, *Availability Analysis*, Prentice Hall, Englewood Cliffs, NJ, 1982.

J. M. Smith and H. C. Van Ness, *Introduction to Chemical Engineering Thermodynamics*, 4th Edition, McGraw Hill Book Co., NY, 1987.

R. Sonntag, C. Borgnakke and G. J. Van Wylen, *Fundamentals of Classical Thermodynamics*, 5th Edition John Wiley & Sons, NY, 1998.

L. M. Sussman, *Elementary General Thermodynamics*, Addison Wesley Publishing Co, 1972.

K. Wark, Advanced *Thermodynamics for Engineers*, McGraw Hill Book Co., NY, 1995.

G. J. Van Wylen, and R. Sonntag, *Fundamentals of Classical Thermodynamics*, 5th Edition, John Wiley & Sons, NY, 1988.

M. Zemansky, *Thermodynamics*, 4th Edition, McGraw Hill Book Co., NY, 1957.

B. STATISTICAL THERMODYNAMICS

J. O. Hirshfelder, C. F. Curtis, and R. B. Bird, *Molecular Theory of Gases and Liquids*, Wiley, NY, 1964.

C. L. Tien, and J. H, Linehard, *Statistical Thermodynamics*, Hemisphere Publishing Corporation, NY, 1979.

C. STATE EQUATIONS, OPTIMIZATIONS AND PROPERTIES

M. Abraham, K. Annamalai, and D. Claridge, *Journal of Energy Resources and Technology*, 119, 236-241, (1997).

K. Annamalai, T. Morris, and D. McNicholls, *AES Vol. 37*, ASME Advanced Energy Systems Division, pp. 239-247, 1997.

K. Annamalai, V.K. Rao and A.V. Sreenath, *Journal of Institute of Fuel,* XLV: 48-51, (1972).

A. A., Bedrassian, and H.Y. Cheh, "Effect of Sodium Acetate on the Vapor-Liquid Equilibrium of Ethanol-water system," *Thermodynamics Data and Correlations*, Ed: D. Zudkevitch and I. Zaremebr, *AIChE Symposium Series*, No 140, Vol. 70, AIChE, NY, 1974.

C. Campo, B. Abushakra, C. Assenlta, K. Annamalai, and T. Lalk, "A Comparison Between the Theoretical and Actual Performance of a Polymer Electrolyte Fuel Cell Using Availability Analysis," ASME-Energy Week, Feb. 2-4, 1998, ASME Paper # ETCE 98-4768.

C. Coimbra and C. Quiroz, *Combustion and Flame*, 101, pp. 209–220, 1995.
H. W. Cooper and J. Goldfrank, *Hydrocarbon Processing*, 46(12), p. 141, 1967.
W. G. Dong, and J. H. Lienhard, *Canadian J. Chem. Eng.*, 64, pp. 158-161, 1986.
R. L. Fosdick and K. R. Rajagopal, *Archive for Rational Mechanics and Analysis*, 81, pp. 317-332, 1983.
E. A. Guggenheim, *J. Chem. Physics*, 13, p. 253, 1945.
J. Joffe and D. Dudkevitch, "Correlation of Liquid Densities of Polar and Nonpolar Compounds," *Thermodynamics Data and Correlations*, Ed: D. Zudkevitch and I. Zaremebr, *AIChE Symposium Series*, No 140, Vol. 70, AIChE, NY, 1974.
J. H. Keenan, F. G. Keyes, P. G. Hill, and J. G. Moore, *Steam Tables*, John Wiley & Sons, NY, 1978.
K. K. Kelley, *U.S. Bur. Mines. Bull.* p. 584, 1960.
S.C. Lau and K. Annamalai, *Transactions of the ASME J. of Energy Resources*, 109:90-95, 1987.
B. I., Lee, and M, G. Kesler, *AIChE Journal*, 21(3), pp. 510-527, 1975.
G. N. Lewis, *Proc. Am. Acad.*, 37, p. 49, 1901.
J. H., Lienhard, N. Shamsundar, and P.O. Biney, *Nucl. Eng. Des.*, 95, pp. 297-314, 1986.
A. L, Lyderson, R. A., Greenkorn, and O. A., Hougen, *Univ. of Wisconsin, Engg. Expt. Sta. Rep* 4, 1955.
JANAF Thermochemical Tables, Dow Chemical Co., 1971.
M. Martin, K. Annamalai, and J. Claridge, "The Effects of Glycol Addition to a Static Water Cooled Thermal Energy Storage system," Proceedings of Renewable and Advanced Energy Systems for the 21st Century, ASME Preprint # RAES99-7674, 10 pages, 1999.
J. W. Miller, Jr., G. R. Schorr, and C. L. Yaws, *Chem. Eng.*, 83(23), p. 129, 1976.
NBS: Selected Values of Chemical Thermodynamic Properties, NBS Technical Note 270-3, 1968, and *API Research Project* 44, Carnegie Press, 1953.
L. B. Pankratz, *U.S. Bur. Mines Bull.*, p. 672, 1982.
K. S. Pitzer, *J. Am. Chem. Soc.*, 77, p. 3439, 1955.
I. K. Puri, *International J. Heat Mass Transfer*, 35, pp. 2571–2578, 1992.
S. Somasundaram, D. Chi, and K. Annamalai, *International Journal of Mechanical Engineering Education*, 16, pp. 141-143, 1988.
S. Somasundaram, D. Chi, and K. Annamalai, *ASHRAE Transactions*, 93, pp. 701-715, 1987.
L. Riedel, *Chem. Ing. Tech.*, 26, p. 259, 1954.
H. M. Spencer, *Ind. Eng. Chem.*, 40, p. 2152, 1948.
R. A. Svehla, *NASA Tech. Report R-132*, Lewis Research Center, Cleveland, OH, 1962.
D. Tootle, K. Annamalai, and A. Kristens, "Work Loss During Compression of a Saturated Mixture in a Compressor", Emerging Energy Technology, 8th Annual International Energy Week, Book V, pp. 433-436, 1997.
P. J. Weinberg, *Nature*, 233, pp 239-241, 1971.

D. OTHER SOURCES

K. Annamalai, W. Ryan, and S.Chandra, *J. of Heat Transfer*, 707-716,115, 1993.
K. Annamalai, and W. Ryan, *Prog. Energy and Combust. Sci.*, 221-295, 18, 1992.
R. B. Bird, W. E. Stewart, and E.N. Lightfoot, *Transport Phenomena*, John Wiley & Sons, NY, 1960.
J. Banhidi, *Radiant Heating Systems*, Pergamon Press, 1991.
R. Feynman, *Lectures on Physics*, Vols. I, II, and III, Addison-Wesley Publishing Co., MA, 1963-65.
A., Kalanidhi, *Engineering Innovations in Pranayama*, Charu-Gayathri Publications, Madras, India.
I. K. Puri, Editor, *Environmental Implications of Combustion Processes*, CRC Press, FL, 1993.

K. Schmidt-Nielsen, *Scaling: Why Is Animal Size So Important?* Cambridge University Press, NY, 1984.
S. S. Zumdahl, *Chemistry*, D.C. Heath and Company, MA, 1986.

E. WEBSITES

http://users.erols.com/sclufer/EmUniv.html (Discussion regarding universe.)
http://thermodex.lib.utexas.edu/ (Website on properties.)
http://www.spacedaily.com/spacecast/news/chandra-00a.html (Discussion regarding universe.)

INDEX